30106017031815
AF746724

# Encyclopedic Handbook of
# Emulsion Technology

# Encyclopedic Handbook of Emulsion Technology


edited by

Johan Sjöblom

*Statoil A/S*
*Trondheim, Norway*



MARCEL DEKKER, INC. NEW YORK • BASEL

**ISBN: 0-8247-0454-1**

This book is printed on acid-free paper.

**Headquarters**
Marcel Dekker, Inc.
270 Madison Avenue, New York, NY 10016
tel: 212-696-9000; fax: 212-685-4540

**Eastern Hemisphere Distribution**
Marcel Dekker AG
Hutgasse 4, Postfach 812, CH-4001 Basel, Switzerland
tel: 41-61-261-8482; fax: 41-61-261-8896

**World Wide Web**
http://www.dekker.com

The publisher offers discounts on this book when ordered in bulk quantities. For more information, write to Special Sales/Professional Marketing at the headquarters address above.

**Copyright © 2001 by Marcel Dekker, Inc. All Rights Reserved.**

Neither this book nor any part may be reproduced or transmitted in any form or by any means, electronic or mechanical, including photocopying, microfilming, and recording, or by any information storage and retrieval system, without permission in writing from the publisher

Current printing (last digit):
10 9 8 7 6 5 4 3 2 1

**PRINTED IN THE UNITED STATES OF AMERICA**

*Cover illustration*: Courtesy of Statoil A/S, Trondheim, Norway.

LEEDS UNIVERSITY LIBRARY

# Preface

In our everyday life we are confronted with different aspects of emulsions—either water- or oil-continuous—with varying amounts of oil, fat, and water. Margarine, butter, milk, and dressings all represent central food emulsions. In specific cases involving food emulsions, it is beneficial for the customer to control the stability of the product, particularly with regard to flocculation, sedimentation/creaming, and coalescence. In other branches of industry, emulsion stability may cause severe problems for the entire process. The most well-known examples are crude-oil–based emulsions, wastewater emulsions, etc. The multiplicity of industrial applications of emulsions has catalyzed a better fundamental understanding of these dispersed systems. The academic research has been successfully interfaced to practical concerns in order to facilitate the solutions to emulsion problems.

The first chapter in this book deals with the fundamental properties and characterization of the water/oil interface. This outstanding contribution by Miller and coworkers is very central in understanding the basics of emulsions from a thermodynamic point of view. The chapter summarizes the newest findings with regard to interfacial processes, theory, and experimental facilities.

In the second chapter, Professor Friberg reviews the use of phase diagrams within emulsion science and technology. Special emphasis is given to emulsion stability, preparation of emulsions, and prediction of structural changes during evaporation. The author has refined to completion the use of phase equilibria (equilibrium conditions for the components) to understand the nature of the processes taking place and the final conditions obtained within the emulsified systems.

The third chapter, by Wasan and Nikolov, discusses fundamental processes in emulsions, i.e., creaming/sedimentation, flocculation, coalescence, and final phase separation. A number of novel experimental facilities for characterization of emulsions and the above-mentioned processes are presented. This chapter highlights recent techniques such as film rheometry for dynamic film properties, capillary force balance in conjunction with differential microinterferometry for drainage of curved emulsion films, Kossel diffraction, imaging of interdroplet interactions, and piezo imaging spectroscopy for drop-homophase coalescence rate processes.

Next, Professor Dukhin et al. contribute a chapter dealing with fundamental processes in dilute O/W emulsions. A basic problem is to couple the processes of coalescence and flocculation by introducing a reversible flocculation, i.e., a process whereby the floc is disintegrated into individual droplets. The authors have utilized video-enhanced microscopy (VEM) to study the emulsified systems and to determine critical time constants for a stepwise flocculation/deflocculation and coalescence.

The Lund group, represented by Professors Wennerström, Söderman, Olsson, and Lindman, addresses the emulsion concept from the perspective of microemulsions. The vast number of studies of microemulsions has contributed to a better understanding of the properties of surfactant films. In this chapter, the contributors discuss the role of surfactant phase behavior (under equilibrium conditions), diffusion properties of both microemulsions and macroemulsions, the implications of the flexible surface model for emulsion stability, and the Ostwald ripening process in a potentially metastable emulsion system.

Dielectric spectroscopy has proved to be an important tool to describe emulsion and microemulsion systems. Professor Yuri Feldman and his Norwegian colleagues review in Chapter 6 the fundamentals of this technique and its plausible applications. The span of applications is very wide and includes flocculation processes in emulsions, diffusion processes and porosity measurements in solid materials, characterization of particulate biosuspensions, and percolation phenomena in microemulsions.

Electroacoustics provide a unique opportunity to estimate both the size of emulsion droplets and the state of the surface (kinetic) charge in a single measurement. The next two chapters, by Hunter and by A. Dukhin, Wines, Goetz and Somasundaran describe in detail the advantages of these techniques and their current development. The latter chapter also describes the application of acoustic techniques to microemulsion systems, revealing interesting structural details.

Professor Dalgleish's chapter contains a comprehensive and detailed review of the important emulsified food systems. The chapter covers the chemistry of the stabilizers (low-molecular-weight and high-molecular-weight stabilizers), the formation of the emulsion during the preparation stage, and the characterization of the formed droplets with respect to particle sizes and size distributions. The author describes the structure of the formed droplets, with details about the stabilizing interfacial layer. Macroscopic stability and destabilization of the systems are also discussed, together with kinetic aspects and the topic of partial coalescence. The author also reviews the concept of multiple emulsions.

Coupland and McClements further elaborate food emulsions in Chapter 10, reviewing the basic theory behind the propagation of ultrasound in emulsified systems and the mechanisms behind the thermal and visco-inertial losses. The pros and cons of different experimental techniques are also reviewed. Crystallization (formation and melting of crystals) and influence of droplet concentration (individual droplets and flocs), as well as droplet size and droplet charge, are all parameters discussed by the authors.

Chapters 11–16 discuss various aspects of measurement techniques as applied to different kinds of emulsifed systems. The chapters review rheology and concentrated emulsions (Princen), the NMR perspective (Balinov and Söderman), surface forces (Claesson, Blomberg, and Poptotshev), microcalorimetry (Dalmazzone and Clausse), video-enhanced microscopy (Sæther), and conductivity (Gundersen and Sjöblom). Some of these experimental techniques represent traditional approaches, while others give new avenues to a physicochemical in-depth interpretation of ongoing processes in emulsions, in both water- and oil-continuous systems.

Professors Nissim Garti and Axel Benichou review formation and preparation of intricate multiple emulsions, of both the water-in-oil-in-water and oil-in-water-in-oil types. Emulsions of this type occur especially when mixtures of both hydrophobic and hydrophilic stabilizers are used. These mixtures can be either commercial or naturally occurring. This chapter is an important complement to the traditional view of "simple emulsions" with only one dispersed phase.

Chapters 18–25 are related in that they discuss aspects of crude oil-based emulsions. The topic of environmental emulsions is covered by Fingas, Fieldhouse, and Mullin. They analyze in depth the emulsification and stabilization processes in oil spills. These processes are crucial because they complicate the removal and treatment of these so-called "mousses" or "chocolate mousses." Natural forces in the form of wind and waves are important mechanisms for the formation of the oil-spill emulsions. Most likely, the stabilization of the formed dispersions is due to naturally occurring components such as asphaltenes and resins. The authors give a comprehensive analysis of different kinds of oil spills with regard to stability and rheological properties.

The next chapter, by Kvamme and Kuznetsova, presents a theoretical approach to molecular-level processes taking place at the W/O interface. The chapter comprises state-of-the-art concepts, experimental results, and atomic-level computer simulations of processes determing the stability of the dispersions. Parallels are drawn to lipid bilayers. A strategy suitable for molecular dynamics simulation of water-in-crude-oil emulsions is presented, with most of its constituents' elements proved by computer simulations of less complex systems.

The processing of extraheavy crude oils/bitumens is extremely important because world reserves amount to 450 billion tons. To date, the mammoth Athabasca deposit in Alberta, Canada, and the Orinoco Oil Belt in western Venezuela are the largest. Together, Canada and Venezuela hold over 40% of the total extraheavy hydrocarbon reserves. The next two chapters by Professors Salager, Briceño, and Bracho from Merida, Venezuela, and Professor Czarnecki from Edmonton, Canada, review the role of emulsions in the processing of these crude oils.

In Venezuela, the Orimulsion concept has been a success in the transportation of extraheavy crudes. These oil-in-water emulsions normally contain about 30% water and about 65% bitumen. Venezuela exports the Orimulsion to countries such as Canada, Japan, China, Denmark, Italy, and Lithuania. In 1998, 4 million tons were exported. The recovery of hydrocarbons from the oil sands is surveyed in detail by Czarnecki, who describes the problems with water-in-oil emulsions during the process steps. A major problem is created by downstream complications due to high amounts of water and high salinity.

The film properties of some selected Chinese and North Sea crude oils are the topics of the next two chapters, by Li, Peng, Zheng, and Wu and by Yang, Lu, Ese, and Sjöblom. Different aspects of interfacial rheology (interfacial shear viscosity) and Langmuir films are explored in depth for the crude oils as such or, alternatively, for selected crude oil components. Correlations between these findings and the macroscopic emulsion stability are pointed out.

Chemical destabilization is the mutual topic for the chapters by Angle and by Sjöblom, Johnsen, Westvik, Ese, Djuve, Auflem, and Kallevik ("Demulsifiers in the Oil Industry"). These contributions present a comprehensive treatise on chemicals used to promote coalescence, as well as their interaction patterns with different indigenous film-forming components in the crude oils. The competition between categories of demulsifiers in the bulk and at the W/O interfaces is central in these chapters. The question of chemical administration to enhance the efficiency of the chemical agents is also raised. The upscaling of the use of the chemicals in actual recovery operations onshore as well as offshore is also reviewed.

Chapters 26–29 all discuss hydrodynamic aspects of emulsified systems. The contribution by Danov, Kralchevsky, and Ivanov presents a very fundamental and thorough survey of different phenomena in emulsions related to dynamic and hydrodynamic motions, such as the dynamics of surfactant adsorption monolayers, which include the Gibbs surface elasticity, and characteristic time of adsorption, mechanisms of droplet–droplet coalescence, hydrodynamic interactions and drop coalescence, interpretation of the Bancroft rule with regard to droplet symmetry, and, finally, kinetics of coagulation in emulsions covering both reversible and irreversible coagulation and the kinetics of simultaneous flocculation and coalescence.

Arntzen and Andresen in their chapter cover the emulsification conditions under different flow conditions in true horizontal gravity separators. The theoretical section includes formation of droplets in turbulent regimes, turbulence-induced coalescence, settling laws in gravity separators, plug velocities and retention times, binary coalescence and hindered settling, and the dispersion layer theory. The authors then describe the choice of internals in order to accelerate the separation of water, oil, and gas, such as foam and mist handling devices, flow distributing devices, and settling enhancing devices. Current models for flow patterns in gravity separators and the impact of various models and internals on the flow of the different phases, including the dispersed phase, are also discussed. The final section in this chapter is devoted to emerging technologies.

The chapter by Urdahl, Wayth, Førdedal, Williams, and Bailey begins by discussing droplet break-up processes under both laminar and turbulent flow conditions and in electrostatic fields. The authors then discuss the droplet coalescence process under normal Brownian motion, under gravity sedimentation, and in laminar shear, including turbulent collisions as well as collisions due to electrostatic forces. The remainder of the chapter is devoted to electrostatic-induced separation of the water-in-oil emulsions and emerging technologies.

Gas hydrate formation is a well-known obstacle in the transport of gas, oil, and water. The formation of such chlatrates and their agglomeration will eventually plug pipes and prevent transport. One way to overcome this problem is to form the gas hydrates in a water-in-oil emulsion. The chapter by Tore Skodvin summarizes some current research at the University of Bergen in this field. It is stated that dielectric spectroscopy is a convenient technique to follow the formation of gas hydrates inside the water droplets, and because of this formation the dielectric properties of water change remarkably. It is also shown that when the gas hydrate particles are emulsified in a water-in-oil matrix one can transport up to about 30 weight% of water without any inhibitors present.

In the final chapter on asphaltene-stabilized crude oil emulsions, Kilpatrick and Spiecker present an extensive survey of naturally occurring crude oil surfactants and their role in the stabilization of the crude oil-based emulsions, based on a large matrix of various crude oils from different parts of the world. The authors also explain the destabilization process by means of a variety of experimental techniques, such as high voltage break-throughs and rheology.

When I began the process of compiling the text in the next 30 chapters and contacted my colleagues all over the world, I was overwhelmed by the extremely positive attitude they all had. This was very important to me, because in editing a volume such as the *Encyclopedic Handbook of Emulsion Technology* it is necessary to ask one's colleagues to give priority — in both time and science — to the project by completing their contributions within the timeframe available. The chapter authors have been very conscientious in this respect. Therefore, I express my most sincere gratitude to all the renowned contributors for their time and effort. I hope that the international community in surface and colloid science/emulsions science and technology will appreciate this volume. I am sure that I speak on behalf of all the authors when I say that all the scientists involved in this project have contributed to a better and deeper understanding of the very complex and intricate emulsified systems.

Finally, I would also like to express my gratitude to my employer Statoil R&D Centre in Trondheim, Norway, for giving me the opportunity to complete this extensive handbook.

*Johan Sjöblom*

# Contents

# Contributors

**Per Arild K. Andresen, Ph.D.** Provida ASA, Oslo, Norway

**Chandra W. Angle, M.Sc.** Natural Resources Canada, Devon, Alberta, Canada

**Richard Arntzen** Kværner Process Systems a.s., Lysaker, Norway

**Inge H. Auflem, M.Sc.** Norwegian University of Science and Technology, Trondheim, Norway

**Adrian G. Bailey, Ph.D., M.I.E.E., F.Inst.P.** University of Southampton, Southampton, Hampshire, England

**Balin Balinov, Ph.D.** Nycomed Imaging AS, Oslo, Norway

**Axel Benichou** Casali Institute of Applied Chemistry, The Hebrew University of Jerusalem, Jerusalem, Israel

**Eva Blomberg, Ph.D.** Royal Institute of Technology and Institute for Surface Chemistry, Stockholm, Sweden

**Carlos Luis Bracho** Universidad de Los Andes, Mérida, Venezuela

**María Isabel Briceño*** Universidad de Los Andes, Mérida, Venezuela

**Per M. Claesson, Ph.D.** Royal Institute of Technology and Institute for Surface Chemistry, Stockholm, Sweden

**Danièle Clausse** Université de Technologie de Compiègne, Compiègne, France

**John N. Coupland, Ph.D.** Pennsylvania State University, University Park, Pennsylvania

**Jan Czarnecki, Ph.D., D.Sc.** Edmonton Research Centre, Syncrude Canada Ltd., Edmonton, Alberta, Canada

**Previous affiliation*: INTEVEP, Los Teques, Venezuela.

**Douglas G. Dalgleish, Ph.D.** Danone Vitapole, Le Plessis-Robinson, France

**Christine S. H. Dalmazzone, Ph.D.** Institut Français du Pétrole, Rueil-Malmaison, France

**Krassimir D. Danov, Ph.D.** University of Sofia, Sofia, Bulgaria

**Jostein Djuve, M.Sc.** University of Bergen, Bergen, Norway

**Andrei S. Dukhin, Ph.D.** Dispersion Technology Inc., Mount Kisco, New York

**Stanislav Dukhin, Ph.D., Dr.Sc.** New Jersey Institute of Technology, Newark, New Jersey

**Marit-Helen Ese, Ph.D.** University of Bergen, Bergen, Norway

**V. B. Fainerman** Institute of Technical Ecology, Donetsk, Ukraine

**Yuri Feldman, Ph.D.** The Hebrew University of Jerusalem, Jerusalem, Israel

**Benjamin G. Fieldhouse, B.Sc.** Environment Canada, Ottawa, Ontario, Canada

**Merv Fingas, Ph.D.** Environment Canada, Ottawa, Ontario, Canada

**Harald Førdedal, Ph.D.** Statoil A/S, Trondheim, Norway

**Stig E. Friberg, Ph.D.** Clarkson University, Potsdam, New York

**Nissim Garti, Ph.D.** Casali Institute of Applied Chemistry, The Hebrew University of Jerusalem, Jerusalem, Israel

**P. J. Goetz** Dispersion Technology Inc., Mount Kisco, New York

**D. O. Grigoriev, Ph.D.** Max-Planck Institute, Berlin, Germany, and St. Petersburg State University, St. Petersburg, Russia

**Stig Are Gundersen, Ph.D.** University of Bergen, Bergen, Norway

**Robert J. Hunter, Ph.D.** University of Sydney, Sydney, New South Wales, Australia

**Ivan B. Ivanov, Ph.D., Dr.Sc.** University of Sofia, Sofia, Bulgaria

**Einar Eng Johnsen, Ph.D.** Statoil A/S, Trondheim, Norway

**Harald Kallevik** Norwegian University of Science and Technology, Trondheim, Norway

**Peter K. Kilpatrick, Ph.D.** North Carolina State University, Raleigh, North Carolina

**J. Krägel** Max-Planck-Institut, Berlin, Germany

**Peter A. Kralchevsky, Ph.D.** University of Sofia, Sofia, Bulgaria

**Tatyana Kuznetsova, Ph.D.*** University of Bergen, Bergen, Norway

**Bjørn Kvamme, Ph.D.** University of Bergen, Bergen, Norway

**D. Y. Kwok** Massachusetts Institute of Technology, Cambridge, Massachusetts

**Mingyuan Li, Ph.D.** University of Petroleum, Changping, Beijing, China

**L. Liggieri, Ph.D.** Istituto di Chimica Fisica Applicata dei Materiali–CNR, Genoa, Italy

**Björn Lindman, Ph.D.** University of Lund, Lund, Sweden

**Wanzhen Lu, Ph.D.** Research Institute of Petroleum Processing, Beijing, China

**A. V. Makievski, Ph.D.** Max-Planck-Institut, Berlin, Germany, and Institute of Technical Ecology, Donetsk, Ukraine

**D. Julian McClements, Ph.D.** University of Massachusetts, Amherst, Massachusetts

**R. Miller, Ph.D.** Max-Planck-Institut, Berlin, Germany

**Joseph V. Mullin, B.O.T.** U.S. Minerals Management Service, Department of the Interior, Herndon, Virginia

**A. W. Neumann, Ph.D.** University of Toronto, Toronto, Ontario, Canada

**Alex D. Nikolov, Ph.D.** Illinois Institute of Technology, Chicago, Illinois

**Ulf Olsson, Ph.D.** University of Lund, Lund, Sweden

**Bo Peng, Ph.D.** University of Petroleum, Changping, Beijing, China

**Evgeni Poptoshev, M.Sc.** Royal Institute of Technology and Institute for Surface Chemistry, Stockholm, Sweden

**H. M. Princen, Ph.D.‡** Mobil Technology Company, Paulsboro, New Jersey

**F. Ravera, Ph.D.** Istituto di Chimica Fisica Applicata dei Materiali–CNR, Genoa, Italy

**Øystein Sæther, Ph.D.** Norwegian University of Science and Technology, Trondheim, Norway

**Jean-Louis Salager** Universidad de Los Andes, Mérida, Venezuela

**Johan Sjöblom, Ph.D.** Statoil A/S, Trondheim, Norway

**Tore Skodvin, Dr. Sc.** University of Bergen, Bergen, Norway

**Olle Söderman, Ph.D.** University of Lund, Lund, Sweden

**P. Somasundaran, Ph.D.** Columbia University, New York, New York

*On leave from Institute of Physics, St. Petersburg University, St. Petersburg, Russia.
‡*Current affiliation*: Consultant, Flemington, New Jersey.

**P. Matthew Spiecker, Ph.D.** North Carolina State University, Raleigh, North Carolina

**Olav Urdahl, Ph.D.** Veslefrikk Operations, Statoil, Sandsli, Norway

**Darsh T. Wasan, Ph.D.** Illinois Institute of Technology, Chicago, Illinois

**Nicholas J. Wayth, Ph.D.** BP Amoco Exploration, Greenford, Scotland

**Håkan Wennerström, Ph.D.** University of Lund, Lund, Sweden

**Arild Westvik** Statoil A/S, Trondheim, Norway

**Trevor J. Williams, Ph.D.** University of Southampton, Southampton, Hampshire, England

**T. H. Wines** Columbia University, New York, New York

**Zhaoliang Wu** University of Petroleum, Changping, Beijing, China

**Xiaoli Yang** Research Institute of Petroleum Processing, Beijing, China

**Xiaoyu Zheng, Ph.D.** University of Petroleum, Changping, Beijing, China

# 1

# Characterization of Water/Oil Interfaces

**R. Miller and J. Krägel**
*Max-Planck-Institut, Berlin, Germany*

**V. B. Fainerman**
*Institute of Technical Ecology, Donetsk, Ukraine*

**A. V. Makievski**
*Max-Planck-Institut, Berlin, Germany, and*
*Institute of Technical Ecology, Donetsk, Ukraine*

**D. O. Grigoriev**
*Max-Planck-Institut, Berlin, Germany, and*
*St. Petersburg State University, St. Petersburg, Russia*

**F. Ravera and L. Liggieri**
*Istituto di Chimica Fisica Applicata dei*
*Materiali–CNR, Genoa, Italy*

**D. Y. Kwok**
*Massachusetts Institute of Technology, Cambridge,*
*Massachusetts*

**A. W. Neumann**
*University of Toronto, Toronto, Ontario, Canada*

## I. INTRODUCTION

The behavior of disperse systems, such as foams and emulsions, is very complex and there have been only few attempts to derive qualitative and quantitative relationships between their stability and physicochemical parameters of the stabilizing adsorption layers. The starting point of most of these approaches is the hydrodynamic theory of thinning of a liquid film between two bubbles or drops according to Reynolds (1) and Levich (2). A simplified picture of the general scenario in an emulsion is the following. When two droplets of equal size approach each other the contact area between the two drops is deformed such that a plane parallel film results (3). The liquid between the two film surfaces flows out until a critical film thickness is reached. In this situation, the two drops can repel each other, form a flock (coagulate), or coalesce to form one larger drop. Coalescence occurs when the film between the drops is not stable enough and ruptures (4). The film thining based on the Reynolds model assumes planar and completely rigid surfaces, which is not the case for films stabilized by adsorption layers of finite dilational elasticity (5).

Barnes (6) and Tadros and Vincent (7) demonstrated the importance of a number of factors on emulsion properties and stability, among them the relative volume of the dispersed phase, i.e., the volume fraction, and the average size of the droplet, the bulk viscosity of each phase, and also the nature and concentration of the emulsifier. The latter must be of vital importance as there are no stable emulsions or foams known without the presence of surface-active compounds. Sometimes this becomes not immediately visible in some systems as stabilizers may be inherent in many natural emulsions or foams.

Some approaches analyzed directly the influence of the stabilizing adsorption layers and concluded that there is a dependence of the stability of an emulsion on the interfacial concentration and the sum of intermolecular interactions (8–10). Murdoch and Leng (11) pointed out the role of bulk and interfacial rheological parameters to describe these processes. This concept was further treated by several authors (12–14). A very comprehensive approach was given by Wasan and co-workers (15,16) who considered the surface shear and dilational rheology, and also some hydrodynamic parameters in their analysis of emulsion films.

In a number of experimental works evidence of the direct effect of adsorption-layer properties on the emulsion (foam) behavior has been discussed. A correlation between film rupture and dilational elasticity for a number of cationic surfactant systems has been shown by Bergeron (17) and also by Espert et al. (18). Dickinson (19) explains that the flocculation behavior of an emulsion requires a deep understanding of how different factors affect the structure and interactions of adsorbed layers, in particular of interfacial protein layers. Various differences in the flocculation are observed depending on the amount of adsorbed surface-active material during emulsification. In emulsion formation, rigid adsorption layers (due to surface-tension gradients, high elasticities) can yield smaller droplets as pointed out by Williams and Janssen (20). The effect of ionic strength and surface charge on emulsion stability has been studied recently (21–23). This might have various ways of action, such as salting out of surfactants and hence changing their surface activity, and changes in the disjoining pressure in the emulsions film. The special effects of ionic surfactants will not be further discussed in this chapter.

The most advanced summary of the importance of the adsorption layer properties on the behavior of an emulsion, i.e., its stability or breakdown, was given recently by Ivanov and Kralchevsky (24). In their review, Ivanov and Kralchevsky demonstrate the importance of the surfactant effect not only qualitatively but also give some general relationships. To evaluate the mass balance for a film under deformation they give the flux at the film surface $z = h/2$ (25):

$$\Gamma_o \frac{1}{r}\frac{\partial(rv_r)}{\partial r} - D_s \frac{\partial \Gamma_o}{\partial c_o}\frac{\partial^2 c}{\partial r^2} = \pm D \frac{\partial c}{\partial z} \tag{1}$$

where $v_r$ is the radial component of the mean mass flow, and $r$ and $z$ are cylindrical coordinates.

The three terms correspond to convection, surface, and bulk diffusion. At very small film thickness $h$ the bulk diffusion term can be neglected in respect to the surface diffusion term. However, this does not take into consideration surfactant flux from the heterogeneous phase, i.e., from inside the emulsion drops.

With respect to the rheological parameters they come to the conclusion that surface elasticity effects are superior to surface viscosity effects. This, however, applies to pure surfactant layers and may be different for pure protein or mixed surfactant/protein adsorption layers. It has been stressed also by Langevin (26), in her review on foams and emulsions, that studies on the dynamics of adsorption and dilational rheology studies for mixed systems, in particular surfactant–polymer systems, are desirable in order to understand these most common stabilizing systems.

The analysis given for the surfactant effect on the thinning rate has shown that a flux from inside the emulsion drops is much less effective than the surfactant present in the homogeneous phase. It will be shown below, however, that almost all surfactants are usually soluble in both liquid phases of an emulsion so that obviously the distribution coefficient will be the parameter which controls the efficiency of a surfactant with respect to film thinning.

In this chapter an overview is given on the possibilities for a quantitative characterization of adsorption layers at liquid/liquid interfaces. After a general introduction to the fundamental thermodynamic relationships and particular ideas on surfactant and protein adsorption, the process of adsorption-layer formation is discussed on the basis of the most frequently used methodology, the measurement of dynamic interfacial tensions. This will also include measurements of extremely low interfacial tensions and the effect of interfacial transfer between the two liquid phases.

Additional information on interfacial layers can be gained from rheological and ellipsometry experiments. There is quite a number of different experimental setups used to determine surface rheological parameters (27). New possibilities to determine surface dilational parameters arise from oscillating-drop experiments. Using axisymmetric drop shape analysis (ADSA) the change in interfacial tension becomes accessible as a function of the drop surface area when the drop deformations are considerably slow. For faster changes, i.e., higher oscillation frequencies, the oscillating-drop technique, as an analog of the pulsating-bubble method (28) has been recently developed (29). Studies of the interfacial shear rheology are described on the basis of a number of experimental methods; however, only a few of the existing techniques are suitable for investigations of liquid/liquid interfacial layers. Special emphasis is placed on torsion pendulum experiments (30).

## II. ADSORPTION ISOTHERMS

### A. Adsorption Isotherms

Interfacial layers at the interface between two immiscible liquid phases are characterized by large gradients in local properties, such as density, tensor of pressure, dielectric permittivity, and concentration of the dissolved components. The profile of the local concentration depends on properties of the dissolved substances. For substances which do not adsorb at the water/oil interface but are soluble in both phases, the concentration in the interfacial layer is between the equilibrium concentrations in the two phases (Fig. 1a).

The presence of a sharp maximum inside the interfacial layer (Fig. 1B) is characteristic of surface-active components. Surface-inactive components can even show a minimum in the local concentration (Fig. 1C). Especially important for the stabilization of emulsions is the adsorption behavior of surfactants at the water/oil interface. To describe such systems makes it necessary to know the adsorption isotherm, to establish relationships between the concentration of a component in the interfacial layer and the bulk phases, and to derive equations of state giving the interfacial tension as a function of the interfacial layer composition. The derivation of these equations is based on the assumption that in equilibrium the temperature and the chemical potentials of any component have identical values in all parts of the system.

### B. Chemical Potentials of Interfacial Layers

The chemical potentials of nonionic components within the interfacial layer $\mu_i^s$ depend on the composition of the layer and its surface tension $\gamma$. The dependence of $\mu_i^s$ on the composition of a surface layer is given by the known relation (31):

$$\mu_i^s = \mu_i^{0s}(T, P, \gamma) + RT \ln f_i^s \tag{2}$$

where $\mu_i^{0s}(T, P, \gamma)$ is the standard chemical potential of component $i$ and depends on temperature $T$, pressure $P$, and surface tension $\gamma$; $f_i$ are the activity coefficients. The standard chemical potential can be presented as a function of pressure and temperature only, if one introduces an explicit dependence of $\mu_i^{0s}(T, P, \gamma)$ on surface tension into Eq. (2) (31). This equation for the chemical potentials is the well-known Butler equation (32):

$$\mu_i^s = \mu_i^0 s + RT \ln f_i^s - \gamma\omega_i \tag{3}$$

where $\omega_i$ is the partial molar area of the $i$th component. In this equation, in contrast to Eq. (2), the standard chemical potential $\mu_i^{0s}(T, P) = \mu_i^{0s}$ is already independent of the surface tension.

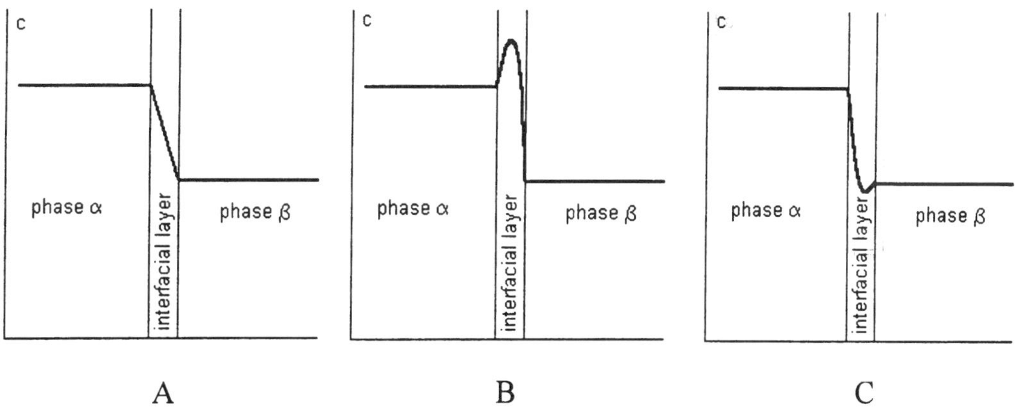


**Figure 1** Change of local concentration in the interfacial layer.

Equations of state for surface layers and adsorption isotherms can be derived by equating the expressions for the chemical potentials at the surface, Eq. (3), to those in the bulk solution:

$$\mu_i^\alpha = \mu_i^{0\alpha} + RT \ln f_i^\alpha x_i^\alpha \tag{4}$$

The standard chemical potentials in Eq. (4), $\mu_i^{0\alpha}$, depend on pressure and temperature. At equilibrium this yields:

$$\mu_i^{0s} + RT \ln f_i^s x_i^s - \gamma\omega_i = \mu_i^{0\alpha} + RT \ln f_i^\alpha x_i^\alpha \tag{5}$$

Note that in equilibrium Eq (4) and (5) are suitable for both bulk phases, water and oil. Now the standard state has to be formulated. For the solvent ($i = 0$) usually a pure component is assumed, and

$$\mu_0^{0s} - \gamma_0\omega_0 = \mu_0^{0\alpha} \tag{6}$$

For the $i$ surface-active components, infinite dilution ($x_i^\alpha \to 0$) as the standard state is experimentally easier to access than the pure state (31, 33). It should be mentioned that setting the activity coefficients equal to unity for infinite dilution is not necessarily consistent with the same unit value of the activity coefficient for pure components. Therefore, an additional normalization of the potentials of the components should be performed. Indicating parameters at infinite dilution by the subscript 0, and those in the pure state by the superscript 0, the two standard potentials are interrelated by

$$\mu_{(0)i} = \mu_i^0 + RT \ln f_{(0)i} \tag{7}$$

for both phases $\alpha$ and the surface phase s. In combination with Eq. (5) this leads to

$$\mu_{0i}^s + RT \ln x_i^s\big|_{x_i^\alpha \to 0} - \gamma_0\omega_i = \mu_{0i}^\alpha + RT \ln x_i^\alpha\big|_{x_i^\alpha \to 0} \tag{8}$$

$$\mu_{0i}^\alpha - \mu_{0i}^s = -\gamma_0\omega_i + RT \ln K_i \tag{9}$$

where $K_i = (x_i^s/x_i^\alpha)_{x_i^\alpha \to 0}$ are the distribution coefficients at infinite dilution. In a similar way it is possible to obtain from Eq. (4) an expression for the distribution coefficient of a component between two volume phases:

$$\mu_{0i}^\alpha - \mu_{0i}^\beta - RT \ln K_i^0 \tag{10}$$

where $K_i^0 = (x_i^\beta/x_i^\alpha)_{x_i^\alpha \to 0}$. For a certain concentration the equilibrium distribution of a component between the oil and water phases (phases $\alpha$ and $\beta$) depends on the activity coefficients:

$$\frac{x_i^\alpha}{x_i^\beta} = K_i^0 \frac{f_i^\beta}{f_i^\alpha} \tag{11}$$

From Eqs (5) and (6) the following relationship results:

$$\ln \frac{f_0^s x_0^s}{f_0^\alpha x_0^\alpha} = -\frac{(\gamma_0 - \gamma)\omega_0}{RT} \tag{12}$$

and from Eqs (5) and (9) one obtains:

$$\ln \frac{f_i^s x_i^s / f_{(0)i}^s}{K_i f_i^\alpha x_i^\alpha / f_{(0)i}^\alpha} = -\frac{(\gamma)_o - \gamma)\omega_i}{RT} \tag{13}$$

The additional (normalizing) activity coefficients introduced in Eq. (7) can be incorporated into the constant $K_i$ which enters Eq. (13). For further derivation it is necessary to express the surface molar fractions, $x_j^s$, in terms of their Gibbs adsorption values $\Gamma_j$. For this we introduce the degree of surface coverage, i.e., $\theta_j = \Gamma_j\omega_j$ or $\theta_j = \Gamma_j\omega_\Sigma$. Here, $\omega_\Sigma$ is the average partial molar area of all components. It is necessary to choose a proper $\omega_0$ and the average partial molar surface area $\omega_\Sigma$ for all components or states. However, the use of a realistic surface demand for $\omega_0$ (approximately 0.1 $nm^2$ for one $H_2O$ molecule) will contradict experimental data. Let us consider first that there is only one dissolved species, and assume that the surface layer and bulk behave ideally. For $\omega_0 = \omega_1$, Eqs (12) and (13) transform into the well-known equations of von Szyszkowski (34) and Langmuir (35)

$$\Pi = \gamma_0 - \gamma = \frac{RT}{\omega_1}\ln(1+K_1 x_1^\alpha) = \frac{RT}{\omega_1}\ln(1 + b_1 c_1) \tag{14}$$

$$\Gamma_1 = \frac{1}{\omega_1}\frac{b_1 c_1}{1 + b_1 c_1} \tag{15}$$

respectively, where the constant $b_1$ is the surface-to-bulk distribution coefficient related to the concentration $c$ rather than to the mole fraction $x$. In order to derive Eq. (14) we have to use a surface-layer model in which the molar surface area of the solvent in Eqs (12) and (13) is chosen equal to the molar surface area of the surfactant. This requirement can be satisfied (36–39) if one chooses the position of the dividing surface in such a way that the total adsorption of the solvent and surfactant are equal to $1/\omega_1$, i.e.,

$$\Gamma_0 + \Gamma_1 = 1/\omega_1 = \Gamma_1^\infty \tag{16}$$

For a saturated monolayer ($\Gamma_1 = 1/\omega_1$), the dividing surface defined by Eq. (16) coincides with the dividing surface of the Gibbs convention, for which $\Gamma_0 = 0$. For $\Gamma_1 = 0$, however, the convention of Eq. (16) shifts the dividing surface towards the bulk solution by the distance $\Delta = (\omega_1 c_0^\alpha)^{-1}$ as compared to the Gibbs convention (40). For large molecules, such as proteins ($\omega \gg \omega_0$), the value of $\Delta$ becomes negligibly small, and therefore for any adsorption the Lucassen-Reynders' dividing surface practically coincides with the Gibbs' dividing surface. Reasons for this choice of the dividing surface have been discussed in (31, 41–45).

For surfactant mixtures or single molecules having several adsorption states within the surface the corresponding values of $\omega_i$ differ and the definition of the dividing surface transforms into a more general relationship:

$$\sum_{i=0}^{n} \Gamma_i = 1/\omega_\Sigma \tag{17}$$

Equations defining an average molecular area demand for all surfactant components of a mixture, taking into account different $\omega_i$, have been proposed by Lucassen-Reynders (38, 39) and Joos and coworkers (33, 41). An example in which the contribution of each component to $\omega_\Sigma$ is determined by its adsorption relative to the other adsorptions (31, 38) is

$$\omega_\Sigma = \left(\sum_{i\geq 1} \Gamma_i \omega_i\right) \Big/ \left(\sum_{i\geq 1} \Gamma_i\right) \tag{18}$$

If dissolved components are ionized, and a separation of charges takes place, resulting in the formation of an electric double layer (EDL), then the electrochemical potential (46–48) has to be used instead of the chemical potential [cf. Eq. (21)]:

$$\mu_i^{\mathrm{S}} = \mu_i^{0\mathrm{S}}(T, P, \gamma) + RT \ln f_i^{\mathrm{S}} x_i^{\mathrm{S}} + z_i F \psi \tag{19}$$

where $F$ is the Faraday constant, $z_i$ is the charge of the ion, and $\psi$ is the electric potential.

Unfortunately, in this case the dependence of the standard potential on the surface tension cannot be excluded, in contrast to the derivation presented for nonionized components. In the solution bulk outside the DEL no charge separation takes place, therefore the chemical potential $\mu_i^\alpha$ for both ionized and nonionized components obeys the same equation [Eq. (4)].

## C. Mixtures of Nonionic Surfactants

Assuming ideality of the bulk solution, and using the surface coverage $\theta_i$ instead of the mole fractions in the form $x_i^{\mathrm{s}} = \theta_i = \Gamma_i \omega_i$ or $\theta_i = \Gamma_i \omega_\Sigma$, the equation of state for a nonideal surface layer can be obtained from Eq. (12):

$$\Pi = -\frac{\mathrm{RT}}{\omega_0}\left[\ln\left(1 - \sum_{i\geq 1} \theta_i\right) + \ln f_0^{\mathrm{s}}\right] \tag{20}$$

and the adsorption isotherm from Eq. (13):

$$K_i c_i = \frac{\theta_i f_i^{\mathrm{s}}}{\left(1 - \sum_{i\geq 1} \theta_i\right)^{n_i} \left(f_0^{\mathrm{s}}\right)^{n_i}} \tag{21}$$

where $c_i$ are the bulk concentrations, and $n_i = \omega_i/\omega_0$. Activity coefficients determined by intermolecular interactions (enthalpic nonideality, $f_i^{\mathrm{sH}}$) can be calculated using the regular solution theory (49–51). Lucassen-Reynders (45) has derived the following expression for the activity coefficient of any surface-layer component for the nonideal entropy of mixing:

$$\ln f_k^{\mathrm{sE}} = 1 - n_k \sum_i (\theta_i/n_i) \tag{22}$$

As the enthalpy and the entropy are active in the Gibbs free energy, this additivity results in

$$f_i^{\mathrm{s}} = f_i^{\mathrm{sH}} \cdot f_i^{\mathrm{sE}} \quad \text{or} \quad \ln \mathrm{f}_i^{\mathrm{s}} = \ln \mathrm{f}_i^{\mathrm{sH}} + \ln \mathrm{f}_i^{\mathrm{sE}} \tag{23}$$

For solutions of two surfactants the substitution of Eqs (22) and (23) into Eqs (20) and (21) leads to (31)

$$\Pi = -\frac{RT}{\omega_0}\left[\ln(1 - \theta_1 - \theta_2) + \theta_1\left(1 - \frac{1}{n_1}\right) + \theta_2\left(1 - \frac{1}{n_2}\right) + a_1\theta_i^2 + a_2\theta_2^2 + a_{12}\theta_1\theta_2\right] \tag{24}$$

$$b_i c_i = \frac{\theta_i}{(1 - \theta_1 - \theta_2)^{n_i}} \exp(-2a_i\theta_i - 2a_{12}\theta_j) \cdot \exp\left[(1 - n_i)\left(a_1\theta_i^2 + a_2\theta_2^2 + a_{12}\theta_1\theta_2\right)\right] \tag{25}$$

where $a_1$, $a_2$ and $a_{12}$ are constants; $b_1 = K_1 \exp(n_i - a_i - 1)$; $i = 1, 2$; and $j - 1, 2 (j \neq i)$. One can easily verify that all known equations describing the interfacial state of solutions of one or two surfactants involving both intermolecular interaction and nonideality of entropy [cf. (33, 36, 37, 52–72)] are limiting cases of Eqs (24) and (25). If the enthalpy of mixing is ideal, i.e., $a_1 = a_2 = a_{12} = 0$, then the following relations result (46, 57, 58):

$$\Pi = -\frac{RT}{\omega_0}\left[\ln(1 - \theta_1 - \theta_2) + \theta_1\left(1 - \frac{1}{n_1}\right) + \theta_2\left(1 - \frac{1}{n_2}\right)\right] \tag{26}$$

$$b_i c_1 = \frac{\theta_i}{(1 - \theta_1 - \theta_2)^{n_i}} \tag{27}$$

If the entropy of mixing for surface-layer components is ideal, that is, $n_1 = n_2 = 1$, then the generalized Frumkin equation of state and adsorption isotherm (36, 37, 52–55) are obtained:

$$\Pi = -\frac{RT}{\omega_0}[\ln(1-\theta_1-\theta_2) + a_1\theta_1^2 + a_2\theta_2^2 + a_{12}\theta_1\theta_2] \tag{28}$$

$$b_i c_i = \frac{\theta_i}{(1-\theta_1-\theta_2})\exp(-2a_i\theta_i - 2a_{12}\theta_j) \tag{29}$$

For the solution of a single surfactant, i.e., for $\theta_2 = 0$ and $c_2 = 0$, these last expressions transform into the usual Frumkin equations (52):

$$\Pi = -\frac{RT}{\omega_0}[\ln(1-\theta) + a\theta^2] \tag{30}$$

$$bc = \frac{\theta}{(1-\theta)}\exp(-2a\theta) \tag{31}$$

Finally, for an ideal surface layer of an $n$-component ideal bulk solution, Eqs (20) and (21) transform into a generalized Szyszkowski–Langmuir equation of state:

$$\Pi = -\frac{RT}{\omega_0}\ln\left(1 - \sum_{i\geq 1}\Theta_i\right) \tag{32}$$

and a generalized Langmuir adsorption isotherm given by Eq. (27) with all $n_i = 1$. A direct consequence of the equality of all $\omega_i$ is that the adsorption ratio of two surfactants remains constant when their concentrations are varied in the same proportion, i.e., at constant $c_1/c_2$. However, for surfactant molecules with different $\omega_i$, Eq. (13) predicts increasingly preferential adsorption of the smaller molecule with increasing surface pressure. This has been shown experimentally (46) and is conveniently illustrated theoretically for ideal surface behavior by the following equation:

$$\frac{\Gamma_1}{\Gamma_2} = \frac{b_1 c_1}{b_2 c_2}\exp\left(\frac{\Pi(\omega_2-\omega_1)}{RT}\right) \tag{33}$$

implying that a smaller molecule will expel a larger one from the surface when their total concentration is increased at constant $c_1/c_2$.

## D. Surface Layers of Surfactants Able to Change Orientation

Equations which describe reorientation of surfactant molecules within the surface layer can be derived from Eqs (20) and (21) (31, 42, 43). Reorientation results in a variation of the partial molar area $\omega_i$. If we assume that the solvent–surfactant and surfactant–surfactant intermolecular interactions do not depend on the state of surfactant molecules at the surface ($\omega_i$), it follows from the regular solution theory [cf. (49–51)] for the convention of Eq. (17) that

$$\ln f_0^{sH} = a\Gamma_\Sigma^2\omega_\Sigma^2 \tag{34}$$

$$\ln f^{sH_i} = a(1-\Gamma_\Sigma\omega_\Sigma)^2,\ i \geq 1 \tag{35}$$

where $a$ is a constant, and $\Gamma_\Sigma = \Sigma_{i\geq 1}\Gamma_i$ is the total adsorption of surfactant in all states. For our choice of the dividing surface, Eq. (22) can be transformed into

$$\ln f_i^{sE} = 1 - \omega_i\sum_{j\geq 0}\Gamma_j = 1 - n_i, \qquad i \geq 1 \tag{36}$$

$$\ln f_0^{sE} = 1 - \omega_0\sum_{j\geq 0}\Gamma_j = 0 \tag{37}$$

It is seen that the convention $\omega_0 = \omega_\Sigma$ means that the entropic contribution to it vanishes. Using Eqs (23) and (34)–(37), from Eqs (30) and (31), and with $K_i = K$ = a constant, one obtains:

$$\Pi = -\frac{RT}{\omega_\Sigma}[\ln(1-\Gamma_\Sigma\omega_\Sigma) + a(\Gamma_\Sigma\omega_\Sigma)^2] \tag{38}$$

$$bc = \frac{\Gamma_i\omega_\Sigma}{(1-\Gamma_\Sigma\omega_\Sigma)^{n_i}}\exp(-n_i)\exp[-2a\Gamma_\Sigma\omega_\Sigma + a(1-n_i)(\Gamma_\Sigma\omega_\Sigma)^2] \tag{39}$$

where $b = K\exp(-a-1)$.

If $K_1 \neq K_2$ [in this case $b_1 = b_2(\omega_1/\omega_2)^\alpha$, where $\alpha$ is a constant (42)], and for surfactant molecules which can adsorb in two states (1 and 2) with different partial molar areas $\omega_1$ and $\omega_2(\omega_1 > \omega_2)$ the adsorption isotherm [Eq. (21)] can be expressed as

$$b_2 c = \frac{\Gamma_2\omega_\Sigma}{(1-\Gamma_\Sigma\omega_\Sigma)^{\omega_2/\omega_\Sigma}} = \left(\frac{\omega_2}{\omega_1}\right)^\alpha \frac{\Gamma_1\omega_\Sigma}{(1-\Gamma_\Sigma\omega_\Sigma)^{\omega_1/\omega_\Sigma}} \tag{40}$$

where $\Gamma_\Sigma = \Gamma_1 + \Gamma_2$ is the total adsorption, and $\omega_\Sigma$ is the mean partial molar area:

$$\omega_\Sigma = \frac{\omega_2 + \omega_1\cdot\beta\cdot\exp[\Pi(\omega_2-\omega_1)/RT]}{1+\beta\cdot\exp[\Pi(\omega_1-\omega_1)/RT]} \tag{41}$$

The parameter $\beta$ in Eq. (41) considers that the adsorption activity of surfactant molecules is larger in state 1 than in state 2 (42, 43):

$$\beta = \left(\frac{\omega_1}{\omega_2}\right)^\alpha \exp\left(\frac{\omega_1-\omega_2}{\omega_\Sigma}\right) \tag{42}$$

An important relationship follows from Eq. (40) for two states 1 and 2 (31, 33, 43):

$$\frac{\Gamma_1}{\Gamma_2} = \beta\exp\left[\frac{\Pi(\omega_2-\omega_1)}{RT}\right] \tag{43}$$

The dynamic surface pressure in the framework of the two-states model is defined by

$$\Pi = -(RT/\omega_\Sigma)\ln(1 - \Gamma_\Sigma\omega_\Sigma) \qquad (44)$$

Let us consider the results obtained for $C_{10}EO_8$ for the water/hexane interface using the pendent-drop method (73). Figure 2 shows the experimental and theoretical interface-tension isotherms.

The theoretical calculations were performed with different models: the reorientation, Langmuir, and Frumkin models. The experimental results are in perfect agreement with the reorientation model and the Frumkin equations, while the Langmuir model is completely invalid. However, for the Frumkin equations a value of $a = -10.8$ for the interaction parameter is obtained which is quite unrealistic. Thus, one can conclude that the model of interacting molecules is inapplicable for $C_{10}EO_8$ at the water/oil interface. The values of geometric parameters ($\omega_1$ and $\omega_2$) for the $C_{10}EO_8$ molecule at the water/air and water/oil interfaces are rather similar to each other, while the $\alpha$ values are quite different: 3.0 at the water/air interface, and 6.5 at the water/oil interface (73). Thus, the adsorption activity of oxyethylene groups at the hexane/water interface is significantly higher than that at the air/water interface. The dependencies of the adsorptions in states 1 and 2 on the interfacial pressure for the two interfaces are shown in Fig. 3.

## E. Models of Interfacial Layer of Ionized Molecules

The Lucassen-Reynders approach considers the surface as a two-dimensional solution described by Eq.

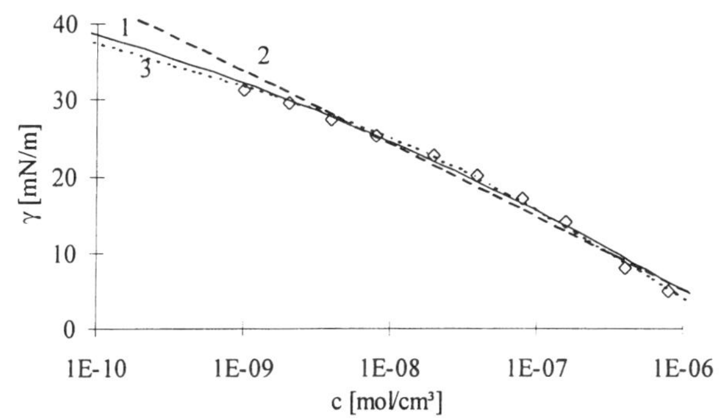


**Figure 2** Surface tension isotherm for $C_{10}EO_8$ at water/hexane interface (Γ); theoretical isotherms:
1 – reorientational model ($\omega_{min} = 4.82 \times 10^5$ m$^2$/mol, $\omega_{max} = 1.09 \times 10^6$ m$^2$/mol, $\alpha = 6.5$);
2 – Langmuir model ($\omega = 5.8 \times 10^5$ m$^2$/mol);
3 – Frumkin model ($\omega = 3.8 \times 10^5$ m$^2$/mol, a = −10.8), according to Ref. 73.

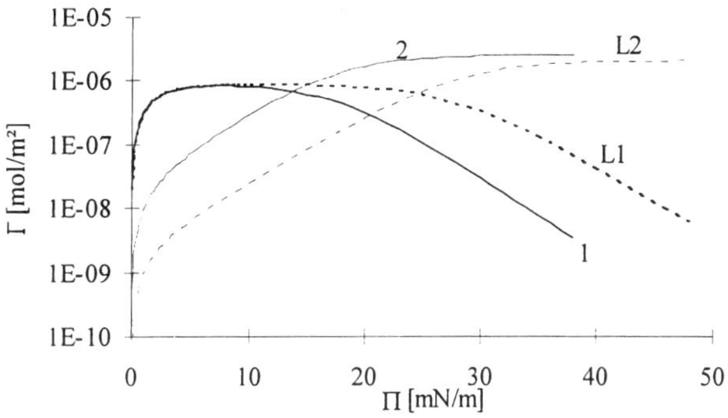


**Figure 3** Dependencies of the adsorption in the state 1 (curves 1 and L1) and 2 (curves 2 and L2) on the interfacial pressure for the two-state model at the water/air (curves 1 and 2) and water/hexane (curves L1, L2) interfaces; according to Ref. 73.

(3) and applied to an electroneutral dividing surface which contains only electroneutral combinations of ions (36, 37, 55). Any additional effects of ionization in this approach should be accounted for in the activity coefficients $f_i$. The surface equation of state is still given by Eq. (12), but the distribution of surfactant between surface and solution bulk is now obtained for electroneutral combinations of ions, say R and X for an anionic surfactant RX (where $R^-$ is the surface-active ion and $X^+$ is the counterion). This means that for both surface and bulk the average ionic product $(c_R c_X)^{-1/2}$ replaces the molar concentration $c_i$ (36–38, 64). This does not make any difference to the adsorption isotherm if there is only one salt RX, but it does when the solution contains in addition an inorganic electrolyte with the same counterion $X^+$ as the surfactant, for example, a salt XY. In such a case it is necessary to take into account the average activity of surface-active ions and all counterions in the solution.

Using the conditions which require the ionic equilibrium to exist in the bulk ($\mu^\alpha_{RX} = \mu^\alpha_R + \mu^\alpha_X$) and surface layer ($\mu^s_{RX} = \mu^s_R + \mu^s_X$), one obtains from Eqs (3) and (4) the relation:

$$\ln\frac{f^s_R f^s_X S^s_R x^s_X}{K f^\alpha_R f^\alpha_X X^\alpha_R X^\alpha_X} = -\frac{(\gamma_o - \gamma)\omega_{RX}}{RT} \qquad (45)$$

As the surface layer is electroneutral, and therefore $x^s_R = X^s_S$, then from Eqs (12) and (40) for nonideal (Frumkin) surface layers and nonideal bulk solutions of one ionic surfactant, with or without additional nonsurface active electrolyte, the adsorption isotherm follows

$$b(c_R - c_{X^+})^{1/2} f_{\pm} = \frac{\Theta}{1-\Theta} \exp(-2a\Theta) \quad (46)$$

where $f_{\pm}$ is the average activity coefficient in the solution bulk, $\Theta = \Gamma_{RX}/\Gamma_{RX\infty}$, $c_{X^+} = c_{RX} + c_{XY}$ and $c_{R^+} = c_{RX}$, and due to the surface-layer electroneutrality $\Gamma_R = \Gamma_X = \Gamma_{RX}/2$. For ideal surface layers ($a = 0$), Eq. (41) is reduced to (6, 7, 25):

$$\theta = \frac{bf_{\pm}(c_R c_X)^{1/2}}{1 + bf_{\pm}(c_R c_X)^{1/2}} \quad (47)$$

corresponding to the following $\Pi(c)$ relationship:

$$\Pi = RT\Gamma_{RX\infty} \ln\left[bf_{\pm}(c_{RX}c_{X^-})^{1/2} + 1\right] \quad (48)$$

A surface-tension isotherm assuming surface-layer nonideality was presented in Refs (36, 37 and 55). For a nonideal surface layer, the unit value which enters the right-hand side of Eq. (48) should be replaced by the activity coefficient of the solvent in the surface layer. It was shown in Refs 36, 37, 55 and 65 that Eq. (48) describes the surface- and interfacial-tension of anionic and cationic surfactant solutions quite well in a wide range of added inorganic electrolyte.

Let us consider now the case when a solution contains a mixture of two anionic (or cationic) surfactants (e.g., homologues $R_1X$ and $R_2X$ with a common counteron $X^+$) with addition of inorganic electrolyte XY. In such systems the counterion concentration $X^+$ is given by the sum of concentrations of $R_1X$, $R_2X$, and XY. After consideration of the surface-to-bulk distribution of both electroneural combinations of ions, the surface-pressure isotherm for ideal surface layers can be written in the form:

$$\Pi = RT\Gamma_{RX\infty} \ln\left[\left((b_1 f_{1\pm})^2 c_{R_1X} c_{X^-} + (b_2 f_{2\pm})^2 c_{R_2X} c_{X^-}\right)^{1/2} + 1\right] \quad (49)$$

One can easily see that Eq. 49 is the straightforward consequence of Eqs (3) and (4) for the compositions of ions within the surface layer $x_i^s = (x_{Ri}^s x_X^s)^{1/2}$ and within the solution bulk $x_i^\alpha = (x_{R_i}^\alpha x_X^\alpha)^{1/2}$, provided that the following conditions are satisfied: $x_{R_1}^s + X_{R_2}^s = x_X^s$ (surface-layer electroneutrality), $x_0^s + x_{R_1}^s + x_{R_2}^s + x_X^s = 1$ (the balance between molar portions of all components within the surface layer), and $\Gamma_0 + \Gamma_{R_1} + \Gamma_{R_2} + \Gamma_X = \Gamma_{RX\infty}$ (dividing surface chosen after Lucassen-Reynders). For nonideal surface layers, the activity coefficient should enter the right-hand side of Eq. (49), instead of unity.

Finally, very large effects on adsorption and surface pressure have been described for mixtures of anionic RX ($R^-X^+$) and cationic RY ($R^+Y^-$) surfactants in the solution. In such systems, the adsorption is represented almost completely by the equimolar composition $R^-R^+$ which has a very high surface activity without any noticeable contribution of $R^-X^+$ and $R^+Y^-$ over a large range of mixing ratios (74). Thus, one can describe the surface tension of this mixture for ideal surface layers by

$$\Pi = RT\Gamma_{R^-R^+\infty} \ln\left[b_{R^-R^+} f_{\pm}(c_{RX}c_{RY})^{1/2} + 1\right] \quad (50)$$

The adsorption equilibrium constant for the composition $R^-R^+$ can be approximated by the constants of adsorption equilibrium of $R^-X^+$ and $R^+Y^-$ in the individual solutions (65):

$$b_{R^-R^+} = (b_{RX} b_{RY})/V \quad (51)$$

where $V$ is the average molar volume of the surfactant.

The advantages of the electroneutral surface-layer model presented above can be regarded also as deficiencies, because this model cannot be used to describe the structure of the surface layer, electric potential of the surface, etc. In addition, no satisfactory treatment of the adsorption of proteins and other polyelectrolytes can be given, if the contribution from DEL into the surface pressure of the adsorption layer is neglected. As no equivalent to the Butler equation exists for the case of ionized layers, the procedure used to derive the equation of state for surface layers should be based on the Gibbs adsorption equation and a model adsorption isotherm equation. The isotherm can also be derived from the theoretical analysis of the expressions for the electrochemical potentials of ions. For the solution of a single ionic surfactant RX, with the addition of inorganic electrolyte XY, starting from Eqs. (19) and (2) one obtains the adsorption isotherm:

$$Kf_R c_R = \frac{\theta}{1-\Theta)} \exp(-2a\Theta) \exp\left(\frac{z_R F\psi}{RT}\right) \quad (52)$$

when $f_R$ is the activity coefficient of the ion R in the solution bulk, $\Theta = \Gamma_R/\Gamma_{R\infty}$. Equation (52) is similar to that derived by Davies (46, 47), and reproduced by other authors (48, 75–80).

For our system, the Gibbs adsorption equation has the form:

$$-d\gamma - \Gamma_R d\mu_R^\alpha + \Gamma_X d\mu_X^\alpha + \Gamma_Y d\mu_Y^\alpha \quad (53)$$

Clearly, the Gibbs dividing surface is used in Eq. (53), where $\Gamma_0 = 0$. The adsorption isotherm [Eq. (52)] involves another definition of the dividing surface (Lucassen-Reynders' surface with $\Gamma_0 \neq 0$), which inevitably introduces some deficiency when a solution of Eqs (52) and (53) is simultaneous used. For a fixed concentration of inorganic electrolyte in ideal bulk solutions Eq. (53) becomes [see (48)]

$$\frac{\partial \gamma}{\partial c_R} = \Gamma_R \frac{\partial \mu_R^\alpha}{\partial c_R} + \Gamma_X \frac{\partial \mu_X^\alpha}{\partial c_R} \tag{54}$$

The values of adsorption for the ions $R^-$ and $X^+$ can be calculated from the integration over the total solution volume, i.e.,

$$\Gamma_i = \int_0^\infty [c_i(y) - c_{i0}]dy \tag{55}$$

where $c_{i0}$ is the concentration of the ions outside the DEL and $y$ is the spatial coordinate. The concentration of ions within the DEL in Eq. (55) can be calculated from the Gouy–Chapman theory. Finally, the relation:

$$\frac{\partial \gamma}{\partial c_R} = RT\frac{\Gamma_R}{c_R} + \frac{2RT}{F}\left(\frac{2\varepsilon RT}{c_R + c_Y}\right)^{1/2} \times \left[\cosh\left(\frac{z_R F \psi^0}{2RT}\right) - 1\right] \tag{56}$$

follows from Eqs (54) and (55) [see (48, 75, 79)], where $\psi^0$ is the electric potential of the surface, and $\varepsilon$ is the dielectric permittivity. Introducing now the value $\Gamma_R$ [cf. adsorption isotherm, Eq. (52)] into Eq. (56) and performing the integration, one obtains the equation of state for surface layers of ionic surfactant solution (46, 48, 75):

$$\Pi = -RT\Gamma_{RX\infty}[\ln(1-\Theta) + a\Theta^2] + \frac{4RT}{F}(2\varepsilon RTc_\Sigma{}^{1/2}[\cosh\varphi - 1] \tag{57}$$

where $c_\Sigma$ is the total concentration of ions within the solution, and $\varphi = z_i F\psi^0/2RT$. It can be thus seen that the inter-ion interaction results in an additional surface-pressure jump. The electric potential is determined by the surface-charge density:

$$\sinh\varphi = z_R\Gamma_R F/(8\varepsilon RTc_\Sigma)^{1/2} \tag{58}$$

It was taken into account in the models developed in Refs 75, 78 and 80 that some portion of the counterions is bound to surface-active ions within the Stern–Helmholtz (S–H) layer, while another (unbound) portion is located within the diffuse region of the DEL. The equivalent relations of Eqs (56)–(58) in this case contain the difference $\Gamma_R - \gamma_X^s$ instead of $\Gamma_R$, where $\Gamma_X^s$ is the adsorption of counterions localized within the monolayer. It follows from the model described by Eqs (56)–(58) that if all counterions are localized within the monolayer (within the S–H layer), then $\Delta\Pi = 0$. However, in this case an additional contribution to the surface pressure $\Delta\Pi$ also exists (81). This contribution is negative, and its value is significantly lower than that given by Eq. (57) for the case of a DEL formation. Examples of a successful application of Eqs (52) and (57) to experimental $\Pi(c_{RX})$ curves are given in Refs 48 and 75.

It was shown in Refs 75 and 80 that the portion of adsorbed surface active 1:1 charged ions which becomes bound to the counterions within the S–H layer is approximately 70–90%, that is, the surface layer is almost electroneutral. These results explain why Lucassen-Reynders' theory can be successfully applied to those systems. It can be shown additionally that for compositions of ion the effect produced by the DEL vanishes. For compositions of ions Eq. (55) can be presented in the form:

$$(\Gamma_R\Gamma_X)^{1/2} = \int_0^\infty [(c_R(y)c_X(y))^{1/2} - (c_{R0}c_{X0})^{1/2}]dy \tag{59}$$

The integration domain on the right-hand side of Eq. (59) can be split into two intervals: 0 to $H$ and $H$ to $\infty$, respectively, where $H$ is the thickness of the S–H layer. For symmetric 1:1 charged electrolytes the contribution to adsorption caused by the diffuse part of the DEL vanishes, and only the contribution of the S–H layer should be considered. Certainly, for a nonsymmetric electrolyte, say a protein, one cannot exclude the contribution by the DEL in the framework of the composition approach, and in these cases the model of a charged monolayer should be preferred.

### F. Adsorption of Proteins

The adsorption isotherm [Eq. 21], and also the equation of stage [Eq. 20] with proper account for Coulomb contributions can be used as a basis to describe adsorption layers of proteins. Here, one has to keep in mind that the subscript $i$ refers to various states of the proten molecule at the surface. The problems arising from a nonideality of the surface layer, the dependence of the $K_i$ values on the state of large molecules at the interface, and the inter-ion interactions within the adsorption layer have been properly considered in Refs 44, 82 and 83.

Assuming an enthalpic contribution of the Frumkin type and taking into account the contribution of the DEL one can transform the equation of state for the surface layer [Eq. 57] into

$$\Pi = -\frac{RT}{\omega_\Sigma}\left[\ln(1-\Gamma_\Sigma\omega\Sigma) + a(\Gamma_\Sigma\omega_\Sigma)^2\right] + \frac{4RT}{F}(2\varepsilon RTc_\Sigma)^{1/2}[ch\varphi - 1] \quad (60)$$

where $\Gamma_\Sigma = \Sigma_{1\geq 1}\Gamma_i$ is the total adsorption of the protein in all states. For protein solutions at high ion concentrations the Debye length $\lambda = (\varepsilon RT/F^2 c_\Sigma)^{1/2}$ is small. This means that for protein solutions the DEL thickness can be smaller than the adsorption layer thickness. Therefore, the concentration of ions in Eqs (59) and (60) is just their concentration within the adsorption layer. It follows from Eqs (59) and (60) that for large $c_\Sigma$ the approximation $\varphi \ll 1$ can be used. After simplification one obtains the following equation of state and adsorption isotherm for nonideally charged surface layers of a protein (83):

$$\Pi = -\frac{RT}{\omega_\Sigma}\left[\ln(1-\Gamma_\Sigma\omega_\Sigma) - a_{el}\Gamma_\Sigma^2\omega_\Sigma^2\right] \quad (61)$$

$$b_1 c = \frac{\Gamma_1\omega_\Sigma}{(1-\Gamma_\Sigma\omega_\Sigma)^{\omega_1/\omega_\Sigma}}. \quad (62)$$

where $a_{el} = z^2F/\omega_\Sigma\,(8\varepsilon RTc_\Sigma)^{1/2}$ and $z$ is the number of unbound unit charges in the protein molecule. The total adsorption amount in these equations can be expressed via the adsorption in state 1:

$$\Gamma_\Sigma = \Gamma_1\sum_{i=1}^{n} i^\alpha \exp\left[-\frac{(i-1)\Pi\omega_1}{RT}\right] \quad (63)$$

Here, $\alpha$ is a constant which determines the variation in surface activity of the protein molecule in the $i$th state with respect to state 1 characterized by a minimum partial molar area $\omega_1 = \omega_{min}$, $b_i = b_1 i^\alpha$; $i$ can be either an integer or fractional, and the increment is defined by $\Delta i = \Delta\omega/\omega_1$. For $\alpha = 0$ one obtains $b_i = b_1$ = a constant, while for $\alpha > 0$ the $b_i$ increase with increasing $\omega_i$. The value of the mean partial molar area for all states, and the adsorption in any $i$th state can be expressed by

$$\omega_\Sigma = \omega_1\left(\sum_{i=1}^{n} i^{(\alpha+1)}\exp\left(-\frac{i\Pi\omega_1}{RT}\right)\right) \Bigg/ \left(\sum_{i=1}^{n} i^\alpha \exp\left(-\frac{i\Pi\omega_1}{RT}\right)\right) \quad (64)$$

$$\Gamma_i = \Gamma_\Sigma\left(i^\alpha \exp\left[-\frac{(i-1)\Pi\omega_1}{RT}\right]\right) \Bigg/ \left(\sum_{i=1}^{n} i^\alpha \exp\left[-\frac{(i-1)\Pi\omega_1}{RT}\right]\right) \quad (65)$$

respectively. The main feature of the theoretical model [Eqs (61–65)] is the self-regulation of both the state of the adsorbed molecules and the adsorption-layer thickness by the surface pressure (82–84). The mechanism of self-regulation is inherent in the Butler equation [Eq. (3)], from which all the main equations are derived.

From Eq. (65) one can calculate the portion of adsorbed molecules which exist in the state $\omega_i$. The dependencies of the distribution function $\Gamma_i/\Gamma_{max}$ on $\omega_i$ and the area per protein molecule in the maximum of the distribution function are shown in Fig. 4. It is seen that the adsorption layer of proteins is characterized by an almost complete denaturation at low surface pressure while at large surface pressures the adsorption layer is composed of molecules in a state with a minimum molecular surface area demand.

A theoretical model for concentrated protein solution was developed in Refs 83 and 85. The calculations performed according to this theory shows good agreement with the experimental data: the adsorption increases significantly while $\Pi$ remains constant. Theoretical studies of the adsorption behavior for mixtures of the globular protein HSA and nonionic surfactants were performed in (86). An anomalous surface tension increase of the mixtures at low surfactant concentrations was found experimentally and explained theoretically.

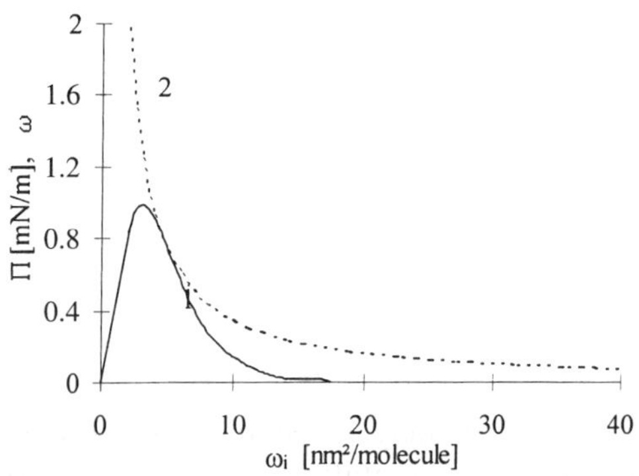


**Figure 4** Distribution of protein adsorption in various states $\Gamma_i$ with respect to the surface area $\omega_i$ covered by the protein molecule in the adsorption layer at $\Pi = 1.2$ mN/m (1), and an area per protein molecule in the maximum of the distribution function as a function of surface pressure (2); parameters used; M = 24,000 g/mol, $\omega_{max} = 40$ nm$^2$/molecule, $\omega_{min} = 2$ nm$^2$/molecule, $\Delta\omega = 1$ nm$^2$/molecule, $a_{el} = 1$ 00, and $\alpha = 1$, according to Ref. 31.

## III. DYNAMIC INTERFACIAL TENSIONS

Numerous methodologies have been developed for the measurement of surface and interfacial tensions as outlined in Refs (87–90). Methods such as the Wilhelmy plate, Du Noüy ring, and capillary-rise techniques are less suitable for liquid/gas interfaces, while a method like the bubble-pressure method is particularly applicable only to a liquid/gas system. Alternative approaches to obtaining liquid–liquid interfacial tension are generally based on drop methods. Overviews of the most frequently used drop methods are given in a monograph, where the pendant-drop (91), drop-volume (92), spinning-drop (93), and drop-pressure methods (94) have been described in detail. This section describes interfacial tension techniques, and gives reference to more details in the literature and experimental examples for a selection of liquid/liquid interfaces.

### A. Axisymmetric Drop Shape Analysis (ADSA)

In essence, the shape of a drop is determined by a combination of interfacial tension and gravity effects. Surface forces tend to make drops spherical whereas gravity tends to elongate a pendant drop or flatten a sessile drop. When gravitational and interfacial tension effects are comparable then, in principle, one can determine the interfacial tension from an analysis of the shape of the drop.

The advantages of pendant and sessile drop methods are numerous. In comparison with a method such as the Wilhelmy plate technique, only small amounts of the liquid are required. Drop-shape methods easily facilitate the study of both liquid–vapor and liquid–liquid interfacial tensions (95, 96). Also, the methods have been applied to materials ranging from organic liquids to molten metals (97) and from pure solvents to concentrated solutions. There is no limitation to the magnitude of surface or interfacial tension that can be measured. The methodology works as well at $10^3$ mJ/m$^2$ as at $10^{-3}$ mJ/m$^2$ as at $10^{-3}$ mJ/m$^2$. Since the profile of the drop may be recorded by photographs or digital image representation, it is possible to study interfacial tensions in dynamic systems, where the properties are time dependent.

In many emulsion or microemulsion systems, the interfacial tension between the oil-rich phase and the aqueous solution is very low (or ultralow), which presents considerable difficulties for many experimental methodologies. The most commonly employed approach for measuring ultralow interfacial tension is the spinning-drop technique (98). However, ADSA has also been used to study these systems and possesses a number of advantages over the spinning-drop technique: higher accuracy, more versatile environmental control (high pressure and temperature), and ability to study time-dependent effects. The problem of extremely low interfacial tensions is discussed in more detail in Sec. IV.

A typical set-up for ADSA is shown in Fig. 5.

In brief, via the CCD camera (1) with objective (2) and the frame grabber (3), an image of the shape of a drop (9) is transferred to a computer, where by using the ADSA software the coordinates of this drop are determined and compared to profiles calculated from the Gauss–Laplace equation of capillarity. The only free parameter in this equation, the interfacial tension $\gamma$, is obtained at optimum fitting of the drop-shape coordinates. The dosing system (7) allows one to change the drop volume and hence the drop surface area. This possibility is used in dilational relaxation experiments as outlined in Sec. VI.

### B. Drop Volume Tensiometry

In recent years the drop-volume method has gained a reputation as a standard technique (99, 100). Its major advantage is that it can be applied to both liquid/gas and liquid/liquid interfaces. Although its experimental conditions and theoretical description are well estab-

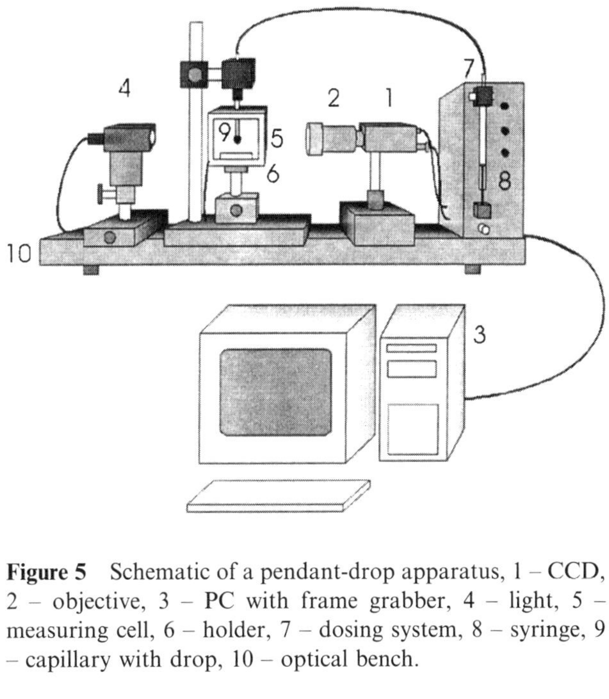


**Figure 5** Schematic of a pendant-drop apparatus, 1 – CCD, 2 – objective, 3 – PC with frame grabber, 4 – light, 5 – measuring cell, 6 – holder, 7 – dosing system, 8 – syringe, 9 – capillary with drop, 10 – optical bench.

lished for a standard range of drop formation times it has only been very recently that a number of peculiarities have been observed and discussed. Commercial instruments based on this principle are widely used in practice now, such as the automatic drop-volume tensiometer TVT1 from Lauda, Germany.

In Fig. 6 the principle of a drop volume apparatus is shown as an example. The motor controller—encoder system (3, 4) linked with a syringe (2) provides a constant and accurate dosing rate while the light barrier (7) is used to detect each detaching drop. Thus, the time in between two drop signals multiplied by the dosing rate gives the drop volume. The dosing system is linked via an interface (8) to the serial port of a PC (9).

The PC software controls complete measurement programs, i.e., drop-volume measurements for a liquid can be performed at different dosing rates. After each measurement the surface tension as a function of time is calculated and plotted as a graph. Other types of automated drop-volume instruments are designed in a similar way.

There are three different measurement modes available with the drop-volume method, which can yield different data. However, taking all peculiarities into consideration, the results obtained by the different procedures are the same. The dynamic version of the drop-volume method is the classical procedure for the measurement of interfacial tensions. This mode consists of

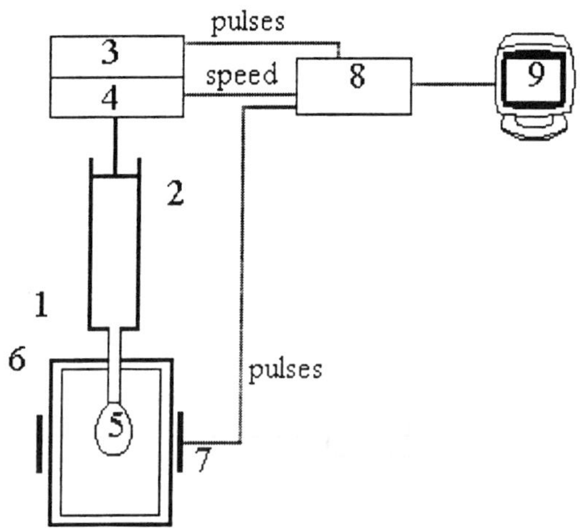


**Figure 6** Principle of an automated drop volume instrument, according to the TVT1 of Lauda, Germany; 1 – capillary, 2 – syringe, 3 and 4-motor controller–encoder system, 5 – drop, 6 – temperature control jacket, 7 – light barrier, 8 – electronic interface, 9 – IBM PC.

creating a continuous formation of drops at the tip of a capillary by means of an accurate dosing system. The interfacial tension is calculated from the average volume measured for several subsequent drops.

The diffusion-controlled adsorption kinetics for the adsorption process is given by the classical Ward and Tordai equation (101) derived more than half a century ago:

$$\Gamma = 2c_0\left(\frac{Dt}{\pi}\right)^{1/2} - 2\left(\frac{D}{\pi}\right)^{1/2}\int_0^t c(0, t-\xi)\mathrm{d}(\xi^{1/2}) \quad (66)$$

where $\Gamma$ is the dynamic adsorption, $D$ is the diffusion coefficient, $c_0$ and $c(0, t)$ are the bulk and subsurface concentrations, respectively, and $\xi$ is a dummy integration variable. The theoretical model to describe the adsorption kinetics of a surfactant at the surface of a continuously growing drop until detachment was first derived by Pierson and Whittaker (102). In analogy with the equation of Ward and Tordai, Eq. (66), the following integral equation was derived (103):

$$\Gamma(t) = 2c_o\sqrt{\frac{3Dt}{7\pi}} - \sqrt{\frac{D}{\pi}}t^{-2/3}\int_0^{3/7t^{7/3}} \frac{c\left(0, \frac{7}{3}\xi^{3/7}\right)}{\left(\frac{3}{7}t^{7/3}-\xi\right)^{1/2}}\mathrm{d}\xi \quad (67)$$

A numerical analysis of this rather complex integral equation showed that the rate of adsorption at the surface of a growing drop with a linear volume increase, as is the case in drop-volume experiments, is about one-third of that at a surface with constant area (92). From experience of adsorption kinetics studies, this approximation for the effective age of one-third of the drop formation time is sufficiently accurate to interpret dynamic interfacial tensions (104–106).

The use of an equation as complex as Eq. (67) requires a lot of numerical calculations so that approximate solutions are very favorable. The first model to describe the adsorption at the surface of a growing drop was derived by Ilkovic in 1938 (107). The boundary conditions were chosen such that the model corresponded to a mercury drop in a polarography experiment. These conditions, however, are not suitable for describing the adsorption of surfactants at a liquid-drop surface. Delahay and coworkers (108, 109) used the theory of Ilkovic and derived an approximation suitable for the description of adsorption kinetics at a growing drop. The relationship was derived only for the initial period of the adsorption process:

$$\Gamma(t) = 2c_o\sqrt{\frac{3Dt}{7\pi}} \quad (68)$$

The relationship already indicates a correlation between the rate of adsorption at a growing drop surface and a stationary interface: the adsorption at a growing drop surface is 3/7 times slower, which is close to 1/3, as discussed before.

The interfacial tension change with time at a growing drop as given by Joos and Van Uffelen (110) has the form:

$$\Delta\gamma(t) = \gamma(t) = \gamma_{eq} = \xi B\sqrt{(2n+1\alpha}\frac{(1+\alpha t)^n - (1-\zeta)}{\sqrt{(1+\alpha t)^{2n+1} - 1}} \quad (69)$$

where $\gamma_{eq}$ is the equilibrium surface tension, $\xi = (\Gamma(t)/\Gamma_e)^2$, $\zeta = (\Gamma_e - \Gamma(t))/\Gamma(t)$,

$$B = \frac{RT\Gamma_{eq}^2}{c_0}\sqrt{\frac{\pi}{4D}} \equiv \left(\frac{d\gamma(t)}{d(t_{ef}^{-1/2})}\right)_{t\to\infty} \quad (70)$$

$R$ is the gas constant, $T$ is the temperature, and $t_{ef} = t/(2\pi+1)$ is the effective adsorption time. Equation (70) follows immediately from Eq. (69) for $\alpha t \gg 1$. The initial drop has a size less than a hemisphere with the radius equal to the capillary radius $r_{cap}$ (68) so that the drop area can be given by $A_o = 1.5\pi r_{cap}^2$. The area of a drop after the break off of the liquid bridge can be described by

$$A(t) = (4\pi)^{1/3}(3V)^{2/3} + A_0 \approx 4.83V^{2/3} + A_0 \quad (71)$$

where $V$ is the drop volume. The value of $\alpha t$ in Eq. (69) can be obtained from Eq. (63):

$$\alpha t = \left(\frac{A(t)}{A_0}\right)^{1/n} - 1 \quad (72)$$

Thus, for any time $t$ the value of $\alpha$ can be calculated from Eqs (71) and (72). If we assume a Langmuir–Szyszkowski adsorption isotherm and interfacial tension equations, the parameters $\xi$ and $\zeta$ can be expressed via the values of the dynamic and equilibrium interfacial pressures, $\Pi(t)$ and $\Pi_{eq}$ as

$$\xi = \left[\frac{1 - \exp(-\Pi(t)RT\Gamma_\infty)}{1 - \exp(-\Pi_{eq}/RT\Gamma_\infty)}\right]^2 \quad (73)$$

$$\xi = \frac{\exp(-\Pi(t)/RT\Gamma_\infty) - \exp(-\Pi_{eq}/RT\Gamma_\infty)}{1 - \exp(-\Pi_{eq}/RT\Gamma_\infty} \quad (74)$$

where $\Pi_{eq} = \gamma_o - \gamma_{eq}$, $\Pi(t) = \gamma_o - \gamma(t)$, $\gamma_o$ is the interfacial tension of pure liquids, and $\Gamma_\infty$ the limiting adsorption value. The respective approximations are useful in order to interpret quantitatively experimental drop-volume results.

## C. Capillary Pressure Tensiometry

The capillary pressure tensiometry (CPT) method has been developed for measuring the interfacial tension of pure liquids and is based on the simple relationship for the capillary pressure:

$$\Delta P = \frac{2\gamma}{R} \quad (75)$$

which is a linear relationship between the capillary pressure $\Delta P$ and the drop curvature $(1/R)$. Thus, from $(\Delta P, 1/R)$ data during the growth of a drop, $\gamma$ can be calculated by fitting a linear relationship. A possible CPT set-up used by Passerone *et al.* (111) is shown in Fig. 7.

The cell is made up of two principal bodies connected by the capillary and containing the two liquids. The cell has been fully constructed in PTFE, PCTFE, and glass to improve the cleaning and filling procedures. The capillary is hand made from a Pyrex glass pipe down on a flame, cut perpendicularly to its axis and then carefully fine grounded. The drop is formed on the inner radius $\alpha$ which is typically in the range 0.25–0.35 mm.

The pressure signal is measured with a pressure transducer placed in contact with the liquid forming the drop. The variation in the pressure difference between the two phases is due to variations in the capillary pressure since all other hydrostatic contributions remain constant. The signal is sampled with a typical frequency of 25 Hz by a PC board.

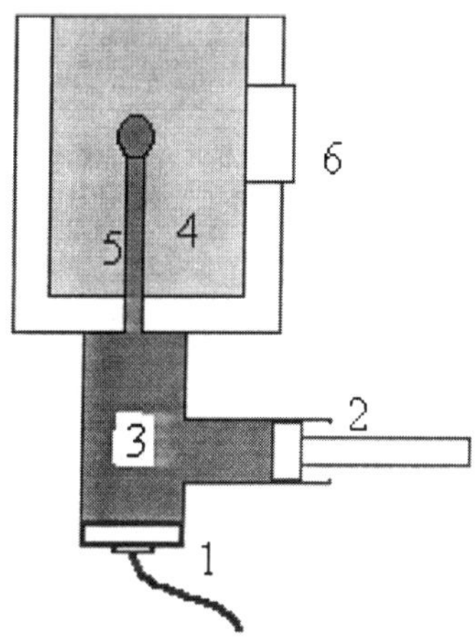


**Figure 7** Sketch of capillary pressure tensiometer for pressure derivative and expanded-drop experiments; 1 – pressure transducer, 2 – injection system, 3 – liquid, 1, 4 – liquid 2; 5 – capillary with drop, 6 – optical window.

## D. Dynamic Interfacial Tensions of Various Systems

When liquid/liquid systems are studied a number of peculiarities have to be considered, the most important of them being the solubility of surfactants in both adjacent liquid phases. There is a striking difference in studies at a liquid surface where only very few surfactants show a comparable phenomenon, the evaporation from the adsorption layer. If the surfactant is soluble in both phases but adsorbs only from one (typically from the aqueous phase) the surfactant is transferred across the interface and desorbs into the oil phase (for details see Sec. V).

In general there are three cases for the adsorption process at a liquid/liquid interface:

1. The surfactant is present in the water phase only.
2. The surfactant is present in the oil phase only.
3. The surfactant is present in both phases with an equilibrium surfactant concentration distribution.

The theoretical solution of models 1 and 2 is a generalized Ward and Tordai equation [Eq. (66)] and was first proposed by Hansen (112):

$$\Gamma(t) = \frac{2}{\sqrt{\pi}}\left[c_{1\mathrm{o}}\sqrt{D_1 t} - \left(\sqrt{D_1} + K\sqrt{D_2}\right)\int_0^{\sqrt{t}} c_1(0, t-\xi)\mathrm{d}\sqrt{\xi}\right] \tag{76}$$

Case 3 is described by the same Ward and Tordai equation; however, $D$ has to be replaced by the effective diffusion coefficient defined as $D_{\mathrm{ef}} = \sqrt{D_1} + K\sqrt{D_2}$.

As an example to demonstrate the solubility of a surfactant in water and also in the adjacent oil phase, here nonane was used, and measurements with Triton X-45 solutions were performed as follows. At the beginning the container in which the drops of the Triton solution were formed contained only pure nonane. The volume of aqueous solution was 300 ml while that of nonane was 10 ml. The experimental results are shown in Fig. 8.

The drops (500 in each run) are formed such that for each flow rate 10 drops are formed and are averaged, starting with the largest flow rate. During the first four runs the obtained $\gamma(t)$ curves change. From the fifth run on no significant changes are observed so that this state refers to the case of adsorption from both adjacent phases, i.e., the equilibrium distribution of the Triton between nonane and water has been reached. Only experiments for this case allow a quantitative interpretation as the experimental conditions can be given in the theoretical model.

Experimental results for solutions of other Tritons have been reported in Ref 113. From these studies it was concluded that the distribution coefficient for Triton X-45 is significantly higher than for Triton X-405. To visualize the significant differences in dynamic surface tensions measured for the three cases discussed above, the results of experiments with Triton X-45 are reported in Fig. 9. It is obvious that for case 3 the adsorption process is the fastest as adsorption takes

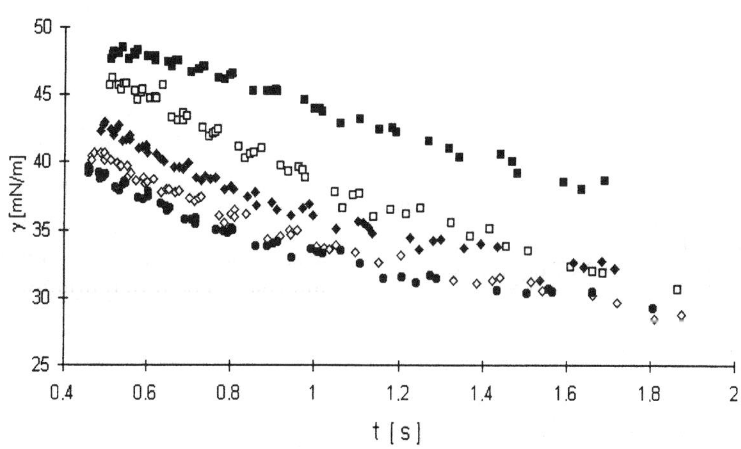


**Figure 8** Dynamic interfacial tension of Triton X-45 solutions as a function of $\sqrt{t}$; $c_{\mathrm{o}} = 1.2 \times 10^{-8}$ mol/cm$^3$, 5 subsequent runs; according to Ref. 113.

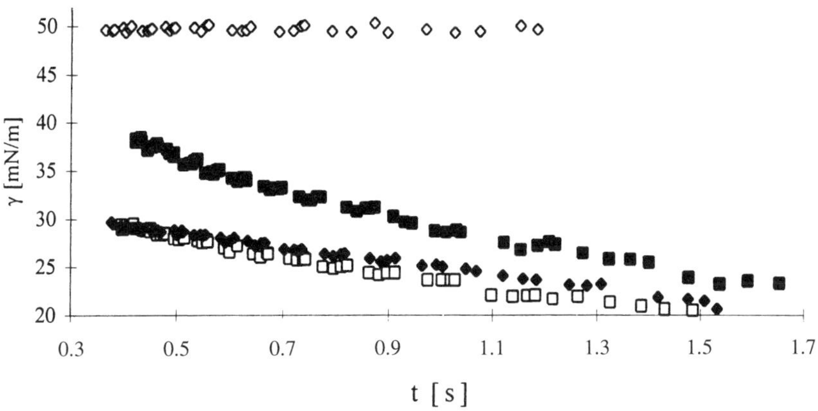


**Figure 9** Dynamic interfacial tension of a Triton X-45 solution as a function of $\sqrt{t}$; $c_o = 2.4 \times 10^{-8}$ mol/cm$^3$ for the three different cases (a) (■), (b) (◆), and (c) (□); and in absence of Triton X-45 (◇), according to Ref. 113.

place from both adjacent liquid phases. In the two other cases (1 and 2) the adsorption is much slower due to adsorption from one phase only and moreover due to the loss of adsorbed molecules via desorption into the second liquid phase. It was emphasized in Refs 113–115 that when dealing with liquid/liquid interfaces one always faces the problem that surfactant molecules are soluble in both adjacent liquids and hence adsorption from one phase generally leads to a transfer across the interface. Experiments particularly dedicated to this transfer are discussed in Sec. V.

## IV. EXTREMELY LOW INTERFACIAL TENSIONS

The measurement of ultralow interfacial tension has been of continued interest (98, 116–128) both in fundamental research and in industrial applications, particularly in surfactant-based (enhanced) oil recovery – an attempt to recover remaining oil reserves by reducing the oil/water interfacial tension through microemulsions (117, 120, 125, 129, 130). Typically, oil is recovered in a primary process by the natural energy of a reservoir. However, as much as 40–60% of the original oil can remain trapped in porous rocks due to capillary retention force. A secondary process of water injection with surfactant is therefore used to facilitate further oil displacement.

Microemulsions are homogeneous mixtures of water and oil with thermodynamically stable oil droplets of diameters ranging from 100 to 1000 Å. The interfacial tension is typically less than 0.001 mJ/m$^2$ and, as a comparison, that for an oil/water system without surfactant is about 50 mJ/m$^2$. The reduction in the interfacial tension decreases the capillary retention force significantly and enables oil droplets to deform and maneuver easily through pores in the rock medium (117).

In the remainder of this section we discuss the thermodynamic consequences and the experimental possibilities for the measurement of ultralow interfacial tensions.

### A. Thermodynamic Consequences

Surface/interfacial tension is a well-defined thermodynamic property (131). It is the energy required to display a unit new interfacial area. From classical Gibbsian thermodynamics, the various modes of energy transfer between the system and the surroundings can be formulated by a relation called a fundamental equation: the fundamental equation (131, 132) of an interface between two bulk phases is given by

$$U^A = U^A(S^A, A, N^A) \tag{77}$$

where $U, S, N,$ and $A$ are, respectively, the internal energy, entropy, total mole number, and interfacial area. The superscript A indicates the property of an interface. The differential of the fundamental equation is given by

$$\mathrm{d}U^A = \left[\frac{\partial U^A}{\partial U^A}\right]_{A,N^A} \mathrm{d}S^A + \left[\frac{\partial U^A}{\partial A}\right]_{S^A,N^A} \mathrm{d}A + \left[\frac{\partial U^A}{\partial N^A}\right]_{S^A,A} \mathrm{d}N^A \tag{78}$$

The intensive parameters, i.e., temperature $T$, interfacial tension $\gamma$, and interfacial chemical potential $\mu^A$, are now defined from the fundamental equation as

$$T = \left[\frac{\partial U^A}{\partial S^A}\right]_{A,N^A} \tag{79}$$

$$\gamma = \left[\frac{\partial U^A}{\partial A}\right]_{S^A,N^A} \tag{80}$$

and

$$\mu^A = \left[\frac{\partial U^A}{\partial N^A}\right]_{S^A,A} \tag{81}$$

Equation (78) can be written in the differential form of a property relation as

$$dU^A = TdS^A + \gamma dA + \mu^A dN^A \tag{82}$$

where the terms $TdS^A$, $\gamma dA$, and $\mu^A dN^A$ correspond respectively to the heat transfer, mechanical work, and chemical work in the system. Here, we look at the mechanical work done ($dW^A$) due to interfacial tension:

$$dW^A = \gamma dA = \left[\frac{\partial U^A}{\partial A}\right]_{S^A,N^A} dA \tag{83}$$

Obviously lowering the value of $\gamma$ can decrease significantly the mechanical work required for a given $dA$. It would be of interest to reduce $\gamma$ as much as possible if the aim is to minimize the surface work. The question then arises as to how low the interfacial tension can get.

Since $U^A$ is the internal energy of an interface, an interface must exist between two bulk phases, for a finite value of interfacial tension. The consequence of Eq. (82) is that a stable interface requires a positive value of interfacial tension, implying that energy must be increased if the interfacial area is to be increased. If this were not the case then, even in the absence of an external agent to apply energy, the interfacial area would keep increasing as this would lead to a decrease in energy – a decrease in energy is always thermodynamically favorable. This would continue until complete molecular dissolution was reached. However, we know that this cannot be true for a stable interface. The implication is that, for a thermodynamically stable system, an experimental value of (ultralow) interfacial tension is always finite and larger than zero.

## B. Experimental Possibilities

Numerous techniques have been developed to measure the interfacial tensions of a liquid/fluid interface (87, 89). Among the commonly used ones, drop-shape methods are very promising for ultralow interfacial tensions: they are based on the idea that the shape of a sessile or pendant drop is determined by the balance betwen surface/interfacial tension and an external force, such as gravity. Two such techniques are axisymmetric drop shape analysis (ADSA) and the spinning-drop technique (SDT). ADSA (133–135) determines the liquid/fluid interfacial tensions from the shape of axisymmetric menisci due to gravitational force. SDT (136–139) employs a similar strategy: instead of gravity, a known centrifugal force is applied for drop deformation. Figure 10 displays a schematic of these effects on drop shape.

Consider a liquid drop immersed in a surrounding liquid medium that is enclosed in a cylindrical glass tube rotating about its horizontal axis (Fig. 10). At zero angular velocity $\varpi$, the droplet behaves as an inverted sessile drop inside the tube. As $\varpi$ increases, the drop starts to rotate about the axis of rotation. The higher the $\varpi$, the more deformed the droplet is. At sufficiently high $\varpi$ the shape of the droplet can be

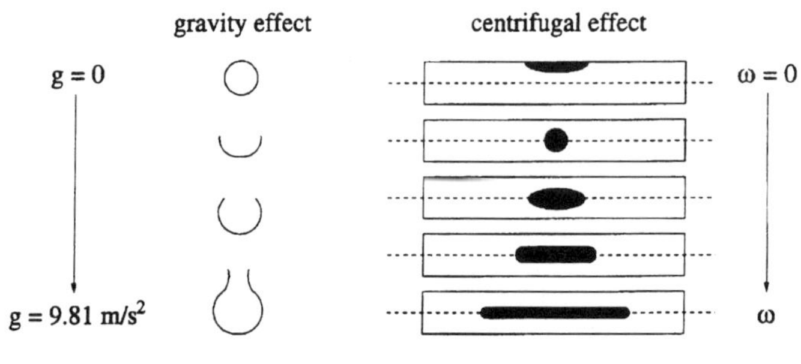


**Figure 10** A schematic of the effects of gravity and centrifugal forces on drop shape.

approximated by a cylinder with rounded ends. Vonnegut (136) has shown that, in the case of a cylindrical droplet, the interfacial tension can be calculated from

$$\gamma = \frac{\Delta\rho\varpi^2 r^3}{4} \tag{84}$$

where $\Delta\rho$ and $r$ are, respectively, the density difference between the liquid/fluid interface and the radius of the deformed cylindrical droplet at high $\varpi$. Thus, knowing $\Delta\rho$, $r$, and $\varpi$ allows the determination of interfacial tension and this is the basic principle of the spinning-drop technique. Equation (84) is often referred to as the Vonnegut equation. A detailed description of ADSA for determining the liquid/fluid interfacial tensions has been described in Sec. III.

## C. Axisymmetric Drop Shape Analysis and Spinning Drop Technique

Both ADSA and SDT have advantages and disadvantages. Here, we summarize them in the following two categories: range of applicability and experimental difficulty.

### 1. Range of Applicability

The range of applicability of ADSA is broad. It has been applied to a variety of studies on the time dependence of liquid/fluid interfacial tensions in the presence of surfactants (140–142, 196), film balance experiments with insoluble (143–145) and soluble films (146, 147), polymer melt experiments (148, 149), pressure (150) and temperature dependence (151) of interfacial tensions, drop size dependence of contact angles and line tension (152, 153), and static (154) and dynamic (155, 156) contact-angle measurements. SDT, on the other hand, is solely restricted to surface/interfacial tension measurements (136,137). It has been applied to polymer melt experiments (157–161) and to situations where the interfacial tension is as low as $10^{-4}$ mJ/m$^2$ (116, 118–120, 125–127). In contrast to the pendant/sessile drop used by ADSA, the "external" centrifugal force can be varied continuously by changing the angular velocity $\omega$ in order to minimize experimental error and to ensure that the system is sufficiently close to gyrostatic equilibrium. A disadvantage of SDT is the fact that only the equilibrium interfacial tension can be obtained.

### 2. Experimental Difficulty

As compared to the pendant-drop arrangement in ADSA, the set-up of SDT is more complex. For example, it can be a very frustrating task to fill the liquid (or bubble) in the matrix inside the cylindrical tube so that it is free of air bubbles and so that escape of volatile components during operation is prevented. This can be a serious problem when dealing with highly viscous polymers (159–162). As the cylindrical drop radius $r$ in Eq. (79) is raised to the third power, reliable experimental results require careful radius calibration. One experimental difficulty of ADSA is that the pendant-drop arrangement may not be appropriate for measurements of ultralow interfacial tension, as the interfacial tension might not withstand the weight of the droplet, i.e., the gravitational effect overpowers that of the interfacial tension. This can be overcome by an inverted sessile-drop arrangement used by Kwok *et al.* (98). Experimental examples are given in the next section.

## D. Experimental Examples

In several instance (98, 127), ADSA ultralow interfacial-tension measurements are available for the same systems for which SDT has been performed. Ultralow interfacial tension measurements by ADSA were by means of an inverted sessile-drop set-up of Aerosol OT (AOT) in NaCl/water solution in an n-heptane matrix solution.

### 1. Aerosol OT (AOT) in Aqueous Solution of NaCl Water and n-Heptane

Figure 11 displays the interfacial-tension results from ADSA, for 0.415 mM AOT in aqueous solution of 0.0513 M NaCl/water and n-heptane. The interfacial tension decreases from about 0.05 mJ/m$^2$ to an equilibrium value of 0.01 mJ/m$^2$ in 12 min. A different result is given in Fig. 12 for 0.420 mM AOT.

It can be seen that the interfacial tension reaches an equilibrium value in a much shorter time, decreasing from about 0.026 to 0.006 mJ/m to 0.006 mJ/m$^2$ in 2 min. Increasing the AOT concentration decreases both the interfacial tension value and the time required to reach equilibrium. The results given in Figs 11 and 12 agree well with those published by Aveyard *et al.* (127) using SDT. The interfacial-tension values reported by Aveyard *et al.*, estimated from their graph, are $\approx 0.01$ and $\approx 0.003$ mJ/m$^2$, respectively, from 0.415 and 0.420 mM AOT. The choice of the method depends on the specific application and is largely a matter of conveni-

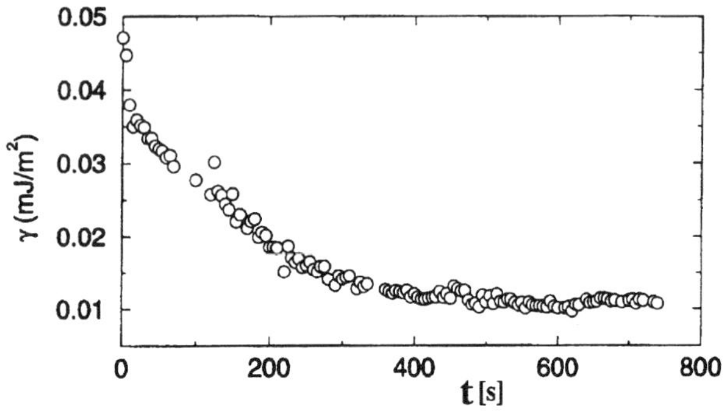


**Figure 11** Interfacial tension vs. time for 0.415 mM AOT in solution of 0.0513 M NaCl/water and n-heptane.

ence, equipment available, and the issues discussed in Sec. IV.B.

## V. SURFACTANT TRANSFER ACROSS THE INTERFACE

The study and description of adsorption processes in liquid–liquid systems deserves some specific consideration because, in most cases, the behavior of these systems is more complex in comparison with that of liquid–air systems.

The principal characteristic of such systems is the solubility of the surfactant in both phases, which is practically never negligible. This implies that the theoretical modeling of the adsorption dynamics needs to consider the transfer of surfactant across the interface during the process and that any experimental study needs careful definition of the initial partition state. Moreover, in most cases the relative volumes of the bulk phases may become an important parameter influencing the adsorption dynamics (114, 115).

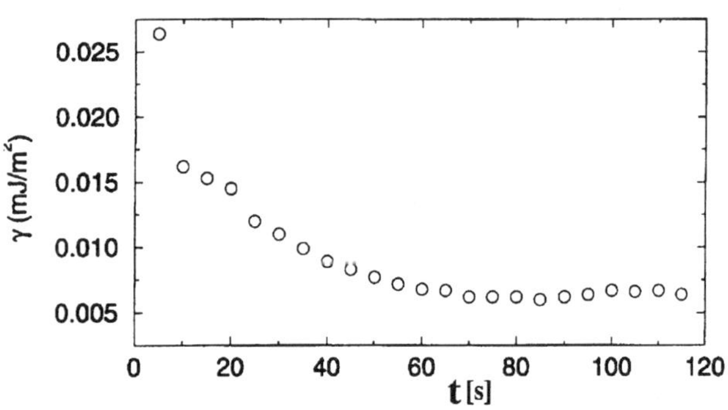


**Figure 12** Interfacial tension vs. time for 0.420 mM AOT in solution of 0.0513 M NaCl/water and n-heptane.

For these reasons knowledge of the partition properties of the systems is a mandatory requirement for characterization of the dynamic behavior of such a system and for an adequate evaluation of the equilibrium adsorption properties.

These properties can be characterized by the distribution coefficient $K_i$ of each adsorbing component $i$. This parameter can be defined as the ratio between the equilibrium concentrations of the surfactant in the two phases, and can be expressed in terms of basic thermodynamic parameters [cf. Eq. (10)]. In fact, considering a solute in two liquids $\alpha$ and $\beta$, under the hypothesis of dilute and ideal solutions, the ratio between the equilibrium concentrations $c_\alpha$ and $\beta$ can be written as (163):

$$K = \frac{c_\alpha}{c_\beta} = \frac{v_\alpha}{v_\beta}\exp\left(-\frac{\mu_\alpha^0 - \mu_\beta^0}{RT}\right) \tag{85}$$

where $K$ is the distribution coefficient for a one-surfactant system, $v_\alpha$ and $v_\beta$ are the molar volumes, $\mu_\alpha^0$ and $\mu_\beta^0$ are the standard chemical potentials, $R$ is the gas constant, and $T$ is the absolute temperature.

Surfactant solutions are typically very diluted, meaning that the hypothesis of the ideal solution is usually satisfied so that, at least at submicellar concentration, $K$ is independent of the concentration, and depends only on the temperature.

Although describing properties of the bulk liquids, in surfactant solutions the value of $K$ strongly influences the adsorption dynamics at the liquid–liquid interface. For example, in adsorption and diffusion processes in water–oil systems, knowledge of the $K$ value is fundamental in interpreting the experimental data (114, 115, 164, 169).

Moreover, the transfer of surfactant between the liquid phases can also have an impact on the interface stability. In fact it has been shown (170, 171) that, depending on the ratio between the diffusion coefficients in the two phases, the transfer of matter across the interface can give rise to interfacial instabilities.

### A. Measurement of the Partition Coefficient

The straightforward way for evaluating $K$ is the measurement of the bulk concentration after equilibration of the immiscible phases.

However, when the partition coefficients of monomeric surfactants have to be evaluated, the utilization of common analytical techniques is very limited—and often impossible—due to the very low values of the concentrations.

The specific surface-active property of the surfactant can be exploited to set up an indirect method for the evaluation of the concentration of a surfactant solution, based on the measurement of the surface tension. In fact, by using the $\gamma$–$c$ isotherm as a calibration curve, $c$ can be evaluated by the equilibrium.

Thus, for a surfactant in an immiscible couple—for example water and oil—$K$ can be measured according to the following methodology (172):

1. First, a $c$–$\gamma$ isotherm is obtained by measuring the equilibrium surface tension of solutions, prepared with oil-saturated water, as a function of the surfactant concentration.
2. A volume $V_w$ of aqueous solution with initial concentration $c_{0w}$ is brought into contact with a volume $V_o$ of pure oil for a time long enough (days) to warrant achievement of the partition equilibrium.
3. The equilibrium surface tension $\gamma_{eq}$ of the aqueous phase is then measured and its concentration $c_w$ is evaluated by using the $\gamma$–$c$ isotherm as a calibration curve.
4. Finally, from the surfactant mass balance, $K$ can be calculated as

$$K = \frac{V_w}{V_o}\left(\frac{c_{w0}}{c_w} - 1\right) \tag{86}$$

The error in this measurement is

$$\Delta K = \frac{V_w}{V_o}\left[\frac{(c_{w0} - c_w)}{c_w}\left(\frac{\Delta V_w}{V_w} + \frac{\Delta V_o}{V_o}\right) + \frac{c_{w0}}{c_w}\left(\frac{\Delta c_{w0}}{c_{w0}} + \frac{\Delta c_w}{c_w}\right)\right] \tag{87}$$

where $\Delta V_w$, $\Delta V_o$, and $\Delta c_{w0}$ are the errors in $V_w$, $V_o$, and $c_{w0}$, respectively, and the error in $c_w$ is given by

$$\Delta c_w = \left|\frac{\partial c_w}{\partial \gamma_{eq}}\right| \Delta \gamma_{eq} \tag{88}$$

which can be calculated by the best fit $\gamma_{eq}$–$c_w$ isotherm.

The meaning and the values of the various terms in Eqs (87) and (88) have been widely discussed in Ref. 172. One of the points resulting from this discussion is that, in order to minimize this error, some experimental parameters, such as the liquid volumes and the concentration range, must be suitably chosen.

The values of $K$ for some surfactants in a water–hexane system are reported in Table 1. As shown, it is easy to evaluate $K$ with an error of the order of 10%. The surfactants listed in Table 1 belong to different classes of nonionic surfactants, and the values of $K$ show that the patitioning is never negligible.

To verify achievement of the partition equilibrium, the ratio between the surfactant concentration in the two phases can be monitored as a function of the time in which the liquids are brought into contact. As shown in Fig. 13, after some time this value reaches a plateau, indicating achievement of the partition equilibrium.

**Table 1** Measured Partition Coefficient $K$ in Water/Hexane System; All Data at 20°C

| Surfactant | K | Range of $c_w$ (mol/cm$^3$) |
|---|---|---|
| $C_8$DMPO | $0.14 \pm 0.02$ | 2e − 7 ÷ 4e − 6 |
| $C_{10}$DMPO | $1.30 \pm 0.05$ | 6 − e8 ÷ 2e − 6 |
| $C_{12}$DMPO | $7.7 \pm 0.3$ | 4e − 8 ÷ 2e − 7 |
| $C_{13}$DMPO | $34.7 \pm 0.6$ | 1e − 8 ÷ 5e − 8 |
| Triton X-405 | $0.098 \pm 0.004$ | 2e − 8 ÷ 2e − 7 |
| Triton X-100 | $0.82 \pm 0.03$ | 2e − 8 |
| $C_{10}E_5$ | $1.65 \pm 0.04$ | 5e − 9 ÷ 4e − 8 |
| $C_{10}E_8$ | $0.49 \pm 0.05$(15°C) | – |
| | $0.85 \pm 0.08$ | 4e − 8 ÷ 1e − 7 |
| | $1.5 \pm 0.1$ (25°C) | – |
| | $1.7 \pm 0.2$ (30°C) | – |
| | $2.3 \pm 0.3$ (35°C) | – |
| $C_{16}E_{20}$ | $0.14 \pm 0.02$ | 1e − 9 ÷ 2e − 9 |

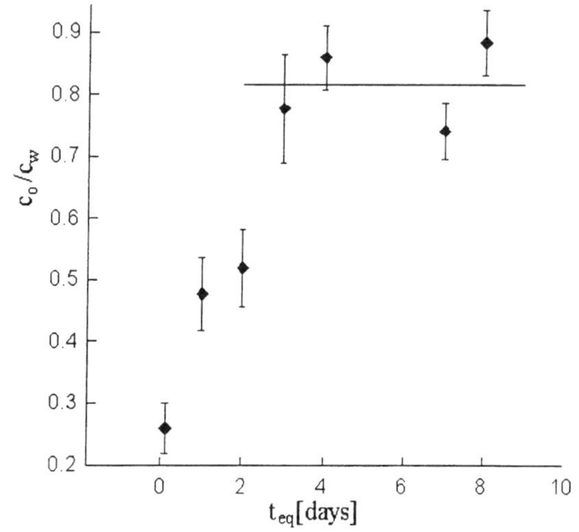


**Figure 13** Dependence of the ratio between the concentration in hexane and in water on the time of contact of the two phases.

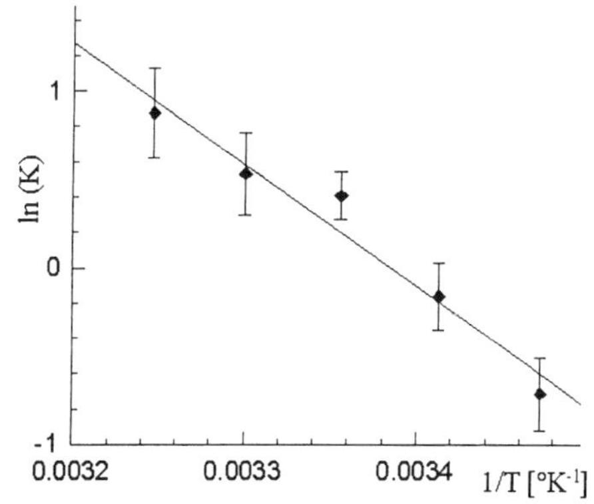


**Figure 14** Logarithm of the partition coefficients of $C_{10}E_8$ in water/hexane vs. the inverse of temperature. The slope of the straight line represents the standard enthalpy of transfer.

According to Eq. (85) the measurement of $K$ as a function of the temperature allows the difference of chemical potential of the surfactant in the two phases and of the transfer enthalpy to be evaluated.

In fact (163), by the reasonable assumption that the exponential term is much more sensitive to the temperature change than to the molar volumes, Eq. (85) leads to

$$\frac{\partial \ln K}{\partial T} = -\frac{1}{R}\frac{\partial}{\partial T}\left(\frac{\mu_{i\alpha}^0 - \mu_{i\beta}^0}{T}\right) \tag{89}$$

By basic thermodynamics arguments the chemical potential and the molar enthalpy are linked by

$$\frac{\partial}{\partial T}\frac{\mu_i}{T} = -\frac{\overline{H}_1}{T^2} \tag{90}$$

Thus, Eq. (88) can be written:

$$\frac{\partial \ln K}{\partial (1/T)} = -\frac{\Delta \overline{H}_i^0}{R} \tag{91}$$

where $\overline{H}_i^0 = \overline{H}_{i\alpha}^0 - \overline{H}_{i\beta}^0$ is the molar standard enthalpy of transfer.

Equation (91) allows the evaluation of the molar standard enthalpy of transfer. For example (173), for $C_{10}E_8$ in water–hexane (Fig. 14), a linear relationship exists between ln($K$) data and $1/T$; thus, it is possible to calculate $\Delta \overline{H}_i^0$ from the slope of the best-fit straight line, which in this case gives $\Delta \overline{H}_i^0 = 5.7\ 10^4$ J/mol.

## B. Effect of Partitioning on Dynamic Adsorption Process

As far as the adsorption dynamics at liquid–liquid interfaces is considered, the partitioning of the surfactants between the two phases has an important role.

In spite of the extensive knowledge of adsorption dynamics at liquid–vapour surfaces, only a few works have been devoted to the study of dynamic and equilibrium propeties of adsorption at liquid–liquid interfaces so that, today, there is an evident lack of data for these systems, and the theoretical approaches developed for liquid–vapour surfaces need to be specified for liquid–liquid interfaces.

For most surfactants it is possible to assume a local equilibrium between the interface and the layer just in contact with it, often called the sublayer. In this case, adsorption is controlled by diffusion since the adsorption $\Gamma$ varies according to the net diffusion flux. Thus, by considering a plane interface betwen two semi-infinites liquid phases 1 and 2, characterized respectively by the diffusion coefficients $D_1$ and $D_2$, it is

$$\frac{d\Gamma}{dt} = -D_1\left.\frac{\partial c}{\partial x}\right|_{x=0} + D_2\left.\frac{\partial c}{\partial x}\right|_{x=0^-} \tag{92}$$

where the $x$ unit vector is taken as perpendicular to the interface and directed towards the liquid 1, and the interface is located in $x = 0$. Owing to the local equilibrium condition, for $t > 0$, the sublayer concentrations $c_{10}(t) = c(0^+, t)$ and $c_{20}(t) = c(0^- t)$ are always at partition equilibrium:

$$c_{20}(t) = Kc_{10}(t) \tag{93}$$

The classical problem of adsorption dynamics is the prediction of the evolution of $\Gamma(t)$ for a "freshly" formed interface—i.e., with $\Gamma(0) = 0$—between two liquids with initial surfactant concentrations:

$$\begin{aligned} c(x,0) &= c_1^\infty \text{ for } x > 0 \\ c(x,0) &= c_2^\infty \text{ for } x < 0 \end{aligned} \tag{94}$$

By solving Eq. (92) with the boundary condition (Eq. (93)] and initial conditions [Eq. (94)], one obtains:

$$\begin{aligned} \Gamma(t) = 2\left[c_1^\infty\sqrt{\frac{D_1 t}{\pi}} + c_2^\infty\sqrt{\frac{D_2 t}{\pi}}\right] \\ - \frac{1}{\sqrt{\pi}}\int_0^t \left[\sqrt{D_1} + K\sqrt{D_2}\right]\frac{c_{01}(\chi)}{\sqrt{t-\chi}}d\chi \end{aligned} \tag{95}$$

which is a generalization of the Ward–Tordai equation (174, 175) derived for a monophasic system.

Owing to the local equilibrium condition, the equilibrium isotherm can be used at any $t$ to describe the relationship beween $\Gamma$ and $c_{01}$, in order to solve Eq. (95).

This straightforward generalization of the monophasic approach to the study of liquid–liquid adsorption dynamics is only possible by assuming local equilibrium conditions. In fact, only in that case are we allowed to use the relationship [Eq. (92)] between the two sublayer concentrations. At present, no theories exist for the description of liquid–liquid adsorption dynamics when the local equilibrium condition is not satisfied. For liquid–vapor systems, some models, often called mixed adsorption dynamics, are available to describe this situation. However, the specification of these models for liquid–liquid systems poses severe problems. Luckily, the local equilibrium condition – and then the diffusion-controlled approach – is suitable for describing adsorption dynamics for most nonionic surfactants. A description of the adsorption dynamics for liquid–liquid systems, considering the presence of energetic adsorption barriers at the two sides of the interface, has been given in Ref. 165.

The models of adsorption from semi-infinite bulk phases predict in any case a monotonic relaxation of the interfacial tension even in the presence of transfer of matter into the second phase. In particular, if the initial bulk concentrations are at partition equilibrium the adsorption asymptotically reaches its equilibrium value, otherwise the system achieves a stationary state.

For some applications, however, the assumption of semi-infinite bulks is not realistic. This can be, for example, the case of the bubbling of drops of surfactant solution in a liquid, which requires one to study the adsorption dynamics in finite volumes.

The effect on adsorption dynamics of the transfer across the interface is particularly remarkable when systems of limited volume are considered which are initially far from the partitioning equilibrium. A first theoretical approach to this problem has been given in Ref. 169, where stirred bulks are considered.

More recently (114, 115, 176), the adsorption dynamics of $C_{13}$ dimethyl phosphine oxide ($C_{13}$DMPO), $C_{12}$DMPO, and $C_{10}$DMPO at freshly formed water–hexane interfaces has been investigated as a function of the initial partition conditions and of the relative volumes between the two liquids. As shown in Table 1 some of these surfactants have large values for the partition coefficients, which enhances the influence of the transfer.

At first, a drop of surfactant aqueous solution was formed in a cell filled with pure hexane, such that a ratio $Q + 10^{-3}$ existed between the volume of the drop (supplying phase) and that of the hexane (recipient phase). The dynamic interfacial tension $\gamma(t)$ was monitored by a computer-enhanced pendant-drop technique. The evolution of $\gamma$ for some initial concentrations of aqueous solution is shown in Fig. 15A–C for $C_{13}$DMPO, $C_{12}$DMPO, and $C_{10}$DMPO, respectively.

Owing to the limited amount of surfactant in the drop, the interfacial tension passes through a minimum when the net number of molecules adsorbing at the interface from the inner phase equates with the net number of molecules desorbing in the external phase. It is important to notice that, in these dynamic conditions, the interfacial tension can reach values which are well below the equilibrium values, which can be relevant for some technological processes such as the control of droplet size or emulsification.

Similar experiments have been run by forming a drop of hexane inside the cell filled with the surfactant aqueous solution, in order to obtain a volume ratio $Q = 1000$ between the supplying and recipient phases. The measured dynamic interfacial tensions for this kind of experiment are shown in Fig 16A–C for $C_{13}$DMPO, $C_{12}$DMPO, and $C_{10}$DMPO, respectively. In this configuration the interfacial tension minima disappear since the internal phase is rapidly saturated and a monotonic relaxation behavior is observed.

A diffusion-controlled model can be applied to describe these experiments, in which a spherical drop of radius $R_1$ is considered embedded in a spherical shell of radius $R_2$ representing the external phase. The volume ratio $Q$ can be adjusted by varying the $R_1/R_2$ ratio.

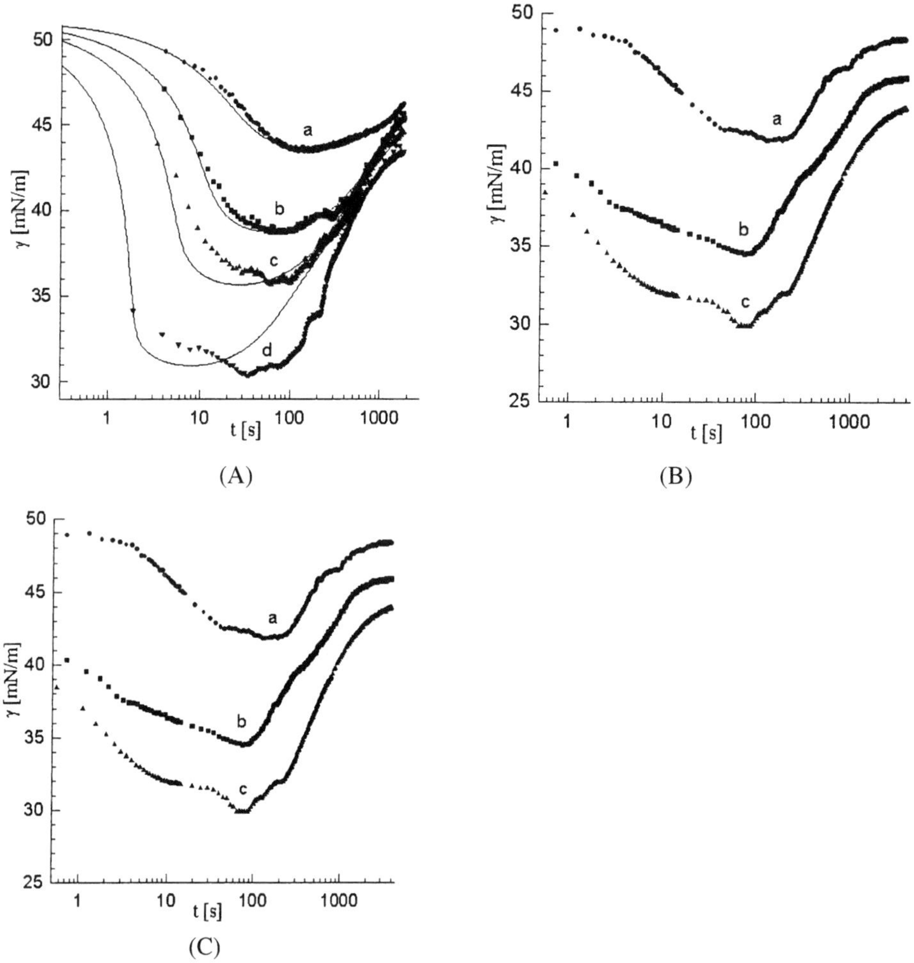


**Figure 15** Dynamic interfacial tension for the adsorption with transfer of surfactant of $C_nDMPO$ at a water/hexane interface. The phase supplying the surfactant is a drop of aqueous solution formed in hexane initially free from surfactant. The water/oil volume ratio is $Q = 10^{-3}$. The solid curves are calculated from the model. The given concentrations are the initial values in water; (A) $C_{13}DMPO$: $C_0 = 1 \times 10^{-8}$ (a), $2 \times 10^{-8}$ (b), $3 \times 10^{-8}$ (c), $5 \times 10^{-8}$ mol/cm$^3$ (d); (B) $C_{12}DMPO$: $C_0 = 2 \times 10^{-8}$ (a), $5 \times 10^{-8}$ (b), $8 \times 10^{-8}$ mol/cm$^3$ (c); (C) $C_{10}DMPO$: $C_0 = 2 \times 10^{-7}$ (a), $3 \times 10^{-7}$ (b), $1 \times 10^{-6}$ mol/cm$^3$ (c).

The model is characterized by the following set of equations:

$$\frac{d\Gamma}{dt} = -D_1 \left.\frac{\partial c}{\partial r}\right|_{r=R_1^-} + D_2 \left.\frac{\partial c}{\partial r}\right|_{r=R_1^+} \tag{91}$$

where $c = c(r, t)$ is the surfactant concentration at time $t$ and at distance $r$ from the origin of the coordinates. This equation is equivalent to Eq. (92) and has to be used as a boundary condition at the interface $r = R_1$ for the diffusion problem in the bulk phases described by the Fick equations:

$$\begin{aligned} \frac{\partial c}{\partial t} &= D_1 \left( \frac{\partial^2 c}{\partial r^2} + \frac{2}{r}\frac{\partial c}{\partial r} \right) \quad \text{for } 0 < r < R_1 \\ \frac{\partial c}{\partial t} &= D_2 \left( \frac{\partial^2 c}{\partial r^2} + \frac{2}{r}\frac{\partial c}{\partial r} \right) \quad \text{for } R_1 < r < R_2 \end{aligned} \tag{96}$$

The initial conditions are:

$$\begin{aligned} c(r, 0) &= c_0 \text{ for } 0 < r < R_1 \\ c(r, 0) &= 0 \text{ for } R_1 < r < R_2 \end{aligned} \tag{97}$$

or exactly the opposite, when the surfactant is initially contained in the external phase. Another boundary

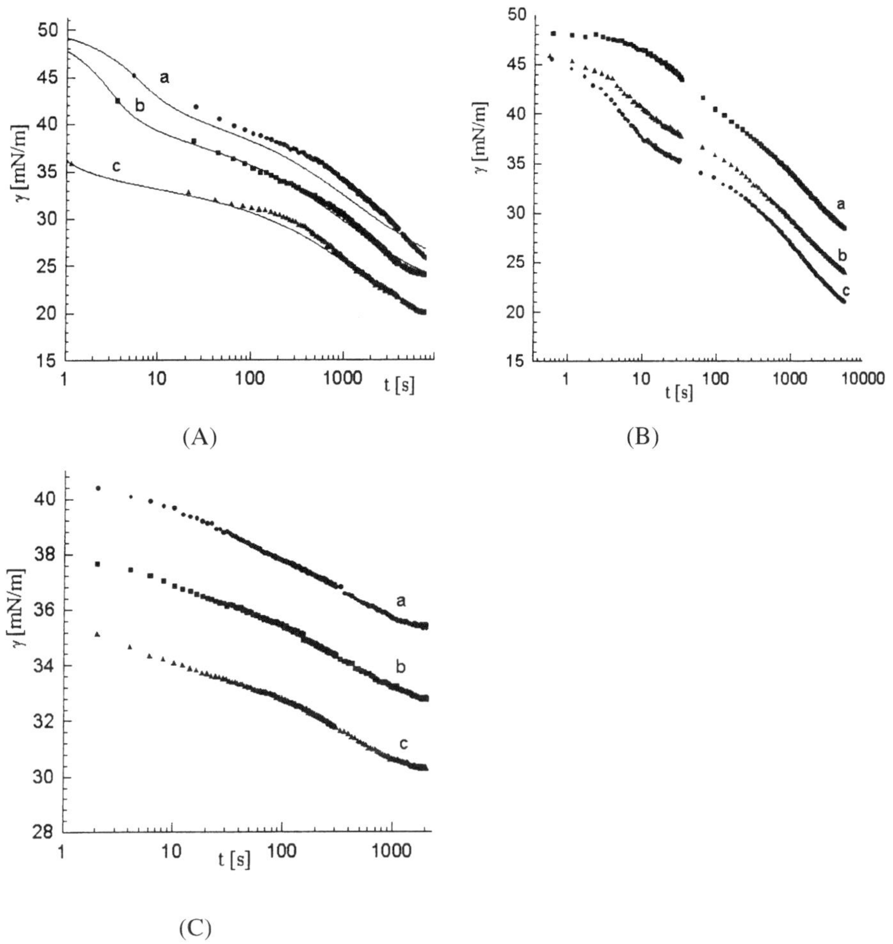


**Figure 16** Dynamic surface tension during the adsorption with transfer of surfactant of $C_nDMPO$ at a water/hexane interface. A drop of hexane initially free from surfactant is formed in the aqueous solution containing the surfactant. The water/oil volume ratio is $Q = 1000$. The solid curves are calculated from the model. The given concentrations are the initial values in water; (A) $C_{13}DMPO$: $C_0 = 1.5 \times 10^{-8}$ (a), $2.3 \times 10^{-8}$ (b), $5.3 \times 10^{-8}$ mol/cm$^3$ (c); (B) $C_{12}DMPO$: $C_0 = 1\ 10^{-8}$ (a) $2 \times 10^{-8}$ (b), $3 \times 10^{-8}$ mol/cm$^3$ (c), (C) $C_{10}DMPO$: $C_0 = 3 \times 10^{-8}$ (a), $5 \times 10^{-8}$ (b), $8 \times 10^{-8}$ mol/cm$^3$ (c).

condition is needed to express the closure of the system:

$$\left.\frac{\partial c}{\partial r}\right|_{r=R_2^-} = 0 \tag{98}$$

Finally, since the interface is considered at local equilibrium with both the adjacent phases, the two boundary concentrations are assumed to be at partition equilibrium:

$$c(R_1^-, t) = k_P c(R_1^+, t)) \tag{99}$$

and the equilibrium relation holds between the boundary concentration $c(R_1^-, t)$ and the adsorption $\Gamma$:

$$\Gamma = \Gamma(c(R_1^-, t)) \tag{100}$$

This set of equations can be solved according to the finite difference scheme given in Ref. 177, by using the values of the isotherm parameters obtained by equilibrium measurements and the values of $K$ reported in Table 1. The model describes the general features observed for the systems and predicts the appearance of the minima in $\gamma(t)$ when $Q < 1$.

Moreover, as shown in Figs 15a and 16a, the calculated $\gamma(t)$ agrees well with the measured dynamic interfacial tension, in paticular at the lower concentrations. For larger concentrations, the deviation increases. It is possible that, at these concentrations, the adoption of a spherical symmetry for the model is no longer adequate, as the drop deformation is no longer negligible owing to the low values reached by $\gamma$.

## VI. INTERFACIAL DILATIONAL RHEOLOGY

Dilational rheological experiments are based on area changes by keeping the shape of the interface constant. Models for the exchange of matter, which sets in after a compression or expansion of the interface, are generally applicable to both harmonic and transient types of relaxations (178). Stress-relaxation experiments may yield results different from those obtained from measurements on small disturbances as the composition of the surface layer can vary (179). Overviews on experimental and theoretical aspects of dilational rheology were given recently in Refs 180–182.

The damping of capillary waves at interfaces is the classic version of all dilational relaxation methods at interfaces. The response of the system is measured in terms of a relative damping of the propagated wave (183, 184). Recent work was focused on modifications of the theoretical background for this technique as well as on experimental improvements (185–187). One of the more recently developed methods to investigate surface relaxations of adsorption layers due to harmonic disturbances is the oscillating-bubble method. The technique involves the generation of radial oscillations of a gas. The theory of pulsating bubbles in surfactant solutions has been further developed (188, 189). Another group of measurements suitable for studying the dilational rheology of interfacial layers are stress-relaxation experiments performed by Joos and coworkers (190–192). All these methods are applicable for studies at liquid/gas interfaces but less suitable for liquid/liquid interfaces. For example, the capillary wave technique can be in principle applied to a water/oil interface; however, the experiment is connected with a number of problems, mainly the huge demand in highly purified oil (193, 194). Also, the overflowing cylinder, one of the effective stress-relaxation experiments for the water/air surface (195), has been successfully used for measurements at the water/oil interface, but again the large amount of solvent needed for an experiment restricts the application of this method.

### A. Pendant-drop Experiments

This technique has been used for relation experiments in the transient as well as harmonic perturbation mode (196–199) and is suitable also for liquid/liquid interfaces (200, 201). The principle set-up of this method has been already described in detail above as a method to investigate the dynamics of adsorption. The computer-controlled motor-driven dosing system can, however, be used to change the volume, and hence its interfacial area in different ways. The most easy area disturbances are step-wise increases or decreases, but also trapezoidal or zig-zag area change can be easily performed. Moreover, the technique allows even harmonic changes of the interfacial area; however, owing to the finite time needed for obtaining the video images only low frequencies can be handled (199, 202).

Due to the changes in the interfacial area a compression or expansion of the adsorption layers is generated which induces a relaxation process in order to re-establish its equilibrium state. By monitoring the evolution of interfacial tension with time the dilational elasticity and the relaxation mechanism can be obtained.

In Fig. 17 some typical interfacial tension changes are shown which have been obtained for a trapezoidal area change of an aqueous protein solution drop in tetradecane. The elasticity can be calculated from the initial jump of the $\gamma(t)$ dependence immediately following an expansion or compression.

The change in interfacial tension during a sinusoidal surface area change for a sunflower oil drop in a protein solution at a comparatively low oscillation frequency is shown in Fig. 18. Due to the time required for image acquisition in the ADSA experiments faster oscillations are not possible without the use of VCR (with a frame frequency of 25 Hz one can perform oscillation experiments at a maximum frequency of 1 Hz).

### B. Drop-oscillation Experiments

The principle set-up of a drop pressure method has been already described in detail above as a method

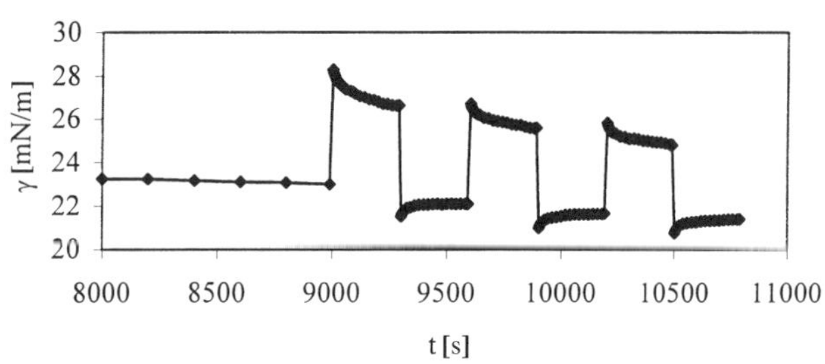


**Figure 17** Dependence of dynamic surface pressure on time for periodic trapezoidal deformations of a solution drop surface ($5 \times 10^{-7}$ M HSA) in tetradecane.

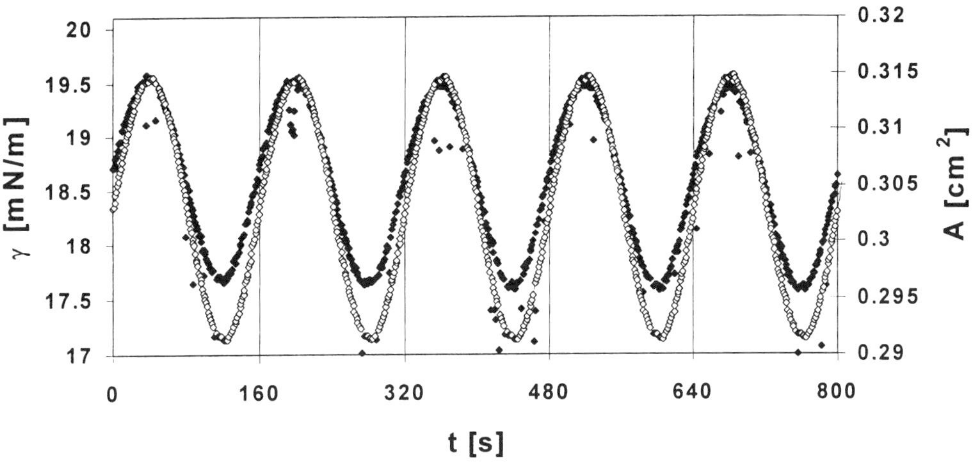


**Figure 18** Harmonic oscillation of the interfacial tension $\gamma$ and drop surface area $A$ of a sunflower oil drop immersed into an aqueous $\beta$-lactoglobulin solution ($10^{-6}$ M/l), frequency $f = 0.00625$ Hz, according to Ref. 201.

to investigate the dynamics of adsorption (203, 204). This set-up can essentially be used for transient relaxation experiments (205).

The drop-shape oscillation technique as developed by Tian *et al.* (206, 207) is another technique suitable for closing the gap in the experimental methods for liquid/liquid interfaces. This method is based on the analysis of drop-shape oscillation modes and yields again the matter-exchange mechanism and the dilational interfacial elasticity. The method is similar to the transient relaxation methods applicable only for comparatively low oscillation frequencies.

A recently developed method allows harmonic changes of the drop surface area in a wide range of frequencies. Figure 19 shows the principle set-up for the oscillating-drop method (29).

The most important components of this set-up are, in analogy to the oscillating-bubble instrument, the pressure sensor, the piezo driver (189), and the capillary. The wetting behavior of the capillary is the key problem as it can control the size of the drop significantly. For the present situation, the capillary inner surface is hydrophobized to be wetted by the oil while the head and outer surface are hydrophilic to be wetted by the aqueous solution under study. The piezo driver and pressure sensor are directly controlled via an interface by the computer. The software allows one to generate a drop oscillation of definite frequency and amplitude while the pressure change is continuously read by the pressure sensor and registered on the computer. From the frequency dependence of the pressure amplitude and phase shift between pressure and drop area change the dilational elasticity and exchange of matter of the interfacial layer are calculated. In Fig. 20 the pressure amplitude measured for a tetradecane drop in water is shown as a function of the oscillation frequency. Up to a frequency of about 100 Hz the expected constant pressure difference due to changes in the radius of curvature of the drop is obtained. At higher frequencies additional contributions from the hydrodynamics of the two liquids arise and have to be considered in the data analysis.

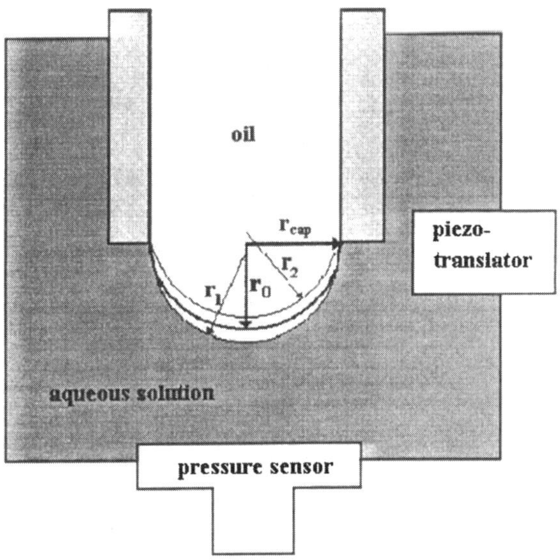


**Figure 19** Schematic of an oscillating drop set-up.

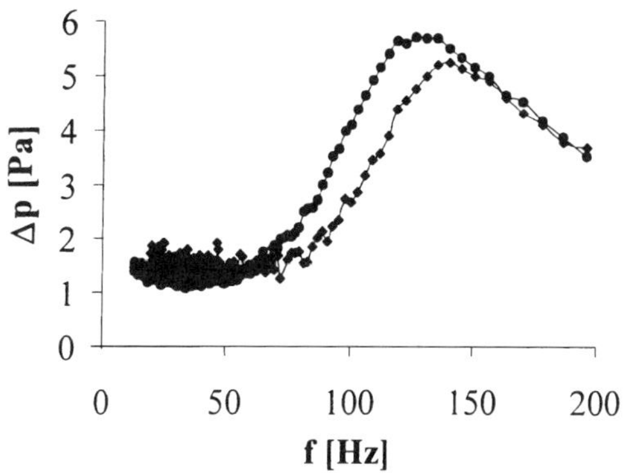


**Figure 20** Frequency dependence of the capillary pressure amplitude for an oil drop in water, two runs.

## C. Emulsion Film Relaxations

In addition to the study of interfacial rheology, studies on thin-film rheology are of particular practical interest. The first ideas were proposed by Kim *et al.* (208). The formation of a foam film can be arranged at the tip of two coaxial capillaries via two independent pumps as described in Ref. 209 (see Fig. 21). Such an assembly was used by Wege *et al.* (210) to exchange the bulk phase of a drop in order to perform penetration experiments at the drop surface. The three stages comprise the formation of an oil drop (dark gray) in water (light gray) as stage 1, and the subsequent formation and increase of a water drop inside this oil drop (stages 2 and 3). The size of the emulsion film and its thickness can be adjusted by actions of the respective pumps.

An oscillation of the foam film can be performed such that the film thickness changes with the area, or can be even kept constant by well-adjusted simultaneous oscillations of both fluid volumes. The coaxial arrangement of two capillaries seems to be an ensemble which allows an easily automated repeat formation of a foam film after an unwanted film rupture. This experimental set-up is under development now.

## D. Exchange of Matter Theory for an Oscillating Drop

The exchange of matter theory for a harmonic interfacial area perturbation:

$$A = A_0(1 + \delta \sin \varpi t) \tag{101}$$

and a diffusional transport of surfactant molecules in the bulk phases given by Fick's equation for a drop of radius $R_0$ in a second infinite liquid:

$$\frac{\partial c}{\partial t} = D_1 \frac{\partial^2 c}{\partial x^2} \text{ at } \infty < x < R_0 \text{ (outside the drop)} \tag{102}$$

$$\frac{\partial c}{\partial t} = D_2 \frac{\partial^2 c}{\partial x^2} \text{ at } 0 < x < R_0 \text{ (inside the drop)} \tag{103}$$

require appropriate initial and boundary conditions. As a useful initial condition one can assume for both cases an equilibrium state of the adsorption layer. The boundary conditions for a bubble and a foam film, however, differ from each other significantly. The boundary condition at the interface reads:

$$\frac{d\Gamma}{dt} + \Theta\Gamma = \frac{\partial c}{\partial x} \text{ at } x = R_0 \tag{104}$$

where $\Theta = d \ln A/dt$ is given by Eq. (101). If one assumes only small oscillation ($\delta < 0.1$) we obtain:

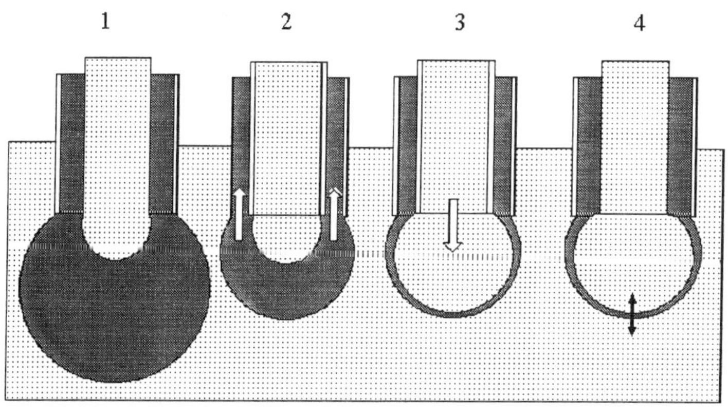


**Figure 21** Steps for the formation of an emulsion film at the tip of a capillary.

$$\Theta = \frac{d}{dt}\frac{\delta \sin \varpi}{1 + \delta \sin \varpi t} \cong \delta \varpi \cos \varpi t \quad (105)$$

The second boundary condition required for the solution of the transport problem is

$$\frac{\partial c}{\partial x} = 0 \ at \ x = 0 \quad (106)$$

To complete the mathematical problem a relationship $\Gamma(c)$, a so-called adsorption isotherm, is needed. For the simple case of bubble or drop oscillations (with the surfactant only outside the drop) a solution was derived in Ref. 189 in analogy to the capillary wave theory (183, 184).

For the emulsion film case, independent of the type of the function $\Gamma(c)$ no analytical solution is available and numerical methods have to be applied. To obtain a link to the experiment, an additional relationship, equivalent to the adsorption isotherm, is required, relating the surface concentration $\Gamma$ with the measured capillary pressure $P = 2\gamma/r$ in the bubble or film pressure of the curved foam film, which in turn is proportional to the surface pressure $\Pi$.

Transient as well as harmonic relaxation experiments give access to the dilational rheology of the studied interface or film (211). The definition of the dilational elasticity $E$ is given by the relation:

$$E = -d\gamma/d\ln\Gamma = d\Pi/d\ln\Gamma \quad (107)$$

From changes of the surface pressure $\Pi$ with time the elasticity can be obtained, while the phase shift between the generation of area oscillations and the pressure oscillation response is a measure of the exchange of matter [introduced by Lucassen as dilational viscosity, cf. (180)].

A systematic analysis of the stability of bubble and drop oscillations in open and closed cells has been performed recently and hydrodynamic limits have been given as a function of the geometry of the bubble and capillary as well as of the bulk properties of the two adjacent liquids (212, 213).

## E. Summary

An experimental technique dedicated to studies of the dynamic and mechanical properties of adsorption layers at the liquid/liquid interface is described with respect to its impact on the characterization of emulsions. A recently developed oscillating-drop technique gives access to the surface rheology of adsorption layers composed of surfactants and/or proteins. The same methodology seems to be suitable for direct investigations of single emulsion-film properties for which a relevant modification of the experimental set-up is proposed.

On the basis of theories for oscillating bubbles, new models for the exchange of matter for surfactant adsorption layers will have to be developed, taking into consideration the effect of surfactant transfer across the interface and the peculiarities of transport in thin emulsion films.

# VII. INTERFACIAL SHEAR RHEOLOGY

The interfacial shear rheological parameters are the analogs of the three-dimensional equivalents of shear elasticity and viscosity, though there are complications for interfaces where material can be exchanged between the interface and the bulk phase during the measurement.

The rate of film thinning in emulsions depends on the interfacial rheology because the flow of fluid is coupled with the flow of interfacial elements. A high interfacial shear viscosity can promote emulsion stability by retarding film thinning and hence the rate of droplet coalescence. The understanding of interfacial rheology in real emulsions is very complicated due to the fact that there are usually a large number of different surface-active components present. This makes it difficult to interpret the rheology of such systems in terms of the respective physicochemical properties of the interface.

The general framework of interfacial rheology has been dealt with systematically by several previous authors: Joly (214), Goodrich (215), Lucassen (216), Edwards *et al.* (217), and Noskov and Loglio (218). Reviews on interfacial rheology in general have been published by Warburton (219) and Miller and coworkers (27, 220). Several researchers have attempted to correlate emulsion stability with interfacial tension and interfacial rheology. The literature up to 1988 has been reviewed by Malhotra and Wasan (221). An introduction to the subject of food emulsions was published by Lucassen-Reynders (222) and more recently by Murray and Dickinson (223) and Murray (224).

## A. Methods

Different techniques for the study of shear rheology of interfacial layers have been developed over the years; however, they are mostly suited for liquid/gas interfaces. The early instruments were constructed to measure the interfacial shear viscosity under constant shear conditions. In more complex systems, nonlinear effects, shear-rate dependencies of the viscosity, and viscoelastic properties are also evident. For measurements at liquid/liquid interfaces two types of rheometers are commonly used—the deep channel and the biconical bob rheometers (both techniques will be discussed later). Warburton (225) proposed an oscillating ring surface rheometer which exploits the phenomenon of mechanical resonance. Benjamins and van Voorst Vader (226) introduced a sensitive method using a concentric ring system which is placed in the interface. The outer ring is driven at a particular frequency and small amplitude, while a torque is applied to the inner ring to keep it stationary.

The oscillatory deep-channel rheometer described by Nagarajan and Wasan (227) can be used to examine the rheological behavior of liquid/liquid interfaces. The method is based on monitoring the motion of tracer particles at an interface contained in a channel formed by two concentric rings, which is subjected to a well-defined flow field. The middle liquid/liquid interface and upper gas/liquid interface are both plane horizontal layers sandwiched between the adjacent bulk phase. The walls are stationary while the base moves. In the instrument described for dynamic studies of viscoelastic interfaces the base oscillates sinusoidally. This movement induces shear stresses in the bottom liquid that are transmitted to the interface. The interfaces are viewed from above through a microscope attached to a rotary micrometer stage which is coaxial to the cylinders.

The interfacial motion is determined from the movement of a small (100 $\mu$m) inert particle placed at the interface. This measurement is sometimes not easy since the particle must be positioned precisely. The measurement is most accurate if the reference is chosen to be the midpoint of the oscillation, where the velocity is maximum. The rheological parameters are calculated from a hydrodynamic analysis for two moving adjacent immiscible liquids incorporating interfacial rheological models. Mechanical considerations restrict the maximum possible frequency of this instrument to about 1 Hz.

Many proposed techniques rely on measuring the rotational motion of a knife-edged disk when placed in the plane of the interface. This arrangement is the two-dimensional equivalent of a Couette viscometer. The biconical disc is often suspended from a torsion wire and different constructions have been devised for monitoring and/or controlling the deflection of the disk in response to the rotation of the disk (228, 229). There is one great advantage of the Couette-type device: the technique can be easily applied to liquid/gas and liquid/liquid interfaces. The sensitivity of this technique is poorer than that of a deep-channel rheometer owing to additional drag on the disk by the bulk phases. However, the biconical disk technique is more widely used, probably because of easier experimental handling. For the biconical disk technique there are some limitations for the interpretation of the results. The shear deformation of the interface can be transferred by a constant low strain-rate experiment or as a very short and small deflection of the biconical disk. In the former experiment the deformation is transferred continuously and sometimes causes the destruction of the interfacial layer. If such a breakdown of the interfacial structure takes place a completely different interfacial layer state will be measured. The latter experimental set-up enables one to prevent the destruction of the interfacial layer. The damped oscillation behavior of the torsion pendulum provides the information on the surface shear rheology; however, for highly nonlinear systems it is difficult to interpret the experimental results.

The biconical bob oscillatory interfacial rheometer of Nagarajan *et al.* (230) is designed to measure the dynamic viscoelastic response of a liquid–liquid interface subjected to a small-amplitude oscillatory shear stress. This instrument is used to examine the rheological behavior of interfaces in the presence of surfactants, in particular of macromolecules. The rheological parameters are calculated from a hydrodynamic analysis incorporating a linear viscoelastic interfacial rheological model. The general response of this instrument has been compared with that of the oscillatory deep-channel interfacial rheometer, which is capable of similar measurements. Measurements of interfacial viscoelasticity for the same liquid–liquid system with the two rheometers are shown to be comparable. This study demonstrates the intrinsic nature and, therefore, the instrument independence of these rheological properties. Accurate measurements of interfacial shear viscoelasticity can be carried out over a wide range of systems by combining the rheometers. A similar experimental set-up was proposed by Miller *et al.* (231) and a schematic is given in Fig. 22.

This torsion pendulum rheometer developed for studies of adsorption layers at the water/air interface been modified in order to allow measurements at the water/oil interface. Instead of a ring with a sharp edge a biconical disk has been used. A detailed description of this device has been given elswhere (232). Basically, a small shear deformation of the interface of the system under study is produced by a freely oscillating, hanging titanium disk. The inter-

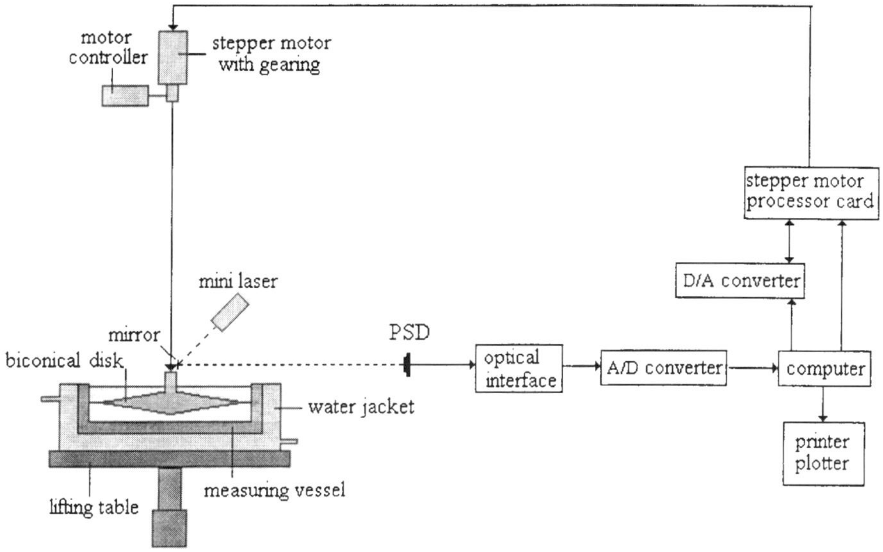


**Figure 22** Schematic view of an interfacial shear rheometer at liquid/liquid interfaces.

facial meniscus is positioned at the edge of the disk, with the interface contained in a concentric glass vessel. The interfacial shear field is generated in the gap between the edge of the disk and the wall of the measuring vessel. Using this device, interfacial shear elasticities and viscosities as a function of adsorption time can be determined by measuring the amplitude ratio and the shift of the eigenfrequency of the pendulum with respect to a surfactant-free interface. A linear viscoelastic model is used to describe the viscoelastic properties of the interfacial layer. For a tungsten wire of 100 $\mu$m diameter and 30 cm length the eigenfrequency is of the order of 0.1 Hz. All rheological experments have to be performed with very small deflection angles ($\leq 2^\circ$) so as to minimize disruption of the interfacial layers. The pendulum experiment, which lasts approximately 30 s, can be repeated every 10 min over a long perod. Unfortunately, this method does not allow large changes in the deformation frequency, as needed for a complete characterization of the viscoelasticity of an interfacial layer. The lower limitation of this instrument is determined by the fact that the bulk phases also exert a drag on the disk. Therefore, the measured effect induced by the interfacial layer must be high enough to be detected. The upper limitation is given by highly nonlinear systems, such as concentrated layers of macromolecules. For such systems it is often difficult to interpret the data obtained from the damped oscillation.

## B. Experimental Results

Interfacial rheological measurements are made on macroscopic interfaces. For the most part, the range of applied stresses, strains, and shear rates do not mirror the vigorous, nonequilibrium conditions of the practical process of emulsification. Nevertheless, interfacial rheology is an efficient and powerful detection technique, which may enhance our knowledge on formation, structure, properties, and behavior of interfacial layers formed in oil/water systems. Lakatos and Lakatos-Szabo (233) determined the interfacial shear rheology of different crude oil/water systems in a wide temperature and shear-rate range in the presence of nonionic emulsifiers (oxyethylated nonylphenols with EO numbers between 10 and 40). They observed that the interfacial viscosity, the nonNewtonian flow behavior, and the activation energy of the viscous flow drastically decrease in the presence of these surfactants. The modification of the rheological properties increase with decreasing EO number, and increasing surfactant concentration and temperature. Opawale and Burgess (234) studied the interfacial shear rheology of different liophilic nonionic surfactants of the sorbitan fatty acid ester type with the aim of selecting appropriate emulsifiers for water-in-oil emulsions under different conditions. For these investigations they used an oscillatory ring surface rheometer. The effects of bulk concentration, temperature, and the presence of salt in the aqueous phase on the interfacial properties of surfactant films were determined. The surfactants exhibited mainly viscoelastic properties. The authors conclude that interfacial association of inverse micelles and/or surfactant multilayer formation are probably responsible for the observed viscoelasticity. The addition of sodium chloride to the aqueous phase and increase in temperature influenced the viscoelastic properties. Mohammed *et al.* (235) studied the effect of demulsifiers on the interfacial rheology and emulsion stability of water-in-crude oil emulsions. The results indicated that the demulsifier used is poor at displacing the naturally occurring asphaltene surfactants from the crude oil/water interface, but if they adsorb at the interface first they prevent the formation of the stable, rigid asphaltene films.

The interaction between polymers and surfactants is one of the most important problems in enhanced oil recoverry. Interfacial shear viscosity is sensitive to surface-active species adsorbed at the oil/water interface. Therefore, interfacial shear viscosity measurements are very useful for investigating the interfacial layer formation by adsorption from mixed polymer–surfactant solutions. Cardenas-Valera and Bailey (236) examined the interfacial rheological properties of spread films of poly(ethylene oxide)/poly(methylmethacrylate) (PEO/PMMA) graft copolymers at toluene/water and toluene–n-heptane/water interfaces. The interfacial shear viscosity was determined from the damped oscillation of a torsion pendulum. The largest viscosity was exhibited by a monolayer spread at the toluene–n-heptane/water interface. Emulsions prepared with this sytem showed the lowest coalescence rate, indicating that at this interface the graft copolymer forms a coherent film which retards interdroplet film drainage. The results show that films with larger values of the interfacial rheological parameters produce a more stable emulsion owing to an increase in the mechanical strength of the interfacial film and its ability to respond to local thickness variations.

Zhang and coworkers (237, 238) studied aqueous solutions of polyacrylamide (PAAM) mixed with three different surfactants at the hexadecane/water interface, using a rotational torsion viscometer. The structure of the interfacial films were shown to be dependent on the shear rate. Based on the experimental results, a mechanism for PAAM–surfactant interactions was proposed. The interfacial viscosity decreased with ionic surfactant concentration. They supposed that the polymer–surfactant interaction model changes with surfactant concentration. For nonionic surfactants the interfacial viscosity was relatively higher and independent of the concentration. The authors did not find an indication that the interaction model changes by increased nonionic surfactant concentration.

In pharmaceutical and food technology, emulsion proteins are often used as stabilizers. The importance of interfacial shear rheological properties on protein-stabilized emulsions was reviewed by Murray and Dickinson (223) and Murray (224). The influence of covalent cross-linking with transglutaminase on the time-dependent surface shear viscosity of adsorbed milk protein films at the n-tetradecane/water interface has been investigated by Faergemand *et al.* (239). They studied the influence of sodium caseinate, $\alpha_{(S1)}$—casein, $\beta$-casein, and $\beta$-lactoglobulin. Proteins were adsorbed from $10^{-3}$ wt % aqueous solutions at pH 7, and apparent surface viscosities were recorded at 40°C in the presence of various enzyme concentrations. Results for casein systems showed a rapid enhancement in surface viscoelasticity due to enzymic cross-linking with a substantially slower development of surface shear viscosity for $\alpha_{(S1)}$-casein than for $\beta$-casein. While adsorbed $\beta$-lactoglobulin showed less relative increase in surface viscosity than the caseins, the results for $\beta$-lactoglobulin showed the presence of a substantial rate of cross-linking of the globular protein in the adsorbed state, whereas in bulk solution $\beta$-lactoglobulin was cross-linked only after partial unfolding in the presence of dithiothreitol. A maximum in shear viscosity at relatively short times following addition of a moderate dose of enzyme was attributed to formation of a highly cross-linked protein film followed by its brittle fracture. Enzymic cross-linking or protein before exposure to the oil–water interface was found to produce a slower increase in surface viscosity than enzyme addition either immediately after interface formation or to the aged protein film.

Williams and Janssen (20) studied the behavior of droplets in a simple shear flow in the presence of a protein emulsifier. The effect of two structurally diverse protein emulsifiers, $\beta$-lactoglobulin and $\beta$-casein, upon the break-up behavior of a single aqueous droplet in a Couette flow field has been studied over a wide range of protein concentrations. It was found that $\beta$-casein and low concentrations of $\beta$-lactoglobulin cause the droplets to be at least as stable as expected from conventional theories based on the equilibrium interfacial tension. In such cases the presence of the emulsifier at the deforming interface is thought to enhance the interfacial elasticity. This effect can be characterized by an effective interfacial tension, which is higher than the equilibrium value. High concentrations of $\beta$-lactoglobulin, on the other hand, have been shown to cause droplets to be less stable than would have been predicted from an equilibrium interfacial tension model. It is thought that an interfacial protein network is formed, which limits the droplet deformation and makes the droplet interface rigid with respect to tangential stresses. As a result, the critical deformation and capillary number are found to be essentially independent of the viscosity ratio. It is proposed that the interfacial structure may be probed using a combination of interfacial shear and dilational rheological measurements. From this type of analysis it may be possible to predict the break-up stability of droplets.

Ogden and Rosenthal (240) studied the influence of solid particles (tristearin crystals) on the stability of protein-stabilized emulsions. A Couette-type torsion-wire surface-shear viscometer was used to measure the apparent interfacial shear viscosity of pH 7 ($I = 0.05$ M) buffered solutions of lysozyme, sodium caseinate, and Tween-40 in contact with either n-tetradecane or purified sunflower oil. When proteins were present in the aqueous phase and tristearin crystals in the oil phase, a synergistic increase in the interfacial shear viscosity was observed. The magnitude of the increase appeared to be independent of the type of protein, but depended on the nature of the oil phase. This increase in the interfacial shear viscosity was not simply due to the presence of protein reducing the interfacial tension and thus affecting the adsorption behavior of the fat crystals. When the aqueous phase contained a small-molecule surfactant (Tween-40) instead of protein, keeping the same interfacial tension, a significantly smaller increase in the interfacial shear viscosity was observed. It therefore seems likely that when proteins are present, hydrophobic peptide residues interact with the tristearin crystals at the interface. More recently, Ogden and Rosenthal (241) studied the interaction of tristearin crystals with $\beta$-casein at the sunflower oil/water interface with the same measuring technique.

An example of shear viscosity measurements on a protein adsorption layer at a water/hexadecane interface, using the biconical disc technique (231), is shown in Fig. 23.

At first, step-by-step increases in the protein concentration results in only rather small increases in the shear viscosity. Above a certain concentration the viscosity increases very strongly. At the water/air interface, in most cases, a maximum in the concentration dependence of the shear viscosity is found, which can be discussed on the basis of conformational changes in the interfacial layer. At the interface between two

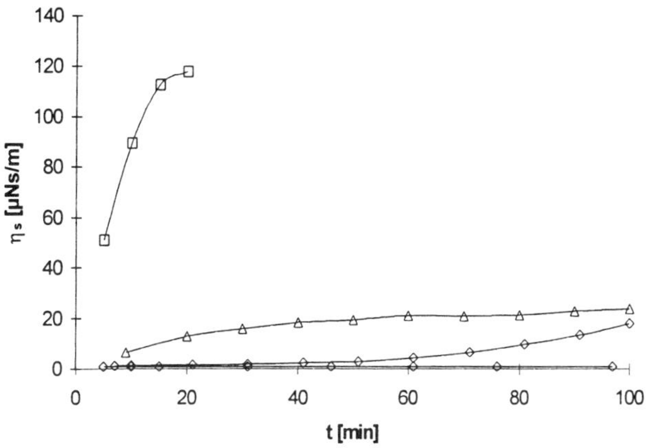


**Figure 23** Interfacial shear viscosity of ABPI (bean protein) at the water/hexadecane interface; protein concentrations: $10^{-4}\%$ (□), $2 \times 10^{-4}\%$ (Δ), $3 \times 10^{-4}\%$ ($\mu$), $4 \times 10^{-4}\%$ (Γ).

liquids the situation seems to be more complicated and qualitatively different results are obtained. The differences may be connected with the additional freedom of adsorbed protein molecules to entangle into the oil phase. In this way also, at higher concentration, an unfolding of adsorbed molecules at the water/oil interface can happen while at the water/air interface, owing to the restricted space, adsorbed molecules will have to remain in their native state as discussed by Wüstneck *et al.* (242).

## VIII. ELLIPSOMETRIC STUDIES

The properties of adsorbed layers at liquid interfaces can be determined either indirectly by thermodynamic methods or directly by means of some particular experimental techniques, such as radiotracer and ellipsometry. For adsorbed layers of synthetic polymers or biopolymers the advantages of the ellipsometry technique become evident as it yields information not only on the adsorbed amount but also on the thickness and refractive index of the layer. The theoretical background of ellipsometry with regard to layers between two bulk phases has been described in literature quite frequently (243). In brief, the principle of the method assumes that the state of polarization of a light beam is characterized by the amplitude ratio $|E_p|/|E_s|$ and the phase difference $(\delta_p - \delta_s)$ of the two components of the electric-field vector $E$. These two components $E_p$ and $E_s$ are parallel (p) and normal (s) to the plane of incidence of the beam and given by

$$E_p = |E_p|e^{i\delta_p}$$
$$E_s = |E_s|e^{i\delta_s} \tag{108}$$

Changes in the state of polarization upon reflection at an interface are given by

$$\tan\Psi = \frac{|E_p^r|/|E_s^r|}{|F_p^i|/|E_s^i|} \tag{109}$$

$$\Delta = (\delta_p^r - \delta_s^r) - (\delta_p^i - \delta_s^i) \tag{110}$$

where $\tan\Psi$ is the change in the amplitude ratio and $\Delta$ is the relative phase change. The superscripts r and i denote the reflected and incident beams. These two parameters are the data usually determined in ellipsometric experiments; however, under special conditions it is possible to estimate only one (244, 245). In terms of the overall reflection coefficients of the parallel $R_p$ and normal $R_s$ components of the beam the total effect caused by the reflection can be written as

$$\rho = \frac{R_p}{R_s} = \frac{E_p^r/E_s^r}{E_p^i/E_s^i} = \tan\Psi e^{i\Delta} \tag{111}$$

This equation is the so-called basic ellipsometric equation. It contains $R_p$ and $R_s$ which depend on the optical properties of the reflecting system, the wavelength of the light $\lambda$, the angle of incidence $\varphi_0$, and the experimentally measurable parameters $\Psi$ and $\Delta$. For the reflection at a clean interface, the $R_p$ and $R_s$ are the Fresnel coefficients (246) of the single uncovered interface. They depend only on the refractive indices of the two adjacent phases and the angle of incidence. For systems that do not absorb light the optical constants of the two bulk phases (ambient and substrate media) are usually obtained from the experimental values of $\Psi_0$ and $\Delta_0$ for the clean interface (denoted by subscript "0") via Eq. (111). For a layer-covered interface, multiple reflections and refractions take place within the layer (Fig. 24).

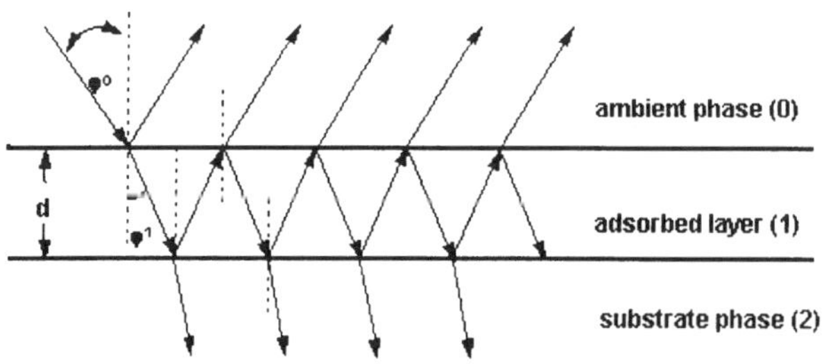


**Figure 24** Multiple reflections and refractions in an interfacial layer.

These effects are enclosed in the overall reflection coefficients as

$$R_j = \frac{r_{01j} + r_{12j}e^{-2i\beta}}{1 + r_{01j}r_{12j}e^{-2i\beta}}, j = p, s \tag{112}$$

Equation (111), in terms of the ellipsometric angles $\Delta$ and $\Psi$, then reads:

$$\rho = \tan\Psi e^{i\Delta}\frac{r_{01p} + r_{12p}e^{-2i\beta}}{1 + r_{01p}r_{12p}e^{-2i\beta}}\frac{1 + r_{01s}r_{12s}e^{-2i\beta}}{r_{01s} + r_{12s}e^{-2i\beta}} \tag{113}$$

In Eqs (112) and (113), $r_{01j}$ and $r_{12j}$ ($j = p, s$) denote the Fresnel reflections coefficients at the 0–1, 1–2 interfaces of the ambient medium (0), layer (1), and substrate medium (2) in the reflecting system; $\beta$ is the phase change of the electromagnetic wave caused by the presence of the interfacial layer.

$$\beta = \frac{2\pi d n_1 \cos\varphi_1}{\lambda} \tag{114}$$

Here, $d$ is the thickness of the layer, $n_1$ is its refractive index, and $\varphi_1$ is the angle of refraction in the layer. When a thin nonadsorbing, plane-parallel, homogeneous, and isotropic layer, with $d \ll \lambda$, is present at an interface between two phases (characterized by $\Delta_0$ and $\Psi_0$) Eq. (113) yields (246):

$$\tan\Psi e^{1\Delta} = \tan\Psi_0 e^{i\Delta_0} \times \left[1 + i\left(\frac{4\pi d\cos\varphi_0 \sin^2\varphi_0 n_2^2 M}{\lambda(n_2^2 - n_0^2)(n_0^2\sin^2\varphi_0 - n_2^2\cos^2\varphi_0)}\right)\right] \tag{115}$$

where $n_0$ and $n_2$ are the refractive indices of ambient and substrate phases, respectively, $\varphi_0$ is the angle of incidence, and $M = n_0^2 + n_2^2 - n_1^2 - n_0^2 n_2^2/n_1$. For systems with a nonabsorbing ultrathin layer the conditions of homogeneity and isotropy of the layer can be invalid (for instance, for insoluble monolayers in the state of a two-dimensional phase transition) (247, 248). In such cases, Eq. (115) can be rearranged to obtain the following relationship (which is exact up to the first-order terms in $d/\lambda$) :

$$\delta\Delta = \Delta - \Delta_0 = -\left(\frac{4\pi d}{\lambda}\right)\left(\frac{\sin\varphi_0 \tan\varphi_0 n_0}{1 - (n_0/n_2)^2\tan^2\varphi_0}\right) \times \left[\frac{n_\perp^2 - n_2^2}{n_0^2 - n_2^2} - \frac{n_0^2}{n_\parallel^2}\left(\frac{n_\parallel^2 - n_2^2}{n_0^2 - n_2^2}\right)\right] \tag{116}$$

In this equation the optical axis of the layer is assumed to be perpendicular to the interface; $n_\perp$ is the real part of the refractive index of the layer perpendicular to its optical axis; $n_{\parallel}$ is the real part of the refractive index of the layer parallel to the optical axis.

The solution of Eq. (113) [or its simplified modification, Eq. (114)] for the calculation of the reflection coefficients is a standard task and is described in the literature, for example, in Ref. 249. The inverse problem, calculation of the refractive index and thickness of the adsorbed layer from the measured ellipsometric angles $\Delta$, $\Delta_0$, $\Psi$, and $\Psi_0$ is unfortunately not so trivial. An analytical solution of these equations is not possible, because the theory does not give explicit expressions for the optical parameters $n_1$ and $d$ of the layer. Therefore, a numerical evaluation, including iteration procedures, is usually applied. Reasonable starting values for the optical properties of the layer are inserted in Eqs (113) or (115) and the iteration process is continued until a satisfactory agreement between the calculated and measured values of $\Delta$ and $\Psi$ has been reached. Modern computers and suitable software make such numeric calculations simple. For anisotropic, ultrathin adsorbed layers even a numerical solution of the basic ellipsometric equation in the respective form Eq. (115) is impossible. These layers have a negligible absorption and therefore the corresponding change in one of the two ellipsometric angles is not measurable ($\delta\Psi \approx 0$). In this connection the basic ellipsometric equation contains only one experimental parameters and three further parameters are to be evaluated. Therefore, it is clear that an infinite number of evaluated parameters can agree with the measured value of $\delta\Delta$.

To proceed and obtain some physical information, additional assumptions about some of the optical properties of the layer are needed (247, 248). These can be derived from sound theoretical considerations, taking into account the structure of the layer and peculiarities of the molecules forming the layer (247, 248).

Although the optical properties of the adsorbed layer by evaluation of the ellipsometric data obtained are quite interesting for its characterization, for interfacial science the information about the amount adsorbed at an interface is especially important. In the calculation of this quantity, however, the problem appears to be of a proper proportionality between the layer properties provided by ellipsometry and the adsorbed amount. Recently, it was shown that for ultrathin adsorbed layers of conventional soluble surfactants ellipsometry is insufficient and additional experimental methods are required (245, 250). Relatively thick layers are also often not homogeneous in the bulk (substrate) normal to the interface. In this case the refractive index and the thickness of the layer calculated from the experimental values of $\delta\Delta$ and $\delta\Psi$ represent mean optical quantities. If, additionally, the refractive index $n_1$ is a linear function of the solute concentration in the layer:

$$n_1 = n_2 + \frac{dn}{dc}(c(z) - c(\infty)) \tag{117}$$

where $c(z)$ is the solute concentration in the layer as a function of the distance from the interface ($z = 0$) to the bulk, and $c(\infty)$ is the solute concentration in the solution. The adsorption $\Gamma$ can be unambiguously calculated from the average layer thickness $d_{av}$ and the average refractive index $n_{1av}$:

$$\Gamma = \frac{d_{av}(n_{1av} - n_2)}{dn/dc} \tag{118}$$

these simplifying assumptions are commonly used in the ellipsometric studies of different layers of synthetic polymers of biopolymers. At the same time it has been accentuated for such applications that the adsorbed amount can be determined more accurately than the layer thickness and refractive index, especially at low interfacial coverages (251).

Ellipsometry is a well-established experimental method for thin-film investigations and nowadays numerous modifications of experimental set-ups exist (252). When ultrafast measurements for monitoring very rapid processes is not necessary, a conventional PCSA null-ellipsometer set-up is often used. The scheme of such an apparatus is shown in Fig. 25. A low-capacity laser serves as light source (beam diameter of about 0.5–1 mm), and the beam passes through the first quarter-wave plate to produce circularly polarized light. The light is then linearly polarized by a Glan–Thompson prism mounted in a rotatable divided circle which can be read with a very high precision. The second quarter-wave plate and the analyzer (a second Glan–Thompson prism) are mounted in a similar manner as a polarizer. A photodiode detector is normally used. Both incidence and reflection arms are motorized and computer controlled; the highly precise motors rotating the polarizer and analyzer are also controlled by the computer.

For ellipsometric investigations of liquid/liquid interfaces numerous measuring cells have been developed. One example is presented schematically in Fig. 26.

For such ellipsometric experiments the cell must be very carefully positioned to place the interface between the two liquids exactly in the ellipsometer axis. The angle of incidence and the angle formed by the two side-walls of the cell must be equal with high accuracy

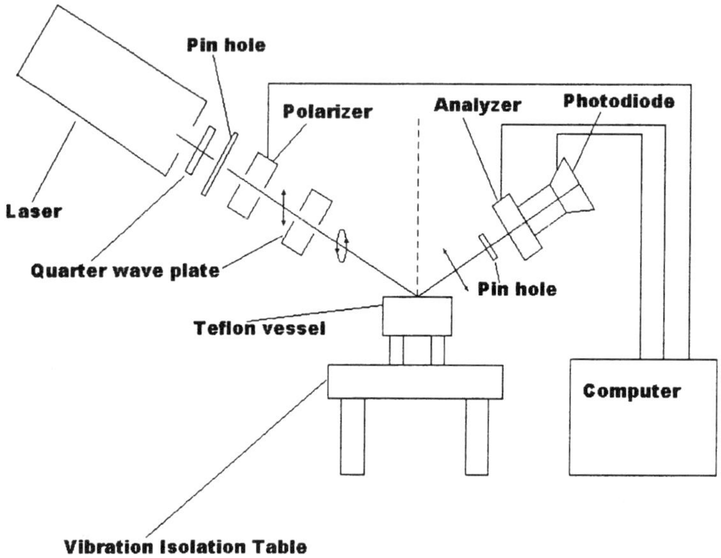


**Figure 25** Principle of an ellipsometer set-up.

to avoid changes in the state of light polarized upon passing through the cell. The measuring procedure by means of the PCSA null-ellipsometer is an experimental routine and typically computerized (253).

Although ellipsometry is well established as an experimental technique for the investigation of adsorbed layers, the number of studies at fluid/liquid interfaces is relatively small. Ellipsometry was used for investigation of the layer thickness between two immiscible liquids near the critical point (254, 255). This technique was also quite often used for in situ studies of the adsorption kinetics at an air/protein solution surface or polymer monolayers at an air/water interface (251, 256). It was also shown that ellipsometric results obtained at the same interface for conventional soluble surfactants or insoluble monolayers cannot be unambiguously interpreted by the standard formalism (244, 245, 247). The application of ellipsometry to the study of coalescence phenomena in emulsion systems was recently reported (257). The newly developed technique of dual-wavelength ellipsometry was used for investigation of the thinning of liquid films between two droplets in an emulsion. A comparison with independent methods shows satisfactory agreement and, hence, ellipsometry can also be applied to such systems.

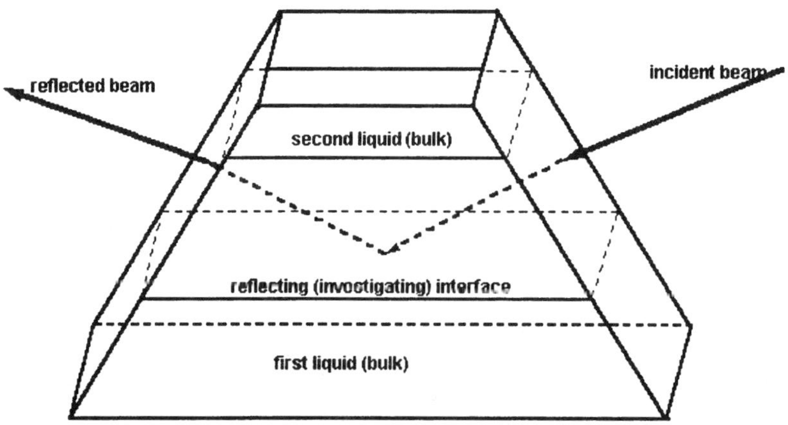


**Figure 26** Measuring cell for ellipsometric studies at liquid/liquid interfaces.

## IX. HLB CONCEPT

Surfactants are the compounds at an interface that reduce the interfacial tension. It follows from thermodynamics, that this property is due to the ability of these compounds to undergo the transfer from within an adjacent fluid (liquid or gas) phase to the interface. In fact, this ability is just the adsorption phenomenon. Clearly, an accumulation at the interface is possible for substances consisting of two parts, each of them separately exhibiting the affinity to one of the contacting phases. Such a property is characteristic of amphiphilic molecules possessing polar (hydrophilic) and nonpolar (lipophilic) parts.

With respect to the properties of polar groups, surfactants can be subdivided into ionic (cation- and anion-active, ampholytic, and zwitterionic) and nonionic surfactants. If the effect produced by the polar group of the surfactant molecule is more significant than that of the lipophilic group, this substance is soluble in water. It is less surface active as compared to any substance characterized by an optimum balance between the activities of hydrophilic and lipophilic groups. Similar conclusions can be drawn also with respect to the solubility in oil: here, the role of the lipophilic group is determining. Clearly, the efficiency of a surfactant is not determined solely by the amphiphilicity, but depends on the hydrophilic/lipophilic balance (HLB) characteristic for this compound. Therefore, this balance is an important characteristic of both the surfactant and the interface.

The first attempt to estimate this hydrophilic/lipophilic balance quantitatively was made by Griffin, who introduced a scale of HLB numbers (258). The initial aim for this classification of surfactants with respect to HLB numbers was to permit the optimum choice of emulsifiers. Subsequently the same method was applied to wetting agents, detergents, etc. The approach proposed by Griffin was further developed and generalized in a number of review papers (259, 260), presenting methods used to determine and calculate HLB numbers for various surfactants. Recently, the HLB concept was analyzed by Rusanov (261) and Kruglyakov (262). In this last book one can also find an extensive bibliography related to Griffin's HLB numbers. The Griffin HLB scale extends from 1 (for extremely lipophilic oleic acid) to 40 (extremely hydrophilic sodium dodecyl sulfate), with a mean HLB value taken to be 10. It is assumed that for the mixture of two or more surfactants, the HLB number additively depends on the HLB numbers of the individual surfactants. Therefore, to determine the HLB number for a surfactant, first the emulsifying ability of this substance is measured, and then that mixture of two surfactants with known HLB numbers is determined which possesses the same emulsifying ability. For the stabilization of oil-in-water emulsions, surfactants with HLB numbers in the range 9–12 are optimal, while to stabilize water-in-oil emulsions, more lipophilic surfactants, possessing HLB numbers in the range from 4 to 6 should be used. As an example, the HLB numbers data are listed below (Table 2) for Tritons X, the substances with the general chemical formula:

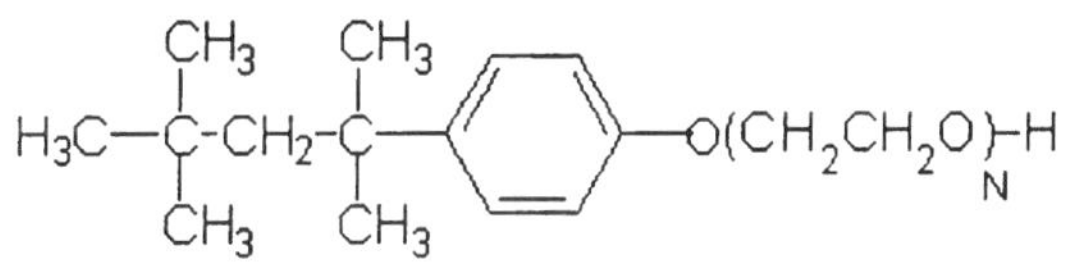

here N is the mean number of oxyethylene groups, and MW is the mean molecular weight.

The stability of an emulsion depends not only on the surfactant type, but also on the nature of the organic phase. To characterize the oil phase, the concept of a necessary (required) HLB number is used. This number is taken to be equal to the HLB number of the surfactant which ensures the best possible emulsification of the oil. Tables of necessary HLB numbers for various oils were published in Ref. 258. For example, with respect to oil-in-water emulsions, the necessary HLB number is 17 for oleic acid, 15 for toluene, 14 for xylene and cetyl alcohol, 10.5 to 12 for mineral oils, 7.5 to 8 for vegetable oils, 5 to 7 for vaseline, and 4 for paraffin. In Refs 263 and 264 the necessary HLB numbers for various oils are compared with the relative dielectric per-

**Table 2** HLB Numbers of Tritons X, According to *Sigma Chemical*

| Triton | N | MW | HLB | Triton | N | MW | HLB |
|---|---|---|---|---|---|---|---|
| X-15 | 1 | 250 | 3.6 | X-102 | 12–13 | 756 | 14.6 |
| X-35 | 3 | 338 | 7.8 | X-165 | 16 | 910 | 15.8 |
| X-45 | 5 | 426 | 10.4 | X-305 | 30 | 1526 | 17.3 |
| X-114 | 7–8 | 536 | 12.4 | X-405 | 40 | 1966 | 17.9 |
| X-100 | 9–10 | 625 | 13.5 | X-705 | 70 | 3286 | 18.7 |

mittivity of the oil $\varepsilon$. In the series of saturated hydrocarbons, a weak inverse dependence between the necessary HLB number and $\varepsilon$ was observed (264); e.g., $\varepsilon = 2.036$ and HLB = 8 for tetradecane, while $\varepsilon = 1.8$ 9 and HLB = 10.5–11.0 for hexane. On the other hand, for various oils a slight increase of the necessary HLB number takes place with increasing dielectric permittivity. All empirical dependence between the necessary HLB number and surface tension or molar volume of oil were proposed, in Ref. 262. It should be noted that the necessary HLB values estimated for the same oil using various methods can differ from each other: for example, the data present in Refs 263 and 264 exceed those listed in Ref. 258.

The attempts to rationalize Griffin's HLB scale from a physicochemical point of view were made in a number of studies. Various correlations were shown to exist between the HLB numbers and the chemical structure or molecular composition of the surfactants. Correlations were also found between the HLB number and physicochemical properties of surfactants and their solutions, for example, surface and interfacial tension, solubility, and heat of solution, spreading and distribution coefficient, dielectric permittivity of the surfactant, cloud point and phase inversion point, critical micelle concentration, foaminess, etc. These studies are reviewed in Ref. 262. However, the correlations found are not generally applicable; moreover, the concept of the additivity of HLB numbers as such for mixtures of surfactants or oils cannot be proven expermentally when the surfactant characteristics are varied over a wider range (265).

An important contribution to the HLB concept was made by Davies (266, 267), where the so-called group numbers were introduced, that is, HLB numbers which correspond not to the molecule as a whole entity, but to the constituting groups (molecular structural units). Once the group numbers $g_i$ are known, one can calculate the HLB number from the chemical formula of a surfactant using the equation:

$$\mathrm{HLB} = 7 + \sum_i g_i \tag{119}$$

For hydrophilic groups $g_i > 0$, while for lipophilic groups $g_i < 0$. The group numbers for some groups calculated by Davies (266, 267) are listed in Table 3.

From analysis of the destruction rates for oil-in-water and water-in-oil emulsions, Davies was able to relate the group numbers with the ratio of coalescence rates for these two types of emulsion. Another correlation was shown to exist to this ratio and the two equilibrium concentrations of the surfactant in the aqueous and oil phases, ($c^{\mathrm{w}}/c^{\mathrm{o}}$—distribution coefficient). Finally, the Davies' theory leads to the relation:

$$\mathrm{HLB} = 7 + 0.36 \ln(c^{\mathrm{w}}/c^{\mathrm{o}}) \tag{120}$$

Comparing this expression with Eq. (10), which determines the difference between the standard chemical potentials of surfactants in the aqueous and oil phases (i.e., by the definition of the free energy of the surfactant transfer from the oil phase into the aqueous phase):

$$\mu_{0i}^{\mathrm{w}} - \mu_{0i}^{\mathrm{o}} = RT \ln K_i^0 \tag{121}$$

where $K_i^0 = (x_i^{\mathrm{o}}/x_i^{\mathrm{w}})_{x_i^{\alpha}\to 0}$. It can be easily seen that the HLB is related to the free energy (or the work of transfer) of the surfactant ($w^{\mathrm{wo}}$) from one phase (water) to the other (oil). Therefore, $w^{\mathrm{wo}} = RT \ln(c^{\mathrm{w}}/c^{\mathrm{o}})$, and, consequently, the energetic interpretation of HLB numbers can be presented as

$$\mathrm{HLB} = 7 + 0.36 w^{\mathrm{wo}}/\mathrm{RT} \tag{122}$$

It is seen from the comparison of Eqs (119) and (122) that the additivity of HLB numbers follows from the additivity of the transfer work, because the group numbers in Eq. (118) are proportional to the partial values of transfer work $w_i^{\mathrm{wo}}$ characteristic to the individual groups which constitute the surfactant molecule:

$$g_i = 0.36 w_i^{\mathrm{wo}}/RT \tag{123}$$

The transfer work can be calculated from the coefficient $K_i^0$ of the surfactant distribution coefficient between the two phases. The values of transfer work for a number of substances are tabulated, for example,

**Table 3** Group Numbers for Some Chemical Groups Calculated by Davies (266, 267)

| | | | |
|---|---|---|---|
| $-SO_4Na$ | 38.7 | $-OH$ | 1.9 |
| $-COOK$ | 21.1 | $-O-$ | 1.3 |
| $-COONa$ | 19.1 | $-(CH_2CH_2-O)-$ | 0.33 |
| $-N^+$ (tert, amine) | 9.4 | $CH_3-$, $-CH_2-$, $-CH=$ | −0.475 |
| $-COOH$ | 2.1 | Benzyl | −1.66 |

in Ref. 262. For a homologous series of a surfactant, the transfer work can be expressed as

$$w^{wo}/RT = a - bn_c \qquad (124)$$

where $a$ and $b$ are constants which correspond to the transfer energy of one hydrophilic group and one methylene group, respectively, and $n_c$ is the number of carbon atoms in the lipophilic part of the molecule. Similar relations are valid also for nonionic surfactants where the specific energy of transfer is calculated per oxyethylene group. If an energetic equilibrium exists between the hydrophilic and lipophilic parts of the surfactant molecule, i.e., $w^{wo}/RT = 0$, then it follows from Eq. (122) that the HLB = 7. For example, with respect to the series of aliphatic acids, according to the condition $a - bc_n = 0$, pentanoic acid is hydrophilic, while hexanoic acid, the next in the series, is lipophilic. Similarly, for the Tritons X homologous series, X-35 is hydrophilic, while X-15 is lipophilic.

While the HLB number concepts proposed by Griffin and Davies certainly facilitate the choice of surfactants and oils with regard to their practical applications, the theoretical deficiencies of these systems are also well known, and were discussed in a number of publications. For example, the additivity principle of the Griffin HLB with respect to mixtures of surfactants is held only within a rather narrow range of surfactants' HLB and necessary HLB numbers for oils (265, 268, 269). No account is taken for the concentration of surfactants, the temperature, or admixtures of electrolytes and other substances. When HLB numbers for nonionic surfactants are calculated from the Griffin equation [see (258)], the hydrophility is estimated only from the mass portion of oxyethylene groups, regardless of their location and the structure of the lyophilic chain. To take account of these effects and the influence of the medium, both for nonionic and ionic surfactants, the concept of effective HLB values was introduced (270, 271). It was shown that the addition of acetone, urea, dioxane, and other substances lead to the increase of the effective HLB numbers for nonionic surfactants, while the addition of glycerine, on the contrary, results in a decrease of the HLB value. The addition of alcohols and polyethylene glycols lead to more complicated changes in the HLB values (270, 271).

In the framework of Davies' concept, the relation [Eq. (119)] is empirical. This relation implies the additivity of group numbers. There is evidence for the fact that the hydrophilicity and lyophilicity of various groups depend on their position in the molecule; this is true, e.g., for isomers. The location of the benzene ring within the alkyl benzene sulfonates affects the distribution coefficient between water and oil, that is, the HLB value. Similar effects were observed with respect to the location of the polar group in the molecules of sodium alkyl sulfates (272). Principal relations of the Davies theory were derived from the analysis of the emulsion coalescence rates, where the Smoluchowsky theory was applied. In this analysis, no account was taken for the difference, which exists between emulsions of low stability, described by this theory, and stable emulsions where the coalescence stage is regulated by the properties of thin liquid films. The assumption of the Davies' model that the repulsive energetic barrier in oil-in-water emulsions does not depend on the lyophilic chain length, while for water-in-oil emulsions this barrier does not depend on the nature of polar groups, contradicts experimental data. Also, some arbitrariness exists in the relation between the behavior of emulsions and HLB numbers. According to Davies, values of HLB > 7 correspond to stable O/W emulsions, while stable W/O emulsions can be obtained for HLB < 7. It was shown experimentally, however, that the formation of stable O/W emulsions is possible for HLBs in the range 2–17, while the HLB interval corresponding to stable W/O emulsions is 2 to 10, see Ref. 262. Also the influence of the concentration of a surfactant on the type and stability of the emulsion remains unclear. For example, at low concentrations O/W emulsions are stable, while when the concentration exceeds 4–6%, a phase inversion takes place, which results in the formation of stable W/O emulsions (267, 273). A comparison between the HLB numbers of Griffin and Davies was made in a number of studies, for example, in Refs 262 and 274). For nonionic surfactants, these numbers are mutually inconsistent.

The HLB scale of Davies is based on the difference between the work of transfer of a surfactant molecule (or its constituents) from the vacuum into aqueous and oil phases. It can be expected that, similar to this difference, the ratio of these values can be used as a measure for the HLB (261). The work of surfactant transfer into a phase from the vacuum can be calculated as $w = w_h + w_l$, where the subscripts "h" and "l" refer to the hydrophilic and lipophilic parts of the surfactant molecule, respectively. This leads to the definition of the HLB indices (261):

$$\chi = w^{w}/w^{o} = (w_h^{w} + w_l^{w})/(w_h^{o} + w_l^{o}) \qquad (125)$$

$$\chi_1 = w_h^{wo}/w_l^{ow} = (w_h^{o} - w_h^{w})/(w_l^{o} - w_l^{w}), \qquad (126)$$

where $w_h^{wo}$ is the work required for the transfer of the hydrophilic group from the aqueous phase into the oil

LEEDS UNIVERSITY LIBRARY

phase, and $w_l^{ow}$ is the work corresponding to the transfer of the lipophilic group from the oil phase into water. If a balance between the hydrophilic and lipophilic group exists, then $\chi = \chi_1 = 1$. The HLB index $\chi$ can be either positive or negative, with lipophilic substances corresponding to $-\infty < \chi < 1$, while for hydrophilic substances $1 < \chi < \infty$. The index $\chi_1$, being the ratio of two positive values, is positive. Using Eq. (117), one obtains from Eqs (125) and (126):

$$\chi = 1 - (RT/w^o)\ln(c^w/c^o) \tag{127}$$

$$\chi_1 = 1 - (RT/w_l^{ow})\ln(c^w/c^o) \tag{128}$$

It is seen that, unlike the scale by Davies, where the distribution coefficient of the surfactant is only necessary to calculate the HLB, to determine $\chi$ and $\chi_1$ one requires some additional information regarding the work necessary for the transfer of the surfactant into the oil phase, or the difference between the works of transfer of lipophilic group from the oil phase into the aqueous one. Also, the ratio:

$$\chi_2 = w_h^w/w_l^o \tag{129}$$

is used as HLB index (275). This value is always possible, being the ratio of two negative values. The disadvantage of the indices $\chi$, $\chi_1$, and $\chi_2$ are compared with Griffin's and Davies' HLB numbers is that they are not additive.

A number of attempts have been made to estimate the HLB from the comparison of the work of surfactant-molecule transfer not between the adjacent bulk phases, but from bulk phases into the surface layer, that is, to use the adsorption work as basis for such estimates. The adsorption work from the aqueous phase is $w(z) - w^w$, while the adsorption work from the oil phase is given by the relation $w(z) - w$. Here, $z$ is the coordinate of the hydrophilic–lipophilic center (HLC) of a surfactant molecule, which corresponds usually to the minimum of $w(z)$ (261). This minimum work of transfer of a surfactant molecule corresponds to the maximum at the plot of the concentration distribution in the surface layer. Usually the location of the minimum of $w(z)$ (i.e., HLC) is displaced from the geometric interface towards the aqueous phase. It is seen from the above relations that the adsorption work, unlike the work of transfer from bulk phases, is not a definite and unambiguous characteristic. Usually the local value of the work $w(z)$ is substituted by the mean (integral) work $w^\sigma$, related to the entire adsorption layer. Due to the inhomogeneity of the interface region, the adsorption work cannot be calculated additively from the work of adsorption for particular groups of the molecule, because these work values depend on the location of the group within the surface layer. Clearly, an interrelation should exist between the adsorption work and HLB characteristics. A trivial approach is based on the comparison of the adsorption work differences. In this case the difference between the adsorption works is just the work necessary for the transfer of the surfactant molecule from one phase into another; this work is the basic value in Davies' concept. The ratio of adsorption works, the so-called hydrophilic–oleophilic ratio (HOR) is often used (261, 262, 276–278):

$$\mathrm{HOR} = w^{0\sigma}/w^{w\sigma} \tag{130}$$

Various expressions for HOR were proposed, expressed via the surfactant distribution coefficient between the phases, and the surfactant's adsorption activity (5). The advantage of HOR as compared to the HLB system is that, for a particular choice of the standard state, this index does not depend on the surfactant concentration, the type of the organic phase, or the presence of various additives soluble in water and oil. Methods were also proposed to determine the HOR for mixtures of surfactants (262). For these systems, however, this index is not additive anymore. The HOR values for mixtures are shifted towards that characteristic for the component which possesses the higher value of the distribution coefficient. Another deficiency of the HOR concept is its suggestiveness: it was mentioned above that this value depends on the coordinate of the HLC. However, for the HOR values other than unity, the HLC position-dependent work of the introduction of a surfactant molecule into the surface layer is uniquely determined by the HOR.

To summarize, among all the proposed characteristics of the HLB, Davies' HLB scale is the most substantiated and most widely used, in spite of the number of deficiencies noted above. Here, the recent publication (279), which relates the electroacoustophoretic behavior of emulsions with Davies' HLB numbers, can be referred to as an example.

## X. CONCLUSIONS

The behavior of emulsions as a particular type of disperse system is controlled by many factors. There is a large number of properties of the corresponding liquid/liquid interface which can be determined by well-established methods, such as dynamic surface tensions, adsorbed amount, exchange of matter across the interface, and dilational and shear rheology. Although first models exist, a general view does not exist yet of how important the individual properties are in respect of emulsion stability or its destabilization. In practice, personal experience and trial and error procedures are most frequently used so far. The access to quantitative methods and extensive studies of model systems will for sure improve the possibilities of designing emulsions with a predefined behavior. This contribution only summarizes the experimental possibilities at extended liquid interfaces rather than providing a link to particular emulsion properties. An overall understanding of real emulsions will certainly require the study of the entire present encyclopedia, and then still questions remain open to be answered in future work.

## ACKNOWLEDGMENTS

The work was financially supported by projects of the European Community (INCO ERB-IC15-CT96-0809), the DFG (Mi418/9-1 and Mi418/7-1), the Fonds der Chemischen Industrie (RM 400429), the German Canadian Agreement on Co-operation in Scientific Research and Technological Development (KAN MPT 22), and the ESA (Topical Team and Fast project).

## NOMENCLATURE

$A$ surface area
$A_0$ surface area at equilibrium
$c$ concentration
$c^w, c^o$ concentration in the aqueous and oil phases
$D$ diffusion coefficient
$dA$ change in interfacial area
$dW^A$ mechanical work due to interfacial tension
$f$ frequency
$g$ gravitational acceleration
HLB hydrophilic/lipophilic balance
$k$ rate constant of transition from state 1 into state 2
$K$ distribution coefficient
MW molecular weight
$N^A$ total interfacial mole number
$P$ pressure
$r$ bubble radius
$R$ gas constant
$r_{cap}$ capillary radius
$S^A$ interfacial entropy
$t$ time
$T$ absolute temperature
$U^A$ interfacial internal energy
$V$ volume
$w^\sigma$ integral work
$x$ direction normal to the interface
$z$ coordinate normal to the surface

*Greek symbols*

$\alpha$ a constant
$\beta = (\omega)1/\omega_2)^\alpha$
$\gamma$ surface tension
$\gamma_0$ surface tension of the pure solvent
$\Gamma = \Gamma_1 + \Gamma_1$ total adsorption
$\delta$ relative oscillation amplitude
$\Delta\overline{H}_i^0$ molar standard enthalpy of transfer
$\Delta\rho$ density difference
$\varepsilon_d$ dilational elasticity
$\eta_d$ dilational viscosity
$\eta_s$ shear viscosity
$\Theta$ relative area change
$\lambda = k/\varpi$ dimensionless rate constant
$\mu^A$ interfacial chemical potential
$\Pi = \gamma_0 - \gamma$ surface tension
$\omega_1, \omega_2$ partial molar areas
$\varpi - 2\pi f$ circular frequency

## REFERENCES


1. O Reynolds. Phil Trans Roy Soc London A177: 1577–165, 1886.
2. VG Levich. Physico-chemical Hydrodynamics. Englewood Cliffs, NJ: Prentice Hall, 1962.
3. AD Scheludko. Adv Colloid Interface Sci 1: 391–430, 1967.
4. IB Ivanov. Pure Appl Chem 52: 1241–1262, 1980.
5. A Sonin, A Bonfillon, D Langevin. J. Colloid Interface Sci 162: 323–330, 1994.
6. HA Barnes. Colloids Surfaces A 91: 89–95, 1994
7. TF Tadros, B Vincent. In: P Becher, ed. Encyclopedia of Emulsion Technology. Vol 1. New York: Marcel Dekker, 1983, p 130.
8. GV Jeffreys, JL Hawksley. AIChE J 11: 413–421, 1965.
9. DR Woods, KA Burrill. J Electroanal Chem 37: 191–203, 1972.
10. AJS Liem, DR Woods. AIChE Symp Ser 70: 8–15, 1974.
11. PG Murdoch, and DE Leng, Chem Eng Sci 26: 1881–1896, 1971.
12. IB Ivanov, TT Traykov. Int J Multiphase Flow 2: 397–410, 1976.
13. AD Barber, S Hartland. Can J Chem Eng 54: 279–288, 1976.

14. Z Zapryanov, AK Malhotra, N Aderangi, DT Wasan. Int J Multiphase Flow 92: 105–129, 1983.
15. DT Wasan, K Sampath, N Aderangi. AIChE Symp Ser 192: 93–97, 1980.
16. RRRP Borwankar, LA Lobo, DT Wasan. Colloids Surfaces A 69: 135–146, 1992.
17. V Bergeron. Langmuir 13: 3474–3482, 1997.
18. A Espert, R von Klitzing, P Poulin, A Collin, R Zana, D Langevin. Langmuir 14: 4251–4260, 1998.
19. E Dickinson. Curr Opinion Colloid Interface Sci 3: 633–638, 1998
20. A Williams, JJM Janssen, A Prins. Colloids Surfaces A 125: 189–200, 1997.
21. E Manev, RJ Pugh. J Colloid Interface Sci 186: 493–497, 1997.
22. JK Angarska, KD Tachev, PA Kralchevsky, A Mehreteab, G Broze. J Colloid Interface Sci 200: 31–45, 1998.
23. D Sentenac, SD Dean. J Colloid Interface Sci 196: 35–47, 1997.
24. IB Ivanov, PA Kralchevsky. Colloids Surfaces A 128: 155–175, 1997.
25. IB Ivanov, DS Dimitrov. In: IB Ivanov, ed. Thin Liquid Films. New York: Marcel Dekker, 1988, p 379.
26. D Langevin. Curr Opinion Colloid Interface Sci 3: 600–607, 1998.
27. R Miller, R Wüstneck, J Krägel, G Kretzschmar. Colloids Surfaces A 111: 75–118, 1996.
28. H Fruhner, KD Wantke. Colloids Surfaces A, 114: 53–60, 1996
29. J Krägel, AV Makievski, VI Kovalchuk, VB Fainerman, R Miller. Colloids Surfaces A (submitted).
30. J Krägel, DO Grigoriev, AV Makievski, R Miller, VB Fainerman, PJ Wilde, R Wüstneck. Colloids Surfaces B 12: 391–397, 1999.
31. VB Fainerman, EH Lucassen-Reynders, R Miller. Colloids Surfaces A 143: 141–165, 1998.
32. JA Butler. Proc Roy Soc Ser A, 138: 348–375, 1932.
33. P Joos. Bull Soc Chim Belg 76: 591–600, 1967.
34. B von Szyszkowski. Z Phys Chem (Leipzig) 64: 385–398, 1908.
35. I Langmuir. J Am Chem Soc 39: 1848–1907, 1917.
36. EH Lucassen-Reynders. J Colloid Sci 19: 584–585, 1964.
37. EH Lucassen-Reynders. J Phys Chem 70: 1777–1785, 1966.
38. EH Lucassen-Reynders. J Colloid Interface Sci 41: 156–167, 1972.
39. EH Lucassen-Reynders. J Colloid Interface Sci 85: 178–185, 1982.
40. EH Lucassen-Reynders. Progr Surface Membrane Sci 10: 253–320, 1976.
41. R Van den Bogaert, P Joos. J Phys Chem 84: 190–194, 1980.
42. VB Fainerman, R Miller, R Wüstneck, AV Makievski. J Phys Chem 100: 7669–7675, 1996
43. VB Fainerman, R. Miller, R. Wüstneck, J Phys Chem 101: 6479–6483, 1997
44. VB Fainerman, R Miller, R Wüstneck. J Colloid Interface Sci 183: 26–34, 1996.
45. EH Lucassen-Reynders. Colloids Surfaces A 91: 79–88, 1994.
46. JT Davies. Proc Roy Soc Ser A 208: 224–231, 1951.
47. JT Davies. Proc Roy Soc Ser A 245: 417–432, 1958.
48. RP Borwankar, DT Wasan. Chem Eng Sci 43: 1323–1337, 1988.
49. I Prigogine. The Molecular Theory of Solutions. Amsterdam: North-Holland, 1968.
50. EA Guggenheim. Mixtures. Oxford: Clarendon Press, 1952.
51. RC Read, JM Prausnitz, TK Sherwood. The Properties of Gases and Liquids. 3rd ed. New York, London, Paris, Tokyo: McGraw-Hill, 1977.
52. AN Frumkin. Z Phys Chem (Leipzig) 116: 466–484, 1925.
53. BB Damaskin, AN Frumkin, SL Djatkina. Izv AN SSSR Ser Chim 2171–2175, 1967.
54. BB Damaskin. Izv AN SSSR Ser Chim 346–351, 1969.
55. EH Lucassen-Reynders. In: EH Lucassen-Reynders, ed. Anionic Surfactants, Physical Chemistry of Surfactant Action. New York-Basel: Marcel Dekker, 1981, p 1.
56. JA Tedoradze, RA Arakeljan, ED Belokolos. Elektrokhimija 2: 563–569, 1966.
57. BB Damaskin. Elektrokhimija 5: 249–255, 1969.
58. BB Damaskin, AN Frumkin, NA Borovaja. Elektrokhimija 8: 807–815, 1972.
59. MJ Rosen, XY Hua. J Colloid Interface Sci 86: 164–172, 1982.
60. XY Hua, MJ Rosen. J Colloid Interface Sci 87: 469–477, 1982.
61. VB Fainerman, SV Lylyk Kolloidn Zh 45: 500–508, 1983.
62. VV Krotov. Kolloidn Zh 47: 1075–1082, 1985.
63. VB Fainerman. Zh Fiz Khim 62: 1003–1010, 1988.
64. VB Fainerman. Zh Fiz Khim 60: 681–685, 1986.
65. VB Fainerman. Kolloidn Zh 48: 512–519, 1986.
66. J Rodakiewicz-Nowak. J Colloid Interface Sci 85: 586–597, 1982.
67. M Karolczak, DM Mohilner. J Phys Chem 86: 2840–2848, 1982.
68. DM Mohilner, H Nakadomari, PR Mohilner. J Phys Chem 81: 244–252, 1977.
69. E Helfand, HL Frisch, JL Lebowitz. J Chem Phys 34: 1037–1044, 1961.
70. R Parsons. J Electroanal Chem 7: 136–144, 1964.
71. E Tronel-Peyroz. J Phys Chem 88: 1491–1496, 1984.
72. H Diamant, D Andelman. J Phys Chem 100: 13732–13742, 1996.
73. R Miller, EV Aksenenko, L Liggieri, F Ravera, VB Fainerman. Langmuir 15: 1328–1336, 1999.

74. EH Lucassen-Reynders, J Lucassen, D Giles. J Colloid Interface Sci 81: 150–162, 1981.
75. VV Kalinin, CJ Radke. Colloids Surfaces A 114: 337–350, 1996.
76. CA MacLeod, CJ Radke. Langmuir 10:3555–3566, 1994.
77. VYa Poberezhnyi, LA Kul'skiy. Kolloidn Zh 46: 735–742, 1984.
78. VYa Poberezhnyi, TZ Sotskova, LA Kul'skiy. Khim i Tekhnol Vody 18: 570–582, 1996.
79. PM Vlahovska, KD Danov, A Mehreteab, G Broze. J Colloid Interface Sci 192: 194–206, 1997.
80. PA Kralchevsky, KD Danov, G Broze, A Mehreteab. Langmuir 15:2351–2365, 1999.
81. VM Muller, BV Derjaguin. J Colloid Interface Sci 61:361–369, 1977.
82. AV Makievski, VB Fainerman, M Bree, R Wüstneck, J Krägel, R Miller. J Phys Chem 102: 417–425, 1998.
83. VB Fainerman, R Miller. In: D Möbius, R Miller, eds. Proteins at Liquid Interfaces. Vol. 7. Amsterdam: Elsevier, 1998, pp 51–102.
84. P Joos, G Serrien. J Colloid Interface Sci, 145: 291–294, 1991.
85. VB Fainerman, R Miller. Langmuir 15: 1812–1816, 1999.
86. R Miller, VB Fainerman, AV Makievski, J Krägel, R Wüstneck. Colloids Surfaces A, 161:151–157, 2000.
87. JF Padday. In: E Matijevic, ed. Surface and Colloid Science. Vol 1. New York: Wiley, 1968, p 101.
88. DS Ambwani, T Fort Jr. In: RJ Good, RR Stromberg, eds. Surface and Colloid Science. Vol 11. New York: Plenum Press.
89. AW Adamson. Physical Chemistry of Surfaces. 5th ed, New York: Wiley, 1990.
90. AI Rusanov, VA Prokhorov. In: D Möbius, R Miller, eds. Studies of Interface Science. Amsterdam: Elsevier, 1996.
91. P Chen, DY Kwok, RM Prokop, OI del Rio, SS Susnar, AW Neumann. In: D Möbius, R Miller, eds. Studies of Interface Science. Vol 6. Amsterdam: Elsevier, 1998, pp 61–138.
92. R Miller, VB Fainerman. In: D Möbius, R Miller, eds. Studies of Interface Science. Vol 6. Elsevier, Amsterdam, 1998, pp 139–186.
93. AM Seifert. In: D Möbius, R Miller, eds. Studies of Interface Science. Vol 6. Amsterdam: Elsevier, 1998, pp 187–238.
94. L Liggier, F Ravera. In: D Möbius, R Miller, eds. Studies of Interface Science. Vol 6. Amsterdam: Elsevier, 1998, pp 239–278.
95. F MacRitchie. Chemistry at Interfaces. San Diego, New York: Academic Press, 1990.
96. JD Andrade. Surface and Interfacial Aspects of Biomaterial Polymers. Vol 2. New York: Plenum, 1985.
97. A Passerone, R Ricci. Bubbles in Interfacial Research. In: D Möbius, R Miller, eds. Studies of Interface Science. Vol 6. Amsterdam: Elsevier, 1998, pp 475–524.
98. DY Kwok, P Chiefalo, B Khorshiddoust, S Lahooti, MA Cabrerizo-Vílchez, OI del Río, AW Neumann. In: R Sharma, ed. Surfactant Adsorption and Surface Solubilization. ACS Symposium Series 615. Washington, DC: American Chemical Society. 1995, ch. 24.
99. JF Padday. In: E Matijevic, ed. Surface and Colloid Science. Vol 1. New York, London, Sydney, Toronto: Wiley-Interscience, 1969, p 39.
100. FC Goodrich. In: E Matijevic, ed. Surface and Colloid Science. Vol 1. New York, London, Sydney, Toronto: Wiley-Interscience, 1969, p 1.
101. AFH Ward, L Tordai. J Phys Chem 14: 453–464, 1946.
102. FW Pierson, S Whittaker. J Colloid Interface Sci 52: 203–213, 1976.
103. R Miller. Colloid Polymer Sci 258: 179–185, 1980.
104. JT Davies, JAC Smith, DG Humphreys. Proc Int Conf Surf Act Subst 2: 281–287, 1957.
105. J Kloubek. J Colloid Interface Sci 41: 1–6, 1972.
106. VB Fainerman. Kolloidn Zh 41: 111–115, 1979.
107. D Ilkovic, J Chim Phys Physicochem Biol 35: 129–135, 1938.
108. P Delahay, I Trachtenberg. J Am Chem Soc 79: 2355–2362, 1957.
109. P Delahay, CT Fike. J Am Chem Soc 80: 2628–2630, 1958.
110. P Joos, M Van Uffelen. J Colloid Interface Sci 171: 297–305, 1995.
111. A Passerone, L Liggier, N Rando, F Ravera, E. Ricci. J Colloid Interface Sci 146: 152–162, 1991.
112. RS Hansen. J Phys Chem 64: 637–641, 1960.
113. VB Fainerman, SA Zholob, R Miller. Langmuir 13: 283–289, 1997.
114. M Ferrari, L Liggieri, F Ravera, C Amodio, R Miller. J Colloid Interface Sci 186: 40–45, 1997.
115. L Liggier, F Ravera, M Ferrari, A Passerone, R Miller. J Colloid Interface Sci 186:46–52, 1997.
116. JL Cayias, RS Schechter, WH Wade. In: KL Mittal, ed. Adsorption at Interface. ACS Symposium Series 8. Washington, DC: American Chemical Society, 1975, pp 234–247.
117. VK Bansal and DO Shah. In: KL Mittal, ed. Micellization, Solubilization, and Microemulsions. Vol 1. New York: Plenum Press, 1977, pp 87–114.
118. JL Cayias, RS Schechter, WH Wade. J Colloid Interface Sci 59: 31–44, 1977.
119. L Cash, JL Cayias, G Fournier, D Macallister, T Schares, RS Schechter, WH Wade. J Colloid Interface Sci 59: 39–44, 1977.
120. EI Franses, JE Puig, Y Talmon, WG Miller, LE Scriven, HT Davis. J Phys Chem 84: 1547–1556, 1980.
121. R Aveyard, BP Binks, TA Lawless, J Mead. J Chem Soc Faraday Trans 1 81: 2155–2168, 1985.

122. R Aveyard, BP Binks, J Mead. J Chem Soc Faraday Trans 1 81:2169–2177, 1985.
123. R Aveyard, BP Binks, J Mead. J Chem Soc Faraday Trans 1 82: 1755–1770, 1986.
124. R Aveyard, BP Binks, S Clark, J Mead. J Chem Soc Faraday Trans 1 82: 125–142, 1986.
125. D Guest, D Langevin. J Colloid Interface Sci 112: 208–220, 1986.
126. R Aveyard, BP Binks, J Mead. J Chem Soc Faraday Trans 1 83: 2347–2357, 1987.
127. R Aveyard, BP Binks, TA Lawless, J Mead. Can J Chem 66: 3031–3037, 1988.
128. R Aveyard, BP Binks, PDI Fletcher, JR Lu. J Colloid Interface Sci 139: 128–138, 1990.
129. WD Harkins, H Zollman. Am Chem Soc 48: 69–81 1926.
130. AM Cazabat, D Langevin, J Meuntier, A Pouchelon. Adv Colloid Interface Sci 16: 175–199, 1982.
131. JW Gibbs. The Scientific Papers of J Willard Gibbs. New York: Dover, 1960.
132. HB Callen. Thermodynamics and an Introduction to Thermostatics. 2nd ed. New York: John Wiley, 1985.
133. Y Rotenberg, L Boruvka, AW Neumann. J Colloid Interface Sci 93: 169–183, 1983.
134. P Cheng, D Li, L Boruvka, Y Rotenberg, AW Neumann. Colloids Surfaces, 43: 151–167, 1990.
135. S Lahooti, OI del Río, P Cheng, AW Neumann. In: JK Spelt, AW Neumann, eds. Applied Surface Thermodynamics. New York: Marcel Dekker, 1996, pp 441–507.
136. B Vonnegut. Rev Sci Instrum 13: 6–9, 1942.
137. HM Princen, IYZ Zia, SG Mason. J Colloid Interface Sci 23: 99–107, 1967.
138. JC Slattery, JD Chen. J Colloid Interface Sci 64: 371–373, 1978.
139. PK Currie, JV Nieuwkoop. J Colloid Interface Sci 87: 301–316, 1982.
140. A Voigt, O Thiel, D William, Z Policova, W Zingg, AW Neumann. Colloids Surfaces 58: 315–326, 1991.
141. R Miller, R Sedev, K-H Schano, C Ng, AW Neumann. Colloids Surfaces 69:209–216, 1993.
142. DY Kwok, MA Cabrerizo-Vílchez, Y Gomez, SS Susnar, IO del Río, D Vollhardt, R Miller, AW Neumann. In: V Pillai, DO Shah, eds. Dynamic Properties of Interfaces and Association Structures. Champaign, IL: American Oil Chemists Society Press, 1996, pp 278–296.
143. DY Kwok, D Vollhardt, R Miller, D Li, AW Neumann. Colloids Surfaces A: Physicochem Eng Aspects 88: 51–58, 1994.
144. DY Kwok, S Tadros, H Deol, D Vollhardt, R Miller, MA Cabrerizo-Vílchez, AW Neumann. Langmuir 12: 1851–1859, 1996.
145. R Wüstneck, P Enders, TH Ebisch, R Miller. Thin Solid Films 298: 39–46, 1997.
146. J Li, R Miller, R Wüstneck, H Möhwald, AW Neumann. Colloids Surfaces A: Physicochem Eng Aspects 96: 295–299, 1995.
147. J Li, R Miller, H Möhwald. Colloids Surfaces A: Physicochem Eng Aspects 114: 113–121, 1996.
148. DY Kwok, LK Leung, CB Park, AW Neumann. Polym Eng Sci, 38: 757–764, 1998.
149. M Wulf, S Michel, K Grundke, OI del Rio, DY Kwok, AW Neumann. J Colloid Interface Sci 210:172–181, 1998.
150. SS Susnar, HA Hamza, AW Neumann. Colloids Surfaces A: Physicochem Eng Aspects 89:169–180, 1994.
151. MA Cabrerizo-Vílchez, Z Policova, DY Kwok, P Chen, AW Neumann. Colloids Surfaces B: Biointerfaces 5: 1–9, 1995.
152. D Duncan, D Li, J Gaydos, AW Neumann. J Colloid Interface Sci 169: 256–261, 1995.
153. A Amirfazli, DY Kwok, J Gaydos, AW Neumann. J Colloid Interface Sci 205: 1–11, 1998.
154. D Li, AW Neumann. J Colloid Interface Sci 148: 190–200, 1992.
155. DY Kwok, CNC Lam, A Li, AW Neumann. J Adhesion, 68: 229–255, 1998.
156. DY Kwok, CNC Lam, A Li, A Leung, R Wu, E Mok, A W Neumann. Colloids Surfaces A: Physicochem Eng Aspects 142: 219–235, 1998.
157. HT Patterson, KH Hu, TH Grindstaff. J Polym Sci C: Polym Symposia 34: 31–43, 1971.
158. J Ryden, PÅ Albertsson. J Colloid Interface Sci 37: 219–222, 1971.
159. JJ Elmendorp, G de Vos. Polym Eng Sci 26: 415–417, 1986.
160. M Heinrich, BA Wolf. Polymer, 33: 1926–1931, 1992.
161. AM Seifert, JH Wendorff. Colloid Polym Sci 270: 962–971, 1992.
162. DD Joseph, MS Arney, G Gillberg, H Hu, D Hultman, C Verdier, TM Vinagre. J Rheol 36:621–662, 1992.
163. J Lyklema. Fundamentals of Interface and Colloid Science. Vol I. London: Academic Press, 1993.
164. R Miller, G Loglio, U Tesei. Colloid Polymer Sci 270: 598–601, 1992.
165. F Ravera, L Liggier, A Passerone, A Steinchen. J Colloid Interface Sci 163: 309–314, 1994.
166. CA MacLeod, CJ Radke. J Colloid Interface Sci 166: 73–88, 1997.
167. VB Fainerman. SA Zholob, R Miller. Langmuir 13: 283–289, 1997.
168. DC England, JC Berg. AIChE J 17: 313–322, 1971.
169. E Rubin, CJ Radke. Chem Eng Sci 35: 1129–1138, 1980.
170. TS Sørensen, M Hennenberg. In: TS Sørensen, ed. Dynamics and Instabilities in Fluid Interfaces, Lecture Notes in Physics 105. Berlin: Springer-Verlag, 1979.

171. M Hennenberg, A Sanfeld, PM Bish. AIChe J 27: 1002–1008, 1981.
172. F Ravera, M Ferrari, L Liggieri, R Miller, A Passerone. Langmuir 13: 4817–4820, 1997.
173. M Ferrari, L Liggieri, F Ravera. J Phys Chem B 102: 10521–10527, 1998.
174. AFH Ward, L Tordai. J Chem Phys 14:453–461, 1946.
175. RS Hansen. J Phys Chem 64: 637–641, 1960.
176. F Ravera, M Ferrari, L Liggieri, R Miller. Progr Colloid Polymer Sci 105: 346–350, 1997.
177. R Miller. Colloid Polymer Sci 259: 375–381, 1981.
178. G Loglio, U Tesei, R Miller, R Cini. Colloids Surfaces A 61: 219–226, 1991.
179. EH Lucassen-Reynders, J Lucassen. Colloids Surfaces A 85: 211–219, 1994.
180. SS Dukhin, G Kretzschmar, R Miller In: D Möbius, R Miller, eds. Studies in Interface Science. Vol 1. Amsterdam: Elsevier, 1995.
181. R Miller, VB Fainerman, J Krägel, G Loglio. Curr Opinion Colloid Interface Sci 2: 578–583, 1997.
182. R Miller, R Wüstneck, J Krägel, G Kretzschmar. Colloids Surface A 111: 75–118, 1996.
183. J Lucassen, M van den Tempel. Chem Eng Sci 27: 1283–1291, 1972.
184. J Lucassen, M van den Tempel. J Colloid Interface Sci 41: 491–498, 1972.
185. SM Sun, MC Shen. J Math Anal Appl 172: 533–566, 1993.
186. BA Noskov, DO Grigoriev, R Miller. J Colloid Interface Sci 188: 9–15, 1997.
187. BA Noskov, DO Grigoriev, R Miller. Langmuir 13: 295–298, 1997.
188. DO Johnson, KJ Stebe. J Colloid Interface Sci 168: 21–35, 1994.
189. KD Wantke, H Fruhner. In: Studies in D Möbius, R Miller, eds. Studies in Interface Science. Vol 6. Amsterdam: Elsevier, 1998, pp 327–365
190. P Joos, M van Uffelen. Colloids Surfaces A 100: 245–253, 1995.
191. M van Uffelen, P Joos. J Colloid Interface Sci 158: 452–459, 1995.
192. T Horozov, P Joos. J Colloid Interface Sci 173: 334–342, 1995.
193. LT Lee, D Langevin, B Farnoux. Physical Review Letters 67: 2678–2681, 1991.
194. A Bonfillon, D Langevin. Langmuir 9: 2172–2177, 1993.
195. A Prins, MA Bos, FJG Boerboom, HKAI van Karlsbeek. In: D Möbius, R Miller, eds. Studies in Interface Science. Vol 7. Amsterdam: Elsevier, 1998, pp 221–265.
196. R Miller, Z Policova, R Sedev, AW Neumann. Colloids Surfaces A 76:179–185, 1993.
197. P Chen, Z Polikova, SS Susnar, CR Pace-Asciak, PM Demin, AW Neumann. Colloids Surfaces A 114: 99–112, 1996.
198. JB Li, R Miller, H Möhwald. Colloids Surfaces A 114: 123–130, 1996.
199. J Benjamins, A Cagna, EH Lucassen-Reynders. Colloids Surfaces A 114: 245–254, 1996.
200. R Miller, J Krägel, AV Makievski, R Wüstneck, JB Li, VB Fainerman, AW Neumann. Proceedings of the Second World Emulsion Congress, Bordeaux, 1997, Vol 4, pp 153–163.
201. R Wüstneck, B Moser, G Muschiolik. Colloids Surfaces A, Colloids Surfaces B 15: 263–273, 1999.
202. AV Makievski, R Miller, VB Fainerman, J Krägel, R Wüstneck. In: E Dickinson, JM Rodíguez Patino, eds. Food Emulsions and Foams: Interfaces, Interfaces, Interactions and Stability, Special Publication No. 227. Cambridge, England: Royal Society of Chemistry, 1999, pp 269–284.
203. R Nagarajan, DT Wasan. J Colloid Interface Sci 159: 164–173, 1993.
204. L Liggieri, F Ravera, A Passerone J Colloid Interface Sci 169: 226–237, 1995.
205. L Liggieri, F Ravera, A Passerone. J Colloid Interface Sci 140: 436–443, 1990.
206. YT Tian, RG Holt, RE Apfel. Theory Phys Fluids 7: 2938–2949, 1995.
207. YT Tian, RG Holt, RE Apfel. J Colloid Interface Sci 187: 1–10, 1997.
208. YH Kim, K Koczo, DT Wasan. J Colloid Interface Sci 187: 29–44, 1997.
209. R Miller, AV Makievski, VB Fainerman, J Krägel, F Ravera, L Liggieri, G Loglio. In: J Banhart, ed., Proceedings of the Workshop "Foams", Leuven, Belgium, 1999.
210. HA Wege, JA Holgado-Terriza, AW Neumann, MA Cabrerizo-Vilchez. Colloids and Surfaces A 156: 509–517, 1999.
211. R Miller, G Loglio, U Tesei, KH Schano. Adv Colloid Interface Sci 37: 73–96, 1991.
212. EK Zholkovskij, VI Kovalchuk, VB Fainerman, G Loglio, J Krägel, R Miller, SA Zholob, SS Dukhin. J Colloid Interface Sci 224: 47–55, 2000.
213. VI Kovalchuk, EK Zholkovskij, J Krägel, R Miller, VB Fainerman, R Wüstneck, G Loglio, SS Dukhin. J Colloid Interface Sci 224: 245–254, 2000.
214. M Joly. In: E Matijevic, ed. Surface and Colloid Science. Vol 5. New York: Wiley-Interscience, 1972, pp 1–93.
215. FC Goodrich. In: KL Mittal, ed. Solution Chemistry of Surfactants. Vol 2. New York: Plenum Press, 1979, pp 733–748.
216. J Lucassen. In: EH Lucassen-Reynders, ed. Anionic Surfactants: Physical Chemistry of Surfactant Action. New York: Marcel Dekker, 1981, pp 217–265.
217. DA Edwards, H Brenner, DT Wasan. Interfacial Transport Processes and Rheology. Boston, MA: Butterworth-Heinemann, 1991

218. B Noskov, G Loglio. Colloids Surf A 143: 167–183, 1998.
219. B Warburton. Curr Opinion Colloid Interface Sci 1:481–486, 1996.
220. R Miller, VB Fainerman, J Krägel, G Loglio. Curr Opinion Colloid Interface Sci 2: 578–583, 1997.
221. AK Malhotra, DT Wasan. In: IB Ivanov, ed. Thin Liquid Films, Surfactant Science Series, Vol 29. New York: Marcel Dekker, 1988, pp 829–890.
222. EH Lucassen-Reynders. Food Struct 12:1–12, 1993.
223. B Murray, E Dickinson. Food Sci Technol Int 2: 131–145, 1996.
224. B Murray. In: D Möbius, R Miller, eds. Proteins at Liquid Interfaces. Amsterdam: Elsevier, 1998, pp 179–220.
225. B Warburton. In: AA Collyer, ed. Techniques in Rheological Measurements. London: Chapman & Hall, 1993, pp 55–95.
226. J Benjamins, F van Voorst Vader. Colloids Surfaces A 65: 161–174, 1992.
227. R Nagarajan, DT Wasan. Rev Sci Instrum 65: 2675–2679, 1994.
228. HO Lee, TS Jiang, KS Avramidis. J Colloid Interface Sci 146: 90–122, 1991.
229. SS Feng, RC MacDonald, BM Abraham. Langmuir 7: 572–576, 1991.
230. R Nagarajan, SI Chung, DT Wasan. J Colloid Interface Sci 204: 53–60, 1998.
231. R Miller, J Krägel, AV Makievski, R Wüstneck, JB Li, VB Fainerman, AW Neumann. Proteins at liquid/liquid interfaces—adsorption and rheological properties. Proceedings of the Second World Emulsion Congress, Bordeaux, 1997, Vol 4, pp 153–163.
232. J Krägel, S Siegel, R Miller, M Born, KH Schano. Colloids Surfaces A 91: 169–180, 1994.
233. I Lakatos, J Lakatos-Szabo. Colloid Polymer Sci 275: 493–501, 1997.
234. FO Opawale, DJ Burgess. J Colloid Interface Sci 197: 142–150, 1998.
235. RA Mohammed, AI Bailey, PF Luckham, SE Taylor. Colloids Surfaces A 91: 129–139, 1994.
236. AE Cardenas-Valera, AI Bailey. Colloids Surfaces A 79: 115–127, 1993.
237. JY Zhang, LP Zhang, JA Tang, L Jiang. Colloids Surfaces A 88: 33–39, 1994.
238. JY Zhang, XP Wang, HY Liu, JA Tang, L Jiang. Colloids Surfaces A 132: 9–16, 1998.
239. M Faergemand, BS Murray, E Dickinson. J Agric Food Chem 45: 2514–2519, 1997.
240. LG Ogden, AJ Rosenthal. J Colloid Interface Sci 191: 38–47, 1997.
241. LG Ogden, AJ Rosenthal. J Am Oil Chem Soc 75: 1841–1847, 1998.
242. R Wüstneck, J Krägel, R Miller, PJ Wilde, DK Sarker, DC Clark. Food Hydrocoll 10: 395–405, 1996.
243. RMA Azzam, NM Bashara. Ellipsmetry and Polarized Light. Amsterdam: North Holland, 1979.
244. R Reiter, H Motschmann, H Orendi, A Nemetz, W. Knoll. Langmuir 8: 1784–1788, 1992.
245. R Teppner, S Bae, K Haage, H Motschmann. Langmuir 15: 7002–7007, 1999.
246. D Ducharme, A Tessier, RM Leblanc. Rev Sci Instrum 58: 571–578, 1987.
247. M Paudler, J Ruths, H Riegler. Langmuir 8: 184–189, 1992.
248. D Ducharme, J-J Max, C Salesse, RM Leblanc. J Phys Chem 94: 1925–1932, 1990.
249. J Lekner. Theory of Reflection of Electromagnetic and Particle Waves, Boston, MA: Martinus Nijhoff 1987.
250. SWH Eijt, MM Wittebrood, MAC Devillers, T Rasing. Langmuir 10: 4498–4502, 1994.
251. JA De Feijter, J Benjamins, FA Veer. Biopolymers 17: 1759–1772, 1978.
252. RW Collins, DE Aspnes, EA Irene, eds. Spectroscopic ellipsometry. Proceedings of the Second International Conference, Charleston, SC, 1997.
253. M Harke, R Teppner, O Schulz, H Orendi, H Motschmann. Rev Sci Instrum 68: 3130–3134, 1997.
254. AI Rusanov, VI Pshenitsyn. Dokl Akad Nauk SSSR 187: 619–629, 1969.
255. A Hirtz, W Lawnik, GH Findenegg. Colloids Surfaces 51: 405–418, 1990.
256. BB Sauer, H Yu, M Yazdanian, G Zografi. Macromolecules 22: 2332–2337, 1989.
257. DG Goodall, ML Gee, G Stevens, J Perera, D Beaglehole. Colloids Surfaces A 143: 41–52, 1998.
258. WC Griffin. J Soc Cosmetic Chem 1: 311–326, 1949; 5: 249–262, 1954.
259. P Beecher. Emulsions, New York: Reinhold, 1965.
260. LA Morris. Manuf Chem Aerosol News 36: 66–69, 1965.
261. AI Rusanov. Chem Rev 27: 1–326, 1997.
262. PM Kruglyakov. Studies in Interface Science. Vol 9. Amsterdam: Elsevier, 2000.
263. WB Borman, GD Hall. J Pharm Soc 52: 442–455, 1964.
264. T Legras. Ann Pharm France 30: 211–232, 1972.
265. N Ohba. Bull Chem Soc Japan 35: 1016–1020, 1962; 35: 1021–1032, 1962.
266. JT Davies. Proceedings of the Second International Congress on Surface Activity. Vol 1. London: Butterworths, 1957, pp 440.
267. JT Davies, EK Rideal. Interfacial Phenomena. New York: Academic Press, 1963.
268. K Shinoda, T Yoneyama, H Tsutsumi. J Disp Sci Technol 1: 1–9, 1980.
269. L Marszall. J Disp Sci Technol 2: 443–454, 1981.
270. L Marszal. J Colloid Interface Sci 60: 570–579, 1977; 65: 589–596, 1978.
271. L Marszal. Tenside Surfactants Detergents 16: 303–311, 1979.

272. IJ Lin, JP Friend, Y Zimmels. J Colloid Interface Sci 45: 378–392, 1973.
273. S Rigelman, G Pichon, Am Perfumer 77: 31–50, 1962.
274. N Schott. J Pharm Soc 60: 648–662, 1971.
275. P Winsor. Solvent Properties of Amphiphilic Compounds. London: Butterworths, 1954.
276. PM Kruglyakov, AF Koretskij. Dokl AN SSSR 197: 1106–1109, 1971.
277. GMM Cook, WR Rodwood, AR Taylor, DA Haydon. Kolloid ZZ Polym 227: 28–37, 1968.
278. SH Ikeda. Adv Colloid Interface Sci 18: 93–125, 1982.
279. OB Ho. J Colloid Interface Sci 198: 249–260, 1998.

# 2

# A Few Examples of the Importance of Phase Diagrams for the Properties and Behavior of Emulsions

**Stig E. Friberg**
*Clarkson University, Potsdam, New York*

## I. INTRODUCTION

The traditional definition of emulsions (1) as consisting of two liquids, of which one is dispersed in the other in the form of macroscopic droplets, was modified by the IUPAC Commission for Nomenclature (2) to include lyotropic liquid crystals. This change was justified by the fact that a large number of commercial emulsions within the areas of foods, pharmaceutics, and personal care contain such structures. Commercial emulsions frequently also contain solid particles, but such systems are usually not called emulsions, but rather emulsions–suspensions to avoid having the term emulsions covering the majority of dispersed systems.

The essential of the emulsion definition is the multiphase feature distinguishing emulsions from microemulsions, which by definition are single-phase liquids (3). This distinction, although not appreciated immediately (4), is essential. With modern mechanical emulsifiers and a judicial choice of components, it certainly is possible to produce liquid dispersions with the dimension of the dispersed phase less than that of a microemulsion of large dimensions, but such emulsions are not microemulsions. They are thermodynamically unstable and, hence, emulsions.

With more than two phases present, phase diagrams become a useful tool to describe the emulsion. It is essential to realize that in an emulsion three compounds may give rise to more than three phases; emulsions are not equilibrium systems.

Phase diagrams are not only a useful tool; they are a necessity in other facets of emulsion applications; this is the case when the application involves evaporation (personal-care formulations, pharmaceutics) or dilution (agricultural emulsions, foods). Some complex model systems will be discussed in the sections designated to these areas; in this introduction only the phase changes in a simple two-phase emulsion (5) will be reviewed to illustrate this point.

The original emulsion is an oil/water (O/W) emulsion with a composition (in percentage) of water/oil/surfactant (W–O–S) 54–40–6. The surfactant is Tween 80, a water-soluble surfactant, and the oil is soybean oil, a liquid triglyceride. This is the simplest case of an emulsion, and the evaporation should in principle consist of the amount of water being reduced causing an inversion from the original O/W emulsion to a W/O emulsion, followed by a slow reduction of the water droplet size and a final disappearance of them to leave an oil phase.

The experimental results show a significantly more complex behavior. After the inversion (between 50 and 80% of the water evaporated) the water droplets (Fig. 1A) appear black, when viewed between crossed polarizers in an optical microscope, as expected. They are formed from an isotropic liquid in another isotropic liquid. However, when 82% of the water is evaporated,

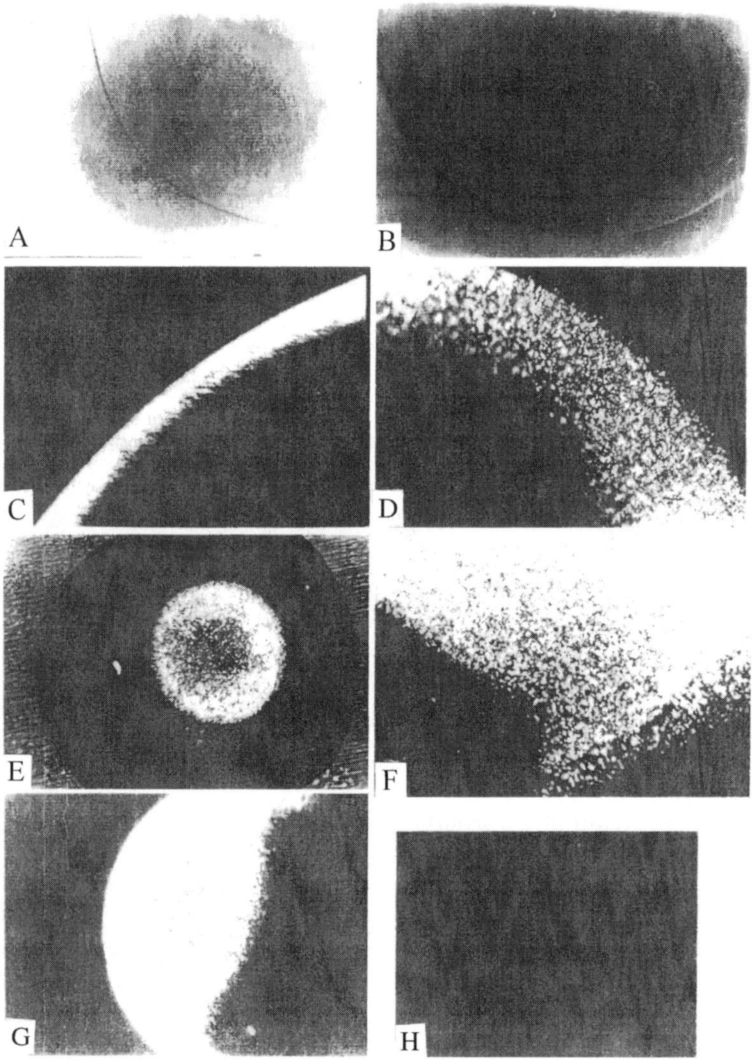


**Figure 1** The optical pattern with sample between crossed polarizers of a simple emulsion during evaporation. (From Colloid & Interface Science with due permission.).

a thin radiant rim in the droplet appears. This radiant part grows with continued evaporation. When 89% of the water has evaporated, the entire droplet is radiant. Subsequently (93% of water evaporated), a nonradiant rim is observed and when 96% of the water has evaporated the droplet is completely black and does not significantly reduce its size thereafter (even after prolonged evaporation.)

Obviously, the experimentally observed changes during evaporation of this simple emulsion are at variance with the two-phase predictions. However, the observations are obvious from the phase diagram (Fig. 2).

The composition of the original emulsion is marked by an 'X' in one of the two-phase regions in the diagram. The tie-line through the total composition (X) ends in the two phases of the emulsion; the oil and an aqueous solution with 9.6% by weight of the surfactant (Fig. 2a).

The evaporation alters the *total* composition along the dashed line emanating from the water corner. The oil phase does not change, but the aqueous phase becomes more concentrated (arrow) when the water evaporates. After the evaporation reaches point B (37% of the water removed) the amount of oil phase is now equal to that of the aqueous phase

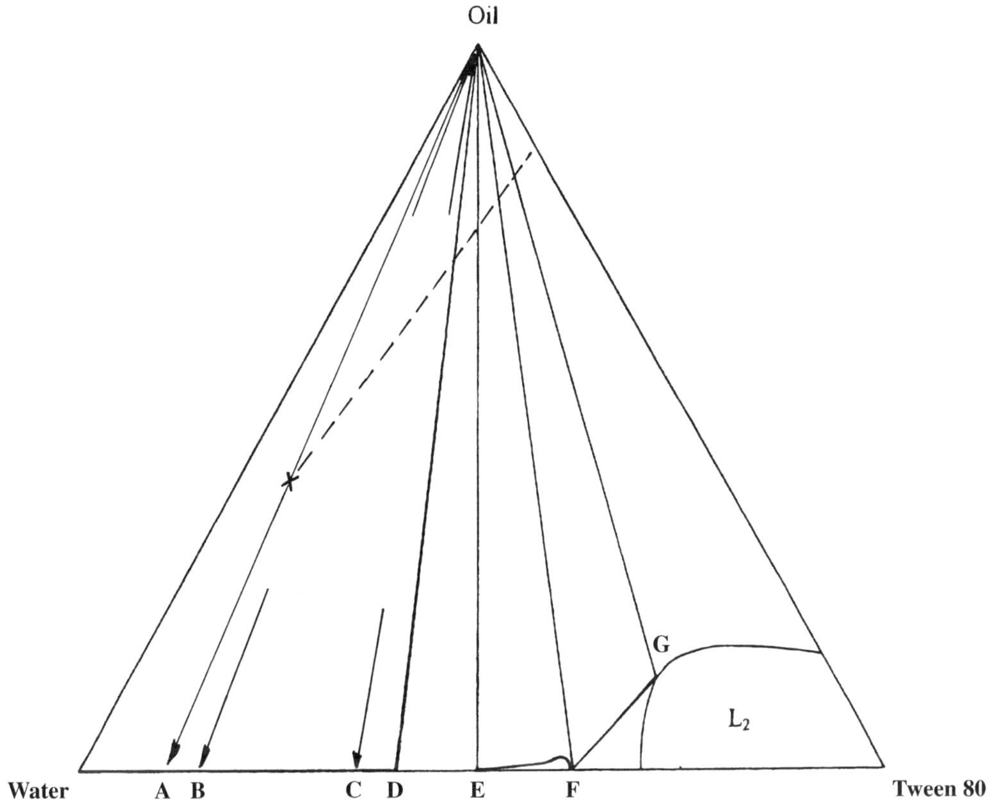


**Figure 2** Phase diagram of the emulsion in Fig. 1 (see text).

and subsequent evaporation causes an inversion in the approximate range B–C (37–79% of the water removed). The evaporation now takes place from the water droplets. When the composition of the aqueous phase reaches point D the isotropic liquid micellar solution becomes saturated and subsequent evaporation leads to the appearance of a new phase; a liquid crystal (LC) of hexagonally close-packed amphiphile cylinders (Fig. 3, bottom). It is easily identified by its characteristic optical pattern when viewed microscopically between crossed polarizers (Fig. 3, top).

The experimental results (Fig. 1) show this phase to be formed and to stay dispersed within the water droplet. In an equilibrium system, this would be interpreted as proof of an interfacial-tension relationship:

$$\gamma_{W/LC} < \gamma_{O/LC} \tag{1}$$

However, the emulsion is not at equilibrium and no conclusion about the relative size of interfacial energies may be drawn from such experimental results.

When the evaporation reaches E the droplets consist entirely of the highly viscous liquid crystal and the emulsion is now transferred to a suspension of almost solid particles (although by definition it is an LC/O *emulsion*). Continued evaporation takes place from the liquid-crystalline particles to point F, when the surfactant liquid G begins to form inside the liquid-crystalline particles. At G all the liquid crystal is changed to the surfactant liquid and with the last water removed an emulsion of the surfactant liquid with 16% triglyceride-in-oil [14% (by weight) surfactant liquid-in-84% oil] is the final state. Owing to the extremely low vapor pressure of the oil, this is the final state for applications.

The results provide an illustration of the need for phase diagrams to be able reliably to predict the behavior of a personal care or pharmaceutical emulsion after topical application and also serves as a strong memento for the evaluation of the effects on skin of different types of formulation.

The opposite phenomenon, the behavior under dilution, is equally important. The following example

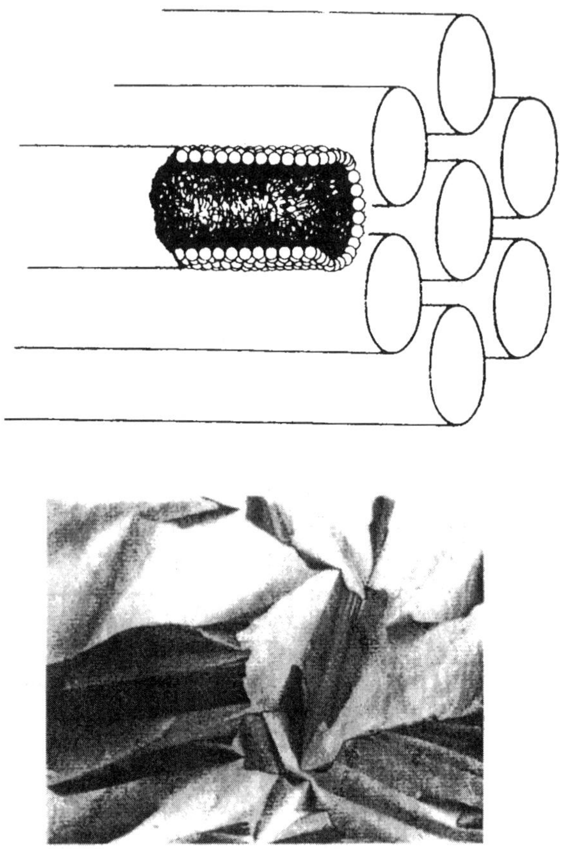

Figure 3 The liquid crystal has a structure of close-packed cylinders (top) as demonstrated by its optical pattern between crossed polarizers (bottom).

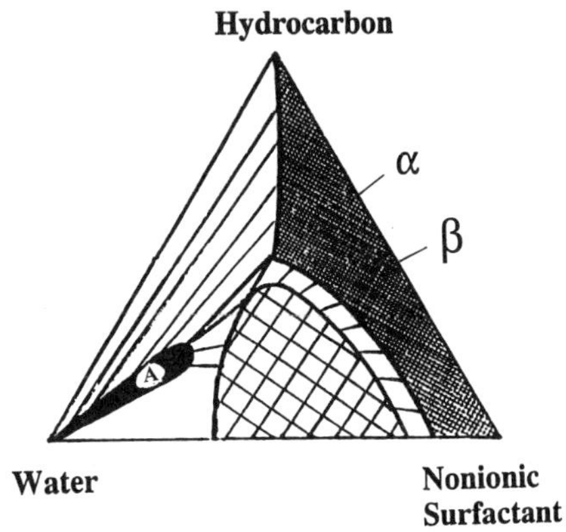


**Figure 4** An infinitely dilutable O/W microemulsion (A).

is from a water-free emulsion that is diluted to 3% in water during application. The formulation may be for agricultural application of herbicides or pesticides, but the example is more general. It gives the condition for obtaining a one-phase system from an emulsion after dilution. Figure 4 shows a typical phase diagram.

The condition to retain a one-phase isotropic liquid during "infinite" dilution is given by the limits for the one-phase region A: an O/W microemulsion. The water-free original formulation is limited to compositions along the "oil"–surfactant axis, and the limit for original compositions to give "infinite" dilutability is marked $\alpha$ and $\beta$ on that axis. Composition $\alpha$ shows the greatest possible amount of "oil" (63%) in the original composition and $\beta$ (47%) the minimum amount. The dilution means a rather complicated array of phase changes, but the final result is a one-phase isotropic liquid A.

These two systems have demonstrated the importance of phase diagrams for emulsions. The following sections will provide a more detailed treatment of some systems from the literature. The discussion will be limited to emulsion stability, and behavior under evaporation.

## II. PHASE DIAGRAMS AND EMULSION STABILITY

Simple emulsions are, of course, two-phase liquid systems with no surfactant present; hydrocarbons or fatty oils do not mix with water. Addition of a surfactant may, depending on the system, lead to the formation of a third phase. The structure of this phase is decisive for the stability of the emulsions.

When the third phase is a liquid (6–15) the emulsion becomes extremely unstable, while, if the third phase is a lamellar liquid crystal, the stability is significantly enhanced (16–18).

In the first case the third phase, a bicontinuous microemulsion, is formed because of temperature-dependent association structures of ethylene oxide adduct surfactants, Fig. 5 (19–21). At low temperatures the surfactant forms micelles in water and the hydrocarbon is solubilized into these micelles (Fig. 5a).

Increasing temperature changes the hydration of the surfactant polar groups, the area per polar group is reduced, and the Ninham $R$ value:

$$R = \nu_H / a_o \ell \qquad (2)$$

($\nu_H$ = volume occupied by the hydrocarbon chain, $a_o$ = area occupied by the polar group, and $\ell$ = the approximate hydrocarbon-chain length) is increased

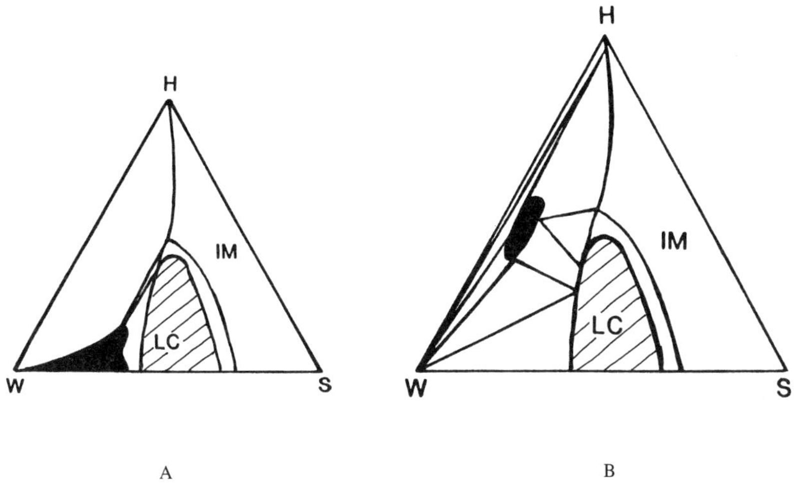


**Figure 5** At low temperatures (A) a nonionic surfactant (S) forms a micellar solution (black) in water (W); it solubilizes a hydrocarbon (H). At enhanced temperatures (B) the micellar solution is changed to a bicontinuous microemulsion (black region).

to the range 0.5–1.0, and a bicontinuous phase is formed (Fig. 5B).

This phase has ultralow interfacial tension both to the oil phase and the aqueous phase and, hence, emulsions at that temperature are extremely unstable (22). The presence of this phase has been used by Lin (23) to obtain efficiently low-energy emulsification.

The opposite effect, the formation of a liquid crystal as a third phase, also depends on the Ninham *R* ratio. It is better discussed using the conditions in water–surfactant systems.

These are of two kinds related to each other by the difference in association structure as illustrated by the temperature variation of surfactant solubility and association. Figure 6 provides a schematic description of the interdependence. At low temperatures the solubility limit of the unimers (s, solid line, Fig. 6) is lower than the limit for amphiphilic association (cmc, dashed line, Fig. 6), and, hence, the latter is not reached and a two-phase equilibrium, aqueous solution of monomers—hydrated surfactant, is established. At temperatures in excess of the Krafft point, $T_K$ (Fig. 6), the association concentration (cmc, solid line, Fig. 6), is now beneath the solubility limit (s, dashed line, Fig. 6). Association takes place and the total solubility (ts, Fig. 6) is drastically increased. Hence, the water–surfactant phase diagram shows a large solubility range for the isotropic liquid solution (unimers plus micelles, Fig. 6) because the association structure, the micelle, is soluble in water. This behavior is characteristic of surfactants with Ninham *R* values less than 0.5.

The other group, those with *R* values in the range 0.5–1.0, also associate at temperatures in excess of the Krafft point, but the molecules are now not spherically packed but rather close to parallel. As a consequence, there is no limit to the size of the association structure, as in the spherical micelles, and a phase separation occurs to form a lamellar liquid crystal. The principle features of the phase diagram in Fig. 6 remain; the Krafft point marks the intersection of unimer solubi-

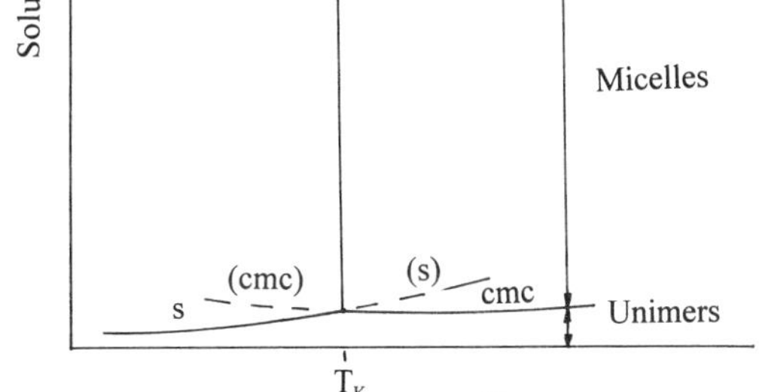


**Figure 6** The temperature-dependent solubility of a micelle-forming surfactant (see text).

lity and association concentration, but the continuous solubility region is not found (Fig. 7).

Surfactants of the latter kind, which are also not significantly soluble in the oil, such as lecithin, give rise to a lamellar liquid crystal at the smallest addition to the emulsion. The water content of the liquid crystal is approximately 50% and the percentage of liquid crystal in the emulsion is easily calculated:

$$P_{\mathrm{LLC}} = 2p_{\mathrm{S}}f_{\mathrm{w}} \tag{3}$$

in which $p_{\mathrm{S}}$ is the percentage of surfactant counted on the water plus surfactant, and $f_{\mathrm{w}}$ is the weight fraction of water in the emulsion.

With partial solubility of the surfactant in the water and in oil the expression for the amount of liquid crystal becomes cumbersome, and the importance of the surfactant concentration is best illustrated by a diagram (Fig. 8). The rapid increase of the lamellar liquid crystal with the surfactant concentration is conspicuous.

The stabilization mechanism by the liquid crystal depends on the emulsification method. With gentle emulsification the liquid crystal forms a "skin" around the droplets, as indicated by Davis (24) and demonstrated in numerous cases (17, 25). Finally, intensive emulsification gives rise to vesicles, which stabilize the emulsion (18).

A very illustrative example of this stabilization has recently been investigated (26) and will be described because of the fact that two kinds of stabilization are experienced within one system. Emulsions (95% water), numbers 1–6 (Fig. 9), are two-phase emulsions with increased amounts of surfactant and reduced amounts of oil (phenethyl alcohol). The surfactant is in the oil phase. Emulsions 7–9 are three-phase emulsions in which the amount of lamellar liquid crystal in the two non-aqueous phases increases with the higher numbers. Emulsions 10–11 are two-phase systems of water plus lamellar liquid crystal.

The enhanced stability for emulsions 7–9 is expected; the amount of liquid crystal is increased, but the greater stability of emulsions 3 and 4 among the two-phase ones needs an explanation. It is due to density matching; the phenethyl alcohol is more dense than water while the surfactant is less dense and for emulsions 3 and 4 the density of the oil phase is close to that of the aqueous phase.

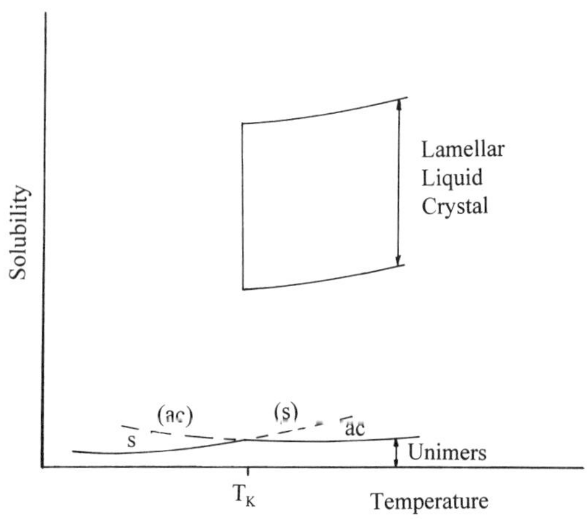


**Figure 7** The temperature-dependent solubility of a surfactant not forming micelles, but for which the primary association is with lamellar liquid crystals (see text).

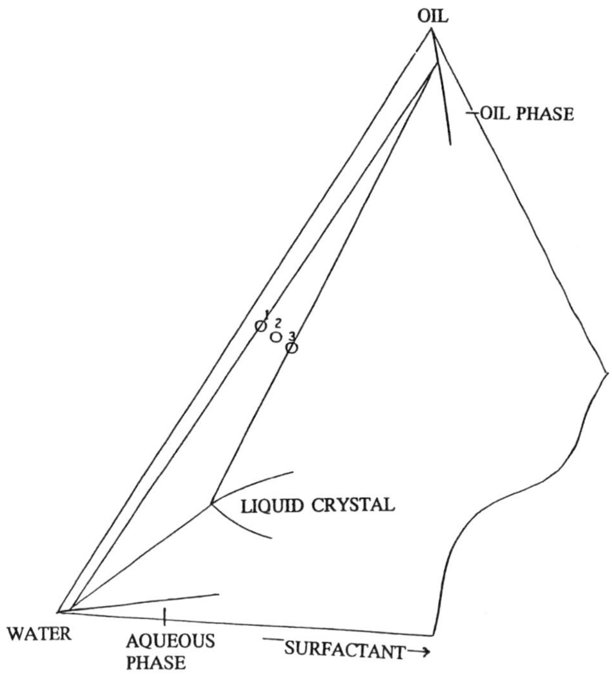


**Figure 8** The amount of liquid crystal varies strongly with the amount of emulsifier:

Percentages in the Three-phase Emulsion

| Emulsion # | Composition: Water | Oil | Emulsifier | Phases: Aqueous | Oil | Liquid crystal |
|---|---|---|---|---|---|---|
| 1 | 48.3 | 49.0 | 2.7 | 48.5 | 51.5 | 0 |
| 2 | 46.8 | 47.3 | 5.9 | 22.7 | 43.3 | 34.0 |
| 3 | 45.3 | 45.7 | 9.0 | 0 | 35.4 | 64.6 |

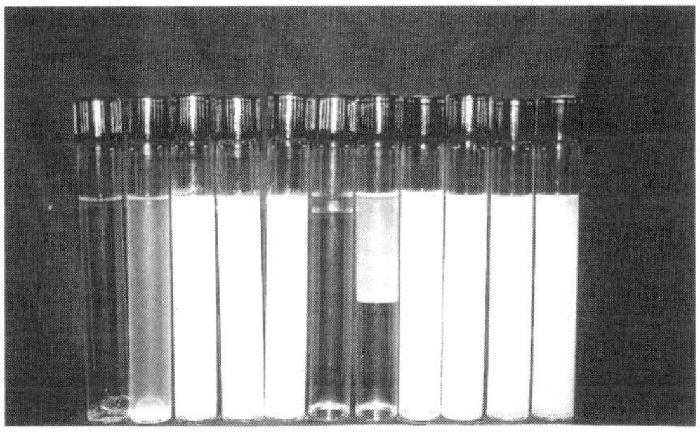


**Figure 9** Variation of stability of phenethyl alcohol emulsions with 95% water and varied emulsifier/oil content:

| Emulsion (%) | 1 | 2 | 3 | 4 | 5 | 6 | 7 | 8 | 9 | 10 | 11 |
|---|---|---|---|---|---|---|---|---|---|---|---|
| Phenethy alcohol | 5 | 4.5 | 4 | 3.5 | 3 | 2.5 | 2 | 1.5 | 1 | 0.5 | 0 |
| Surfatant | 0 | 0.5 | 1 | 1.5 | 2 | 2.5 | 3 | 3.5 | 4 | 4.5 | 5 |

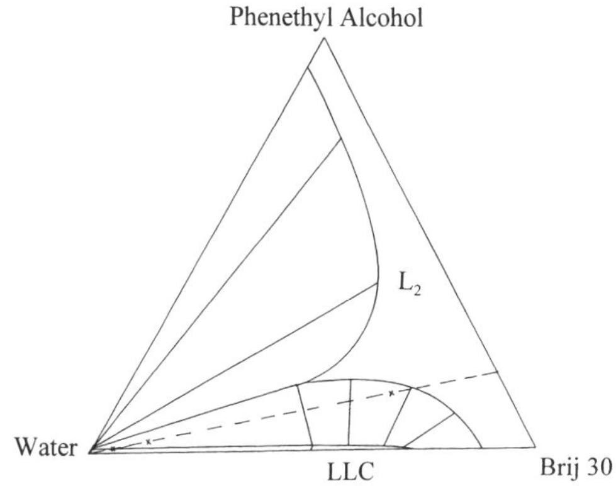


**Figure 10** Phase diagram of a simple fragrance compound system (phenethyl alcohol) with water and a commercial nonionic surfactant (Brij 30) (see text).

## III. PHASE DIAGRAMS AND EVAPORATION FROM EMULSIONS

This section will use examples from two areas of emulsion technology to illustrate the advantages and limitations in the use of phase diagrams to evaluate the structural changes during evaporation. As the first example, a system with a solubilized fragrance compound will be chosen, because of the relationship between the vapor pressure and the different phases.

The phase diagram (Fig. 10) will first be described in order to relate the structural changes during evaporation to the equilibrium structures. The solubility of water in the phenethyl alcohol was 2% by weight and that of the surfactant was close to zero. The surfactant dissolved up to 12% of water and between 27 and 50% of lamellar liquid crystal. The phenethyl alcohol solubilized in liquid crystal was a modest 2% by weight. The two-phase region between the phenethyl alcohol surfactant solution with solubilized water and the lamellar liquid crystal ranged from zero for phenethyl alcohol to a phenethyl alcohol/surfactant ratio of 3/7; also, the point of maximum water content was ≈ 45%. This composition was in equilibrium both with an aqueous solution containing 1.3% phenethyl alcohol and with a lamellar liquid crystal of high water content (49.5%) and a small amount of phenethyl alcohol (1.1%). A small two-phase region existed along the water/surfactant axis between the water and the liquid-crystal region. A second two-phase region consisted of an aqueous solution of phenethyl alcohol dissolved in the range 1.3–2% and a water-saturated limit for the phenethyl alcohol/surfactant solution with phenethyl alcohol/surfactant ratios in excess of 3/7. Up to 7.5% of water dissolved in phenethyl alcohol, and the phenethyl alcohol and the surfactant are mutually completely soluble.

The vapor pressure of phenethyl alcohol in these phases was used to estimate its variation during vaporation (27) and was evaluated in some detail (28).

The influence of the interaction with water is shown in Fig. 11. The right-hand part of the figure shows vapor pressure with dissolved water; the values are slightly in excess of those for an ideal solution. The increase in vapor pressure, with phenethyl alcohol added to water, is extremely high and of a different magnitude to those for an ideal solution (insert to Fig. 11).

The phenethyl alcohol pressure in the phenethyl alcohol/surfactant solutions (Fig. 12) initially has slightly lower values than those for an ideal solution (hatched curve), while in the surfactant-rich part the vapor pressures are slightly higher. The vapor pressures for compositions at the limit of water solubility in the phenethyl alcohol/surfactant region (Fig. 13), follows those for an ideal solution except in the range of high water content, in which they are significantly in excess.

These values have one feature of interest for application purposes when combined with the results from stability determinations (see Sec. II). The vapor pressure of phenethyl alcohol in the liquid crystal with approximately 1% of it solubilized is equal to that of the liquid with 16% alcohol. The consequence is extre-

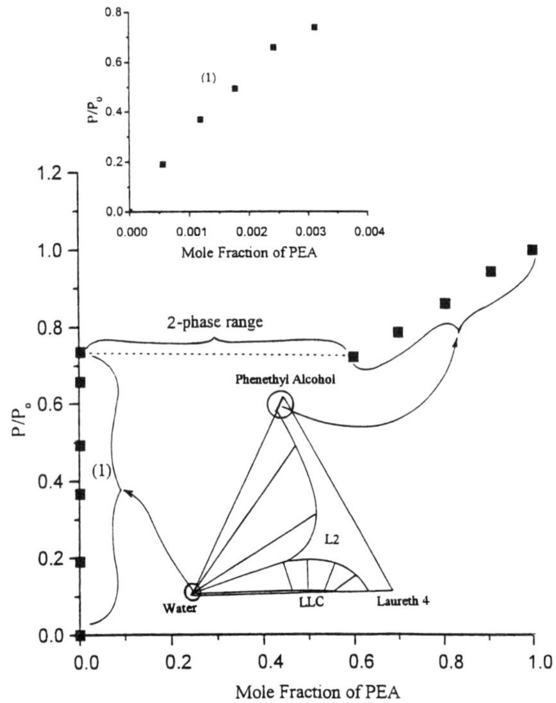


**Figure 11** Vapor pressure of water–phenethyl alcohol solutions: $P$ = partial vapor pressure of the alcohol; $p_o$ = vapor pressure of the pure alcohol.

mely important for formulation purposes; a composition with a 10 times lower amount of alcohol has identical vapor pressure! Relating this information to the fact that the fragrance is the most expensive part of a personal-care product, the conclusions are obvious.

With this fact established a description of the evaporation path and the concomitant vapor pressure variation is of interest.

The initial direction of the evaporation path is determined by the vapor pressure of water and fragrance (Fig. 14) and for an emulsion the vapor pressure of the oil is to a first approximation monitored by the volume ratio of water to surfactant in the aqueous phase (29). The composition on the water–fragrance axis of the vapor is calculated as the weight fraction of water $F_F^1$ in the vapor:

$$F_F^1 = P_F M_F/(P_F M_F + P_W \cdot 18) \tag{4}$$

in which $P_w$ and $P_F$ are the vapor pressures of water fragrance, respectively, while $M_F$ is the molecular weight of the fragrance compound. Simplifying the expression by putting $P_W = 20$ mmHg and $M_F = 180$ one obtains:

$$F_F^1 = (1 + 2/P_F)^{-1} \tag{5}$$

The vaporation trajectory is a straight line through points 1 and 2 (Fig. 14):

$$F_F - F_F^1 = (F_F^0 - F_F^1)F_S/F_S^0 \tag{6}$$

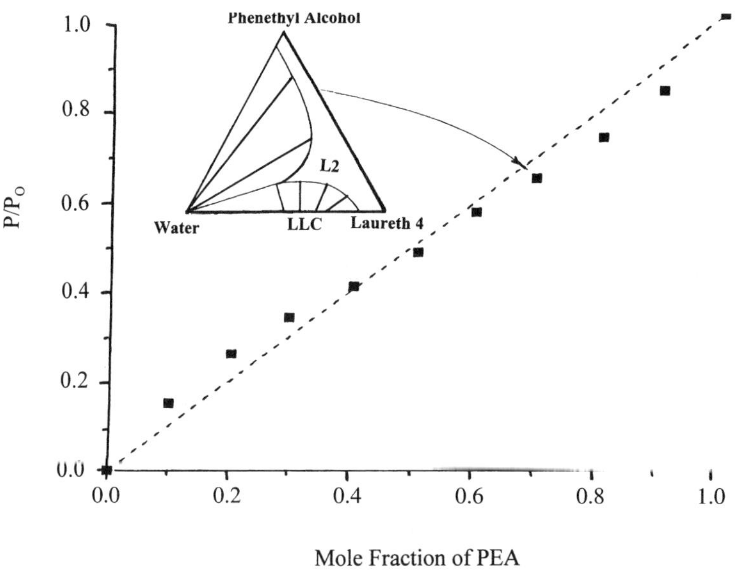


**Figure 12** Vapor pressure of surfactant–phenethyl alcohol solutions: $p$ = partial vapor pressure of the alcohol; $p_o$ = vapor pressure of the pure alcohol.

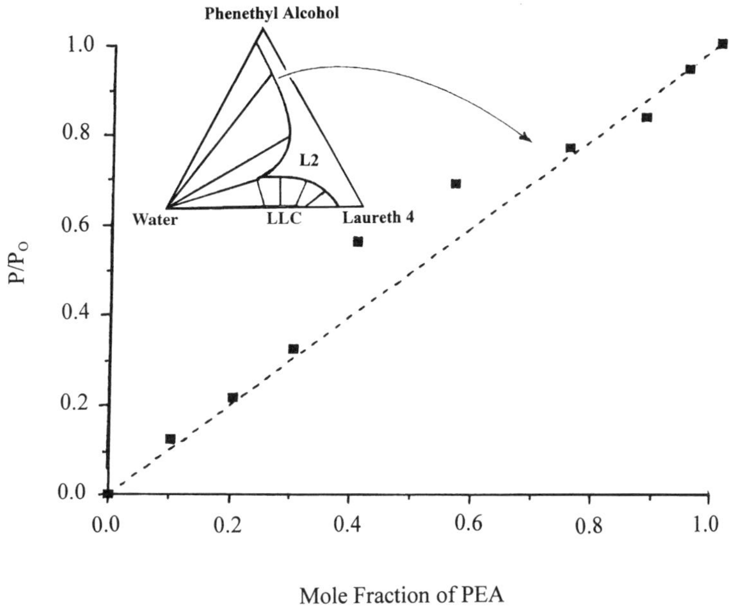


**Figure 13** Vapor pressure of water–phenethyl alcohol solutions saturated with water: $p$ = partial vapor pressure of the alcohol; $p_o$ = vapor pressure of the pure alcohol.

in which $F_F^0$ and $F_S^0$ are the initial weight fractions of fragrance compound and surfactant, respectively.

Putting $F_F = 0$ gives the weight fractions of surfactant left when the fragrance is evaporated, assuming a straight-line dependence through the entire process. An assumption that gives too large a value for $F_S$ (30, 31).

Putting $F_F^0 = \alpha F_F^1$ one obtains

$$F_S = F_S^0/(1 - \alpha) \tag{7}$$

For $\alpha = 1$, $F_S = \infty$ as expected and, furthermore, to obtain the pure surfactant directly as the end product:

$$\alpha = 1 - F_S^0 \tag{8}$$

In addition, Eq. 6 gives the weight fraction of surfactant at zero water content.

$$F_S = F_S^0(1 - F_F^1)/\{F_F^1(\alpha - 1) + F_S^0\} \tag{9}$$

where $\alpha = 1$ gives, as expected

$$F_S = 1 - F_F^1 \tag{10}$$

and the influence by an increase of $\alpha$ beyond 1 is found by putting $\alpha = 1 + \delta$. Now

$$F_S = F_S^0(1 - F_F^1)/(F_F^1\delta + F_S^0) \tag{11}$$

demonstrating the influence of $\alpha$ in reducing the amount of surfactant in the final composition.

After these general evaluations a comparison is of interest between experimental data for vapor pressures

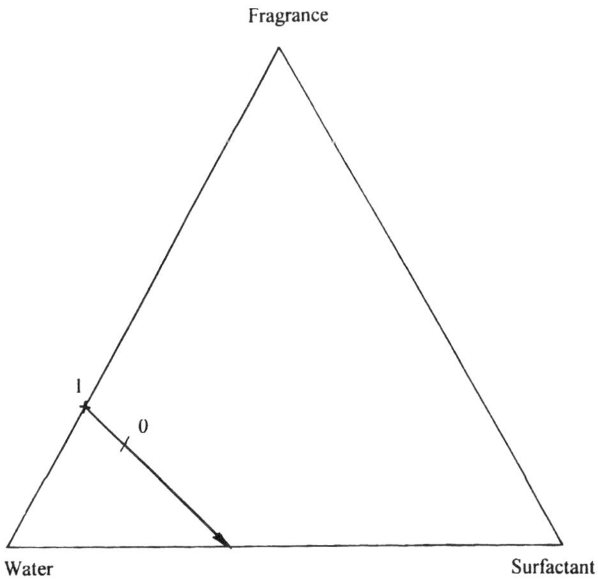


**Figure 14** The direction of the initial path of evaporation for a sample (0) in the water–surfactant–fragrance system is determined by the vapor composition (I) as obtained from the vapor pressures.

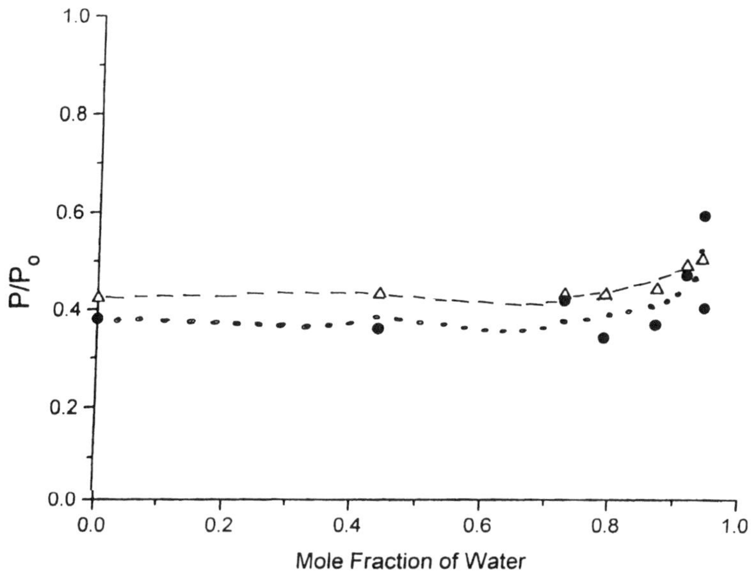


**Figure 15** The vapor pressure of phenethyl alcohol over a complex emulsion (•) during evaporation and the values calculated from phase equilibria (Δ), Fig. 10.

during evaporation and those estimated from static measurements in different areas.

The evaporation path shown in Fig. 10 is very close to a straight line owing to the low vapor pressure of phenethyl alcohol. At first the structural changes during evaporation will be described followed by a comparison between the measured vapor pressures and those estimated from the static ones.

The vapor pressure above the fragrance emulsion for different water contents is shown in Fig. 15. During the evaporation process, the vapor pressures of the fragrance varied only to a small extent. The figure also shows the vapor pressure for corresponding compositions calculated from earlier measurements (32) of vapor pressure in the entire system in Fig. 10.

The consequence of these results is a strong indication that a determination of the vapor pressures of the different phases in a complex emulsion system is sufficient to give a reasonable prediction of the variation in vapor pressure during evaporation of any formulation built on the components of the phase diagram. From an application point of view, the discrepancy between the data in Fig. 15 is not important; the measured values are lower than the estimated ones, but not significantly so. Evaporation form more complex systems has also been investigated (33). Surprisingly, even for such systems the predictions from static measurements gave a reasonable picture of the evaporation phenomenon.

The phase diagrams are also useful for estimating the decisive elements in the interaction between skin-care formulations and the skin. It is essential to realize that the initial formulation is not important for the action on the skin; its influence is limited to the esthetics, the feel, and the fragrance perception at application. These are, of course, decisive for customer selection *at the first purchase*; repeat customers depend also on the perceived action on the skin.

This means that there is no justification for analysis of the entire phase diagram; not least because such diagrams tend to be rather complicated as demonstrated by Fig. 16, which shows the complete diagram for a system of water, an $\alpha$-hydroxy acid (glucolic acid), a white oil, and a nonionic surfactant (Laureth 4) (34).

For the interaction with the skin the non-aqueous part is essential and as shown by Fig 16b, the system now becomes a simple two-phase one with glycolic acid dispersed in the white oil surfactant liquid. Hence, Fig. 16b contains the essential information; all the complex relations in the remaining parts of the figure are not directly useful for this particular aspect.

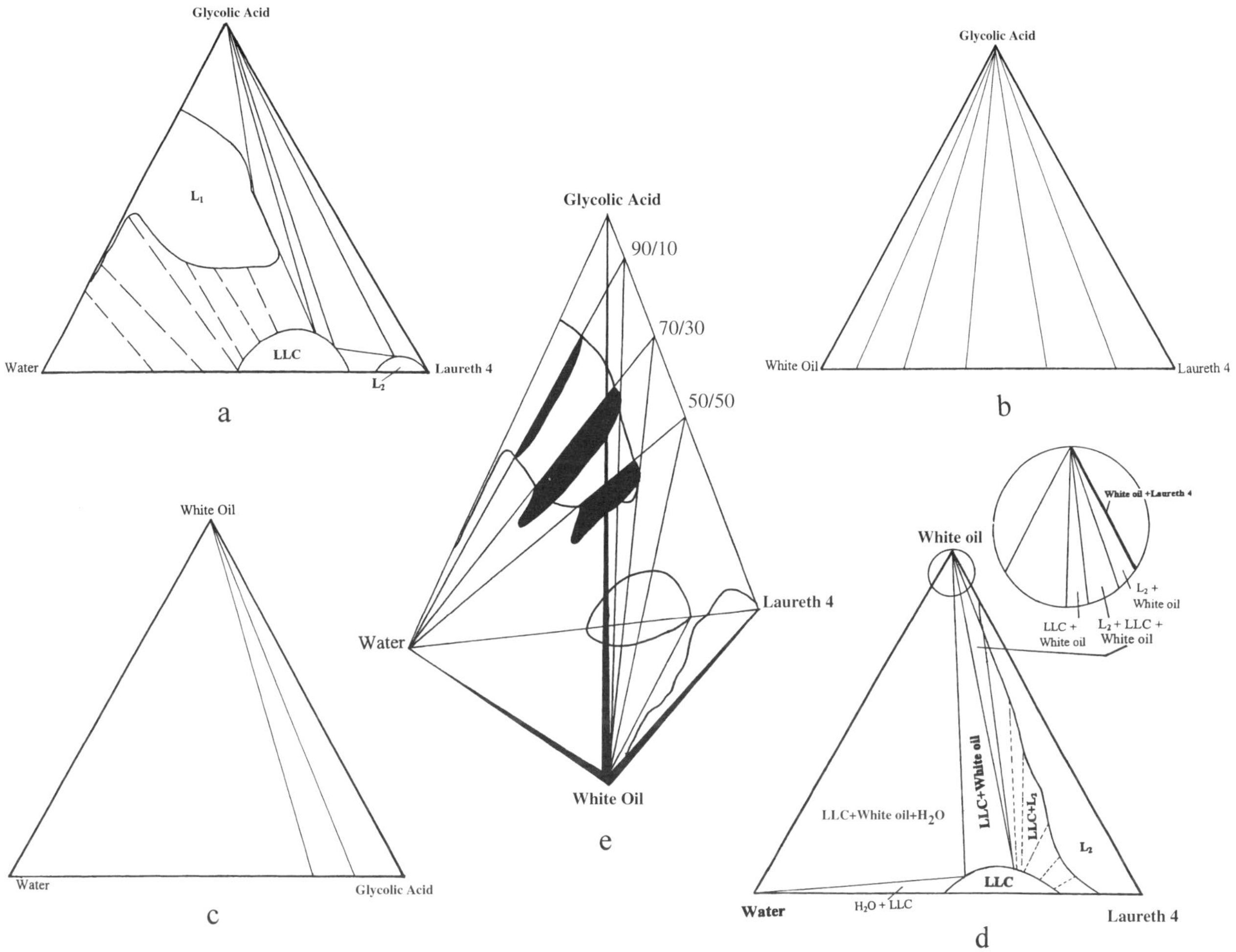


**Figure 16** The phase diagram of water–glucolic acid–white oil-Laureth 4 is complicated. (From Ref. 34.)

## IV. SUMMARY

Some pertinent examples have been given of the importance of phase diagrams for emulsion systems. Their importance in judging the essential structures for stabilization and for changes during evaporation has been emphasized.

## REFERENCES

1. P Sherman, ed. Emulsion Science. New York: Academic Press, 1968.
2. International Union of Pure and Applied Chemistry. Manual on Colloid and Surface Science. London: Butterworths, 1972.
3. B Lindman, I Danielsson. Colloids Surf 3: 391, 1981.
4. SE Friberg. Colloids Surf 4: 201, 1982.
5. SE Friberg, T Huang, PA Aikens. Colloids Surf 121: 1, 1997.
6. K Shinoda, H Arai. J Phys Chem 68: 3485, 1964.
7. K Shinoda, J Colloid Interface Sci 24: 4, 1967.
8. K Shinoda. J Colloid Interface Sci 34: 278, 1970.
9. K Shinoda, H Kunieda. J Colloid Interface Sci 42: 381, 1973.
10. S Friberg, I Lapczynska. Prog Colloid Polym Sci 56: 16, 1975.
11. S Friberg, I Lapczynska, G Gillberg. J Colloid Interface Sci 56: 19, 1976.
12. M Kahlweit. J Colloid Interface Sci 90: 197, 1982.
13. M Kahlweit, E Lessner, R Strey. J Phys Chem 87: 5032, 1983.
14. PG Nilsson, B Lindman. J Phys Chem 86: 271, 1982.
15. PG Nilsson, B Lindman. J Phys Chem 87: 4756, 1983.

16. S Friberg, L Mandell, K Fontell. Acta Chem Scand 23: 1055, 1969.
17. J Yang, SE Fribreg. In: J Sjöblom, ed. Emulsion and Emulsion Stability. New York: Marcel Dekker, 1996, p 1.
18. B Sjöström, K Westesen, B Bergenståhl. Int J Pharm 94: 89, 1993.
19. SE Friberg. In: HF Eicke, GD Parfitt, eds. Interfacial Phenomena in Apolar Media. New York: Marcel Dekker, 1987, p 93.
20. SE Friberg. Adv Colloid Interface Sci 32: 167, 1990.
21. SE Friberg. Langmuir 8: 8, 1992.
22. K Mandani, SE Friberg. Progr Colloid Polym Sci 65: 165, 1978.
23. KJ Lin. Macromolecules 1: 213, 1968.
24. JT Davies. Recent Progr Surface Sci 2: 129, 1964.
25. SE Friberg, L Mandell, M Larsson. J Colloid Interface Sci 29: 155, 1969.
26. Z Zhang, T Denler, SE Friberg, PA Aikens. Int J Cosmet Sci, 22: 105, 2000.
27. SE Friberg, M Szymula, L Fei, J Barber, A Al-Bawwab. Int J Cosmet Sci 19: 259, 1997.
28. SE Friberg, T Young, R Mackay, J Oliver, M Breton. Colloids Surf 100: 83, 1995.
29. JM Behan, KD Perring. Int J Cosmet Sci 9: 261, 1987.
30. SE Friberg, B Yu, J Lin, E Barni, T Young. Colloids Polym Sci 271: 152, 1993.
31. SE Friberg, T Young, R Mackay, J Oliver, M Breton. Colloids Surf 100: 83, 1995.
32. SE Friberg, T Huang, L Fei, SA Vona Jr, PA Aikens. Progr Colloid Polym Sci 101: 18–22, 1996.
33. SE Friberg, Q Yin, PA Aikens. Int J Cosmet Sci 20: 335, 1998.
34. SE Friberg, A Al-Bawab, JL Barber, PA Aikens. J Disp Sci Technol 19: 399, 1998.

# 3

# Structure and Stability of Emulsions

**Darsh T. Wasan and Alex D. Nikolov**
*Illinois Institute of Technology, Chicago, Illinois*

Emulsion stability is characterized in different ways—creaming or sedimentation, flocculation of drops, coalescence between drops, or phase separation. A number of novel experimental techniques have been developed in our laboratory to examine both the structure and stability of emulsions. This chapter highlights our more recent experimental methods which include: (Sec. I) film rheometry for dynamic film properties; (Sec. II) capillary force balance in conjunction with differential microinterferometry for drainage of curved emulsion films; (Sec. III) back-light scattering (Kossel diffraction) for structure factor; (Sec. IV) direct imaging for effective interdroplet interactions; and (Sec. V) piezo imaging spectroscopy for drop-homophase coalescence-rate processes. These experimental techniques are being used by us to gain a mechanistic understanding of both the structure and stability of polydisperse emulsion and foam systems (1–7).

## I. FILM RHEOMETRY

The stability of any emulsion is largely due to the nature of the film that is formed between two approaching droplets. Coalescence of drops in any emulsion system is a dynamic process. The rheological behavior of emulsions depends on the response of the thin liquid films and the plateau borders during shear and dilation. In real emulsions, the size and distribution of the drops is generally polydisperse. Hence, thin liquid films formed between drops are typically not flat, as in a homogeneous dispersion, but have a spherical, curved shape due to the capillary pressure difference between drops of unequal size.

A versatile interfacial and film rheometer has been developed in our laboratory (7–10). In this technique, a curved, spherical cap-shaped fluid interface or liquid film is formed at a capillary tip and the interfacial tension (IFT) of the single interface or the film tension of the film can be determined by measuring the capillary pressure of the interface or film (Fig. 1). The IFT or film tension is related to the capillary pressure and the radius of the interface or film curvature by the Young–Laplace equation. The IFT and film tension can be measured not only in equilibrium, but also in dynamic conditions as well. The automated apparatus makes it possible to change the interfacial or film area in virtually any mode (expansion or contraction) at various rates (Fig. 2). This instrument is now made available through our laboratory.

The flocculation and coalescence processes of a polydispersed lamella or film can be divided into two processes: film drainage and film rupture. To model the film-rupture process of polydispersed emulsions, film stress-relaxation experiments were carried out. In these experiments, the film was quickly expanded and then the relaxation of the film was measured. To characterize the film-drainage process, dynamic film-tension measurements were conducted in which the film was continuously and slowly expanded while the film tension was monitored. Single interfaces were also studied by forming a drop at the capillary (7).

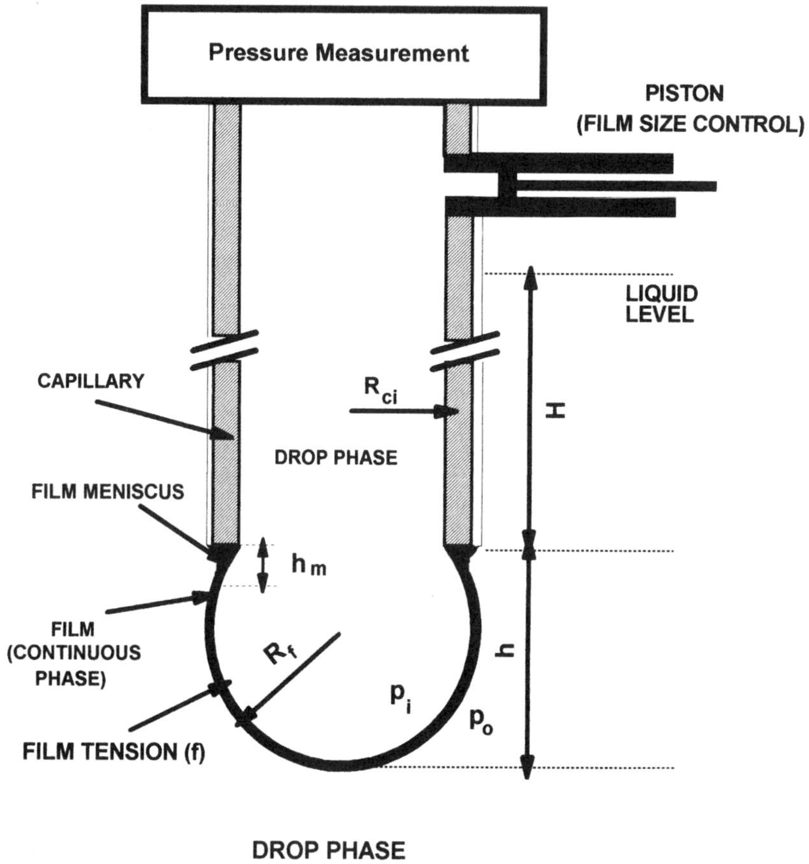


$$P_C = \frac{4f}{R_c}$$

**Figure 1** Principle of studying liquid film formed at the tip of a capillary.

Figure 3 shows a film stress-relaxation experiment with an aqueous emulsion film formed between dodecane drops. The film was suddenly expanded by 22% in area and then the film size was kept constant. The stress-relaxation curve provides information about the kinetics of emulsifier adsorption on the film surfaces. Figure 3 also shows that the reproducibility of the film stress-relaxation experiment was very good.

In the dynamic film-tension experiments, the film area is continuously increased by a constant rate and the dynamic film tension is monitored. The measured film tensions were compared with the interfacial tensions of the oil/water interfaces. It was found that under dynamic conditions, the film tension is higher than twice the single interfacial tension (Fig. 4). These results have important implications for the stability and rheology of emulsions with high disperse phase ratios (polyhedral structure).

The initial (maximum) film tension after the expansion in the film stress-relaxation experiments can also be used to determine the film elasticity (7). A plot of the initial film tension versus the logarithm of the relative film expansion is shown in Fig. 5. For comparison, the initial single interfacial tensions obtained in the experiments with the respective single oil/water interface are also plotted. The film elasticity obtained from the top of the curve is equal, within experimental error, to twice the interfacial elasticity of the single interface.

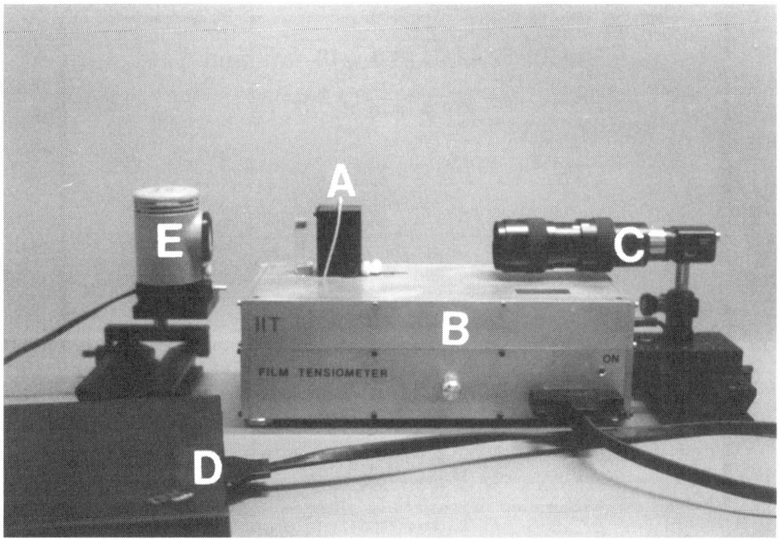


**Figure 2** Photograph of our new film rheometer: (A) capillary with capillary holder; (B) operation control; (C) zoom objective with camera; (D) computer; (E) light illuminator.

## II. CAPILLARY FORCE BALANCE

We have constructed a new surface-force apparatus capable of measuring the capillary pressure and structural disjoining pressure of the thinning curved emulsion film as a function of time and film thickness (11–16). This apparatus is equipped with Max Zhender differential interferometry (DI) which is used to measure the film curvature (2). A sketch of the surface-force balance experimental set-up is shown in Fig. 6. For measuring oil-in-water emulsions the inner capillary of the cell is filled with oil phase, the bottom part of the outer capillary is filled with water phase, and the top part of the outer capillary if filled with oil phase. A

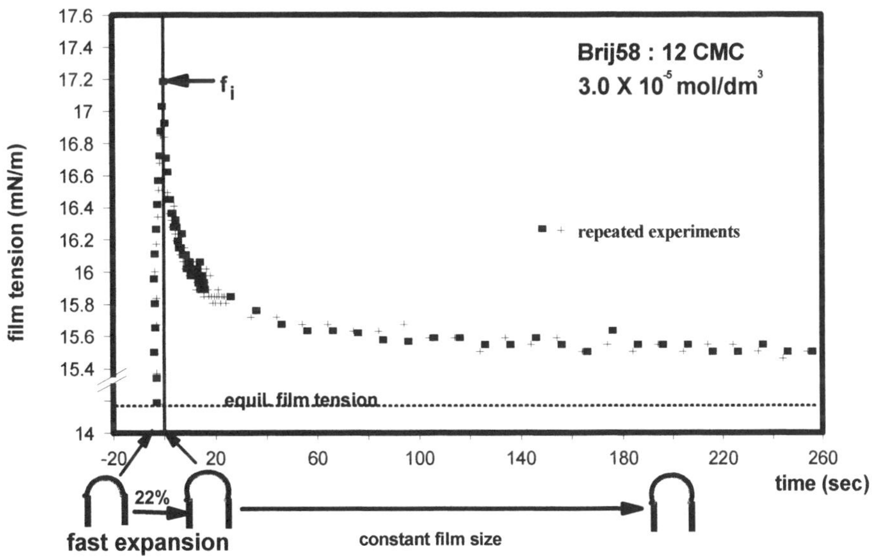


**Figure 3** Reproducibility of the film stress-relaxation experiment for the aqueous emulsion film stabilized with $3.0 \times 10^{-5}$ mol/dm$^3$ (12 CMC) Brij 58.

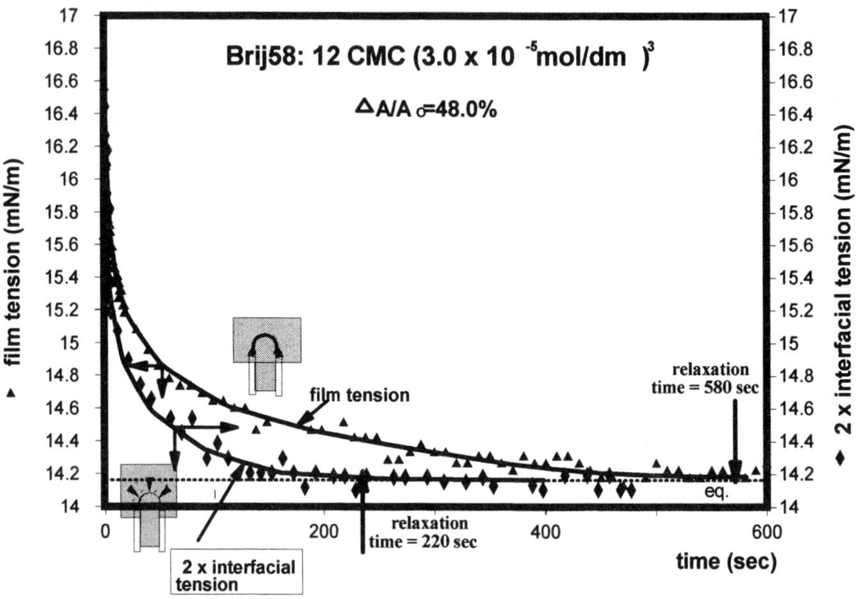


**Figure 4** Comparison of the relaxation of film tension and single interfacial tension (IFT) with 48% relative expansion of the emulsion system with $3.0 \times 10^{-5}$ mol/dm$^3$ Brij 58 (< 1 CMC).

curved film is formed by drawing the oil phase from the inner capillary, using a piston pump. The film curvature can be varied by changing the ratio of the outer to inner capillary diameter.

Also, we have recently used the capillary force balance in conjunction with reflected-light microinterferometry to study stratification (i.e., micelles ordered in layers) phenomena inside emulsion films, which is one of the key mechanisms controlling film stability (4, 11–16). Figure 7 shows a photocurrent versus time interferogram of the film-thinning process in a microscopic horizontal film (film diameter $3.6 \times 10^{-2}$cm) stabilized

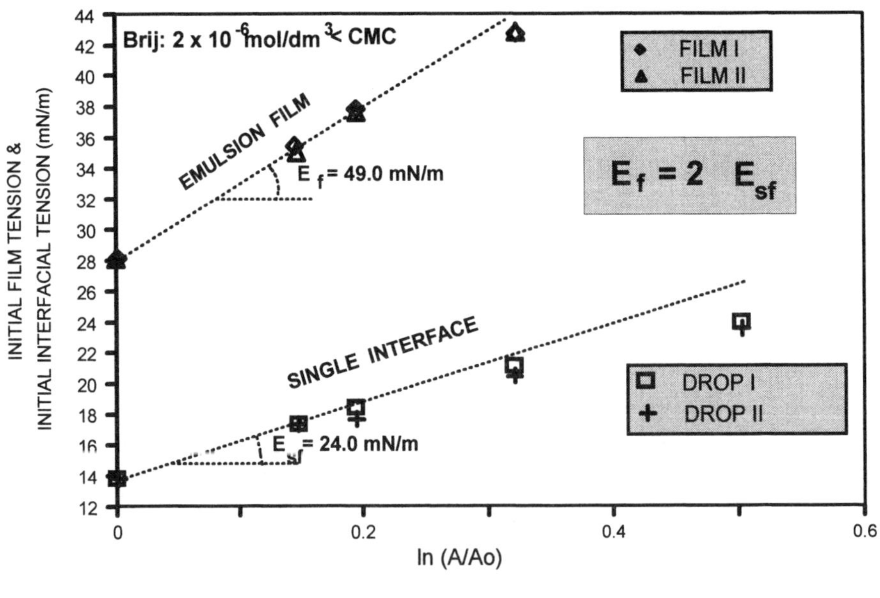


**Figure 5** Initial film tension and IFT in the stress-relaxation experiments as a function of $\ln(A/A_0)$ for the emulsion system, in the presence of $2.0 \times 10^{-6}$ mol/dm$^3$ Brij 58 (< 1 CMC).

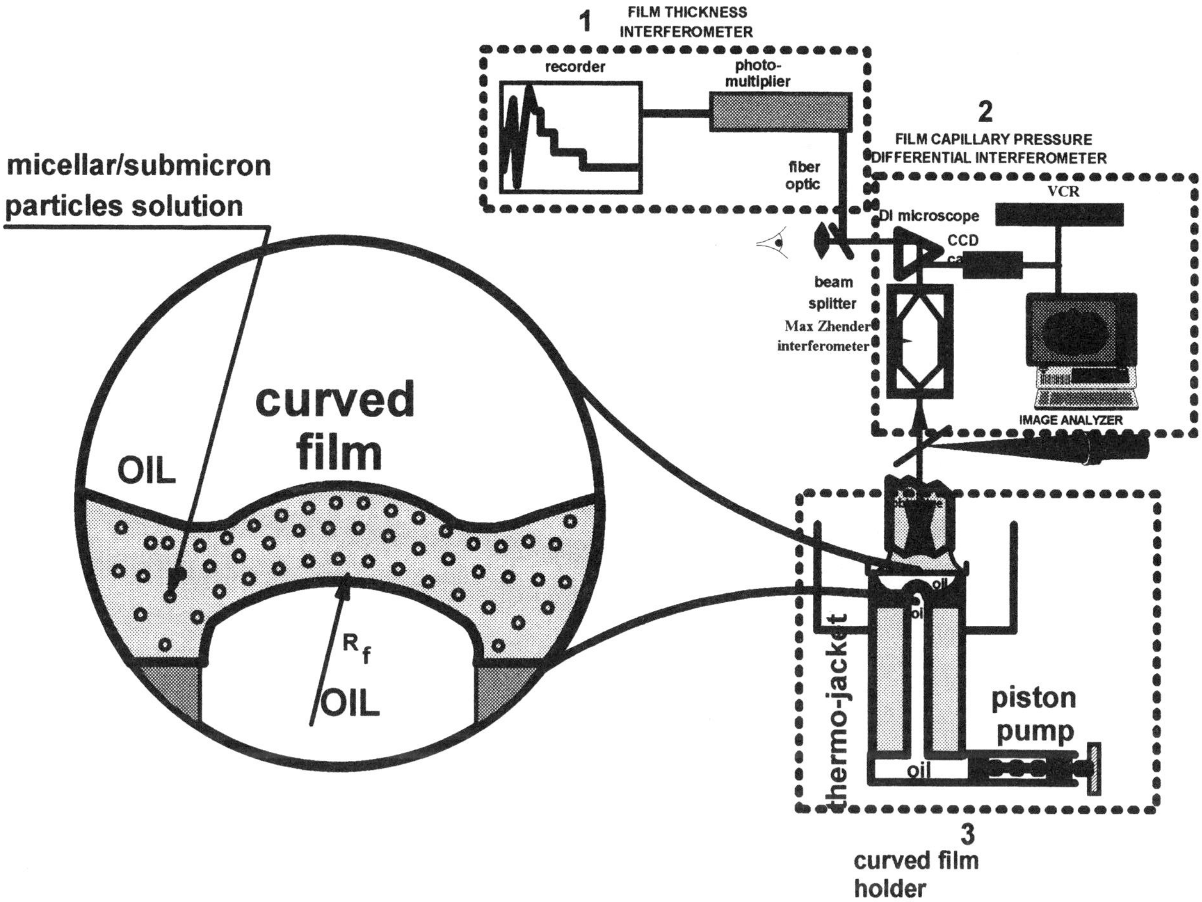


**Figure 6** Capillary force balance.

by sodium dodecyl sulfate ($6 \times 10^{-2}$M). The thickness at which the stepwise transition begins is marked with an arrow. The system consists of an oil-in-water type emulsion. In this case, the water film changes its thickness only once, and at this thickness, the film contains three layers of micelles. The final film contains two layers of micelles and is stable.

Figure 8 shows photomicrographs of the various stages of stepwise thinning of a microscopic, horizontal oil film stabilized by asphaltene particles (7 vol%) in a 1:1 volume mixture of n-heptane and toluene. At a film thickness greater than about 300-nm, the asphaltene particles inside the film form a random structure which causes the white and dark interference patterns produced in reflected monochromatic light to form a mosaic structure (Fig. 8a). The film is irregular. After a while, a white expanding spot surrounded by a dark rim appears inside the film with a thickness of about 100-nm (Fig. 8b). Here, one can see that the film thickness at the spot area appears to be much more regular than the surrounding film. Subsequently, the spot expands (Fig. 8c, d) and, finally, the white spot occupies the whole film.

We have also observed the dynamic film-thickness transition phenomenon (i.e. stratification), inside an ice-cream emulsion film, caused by layering of caseinate submicelles inside it (13).

The investigations in our laboratory showed that the film microlayering was a universal phenomenon (17, 18) which fundamentally differed from the classical film-thinning mechanism by the common black film/Newton film transition. The particles may be any kind of isotropic structures in the 10–100-nm range including micelles, fine solid particles, globular protein molecules, or random coil-shaped polysaccharide molecules or protein aggregates such as caseinate submicelles. This ordering occurs because highly charged Brownian particles (micelles) interact via repulsive

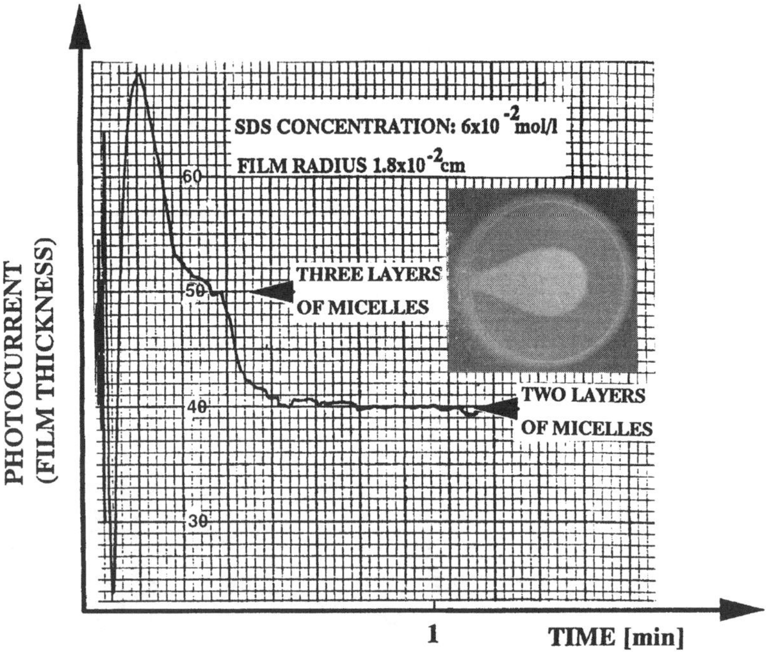


**Figure 7** Photocurrent vs. time interferogram of thinning of the emulsion film of sodium dodecyl sulfate. The photomicrograph depicts the moment of film thickness transition from three to two micellar layers.

forces inside the restricted volume of the film. The classical Derjaguin, Landau, Verwey, Overbeck (DLVO) theory of colloid stability, which explains order in colloidal systems as a balance of van der Waals attractive forces and electrostatic forces, cannot be used here because the intermicellar distances are too large for the van der Waals forces to be sufficiently significant to balance the respulsive forces (19, 20).

We have investigated theoretically film-thickness stability and structure formation inside a liquid film by Monte Carlo numerical simulations and analytical methods, using the Ornstein–Zernicke (O–Z) statistical mechanics theory (21–24). The formation of long-range, ordered microstructures (giving rise to an oscillating force) within the liquid film leads to a new mechanism of stabilization of emulsions (3, 4, 25). In addition to the effective volume of micelles or other colloidal particles and polydispersity in micelle size, the film size is also found to be the main parameter governing emulsion stability (15).

## III. BACK-LIGHT SCATTERING—KOSSEL DIFFRACTION

This optical technique can be used to investigate the structure and texture of emulsions. In the method, the emulsion, in a transparent vessel, is illuminated by a collimated laser beam. A portion of the light rays are scattered from the emulsion droplets through the wall of the vessel and form a concentric interference pattern (26). The back-scattering phenomenon is analogous to the operation of diffraction gratings. The measurement can be used to characterize the packing structure of the emulsion. The average pair potential (potential of mean force), which is the potential (free) energy of a pair of droplets in the presence of other droplets, can be calculated from the radial distribution function.

Figure 9 shows the structure factor as a function of the light-scattering vector depicting the fat particle structure inside the food emulsion. There are two samples shown in this figure. The two samples included the same fat concentration (5.14 wt%) except that the caseinate concentration inside sample 4 was half of that inside sample 1. The first peak height of structure factor $S(a)$) of the sample with higher caseinate concentration was higher, indicating that the addition of caseinate facilitates fat-particle structure formation. This could be explained by the stabilization mechanism of caseinate submicelles in the aqueous phase (13). Results for a binary system consisting of fat particles and caseinate submicelles were calculated by us from the O–Z equation (Fig.

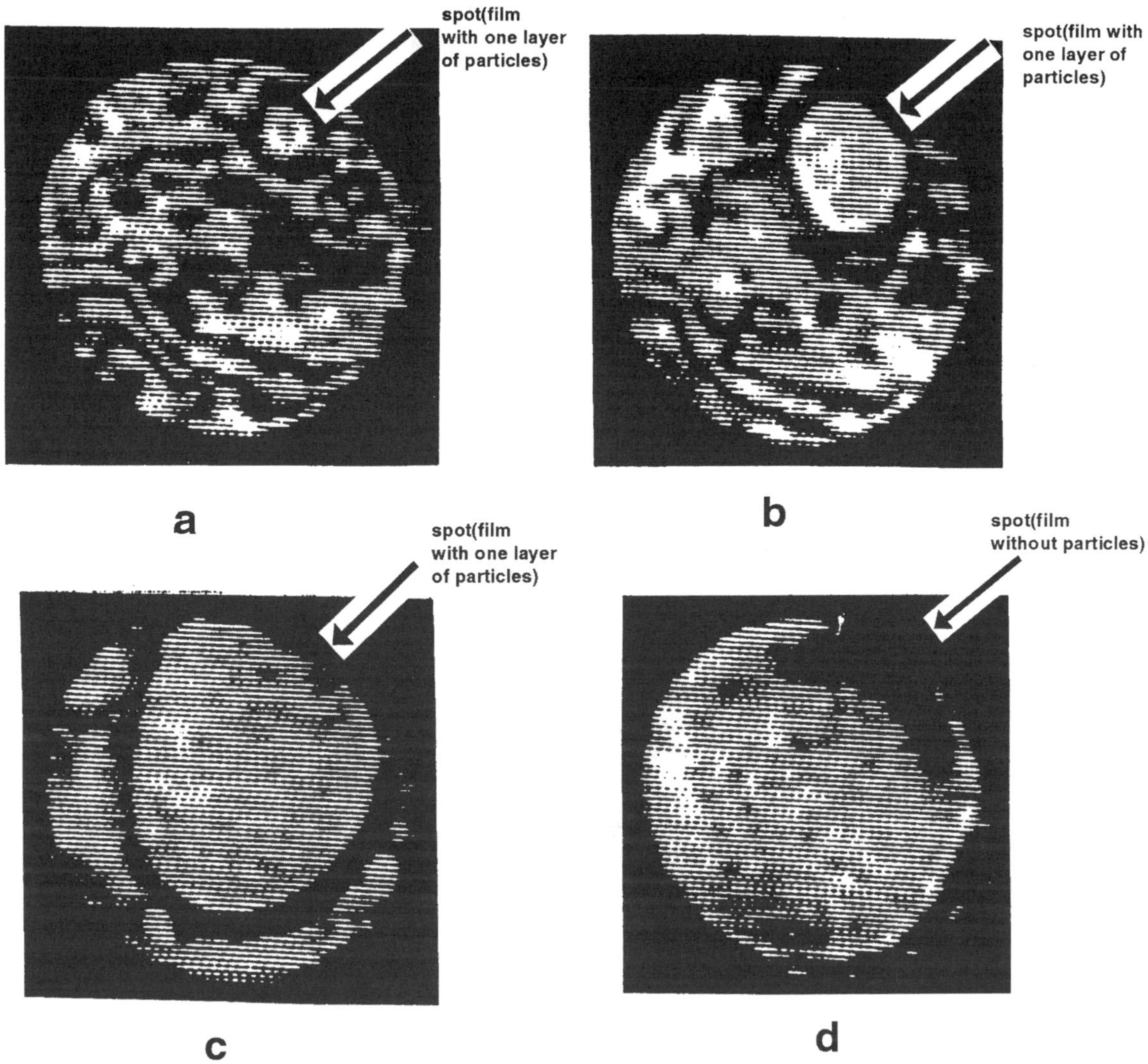


**Figure 8** Sequence of photomicrographs depicting the stages of stepwise thinning of an oil emulsion film in the presence of 7 vol% asphaltene.

10). The parameters used in these calculations were $D$ (large fat)/$d$ (small caseinate) = 20 and the volume fraction of large particles was equal to 5 wt%. We observed that with increasing caseinate concentration the structure barrier between large fat particles increased rapidly; when the caseinate concentration reached 20 vol% the structure energy barrier was larger than 3 kT. Such a high energy barrier was enough to prevent large fat-particle aggregation; therefore, the emulsion became stable. In sample 1, the caseinate submicelle concentration was estimated to be around 20 vol%. Therefore, microlayering stabilization went into effect.

We have also used the back-light scattering technique to investigate the effect of shear rate on emulsion structure. The microstructure distortions occurred at high shear rates (25).

## IV. DIRECT IMAGING

This technique is particularly useful for highly concentrated emulsion systems. We have used the digitized optical imaging technique to study the microstructure of a number of different emulsion systems (27). In this method, an emulsion sample was taken under a micro-

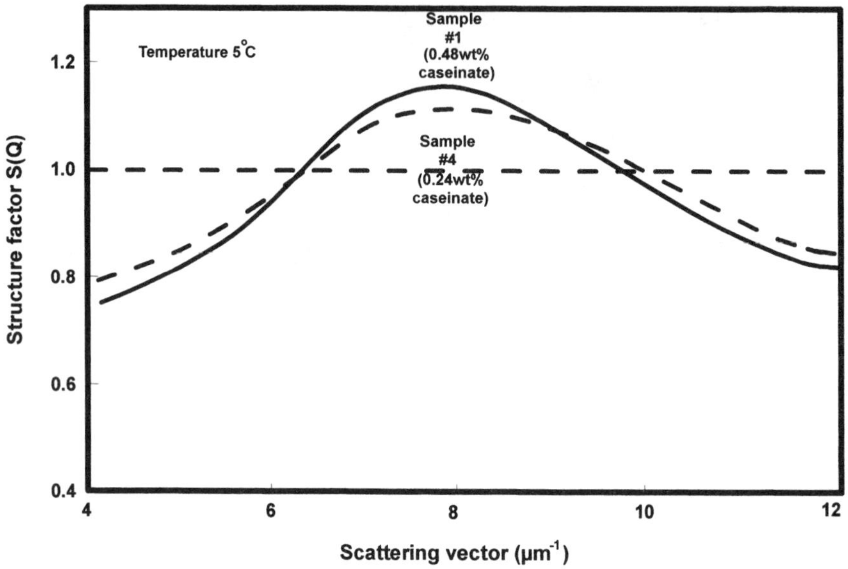


**Figure 9** Effect of caseinate on fat-particle structure inside food emulsion (fat concentration: 5.14 wt%).

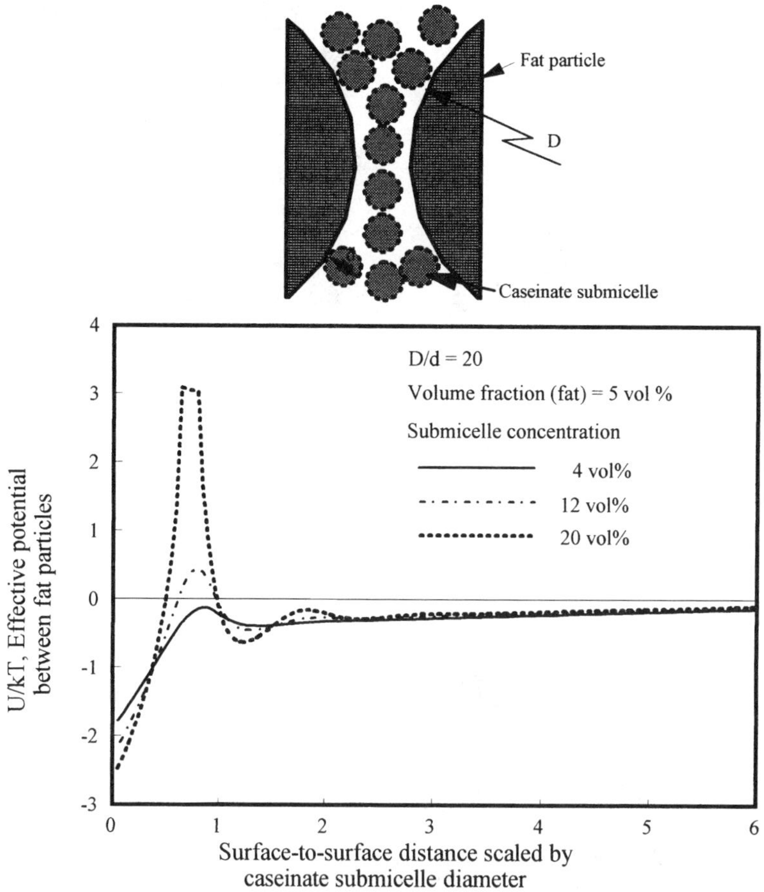


**Figure 10** Stabilization mechanism of caseinate submicelle microlayering: calculated results using O–Z method (particle size ratio: 20, and fat concentration: 5 wt%).

scope to record a microstructural image. This image was recorded using a video camera with imaging software (Image Pro) attached to the microscope. The microstructural image was magnified and then the analysis was done to measure the interdroplet distance. This acquired data was processed in MATLAB to calculate the radial distribution function (RDF) and structure factor. The RDF, $g(r)$, measured the probability of finding an emulsion droplet center at a distance, $r$, from a reference droplet. It is oscillatory in nature and tends to unity as the distance from the reference droplet tends to infinity, implying that the probability of finding a droplet at infinity is the same as that in the bulk. It typically has a maximum at a distance of one droplet diameter for a monodisperse phase system.

Figure 11 shows the $g(r)$ for two emulsion samples. The emulsion samples had the same composition, except one had sucrose ester (0.1 wt%) as the water-soluble surfactant and the other had sucrose oleate (0.1 wt%). The fat content was 40 wt%, the protein (sodium caseinate) was 4 wt%, and the water content was 56 wt%. The RDF shows that the corresponding effective pair potential of interaction between fat particles is also oscillatory. The periodicity of the curve is nearly the size of the particles. The structure factor $S(\sigma)$ for these samples is shown in Fig. 12. The first peak height of the structure factor of the sucrose oleate sample is higher, indicating that the addition of sucrose oleate facilitates the fat-particle structure formation. Thus, the fat-particle structure in the sucrose oleate sample is much better organized than the particle structure in the sucrose ester sample. This leads to higher fat-particle flocculation in the sucrose stearate sample and the emulsion is less stable.

In summary, the nondestructive digitized imaging technique is very useful for studying structure formation in oil–water emulsion systems.

## V. PIEZO IMAGING SPECTROSCOPY

This technique is based on using a piezo-transducer to monitor the process of coalescence of a drop at a liquid–liquid interface (28). A drop is formed at the tip of a capillary; the drop causes the interface to oscillate, and the oscillations of the interface are traced on a digital storage oscilloscope. The typical response of the piezo-transducer consists of an initial high-frequency (10 Hz), low-amplitude damped oscillation followed by a relatively low frequency (2 Hz), higher amplitude, highly damped oscillation (Fig. 13). The high-frequency part of the signal is attributed to film rupture while the low-frequency part is due to the formation of a jet during film drainage. Both the frequency of the signal and the damping factor are calculated from these measurements (28). This novel technique for monitoring drop/homophase coalescence in liquid–liquid dispersions having an opaque or turbid dispersion medium has been developed by us. The effects of interfacial tension, homophase viscosity, and surfactant concentration in the dispersion on the coalescence process have been studied. Several

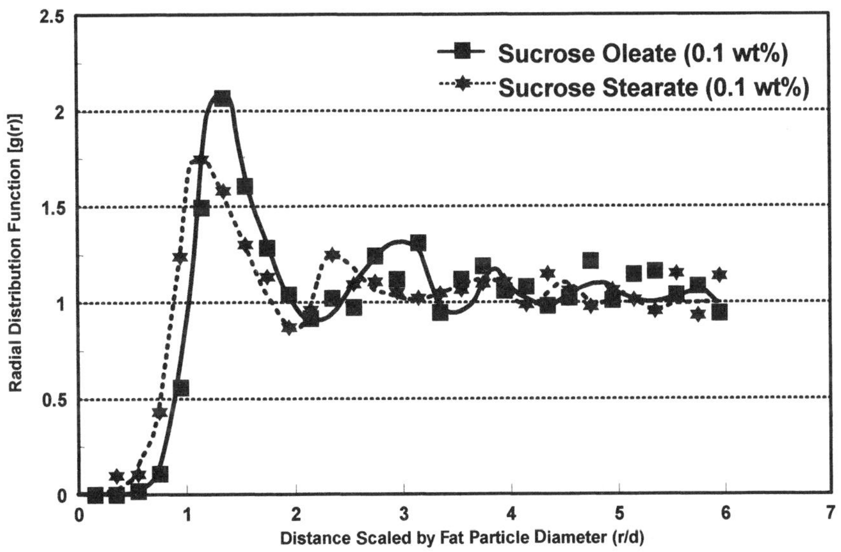


**Figure 11** Effect of surfactant on radial distribution function of fat particles in oil-in-water emulsion.

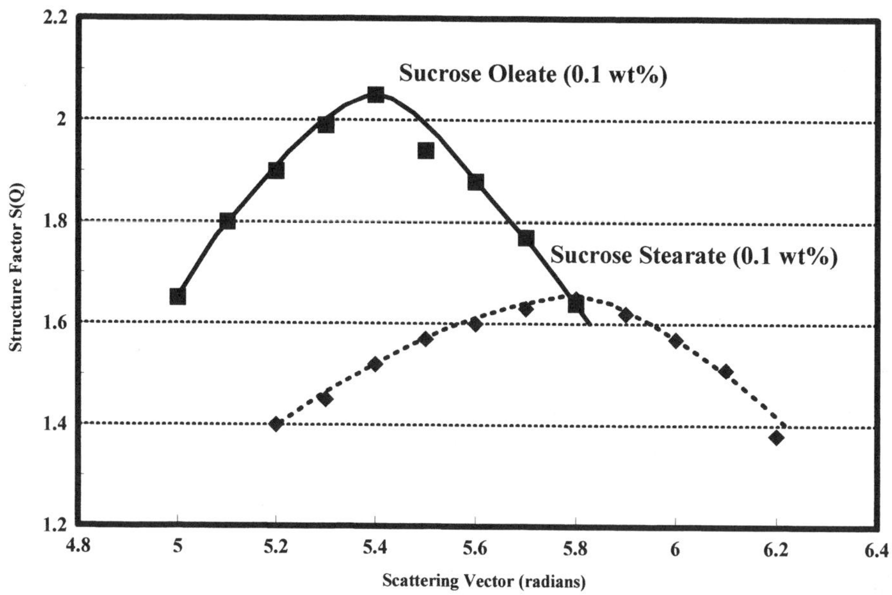


**Figure 12** Effect of surfactant on structure factor of fat particles in oil-in-water emulsion.

other potential applications of this technique such as measuring the rate of phase separation are also discussed by us elsewhere (28).

## VI. SUMMARY

The research discussed above is based on the work performed in our laboratory in the subject area of emulsion microstructure and stability. A critical thrust of our on-going research program has been the development of instrumental techniques for understanding the mechanism of emulsion stability in various systems including food, pharmaceutical, cosmetic, and petroleum emulsions. The development of reliable measurement techniques has been followed up by us in a series of studies, both theoretical and experimental, which were aimed at understanding the role of dynamic interfacial prop-

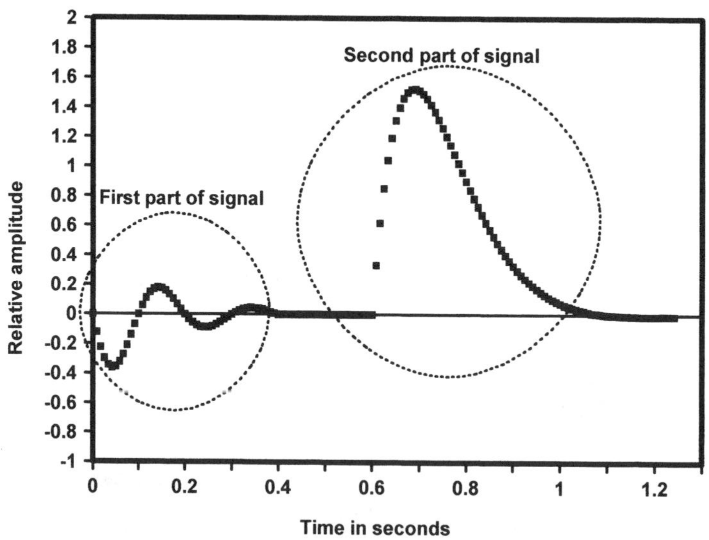


**Figure 13** Typical transducer response to droplet–fat interface coalescence.

erties in the stability of thin liquid films associated with drops and bubbles, as in emulsion and foam systems (29).

We have developed a capillary force balance to study the phenomenon of nano-sized particle/micelle structuring inside the thin liquid films, the so-called emulsion and foam films, and discovered the presence of long-range (nonDLVO) oscillatory structural forces (oscillatory disjoining pressures) induced by the confined boundaries of the film with fluid surfaces. We carried out a theoretical analysis of these forces using the statistical mechanics approach and Monte Carlo simulations. At low micelle/particle concentrations, the long-range oscillatory structural force leads to an attractive depletion effect which gives rise to phase separation in emulsions and other colloidal dispersions. However, at high micelle/particle concentrations, the oscillatory structural force induces micelle/particle structural transitions inside the film and the formation of two-dimensional crystalline layers with hexagonal interplanar ordering which offers a new mechanism for stabilizing emulsions, particle dispersions and foams. We have used both nondestructive back-light scattering and direct optical imaging techniques to characterize quantitatively these long-range structural forces in supramolecular fluids such as concentrated suspensions of nano-sized particles, surfactant micellar solutions and microemulsions, and in systems of fat particles, emulsifiers, and gums (hydrocolloids). This discovery of oscillatory structural forces with a period of oscillation equal to the effective size of the micelle/particle arising from the self-organization of nano-sized particles has opened up new vistas in emulsion/dispersion science and technology. Our work on the thinning of emulsion and foam films provides a theoretical link between oscillatory disjoining pressure in thin films and oscillatory structural and depletion forces in concentrated suspensions.

## ACKNOWLEDGMENTS

The authors gratefully acknowledge the financial support provided by the National Science Foundation and the U.S. Department of Energy in addition to a number of industrial organizations.

## REFERENCES

1. D Edwards, H Brenner, DT Wasan. Interfacial Transport Processes and Rheology. Boston: Butterworth-Heinemann, 1991, pp 1–558.
2. AD Nikolov, DT Wasan. In: AT Hubbard, ed. Handbook of Surface Imaging and Visualization. Boca Raton, FL: CRC Press, 1995, pp 209–214.
3. PJ Breen, DT Wasan, YH Kim, AD Nikolov, CS Shetty. In: J. Sjöblom, ed. Emulsions and Emulsion Stability. New York: Marcel Dekker, 1996, pp 237–286.
4. DT Wasan, AD Nikolov. Emulsion Stability Mechanisms. Proceedings of the First World Congress on Emulsions, Paris, 1993, pp 93–112.
5. YH Kim, DT Wasan, PJ Breen. Colloids Surfaces 95: 235–247, 1995.
6. YH Kim, AD Nikolov, DT Wasan, H Diaz-Arauzo, CS Shetty. J Dispersion Sci Technol 17: 33–53, 1996.
7. YH Kim, K Koczo, DT Wasan. J Colloid Interface Sci 187: 29–44, 1997.
8. R Nagarajan, DT Wasan. J Colloid Interface Sci 159: 164–173, 1993.
9. JM Soos, K Koczo, E Erdos, DT Wassan. Rev Sci Instrum 65: 3555–3562, 1994.
10. R Nagarajan, K Koczo, E Erdos, DT Wasan. AIChE J 41: 915–923, 1995.
11. ED Manev, SV Sazdanova, DT Wasan. J Dispersion Sci Technol 5: 111, 1984.
12. AD Nikolov, DT Wasan. J Colloid Interface Sci 133: 1–12, 1989.
13. K Koczo, AD Nikolov, DT Wasan, RP Borwankar, A Gonsalves. J Colloid Interface Sci 178: 694–702, 1996.
14. AD Nikolov, DT Wasan. Powder Technol 88: 299–304, 1996.
15. AD Nikolov, DT Wasan. Colloids Surfaces 123/124: 375–381, 1997.
16. AD Nikolov, DT Wasan. Colloids Surfaces 128: 243–253, 1997.
17. DT Wasan. Chem Eng Ed. Spring Issue: 104, 1992.
18. DT Wasan, AD Nikolov, P Kralchevsky, IB Ivanov. Colloids Surfaces 67: 139–145, 1992.
19. AD Nikolov, PA Kralchevsky, IB Ivanov, DT Wasan. J Colloid Interface Sci 133: 13, 1989.
20. AD Nikolov, DT Wassan. Langmuir 8: 2985–2994, 1992.
21. XL Chu, AD Nikolov, DT Wasan. Langmuir 10: 4403–4408, 1994.
22. XL Chu, AD Nikolov, DT Wasan. Langmuir 12: 5004–5010, 1996.
23. XL Chu, AD Nikolov, DT Wasan. J Chem Phys 103: 6653–6661, 1995.
24. DT Wasan, AD Nikolov In: S Manne and G Warr, eds. Supramolecular Structure in Confined Geometries. ACS Symposium Series No. 736, 1999, pp 40–53.
25. W Xu, AD Nikolov, DT Wasan, A Gonsalves, R Borwankar. J Food Sci 63: 183–188, 1998.

26. W Xu, AD Nikolov, DT Wasan. J Colloid Interface Sci 191: 471–481, 1997.
27. K Kumar, AD Nikolov, DT Wasan. In: K Mittal and P Kumar, eds. Emulsions, Foams and Thin Films, New York: Marcel Dekker, 2000, pp 87–104.
28. J Chatterjee, AD Nikolov, DT Wasan. Ind Eng Chem Res 35: 2933–2938, 1996.
29. DT Wasan. In: K Mittal and P Kumar, eds. Emulsions, Foams and Thin Films, New York: Marcel Dekker, 2000, pp 1–30.

# 4

# Coupling of Coalescence and Flocculation in Dilute O/W Emulsions

**Stanislav Dukhin**
*New Jersey Institute of Technology, Newark, New Jersey*

**Øystein Sæther**
*Norwegian University of Science and Technology, Trondheim, Norway*

**Johan Sjöblom**
*Statoil A/S, Trondheim, Norway*

## 1. GENERAL

### A. Kinetic and Thermodynamic Stability in Macroemulsions and Miniemulsions

The majority of emulsion-technology problems relate to the stabilization and destabilization of emulsions (1–7). Despite the existence of many fundamental studies related to the stability of emulsions, the extreme variability and complexity of the systems involved in any specific application often pushes the oil industry to achieve technologically applicable results without developing a detailed understanding of the fundamental processes. Nevertheless, since in most cases technological success requires the design of emulsions with a very delicate equilibrium between stability and instability, a better understanding of the mechanisms of stabilization and destabilization might lead to significant breakthroughs in technology.

Notwithstanding their thermodynamic instability, many emulsions are kinetically stable and do not change appreciably for a prolonged period. These systems exist in the metastable state (8–15). The fundamentals of emulsion stability (destabilization) comprise emulsion surface chemistry and physicochemical kinetics.

In contrast to the large success in industrial applications of emulsion surface chemistry the potential of physicochemical kinetics as a basis for emulsion dynamics modeling is almost never used in emulsion technology. This situation has started to change during the last decade. Although the coupling of the subprocesses in emulsion dynamics modeling (EDM) continues to represent a large problem not yet solved, models are elaborated for (1) macroemulsions (10, 16–22); and (2) miniemulsions (23–30), for long and short lifetimes of thin emulsion films.

1. For large droplets (larger than 10–30 μm) in macroemulsions the rate of thinning of the emulsion film formed between two approaching droplets is rather low, and correspondingly, the entire lifetime of an emulsion need not be short, even without surfactant stabilization of the film. For this case the notion of *kinetic stability* is introduced (10, 16–19) to denote the resistance of the film against *rupture during thinning*. The droplet deformation and flattening cause this strong resistance, which is described by the Reynolds equation (31, 32). According to theory, the role of this

deformation (33–35) decreases rapidly with decreasing droplet dimension.

2. For small droplets (smaller than 5–10 μm) in miniemulsions droplet deformation can be neglected, because the Reynolds drainage rate increases as $R_d^{-5}$ (10, 36) ($R_d$, the Reynolds film radius) and because the smaller the droplets, the smaller is their deformation (33–35).

In distinction from macroemulsions, where the kinetic stability is the manifestation of droplet–droplet hydrodynamic interaction and droplet deformation, in miniemulsions the kinetic stability is the manifestation of the interplay between surface forces and Brownian movement (23). As the molecular forces of attraction decrease linearly with decreasing droplet dimension, namely, approximately 10 times at the transition from macroemulsions to miniemulsions, the potential minimum of droplet–droplet interaction (secondary minimum) decreases, and for miniemulsions this depth can be evaluated as 1—5 kT (12, 37). At this low energy, Brownian movement causes droplet doublet disaggregation after a short time (the doublet fragmentation time, $\tau_d$). If this time is shorter than the lifetime of the thin film, rapid decrease in the total droplet concentration (t.d.c.) is prevented (restricted by the coalescence time, $\tau_c$), i.e., stability is achieved due to this kinetic mechanism (23).

### B. Current State of Emulsion Stability Science

A large disparity exists between knowledge concerning kinetic stability and thermodynamic stability. The main attention has been paid to *kinetic stability* for both macroemulsions (16–22) and miniemulsions (23–30). As a result, the droplet–droplet interaction and the collective processes in dilute emulsions are quantified (38, 39) and important experimental investigations are made (27, 28, 40). Some models are elaborated for the entire process of coalescence in concentrated emulsions as well (41, 42). Given thermodynamic stability, a thin interdroplet film can be metastable.

In contrast to the large achievements in investigations of kinetic stability, modest attention has been paid to the fundamentals of thermodynamic stability in emulsions, especially regarding the surfactant adsorption layer's influence on the coalescence time. There are several investigations devoted to the surface chemistry of adsorption related to emulsification and demulsification. However, the link between the chemical nature of an adsorption layer, its structure, and the coalescence time is not yet quantified.

A premiss for such quantification is the theory of a foam-bilayer lifetime (43). The main notions of this theory are similar to the theory of Derjaguin and coworkers (44, 45). However, the theory (43) is specified for amphiphile foam films, it is elaborated in detail, and is proven by experiment with water-soluble amphiphiles, such as sodium dodecyl sulfate (47). As the dependence of the rupture of the emulsion film on surfactant concentration is similar to that for a foam film, the modification of theory with respect to emulsions may be possible. Although this modification is desirable the specification of a theory for a given surfactant will not be trivial, since the parameters in the equation for the lifetime (45) are unknown and their determination is not easy. As the theory (43, 47) is proposed for amphiphiles and since a wider class of chemical compounds can stabilize emulsions, the film-rupture mechanism (44) is not universal regarding emulsions.

Thus, in contrast to the quantification of kinetic stability, the empirical approach continues to predominate regarding thermodynamic stability. Meanwhile, thermodynamic stability provides greater opportunity for long-term stabilization of emulsions, than does kinetic. This means that the experimental characterization of thermodynamic stability, i.e., the measurement of coalescence time, is of major importance.

### C. Specificity of Emulsion Characterization

Generalized emulsion characterization, i.e., measurement of droplet size distribution, electrokinetic potentials, Hamaker constant, etc., is not always sufficient. Thermodynamic stability with respect to bilayer rupture cannot be quantified with such a characterization procedure alone. Consequently, measurement of the coalescence time $\tau_c$ is of major importance for an evaluation of emulsion stability; it is an important and specific parameter of emulsion characterization.

The current state of miniemulsion characterization neglects the importance of $\tau_c$ measurement. A practice for $\tau_c$ measurement is practically absent with the exception of only a few papers considered in this chapter. Meanwhile, many papers devoted to issues more or less related to emulsion stability do not discuss $\tau_c$ measurement. One reason for this scientifically and technologically unfavorable situation in which emulsions are incompletely characterized may originate from a lack of devices enabling $\tau_c$ measurements to be made.

## D. Scope of the Chapter

This chapter is focused on kinetic stability in miniemulsions with emphasis on the coupled destabilizing subprocesses, in distinction from other chapters in the *Encyclopedia* describing other aspects of emulsion stability.

In general there are three coupled subprocesses that will influence the rate of destabilization and phase separation in emulsions. These are aggregation, coalescence, and floc fragmentation. Often, irreversible aggregation is called coagulation and the term flocculation is used for reversible aggregation (13, 48). Ostwald ripening (49, 50) coupled (24) with aggregation and fragmentation is a separate topic which will be not considered here.

A simplified theory is available for the coupling of coalescence and flocculation in emulsions void of larger flocs. This theory is considered in Sec. II and will assist in the consideration of the more complicated theory of coupling of coalescence and coagulation (Sec. III). The experimental investigations are described in parallel. Section IV is devoted to the theory of doublet fragmentation time and its measurement, as this characterizes an emulsion regarding fragmentation and because its measurement is an important source of information about surface forces and the pair interaction potential. The discrimination between conditions for coupling of coalescence with coagulation or with flocculation is considered in Sec. V. The quantification of kinetic stability creates new opportunities for long-term prediction of miniemulsion stability, for stability optimization, and for characterization with thc standardization of $\tau_c$ and $\tau_d$ mcasurcments. This forms the basis for emulsion dynamics modeling (Sec. VI).

## II. COUPLING OF COALESCENCE AND FLOCCULATION

### A. Singlet–Doublet Quasiequilibrium

Each process among the three processes under consideration is characterized by a characteristic time, namely, $\tau_{Sm}$, $\tau_d$, and $\tau_c$. The Smoluchowski time (51), $\tau_{Sm}$, gives the average time between droplet collisions. If the time between two collisions is shorter than $\tau_d$ a doublet can transform into a triplet before it spontaneously disrupts. In the opposite case, i.e., at

$$\tau_{Sm} \gg \tau_d \tag{1}$$

the probability for a doublet to transform into a triplet is very low because the disruption of the doublet occurs much earlier than its collision with a singlet. The rate of multiplet formation is very low for

$$\text{Rev} = {}^{\tau_d}\big/_{\tau_{Sm}} \ll 1 \tag{2}$$

where we introduce the notation "Rev" for small values of the ratio corresponding to the reversibility of aggregation and a singlet–doublet quasiequilibrium.

The kinetic equation for reversible flocculation in a dilute monodisperse o/w emulsion when neglecting coalescence is [52–54]

$$\frac{dn_2}{dt} = \frac{n_1^2}{\tau_{Sm}} - \frac{n_2}{\tau_d} \tag{3}$$

where $n_1$ and $n_2$ are the dimensionless concentrations of doublets and singlets, $n_1 = N_1/N_{10}$, $n_2 = N_2/N_{10}$, $N_1$ and $N_2$ are the concentrations of singlets and doublets, and $N_{10}$ is the initial concentration, and

$$\tau_{Sm} = \left(\frac{4kT}{3\eta}N_{10}\right)^{-1} = (K_f N_{10})^{-1} \tag{4}$$

where $k$ is the Boltzmann constant, $T$ is the absolute temperature, and $\eta$ is the viscosity of water. For aqueous dispersions at 25°C, $K_f = \dfrac{4kT}{3\eta} = 6 \times 10^{-18} m^3 s^{-1}$. The singlet concentration decreases with time due to doublet formation, while the doublet concentration increases. As a result, the rates of aggregation and floc fragmentation will approach each other. Correspondingly, the change in the number of doublets $dn_2/dt = 0$. Thus, a dynamic singlet–doublet equilibrium (s.d.e.) is established:

$$n_{2eq} = \frac{\tau_d}{\tau_{Sm}} n_{1eq}^2 \tag{5}$$

Under condition (2) it follows from Eq. (2.5) that

$$n_{1eq} \cong 1, N_{1eq} = N_{10} \tag{6}$$

$$N_2 = \text{Rev} \cdot \text{N}_{10} \text{ or } n_2 \ll 1 \tag{7}$$

Thus, at small values of Rev the s.d.e. is established with only small deviations in the singlet equilibrium concentration from the initial concentration [Eq. (6)]. The doublet concentration is very low compared to the singlet concentration, and the multiplet concentration is very low compared to the doublet concentration. The last statement follows from a comparison of the production rates of doublets and triplets. The doublets appear due to singlet–singlet collisions, while the triplets appear due to singlet–doublet collisions. The latter rate is lower owing to the low doublet concentration. The ratio of the number of singlet–doublet collisions to the number of singlet–singlet collisions is proportional to Rev.

### B. Kinetic Equation for Coupling of Flocculation and Intradoublet Coalescence in Monodisperse Emulsions

Both the rate of doublet disaggregation and the rate of intradoublet coalescence are proportional to the momentary doublet concentration. This leads (23, 29) to a generalization of Eq. (3):

$$\frac{dn_2}{dt} = \frac{n_1^2}{\tau_{Sm}} - n_2\left(\frac{1}{\tau_d} + \frac{1}{\tau_c}\right) \tag{8}$$

There are two unknown functions in Eq. (8), so an additional equation is needed. This equation describes the decrease in the droplet concentration caused by coalescence:

$$\frac{d}{dt}(n_1 + 2n_2) = -\frac{n_2}{\tau_c} \tag{9}$$

The initial conditions are

$$n_2\bigg|_{t=0} = 0 \tag{10}$$

$$\frac{dn_2}{dt}\bigg|_{t=0} = \frac{n_{10}}{\tau_{Sm}} \tag{11}$$

Condition (11) follows from Eqs (8) and (9). The solution of the set of Eqs (8) and (9), taking into account boundary conditions (10) and (11), is a superposition of two exponents (23, 29). In the case

$$\tau_c \gg \tau_d \tag{12}$$

the solution simplifies (23, 29) to

$$n_2(t) = \frac{\tau_d}{\tau_{Sm}}\left[\exp\left(-\frac{2\tau_d t}{\tau_{Sm}\tau_c}\right) - \exp\left(-\frac{t}{\tau_d}\right)\right] \tag{13}$$

Equation (13), as compared to Eqs (5) and (6), corresponds to the s.d.e. if the expression in the second brackets equals unity. In the time interval:

$$\tau_d \lessdot t \lessdot \tau \tag{14}$$

where

$$\tau = \tau_c \frac{\tau_{Sm}}{2\tau_d} \tag{15}$$

the first term in the second brackets is approximately equal to 1, while the second one decreases from 1 to a very small value. Thus, the s.d.e. is established during the time $\tau_d$ and is preserved during the longer time interval [Eq. (14)].

For times longer than $\tau$ there is no reason to apply Eq. (13) since the condition to linearize Eq. (8) is no longer valid with the concentration decrease. At the beginning of the process the doublet concentration increases, while later coalescence predominates and the doublet concentration decreases. Thus, function (13) has a maximum (23, 29).

### C. Coalescence in a Singlet–Doublet System at Quasiequilibrium

After a time $t_{max}$ a slow decrease in the doublet concentration takes place simultaneously with the more rapid processes of aggregation and disaggregation. Naturally, an exact singlet–doublet equilibrium is not valid owing to the continuous decrease in the doublet concentration. However, the slower the coalescence, the smaller is the deviation from the momentary dynamic equilibrium with respect to the aggregation–disaggregation processes.

It is reasonable to neglect the deviation from the momentary doublet–singlet equilibrium with the condition:

$$\frac{dn_2}{dt} \ll \frac{n_2}{\tau_d} \tag{16}$$

Indeed, for this condition the derivative in Eq. (3) can be omitted, which corresponds to s.d.e. characterized by Eq. (5).

It turns out (23, 27–29) that the deviation from s.d.e. is negligible as the condition (13) is valid, i.e., for conditions (2) and (12). For these conditions the fragmentation of flocs influences the coalescence kinetics which can be represented as a three-stage process, as illustrated in Fig. 1. During a rather short time $\tau_d$ the approach to s.d.e. takes place, i.e., a rather rapid increase in the doublet concentration (stage 1). During the next time interval $\tau_d < t < t_{max}$ the same process continues. However, the rate of doublet formation declines due to coalescence (stage 2). The exact equilibrium between the doublet formation and their disappearance due to coalescence takes place at the time $t_{max}$ when the doublet concentration reaches its maximum value $n_2(t_{max})$. During the third stage, when $t > t_{max}$, the rate of doublet fragmentation is lower than the rate of formation, because of the coalescence within doublets. This causes a slow monotonic decrease in the concentration. Taking

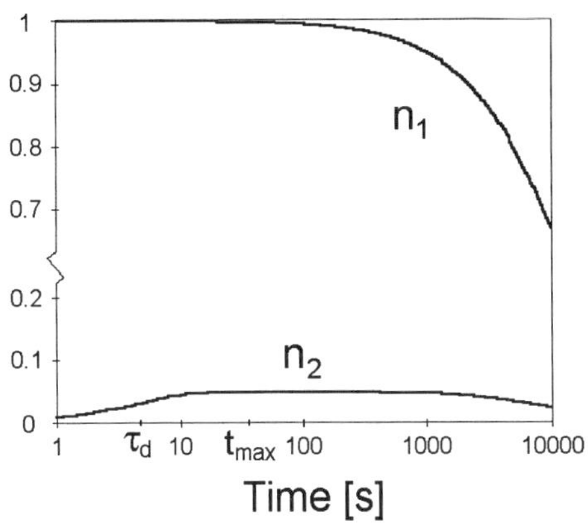


**Figure 1** Three stages in the coupling of aggregation, fragmentation, and coalescence under the condition $\tau_d << \tau_{Sm} << \tau_c$. Initially, the doublet concentration $n_2$ is very low and the rates of doublet fragmentation and of coalescence are correspondingly low compared to the rate of aggregation (first stage, no coupling). Owing to increasing $n_2$ the fragmentation rate increases and equals the aggregation rate at $t_{max}$ (exact s.d.e.). The growth in $n_2$ stops at $t_{max}$ (second stage, coupling of aggregation and fragmentation). Intradoublet coalescence causes a slight deviation from exact s.d.e. to arise at $t > t_{max}$, and the singlet concentration $n_1$ and the doublet concentration decrease due to intradoublet coalescence (third stage, coupling of aggregation, fragmentation, and coalescence); $n_1$ and $n_2$ are dimensionless, $n_1 = N_1/N_{10}$; $n_2 = N_2/N_{10}$; $N_{10}$ is the initial singlet concentration. (From Ref. 23.)

into account the s.d.e. [Eqs (5) and (7)], Eq. (9) can be expressed as

$$\frac{dn_1}{dt} = -\frac{\tau_d}{\tau_{Sm}\tau_c} n_1^2 \tag{17}$$

The result of the integration of Eq. (17) can be simplified to

$$n_1(t) = \frac{n_1(t_{max})}{1 + n_1(t_{max})\dfrac{t}{2\tau}} \cong \left(1 + \frac{t}{2\tau}\right)^{-1} \tag{18}$$

with a small deviation in $n_1(t_{max})$ from unity. As opposed to the preceding stages when the decrease in droplet concentration caused by coalescence is small, a large decrease is now possible during the third stage. Thus, this is the most important stage of the coalescence kinetics.

## D. Reduced Role of Fragmentation with Decreasing $\tau_c$

With decreasing $\tau_c$, condition (12) is violated and new qualitative features of the destabilization process, not discussed in Refs 27–29, arise. As the ratio $\tau_c/\tau_d$ diminishes and

$$\tau_c < \tau_d \tag{19}$$

the s.d.e. is violated because a larger part of the doublets disappear due to coalescence. Correspondingly, the smaller the ratio $\tau_c/\tau_d$ the smaller is the fragmentation rate in comparison with the aggregation rate, i.e., the larger the deviation from s.d.e. In the extreme case:

$$\tau_c \ll \tau_d \tag{20}$$

the fragmentation role in s.d.e. can be neglected. This means that almost any act of aggregation is accompanied by coalescence after the short doublet lifetime. Neglecting this time in comparison with $\tau_{Sm}$, in agreement with condition (1), one concludes that any act of aggregation is accompanied by the disappearance of one singlet:

$$\frac{dn_1}{dt} = -\frac{n_1^2}{\tau_{Sm}} \tag{21}$$

This leads to a decrease in the singlet concentration described by an equation similar to the Smoluchowski equation for rapid coagulation:

$$n_1(t) = (1 + t/\tau_{Sm})^{-1} \tag{22}$$

The Smoluchowski equation describing the singlet time evolution does not coincide with Eq. (22). The peculiarity of Eq. (22) is that it describes the kinetics of coupled aggregation and coalescence with a negligible fragmentation rate. Due to fragmentation, doublet transformation into multiplets is almost impossible under condition (1).

The coupling of aggregation, fragmentation, and coalescence in the more general case described by condition (19) leads to equation:

$$n_1(t) = \frac{n_1(t_{max})}{1 + n_1(t_{max})t/2\tau_g} \cong (1 + t/2\tau_g)^{-1} \quad (t > \tau_d + \tau_c) \tag{23}$$

with a small deviation of $n_1(t_{max})$ from unity and

$$\tau_g = \frac{\tau_{Sm}(\tau_d + \tau_c)}{2\tau_d} \tag{24}$$

Under conditions (1) and (12) $\tau_g \approx \tau$ and Eq. (23) transforms into Eq. (18). Under conditions (1) and (20) $\tau_g \approx \tau_{Sm}$ and Eq. (23) transforms into Eq. (22). Equation (24) demonstrates the reduction of the role of fragmentation with decreasing $\tau_c$. It is seen that at the transition from condition (19) to condition (20) $\tau_d$ cancels out in Eq. (24), i.e., the fragmentation role diminishes.

## E. Experimental

### 1. Application of Video-enhanced Microscopy Combined with the Microslide Technique for Investigation of Singlet–Doublet Equilibrium and Intradoublet Coalescence (27–29)

Direct observation of doublets in the emulsion bulk is difficult because the doublets tend to move away from the focal plane. The microslide preparative technique can, however, be successfully applied, providing pseudobulk conditions. A microslide is a plane-parallel glass capillary of rectangular cross-section. The bottom and top sides of the capillary are horizontal, and the gravity-induced formation of a sediment or cream on one of the inner normal surfaces is rapidly completed owing to the modest inner diameter of the slide. If both the volume fraction of droplets in an emulsion and the capillary height are small, the droplet coverage on the inside surface amounts to a few per cent, and the analysis of results is rather simple. It can be seen through the microscope that the droplets which have sedimented on to the capillary surface participate in chaotic motion along the surface. This indicates that a thin layer of water separating the surface of the microslide from the droplets is preventing the main portion of droplets from adhering to the microslide surface, an action which would stop their Brownian motion.

During diffusion along the microslide ceiling the droplets collide. Some collisions lead to the formation of doublets. Direct visual observation permits evaluation of the doublet-fragmentation time which varies in a broad range (25). Another approach to doublet-fragmentation time determination is based on evaluation of the average concentration of singlets and doublets and using the theory outlined above.

Application of the microslide preparative technique combined with video microscopy is promising and has allowed the measurement of the coupling of reversible flocculation and coalescence (27, 29). However, some experimental difficulties were encountered: droplets could sometimes be seen sticking to the glass surface of the microslide.

### 2. Improving the Experimental Technique with Use of Low-density Contrast Emulsions (28)

The sticking of droplets indicates a droplet-wall attraction and the existence of a secondary potential pit as that for the droplet–droplet attraction in a doublet. The droplet concentration within the pit is proportional to the concentration on its boundary. The latter decreases with a decrease in the density contrast.

The electrostatic barrier between the potential pit and the wall retards the rate of sticking. The lower the droplet flux through this barrier, the lower is the potential pit occupancy by droplets. Thus, an essential decrease in the rate of sticking is possible with decreasing density contrast.

Oil/water emulsions were prepared (28) by mixing dichlorodecane (DCD, volume fraction 1%) into a $5 \times 10^{-5}$ M sodium dodecyl sulfate (SDS) solution with a Silverson homogenizer. The oil phase was a 70:1 mixture of DCD, which is characterized by an extremely low density contrast to water, and decane.

The droplet distribution along and across the slide was uniform (28). This indicates that there was no gravity-induced rolling either. One slide among four was examined for two weeks without any sticking being observed (28). The absence of the rolling and sticking phenomena allowed acquisition of quite accurate data concerning the time dependence of the droplet size distribution.

### 3. The Measurement of Coalescence Time and Doublet-fragmentation Time

The doublet-fragmentation time was measured by direct real-time observation of the doublets on the screen and by analysis of a series of images acquired at 1–3 min intervals (25). The formation and disruption/coalescence of a doublet could thus be determined.

The general form of the concentration dependence agrees with the theory. At $C \sim 3 \times 10^{-3}$ M, both theory and experiment yield times of about 1 min; at $C = 9 \times 10^{-3}$ M, these times exceed 10 min. For calculation of the doublet-fragmentation time the electrokinetic potential was measured (29, 46).

In experiments with different droplet concentrations it was established that the higher the initial droplet concentration, the higher the doublet concentration. This corresponds to the notion of singlet–doublet equilibrium. However, if the initial droplet concentration exceeds 200–300 per observed section of the microslide, multiplets predominate. Both the initial droplet concentration and size affect the rate of decrease in the droplet concentration. The larger the droplets, the smaller the concentration sufficient for the measurement of the rate of decrease in the droplet concentration. This agrees with the theory of doublet-fragmentation time which increases with droplet dimension. Correspondingly, the probability for coalescence increases. These first series of experiments (27, 29) were accomplished using toluene-in-water emulsions without the addition of a surfactant and decane-in-water emulsions stabilized by SDS. The data obtained, concerning the influence of the electrolyte concentration and surface charge density, were in agreement with the existing notions about the mechanism of coalescence. With increasing SDS concentration, and correspondingly increasing surface potential, the rate of decrease in the droplet concentration was reduced.

Two methods were used for the measurement of the coalescence time (28, 29). Measurement of the time dependence for the concentrations of singlets and doublets and a comparison with Eq. (9) enables an evaluation of the coalescence time to be made. Further, information about the time dependence for singlets and the doublet-fragmentation time may be used as well. These results, in combination with Eqs (15) and (18), determine the coalescence time. The good agreement between results obtained by these very different methods indicates that the exactness of the theory and experiments is not low.

In recent years several research groups have improved significantly the theoretical understanding of coalescence of droplets or bubbles. The newer results (53–57), together with results of earlier investigations (58–62), have clarified the role of double-layer interaction in the elementary act of coalescence.

DLVO theory was applied (63, 64) for the description of "spontaneous" and "forced" thinning of the liquid film separating the droplets. These experimental results and DLVO theory were used (63) for the interpretation of the reported visual study of coalescence of oil droplets 70–140 μm in diameter in water over a wide pH interval. A comparison based on DLVO theory and these expermental data led the authors to condlue (63) that "if the total interaction energy is close to zero or has a positive slope in the critical thickness range, i.e., between 30 and 50 nm, the oil drops should be expected to coalesce." In the second paper (64), where both ionic strength and pH effects were studied, coalescence was observed at constant pH values of 5.7 and 10.9, when the Debye thickness was less than 5 nm. The main trend in our experiments and in Refs 63 and 64 were in accordance, because it was difficult to establish the decrease in t.d.c at NaCl concentrations lower than $5 \times 10^{-3}$ M, i.e., double-layer (DL) thicknesses larger than 5 nm. An almost quantitative coincidence in the double-layer influence on coalescence, established in our work for micrometer-sized droplets and in Refs 63 and 64 for almost 100 times larger droplets, is important for general knowledge on coalescence.

### F. Perspective for Generalization of the Theory for Coupling of Coalescence and Flocculation

The proposed theory for coupling of coalescence and flocculation at s.d.e. permits the proposal of some important applications (Sec. VI). At the same time generalized theory is necessary, since the role of multiplets increases after a long time or with a higher initial concentration. At least two approaches to this difficult task are seen.

According to our videomicroscopic observations there are large peculiarities in the structure and behavior of multiplets arising at conditions near to s.d.e. These peculiarities can be interpreted as the manifestation of quasiequilibrium, comprising singlets, doublets, and multiplets. Similar to doublets, the lifetime of triplets, tetraplets, etc., can be short due to fragmentation and coalescence. This can be valid for multiplets with an "open" structure, in distinction from another structure which can be called "closed." In open multiplets any droplet has no more than one or two contacts with other droplets, which corresponds to a linear chain-like structure. This causes easy fragmentation, especially for the extreme droplets within a chain. The "closed" aggregates have a more dense and isometric structure, in which droplets may have more than two contacts with neighboring droplets. As result, fragmentation is more difficult and the frequency is lower.

Progress in the theory of aggregation with fragmentation (65–69) for a suspension creates a premiss for a theoretical extension towards emulsions. However, the necessity in accounting for coalescence makes this task a difficult one.

## III. COUPLING OF COALESCENCE AND COAGULATION

### A. General

For emulsion characterization the notation $n_1$ represents the number density of single droplets and $n_i$ the number density of aggregates comprising $i$ droplets

($i = 2, 3 \ldots$). The total number density of single droplets and all kinds of aggregates is given by

$$n = \sum_{i=1}^{\infty} n_i \tag{25}$$

This characterization corresponds with Smoluchowski theory (51). To characterize coalescence, the total number of individual droplets moving freely, plus the number of droplets included in all kinds of aggregates, $n_T$:

$$n_T = \sum_{i=1}^{\infty} i n_i \tag{26}$$

is introduced as well.

As distinct from the Smoluchowski theory for suspensions, which predicts the time dependence of the concentration of all kinds of aggregates, the time dependence for the total droplet number can be predicted at the current state of emulsion dynamics theory.

The quantification of coagulation within the theory of coupled coagulation and coalescence (CCC theory) is based on the Smoluchowski theory of perikinetic coagulation. Correspondingly, all restrictions inherent in the Smoluchowski theory of Brownian coagulation are preserved in the CCC theory. This means that creaming and gravitational coagulation are not accounted for. A variant of the Smoluchowski theory specified with regard for gravitational coagulation is well known (70). However, its application is very difficult because the rate constant of collisions induced by gravity depends on droplet dimension (12). Owing to the weak particle (aggregate) dimension dependence of the rate constants for Brownian collisions the Smoluchowski theory is valid for polydisperse suspensions and remains valid as polydisperse aggregates arise. Unfortunately, this advantage of the Smoluchowski theory can almost disappear when combined with the coalescence theory, because the coalescence rate coefficients are sensitive to droplet dimension. Thus, droplet and aggregate polydispersity does not strongly decrease the exactness of the description of coagulation in the CCC theory, while the exactness of coalescence description can be severely reduced.

Although the coalescence influence on the Brownian coagulation rate coefficient can be neglected, its influence on the final equations of the Smoluchowski theory remains. It can be shown that Smoluchowski's equation for the total number of particles:

$$n(t) = \left(1 + \frac{t}{\tau_{Sm}}\right)^{-1} \tag{27}$$

remains valid, while in parallel the equations for the singlet and aggregate concentrations cannot be used to account for coalescence. Regarding coupled coagulation and coalescence, the Smoluchowski equation for $n_1(t)$ is not exact because it does not take into account the singlet formation caused by coalescence within doublets.

The coalescence within an aggregate consisting of $i$ droplet is accompanied by the aggregate transforming into an aggregate consisting of $(i-1)$ droplets. As coalescence changes the aggregate type only, the total quantity of aggregates and singlets does not change. This means that the Smoluchowski function $n(t)$ does not change during coalescence, since Smoluchowski defined the total quantity of particles as consisting of aggregates and singlets.

## B. Average Models

Average models do not assign rate constants to each possibility for coalescence within the aggregates, but deal with certain averaged characteristics of the process. The models in Refs 38 and 71 introduce the average number of drops in an aggregate $m$, because the number of films in an aggregate $n_f$ and $m$ are interconnected. For a linear aggregate:

$$n_f = m - 1 \tag{28}$$

As the coalescence rate for one film is characterized by $\tau_c^{-1}$, the decrease in the average droplet quantity in an aggregate is $n_f$ times larger. This is taken into account in the model of van den Tempel (71) for simultaneous droplet quantity increase due to aggregation and decrease due to coalescence. Van den Tempel formulates the equation which describes the time dependence for the average number of droplets in an aggregate as

$$\frac{dm}{dt} = K_f N_{10} - \tau_c^{-1}(m-1) \tag{29}$$

where the first term is derived using Smoluchowski theory.

The total number of droplets $n_T$ is the sum of single droplets $n_1(t)$ and the droplets within aggregates:

$$n_T(t) = n_1(t) + n_v(t) m(t) \tag{30}$$

where $n_v$ is the aggregate number. The latter can be expressed as

$$n_v(t) = n(t) - n_1(t) \tag{31}$$

Both terms are expressed by Smoluchowski theory. The integration of Eq. (29) and the substitution of the result into Eq. (30) yields the time dependence $n_T(t)$ according to the van den Tempel model.

### 1. The Model of Borwankar *et al.*

In Ref 38 the van den Tempel model is criticized and improved through the elimination of Eq. (29). The authors point out that the "incoming" aggregates which cause the increase in $m$ have themselves undergone coalescence. This is not taken into account in the first term on the right-hand side of Eq. (29). Instead of taking a balance on each aggregate (as van den Tempel did) Borwankar *et al.* took an overall balance on all particles in the emulsion. For linear aggregates, the total number of films in the emulsion is given by

$$n_f n_v = (m-1)n_v \tag{32}$$

Thus, instead of Eq. (29) the differential equation for $n_T$ follows:

$$-\frac{dn_T}{dt} = \tau_c^{-1}(m-1)n_v \tag{33}$$

where $m$ can be expressed through $n_T$ using Eq. (30). The advantage of this equation in comparison with Eq. (29) is obvious. However, there is a disadvantage common to both theories, caused by the use of the Smoluchowski equation for $n_1(t)$. Coalescence does not change the total particle concentration $n(t)$, but changes $n_1(t)$ and correspondingly $n_v(t)$, according to Eq. (31).

The application of Smoluchowski theory in the quantification of the coupling of coalescence and coagulation has to be restricted with the use of the total particle concentration $n(t)$ only. The average models of van den Tempel and Borwankar *et al.* (38) do not meet this demand.

The theory of Danov *et al.* (39) does not contradict this demand, which makes it more correct than the preceding theories. Among the Smoluchowski results the function $n(t)$ is only present in the final equations of this theory. Although the exactness of averaged models is reduced due to the violation of the restriction in the use of Smoluchowski theory, results for some limiting cases are not erroneous.

### 2. The Limiting Cases of Fast and Slow Coalescence

Two limiting cases can be distinguished: the rate of coalescence is much greater than that of flocculation (rapid coalescence):

$$\tau_c^{-1} \gg \tau_{Sm}^{-1} \tag{34}$$

and the rate of flocculation is much greater than that of coalescence (slow coalescence):

$$\tau_c^{-1} \ll \tau_{Sm}^{-1} \tag{35}$$

According to the general rules of physicochemical kinetics the slowest process is rate controlling. If the coagulation step is rate controlling, namely, when condition (34) is valid, then the coalescence is rapid and the general equation of the theory in Ref. 38 is reduced to second-order kinetics, i.e., to Smoluchowski's equation [Eq. (27)]. Flocs composed of three, four, etc., droplets cannot be formed, because of rapid coalescence within the floc. In this case the structure of the flocs becomes irrelevant.

At first glance the coagulation rate has not manifested itself in the entire destabilization process in the case of slow coalescence [condition (35)]. At any given moment the decrease in the total droplet concentration is proportional to the momentary total droplet concentration (first-order kinetics), which causes an exponential decrease with time:

$$n = \exp\left(-\frac{t}{\tau_c}\right) \tag{36}$$

However, this equation cannot be valid for an initial short period, because at the initial moment there are no aggregates and their quantity continues to be low during a short time. This means that the coagulation is limiting during an initial time at any slow coalescence rate. This example illustrates the necessity of a more exact approach than that which uses average models. This was done by Danov *et al.* (39).

## C. DIGB Model for the Simultaneous Processes of Coagulation and Coalescence

This kinetic model, proposed by Danov, Ivanov, Gurkov, and Borwankar, is called the DIGB model for the sake of brevity. Danov *et al.* (39) generalized the Smoluchowski scheme (Fig. 2a) to account for droplet coalescence within flocs. Any aggregate (floc) composed of $k$ particles can partially coalesce to become an aggregate of $i$ particles ($1 < i < k$), with the rate constant being $K_c^{k,i}$ (Fig. 2b). This aggregate is further involved in the flocculation scheme, which makes the flocculation and coalescence processes interdependent. Therefore, the system exhibiting both flocculation and coalescence is described by a combination of schemes 1 and 2.

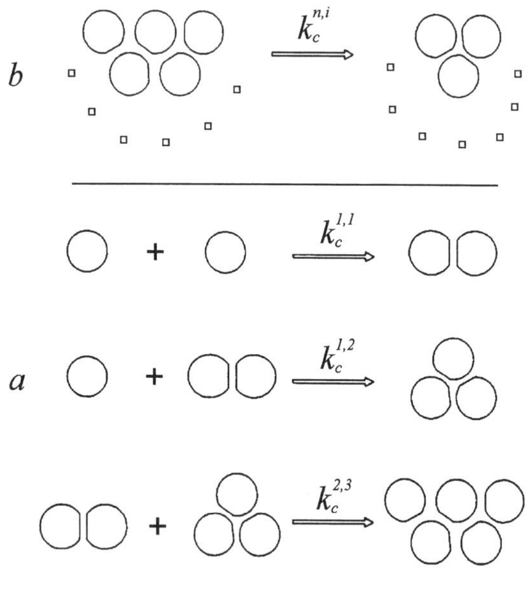


**Figure 2** (a) Model of flocculation according to the Smoluchowski scheme; (b) coalescence in an aggregate of $k$ particles to become an aggregate of $i$ particles, with a rate constant $K_c^{k,i}$, $1 < i < k$. (From Ref. 39.)

$$\frac{dn_k}{dt} = \sum_{i=1}^{k-1} K_f^{i,k-i} n_i n_{k-i} - 2\sum_{i=1}^{\infty} K_f^{k,i} n_k n_i + \sum_{i=k+1}^{\infty} K_c^{i,k} n_i - \sum_{i=1}^{k-1} K_c^{k,i} n_k \quad (37)$$

Equation (37) is multiplied by $k$ and summed up for all $k$; this yields the equation for $n_T$ which is expressed through double sums. The change of the operation sequence in these sums leads to the important and convenient equation:

$$\frac{dn_T}{dt} = \sum_{k=1}^{\infty} k \sum_{i=k+1}^{\infty} K_c^{i,k} n_i - \sum_{i=2}^{\infty} k \sum_{i=1}^{k-1} K_c^{k,i} n_k \quad (38)$$

Afterwards, a total rate coefficient referring to complete coalescence of the $i$th aggregate:

$$K_{c,T}^{i} = \sum_{k=1}^{i-1} (i-k) K_c^{i,k} \qquad i = 2, 3 \ldots \quad (39)$$

is introduced. For linearly built aggregates the following expression is derived:

$$K_{c,T}^{i} = K_c^{2,1}(i-1) \quad (40)$$

With the expression for $K_{c,T}^{i}$, using also Eqs (25) and (26), Eq. (40) is transformed into

$$\frac{dn_T}{dt} = -K_c^{2,1}(n_T - n) \quad (41)$$

The integration result of this first-order linear differential equation is well known and is represented in general form without specification of $n(t)$ [Eq. (18) in Ref. 39]. An interesting peculiarity of this important derivation is the disappearance of terms, related to coagulation at the transition, from the equation set [Eq. (37)] to the main Eq. (38). This corresponds to the fact that the total quantity of droplets does not change due to coagulation; it decreases due to coalescence only.

The coagulation regularity manifests itself in the $n(t)$ dependence, arising in Eq. (41). It creates the illusion that Eq. (41) can be specified for any $n(t)$ function corresponding to any subprocess affecting the droplet aggregate distribution. For example, the gravitational coagulation theory leads to a function $n_g(t)$ (70), but it does not create the opportunity to describe the gravitational coagulation coupling with coalescence by means of substituting $n_g(t)$ into the integral of Eq. (41). As the coalescence influences the gravitational coagulation another function has to be substituted into Eq. (41) instead of $n_g(t)$. This function has to be derived accounting for the coupling of coalescence and coagulation. One concludes that Eq. (41) cannot be used, because its derivation assumes that the coupling of gravitational coagulation (or another process) and coalescence is already quantified.

A happy exception is Brownian coagulation and its modeling by Smoluchowski with the coagulation rate coefficients, of which sensitivity to aggregate structure and coalescence is low. The substitution of function (27) into the integral of Eq. (41) yields the equation characterizing the coupling of coalescence and Brownian coagulation (39).

In fractal theory (72) it is established that diffusion-limited aggregates and diffusion-limited cluster–cluster aggregates are built up linearly. This can simplify application of the DIGB model. However, the diffusivity of fractal aggregates (73) cannot be described by simple equations and Smoluchowski theory. This will cause coagulation-rate coefficient dependence on aggregate structure, decreasing the exactness of Eq. (41) when applied to fractal aggregates. However, there is no alternative to the DIGB model, which can be used as a crude but useful approximation in this case as well. In the absence of an alternative the DIGB model can be recommended for evaluation in the case of gravitational coagulation.

Danov *et al.* (39) compares their theory with the predictions of averaged models for identical conditions. It turns out that if coalescence is much faster than flocculation, the predictions of the different models coincide. Conversely, for slow coalescence the results of the averaged models deviate considerably from the exact solution. These two results of the comparison are in agreement with the qualitative considerations in Sec. III.B.

Data for the relative change in the total number of droplets as a function of time are presented in Fig. 3 from Ref. 39. Figure 3 a–c refers to $K_F N_{10} = 0.1\ s^{-1}$ and the coalescence constant $K_c^{2,1}$ varies between 0.1 $s^{-1}$ (a) and 0.001 $s^{-1}$ (c). It is seen that the agreement between the Danov *et al.* and Borwankar *et al.* models is better the faster the coalescence, as was explained qualitatively above. The van den Tempel curves deviate considerably from the other two solutions.

For very long times, and irrespective of the values of the kinetic parameters, the model of Borwankar *et al.* (38) is close to the numerical solution. This is probably because the longer the time, the smaller is the concentration of single droplets. In this extreme case the error caused in the average models due to the influence of coalescence on the singlet concentration [not taken into account in the equation for $n(t)$] is negligible.

The shortcomings of the averaged models (38, 71) and the advantages of the DIGB model are demonstrated in Ref. 39. However, the range of applicability of this model is restricted by many simplifications and the neglect of other subprocesses (see Sec. II.A). An efficient analytical approach was made possible due to the neglect of the coalescence rate coefficient's dependence on the dimensions of both interacting droplets.

The model of Borwankar *et al.* was examined experimentally in Ref. 40. The emulsions were oil-in-water with soybean oil as the dispersed phase, volume fraction 30% and number concentration $10^7$–$10^{10}$ $cm^{-3}$. The emulsions were gently stirred to prevent creaming during the aging study. A sample was placed on a glass

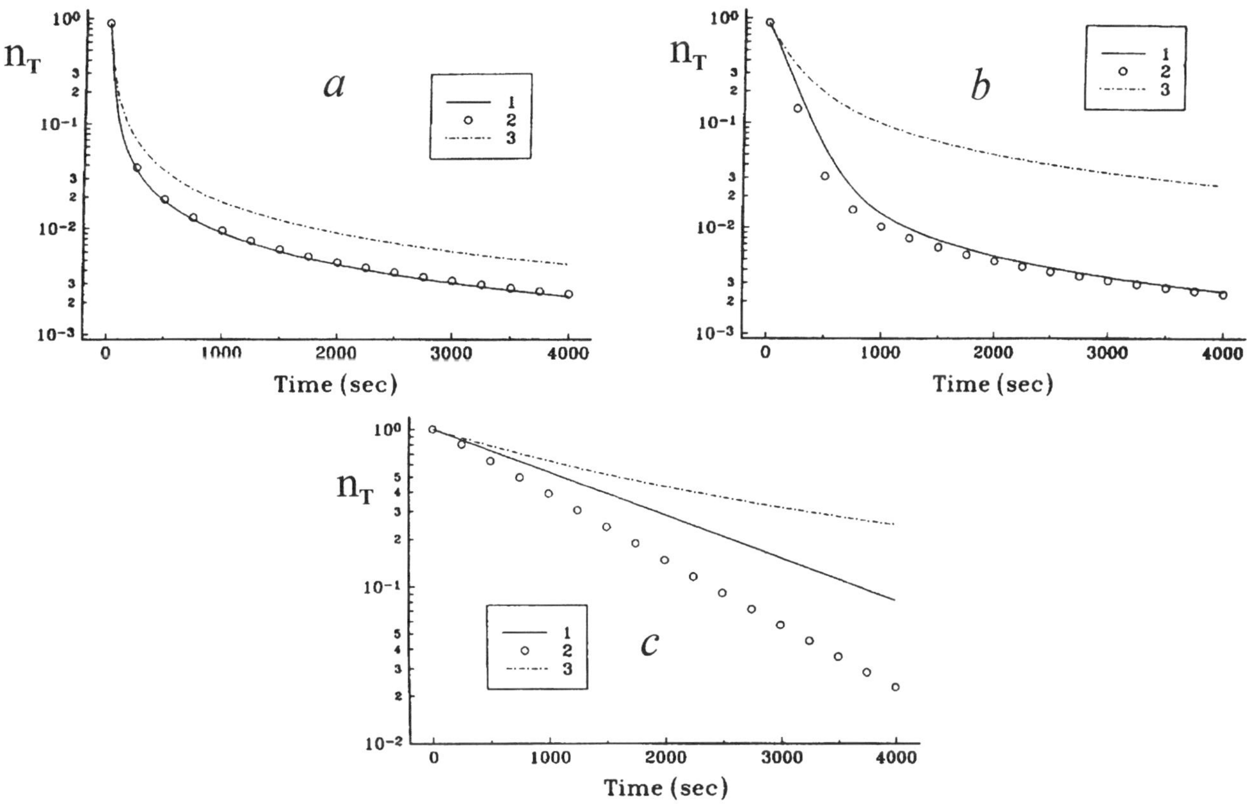


**Figure 3** Relative change in the total number of droplets vs. time: initial number of primary particles $N_{10} = 1 \times 10^{10}\ cm^{-3}$; flocculation rate constant $K_f = 1 \times 10^{-11}\ cm^3/s$; curve 1, the numerical solution of the set Eq. (37); curve 2, the model of Borwankar *et al.* (38) for diluted emulsions; curve 3, the model of van den Tempel (71): (a) coalescence rate constant $K_c^{2,1} = 1 \times 10^{-1}\ s^{-1}$; (b) $K_c^{2,1} = 1 \times 10^{-2}\ s^{-1}$; (c) $K_c^{2,1} = 1 \times 10^{-3}\ s^{-1}$. (From Ref. 39.)

slide, all aggregates were broken up, and the size of the individual droplets was measured. A quite good agreement with the theory was established. However, the fitting of the experimental data was accomplished using two model parameters, namely, the coalescence and coagulation rate coefficients. For the latter coefficient optimal values (different for two emulsions) were obtained, strongly exceeding the Smoluchowski theory value (Sec. II.A). An interpretation is that orthokinetic and perikinetic coagulation took place simultaneously as a result of stirring. Several experiments are known (discussed in Ref. 54) which demonstrate better agreement with the value for the coagulation constant predicted in Smoluchowski theory.

## IV. DOUBLET-FRAGMENTATION TIME

### A. Theory of Doublet-fragmentation Time

A doublet fragmentation was described by Chandrasekhar (74) as the diffusion of its droplets from the potential minimum, characterizing their attraction. The time scale for this process takes the form (75):

$$\tau_d = \frac{6\pi\eta a^3}{kT}\exp(-U_{min}/kT) \tag{42}$$

where $U_{min}$ is the depth of the potential minimum, $k$ is the Boltzmann constant, and $n$ is water dynamic viscosity.

To derive the formula for the average lifetime of doublets, Muller (76) considered the equilibrium in a system of doublets and singlets: that is, the number of doublets decomposing and forming are equal. Both processes are described by the standard diffusion flux $J$ of particles in the force field of the particle that is regarded as central.

Each doublet is represented as an immovable particle with the second singlet "spread" around the central one over a spherical layer, which corresponds to the region of the potential well. The diffusion flux $J$ of "escaping" particles is described by equations used in Fuchs' theory of slow coagulation. The first boundary condition corresponds to the assumption that the escaping particles do not interact with other singlets. The second condition reflects the fact that the potential well contains exactly one particle.

At small separation between the droplets in a doublet the droplet diffusivity reduces because of the increasing hydrodynamic resistance during the droplet approach. A convenient interpolation formula was used (76) for the description of the influence of hydrodynamic interaction on the mutual diffusivity. The difference between the more exact Muller equation and Eq. (42) is caused mainly on account of this hydrodynamic interaction.

### B. Doublet-fragmentation Time of Uncharged Droplets

In this section we consider a doublet consisting of droplets with a nonionic adsorption layer. The closest separation between two droplet surfaces $h_0$ exceeds the double thickness of the adsorption layer ($2h_a$). As a crude approximation $h_0$ can be identified with $2h_a$. In the case of small surfactant molecules $2h_a \approx 2$ nm.

In this case, the potential well has a sharp and deep minimum. This means that the vicinity of this minimum determines the value of the integral in the more exact Muller (76) equation. For examination of this assumption, this integral was calculated numerically and according to the approximate equation (26):

$$\int_{\alpha}^{\beta} \varphi(t)\exp[f(t)]dt = \left[\frac{2\pi}{f''(t_m)}\right]^{1/2} \varphi(t_m)\exp[f(t_m)] \tag{43}$$

where $t_m$ corresponds to the potential well minimum.

The difference in results was small and enabled application of Eq. (43) to the calculation and substitution of the asymptotic expression (11, 14):

$$U(h) = -\frac{A}{12}\cdot\frac{a}{h} \tag{44}$$

which is valid at small distances to the surface. The result of calculations according to Eqs (43) and (44) (the Hamaker constant $A = 1.3\times10^{-20}$ J) are shown in Fig. 4. The chosen value of the Hamaker constant is consistent with those reported elsewhere (77, 78). In addition to the value of $A = 1.3\times10^{-20}$ J, we mention other values of the Hamakar constant which were employed elsewhere. For example, in food emulsions (78) the Hamakar constant lies within the range $3\times10^{-21}$–$10^{20}$ J. The results of calculations for smaller Hamaker constants are also presented in Fig. 4.

The influence of the adsorption layer thickness on doublet lifetime is shown in Fig. 5 for one value of the Hamaker constant. There is high specificity in the thickness of a polymer adsorption layer. β-Casein adsorbed on to polystyrene latex causes an increase in the radius of the particle of 10–15 nm (79). A layer of β-lactoglobulin appears to be in the order of 1–2 nm thick, as compared to 10 nm for the caseins (80).

When adsorbed layers of a hydrophilic nature are present the repulsive hydration forces must be taken

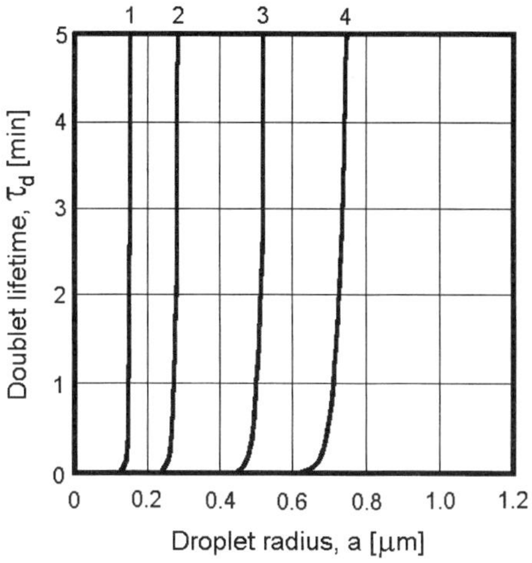


**Figure 4** Dependence of doublet lifetime on droplet dimension at different values of the Hamaker constant $A$ : (1) $A = A_1 = 1.33 \times 10^{-20}$ J; (2) $A = 0.5 \times A_1$; (3) $A = 0.35 \times A_1$; (4) $A = 0.25 \times A_1$; (5)$A = 0.1 \times A_1$. The shortest interdroplet distance is 2 nm. (From Ref. 26.)

into account. At low ionic strengths, the repulsion follows the expected exponential form for double-layer interaction:

$$U(h) = K_s e^{-(h/h_s)} \tag{45}$$

In Ref. 81 the authors emphasize that the surface charge in food emulsions is low, electrolyte concentrations are high, and hence, the DL is not responsible for emulsion stability. The stabilization can be caused by hydration forces. However, flocculation to the secondary minimum remains. Meanwhile, this conclusion must be specified to account for droplet dimension.

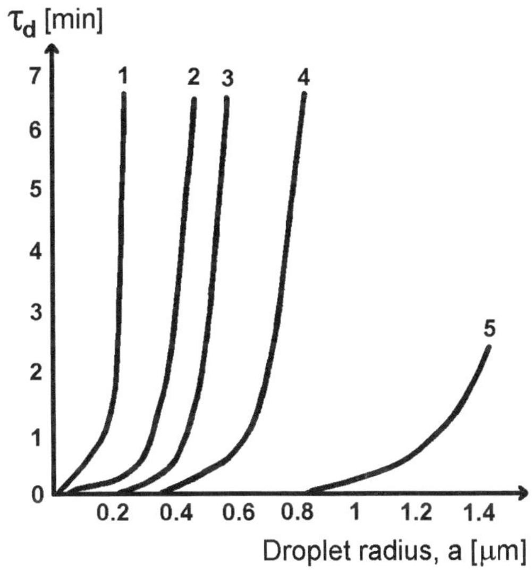


**Figure 5** Influence of adsorption layer thickness on the droplet lifetime of an uncharged droplet. Adsorption layer thickness: (1) $h_0 = 1$ nm; (2) $h_0 = 2$ nm; (3) $h_0 = 4$ nm; (4) $h_0 = 6$ nm. $A = 1.33 \times 10^{-20}$ J. (From Ref. 26.)

### C. Lifetime of a Doublet of Charged Droplets and Coagulation/Flocculation

As seen in Fig. 1 of Ref. 37 the coordinates of the secondary minimum corresponds to $\kappa h_{min} = 5$–12 nm. Owing to this rather large distance the frequency dependence of the Hamaker constant may be of importance, and the Hamaker function $A(h)$ characterizing molecular interaction should be introduced.

In Ref. 82 the distance-independent interaction at zero frequency and interaction at nonzero frequency is considered separately:

$$A(h) = A_o + [A(h) - A_o] \tag{46}$$

The result from 36 systems in Ref. 82 are in quite good accordance with the calculations of other papers. According to Churaev, the system polystyrene–water–polystyrene can be used to estimate the Hamaker function for oil–water systems. However, with increasing droplet separation the importance of $A_0$ increases on account of $[A(h) - A_0]$. The component $A_0$ is screened in electrolyte concentrations, because of dielectric dispersion (83–85). At a distance of $\kappa h_{min} \approx 3$–5 nm the authors (84) found that molecular interaction disappeared at zero frequency. Experimental evidence concering this statement is discussed in Ref. 14. When evaluating the secondary minimum coagulation, $A_0$ can be omitted, as illustrated in Ref. 85.

For illustration of the influence of electrolyte concentration, Stern potential, and particle dimension some calculations of doublet lifetime are made and their results are presented in Fig. 6. The potential well depth increases and in parallel doublet lifetime increases with increasing particle dimension and electrolyte concentration and decreasing surface potential.

## V. COALESCENCE COUPLED WITH EITHER COAGULATION OR FLOCCULATION IN DILUTE EMULSIONS

Limited attention is paid to the role of fragmentation in emulsion science. A comparison of the prediction of coalescence with and without accounting for fragmentation (Sec. II and III) enables evaluation of the frag-

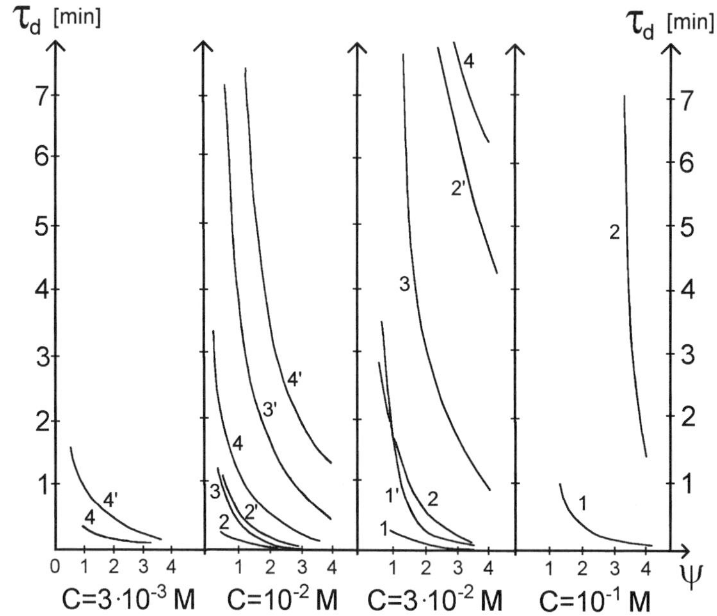


**Figure 6** Dependence of doublet lifetime on the Stern potential for different electrolyte concentrations and droplet dimensions. Numbers near curves correspond to droplet radius. (1) Curves 1–4 without account for retardation of molecular forces of attraction, $\Psi - e\psi/kT$; (2) curves 1′–4′ with account for retardation. (From Ref. 26.)

mentation significance to be made. This comparison will be carried out in Sec. V. A.

The theories (39) and (23) have different areas of applicability (not specified in the papers) and are complementary. Naturally, this complicates the choice between these theories when taking into account the concrete conditions for experiments. An approximate evaluation of the aforesaid areas of applicability is given in Sec. V.B.

### A. Fragmentation of Primary Flocs in Emulsions and the Subsequent Reduction of Coalescence

Floc fragmentation decreases the quantity of inter-droplet films and correspondingly reduces the entire coalescence process. This reduction can be characterized by comparison of Eq. (18) with theory (39), which neglects fragmentation. The longer the time, the greater the reduction, which allows the use of the simpler theory (38) for comparison. The results for longer times coincide with the predictions of the more exact theory (39).

The results of theory (39) concering slow coalescence are illustrated by curve 1 in Fig. 3c in Ref. 39, which is redrawn in Fig. 7a. It can be seen that for a low value of the coalescence rate constant, the semi-logarithmic plot is linear, indicating that the process follows a coalescence rate-controlled mechanism according to Eq. (36). As opposed to the simple exponential time dependence in Eq. (36), second-order kinetics dominate at rapid doublet fragmentation, even if coalescence is very slow. The physical reason becomes clear when considering how Eq. (18) is derived. As seen from Eq. (17) the rate of decline in the droplet concentration is proportional to the doublet concentration. The latter is proportional to the square of the singlet concentration at s.d.e., which causes second-order kinetics. Thus, at slow coalescence the disaggregation drastically changes the kinetic law of coalescence, i.e., from the exponential law to second-order kinetics.

In the second stage, coagulation becomes the rate-controlling process because of the decrease in the collision rate accompanying the decrease in the droplet concentration. Thus, at sufficiently long times, second-order kinetics characterize both reversible and irreversible aggregation. Nevertheless, a large difference exists even when identical functions describe the time dependence, as the characteristic times are expressed through different equations for irreversible and reversible aggregation. In the first

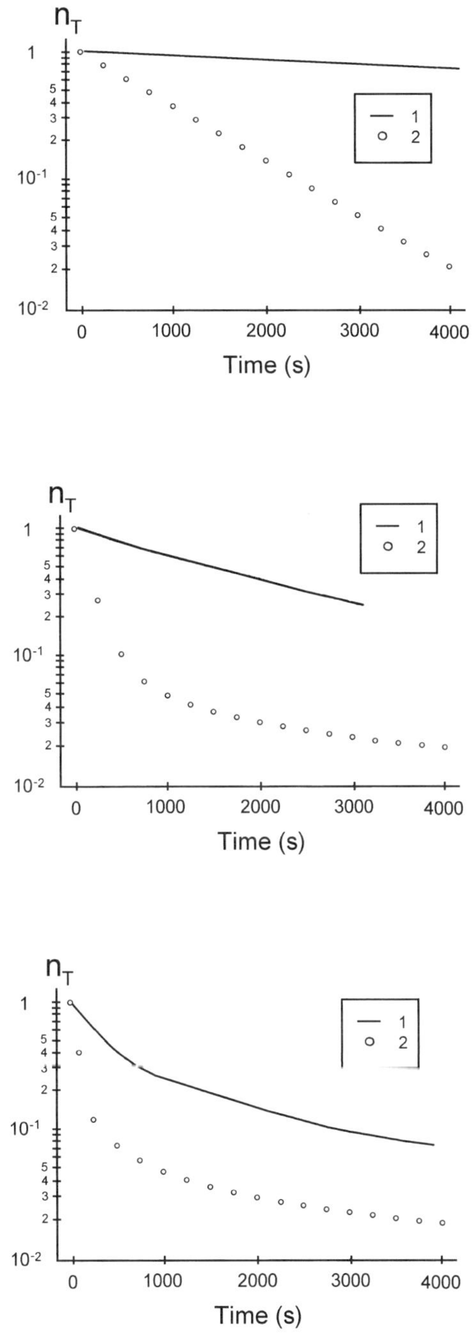


Figure 7 Relative change in the total number of droplets vs. time; initial number of droplets $N_{10} = 1 \times 10^{10}$ cm$^{-3}$; flocculation rate constant $K_f = 1 \times 10^{-11}$ cm$^3$s$^{-1}$; curve 1 – calculations according to Eq. (18); curve 2 – the model of Borwankar *et al* [38] for dilute emulsions, coalescence rate constant (a) $K_c^{2,1} = 1 \times 10^{-1}$ s$^{-1}$, (b) $K_c^{2,1} = 1 \times 10^{-2}$ s$^{-1}$, (c) $K_c^{2,1} = 1 \times 10^{-3}$ s$^{-1}$. Coalescence time $\tau_c = 10^3$ s (a); $\tau_c = 10^2$ s (b); $\tau_c = 10$ s (c). Smoluchowski time $\tau_{Sm} = 10$ s. Doublet lifetime $\tau_d = 0.5$ s; $n_T$ is the dimensionless total droplet concentration, $n_T = N_T/N_{10}$. (From Ref. 23.).

case it is the Smoluchowski time, in the second case it is the combination of three characteristic times, i.e., Eq. (18).

Let us now try to characterize quantitatively the reduction in coalescence caused by doublet distintegration. For this purpose the calculations are performed according to Eq. (18) at $\tau_{Sm} = 10$ s and $\tau_c = 10^3$ s (Fig. 7a), $10^2$ s (Fig. 7b), and 10 s (Fig. 7c). For all figures the same value of the ratio $2\tau_d/\tau_{Sm} = 0.1$ is accepted, satisfying condition (2). In all these figures the calculations according to Eq. (18) are illustrated by curve 1.

The comparison of curves 1 and 2 characterizes the reduction of coalescence caused by doublet disintgration; the lower the Rev values, the stronger the reduction. The simple curve 1 in Fig. 7a can be used also for higher $\tau_c$ values, because then the condition of Eq. (31) is even better satisfied. Thus, if $\tau_{c1}$ and $t_1$ correspond to the data in Fig. 7a, and $\tau_{c2} = m\tau_{c1}$ with $m \gg 1$, the identity:

$$\tau_{c2} t_2 = \tau_{c1} m \frac{t_1}{m} \tag{47}$$

is useful. This means that

$$\frac{n_T}{n_o}\left(\tau_{c1} m, \frac{t_1}{m}\right) = \frac{n_T}{n_o}(\tau_{c1}, t_1) \tag{48}$$

i.e., $t_2 = t_1/m$ where the right-hand side of Eq. (48) is drawn in Fig. 8. For example, Fig. 8 is similar to Fig. 7a and can be used for a 100-fold longer time, as shown on the abscissa. The increase in $\tau_c$ enables us to increase $\tau_{Sm}$ without violating condition (35) and with Eq. (36) valid. Thus, $\tau_{Sm} = 1000$ s or lower can be chosen as the condition for Fig. 8. Curve 1, characterizing the rate of doublet disintegration, is pre-

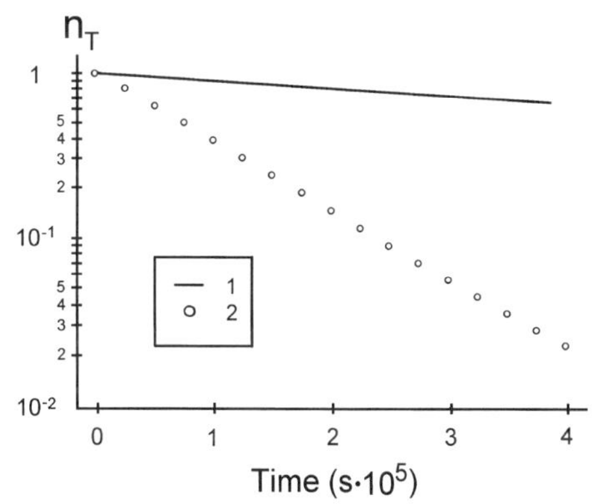


Figure 8 Similar to Fig. 7, with other values for the characteristic times. Coalescence time $\tau_c = 10^5$ s; Smoluchowski time $\tau_{Sm} = 10^3$ s; doublet-fragmentation lifetime $\tau_d = 50$ s. (From Ref. 23.)

served as well if the value of $2\tau_d/\tau_{Sm} = 0.1$ remains; now it corresponds to a higher $\tau_d$ value of 5 s.

### B. Domains of Coalescence Coupled Either with Coagulation or with Flocculation

The condition:

$$\text{Rev} \gg 1 \tag{49}$$

corresponds to coagulation. The theory for the intermediate case:

$$\text{Rev} \sim 1 \tag{50}$$

when part of the droplets participate in flocculation and another coagulate is absent. To specify the conditions (2) and (49) the doublet lifetime must be expressed through surface-force characteristics, namely, through the surface electric potential, the Hamaker function, and droplet dimension, as was described in Sec. IV.

In the equation for the Smoluchowski time [Eq. (4)] the droplet numerical concentration $N_{10}$ can easily be expressed through the droplet volume fraction $\varphi$ and the average droplet radius $a$ (we replace a polydisperse emulsion by an "equivalent" monodisperse emulsion). The resulting analysis respective to $a$ and $\varphi$ is easier than relating to $N_{10}$ because the boundary of application of different regularities are usually formulated with respect to $a$ and $\varphi$. The Smoluchowski time is

$$\tau_{Sm} = \kappa_F^{-1}\varphi^{-1}\frac{4}{3}\pi a^3 \tag{51}$$

We exclude from consideration a special case of extremely dilute emulsions. Comparing Fig. 6 with the results of calculations according to Eq. (51) one concludes that condition (49) is mainly satisfied. It can be violated if simultaneously the droplet volume fraction and the droplet dimension are very small. This occurs if $\varphi < 10^{-2}$ and $a <$ (0.2–0.3) μm. Discussing this case we exclude from consideration the situation when $a <$ 0.1 μm, corresponding to microemulsions and $\varphi \ll 10^{-2}$. With this exception one concludes that, for uncharged droplets, flocculation is almost impossible because condition (2) cannot be satisfied. A second conclusion is that at

$$a < (0.2\text{–}0.3)\ \mu\text{m} \tag{52}$$

theory (39) cannot be applied without some corrections made necessary by the partially reversible character of the aggregation. The main conclusion is that when

$$a > 0.2\ \mu\text{m and } \phi > 10^{-2} \tag{53}$$

theory (39) does not need corrections in respect of the reversibility of flocculation. However, this conclusion will change at the transition to a thicker adsorption layer. As described in Sec. IV, the thicker the adsorption layer, the shorter is the doublet-fragmentation time.

The electrostatic repulsion decreases the depth of the potential well and correspondingly decreases the doublet lifetime. As a result, flocculation becomes possible for submicrometer droplets as well as for micrometer-sized droplets, if the electrolyte concentration is not too high, the surface potential is rather high, and the droplet volume fraction is not too high. This is seen from Fig. 6.

The reversibility criterion depends on many parameters in the case of charged droplets. To discriminate and to quantify the conditions of coagulation and flocculation let us consider Rev values lower than 0.3 as low and values higher than 3 as high. In other words, coagulation takes place when Rev $> 3$, while at Rev $< 0.3$ there is flocculation; that is, the conditions:

$$\text{Rev} = \frac{\tau_d(c_o, \varphi, a)}{\tau_{Sm}(a, \varphi)} > 3 \tag{54}$$

$$\text{Rev} < 0.3 \tag{55}$$

determine the boundaries for the domains of coagulation and flocculation. These domains are characterized by Fig. 9 and correspond to fixed values of the droplet volume fraction. In addition, a definite and rather large droplet dimension $2a = 4$ μm is fixed. After fixation of the values of volume fraction and droplet dimension the domains are characterized in coordinates $\Psi$ and $C$.

In Fig. 9, the domain of flocculation is located above and to the left of curve 2; the domain of coagulation is located beneath and to the right of curve 1. To characterize the sensitivity of the domain boundaries to the Hamaker function value, curves 1′ and 2′ are calculated using values twice as high as those of curves 1 and 2.

As distinct from uncharged droplets, flocculation in the range of micrometer-sized droplets is possible. As seen in Fig. 9, even rather large droplets (4 μm) aggregate reversibly if the electrolyte concentration is lower than $(1-5)\times 10^{-2}$ M and the Stern potential is higher than 25 mV. For smaller droplets the domain of flocculation will extend while the domain of coagulation will shrink. For submicrometer droplets, flocculation takes place even at high electrolyte concentrations (0.1 M).

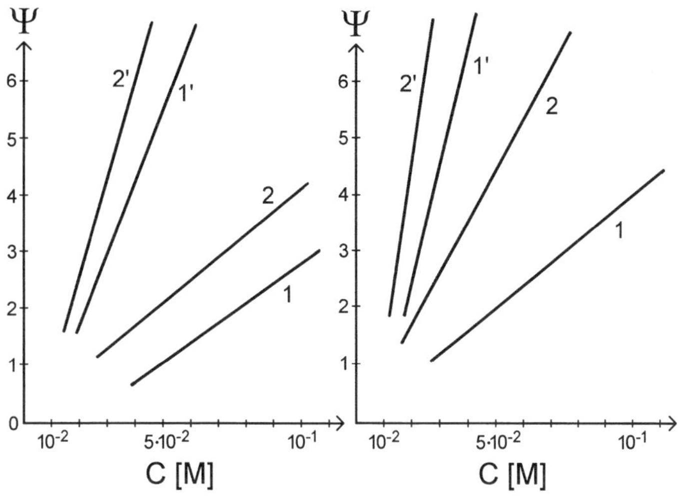


**Figure 9** Domains of coagulation and flocculation. Curves 1 and 2 are calculated with the Rabinovich–Churaev Hamaker function; a twice higher value is used for calculation of curves 1′ and 2′. The domain of flocculation is located above curve 1, while the domain of coagulation is located beneath curve 2. Volume fraction $\phi = 0.01$ (a); $\phi = 0.1$ (b). Particle dimension $2a = 4$ μm. (From Ref. 26.)

## C. Hydration Forces Initiate Flocculation

Due to the similarity of the exponential distance ($h$) dependence of hydration forces and that for the electrostatic interaction the decrease in the doublet lifetime caused by hydration forces of repulsion can be calculated on account of this similarity. It is sufficient to use the substitution $h_s$ for $\kappa^{-1}$ and $K_s$ for

$$16\varepsilon\left(\frac{kT}{e}\right)^2 \tanh\left(\frac{e\psi}{kT}\right)^2 \tag{56}$$

where $k$ is the Boltzmann constant, $T$ is the absolute temperature, $\varepsilon$ is the dielectric permittivity of water, and $e$ is the elementary charge. The doublet lifetime can be determined with use of the results presented in Fig. 6. For the sake of brevity, a similar figure with $K_s$ and $h_s$ as the abscissa is not shown. It turns out that the decrease in $\tau_d$ caused by hydration forces leads to reversible aggregation of submicrometer droplets. As to micrometer-sized droplets, coagulation takes place except for the case when both $h_s$ and $K_s$ are rather large.

## VI. APPLICATIONS

The restrictions in Eqs (1) and (12) corresponding to strong retardation of the rate of multiplet formation and slow intradoublet coalescence are not frequently satisfied. Nevertheless, these conditions are important because they correspond to the case of very stable emulsions. As the kinetics of the retarded destabilization of fairly stable emulsions is of interest, attention has to be paid to provide these conditions and thus the problem of coupled coalescence and flocculation arises.

There are large qualitative distinctions in the destabilization processes for the coupling of coalescence and coagulation, and coalescence and flocculation. In the first case, rapid aggregation causes rapid creaming and further coalescence within aggregates. In the second case, the creaming is hampered owing to the low concentration of multiplets, and coalescence takes place both before and after creaming. Before creaming, singlets predominate for a fairly long period of gradual growth of droplet dimensions due to coalescence within doublets. The discrimination of conditions for coupling of coalescence with either flocculation or coagulation is accomplished in Ref. 26.

The creaming time is much shorter in the coagulation case and correspondingly the equation describing the coupling of coalescence and flocculation preserves its physical sense for a longer time than is the case for coagulation. One concludes that the theory of the coupling of coalescence and flocculation provides a new opportunity for long-term prediction of emulsion stability, although creaming restricts the application of this theory as well. Note that this restriction weakens in the emulsions of low-density contrast and in W/O emulsions with a high viscosity continuum.

Long-term prediction is a two-step procedure. The first step is the determination of whether an emulsion exhibits coagulation or flocculation. It means that the characteristic time $\tau_d$ must be measured and compared with $\tau_{Sm}$, the value of which is easily evaluated with account taken for the measured concentration using Eq. (4). A comparison of these times allows a choice between condition (1) and the opposite condition ($\tau_{Sm} \ll \tau_d$). The second step is the prediction of the evolution in time for the t.d.c. If condition (1) is valid, Eq. (18) has to be used for the prediction; $\tau$ in Eq. (18) has to be specified in accordance with Eq. (15). In the opposite case DIGB theory must be used.

### A. Long-term Prediction of Emulsion Stability

It is possible, in principle, to give a long-term prediction of emulsion stability based on the first indications of aggregation and coalescence. The next example clarifies the principal difficulty in a reliable long-term prediction if a dynamic model of the emulsion is not available.

The first signs of aggregation and coalescence can always be characterized by a linear dependence, if the investigation time $t$ is small in comparison with a characteristic time $\tau$ for the evolution of the total droplet concentration $n(t)$:

$$n(t) = n_0\left(1 - \frac{t}{\tau}\right) \tag{57}$$

This short time asymptotic corresponds to many functions, for example, to Eq. (18) or (36). The first can arise in the case of coalescence coupled with coagulation (39), while the second can arise for coalescence coupled with flocculation (29). The discrimination between irreversible and reversible aggregation is only one component of emulsion dynamics modeling (EDM) and it is seen that without this discrimination the difference in the prediction of the time necessary for a droplet concentration decrease, for example, 1000 times, can be $7\tau$ and $1000\tau$.

### B. Perfection of Methods for Emulsion Stabilization (Destabilization) by Means of the Effect on Both Coalescence and Flocculation

Stability (instability) of an emulsion is caused by the coupling of coalescence and flocculation. Meanwhile, for emulsifiers (or demulsifiers) the elaboration of their influence on the elementary act of coalescence only is mainly taken into account. The coupling of coalescence and flocculation is reflected in Eq. (15) and one concludes that it follows the multiplicativity rule and not the additivity rule. This means that the total result of the application of a stabilizer (destabilizer) depends very much on both flocculation and fragmentation. The development of a more efficient technology for emulsion stabilization (destabilization) is possible by taking into account the joint effect on both the coalescence and the aggregation (disaggregation) processes.

#### 1. Combining Surfactants and Polymers in Emulsion Stabilization

The coalescence rate depends mainly on the thin (black) film stability and correspondingly on the short-range forces. The flocculation depends on the long-range surface forces. Owing to this large difference, synergism in the dependence of these processes on the different factors can be absent.

The use of one surfactant only may not provide both the optimal fragmentation and optimal stability of an emulsion film. Probably the use of a binary surfactant mixture with one component which provides the film stability, and a second one which prevents the flocculation may provide perfect emulsion stabilization. Naturally their coadsorption is necessary. For such an investigation a measurement method for both the doublet-fragmentation time and the coalescence time is necessary.

#### 2. Strong Influence of Low Concentrations of Ionic Surfactant on Doublet-fragmentation Time and Coalescence Time

Let us consider the situation when an emulsion is stabilized against coalescence by means of an adsorption layer of nonionic surfactant and is strongly coagulated because of the subcritical value of the Stern potential that is usual for inorganic electrolytes (46) at moderate pH. In a large floc any droplet has many neighbors, meaning a rather high number of interdroplet films per droplet. The coalescence rate is proportional to the total number of films and can be quite high. It can be strongly decreased by adding a low concentration of an ionic surfactant. This can be sufficient to provide a supercritical Stern potential value that will be accompanied by a drastic decrease in the doublet lifetime compared to that of weakly charged droplets.

At shorter doublet lifetimes flocculation can become reversible and it can stop at the stage of singlet–doublet equilibrium. It will provide a strong decrease in the coalescence rate because coalescence occurs within doublets only and their concentration can be very low.

Thus, a small addition of an ionic surfactant to a higher concentration of a nonionic surfactant, sufficient to provide an almost saturated adsorption layer, can make the overall emulsion stabilization more efficient. The nonionic surfactant suppresses coalescence but cannot prevent flocculation, while the ionic surfactant retards the development of flocculation.

We can give an example when both coalescence and flocculation are affected by an ionic surfactant (SDS). In Ref 86 it is established that coalescence is suppressed at SDS concentrations exceeding $6 \times 10^{-5}$ M. Meanwhile, the CCC is $2 \times 10^{-2}$ M NaCl at $10^{-6}$ M SDS. Thus, SDS concentrations slightly above $10^{-6}$ M are sufficient to retard flocculation. In this example it is essential that the concentrations needed to retard flocculation are very low compared to those needed to prevent coalescence.

It is noteworthy that low concentrations of an ionic surfactant can increase emulsion stability as a result of the simultaneous manifestation of three mechanisms. First, the depth of the secondary potential minimum decreases owing to the electrostatic repulsion that is accompanied by a $\tau_d$ decrease. Second, the transition from the secondary minimum through an electrostatic barrier and into the primary minimum extends the coalescence time. Third, the time of true coalescence, i.e., the time necessary for thin-film rupture increases because of electrostatic repulsion as well (27, 63).

## C. Standardization of the Measurement of $\tau_c$ and $\tau_d$

Direct investigation of the coalescence subprocess in emulsions is difficult. Instead, the entire destabilization process is usually investigated. Meanwhile, the rate of the destabilization process depends on the rates of both flocculation and disaggregation and on the floc structure as well. All these characteristics vary in a broad range. At a given unknown value for the time of the elementary act of coalescence $\tau_c$ the different times can be measured for the integrated process and different evaluations of $\tau_c$ are possible.

The rate of coalescence in an aggregate essentially depends on the number of droplets within it and the packing type, i.e., on the number of films between the droplets. This complication is absent when considering the case of the s.d.e.

The possible advantage of $\tau_c$ measurement at s.d.e. is in avoiding the difficulty caused by polydispersity of droplets appearing during preceding coalescence within large flocs. At the s.d.e. the initial stage of the entire coalescence process can be investigated when the narrow size distribution of an emulsion is preserved.

At s.d.e. the determination of the time dependence of the t.d.c. is sufficient for the investigation of coalescence. In Refs 27 and 28 this was accomplished through direct visual observation. By using video-enhanced microscopy and computerized image analysis the determination of t.d.c. can be automated. Such automated determination of total droplet number in a dilute DCD-in-water emulsion at the s.d.e. can be recommended as a standard method for the characterization of the elementary act of coalescence.

In parallel, the second important characteristic, namely, the doublet-fragmentation time is determined by the substitution of $\tau_c$, $\tau_{Sm}$, and measured $\tau_d$ into Eq. (18).

## D. Experimental–Theoretical Emulsion Dynamics Modeling

### 1. General

To predict the evolution of the droplet (floc) size distribution is the central problem in emulsion stability. It is possible, in principle, to predict the time dependence of the distribution of droplets (flocs) if information concering the main subprocesses (flocculation, floc fragmentation, coalescence, creaming), constituting the whole phenomenon, is available. This prediction is based on consideration of the population balance equation (PBE).

The PBE concept was proposed by Smoluchowski (70). He specified this concept for suspensions and did not take into account the possibility of floc fragmentation. Even with this restriction he succeeded in the analytical solution, neglecting gravitational coagulation and creaming, and obtained the analytical time dependence for a number of aggregates $n_i$ comprising $i$ particles ($i = 2, 3 \ldots$).

In the most general case the equation for the evolution of the total droplet number takes into account the role of aggregation, fragmentation, creaming, and coalescence. There is no attempt to propose an algorithm even for a numerical solution to such a problem.

The usual approach in the modeling of an extremely complicated process is the consideration of some extreme cases with further synthesis of the results obtained. The next three main simplifications are inherent in the current state of emulsion dynamics modeling: the neglect of the influence of the gravitational field, i.e., neglect of creaming/sedimentation; in a first approximation it is possible to consider either coagulation or flocculation; finally, the neglect of the rate-constant dependence on droplet dimension.

## 2. Combined Approach in Investigations of Dilute and Concentrated Emulsions

The modeling of collective processes in concentrated emulsions is extremely complicated. Recently, the efficiency of computer simulation in the systematic study of aggregates, gels, and creams has been demonstrated (48). Monte Carlo and Brownian dynamics are particularly suited to the simulation of concentrated emulsions. However, information about droplet–droplet interaction is necessary. The reliability of this information is very important in providing reasonable results concerning concentrated emulsions. In other words, the assumption concerning pair additive potentials for droplet/droplet interaction and the thin emulsion film stability must be experimentally confirmed. The extraction of this information from experiments with concentrated emulsions is very difficult. On the other hand, measurement of the doublet-fragmentation time in dilute emulsions is a convenient method for obtaining information about pair additive potentials. Information about pair potentials and the elementary act of coalescence obtained in experiments with dilute emulsions preserves its significance for concentrated emulsions as well.

One concludes that modeling of concentrated emulsions becomes possible by combining experimental investigation of the simplest emulsion model system with computer simulation accounting for the characteristics of a concentrated emulsion (high droplet-volume fraction, etc.).

## 3. Kernel Determination Is the Main Task that Must Be Solved to Transform the PBE in an Efficient Method for Emulsion Dynamics Modeling

The levels of knowledge concerning kernels describing different subprocesses differ strongly. There exists a possibility for quantification of kernels related to aggregation and fragmentation (12, 37, 76). On the other hand, the current state of knowledge is not sufficient for prediction of the thin film disruption time.

The deficit in knowledge of thin-film stability makes purely theoretical modeling of emulsion dynamics impossible. As a result a complex semi-theoretical approach to EDM is necessary. The PBE is the main component of both the experimental and the theoretical stages of this approach. In the experimental stage the PBE, simplified with regard to s.d.e., provides the background for determination of the coalescence kernels with use of the experimental data (27, 28).

For determination of the coalescence kernels the more complicated reverse task must be solved, namely, their determination based on comparison of the experimental data on the emulsion evolution in time with the PBE solution. In the absence of an analytical solution the reverse task is usually very difficult. The most efficient way to overcome this difficulty is experimental realization with the use of the universally simplest conditions for emulsion time evolution, which can be described analytically.

## 4. Singlet–Doublet Quasiequilibrium with Slow Coalescence within Doublets Is the Simplest Emulsion State for which Investigation Can Provide Information about Coalescence

The simplest singlet–doublet emulsion can exist at singlet–doublet quasiequilibrium and with slow coalescence within doublets. Its simplicity results in a very simple kinetic law for the entire kinetics of coupled flocculation and coalescence, namely, Eq. (18). Thus, s.d.e. provides the most convenient conditions for investigations of the elementary act of coalescence and the doublet-fragmentation time.

The main simplification in all existing models for emulsion dynamics (23, 39) is the neglect of the coalescence time dependence on droplet dimensions. This simplification is not justified and decreases very much the value of the prediction, which can now be made with use of the PBE. For elimination of this unjustified simplification it is necessary to determine the coalescence time for emulsion films between droplets of different dimensions $i$ and $j$, namely, $\tau_{cij}$, similar to the existing analytical expressions for the doublet-fragmentation time, $\tau_{dij}$ (12). The determination of a large set of $\tau_{cij}$ values by means of a comparison of experimental data obtained for an emulsion consisting of different multiplets and the PBE numerical solutions for it is impossible. On the other hand, this paramount experimental–theoretical task can be solved for a dilute emulsion at s.d.e. and slow intra-doublet coalescence.

### 5. Substitution of the Coalescence Kernels Makes the PBE Equation Definite and Ready for Prediction of Emulsion Time Evolution with the Restriction of Low-density Contrast and without Account for Gravitational Coagulation and Creaming

With application of the scaling procedure for the representation of the kinetic rate constants for creaming and gravitational coagulation the PBE is solved analytically in Ref. 87. This scaling theory creates a perspective for the incorporation of creaming in the emulsion dynamics model in parallel with coalescence, aggregation, and fragmentation.

## VII. SUMMARY

The mechanisms of kinetic stability in macroemulsions and miniemulsions are completely different. The strong droplet deformation and flattening in a macroemulsion cause the Reynolds mode of drainage which prolongs the life of the emulsions. This mechanism is not important for miniemulsion droplet interaction, because either the deformation and flattening are weak (charged droplets) or the Reynolds drainage is rapid owing to the small dimensions of the interdroplet film (uncharged droplets). The kinetic stability of a miniemulsion can be caused by floc fragmentation if the electrokinetic potential is not too low and the electrolyte concentration is not too high, corresponding to some electrostatic repulsion.

The potential strength of physicochemical kinetics with respect to emulsions is the PBE, allowing prediction of the time evolution of the droplet size distribution (d.s.d.) when the subprocesses [including droplet aggregation, aggregate fragmentation, droplet coalescence, and droplet (floc) creaming] are quantified. The subprocesses are characterized in the PBE by the kinetic coefficients. The coupling of the four subprocesses, the droplet polydispersity, and the immense variety of droplet aggregate configurations causes extreme difficulty in EDM. The processes of aggregation, fragmentation, and creaming can be quantified. In contrast, only the experimental approach is now available for efficient accumulation of information concering emulsion-film stability and coalescence kernel quantification for EDM.

Correspondingly, EDM may be accomplished by combining experiment and theory: (1) determination of coalescence and fragmentation kernels with the use of emulsion stability experiments at low-density contrast (l.d.c.) and s.d.e., because this permits the omittance of creaming and gravitational terms in PBE, simplifying it and making solution of the reverse task possible; and (2) prediction of the droplet size evolution with time by means of solution of the PBE, specified for the determined coalescence and fragmentation kernels. This mathematical model has to be based on the PBE supplemented by terms accounting for the role of creaming and gravitational coagulation in the aggregation kinetics.

EDM with experiments using l.d.c. emulsions and s.d.e. may result in: (1) the quantification of emulsion film stability, namely, the establishment of the coalescence time dependence on the physicochemical specificity of the adsorption layer of a surfactant (polymer), its structure, and the droplet dimensions. This quantification can form a basis for the optimization of emulsifier and demulsifier selection and synthesis for emulsion technology applications, instead of the current empirical level applied in this area; and (2) the elaboration of a commercial device for coalescence-time measurement, which in combination with EDM will represent a useful approach to the optimization of emulsion technology with respect to stabilization and destabilization.

## ACKNOWLEDGMENTS

The technology program FLUCHA, financed by the oil industry and the Norwegian Research Council, is acknowledged for financial support.

## REFERENCES

1. P Becher, ed. Encyclopedia of Emulsion Technology. Vol 1. New York: Marcel Dekker, 1983.
2. P Becher, ed. Encyclopedia of Emulsion Technology. Vol 2. New York: Marcel Dekker, 1985.
3. J Sjöblom, ed. Emulsions – A Fundamental and Practical Approach. Dordrecht: Kluwer Academic, 1992, p 1.
4. RA Mohammed, AI Bailey, PF Luckham, SE Taylor. Colloids Surfaces 80: 223 1993; 80: 237, 1993; 83: 261, 1994.
5. CI Chitewelu, V Hornof, GH Neale, AE George. Can J Chem Eng 72: 534, 1994.
6. L Bertero, AD Lullo, A Lentini, L Terzi. SPE 28543. Society of Petroleum Engineers, Italy, 1994.
7. JF McCafferty, GG McClaflin. SPE 24850. Society of Petroleum Engineers, 1992.
8. JA Kitchener, PR Musselwhite. In: IP Sherman, ed. Emulsion Science. New York: Academic Press, 1968.

9. J Sjöblom, ed. Emulsions and Emulsion Stability. New York: Marcel Dekker, 1996.
10. IB Ivanov, PA Kralchevsky. Colloids Surfaces 128: 155, 1997.
11. BV Derjaguin. Theory of Stability of Colloids and Thin Films. New York: Plenum Press, 1989.
12. SS Dukhin, J Sjöblom. In: J Sjöblom, ed. Emulsions and Emulsion Stability. New York: Marcel Dekker, 1996, p 41.
13. I Gregory. Crit Rev Environ Control 19: 185, 1989.
14. JN Israelachvili. Intermolecular and Surface Forces. London: Academic Press, 1991.
15. J Lyklema. Fundamentals of Interface and Colloid Science. Vol 1. London: Academic Press, 1993.
16. PA Kralchevsky, KD Danov, ND Denkov. Handbook of Surface and Colloid Chemistry. London: CRC Press, 1996.
17. PA Kralchevsky, KD Danov, ND Denkov. Handbook of Surface and Colloid Chemistry. London: CRC Press, 1996, Chap. 11.4.
18. RK Prud'homme, ed. Foams: Theory, Measurement and Applications. New York: Marcel Dekker, 1995.
19. PA Kralchevsky, KD Danov, IB Ivanovv. In: RK Prud'homme, ed. Foams: Theory, Measurement and Applications. New York: Marcel Dekker, 1995, pp 1–97.
20. SAK Jeelany, S Hartland. J Colloid Interface Sci 164: 296, 1994.
21. PDI Fletcher. In: D Mobius, R Miller, eds. Drops and Bubbles in Interface Research. Vol. 6. New York: Elsevier, 1998.
22. PJ Breen, DT Wasan, YH Kim, AD Nikolov, CS Shetty. In: J Sjöblom, ed. Emulsions and Emulsion Stability. New York: Marcel Dekker, 1996.
23. SS Dukhin, J Sjöblom. J Disp Sci Technol 19: 311, 1998.
24. Ø Holt, Ø Sæther, J Sjöblom, SS Dukhin, NA Mishchuk. Colloids Surfaces 123/124: 195, 1997.
25. Ø Sæther, SS Dukhin, J Sjöblom, Ø Holt. Colloid J 57: 836, 1995.
26. SS Dukhin, J Sjöblom, DT Wasan, O Sæther. Colloid and Surfaces (in press).
27. Ø Sæther, J Sjöblom, SV Vrebich, NA Mischchuk, SS Dukhin. Colloids Surfaces 142: 189, 1998.
28. Ø Sæther, J Sjöblom, SV Verbich, SS Dukhin. J Disp Sci Technol 20: 295, 1999.
29. Ø Holt, Ø Sæther, J Sjöblom, SS Dukhin, NA Mischuk. Colloids Surfaces 141: 269, 1998.
30. SS Dukhin, Ø Sæther, J Sjöblom. Program and Abstracts, 72nd ACS Colloid and Surface Science Symposium, Pennsylvania State University, University Park, PA, June 21–24, 1998 (abstr 401).
31. O Reynolds. Phil Trans R Soc (London), A177: 157, 1886.
32. A Scheludko. Adv Colloid Interface Sci 1: 391, 1967.
33. KD Danov, DN Petsev, ND Denkov, R Borwankar. J Chem Phys 99: 7179 1993.
34. KD Danov, ND Denkov, DN Petsev, R Borwankar. Langmuir 9: 1731, 1993.
35. DN Petsev, ND Denkov, PA Kralchevsky. J Colloid Interface Sci 176: 201, 1995.
36. IB Ivanov, DS Dimitrov. In: IB Ivanov, ed. Thin Liquid Films. New York: Marcel Dekker, 1998.
37. NA Mishchuk, J Sjöblom, SS Dukhin. Colloid J 57: 785 1995.
38. RP Borwankar, LA Lobo, DT Wasan. Colloids Surfaces 69: 135, 1992.
39. KD Danov, IB Ivanov, TD Gurkov, RP Borwankar. J Colloid Interface Sci 167: 8, 1994.
40. LA Lobo, DT Wasan, M Ivanova. In: KL Mittal, DO Shah, eds. Surfactants in Solution. New York: Plenum Press, 1992, Vol 11. p 395.
41. L Lobo, I Ivanov, D Wasan. Mater Interfaces Electrochem Phenomena. 39: 322, 1993.
42. G Narsimhan, E Ruckenstein. In: RT Prud'homme, ed. Foams: Theory, Measurement and Applications. New York: Marcel Dekker, 1995, p 99.
43. D Exerova, D Kaschiev, D Platikanov. Adv Colloid Interface Sci 40: 201, 1992.
44. BV Derjaguin, YuV Gutop. Kolloidn Zh 24: 431, 1962.
45. AV Prokhorov, BV Derjaguin. J Colloid Interface Sci 125: 111, 1988.
46. SV Verbich, SS Dukhin, A Tarovsky, Ø Holt, Ø Sæther, J Sjöblom. Colloids Surfaces 23: 209, 1997.
47. D Kashchiev, D Exerowa. J Colloid Interface Sci 203: 146, 1998.
48. E Dickinson, SR Euston. Adv Colloid Interface Sci 42: 89, 1992.
49. P Taylor. Adv Colloid Interface Sci 75: 107, 1998.
50. HW Yarranton, YH Masliyah. J Colloid Interface Sci 196: 157, 1997.
51. M Smoluchowski. Phys Zeit 17: 557, 1916.
52. GA Martynov, VM Muller. Kolloidn Zh 36: 687, 1974.
53. VM Muller. Kolloidn Zh 40: 885, 1978.
54. H Sonntag, K Strenge. Coagulation Kinetics and Structure Formation. Berlin: VEB Deutcher Verlag der Wissenschaften, 1987.
55. JD Chan, PS Hahn, JC Slattery. AIChE J 34: 140, 1988.
56. SAK Jeelani, S Hartland. J Colloid Interface Sci 156: 467, 1993.
57. RW Aul, WI Olbriht. J Colloid Interface Sci 115: 478, 1991.
58. A Scheludkdo D Exerova. Colloid J 168: 24, 1960.
59. A Scheludko, Proc. K. Ned. Akad. Wet. Ser. B. 65, 87, 1962.
60. D Platikanov, F Manev. Proceedings of the 4th International Congress of Surface Active Substances, New York, Plenum Press, 1964, p 1189.
61. KA Burrill and DR Woods. J Colloid Interface Sci 42: 15, 1973.
62. KA Burrill, DR Woods. J Colloid Interface Sci 42: 35, 1973.

63. SR Deshiikan, KD Papadopoulos. J Colloid Interface Sci 174: 302, 1995.
64. SR Deshiikan, KD Papadopoulos. J Colloid Interface Sci 174: 313, 1995.
65. ME Costas, M Moreau, L Vicente. J Phys A: Math Gen 28: 2891, 1995.
66. F LeBerre, G Chauveteau, E Pefferkorn. J Colloid Interface Sci 199: 1, 1998.
67. BJ McCoy, G Madras. J Colloid Interface Sci 201: 200, 1998.
68. J Widmaier, E Pefferkorn. J Colloid Interface Sci 203: 402, 1998.
69. IM Elminyawi, S Gangopadhyay, CM Sorensen. J Colloid Interface Sci 144: 315, 1991.
70. M. von Smoluchowski. Phys Chem 92: 129, 1917.
71. M van den Tempel. Rec Trav Chim 72: 419, 1953, 72: 433, 1953.
72. P Meakin. Adv Colloid Interface Sci 28: 249, 1988.
73. R Nakamura, Y Kitada and T Mukai. Planet Space Sci 42: 721, 1994.
74. S. Chandrasekhar. Rev Mod Phys 15: 1, 1943.
75. WB Russel, DA Saville, WR Schowalter. Colloidal Dispersions. New York: Cambridge Univrsity Press, 1989.
76. VM Muller. Colloid J 58: 598, 1996.
77. YaI Ravinovich, AA Baran. Colloids Surfaces 59: 47, 1991.
78. P Walstra. In: GO Philips, PA Williams, DJ Wedlock, eds. Gums and Stabilizers for the Food Industry. Vol 4. Oxford: IRL, 1988, p 233.
79. DG Dalgleish. Colloids Surfaces 46: 14, 1990.
80. DG Dalgleish, Y Fang. J Colloid Interface Sci 156: 329, 1993.
81. B Bergenstahl, PM Claesson. In: K Larson, S Friberg, eds. Food Emulsions. New York: Marcel Dekker, 1989, p 41.
82. YaI Rabinovich, NV Churaev. Kolloidn Zh 46: 69, 1984; 52: 309, 1990.
83. DJ Mitchell, P Richmond. J Colloid Interface Sci 46: 128, 1974.
84. VN Gorelkin, VP Smilga. Kolloidn Zh 34: 685, 1972.
85. VN Gorelkin, VP Smilga. In: BV Derjaguin, ed. Poverkhnostnye Sily v Tonkikh Plenkakh i Ustoichivost Kolloidov. Moskow: Nauka, 1974, p 206.
86. S Usui, Y Imamura, E Barough. J Disp Sci Technol 8: 359, 1987.
87. SB Grant, C Poor, S Relle. Colloids Surfaces 107; 155, 1996.

# 5

# Macroemulsions from the Perspective of Microemulsions

**Håkan Wennerström, Olle Söderman, Ulf Olsson, and Björn Lindman**
*University of Lund, Lund, Sweden*

## I. INTRODUCTION

Microemulsions and surfactant-stabilized (macro) emulsions are distinctively different with respect to thermodynamic stability and, therefore, while most significant for both types of systems, the role of studies of phase behavior is different in the two cases. For emulsions we are concerned with two- or multi-phase regions in the phase diagrams, and for microemulsions with one-phase regions. Because of that microemulsion studies are closely related to studies of other thermodynamically stable phases, notably liquid crystalline phases and micellar solutions. Structural models of microemulsions have to a considerable extent been advanced on the basis of our understanding of other stable phases; the formation and stability of a microemulsion phase for a certain surfactant results from the comopetition with alternative phases. The principal differences between microemulsions and emulsions, together with the related nomenclature, is bound to lead to considerable confusion; for example, the persistence in literature of emulsion-based structural pictures of microemulsions can be traced to the related names. However, the term microemulsions is kept for historical reasons.

There are also many similarities between micro- and macro-emulsions, such as the presence of surfactant films, and studies of microemulsions have resulted in a detailed understanding of the properties of surfactant films. Progress has been made both with respect to experimental methods and theoretical concepts.

In this chapter, which is a slight elaboration of a recent publication (1), we will discuss how some of our knowledge of microemulsions can be used in emulsion studies. We discuss in particular the role of phase behavior; diffusion studies of both microemulsions and macroemulsions; the implications of the flexible surface model for emulsion stability; and the Ostwald ripening process in a potentially metastable emulsion system. By way of these examples it is demonstrated how microemulsion research has helped to develop our current understanding of macroemulsions.

Already at an early stage in the development of surface and colloid science emulsions received a lot of attention. It was primarily clear that the properties of foodstuffs like milk, butter, and sauces contained dispersions of one liquid phase in another; either oil (as fat) in water or water in oil (fat). Pioneering studies were performed by Bancroft in the first quarter of this century and his studies were followed up by Harkins and others, leading to a phenomenological knowledge that is still relevant today. Attempts were also made to understand and rationalize the experimental findings, but in this aspect progress was much slower.

The majority of emulsions are stabilized by surfactants and in this case Bancroft understood that the stability of an emulsion was related to the properties of the surfactant film. However, at the time virtually the only way to study such films experimentally was in the surface balance obtaining area – pressure isotherms. Such measurements, although informative, did not provide a sufficient characterization of the surfactant film, and the stability problem was left unsolved. With the emergence of the DLVO theory some aspects of emulsion stability could be discussed in a theoretically coherent way, but it was also clear that the DLVO picture was irrelevant under many important circumstances.

During the 1960s a new generation of surfactant scientists appeared. They primarily focused their interest on the properties of bulk phases. For bulk phases many more experimental techniques could be applied and a better understanding of the molecular properties of surfactant films emerged. Together with Kozo Shinoda, Stig Friberg pioneered the approach of connecting bulk properties like phase equilibria with molecular properties of surfactant films and ultimately with emulsion stability (2). They particularly pointed out the importance and relevance of the appearance of lamellar phases for long time emulsion stability.

In the same period, Schulman *et al.* (3) and Winsor (4) observed that certain mixtures of oil, water, surfactant, and a cosurfactant yielded seemingly clear one-phase systems even without substantial input of mechanical energy. These systems typically contained small droplets of one liquid in the other and they were called microemulsions. It was in particular Friberg and his group who stressed that the long-term stability of these microemulsions was in fact a result of their thermodynamic stability (5). A fact that was used as a characterizing dividing line between microemulsions and emulsions, where the latter now are termed macroemulsions in the cases where one wants to distinguish them explicitly from the microemulsions. Friberg *et al.* also showed (6), by way of example, that a prerequisite for studies of the properties of microemulsions is a knowledge of the phase equilibria, a view that is now finally accepted in the field.

In the 1970s and 1980s there was an explosion in the scientific interest in microemulsions, which by themselves posed many intriguing challenges. The original connection to the, technically more important, emulsion problem was largely forgotten. However, one very important result of the microemulsion studies is a much improved understanding of the properties of surfactant films at the water–oil interface both in qualitative and quantitative terms.

In the spirit of Friberg's pioneering work we review in this paper some recent progress in the understanding of emulsions which explicitly build on studies of microemulsions. We start by discussing phase diagrams showing both microemulsion areas and two- or three-phase areas where emulsions may potentially be formed. Self-diffusion measurements has turned out to be a versatile method for studying microemulsion structure. In contrast to macroemulsions, microemulsions not only exhibit a droplet structure but there is also a transition to a bicontinuous structure which is clearly revealed in diffusion studies. Similarly, self-diffusion measurements are effective for the characterization of macroemulsions as discussed in the following section. We then summarize the present understanding of microemulsions which have the focus on the properties of the surfactant film. It is then outlined how this knowledge can be transformed to discuss microemulsion stability. The paper is concluded with a discussion of the transition from microemulsion to macroemulsion.

## II. PHASE BEHAVIOR

To form a microemulsion three ingredients are necessary: polar solvent (water), apolar solvent (oil), and surfactant. Since typical microemulsions only occur under rather selective circumstances it is in practice necessary to have an additional tuning variable that can be adjusted to obtain optimal conditions for microemulsion formation. In the early studies of Schulman *et al.* (3) the amount of cosurfactant was used to tune the systems in addition to the salt concentration. This introduces a fourth (cosurfactant) and sometimes a fifth (salt) component, making the thermodynamic description nearly intractable. Below we illustrate the basic principles by staying with three-component systems, using the temperature as the tuning variable. This situation is most easily realized in practice with nonionic surfactants of the $C_mE_n$ type, where E denotes an ethylene oxide unit.

In practice the phase equilibria are determined at different constant temperatures. An illustrative Gibbs triangle is shown in Fig. 1. Data from different temperatures are collected into a phase prism as illustrated in Fig. 2a. We have three independent intensive variables and since two-dimensional projections are most illustrative we have to make cuts in the three-dimensional prism. The structural transitions are best illustrated in a so-called Shinoda cut (7), where the temperature and oil–water ratio is varied at a constant surfactant concentration, as shown in Fig. 2, which also displays schematic drawings of the different aggregate structures. As will be discussed in more detail later in the paper the temperature changes the properties of the surfactant film, resulting in structural changes. At low temperatures, oil-swollen micelles or oil microemulsion droplets occur. An increase in temperature leads to the

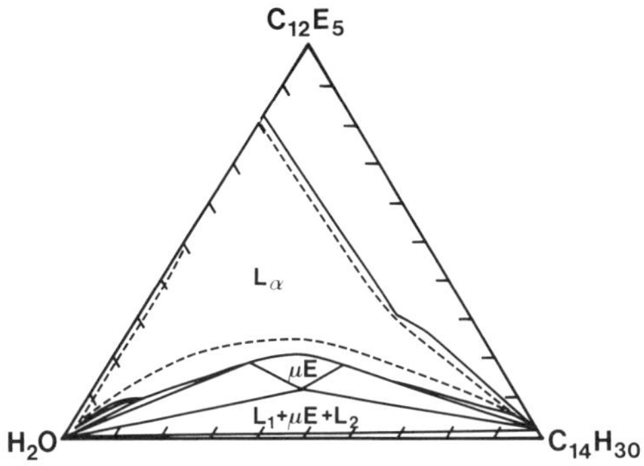


**Figure 1** Isothermal phase diagram of the $C_{12}E_5/H_2O$/tetradecane system at 47.8°C, which is near the balance temperature. The microemulsion phase (μE) forms a narrow island near equal amounts of water and oil. At low surfactant concentrations the microemulsion is in simultaneous equilibrium with excess of oil ($L_2$) and water ($L_1$). At higher concentrations a lamellar phase ($L_\alpha$) is stable, which also extends to the oil and water corners. (Figure redrawn from Ref 46.)

gradual formation of a bicontinuous phase, which at even higher temperatures goes over to a water-droplet phase. At high temperature and high water content, or low temperature and low water content, there is a sponge phase showing a bicontinuous structure containing a bilayer network structure.

The importance of the amount of surfactant is best illustrated in the so-called Kahlweit fish plot (8), where the oil-to-water ratio is kept constant, typically at 1:1, as illustrated in Fig. 3. The diagram is virtually symmetrical with respect to a reflection at a temperature $T_0$, representing the temperature where the surfactant film is balanced, that is, exhibits a zero mean curvature. Increasing the surfactant concentration from zero we first encounter a narrow two-phase area of water and oil, which, with only small additions of surfacant, changes into a Winsor three-phase area, which in addition to water and oil also has a (bicontinuous) microemulsion as the third phase. With a further increase in surfactant concentration we enter the one-phase microemulsion at the apex of a three-phase triangle. The more effective the surfactant the smaller is the concentration, $\phi_s^*$, at which we encounter the one-phase area. The microemulsion region has a relatively small width in the surfactant concentration direction, and one enters a narrow microemulsion–lamellar two-phase area followed by a single lamellar phase, which prevails to high surfactant concentrations.

A third way of representing the phase behavior is to prepare a plot at a constant ratio of surfactant to oil, which is particularly relevant for temperatures below the balanced temperature $T_0$. The diagram shown in Fig. 4 provides an alternative illustration of the structural transitions that occur with an increase in temperature that we discussed in connection with Fig. 2b. A new aspect is that for an oil-in-water droplet there is a given surfactant-to-oil ratio and a decrease in water content, which should simply produce more droplets (per unit volume), but with the same properties until droplet–droplet interactions become significant (9, 10). The horizontal lower boundary of the droplet microemulsion phase in the phase diagram is a clear illustration of this effect.

Above we have focused attention on the different one-phase areas in Figs 1–4. There are also a number of two- and three-phase areas, which typically contain virtually pure oil or water as one of the phases. Under these multiphase conditions emulsions can be formed by mechanical agitation of some kind. Interfacial tensions are typically small to very small, which facilitates the formation of an emulsion. One of the classical questions in emulsion science is the relation between emulsifier and the type of emulsion, which is formed on mechanical agitation. Early on, Bancroft formulated a rule that the liquid in which the solubility of the surfactant is highest becomes the continuous phase. This rule has a substantial empirical foundation but the rule that emerges from the emulsions formed from the systems shown in Figs 1–4 is somewhat different. Below the balanced temperature $T_0$ we obtain oil-in-water emulsions and above it the emulsions are of the water-in-oil type irrespective of the solubility properties of the surfactant, which also change with temperature, but not in the same dramatic way (11). This illustrates how a knowledge of microemulsion phase behavior leads to a straightforward correlation with macroemulsion properties and how studies of microemulsions can be utilized to obtain a deeper understanding of macroemulsions. This is the theme of this review.

## III. SELF-DIFFUSION IN MICROEMULSIONS

It is necessary, but not sufficient, to characterize microemulsion systems thermodynamically in terms of phase equilibria. To obtain an understanding on the molecular level we have also to study the molecular behavior. One of the early scientific challenges in the microemulsion field was to understand the transition from an oil-in-water to a water-in-oil droplet structure. The question was clearly formulated by Friberg, but it turned

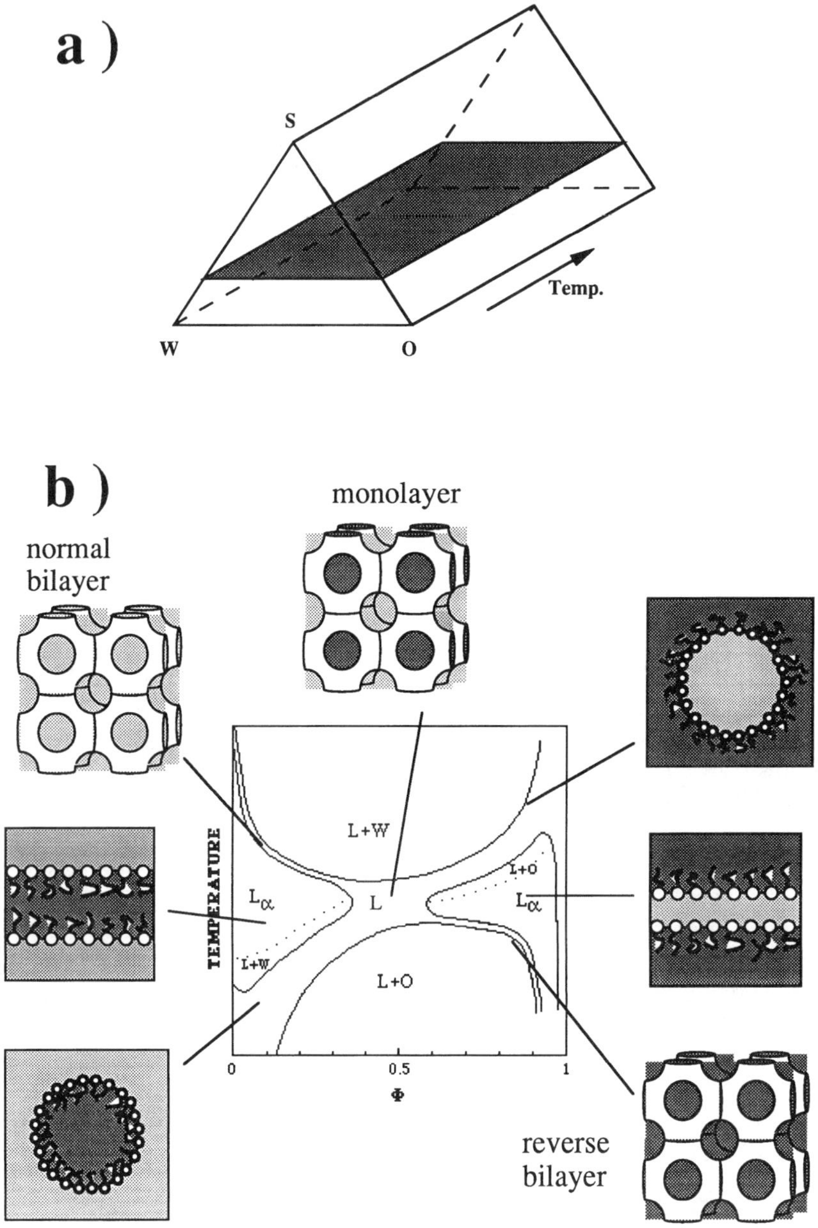


**Figure 2** (a) A schematic phase diagram cut at constant surfactant concentration through the temperature–composition phase prism of a ternary system with nonionic surfactant (Shinoda cut) showing the characteristic X-like extension of the isotropic liquid phase (L). (b) Schematic drawings of the various microstructures are also shown.

out to be difficult to find an experimental method that was suitable for settling the question. Scriven (12) suggested that one should consider bicontinuous structures and it was subsequently demonstrated that measurements of molecular self-diffusion was the method of choice to demonstrate bicontinuity on the molecular level (13–15). In some early studies tracer diffusion was used, but with the emerging refinements of the pulsed field gradient NMR method (16, 17), it soon became the dominant method. A great advantage is that one can determine the diffusion of all three components water, oil, and surfactant in one experiment. This is not only convenient but it is also necessary for demonstrating bicontinuity.

Figure 5 shows data from a typical crucial experiment (7, 18), where the diffusion behavior is monitored

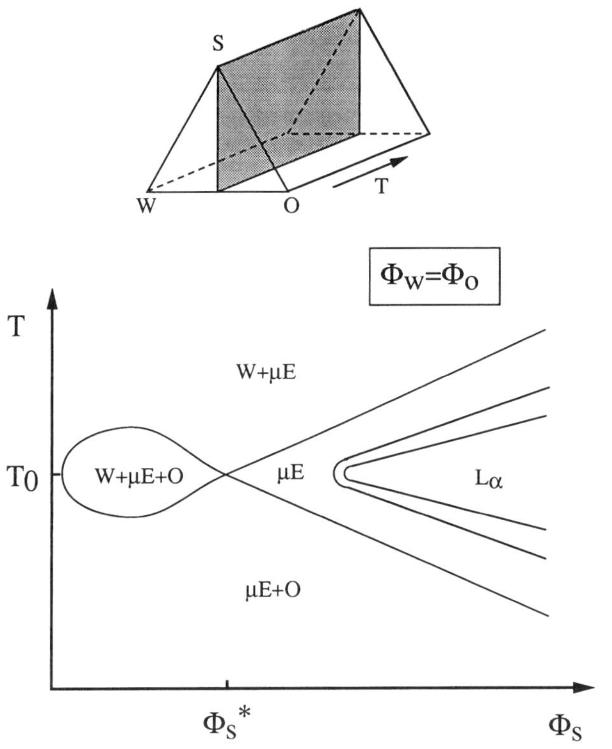


**Figure 3** A schematic phase diagram cut at equal amounts of water and oil (Fish plot), plotted as temperature vs. surfactant concentration. $T_0$ is the balance temperature and $\Phi_s^*$ is the minimum surfactant concentration needed to mix equal amounts of water and oil at $T_0$. The lower $\Phi_s^*$, the more "efficient" is the surfactant.

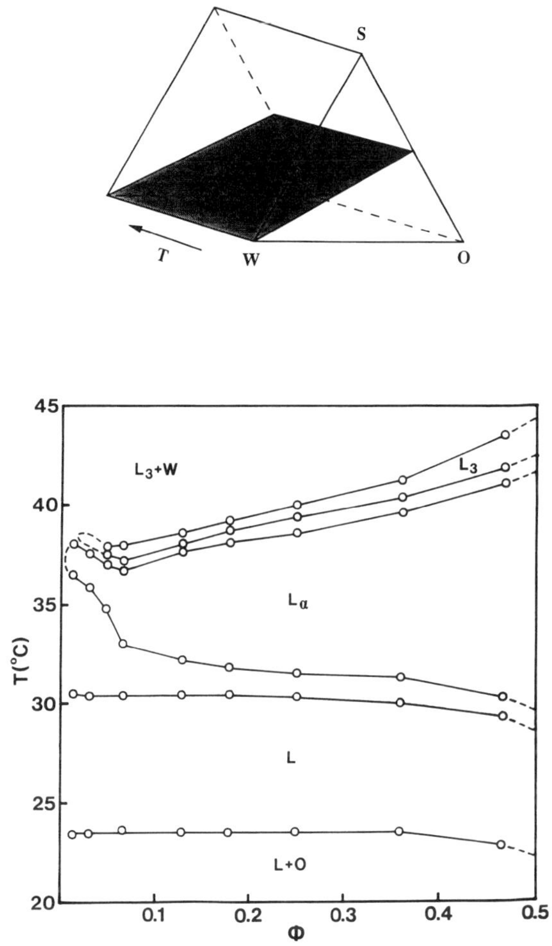


**Figure 4** Illustration of the section through the phase prism defined by a constant surfactant-to-oil ratio and a partial phase diagram of the $C_{12}E_5/D_2O$/decane system at constant ratio $\Phi_s/\Phi_O = 0.815$. (Figure redrawn from Ref. 9.)

along a path in the microemulsion channel of the Shinoda cut in Fig. 1. In the transition region, close to the balanced temperature $T_0$, diffusion is rapid for both oil and water with only a reduction to approximately 60% of the value in the neat liquid. At this point the phase structure generates the same obstruction effect for both liquids and in addition is the obstruction effect close to the ideal value of 67% (2/3) obtained for an ordered cubic bicontinuous structure, where one can neglect the thickness of the film (19). Furthermore, the surfactant diffusion shows a pronounced maximum at $T_0$, also indicating that the surfactant film has a bicontinuous structure allowing for molecular diffusion over macroscopic distances. Measurements of this type have provided strong experimental evidence in favor of a disordered bicontinuous structure of balanced microemulsions; a view that is now firmly established.

Having established the existence of oil-in-water droplets at low temperatures, an unbiased bicontinuous structure at the balance point and water-in-oil droplet structure at high temperatures there still remains to establish how these rather disparate structures continuously evolve into one another. Also in this case self-diffusion measurements are very informative. In Fig. 6 we show how the self-diffusion coefficients for cyclohexane, hexadecane, and $C_{12}E_5$ vary with temperature for a microemulsion system with water as the major component and a mixed oil (20). At the low temperature end, the $D$ values are the same for the three components and the absolute value is the one for a sphere of the expected size. An increase in temperature leads after a couple of degrees to a decrease in the diffusion, signaling aggregate growth, but this is followed by a rapid increase in the diffusion. The self-diffusion coefficient increases by an order of magnitude within half a degree, which is exceptional

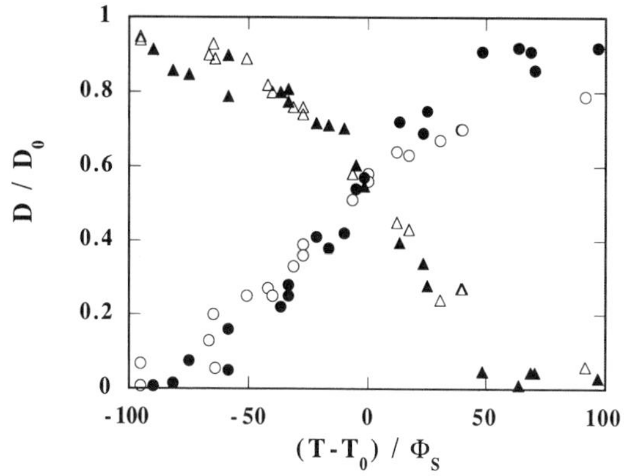


**Figure 5** Relative diffusion coefficients ($D/D_0$) of water (triangles) and oil (circles) plotted as a function of a reduced temperature $(T - T_0)/\Phi_s$, where $T_0$ is the balance temperature and $\Phi_s$ is the surfactant volume fraction, in the main microemulsion channel of a Shinoda-cut diagram. Open symbols refer to the $C_{12}E_5$/water/cyclohexane/tetradecane (equal weights of cyclohexane and hexadecane) system (20) and filled symbols to the $C_{12}E_5$/water/tetradecane system (18). The data illustrate the symmetric inversion of the microstructure around $T_0$. (For a further discussion about the chosen reduced temperature, see Ref. 7.)

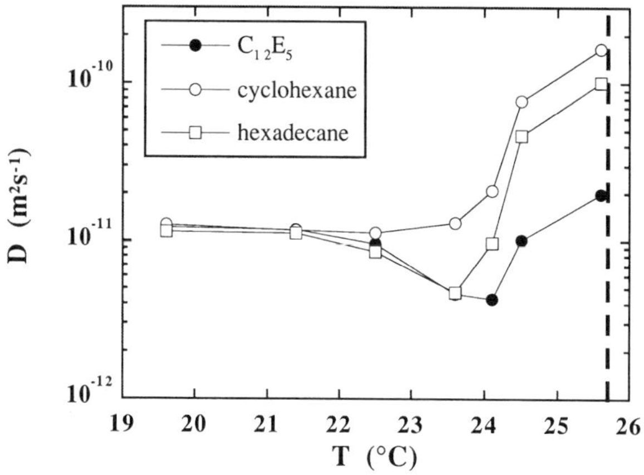


**Figure 6** Double oil-diffusion experiment with nonionic surfactant: self-diffusion coefficients as a function of temperature in a water-rich microemulsion with nonionic surfactant. A transition from oil-in-water droplets to a bicontinuous microstructure occurs with increasing temperature (decreasing spontaneous curvature of the $C_{12}E_5$ surfactant film). Note that the initial decrease of the self-diffusion coefficients shows that the droplets grow in size before the bicontinuous transition. (Data from Ref. 20.)

for a one-phase system. There is a dramatic change in the connectivity in the system.

We obtain further insight by observing that the changes in diffusion of cyclohexane and hexadecane are slightly out of phase. Figure 7 shows how the ratio of the two diffusion coefficients change with temperature. At low temperatures the ratio is unity since both molecules travel in the same aggregate. At the high-temperature end, on the other hand, the ratio is 1.6, which is the same value as found in the bulk oil mixture, and also in the developed bicontinuous structure. In the intermediate range there is a peak in the ratio, indicating that the diffusion process involves a step across a region of nonbulk properties, resulting in molecular selectivity. These studies demonstrate that dramatic structural and dynamic changes can occur over a very narrow temperature range and that NMR self-diffusion measurements is a sensitive and versatile method for monitoring such changes.

## IV. NMR SELF-DIFFUSION STUDIES OF MICROEMULSIONS

Above we described the application of the NMR self-diffusion technique to microemulsions, where, as noted above, it yields invaluable information with respect to the microstructure of the systems under study. One of the key features of the method is the fact that it measures the transport of molecules over a time (usually termed Δ) which we are free to choose at our own will in the range of from a few milliseconds to several seconds. This means that the length scale over which we

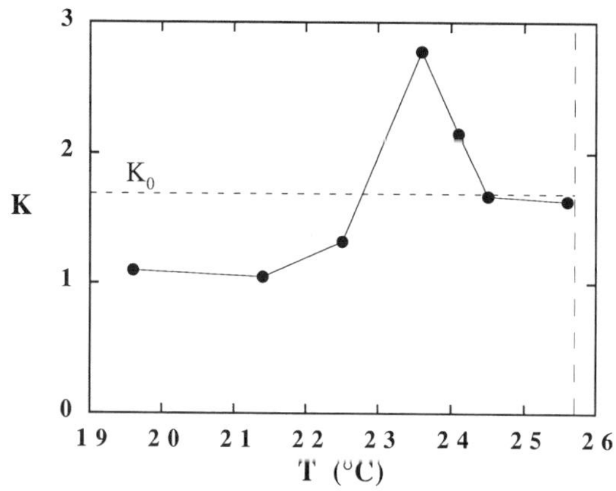


**Figure 7** Diffusion coefficient ratio, $K$, as a function of temperature in the same system as that in Fig. 6. The maximum in $K$ indicates that an attractive interaction between the micelles is operating prior to the formation of a bicontinuous structure. (Data from Ref. 20.)

are measuring the molecular transport is the micrometer regime for low molecular weight liquids. For a microemulsion the molecules do not experience any boundaries which halts their diffusion on such length scales. For the dispersed phase in a macroemulsion the situation is different, for here the molecules may experience a boundary during their diffusion $\Delta$, a situation commonly referred to as restricted diffusion. As a consequence, the molecular displacement is lowered as compared to free diffusion, and the outcome of the experiment becomes drastically changed (21–23). In fact, for restricted diffusion, the experiment can be used for structure characterization.

The obvious application of the method is to determine emulsion droplet sizes. The NMR sizing method, which was apparently first suggested by Tanner in Ref. 24, has been applied to a number of different emulsions ranging from cheese to crude-oil emulsions (25–31).

When applied to a real emulsion one has to consider the fact that the emulsion droplets in most cases are polydisperse in size. This effect can be accounted for if the molecules confined to the droplets are in a slow exchange situation, meaning that their lifetime in the droplet must be longer than $\Delta$. For such a case, the echo attenuation is given by an appropriately weighted sum of the contributions from the different droplet sizes. The approach is to assume a certain droplet size distribution function and to derive the parameters of the distribution from the raw NMR data. An often used such distribution function is the lognormal form. Given in Fig. 8 is an example of such a study, where a low calorie spread ("margarine") has been investigated. Shown is the raw NMR data and the lognormal size distribution function derived from the data. One decisive advantage of the method is that it can be used to investigate macroemulsion stability from the point of view of the time evolution of the droplet size distribution. Since it is nondisturbing the *same* sample can be studied over an extended period and the development of the size distribution can be monitored. This feature is important if one wants to study long-term stability or the effect of certain additives on the droplet size. Such data are important when theories for describing various aspects (see below) of emulsion stability are being developed.

To conclude this section we summarize the main advantages of the NMR diffusion method as applied to emulsion droplet sizing. It is nonperturbing, requiring no sample manipulation (such as dilution with the continuous phase) and, as noted above, it is nondestructive. It is insensitive to the physical appearance of the sample, and can be applied to nontransparent samples. It requires small amounts of sample (typically of the order of a few hundred milligrams) and is normally quite rapid (of the order of 10 min per sample).

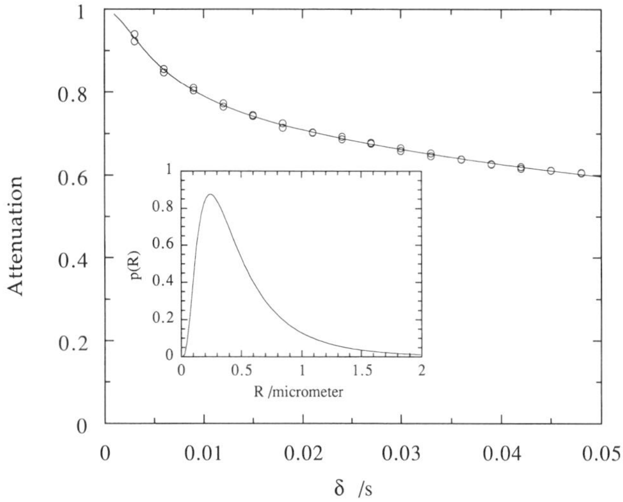


**Figure 8** Echo intensity for the entrapped water in droplets formed in a low-calorie spread containing 60% fat vs. the experimental parameter $\delta$. The solid line corresponds to the fit of a model of water entrapped in droplets of varying size. The size distribution obtained from the fit is given as an insert. (Figure adapted from Ref. 25.)

The NMR sizing method requires that the time of residence of the dispersed phase in the emulsion droplets is long relative to $\Delta$. The situation is different for cases where the lifetimes in the droplets are shorter or equal to $\Delta$; such a situation is possible if the dispersed phase actually crosses the film separating the droplets by some mechanism the detailed nature of which need not concern us here. We are then dealing with a system with permeable barriers (on the relevant time scale), and the system can now be regarded as belonging to the general class of porous systems.

For the case when the lifetime is short with respect to $\Delta$, the NMR experiment enables one to determine the long-term diffusion coefficient for the dispersed phase (or for molecules dissolved in the dispersed phase). An example of such a case is given by an emulsion composed of 96 wt % brine (of concentration 0.17 M with respect to NaCl), 2.3 wt % heptane, 1.4 wt % tetraethylene glycol dodecyl ether ($C_{12}E_4$), and 0.3 wt % soybean phosphatidyl choline. The echo attenuation for three different values of $\Delta$ for this system is presented in Fig. 9. The data set in Fig. 9 is not compatible with diffusion within a closed droplet. Rather, the data is compatible with a Gaussian diffusion and one can obtain a common diffusion coefficient the value of which is $2.25 \times 10^{-10}\,\mathrm{m^2\,s^{-1}}$. This value should be compared with the bulk value of water diffusion which is $2.23 \times 10^{-9}\,\mathrm{m^2\,s^{-1}}$ at the relevant temperature. Thus, the water diffusion is Gaussian in nature, albeit with a reduced value of the diffusion coefficient, indicating that the droplets are semipermeable. There are essentially two parameters governing this process, viz. the lifetime in the droplets, $\tau$, and the size of the droplets, $R$. It is of great interest to determine the values of these parameters for these systems. However, this is not possible when the residence lifetime is shorter than the value of $\Delta$ used.

There is a simple physical reason for this state of affairs. The diffusion process at long times is essentially a random walk of step length $2R$, where $R$ is the droplet radius. For such a case the diffusion coefficient is given by $(2R)^2/(6\tau)$, where $\tau$ is the lifetime of the walker in each droplet. Clearly, an infinite number of combinations of $R$ and $\tau$ yield the same value for the long-term diffusion coefficient. An increase in the lifetime in the droplets can be compensated for by an increase in the step size. However, if additional information is available one may of course separate the parameters $R$ and $\tau$. The system described here has been studied by Kunieda *et al.* (32) and they report a value of the radius equal to 4 μm. Using this value for $R$ one obtains a value for the lifetime $\tau$ of 47 ms.

We now turn to the case where the lifetime of the dispersed phase is of the same order of magnitude as $\Delta$. For this case one may actually separate the dynamic and structural information. Under these conditions, one may in some cases obtain a peak in the plot of the echo amplitude versus the gradient pulse duration, $\delta$. This is a surprising result at first sight, as we are accustomed to observe a monotonic decrease in the echo amplitude with the relevant experimental parameter, but it is actually a manifestation of the fact that the diffusion is no longer Gaussian. Such peaks can be rationalized within a formalism related to the one used to treat diffraction effects (33), and the analysis of the data may yield important information regarding not only the size of the droplets but also the permeability of the dispersed phase through the thin films as well as the long-term diffusion behavior of the dispersed phase. We show in Fig. 10 an example of such a diffraction-like effect in a concentrated emulsion system (34). The particular example pertains to a concentrated emulsion based on a fluorinated nonionic surfactant, where the continuous medium is a perfluorinated oil.

The data in Figs 9 and 10 are presented with the value of the quantity $q$ on the abscissa. This quantity has the dimension of inverse length and it is related to the scattering vector in scattering techniques. In fact,

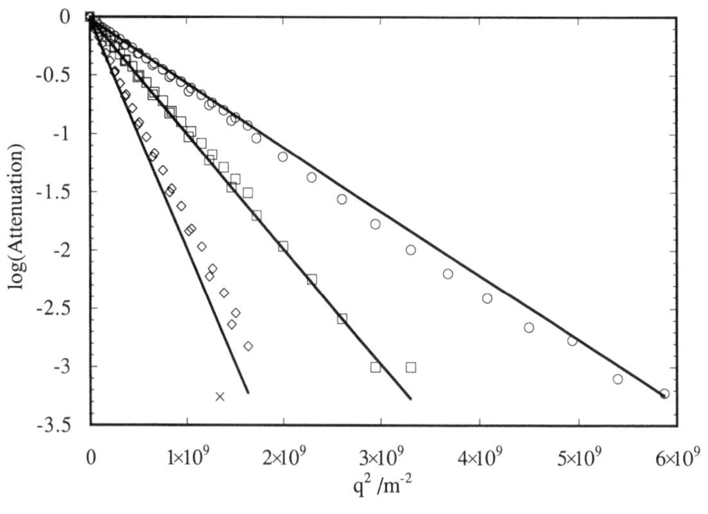


**Figure 9** Experimental echo attenuation curves vs. the parameter $q^2$ for $\Delta = 140$ ms (right curve), 250 ms (middle curve), and 500 ms (left curve) (see text for details). A global fit to the three attenuation curves gives $D = (2.25 \pm 0.022)\ 10^{-10}\ \mathrm{m^2\ s^{-1}}$. (Figure adapted from Ref. 47.)

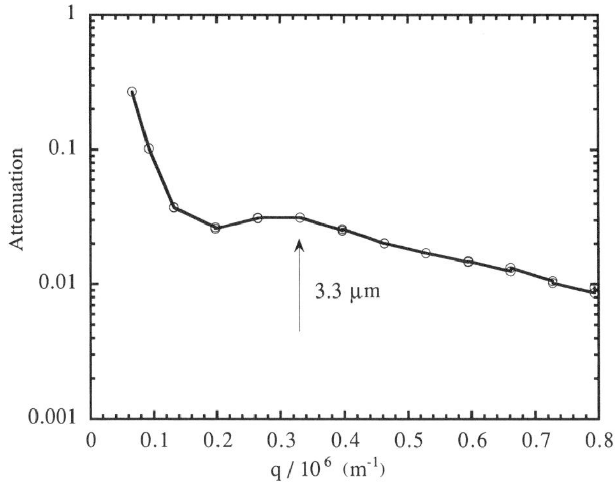


**Figure 10** Echo intensity vs. the parameter $q$ for the water in a concentrated W/O emulsion consisting of a partially fluorinated surfactant, perfluorodecaline, and water. (Figure adapted from Ref. 34.)

the inverse of the value of the position of the peak can be related to the center-to-center distance of the droplets. In the example given in Fig. 10 this value is 3.3 μm which is in good agreement with twice the droplet radii as judged from microscope images taken of the emulsion. To obtain the droplet lifetime one needs access to a theory for diffusion in these porous systems. Alternatively, one may use an approach based on Brownian dynamics simulations for a relevant model system.

We end this section by summarizing the areas where we feel that the NMR diffusion method will prove important in future studies of emulsions and refer to a more detailed account presented in Chapter 10 of this book. As theories describing emulsion stability become more refined, there will be a need for data on droplet size distribution and also on total emulsion droplet area and how these quantities evolve with time. As outlined above, NMR is capable of providing such data. Another important question pertains to the microstructure of the continuous phase, which can be studied both in the emulsion phase and also in the phase-separated systems which yield the emulsion. Finally, we note that one important class of emulsions, namely, multiple emulsions, is practically virgin territory with regard to NMR studies. In the characterization and understanding of important features of these systems NMR will most likely play an important role.

## V. THE FLEXIBLE SURFACE MODEL OF MICROEMULSIONS

The most lucid way of conceptually and quantitatively understanding the rich structural variation and structural transitions of microemulsions is to use the framework of the flexible surface model (35). The basic assumption in this model is to describe a surfactant monolayer or bilayer as a mathematical surface dividing space into two or more separate regions. With each configuration of the surface one associates a curvature (free) energy $G_c$ obtained as a surface integration of a local curvature free-energy density $g_c$:

$$G_c = \int dA g_c \tag{1}$$

The total free energy is then obtained from the partition function containing an integration of the Boltzmann factor, $\exp(-G_c/k_B T)$, over all allowed configurations.

It is customary to expand the curvature free-energy density to second order in the curvatures of the surface (36):

$$g_c = 2\kappa(H - H_0)^2 + \bar{\kappa}K \tag{2}$$

where $H$ is the mean and $K$ the Gaussian curvature of the surface. Equation (2) contains three parameters characteristic for a system; the bending rigidity $\kappa$, the saddle splay constant $\bar{\kappa}$, and the spontaneous curvature $H_0$. In particular, the spontaneous curvature is of crucial importance for the behavior of a microemulsion system since it determines to what extent and in what direction a surfactant film prefers to curve. For a balanced microemulsion $H_0 = 0$ by definition.

For the case of a balanced microemulsion, Eqs (1) and (2) combine to give a partition function that is scale invariant, and Porte *et al.* (37) showed that this leads to a free energy that varies with the third power of the surfactant volume fraction. Daicic *et al.* (38) argued that one has to extend the model to account for the observed microemulsion phase behavior and by expanding to fourth order in Eq. (2) they could reproduce the observed phase behavior nearly quantitatively. Figure 11 shows a calculated phase diagram to be compared with the experimental one of Fig. 1. Two important effects that the model captures is that when the surfactant concentration is increased there is a transition from a microemulsion to a lamellar phase owing to the increased importance of higher order terms in Eq. (2). If one, on the other hand, changes the oil–water ratio the microemulsion also turns into a lamellar phase, in this case caused by an inability of the

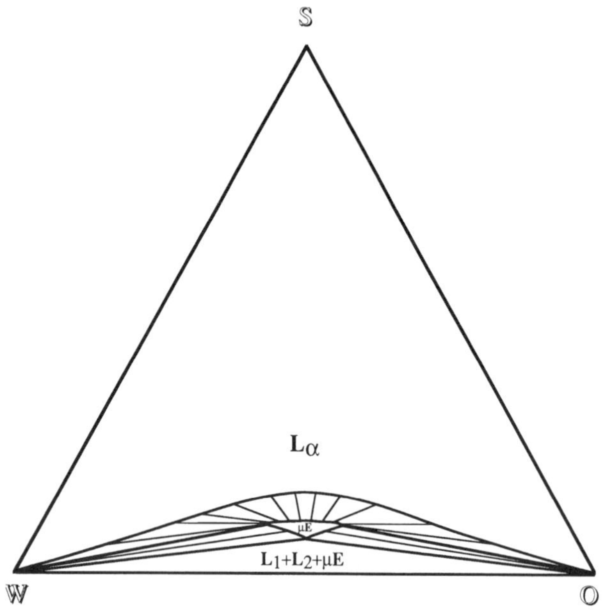


**Figure 11** Calculated phase diagram (38) for a surfactant/water/oil system for zero spontaneous curvature showing the microemulsion equilibria with excess of solvents and the lamellar phase.

bicontinuous network to sustain a close to zero mean curvature with either water or oil in excess of the other.

Based on these studies and on many others with different techniques [see, e.g., (39)] one has obtained a quite detailed knowledge of the properties of the surfacant film separating the oil and water domains. In particular, the value of $\kappa$ is relatively well known ($5$–$10 \times 10^{-21}$) J) while for $\bar{\kappa}$ the picture is still unclear. From studies of the temperature variation of the properties of microemulsions the temperature dependence of $H_0$ has been characterized for $C_mE_n$ films. It can be written as

$$H_0(T) = c(T - T_0) \tag{3}$$

where $T_0$ is the balanced temperature, which depends on the type of surfactant, nature of the oil, and on possible additives to either solvent. The coefficient $c$ is approximately $-10^7\,\mathrm{m}^{-1}\mathrm{K}^{-1}$ for $C_{12}E_5$ (40, 41), where the sign is positive for curvature towards oil. Based on this value of $c$ we readily understand why in Fig. 2 the swollen micelles of radius 10 nm changed to reversed aggregates over a temperature range of 20°C. In the subsequent section of the paper we explore how this information on the physical properties of the surfactant film can be used to interpret the stability behavior of emulsions.

## VI. EMULSION COALESCENCE

In the final stage of emulsion-droplet coalescence the two droplets have made contact, the protecting surfactant films have fused, and the droplets are connected by a narrow neck. In this situation the neck can either grow spontaneously at all sizes or there is a critical neck size that has to be exceeded before growth occurs spontaneously. The driving force for the coalescence process is a decrease in surface area with a concomitant gain in surface free energy. One could then expect that the process is slower the smaller the surface tension. As the analysis presented below demonstrates, this intuitive expectation is erroneous (42).

Consider two emulsion droplets joined by a neck, as illustrated in Fig. 12. The surfactant film is highly curved in the region of the neck while it is virtually planar elsewhere. As the neck grows or shrinks there is a change in both the area of the film and in its curvature. We can write the total free energy of the film, $G_F$, as a sum of three terms:

$$G_F = W_1 + W_2 + W_3 \tag{4}$$

Here, $W_1$ is the surface free energy:

$$W_1 = \gamma\Delta A = 2\pi\gamma(\pi - 3)b^2 + (\pi - 2)ab - a^2 \tag{5}$$

where $\gamma$ is the surface tension, $a$ is the hole radius, and $2b$ is the liquid film thickness as shown in Fig. 12. The mean curvature energy is obtained from Eq. (2) with the planar state as reference:

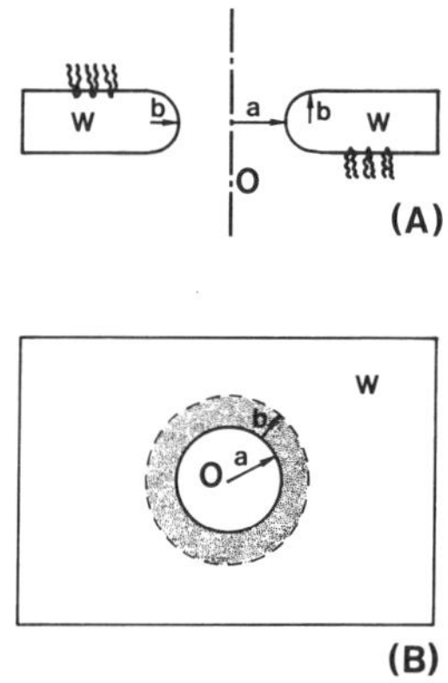


**Figure 12** Illustration of hole nucleation in a liquid (water) film originally separating two oil droplets (whose radius is assumed to be much larger than the water film thickness which then can be thought of as being flat with constant thickness). (A) Viewed perpendicular to the hole axis where $a$ is the hole radius and $b$ is the film half thickness; (B) Viewed along the hole axis.

$$W_2 = \kappa \int dA\left(H^2 - 2H_0H\right) = 2\pi\kappa\left\{2\pi H_0 a + \frac{2(a+b)^2}{b\sqrt{a(a+2b)}} \arctan \sqrt{\frac{a+2b}{a}} + 2(\pi-4)bH_0 - 4\right\} \qquad (6)$$

where the second equality involves some elaborate calculations. The third term in Eq. (4) involves the Gaussian curvature energy. Once the neck has been formed there is no change in topology, so, by the Gauss–Bonnet theorem this energy is constant:

$$W_3 = -4\pi\bar{\kappa} \qquad (7)$$

and hence has no effect on the growth of the neck.

Of the parameters entering Eqs (5)–(7) we have a knowledge of $H_0$ and $\kappa$ from the microemulsion studies, while $\bar{\kappa}$ is irrelevant when considering changes in $G_F$. Surface tensions have been extensively studied for microemulsions and it has been demonstrated that $\gamma$ has a deep minimum at balanced conditions and that its value is dependent on the spontaneous curvature of the film. Under droplet microemulsion conditions:

$$\gamma = \gamma_0 + 4H_0^2 \frac{\kappa^2}{2\kappa + \bar{\kappa}} \qquad (8)$$

where the constant $\gamma_0$ has a low value of order $10^{-6}$ N/m or less.

The surface-energy term $W_1$ goes to high negative values for sufficiently large values of the radius $a$ of the neck, but in an intermediate range where $a$ and $b$ are of similar magnitude the term $W_1$ for the surface free energy and the term $W_2$ for the curvature energy are of similar magnitude. The $W_1$ term is insensitive to the sign of $H_0$ but that is not true for the curvature energy. In the neck there is a region of the surfactant film that tends to be strongly curved towards the continuous liquid medium. For an oil-in-water emulsion this implies substantial positive curvature energy contribution when $H_0$ is positive but no longer when $H_0$ is negative. Detailed calculations using Eqs (4) to (8) reveal that this factor has a profound influence on the energetics of neck growth. Figure 13 shows how the calculated barrier to neck growth varies with temperature in the vicinity of the balanced state.

We identify three crucial features of the droplet-coalescence process based on this calculation:

1. When the spontaneous curvature has a sign that tends to promote curvature away from the continuous medium, there is a substantial barrier towards growth of a neck that has formed. On the other hand, for the

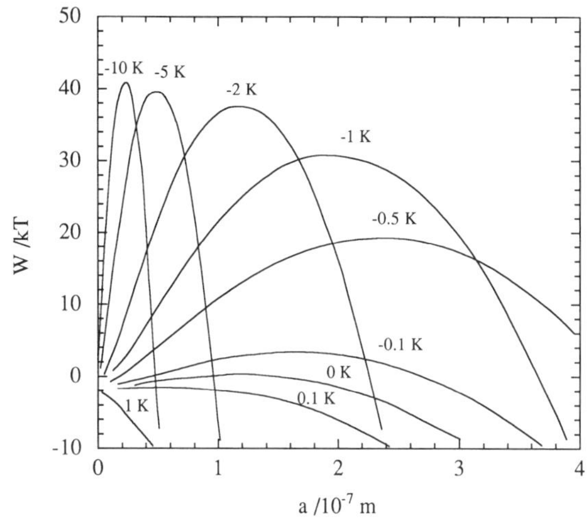


**Figure 13** Calculated barrier to neck growth, $W$, plotted as a function of neck radius, $a$ (see Fig. 12). Each curve corresponds to a specific temperature difference from the balance temperature, $T_0$. (Figure redrawn from Ref. 42.)

opposite sign of $H_0$, neck growth is spontaneous. This provides a detailed mechanistic interpretation of the modified Bancroft rule discussed in Sec. I.

2. The transition from a stable to an unstable system occurs over a remarkably narrow temperature range of one degree or less. In careful experiments, Kabalnov and Weers (43) have demonstrated that this prediction is practically quantitatively accurate.

3. In the stable region the barrier appears to be constant except close to the balanced point. We expect that the picture changes in this respect when one also consider direct interactions between different patches of the film.

## VII. NUCLEATION AND OSTWALD RIPENING

In addition to coalescence, emulsions can also break through an Ostwald ripening process, where a few emulsion droplets grow in size at the cost of the majority shrinking and disappearing. Microemulsions can also be used to advantage to study this process. In Fig. 4 we show a partial phase diagram, where the surfactant/oil ratio was constrainted to a constant value. At approximately 24°C there is a virtually horizontal phase boundary separating at higher temperature a microemulsion-droplet phase from a two-phase area with somewhat smaller droplets and excess of oil. At either side of the phase boundary the droplet size is determined by the spontaneous curvature at the temperature in question; thus, the lower the temperature the smaller the droplets. This region of the phase diagram provides a neat illustration of the difference between a microemulsion and an emulsion. In the one-phase area we have a microemulsion while in the two-phase area we can potentially make an emulsion with oil droplets in water. In fact such an emulsion is easily prepared by taking the microemulsion and cooling it rapidly into the two-phase area. For a sufficiently rapid quench we have an emulsion with exactly the same molecular structure as the microemulsion.

An emulsion prepared in this way can serve as a useful model system for studies of the Ostwald ripening process. Right after the quench we have a system of small, nearly monodisperse emulsion droplets. In an Ostwald ripening process some of these should grow at the expense of the others. It is normally assumed that such a fluctuation is unstable and that the growth rate is only limited by material transport. However, for a monodisperse droplet system this is not trivially true. Since we have a detailed knowledge of the properties of the surfactant film in the present case it is possible to make an analysis of the growth process. This analysis reveals that there are both locally stable and unstable situations. In the former case a nucleation process is needed to trigger a further Ostwald ripening (44, 45).

## VIII. CONCLUSIONS

In this paper, devoted to the research efforts of Stig Friberg, we have tried to demonstrate how research on equilibrium microemulsion systems can be used to deepen the knowledge on nonequilibrium emulsions. We have discussed both experimental methodology in terms of NMR self-diffusion measurements and phase-equilibrium studies as well as conceptual advances such as the flexible surface model combined with a quantitative determination of the relevant parameters. We feel that we are following a tradition set by Stig Friberg and Kozo Shinoda in considering the surfactant self-association process as a fundamental phenomenon governing both the equilibrium properties such as, for example, phase behavior, and the emulsion formation and coalescence.

## ACKNOWLEDGMENTS

A major part of the work presented in this paper has been done in collaboration with several colleagues. In particular we would like to thank David Anderson, Balin Balinov, John Daicic, Alexey Kabalnov, Marc Leaver, Jane Morris, and Kozo Shinoda. Financial support from the Swedish Natural Science Research Council (NFR), the Swedish Board for Industrial and Technical Development (Nutek), and the Swedish Research Council for Engineering Sciences (TFR) is kindly acknowledged.

## REFERENCES

1. H Wennerström, O Söderman, U Olsson, B Lindman. Colloids Surfaces A 123–124: 13–26, 1997.
2. K Shinoda, SE Friberg. Emulsions and Solubilization. New York: Wiley-Interscience, 1986.
3. JH Schulman, W Stoeckenius, LM Prince, J Phys Chem 63: 1677–1680, 1959.
4. PA Winsor. Chem Rev 68: 1–40, 1963.
5. I Danielsson, B Lindman. Colloids Surfaces 3: 391–392, 1981.
6. S Friberg, L Mandell, K Fontell. Acta Chem Scand 23: 1055–1057, 1969.
7. U Olsson, H Wennerström. Adv Colloid Interface Sci 49: 113–146, 1994.
8. M Kahlweit, R Strey, D Haase, P Firman. Langmuir 4: 785–790, 1988.
9. U Olsson, P Schurtenberger. Langmuir 9: 3389–3394, 1993.
10. MS Leaver, U Olsson, H Wennerström, R Strey. J Phys II 4: 515–531, 1994.
11. BP Binks. Langmuir 9: 25–28, 1993.
12. LE Scriven. Nature 263: 123–125, 1976.
13. T Bull, B Lindman. Mol Cryst Liq Cryst 28: 155–160, 1974.
14. B Lindman, N Kamenka, T Kathopoulis, B Brun, PG Nilsson. J Phys Chem 84: 2485–2490, 1980.
15. B. Lindman, K Shinoda, U Olsson, D Anderson, G Karlström, H Wennerström. Colloids Surfaces 38: 205–224, 1989.
16. PT Callaghan. Aust J Phys 37: 359–387, 1984.
17. P Stilbs. Prog Nucl Magn Reson Spectrosc 19: 1–45, 1987.
18. U Olsson, K Shinoda, B Lindman. J Phys Chem 90: 4083–4088, 1986.
19. DM Anderson, H Wennerström, J Phys Chem 94: 8683–8694, 1990.
20. U Olsson, K Nagai, H Wennerström. J Phys Chem 92: 6675–6679, 1988.
21. PT Callaghan, A Coy. In: P Tycko, ed. NMR Probes of Molecular Dynamics. Dordrecht: Kluwer Academic, 1993.
22. PT Callaghan. Principles of Nuclear Magnetic Resonance Microscopy. Oxford: Clarendon Press, 1991.
23. PT Callaghan, A Coy, TPJ Halpin, D MacGowan, JK Packer, FO Zelaya, J Chem Phys 97: 651–662, 1992.

24. JE Tanner. PhD thesis, University of Wisconsin, Madison, WI, 1966.
25. B Balinov, O Söderman, T Wärnheim. J Am Oil Chem Soc 71: 513–518, 1994.
26. B Balinov, O Urdahl, O Söderman, J Sjöblom. Colloids Surfaces. 82: 173–181, 1994.
27. PT Callaghan, KW Jolley, R Humphrey. J Colloid Interface Sci 93: 521–529, 1983.
28. X Li, JC Cox, RW Flumerfelt. AIChE J 38: 1671, 1992.
29. I Löennqvist, A Khan, O Söederman, J Colloid Interface Sci 144: 401-411, 1991.
30. KJ Packer, C Rees. J Colloid Interface Sci 40: 206–218, 1972.
31. JC Van den Enden, D Waddington, H Van Aalst, CG Van Kralingen, KJ Packer, J Colloid Interface Sci 140: 105–113, 1990.
32. H Kunieda, N Yano, C Solans. Colloids Surfaces 36: 313–322, 1989.
33. PT Callaghan, A Coy, D MacGowan, KJ Packer, FO Zelaya. Nature (London) 351: 467–469, 1991.
34. B Balinov, O Söderman, JC Ravey. J Phys Chem 98: 393, 1994.
35. H. Wennerström, U Olsson. Langmuir 9: 365–368, 1993.
36. W Helfrich. Z Naturforsch 28c: 693–703, 1973.
37. G Porte, J Appell, P Bassereau, L Marignan, J Phys France 50: 1335–1347, 1989.
38. J Daicic, U Olsson H, Wennerström. Langmuir 11: 2451–2458, 1995.
39. LT Lee, D Langevin, J Meunier, K Wong, B Cabane. Progr Colloid Polym Sci 81: 209–214, 1990.
40. R Strey. Colloid Polymer Sci 272: 1005–1019, 1994.
41. V Rajagopalan, H Bagger-Jörgensen, K Fukuda, U Olsson, B Jönsson. Langmuir 12: 2939–2946, 1996.
42. A Kabalnov, H Wennerström. Langmuir 12: 276–292, 1996.
43. A Kabalnov, J Weers. Langmuir 12: 1931–1935, 1996.
44. JM Morris, U Olsson, H Wennerström. Langmuir 13: 606–608, 1997.
45. H Wennerström, J Morris, U Olsson. Langmuir 13: 6972–6979, 1997.
46. H Kunieda, K Shinoda. J Disp Sci Technol 3: 233–244, 1982.
47. B Balinov, P Linse, O Söderman. J Colloid Interface Sci 182: 539–548, 1996.

# 6

# Dielectric Spectroscopy on Emulsion and Related Colloidal Systems—A Review

**Yuri Feldman**
*The Hebrew University of Jerusalem, Jerusalem, Israel*

**Tore Skodvin**
*University of Bergen, Bergen, Norway*

**Johan Sjöblom**
*Statoil A/S, Trondheim, Norway*

## I. INTRODUCTION

In general, there seems to be a shortage in modern process industry of elegant and nonintrusive techniques to describe accurately the process with regard to yield, competing reactions, external parameters, contaminations, and formation of disturbing colloidal states. It goes without saying that the development of such techniques will highly secure the success of many uncertain processes in operation today. In reviewing possible techniques one can point out dielectric (or capacitance), ultrasound, and analytical techniques, together with spectroscopic alternatives.

In this review we have chosen to highlight the pros and cons of dielectric spectroscopy applied to emulsified and related systems. The reason for the choice of these systems is that they represent multiphase systems containing phase boundaries or interfaces. It is shown that dielectric spectroscopy scanning over large frequency intervals (from some kilohertz up to several gigahertz) is extremely sensitive towards interfacial phenomena and interfacial polarization. Hence, there exist many possibilities for this technique to map processes taking place in the bulk phases as well as at the phase boundaries. This is one of several reasons why the technique is so powerful and frequently used in the study of heterogeneous systems.

In a large number of processes, one factor determining failure or success may be the state of colloids present in the system at some stage in the process. This is true whether the process is the manufacture of pharmaceuticals, application of paint, separation of water from crude oil, blood flow, preparation of carrier matrices for catalysts, preparation of novel materials, food production, etc. (the list could be expanded several times without becoming complete). Even though the science of colloids has reached a high level of sophistication, new or improved techniques that can shed light on the complex and highly dynamic interactions taking place in colloidal systems are still needed.

In the following we combine dielectric spectroscopy and colloidal systems. We believe this to be a fruitful combination of two highly important and current topics. Dielectric measurements have roots from over a hundred years ago; in the beginning of course only simple capacitance measurements were made.

However, the findings made especially on more complex systems encouraged scientists to analyze the physics and chemistry of what we call heterogeneous systems. A prediction of the dielectric properties of such systems was a main concern of scientists such as Maxwell, Wagner, Debye, and Sillars. Historically, we find a genuine interest in understanding dielectric properties of chemically very intricate systems. Depending on the problem, either the theory or the experiments have been in the lead.

Today, both dielectric measurements and colloidal systems (as a representative of heterogeneous systems) are of great interest both with regard to basic as well as applied science.

Dielectric measurements have developed from cumbersome Wheatstone-bridge measurements to an efficient, precise, and rapid spectroscopic technique. The new technique dielectric spectroscopy soon found interesting applications within the field of colloid chemistry. The technique can be applied as a precision method in a thorough mapping of the static and dynamic properties of colloidal systems. Industrially, dielectric measurements can be utilized in the on-line characterization of such complex systems.

First, we give an introduction to the basic concepts underlying the measuring technique, before the technique itself is reviewed. Finally, experimental work is presented. The experimental part has three main sections, i.e., one section covers equilibrium systems, in the second non-equilibrium systems are regarded and the last one is on the application of dielectric spectroscopy to biological systems.

In the writing of this review we have not sought to cover every aspect of the dielectric properties of colloidal systems. Our aim has rather been to demonstrate the usefulness of dielectric spectroscopy for such systems, using the application to selected systems as illustrations.

## II. DIELECTRIC POLARIZATION — BASIC PRINCIPLES

Recently a new field, mesoscopic physics, has emerged. It is interesting to understand the physical properties of systems that are not as small as a single atom, but small enough that the properties can be dramatically different from those in a larger assembly. All these new mesoscopic phenomena can easily be observed in the dielectric properties of colloid systems. Their properties strictly depend on the dimensional scale and the time scale of observation. Self-assembling systems such as micellar surfactant solutions, microemulsions, emulsions, aqueous solutions of biopolymers, and cell and lidposome suspensions all together represent this population of complex liquids that have their almost unique dynamic and structural properties emphasized on the mesoscale.

Consequently, let us consider different types of polarization that can take place in such structures. There can be two considerations in this case. One from a phenomenological point of view, in terms of the macroscopic polarization vector, another from the molecular point of view, taking into account all possible contributions to the macroscopic polarization vector of unit sample volume.

### A. Dielectric Polarization in Static Electric Fields

Being placed in an external electric field, a dielectric sample acquires a nonzero *macroscopic dipole moment*. This means that the dielectric is polarized under the influence of the field. The polarization $\boldsymbol{P}$ of the sample, or dipole density, can be presented in a very simple way:

$$\boldsymbol{P} = \frac{\boldsymbol{M}}{V} \tag{1}$$

where $\boldsymbol{M}$ is the macroscopic dipole moment per unit volume, and $V$ is the volume of the sample. In a linear approximation the polarization of the dielectric sample is proportional to the strength of the applied external electric field $\boldsymbol{E}$(1):

$$\boldsymbol{P} = \chi \boldsymbol{E} \tag{2}$$

where $\chi$ is the *dielectric susceptibility* of the material. If the dielectric is isotropic, $\chi$ is scalar, whereas for an anisotropic system $\chi$ is a tensor.

In the Maxwell approach, in which matter is treated as a continuum, we must in many cases ascribe a dipole density to matter. Let us compare the vector fields $\boldsymbol{D}$ and $\boldsymbol{E}$ for the case in which only a dipole density is present. Differences between the values of the field vectors arise from differences in their sources. Both the external charges and the dipole density of the sample act as sources of these vectors. The external charges contribute to $\boldsymbol{D}$ and $\boldsymbol{E}$ in the same manner (2). The electric displacement (electric induction) vector $\boldsymbol{D}$ is defined as

$$\boldsymbol{D} = \boldsymbol{E} + 4\pi \boldsymbol{P} \tag{3}$$

For a uniform isotropic dielectric medium, the vectors $\boldsymbol{D}, \boldsymbol{E}, \boldsymbol{P}$ have the same direction, and the susceptibility is coordinate independent, therefore

$$\boldsymbol{D} = \boldsymbol{E}(1 + 4\pi\chi)\boldsymbol{P} = \varepsilon\boldsymbol{E} \tag{4}$$

where $\varepsilon$ is *the dielectric permittivity*. It is also called the dielectric constant, because it is independent of the field strength. It is, however, dependent on the frequency of the applied field, the temperature, the density (or the pressure), and the chemical composition of the system.

### 1. Types of Polarization

For isotropic systems and static linear electric fields, we have

$$\boldsymbol{P} = \frac{\varepsilon - 1}{4\pi}\boldsymbol{E} \tag{5}$$

The applied electric field gives rise to a dipole density through the following mechanisms:

**Deformation polarization** — This can be further divided into two independent types:

*Electron polarization* — the displacement of nuclei and electrons in the atom under the influence of the external electric field. As electrons are very light they have a rapid response to the field changes; they may even follow the field at optical frequencies.

*Atomic polarization* — the displacement of atoms or groups of atoms in the molecule under the influence of the external electric field.

**Orientation polarization** — The electric field tends to direct the permanent dipoles. The rotation is counteracted by the thermal motion of the molecules. Therefore, the orientation polarization is strongly dependent on the temperature and the frequency of the applied electric field.

**Ionic polarization** — In an ionic lattice, the positive ions are displaced in the direction of an applied field while the negative ions are displaced in the opposite direction, giving a resultant dipole moment to the whole body. The mobility of ions also strongly depends on the temperature, but contrary to the orientation polarization the ionic polarization demonstrates only a weak temperature dependence and is determined mostly by the nature of the interface where the ions can accumulate. Most of the cooperative processes in heterogeneous systems are connected with ionic polarization.

To investigate the dependence of the polarization on molecular quantities it is convenient to assume the polarization $\boldsymbol{P}$ to be divided into two parts: the induced polarization $\boldsymbol{P}_\alpha$, caused by translation effects, and the dipole polarization $\boldsymbol{P}_\mu$, caused by orientation of the permanent dipoles.

$$\frac{\varepsilon - 1}{4\pi}\boldsymbol{E} = \boldsymbol{P}_\alpha + \boldsymbol{P}_\mu \tag{6}$$

We can now define two major groups of dielectrics: polar and nonpolar. A *polar dielectric* is one in which the individual molecules possess a dipole moment even in the absence of any applied field, i.e., the center of positive charge is displaced from the center of negative charge. A *nonpolar dielectric* is one where the molecules possess no dipole moment, unless they are subjected to an electric field. The mixture of these two types if dielectrics is common in the case of complex liquids and the most interesting dielectric processes are going on at their phase borders or at liquid–liquid interfaces.

Owing to the long range of the dipolar forces an accurate calculation of the interaction of a particular dipole with all other dipoles of a specimen would be very complicated. However, a good approximation can be made by considering that the dipoles beyond a certain distance, say some radius $a$, can be replaced by a continuous medium having the macroscopic dielectric properties of the specimen. Thus, the dipole whose interaction with the rest of the specimen is calculated may be considered as surrounded by a sphere of radius $a$ containing a discrete number of dipoles, beyond which there is a continuous medium. To make this a good approximation the dielectric properties of the whole region within the sphere should be equal to those of a macroscopic specimen, i.e., it should contain a sufficient number of molecules to make fluctuations very small (3, 4). This approach can be successfully used for the calculation of dielectric properties of ionic self-assembled liquids. In this case the system can be considered to be a monodispersed system consisting of spherical water droplets dispersed in the nonpolar medium (5).

Inside the sphere where the interactions take place, the use of statistical mechanics is required. This is a second method for calculating the polarization from molecular parameters that allows us to take into account the short range of dipole–dipole interaction (1, 2). Here, we may write for the dipole density $\boldsymbol{P}$ of a homogeneous system:

$$\boldsymbol{P}V = \langle \boldsymbol{M} \rangle \tag{7}$$

where $V$ is the volume of the dielectric under consideration and $\langle \boldsymbol{M} \rangle$ is its average total dipole moment and $\langle \rangle$ denote a statistical mechanical average. For an isotropic system, using Eq. (4), we find:

$$(\varepsilon - 1)\boldsymbol{E} = \frac{4\pi}{V}\langle \boldsymbol{M} \rangle \tag{8}$$

Further calculations, using all necessary statistical mechanical approaches for the appropriate border conditions, allow us to obtain the well-known Kirkwood–Frölich relationship for static dielectric permittivity:

$$g\mu = \frac{9kT}{4\pi}\frac{(\varepsilon - \varepsilon_\infty)(2\varepsilon + \varepsilon_\infty)}{\varepsilon(\varepsilon_\infty + 2)^2} \tag{9}$$

where $\mu$ is the dipole moment of the molecule in the gas phase, $k$ is the Boltzmann constant, $T$ is the absolute temperature, $\varepsilon_\infty$ is the high-frequency limit of complex dielectric permittivity, and $g$ is a correlation factor. Kirkwood introduced the correlation factor in his theory in order to take into account the short-range order interactions in associated polar liquids.

An approximate expression for the Kirkwood correlation factor can be derived by taking into account only the nearest-neighbors' interactions. In this case the sphere is reduced to contain only the $j$th molecule and its $z$ nearest neighbors. For this definition it is possible to derive the following relationship for the parameter $g$:

$$g = 1 + z\langle \cos\theta_{ij} \rangle \tag{10}$$

where $\langle \cos\theta_{ij} \rangle$ gives the average angle of orientations between the $i$th and $j$th dipoles in the sphere of short-range interactions. From Eq. (10) it is obvious that when $\langle \cos\theta_{ij} \rangle \neq 0$, $g$ will be different from 1. It means that the neighboring dipoles are correlated between themselves. The parallel orientation of dipoles leads to a positive value of the average cosines and $g$ larger than 1. When the antiparallel orientation of dipoles can be observed, $g$ will be smaller than 1. Both cases are observed experimentally (1, 2, 4). This parameter will be extremely useful in the understanding of the short-range molecular mobility and interaction in self-assembled systems.

### B. Dielectric Polarization in Time-dependent Electric Fields

When an external field is applied the dielectric polarization reaches its equilibrium value, not instantly, but over a period of time. By analogy, when the field is removed suddenly, the polarization decay caused by thermal motion follows the same law as the relaxation or decay function:

$$\alpha(t) = \frac{\boldsymbol{P}(t)}{\boldsymbol{P}(0)} \tag{11}$$

The relationship, Eq. (4), for the displacement vector in the case of time-dependent fields may be written as follows (1, 3):

$$\boldsymbol{D}(t) = \varepsilon_\infty \boldsymbol{E}(t) + \int_{-\infty}^{t} \boldsymbol{E}(t')\dot{\Phi}(t - t')\mathrm{d}t' \tag{12}$$

where $\boldsymbol{D}(t) = \boldsymbol{E}(t) + 4\pi\boldsymbol{P}(t)$, $\varepsilon_\infty$ is the high-frequency limit of the complex dielectric permittivity $\varepsilon^*(\omega)$, and $\dot{\Phi}(t)$ is the dielectric response function ($\dot{\Phi}(t) = (\varepsilon_s - \varepsilon_\infty)[1 - \alpha(t)]$ where $\alpha(t)$ is the relaxation function or the decay function of dielectric polarization). ($\varepsilon_s$ is the low-frequency limit of the permittivity, commonly denoted the static permittivity.) The point denotes the time derivative in Eq. (12). The complex dielectric permittivity $\varepsilon^*(\omega)$ is connected with the relaxation function by a very simple relationship:

$$\frac{\varepsilon^*(\omega) - \varepsilon_\infty}{\varepsilon_s - \varepsilon_\infty} = L\{-\dot{\alpha}(t)\} \tag{13}$$

where $L$ is the operator of the Laplace transform, which is defined for the arbitrary time-dependent function $f(t)$ as:

$$L\{f(t)\} \equiv F(\omega) = \int_0^\infty \mathrm{e}^{-s}f(t)\mathrm{d}t, \quad s = i\omega + x, \text{ where } x \to 0 \tag{14}$$

If

$$\alpha(t) \sim \exp(-t/\tau_m) \tag{15}$$

where $\tau_m$ represents the dielectric relaxation time, then the relation first obtained by Debye, is true for the frequency domain (1, 3, 4):

$$\frac{\varepsilon^*(\omega) - \varepsilon_\infty}{\varepsilon_s - \varepsilon_\infty} = \frac{1}{1 + i\omega\tau_m} \tag{16}$$

For most of the systems being studied such a relation does not sufficiently describe the experimental results. This makes it necessary to use empirical relations which formally take into account the distribution of relaxation times with the help of various parameters $(\alpha, \beta)$ (3). In the most general way such nonDebye dielectric behavior can be described by the so called Havriliak–Negami relationship (3, 4, 6):

$$\varepsilon(\omega) = \varepsilon_\infty + \frac{\varepsilon_s - \varepsilon_\infty}{[1 + (i\omega\tau)^\alpha]^\beta}, \qquad \alpha, \beta > 0 \tag{17}$$

The specific case $\alpha = 1, \beta = 1$ gives the Debye relaxation law, $\beta = 1, \alpha \neq 1$ corresponds to the so-called Cole–Cole equation, whereas the case $\alpha = 1, \beta \neq 1$ corresponds to the Cole–Davidson formula. Recently, some progress in the understanding of the physical meaning of the empirical parameters $(\alpha, \beta)$ has been made (7, 8). Using the conception of a self-similar relaxation process it is possible to understand thenature of a nonexponential relaxation of the Cole–Cole, Cole–Davidson, or Havriliak–Negami type.

An alternative approach is to obtain information on the dynamic molecular properties of a substance directly in the time domain. Relation Eq. (13) shows that the equivalent information on the dielectric relaxation properties of a sample being tested can be obtained both in the frequency and the time domains. Indeed, the polarization fluctuations caused by thermal motion in the linear response case are the same as for the macroscopic reconstruction induced by the electric field (9, 10). This means that one can equate the relaxation function $\alpha(t)$ and the macroscopic dipole correlation function $\Gamma(t)$ :

$$\alpha(t) \cong \Gamma(t) = \frac{\langle \boldsymbol{M}(0)\boldsymbol{M}(t)\rangle}{\langle \boldsymbol{M}(0)\boldsymbol{M}(0)\rangle} \tag{18}$$

where $\boldsymbol{M}(t)$ is the macroscopic fluctuating dipole moment of the sample volume unit which is equal to the vector sum of all the molecular dipoles. The rate and laws govering the decay function $\Gamma(t)$ are directly related to the structural and kinetic properties of the sample and characterize the macroscopic properties of the system studied. Thus, the experimental function $\Phi(t)$ and hence $\alpha(t)$ or $\Gamma(t)$ can be used to obtain information on the dynamic properties of a dielectric in terms of the dipole correlation function.

## C. Dielectric Polarization in Heterogeneous Systems

Complex fluids composed of several pseudophases with a liquid–liquid interface (emulsions, macroemulsions, cells, liposomes) or liquid–solid interface (suspensions of silica, carbon black, latex, etc.) can, from a dielectric point of view, be considered as classical heterogeneous systems. Several basic theoretical approaches have been developed in order to describe the dielectric behavior of such systems. Depending on the concentration, the shape of the dispersed phase, and the conductivity of both the media and disperse phase, different mixture formulas can be applied to describe the electric property of the complex liquids (11–15).

Two general theoretical approaches have been applied in the analysis of heterogeneous materials. The macroscopic approach, in terms of classical electrodynamics, and the statistical mechanics approach, in terms of charge-density calculations. The first is based on the application of the Laplace equation to calculate the electric potential inside and outside a dispersed spherical particle (11, 12). The same result can be obtained by considering the relationship between the electric displacement $\boldsymbol{D}$ and the macroscopic electric field $\boldsymbol{E}$ in a disperse system (12, 13). The second approach takes into account the coordinate-dependent concentration of counterions in the diffuse double layer, regarding the self-consistent electrostatic potential of counterions via Poisson's equation (5, 16, 17). Let us consider these approaches briefly.

The first original derivation of mixture formula for spherical particles was performed by Maxwell (18) and was later extended by Wagner (19). This Maxwell–Wagner (MW) theory of interfacial polarization usually can be successfully applied only for dilute dispersions of spherical particles. The dielectric permittivity of such a mixture can be expressed by the well-known relationship:

$$\varepsilon = \varepsilon_1 \frac{(2\varepsilon_1 + \varepsilon_2) - 2\phi(\varepsilon_1 - \varepsilon_2)}{(2\varepsilon_1 + \varepsilon_2) + \phi(\varepsilon_1 - \varepsilon_2)}, (\phi \ll 1) \tag{19}$$

Here, $\varepsilon_1$ and $\varepsilon_2$ are the dielectric permittivities of the continuous phase and inclusions, respectively, and $\phi$ is the volume fraction of inclusions.

Fricke and Curtis (20) and then Sillars (21) generalized the MW theory for the case of ellipsoidal particles. Bruggemann (22) and Hanai (23) in his series of papers extended the theory also for the case of concentrated disperse systems. For a system containing homogeneously distributed spherical particles Hanai found that the complex permittivity of the system is given by

$$\left(\frac{\varepsilon_2^* - \varepsilon^*}{\varepsilon_2^* - \varepsilon_1^*}\right)\left(\frac{\varepsilon_1^*}{\varepsilon^*}\right)^{1/3} = 1 - \phi \tag{20}$$

Boned and Peyrelasse (24) and Boyle (25) made extensions of the MW theory to include nonspherical droplets and concentrated emulsions. In the case of disperse systems containing spheroidal particles aligned in parallel, the total permittivity according to Boyle may be found from (25):

$$\left(\frac{\varepsilon_2^* - \varepsilon^*}{\varepsilon_2^* - \varepsilon_1^*}\right)\left(\frac{\varepsilon_1^*}{\varepsilon^*}\right)^{A_a} = 1 - \phi \tag{21}$$

where $A_a$ is a depolarization factor dependent on the axial ratio $a/b$ describing the spheroids. When $A_a = 1/3$, Eq. (21) is reduced to Eq. (20).

In the cases where spheroidal particles are randomly oriented, the Boned–Peyrelasse equation may be applied (24):

$$\left(\frac{\varepsilon_1^*}{\varepsilon^*}\right)^{3d}\left(\frac{\varepsilon_2^*-\varepsilon^*}{\varepsilon_2^*-\varepsilon_1^*}\right)\left(\frac{\varepsilon_1^*(1+3A)+\varepsilon_2^*(2-3A)}{\varepsilon^*(1+3A)+\varepsilon_2^*(2-3A)}\right)^{3K} = 1-\phi \tag{22}$$

Here, $A = (1 - A_a)/2$ is the depolarization factor while $d$ and $K$ are functions of $A$. Also in this case, when all three axis are equal, i.e., the particles are spherical, $A = 1/3$ and Eq. (22) simplifies to Eq. (20).

The main assumption in all these approaches is that the characteristic sizes of the single-phase regions are much larger than the Debye screening length (26). Provided that the dielectric permittivity and electric conductivity of the individual phases are known, the MW models enable us to calculate the total frequency-dependent permittivity of the system.

An essential feature of the MW polarization effect in complex liquids with liquid–liquid interfaces is the appearance of an accumulated charge at the boundaries between differing dielectric media as a result of ionic migration processes (12, 13). It is further assumed in the MW theory that the conductivities of each phase are uniform and constant. However, the presence of spatial charges in the suspending medium alters the potential gradient near the interface. Thermal motions and ion concentration effects result in this surface-charge layer. In emulsion systems, where the water droplet also contains counterions, a considerable part of the dielectric response to an applied field originates from the redistribution of counterions (27). It is known (12, 28–31) that counterions near the charged surface can be distributed into two regions: the Stern layer and the Gouy–Chapman diffuse double layer. Counterions in the Stern layer are considered fixed on the inner surface of the droplet and are unable to exchange position with the ion in the bulk. The distribution of counterions in the diffuse double layer is given by the Boltzmann distribution in terms of the electrostatic potential. This potential is given, self-consistently, in terms of these counterion charges by Poisson's equation instead of Laplace's equation just as it was performed in the models of interfacial polarization.

The distribution of counterions is essentially determined by their concentration and the geometry of the water core. Thus, in the case of large droplets, the assumption can be made that the droplet radius is much larger than the length of the diffuse double layer as measured by the Debye screening length. In such an approximation, the counterions can be considered to form a thin layer near the inner surface of the droplet. The difference between the Stern layer and the diffuse double layer peters out and the polarization can be described by the Schwarz model (32). In the Schwarz model the mechanism controlling the relaxation is the diffusion of the counterions along the surface. This model is more relevant to the dielectric behavior of macroemulsions than to microemulsions, for example. The above approach (unlike the "classical" MW) institutes the dependence of the dielectric properties on the characteristic size of mesoscale structures (e.g., on the radius of dispersed particle), and the ionic strength or dissociation ability of substances.

## III. BASIC PRINCIPLES OF DIELECTRIC SPECTROSCOPY

The dielectric spectroscopy (DS) method occupies a special place among the numerous modern methods used for physical and chemical analysis of material, because it allows investigation of dielectric relaxation processes in an extremely wide range of characteristic times ($10^4$–$10^{-12}$ s). Although the method does not possess the selectivity of NMR or ESR it offers important and sometimes unique information on the dynamic and structural properties of substances. DS is especially sensitive to intermolecular interactions, and cooperative processes may be monitored. It provides a link between the properties of the individual constituents of a complex material and the characterization of its bulk properties (see Fig. 1).

However, despite its long history of development, this method is not widespread for comprehensive use, because the wide frequency range ($10^{-5}$–$10^{12}$ Hz), overlapped by discrete frequency domain methods, have required a great deal of complex and expensive equipment. Also, for different reasons, not all the ranges have been equally available for measurement. Investigations of samples with variable properties over time (e.g., nonstable emulsions or biological systems) have thus been difficult to conduct. The low-frequency measurements of conductive systems had a strong limitation due to electrode polarization. All the above-mentioned reasons led to the fact that information on dielectric characteristics of a substance could only be obtained over limited frequency ranges. As a result the investigator had only part of the dielectric spec-

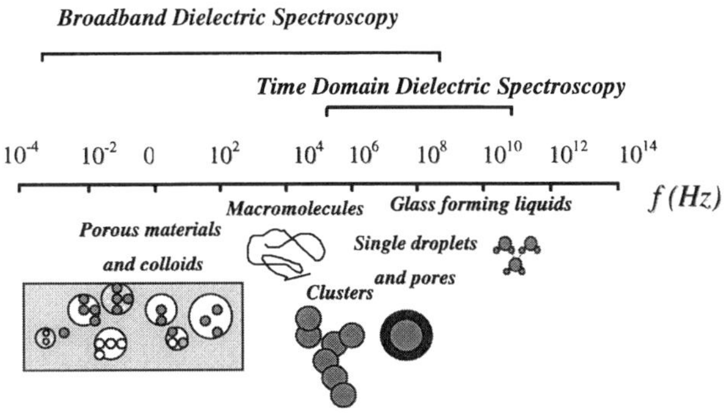


**Figure 1** Electromagnetic wave scale of DS applicability for complex materals.

trum at his/her disposal to determine the relaxation parameters.

The successful development of the time-domain dielectric spectroscopy method (generally called time-domain spectroscopy – TDS) (33–38) and of broadband dielectric spectroscopy (BDS) (39–41) have radically changed attitudes toward DS, making it an effective tool for the investigation of solids and liquids, on the macroscopic, microscopic, and mesoscopic levels.

The basic approaches, the principles of experimental realization, sample holders for different applications, data treatment and presentation, and different TDS methods, which enable one to obtain the complete spectrum of $\varepsilon^*(\omega)$ in the frequency range $10^5$–$10^{10}$ Hz, are given below.

## A. Basic Principles of the TDS Method

TDS is based on transmission-line theory in the time domain that aids in the study of heterogeneities in coaxial lines according to the change in shape of a test signal (33–38). As long as the line is homogeneous the shape of this pulse will not change. However, in the case of a heterogeneity in the line (the inserted dielectric, for example) the signal is partly reflected from the air-dielectric interface and partly passes through it. Dielectric measurements are made along a coaxial transmission line with the sample mounted in a cell that terminates the line. A simplified block diagram of the set-up common for most TDS methods (except transmission techniques) is presented in Fig. 2. Differences mainly include the construction of the measuring cell and its position in the coaxial line. These lead to different kinds of expressions for the values that are registered during the measurement and for the dielectric characteristics of the objects under study.

A rapidly increasing voltage step $V_0(t)$ is applied to the line and recorded, along with the reflected voltage $R(t)$ returned from the sample and delayed by the cable propagation time (Fig. 2). Any cable or instrument artifacts are separated from the sample response as a result of the propagation delay, thus making them easy to identify and control. The entire frequency spectrum is captured at once, thus eliminating drift and distortion between frequencies.

The complex permittivity is obtained as follows: for nondisperse materials (frequency-independent permittivity), the reflected signal follows the RC exponential response of the line–cell arrangement; for disperse materials, the signal follows a convolution of the line–cell response with the frequency response of the sample. The actual sample response is found by writing the total voltage across the sample:

$$V(t) = V_0(t) + R(t) \tag{23}$$

and the total current through the sample (38, 42, 44):

$$I(t) = \frac{1}{Z_0}[V_0(t) - R(t)] \tag{24}$$

where the sign change indicates direction and $Z_0$ is the characteristic line impedance. The total current through a conducting dielectric is composed of the displacement current $I_D(t)$, and the low-frequency current between the capacitor electrodes $I_R(t)$. Since the active resistance at zero frequency of the sample-containing cell is (38) (see Fig. 3):

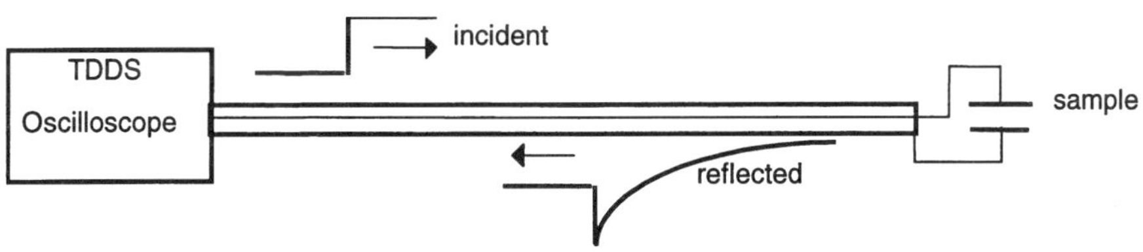


**Figure 2** Basic TDS set-up.

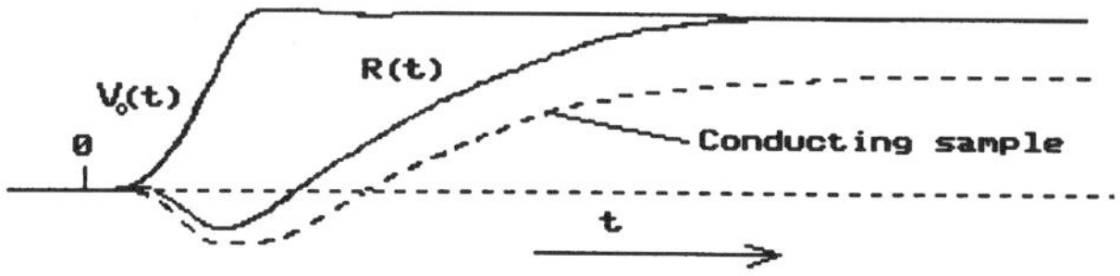


**Figure 3** Characteristic shape of the signal recorded during a TDS experiment; $V_0(t)$: incident pulse, $R(t)$: reflected signal.

$$r = \lim_{t\to\infty} \frac{V(t)}{I(t)} = Z_0 \lim_{t\to\infty} \frac{V_0(t) + R(t)}{V_0(t) - R(t)} \tag{25}$$

the low-frequency current can be expressed as:

$$I_R(t) = \frac{V(t)}{r} = \frac{V_0(t) + R(t)}{Z_0} \lim_{t\to\infty} \frac{V_0(t) - R(t)}{V_0(t) + R(t)} \tag{26}$$

Thus, Eq. 24 can be written as:

$$I_D(t) = \frac{1}{Z_0}\left\{[V_0(t) = R(t)] - [V_0(t) + R(t)] \lim_{t\to\infty} \frac{V_0(t) - R(t)}{V_0(t) + R(t)}\right\} \tag{27}$$

Relations (23) and (27) represent the basic equations that relate $I(t)$ and $V(t)$ to the signals recorded during the experiment. In addition, Eq. (27) shows that TDS permits one to determine the low-frequency conductivity $\sigma$ of the sample directly in the time domain (36–38):

$$\sigma = \frac{\varepsilon_0}{Z_0 C_0} \lim_{t\to\infty} \frac{V_0(t) - R(t)}{V_0(t) + R(t)} \tag{28}$$

where $\varepsilon_0 = 8.85 \times 10^{-12}$ F/m, and $C_0$ is the electric capacity of the coaxial sample cell terminated to the coaxial line. Using $I(t)$ and $V(t)$ or their complex Laplace transforms $i(\omega)$ and $v(\omega)$ one can deduce the relations that will describe the dielectric characteristics of a sample being tested either in frequency or time domain. The final form of these relations depends on the geometric configuration of the sample cell and its equivalent presentation (33–38).

The sample admittance for the sample cell terminated to the coaxial line is then given by

$$Y(\omega) = \frac{i(\omega)}{v(\omega)} \tag{29}$$

and the sample permittivity can be presented as follows:

$$\varepsilon(\omega) = \frac{Y(\omega)}{i\omega C_0} \tag{30}$$

where $C_0$ is the geometric capacitance of the empty sample cell. To minimize line artifacts and establish a common time reference, Eq. (29) is usually rewritten in differential form, to compare reflected signals from the sample and a calibrated reference standard and thus eliminate $V_0(t)$ (33–38).

If one takes into account the definite physical length of the sample and multiple reflections from the air–dielectric or dielectric-air interfaces, Eq. (30) must be written in the following form (33–36, 44):

$$\varepsilon^*(\omega) = \frac{c}{i\omega(\gamma d)} Y(\omega) X \cot X \tag{31}$$

where $X = (\omega d/c)\sqrt{\varepsilon^*(\omega)}$, $d$ is the effective length of the inner conductor, $c$ is the velocity of light, and $\gamma$ is the ratio between the capacitance per unit length of the cell and that of the matched coaxial cable. Equation (31), in contrast to Eq. (30), is a transcendental one, and its exact solution can be only obtained numerically (33–37, 42). The essential advantage of TDS methods in comparison with frequency methods is the ability to obtain the relaxation characteristics of a sample directly in the time domain. Solving the integral equation one can evaluate the results in terms of the dielectric response function $\Phi(t)$ (38, 43, 44). It is then possible to associate $\varphi(t) = \Phi(t) + \varepsilon_\infty$ with the macroscopic dipole correlation function $\Gamma(t)$ (9, 46) in the framework of linear-response theory.

## B. Experimental Tools

### 1. Hardware

The standard time-domain reflectometers used to measure the inhomogeneities of coaxial lines (38, 42, 47, 48) are the basis of the majority of modern TDS set-ups. The reflectometer consists of a high-speed voltage step generator and a wide-band registering system with a single- or double-channel sampling head. In order to meet the high requirements of TDS measurements such commercial equipment must be considerably improved. The main problem is due to the fact that the registration of incident $V(t)$ and reflected $R(t)$ signals is accomplished by several measurements. In order to enhance the signal-to-noise ratio one must accumulate all the registered signals. The high level of drift and instabilities during generation of the signal and its detection in the sampler are usually inherent to serial reflectometry equipment.

The new generation of digital sampling oscilloscopes (36, 45) and specially designed time-domain measuring set-ups (TDMS) (38) offer comprehensive, high precision, and automatic measuring systems for TDS hardware support. They usually have a small jitter factor (< 1.5 ps), important for rise time, a small flatness of incident pulse (< 0.5% for all amplitudes), and in some systems a unique option for parallel-time nonuniform sampling of the signal (38).

The typical TDS set-up consists of a signal recorder, a two-channel sampler, and a built-in pulse generator. The generator produces 200-mV pulses of 10 μs duration and short rise time (~ 30 ps). Two sampler channels are characterized by an 18 GHz bandwidth and 1.5 mV noise (RMS). Both channels are triggered by one common sampling generator that provides their time correspondence during operation. The form of the voltage pulse thus measured is digitized and averaged by the digitizing block of TDMS. The time base is responsible for major metrology TDMS parameters. The block diagram of the described TDS set-up is presented in Fig. 4 (38).

*a. Nonuniform Sampling*

In highly disperse materials, as described in this review, the reflected signal $R(t)$ extends over wide ranges in time and cannot be captured on a single time scale with adequate resolution and sampling time. In an important modification of regular TDS systems, a nonuniform sampling technique (parallel or series) has been developed (38, 49).

In the series realization, consecutive segments of the reflected signal on an increasing time scale are registered and linked into a combined time scale. The combined response is then transformed using a running Laplace transform to produce the broad frequency spectra (49).

In the parallel realization, a multiwindow time scale of sampling is created (38). The implemented time scale is the piecewise approximation of the logarithmic scale. It includes $nw \subseteq 16$ sites with the uniform discretization step determined by the following formula:

$$\delta_{nw} = \delta_1 \times 2^{nw} \quad (32)$$

where $\delta_1 = 5$ ps is the discretization step at the first site, with the number of points in each step except for the first one being equal to $npw = 32$. At the first site, the number of points $npw_1 = 2npw$. The doubling of the number of points at the first site is necessary in order to have the formal zero-time position, which is impossible in the case of the strictly logarithmic structure of the scale. In addition, a certain number of points located in front of the zero-time position is added. They serve exclusively for the visual estimation of the stability of the time position of a signal and are not used for the data processing.

The structure of the time scale described allows the overlapping of the time range from 5 ps to 10 μs during one measurement, which results in a limited number of registered readings. The overlapped range can be shortened, resulting in a decreasing number of registered points and thus reducing the time required for data recording and processing.

The major advantage of the multiwindow time scale is the ability to obtain more comprehensive information. The signals received by using such a scale contain information within a very wide time range and the user merely decides which portion of this information to use for further data processing. Also, this scale provides for the filtration of registered signals close to the optimal one already at the stage of recording.

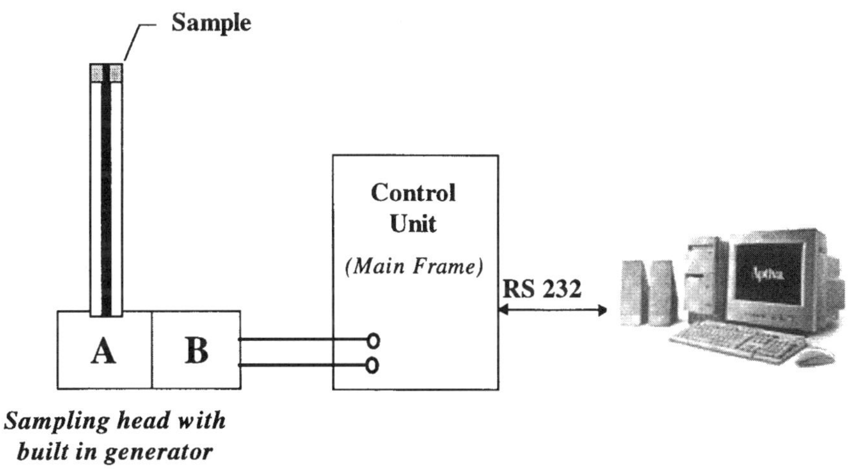


**Figure 4** Circuit diagram of a TDS set-up.

*b. Sample Holders*

A universal sample holder that can be used for both liquid and solid samples in both the low- and high-frequency regions of the TDS method is unfortunately not yet available. The choice of its configuration depends on the measurement method and data-treatment procedure. In the framework of the lumped capacitor approximation one can consider three general types of sample holders (38, 44) (Fig. 5a): a cylindrical capacitor filled with sample. This cell (a cut-off cell) can also be regarded as a coaxial line segment with the sample having an effective $\gamma d$ length characterized in this case by the corresponding spread parameters. This makes it possible to use practically identical cells for various TDS and BDS method modification (50). For the total-reflection method the cut-off cell is the most frequent configuration (33–37, 42). The theoretical analysis of the cut-off sample cell (Fig. 5a) showed that a lumped-element representation enables the sample-cell properties to be accurately determined over a wide frequency range (50). Another type of sample holder that is frequently used is a plate capacitor terminated to the central electrode on the end of the coaxial line (Fig. 5b) (38, 44, 51, 52). The most popular now for different applications is an open-ended coaxial line sensor (Fig. 5c) (53–59). In lumped-capacitance approximation the configurations in Fig. 5a,b have high-frequency limitations, and for highly polar systems one must take into account the finite propagation velocity of the incident pulse or, in other words, the spread parameters of the cell (34–38). The choice of cell shape is determined to a great extent by the aggregate condition of the system studied. While cell (a) is convenient for measuring liquids (see Fig. 6a), configuration (b) is more suitable for the study of solid disks and films (Fig. 6b). Both cell types can be used to measure powder samples. While studying anisotropic systems (liquid crystals, for instance) the user may replace a coaxial line by a strip line or construct a cell with the configuration providing the measurements under various directions of the applied electric field (35, 36). The (c) type cell (see Fig. 5c) is used only when it is impossible to place the sample in the (a) or (b) type cells (38, 54–61). The fringing capacity of the coaxial-line end is the working capacity for such a cell. This kind of cell is widely used now for investigating the dielectric properties of biological materials and tissues (56–58), petroleum products (58), constructive materials (45), soil (60), and numerous other nondestructive

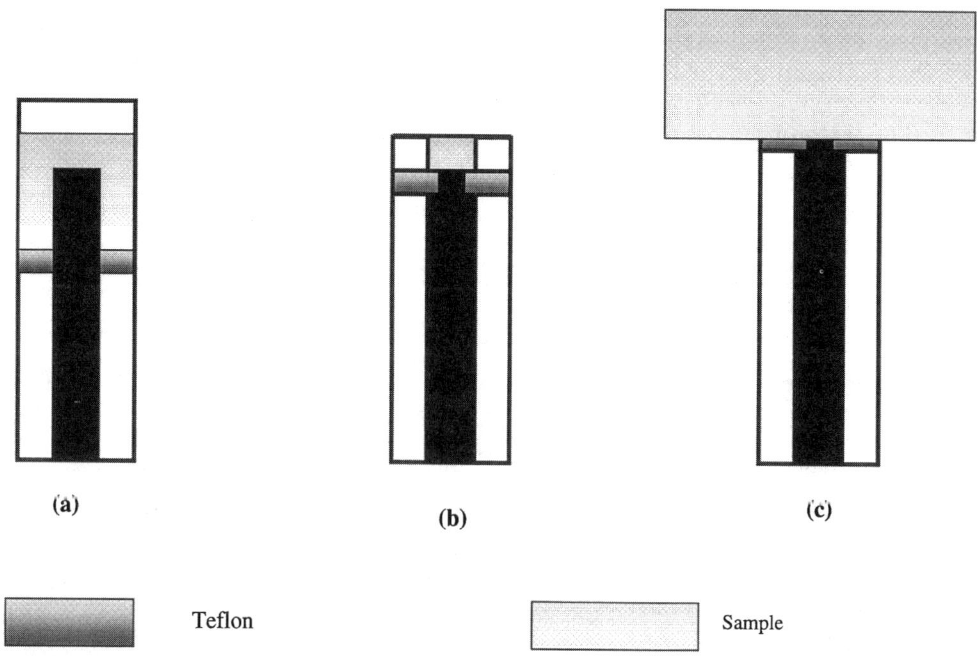


**Figure 5** Simplified drawings of sample cells: (a) open coaxial line cell; (b) lumped capacitance cell; (c) end capacitance cell.

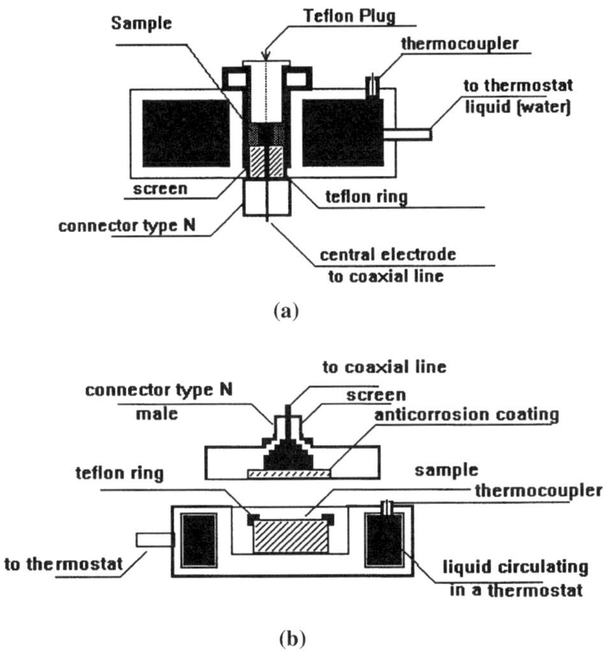


**Figure 6** (a) Sample cell for high-frequency measurements of liquids; (b) sample cell for low-frequency measurements of liquids, solids, and powders. (From Ref. 38. With permission from American Institute of Physics.)

permittivity and permeability measurements. The theory and calibration procedures for such open-coaxial probes are well developed (61, 62) and the results are meeting the high standards of modern measuring systems.

### 2. Software

Measurement procedures, registration, storage, time referencing, and data analyses are carried out automatically in modern TDS systems. The process of operation is performed in on-line mode and the results can be presented both in frequency and time domains (34, 36, 38, 45, 49). There are several features of the modern software that control the process of measurement and calibration. One can define the time windows of interest that may be overlapped by one measurement. During the calibration procedure precise determination of the front-edge position is carried out and the setting of the internal autocenter on this position applies to all the following measurements. The precise determination and settings of horizontal and vertical positions of calibration signals are also carried out. All parameters may be saved in a configuration file, allowing for a complete set of measurements, using the same parameters and without additional calibration.

A typical flow chart of the data-processing software is presented in Fig. 7. It includes the options of signal corrections, correction of electrode polarization and d.c. conductivity, and different fitting procedures both in time and frequency domains.

## C. Electrode Polarization Corrections

Many dielectric materials are conductive. This complicates the TDS study of conductive samples, and the effect of low-frequency conductivity needs to be corrected for. Usually, for a low-conductivity system the value of the d.c. conductivity can be evaluated as described in Sec. III.A.

One of the greatest obstacles in TDS measurements of conductive systems is the parasitic effect of electrode polarization. This accumulation of charge on electrode surfaces results in the formation of electric double layers (63–67). The associated capacitance and complex impedance due to this polarization is so large that the correction for it is one of the major requisites in obtaining meaningful measurements on conductive samples, especially in aqueous biological and colloidal systems (65–71). The details of electrode polarization depend microscopically upon the electrode surface topography and surface area, as well as upon the surface chemistry (reactive surface groups or atoms) and the interactions with the dielectric material or sample being examined. In the case of complex conductive liquids, the principal motivation of this work, surface ionization and ion-exchange processes in the electric double layer, can depend critically upon the chemical nature of the sample being investigated as well as upon the chemical and physical nature of the electrodes used. Because these many effects can be so diverse, no simple correction technique has been widely accepted. Several equivalent circuits have been proposed for describing the essential elements of a sample cell containing electrolyte solution (15, 65, 68), and the most generally accepted approach is shown in Fig. 8. Under the assumption that the electrodes are blocking with respect to Faradaic electron transfer, the polarization impedance of the electrodes $Z$, may be expressed as

$$Z_p = R_p + \frac{1}{i\omega C_p} \tag{33}$$

where $i = \sqrt{-1}$. Both $C_p$ and $R_p$ vary with frequency and $Z_p$ is often considered negligible at sufficiently high frequencies. This high-frequency limit for electrode polarization has been estimated in different

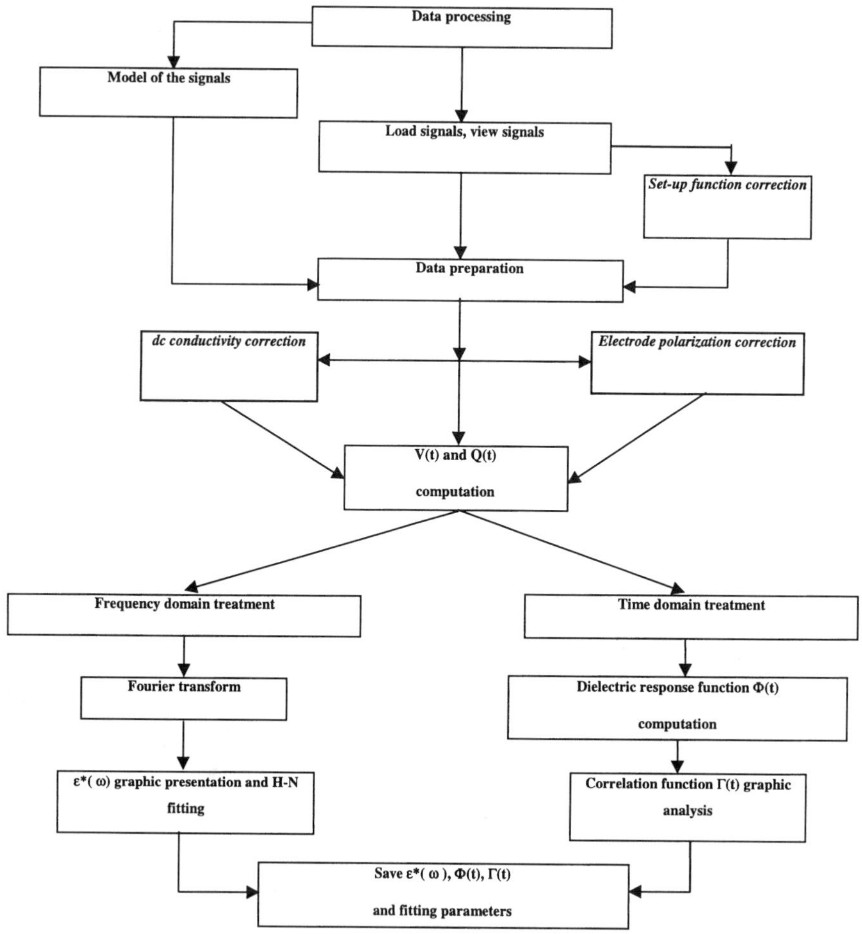


**Figure 7** Flow chart of data-treatment software. (From Ref. 38. With permission from American Institute of Physics.)

ways, depending on the particular type of electrode assembly and DS experiment, but has generally fallen in the interval 100–500 kHz. Owing to the diverse nature of the processes that can contribute to electrode polarization, such as sample-dependent chemical processes alluded to above, it is difficult to estimate an upper bound for this frequency limit.

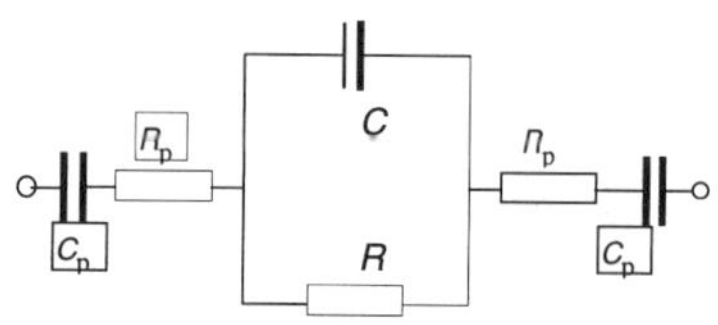


**Figure 8** Equivalent circuits for a conductive dielectric sample with electrode polarization impedance described by $C_p$ and $R_p$.

Two different approaches (71, 72) have been developed to correct for this phenomenon in TDS measurements directly in the time domain. One of them is applicable to weak electrolytes with a small level of low-frequency conductivity and hence a comparatively small effect of electrode polarization (72). In the other approach, applied to very conductive systems, the fractal nature of electrode polarization is considered (71). Let us consider these approaches.

In the case of TDS we can present a double layer with a capacitance $C_p$ that is connected in series to the sample cell filled with the conductive material (Fig. 9). The characteristic charge time of $C_p$ is much larger than the relaxation time of the measured sample. This allows us to estimate the parameters of parasitic capacitance in the long-time window where only the parasitic electrode polarization takes place. Considering the relationship for the current $i(s)$

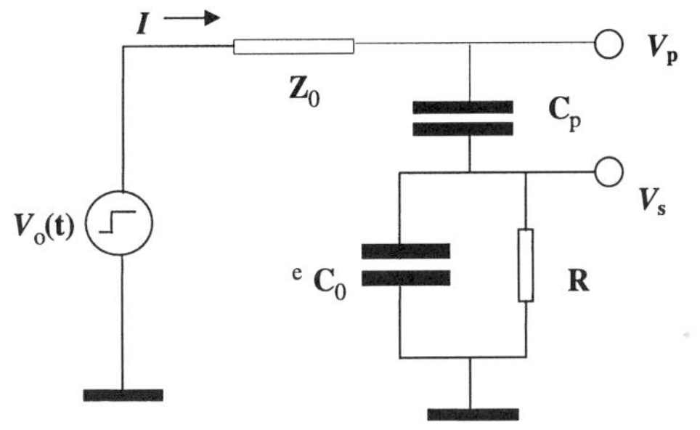


**Figure 9** Equivalent circuit accounting for the electrode-polarization effect. $V_0(t)$ is a rapidly increasing voltage step; $I(t)$ is a current; $Z_0$ is the coaxial line impedance; $C_p$ is the capacitance of electrode polarization; $C_0$ is an empty cell capacitance filled with a dielectric sample of permittivity $\varepsilon$ and conductivity $1/R$; $V_p(t)$ and $V_s(t)$ are the voltages at the appropriate parts of the circuit. (From Ref. 72. With permission from Elsevier Science B.V.)

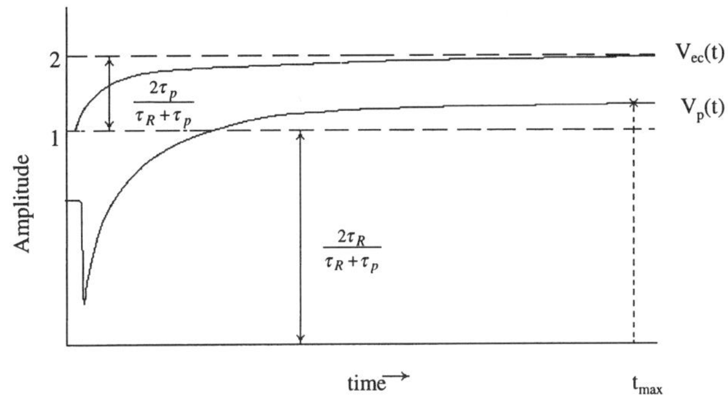


**Figure 10** Schematic presentation of signal from a sample with conductivity [signal $V_p(t)$] and the correction exponential function described by the electrode polarization curve $Vec(t)$. (From Ref. 72. With permission from Elsevier Science B.V.)

($s = \gamma + i\omega$, $\gamma \to 0$ is a generalized frequency in the Laplace transform) and voltage $v_p(s)$ in the frequency domain (Fig. 9) and making an inverse Laplace transform in the limit $t \to \infty$, one can obtain the analytical expression for the electrode polarization correction function $V_{ec}(t)$ as follows (72):

$$V_{ec}(t) = \frac{2\tau_p}{\tau_R + \tau_p}(1 - e^{-t/\tau_2}) \tag{34}$$

where $\tau_p = Z_0 C_p$, $\tau_R = RC_R$, and $\tau_0 = C_0 R = C_0/\sigma$. The parameter $\tau_2$ may be obtained from the tail of the signal where only the electrode polarization effect takes place:

$$\tau_2 = \lim_{t\to\infty} \frac{2 - V_p(t)}{V_p(t)} \tag{35}$$

In order to eliminate the influence of the polarization capacitance it is necessary to subtract the exponential function with the appropriate parameters from the raw signal of the conductive sample. The exponential function $V_{ec}(t)$ of electrode polarization correction can easily be fitted to the real signal (see Fig. 10).

The TDS measurements on aqueous solutions of proteins and cell suspensions at up to several gigahertz (70–73) have shown that electrode polarization has to be taken into account even at frequencies as high as several hundred megahertz. Schwan noted the porous nature of electrode polarization phenomena (67). We nowadays often attribute such porosity to fractality and characterize porous tortuosity in terms of fractal dimension. Schwan also mentioned the increasing magnitude of this effect with increasing frequency. While the fractal nature of *electrode surfaces* is now well appreciated (74–79), no applications for making polarization corrections, capitalizing upon the fractal nature of *electrode polarization*, appear to have been developed previously.

A general form for depicting the fractal nature of an electrode double-layer impedance is given by

$$Z_v(j\omega) = A(v)(j\omega)^{-v} \tag{36}$$

where $0 < v < 1$, $A(v)$ is an adjustable paramter, and the frequency $\omega$ is located in a certain range $\omega_{min} \le \omega \le \omega_{max}$ due to the self-similar electrode-polarization properties of the electrode surface (74–78). At sufficiently high and low frequencies the self-similarity of the electrode polarization disappears.

The exponent $v$ has often been connected with the fractal dimension of the electrode surface, but this connection is not necessary. Pajkossy and coworkers (76, 77) have shown, however, that specific adsorption effects in the double layer necessarily do appear for such a dispersion. We can connect the exponent to the fractality of the dynamical polarization and show that the polarization is self-similar in time, in contrast to the self-similar geometrical structure.

The specific frequency dependence in Eq. (36) is known as the *constant phase angle* (CPA) dependence (78–80). This impedance behavior occurs for a wide class of electrodes (75–77) and suggests the introduction of a new equivalent circuit element with impedance characteristics similar to those of Eq. (36). We call this element a *recap* element (derived from *re*sistance and *cap*acitance). The electric and fractal properties for this recap complex impedance $C_v(s)$ are given by:

$$C_v(s) = R(RC)^{-v}s^{-v} \qquad 0 < v < 1 \tag{37}$$

This circuit element occupies an intermediate position between $R_p(v=0)$ and $C_p(v=1)$ and expresses its impedance in the finite range of frequencies articulated above. Using the definition of Eq. (37) we can rearrange the equivalent circuit of the measuring sample cell filled with electrolyte solution, as shown in Fig. 11.

The impedance of the sample cell (electrodes) containing an electrolyte solution can then be derived using one of these recap elements for the impedance at each electrode. Since the electrodes in general are not perfectly identical, each can be defined according to Eq. (37) to give

$$C_{v_i}(s) \equiv C_{v_i} \cdot s^{-v_i} = R_i(R_iC_i)^{-v_i}s^{-v_i} \tag{38}$$

with $i = 1$ and 2. In this fractal representation the measuring cell is defined in terms of the fractal impedance of the electrode polarization.

In order to derive the current–voltage relationship obtained for the equivalent circuit of Fig. 10, it is convenient to note the following identity (81):

$$s^{-v}i(s) = \frac{1}{\Gamma(v)}\int_0^t (t-t')^{v-1}I(t')\mathrm{d}t' \tag{39}$$

where $I(t)$ is a time-dependent current, $\Gamma(v)$ is the gamma function, $s$ is a complex frequqency, and $i(s)$ is the Laplace transformation of $I(t)$. This resulting current–voltage relationship:

$$\begin{aligned} V(t) = {}& \frac{C_{v_1}}{\Gamma(v_1)}\int_0^t (t-t')^{v_1-1}I(t')\mathrm{d}t' \\ & + \frac{C_{v_2}}{\Gamma(v_2)}\int_0^t (t-t')^{v_2-1}I(t')\mathrm{d}t' \\ & + \frac{1}{C}\int_0^t \exp[-(t-t')/\tau_c]I(t')\mathrm{d}t' \end{aligned} \tag{40}$$

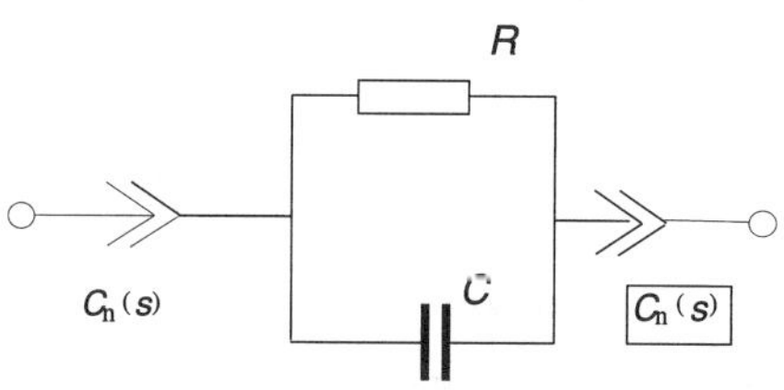


**Figure 11** Equivalent circuits for a conductive dielectric sample with electrode-polarization impedance described by recap $C_V(s)$. (From Ref. 71. With permission from American Physical Society.)

has two terms describing the polarization of the respective electrodes and a third term describing the contribution from the bulk sample, where $\tau_c = RC$. Equation (40) shows how $V(t)$ and $I(t)$ are related for different kinds of measuring cells (i.e., different $C_{v_i}, v_i$) containing conductive solutions that polarize electrodes in conformity with the equivalent circuit illustrated in Fig. 11.

When $I(t)$ is constant, as will be the case at long times after all the transients associated with sample relaxation have died out, Eq. (40) reduces to

$$V(t) \propto \frac{C_{v_1}}{v_1\Gamma(v_1)}t^{v_1} + \frac{Cv_2}{v_2\Gamma(v_2)}t^{v_2} + R[1-\exp(-t/\tau_c)] \tag{41}$$

where $\tau_c$ defines the time scale wherein the recap elements (the electrode polarization) affect the $V(t)$ measured for the sample cell containing a conductive solution. If we make the simplifying assumptions that (1) both electrodes of the sample cell have the same (or equivalent) fractal polarization ($v_1 = v_2$; $C_{v_1} = C_{v_2}$); and (2) that there is no dispersion of the conductive solution (sample) in the time window defined by $\tau_c(> t_{\min} = 1/\omega_{\max})$, Eq. (41) can be rewritten in the following way:

$$V(t) = Bt^v + Rt/\tau_c \qquad \text{for } t/\tau_c \ll 1 \tag{42}$$

and

$$V(t) = Bt^v + R \qquad \text{for } t/\tau_c \gg 1 \tag{43}$$

Equations (41) and (42) are particularly useful in illustrating how the contribution of electrode polarization $Bt^v$ should be substracted from $V(t)$.

The voltage $V(t)$ and the current $I(t)$ observed at the sample cell [plate or cylindrical capacity (72, 73)] at the end of a coaxial line are presented by Eqs (23) and (24), respectively. In the case of conductive solutions, Eqs. (23) and (24) show that both the voltage and the current flow are influenced by electrode polarization. The observed voltage $V(t)$ monotonically increases in the TDS time window of observation and the current $I(t)$ monotonically decays. The electrode polarization *correction* is then obtained by subtracting the function $Bt^v$ from $V(t)$. The incident pulse $V_o(t)$ generally is an approximation to a step function with zero long-time slope, and the monotonically increasing behavior of $V(t)$, associated with the correction $Bt^v$, is a component of the reflected pulse $R(t)$. Since this component is subtracted in Eq. (23) from $R(t)$, it needs to be added $(+Bt^v/Z_o)$ to $I(t)$ in Eq. (24) and only after this can the correction of conductivity contribution by relation (27) be taken into account.

The experimental signal, $V(t)$, for a strong electrolyte solution of NaCl in water at 25°C (adjusted to pH 1.25 with 0.05 M HCl and with a low-frequency conductivity of 1.57 S $m^{-1}$) is illustrated on a log–log scale in Fig. 12 (71). The TDS multiwindow measurement (38) allow a long-time (up to 10 μs) registration of signal tails, where the signal from a dielectric with low-frequency conductivity has completely relaxed (decayed) and only the signal from the electrode polarization remains. The fractality index $v$ was determined to be 0.785 with $B = 0.593$ by linear regression of the asymptotically linear portion. The fractality is strictly dependent on the electrode material (82), the electrode polishing (71), and the chemistry related to the interactions between aqueous electrolyte and the electrode surfaces.

The electrode polarization correction applicable for TDS measurements of conductive colloidal samples can be summarized in the following way. A refrence sample with an electrolyte composition and conductivity equal to the continuous phase is measured, and these data are subsequently used to fit the parameters $v$ and $B$ (as was illustrated above for the data in Fig. 12; then, as long as the conductivity and electrolyte composition of the continuous phase remain equal to that of the reference, the sample may be measured with the same electrodes. The polarization correction embodied in the $v$ and $B$ obtained earlier is applied by substraction of $Bt^v$ from $V(t)$ and by adding this same function (scaled by $Z_0$) to $I(t)$. These corrected signals are then Fourier transformed (71) to obtain the resulting dielectric spectra, $\varepsilon'(\omega)$ and $\varepsilon''(\omega)$. In this exposition of the correction procedure we have illustrated the method, focusing upon the polarization properties of a simple two-component electrolyte solution. After correction in this case we are left with the properties of the neat solvent, water.

A comparison between the dielectric spectra of a simple electrolyte solution (pH = 1.25) with and without electrode polarization correction is presented in Fig. 13. The permittivity before and after the correction is illustrated in Fig. 13a. The uncorrected permittivity exhibits an anomalously large value at low

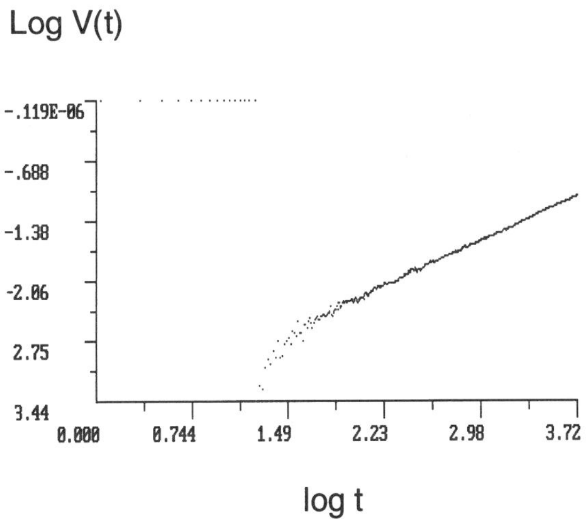


**Figure 12** Voltage $V(t)$ applied to the electrolyte solution; $v = 0.785$, $V = 0.593$ (with polished stainless-steel electrodes). (From Ref. 71. With permission from American Physical Society.)

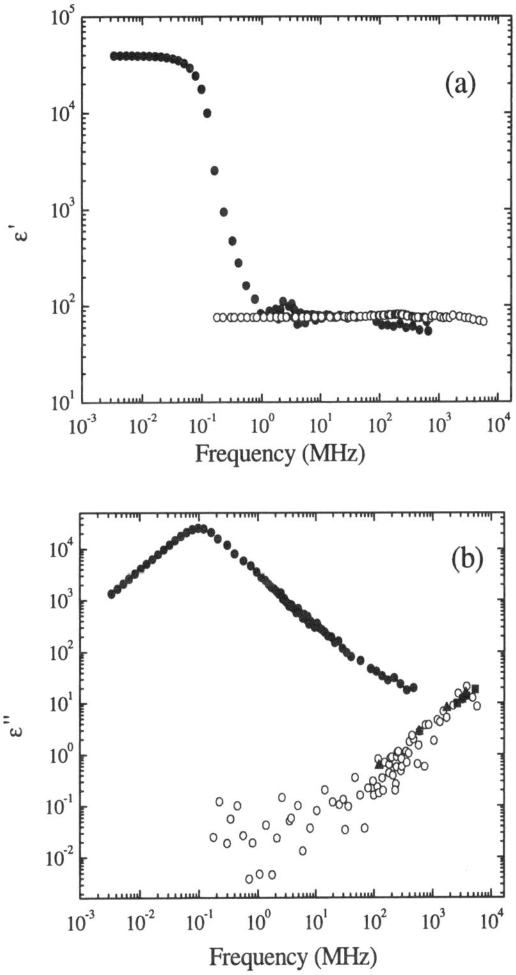


**Figure 13** Fractal electrode-polarization correction for dielectric spectrum (a) $\varepsilon'$ and (b) $\varepsilon''$ of simple electrolyte solution; 0.1 M NaCl at pH 1.25; ● – uncorrected; ○ – corrected; ▲ – pure water at 20°C (from Ref. 83); ■ – pure water at 25°C. (From Ref. 84.)

frequencies and seemingly undergoes some sort of dispersion process in the 100 kHz to 1 MHz region. This dispersion is highlighted as a peak (at 100 kHz) in the uncorrected dielectric loss spectrum of Fig. 13b. This dispersion essentially completely vanishes after the correction for electrode polarization has been made. The corrected permittivity in Fig. 13a is essentially independent of frequency at about 78. The dielectric loss, after correction for electrode polarization, becomes very small in the 100 kHz–100 MHz range, but shows a steady increase with increasing frequency. This increase is consistent with the well-known dielectric loss maximum of water in the region of 3 to 6 GHz, and compares well with the experimental dielectric loss values for water reported at 20°C (83) and at 25°C (84), and illustrated in Fig. 13b for comparison.

The data derived after the electrode polarization correction are in good agreement with previously published data on water. These data show unequivocally the need for electrode polarization correction at frequencies in excess of 100 MHz, and that such correction can be effected by explicitly considering the fractality of electrode polarization.

## D. External Fields

The dielectric properties of a sample may be strongly influenced by its environment, and measurements of the dielectric behavior as a function of, for instance, temperature or pressure (2) are performed on a routine basis. In addition to the temperature and pressure effects, the TDS method also allows for subjecting the samples to externally applied electric or magnetic fields.

### 1. Electric (High-voltage Measurements)

High electric fields have been used in order to study the effect on W/O emulsions (52). In investigations on liquid crystals, external electric fields are applied in order to ensure the desired orientation of the crystals (85).

The TDS sampling heads are very vulnerable towards high voltages and electric currents. Thus, the signal applied to a sample and consequently recorded by the TDS equipment normally cannot exceed 0.2 V. In order to attain sufficiently high voltages between the electrodes to induce any changes in the sample and, at the same time and through the same line, transmit the low-voltage step pulse used in the characterization of the sample, special equipment has to be used. Thus, a broadband coaxial bias-tee is inserted into the transmission line. The bias-tee is designed so that it will let the fast rise-time pulses pass through with negligible distortion of the waveform, while the high-voltage d.c. is effectively blocked from reaching the part of the transmission line that is connected to the sampling head.

The modifications of the standard experimental set-up needed to include this method are schematically depicted in Fig. 14. A d.c. voltage supply is connected to the coaxial line via a bias-tee (Picosecond Pulse Labs, 5530A), as described above. In this way a potential difference can be applied between the cell electrodes. Owing to the short distance between the electrodes (controlled by the spacer, see Fig. 15) strong electric field result, even from moderate voltages. With a spacer thickness of 120 μm and a potential difference of 60 V, the electric field applied to the sample will be 5 kV/cm. This field strength is sufficiently strong to lead to a marked distortion of the shape of water droplets in an emulsion, and in many cases the electric field induces the coalescence of emulsion droplets. A d.c. block inserted in the line between the bias-tee and the sampling head may be applied as an additional protection of the sampling head from the high voltage; only the step pulse is allowed to travel through. The blocking of a.c. fields is more difficult to accomplish; thus, at present, this method is limited to external d.c. fields.

### 2. Magnetic Fields

The orientation of a sample (on a molecular or aggregate level) by the action of a magnetic field may be achieved, using a set-up as illustrated in Fig. 16 (86). The dielectric cell shown in Fig. 16a is of the open-ended coaxial sensor type, and the electrical length is found to be 0.027 mm. The magnetic field is created using rod magnets with the magnetic poles placed on either side of the dielectric cell. An alternative set-up used for measuring the sedimentation profile of suspensions containing magnetic particles is shown in Fig. 16b. In this case a magnetic field up to approximately 0.4 T is created by an electromagnet. Also, the dielectric cell is modified in order to increase the functionality of the experimental set-up (Fig. 16b).

## IV. DIELECTRIC PROPERTIES OF MICROEMULSIONS

Microemulsions are thermodynamically stable, clear fluids, composed of oil, water, surfactant, and sometimes cosurfactant, that have been widely investigated

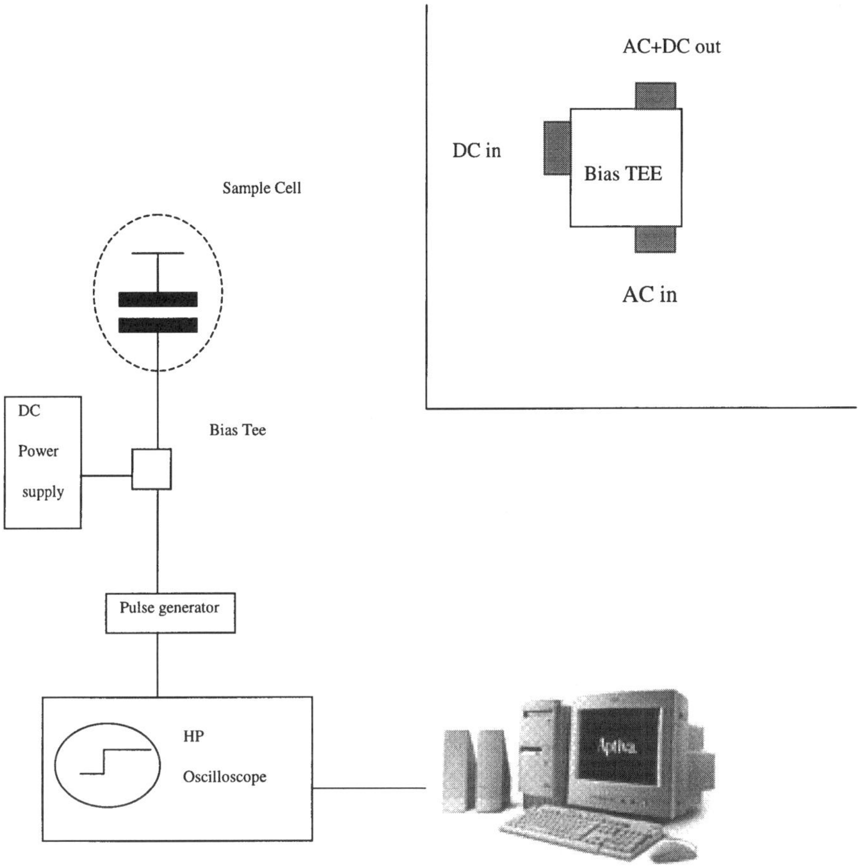


**Figure 14** TDS set-up for high external electric field measurements.

during recent years because of their numerous practical applications. The chemical structure of surfactants may be of low molecular weight as well as being polymeric, with nonionic or ionic components (87–90). In the case of an oil-continuous (W/O) microemulsion, at low concentration of the dispersed phase, the structure is that of spherical water droplets surrounded by a monomolecular layer of surfactant molecules whose hydrophobic tails are oriented towards the continuous oil-phase. When the volume fractions of oil and water are both high and comparable, random, bicontinuous structures are expected to form. There are also microemulsions in which the minor component forms disks, sheets, or rods, as well as mixtures, which are microlamellar (89–92). It was found that alcohol added as cosurfactant could affect the solubilization of water in microemulsions (93, 94). The alcohol molecules can reside in both the aqueous and oil phases, and/or in the amphiphilic monolayer at the interface. A clear understanding of the role that the alcohol plays in the organization of the morphology of microemulsions has not yet been obtained and the problem of estimating the amount of alcohol participating at the interface of the microemulsion and in the bulk is not yet completely resolved.

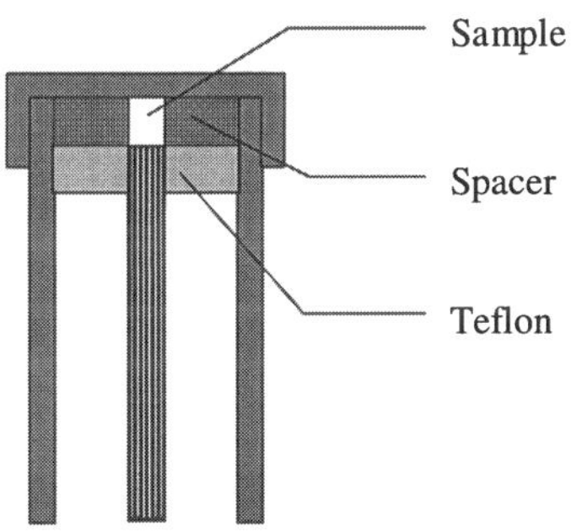


**Figure 15** TDS sample cell for high-frequency for high external electric field measurements.

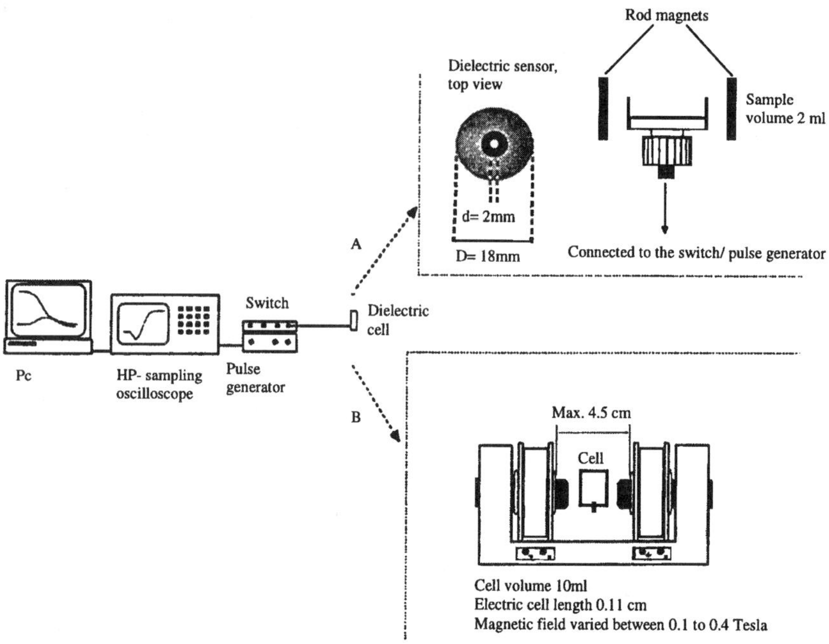


**Figure 16** TDS set-up for external magnetic field measurements.

The structure of the microemulsion depends on the interaction between droplets. In the case of repulsive interaction, the collisions of the droplets are short and no overlapping occurs between their interfaces. However, if the interactions are attractive, transient droplet clusters are formed. The number of such clusters increases, when the water fraction, the temperature, the pressure, or the ratio of water to surfactant is increased, leading to a percolation in the system (95–101).

The majority of the different chemical and physical properties, as well as the morphology of microemulsions, is determined mostly by the microBrownian motions of its components. Such motions cover a very wide spectrum of relaxation times ranging from a few picoseconds to tens of seconds. Given the complexity of the chemical make up of the microemulsions, there are many various kinetic units in the system. Depending on their nature, the dynamic processes in the microemulsions can be classified into three types.

The *first type* of relaxation process reflects characteristics inherent to the dynamics of single droplet components. The collective motions of the surfactant molecule head groups at the interface with the water phase can also contribute to relaxation of this type. This type can also be related to various components of the system containing active dipole groups, such as cosurfactant, and bound and free water. The bound water is located near the interface, while "free" water, located more than a few molecule diameters away from the interface, is hardly influenced by the polar or ionic groups. In the case of ionic microemulsions, the relaxation contributions of this type are expected to be related to the various processes associated with the movement of ions and/or surfactant counterions relative to the droplets and their organized clusters and interfaces.

For percolating microemulsions, the *second* and the *third types* of relaxation processes are pertinent, characterizing the collective dynamics in the system and having a *cooperative nature*. The dynamics of the second type may be associated with the transfer of an excitation caused by the transport of electrical charges within the clusters in the percolation region. The relaxation processes of the third type are caused by rearrangements of the clusters and are associated with various types of droplet and cluster motions, such as translations, rotations, collisions, fusion, and fission.

MicroBrownian dynamics of microemulsions can be studied by various techniques including dynamic–mechanical, dielectric, ultrasonic and NMR relaxation, ESR, volume, enthalpy and specific heat relaxation, quasielastic light and neutron scattering, fluorescence-depolarization experiments, and many other methods (90, 102–107). The information thus acquired provides an opportunity to clarify the organizational structure and the dynamic behavior of such systems.

Dielectric spectroscopy may be successful in providing unique information about the dynamics and structure of microemulsions on various spatial and temporal scales. Being sensitive to percolation, DS is expected to provide unambiguous conclusions concerning the stochastic type, the long time scale cooperative dynamics, and the imposed geometric restrictions of molecular motions before, during, and after the percolation threshold in microemulsions. It also can give valuable information about fractal dimensions and sizes of the percolation clusters. On the other hand, an analysis of the dynamics on the short time scale can provide an understanding of the relaxation mechanisms in microemulsions on a geometrical scale of one microdroplet or the dynamics of surfactant or cosurfactant molecules in the interface. This is important, as it can give quantitative information about amounts of alcohol residing both in the interface and in the bulk and thus enables one to calculate the amount of bound water in the system.

The purpose of this chapter is to describe how DS can be applied to the investigation of microemulsions and how information about molecular mobility and structure can be extracted. We will show that an experimentally monitored temperature-dependent increase in $\varepsilon$ can be explained by the temperature-dependent growth of the mean-square fluctuation dipole moment of a droplet. Analysis of the dynamic features of the known ionic and nonionic microemulsions on various time and geometrical scales will provide knowledge on both the components of the system and microdroplets as a whole. For instance, by choosing an ionic AOT/water/decane microemulsion near the percolation threshold, we can investigate the cooperative relaxation associated with charge transport in the system. By investigating a series of quaternary oil/surfactant/cosurfactant/water microemulsions prepared with the nonionic surfactants $C_{18:1}(EO)_{10}$ or $C_{12}(EO)_8$, we can calculate the amount of alcohol residing in the interface and in the bulk phases as well as the amount of bound and free water in the system. The bound water is located near the interface and is hydrogen bonded to the hydrophilic head groups of surfactant and alcohol molecules.

## A. Dielectric Spectroscopy of Ionic Microemulsions Far Below Percolation

The microemulsions formed with the surfactant, sodium bis(2-ethylhexyl) sulfosuccinate (AOT), water, and oil are widely investigated systems whose dynamics, phase behavior, and structure are well known (87–90). These microemulsions reside in the $L_2$ phase over a wide temperature range, i.e., the microemulsions consist of nanometer-sized spherical droplets with water in the central core surrounded by a layer of surfactant molecules. In this phase, the surfactant molecules have their hydrophilic head groups facing the water and their hydrophobic tails oriented towards the continuous oil phase. Molecules of AOT can dissociate into anions containing negatively charged head groups, $SO_3^-$, staying at the interface and positive counterions, $Na^+$, distribution in the droplet interior. There is a characteristic feature for these systems, such as a small droplet radius, comparable to the thickness of the electric double layer as measured by the Debye length. The thickness of the double layer and the distribution of mobile counterions within it can be calculated from the Poisson–Boltzmann equation (108, 109).

The electrical conductivity and dielectric permittivity of the ionic water-in-oil microemulsions show quite remarkable behavior when the temperature, the water fraction, pressures, or ratio of water to surfactant is varied (95–101). In our prior research (96, 97, 107), the dielectric relaxation, electrical conductivity, and diffusion properties of the ionic microemulsions were investigated in a broad temperature region. In particular, the investigation showed that ionic microemulsions start to exhibit percolation behavior that is manifested by a rapid increase in the static dielectric permittivity $\varepsilon$ and electrical conductivity $\sigma$ when the temperature reaches the percolation onset $T_{on}$ (Fig. 17). The appearance of the percolation reveals that in the region $T > T_{on}$ the droplets from transient clusters. When the system approaches the percolation threshold $T_p$, the characteristic size of such clusters increases, leading to the observed increase in $\sigma$ and $\varepsilon$. This increase in the percolation temperature region is governed by scaling laws:

$$\sigma \sim \begin{cases} (T_p - T)^{-s} & T < T_P \\ (T - T_p)^t & T > T_p \end{cases} \tag{44}$$

and

$$\varepsilon \sim (T_p - T)^{-s} \qquad T < T_p \tag{45}$$

characterized by critical exponents $s$ and $t$. Experimentally, the critical exponents are found to have the values $s \approx 1.2$ and $t \approx 1.9$ (96). We define the *percolation onset* as the temperature at which the microemulsion starts to display a scaling behavior for

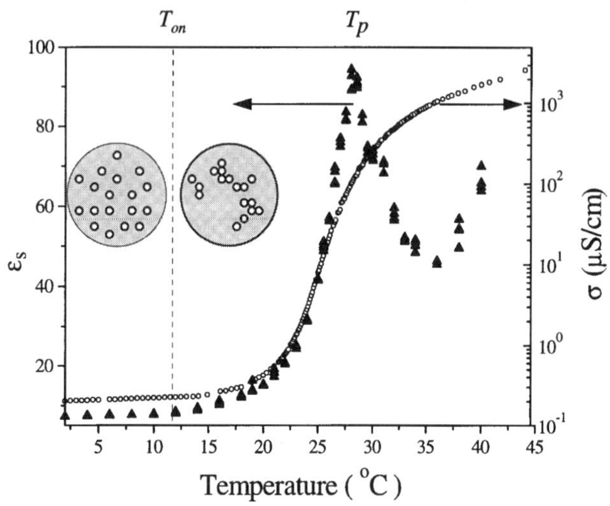


**Figure 17** Schematic illustration of the structures and temperature dependence of static dielectric permittivity and conductivity for the AOT–water–decane microemulsion (17.5:21.3:61.2). (From Ref. 5. With permission from Elsevier Science B.V.)

conductivity [Eq. (44)] and for dielectric permittivity [Eq. (45)].

Below the percolation onset, both the conductivity $\sigma$ and static dielectric permittivity $\varepsilon$ of the microemulsions increase as a function of the volume fraction of droplets $\varphi$ and/or temperature $T$(95–102). However, this increase is not significantly essential, as it is within the percolation region.

The increase of the conductivity versus temperature and volume fraction of droplets below the percolation onset can be described by the charge-fluctuation model (110, 111). In this model the conductivity is explained by the migration of charged aqueous noninteracting droplets in the electric field. The droplets acquire charges owing to the fluctuating exchange of charged surfactant heads at the droplet interface and the oppositely charged counterions in the droplet interior. The conductivity is then proportional to $\varphi$ and $T$:

$$\sigma = \frac{\varepsilon_0 \varepsilon k_B}{2\pi\eta R_d^3} T\varphi \tag{46}$$

where $\eta$ is the solvent viscosity and $R_d$ is the droplet radius.

Unlike the mechanism of increasing conductivity below the percolation onset as a function of temperature, the temperature behavior of static dielectric permittivity has been hitherto puzzling. The static dielectric permittivity of dipolar liquids $\varepsilon$ is proportional to the macroscopic mean-square dipole moment $\langle M^2 \rangle$ of the system unit volume and inversely proportional to the temperature $T$, as $\varepsilon \sim \langle M^2 \rangle / T(1)$. However, the ionic microemulsions exhibit a growth of the dielectric permittivity as a function of temperature in the whole temperature interval.

Given the complexity of the chemical composition of the microemulsions, there are several sources of the dielectric polarization in the system. The contributions are expected to be related to the various processes connected with interfacial polarization, counterion polarization, and the motions of the anionic head groups of the surfactant molecules at the interface with the water phase (39, 96, 97, 112, 113). The contribution in polarization can also be related to various components of the system containing dipole groups, such as bound and free water (114).

The experimentally observed (95–101) increase in the dielectric polarization in the microemulsions in the nonpercolating region can qualitatively be imputed to two mechanisms. The first mechanism attributes the increase in $\varepsilon$ below percolation to an aggregation of the spherical droplets with polarizability that is independent of temperature (115). However, an aggregation of droplets seems to be very unlikely at temperature far below the percolation region. An alternative mechanism is related to the temperature dependence of the fluctuation dipole moment of noninteracting and therefore nonaggregating droplets dispersed in oil. In order to provide the experimentally monitored temperature increase in the dielectric permittivity of a monodispersed system consisting of spherical droplets at a constant volume fraction, the value of the mean-square dipole moment $\langle \mu^2 \rangle$ of the droplet must grow faster than the linear function of temperature.

It has been argued that the interaction of the droplets can be modulated by changing the length of the oil chain (116). In light of this, a direct approach to elucidation of the mechanism responsible for the increase in dielectric permittivity would be to investigate the temperature dependence of $\varepsilon$ of the microemulsions built up with various oils. An understanding of the mechanisms leading to the temperature dependence of the fluctuation dipole moment, and the development of a model for the dielectric permittivity of ionic microemulsions, is also important since it will provide insight into dielectric polarization and relaxation mechanisms of such systems.

The purpose of this part is to show how the controversy was resolved concering the main mechanism that provides the temperature dependence of the dielectric permittivity in ionic water-in-oil microemulsions far below the percolation region. It was shown that dielectric permittivity does not depend on the length of the oil chains and, therefore, aggregation of droplets cannot be responsible for the observed temperature dependence of $\varepsilon$ (5, 117).

In order to explain the temperature behavior of $\varepsilon$ far below the percolation onset a simple statistical model of polarization of nanometer-sized droplets containing negatively charged ions at the interface and positive counterions distributed in the droplet interior was developed (5, 117). In the framework of this model, when the values of the droplet size and the constant of dissociation of ionic surfactant are both small, an experimentally monitored temperature increase in $\varepsilon$ can be explained by the temperature growth of the mean-square fluctuation dipole moment of a droplet.

### 1. Effect of Oil Chain Length on the Microemulsions

The dependence of static permittivity of the microemulsions as a function of temperature and volume fraction is shown in Fig. 18. This behavior over the measured temperature region can be analyzed in two separate intervals: below the onset of a percolation region $T_{on}$ and above it. At the onset of percolation, the microemulsion starts to display a scaling behavior

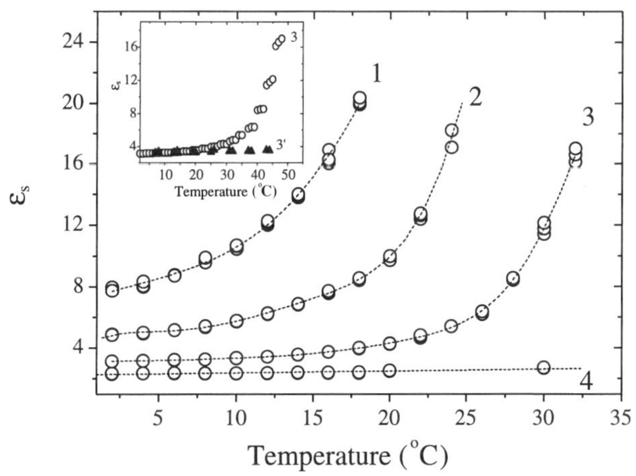


**Figure 18** Static dielectric permittivity vs. temperature for the AOT–water–decane microemulsions for various volume fractions $\phi$ of the dispersed phase: 0.39 (1); 0.26 (2); 0.13 (3); 0.043 (4). In the inset a similar plot for the AOT–water–decane (3, ○) and AOT–water–hexane (3′, $\sigma$) microemulsions for $\phi = 0.13$. The value $W =$ [water]/[AOT] is kept constant at 26.3 for all the microemulsions. The lines are drawn as a guide for the eye. (From Ref. 5. With permission from Elsevier Science B.V.)

of conductivity and dieletric permittivity due to droplet aggregation. For the most concentrated microemulsion, $\varphi = 0.38$, a temperature of the percolation onset of $T_{on} \approx 12°C$ was determined (98); this temperature has to be significantly higher for the more diluted microemulsions ($\varphi = 0.26, 0.13$, and 0.043).

In the percolation region ($T > T_{on}$) the main origin of the steep increase in permittivity as a function of temperature (Fig. 17) is the clustering of droplets (29). We estimated (97) that at the onset of the percolation region ($T \approx T_{on}$) on the droplets have a tendency to form small dynamic, weakly bound aggregates consisting of $10 \pm 5$ droplets. However, far below the percolation onset the ionic microemulsion could be assumed to consist of separated noninteracting water droplets. Thus, the weak increase in $\varepsilon$ far below the percolation onset might be explained either by the clustering of droplets or by another "unknown" mechanism. The mechanism should bring about the temperature increase in $\varepsilon$ in the system in which the droplets are considered to be separated from one another and noninteracting. The hypothesis that the interaction between droplets increases with the oil chain length (116) can be examined by measuring the permittivity of two microemulsions of identical droplet size but made from different oils. According to the clustering mechanism, the effect of droplet aggregation should be more pronounced (at a given temperature) in a microemulsion containing decane rather than hexane, thus resulting in higher values of $\varepsilon$ for the decane-containing microemulsion.

The dielectric measurements performed for the AOT/water/decane and AOT/water/hexane micro emulsions at the volume fraction of the dispersed phase of $\varphi = 0.13$ demonstrate the significant shift of the percolation region to the direction of high temperatures when the oil chain length decreased (Fig. 18) (118, 119). However, the values of $\varepsilon$ for both the microemulsions are the same at low temperatures, i.e., below the percolation onset. Thus, those results do not support the hypothesis that the clustering can be responsible for the temperature behavior of the static dielectric permittivity at $T < T_{on}$, and it must be the internal processes within a droplet that determine the behavior of the dielectric polarization in the system.

### 2. Analysis of the Dielectric Relaxation Behavior Far Below Percolation

In order to ascertain the origins and mechanisms responsible for the observed temperature behavior of the static dielectric permittivity, let us analyze the total dielectric relaxation behavior of ionic microemulsions. Dynamic aspects of the dielectric polarization can be taken into account by considering the macroscopic dipole correlation function (DCF) $\psi(t)$ in the time domain, or the complex dielectric permittivity $\varepsilon^*(\omega)$ in the frequency domain (38). In the microemulsions, DCF is associated with the relaxation of the entire induced macroscopic fluctuation dipole moment $\overline{M}(t)$, which is equal to the vector sum of all the dipole moments of the system (see Eq. 18).

It was shown (96) that ionic microemulsions exhibit a complex nonexponential behaviour that is strongly dependent on temperature. Far below the percolation onset ($T < T_{on}$) where the microemulsion has a structure of single spherical droplets, the main contribution in the relaxation mechanism comes from the fast relaxation processes with characteristic relaxation times distributed in the range from dozens of picoseconds to a few nanoseconds (96). These processes are inherent to the dynamics of the single droplet components and interfacial polarization. In the percolation region ($T > T_{on}$), transient clusters of droplets are formed as a result of attractive interactions between the droplets. The dielectric dispersion related to the clustering describes the collective dynamics in the system and has a *cooperative character*. The relaxation processes of this type are associated with the transfer of an excitation of a fluctuating dipole moment caused by the transport of electrical charges within the droplets and clusters as well as the rearrangement of the clusters. The characteristic time scale of these processes is within tens and hundreds of nanoseconds (96).

Let us consider the dynamics of microemulsions in the region $T < T_{on}$. The dependence of the macroscopic dipole correlation function $\psi(t)$ for the AOT/water decane microemulsion versus time at different temperatures is presented in Fig. 19. One can see that the dipole correlation function has a complex nonexponential behavior. In the first approximation it can be presented by the formal sum of $N$ Debye relaxation processes as

$$\psi(t) = \sum_{i=1}^{N} A_i \exp(-t/\tau_c) \qquad (47)$$

where $\tau_i$ are the relaxation times and $A_i$ are the amplitudes of the processes. We note that $\sum_{i=1}^{N} A_i = 1$.

The interpolation of the experimental data was carried out by a least-squares fitting procedure of the DCF values. The most appropriate number of elementary Debye processes involved is determined by the minimum of the standard deviation $\chi^2$. The dielectric

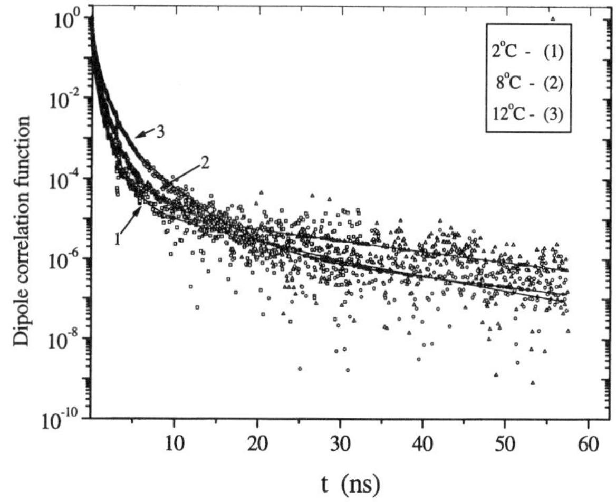


**Figure 19** Time dependence of the macroscopic dipole correlation functions for AOT–water–decane microemulsion: $\phi = 0.38$ at $T = 2°C$ (1); $T = 6°C$ (2); and $T = 10°C$ (3). (From Ref. 5. With permission from Elsevier Science B.V.)

response obtained reflects some properties inherent in single particle dynamics. The best-fit curves of the experimental data are reported in Fig. 19 (solid lines) together with the experimental data. In the temperature range 2°–12°C the best fitting gave the four elementary exponential relaxation processes (5).

The longest relaxation process with the characteristic time $\tau_1$ has a very small amplitude ($\sim 2\%$). An experimental value of $\tau_1$ is near 10 ns. The experimental relaxation times $\tau_2$ and $\tau_3$ are within the ranges 1.2–1.6 and 0.2–0.3 ns, respectively. The amplitudes $A_2$ and $A_3$ of the second and third relaxation processes increase with the temperature. Since the amplitudes $A_2$ and $A_3$ have a similar temperature behavior, it is reasonable to associate them with the same relaxation mechanism. The fourth process seems to be distributed around 50 ps. Its experimental amplitude and relaxation time decrease with the temperature.

In order to understand this complex relaxation behavior of the microemulsions, it is necessary to analyze dielectric information obtained from the various sources of the polarization. For a system containing more than two different phases the interfacial polarization mechanism has to be taken into account. Since the microemulsion is ionic, the dielectric relaxation contributions are related to the movement of surfactant counterions relative to the negatively charged droplet interface. A reorientation of AOT molecules, and of free and bound water molecules, should also be mentioned in the list of polarization mechanisms. In order to ascertain which mechanism can provide the experimental increase in dielectric permittivity, let us discuss the different contributions.

The contribution from interfacial polarization can be estimated by using one of the shell models (14, 23, 27). On the basis of the Maxwell-Wagner approach, these models describe the dielectric propreties of monodispersed suspension of coated spherical particles dispersed in a continuous medium. In all cases, the assumption that the particle radius is much larger than the Debye screening length is also made. Thus, the Laplace equation is applied in order to describe the electric potential in all three phases (core, shell, and continuous medium). If the dielectric constants and electrical conductivities of the phases are known, the shell models enable us to calculate the interfacial polarization. In particular, when the electrical conductivity of the components is negligibly small, the shell models are reduced to the dielectric mixture models (14, 27, 114, 115). The numerical calculations of the contribution of interfacial polarization to the dielectric permittivity, performed on the basis of various shell models, give similar results for all the models (96, 117). For instance, in the case of the most concentrated microemulsion, the increment of the dielectric permittivity associated with interfacial polarization, determined with the help of various shell models, ranges between 3.4 and 3.7. Note that since the polarization of both water and surfactant is inversely proportional to $T$, the interfacial polarization will also provide a weak temperature behavior of permittivity which is inversely proportional to $T$, and thus does not explain the monitored increase in permittivity.

In AOT microemulsions, where the aqueous core of the droplets also contains counterions, a considerable part of the dielectric response to the applied fields originates from the redistribution of the counterions. As mentioned in Sec. II, the counterions near th charged surface can be distributed between the Stern layer and the Gouy–Chapman diffuse double layer (28–31). The distribution of counterions is essentially determined by their concentration and the geometry of the water core. Thus, for very large droplets the diffuse double layer peters out and the polarization can be described by the Schwarz model (32). However, as already mentioned, this approach is more relevant to the dielectric behavior of emulsions than to that of microemulsions.

In the opposite case of very small droplets, AOT-hydrated micelles can be considered. The high-frequency dielectric response of very small AOT reverse micelles has been analyzed (118, 119) at a molar ratio of water to surfactant of $W < 10$. The avrage radius $R_w$ of the water core is related to $W$ by the semiempirical relation $R_w = (1.25\ W + 2.7)$ Å (13, 36). For almost dehydrated reverse micelles $R_w < 5$ Å, one can expect that nearly all the counterions are bound in the surfactant layer structure and immobilized. The dynamics of such a dehydrated system with a characteristic relaxation time of a few nanoseconds has been described in terms of the rotational diffusion of the whole micelle, which represents a nearly rigid structure. On increasing $R_w$ to 15 Å ($W < 10$), an increasing number of AOT ion pairs can achieve sufficient mobility to contribute separately to the dielectric relaxation with a characteristic relaxation time of hundreds of picoseconds. The authors restricted their consideration to the case of small droplets where most of the water and counterions are considered to be bound. They associated the dynamics in the range of hundreds of picoseconds with the rotation of completely hydrated surfactant ion pairs and neglected all bulk diffusion effects of "free" counterions in the double layer. The high-frequency dynamics detected in Refs 35 and 36, with a characteristic relaxation time significantly shorter than 100 ps, were attributed to water relaxation. We note here that the dynamics associated with reorientation of water and surfactant molecules were independent of the droplet size. Therefore, we can assume that the distributed process, found from the fitting, with a characteristic relaxation time $\tau_4$, can be associated with the described mobility of bound water and AOT.

In the intermediate case ($W > 10$), the radius of the droplet is large enough to cause the water molecules to form a pool of free water (120). The rotational diffusion of whole droplets of our microemulsions is expected to be in the range 250–300 ns (96). We did not observe any relaxation with these characteristic times, perhaps because of the very small amplitude (<1%) of this process. One can expect that in the case of systems with intermediate droplet sizes there can be two contributions to polarization caused by counterions, one stemming from bound counterions in the Ster layer and another one from concentration polarization in the diffuse part of the double layer. It was found from counterion $^{23}$Na spin-relaxation measurements that, in the intermediate region of $10 < W < 70$, the diffusion motion of counterions was in fact three dimensional rather than two dimensional (121). In particular, in the case of $W = 26.3$ the magnitude of the relative $Na^+$ diffusion $D_S/D_0 \approx 0.2$, where $D_S$ is the diffusion coefficient in the surface layer of the microemulsion and $D_0$ is the $Na^+$ bulk diffusion coefficient. The orientation of the fluctuating dipole moment of the droplets caused by the applied electric field is determined by the diffusion of the counterions both along the inner surface of the droplets and along the radial direction of the double layer. The characteristic relaxation time corresponding to the surface diffusion is $\tau_s = R_W^2/2D_s$. By the same token, the characteristic relaxation time for the radical diffusion of counterions can be estimated as $\tau_0 = l_D^2/6D_0$, where $l_D$ is the Debye length. For our microemulsions we can set $l_D$ to be of the order of the radius of a water core $R_w = 37$ Å. Taking $D_0$ as in the bulk water ($2 \times 10^{-5}$ cm$^2$/s) and $D_S \approx 0.2D_0 = 0.4 \times 10^{-5}$ cm$^2$/s gives characteristic diffusion times of 12 and 1 ns for the surface and radial diffusion, respectively. These characteristic times agree with the observed relaxation times $\tau_1$ and $\tau_2$, respectively. Since the amplitude of the longest relaxation process $A_1$ is very small ($<2\%$) the process associated with the diffusion along the surface is not essential in the polarization. On the other hand, the amplitudes $A_2$ and $A_3$ both increase with the temperature. Thus, we assume that the three-dimensional counterion movement in the diffuse layer is responsible for the temperature rise of static permittivity at $T < T_{on}$. Hence, this diffusion movement yields the experimentally observed second and third relaxation processes.

### 3. Model of Dielectric Polarization of Ionic Microemulsions

#### *a. Static Permittivity of Ionic Microemulsions*

The dielectric permittivity of an ionic microemulsion can be calculated in a general way by treating it as a monodispersed system consisting of spherical water droplets dispersed in the oil medium. In this way the system is considered as a homogeneous specimen consisting of a number of charges and/or dipoles, each of which is described in terms of its displacement from the position of its lowest energy level. The permittivity $\varepsilon$ of the system can be derived (1, 5, 117) in terms of the dielectric polarization $P$ and/or the total electric dipole moment $M$ of some macroscopic volume $V$ in the presence of the macroscopic electric field $E$ as

$$\varepsilon - 1 = \frac{4\pi P}{E} = \frac{4\pi M}{VE} \tag{48}$$

For ionic microemulsions the total electric dipole moment $M$ can be represented as a sum of the two contributions. One is associated with the moment, $M_E$, due to displacements of mobile ions in the diffuse double layer, which follow the laws of statistical mechanics, and another with the moment, $M_{mix}$, resulting from all other displacements in the mixture (5, 117). As was discussed above, $M_{mix}$ is related to the contributions from various processes of polarization in the microemulsion, which is treated as a heterogeneous sytem of various components.

It was shown (5, 117) that the permittivity $\varepsilon$ can be described by the relationship:

$$\frac{(\varepsilon - \varepsilon_{mix})(2\varepsilon + \varepsilon_{mix})(2\varepsilon + \varepsilon_w)}{\varepsilon^2} = \frac{9\varphi\langle\mu^2\rangle}{R_d^3 k_B T} \tag{49}$$

where $\langle\mu^2\rangle$ is the mean-square dipole moment of a droplet, $\varphi$ is the volume fraction of the droplets, $R_d$ is the radius of the surfactant-coated water droplet, $T$ is the temperature, $k_B$ is the Boltzmann constant, and $\varepsilon_{mix}$ is the permittivity due to polarization of the heterogeneous system. Since each droplet consists of a water core surrounded by a surfactant layer in a continuous phase prepared from oil, the effect of the interfacial polarization can be accurately regarded by using a Maxwell-Wagner mixture formula [one-shell model (14)].

Equation (49) establishes a dependence of the permittivity of a microemulsion on the temperature $T$, volume fraction of droplets $\varphi$, and apparent dipole moment of a droplet $\mu_a = (\langle\mu^2\rangle)^{\frac{1}{2}}$.

#### *b. Fluctuating Dipole Moment of a Droplet*

For calculation of the mean-square dipole moment of a droplet, $\langle\mu^2\rangle$, the theoretical development is carried out within the framework of the following assumptions (5, 117):

1. The droplets are considered identical and the interaction between them is neglected.
2. A nanodroplet contains $N_a$ *surfactant* molecules, $N_s$ of which are dissociated. Due to electroneutrality the numbers of the negatively charged surfactant molecules, $N^-$, and the number of positively charged counterions, $N^+$, are equal, i.e., $N^+ = N^- = N_s$.
3. The ions are treated as point charges.
4. The average spatial distribution of counterions inside the droplets is continuous and governed by the Boltzmann distribution law.
5. All the negatively charged surfactant molecules are assumed to be located in the interface at the spherical plane of radius $R_w$, corresponding to the radius of the droplet water pool.

In the model a single droplet is described by the spherical coordinate system shown in Fig. 20. The dipole moment of a single droplet is given by

$$\mu = e\sum_{i=1}^{N_s}(\boldsymbol{r}_i^+ - \boldsymbol{r}_i^-) \tag{50}$$

where $\boldsymbol{r}_i^+$ and $\boldsymbol{r}_i^-$ are the radius vectors of the positively charged counterion and negatively charged surfactant head, respectively; $e$ is the magnitude of the ion charge.

The quantity of interest is the mean-square dipole moment $\langle\mu^2\rangle$ of a droplet. It can be expressed in terms of the mean-squared fluctuations of the dipole moment $\mu$ by

$$\langle\mu^2\rangle = \langle(\Delta\mu)^2\rangle + \langle\mu\rangle^2 \tag{51}$$

where $\langle(\Delta\mu^2)\rangle = \langle(\mu - \langle\mu\rangle)^2\rangle$. As mentioned above, the mean-square dipole moment of a droplet $\langle\mu^2\rangle$ is calculated in the equilibrium state in the absence of an electric field. In this case the calculation of the electric polarization is retained in the framework of the linear theory of the electric field one can assume a spherical symmetry of the distribution of charges within a droplet. That means $\langle\mu\rangle = 0$. Hence, the apparent dipole moment in the system has a fluctuation ntaure (5), i.e.,

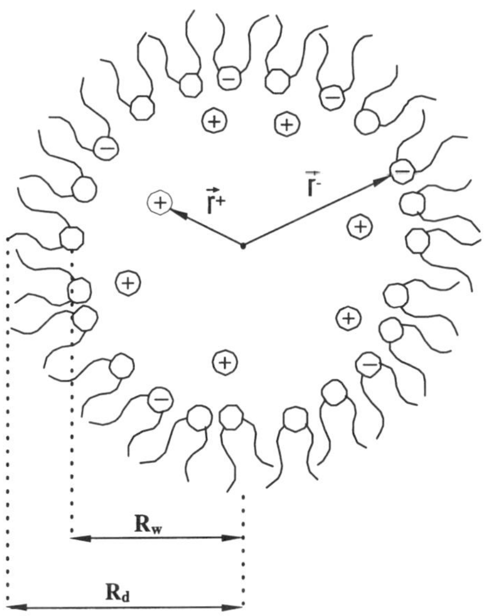


**Figure 20** Schematic picture of the spherical water–surfactant droplet. The reference point is chosen at the center of the droplet. The $i$th ion–counterion pair is represented by radius vectors of ion and counterion, $\boldsymbol{r}_i^+$ and $\boldsymbol{r}_i^-$. (From Ref. 5. With permission from Elsevier Science B.V.)

$$\langle\mu^2\rangle \approx \langle(\Delta\mu)^2\rangle \tag{52}$$

In order to calculate the value of $\langle\mu^2\rangle$, we square the left- and right-hand sides of Eq. (50) and average the result by the ensemble of the realizations of random positions of ions. Retaining the main terms in the quadratic form, we then obtain:

$$\langle(\Delta\mu)^2 \approx e^2\sum_{i=1}^{N_s}\left[\langle(\boldsymbol{r}_i^+)^2\rangle + \langle(\boldsymbol{r}_i^-)^2\rangle\right] \tag{53}$$

A calculation of the terms $\langle(\boldsymbol{r}_i^+)^2\rangle$ and $\langle(\boldsymbol{r}_i^-)^2\rangle$ entering Eq. (53) can be performed by using the one-particle distribution functions $W_1^+(\boldsymbol{r}^+)$ and $W_1^-(\boldsymbol{r})$ that are proportional to the ion density:

$$W_1^+(\boldsymbol{r}^+) = W_1^+(|\boldsymbol{r}^+|) = \frac{1}{N_S}c(r) \tag{54}$$

$$W_1^-(\boldsymbol{r}^-) = \frac{1}{4\pi R_W^2}\delta(|\boldsymbol{r}^-| - R_W) \tag{55}$$

where $R_w$ is the radius of the water core, $c(r)$ is the density of the counterions at the distance $r = |\boldsymbol{r}^+|$ from the center of the droplet, and $N_s$ is the total number of the counterions in the droplet interior:

$$N_s = 4\pi\int_0^{R_W} r^2c(r)\mathrm{d}r \tag{56}$$

By taking into account Eqs (54) and (55), Eq. (53) reads

$$\langle(\Delta\mu)^2\rangle = e^2\left\{4\pi\int_0^{R_W} r^4c(r)\mathrm{d}r + N_S R_W^2\right\} \tag{57}$$

According to Eq. (52), relation (57) allows us to calculate the apparent dipole moment $\mu_a = (\langle(\Delta\mu)^2\rangle)^{\frac{1}{2}}$ of a droplet.

The distribution of the counterions in the droplet interior is assumed (27, 122, 123) to be governed by the Poisson–Boltzmann equation:

$$\Delta\Psi = -\frac{4\pi ec(r)}{\varepsilon_W} = -\frac{4\pi ec_0}{\varepsilon_W}\mathrm{e} - \frac{e(\Psi(r) - \Psi(0))}{k} \tag{58}$$

Where $\Psi$ is the electrostatic potential, and $\varepsilon_W$ is the dielectric permittivity of the water core. Here, the reference point is chosen at the center ($r = 0$) of the spherical droplet, where the counterion density is $c_0$ and the electric potential is $\Psi(0)$.

Equation (58) reads in the dimensionless form as

$$\psi'' + \frac{2}{x}\psi' = -e^{-\psi} \tag{59}$$

with the boundary conditions:

$$\psi(0) = 0 \tag{60}$$

and

$$\psi'(0) = 0 \tag{61}$$

where $\psi$ and $x$ are the dimensionless potential $\psi = e[\psi - \psi(0)]/k_B T$ with respect to the center and the dimensionless distance $x = r/l_D$, respectively. Here, the characteristic thickness of the counterion layer near the surface of the water core:

$$l_D = \left(\frac{\varepsilon_w k_B T}{4\pi e^2 c_0}\right)^{1/2} \tag{62}$$

is the Debye screening length.

The Poisson–Boltzmann equation [Eq. (58)] can be solved by numerical integration (5, 123) or by expansion of the potential $\psi$ in the radical coordinate $x$ (117, 122). The theoretical details of the calculations were published elsewhere (117).

It was shown (5, 117) that the total number of the counterions in the droplet interior $N_s$ can be presented in the following way:

$$N_s = 4\pi l_D^3 c_0 \sum_{j=0}^{\infty} \frac{a_j}{2j+3} x_R^{2j+3} \tag{63}$$

where $x_R = R_w/l_D$ and $a_j$ are coefficients of the logarithm of the power series in the Poisson–Boltzmann equation solution. Furthermore, using Eq. (62) and a result for the charge density $c(x)$, the relationship for the mean-square dipole moment $\langle\mu^2\rangle$ of a droplet can be written as follows (117):

$$\langle\mu^2\rangle = 4\varepsilon_w k_B T R_w^3 \sum_{j=0}^{\infty} \frac{a_j(j+2)}{(2j+3)(2j+5)} x_R^{2(j+1)} \tag{64}$$

In order to find the counterion density at the center of a droplet $c_0$ and entering Eq. (62) for the Debye length $l_D$, the counterion concentration $c(r)$ must be related to the dissociation of the surfactant molecules in the water core of the droplet. The dissociation of the surfactant molecules is described by the equilibrium relation (122, 123):

$$K_s(T) = c_0 e^{\psi\left(\frac{R_0}{D}\right)} \frac{N_0}{N_a - N_s} \tag{65}$$

where $N_a$ is the micelle aggregation number, $K_s$ is the equilibrium dissociation constant of the surfactant, and $\psi(x_R)$ is the dimensionless electrical potential near the surface of the water core (i.e., at $r = R_w$). Substituting $\psi(x_R)$ in Eq. (65) with the logarithm of a power series (124) $\psi(x) = -\ln\sum_{j=0}^{\infty} a_j x^{2j}$ it is possible to obtain:

$$K_S(T) = \frac{c_0 N_S}{N_a - N_S} \sum_{j=0}^{\infty} a_j x_R^{2j} \tag{66}$$

The system of coupled equations [Eqs (63), (64), and (66)], along with the recurrence formula for $a_j$ (119) and Eq. (62) for $l_D$, constitute the model describing the temperature and geometry dependence of the mean-square dipole moment of the droplet $\langle\mu^2\rangle$. Furthermore, by inserting the calculated values of the mean square dipole moment into Eq. (49), we can obtain the equation:

$$\frac{(\varepsilon - \varepsilon_{mix})(2\varepsilon + \varepsilon_{mix})(2\varepsilon + \varepsilon_w)}{\varepsilon^2} = 36\varphi\varepsilon_w \left(\frac{R_w}{R_d}\right)^3 \sum_{j=0}^{\infty} \frac{a_j(j+2)}{(2j+3)(2j+5)} x_R^{2(j+1)} \tag{67}$$

This enables us to calculate the values of the dielectric permittivity $\varepsilon$ of the system.

### c. *Approximate Relationships for the Fluctuation Dipole Moment of a Droplet and the Permittivity of Ionic Microemulsions*

An adequate approximate relationship for the calculation of the mean-square dipole moment $\langle\mu^2\rangle$ in the case of a small droplet and/or the small dissociation of surfactant ($R_w \leq l_D$) can be obtained as a first approximation, by taking into account the first term ($j = 0$) only, in the series of expressions, Eqs (64)–(66). In this approximation, by using the relationship $x_R = R_w/l_D$, and taking into account Eq. (62), we obtain for the mean-square dipole moment:

$$\langle\mu^2\rangle \approx \frac{8\varepsilon_w k_B T R_w^5}{15 l_D} = \frac{32}{15}\pi e^2 c_0 R_w^5 \tag{68}$$

An approximate relationship for the counterion density at the droplet center $c_0$ can be obtained by using Eqs (63) and (65) in the first approximation, which reads:

$$c_0 \approx \left(\frac{3K_s}{R_w A_s}\right)^{1/2} \tag{69}$$

where $A_s = 4\pi R_w^2/N_a$ is the average area (cm$^2$/molecule) on the surface of a water core associated with one surfactant molecule. After combining Eqs (68) and (69), the mean-square fluctuation dipole moment of a droplet becomes:

$$\langle \mu^2 \rangle \approx \frac{32\sqrt{3}}{15}\pi e^2 \left( \frac{R_w^9 K_s}{A_s} \right)^{1/2} \tag{70}$$

In order to obtain a tractable relationship for the dielectric permittivity $\varepsilon$, we can further simplify Eq. (49) by taking into account the relative magnitudes of $\varepsilon$, $\varepsilon_{mix}$, and $\varepsilon_w$. For $\varepsilon_w \approx 78$ and $\varepsilon \sim \varepsilon_{mix} \ll \varepsilon_w$, we approximate in Eq. (49) the term $(2\varepsilon + \varepsilon_w)$ by $\varepsilon_w$ and $(2\varepsilon + \varepsilon_{mix})$ by $3\varepsilon$; then, by substituting Eq. (70) into Eq. (49), we obtain:

$$\varepsilon \approx \frac{\varepsilon_{mix}}{1 - X} \tag{71}$$

where

$$X = \frac{32\sqrt{3}\pi\varphi e^2}{5R_d^3 k_B T \varepsilon_w} \left( \frac{R_w^9 K_s}{A_s} \right)^{1/2} \tag{72}$$

It is easy to show that, for small droplet concentrations, $X \ll 1$; thus, an approximate relationship for $\varepsilon$ is

$$\varepsilon \approx \varepsilon_{mix}(1 + X) \tag{73}$$

In order to explain the experimental temperature behavior of the dielectric permittivity of the system we have to consider the temperature behavior of the dissociation constant of the surfactant $K_s$, which has an Arrhenius behavior (108):

$$K_s(T) = K_0 \exp\left( -\frac{\Delta H}{k_B T} \right) \tag{74}$$

where $\Delta H$ is the apparent activation energy of dissociation of the surfactant in the water pool of a droplet, and $K_0$ is the pre-exponential factor. It is easy to show that the permittivity of microemulsions obtained from Eq. (71) or (73) for the Arrhenius behavior of the dissociation constant is the growing function of temperature in the temperature range $T < \frac{\Delta H}{2k_B}$. This is always fulfilled in the measured temperature interval for any reasonable value of the activation energy.

For numerical evaluations of the model we have to set the values of the parameters matching the studied systems. The value of the dielectric permittivity of water was assumed to be equal to that of bulk water at the corresponding temperature throughout all the calculations, i.e., $\varepsilon_w = 87.74 - 0.40008t + 9.398 \times 10^{-4}t^2 - 1.41 \times 10^{-6}t^3$ (83), where $t$ is the temperature in degrees Celsius. The value of 2 for the dielectric permittivity of decane was adopted in the present calculations. The effective value of 8.5 was used (125) for the dielectric permittivity of AOT. The value of $\varepsilon_{mix}$ was calculated by using the one-shell model (14). The aggregation number $N_a$ was estimated to be 244 molecules per droplet. The value of $A_s = 65$ Å$^2$ was adopted for the average area on the surface of the water core associated with one AOT molecule (115).

Note that the developed model can only be applied within the special ranges of the droplet radius and ionic dissociation of the surfactant. From one side, the droplets cannot be too small. The radius of the water core must be larger than 15 Å (water-to-surfactant ratio $W > 10$) to ensure that a core of "free" water exists (118). On the other hand, the droplets cannot be too large, since the applicability range of the solution Eq. (68) is restricted by the condition $R_w < 3.27 l_D$. This condition may also be expressed in terms of the strength of the electrolyte in the droplet interior $pK_s$ ($pK_s \equiv -\log K_s$) and/or by the degree of dissociation of surfactant $\alpha = N_s/N_a$ (5, 119). Regarding $K_s$ of the surfactant AOT, little is known and we did not find any reliable experimental data for it in the literature. Thus, $K_s$ can be considered as an adjustable parameter of the theory which can be calculated from the inverse problem, i.e., we can determine $K_s$ from a knowledge of the experimentally measured permittivity of the studied microemulsions. The equilibrium dissociation constant $K_s$ can be calculated by using Eqs (62), (66), and (67) or, in the case of small droplets and/or a low degree of surfactant dissociation, by the approximate equation (71).

Figure 21 compares the values of the experimental apparent dipole moment $\mu_a$ of the studied microemulsions, obtained from Eq. (49), together with the theoretical values obtained on the basis of Eq. (67). One can see that the apparent dipole moment $\mu_a = (\langle \mu^2 \rangle)^{1/2}$ of the microemulsions increases versus temperature. For all the microemulsions studied the magnitude of the dipole moments for various volume fractions $\varphi$ of the dispersed phase does not depend on $\varphi$ within a degree of accuracy better than 10%, which confirms the assumption of the model that droplets in the system can be considered as noninteracting for such concentrations of droplets.

As a final comment, let us briefly discuss the permittivity of the studied microemulsions. Figure 22 shows the temperature dependencies of the experimental permittivity and the results of the calculations on the basis of the developed model performed by using Eqs (67) and (71). The difference between the values of $\varepsilon$ obtained from these formulas can only be observed at high $\varphi$. The calculated values of $\varepsilon$ agree well with the experimental data in the region far below the onset

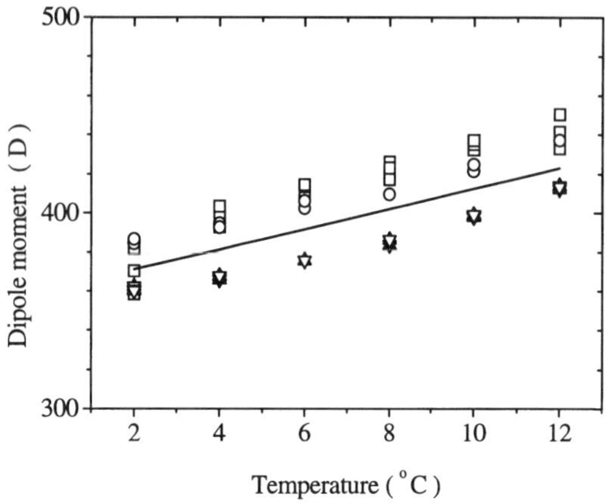


**Figure 21** Temperature dependence of experimental [and calculated on the basis of Eq. (64)] macroscopic apparent dipole moments of a droplet of the AOT/water/decane microemulsions. Experimental values for the dipole moment are shown for various volume fractions $\phi$ of the dispersed phase: 0.043 (□ ; 0.13 (○); 0.26 (Δ); and 0.39 (∇). Calculated values are shown by the solid line. (From Ref. 5. With permission from Elsevier Science B.V.)

of percolation ($T < T_{on)}$, where the assumptions of the model are fulfilled. At temperatures close to the percolation onset $T_{on}$ and beyond it, deviations in the theoretical values from experimental data are observed. These deviations indicate the structural changes in the system that appear at percolation.

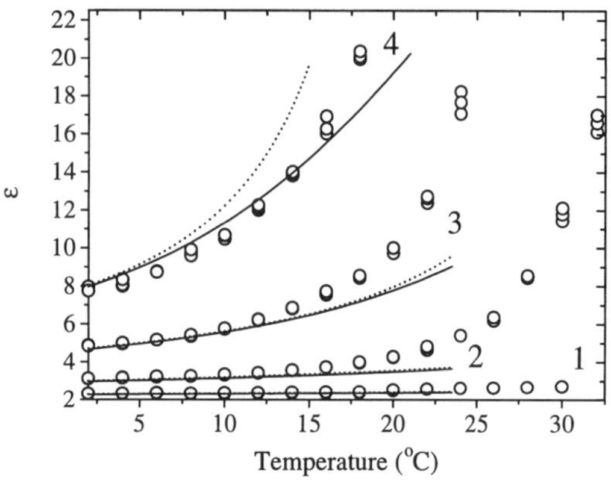


**Figure 22** Experimental (○) and calculated static dielectric permittivity vs. temperature for the AOT–water–decane microemulsions for various volume fractions $\phi$ of the dispersed phase: 0.39 (curve 1); 0.26 (curve 2); 0.13 (curve 3); 0.043 (curve 4). The calculations were performed by using the formulas: Eqs. (67) and (71) (dashed line). (From Ref. 117. With permission from American Physical Society.)

## B. Dielectric Properties of Ionic Microemulsions at Percolation

A percolation phenomenon was found in ionic microemulsion droplets when the water fraction, the temperature, the pressure, the strength of the electric field, or the ratio of water to the surfactant was varied (95–98, 101). Basically, the percolation behavior is manifested by the rapid increase in electrical conductivity $\sigma$ and static dielectric permittivity $\varepsilon$ as the system approaches the percolation threshold (Fig. 17).

The dielectric-relaxation properties in sodium bis(2-ethylhexyl) sulfosuccinate (AOT)/water/decane microemulsion near the percolation temperature threshold have been investigated in a broad temperature region (96, 97, 107). It was found that the system exhibits a complex nonexponential relaxation behavior that is strongly temperature dependent. The time–decay behavior of the dipole correlation function of the system $\Psi(t)$ was deconvoluted into normal modes and represented as a sum of a few Kohlrausch–Williams–Watts (KWW) terms, $\exp[-(t/\tau_M)^V]$, each with characteristic macroscopic relaxation times, $\tau_M$, and stretched exponents, $v$, respectively (96). It was shown that, in the percolation region, transient clusters of a fractal nature are formed because of attractive interactions between droplets. An interpretation of the results was carried out in the framework of the dynamic percolation model (126). According to this model, near the percolation threshold, in addition *to the fast relaxation* related to the dynamics of droplet components, there are at least two much longer characteristic time scales. The *longest process* has characteristic relaxation times greater than a few microseconds and should be associated with the rearrangements of the typical percolation cluster. The temporal window of the intermediate process is a function of temperature. This *intermediate process* reflects the *cooperative relaxation phenomenon* associated with the transport of charge carriers along the percolation cluster (126–128).

For a description of the mechanism of cooperative relaxation, Klafter, Blumen, and Shlesinger (KBS) (129, 130) considered a transfer of the excitation of donor molecule to the acceptor molecule through many parallel channels in various condensed media. The KBS theory might be modified for describing the process of the charge transfer in colliding droplets forming a cluster and giving rise to the relaxation of the entire fluctuation dipole moment. The normalized decay function $\alpha(t)$ in the microemulsions is associated with the relaxation of the entire induced macroscopic fluctuation dipole moment $\overline{M}(t)$ of the sample of unit volume, which is equal to the vector sum of all the fluctuation dipole moments of droplets.

The relaxation of the fluctuational dipole moment of a droplet is related to the transfer of the excessive charge (excitation) within two colliding droplets from a charged droplet (donor in the KBS model) to a neutral droplet (acceptor). The theoretical details of the model has been published elsewhere (97). This model of cooperative relaxation can be applied to fractal media such as the ionic microemulsion represented in the percolation region.

In the framework of the theory of cooperative relaxation in fractal media it is shown (97) that the macroscopic dipole correlation function $\psi(t)$ of the system is given by

$$\Psi(t) = \exp\left\{\Gamma\left[g\left(\frac{t}{\tau}\right)N, k, v\right]\left(\frac{t}{\tau}\right)^{v} + B(v)\left(\frac{t}{\tau}\right)\right\} \qquad (75)$$

where the coefficient $\Gamma[g(t/\tau), N, k, v]$ depends on the microscopic relaxation function $g(t/\tau)$ describing the elementary act of a charge transfer along the percolation cluster on the scaling parameters $k$, characterizing the type of the fractal similarity, and on the number of stages of self-similarity of the clusters $N$ (Fig. 23). The $\tau$ is the microscale relaxation time describing the charge transfer between two neighboring droplets. The coefficient $B(v)$ is a correction for the KWW function at large times. The parameter $v$ in Eq. (75) characterizes the cooperative dynamics and structure of the fractal clusters. The relationship between the exponent $v$ and the fractal dimension $D_f$ is given by $D_f = 3v$ (97).

As noted above, the dynamical processes in microemulsions can be classified into three types. The first

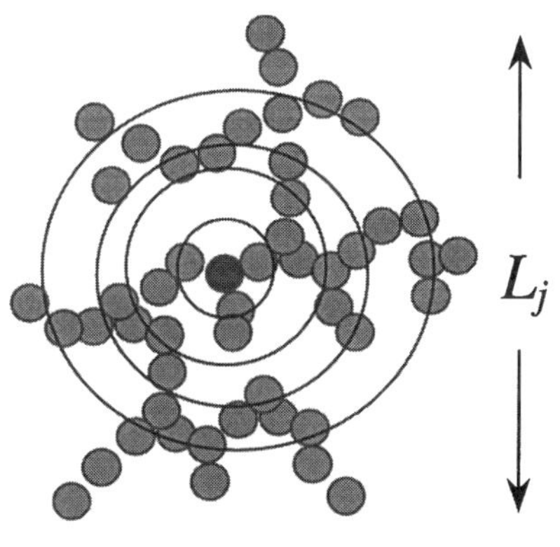


**Figure 23** Schematic picture of the excitation transfer via parallel relaxation channels in the fractal cluster of droplets in the percolating microemulsions.

type of relaxation process reflects characteristics inherent to the dynamics of the single-droplet components. The second and third types of relaxation processes characterize the collective dynamics in the system and have a cooperative nature.

A detailed analysis and estimations of the relaxation-time values show (5, 96) the following hierarchy of processes on the time scale: the relaxation processes of the first type $\tau_1$ are the fastest, with an order of hundreds of picoseconds, when compared with the time $\tau_c$, needed to "explore" the cluster and with the rearrangement time, $\tau_R$. The rearrangements occur on timescales of microseconds (128) and are considered the slowest process. The intermediate process ($\tau_1 < \tau_c < \tau_R$), relating to the cooperative transport of charge carriers along the clusters, has a temporal window depending on temperature. The minimal time boundary is of the order of hundreds of picoseconds, whereas the maximal time boundary has a value of tens of nanoseconds at the beginning of the percolation region and reaches 700 ns at $T_p$. All these contribute to a complex behavior of the dielectric correlation function. We can suppose, therefore, that near the percolation threshold the main contribution to the dynamics results from the cooperative effect related to the transfer of charge carriers along the percolation clusters, as given by Eq. (75).

The typical decay behavior of the dipole correlation function of the microemulsion in the percolation region is presented at Fig. 24. Figure 25 shows the temperature dependence of the effective relaxation time, $\tau_{eff}$, defined within the fractal parameters, and corresponding to the macroscopic relaxation time $\tau_M$ of the KWW model. In the percolation threshold $T_p$, the $\tau_{eff}$ exhibits a maximum and reflects the well-known *critical slowing down* effect (131).

The stretched exponent $v$ depends essentially on the temperature (Fig. 26). At 14°C, $v$ has a value of 0.5. However, when the temperature approaches the percolation threshold $T_p = 27$°C, $v$ reaches its maximum value of 0.8, with an error margin of less than 0.1. Such rapid decay of the KWW function at the percolation threshold reflects the increase of the cooperative effect of the relaxation in the system. At temperatures above the $T_p$, the value of the stretched exponent $v$ decreases, and indicates that the relaxation slows down in the interval 28°–34°C. At temperatures above 34°C, the increase in $v$ with the rise in temperature suggests that the system undergoes a structural modification. Such a change implies a transformation from an $L_2$ phase to lamellar or bicontinuous phases (132, 133). On the other hand, the temperature beha-

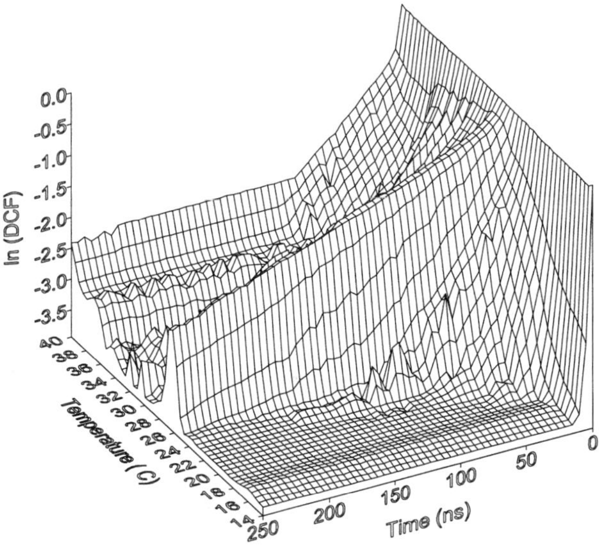


**Figure 24** Three-dimensional plots of time and temperature dependence of the macroscopic dipole correlation function for the AOT–water-decane microemulsion. (From Ref. 97. With permission from American Physical Society.)

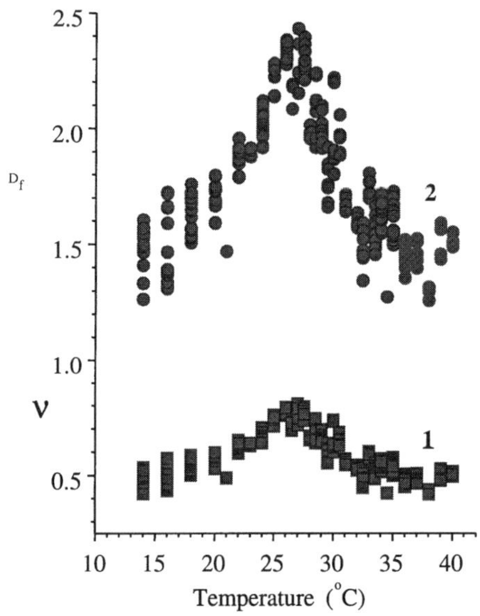


**Figure 26** Temperature dependence of the stretching parameter $v$ (□) and fractal dimension $D_f$ (O). (From Ref. 97. With permission from American Physical Society.)

vior of the fractal dimension $D_f$ (Fig. 26) shows that below the percolation threshold $D_f < 2$. This corresponds to a system of small clusters dispersed in space and can be described by the model of unbounded fractal sets with a $D_f$ of less than 2 (134, 135).

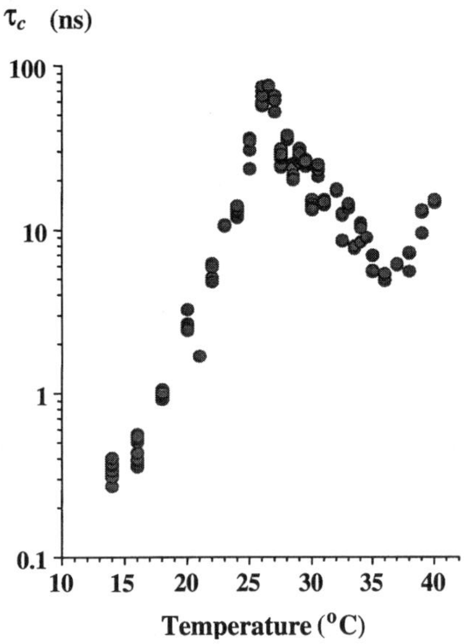


**Figure 25** Temperature dependence of the macroscopic effective relaxation time $\tau_{eff}$. (From Ref. 97. With permission from American Physical Society.)

At the percolation threshold $D_f = 2.4 \pm 0.2$, satisfactorily concurring with the literature value of 2.5 (136). Above the percolation threshold $D_f$ decreases, which can be explained by reorganizations of the system with corresponding structural changes. A structural modification of the system at temperatures above 34°C and the appearance of more prolonged and/or ordered regions in the microemulsion leads to a new observable increase in $D_f$.

It was shown that the effective length of the clusters increases sharply (97, 136) and diverges in the percolation threshold in accordance with the percolation scaling law $L_N \sim (T - T_p) - v_g$ where $v_g$ is the geometrical exponent (131). However, dispersion of the data obtained from the fitting does not enable one to estimate precisely a critical exponent $v_g$ of this growth.

The typical number of droplets $S$ in the aggregates may be estimated according to the relationship given by

$$S \sim (L_N/d_{drop})^{1/D_f} \tag{76}$$

where $d_{drop}$ is the diameter of the surfactant-coated water droplet, estimated to be 100 Å. The temperature dependence of the number of droplets in the typical fractal cluster $S$ is presented in Fig. 27.

Analysis of the temperature behavior of the calculated parameters shows that at the onset of the percolation region the droplets have a tendency to form

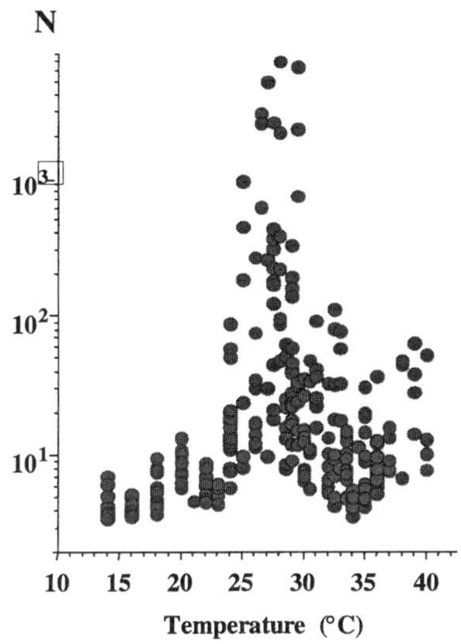


**Figure 27** Temperature dependence of the number of droplets in a typical percolation cluster. (From Ref. 97. With permission from American Physical Society.)

small dynamic aggregates consisting of $10 \pm 5$ droplets that are weakly bound. The characteristic length of such aggregates changes in the interval $L_N \sim 600$–1000 Å. The fractal dimension at these temperatures has a value of less than 2, indicating that the aggregates are surrounded by empty spaces, i.e., separated from another. We note that each of these aggregates participates in the relaxation as independent objects with no correlation between them. At the percolation threshold, the aggregates tend to form a large percolation cluster, which participates in the cooperative relaxation as a whole object.

The fractal model described above enables us to relate the characteristics of a total cooperative macroscopic relaxation function with the parameters of the fractal medium (fractal dimension $D_f$ and the degree of the development of fractality which is expressed by the number of self-similarity stages $N$). Herewith, the fractal medium has been represented by the single "effective" self-similar geometric structure. The recursive model neither describes the cluster polydispersity (cluster size distribution), nor the relaxation of the individual cluster. Therefore, it is expedient to use the complementary model in order to estimate cluster statistics and dynamics in more detail. Such a model might be developed in the framework of the general statistical description. In this description, the DCF of a complex system is represented as a result of averaging of the dynamic relaxation functions corresponding to the different random *states* of the system (99, 137):

$$\Psi(t, s_m) = \int G(t, s) W(s, s_m) \mathrm{d}s \tag{77}$$

where $W(s, s_m)$ is the state probability density distribution function for the state variable $s$, depending on the effective maximal scale $s_m$, and $G(t, s)$ is the dynamic relaxation function in state $s$. The integration in Eq. (77) is performed over all possible states of the system.

For the percolation case, the function $W(s, s_m)$ was taken in the form of the "generalized exponential distribution" or "main statistical distribution" as follows (138–140):

$$W(s) \sim s^{\gamma} \exp[(s/s_m)]^{\eta} \tag{78}$$

where $\gamma$ is the polydispersity index, $s_m \gg 1$ is the cut-off cluster size, and $\eta > 0$ is the cut-off rate index. The numerical values of these parameters depend on the distance from the percolation threshold. Above the percolation threshold, i.e., at $u > 0$, the function $W(s)$ describes the size distribution of all clusters except for the infinite cluster. In the continuum limit, the function $W(s)$ is the cluster size probability density distribution function. Herewith, the state variable $s$ is just the cluster size (number of monomers in the cluster). The form of $G(t, s)$ was chosen with the help of the hypothesis of dynamical scaling as $G(t, s) = \exp[-t/\tau(s)]$, where $\tau(s) = \tau_1 \cdot s^{\alpha}$, $\tau_1$ is the minimal time associated with the monomer relaxation, and the parameter $\alpha$ is the scaling index establishing a correspondence between the size of a cluster and its relaxation time. The tentative application (140) of the statistical fractal model validated the usefulness of the statistical approach. The treatment performed on the experimental data suggested that the polydispersity index, the dynamic scaling exponent, and the $\sigma$ exponent are not universal quantities and that they might depend on the specific interactions in the system. Further development of that approach will give an effective tool for investigation of the polydispersity index, the dynamic scaling exponent, and other structural and dynamical parameters of microemulsions in percolation.

## C. Dielectric Properties of Nonionic Microemulsions

Analysis of the dynamics on short time scales can unravel the nature of the relaxation processes and provide information about the partition of the water and alcohol between bulk and interface. It was shown

(141–143) that nonionic microemulsions usually have a very complicated dielectric behavior. From the Cole–Cole plot, for example, Fig. 28, it can be seen that the complete dielectric spectrum cannot be described by a single relaxation process. The wide nonuniform time window used (38) allows one to deconvolute the relaxation spectrum to distinguish three or four relaxation processes (dependent on water content and temperature) distributed in the interval between $10^{-11}$ and $10^{-9}$ s. Given the complexity of the chemistry of microemulsions, a precise interpretation of the dielectric relaxation mode is difficult. Dielectric relaxation contributions in such systems are expected to be related to bound and free water, alcohol molecules, and active dipole groups belonging to the surfactant molecules and associates.

### 1. Evaluation of the Amount of Alcohol at the Interface

Two main dielectric relaxation processes can be considered in nonionic microemulsions. The longest one, with characteristic time changes between 1 and 2 ns down to hundreds of picoseconds with increasing water content in the system, and the shorter relaxation rpocess that is characterized by time changes from 100 ps down to dozens of picoseconds. The long-term behavior has been correlated with the self-associated state of the alcohol and associated with the break-up of linear alcohol complexes. The relaxation time is related to relaxation of the alcohol in microemulsions at various water contents and can be juxtaposed with that of oil–alcohol mixtures for various concentrations of alcohol in the system.

The two relaxation processes were evaluated both for the alcohol/dodecane mixtures (Fig. 29) and for the microemulsions (Fig. 30). In order to understand the way to calculate the amount of alcohol at the interface and in the oil phase, let us first analyze the alcohol/dodecane binary mixture.

#### a. *The Alcohol/Dodecane Binary Mixture*

In such mixtures the long relaxation time $\tau_1$ increases with increasing concentration of butanol in dodecane (Fig. 29a). In other words, diluting butanol with oil leads to a long relaxation-time reduction. The decrease in $\tau_1$ with the oil concentration is in good agreement with the literature (144–146). On the other hand, the short relaxation time $\tau_2$ is hardly affected by the presence of dodecane. The same behavior between the alcohol concentration and the relaxation times was detected in a pentanol/dodecane mixture (Fig. 29b).

Aliphatic alcohols are known to be strongly associated polar liquids. Alcohol molecules in the liquid

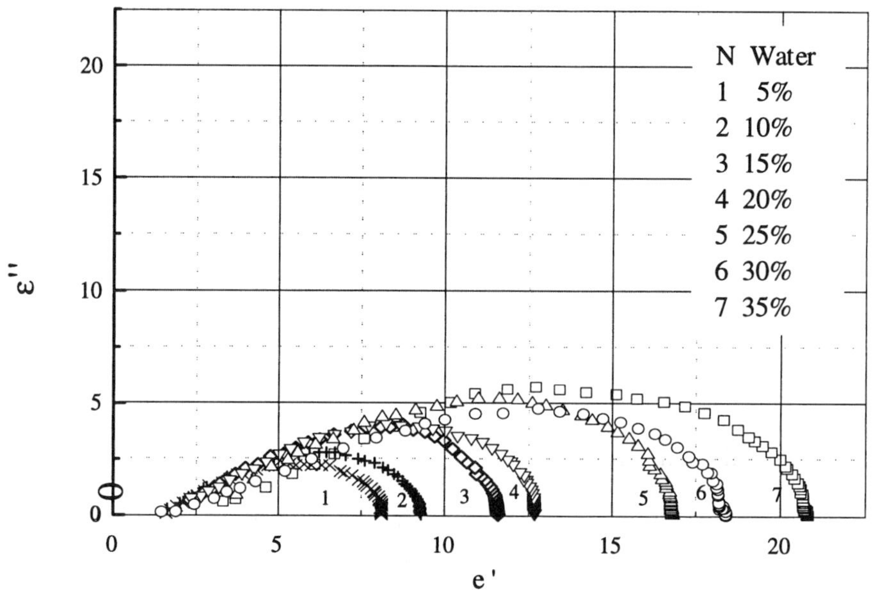


**Figure 28** Cole–Cole diagram for seven mixtures of W/O dodecane/butanol/Brij 97/water microemulsions at 20°C for various water contents: 5% (1); 10% (2); 15% (3); 20% (4); 25% (5); 30% (6); 35% (7). (From Ref. 143. With permission from Elsevier Science.)

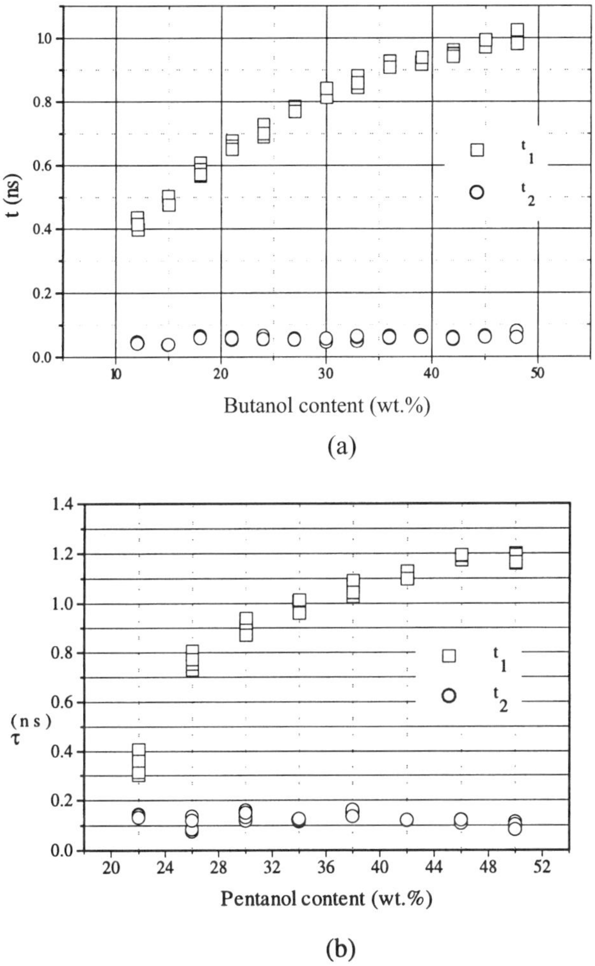


**Figure 29** Dielectric relaxation times of dodecane/alcohol mixtures at 10°C for different alcohol contents (a) dodecane/butanol; (b) dodecane/pentanol. (From Ref. 141. With permission from Elsevier Science.)

phase are linked into oligomer chains by intermolecular hydrogen bonds (H-bonds) (147–149). The alcohol molecules are able to rotate around the H-bonds, resulting in the associated species existing mainly in the convoluted form. In the general case, rotation around the H-bonds is partially hindered, and the longer the hydrocarbon radical, the stronger is the hindering.

As the nonpolar dodecane is dispersed in the alcohol medium, the alcohol chains are disrupted and broken into smaller clusters. As more dodecane is introduced into the mixture, alcohol complexes become smaller until the alcohol appears as monomers, i.e., nonclustered alcohol (144–146).

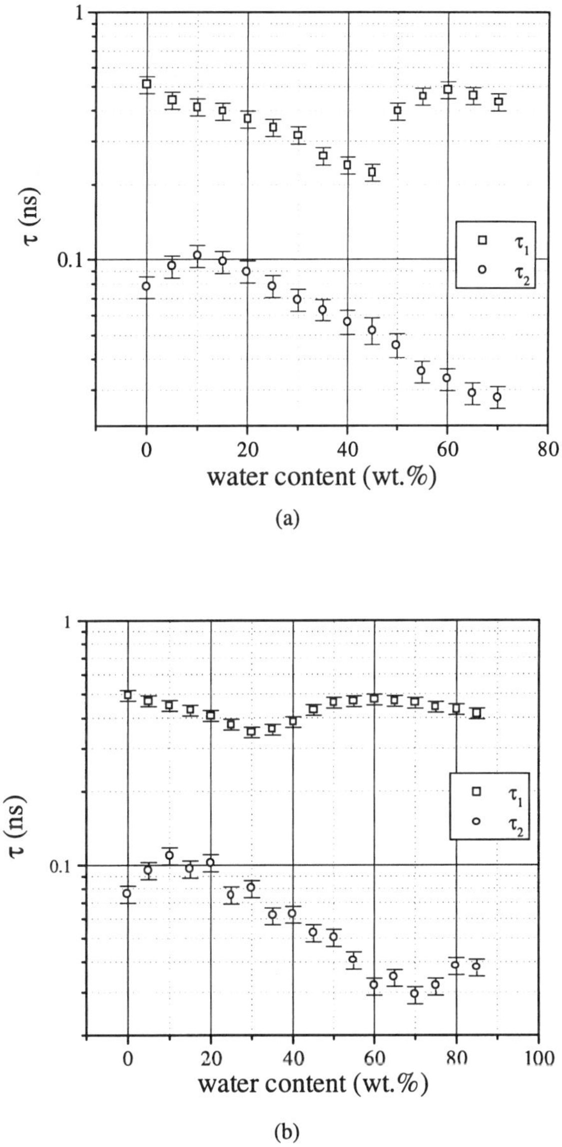


**Figure 30** Dielectric relaxation times at 10°C of (a) dodecane/butanol/Brij 97/water microemulsion (system 1) and (b) dodecane/pentanol/$C_{12}(EO)_8$/water microemulsions (system 2), for different water contents. (From Ref. 141. With permission from Elsevier Science.)

The long dielectric relaxation process (presented here by $\tau_1$) has been correlated with the self-associated state of the alcohols. It is caused by a cooperative relaxation of the average dipole moment of the whole alcohol aggregate in the pure alcohol or in the alcohol/dodecane mixture (148). Therefore, the long relaxation process will exist only in the presence of alcohol aggregates. It was shown (149) that the long relaxation time of the alcohol is related to the average number of monomers in the alcohol cluster $n$, and to the relaxation time of the alcohol monomers' isotropic rotation $\tau'$ by the relation:

$$\tau = 6n\tau' \tag{79}$$

As a result of adding the hydrocarbon, smaller units of alcohol aggregate are created, and according to Eq. (79) this will lead to a decrease of the long relaxation time $\tau_1$.

The short dielectric relaxation process (presented here by $\tau_2$) is associated with the anisotropic motion of the monomer alcohol species in a chain cluster (149). In microemulsions, the short process is the superposition of several dielectric relaxation processes, which have similar relaxation times such as movement or rotation of the alcohol monomers, hydrate water, and surfactant polar head groups. The short relaxation time is barely affected by the alcohol concentration in the mixture since it is less sensitive to the aggregation process.

*b. The Microemulsion Systems*

As previously discussed, the alcohol in the studied microemulsion is located in the oil phase or in the water aggregate's interfacial film. Thus, the continuous phase of the microemulsion is actually an alcohol/dodecane binary mixture (neglecting a very small amount of water). Since the alcohol located at the interface exists as monomers, the long relaxation time observed in the microemulsion is associated with an alcohol aggregate that is only present in the continuous phase with the dodecane. In Fig. 30 it can be seen, both for Brij 97 and the $C_{12}$ $(EO)_8$ systems, that in the range of the $L_2$ phase (0–50% of water), as water is added to the microemulsion, the long relaxation time $\tau_1$ decreases. This phenomenon can be explained as follows: as more water is introduced into the system some of the alcohol (butanol or pentanol) migrates from the continuous oil phase to the interface. As mentioned above, when the alcohol concentration in the dodecane phase decreases, the alcohol clusters became smaller, reflected by a reduction of the long relaxation time. The same phenomenon of decreasing the long relaxation time with increasing water content was observed in systems containing Brij 97 with butanol as cosurfactant (Fig. 30a) and $C_{12}$ $(EO)_8$ with pentanol as cosurfactant (Fig. 30b).

The average association number $n$ of the alcohol molecules is determined by the alcohol concentration in the oil. The dependence of the association number $n$ of butanol and pentanol as a function of the alcohol concentration in binary mixtures is presented in Fig. 31. The association number was calculated from the long dielectric relaxation time of the binary mixture using Eq. (77) and taking $\tau'$ as 7.1 ps for butanol and 10.4 ps for pentanol (146).

The oil phase of the microemulsion can be considered as a binary mixture of dodecane and alcohol. In this case, by using Eq. (79), it is possible to calculate the association number of the alcohol molecules in the oil phase from the long relaxation time of the microemulsion (Fig. 32). For the $C_{12}(EO)_8$/water/dodecane/pentanol microemulsions the association number varies in the range from 4 to 6 (Fig. 32a). From Fig. 31 it can be seen that these $n$ values are in the range of 21 to 24% of pentanol in dodecane. Thus, we estimate that the pentanol concentration in the oil phase of $C_{12}(EO)_8$/water/dodecane/pentanol microemulsion decreases just slightly in the range (21–24%) when water is added. Since the total amount of each component in the microemulsion is known, it is easy to calculate the partition of the alcohol between the oil phase and the interface. From such calculations we can evaluate that, for the microemulsions with high water content (60%), 70% of the pentanol is located at the interface and 30% is dissolved in the oil. As mentioned before, this ratio changes slightly when the water content in the microemulsion is changed.

Adding water to the microemulsion can in some cases cause an inversion from the $L_2$ to the $L_1$ phase.

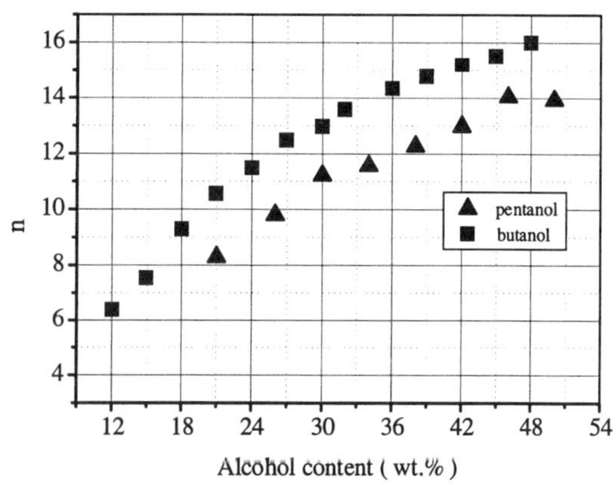


**Figure 31** Association number $n$ at 10°C, of alcohol molecules vs. its concentration in alcohol/dodecane mixtures. The association number was calculated from the long dielectric relaxation time of the binary mixture using Eq. (77). (■) Butanol/dodecane; (▲) pentanol/dodecane. (From Ref. 141. With permission from Elsevier Science.)

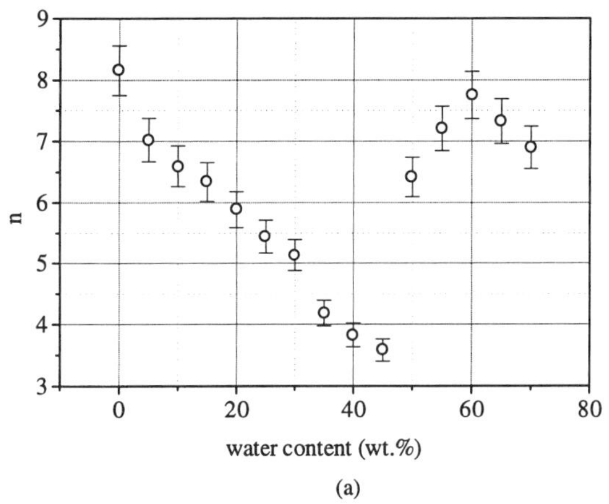


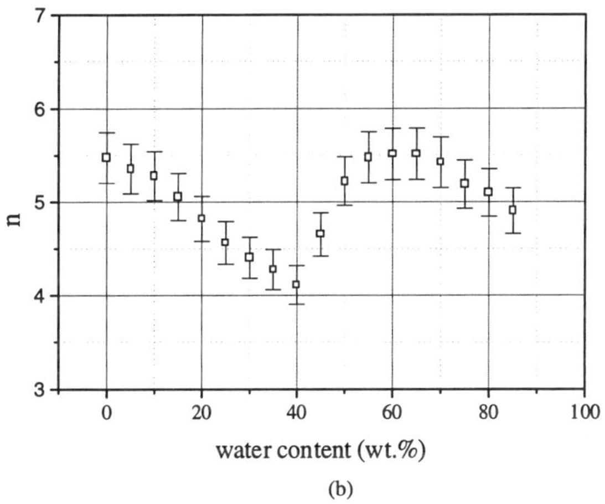


**Figure 32** Association number $n$ at 10°C, of alcohol molecules in the oil phase of the microemulsions with different water contents. The association number was calculated from the long dielectric relaxation time of microemulsions. (a) (System 1)dodecane/butanol/Brij 97/water microemulsions; (b) (system 2) dodecane/pentanol/$C_{12}(EO)_8$/water microemulsions. (From Ref. 141. With permission from Elsevier Science.)

In our system the inversion was found to be in the range 45–60 wt % water. In Fig. 32 it can be seen that the dielectric long relaxation process $\tau_1$ and the association number $n$ increase abruptly at 50% water concentration. This "kink" is more evident in the Brij 97–butanol system. It is reasonable to believe that this kink, which occurs in the inversion area, is due to expulsion of part of the alcohol from the interface to the oil phase. As it happens, the concentration of the alcohol in the oil phase increases and the long relaxation time increases. Hence, the inversion can be detected by the TDS method.

## 2. Evaluation of the Amount of Bound Water in Microemulsions

Information about the bound water fraction in some colloid systems, silica gels, and biological systems is usually inferred on the basis of the frequency- and time-domain DS measurements from the analysis of the dielectric decrements or the relaxation times (64, 150–152). However, the nonionic microemulsions are characterized by a broad relaxation spectrum as can be seen from the Cole–Cole plot (Fig. 33). Thus, these dielectric methods fail because of the difficulties of deconvoluting the relaxation processes associated with the relaxations of bound water and surfactant occurring in the same frequency window.

In microwave dielectric measurements (> 30 GHz) the dielectric permittivity and dielectric losses for bound and free water show significantly different magnitudes. Thus, in measurements at high microwave frequencies the contribution from bound water in the dielectric losses will be negligibly small, and the contribution from the free water fraction can be found. In contrast to the above-mentioned procedures used for calculation of bound water from the relaxation spectrum analysis, this approach will not involve analyses of overlapping relaxation processes and can thus easily be applied to microemulsions having a complex relaxation spectrum.

It is necessary to choose a model that will adequately describe the dielectric properties of the microemulsion. Most of the existing theories (106, 153) operate with a system consisting of well-defined geometrical structures such as spherical or ellipsoidal

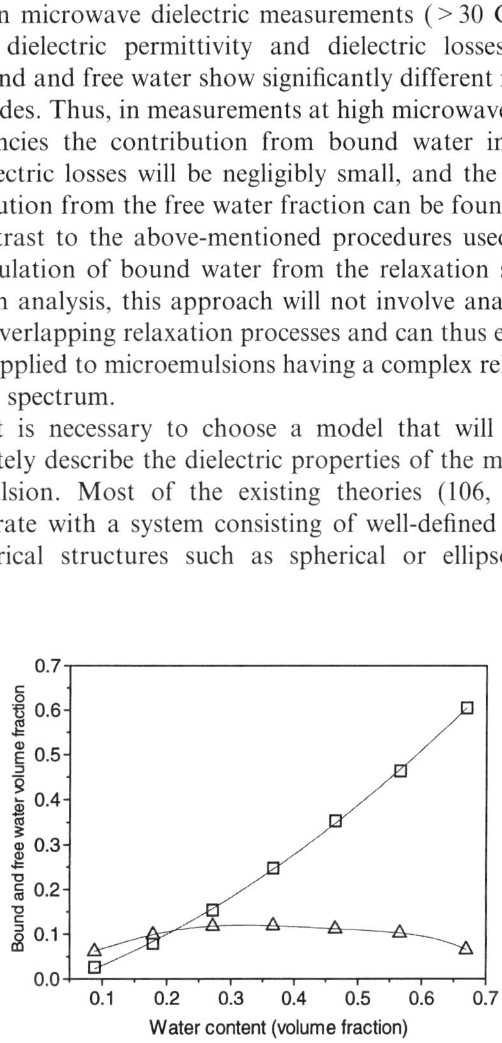


**Figure 33** Relative amounts of free (□) and bound (△) water contents of the investigated microemulsion vs. total water content. (From Ref. 141. With permission from Elsevier Science.)

droplets, cylindrical or flat lamellars, etc. Since it was not known how the morphology of the microemulsion system changes with the addition of water, an approach suggested by Kraszewski and coworkers (154, 155) for the prediction of the permittivity mixture was applied. In this case, the relationship relating complex permittivity of the microemulsion with those of its components and their volume fractions may be written:

$$\sqrt{\varepsilon^*} = \varphi_m\sqrt{\varepsilon_m^*} + \varphi_f\sqrt{\varepsilon_f^*} + \varphi_b\sqrt{\varepsilon_b^*} \tag{80}$$

where $\varphi_m$ denotes the volume fraction of the system occupied by the microemulsion without added water (i.e., "dry" microemulsion consisting of dodecane, butanol, and Brij 97), and $\varphi_f$ and $\varphi_b$ are the volume fractions of free and bound water, respectively. The volume fractions are calculated per unit volume of microemulsion. The complex permittivities of the entire system, the microemulsion with no added water, and free and bound water, are represented by $\varepsilon^*$, $\varepsilon_m^*$, $\varepsilon_f^*$, and $\varepsilon_b^*$, respectively.

Since the volume fractions of free, $\varphi_f$, and bound, $\varphi_b$, water are both unknown, it is convenient to measure the dielectric permittivity in a frequency range where the dielectric loss of bound water may be safely neglected. The relaxation spectrum of free and bound water for our systems will safely satisfy this requirement at the measurement frequency of 75 GHz. In this case, the complex permittivity of the bound water is equal to its real part, i.e., $\varepsilon_b^* = \varepsilon_b' + i0$. For the sake of simplicity, calculations of the complex dielectric constant can be written in polar coordinates, $\varepsilon^* = R\exp(i\theta)$, where $R = (\varepsilon'^2 + \varepsilon''^2)^{1/2}$ and $\tan\theta = \varepsilon''/\varepsilon'$. Now $\exp(i\theta) = \cos\theta + i\sin\theta$; therefore, substituting the corresponding terms in polar coordinates in Eq. (80) for the "dry" microemulsion, free and bound water, respectively, then rearranging and separating the imaginary parts, one can obtain the following relationship for the calculation of the bound-water fraction (156):

$$\varphi_b = \frac{\sin\left(\frac{\theta_m}{2}\right)\sqrt{R_m} - \sin\left(\frac{\theta_e}{2}\right)\sqrt{R_e} - \sin\left(\frac{\theta_w}{2}\right)\sqrt{R_w} - \varphi_m\left[\sin\left(\frac{\theta_m}{2}\right)\sqrt{R_m} - \sin\left(\frac{\theta_f}{2}\right)\sqrt{R_f}\right]}{\sin\left(\frac{\theta_f}{2}\right)\sqrt{R_f}} \tag{81}$$

Figure 33 shows the volume fraction of free, $\varphi_f$, and bound water, $\varphi_b$, in the microemulsion, calculated from Eq. (81), as a function of the total water content, $\varphi_w$. As long as there are unoccupied binding sites at the interface on the surfactant head groups and on the cosurfactant molecules, the increase in $\varphi_w$ leads to an increase in $\varphi_b$. The bound-water fraction shows a maximum between 0.30 and 0.40 of the total water volume fraction, corresponding to 0.12 of bound-water volume fraction. Further addition of water dilutes the alcohol and surfactant, decreasing the number of active centers, which are able to absorb water molecules. Thus, the bound water fraction $\varphi_b$ decreases after reaching the maximum.

Figure 34 plots the volume ratio between bound water and the total water amount versus water contents. The fraction of bound water decreases with increasing total water content in the system. As more water is introduced into the system, the interface approaches saturation and it is less favorable for additional water to be bound. The dilution effect described above also leads to this decrease.

Water molecules can be bound to both ethylene oxide (EO) groups of Brij 97 and to the hydroxyl (OH) groups of butanol. Unfortunately, it is not known for this system how many molecules are bound to either alcohol or surfactant. Therefore, two curves are presented in Fig. 35. The curves demonstrate the number of bound water molecules calculated per EO group and per EO + OH group versus the total water content, respectively. As the water content increases, the number of water molecules bound to EO or EO + OH groups increases and reaaches a maximum at ~ 2.5 molecules for one curve and ~ 1.5 for the other. This increase means that the water-binding

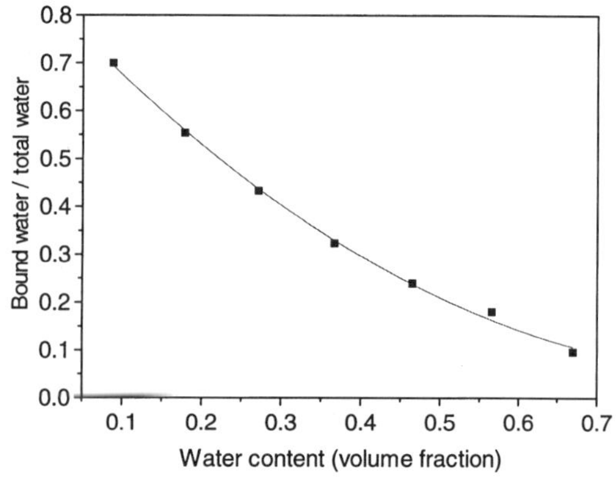


**Figure 34** Ratio of bound water-to-total water content (volume of bound water divided by volume of total water) vs. water content. (From Ref. 141. With permission from Elsevier Science.)

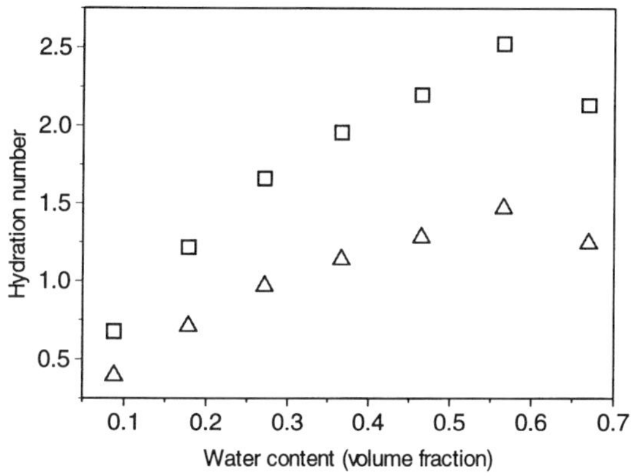


**Figure 35** Number of bound water molecules for the investigated microemulsion calculated per ethylene oxide (EO) group (□) and per EO + OH group (△) vs. the total water content. (From Ref. 141. With permission from Elsevier Science.)

centers do not become saturated until ~ 0.6 of the total water volume fraction is obtained. The decrease in the number of bound water molecules per EO or EO + OH group can be due to changes in the morphology of the system above 0.6 of the water volume fraction content. The number of bound water molecules per EO group at saturation obtained from the analysis of the dielectric properties, 2.5, is in good agreement with that obtained by differential scanning calorimetry (DSC) (157), 2.8, for a similar system. However, DSC measurements may be performed only after the samples have been cooled to a very low temperature, while the microwave dielectric method can be applied at any desired temperature.

In this section we have demonstrated the potential of time-domain dielectric spectroscopy in obtaining information about both the structure and dynamics of ionic and nonionic microemulsions on different temporal scales.

## V. NONEQUILIBRIUM COLLOIDAL SYSTEMS

Emulsions and suspensions of solid particles are common examples of colloidal systems that are not in the equilibrium state. As the systems destabilize and approach the equilibrium state, several processes will be involved. Typically, flocculation, sedimentation, and coalescence, etc., take place simultaneously at more or less well-defined rates, continuously changing the properties of the system.

Flocculation leads to strongly reduced interparticle distances. Also, the sedimentation process alters the particle distribution throughout the system, leading to a particle density gradient. The size distribution will change as a result of coalescence or coagulation.

There is a constant challenge for improved techniques in order to make accurate predictions on the colloidal stability of various sytems. In this section we demonstrate how dielectric spectroscopy can be applied as a technique to follow the breakdown of water-in-oil emulsions and to monitor the sedimentation of particle suspensions. Dielectric spectroscopy, combined with statistical test design and evaluation, seems to be an appropriate technique for the study of these problems. However, one should continue to seek satisfactory theoretical models for the dielectric properties of inhomogeneous systems.

### A. Dielectric Properties of Emulsions

We have restricted this presentation to involve only oil-continuous emulsions. The reason for excluding water-continuous systems is that the O/W emulsions are usually stabilized by means of ionic surfactants (or surfactant mixtures) and consequently the electric double-layer effects can be very large. The electrode polarization will normally also be very strong in many O/W systems. For the TDS technique water-in-oil emulsions stabilized by means of nonionic surfactants are very good model systems.

#### 1. Effect of Flocculation

Theoretical models for the dielectric properties of heterogeneous mixtures [for instance, Eq. (20), or extensions of this model] are commonly applied in order to explain or predict the dielectric behavior also of emulsions (106, 158). However, in the present theories a homogeneous distribution of the dispersed phase is required. This requirement is rarely fulfilled in a real emulsion system where the inherent instability makes the emulsions go through different stages on the way towards complete phase separation. Processes like sedimentation, flocculation, and coalescence continuously alter the state of the system (Fig. 36). These processes also influence the dielectric properties (159–162). Thus, the dielectric properties of one given sample may vary considerably over a period of time (160), depending on the emulsion rate.

The effect of flocculation on the dielectric properties of disperse systems is well documented, both when it comes to suspensions of solid particles and emulsions.

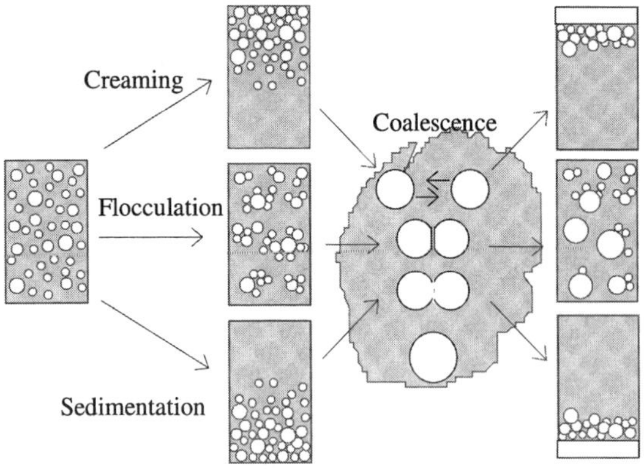


**Figure 36** Schematic view of some of destabilizing processes that may take place in an emulsion, eventually leading to complete phase separation. Similar processes also take place in other types of disperse systems.

**Table 1** Dielectric Parameters for a W/O Emulsion Containing 50 vol % water, Stabilized by 1% Berol 26. The Aqueous Phase Contains 5 wt% NaCl. Measurements Have Been Performed Both during Emulsification (Using Different Methods) and at Rest (Immediately after Emulsification).

| Device | Rotation speed (r.p.m.) | $\varepsilon_s$ During mixing | $\varepsilon_s$ At rest | $\varepsilon_\infty$ During mixing | $\varepsilon_\infty$ At rest |
|---|---|---|---|---|---|
| Laboratory mixer | $\approx 2500$ | 15.6 | 24.2 | 9.4 | 11.0 |
| | $(\approx 500)$ | 16.9 | 20.9 | 10.4 | 11.3 |
| Vortex mixer | | 15.4 | 23.9 | 11.0 | 12.9 |

*Source*: Ref. 164.

Genz *et al.* (163) found that the conductive and dielectric properties of suspensions of carbon black in mineral oil was highly dependent on shear. When a shear force was applied to the suspensions, flocculated aggregates were torn apart and a reduction in the permittivity levels was observed. When the shear stopped, the permittivity rose to previous levels. Thus, the reversibility of the shear-induced floc disintegration could be followed by means of dielectric spectroscopy.

Also, Hanai (11) exposed his systems to shear. In order to verify his theory on the dielectric properties of concentrated emulsions (161, 62), dielectric measurement on W/O emulsions were performed at rest and under influence of shear forces. At rest the static permittivities by far exceeded the values predicted from Eq. (20). However, when modestly high shear forces were applied, Hanai found good agreement between measured and predicted values. Also, in this case the reduction in permittivity upon shear was ascribed to a disintegration of flocs.

Table 1 gives the results from a simple experiment visualizing the effect of stirring on the dielectric properties of emulsified systems (164). It is seen that the difference in the static permittivity between the emulsions during stirring and at rest can be as high as 50%. In this case the volume fraction of water is constant, so the only difference between the measurements is in the state of the emulsions. Visually, no phase separation could be detected, so the difference must be due to flocculation or sedimentation, leading to a reduced droplet–droplet distance. However, the variations cannot be so large that they promote coalescence, since this was not visually observed. In a study of model W/O emulsions containing different amounts of water (Table 2) (165) it was found that dielectric models based upon spherical, noninteracting droplets could

**Table 2** Dielectric Parameters Obtained Experimentally for Different Water-in-Crude Oil Emulsions, Together with Corresponding Shape Factors Found from Comparison with Theoretical Parameters (Eq. 19)

| Experimental values | | | | Oblate spheroids $(0 < A < 1/3)$ | | | Prolate spheroids $(1/3 < A < 1/2)$ | | |
|---|---|---|---|---|---|---|---|---|---|
| $\phi$ | $\varepsilon_s$ | $\varepsilon_\infty$ | $\tau$ (ns) | $A^{\varepsilon s}$ | $A^{\varepsilon\infty}$ | $A^{\tau}$ | $A^{\varepsilon s}$ | $A^{\varepsilon\infty}$ | $A^{\tau}$ |
| 0.17 | 17.5 | 6.5 | 0.8 | 0.06 | 0.06 | 0.06 | 0.48 | 0.49 | 0.48 |
| 0.23 | 22.4 | 7.4 | 1.0 | 0.08 | 0.09 | 0.07 | 0.48 | 0.48 | 0.48 |
| 0.28 | 28.2 | 9.6 | 1.0 | 0.09 | 0.09 | 0.08 | 0.47 | 0.48 | 0.47 |
| 0.38 | 35.7 | 13.2 | 1.3 | 0.12 | 0.12 | 0.10 | 0.46 | 0.47 | 0.47 |
| 0.44 | 46.0 | 13.5 | 1.2 | 0.14 | 0.14 | 0.14 | 0.46 | 0.45 | 0.46 |
| 0.50 | 60.0 | 14.0 | 1.3 | 0.15 | 0.16 | 0.13 | 0.45 | 0.43 | 0.46 |

*Source*: Ref. 165.

not predict the permittivities found. In order to obtain agreement between measured values and predicted ones the traditional models on heterogeneous mixtures were modified, taking a flocculated state of the emulsions into account.

*a. Linear flocs*

In a first approach, it is assumed that the flocculated aggregates can be considered as spheroids, as depicted in Fig. 37. By further assuming that the volume of the floc equals the volume of the droplets, and neglecting any interactions over the thin films separating the droplets, the overall permittivity can be predicted by use of Eq. (21) or (22). The flocs are in this case characterized by one single variable, namely a shape factor, $A$. The easiest way to interpret $A$ is to consider a linear flocculation (see Fig. 37) and to define an axial ratio between the major and minor axes. When doing so, corresponding axial ratios were found to be in the range from 1:3 to 1:10 (165). However, a small linear flocculation is not entirely consistent with the complex structures a flocculated system can build up. Other more complex flocculation models have also been developed.

*b. Flocs as Subsystems*

An alternative approach is where the aggregates, formed as a result of flocculation, are treated as subsystems of the emulsions (166). The dielectric properties of the subsystems will be decisive for the dielectric properties of the overall system.

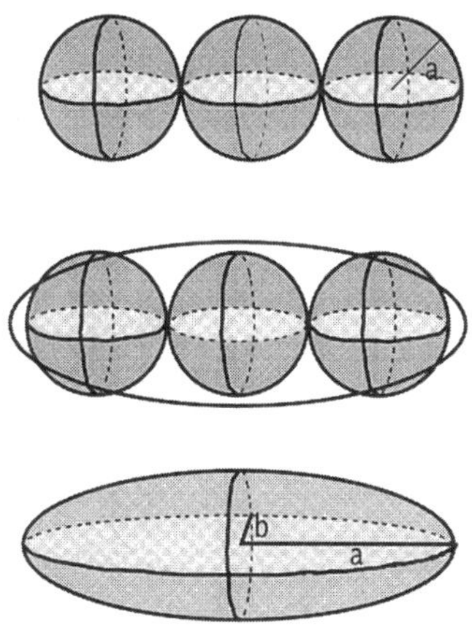


**Figure 37** Linear flocs: if the droplets form small linear aggregates, the floc aggregates can be treated as spheroids and can as such be characterised by the axial ratio $a/b$ (or consequently by the shape factor $A$) as introduced in Eq. (20).

In floc aggregates some of the continuous phase is entrapped between the emulsion droplets. Thus, the volume occupied by a floc will exceed the volume of the individual droplets building up the floc (Fig. 38). The volume fraction occupied by flocs, $\phi_f$, can be expressed as

$$\phi_f = \frac{V_f}{V} = \frac{V_f V_d}{V_d} \Big/ V = \phi \frac{V_f}{V_d} = \frac{\phi}{\phi_{d,f}}, \quad (\phi < \phi_{d,f} < 1) \tag{82}$$

where $V_f$ and $V_d$ are the volumes of the flocs and droplets, respectively, $V$ is the total volume of the system; $\phi$ is the volume fraction of disperse phase with regard to the total volume, whereas $\phi_{d,f}$ is the volume fraction of disperse phase with regard to the floc volume. The factor $\phi_{d,f}$ is a measure on the packing of the droplets in the flocs, i.e., a small value of $\phi_{d,f}$ means a loose structure, and a large value represents a densely packed floc.

Assuming that the droplets retain a spherical shape, the permittivity for the flocs, $\varepsilon_f^*$, can be calculated by using Eq. (20):

$$1 - \phi_{d,f} = \left(\frac{\varepsilon_f^* - \varepsilon_1^*}{\varepsilon_2^* - \varepsilon_1^*}\right)\left(\frac{\varepsilon_2^*}{\varepsilon_f^*}\right)^{1/3} \tag{83}$$

We can then proceed to calculate the total permittivity $\varepsilon^*$ for the system, now treating the flocs as the disperse phase, at a concentration that equals $\phi_f$ :

$$1 - \phi_f = \left(\frac{\varepsilon^* - \varepsilon_f^*}{\varepsilon_2^* - \varepsilon_f^*}\right)\left(\frac{\varepsilon_2^*}{\varepsilon^*}\right)^{3d,f} \times \left(\frac{\varepsilon_2^*(1+3A_f) + \varepsilon_f^*(2-3A_f)}{\varepsilon^*(1+3A_f) + \varepsilon_f^*(2-3A_f)}\right)^{3K,f} \tag{84}$$

By use of Eq. (84) we allow for a spheroidal shape of the flocs, as expressed through the parameters $A_f$, 3d, f and $3K$, f. In this model the flocculated aggregates are characterized by the packing density of the droplets in the floc and a shape factor.

*c. Two-component Model with a Partial Flocculation*

In the approaches above (i.e., Sec. V.A.1.a and V.A.1.b) a complete flocculation, i.e., that all droplets are part of a floc, is taken for granted. This is seldom the case, and the situation sketched in Fig. 39 is more likely to occur. The system properties are then dependent both on the permittivity and volume fraction of the free droplets as well as on the floc permittivity and $\phi_{p,f}$.

The following expression, derived by Hanai and Sekine (167) gives the permittivity of a system where

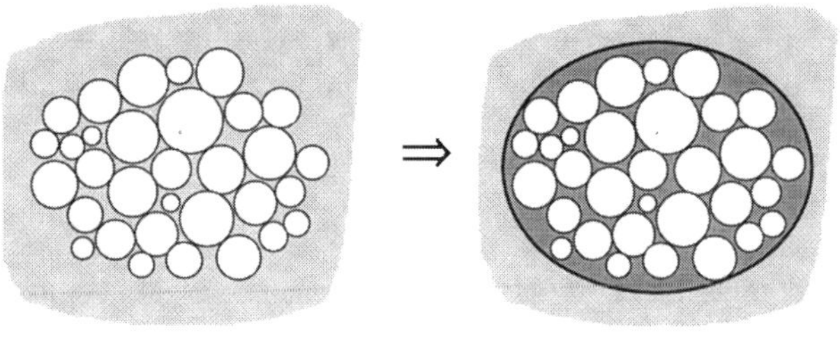

**Figure 38** Flocs treated as subunits: the flocs are treated as subunits, with a volume fraction of droplets in floc (or the floc density), $\phi_{df}$, that is higher than the overall volume fraction $\phi$. Some of the continuous phase will be entrapped between the droplets in the floc. Thus, the volume of a floc exceeds the total volume of the droplets in it. The floc shapes are assumed to be spheroidal.

two different types of spherical particles are dispersed in the same system:

$$\ln(1-\phi) = \frac{1}{3}[\log \varepsilon_2^* - \log \varepsilon^*] + A[\log(\varepsilon^* - \alpha) - \log(\varepsilon_2^* - \alpha)] + B[\log(\varepsilon^* - \beta) - \log(\varepsilon_2^* - \beta)] \tag{85}$$

where $A$, $B$, $\alpha$, and $\beta$ are functions of the permittivities $\varepsilon_{1,j}$ and $\varepsilon_{1,k}$ and of the ratio $\phi_k/\phi_j$, where the subscripts $j$ and $k$ represent the two different kinds of particles; $\varepsilon_2$ is the dielectric permittivity of the continuous phase. In Eq. (85) "ln" denotes the natural logarithm of a real number whereas "log" is the principal value of the complex logarithm; $\phi$ is the total volume fraction particles in the system, i.e., $\phi = \phi_j + \phi_k$. [Similar expressions for multicomponent systems with randomly oriented spheroids have been derived by Boned and Peyrelasse (24).]

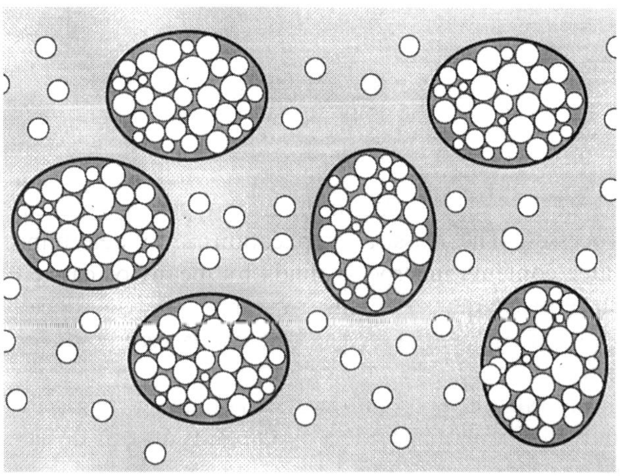

**Figure 39** Model for partial flocculation.

If we treat the free water droplets as one type of particle and the flocs as the other type, Eq. (85) may be used to find the dielectric properties of a partially flocculated emulsion.

The dispersed water, present at a concentration $\phi$, is distributed between two states, either as free droplets, or bound in the flocs. We can write

$$\phi = \phi_w^{\text{Free}} + \phi_w^{\text{Bound}}$$

The volume fraction of flocs in the system ($\phi_f$) is related to $\phi_w^{\text{Bound}}$ through

$$\phi_f = \phi_w^{\text{Bound}} \phi_{d,f}$$

where $\phi_{d,f}$ is the volume fraction droplets in the flocs, as in the previous section. To apply Eq. (85) the total volume fraction occupied by the two "components" is needed and

$$\phi_{\text{Total}} = \phi_w^{\text{Free}} + \phi_f = (\phi - \phi_w^{\text{Bound}}) + \frac{\phi_w^{\text{Bound}}}{\phi_{d,f}}$$

Thus, for a given $\phi$ we can have a wide range of different systems, as we can freely vary $\phi_w^{\text{Free}}$ and $\phi_{d,f}$, within the limits $\{0 < (\phi - \phi_w^{\text{Bound}}) < \phi$ and $\theta < \theta_{d,f} < 1\}$.

Common for all models presented is that, by the right choice of parameters, one can theoretically predict static permittivities that match experimental values for W/O emulsions, and that are higher than those predicted from the models presuming a homogeneous distribution of spherical emulsion droplets. However, there are more features to the dielectric spectrum than just the static permittivity, and for the models to be deemed useful in characterizing emulsion properties they should be able to reproduce also the frequency dependence of permittivity. In other words, the complete relaxation process must be satisfactorily accounted for. When the models above have been fitted to experimental data we have experienced only a limited success in finding sets of floc parameters that yield theoretical spectra in accordance with the experimental ones (164, 165, 168, 169). Taking into account the multitude of possible configurations of the systems after flocculation, this is not unexpected, and an effort still has to be made before the dielectric properties of flocculated systems can readily be predicted.

## 2. Electrocoalescence

In a strong electric field the aqueous droplets in a W/O emulsion become polarized due to ion separation in the droplets. This polarization will create an opposite field inside the water droplets. As a consequence we will have a flocculation of aqueous droplets, a so-called bridging, between the electrodes. The potential difference between droplets may be written as:

$$V_{pot} \propto Er\cos\theta \qquad (86)$$

where $E$ is the applied field, $r$ is the droplet dimension, and $\theta$ is the angle to the applied field. With a droplet radius of 10 μm, an applied field of 0.5 kV/cm, and droplets parallel with the external field, the potential will be of the order of 0.5 to 1.0 V. Under these conditions the field over the surfactant membrane ($\approx 2l_{surfactant}$) separating two aqueous droplets is of the order of $10^6$ V/cm. This is a very high local field that can create an ion transport over the surfactant film. When the ions start to cross the film a coalescence of adjacent droplets will take place. However, this transport requires a certain level, $V_{crit}$, before it will take place; $V_{crit}$ can hence be taken as a measure on the emulsion stability (52). Below $V_{crit}$ different kinds of phenomena can take place. For a single-droplet assembly one can predict a substantially increased level of flocculation, i.e., the static permittivity should increase under these conditions. If the emulsion is already in a flocculated state the applied voltage may first of all alter the droplet size distribution towards larger droplets and hence also the size of the flocs. As a consequence a drop in $\varepsilon_s$ can be predicted. However, when the applied field is switched off the original floc size and structure will be slowly obtained and a relaxation process towards the original $\varepsilon_s$ can be observed. The level of irreversible distortions in the droplet sizes can determine a difference in the initial $\varepsilon_s$ and the $\varepsilon_s$ after relaxation.

*a. Percolation Phenomena in W/O Emulsions in High Electric Fields*

Figure 40 shows the static permittivity of an asphaltene-stabilized model W/O emulsion versus the applied external electric field (170). The static permittivity increases with the applied electric field. This is an indication of a low degree of attraction between the aqueous droplets and, initially, a low level of flocculation. The applied electric field will induce a flocculation that consequently leads to an increase in $\varepsilon_s$. First to a small extent only, later more pronounced as the critical voltage is approached. However, when the critical electric field is exceeded, a steep decline in the static permittivity to a value of 5–7, is observed. This can be viewed as a percolation phenomenon according to a static model (171, 172).

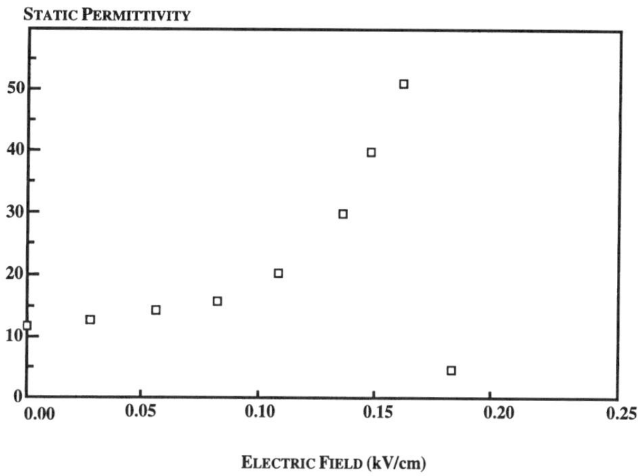


**Figure 40** Static permittivity of an asphaltene-stabilized model emulsion vs. the applied external electric field. The percolation behavior is clearly seen. (From Ref. 170.)

From studies of the percolation phenomena in W/O microemulsions it has been proposed that permittivity follows a scaling law with regard to temperature (96). However, when we induce the percolation by applying a high electric field, the following scaling law may apply:

$$\varepsilon_S \sim (E_{cr-E})^{-S} \qquad (87)$$

where $E < E_{cr}$. From linear regression analysis, the critical exponent $s$ can be estimated from the slope. The data from Fig. 40 will give a slope of 0.52, which is close to the theoretical value expected from a static percolation model ($s = 0.6$) containing bicontinuous oil and water structures (172).

## 3. Dielectric Spectroscopy on Technical Emulsions (Gas Hydrate Formation)

Literature reports on a variety of applications of dielectric measurements in different types of technical processes. The classical application is to determine water contents in process fluids by means of capacitance measurements. This technique has also been extended to higher frequencies by Wasan and coworkers (173, 174). In the following we present a technically very important problem that combines a controlled reaction inside a W/O emulsion and dielectric spectroscopy as a process on-line instrumentation. This problem concerns the formation and transport of gas hydrates in pipelines.

Gas hydrates (clathrates) may technically be considered as an alternative form of ice that has the ability to entrap relatively large volumes of gas within cavities in the hydrate crystal matrix. The entrapped guest molecules (gas) stabilize the structure by means of van der Waals interactions, and combinations of the different unit cells give rise to structures I, II (175–177), and H (178). The most common gas to form gas hydrates is methane, but ethane, propane, butane, carbon dioxide, nitrogen, and many other types of gases may also give rise to gas hydrates.

When gas hydrates are formed in a W/O emulsion, the emulsified water is converted into clathrate structures and thus the volume fraction of free water in the emulsion droplets decreases. The formation of these new structures will alter the overall dielectric properties of the emulsion, and dielectric measurements can thus be used to follow the gas hydrate formation. In Fig. 41 this is illustrated with $CCl_3F$ as the gas hydrate forming species in a nonionic W/O emulsion (179, 180). The decrease in the overall permittivity of the system is due to the transition of water from the liquid to the solid state (i.e., from free water to water bound in the hydrates). From the dielectric measurements we can extract information on the onset temperature for hydrate formation, induction time for the process, and so on. The permittivity level after the hydrate formation is completed is indicative of the ratio between liquid water and hydrate water, i.e., the amount of water converted into hydrate water can be readily found (179, 180).

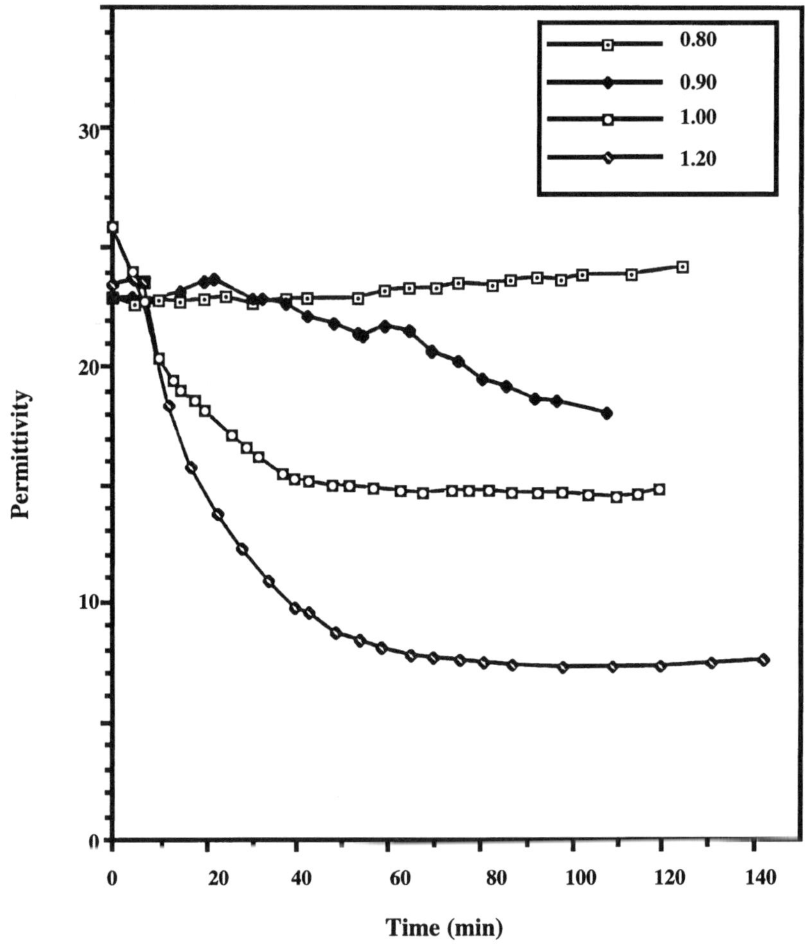


**Figure 41** Static permittivity vs. time for clathrate hydrate formation in model W/O emulsions. The concentrations of $CCl_3F$ in $H_2O$ were 0.80, 0.90, 1.00, and 1.20, respectively, times the theoretical molar fraction 1:17 expected to be found in the hydrates. (From Ref. 179.)

If the aqueous phase contains electrolytes, a relaxation due to the Maxwell–Wagner–Sillars effect will be observed. Since the electrolyte is not incorporated in the clathrate structures, an increased electrolyte concentration in the remaining free water will result, thus changing the dielectric relaxation mode. In Fig. 42 we note that the relaxation time $\tau$ decreases from the initial 1000±100 ps to a final level of 200±20 ps during hydrate formation. The experimental value of 200 ps corresponds roughly to a 3% (w/v) NaCl solution, as compared with the initial salt concentration of 1% (w/v).

## B. Suspensions

### 1. Sedimentation in Particle Suspensions

During sedimentation the volume fraction of suspended particles will increase near the bottom of the sample while the concentration of particles in the top layer will be correspondingly smaller. Information on the sedimentation rate and structure of the sediment layer can thus be extracted from dielectric measurements carried out at different levels of the sample.

The application of the TDS method for sedimentation studies may be illustrated with the studies carried out on two model systems, i.e., aqueous suspensions of $SiO_2$ and $Al_2O_3$, respectively. Figure 43 shows the permittivity spectra of $Al_2O_3$ in water as a function of the volume fraction of particulate matter. The most notable changes are in the static permittivity, which is also focused on in the following figures. Figure 44a shows the variation in static permittivity as a function of the volume fraction of $SiO_2$. The Hanai equation, Eq. (20), gives good estimates of the observed behavior. Figure 44b presents the corresponding data for $Al_2O_3$ and also in this case the Hanai model assuming spherical particles seems to describe the experimental data fairly well (181, 182).

In order to alter the sedimentation rate the surfaces of the $SiO_2$ and $Al_2O_3$ particles were modified through the adsorption of surfactants or polymers. The nonionic surfactant $C_9PhEO_6$ and the nonionic polymer ethyl hydroxyethyl cellulose (EHEC) were used. Figure 45 gives the influence of pH on the sedimentation of $Al_2O_3$ in water for pure particles (Fig. 45a) and EHEC-coated particles (Fig. 45b). At pH 8.5 (i.e., close to the isoelectric point (IEP) for alumina) the sedimentation is fast and gives rise to a porous sediment (leading to a high final level of $\varepsilon$). In contrast to this one can follow the slow sedimentation at pH 3.5 where we end up with a denser sediment and consequently a lower $\varepsilon$. For the coated particles the effect of pH is rather small (182).

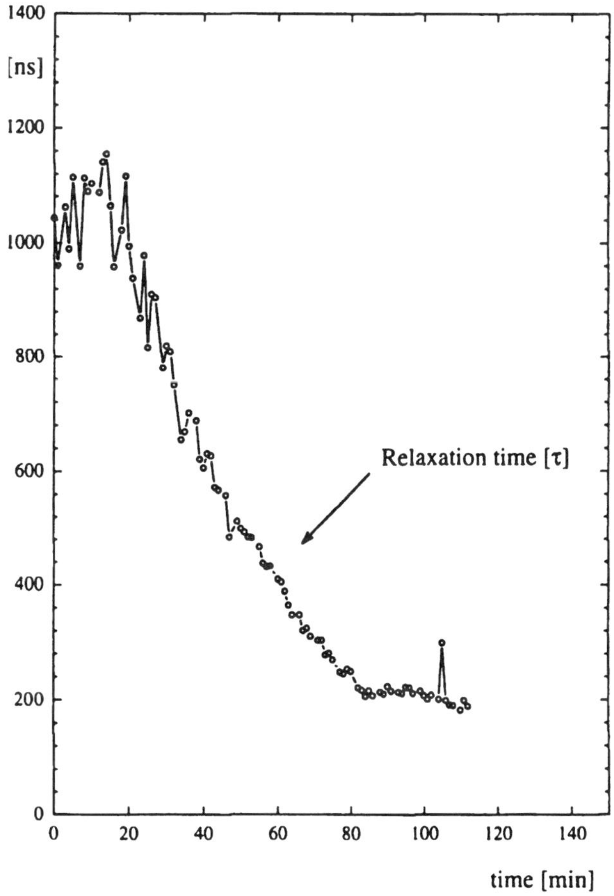


Figure 42 Relaxation time vs. time during clathrate hydrate formation. (From Ref. 179.)

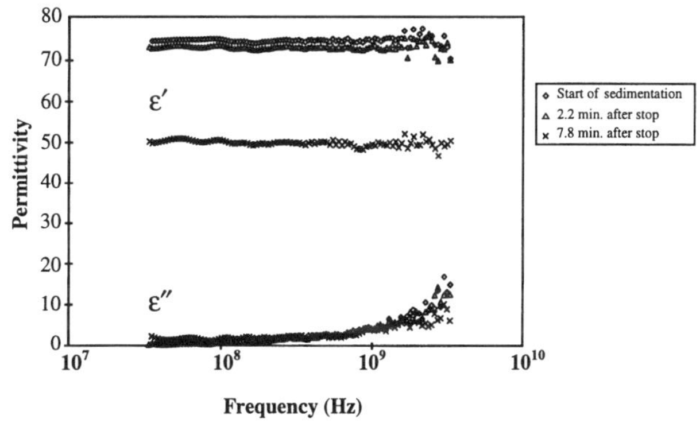


Figure 43 Dielectric spectra from an aqueous alumina particle suspension, recorded during sedimentation. (From Ref. 181.)

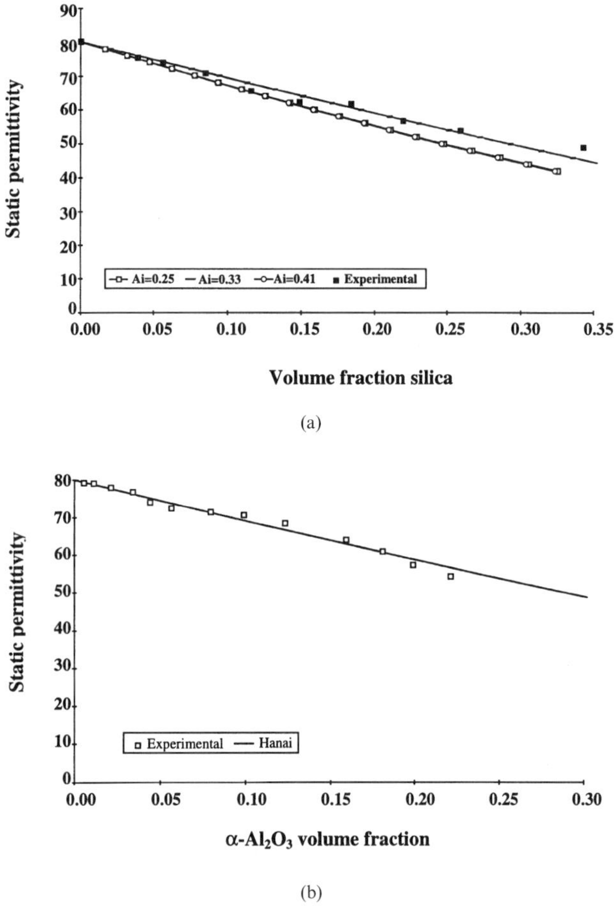


**Figure 44** Static permittivity vs. volume fraction particles in aqueous suspensions: (a) silica particles, (b) alumina particles. (From Ref. 181.)

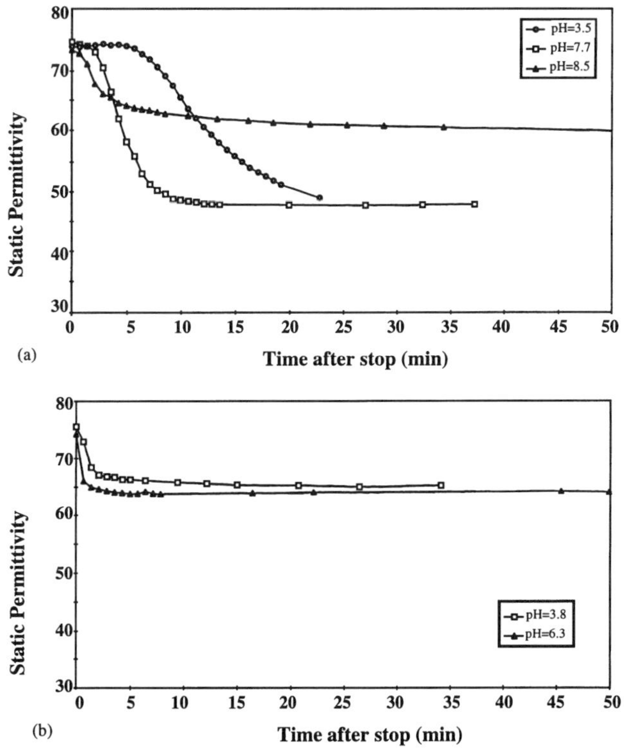


**Figure 45** Static permittivity of particle suspensions vs. time, recorded during sedimentation. Measurements are performed in the bottom of the sedimentation cell, and at different pH conditions. (a) Alumina particles; (b) alumina particles coated with EHEC. (From Ref. 182.)

## 2. Magnetic Particles

Magnetically active monodisperse organic particles have found many technical applications especially within the separation and isolation of live cells and microorganisms. The magnetic particles are also almost perfect for sedimentation tests in applied magnetic fields, in as much as the degree of flocculation may be controlled quite accurately. In Fig. 16 an experimental design for the dielectric study of aqueous suspensions of magnetic particles is displayed (86). In this survey, monodisperse polystyrene particles (2.8–4.5 μm) containing 25% magnetic oxides have been used. When the particles are subjected to a magnetic field an induced flocculation takes place (Fig. 46). The next figure (Fig. 47) shows how the static permittivity changes with time (due to sedimentation) and with the strength of the applied magnetic field. The strength of the magnetic field will dictate the degree of flocculation. For single-particle sedimentation with a linear decrease in the static permittivity as a function of concentration, one would expect proportionality between the derivatives of volume fraction and permittivity versus time. From Fig. 47 we can observe that this is more or less always the case when the sedimentation occurs without the presence of a strong magnetic field, at least up to a period between 0 and maximally 5 min. In the presence of an external field this period is substantially shorter for low-volume fractions and substantially extended for high-volume fractions.

Two major effects will dictate the suspension behavior, i.e., the gravity-induced sedimentation and the bridging induced by the magnetic field. At lower volume fractions the distance between the particles is rather large, and the hindrance of settling is low. When a magnetic field is applied across the suspension there

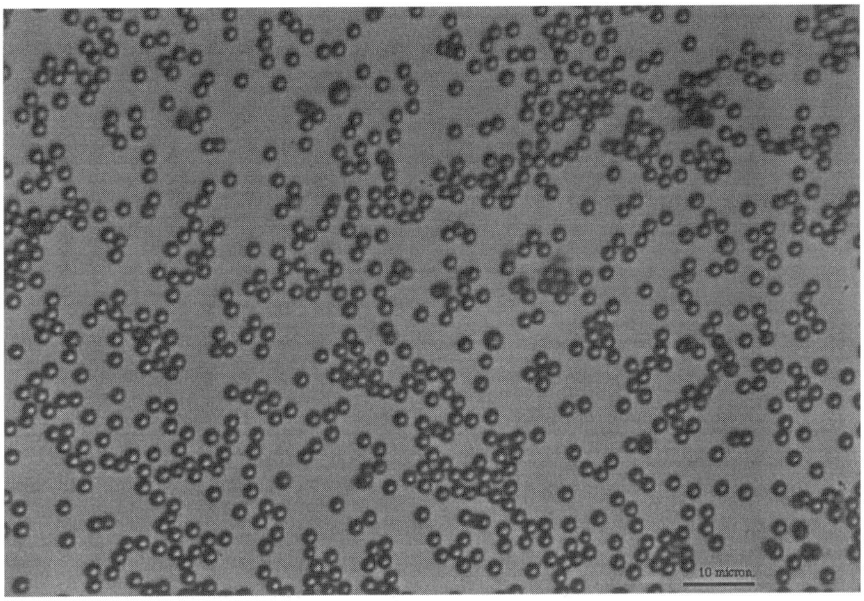


(a)

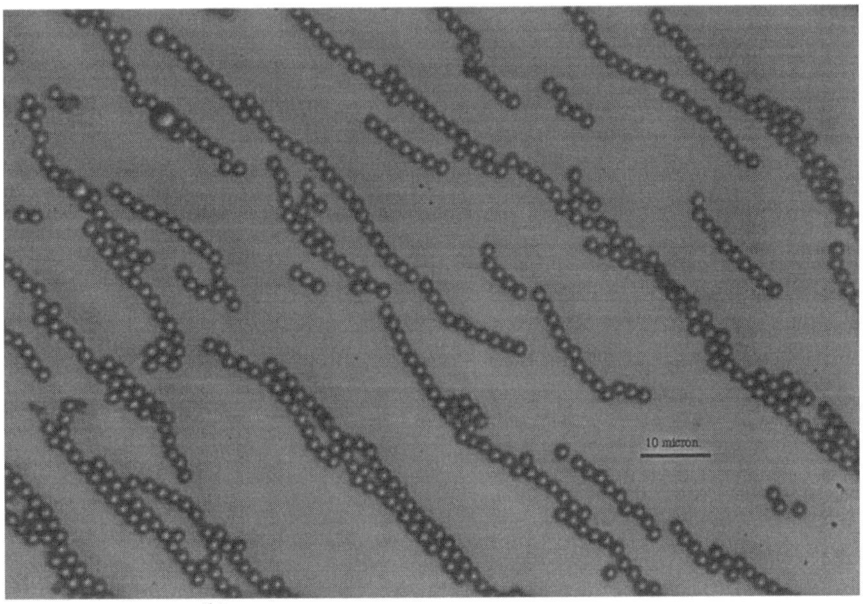


(b)

**Figure 46** (a) Micrograph of polystyrene-based magnetic monodisperse particles; the particle diameter is 2.8 μm. (b) The particles are subjected to an external magnetic field. (From Ref. 86.)

will be an instantaneous sedimentation of small flocs (as seen from Fig 47 with $\phi = 0.02$ and 0.05). At higher $\phi$ values the initial distance between the particles is much smaller and the bridging induced by the magnetic field may lead to a network formation resulting in a hindered sedimentation. This can be observed for $\phi = 0.20$ and 0.25. For a field of 0.4 T it takes about 5 min before the sedimentation sets in while a weaker field (0.1 T) can delay the sedimentation for 3.5 min. For the highest volume fraction ($\phi = 0.25$) no significant sedimentation is observed when the magnetic field is applied.

When the external magnetic field is applied the particles will, to a greater degree, flocculate and settle as

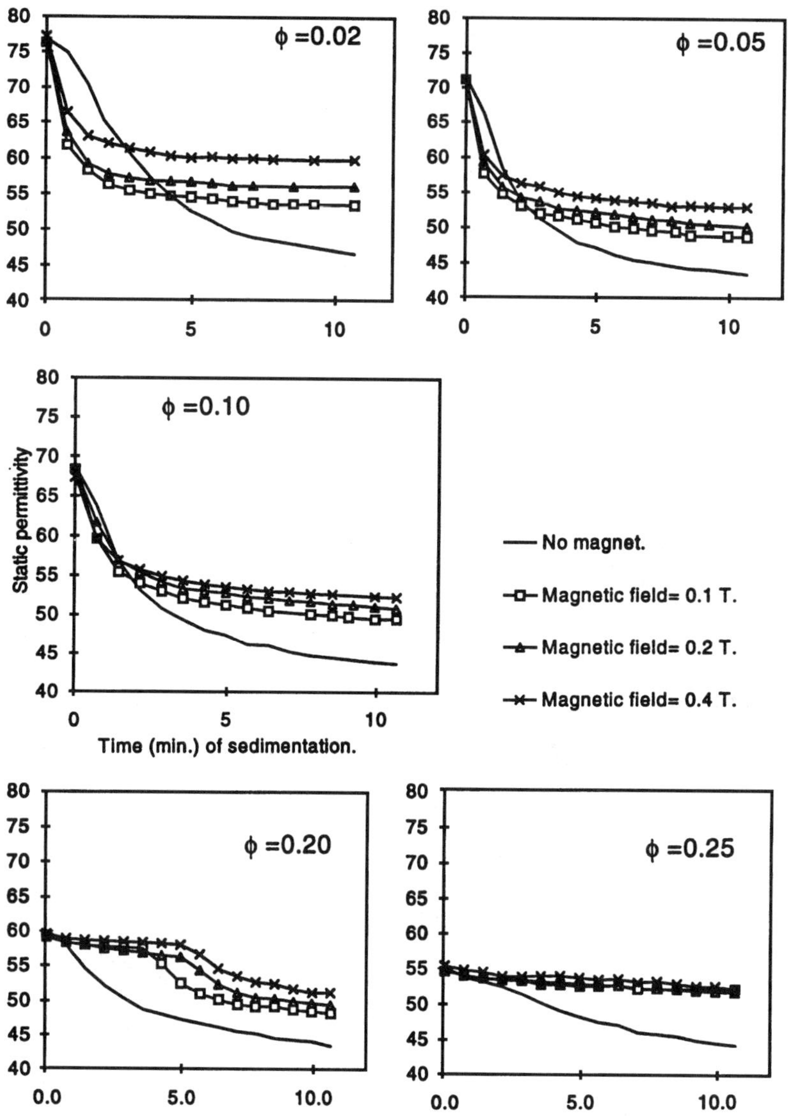


**Figure 47** Static permittivity vs. time for suspensions containing the particles from Fig. (46) recorded during sedimentation. Different magnetic field strengths were applied, and the volume fraction particles were varied according to the legends. (From Ref. 86.)

flocs, giving a more porous bottom layer, as illustrated in Fig. 48. When the particles sediment individually the result is a more dense sediment. The porosity of the sediment is reflected in the level of the permittivity, a low permittivity in the sediment layer indicating a close packing of the particles.

By monitoring the dielectric properties of nonequilibrium systems a wide range of parameters describing the system may be deduced. However, more work of both a theoretical and experimental nature needs to be performed before we have a complete understanding of how the nonideal situations influence the dielectric parameters.

## VI. DIELECTRIC STUDY OF HUMAN BLOOD CELLS

One of the important subjects in biophysics is the investigation of the dielectric properties of cells and the structural parts of the cells (i.e., membrane, cytoplasm, etc.). These can provide valuable knowledge

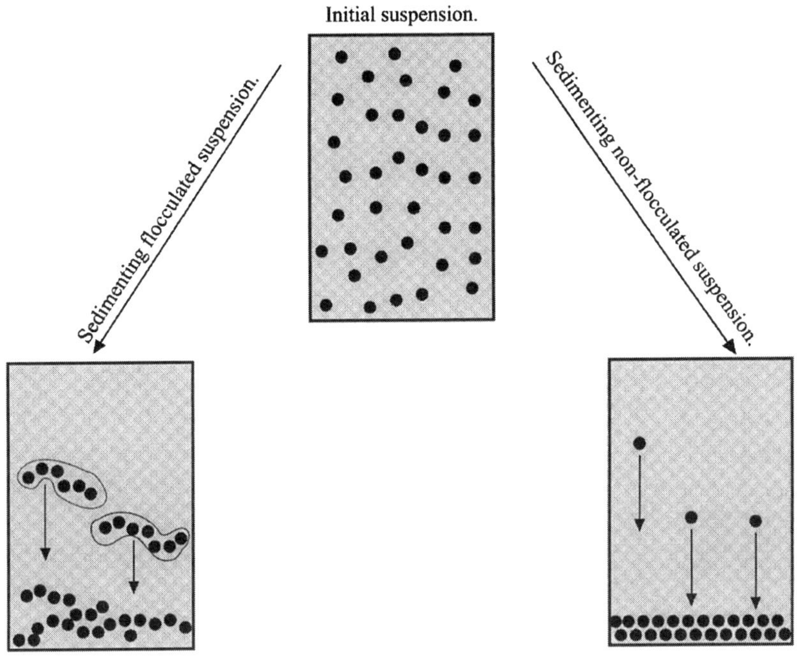


**Figure 48** Schematic view on how flocculation may influence the nature of the sediments.

about different cell structures, their functions, and metabolic mechanisms.

The cell-suspension spectra are known to show a so-called $\beta$-dispersion (183), which is observed in the frequency range 100 kHz–10 MHz and can be interpreted as the interface polarization. This dispersion is usually described in the framework of different mixture formulas and shelled models of particles (14, 70, 72, 183, 184). In the example of biological cells, the interface polarization is connected to the dielectric permittivity and conductivity of the cell structural parts.

Before time-domain spectroscopy (TDS) methods were developed, investigations by frequency-domain methods were restricted by lack of techniques which allowed quick determination of the dielectric spectrum within the frequency range $10^5$–$10^{10}$ Hz (14, 64). It was quite difficult to ensure the stability of biological materials because of the long duration of the experiment. TDS allows one to obtain information on dielectric properties in a wide frequency range during a single measurement and hence to study unstable biological systems properly. This is also a much less time-consuming method, which requires very few devices to work with. Moreover, the sensitivity of the system allows one to study dilute solutions or suspensions of cells (volume fraction $< 10\%$).

## A. Cell and Cell-suspension Dielectric Models

For analysis of the dielectric properties of blood-cell suspensions, several classical models are usually used (11, 14, 185–201). For small volume fractions of cells the Maxwell-Wagner model is used, while for larger ones (see Sec. II) the Hanai formula would be preferable (14, 186). It was shown (70, 72) that for dilute suspensions of human blood cells the dielectric spectra of a single cell can be successfully calculated from the Maxwell model of suspension, according to the mixture formula [Eq. (19)]:

$$\varepsilon^*_{\text{mix}} = \varepsilon^*_{\text{sup}} \frac{(2\varepsilon^*_{\text{sup}} + \varepsilon^*_{\text{c}}) - 2p(\varepsilon^*_{\text{sup}} - \varepsilon^*_{\text{c}})}{(2\varepsilon^*_{\text{sup}} + \varepsilon^*_{\text{c}}) + p(\varepsilon^*_{\text{sup}} - \varepsilon^*_{\text{c}})} \tag{88}$$

where $p$ is the cell volume fraction, $\varepsilon^*_{\text{mix}}$ is the effective complex dielectric permittivity of the whole mixture (suspension), $\varepsilon^*_{\text{sup}}$ is the complex dielectric permittivity of the supernatant, $\varepsilon^*_{\text{sup}}(\omega) = \varepsilon_{\text{s,sup}} - i\frac{\sigma_{\text{sup}}}{\omega}$, $\sigma_{\text{sup}}$ is its conductivity, and $\varepsilon^*_{\text{c}}$ is the effective complex dielectric permittivity of the average cell.

### 1. Single-shell Model of the Erythrocytes

The dielectric properties of the spherical erythrocyte can be described by the single-shell model (14, 70,

186). In this model the cell is considered as a conducting homogeneous sphere (or ellipsoid) covered with a thin shell, much less conductive than the sphere itself (14, 186–189).

The expression for the complex dielectric permittivity [$\varepsilon_c^*(\omega)$] in the single-shell model contains five parameters: dielectric permittivity and conductivity of the cell membrane ($\varepsilon_m$ and $\sigma_m$); dielectric permittivity and conductivity of the cell interior (cytoplasm) ($\varepsilon_{cp}$ and $\sigma_{cp}$); and a geometrical parameter $v = (1 - d/R)^3$ where $d$ is the thickness of the cell membrane and $R$ is the radius of the cell. This expression is as follows (188):

$$\varepsilon_c^*(\omega) - \varepsilon_m^*(\omega)\frac{2(1-v) + (1+2v)E(\omega)}{2 + v + (1-v)E(\omega)} \qquad (89)$$

where $E(\omega) = \dfrac{\varepsilon_{cp}^*(\omega)}{\varepsilon_m^*(\omega)}$, $\varepsilon_{cp}^*(\omega) = \varepsilon_{cp} - i\dfrac{\sigma_{cp}}{\varepsilon_0\omega}$, $\varepsilon_m^*(\omega) = \varepsilon_m - i\dfrac{\sigma_m}{\varepsilon_0\omega}$, and $\varepsilon_0$ is the vacuum permittivity.

It is possible to show that the single-shell model equation can be presented as the sum of a single Debye process and a conductivity term:

$$\varepsilon_c^* = \varepsilon_\infty + \frac{\Delta\varepsilon}{1 + i\omega\tau} + \frac{\sigma}{i\omega\varepsilon_0} \qquad (90)$$

The last equation depends only on four parameters ($\Delta\varepsilon$ – dielectric strength, $\tau$ – relaxation time, $\varepsilon_\infty$ – dielectric permittivity at high frequency, and $\sigma$ – the conductivity), which are functions of the dielectric and geometrical parameters of the single-shell models. This connection between the experimental presentation [Eq. (90)] and the single-shell model [Eq. (88)] formulas allows one to conclude that only four independent parameters are required to describe the spectrum of a single cell. The fifth parameter can be expressed in terms of these four parameters. Using the high- and low-frequency limits one can derive the following relations:

$$\varepsilon_{c_0}^* \cong \varepsilon_m\left(\frac{1+2v}{1-v}\right) - j\frac{\sigma_m}{\omega\varepsilon_0}\left(\frac{1+2v}{1-v}\right)\omega \to 0 \qquad (91)$$

$$\varepsilon_{c\infty}^* \cong \varepsilon_{cp} - j\frac{\sigma_{cp}}{\omega\varepsilon_0}\omega \to \infty \qquad (92)$$

By fitting the single-shell model to the experimental spectrum the following four parameter combinations can be obtained:

$$\varepsilon_m\left(\frac{1+2v}{1-v}\right);\ \sigma_m\left(\frac{1+2v}{1-v}\right);\ \varepsilon_{cp} \text{ and } \sigma_{cp} \qquad (93)$$

The first two terms indicate that one of the three parameters ($\varepsilon_m$, $\sigma_m$, or $v$) has to be obtained by an independent method and to be kept fixed during the fitting procedure, whereas the other two parameters can be fitted. Usually, the geometrical parameter $v$ is fixed, since it can be calculated by using the values of the cell radius ($R$) and the cell membrane thickness ($d$), which are evaluated independently. The parameters $\varepsilon_{cp}$ and $\sigma_{cp}$ in set (93) can be calcualted directly from the fitting procedure.

It can be shown analytically that the expression for the model spectrum of a cell suspension represented by the combination of the Maxwell–Wagner formula and the single-shell model of cells can be rewritten as the sum of two Debye proceesses and a conductivity term. This is in contrast to the conclusions (190) that every interface of a shelled particle gives a single Debye-type dispersion. In the case of a one-shell particle suspension there are two interfaces. One of them is the interface between the cytoplasm and the cell membrane and the other one is the interface between the cell membrane and the suspending medium. However, Pauly and Schwan (202) have proven that for biological cells these two dispersions degenerate to only one relaxation process. Indeed, the numerical model experiment (203) has shown that the dielectric strength of the high-frequency process is about 2–3 orders of magnitude smaller than the dielectric strength of the low-frequency process and can therefore be neglected.

## 2. Double Shell Model of the Lymphocytes

It is well known that lymphocytes are spherical, and have a thin cell membrane and a spherical nucleus (surrounded by a thin nuclear envelope) that occupies about 60% of the cell volume (188). Therefore, the dielectric properties of lymphocytes can be described by the double-shell model (188, 190, 191) (see Fig. 49). In this model the cell is considered to be a conducting sphere covered with a thin shell, much less conductive than the sphere itself, in which a smaller sphere with a shell (i.e., the nucleus) is incorporated. In addition, one assumes that every phase has no dielectric losses and the complex dielectric permittivity can thus be written as:

$$\varepsilon_i^*(\omega) = \varepsilon_i - j\frac{\sigma_i}{\varepsilon_0\omega} \qquad (94)$$

where $\varepsilon_i$ is the static permittivity and $\sigma_i$ is the conductivity of every cell phase. The subscript $i$ can denote "m" for membrane, "cp" for cytoplasm, "ne" for nuclear envelope, or "np" for nucleoplasm.

The effective complex dielectric permittivity of the whole cell ($\varepsilon_c^*$) is represented as a function of the phase

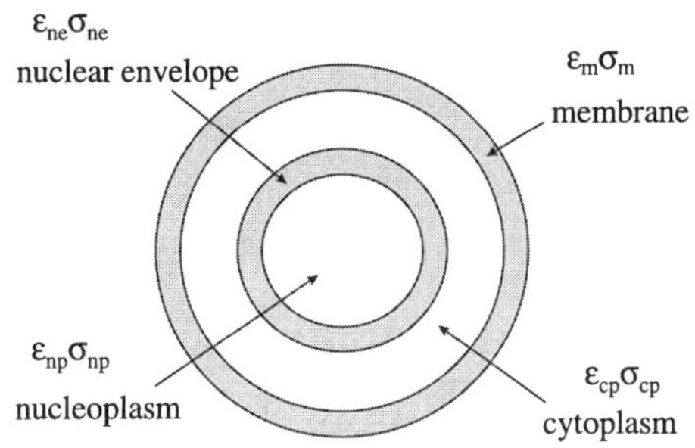


**Figure 49** Schematic picture of the double-shell dielectric model of the cell. Every phase of the cell is described by the corresponding dielectric permittivity ($\varepsilon$) and conductivity ($\sigma$). (From Ref. 72. With permission from Elsevier B.V.)

parameters, i.e., the complex permittivities of cell membrane ($\varepsilon_m^*$), cytoplasm ($\varepsilon_{cp}^*$), nuclear envelope ($\varepsilon_{ne}^*$), and nucleoplasm ($\varepsilon_{np}^*$) :

$$\varepsilon_c^* = \varepsilon_m^* \frac{1(1 - v_1) + (1 + 2v_1)E_1}{(2 + v_1) + (1 - v_1)E_1} \tag{95}$$

where the geometrical parameter $v_1$ is given by $v_1 = (1 - d/R)^3$, where $d$ is the thickness of plasma membrane and $R$ is the outer cell radius. The intermediate parameter, $E_1$, is given by

$$E_1 = \frac{\varepsilon_{cp}^*}{\varepsilon_m^*} \frac{2(1 - v_2) + (1 + 2v_2)E_2}{(2 + v_2) + (1 - v_2)E_2} \tag{96}$$

where $v_2 = [R_n/(R - d)]^3$, $R_n$ being the outer radius of the nucleus. Finally, $E_2$ is given by

$$E_2 = \frac{\varepsilon_{nc}^*}{\varepsilon_{cp}^*} \frac{2(1 - v_3) + (1 + 2v_3)E_3}{(2 + v_3) + (1 - v_3)E_3} \tag{97}$$

where $v_3 = (1 - d_n/R_n)^3$, $E_3 = \varepsilon_{np}^*/\varepsilon_{ne}^*$, $d_n$ being the thickness of the nuclear envelope.

In the double-shell model every structural part of the cell (cell membrane, cytoplasm, nuclear envelope, and nucleoplasm) can be described by two parameters – permittivity and conductivity. Therefore, a cell is described by eight dielectric phase parameters, and there are three geometrical parameters that are a combination of thickness of outer and internal membranes with radii of nucleus and cell. Thus, as it seems, we could obtain all 11 parameters from the fitting of a one-cell spectrum by the double-shell equation. However, it can be demonstrated that some of the parameters are not independent in this model.

In the multishell model every shell gives rise to an additional Debye process with two extra parameters (190). Thus, in the case of the double-shell model, which contains eleven parameters, the spectrum of a suspension can be written as a sum of two Debye processes with conductivity, which includes only six parameters. This means that only six parameters from the total 11 in the double-shell model are arguments that can be fitted. The other five parameters have to be measured by independent methods and have to be fixed in the fitting process. In the fitting procedure, described elsewhere (72, 203), the radius of a cell, the thickness of both membranes, and the permittivity of cytoplasm and nucleoplasm were all fixed.

The specific capacitance of the cell membrane can be calculated directly from the cell suspension spectrum by using the following formula, derived from the Hanai–Asami–Koisumi model (192):

$$C_m = \frac{2}{3} \frac{\varepsilon_0}{R} \frac{\varepsilon_{mix(low)}}{(1 - (1 - p)^{3/2})} \tag{98}$$

where $\varepsilon_{mix(low)}$ is the low-frequency limiting value of the dielectric permittivity (static permittivity) of the suspension.

## B. Protocol of Experiment, Fitting Details, and Statistical Analysis

The whole procedure of the human blood-cell suspension study is presented schematically in Fig. 50. The TDS measurements on the cell suspension, the volume-fraction measurement of this suspension, and measurements of cell radius are excecuted during each experiment on the sample. The electrode-polarization correction (see Sec. II) is performed at the stage of data treatment (in the time domain) and then the suspension spectrum is obtained. The single cell spectrum is calculated by the Maxwell–Wagner mixture formula [Eq. (88)], using the measured cell radius and volume fraction. This spectrum is then fitted to the single-shell model [Eq. (89)] in the case of erythrocytes or to the double-shell model [Eqs (94)–(98)] to obtain the cell-phase parameters of lymphocytes.

In the fitting procedure for lymphocytes, the following parameters were fixed: the radius of the measured cell; the cell membrane thickness ($d = 7$ nm) (188); the nuclear envelope thickness ($d_n = 40$ nm) (188, 191); and the ratio of the nucleus radius to the cell radius ($R_n/R = (0.6)^{1/3}$). These four values represent only three geometrical parameters of the double-shell model (188, 190, 191). Two other fixed parameters were the dielectric permittivity of the cytoplasm ($\varepsilon_{cp} = 60$) (188) and the dielectric permittivity of the nucleoplasm ($\varepsilon_{np} = 120$); $\varepsilon_{np}$ was chosen as the middle value from the range presented in other papers (188,

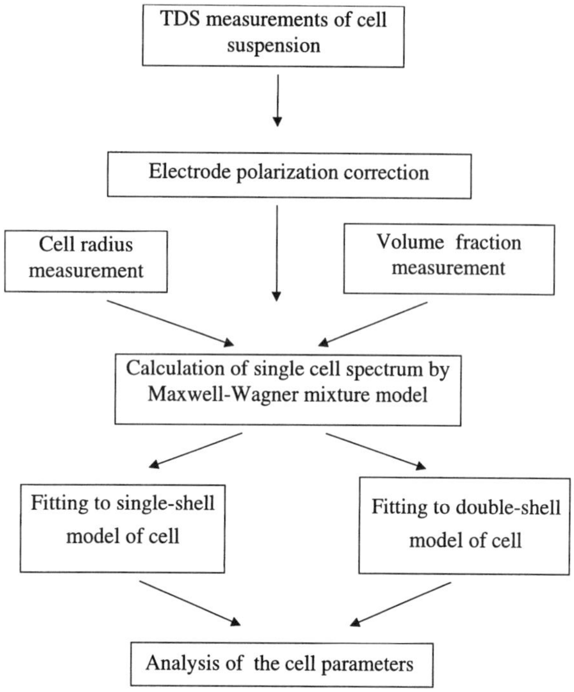


**Figure 50** Flow chart of the whole protocol of human blood-cell suspension study.

191). Moreover, numerical evaluations and evidence in Refs 188 and 191 have shown that the suspension spectrum is almost insensitive to changes in the $\varepsilon_{cp}$ and $\varepsilon_{np}$ parameters from 30 up to 300. Students' $t$-test (201) was usually applied to analyze the results of the fittings.

## C. Erythrocytes and Ghosts

The erythrocyte and erythrocyte ghost suspensions are very similar systems. They differ in their inner solution (in the case of erythrocytes it is an ionic hemoglobin solution; in the case of ghosts it is almost like the surrounding solution they were in while they were sealed). The cell sizes in a prepared suspension depend both on the ion concentration in the supernatant and in the cell interior (70). Thus, the dielectric spectra of erythrocytes and erythrocyte ghost suspensions have the same shape, which means that there are no additional (except Maxwell–Wagner) relaxation processes in the erythrocyte cytoplasm; thus, the single-shell model (Eq. 89) can be applied.

The frequency dependence of the effective dielectric permittivity of a single "average" erythrocyte was calculated for both samples, according to Eq. (88). The spectra obtained were almost identical, as was expected since both samples were prepared from erythrocytes of the same blood sample. The normalized dielectric permittivity spectrum of one of them (No. 1) is shown in Fig. 51 by points. The relaxation process for a single erythrocyte can be described by the Debye equation. The effective dielectric permittivity spectra of an "average" erythrocyte were fitted to the single-shell model equation [Eq. (89)] and the phase parameters of the cell were found. The fitted normalized dielectric spectrum is given in Fig. 51 by a solid line.

The membrane dielectric permittivity $\varepsilon_m$ was evaluated from fitting with an accuracy of $\pm 0.05$ (70). The membrane conductivity could not be estimated from the fitting as far as it was less than the accuracy limits, so it was set to zero. Insufficient accuracy of the inner phase dielectric permittivity $\varepsilon_S$ allows only estimates of the lower and upper limits of its value. The inner phase conductivity $\sigma_i$ was found with an accuracy of $\pm 0.01$ S/m. The dimensionless parameter $v$ was found with an accuracy of $\pm 1 \times 10^{-4}$.

As regards the ghosts, their membrane dielectric permittivity $\varepsilon_m$ is that of the erythrocytes they were prepared from. The inner phase conductivity $\sigma_i$ of

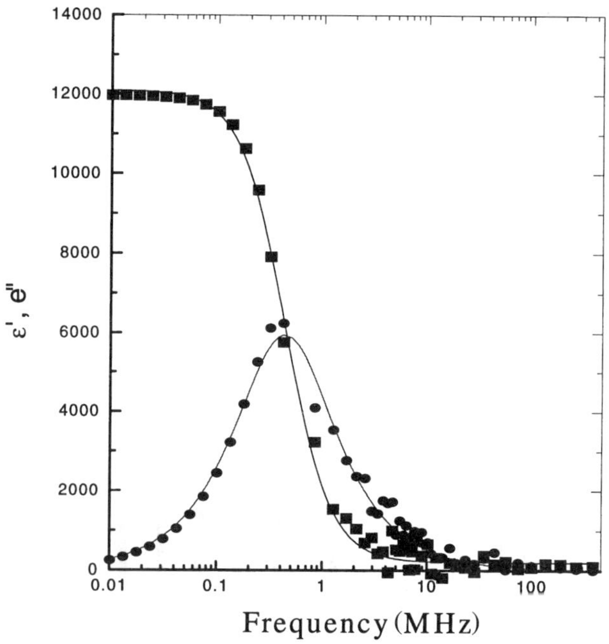


**Figure 51** Dielectric permittivity spectrum of the "average" erythrocyte fitted to the single-shell model. Experimental results are figured by points, and the fitting results by a solid line. (From Ref. 70. With permission from Elsevier Science B.V.)

ghosts varied in correspondence with the conductivity of the supernatant they were placed in. The parameter $v$ depends both on the supernatant conductivity and on the ghost preparation method.

The limits of the phase parameter variation of different erythrocytes and ghosts at room temperature are given in Table 3. The dielectric permittivities of the erythrocyte membranes are found to be distributed near 5 (Fig. 52). The erythrocyte membrane thickness was found to be 3.1 nm, assuming an average radius of 2.7 μm. This result is in good agreement with that obtained by Fricke (193, 194).

Thus, it was shown that one could apply the mixture equation [Eq. (88)] and the single-shell model [Eq. (89)] to dilute solutions of ghosts and erythrocytes. The membrane dielectric permittivity, cytoplasm conductivity, and geometric parameter $v$ for each kind of cell were calculated. The next step was to use this approach for the dielectric property study of normal and pathological blood cells.

**Table 3** Limits of Phase Parameters' Variation for Different Erythrocytes and Ghosts at Room Temperature

| | $\varepsilon_m$ | $v$ $\left(1-\frac{d}{R}\right)^3$ | $d$ (nm) Independent measurement ($R = 2.7$ μm) |
|---|---|---|---|
| Erythrocytes | ~ 5 | ~ 0.9967 | ~ 3.1 |
| Ghosts | ~ 4.8 | ~ 0.9967 | ~ 3.1 |

*Source*: Ref. 70 (with permission from Elsevier Science B.V.).

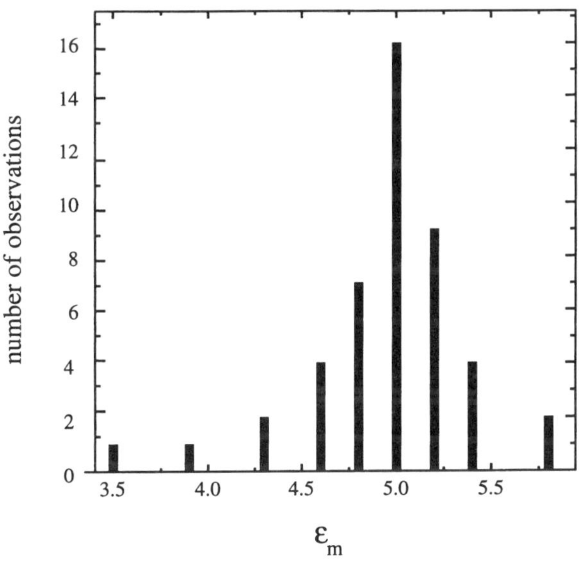


**Figure 52** Distribution of the observed values of the erythrocyte membrane dielectric permittivity. (From Ref. 70. With permission from Elsevier Science B.V.)

## D. Lymphocytes

### 1. Description of Measured Cell Samples

Nine cell populations were investigated (72): normal perpheral blood T cells, normal tonsillar B cells, peripheral blood B cells, which were transformed by infection with Epstein–Barr Virus (EBV) (Magala line), malignant B cell lines (Farage, Raji, Bjab, and Daudi) and malignant T cell lines (Peer and HDMAR). The sizes of the cells were determined by using a light microscope. The typical size distribution is shown in Fig. 53. The volume fractions were measured with a micro-centrifuge (Haematocrit) and corrected for the intercellular space, which was determined with Dextran Blue, and found to be 20.2 ± 3.2% of the volume of the pellet (204, 205). The cell populations with corresponding names of diseases, the mean value of cell radii and the volume fractions are presented in Table 4.

### 2. Dielectric Properties of Measured White Blood Cells

Typical examples of single cell spectra obtained for studied cell lines by the procedure described above are presented in Fig. 54. One can see that the spectra of the various cell lines are different. It should be mentioned that the transition from the suspension spectrum to a spectrum of a single average cell leads to a noise increase that is especially noticeable at high frequencies. This phenomenon is the result of the nonlinearity of Eq. (88), which was used for this calculation. In particular, the parameters for the nucleus envelope, cytoplasm, and nucleoplasm (presented in Table 5) are connected with the high frequency of the cell spectrum; therefore, these parameters were obtained with a relatively low accuracy.

The specific capacitance of a cell membrane was estimated from the value of the state permittivity of the suspension spectra by the relationship Eq. (94). This value is proportional to the ratio of the cell membrane permittivity to the membrane thickness $\varepsilon_m/d$, according to the formula of a plate capacitor, i.e., $C_m = \varepsilon_m\varepsilon_o/d$. Both the capacitance $C_m$ and ratio, $\varepsilon_m/d$, are presented in Fig. 55. Note that in

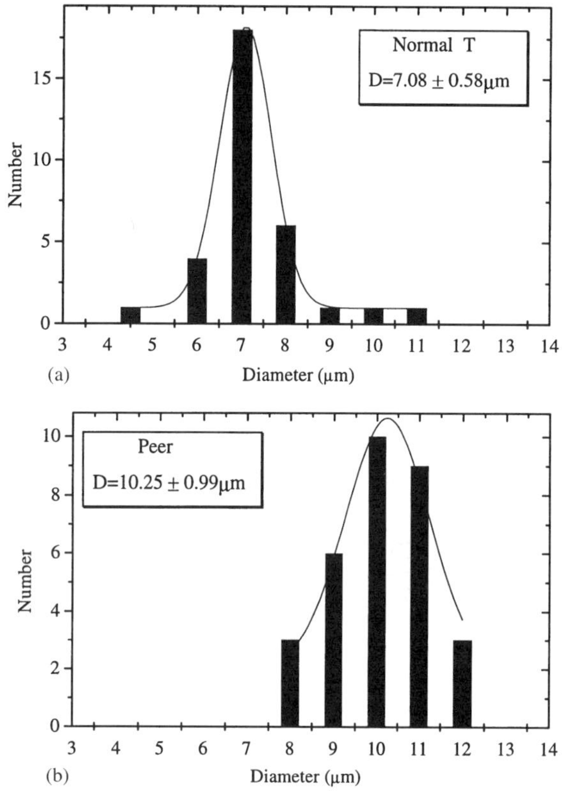


**Figure 53** Typical size distributions for normal (a) and malignant (b) lymphocytes. The results of fitting by Gauss distribution function are shown by the solid line; *D* is the mean diameter of cells. (From Ref. 72. With permission from Elsevier Science B.V.)

this study we are not able to evaluate the permittivity and the thickness of the cell membrane independently.

One can see (Table 5; Fig. 55) that the membrane capacitance (or the similar parameter $\varepsilon_m/d$) has different values for the different cell populations. These parameters for B-normal cells exceed by 11% the values for the T-normal ones. Even more dramatic is the difference between the value of the membrane capacitance of the normal cells and that for all the malignant cells.

For B lymphocytes the capacitance of the normal cells is higher than that of all the malignant cells. The same parameter for the EBV-transformed line (Magala) is intermediate between the values for normal and malignant cells. According to statistical analysis by the *t*-test, the difference between transformed (Magala) and malignant lines is statistically significant (*t*-test gives the probability $0.01 < Pr < 0.02$), whereas there is no statistically significant difference between the nondividing (B normal) and transformed (Magala) populations ($Pr > 0.2$).

As for the T-cell population, the membrane capacitance of the malignant cells (see Fig. 55) was smaller than that of the normal T cells. However, this difference was borderline statistically significant ($0.05 < Pr < 0.1$).

As can be seen in Fig. 56 and Table 5 the membrane conductivities of normal cells of both the B and

**Table 4** List of Cell Populations Studied

| Cells | Radius (μm) | Number of experiments | Volume fraction of suspension (%) | Disease |
|---|---|---|---|---|
| *B cells* | | | | |
| B-normal[a] | 3.3 | 3 | 1.7–3.2 | – |
| Magala | 5.3 | 2 | 5.2–7 | EBV immortalized |
| Farage | 5.2 | 5 | 4.4–8.56 | Non-hodgkin's lymphoma |
| Raji | 6.4 | 4 | 4.4–6.96 | Burkitt's lymphoma |
| Bjab | 6.3 | 3 | 2–6.4 | Burkitt's lymphoma |
| Daudi | 6.8 | 4 | 2.8–6.2 | Burkitt's lymphoma |
| *T cells* | | | | |
| T normal[b] | 3.4 | 2 | 4 | – |
| Peer | 5.1 | 4 | 2.8–6.96 | Acute lymphocytic leukemia |
| HDMAR | 5.9 | 2 | 2.8–5.2 | Hodgkin's lymphoma |

[a]Tonsillar B cells.
[b]Peripheral blood T cells.
*Source*: Ref. 72 (with permission from Elsevier Science B.V.).

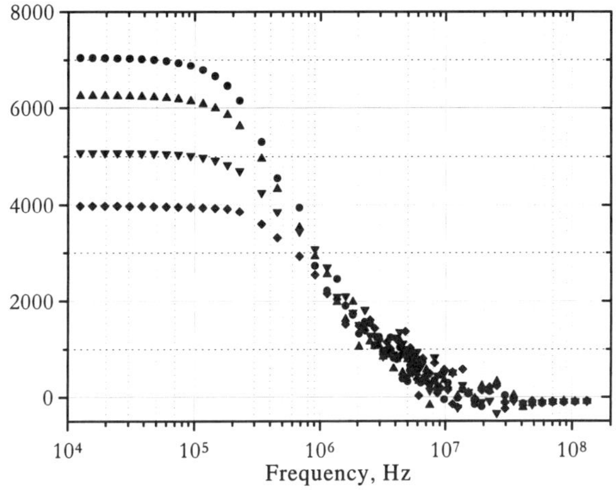


**Figure 54** The real part of complex dielectric permittivity for differnt cell populations calculated from experimental suspension spectra by Maxwell–Wagner mixture model: (●) Magala; (▲) Raji; (▼) Bjab; (◆) HDMAR. (From Ref. 72. With permission from Elsevier Science B.V.)

T populations were significantly higher than for that of malignant and transformed cells. In the B-cell group, the membrane conductivity of normal cells was about six times larger than that of the average value of the malignant cell lines and the transformed cells. This difference is statistically highly significant ($Pr < 0.01$). No significant difference in the conductivity between the transformed and malignant cells in the B population were found ($Pr > 0.2$). Concerning the conductivity of T cells, the difference between normal and malignant cells was not so large as for B cells, but it was statistically significant ($0.01 < Pr < 0.02$).

The $\varepsilon_{ne}$ value of normal B cells is about 1.3 times larger than the average value of other lines of this group (Table 5). Statistically, the difference is borderline significant ($0.05 < Pr < 0.1$).

There is a difference in the B group between normal B cells and the malignant cell lines, which is statistically high significant ($0.01 < Pr$). The normal T cells value is more than twice the average of the malignant cell lines (Table 5) and the difference is statistically significant ($0.01 < Pr < 0.02$).

The conductivity of normal B cells is larger by a factor of 1.8 than that of the average of the malignant cell lines (Table 5). This difference is statistically highly significant ($Pr < 0.01$). There is no difference in this parameter between the average of the malignant cell lines and that of the transformed cell lines. There is almost no difference between normal T cells and malignant T cells.

The normal B-cell value is larger by a factor of 1.6 or more than that of the average value of the malignant cells (Table 5), and it is statistically significant ($0.01 < Pr < 0.02$). There is a small difference between normal T cells and the average malignant cells, but this difference is statistically not significant ($0.2 < Pr$).

It is interesting to note that the conductivity of nucleoplasm is about twice that of the cytoplasm for almost all cell populations.

**Table 5** Dielectric Parameters[a] of Cell Structural Parts for All Cell Populations Studied

| | $\varepsilon_m$ | $\sigma_m$ ($\times 10^{-6}$S/m) | $\varepsilon_{ne}$ | $\sigma_{ne}$ ($\times 10^{-3}$S/m) | $\sigma_{ep}$ (S/m) | $\sigma_{np}$ (S/m) |
|---|---|---|---|---|---|---|
| B normal | $12.8 \pm 1.6$ | $56 \pm 29$ | $106 \pm 35$ | $11.1 \pm 72$ | $1.31 \pm 0.08$ | $2.04 \pm 0.29$ |
| Magala | $11.4 \pm 2.4$ | $8.8 \pm 0.7$ | $72.5 \pm 11.6$ | $3.7 \pm 0.9$ | $0.55 \pm 0.2$ | $1.08 \pm 0.03$ |
| Farage | $9.8 \pm 1.1$ | $9.1 \pm 1.4$ | $60.3 \pm 22.6$ | $4.4 \pm 2.5$ | $0.48 \pm 0.14$ | $1.07 \pm 0.43$ |
| Raji | $8.8 \pm 1.1$ | $8.2 \pm 0.6$ | $79.9 \pm 34.4$ | $4.0 \pm 1.6$ | $0.58 \pm 0.02$ | $1.02 \pm 0.25$ |
| Bjab | $8.0 \pm 0.7$ | $11.0 \pm 5.3$ | $108 \pm 35$ | $2.1 \pm 0.7$ | $0.88 \pm 0.11$ | $1.39 \pm 0.54$ |
| Daudi | $7.2 \pm 0.7$ | $9.5 \pm 1.4$ | $66.1 \pm 7.5$ | $2.7 \pm 0.3$ | $0.85 \pm 0.09$ | $1.44 \pm 0.35$ |
| *T cells* | | | | | | |
| T normal | $11.1 \pm 1.4$ | $27.4 \pm 6.2$ | $85.6 \pm 16.7$ | $8.8 \pm 0.6$ | $0.65 \pm 0.13$ | $1.26 \pm 0.27$ |
| Peer | $9.5 \pm 0.7$ | $12.9 \pm 3.6$ | $61.6 \pm 17$ | $2.1 \pm 0.6$ | $0.81 \pm 0.09$ | $1.42 \pm 0.2$ |
| HDMAR | $7.4 \pm 1.2$ | $14.5 \pm 4$ | $101.2 \pm 55.3$ | $3.0 \pm 0.2$ | $0.88 \pm 0.25$ | $1.58 \pm 0.28$ |

[a]Fitting procedure was performed by fixing the following parameters: $\varepsilon_{ep} = 60$; $\varepsilon_{np} = 120$; $d = 7$ nm; $d_n = 40$ nm, $R_n = R(0.6)^{1/3}$
*Source*: Ref. 72 (with permission from Elsevier Science B.V.).

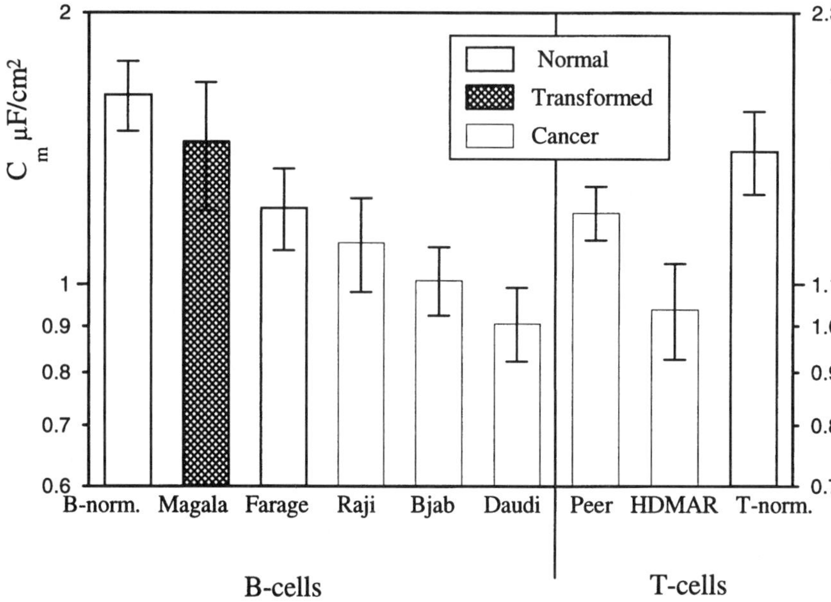


**Figure 55** Capacitance of cell membrane and ratio of dielectric permittivity of the cell membrane thickness ($\varepsilon_m/d$) for all cell populations under investigation. Shown are the mean values of the results ±SD. (From Ref. 72. With permission from Elsevier Science B.V.)

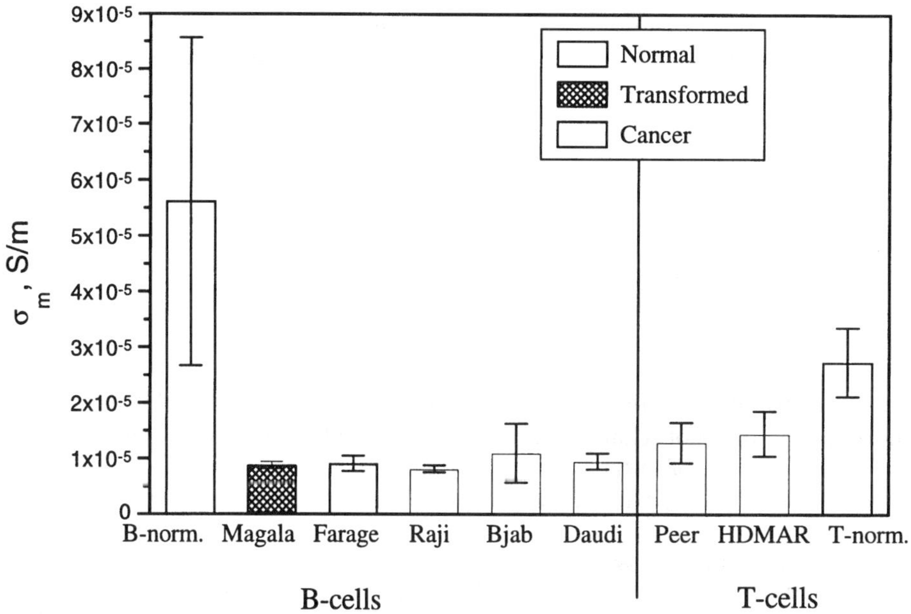


**Figure 56** Conductivity of cell membrane for all populations under investigation. Shown are the mean values of the results ±SD. (From Ref. 72. With permission from Elsevier Science B.V.)

## 3. Main Features of Dielectric Response of Pathological Cells

The various dielectric parameters of the cell populations presented in Table 5 were obtained with cells suspended in a low-conductivity medium, in which the major components were sucrose and glucose rathr than salts. The reason for using this medium was to decrease the electrode polarization. A priori, this choice of medium raises questions about the state of the cells as compared to their native state, in which they are immersed in a solution containing salts and proteins. One can expect at least two kinds of changes when cells are transferred to a medium of low ionic strength. First, a direct change in the cell membrane integrity, and second, changes in the ionic environment within the cell due to disturbances in the cybernetic mechanism of ion regulation. Another question is whether different normal and malignant cell populations will react similarly to an alteration of the ionic strength of the environment. If changes do occur in the cells, it is important to determine their rate.

Glascoyne et al. (206) followed the change in conductivity of the medium and the leakage of $K^+$ after suspending various cells in low-conductivity medium. The half-time of the increase in conductivity (90% of the cation flux was of $K^+$ ions) was roughly 20 to 30 min, which corresponds to the value obtained by Hu et al. (200) with murine B and T lymphocytes.

In our experiments the complete measurement cycle (three repetitions) for each cell line took about 10 min and was made as soon as cells were suspended in the low-conductivity medium. Results of three successive measurements of the conductivity show very small change (noise level) of that parameter for both normal and malignant cell suspensions (72). This means that neither the ionic composition of the cells nor the cell membrane integrity changed considerably during the time required for the measurements (about 10 min) (72). This was consistent with the results of More et al. (207), Hu et al. (200), and Gascoyne et al. (206).

Furthermore, the size distribution of the cells and the microscopic morphology did not change in a noticeable way, even after an hour of suspension in the low ionic strength medium (72). It was the same as that of cells kept in their growth medium. This implies that no large changes in the intracellular ion composition occurred as a result of the suspension (208).

Our findings are consistent with the microscopic observations made by Hu et al. (200) on B and T murine lymphocytes. They also tested possible damage by the low ionic strength medium by determining the proportion of the cells that were permeable to propidium iodide. Their findings indicated that the resuspension of normal and malignant cells for not more than 10 min did not seem to effect in a considerable way the state of neither the normal nor the malignant cells.

Normal, EBV-transformed and malignant lymphocytes were recently investigated (72). Normal lymphocytes do not live for a prolonged time in culture and do not divide without addition of mitogenic stimuli. One of the methods used to immortalize lymphoid cells, so that they can live and divide under culture conditions, is to transform them with a virus such as the EBV virus. The Magala B cell line was developed in this way. Transformed cells and malignant cells share the capacity to divide in culture. Malignant cells differ from normal and transformed cells by numerous additional features. In our experiments both B- and T-cell populations could be characterized by two positive/negative properties – malignant or nonmalignant nature of the cells (cancer/noncancer), and the capacity or lack of capacity to divide (dividing/nondividing). Thus, the EBV-transformed Magala cells are noncancerous but dividing, while Farage, Raji, Bjab, Daudi, Peer, and HDMAR lines are malignant and can divide in culture. Classifying the cells in this manner and by using the *t*-test, the analysis of cell parameters was carried out,

Analysis of cell membrane capacitance (see Fig. 55) has shown that this parameter is different for various cell populations. For the B lymphocytes the capacitance of membranes of normal cells is higher than that of all malignant cells. The same parameter for the EBV-transformed line (Magala) is intermediate between the values for normal and malignant cells. This probably reflects the fact that this transformed line possesses the same dividing feature as cancer cells, but it is really noncancerous. The difference between transformed (Magala) and malignant lines is statistically significant, whereas there is no statistically significant difference between the nondividing (normal) and transformed (Magala) populations. Thus, it is reasonable to assume that in the B-cell population the decrease in specific capacitance of the cell membrane is more strongly correlated with cancer than with the dividing feature. As for the T-cell population, the membrane capacitance of the malignant cells (see Fig. 55) was smaller than that of the normal T cells. This difference was of borderline statistical significance. However, the trend of decrease in cell membrane capacitance associated with malignancy is the same as within the B-cell group. Yet, at present, it is not possible to draw strong conclusions as in the case of the B-cell group; also, there is a lack of measurements on transformed T cells. This difficulty is true also for other parameters, even if the difference between normal T cells and the malignant ones is statistically significant.

Now let us consider the conductivity of the cell membranes (72). One can see in Fig. 56 that the membrane conductivity of normal cells of both the B and T populations was significantly higher than for that of malignant and transformed cells. There was no notable difference in the conductivity between the transformed and malignant cells in the B-cell population. Thus, in the B-cell population the decrease in cell membrane conductivity seemed to result from the dividing properties of the cells rather than from acquisition of malignant properties. Concerning the conductivity of T cells, the difference between normal and malignant cells was not so large as for B cells, but was statistically significant. Note again that no measurements for T transformed cells were performed.

For normal B cells (see Table 5) $\sigma_{ne}$, $\sigma_{cp}$, and $\sigma_{np}$ are higher than for the other cell populations. Concerning the parameters of transformed cells they are in the value range of cancer cells. Thus, the dividing feature which is imminent to cancer cells as well as to transformed cells seems to be responsible for the change in these parameters in B populations.

It would be important to understand in molecular terms the mechanisms underlying the difference in the various dielectric parameters that were measured. Parameters such as conductivity and permittivity reflect the distribution of free and bound charges, in membranes, respectively, and also the polarizability of various components.

The average ionic composition of T and B human lymphocytes in the quiescent state is $K^+$: 130–150 mM; $Na^+$: 15–30 mM; $Cl^-$: 70–90 mM; and $Ca^{2+}$ $\sim$ 0.1 μM. This composition has also been found in B-CLL lymphocytes (chronic lymphocytic leukemia) (209, 210). Averaging the conductivity of the cytoplasmic $\sigma_{cp}$ and the nucleoplasmic $\sigma_{np}$ (Table 5) of the normal B lymphocytes it was shown to have a value of 130 mM for KCl (72). Thus, as a first approximation, data on the conductivity of the inner cell solution are consistent with an assumption that the conductivities of the cytoplasm and the nucleoplasm are mainly as a result of the ionic species $K^+$, $Cl^-$, and $Na^+$ in aqueous media.

We want to note the fact that the conductivity of the cytoplasm is consistently lower by a factor of $\sim$ 2 than that of the nucleoplasm. This was invariably found in all B- and T-cell lines, both normal and malignant (Table 5). If the conductivity of either the cytoplasm or nucleoplasm indeed expresses mainly the transport of small ions, such as $K^+$, $Cl^-$, and $Na^+$ in an aqueous environment, then there must be a selective barrier between the above two phases, which is very likely the nuclear envelope (NE).

The NE of eukaryotic cells is made up of two concentric membranes, the inner and the outer envelope membrane (211). They are separated by the perinuclear space, but fuse at specific points, where they form the nuclear pore complex. The nuclear pore is composed of up to 100 different proteins, arranged in two octagonal arrays. The nuclear pores are believed to regulate the bidirectional nucleocytoplasm transport of macromolecules, such as mRNA, transcription factors, proteins, etc., which is a process that requires metabolic energy. This means that they indeed act as diffusion barriers. The open inner diameter of the pore was shown to be approximately 90 Å. The large diameter of nuclear pores has led to the conclusion that the nuclear pores are unable to regulate fluxes of ions (diameter of 3 to 4 Å) or maintain a gradient of ions across the NE. Studies, using the "patch-clamp" technique, have detected ion-channels' activity at the NE (212, 213). These findings put into focus the regulation of nuclear ions by the NE. Several classes of $K^+$-selective ion channels were recorded by patch-clamp techniques, with conductances of 100 to 550 pS in nuclei of different cells. Also a large, cation-selective ion channel with a maximum conductance of 800 pS in 100 mM KCl was detected. This channel is a possible candidate for the open nuclear pore, as its conductance is consistent with the geometrical dimension of the open pore. It is important to notice that this channel is open only for short times, and is mainly in the closed mode. These findings are consistent with older data (213) on microelectrode studies with in-situ nuclei, which claimed to measure a potential difference across the NE and reported low electrical conductance of that membrane complex.

The findings that the NE acts as a diffusion barrier can explain the fact that the nucleoplasm and the cytoplasm can have a different steady-state conductivity and probably also different compositions.

As shown in Table 5, the electrical conductivity of the NE is larger by two orders of magnitude than that of the cell membrane. We would then expect that the numbers of ionic channels and their nature, including their gating mechanism, would then be very different in these two types of cellular membranes.

The ability of the noninvasive TDS technique to analyze in situ the properties of the intracellular structures is very important. If indeed the two-shell model represents also the NE and the nucleoplasm, then it describes the very regions of the cell where the putative control of cell division resides, and also the region where the gene expression takes place. Electrical methods are relatively easy to apply (72, 214) and can lead the molecular biologist to decipher more rationally the control mechanism of cell growth in situ.

## ACKNOWLEDGMENTS

The University of Bergen (Foundation for Senior Research Fellowship) is thanked for a grant to Professor Yuri Feldman, facilitating his contribution to this review during his stay in Bergen. The Flucha program, financed by the Norwegian Research Council (NFR) and the oil industry, is also acknowledged for financial support.

## REFERENCES

1. H Fröhlich. Theory of dielectrics. Dielectric constant and Dielectric Loss. 2nd ed. Oxford: Clarendon Press, 1958.
2. CJF Böttcher. Theory of Electric Polarization, Vol 1. 2nd ed. Amsterdam: Elsevier Science B.V., 1993.
3. CJF Böttcher, P Bordewijk. Theory of Electric Polarization, Vol 2. 2nd ed. Amsterdam: Elsevier Science B.V., 1992.
4. NE Hill, WE Yaughan, AH Price, M Davis. Dielectric Properties and Molecular Behavior. London: Van Nostrand, 1969.
5. N Kozlovich, Yu Alexanadrov, A Pusenko, Yu Feldman. Colloids Surfaces A 140: 229–312, 1998.
6. S Havriliak, SJ Havriliak. Polymer 37: 4107, 1996.
7. RR Nigmatullin, YE Ryabov. Phys Solid State. 39: 87, 1997.
8. L Nivanen, R Nigmatullin, A LeMehaute. Le Temps Irreversible a Geometry Fractale. Paris: Hermez, 1998.
9. RH Cole. Dielectric Polarization and Relaxation. NATO ASI Ser., Ser. C, Vol 135. Molecular Liquids, 1984, p 59–110.
10. Yu D Feldman, VV Levin. Chem Phys Lett 85: 528, 1982.
11. T Hanai. Kolloidn Zh. 177: 57, 1960.
12. SS Dukhin, VN Shilov. Dielectric Phenomena and the Double Layer in Disperse Systems and Polyelectrolytes. trans., D Lederman. New York: Halsted Press, 1974.
13. R Pethig. Dielectric and Electronic Properties of Biological Materials. Chichester: John Wiley, 1979.
14. S Takashima. Electric Properties of Biopolymers and Membranes. Bristol: Adam Hilger, 1989.
15. B Nettelblad, GA Niklasson. J Mater Sci 32: 3783, 1997.
16. EM Trukhan. Soviet Phys – Solid State. 4: 2560, 1963.
17. TS Sørensen. J Colloid Interface Sci 168: 437, 1994.
18. JC Maxwell. A Treatise of Electricity & Magnetism, articles 310–314. New York: Dover, 1954 (originally Clarendon Press, 1891).
19. KW Wagner. Arch Electrotech 2: 371, 1914.
20. H Fricke, HJ Curtis. J Phys Chem 41: 729, 1937.
21. RW Sillars. J Inst Electr Engrs 80: 378, 1937.
22. DAG Bruggemann. Ann Phys Lpz 24: 636, 1935.
23. T Hanai. In: P Sherman, ed. Emulsion Science. New York: Academic Press, 1968.
24. C Boned, J Peyrelasse. Colloid Polymer Sci 261: 600, 1983.
25. MH Boyle. Colloid Polymer Sci 263: 51, 1985.
26. P Debye, E Hückel. Phys Z 24: 185, 1923.
27. J Sjöblom, B Jönsson, C Nylander, I Lundström. J Colloid Interface Sci 96: 504, 1983.
28. J Lyklema, SS Dukhin, VN Shilov. J Electroanal Chem 143: 1, 1983.
29. JF Rathman, JF Scamehorn. J Phys Chem 88: 5807, 1984.
30. C Grosse, KR Foster. J Phys Chem 91: 3073, 1987.
31. R Barchini, DA Saville. J Colloid Interface Sci 173: 86, 1995.
32. G Schwarz. J Phys Chem 66: 2636, 1962.
33. RH Cole, JG Berberian, S Mashimo, G Chryssik, A Burns, E Tombari. J Appl Phys 66: 793, 1989.
34. S Mashimo, T Umehara, T Ota, S Kuwabara, N Shinyashiki, S Yagihara. J Molec Liquids 36: 135, 1987.
35. D Bertolini, M Cassettari, G Salvetti, E Tombari, S Veronesi. Rev Sci Instrum 61: 450, 1990.
36. R Nozaki, TK Bose. IEEE Trans Instrum Meas 39: 945, 1990.
37. B Gestblom, E Noreland, J Sjöblom. J Phys Chem 91: 6329, 1987.
38. Yu Feldman, A Andrianov, E Polygalov, G Romanychev, I Ermolina, Yu Zuev, B Milgotin. Rev Sci Instrum 67: 3208, 1996.
39. A Schönhals, F Kremer, E Schlosser. Phys Rev Lett 67: 999, 1991.
40. U Kaatze, V Lonneckegabel, R Pottel. Z Phys Chem (Int J Res Phys) 175: 165, 1992.
41. K Folgerø, T. Friisø, J Hilland, T Tjomsland. Meas Sci Technol 6: 995, 1995.
42. JG Berberian. J Molec Liquids 56: 1, 1993
43. B Gestblom, P Gestblom. Macromolecules 24: 5823, 1991.
44. YuD Feldman, YuF Zuev, EA Polygalov, VD Fedotov. Colloid Polymer Sci 270: 768, 1992.

45. N Miura, N Shinyashiki, S Yagihara, M Shiotsubo. J Am Ceram Soc 81: 213, 1998.
46. Yu Feldman, V Levin. Chem Phys Lett 85: 528, 1982.
47. AM Bottreau, A Merzouki. IEEE Trans Instrum Meas 42: 899, 1993.
48. EK Miller, ed. Time Domain Measurements in Electromagnetics. New York: Van Nostrand Reinhold, 1986.
49. NE Hager III. Rev Sci Instrum 65: 887, 1994.
50. O Götmann, U Kaatze, P Petong. Meas Sci Technol 7: 525, 1996.
51. JG Berberian, RH Cole. Rev Sci Instrum 63: 99, 1992.
52. B Gestblom, H Førdedal, J Sjöblom. J Disp Sci Technol 15: 449, 1994.
53. T Skodvin, T Jakobsen, J Sjöblom. J Disp Sci Technol 15: 423, 1994.
54. GQ Jiang, WH Wong, EY Raskovich, WG Clark, WA Hines, J Sanny. Rev Sci Instrum 64: 1614, 1993.
55. T Jakobsen, K Folgerø. Meas Sci Technol 8: 1006, 1997.
56. YJ Lu, HM Cui, J Yu, S Mashimo. Bioelectromagnetics 17: 425, 1996.
57. S Naito, M Hoshi, S Mashimo. Rev Sci Instrum 67: 3633, 1996.
58. JZ Bao, ST Lu, WD Hurt. IEEE Trans Microwave Theory Technol 45: 1730, 1997.
59. T Skodvin, J Sjöblom. Colloid Polym Sci 274: 754, 1996.
60. MS Seyfried, MD Murdock. Soil Sci 161: 87–98, 1996.
61. JM Anderson, CL Sibbald, SS Stuchly. IEEE Trans Microwave Theory Technol 42: 199, 1994.
62. DV Blackham, RD Pollard. IEEE Trans Instrum Meas 46: 1093, 1997.
63. HP Schwan. Ann NY Acad Sci 148: 191, 1968.
64. EH Grant, RJ Sheppard, GP South. Dielectric Behavior of Biological Molecules in Solution. Oxford: Clarendon Press, 1978.
65. J Lyklema. Fundamentals of Interface and Colloid Science. Vol II. Solid–Liquid Interfaces. London: Academic Press, 1995.
66. J O'M Bockris, AKN Reddy. Modern Electrochemistry. Vol 2. New York: Plenum Press, 1977.
67. HP Schwan. Ann Biomed Eng 20: 269, 1992.
68. HP Schwan. Physical Techniques in Biological Research. Vol 6. New York: Academic Press, 1963.
69. CL Davey, GH Marks, DB Kell. Eur Biophys J 18: 255, 1990.
70. R Lisin, B Ginzburg, M Schlesinger, Yu Feldman. Biochim Biophys Acta 1280: 34, 1996.
71. Yu Feldman, R Nigmatullin, E Polygalov, J Texter. Phys Rev E 58: 2179, 1998.
72. Yu Polevaya, I Ermolina, M Schlesinger, B-Z Ginzburg, Yu Feldman. Biochim Biophys Acta 1419: 257–271, 1999.
73. I Ermolina, V Fedotov, Yu Feldman. Phisica A 249: 347, 1998.
74. RR Nigmatullin, in The Proceedings of Bordeaux Summer School–Geometrie Fractale et Hyperbolique. Derivation Fractionnaire et Fractale 3–8 July 1994. Fractals, Solutions and Chaos 11: 1994.
75. T Pajkossy, L Nyikos. Phys Rev B 42: 709, 1990.
76. T Pajkossy. J Electroanal Chem 364: 111, 1994.
77. T Pajkossy, AP Borosy, A Imre, SA Martemyanov, G Nagy, R Schiller, L Nyikos. J Electroanal Chem 366: 69, 1994.
78. SH Liu, Phys Rev Lett 55: 529–532, 1985.
79. T Kaplan, LJ Gray. Phys Rev B 32: 7360, 1985.
80. RM Hill, LA Dissado, RR Nigmatullin. J Phys Condens Matter 3: 9773, 1991.
81. K Oldham, J Spanier. The Fractional Calculus. New York: Academic Press, 1974.
82. M Stromme, A Niklasson, CG Granqvist. Phys Rev B 52: 14192, 1996.
83. JB Hasted. Aqueous Dielectrics. London: Chapman and Hall, 1973.
84. EH Grant, TJ Buchanan, HF Cook. J Chem Phys 26: 156, 1957.
85. B Gestblom, S Urban. Z Naturforsch 50a: 595, 1995.
86. B Pettersen, T Skodvin, J Sjöblom. Colloids Surfaces A 143: 323, 1998.
87. SE Friberg. Microemulsions: Structure and Dynamics. Boca Raton, FL: CRC Press, 1987.
88. B Lindman, ed. Surfactants, Adsorption, Surface Spectroscopy and Disperse Systems. Darmstadt: Steinkopff, 1985.
89. D Langevin, Annu Rev Phys Chem 43: 341, 1992.
90. J Sjöblom, R Lindberg, SE Friberg. Adv Colloid Interface Sci 95: 125, 1996.
91. U Olsson, K Shinoda, B Lindman. J Phys Chem 90: 4083, 1986.
92. PG DeGennes, C Touplin. J Phys Chem 86: 2294, 1982.
93. R Strey, MJ Jonströmer. J Phys Chem 96: 4537, 1992.
94. R Kahlweit, R Strey and G Busse. J. Phys. Chem., 95, 5344, 1991.
95. MA Dijk, G Casteleijn, JGH Joosten, YK Levine. J Chem Phys 85: 626, 1986.
96. Yu Feldman, N Kozlovich, I Nir, N Garti. Phys Rev E 51: 478, 1995.
97. Yu Feldman, N Kozlovich, Y Alexandrov, R Nigmatullin, Y Ryabov. Phys Rev E 54: 5420, 1996.
98. C Cametti, P Codastefano, P Tartaglia, S Chen, J Rouch. Phys Rev A 45: R5358, 1992.
99. F Bordi, C Cametti, J Rouch, F Sciortino, P Tartaglia. J Phys: Condens Matter A19: 8, 1996.
100. C Boned, J Peyrelasse, Z Saidi. Phys Rev E 47: 468, 1993.
101. A Ponton, TK Bose, G Delbos. J Chem Phys 94: 6879, 1991.
102. A D'Aprano, G D'Arrigo, A Paparelli. J Phys Chem 97: 3614, 1993.
103. M Wasserman. Russ Chem Rev 63: 373, 1994.

104. F Mallamace, N Micali, C Vasi, G D'Arrigo. Phys Rev A 43: 5710, 1991.
105. F Sicoli, D Langevin. J Phys Chem 99: 14819, 1995.
106. M Clausse. In: P Becher, ed. Encyclopedia of Emulsion Technology. Basic Theory. Vol 1. New York: Marcel Dekker, 1983.
107. Yu Feldman, N Kozlovich, I Nir, N Garti, V Archipov, Z Idiyatullin, Y Zuev, V Fedotov. J Phys Chem 100: 3745, 1996.
108. JA Beunen, E Ruckenstein. J Colloid Interface Sci. 96: 469, 1983.
109. E Ruckenstein, JA Beunen. Langmuir 4: 77, 1988.
110. HF Eicke, M Bercovec, B Das-Gupta. J Phys Chem 93: 314, 1989.
111. DG Hall, J Phys Chem 94: 429, 1990.
112. SI Chou, DO Shah. J Phys Chem 85: 1480, 1981.
113. C Cametti, P Codastefano, P Tartaglia. Ber Bunsenges Phys Chem 94: 1499, 1990.
114. J Peyrelasse, C Boned. J Phys Chem 89: 370, 1985.
115. MA Dijk, JGH Joosten, YK Levine, D Bedeaux. J Phys Chem 93: 2506, 1989.
116. J Lang, A Jada, A Malliaris. J Phys Chem 92: 1946, 1988.
117. N Kozlovich, A Puzenko, Yu Alexandrov, Yu Feldman. Phys Rev E 58: 2179, 1998.
118. M D'Angelo, D Fioretto, G Onori, L Palmieri, A Santucci. Phys Rev E 52: R4620, 1995.
119. M D'Angelo, D Fioretto, G Onori, L Palmieri, A Santucci. Phys Rev E 54: 993, 1996.
120. G Giammona, F Goffrede, V Turco Liveri, G Vassallo. J Colloid Interface Sci 154: 411, 1992.
121. P Huang Kenez, G Carlström, I Furo, B Halle. J Phys Chem 96: 9524, 1992.
122. M Tomic, N Kallay. J Phys Chem 96: 3874, 1992.
123. P Karpe, E Ruckenstein. J Colloid Interface Sci 137: 408, 1990.
124. M Boas. Mathematical Methods in Physical Sciences. New York: John Wiley, 1983, p 12.
125. M Belletete, M Lachapelle, G Durocher. J Phys Chem 94: 5337, 1990.
126. ALR Bug, SA Safran, GS Grest, I Webman. Phys Rev Lett 55: 1896, 1985.
127. G Grest, I Webman, S Safran, A Bug. Phys Rev A 33: 2842, 1986.
128. C Cametti, P Codastefano, A Di Biasio, P Tartaglia, S Chen. Phys Rev A 40: 1962, 1989.
129. J Klafter, A Blumen. Chem Phys Lett 119: 377, 1985.
130. J Klafter, MF Shlesinger. Proc Natl Acad Sci USA 83: 848, 1986.
131. D Stauffer, A Aharony. Introduction to Percolation Theory. Revised 2nd edn. London: Taylor & Francis, 1994.
132. D Vollmer, J Vollmer, HF Eicke. Europhys Lett 26: 389, 1994.
133. D Vollmer. Europhys Lett 27: 629, 1994.
134. BB Mandelbrot. The Fractal Geometry of Nature. New York: Freeman, 1982.
135. F Feder. Fractals. New York: Plenum Press, 1988.
136. S-H Chen, J Rouch, F Sciortino, P Tartaglia. J Phys: Condens Matter 6: 10855, 1994.
137. A Bunde, S Havlin, J Klafter, G Graff, A Shehter. Phys Rev Lett. 78: 3338, 1997.
138. LH Peebles, Jr. Molecular Weight Distributions in Polymers. New York: Wiley-Interscience, 1971.
139. RR Nigmatullin. J Appl Magn Reson 14: 601, 1998.
140. Yu Alexandrov. Dielectric Spectroscopy Investigation of Dynamic and structural properties of microemulsions. Thesis. Jerusalem, 1998.
141. Yu Feldman, N Kozlovich, I Nir, N Garti. Colloids Surfaces A 128: 47, 1997.
142. J Sjöblom, B Gestblom. J Colloid Interface Sci 115: 535–544, 1987.
143. Yu Feldman, N Kozlovich, I Nir, A Aserin, S Ezrahi, N Garti. J Non-Crystal Solids 172–174: 1109, 1994.
144. SC Mehrotra, B Gestblom, J Sjöblom. Finn Chem Lett 34: 87, 1985.
145. E Noreland, B Gestblom, J Sjöblom. J Solution Chem 18: 303, 1989.
146. MK Kreoger. J Molec Liquids 36: 101, 1987.
147. PC Brot. Ann Phys 2: 714, 1957.
148. SK Garg, CP Smyth. J Phys Chem 69: 1294, 1965.
149. VV Levin, YuD Feldman. Chem Phys Lett 87: 162, 1982.
150. M Fukuzaki, U Toshihiro, D Kurita, S Shioya, M Haida, S Mashimo. J Phys Chem 96: 1087, 1992.
151. Y Marcus. Cell Biochem Function 13: 157, 1995.
152. U Kaatze. J Solution Chem 28: 1049, 1997.
153. J Sjöblom, B Gestblom. In: SE Friberg, B Lindman, eds. Organized Solutions. Surfactants in Science and Technology. New York: Marcel Dekker, 1992, p 193.
154. A Kraszewski. J Microwave Power 12: 215, 1977.
155. A Kraszewski, S Kulinski, M Matuszewski. J Appl Phys 47: 1275, 1976.
156. N Kozlovich, I Nir, B Tsentsiper, VI Zhuravlev, A Aserin, S Ezrahi, N Garti, Yu Feldman. J Surfc Sci Technol. in press.
157. N Garti, A Aserin, S Ezrahi, I Tiunova, G Berkovic. J Colloid Interface Sci 178: 60, 1996.
158. K Asami, T Hanai. Colloid Polymer Sci 270: 78, 1992.
159. ID Chapman. J Phys Chem 75: 537, 1971.
160. B-M Sax, G Schön, S Paasch, MJ Schwuger. Progr Colloid Polymer Sci 77: 109, 1988.
161. T Hanai. Kolloidn Zh 171: 23, 1960.
162. T Hanai. Kolloidn Zh 175: 62, 1961.
163. U Genz, JA Helsen, J Mewis. J Colloid Interface Sci. 165: 212, 1994.
164. J Sjöblom, T Skodvin, T Jakobsen, SS Dukhin. J Disp Sci Technol 15: 401, 1994.
165. T Skodvin, J Sjöblom, JO Saeten, O Urdahl, B Gestblom. J Colloid Interface Sci 166: 43, 1994.

166. T Skodvin, J Sjöblom. J Colloid Interface Sci 182: 190, 1996.
167. T Hanai, K Sekine. Colloid Polymer Sci 265: 888, 1986.
168. T Skodvin, J Sjöblom, JO Saeten, T Wärnheim, B Gestblom. Colloids Surfaces A 83: 75, 1994.
169. J Sjöblom, T Skodvin, T Jakobsen. J Disp Sci Technol 15: 423, 1994.
170. H Førdedal J Sjöblom. J Colloid Interface Sci 181: 1996, 589–594.
171. G Capuzzi, P Baglioni, CMC Gambi, EY Sheu. Phys Rev E 60: 792, 1999.
172. J Peyrelasse, M Moha-Ouchane, C Boned. Phys Rev A 38: 904, 1988.
173. C Thomas, JP Perl, DT Wasan. J Colloid Interface Sci 139: 1, 1990.
174. JP perl, HW Bussey, DT Wasan. J Colloid Interface Sci 108: 528, 1985.
175. M von Stackelberg. Naturwissenschaften 11: 327, 1949.
176. M von Stackelberg, HR Müller. Z Elektrochem 58: 25, 1954.
177. M von Stackelberg, W Jahns. Z Elektrochem 58: 162, 1954.
178. JA Ripmeester, JS Tse, CI Ratcliffe, BM Powell. Nature 325: 135, 1987.
179. T Jakobsen, J Sjöblom, P Ruoff. Colloids Surfaces A 112: 73, 1996.
180. J Sjöblom, H Førdedal, T Jakobsen, T Skodvin. In: KS Birdi, ed. Handbook of Surfactants. Boca Raton, FL: CRC Press, 1997.
181. B Pettersen, J Sjöblom. Colloids Surfaces A 113: 175, 1996.
182. B Pettersen, L Bergflødt, J Sjöblom. Colloids Surfaces A 127: 175, 1997.
183. HP Schwan. Adv Biol Med Phys 5: 147, 1957.
184. F Bordi, C Cametti, A Rosi, A Calcabrini. Biochim Biophys Acta 1153: 77–88, 1993.
185. H Looyenga. Physica 31: 401, 1965.
186. T Hanai, K Asami, N Koizumi. Bull Inst Chem Res Kyoto Univ 57: 297, 1979.
187. K Asami, T Hanai, N Koizumi. Jap J Appl Phys 19: 359, 1980.
188. K Asami, Y Takahashi, S Takashima. BBA 1010: 49, 1989.
189. K Asami, A Irimajiri. Biochim Biophys Acta 778: 570, 1984.
190. A Irimajiri, T Hanai, A Inouye. J Theor Biol 78: 251, 1979.
191. A Irimajiri, Y Doida, T Hanai, A Inouye. J Mem Biol. 38: 209, 1978.
192. K Asami, Y Takahashi, S Takashima. Biophys J. 58: 143, 1990.
193. H Fricke. Nature 4381: 731, 1963.
194. H Fricke. J Phys Chem 57: 934, 1953.
195. R Pethig, DB Kell. Phys Med Biol 32: 933, 1987.
196. A Surowiec, SS Stuchly, C Izaguirre. Phys Med Biol 31: 43, 1986.
197. FF Becker, X-B Wang, Y Huang, R Pethig, J Vykoukal, PRC Gascoyne. Proc Natl Acad Sci USA 92: 860, 1995.
198. A Irimajiri, K Asami, T Ichinowatari, Y Kinoshita. Biochim Biophys Acta 869: 203, 1987.
199. A Irimajiri, K Asami, T Ichinowatari, Y Kinoshita. Biochim Biophys Acta 896: 214–223, 1987.
200. X Hu, WM Arnold, U Zimmerman. Biochim Biophys Acta 1021: 191–200, 1990.
201. GW Snedecor, WG Cochran. Statistical methods. Ames, IA: The Iowa State College Press, 1956.
202. H Pauly, P Schwan. Z Naturforsch 14b: 125, 1959.
203. I Ermolina, Yu Polevaya, Yu Feldman. Eur Biophys J Biophy 29: 141–145, 2000.
204. M Ginzburg, B-Z Ginzburg. Biochim Biophys Acta 584: 398, 1979.
205. A Zmiri, Z Ginzburg. Plant Sci Lett 30: 211, 1983.
206. PRC Gascoyne, R Pethig, JPH Burt, FF Becker. Biochim Biophys Acta 1149: 119–120, 1993.
207. R More, I Yron, S Ben-Sasson, DW Weiss. Cellular Immunol 15: 382, 1975.
208. BCK Rossier, K Geeringh, JP Kraehenbuhl. Trends Biochem Sci 12: 483, 1987.
209. GB Segal, MA Lichtman. J Cell Physiol 93: 277, 1977.
210. S Grinstein, J Dixon. Physiol Rev 69: 417, 1989.
211. C Dingwell, R Laskey. Science 258: 942, 1992.
212. AJM Matzke, AM Matzke. Bioelectrochem Bioenergetics 25: 357, 1991.
213. JO Bustamente. J Mem Biol 138: 105, 1994.
214. MT Santini, C Cametti, PL Indovina, G Morelli, G Donelli. J Biomed Mater Res 35: 165, 1997.

# 7

# Electroacoustic Characterization of Emulsions

**Robert J. Hunter**
*University of Sydney, Sydney, New South Wales, Australia*

## I. INTRODUCTION

Electroacoustics provides a unique opportunity to estimate both the size of emulsion droplets and the state of the surface (kinetic) charge in a single measurement. There are two principal methods used: the colloid vibration potential (or current) (CVP or CVC) and the electrokinetic sonic amplitude (ESA). The CVP and CVC have mostly been used in the kilohertz region and the ESA method in the megahertz region. The theoretical developments are somewhat different in these two regimes, although complete theories should yield identical data from the two sources. Measurements in the kilohertz range have, for the most part, been limited to a single frequency which can only provide measurements of the zeta potential (if the particle size is known). When measurements are taken over a range of megahertz frequencies (0.3–20 MHz, say), as has been done in the ESA mode, the possibility emerges of determining both the particle size distribution and the zeta potential simultaneously. Such measurements, sometimes referred to as *electroacoustic spectroscopy* (1), can be made in concentrated emulsion systems with provision of useful data up to concentrations in excess of 60% by volume. The results can also be used to investigate details of the fluid flow in the neighborhood of the particle surface in the presence, for example, of adsorbed polymer molecules.

Measurement of the stability and anticipated rheological behavior of emulsions has, in the past, been limited by the difficulty of determining the surface charge on emulsion droplets at the normal concentrations at which they occur in industrial and biological systems. Even the relatively modest concentration of fat droplets in milk (about 3%) produces a fluid which is optically opaque and not measurable by the normal procedures of electrophoresis. More recent developments, in which light-scattering methods are coupled with optical fibers to introduce a light beam into the sample and extract the scattered beam, still suffer from problems of interpretation. The alternative procedure, of diluting the emulsion before measurement, is far from satisfactory. Even if one were able to find the correct diluent (to duplicate the electrolyte solution which bathes the droplets) the dilution process itself changes the phase-volume ratio and hence alters the distribution of any component that is soluble in the oil and water phases. The problem is compounded by the fact that one can never be sure that such redistribution processes are not contributing to the end result (especially if some unusual behavior is being investigated).

The possibility of making measurements directly on an emulsion of essentially any concentration up to around 60% is therefore a very appealing one. When the same measurement can yield a consistent measure of both the size and the electrokinetic charge on the particles, then the method becomes of unique value. That is the present situation with electroacoustic measurements of the ESA effect in the megahertz frequency range.

The two principal electroacoustic processes are the CVP and the ESA effect. The CVP is an outgrowth of the first proposed electroacoustic effect, namely, the *ion vibration potential* (IVP) which was investigated theoretically by Debye in 1933 (2) as a possible method for estimating the hydration numbers of the various ions. Debye showed that if a sound wave were passed through a salt solution it would disturb the ions and their surrounding atmospheres. This would create an array of tiny dipoles which would give rise to a macroscopically measurable potential difference between the peak and trough of the sound wave. Although there were formidable experimental difficulties to overcome (3) before the theory could be adequately tested, some progress was made in this area in the 1960s. It has recently proved possible to develop improvements in the theoretical treatment, which are giving more consistent results (4), although there remain some unresolved inconsistencies in the experimental data.

It was recognized very early (5), however, that the corresponding effect in a colloidal suspension (the CVP) should be much larger and easier to use since in that case the dipole would be created by the particle and its surrounding double layer. That has proved to be the case and the application of the method to the investigation of polyelectrolytes and proteins has been reviewed by Zana and Yeager (3). More recently, the application of the CVP method to colloidal suspensions and emulsions has been reviewed by Marlow *et al.* (6) and the more recent developments will be discussed in the next chapter. Here, we will concentrate on the developments in the alternate electroacoustic procedure, the ESA effect. This refers to the production of a sound wave when a high-frequency electric field is applied to a suspension or emulsion.

The ESA effect was first recognized in the early 1980s by engineers at Matec Applied Sciences (Hopkinton, MA, USA) who patented the application of the method (7) to colloidal systems. The result was an instrument (the ESA-8000) which operated at a single frequency (around 1 MHz) and was able to do little more that determine the isoelectric point of a suspension, i.e., the point at which the zeta potential ($\zeta$) passes through zero when the suspension is titrated with a reagent capable of changing the effective surface charge. The subsequent development by O'Brien (8) of an adequate theory for the ESA effect made it possible to estimate the zeta potential from the measured ESA signal if the particle size were known. More importantly, the theory showed that if the measurements were made over a range of frequencies and one could accurately measure both the magnitude of the sound signal and the phase relationship between the applied field and the resulting sound response then the method could be used to measure both the zeta potential and the size simultaneously. An instrument which does that, called the AcoustoSizer (Colloidal Dynamics Inc., Warwick, RI), has been available commercially since 1994. Developments in the measurement of the ESA effect have been reviewed elsewhere (9).

A minimal arrangement for observing the electroacoustic effects is displayed in Fig. 1. The application of the electric field to an emulsion of charged particles causes the droplets to oscillate backwards and forwards with the same frequency as the field. The drops are driven in one direction by the field and the surrounding double layer is driven in the opposite direction. This motion causes the formation of an acoustic dipole at each droplet, but in the body of the emulsion the dipoles cancel one another. Near the electrodes the cancellation does not occur and the dipoles reinforce one another to create a sound wave which emerges from the emulsion and travels down the delay rod to the transducer. A second signal, arising from the left-hand electrode, travels through the emulsion and then down the delay rod to the transducer, arriving a few microseconds later.

It must be borne in mind that these effects can also occur in salt solutions, so the resulting signal is the sum of signals derived from the emulsion droplets and the surrounding electrolyte. Normally, however, the size of the electrolyte signal is much smaller than that of the droplets, and can be ignored, unless the charge on the drops is very low.

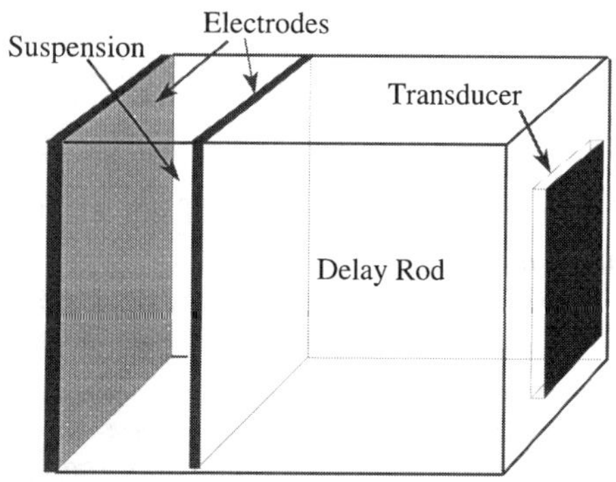


**Figure 1** Minimal arrangement for observation and measurement of the ESA or CVP effect.

## II. ELECTROACOUSTIC THEORY

O'Brien's theoretical analysis (8, 10) is for a suspension of solid particles, but the evidence to date indicates that emulsion droplets behave in the same way as solid particles at the frequencies involved in the ESA effect. This is understandable on a number of counts. First, it is usually observed that surfactant-stabilized emulsion droplets in a flow field do not behave as though they were liquid. The presence of the stabilizing layer at the interface restricts the transfer of momentum across the phase boundary so that there is little or no internal motion in the drop. Also, the motions which are involved are extremely small (involving displacements of the order of fractions of a nanometer) so the perturbations are small compared to the size of the drop. Finally, O'Brien has shown in some unpublished calculations that if the surface is unsaturated, so that the surfactant groups can move under the influence of the electric field, then the effect on the electroacoustic signal would depend on the quantity $d\gamma/d\Gamma$, where $\gamma$ is the surface tension and $\Gamma$ is the surface excess of the surfactant. We have not been able to find any evidence for such an effect, if it exists, so we will assume that the analysis for a solid particle holds also for emulsions.

O'Brien showed in his initial analysis (8) that there was a reciprocal relationship between the CVP and the ESA effects so that essentially the same information could be obtained from either. However, it transpires that the information is easier to obtain from the ESA effect because it appears directly. The same information can be obtained from the CVP only if one knows the complex conductivity of the system. This limitation can, however, be overcome by measuring the CVC. O'Brien's initial analysis was confined to dilute systems, but was subsequently extended to systems of arbitrary concentration as long as the particles were small compared to the wavelength of the generated sound (10). This condition is always fulfilled in practice for the normal emulsion sizes and for frequencies up to 20 MHZ for which the wavelength is of order 100 μm. The reciprocal relationship between CVP and ESA has been demonstrated for solutions of polyelectrolytes by O'Brien *et al.* (11).

The particle property which is extracted from the measured ESA response is the *dynamic mobility*, $\mu_d$ of the drops. This is a complex quantity, having a magnitude and a phase angle (just as the ESA signal is a complex quantity). The magnitude of $\mu_d$ is analogous to the electrophoretic mobility obtained in, say, an electrophoresis experiment, where a d.c. field is applied. It is essentially determined by the electrokinetic charge (or the zeta potential) on the drops. As the frequency of the electric field is increased, the particles are able to follow the field quite well up to frequencies in the kilohertz range, but the magnitude of the mobility gradually decreases with increase in frequency. The effect is small for small drops but larger for larger drops. At the same time, the lag between the applied field and the resulting sound signal increases with frequency and this is reflected in the phase angle of the dynamic mobility. It, too, increases from zero for small particles to a value of around 45° for larger particles and/or higher frequencies. It is this effect which enables the size to be obtained from the measured signal. Since both the magnitude and the phase angle depend upon the frequency for drops of a given size, it is possible to use the two effects to obtain a more reliable assessment of both the zeta potential and the size of the particles from a measurement over a range of frequencies.

To determine the precise relationship between the ESA signal and the dynamic mobility one must solve the set of differential equations given by O'Brien in his 1990 paper (10). For the AcoustoSizer that problem is simplified by the geometry because the electrode dimensions and separation are both large compared to the wavelength of the sound (of millimeter order at the frequencies used). In that case the relation is given by O'Brien *et al.* (12) as:

$$\mathrm{ESA} = A(\omega)\phi(\Delta\rho/\rho)\mu_d \cdot [Z_e Z_r/(Z_e + Z_r)] \quad (1)$$

where $A(\omega)$ is an instrument function [which depends, among other things, on frequency ($\omega = 2\pi\times$ frequency in hertz) and conductivity], $\phi$ is the volume fraction of the emulsion, and $\Delta\rho$ is the difference in density between the drops and the surrounding medium (density $\rho$). The $Z$ functions are acoustic impedances of the emulsion (e) and the delay rod (r), respectively.

The acoustic impedance function measures how effectively the sound signal is transferred from the emulsion to the delay rod. The value of $Z_r$ is well defined and constant (equal to the product of the density and the sound-wave velocity in the medium), and in dilute solutions the value of $Z_e$ is little different from that of the suspension medium. In that case the impedance factor is constant and can be incorporated into the function $A$ which is an instrument calibration factor. That procedure is used in the Matec ESA-8000 device. For more concentrated emulsions, $Z_e$ depends on the density and volume fraction of the drops so it is necessary to monitor the acoustic impedance directly. That is done in the AcoustoSizer so that it is possible to measure the dynamic mobility accurately up to concentrations of order 60% by volume.

## A. Relationship Between Dynamic Mobility and Particle Properties for Dilute Systems

The analysis of the relationship between the dynamic mobility and the particle properties has been made possible by the development of special procedures for dealing with systems in which the double layer around the particle or droplet is thin compared to the radius of curvature. The double-layer thickness is measured by the Debye–Hückel parameter $\kappa$ which is related to the ionic strength of the electrolyte (13). For a 1 mM solution of a 1:1 electrolyte, the double-layer thickness, $\kappa^{-1}$, is about 10 nm and it decreases as the square root of the concentration, so for a 0.1 M solution it would be about 1 nm. The double layer is regarded as thin if the ratio of radius to thickness ($\kappa a$) exceeds about 20 and that will be the case for most normal emulsions at most electrolyte concentrations.

O'Brien has shown (10) that for a dilute suspension of spherical particles (less than about 4% by volume, say) with thin double layers, the dynamic mobility is related to the particle properties as follows:

$$\mu_{\rm d} = \frac{2\epsilon\zeta}{3\eta} G\left(\frac{\omega a^2}{v}\right)[1 + f(\lambda, \omega')] \tag{2}$$

where $\epsilon$ is the dielectric permittivity and $\eta$ is the viscosity of the medium, $a$ is the particle radius, and $v$ is the kinematic viscosity of the suspension medium ($= \eta/\rho$). The functions $G$ and $f$ are both complex and measure the effects of inertia and the bending of the electric field around the particles or droplets, respectively. They are given by the following expressions:

$$G(\alpha) = \frac{1 + (1+i)\sqrt{(\alpha/2)}}{1 + (1+i)\sqrt{(\alpha/2)} + i\dfrac{\alpha}{9}\left(3 + 2\dfrac{\Delta\rho}{\rho}\right)} \tag{3}$$

$$f(\lambda, \omega') = \frac{1 + i\omega' - \left[2\lambda + i\omega'\dfrac{\epsilon_p}{\epsilon}\right]}{2(1 + i\omega') + \left[2\lambda + i\omega'\dfrac{\epsilon_p}{\epsilon}\right]} \tag{4}$$

where $\omega' = \omega\epsilon/K^{\infty}$ and $\alpha = \omega a^2/v$. Here, $\epsilon_{\rm p}$ is the particle permittivity and $K^{\infty}$ is the bulk conductivity of the electrolyte suspension. The parameter $\lambda$ will be described shortly.

The $G$ factor is determined by the size of the droplets, $a$; the variation of $G$ with the parameter $\alpha$ is shown in Fig. 2. For small particle sizes and/

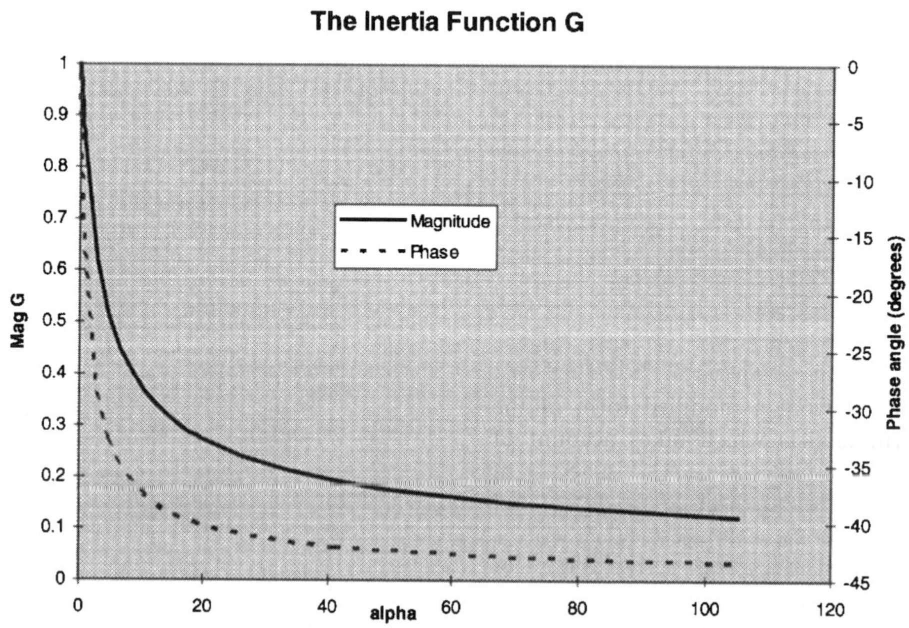


**Figure 2** Inertia function $G(\alpha)$ where $\alpha = \omega a^2/v$. At 1 MHz in water at room temperature $\alpha \approx 6a^2$ for $a$ in μm.

or low frequencies ($\alpha \to 0$) the value of $G$ is 1 and it has a zero phase angle. The inertia effect is then negligible and the particles essentially behave as they do in a d.c. field. As $\alpha$ increases, the magnitude of $G$ decreases to zero and the phase angle becomes more negative, ultimately reaching a value of 45°. For a 1-μm droplet in water at 20°C $\omega a^2/v = 1$ at a frequency of 0.15 MHz. For a 0.1-μm droplet the corresponding frequency is 15 MHz. The frequency range of the AcoustoSizer (0.3–11 MHz) thus corresponds to a size range (for dilute systems) from 0.1 to 10 μm diameter. The range is, however, shifted to larger values for more concentrated systems. There is some contribution (mostly positive) to the phase angle from the function $f$, especially for higher values of the $\zeta$-potential and lower electrolyte concentrations. In principle, the phase angle of $\mu_d$ alone could be used to determine the drop size, and the magnitude then used to determine the $\zeta$-potential. In practice it turns out to be better to use the variation of both elements of $\mu_d$ to determine both parameters simultaneously.

The factor $(1+f)$ in Eq. (2) measures the tangential electric field at the particle surface. It is this component which generates the electrophoretic or electroacoustic motion. For a fixed frequency, it can be seen from Eq. (4) that $(1+f)$ depends on the permittivity of the particles and on the function $\lambda = K_s/K^{\infty}a$, where $K_s$ is the surface conductance of the double layer; $\lambda$ measures the enhanced conductivity due to the charge at the particle surface. It is usually small unless the zeta potential is very high, so for most emulsions with large $\kappa a$, $\lambda$ has a negligible effect. The ratio $\epsilon_P/\epsilon$ is also small for oil-in-water emulsions. Equation (4) can then be reduced to $f = 0.5$ and hence the dynamic mobility becomes:

$$\mu_d = (\epsilon\zeta/\eta)G(\omega a^2/v) \tag{5}$$

Now the frequency dependence and the phase lag are determined entirely by the inertia term $G$, and the zeta potential is calculated from a modified form of the Smoluchowski formula (41) which takes account of the inertia effect for the larger particles, especially at the higher frequencies. The determination of size and charge is particularly simple in this case.

Figure 3 gives a good indication of the validity of the theory, for solid submicrometer spherical particles of silica. The spherical nature of emulsion drops means that particle shape is not likely to be a problem under most circumstances.

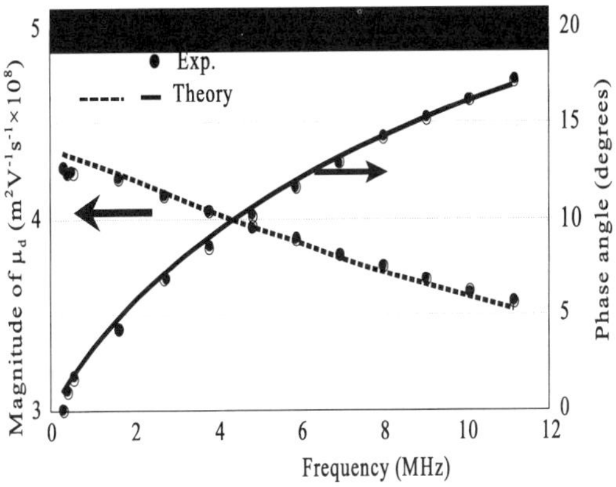


**Figure 3** Comparison of theoretical (10) and experimental values of the magnitude and phase of the dynamic mobility for a silica sol of radius 300 nm.

## III. INSTRUMENTATION

### A. The Matec ESA-8000

A number of descriptions of this instrument have already appeared in the literature (14, 15), including a full description of its main features by Cannon (16). Sayer (17) also provides a general block diagram. The instrument comes in a number of different configurations: a flow-through cell (the PPL-80 sensor) and a dip-type probe (SPP-80). The flow-through cell has flat electrodes separated by a few millimeters, while the dip probe (Fig. 4) has a circular disk electrode and a thin-bar counter-electrode again a few millimeters away. In

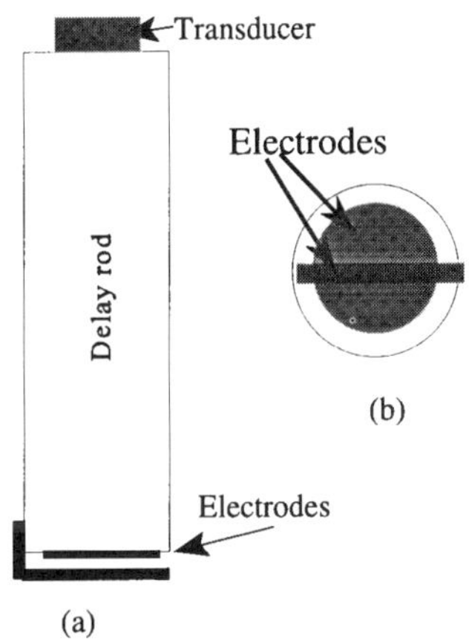


**Figure 4** Dip-type probe for the ESA-8000 device. The electrode spacing is a few millimeters, corresponding to 1.5 wavelengths of sound in the liquid medium.

each case the electrode spacing is designed to establish a resonance condition (at 3/2 times the wavelength) in the space between the electrodes, so that the sensitivity is enhanced.

The ESA-8000 can be used for both CVP and ESA measurements, depending on whether the electric field (in the form of a short pulse, or more strictly a "tone-burst") is applied to the transducer or to the electrodes of the cell. In the latter case (the ESA mode) the generated sound wave travels along the delay rod to the transducer and the resulting voltage pulse is then sent to the signal processor unit for estimation of the magnitude and phase angle, using quadrature detection (16). The delay rod is essential because the application of the field to the cell electrodes results in an immediate signal in the transducer due simply to electromagnetic coupling. The delay rod must be long enough so that that "noise" has died away before the sound signal arrives for measurement.

When the ESA-8000 was first introduced there was no adequate theory on which to base interpretation of the signals. Since the instrument measures at only one frequency it does not provide enough information to estimate the size. It seemed reasonable to assume, however, that the magnitude of the sound signal was related to the amount of charge on the particles and this could be calibrated to some extent by using a standardizing colloid of known charge (or $\zeta$-potential). For this purpose, a commercial nanometer-sized silica sol (Ludox) was normally used. After O'Brien's theory became available (10) it became possible to obtain quantitative estimates of zeta potential. For the larger particles (above about 1 μm) the zeta potential depends strongly on size and a suitable average value must be provided to enable a valid estimate of $\zeta$ to be made. A spreadsheet program (in Lotus-123 or Excel 5) is now available (18) for estimating the appropriate average size from data provided by some other size measuring method, such as light scattering.

The ESA-8000 can make accurate estimates of zeta potential in both aqueous and nonaqueous environments, but it has a number of limitations. Since it measures at only one frequency, it cannot determine both the size and the charge. It is also unsuitable for handling concentrated systems since it has no provision for estimating the acoustic impedance of the suspension which is required to obtain $\mu_d$ from Eq. (1). Determining the acoustic impedance is relatively easy, but estimating the phase angles with the necessary precision (about 1°) is quite difficult. Both those problems were addressed in the design of the AcoustoSizer.

## B. The AcoustoSizer

The AcoustoSizer is designed to measure in the ESA mode over a range of frequencies around 1 MHz. The original version of the instrument performed measurements at 13 frequencies from 0.3 to 11.2 MHz which gave a size range from 0.1 to 10 μm (diameter). More recent versions have extended both the hardware (to 20 MHz) and the software to expand the range from 0.07 to 15 μm. That range is shifted upwards somewhat in systems at higher concentration.

The cell of the AcoustoSizer is made of a highly chemically resistant epoxy resin and has a capacity of about 400 mL. Its contents can be stirred by an overhead propeller/impeller stirrer with a variable speed drive. Probes dip into the cell to measure the temperature, electrical conductivity, and pH. Provision is also made for conducting pH and other titrations using built-in, computer-controlled microburets of high (0.1 μL) precision. The electrodes in this case are embedded in the cell walls and are about 5 cm apart so that there is no resonance in the cell and the signals from the two electrodes are quite separate. When the electric field is applied across the cell (again as a short pulse lasting a few microseconds) the droplets of the emulsion will oscillate backwards and forwards. As we noted above, the motions induced by the applied field are extremely small. In a typical field of around 40 V/cm the particles will oscillate through distances of less than 0.1 nm, which is less than the size of a single atom. The amplitude of the sound wave moving along the right-hand delay rod is therefore very small and its effects must be greatly amplified before processing. The complex Fourier transform of the signal is first calculated (to determine what the response would have been to a continuous sine wave rather than a pulse of limited duration). The result can then be compared with O'Brien's equation [Eq. (1)], which is derived for a continuous sinusoidal field (8). To do that we also need, in the general case, the acoustic impedance of the suspension. That is obtained using the transducer on the left-hand side of Fig. 5. In this case the field is applied to the transducer and the resulting sound wave travels down the delay rod and is reflected at the interface with the emulsion. The ratio of the (complex) amplitude of the reflected wave to that of the incident wave is the reflection coefficient, and by comparing the reflection coefficient of an empty cell with that from the cell containing an emulsion one can determine the function $Z_e$ (12).

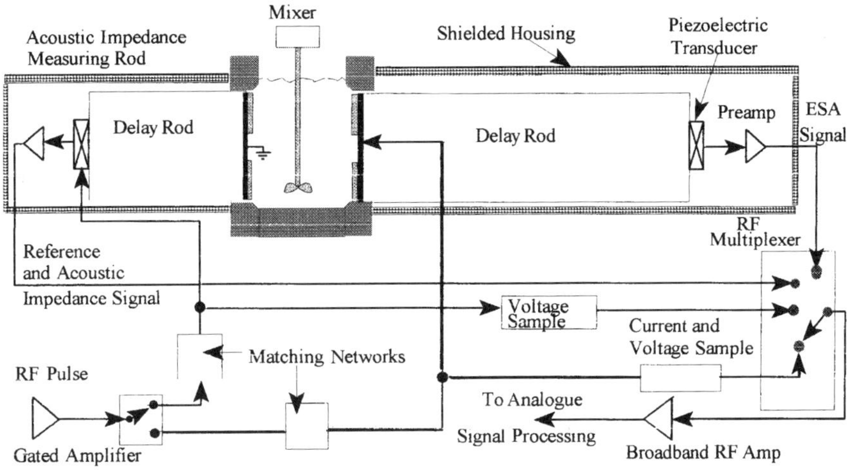


**Figure 5** Schematic diagram of the arrangement of electrodes and transducers in the AcoustoSizer. The ESA signal is taken from the right-hand side transducer while the left-hand side is used for determination of the acoustic impedance.

## IV. CALIBRATION

### A. Determining the Function $A(\omega)$

Before the dynamic mobility can be obtained from Eq. (1) we need to be able to determine the function $A(\omega)$, which depends on the length of the delay rods, on the transducer characteristics, and on the amplifier settings in the signal-processor scheme. It also depends to some extent on the electrical conductivity of the emulsion, especially at low and high conductivities. (At very high conductivities, the current required to establish the standard field strength may exceed the capacity of the driving amplifiers so the field decreases in magnitude. At low conductivities the field lines in the cell become altered because it is then possible for some of the field to leak into the plastic walls of the cell. The field arrangement inside the cell is very complicated because of all the probes in there so any alteration to the disposition of the field alters the particle response.)

The calibration is performed using a special salt solution. As noted above all salt solutions give an ESA signal, but usually it is small compared to the signal from colloid-sized particles or droplets. There are, however, some salts for which the ESA signal is quite large because there is a large difference in the sizes of the cation and anion. The one used for the AcoustoSizer (12) is the potassium salt of $\alpha$-dodecatungstosilicic acid. The octadecahydrate ($K_4[SiW_{12}O_{40}].18H_2O$) tends to lose some water of crystallization, but is still an effective standard so long as it is very pure (i.e., has no extraneous ions). The loss of water from the crystal is unimportant because the anticipated ESA signal can be calculated from the electrical conductivity of the salt. Changes in the salt concentration due to efflorescence are therefore taken into account by the measured conductivity.

The details of the calculation are given by O'Brien et al. (12) so we will not repeat them here but merely quote the value of $A(\omega)$:

$$A(\omega) = \frac{\text{ESA}}{M(S_a - S_s)} \quad \text{with} \quad M = -3.02 \times 10^{-9} K \tag{6}$$

where $K$ is the (low frequency) conductivity measured in SI units, $S$ is the measured reflection coefficient, and subscripts "s" and "a" refer to the salt and the empty (air-filled) cell, respectively. [Note that in the original paper (12) the constant in the expression for $M$ in Eq. (6) was misquoted.] The calibration procedure has been shown to be consistent with the independent method developed by James et al. (19), who used a colloidal dispersion to calibrate the ESA-8000.

It would be hard to overestimate the significance of this new procedure, however. There is a great deal of difficulty attached to the problem of finding a suitable standard material, especially for the zeta potential. Different manufacturers and standardizing bodies have produced different materials: the National Institute of Standards and Technology in Washington, for example, provides a standard iron oxide which, if made up to a defined recipe, is reported to give reproducible results for $\zeta$. Here, we have a procedure which couples the zeta potential back to the classical methods of measuring the transport numbers of ions in solution and the electrical conductivity of a simple salt. It may well prove to be a more viable and robust standard than any other currently available in electrokinetics.

## B. Behavior of Polydisperse Systems

In a polydisperse system, the ESA signal is related to the volume average dynamic mobility of the particles, $\langle \mu_d \rangle$ which is defined as:

$$\langle \mu_d(\omega) \rangle = \int_0^{\infty} \mu(\omega, a) p(a) \mathrm{d}a \tag{7}$$

where $p(a)\mathrm{d}a$ is the mass fraction of particles with radius between $a + \mathrm{d}a/2$ and $a - \mathrm{d}a/2$. For a dilute system, Eq. (2) can be used for $\mu_d(\omega)$ for each value of $a$ provided only that the double layer is thin. The only unknown terms in Eq. (7) are then $\zeta$ and the function $p(a)$ which must be adjusted until the best fit is found to the dynamic mobility spectrum.

The AcoustoSizer software assumes that the size follows a lognormal distribution and adjusts the median and spread of the distribution, along with the zeta potential, to give the best fit to the mobility spectrum, by minimizing the relative root mean square error (superscript "th" is the theoretical):

$$\sqrt{\frac{\sum_1^{13} |\mu_j - \mu_j^{\mathrm{th}}|^2}{\sum_1^{13} |\mu_j|^2}}$$

O'Brien et al. (12) show that good agreement can be achieved between the ESA lognormal size distribution and the 'true' distribution for a pair of ground quartz standards supplied by the Bureau of Common Reference of the EEC. In the same paper they also describe the results on a variety of industrial samples of ceramics, paper coatings, and pigments, indicating good agreement between sizes obtained by the ESA method and by an alternative sedimentation technique (the Horiba Capa-700).

# V. ELECTROACOUSTICS OF EMULSIONS

One of the earliest publications referring to the use of the ESA-8000 apparatus in nonaqueous media was that of Isaacs et al. (20) who successfully monitored the coalescence of water-in-oil emulsions in the dilute concentration regime. There has been little further published work on emulsions with the ESA-8000, although Washington (21) reported a preliminary study of the zeta potential of "Intralipid" (a proprietary phospholipid-stabilized emulsion used for intravenous feeding of postoperative patients). A more detailed study of that system is discussed below. Lopez *et al.* (22) reported on the value of the ESA method for investigating bitumen emulsions (see below) and Do Carmo Marques *et al.* (23) were able to establish a direct relationship between the ESA signal and the asphaltene content of a toluene-in-water miniemulsion. A paper by Goetz and El-Aasser (24), on the behavior of concentrated miniemulsions, will be discussed when we treat the problem of concentrated systems.

A significant amount of ESA work has been in progress in various industrial laboratories but that has not appeared in the general literature until recently. Ho (25), for example, has given a very interesting account of the use of the ESA-8000 for studying the efficacy of various ionic and zwitterionic surfactants as emulsion stabilizers. He prepared hexane-in-water emulsions at about 10% concentration (by weight) and studied the electroacoustic behavior as a function of the stabilizing surfactant. He looked at some 30 surfactants, mostly cationic, but with some anionic and some zwitterionics. The pH behavior was unsurprising, with the zwitterionics showing an isoelectric point (IEP) at some intermediate pH values and the weak-base types increasing in charge at low pH. The inability to measure droplet size made interpretation of some of the results problematic and there would clearly be an advantage in repeating this kind of study using electroacoustic spectroscopy where the size could be determined. With the AcoustoSizer one would also have the opportunity to eliminate any artifacts created by differences in acoustic impedance which were not able to be accounted for in the ESA-8000 study. Nevertheless, Ho was able to make some very interesting findings. The plot of the ESA signal versus concentration of surfactant, $c$, is very like the typical high-affinity Langmuir isotherm, with a well-defined plateau in most cases. Strictly linear plots were obtained of ESA/$c$ against $c$ and the initial slope of these plots could be related to the number, $N$, of $CH_2$ groups in the alkyl chain of the surfactant. The loglinear relation between slope and $N$ is reminiscent of Traube's rule relating various surfactant-micellization characteristics to chain length. Ho proposes these ESA plots as a means of rapidly assessing the hydrophile/lipophile balance (HLB) values for different ionic surfactants in accordance with the Davies HLB scale (26).

Carasso *et al.* (27) used the AcoustoSizer to determine the variation in droplet size and zeta potential for an intravenous emulsion (Intralipid, Kabi Pharmacia) as a function of the pH and other variables. This material is a 20% suspension (in water) of a triglyceride fat, stabilized by egg lecithin, and they were able to characterize it successfully, without dilution, using a theoretical development which is discussed below. They were able to show that it is very stable with respect to pH, showing essentially reversible zeta-potential behavior over the pH range 4–10. The zeta potential varied from −14 to −46 mV over that range and was −24 mV at the natural pH of 7. The diameter of the particles was essentially constant, at 0.23 ± 0.02 μm, over the whole pH range. Calcium and sodium salts are often added to these emulsions along with other essential nutrients, and it is important to know when such additions are likely to destabilize the emulsion. Carasso *et al.* (27) were able to show that calcium ions rapidly decreased the magnitude of the zeta potential and produced a reversal of sign at about 5 to 7 mM (Fig. 6). At this point the droplet size appeared to increase, though it returned to smaller values at higher concentrations of calcium when the zeta potential became sufficiently positive. Sodium ions at higher concentrations produced some reduction in the magnitude of zeta, but were not able to reverse the sign. Thus, sodium ions would be classed as "indifferent" and calcium ions as "specifically adsorbed" at this interface. These results were consistent with optical microscopic observations of the emulsion and are to be expected on general double-layer theory grounds. They are clearly relevant to deciding the levels of mineral nutrients which may be added to the emulsion before injection or perfusion.

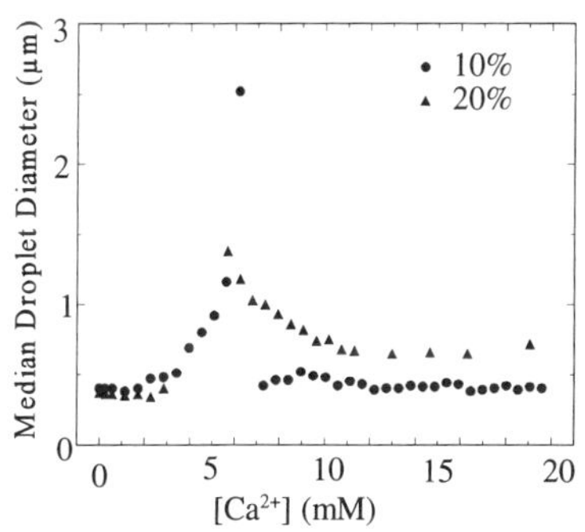


**Figure 6** The size of Intralipid emulsion drops as a function of calcium ion concentration.

Of even more significance, perhaps, was some related work by Lilley *et al.* (28) on the destabilizing effect of adding anesthetic drugs to an injectable emulsion (Propofol). The normal injection dose is 20 mL, and this study showed that up to 10 mg of the anesthetic lignocaine could be added to 20 mL of propofol, but any increase beyond that caused the magnitude of the zeta potential to fall below 15 mV and the droplet size to increase dramatically. This would seem then to be the maximum dosage permitted by this route. A related study (29) was carried out on a mixture of anesthetic, opiate, and muscle-relaxant drugs to determine their mutual compatibility in terms of the stability of the mixed emulsion, using the AcoustoSizer to assess the zeta potential and size of various mixtures, both immediately after preparation and after storage under various conditions. The significance of these studies lies in the fact that the stability can be assessed at the normal emulsion concentration.

Of more general interest are the emulsions natural to the dairy industry, such as milk and cream and their various products. Wade and Beattie (30) have studied such systems using the AcoustoSizer with some interesting results. They examined the fat emulsion separated from homogenized milk (about 4% concentration) and natural cream (about 38% concentration) and also an artificial milk and cream produced by dispersing anhydrous fat in skim milk. The milk emulsions are "dilute" (in the sense that the hydrodynamic interactions between droplets are unimportant). Both the commercial cream and the reconstituted cream were studied at this same concentration (4%) at the natural pH (6.7). The behavior of undiluted cream will be discussed when we deal with the problem of concentrated systems.

The fat droplets in raw milk are stabilized by a thin protective layer known as the milk fat globular membrane (MFGM). Cream produced by simple separation from the raw milk should retain that membrane intact. Milk which has been homogenized will have smaller droplet size and a larger surface area so the membrane will only partially cover the droplet surface and the exposed surface will become covered with a protein mixture from the milk plasma. When artificial milk and cream are prepared from anhydrous milk fat, there would be little, if any, MFGM and the entire surface would be expected to be covered by proteins from the plasma. One would expect therefore that the zeta potential would show significant differences between the surfaces of these products. Table 1 shows the results obtained by Wade and Beattie (30) for the homogenized and reconstituted milk and for

**Table 1** Particle Size and Zeta Potential of Natural and Reconstituted Milk Fat Emulsions; $d_{50}$, $d_{15}$, and $d_{85}$ Are the Median, and the 15th and 85th Percentiles of the (Lognormal) Size Distribution. All Samples Measured at 4% Volume Fraction and Natural pH (6.7).

| Emulsion | Zeta (mV) | $d_{50}$ | $d_{15}$ | $d_{85}$ |
|---|---|---|---|---|
| Commercial cream | −27 | 3.4 | 3.1 | 3.8 |
| Recombined cream | −24 | 2.5 | 2.3 | 2.8 |
| Commercial milk | −18 | 1.0 | 0.9 | 1.1 |
| Recombined milk | −35 | 1.0 | 0.9 | 1.1 |

*Source*: Ref. 30.

the dilute samples of the natural and reconstituted cream. To obtain these results the total ESA signal must be corrected for the contributions from the salts, the serum proteins, and the casein micelles. The zeta potential obtained for the homogenized milk sample agrees well with that obtained by Dalgleish (31) using laser Doppler electrophoresis, but the $\zeta$ value for the cream emulsion is significantly different. Dalgleish observed some time dependence, but the $\zeta$ value settled to −10 mV after about 10 min. There are considerable problems associated with dilution of these very complex systems and since both of them were diluted before measurement we will suspend judgment for the moment on which is the more reliable of these two estimates.

Another interesting application of the electroacoustic procedure is given by Hunter and O'Brien (32) in a study of a highly charged emulsion system. The emulsion droplets were produced by stabilizing perfluorodecalin droplets with sodium dodecyl sulfate (SDS) and passing the resulting (rather unstable) emulsion several times through an homogenizer. This has a very small orifice which so constricts the flow that the oil droplets are drawn out and broken down to sizes in the submicrometer range. When measured with the AcoustoSizer after several passes through the homogenizer, the emulsion showed a droplet diameter of about 0.7 μm and a zeta potential of about −175 mV at a solution concentration of around $1.4 \times 10^{-3}$ M, corresponding to a $\kappa a$ value of about 43. Such very high zeta potentials have seldom been reported previously and Fig. 7 shows why this is so. The computer calculations of O'Brien and White (33) show that when the d.c. electrophoretic mobility is plotted as a functon of zeta potential, for $\kappa a$ values around 50, there is a pronounced maximum in the curve. In a d.c. measurement yielding a reduced mobility of about 4.7 one would be unable to determine

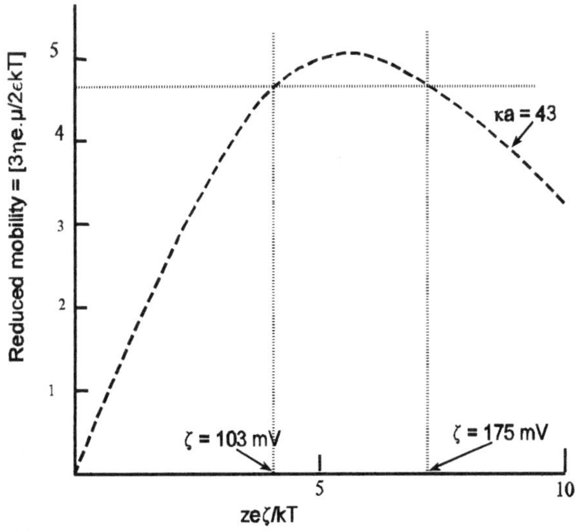


**Figure 7** Dimensionless d.c. mobility as a function of dimensionless $\zeta$ potential according to the numerical calculations of O'Brien and White (33) for the value of $\kappa a$ relevant to the highly charged emulsion system (see text).

whether the appropriate zeta potential was −103 or −175 mV. Figure 8 shows that there is no such ambiguity in the dynamic mobility, for which both the magnitude and the phase angle clearly indicate −175 mV rather than the lower value. Note, however, that the magnitude curves in both cases appear to converge to the same low frequency value as would be expected from Fig. 7.

## VI. EFFECT OF CONCENTRATION OF PARTICLES

### A. Acoustic Impedance

One immediate effect of increasing the particle concentration in the emulsion is that the acoustic impedance, $Z_e$, can no longer be approximated as equal to that of the dispersion medium. Since Eq. (1) remains valid at all concentrations commonly encountered, it is important that the correct value of $Z_e$ is used, so that the correct value of the dynamic mobility is obtained from the measured ESA signal. In principle, the value of $Z_e$ for the emulsion could be a complex function of the frequency and the properties of the suspension, but the exact behavior is of little consequence for measurements with the AcoustoSizer, since it measures the value at each frequency before calculating $\mu_d$ from ESA signal.

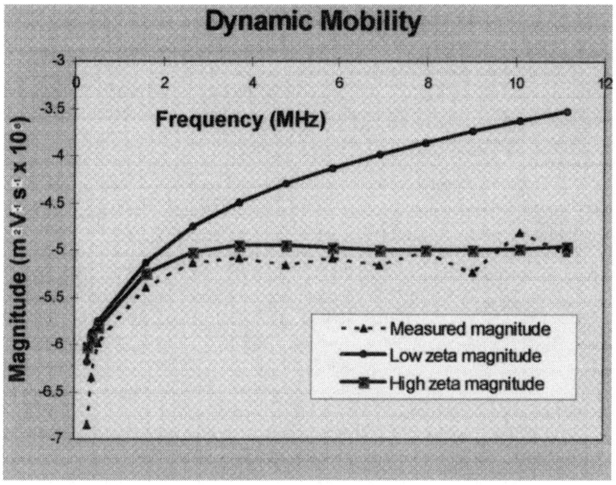


(a)

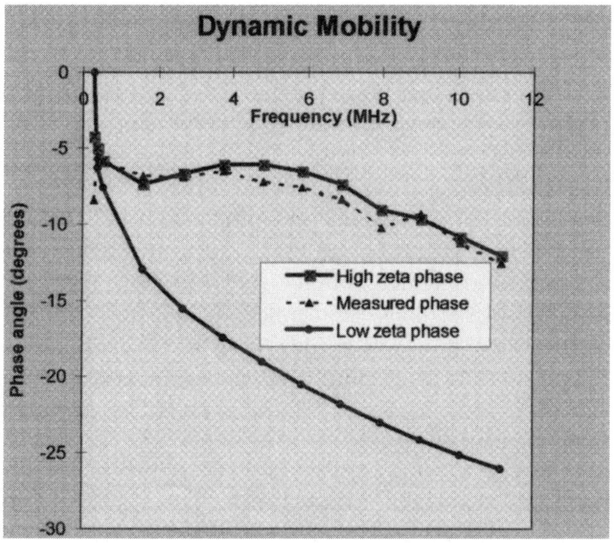


(b)

**Figure 8** (a) Comparison of the magnitude of the dynamic mobility of the emulsion with the calculated values for low (103 mV) and high (175 mV) zeta potential; (b) the same for the phase angles.

## B. Effect of Concentration on Dynamic Mobility

Equation (1) shows that the ESA effect should be proportional to the volume fraction $\phi$ for dilute systems, and the measurements of Klingbiel *et al.* (15) suggest that this holds for at least some spherical particles up to concentrations of order 5%. Texter's nonspherical particles were linear only up to about 2% by volume (34) but one would expect nonspherical particles to show departures from the simple relation at lower concentrations than for spherical particles. Departures from sphericity will not be important in emulsion systems until one reaches very high concentrations indeed so we may reasonably assume that the dilute formula should hold up to at least 5% by volume. Even this value is a great deal higher than the normal concentrations at which d.c. electrophoresis is conducted* but it is at the lower end of the range of the ESA method.

A limited number of studies have been carried out on more concentrated systems using variations of the traditional electrophoretic method, e.g., the tracer and mass-transport methods. Reed and Morrison (35) have shown that, for d.c. fields, even in highly concentrated systems, the hydrodynamic and electrostatic interactions cancel one another when the double layers are thin, and the only effect which must be taken into account is the reverse flow of fluid displaced by the moving particles. Zukoski and Saville (36), using red blood cells mixed with ghosts, have verified that this is so and that the d.c. mobility, $\mu_c$, of a concentrated system of volume fraction $\phi$ is given by the simple relation:

$$\mu_c = \mu_0[1 - g\phi] \tag{8}$$

where $\mu_0$ is the mobility at infinite dilution, and $g$ is, within the limits of experimental error, equal to unity.

Marlow and Rowell (37) working with coal/water slurries and using the CVP technique have shown that, at the frequencies of their measurements (200 kHz), the effect of particle concentration can be adequately described by introducing a factor $(1 - g\phi)$ into their equivalent of Eq. (1) where again, $g$ was very close to unity. In their review article Marlow *et al.* (6) discuss the way the cell model of Levine and coworkers (38, 39) is introduced into the CVP theory and show that, for thin double layers, the result is that the hydrodynamic and electrostatic interactions essentially cancel one another and one is left with only the factor $(1 - \phi)$ to take account of the backflow of liquid caused by the particle motion.

*Traditional methods of determining the electrophoretic mobility in a d.c. electric field have involved particle concentrations with $\phi \ll 0.001$. This provides the infinite dilution limiting value, and the appropriate theoretical analysis is for an isolated particle in an infinite volume of electrolyte.

Unfortunately, no such simple solution is available for the ESA effect at the high frequencies at which it is currently used. Goetz and El-Aasser (24) attempted to compare the electroacoustic and electrophoretic behavior of concentrated miniemulsion systems of toluene in water, stabilized by cetyl alcohol and sodium lauryl sulfate. They concluded that the simple correction which works well for CVP does not produce a similar reconciliation in the case of the ESA effect. Their conclusions are, however, suspect because of uncertainties arising from the dilution of the emulsion system; this can so easily lead to changes in surface properties, no matter how carefully it is done. Texter's results (34) referred to above are perhaps more definitive in this case. He showed that in the range from 2 to 5% by volume where his particles showed a nonlinear dependence of the ESA signal on volume fraction, the Levine and Neale model (38) was unable to account for the nonlinearity. His particles were, however, nonspherical and that may at least partially explain the discrepancy.

Nonetheless, there are good reasons to believe that the concentration correction for the ESA method is not as simple as Eq. (8) would suggest. The Levine approach uses Kuwabara's "zero vorticity" model (40) in which the vorticities of both the hydrodynamic and the electric fields are zero on the defining surface of the cell which encloses each particle (41). At frequencies in the megahertz range, the vorticity of the flow field stretches out beyond the confines of the cell so that hydrodynamic interactions between the particles are very much more significant. O'Brien *et al.* (42) showed that the Levine cell model drastically underestimates the effect of concentration on both the magnitude and the phase angle of the dynamic mobility in the range 0.5–11 MHz. It should be noted that the reciprocal relationship makes it clear that precisely the same limitations would apply to the CVP in the same frequency range. Those experimental studies should be borne in mind in considering Ohshima's calculations of the concentration effect using the Kuwabara model (43). He gives the results of his numerical calculations of the magnitude of the dynamic mobility (but not the phase) for various $\kappa a$ values from near zero ($10^{-3}$) to infinity ($10^{3}$) for values of $\alpha = \omega a^2/\nu$ from 0.1 to 100 and for $\phi$ values from 0 to 0.7 assuming that $\zeta$ is small. He also provides an approximate analytical solution valid for low $\zeta$ potentials and insulating particles ($\epsilon_p = 0$). The experimental results (42) would suggest that the frequency range over which those results can be used is rather limited.

Rider and O'Brien (44) have extended the dilute-solution theory to incorporate the order $\phi$ correction which allows one to describe the ESA behavior up to particle concentrations of order 10% by volume. For higher concentrations, the corrections depend to a considerable extent on the density difference between the disperse phase and the dispersion medium. Fortunately, in the case of the emulsion systems, where that density difference is usually relatively small, O'Brien has provided an approximate analytical solution to the problem which appears to be very effective. For near-neutrally buoyant particles it is only necessary to take account of the near-neighbor hydrodynamic interactions. The effect of the particles in modifying the electric field experienced by each particle can also be relatively easily taken into account, using the Clausius–Mosotti approach, familiar from the theory of dielectric permittivity. Using the Percus–Yevick approximation (47) to estimate the distribution function for the nearest neighbors [$g(r)$] and assuming additivity of the contributions from each particle in the vicinity of the central particle, O'Brien *et al.* have shown (45) that the dynamic mobility is given by:

$$\mu_d = \frac{\epsilon\zeta}{\eta} H \left[ \frac{1 - \dfrac{2\beta^2}{(2+\phi)(3+3\beta+\beta^2)} - \dfrac{3\phi}{2+\phi} F}{\left(1 + \phi \dfrac{\Delta\rho}{\rho} HF\right)} \right] \tag{9}$$

where the factors $H$ and $F$ are defined as:

$$H = \frac{3+3\beta+\beta^2}{3+3\beta+\dfrac{1}{3}\beta^2\left(1+2\dfrac{\rho_p}{\rho}\right)} \tag{10}$$

$$F = \frac{2}{3}\left[(4\beta^2 I + \left(1+2\beta)e^{-2\beta}\right)J^2 + \frac{1}{2}\right] \tag{11}$$

Here

$$J = \frac{e^{\beta}}{1+\beta+\dfrac{\beta^2}{3}} \quad \text{and} \quad I = \int_1^{\infty} (g(r)-1) r e^{-2\beta r}\, dr \tag{12}$$

where $\beta = (1+i)\sqrt{(\omega a^2/2\nu)}$.

The results of the theory are shown in Fig. 9 where it is apparent that increasing particle concentration reduces the variation of the signal with frequency (both in magnitude and in phase angle). The effect is to make the particles appear smaller in size as the concentration is increased. It also makes the measurement

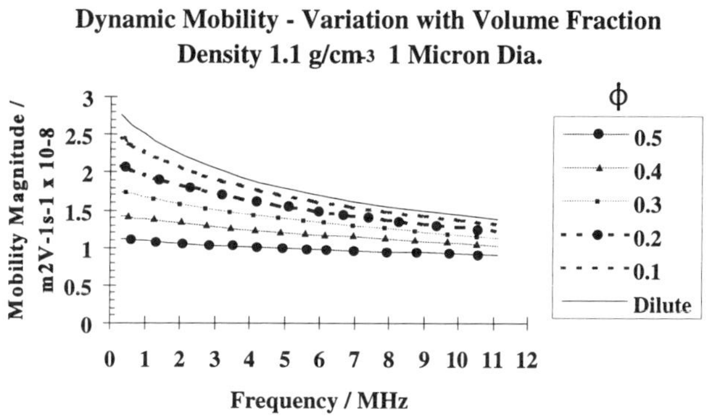


(a)

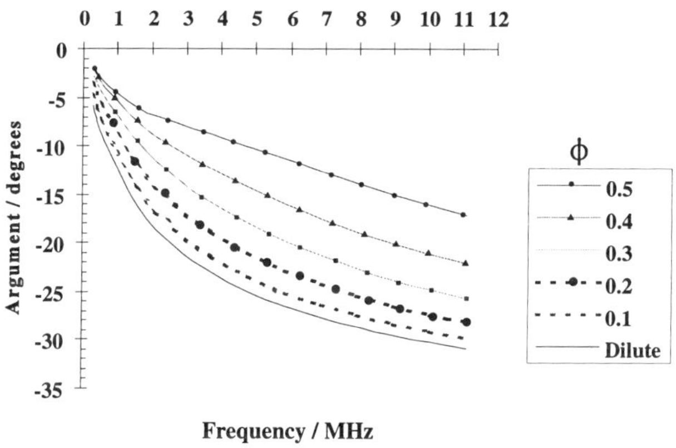


(b)

**Figure 9** (a) Magnitude of the dynamic mobility as a function of frequency for various volume fractions for a particle of radius 1 μm; (b) phase angles for the same conditions as in (a).

of their size dependent on increasingly precise measurement of both the magnitude and the phase. Fortunately, the magnitude of the ESA signal increases with $\phi$ so the signal-to-noise ratio is improved, though there is no doubt that there are limits to the concentrations at which sizing will be successful. Only one of Ohshima's sets of numerically calculated results (43) (his Fig. 9) is in a region where it can be compared with the O'Brien calculation ($\kappa a = 50$) and there appears to be little or no correlation between the two calculations.

The efficacy of O'Brien's analysis is demonstrated by the data in Table 2 which shows a comparison of the estimated values of zeta potential and of size using the dilute-solution theory and the more elaborate theory of Eqs. (9) to (12). One of the systems used in this study was the same parenteral/intravenous emulsion used by Carasso *et al.* and referred to above (27). This is a very stable material and it is supplied as a 20 or 10% (w/v) emulsion which was carefully diluted with the suspending fluid and measurements made at varying particle concentrations. It is clear from Table 2 that using the concentrated-suspension theory gave rise to almost identical zeta and size values at all dilutions whereas the dilute theory would suggest rather unlikely variations of order 15% in both size and zeta potential.

The other systems used in the study were somewhat more variable in composition. Some were standard examples of common dairy products (30): full cream and reconstructed cream (made by mixing cream with skim milk). They too gave much more consistent results when analyzed using the concentrated formula than were obtained with the dilute formula [Eq (2)]. It should also be noted that the zeta potential and size data obtained for the concentrated systems (cream and reconstituted cream) before and after dilution are reasonably consistent (comparing Tables 1 and 2). Both show almost the same size, and the zeta potentials differ by only 5 to 6 mV, which suggests that the dilution procedure used in preparing the data for Table 1 is more satisfactory than the alternatives but that one should still favor the results obtained on systems which have not been diluted at all.

The bitumen cited in Table 2 was an "emulsion" prepared industrially by mixing hot (140°C) bitumen with surfactant and water (~ 20°C) to produce an emulsion (at ~ 90°C) and then cooling it to room temperature. The disperse phase in that case had a very high viscosity and behaved essentially as a solid.

## VII. CONCLUSIONS

Electroacoustic spectroscopy offers the prospect of studying the size distribution, and electrokinetic and stability behavior of emulsion systems while avoiding the very real problems associated with dilution of such systems. Studies are as yet in their infancy but they have already revealed new insights into electrokinetic processes, especially for the very highly charged systems used in industry. The possibility of studying polymer adsorption on emulsion systems, as an extension of the work already performed at the solid-solution interface (46) opens up entirely new prospects for the examination of both biological and technological emulsion systems.

**Table 2** Particle Size Distributions and Zeta Potentials Calculated from Dynamic Mobility Data According to O'Brien's Concentrated Formulas [Eqs (9)–(12)] Compared to the Dilute Relation [Eq. (2)]. Particle Sizes are Given as Diameters in μm, Where $d_{50}$, $d_{15}$, and $d_{85}$ Represent the Median, and 15th and 85th Percentiles of a Lognormal Distribution. Zeta Potentials Are in mV.

| Emulsion | Volume fraction | Concentrated formula | | | | Dilute formula | | | |
|---|---|---|---|---|---|---|---|---|---|
| | | $\zeta$ | $d_{50}$ | $d_{15}$ | $d_{85}$ | $\zeta$ | $d_{50}$ | $d_{15}$ | $d_{85}$ |
| 1. Reconstituted | 0.38 | −36.8 | 1.49 | 1.04 | 2.12 | −19.7 | 0.82 | 0.74 | 0.90 |
| dairy cream[a] | 0.27 | −37.6 | 1.38 | 1.13 | 1.68 | −25.0 | 0.95 | 0.86 | 1.05 |
| | 0.16 | −40.0 | 1.33 | 1.20 | 1.47 | −32.0 | 1.08 | 0.98 | 1.20 |
| | 0.11 | −40.5 | 1.32 | 1.17 | 1.46 | −35.0 | 1.15 | 1.05 | 1.28 |
| 2. Dairy cream[b] | 0.38 | −23.0 | 3.08 | 2.79 | 3.41 | −12.9 | 1.86 | 1.68 | 2.06 |
| | 0.04 | −25.3 | 3.14 | 2.84 | 3.48 | −24.2 | 3.05 | 2.75 | 3.37 |
| 3. Intravenous fat | 0.217 | −40.8 | 0.40 | 0.36 | 0.44 | −29.4 | 0.28 | 0.26 | 0.31 |
| emulsion | 0.163 | −40.3 | 0.40 | 0.36 | 0.44 | −31.9 | 0.31 | 0.28 | 0.35 |
| (Intralipid | 0.109 | −40.2 | 0.40 | 0.30 | 0.53 | −34.2 | 0.35 | 0.31 | 0.38 |
| 20% w/v) | 0.054 | −40.6 | 0.40 | 0.26 | 0.60 | −37.5 | 0.39 | 0.28 | 0.53 |
| 4. Intralipid[c] | 0.109 | −40.4 | 0.33 | 0.23 | 0.48 | −34.4 | 0.28 | 0.21 | 0.38 |
| 5. Bitumen A1 | 0.40 | 67 | 5.9 | 2.6 | 13 | 32 | 2.9 | 1.5 | 5.4 |
| Bitumen A2 | 0.40 | 65 | 5.3 | 2.5 | 11 | 31 | 2.6 | 1.5 | 4.5 |
| Bitumen A3 | 0.40 | 45 | 6.6 | 2.0 | 21 | 21 | 3.2 | 1.2 | 9.1 |
| 6. Bitumen B | 0.60 | 43 | 3.5 | 2.6 | 4.7 | 12 | 1.0 | 0.9 | 1.1 |
| Diluted B | 0.40 | 55 | 3.9 | 2.2 | 6.7 | 26 | 1.9 | 1.4 | 2.5 |

[a] In commercial skimmed milk.
[b] In a centrifugate.
[c] 10% (w/v) as supplied.

## REFERENCES


1. VA Hackley, J Texter. J Res Nat Inst Stand and Tech 103(2), 1998.
2. P Debye. J Phys Chem 1: 13, 1933.
3. R Zana, E Yeager. Ultrasonic Vibration Potentials. In: J O'M Bockris, BE Conway, eds. Modern Aspects of Electrochem. Plenum, New York: 14: 1–61, 1982.
4. S Durand Vidal, JP Simonin, P Turq, O Bernard. J Phys Chem 99: 6733–6738, 1995.
5. J Hermans. Philos Mag 25: 426; 26: 674, 1938.
6. BJ Marlow, D Fairhurst, HP Pense. Langmuir, 4: 611–626, 1988.
7. T Oja, D Cannon, GL Petersen. US Patent 4 497 208, 1985.
8. RW O'Brien J Fluid Mech 190: 71–86, 1988.
9. RJ Hunter. Colloids Surfaces A: Physicochem Eng Aspects 141: 37–66, 1988.
10. RW O'Brien. J Fluid Mech 212: 81, 1990.
11. RW O'Brien, P Garside, RJ Hunter. Langmuir 10: 931–935, 1994.
12. RW O'Brien, D Cannon, WN Rowlands. J Colloid Interface Sci 173: 406–418, 1995.
13. RJ Hunter. Foundations of Colloid Science. Vol I. Oxford: Oxford University Press, 1987, p 332.
14. A Babchin, RS Chow, RP Sawatsky. Adv Colloid Interface Sci 30: 111–151, 1989.
15. RT Klingbiel, H Coll, RO James, J Texter. Colloids Surfaces 68: 103–109, 1992.
16. DW Cannon. In: SB Malghan, ed. Electroacoustics for Characterization of Particulates and Suspensions. NIST Special Publication 856. Washington, DC: National Institute of Standards and Technology, 1993, pp 40–66.
17. TSB Sayer. Colloids Surfaces A: Physicochem Eng Aspects 77: 39–47, 1993.
18. RJ Hunter. Spreadsheet program. Available from the author at hunter_r@chem.usyd.edu.au.
19. RO James, J Texter, PJ Scales. Langmuir 7: 1993–1997, 1991.
20. EE Isaacs, H Huang, AJ Babchin, RS Chow. Colloids Surfaces 46: 177–192, 1990.
21. C Washington. Int J Pharm 87: 167–174, 1992.
22. FJ Lopez, H Rivas, RE Lujano, Proceedings of Seventh International Conference on Surface and Colloid Science, Compiegne, 1991, Sect. B4, p 59. (Quoted in Ref. 24.)
23. LC Do Carmo Marques, JF De Oliveria, G Gonzalez. J Dispersion Sci 18: 477–488, 1997.
24. RJ Goetz, MS El-Aasser. J Colloid Interface Sci 150: 436–452, 1992.

25. OB Ho. J Colloid Interface Sci 198: 249–260, 1998.
26. JT Davies, EK Rideal. Interfacial Phenomena. 2nd ed. London: Academic Press, 1963, p 371.
27. ML Carasso, WN Rowlands, RA Kennedy. J Colloid Interface Sci 174: 405–413, 1995.
28. EM Lilley, PR Isert, ML Carasso, RA Kennedy. Anaesthesia 52: 288, 1997.
29. PR Isert, D Lee, D Naidoo, ML Carasso, RA Kennedy. J. Clin Anaesthesia 8: 329–336, 1996.
30. T Wade, JK Beattie. Colloids Surfaces B: Biointerfaces 10: 73–85, 1997.
31. DG Dalgleish. J Dairy Res 51: 425, 1984.
32. RJ Hunter, RW O'Brien. Colloids Surfaces A: Physicochem Eng Aspects 126: 123–128, 1997.
33. RW O'Brien, LR White. J Chem Soc Faraday II 74: 1607, 1978.
34. J Texter. Langmuir 8: 291, 1992.
35. LD Reed, FA Morrison. J Colloid Interface Sci 54: 117, 1976.
36. CF Zukoski, DA Saville. J Colloid Interface Sci 115: 422–436, 1987.
37. BJ Marlow, RL Rowell. J Energy Fuels 2: 125–131, 1988.
38. S Levine, G Neale. J Colloid Interface Sci 47: 520; 49: 332, 1974.
39. S Levine, G Neale, J Epstein. J Colloid Interface Sci 57: 424, 1976.
40. S Kuwabara. J Phys Soc Japan 14: 527, 1959.
41. RJ Hunter. Zeta Potential in Colloid Science. London: Academic Press, 1981, 386 pp.
42. RW O'Brien, WN Rowlands, RJ Hunter. In: SB Malghan, ed. Electroacoustics for Characterization of Particulates and Suspensions. NIST Special Publication 856. Washington, DC: National Institute of Standards and Technology, 1993, pp 1–22.
43. H Ohshima. J Colloid Interface Sci 195: 137–148, 1997.
44. P Rider, RW O'Brien. J Fluid Mech 257: 607–636, 1993.
45. RW O'Brien, TA Wade, ML Carasso, RJ Hunter, WN Rowlands, JK Beattie. In: T Provder, ed. Proceedings of the American Chemical Society Symposium, Orlando, FL, August 1996. ACS Symposium Series 693. Washington, DC: American Chemical Society, 1998, p 311.
46. ML Carasso, WN Rowlands, RW O'Brien. J Colloid Interface Sci 193: 200–214, 1997.
47. RJ Hunter. Foundations of Colloid Science. Vol. II. Oxford: Oxford University Press, 1989, p 701.

# 8

# Acoustic and Electroacoustic Spectroscopy for Characterizing Emulsions and Microemulsions

**Andrei S. Dukhin and P. J. Goetz**
*Dispersion Technology Inc., Mount Kisco, New York*

**T. H. Wines and P. Somasundaran**
*Columbia University, New York, New York*

## I. INTRODUCTION

The widespread acceptance and commercialization of acoustic spectroscopy has been slow to develop. This technique has been overlooked by many in academia and industry in the past, but has recently been showing increased levels of accptance. This powerful method of characterizing concentrated heterogeneous systems has all the capabilities for being successful. The first hardware for measuring acoustic properties of liquids was developed more then 50 year ago at the Massachusetts Institute of Technology (1) by Pellam and Galt. The first acoustic theory for heterogeneous systems was created by Sewell 90 years ago (2). The general principles of the acoustic theory were formulated 47 years ago by Epstein and Carhart (3). There is a long list of applications and experiments based on acoustic spectroscopy – see reviews (4, 5). Despite all of these developments, however, acoustic spectroscopy is rarely mentioned in modern handbooks on colloid science (6, 7).

Acoustics is able to provide reliable particle size information for concentrated dispersions without any dilution. There are examples when acoustics yields size information at volume fractions above 40%. This in-situ characterization of concentrated systems makes the method very useful and unique in this capability compared to alternative methods, including light scattering, where dilution is required. Acoustics is also able to deal with low dispersed phase volume fractions and in some systems can characterize down to below 0.1% vol. This flexibility for concentration range provides an overlap with classical methods for dilute systems. In the overlap range, acoustics size characterization has been found to have excellent agreement with these other techniques.

Acoustics is not only a particle sizing technique, but also provides information about the microstructure of the disperesed system. The acoustic spectrometer can be considered as a microrheometer. In acoustics, stresses are applied in the same way as with regular rheometers, but over very short distances on the micrometer scale. In this way, the microstructure of the dispersed system can be sensed. Currently, this feature of acoustics is only beginning to be exploited, but it is certainly very promising.

Many people have perceived acoustics to have a high degree of complexity. The operating principles are in fact quite straightforward. The acoustic spectrometer generates sound pulses that pass through a sample system and are then measured by a receiver. The passage through the sample system causes the sound energy to change in intensity and phase. The acoustic instrument measures the sound energy losses (attenuation) and the sound speed. The sound attenuates due to interaction with the particles and liquid in the sample system. Acoustic spectrometers generally operate in the frequency range 1–100 MHz. This is a much higher sound frequency than the upper limit of our hearing which is only 0.02 MHz.

While the operating principles are relatively simple, the analysis of the attenuation data to obtain particle size distributions does involve a degree of complexity in fitting experimental results to theoretical models based on various acoustic loss mechanisms. The advent of high-speed computers and the refinement of these theoretical models has made the inherent complexity of this analysis of little consequence. In comparison, many other particle sizing techniques such as photon-correlation spectroscopy also rely on similar levels of complexity in analyzing experimental results.

Acoustics has a related field that is usually referred to as "electroacoustics" (8). Electroacoustics can provide particle size distribution as well as zeta potential. This relatively new technique is more complex than acoustics because an additional electric field is involved. As a result, both hardware and theory become more complicated. There are even two different versions of electroacoustics depending on what field is used as a driving force. Electrokinetic sonic amplitude (ESA) involves the generation of sound energy caused by the driving force of an applied electric field. Colloid vibration current (CVC) is the phenomenon where sound energy is applied to a system and a resultant electric field or current is created by the vibration of the colloid electric double layers.

Returning to acoustics, its lack of widespread acceptance may be related to the fact that it yields too much, sometimes overwhelming, information. Instead of dealing with interpretation of the acoustic spectra it is often easier to dilute the system of interest and apply light-based techniques. It was often naïvely assumed that the dilution had not affected the dispersion characteristics. Lately, many researchers are coming to the realization that dispersed systems need to be analyzed in their natural concentrated form, and that dilution destroys a number of useful and important properties.

We are optimistic about the future of acoustics in colloid science. It is amazing what this technique can do especially in combination with electroacoustics for characterizing electric surface properties. We hope that this review will allow you to taste the power and opportunities related to these sound-based techniques.

## II. THEORETICAL BACKGROUND

There are six known mechanisms for the ultrasound interaction with a dispersed systsem: (1) viscous; (2) thermal; (3) scattering; (4) intrinsic; (5) structural; and (6) electrokinetic. Here, we give only short qualititive descriptions, omitting complicated mathematical models.

1. The viscous losses of the acoustic energy occur due to the shear waves generated by the particle oscillating at the acoustic pressure field. These shear waves appear because of the difference in the densities of the particles and medium. This density contrast causes particle motion with respect to the medium. As a result, the liquid layers in the particle vicinity slide relative to each other. This sliding nonstationary motion of the liquid near the particle is referred to as the "shear wave". Viscous losses are dominant for small rigid particles with sizes below 3 μm, such as oxides, pigments, paints, ceramics, cement, graphite, etc.
2. The reason for the thermal losses is the temperature gradients generated near the particle surface. These gradients are a result of the thermodynamic coupling between pressure and temperature. This mechanism is dominant for soft particles, including emulsion droplets and latex beads.
3. The mechanism of the scattering losses is quite different than the viscous and thermal losses. Acoustic scattering does not produce dissipation of acoustic energy. This mechanism of scattering is similar to that of light scattering. Particles simply redirect a part of the acoustic energy flow and as a result this portion of the sound does not reach the sound transducer. This mechanism is important for larger particles (> 3 μm) and high frequency (> 10 MHz).
4. The intrinsic losses of the acoustic energy occur due to the interaction of the sound wave with the materials of the particles and medium as homogenous phases on a molecular level.
5. Structural losses are caused by the oscillation of a network of particles that are interconnected. Thus, this mechanism is specific for the given type of structured system.
6. Electrokinetic losses are caused by the oscillation of charged particles in an acoustic field, which leads to the generation of an alternating electrical field, and consequently to an alternat-

ing electric current. As a result, a part of the acoustic energy is transformed into electrical energy and then irreversibly to heat.

Only the first four loss mechanisms (viscous, thermal, scattering, and intrinsic) make a significant contribution to the overall attenuation spectra in most cases. Structural losses are significant only in structured systems that require a quite different theoretical framework. These four mechanisms form the basis for acoustic spectroscopy.

The contribution of electrokinetic losses to the total sound attenuation is almost always negligibly small (9) and will be neglected. This opens an opportunity to separate acoustic spectroscopy from electroacoustic spectroscopy because acoustic attenuation spectra are independent of the electric properties of the dispersed system.

Following this distinction between acoustics and electroacoustics, the corresponding theories will be considered separately.

## III. THEORY OF ACOUSTICS

The most well known acoustic theory for heterogeneous systems was developed by Epstein and Carhart (3), and Allegra and Hawley (10). This theory takes into account the four most important mechanisms (viscous, thermal, scattering, and intrinsic) and is termed the "ECAH theory." This theory describes attenuation for a monodisperse system of spherical particles and is valid only for dilute systems.

The term "monodisperse" assumes that all of the particles have the same diameter. Extensions of the ECAH theory to include polydispersity have typically assumed a simple linear superposition of the attenuation for each size fraction. The term "spherical" is used to denote that all calculations are performed assuming that each particle can be adequately represented as a sphere.

Most importantly, the term "dilute" is used to indicate that there is no consideration of particle–particle interactions. This fundamental limitation normally restricts the application of the resulting theory to dispersions with a volume fraction of less than a few volume per cent. However, there is some evidence that the ECAH theory, in some very specific situations, does nevertheless provide a correct interpretation of experimental data, even for volume fractions as large as 30%.

An early demonstration of the ability of the ECAH theory was provided by Allegra and Hawley. They observed almost perfect correlation between experiment and dilute case ECAH theory for several systems: a 20% by volume toluene emulsion; a 10% by volume hexadecane emulsion; and a 10% by volume polystyrene latex. Similar work with emulsions by McClements (11, 12) has provided similar results. The recent work by Holmes *et al.* (13, 14) shows good agreement between ECAH theory and experiments even for 30% by volume polystyrene latex.

A surprising absence of particle–particle interaction was observed with neoprene latex (15). This experiment showed that attenuation is a linear function of volume fractions up to 30% for this particular system (Fig. 1). This linearity is an indication that each particle fraction contributes to the total attenuation independently of other fractions, and is a superposition of individual contributions. Superposition works only when particle–particle interaction is insignificant.

It is important to note that the surprising validity of the dilute ECAH theory for moderately concentrated systems has only been demonstrated in systems where the "thermal losses" were dominant, such as emulsions and latex systems. In contrast, a solid rutile dispersion exhibits nonlinearity of the attentuation above 10% by volume (Fig. 2).

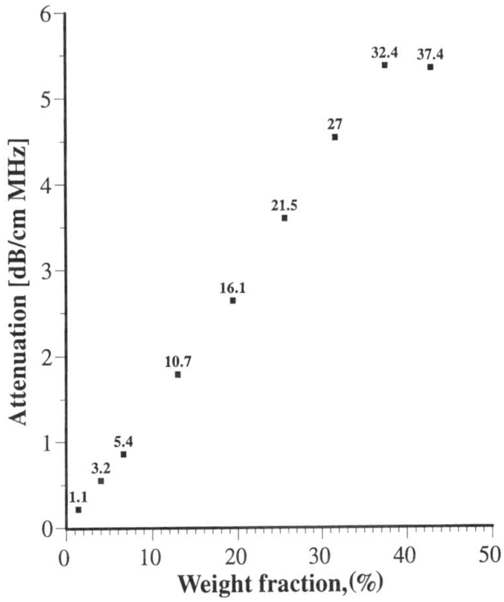


**Figure 1** Dependence of attenuation in the neoprene latex at a frequency of 15 MHz) on the dispersed system weight fraction. Corresponding volume fractions in % are shown as the data points labels.

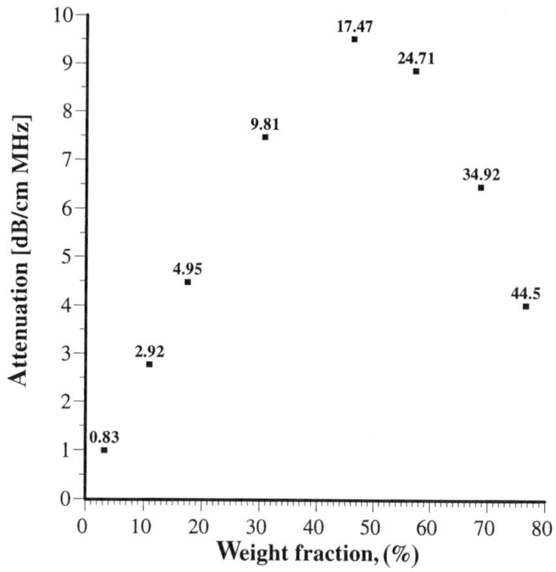


**Figure 2** Dependence of attenuation in the rutile dispersion (rutile R-746, DuPont), at a frequency of 15 MHz, on the dispersed system weight fraction. Corresponding volume fractions in % are shown as the data points labels.

The difference between the "viscous depth" and the "thermal depth" provides an answer to the observed differences between emulsions and solid particle dispersions. These parameters characterize the penetration of the shear wave and thermal wave, respectively, into the liquid. Particles oscillating in the sound wave generate these waves which damp in the particle vicinity. The characteristic distance for the shear wave amplitude to decay is the "viscous depth" $\delta_v$. The corresponding distance for the thermal wave is the "thermal depth" $\delta_t$. The following expressions give these parameter values in dilute systems:

$$\delta_v = \sqrt{\frac{2v}{\omega}} \tag{1}$$

$$\delta_t = \sqrt{\frac{2_\tau}{\omega \rho C_p}} \tag{2}$$

where $v$ is the kinematic viscosity, $\omega$ is the frequency, $\rho$ is the density, $\tau$ is the heat conductance, and $C_p$ is the heat capacity at constant pressure.

The relationship between $\delta_v$ and $\delta_t$ has been considered before. For instance, McClements plots "thermal depth" and "viscous depth" versus frequency (4). It is easy to show that "viscous depth" is 2.6 times more than "thermal depth" in aqueous dispersions (15). As a result, the particle viscous layers overlap at a volume fraction lower than that of the particle thermal layers. Overlap of the boundary layers is a measure of the corresending particle–particle interaction. There is no particle interaction when corresponding boundary layers are sufficiently separated.

Thus, an increase in the dispersed volume fraction for a given frequency first leads to the overlap of the viscous layers because they extend further into the liquid. Thermal layers overlap at higher volume fractions. This means that the particle hydrodynamic interaction becomes more important than the particle thermodynamic interaction at the lower volume fractions.

The 2.6 times difference between $\delta_v$ and $\delta_t$ leads to a large difference in the volume fractions corresponding to the beginning of the boundary layers' overlap. The dilute case theory is valid for volume fractions smaller than the critical volume fractions $\varphi_v$ and $\varphi_t$. These critical volume fractions are functions of the frequency and particle size. These parameters are conventionally defined from the condition that the shortest distance between particle surfaces is equal to $2\delta_v$ or $2\delta_t$. This definition yields the following expression for the ratio of the critical volume fractions in aqueous dispersions:

$$\frac{\varphi_v}{\varphi_t} = \left(\frac{a\sqrt{\pi f} + 2.6^{-1}}{a\sqrt{\pi f} + 1}\right) \tag{3}$$

where $a$ is the particle radius in micrometers, and $f$ is the frequency is in megahertz.

The ratio of the critical volume fractions depends on the frequency. For instance, for neoprene latex, the critical "thermal" volume fraction is 10 times higher than the critical "viscous" volume fraction for 1 Mhz and only three times higher for 100 Mhz.

It is interesting that this important feature of the "thermal losses" works for almost all liquids. We have more than 100 liquids with their properties in our database. The core of this database is the well known paper by Anson and Chivers (16). We can introduce a parameter referred to as the "depth ratio":

$$depth\ ratio = \frac{\delta_v}{\delta_t}$$

This parameter is 2.6 for water, as was mentioned before. Figure 3 shows values of this parameter for all liquids from our database relative to the viscous depth of water. It is seen that this parameter is even larger for many liquids.

Therefore, "thermal losses" are much less sensitive than "viscous lossess" to the particle–particle interaction for almost all known liquids. It makes ECAH

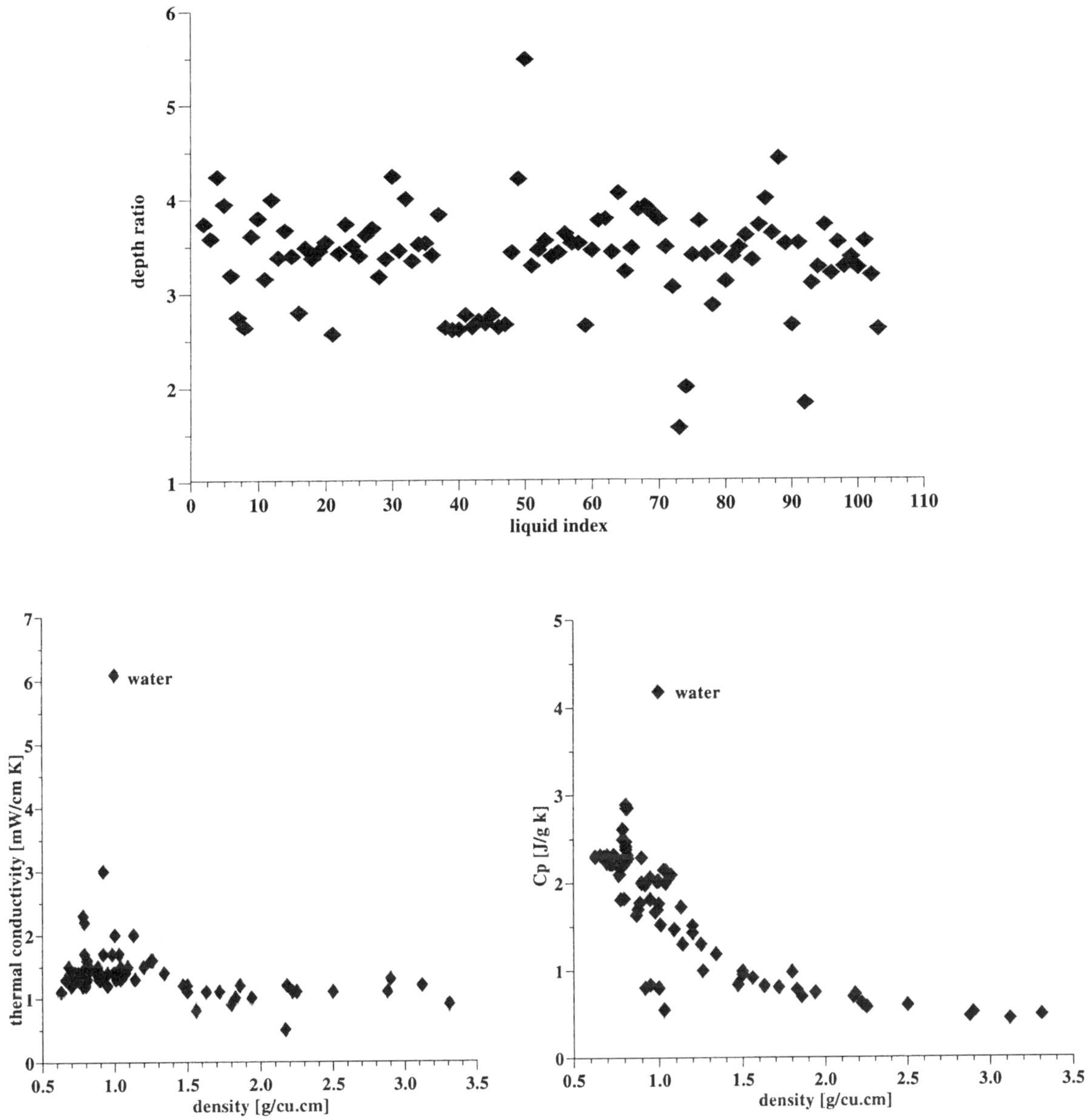


**Figure 3** Thermal properties of various liquids.

theory valid in a much wider range of emulsion volume fractions than one would expect.

There is one more fortunate fact for ECAH theory that follows from the values of the liquid's thermal properties. In general, ECAH theory requires information about three thermodynamic properties: thermal conductivity $\tau$, heat capacity $C_p$, and thermal expansion $\beta$. It turns out that $\tau$ and $C_p$ are almost the same for all liquids except water. Figure 3 illustrates the variation of these parameters for more than 100 liquids from our database. This reduces the number of required parameters to one–thermal expansion. This parameter plays the same role in "thermal losses" as density does in "viscous losses."

ECAH theory has the great disadvantage of being mathematically complex. It cannot be generalized for particle–particle interactions. This is not important, as we have found for emulsions, but may be important for latex systems, and is certainly very important for high-density contrast systems. There are two ways to simplify this theory by using a restriction on the frequency and particle size. The first one is the so-called "long wave requirement" (10) which requires the wavelength of the sound wave $\lambda$ to be larger than particle radius $a$. This "long-wave requirement" restricts the particle size for a given set of frequencies. Our experience shows that particle size must be below 10 μm for the frequency range 1–100 MHz. This restriction is helpful for characterizing small particles.

The long-wave requirement provides a sufficient simplification of the theory for implementing particle–particle interactions. It has been done in the work in Ref. 20 on the basis of the "coupled phase model" (18, 19). This theory (19) works up to 40% volume even for heavy materials including rutile.

There is another approach to acoustics which employs a "short-wave requirement." It was introduced by Riebel et al. (20). This approach works only for large particles (above 10 μm, but requires only limited input data on the sample. The theory may provide an important advantage in the case of emulsions and latex systems when the thermal expansion is not known.

There is opportunity in the future to create a mixed theory that could use a polynomial fit merging together "short" and "long" wave range theories. Such a combined theory will be able to cover a complete particle size range from nanometers to millimeters for concentrated systems.

There are two recent developments in the theory of acoustics which deserve to be mentioned here. The first one is a theory of acoustics for flocculated emulsions (21). It is based on ECAH theory, but it uses an addition an "effective medium" approach for calculating thermal properties of the flocs. The success of this idea is related to the feature of the thermal losses that allows for insignificant particle – particle interactions even at high volume fractions. This mechanism of acoustic energy dissipation does not require relative motion of the particle and liquid. Spherical symmetrical oscillation is the major term in these kinds of losses. This provides the opportunity to replace the floc with an imaginary particle, assuming a proper choice of the thermal properties.

Another significant development is associated with the name of Samuel Temkin. He offers in his papers (22, 23) a new approach to acoustic theory. Instead of assuming a model dispersion consisting of spherical particles in a Newtonian liquid, he suggests that the thermodynamic approach be explored as far as possible. This very promising theory operates with notions of particle velocities and temperature fluctuations, and yields some unusual results (22, 23). It has not yet been used, as far as we know, in commercially available instruments.

## IV. THEORY OF ELECTROACOUSTICS

Whereas acoustic spectroscopy describes the combined effect of the six separate loss mechanisms, electroacoustic spectroscopy, as it is presently formulated, emphasizes only one of these interaction mechanisms, the electrokinetic losses.

In acoustic spectroscopy sound is utilized as both the excitation and the measured variable, and therefore there is but one basic implementation. In contrast, electroacoustic spectroscopy deals with the interaction of electric and acoustic fields and therefore there are two possible implementations. One can apply a sound field and measure the resultant electric field which is referred to as the colloid vibration potential (CVP), or conversely one can apply an electric field and measure the resultant acoustic field which is referred to as the ESA.

First, let us consider the measurement of CVP. When the density of the particles $\rho_p$ differs from that of the medium $\rho_m$, the particles move relative to the medium under the influence of an acoustic wave. This motion causes a displacement of the internal and external parts of the double layer (DL). The phenomenon is usually referred to as a polarization of the DL (6). This displacement of opposite charges gives rise to a dipole moment. The superposition of the electric fields of these induced dipole moments over the collection of particles gives rise to a macroscopical electric field which is referred to as the colloid vibration potential (CVP). Thus, the fourth mechanism of particles' interaction with sound leads to the transformation of part of the acoustic energy to electrical energy. This electrical energy may then be dissipated if the opportunity for electric current flow exists.

Now let us consider the measurement of ESA which occurs when an alternating electric field is applied to the disperse system (7). If the zeta potential of the particle is greater than zero, then the oscillating electrophoretic motion of the charged dispersed particles generates a sound wave.

Both electroacoustic parameters CVP and ESA can be experimentally measured. The CVP or ESA spectrum is the experimental output from electroacoustic spectroscopy. Both of these spectra contain information on the zeta potential and particle size distribution (PSD); however, only one of the electroacoustic spectra is required because both of them contain essentially the same information about the dispersed system.

The conversion of electroacoustic spectra into PSD requires a theoretical model of the electroacoustic phenomenon. This conversion procedure is much more complicated for electroacoustics than for acoustics. The reason for the additional problems relates to the additional field involved in the characterization, i.e., the electric field. The theory becomes much more complicated because of this additional field.

For some time, O'Brien's theory (24, 25) has been considered as a basis for electroacoustics, including concentrated systems. For instance, the review of electroacoustics published by Hunter (7) mentions a somewhat modified version of O'Brien's theory for the electroacoustic characterization of emulsions. However, a few papers have appeared recently (26–28) which express some doubts in O'Brien's theory. It is shown in these papers that O'Brien's theory contradicts the Onsager principle if applied to concentrated systems. There is also a large discrepancy between this theory and experiment. These newpapers offer a different electroacoustic theory which is supposed to be valid in concentrates. In particular, this theory gives correct interpretation of the two equilibrium dilution tests with small silica (30 nm) and larger rutile (300 nm) particles. In both cases the theory works with concentrates (up to 40% by volume) of rutile, for instance.

So far, this new electroacoustic theory has been tested with rigid heavy particles only. It is not clear yet how it will work with emulsions as there were no experimental data for emulsions available. This concern is related to the fact that this theory as well as O'Brien's theory neglect thermodynamic effects. It is rather surprising, keeping in mind that the thermodynamic effect of "thermal losses" is dominant for the acoustics of emulsions. It is not yet clear why electroacoustics is so different from acoustics in that thermodynamic effects are not important.

We offer one simple hypothesis that might explain this difference. Electroacoustics is related to the displacement of the electric charges in the DL. This displacement is characterized by dipole symmetry. At the same time "thermal losses" measured by acoustics are associated mostly with spherical symmetry. They are caused by oscillation of the particle's volume in the sound wave. It is clear that such a spherically symmetrical oscillation does not cause displacement of electric charges in DLs with dipole structure.

This is a hypothesis and a fundamental theory that will take into account thermodynamic effects in addition to electrodynamic and hydrodynamic effects which should resolve the question. The electroacoustic theory of emulsions will not be complete unless such a theory is developed. Nevertheless, electroacoustics, even at the present stage, can yield very important information about electric surface properties of emulsions as it will be shown below.

## V. BUBBLES PROBLEM

One of the experimental problems that may affect acoustics is the presence of air bubbles during measurements. While bubbles will affect sound attenuation and speed, it is worth considering how much of an effect they really have and whether the bubbles will detract from the acoustic techniques:

1. It has been determined that acoustic spectra are affected by bubbles. An acoustic theory describing sound propagation through bubbly liquid was developed by Foldy in 1944 (29), and confirmed experimentally in the 1940s and 1950s (30, 31).
2. The contribution of bubbles to sound speed and attenuation depends on the bubble size and sound frequency. For instance, a 100-μm bubble has a resonance frequency of about 60 kHz. This frequency is reciprocally proportional to the bubble diameter. A bubble of 10 μm diameter will have a resonance frequency of about 0.6 MHz.
3. Acoustic spectroscopy of dispersed systems operates with frequencies above 1 MHz and usually up to 100 MHz. The size of the bubbles must be well below 10 μm in order to affect the complete frequency range of the acoustic spectrometer.
4. Bubbles with sizes below 10 μm are very unstable as is known from general colloid chemistry and the theory of flotation. "Colloid-sized gas bubbles have astonishingly short lifetimes, normally between 1 μs and 1 ms" (32). They simply dissolve in liquid because of their high curvature.

Bubbles can only affect the low-frequency part of the acoustic spectra (below 10 MHz). The frequency range 10–100 MHz is available for particle characterization even in the bubbly liquids. Acoustic spectrometers can both sense bubbles and characterize particle size. We can confirm this conclusion with thousands of measurements performed with hundreds of different systems. Sensitivity to bubbles, in fact, is an important advantage of acoustics over electroacoustics. The presence of bubbles may affect the properties of the solid dispersed phase. For instance, bubbles can be centers of aggregation, which makes them an important stability factor.

## VI. MEASURING TECHNIQUE

Currently, there are three acoustic spectrometers on the market: Ultrasizer from Malvern, Opus of Sympatec, and DT-100 from Dispersion Technology. All of them are claimed to be able to characterize emulsions in the wide droplet size range. There are some major differences between them. For instance, Opus was designed initially for large particles only because it employs the "short wavelength requirement" (21).

There are also two electroacoustic spectrometers on the market: the AcoustoSizer from Colloidal Dynamics and the DT-200 from Dispersion Technology. There is only one instrument which provides both features, acoustics and electroacoustics together, and this is the DT-1200 Acoustic and Electroacoustic Spectrometer from Dispersion Technology.

Comparison of the different instruments lies beyond the scope of this review. The Dt-1200 was used for all experiments described in this work. A description of this instrument is given below.

The DT-1200 has two separate sensors for measuring acoustic and electroacoustic signals separately. Both sensors use the pulse technique. The acoustic sensor has two piezo crystal transducers. The gap between the transmitter and receiver is variable in steps. In default mode, the gap changes from 0.15 mm up to 20 mm in 21 steps. The basic frequency of the pulse changes in steps as well. In default mode, the frequency changes from 3 to 100 MHz in 18 steps. The number of pulses collected for each gap and frequency is automatically adjustable in order to reach the target signal-to-noise ratio.

The variable-gap technique is an essential feature of the acoustic spectrometer. This makes it possible to cover a wide dynamic range of possible attenuations. For instance, pure water is almost transparent to ultrasound at a low frequencies (below 10 MHz). The attenuation of water reaches only 5 dB/cm at 50 MHz. Therefore, the attenuation of water should be measured at large gaps, as little information is obtained from small gaps.

Water is the least attenuating liquid known to us. In contrast, a 40% by weight water-in-cyclo methicone emulsion attenuates ultrasound very strongly (Fig. 4). This attenuation reaches 450 dB/cm 50 MHz. This

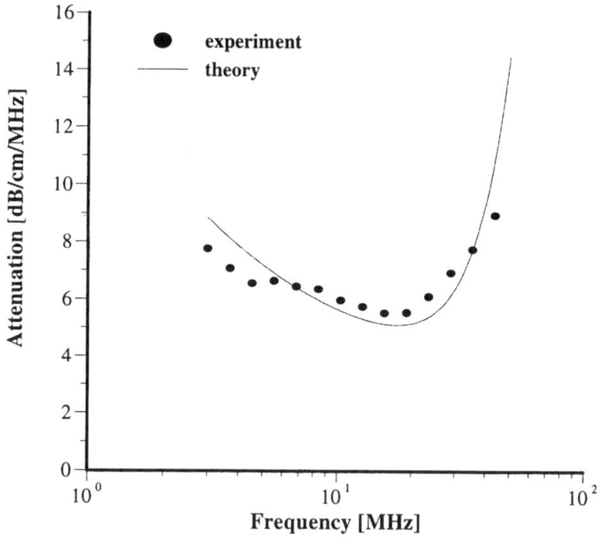


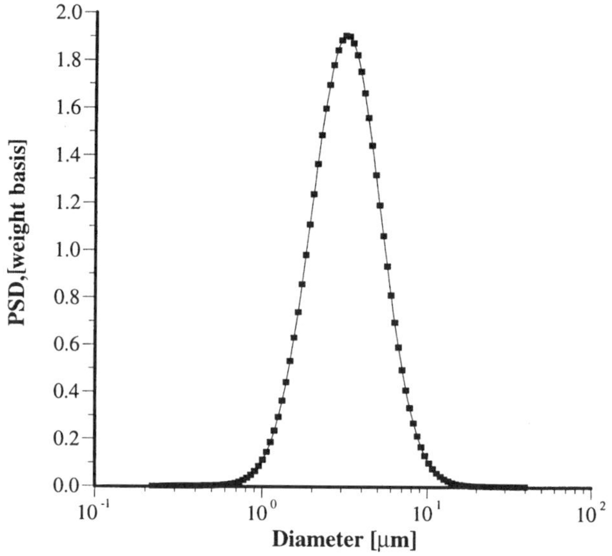


**Figure 4** Attenuation spectra and droplet size distribution of 40% water in cyclo methicone emulsion.

occurs because cyclo methicone has a very high thermal expansion coefficient of 14.5 $10^{-4}$ $1K^0$. The acoustic attenuation results for this system can only be obtained at small gaps. At larger gap size, the ultrasound signal simply does not pentrate to the receiver because of the high attenuation.

The acoustic sensor measures also sound speed at a single chosen frequency. The sound speed is measured by recording the time it takes for a pulse to arrive at the receiver. The instrument automatically adjusts the pulse sampling, depending on the value of the sound speed. An accurate knowledge of the sound speed is necessary for eliminating possible artifacts such as excessive attenuation at low frequencies.

Sound-speed measurement is especially critical for characterizing emulsions. There are considerable data indicating that the sound speed of various liquids is extremely dependent on small traces of contamination. Examples of such complex behavior of different mixed liquids is given in the literature (33). Figure 5 illustrates one of the examples from this book. Therefore, the only reliable way to obtain the sound speed of a given emulsion is by experimental measurement. This is important to keep in mind when evaluating different acoustic instrument models as an instrument without the capability to measure sound speed is very limited.

The DT-1200 instrument is able to measure sound speed quite reliably, which is is illustrated in Fig. 6 showing the results of a dilution test with silica Ludox TM. The experimentally measured sound speed was found to be very close to theoretical calculations.

The electroacoustic sensor measures the magnitude and phase of the CVC at 1.5 and 3 MHz. It has a piezo crystal sound transmitter and a specially designed electric antenna. The distance between the transmitter and antenna is 5 mm. There is a provision for automatic correction of the sound speed and attenuation measured with the acoustic sensor.

There is a special analysis program which calculates (PSD) from attenuation spectra and zeta potential from the CVC. This program uses the ECAH theory for calculating "thermal losses", the Waterman–Truell theory (34) for calculating scattering losses, and the theory described in the Ref. 19 for calculating viscous losses. The electroacoustic theory used with the new version of the instrument is described in Ref. 28.

This program tests lognormal, bimodal, and modified lognormal (35) PSDs. It uses an error analysis in order to search for the best PSD. The goal of the optimization procedure is to minimize the error of the theoretical fit to the experimental attenuation spectra.

The analysis program takes into account the PSD correction when it calculates zeta potential. It uses either PSD calculated from the attenuation spectra or a priori known PSD. The analysis routine also makes a correction for attenuation of the sound pulse.

The total required sample volume is about 100 ml. There is a special magnetic stirrer preventing sedimentation and promoting mixing of chemicals during titration. The instrument has two burets and appropriate software for automatic titration. Conductivity and temperature probes are also available.

One attenuation spectra measurement with a default set-up takes from 5 to 10 min. A user can speed up the

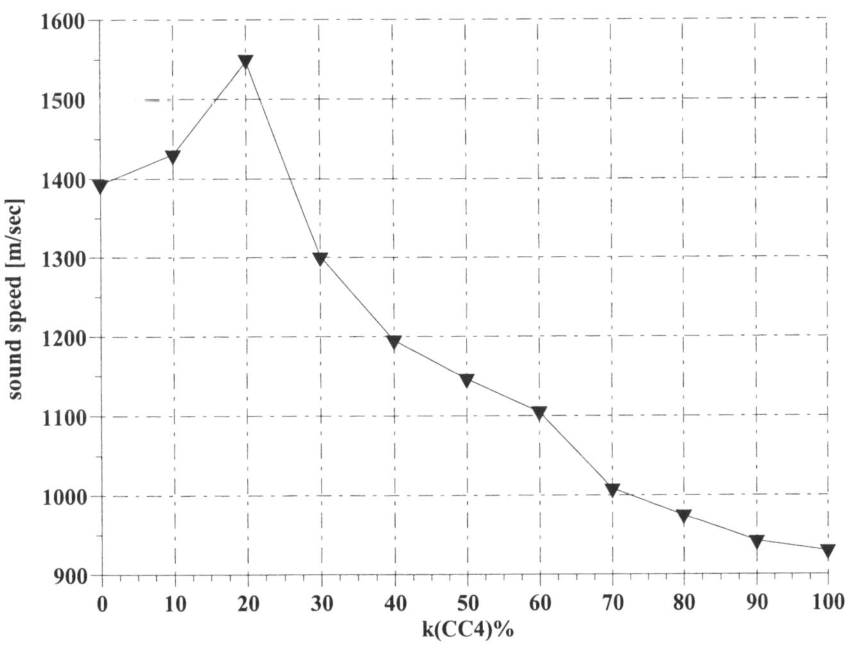


**Figure 5** Sound speed of water—$CC_4$ mixture.

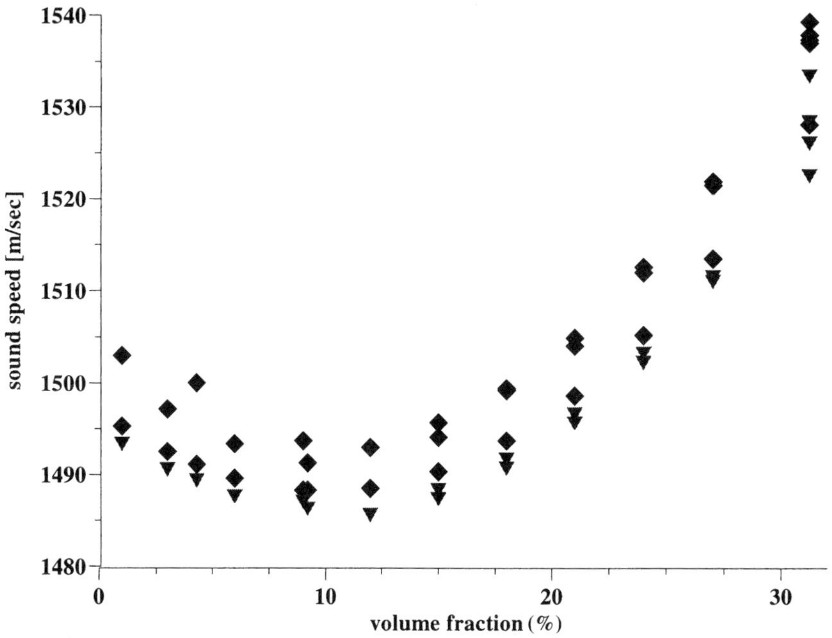


**Figure 6** Sound speed for silica Ludox TM vs. volume fraction. Equilibrium dilution using dialysis. Theory (triangles) according to the Wood expression; experiment diamonds.

measurement by changing set-up parameters. One CVC measurement takes from 10 s to 1 min depending on the system properties.

The precision and accuracy of the DT-1200 for emulsions are described below. We start here with solid particles because it is much easier to test the reproducibility and accuracy with a stable dispersion of solid particles (36). Possible variation of the emulsion droplets can affect this test.

The attenuation spectra in Fig. 7 provide an example of the acoustic sensor precision. These attenuation spectra were measured using alumina Sumitomo AA-2 and silica Ludox TM. The alumina sample was measured 10 times repeatedly while the silica sample was measured 11 times repeatedly. The corresponding median particle size results are given in Table 1.

The absolute variation of the median particle size was 0.9% for alumina and 1.5% for silica. These values show the precision of the acoustic sensor.

Figures 8 and 9 illustrate the precision of the electroacoustic sensor. Figure 8 provides results for 51 continuous CVC measurements on silica Ludox. The precision measured as the absolute variation of the zeta potential measurement is a fraction of a millivolt. Figure 9 shows titration curves for two different silica samples.

The accuracy characterizes correlation between real and measured values. The accuracy of PSD measurements is a measure of the adequacy of the measured PSD. In order to determine the accuracy of the PSD, one needs a standard system with a known particle size distribution. BCR silica quartz was chosen as a standard with a median size of about 3 μm. This system was chosen because it is a well-known PSD standard in Germany.

Figure 10 shows the standard particle size distribution and PSD measured with the DT-1200. The difference in the median particle size between the standard and experimental results obtained with the DT-1200 was less than 1%. The PSD was also found to contain a higher percentage of the smaller particles and this gave an accuracy of 5% for the standard deviation (measure of polydispersity).

A test of the measurement accuracy of zeta potential is much more complicated because there is no zeta potential standard for concentrated systems. The absence of electroacoustic theory for concentrated systems creates additional complexity. Our experience is that CVC makes it possible to measure $\zeta$ with almost the same accuracy as microelectrophoresis.

We have also tested the accuracy and precision of the D-1200 Acoustic Spectrometer using Standard Dow Latex with an expected median particle size of 0.083 μm. The results are shown in Table 2.

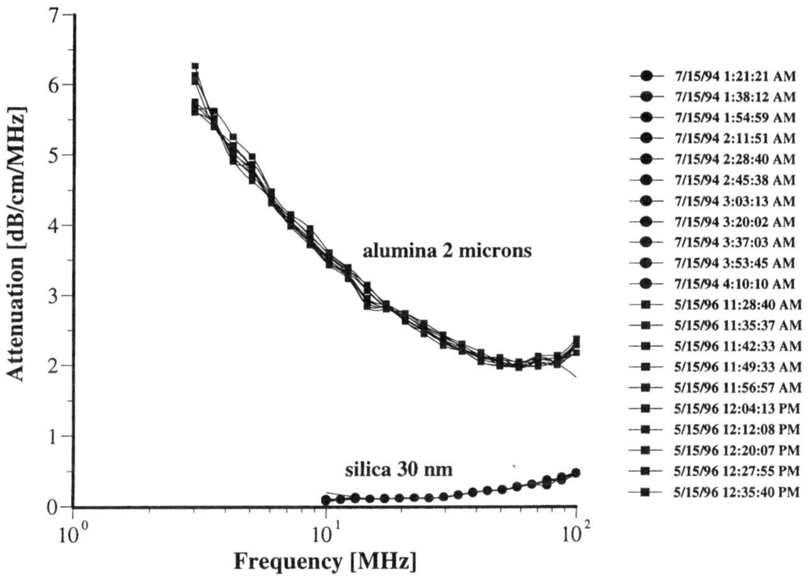


**Figure 7** Attenuation of the multiple measurements with alumina Sumitomo AA-2 and silica Ludox TM at 10% wt.

**Table 1** Median Particle Size (μm)

| | | | | | | | | | | | |
|---|---|---|---|---|---|---|---|---|---|---|---|
| Alumina | 2.015 | 2.076 | 2.057 | 2.092 | 2.065 | 2.047 | 2.035 | 2.075 | 2.039 | 2.085 | |
| Silica | 0.03 | 0.029 | 0.029 | 0.029 | 0.029 | 0.03 | 0.03 | 0.029 | 0.03 | 0.03 | 0.03 |

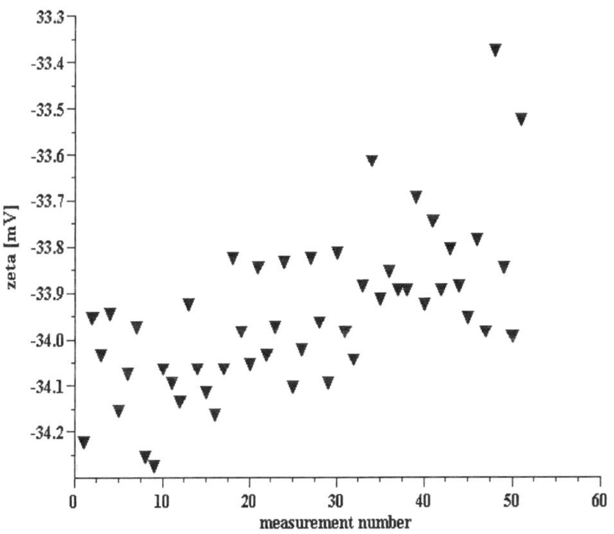


**Figure 8** Multiple ζ-potential measurements of 10% wt silica Ludox.

## VII. APPLICATIONS AND EXPERIMENTS

### A. Emulsions

There are many instances of successful characterization of the PSD and zeta potential of emulsion droplets. There are two quite representative reviews of these experiments published by McClements (4) (acoustics) and Hunter (8) (electroacoustics).

Some results of our recent investigation are presented that were not published before. Various factors that affected stability, size, and zeta potential of the emulsion droplets were investigated.

The first experiment was a repetition to some extent of McClements' work with hexadecane-in-water emulsions. An emulsion was prepared following McClements work (37), containing 25% by weight of hexadecane in water. The measured attenuation spectra (Fig. 11) exhibited a pronounced time dependence. The sound attenuation was found to increase in magnitude as time elapsed. This increase in the attenuation corresponded to the droplet population becoming smaller in size. The median droplet size was reduced

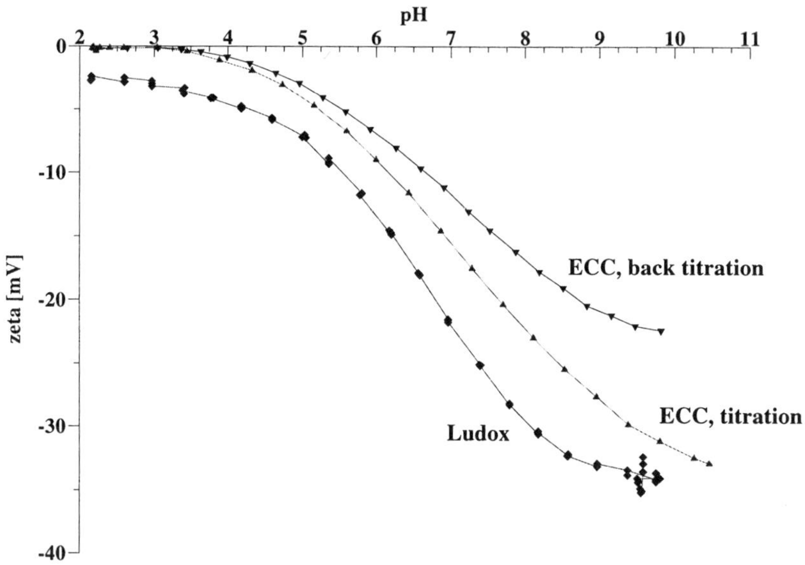


**Figure 9** Titration of silica Ludox TM at 10% wt and chemical–mechanical polishing silica ECC.

by almost half during a half-hour experiment. This reduction in droplet size was caused by the shear induced by a magnetic stirrer used in the sample chamber of the DT-1200 instrument. As the emulsion was stirred, the larger drops were fragmented into smaller droplets.

Another important parameter affecting emulsions is the surfactant concentration that affects surface chemistry. This factor was tested for reverse water-in-oil emulsion. The oil phase was simply commercially available car-lubricating oil diluted twice with paint thinner in order to reduce the viscosity of the final sample. Figure 12 illustrates results for emulsions prepared with 6% by weight of water.

This figure shows the attenuation spectra for three samples. The first sample was a pure oil phase and exhibited the lowest attenuation. It is important to measure the attenuation of the pure dispersion medium when a new liquid is evaluated. In this particular case, the intrinsic attenuation of the oil phase was almost 150 dB/cm at 100 MHz which is more than seven times

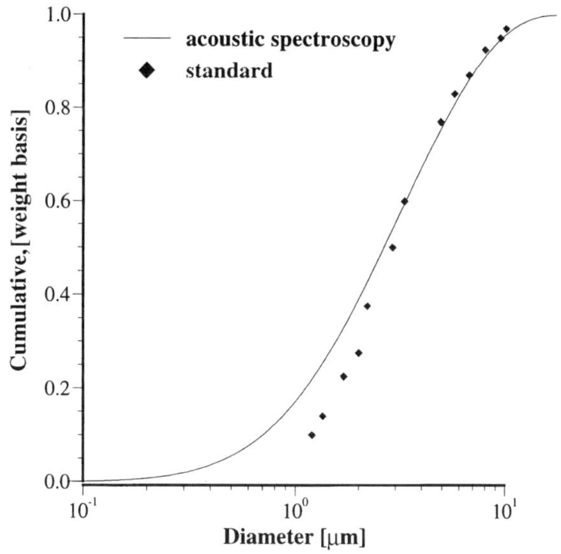


**Figure 10** Particle size distribution (PSD) of the silica quartz BCR. Acoustic measurement has been performed with an 11% wt in ethanol.

**Table 2** Median Diameter Measured for 10% wt Latex Serva 44405, Standard Dow Latex with Expected Median Size of 0.083 μm

| | 1 | 2 | 3 | 4 | 5 | 6 | 7 | 8 | 9 | 10 |
|---|---|---|---|---|---|---|---|---|---|---|
| Median diameter μm | 0.0767 | 0.0765 | 0.0867 | 0.0823 | 0.0775 | 0.0737 | 0.0713 | 0.081 | 0.076 | 0.075 |

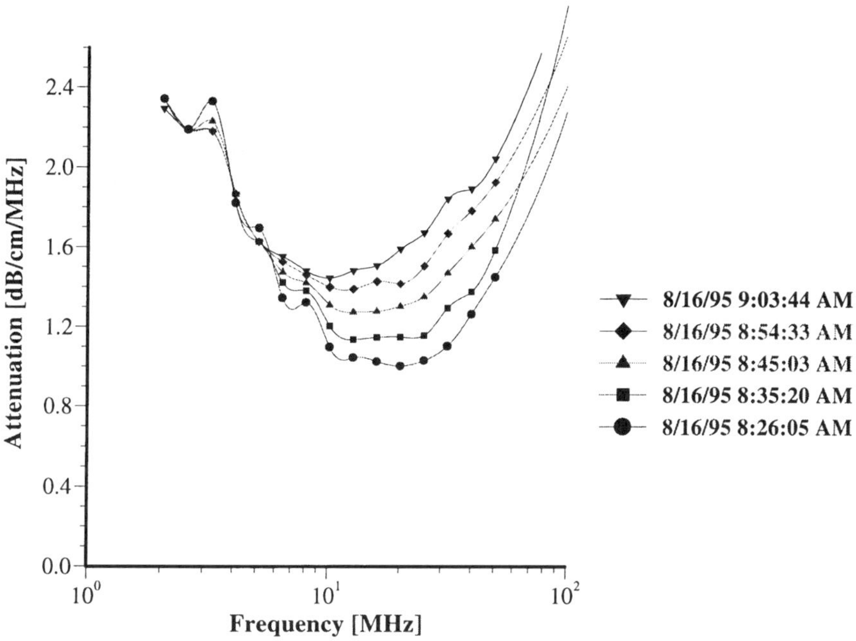


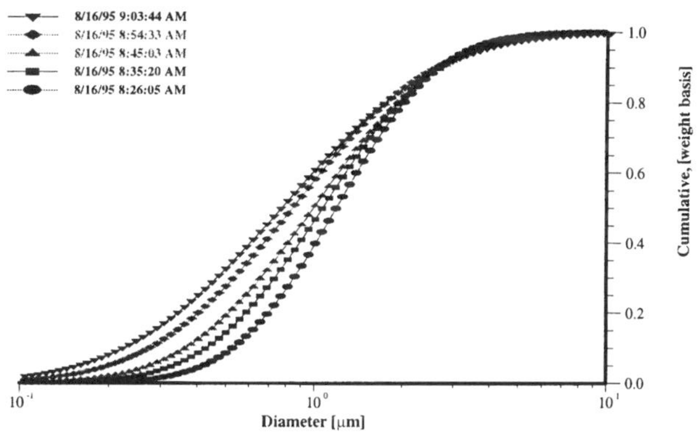


**Figure 11** Attenuation and corresponding PSD of 25% wt hexadecane-in-water emulsion.

higher than for water. This intrinsic attenuation is a very important contribution to the attenuation of ultrasound in emulsions. It is the background for characterizing emulsion systems.

The emulsion without added surfactant was measured twice with two different sample loads. As the water content was increased the attenuation became greater in magnitude. For this system, the attenuation was found to be quite stable with time. Addition of 1% by weight of AOT [sodium bis (2-ethylhexylsulfosuccinate] changed the attenuation spectra dramatically. This new emulsion with modified surface chemistry was measured twice in order to show reproducibility. The corresponding PSD is shown in Fig. 12 and indicates that the AOT converted the regular emulsion into a microemulsion as one could expect.

These experiments proved that the acoustic technique is capable of characterizing the PSD of relatively stable emulsions. In many instances, emulsions are found that are not stable at the dispersed volume concentration required to obtain sufficient attenuation signals (usually above 0.5%). Hazy water in fuel emulsions (diesel, jet fuel, gasoline) may exist at low water concentrations of only a few 100 ppmv (0.01%) of dispersed water. Attempts at characterizing these systems without added surfactant resulted in unstable

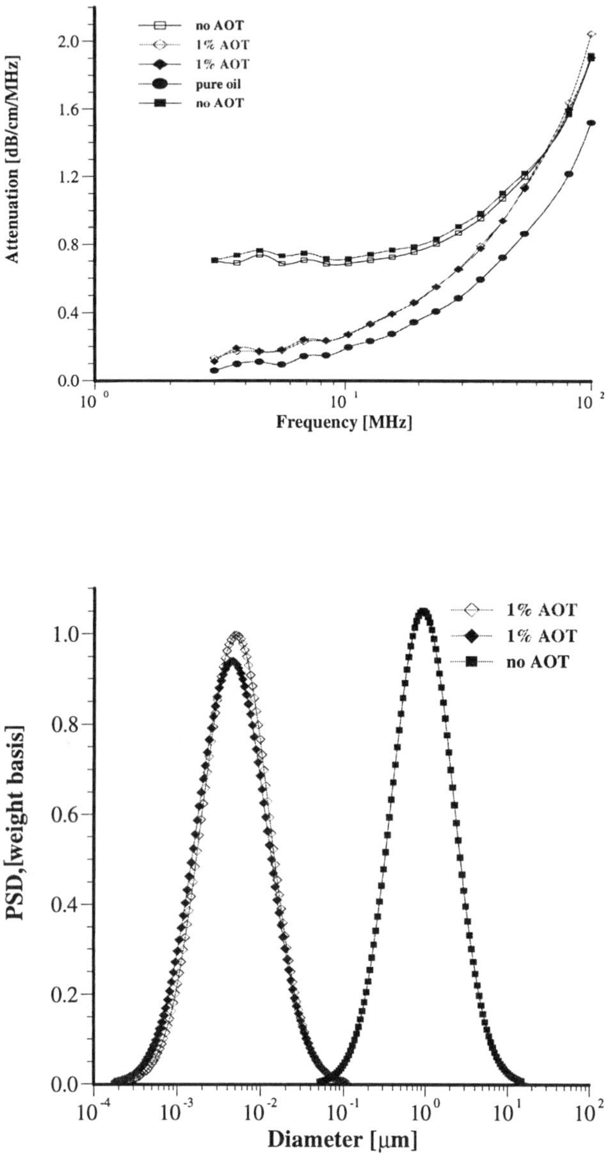


**Figure 12** Attenuation and corresponding PSD of 6% wt water-in-car-oil emulsion and microemulsion caused by AOT.

attenuation spectra, and water droplets were discovered to separate from the bulk emulsion and settle out on the chamber walls. This problem is less important for thermodynamically stable microemulsions.

## B. Microemulsions

The mixture of heptane with water and AOT is a classic three-component system. It has been widely studied due to a number of interesting features it exhibits. This system forms stable reverse mciroemulsions (water in oil) without the complication introduced by additional cosurfactant. Such a cosurfactant (usually alcohol) is required by many other reverse microemulsion systems. This simplification makes the alkane/water/AOT system a model for studying reverse microemulsions.

There have been many studies devoted to characterization of these practically important systems. Reverse emulsion droplets have been used as chemical microreactors to produce nanosize inorganic and polymer particles with special properties that are not found in the bulk form (38–42). These microemulsion systems have also been a topic of research for biological systems and the AOT head groups have been found to influence the conformation of proteins and increase enzyme activity (43–46). The unique environment created in the small water pools of swollen reverse micelles allows for increased chemical reactivity. The increase in surface area with decreased in size of the droplets also can significantly increase reactivity by allowing greater contact of immiscible reactants.

There have been many attempts to measure the droplet size of ths microemulsion. Several different techniques were used: PCS (47–52), classic light scattering (49, 51, 53), neutron scattering (SANS) (54–56), X-ray scattering (SAXS) (48, 57, 58), ultracentrifugation (46, 50, 53), and viscosity (48, 50, 53). It was observed that the heptane/water/AOT microemulsions have water pools with diameters ranging from 2 nm up to 30 nm. The water drops are encapsulated by the AOT surfactant so that virtually all of the AOT is located at the interface shell. The size of the water droplets can be conveniently altered by adjusting the molar ratios of water to surfactant designated as $R$ ([H2O]/[AOT]). At low $R$ values ($\leq 10$) the water is strongly bound to the AOT surfactant polar head groups and exhibits unique characteristics different from bulk water (53). At higher water ratios ($R > 20$), free water is predominant in the swollen reverse micellular solutions, and at approximately $R = 60$, the system undergoes a transition from a transparent microemulsion into an unstable turbid macroemulsion. This macroemulsion separates on standing into a clear upper phase and a turbid lower phase.

The increase in droplet size and phase boundary can also be achieved by raising the temperature up to a critical value of 55°C. In adition, this system has been found to exhibit an electrical percolation threshold whereby the conductivity increases by several orders of magnitude by either varying the $R$ ratio or increasing the temperature (56, 57, 59, 60). Despite all these efforts, there still remain questions regarding the polydispersity of the water droplets, and few studies are available above the $R$ value of 60 where a turbid macroemulsion state exists.

Acoustic spectroscopy offers a new opportunity for characterizing such complicated systems. Details of this experiment are presented in the Ref. 61. The reverse microemulsions were prepared by first making a 0.1 M solution of AOT in heptane (6.1% wt AOT). The heptane was obtained from Sigma as HPLC grade (99+ % purity). Known amounts of 18 MΩ· cm water were added to the AOT–heptane solution using a 1 ml total volume graduated glass syringe and then shaken for 30 s in Teflon-capped glass bottles. The shaking action was required to overcome an energy barrier to distribute the water into the nanosized droplets, as it could not be achieved using a magnetic stirrer.

In all cases, the reported $R$ values was based on the added water, and were not corrected for any residual water that may have been in the dried-AOT or heptane solvent. Karl Fischer analysis of the AOT–heptane solutions before the addition of water resulted in an $R$ value of 0.4. This amount was considered to be negligible.

Measurements were made starting with the pure water and heptane and then the AOT–heptane sample with no added water ($R = 0$). The sample fluid was removed from the instrument cell and placed in a glass bottle with a Teflon cap. Additional water was titrated and the microemulsion was shaken for 30 s before being placed back in the instrument cell. The sample cell contained a cover to prevent evaporation of the solvents. The samples were visually inspected for clarity and rheological properties for each $R$ value. These steps were repeated for increasing water weight fraction or $R$ ratios up to 100. At $R \geq 60$ the microemulsions became turbid. At $R > 80$, the emulsions became distinctly more viscous.

The weight fractions of the dispersed phase were calculated for water only, without including the AOT. Each trial run lasted approximately 5–10 min with the temperature varied from 25–27°C. A separate microemulsion sample for $R = 4-$ was made up a few days before the first study. For the $R = 70$ sample, a second acoustic measurement was carried out with the same sample used for the first study. The complete set of experiments for water, heptane, and the reverse microemulsions from $R = 0$ to 100 was repeated to evaluate the reproducibility.

Attenuation spectra measured in the first run up to $R = 80$ are presented in Fig. 13. The results for $R = 90$ and $R = 100$ are not reported because they were found

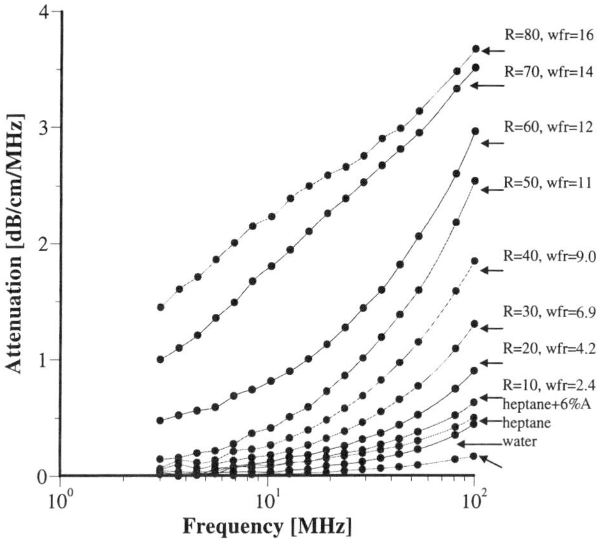


**Figure 13** Acoustic attenuation spectra measured for water/AOT/heptane system for different water-to-AOT ratios $R$.

to vary appreciably. As the water concentration is increased, the attenuation spectrum rises in intensity and there is a distinct jump in the attenuation spectrum from $R = 5-$ to $R = 60$ in the low-frequency range. This discontinuity is also reflected in the visual appearance, as at $R = 60$ the system becomes turbid. The smooth shape of the attenuation curve also changes at $R > 60$. The stability and reproducibility of the system was questioned owing to the irregular nature of the curve, so the experiment at $R = 70$ was repeated and gave almost identical results. An additional experiment was run at $R = 40$ for a separate microemulsion prepared a few days earlier. This showed excellent agreement with the results for freshly titrated microemulsion.

For $R$ values > 70, an increase in the viscosity and a decrease in the reproducibility of the attenuation measurement were observed. This could be due to the failure of the model of this system as a collection of separate droplets at high $R$ values.

A second set of experiments was run to check the reproducibility. The results of both sets of experiments up to $R = 60$ are given in Fig. 14. It can be seen that the error related to the reproducibility is much smaller than the difference between attenuation spectra for the different $R$ values. This demonstrates that the variation of attenuation reflects changes in the sample properties of water weight fraction and droplet size. The sound attenuation at $R$ values above 60 were not as reproducible, but did give the

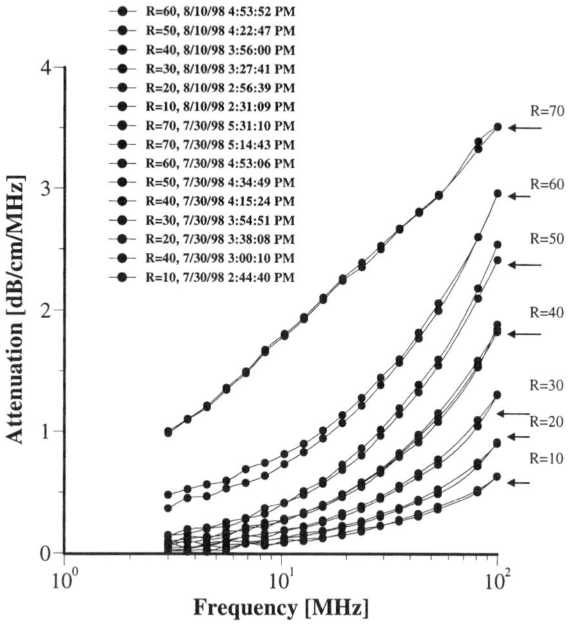


**Figure 14** Reproducibility test of the attenuation measurement.

same form of a bimodial distribution as the best fit for the experimental data.

The two lowest attenuation curves correspond to the attenuation in the two pure liquids: water and heptane. This attenuation is associated with oscillation of liquid molecules in the sound field. If these liquids are soluble in each other, the total attenuation of the mixture would lie between these two lowest attenuation curves. However, it can be seen that the attenuation of the mixture is much higher than that of the pure liquids. The increase in attenuation is, therefore, due to this heterogeneity of the water in the heptane system. The extra attenuation is caused by motion of droplets, not separate molecules. The scale factor (size of droplets) corresponding to this attenuation is much higher than that for pure liquids (size of molecules).

The current system contains a third component – AOT. A question arises on the contribution of AOT to the measured attenuation. In order to answer this question, measurements were performed on a mixture of 6.1% by weight. AOT in heptane ($R = 0$). It is the third smallest attenuation curve in Fig. 13. It is seen that attenuation increases somewhat due to AOT. However, this increase is less than the extra attenuation produced by water droplets. The small increase in attenuation is attributed to AOT micelles. Unfortunately the thermal properties of AOT as a liquid phase are not known and the size of these micelles could not be calculated.

The PSDs corresponding to the measured attenuation spectra are prsented in Fig. 15. It can be seen that the distribution becomes bimodal for $R \geq 60$, which coincides with the onset of turbidity. It is to be noted that such a conclusion could not easily be arrived at with other techniques. However, Fig. 15 illustrates a peculiarity of this system that can be compared with independent data from the literature (54, 55): mean particle size increases with $R$ in an almost linear fashion. This dependence becomes apparent when mean size is plotted as a function of $R$ as in Fig. 16.

It is seen that mean particle sizes measured used acoustic spectroscopy are in good agreement with those obtained independently using the SANS and SAXS techniques (43, 48, 54) for $R$ values ranging from 20 to 60. A simple theory based on equipartition of water and surfactant (36) can reasonably explain the observed linear dependence.

At $R = 10$ the acoustic method gave a slightly larger diameter than expected. This could be as a result of the constrained state of the "bound water" in the swollen reverse micelles. The water under these conditions may exhibit different thermal properties from those of the bulk water used in the particle size calculations. Also, at the low $R$ values ($R \leq 10$ or $\leq 2.4\%$ water), the attenuation spectrum is not very large as compared to the background heptane signal. The contribution

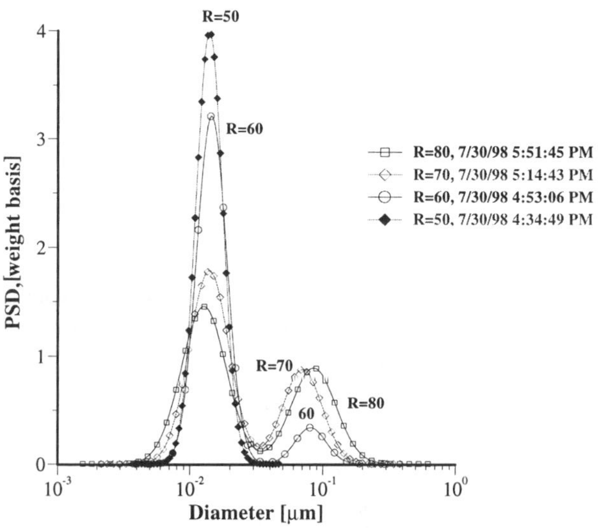


**Figure 15** Drop size distribution for varying $R$ [H20]/[AOT] from 10 to 50 and from 50 to 80.

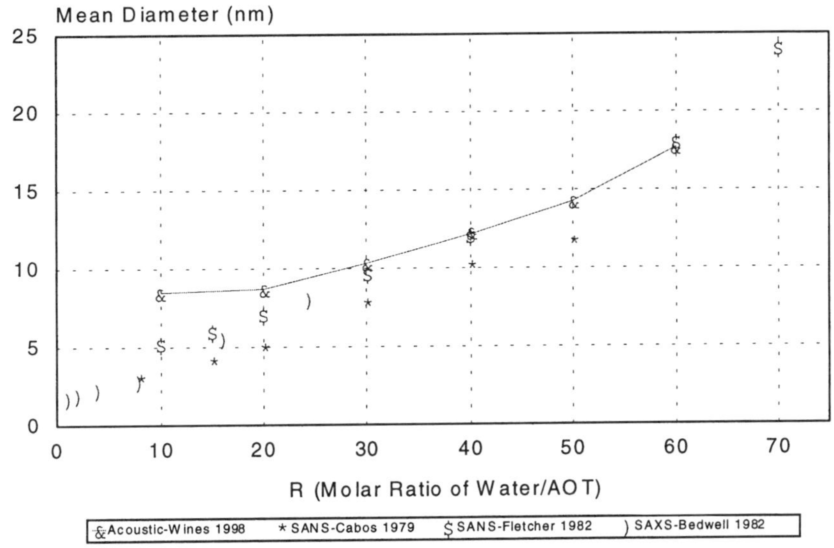


**Figure 16** Comparison of mean droplet size measured using acoustic spectroscopy, neutron scattering, and X-ray scattering.

of droplets to the attenuation spectrum then may become too low to be reliably distinguished from the background signal coming from heptane molecules and AOT micelles.

In addition to particle size, the CVC was also measured for calculating zeta potential. The results are presented in Fig. 17, and $\zeta$ was found to depend on the water content. An increased concentration of water resulted in higher zeta potentials. However, the water content was not the most important factor. This experiment was performed at two different AOT concentrations and the ratio of water to AOT ($R$) was discovered to be the key parameter. When $\zeta$ wasplotted versus the $R$ values, the same curve was obtained for both AOT concentrations. This demonstrates that the zeta potential depends on the degree of the water surface coverage by AOT molecules.

This experiment allows us to suggest a mechanism for electric-charge formation on the surface of the water droplets in the oil phase. This is a field of great interest in modern emulsion science. According to our experiment, the zeta potential appears when there is a deficit of AOT molecules for complete coverage of the water droplets. As more elements of the water phase become exposed to the oil, higher values of $\zeta$ are measured. The water phase also contains a considerable concentration of sodium ions that originate from the AOT and serve as counterions to the negatively charged sulfosuccinate head groups. As a result of decreased surface coverage, the water droplets gain surface charge when they are in contact with oil. This surface charge can appear because of ion exchange between the water and oil phases caused by the difference in standard chemical potentials in each phase. Molecules of AOT do not create surface charge, but conversely screen the surface charge of the initial water droplets. At the same time these AOT molecules change the interfacial tension, creating conditions for a thermodynamically stable microemulsion. This is only a hypothesis so far and further investigation is required for confirmation.

## C. Latex Systems

There have been many successful experiments that have characterized latex systems by using both acoustics and electroacoustics. For instance, Allegra and Hawley (10) measurd polystyrene latex. We measured Standard Dow latex which is also polystyrene in nature (see above). There is another successful application, this time with neoprene latex, which is described in Ref. 15.

This low-density latex dispersion (Neoprene Latex 735A) is designed by DuPont as a wet-end additive to fibrous slurries. The fraction of the latex in the initial dispersion is 42.8% by weight (37.3% by volume). The pH value at 25°C is 11.5. The physical properties of the neoprene (slow crystallizing polychloroprene homopolymer) have been measured in

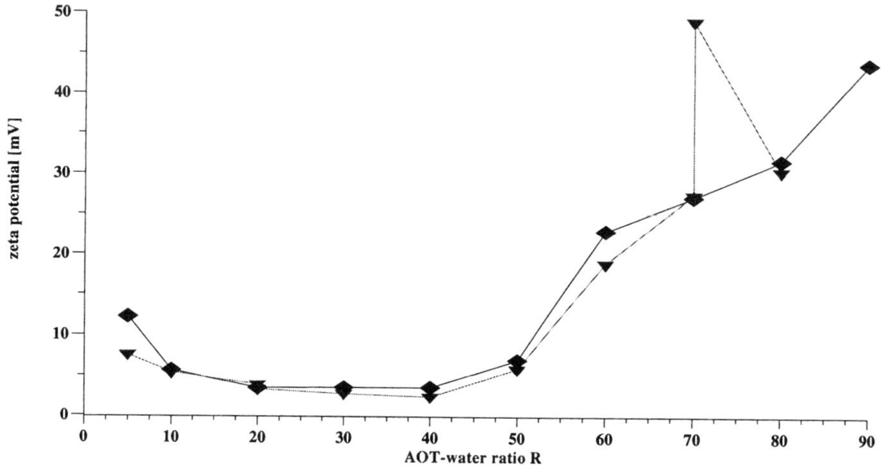


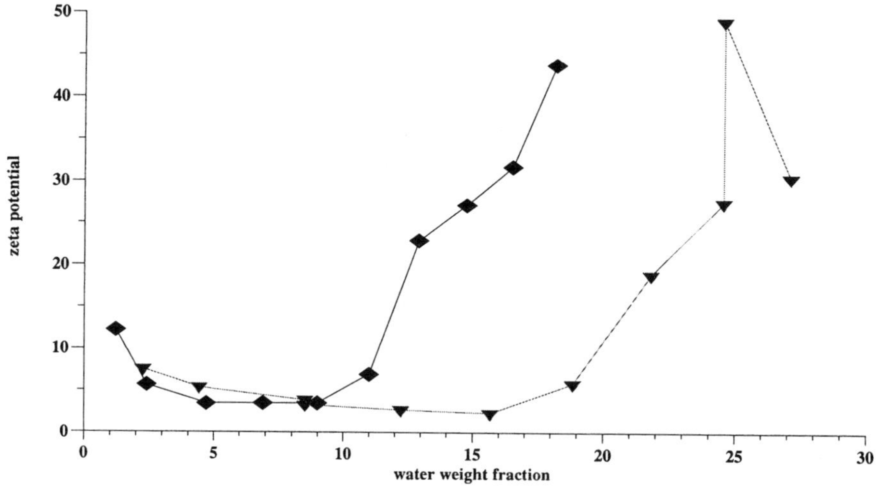


**Figure 17** Zeta potential measured electroacoustically for water droplets covered with AOT in heptane vs. water content.

the DuPont laboratories many years ago. These data are summarized in the monograph "The Neoprenes" (62).

A dilution test was performed on this latex by using distilled water with a pH adjusted to 11.5 with 1 N potassium hydroxide. The samples were prepared with various dispersed concentrations (1.4, 4, 6.6, 13, 19.4, 25.6, 31.6, and 37.5 % by weight) by adding diluting solution to the initial neoprene latex.

Interpretation of the attenuation spectra requires information on the entire particle size spectra. A log-normal approximation was used with a median size of 0.16 μm and a standard deviation of 6% for the PSD measured with hydrodynamic chromatography.

The experimental data collected by the acoustic method with the neoprene latex provided an opportunity to check the validity of the ECAH theory when thermal losses were the dominant mechanism of the sound attenuation (see Fig. 18). In order to calculate the theoretical attenuation spectra, information is required about the particle size, thermodynamic properties of the dispersed phase, and dispersion medium materials as well as "partial intrinsic attenuations." Fortunately, all of the required parameters are available in this case. The approximate thermodynamic properties of the neoprene are known from the independent investigation performed by DuPont.

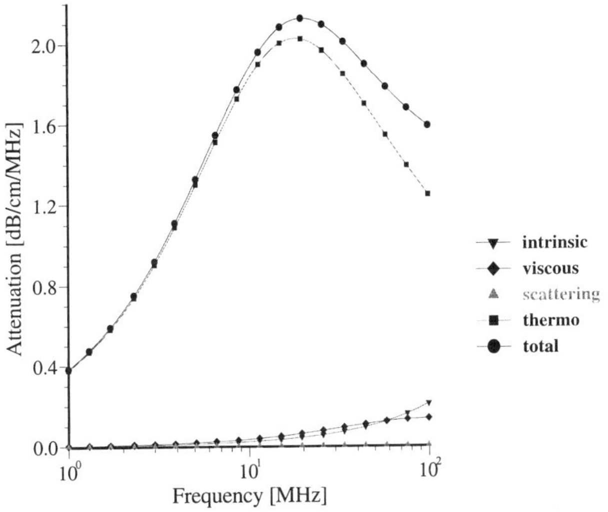


**Figure 18** Theoretical attenuation spectra for the various mechanisms of the acoustic energy losses. Volume fraction is 10% vol; particle size 0.16 μm.

Figure 19 shows experimental and theoretical attenuation spectra for all the measured volume fractions. It is seen that the correlation between theory and experiment is very good up to 37.5% by weight (32.4% by volume).

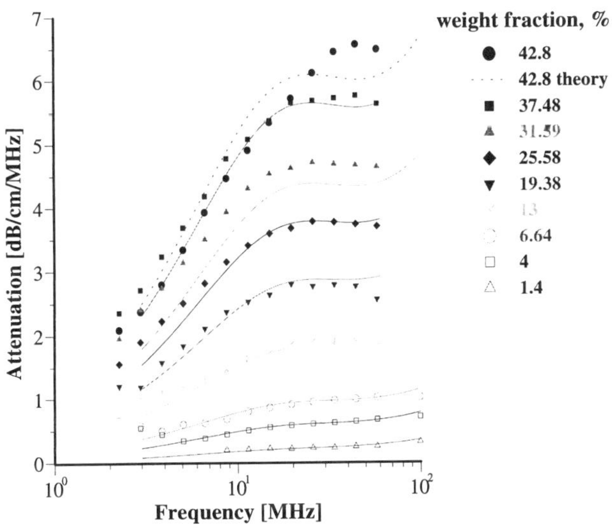


**Figure 19** Experimental and theoretical attenuation spectra for neoprene latex for the weight fractions indicated in the legends.

These successful examples of characterizing latex systems are possible only when thermal expansion coefficients are known. Unfortunately, this parameter is not known for many latex polymers. This problem becomes even more complicated for latex systems than for emulsions because the value of the thermal expansion depends strongly on the chemical composition of the polymer. Figure 20 illustrates this fact for several ethylene copolymers with different ethylene contents. Variation of the ethylene content from 5 to 10% was found to cause significant change in the attenuation spectra. This change is associated with the thermal expansion coefficient, but not the particle size.

The uncertainty related to the thermal expansion coefficient makes latex systems the most complicated systems for acoustics. This is important to keep in mind for testing a particular model of an acoustic instrument. Latex dispersions that are used as standards for light-based methods should be used with caution as in many cases the thermal-expansion properties of these standards are not well known.

## VIII. CONCLUSIONS

We hope that we have proved with this short review that acoustics and electroacoustics can be extremely helpful in characterizing particle size, zeta potential, and some other properties of concentrated emulsions, microemulsions, and latex systems. Both methods are commercially available already. There are still some problems with the theoretical background for electroacoustics, but analysis of the literature shows gradual improvement in this field.

The combination of acoustic and electroacoustic spectroscopy provides a much more reliable and complete characterization of the disperse system than either one of those techniques separately. Electroacoustic phenomena are more complicated to interpret than acoustic phenomena because an additional field (electric) is involved. This problem becomes even more pronounced for a concentrated system. It makes acoustics favorable for characterizing particle size, whereas electroacoustics yields electric surface properties.

We believe that these ultrasound-based techniques provide a very valuable addition to the traditional colloid chemical arsenal of tools designed for characterizing surface phenomena.

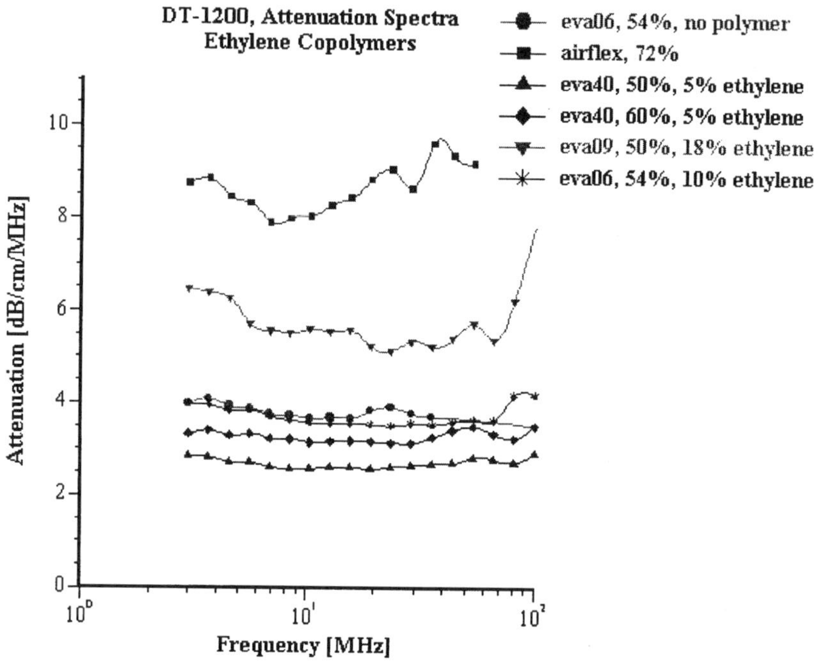


**Figure 20** Attenuation spectra for latex dispersions with different contents of ethylene.

## REFERENCES


1. JR Pellam, JK Galt. J Chem Phys 14: 608–613 (1946).
2. CTJ Sewell. Phil Trans Roy Soc, London, 210: 239–270, 1910.
3. PS Epstein, RR Carhart. J Acoust Soc Am 25: 553–565, 1953.
4. DJ McClements. Adv Colloid Interface Sci 37: 33–72, 1991.
5. VA Hackley, J Texter, eds. Ultrasonic and Dielectric Characterization Techniques for Suspended Particulates. New York: The American Chemical Society, 1998.
6. J Lyklema. Fundamentals of Interface and Colloid Science. Vol 1. New York: Academic Press, 1993.
7. RJ Hunter. Foundations of Colloid Science. Oxford: Oxford University Press, 1989.
8. RJ Hunter. Colloids Surfaces 141: 37–65, 1998.
9. TA Strout. Attenuation of Sound in High Concentration Suspensions: Development and Application of an Oscillatory Cell Model. Thesis. The University of Maine, 1991.
10. JR Allegra, SA Hawley. J Acoust Soc Am 51: 1545–1564, 1972.
11. JD McClements. Colloids Surfaces 90; 25–35, 1994.
12. DJ McClements. J Acoust Soc Am 91: 849–854, 1992.
13. AK Homes, RE Challis, DJ Wedlock. J Colloid Interface Sci 156: 261–269, 1993.
14. AK Holmes, RE Challis, DJ Wedlock. J Colloid and Interface Sci 168: 339–348, 1994.
15. AS Dukhin, PJ Goetz, CW Hamlet. Langmuir 12: 4998–5004, 1996.
16. LW Anson, RC Chivers. Ultrasonic 28: 16–25, 1990.
17. AH Harker, JAG Temple. J Phys D Appl Phys 21: 1576–1588, 1988.
18. RL Gibson, MN Toksoz. J Acoust Soc Am 85: 1925–1934, 1989.
19. AS Dukhin, PJ Goetz. Langmuir 12: 4987–4997, 1996.
20. U Riebel, et al. Part Part Syst Charact: 135–143, 1989.
21. R Chanamai, JN Coupland, DJ McClements. Colloids Surfaces 139: 241–250, 1998.
22. S Temkin. Phys Fluids 4: 2399–2409, 1992.
23. S Temkin. J Acoust Soc Am 103: 838–849, 1998.
24. RW O'Brien. J Fluid Mech 190: 71–86, 1988.
25. RW O'Brien. Determination of particle Size and Electric Charge. US Patent 5 059 909, 1991.
26. AS Dukhin, VN Shilov, Yu Borkovskaya. Langmuir 15 (10): 3452–3457, 1999.
27. AS Dukhin, H Ohshima, VN Shilov, PJ Goetz. Langmuir 15 (10): 3445–3451, 1999.
28. AS Dukhin, VN Shilov, H Ohshima. Langmuir 15 (20): 6692–6706, 1999.
29. LL Foldy. Propagation of Sound Through a Liquid Containing Bubbles. OSRD Report No. 6.1-sr1130–1378, 1944.

30. EL Carnstein, LL Foldy. J Acoust Soc Am 19: 481–499, 1947.
31. FE Fox, SR Curley, GS Larson. J Soc Am 27: 534–539, 1957.
32. S Ljunggren, JC Eriksson. Colloids Surfaces 129/130: 151–155, 1997.
33. W Schaaffs. In: K Hellwege, ed. Molecular acoustics. Vol 5. Atomic and Molecular Physics. Berlin, New York: Landolt-Bornstein, 1967.
34. PS Waterman, RJ Truell. Math Phys 2: 512, 1961.
35. RR Irani, CF Callis. Particle Size: Measurement, Interpretation and Application. New York, London: John Wiley, 1971.
36. AS Dukhin, PJ Goetz. Colloids Surfaces 144: 49–58, 1998.
37. E Dickinson, DJ McClements, MJW Povey. J Colloid Interface Sci 142: 103–110, 1991.
38. JP Wilcoxon, RL Williamson. In: CR Safinya, SA Safran, PA Pincus, eds. Material Research Society Symposium Proceedings, Macromolecular liquids. Vol 177, Pittsburgh, PA: Materials Research Society, 1990.
39. F Candau. In: CR Safinya, SA Safran, PA Pincus, eds. Material Research Society Symposium Proceedings: Macromolecular Liquids. Vol 177. Pittsburgh, PA: Materials Research Society, 1990.
40. L Motte, A Lebrun, MP Pileni. Progr Colloids Polym Sci 89: 99, 1992.
41. D Ichinohe, T Arai, H Kise. Synth Metals 84: 75, 1997.
42. KV Schubert, KM Lusvardi, EW Kaler. Colloids Polym Sci 274: 875, 1996.
43. FM Menger, K Yamada. J Am Chem Soc 101: 6731, 1979.
44. D Chatenay, W Urbach, AM Cazabat, M Vacher, M Waks. Biophys J 48: 893, 1985.
45. GS Timmins, MJ Davies, BC Gilbert, H Caldarau. J Chem Soc Faraday Trans 90: 2643, 1994.
46. AV Kabanov. Makromol Chem, Macrormol Symp. 44: 253, 1991.
47. V Crupi, G Maisano, D Majolino, R Ponterio, V Villari, E Caponetti. J Mol Struct 383: 171, 1996.
48. B Bedwell, E Gulari. In: KL Mittal, ed. Solution Behavior of Surfactants. Vol 2. New York: Plenum Press, 1982.
49. E Gulari, B Bedwell, S Alkhafaji. J Colloid Interface Sci 77: 202, 1980.
50. M Zulauf, HF Eicke. J Phys Chem 83: 480, 1979.
51. HF Eicke. In: ID Rob, ed. Microemulsions. New York: Plenum Press, 1982, p 10.
52. JD Nicholson, JV Doherty, JHR Clark. In: ID Rob, ed. Microemulsions. New York: Plenum Press, 1982, p 33.
53. HF Eicke, J Rehak. Helv Chim Acta 59: 2883, 1976.
54. PDI Fletcher, BH Robinson, F Bermejo-Barrera, DG Oakenfull, JC Dore, DC Steytler. In: ID Rob, ed. Microemulsions. New York: Plenum Press, 1982, p 221.
55. PC Cabos, P Delord, J App Cryst. 12: 502, 1979.
56. S Radiman, LE Fountain, C Toprakcioglu, A de Vallera, P Chieux. Progr. Colloids Polym Sci 81: 54, 1990.
57. JP Huruguen, T Zemb, MP Pileni. progr Colloids Polym Sci 89: 39, 1992.
58. MP Pileni, T Zemb, C Petit. Chem Phys Lett 118: 414, 1985.
59. W Sager, W Sun, HF Eicke. Progr Colloids Polym Sci 89: 284, 1992.
60. SA Safran, GS Grest, ALR Bug. In: HL Rosano, M Clause, eds. Microemulsion Systems. New York: Marcel Dekker, 1987, p 235.
61. TH Wines, AS Dukhin, P Somasundaran. Colloids Surfaces, 2000.
62. NL Catton. The Neoprenes. Monograph. Rubber Chemical Division, EI Du Pont De Nemours, Wilmington, DE, 1953.

9

# Food Emulsions

**Douglas G. Dalgleish**
*Danone Vitapole, Le Plessis-Robinson, France*

## I. INTRODUCTION

Food emulsions are familiar to almost everyone. Unskimmed milk and cream are emulsions, as are, for example, butter, margarine, spreads, mayonnaises and dressings, coffee creamers, cream liqueurs, some fruit drinks, processed cheeses, ice creams, and whippable toppings. This wide variety of products is formulated and stabilized using many ingredients, although naturally they are governed by the same physical principles as are other emulsion systems. Specific constraints on the formulation of the emulsions are the requirement of stability over extended periods (i.e., they may need to have a shelf-life of several months or longer) and that the emulsions must be edible (i.e., they should contain only ingredients which are approved as safe for human consumption). Added to this is the essential requirement that they need to be safe microbiologically, even after extended storage, so that they must be prepared in such a way that neither pathogenic nor spoilage organisms are present. Clearly, all of these requirements place restrictions on the ways in which the emulsions can be formulated and produced. A further potential constraint arises from the increasing public interest in the availability of foods based on what are perceived to be "natural" or "healthy" ingredients; thus, it is important to maximize the use of naturally available materials such as proteins or phospholipids in formulating the final product.

Therefore, to understand the behavior of food emulsions, we need to know as much as possible about these types of emulsifiers, because they may not behave exactly similarly to "classical" small-molecule emulsifiers. For example, phospholipid molecules can interact with each other to form lamellar phases or vesicles; they may interact with neutral lipids to form a mono- or multi-layer around the lipid droplets, or they may interact with proteins which are either adsorbed or free in solution. Any or all of these interactions may occur in one food emulsion. The properties of the emulsion system depend on which behavior pattern predominates. Unfortunately for those who have to formulate food emulsions, it is rarely possible to consider the emulsion simply as oil coated with one or a mixture of surfactants. Almost always there are other components whose properties need to be considered along with those of the emulsion droplets themselves. For example, various metal salts may be included in the formulation (e.g. $Ca^{2+}$ is nearly always present in food products derived from milk ingredients), and there may also be hydrocolloids present to increase the viscosity or yield stress of the continuous phase to delay or prevent creaming of the emulsion. In addition, it is very often the case, in emulsions formulated using proteins, that some of the protein is free in solution, having either not adsorbed at all or been displaced by other surfactants. Any of these materials (especially the metal salts and the proteins) may interact with the molecules which form the adsorbed surface layer on the emulsion droplets, and cause instability (flocculation, gelation) of the emulsions. This is likely to be especially true at high temperatures, and it is again a feature of many food products that they have to undergo a heating process so as to reduce or eliminate the bacterial load. These heat treatments range from simple pasteurization (72°C for 15 s) to retort sterilization (120°C for 10 min) to ultrahigh-temperature (UHT) treatments (140°C for a few seconds). During heating, proteins present in the product may become denatured and may cause immediate or delayed flocculation or gelation, which are generally undesirable. In some cases, however, the heat treatment is a factor in texturizing the product (as in cream cheeses).

The basic lipid material used in food emulsions is almost always triglycerides, which also play a part in determining the emulsion properties, mainly because many of them are partly crystalline at room temperature. There is little evidence that the particular fatty acids in the oil significantly alter adsorption of surfactant (although there are differences between triglyceride oils and simple hydrocarbons in this respect), but the crystallinity of the oil affects the homogenization process and is also critical in processes such as partial coalescence, providing texture to ice creams and whipped toppings. Therefore, when a product is being formulated it is important to define a melting profile of the oil which is suitable for the product. Although emulsions can be made using fat or oil (basically reflecting the source of triglycerides), no distinction will be made in this chapter between the two terms; the term "oil" will be used throughout to mean liquid or solid triglycerides of animal or vegetable origin.

Emulsifiers perform at least two functions. First and predominantly, they must stabilize the dispersion of oil which is produced. However, they have a second function which is to modify the environment of the oil. The emulsifiers are not necessarily all on the interface, since the excess over that required simply to emulsify the oil will be dissolved either in the oil or in the aqueous phase. This may in turn modify the properties of the phase in which they are dissolved. Moreover, the type of emulsifier used affects the behavior of the emulsion, not only from the point of view of stability, but because it may promote or hinder interactions between the emulsion droplets and other ingredients of the food system. Therefore, in selection of an emulsifier, it is important to have a clear idea of which functions it is expected to perform.

Oil-in-water (O/W) emulsions (e.g., creams, coffee creamers, cream liqueurs, and mayonnaise) are mainly fluid, although they may have partly crystalline oil phases. Stability of these emulsions may be maintained by adsorption of small-molecule emulsifiers, protein molecules, or aggregates of protein molecules (casein micelles, egg yolk granules), or by mixtures of these. Not all of the O/W emulsions used in foods are required to have very high long-term stability; indeed, for whipped toppings and ice-cream mixes, the emulsion needs to be stable for some time, but then must be capable of being destabilized to form the final product. On the other hand, emulsions such as evaporated milks and mayonnaises are required to be stable to flocculation, creaming, and coalescence for long periods so as to maintain the product in an acceptable form for consumption. Although stability may be enhanced by the presence of materials giving viscosity or a yield stress to the continuous phase (e.g., gums), it is not good practice to rely solely upon these effects to stabilize the emulsions. However, the gums are often necessary to provide increased organoleptic properties in low-fat emulsion types.

Water-in-oil (W/O) emulsions (butter, margarines, spreads) are not stabilized simply by forming an adsorbed layer of surfactant which minimizes the effects of interparticle collisions. Important factors in these emulsions are the crystallinity of the oil phase, the presence of rigid surfactants on the O/W interface, and the presence of agents which increase the viscosity of the aqueous phase in the droplets. A degree of mechanical stabilization is, therefore, more important in W/O than in O/W emulsions. In some cases, it is possible to make water-in-oil-in-water (W/O/W) multiple emulsions by homogenizing a W/O emulsion in the presence of suitable surfactants (1), although at the time of writing there have been few of these types of emulsions used in foods.

This chapter will be concerned almost exclusively with liquid O/W emulsions, mainly because they are the most exhaustively studied and the principles for their behavior are the most thoroughly established, not necessarily because they are the most important of the emulsions. For example, no description of emulsions in meat products or in bread mixes and cake batters is given, as these are less undrstood from a fundamental point of view than are the more simple O/W emulsions. What is attempted here is a description of the structures of emulsion droplets and how these affect the properties of the emulsion.

## II. SURFACTANTS

### A. Small-molecule Surfactants

There is a range of surfactants which are permitted for use in the formulation of food emulsions. Small-molecule surfactants (monoglycerides and diglycerides, sorbitan esters of fatty acids, polyoxyethylene sorbitan esters of fatty acids, phospholipids, and many others) generally contain long-chain fatty acid residues, which provide the hydrophobic group which binds to the lipid phase of the oil–water interface and causes adsorption. The head groups of these emulsifiers are more varied (Fig. 1), ranging from glycerol (in monoglycerides and diglycerides) and substituted phosphoglyceryl moieties (in phospholipids) to sorbitan highly substituted with polyoxyethylene chains (2). The differences between these emulsifiers are generally expressed in terms of the hydrophile–lipophile balance (HLB), so that the more oil-soluble surfactants have low HLB numbers and the more water-soluble ones have high numbers, with the value of 7 being treated as "neutral" (3). The presence of emulsifiers of a low HLB number favors the formation of W/O emulsions, and a high HLB number promotes O/W emulsions, although this is not a completely hard and fast rule (4).

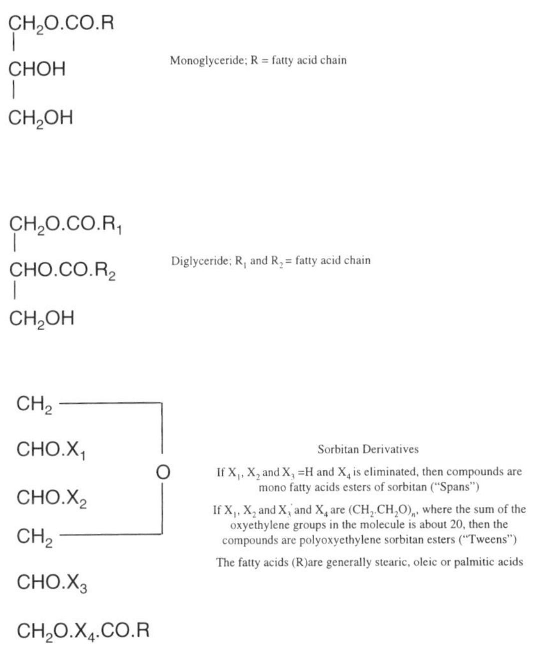


**Figure 1** Structures of some small-molecule emulsifiers used in foods.

Because these molecules adsorb strongly to the oil–water interface and have few steric constraints to prevent them from packing closely, they generate low interfacial tensions (5) and lower the Gibbs interfacial energy. However, they do not generally give highly cohesive or viscous surface layers, so that adsorbed layers of these small molecules may be quite easily disrupted (relative to adsorbed proteins, see below). This property is indeed used in certain types of emulsions, where limited stability to coalescence is required.

Of these small surfactant molecules, the phospholipids (lecithins) behave somewhat differently from the others, because they are capable of forming particular structures (e.g., bilayers and vesicles), which, in turn, may interact with the O/W interface. Experience suggests that these materials do not behave as simply as other small molecules on the interface, and this will be discussed in a later section.

### B. Proteins

Proteins, on the other end of the scale of molecular complexity, act as emulsifiers but behave differently from the small molecules, because of their individual molecular structures, and, indeed, it is the particular proteins present which give many food emulsions their characteristic properties. Most, if not all, proteins in their native states possess specific three-dimensional structures which are maintained in solution, unless they are subjected to disruptive influence such as heating (6). When they adsorb to an oil–water interface, it is unlikely that the peptide chains of proteins dissolve significantly in the oil phase,* as they are quite hydrophilic as a result of the presence of carboxyl or amido groups; it is more likely that the major entities penetrating the interface are the side chains of the amino acids (Table 1). It is possible, for example, for an $\alpha$-helical portion of a protein to have a hydrophobic side, created by the hydrophobic side chains which lie outside the peptide core of the helix. However, even proteins lacking such regular structures possess amino acids with hydrophobic side chains which will adsorb to the oil–water interface. When a protein is adsorbed, the structure of the protein itself will prevent close

*This statement refers to the proteins generally used to produce food emulsions; it is not true for the types of protein which are found naturally in transmembrane locations in tissue cells, where the particular sequences of the proteins allow them to make hydrophobic pockets into regions of lipid.

**Table 1** Hydrophobic and Hydrophilic Amino Acids[a] in the Side Chains of Protein Molecules

| Hydrophobic | Hydrophilic |
|---|---|
| Tryptophan | Glutamic Acid |
| Isoleucine | Aspartic Acid |
| Tyrosine | Histidine |
| Phenylalanine | Serine |
| Proline | Threonine |
| Leucine | Arginine |
| Valine | |

[a]Both lists are in descending order; amino acids not mentioned are of indeterminate nature.
*Source*: Ref. 7.

packing of the points of contact with the interface, and, as a result, the adsorption of a protein reduces the interfacial tension less than does the adsorption of small molecules (Table 2). Although some proteins are excellent emulsifiers, not all proteins can adsorb strongly to an O/W interface, either because their side chains are strongly hydrophilic or because they possess rigid structures which do not allow the protein to adapt to the interface. Examples of such proteins are gelatin, which forms poor emulsions because it has a hydrophilic character and is a large, rather rigid, molecule (9), and lysozyme, which although it does adsorb to O/W interfaces, tends (presumably because of its relatively inflexible structure) to be a poor emulsifier (10).

Because adsorption of proteins occurs via the hydrophobic side chains of amino acids, it is often suggested that a measurement of surface hydrophobicity (11) should allow prediction of the emulsifying power of a protein (12). However, surface hydrophobicity, as determined by the binding of probe molecules

**Table 2** Reduction of Interfacial Tension by Proteins and Surfactants at the Oil–Water Interface

| Molecule | $(\gamma_0 - \gamma)$ (mN m$^{-1}$) | Oil phase | Ref. |
|---|---|---|---|
| $\beta$-Casein | 25 | Tetradecane | 14 |
| $\beta$-Lactoglobulin | 21 | Tetradecane | 111 |
| Gelatin | 15 | Tetradecane | 9 |
| Phosvitin | 12 | Tetradecane | 14 |
| Lysozyme | 17 | Toluene | 35 |
| $C_{12}E_2$ surfactant | 35 | Tetradecane | 8 |
| Glycerol monostearate | 28 | Sunflower oil | 5 |

$\gamma_0$ and $\gamma$ are the interfacial tensions in presence and absence of surfactants, respectively.

to the protein in solution (13), may be an unreliable measure because of the capacity of the protein to change conformation during adsorption or after it has adsorbed (see below), with the effect that adsorption becomes stronger with time. Equally, even apparently hydrophilic proteins may adsorb strongly, as shown by the egg protein, phosvitin, which is a surprisingly good emulsifier (14–17) despite having more than 50% of its residues composed of phosphoserine (18), an amino acid which is charged and, of course, hydrophilic. Even the relatively few hydrophobic residues in the protein are sufficient to cause adsorption.

### C. Adsorption and Protein Conformation

Much research has been aimed at determining the mechanism of protein adsorption, and it is likely that most of the proteins which adsorb well to interfaces are capable of changing conformation either as they adsorb or shortly afterwards. Surface denaturation is an established concept (19, 20), and it emphasizes the probability that unfolding of the protein after adsorption can maximize the amount of hydrophobic contacts with the oil interface (Fig. 2).

A number of methods have confirmed that proteins change their conformation when they adsorb to liquid or solid interfaces. Spectroscopic studies of lysozyme, for example, show a decrease in the secondary structure caused by adsorption to polystyrene latex (21), and it is possible that the protein goes through a number of conformational states as the adsorption process continues (22). The Fourier Transform Infra Red (FTIR) spectra of both $\alpha$-lactalbumin and $\beta$-lactoglo-

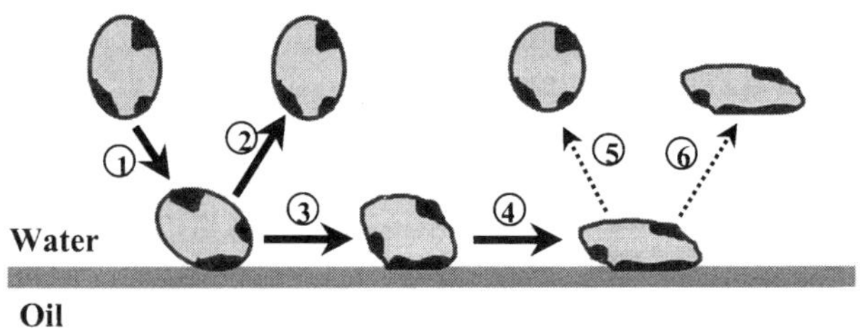


**Figure 2** Presumed mechanism of adsorption of protein. The protein approaches the interface and starts to adsorb via hydrophobic areas on the surface (1). At this point, rapid desorption without change of conformation can occur (2). The adsorbed protein changes conformation (3 and 4) to expose more hydrophobic residues to the oil surface. Desorption of the protein at this stage yields a structure which is near to native, although slightly altered (5), or one which is denatured extensively (6).

bulin have been shown to differ significantly from those of the native proteins (23, 24). Proteins absorbed to a surface and subsequently desorbed by the action of small molecules have been found to possess an altered conformation (25), confirming that the changes caused by adsorption may be irreversible; for example, lysozyme and chymosin lose their enzymic activity on adsorption, and do not regain it after being desorbed from the interface (26). Although we may intuitively expect that the change in the conformation during adsorption will cause the destruction of secondary and tertiary structures of proteins, this is not always the case; it is possible to increase the ordered structure in some cases (27), perhaps by the formation of amphipathic helices. Adsorbed proteins are capable of interacting chemically by formation of intermolecular disulfide bonds to form oligomers, as has been shown for adsorbed $\beta$-lactoglobulin and $\alpha$-lactalbumin (28), although such reactions do not occur in solution unless the proteins are denatured by heating (29). Recently, several studies have demonstrated that the heat of denaturation of adsorbed proteins, a measured by differential scanning calorimetry (DSC), is very much diminished, against suggesting that unfolding on the surface has occurred (30–32). In some cases (e.g., lysozyme and $\alpha$-lactalbumin), this surface denaturation appears to be at least partially reversible, but in others (e.g., $\beta$-lactoglobulin), adsorption causes irreversible changes in the protein molecules (32).

The casein proteins tend to be a special case. Because these proteins appear not to contain much rigid secondary structure ($\alpha$-helix or $\beta$-pleated sheet) (33) and because they possess considerable numbers of hydrophobic residues (34), they adsorb well (35, 36). However, because of the lack of definition of their original native structures, it is impossible to determine whether conformational changes occur during adsorption, as neither spectroscopic changes nor DSC are capable of demonstrating conformational changes in these proteins.

It is worth remembering that the reactions between adsorbed protein molecules in emulsions (for instance, disulfide-bridging interactions such as those mentioned above) will be encouraged by the very high local concentration of protein within the adsorbed interfacial layers. Generally, we know (37, 38) that for proteins the interfacial concentration (surface excess, $\Gamma$) is between 1 and 3 mg m$^{-2}$ and that the adsorbed layers are generally less than 5 nm thick (39), so it is simple to calculate that in the interfacial region a monolayer of protein has a concentration of about 500 mg ml$^{-1}$, that is, 50%. This is much higher than can be achieved by attempting to dissolve the proteins directly in solution because of the extremely high viscosity which is generated, so direct comparisons between adsorbed and unadsorbed proteins at equal effective concentrations are not possible. However, the protein in the adsorbed layer may be in a favorable position for intermolecular interactions, because the molecules are very close to one another and adsorption holds them in position, so that diffusion is slow. It is probable that the adsorbed layer of protein is more like a gel than a solution; this is at least partly the reason why many adsorbed proteins form highly viscous interfacial layers. It must be remembered that these will be essentially two-dimensional gels, with each molecule occupying about 11 nm$^2$ of interface [calculated on the basis of a molecule of 20,000 Da and a surface coverage of 3 mg m$^{-2}$; this agrees well with the expected dimensions of a globular protein of this weight (40) and is much larger than the 0.5–2.5 nm$^2$ per molecule which has been found for adsorbed modified monoglycerides (41)]. It is, therefore, not surprising that adsorption can alter the behavior of proteins. The formation of such concentrated layers has relatively little to do with the overall bulk concentration of the protein in solution, which may give stable emulsions at relatively low bulk concentrations (although this depends on the amount of oil and the interfacial area to be covered). Caseins form extended layers about 10 nm thick, and even at a $\Gamma$ of 3 mg m$^{-2}$ have a "concentration" of about 300 mg ml$^{-1}$. Conversely, whey proteins form much thinner layers (about 2 nm thick) and will have to begin to form multilayers if $\Gamma$ is more than about 2 mg m$^{-2}$, as there is no further space available for monolayer adsorption beyond that point.

One factor which can have considerable importance in the emulsifying properties of proteins is their quaternary structure. For example, in milk the caseins exist in aggregates of considerable size (casein micelles) containing between about 500 and 10,000 individual protein molecules (42). These particles act as the surfactants when milk is homogenized (43). On the other hand, sodium caseinate prepared from milk exists in a much less aggregated state (44) and is much superior to the micelles in emulsifying properties (45) (i.e., the amount of oil which can be stabilized by a given weight of casein, i.e., either of the two forms). This effect is probably simply because the effective concentration of emulsifier is much less when the casein is in its micellar form, which is relatively resistant to destruction. Therfore, during homogenization, the nonmicellar casein will arrive at the interface more readily than the micelles. Molecules such as $\beta$-lactoglobulin also show changes in quaternary structure as a function of pH (46), and these may be related to the changes in the protein's surfactant properties at different pH values (47). The denaturation of $\beta$-lactoglobulin by heat causes the protein to aggregate, and this decreases the emulsifying power to a considerable extent (48).

With a few exceptions, most of the detailed research has been performed on relatively few proteins. Of these, the caseins ($\alpha_{s1}$, $\alpha_{s2}$, $\beta$, and $\kappa$) and whey proteins ($\alpha$-lactalbumin and $\beta$-lactoglobulin) predominate. This is principally because these proteins are readily available in pure and mixed forms in relatively large amounts; they are all quite strongly surfactant and are already widely used in the food industry, in the form of caseinates and whey protein concentrates or isolates. Other emulsifying proteins are less amenable to detailed study by being less readily available in pure form (e.g., the proteins and lipoproteins of egg yolk). Many other available proteins are less surface active than the milk proteins, for example, soya isolates (49), possibly because they exist as disulfide-linked oligomeric units rather than as individual molecules (50). Even more complexity is encountered on the phosphorylated lipoproteins of egg yolk, which exist in the form of granules (51), which themselves can be the surface-active units (e.g., in mayonnaise) (52).

Therefore, in the detailed descriptions of model emulsions given below, nearly all of them (especially where the surfactant proteins are considered at the molecular level) concern themselves with the milk proteins.

## III. EMULSIFYING ACTIVITY OF SURFACTANTS

It is essential to estimate the potential of given surfactants for forming emulsions. Ideally, we need techniques which are method independent, that is, which give absolute results, or at least give results applicable to specific methods for preparing emulsions. The two most widely used methods, namely, emulsifying activity index (EAI) and emulsifying capacity (EC) do not fit these criteria, although they are simple to apply. In the second of those, a known quantity of surfactant is dissolved in water and then oil is added to it in a blender. This forms a crude emulsion, and further aliquots of oil are added until the emulsion inverts or free oil is seen to remain in the mixture. This ostensibly gives the weight of oil which can be emulsified by the defined weight of protein. It is evident that this method is dependent on the particular blender because what is important in emulsion formation is not the weight of oil but its interfacial area. Thus, if the emulsion is made of large droplets, it will consume less protein than if small droplets are present. The conditions of emulsion formation are therefore critical to the method, as it is possible to obtain different results at different blender speeds. Therefore, the method may not be generally transferable between different laboratories, nor is it in any sense an absolute measure. As a quality-control measure, or as an internal method in a single laboratory, it can be, however, useful.

To measure the EAI, an emulsion is made and the particle size is estimated, usually by turbidimetric methods (53), although the particle sizes of the emulsion droplets can be measured by other means. The assumption is then made that all of the protein is adsorbed to the interface, and so a measure of emulsifying potential can be measured. Although it provides more information than EC, the method has two major defects: first, it is by no means certain that all of the available protein is adsorbed, or that it is adsorbed as a monolayer. Indeed, it is known that at concentrations of protein of more than about 0.5% (with oil concentration of 20%), some of the protein remains unadsorbed, even after powerful homogenization where the concentration of protein is the limiting factor in the determination of the sizes of the droplets (54, 55). If homogenization is less extensive, then the proportion of protein which is adsorbed decreases. The second major problem in interpretation of EAI is simply the difficulty of determining the particle sizes and their distribution. There are a variety of methods for measuring the size distribution of suspended particles, and care must be taken to avoid error in this measurement. Traditionally, the particle sizes in determinations of EAI are measured by determining the turbidity of diluted suspensions of the emulsions, which is a method much subject to error.

Ideally, to describe the emulsifying capacity of a protein, the particle size distribution and the amounts of individual surfactants adsorbed to the oil–water interface need to be measured. It is possible to measure the amount of adsorbed protein by centrifuging the emulsion so that all of the fat globules form a layer above the aqueous phase, and measuring the concentration of surfactant left in this phase by ion-exchange of reverse-phase chromatography. Alternatively, the fat layer after centrifugation can itself be sampled, and the adsorbed protein can be desorbed from the interface by the addition of sodium dodecyl sulfate (SDS) and quantified by electrophoresis on polyacrylamide gels. In addition, although this is more difficult to determine, it is desirable to know the state of the adsorbed material (e.g., its conformation, which parts of the adsorbed molecules protrude into solution and are available for reaction, etc.). This represents an ideal which is rarely possible to achieve, but the explanation of the behavior of emulsions, and perhaps the design of new ones, may depend on this knowledge.

## IV. FORMATION OF EMULSIONS

Food O/W emulsions are generally produced using colloid mills or high-pressure homogenization. In the former method, the oil–water–surfactant mixture is passed through a narrow gap between a rotor and a stator, in which the stresses imposed on the mixture are sufficient to break up the oil into droplets, to which the surfactant adsorbs (3). This method tends to produce droplets of emulsion which are larger than those produced by high-pressure homogenization, being of the order of 2 μm in diameter. The technique is used to manufacture mayonnaises and salad creams, in which stability depends less on the presence of very small particles than on the overall composition and high viscosity of the preparation. In liquid emulsions, however, smaller particles are required to prevent creaming and possible coalescence.

High-pressure homogenization is used to produce these smaller droplets. First, a coarse emulsion of the ingredients is formed by blending, and this suspension is then passed through a homogenizing valve, at pressures which are generally in the region 6.8–34 MPa (1000–5000 psi). The high-pressure flow through the valve creates turbulence, which pulls apart the oil droplets, during and after which the surfactant molecules adsorb to the newly created interface (56). If the adsorption is not rapid, or if there is insufficient surfactant present, then recoalescence of the oil droplets occurs (57). Apart from the mechanical design of the homogenizer, the sizes of the emerging droplets depend on, among other factors, the homogenization pressure (Fig. 3), the viscosity of the suspension, the number of passes (58), and the amount of surfactant present (Fig. 4) (54). When there is a large excess of surfactant present, the particle size is limited by the characteristics of the homogenizer and of the suspension; on the other hand, if only small amounts of surfactant are present, the surfactant concentration limits the sizes of the particles, since insufficiently covered emulsion droplets will recoalesce. It is therefore likely that, as the compositions of products are reformulated, the sizes of the emulsion droplets in them will change.

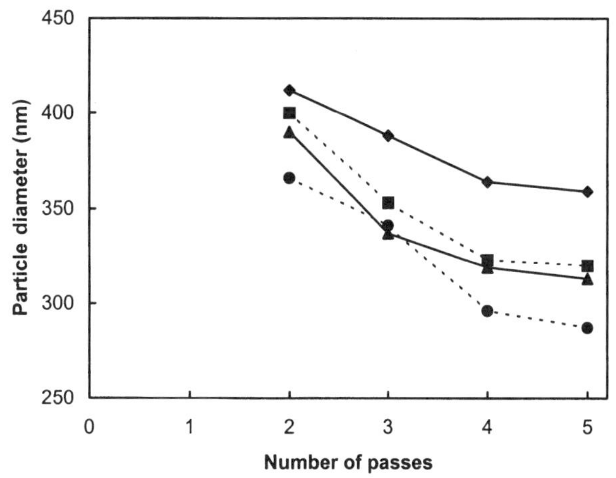


**Figure 3** Average diameters of particles in homogenized milk, as a function of the number of passes through the homogenizer (Microfluidizer), and the pressure. Diamonds and filled line: homogenization pressure of 14 MPa; squares and broken line: pressure of 21 MPa; triangles and full line: pressure of 28 MPa; circles and broken line: pressure of 35 MPa.

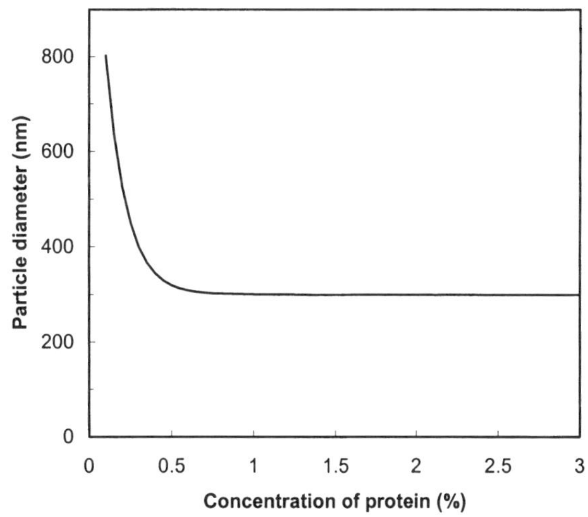


**Figure 4** Average particle diameter as a function of protein concentration for emulsions made using 20% soya oil and caseinate at different concentrations. Below a concentration of about 0.5% caseinate, the size of droplets is dependent upon the concentration of casein; above this, the concentration depends on the conditions of homogenization.

In addition to recoalescence and increased droplet size in the presence of insufficient surfactant, the phenomenon of bridging flocculation, in which the emulsion droplets form clusters during homogenization, can be observed. For the bridging to occur, it is necessary to have macromolecular surfactants with at least two sites by which they can adsorb to interfaces, and at low surfactant concentrations such molecules can become adsorbed to two separate oil droplets. Proteins can form bridges in this way (59), and even more commonly, natural aggregates of proteins such as casein micelles can induce clustering of the oil droplets (60). Bridging flocculation may be reversed by incorporating more surfactant (which need not in this case be macromolecular) so as to provide enough material to cover the nascent interface. In the case of clustering by particles which themselves can be broken up, a second-stage homogenization at lower pressure (3.4 MPa, 500 psi) can also be sufficient to break down the bridging aggregates and to separate the clustered fat globules. Clearly, however, such treatment will be inapplicable to clusters bridged by single macromolecules, which cannot be broken up in this way.

## V. MEASUREMENT OF PARTICLE SIZES AND SIZE DISTRIBUTIONS IN EMULSIONS

Once an emulsion has been formed by using homogenization or other means, it is generally necessary to characterize it, specifically in terms of its size distribution. This is important in a number of respects: knowledge of the size distribution provides information on the efficiency of the emulsification process, and the monitoring of any changes in the size distribution as the emulsion ages gives information on the stability of the system. Thus, measurement of particle size should be part of a quality-control operation. Also, the definition of size distribution may be important when emulsion systems or processes are patented. However, the measurement of true size distributions or even the average sizes of emulsion droplets is not simple, despite the existence of a number of potentially useful and apparently simple methods.

The most direct method, and one which is perhaps least subject to errors, is electron microscopy (61). This technique can be used to determine the number-average size distribution, providing that (1) a fully representative sample of the emulsion is prepared, fixed, mounted, sectioned, and stained without distortion; (2) a sufficient number of particles is measured to ensure statistical accuracy of the distribution; and (3) proper account is taken of the effects of sectioning on the apparent size distribution. All of this requires considerable time, effort, and calculation, so that the technique cannot be used routinely to determine size distributions. It may be used as a standard against which to compare other methods, and also finds a use in measuring systems where dilution causes changes in the particle sizes, as in microemulsions (62). The technique gives a number-average distribution of the particles, since every particle is accorded the same weighting.

The more rapid methods tend to emphasize the large end of the size distribution and, therefore, to give biased results. Measurement of the particle sizes by the now somewhat outdated method of conductivity (Coulter counter), for example, is insensitive to the presence of small particles, with sizes below about 0.8 μm (63, 64). For fine emulsions, therefore, this technique leads to underestimation of the population of small particles. Techniques involving light scattering, which are probably the most widely used, also tend to emphasize large particles in the distribution. The simplest of these methods depends on the measurement of turbidity at one or a number of wavelengths (53, 65). Turbidity, or apparent absorbance of light, is a measure of the total amount of light scattered as it passes through a cuvette containing diluted emulsion. Although the method is rapid and may be performed in any laboratory possessing a spectrophotometer, it cannot be used to give the true distribution of particle sizes, but at best to give an average. If required, it can be assumed that the particles form a distribution of known shape, but this, of course, assumes that the distribution is known beforehand. The turbidimetric method also emphasizes the larger particles because they scatter more light, and so the method tends to underestimate the contribution of smaller particles.

A number of commercial instruments measure the distributions of particle sizes by determining the intensity of light scattered from a highly diluted sample at specific scattering angles in the range 0°–30° (integrated light scattering, ILS). With knowledge of the scattering properties (i.e., the Mie scattering envelope) of the particles (66), software is used to calculate the most probable distribution of particle sizes.

This does not always yield the true absolute distribution, for two main reasons. The first of these is that the angular range is generally too restricted to allow measurement of small particles of diameters less than about 50 nm; these scatter almost isotropically, whereas large particles preferentially scatter in the forward directions, so that to measure the distribution accurately, a larger span of scattering angles between 0° and 150° is essential (67). The result of the limitation of the angular range is that once again the contribution of small particles tends to be underestimated, and the instruments generally are inaccurate in their estimates of the contribution of particles smaller than 0.1 μm. Unfortunately, many food emulsions contains particles ranging from 50 nm to 1 μm, and this is precisely where many instruments are least accurate. A further problem with any light-scattering method is that the accuracy of the calculated distribution depends on how well the optical properties of the emulsion droplets (i.e., their real and imaginary refractive indices) can be defined. Also, in all cases, the droplets are assumed to be spherical, but it may be necessary to make assumptions about the structures of the interfacial layers. An emulsion droplet is essentially a coated sphere (68), which is characterized by refractive indices of the core and the coat, and these need not be equal; in fact, they may be quite different. Calculations of the scattering behavior of emulsion droplets may, therefore, depend on the presumed structures of the particles.

If the emulsion is unstable, the particle size distribution will of course change, but a simple measure of light scattering cannot distinguish between droplets which are flocculated and those which have coalesced, and other methods of measurement are needed to define which type of instability has occurred. Flocculation introduces another inaccuracy, since it produces particles which are neither homogeneous nor spherical. To determine the type of instability which has occurred, it is often possible to use a light microscope, or alternatively the destabilized emulsion can be treated with SDS to dissociate any flocs. A second measurement of the particle size will show no change if the emulsion has coalesced, but will revert to the original particle size distribution if the destabilization has been by flocculation.

Dynamic light scattering (DLS) offers an alternative means of measurement (69). This technique does not measure the total amount of light scattered, but the dynamics of the scattered light over very short time periods. Usually, the light scattering is measured at a fixed angle of 90°, and a correlation function is measured. This is essentially a weighted sum of exponentials, which depend on the diffusion coefficients of the particles through the aqueous medium. As with ILS, the calculation of the true size distribution depends on the knowledge of the detailed light-scattering properties of the emulsion droplets. In addition, the fit of theory to the true correlation function is ill-conditioned (70), so that the size distribution which is obtained can depend on the technique used to fit the correlation function. This technique usually overemphasizes the contribution of larger particles to the size distribution. This is partly because they tend to scatter more (i.e., have higher weighting factors), but also because of the nature of the correlation function itself, as the information about the small particles is contained only in the short-time part of the function, whereas information about the large particles is contained at all points.

In experiments which compared various methods of measuring the particle size distribution in an emulsion, as determined by different methods, it was evident that a custom-built ILS spectrometer measuring between angles of 4° and 145° was capable of demonstrating that a considerable population of particles with diameters $< 0.1$ μm was present in the mixture, and that none of the other methods could do so; the existence of these small particles was confirmed by electron microscopy (71). The DLS methods were dependent on the configuration of the instrument, but provided some information about the small particles. These measurements also illustrated the general principle that the more accurate the answer required, the more laborious and time consuming is the experimental procedure, which makes it difficult to use the technique routinely.

Although it is not particularly serious for simple stable emulsions, it is necessary to ensure that no dissociation of particles is caused by the high dilution which is required for accurate light-scattering experiments. This may lead to dissociation of flocculated material, or the breakdown of complex interfacial layers (e.g., those formed by casein micelles on the oil–water interfaces in homogenized milk). Although a method for measuring light scattering in concentrated solutions exists (72), very few experiments involving food emulsions have used the technique, and its full effectiveness remains to be demonstrated.

Despite all of these problems, light-scattering methods are at present the most effective means of obtaining information about the size distributions of particles in emulsion systems. In particular, these may be used in a comparative mode, to measure changes which occur in the suspensions. All of the methods can detect whether aggregation is occurring, so that they may all be used to detect instability of emulsions. It is under these circumstances that the light-scattering methods come into their own, as they are almost unique in allowing the kinetics of aggregation to be studied on a real-time basis (73). Simply, the fact that the particle size is increasing can be determined without any particular attributes of the particles needing to be known.

Light scattering is most reliably used in diluted solutions, and the act of dilution of the samples may cause changes in the particles (e.g., flocs may be dissociated). Recently, techniques based on ultrasonic acoustic spectroscopy have been demonstrated to provide information on the size distribution of emulsion droplets in a dispersion. The measurement is based on the fact that the attenuation of ultrasound of a defined frequency through a suspension depends on the size of the particles. By measuring the attenuation of sound of a series of different frequencies through the sample, it is possible to calculate the size distribution of the particles in the suspension (74). Instruments to perform the measurements are available, and they are capable of comparing well with the information available from light-scattering measurements. Just as the scattering of light depends on the relative refractive indices of the dispersed phase and the continuous phase, so particle sizing by ultrasound also depends on the physical properties of the dispersed phase, for example, the density, viscosity, thermal conductivity, and specific heat are all required to be known to permit the true size distribution to be determined (75). In the absence of these factors, ultrasonic attenuation spectroscopy can give only relative size distributions. However, because of the applicability of the method to real (undiluted) emulsion systems, it is likely that the method will increase in its usage.

A related method which is also finding uses is electroacoustic spectroscopy (76). Because the passage of ultrasound through a dispersion disturbs the double layer which surrounds the dispersed particles, an electric current is set up. Measurement of this current allows the calculation also of the $\zeta$-potential of the particles as well as their size distribution. The measurement may also be performed in reverse; an oscillating electric field causes the emission of ultrasound, which in turn permits the size distribution and the $\zeta$-potential to be measured (77). Like the more simple ultrasonic spectroscopy, these methods require precise knowledge of the physical characteristics of the medium and the dispersed phase to give absolute, rather than relative, results.

## VI. STRUCTURES OF EMULSION DROPLETS

The structures of the interfacial layers in emulsion droplets might be expected to be simple when small-molecule emulsifiers are used, but this is not necessarily the case, especially when not one but a mixture of surfactant molecules is present. Although simple interfacial layers may be formed where the hydrophobic moieties of the surfactants are dissolved in the oil phase, and the hydrophilic head groups are dissolved in the aqueous phase, it is also possible to form multilayers and liquid crystals close to the interface (78). These, of course, depend on the nature and the concentrations of the different surfactants. Interactions between surfactants generally enhance the stability of the emulsion droplets, because more rigid and structured layers tend to inhibit coalescence. Also, mixtures of different surfactants having different HLB numbers appears to provide structured interfacial layers, presumably because of the different affinities of the surfactants for the oil–water interface (79).

Specifically, phospholipids may form multilamellar structures around the oil–water interface, and presumably these layers will have different spacing depending on the amount of hydration (80). So, although the major adsorption of phospholipids at low concentration is likely to be in the form of monolayers, it may be possible to produce more complex structures when large amounts of phospholipid are present or when other surfactants are coadsorbed to the interface.

Undoubtedly, the most complex structures are produced when the surfactants are proteins, because of the great range of conformational states which is (at least potentially) accessible to such molecules (Fig. 5). This is of interest because of the implications of conformational change on the reactivity and functionality of the proteins (e.g., it appears that adsorbed $\beta$-lactoglobulin cannot form disulfide bonds with $\kappa$-casein when heated, although this reaction is known to occur between the proteins in solution and in heated milk). Flexible molecules such as caseins may be considered to adsorb as if they were heteropolymers (81) because of their presumed high conformational mobility (33). Certainly, adsorbed $\beta$-casein exhibits different susceptibility to attack by proteolytic enzymes, compared with the protein in solution (82). Indeed, adsorption to different materials cause differences in the conformation of the adsorbed molecule; for example, the protein seems to have somewhat different conformations when adsorbed to n-tetradecane–water and soya oil–water interfaces (83). As a result of these measurements, it has been demonstrated that model hydrocarbon–water systems are not necessarily suitable for describing triglyceride–water systems.

Nevertheless, much is known about the structure of adsorbed $\beta$-casein, certainly more than is known for any other food protein, and various techniques have been used to study the adsorbed protein. The first evidence from DLS showed that $\beta$-casein adsorbed to a polystyrene latex caused an increase in the radius of the particle by 10 to 15 nm (84). Later studies using small-angle X-ray scattering confirmed this and

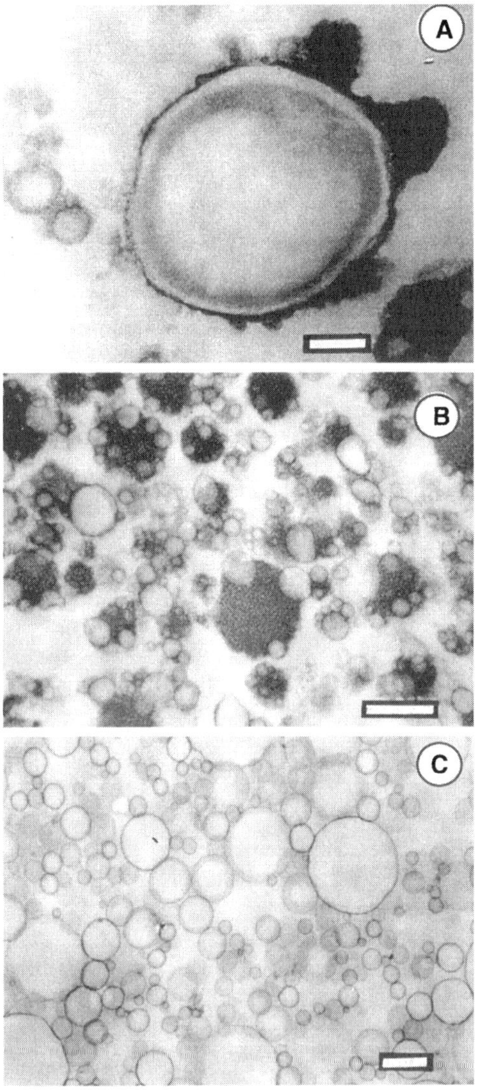


**Figure 5** Electron micrographs (transmission) of different emulsion particles. (A) Large fat globule from homogenized milk, showing nonhomogeneous distribution of protein (fragmented casein micelle) on the surface; (B) smaller fat globules from homogenized (Microfluidized) milk, showing bridging flocculation to form clusters, and the association of small fat globules with large aggregates (micelles) of caseins; (C) caseinate-stabilized soya oil-in-water emulsion showing polydisperse distribution of oil droplets and a thin layer of casein on the surface of the globules. Scale bar is 100 nm in (A), 200 nm in (B), and 300 nm in (C).

showed, in addition, that the bulk of the mass of the protein was close to the interface, so the interfacial layer was not of uniform density throughout (85). Neutron-reflectance studies also showed that most of the mass of protein was close to the interface (86). Only a relatively small portion of the mass of the adsorbed protein extends from the tightly packed interface into the solution, but it is this part which determines the hydrodynamics of the particle and which is almost certainly the source of the steric stabilization which the $\beta$-casein affords to emulsion droplets (84). It is to be noted that all of the studies just described were performed on latex particles or on planar interfaces; however, it has also been demonstrated that the interfacial structures of $\beta$-casein adsorbed to emulsion droplets resemble those of the model particles (39, 85). Although detailed control of emulsion droplets during their formation in a homogenizer is impossible, it is possible to break down the surface layers by using proteolytic enzymes (87), and by comparison with the behavior of model systems under similar conditions, it is possible to demonstrate that the proteins seem to have similar conformations in the model systems and emulsions (88).

It is also possible to use proteolytic enzymes to demonstrate which part of the $\beta$-casein is protruding into the solution. There are many sites in the molecule where, in principle, trypsin can attack the protein, and these are found to be almost equally susceptible to attack when the protein is in solution. In the adsorbed protein, sites close to the N-terminal group are most readily attacked, suggesting that they are the most accessible, presumably because they form the part of the adsorbed layer which protrudes into the solution (82). This region of the molecules is the one which would be expected to have this function, as it is the most hydrophilic and highly charged part of the protein; thus, for $\beta$-casein, it is possible to predict the conformation of the protein in the adsorbed state from a study of its sequence (82). It seems that $\beta$-casein is perhaps the only protein for which this kind of prediction can be done; most other proteins (even the other caseins) have much less distinctive hydrophilic and hydrophobic regions and, therefore, have conformations which are more difficult to predict (88). From studies of the structure of adsorbed $\alpha_{s1}$-casein in model systems and in emulsions, it is established that neither the most accessible sites for trypsinolysis (89) nor the extent of protrusion of adsorbed protein into the solution (90) can be readily predicted.

It is possible to calculate from statistical mechanical principles the approximate conformations of the adsorbed caseins, by assuming that they are flexible, and composed of chains of hydrophilic and hydrophobic amino acids (91). The calculations of these model systems show many of the features of the actual measured properties, especially the tendency of the adsorbed $\beta$-casein to protrude further from the interface than the $\alpha_{s1}$-casein (92). These calculations have in turn been used to explain the differing stability of the two different types of emulsions (93). These calculations have considerable success in explaining both the structure and stability of casein-coated emulsions, but are less adaptable to explain the behavior of more rigid protein surfactants. However, the same principles have been used to explain the apparently anomalous adsorption of phosvitin (16).

The difficulty of ascertaining the structure of the adsorbed protein is greater for globular proteins. In these cases, the adsorbed layer is much thinner than it is for the caseins, so that layers of $\beta$-lactoglobulin appear to be of the order of 1 to 2 nm thick instead of about 10 nm measured for the caseins (39, 90, 94). From hydrodynamics and scattering experiments, it is even possible to suggest that the thickness of the adsorbed layers is smaller than would be expected from the protein in its natural conformation, so that these measurements of the size of the adsorbed protein suggest that adsorption causes it to change conformation. This may be confirmed by other techniques. For example, it is known that adsorbed $\beta$-lactoglobulin forms intermolecular disulfide bonds (28), which does not occur when the molecules are in their native conformations in solution [although the high concentration of protein in the adsorbed layer (see above) will certainly enhance any tendency that the molecules have to aggregate]. In addition, detailed studies of the DSC of emulsions containing $\beta$-lactoglobulin (32) have shown that (1) the protein when adsorbed to the oil–water interface in emulsions loses its heat of denaturation (i.e., shows no intake of heat which can be associated with denaturation, presumably because the protein is already surface denatured); and (2) if the protein is desorbed from the interface by treatment with detergent (Tween-20), it can be seen to be denatured irreversibly (i.e., no recovery of the denaturation endotherm is seen). This may be contrasted with the behavior of $\alpha$-lactalbumin, which loses its heat of denaturation when adsorbed, but recovers its original thermal behavior when the protein is competitively desorbed by Tween (22, 32). These studies confirm that different proteins show quite different degrees of denaturation when they are adsorbed to oil–water interfaces. This is further confirmed by studies of the infrared spectra of the proteins (Fig. 6) (23, 24).

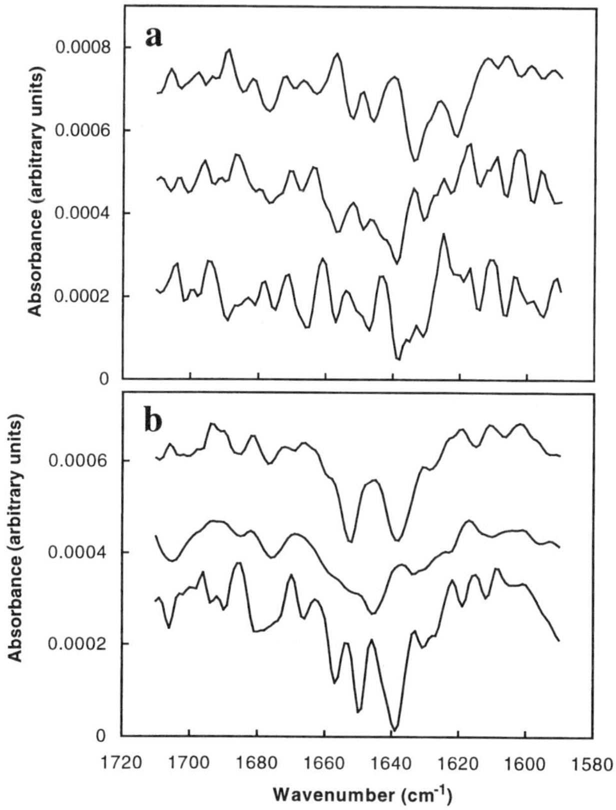


**Figure 6** FTIR second derivative spectra of the amide-II region of the spectra of $\beta$-lactoglobulin (a) and $\alpha$-lactalbumin (b), showing that changes in spectra, and hence the conformation, occur during adsorption. In the two parts of figures, the top spectrum is of native protein, the middle spectrum is of adsorbed protein, and the bottom spectrum is of protein displaced from the interface by excess of Tween-60. For both proteins, adsorption gives a change in the spectrum which is not reversed by subsequent desorption (the spectra of the native proteins are not altered by the presence of Tween).

From a combination of these studies, it is possible to conclude that the adsorption of proteins during emulsion formation leads to at least partial and sometimes complete denaturation of the molecules, which may be partially or competely irreversible. It is potentially possible (although rare in practice) to define the structure of the interfacial layer in terms of the extent of denaturation of the adsorbed proteins and whether they form an extended monolayer. To this may be added two complicating factors: the possibility that multilayers, rather than monolayers, are formed and the possibility that specific proteins may exhibit variable behavior depending on the conditions. Caseins are capable of the latter behavior; for example, it is possible to prepare stable emulsions containing 20% (w/w) of soya oil, with as little as 0.3% casein, and in these the surface coverage has been measured to be a little less than 1 mg m$^{-2}$. The hydrodynamic thickness of the adsorbed layer in these emulsions is about 5 nm

(Fig. 7). In emulsions prepared with larger amounts of casein (1–2%), the surface coverage is increased to 2–3 mg $m^{-2}$ and the thickness of the adsorbed layer is about 10 nm (i.e., about twice that of the layer at lower surface coverage) (54). It has been suggested that this is a result of the adsorbed casein molecules adopting two different conformations: one at low coverage, where the proteins have to cover a maximum area of surface (about 48 $nm^2$ per molecule), and one at high coverage, where the molecules are more closely packed (about 13 $nm^2$ molecule). It is not thought that the caseins form multilayers in these emulsion particles, particularly because the binding curve is smooth as the concentration of casein is increased and does not show steps such as are typical of the formation of multilayers (55). The addition of more protein to the aqueous phase of emulsions made with low concentrations of caseinate results in adsorption of some of the added protein and a corresponding increase in the thickness of the adsorbed layers. This is true, even if the added protein is not casein, and illustrates that, for caseins at least, the proteins on the surface possess sufficient mobility to be moved as other proteins adsorb.

However, it seems clear that under some conditions, caseins and other proteins can form multilayers; this has been demonstrated for adsorption to planar interfaces, where large surface excesses are easily generated

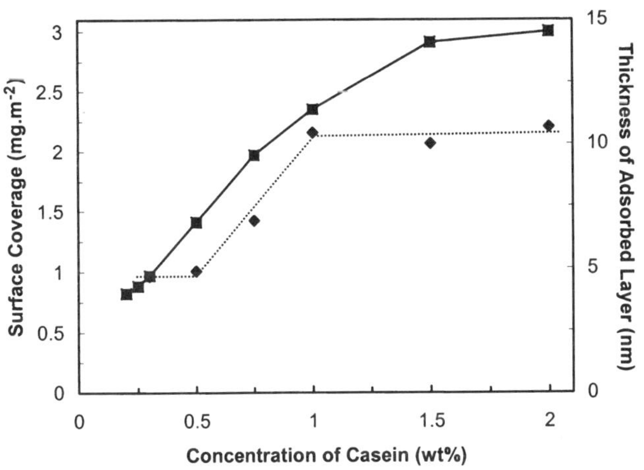


**Figure 7** Surface load of caseinate on droplets in emulsions containing 20% soya oil and different amounts of caseinate. Solid line and left-hand axis: surface load of caseinate; dotted line and right-hand axis: thickness of adsorbed layer of casein as measured by photon-correlation spectroscopy, showing the tendency for the layer to be thin or extended depending on the amount of protein adsorbed.

and multilayers are formed (35). There is less evidence for this in emulsions, although some high surface coverages (up to about 10 mg $m^{-2}$) have been measured (95), which must demand that there is more than a single layer on the oil–water interface because it would be impossible to pack this amount of protein into a monolayer. It is not clear why in some cases monolayers and in some cases multilayers are formed, although it is likely that the physical conditions of homogenization may be important. Also, differences in the methods of preparing the caseins may be relevant; at neutral pH values, highly purified caseins have not generally been associated with multilayer formation, which seems generally to be associated with the use of commercial sodium caseinates.

Multiple layers seem to be more easily formed in emulsions containing whey proteins than with caseins. Perhaps because these proteins are originally globular and form thinner layers, which cannot extend far into the solution from the interface, they are forced to form multilayers. Because the whey proteins project less into solution than do the caseins, they may be less effective at sterically preventing the approach of additional molecules which go to form the multilayers. Finally, because they change conformation when they adsorb, they may offer new possibilities for interaction with incoming whey proteins from solution. There is evidence for multiple layers of whey proteins from both planar interfaces and emulsion droplets (55, 96). However, although these multiple layers exist, there is no definite evidence which links them to changes in the functional properties of the emulsions. Nor is it well determined how stable the multiple layers are, compared with a monolayer. It is very difficult, or perhaps impossible, to simply wash adsorbed proteins from adsorbed monolayers formed on the interfaces of oil droplets (97). However, the outer portion of multilayers may be more readily displaced because it is held in place by protein–protein interactions only, which may be weaker than the forces which lead to adsorption. Generally, however, the properties of the outer parts of multilayers have been little studied.

As a final degree of complexity, food emulsions may be stabilized by particles (Fig. 5). Perhaps the most common are the protein "granules" from egg yolk which play a role in the stabilization of mayonnaise (52), and casein micelles in products such as homogenized milk. Both of these emulsifiers are known to be adsorbed to the oil–water interface as complex particles, which do not dissociate completely to their individual proteins either during or after adsorption (98, 99). During the homogenization of milk, casein micelles are partially disrupted at the oil–water interface so that they adsorb either whole or in fragments. Indeed, once a micelle has adsorbed, it appears to be able to spread over an area of the interface (61, 100, 101). Thus, the fat droplets in homogenized milk are surrounded by a membrane which must contain some of the original fat globule membrane [phospholipid and protein (102)] but is primarily constituted of semi-intact casein micelles. Likewise, the oil–water interface in mayonnaise is partly coated by the granular particles formed from the phosphoprotein and lipoprotein constituents of egg yolk (103). In this high-lipid product the granules also may act to keep the oil droplets well separated and prevent coalescence.

Emulsion formation by means of these aggregates of protein is generally less efficient than by the proteins when they are present in the molecular state, simply because the efficient formation of the emulsion depends on rapid coverage of the newly formed oil surface in the homogenizer. A particle containing many molecules of protein will encounter a fat surface less frequently than an equivalent amount of molecular protein. Thus, although it is possible to prepare homogenized milk with the proportions of casein and fat which occur naturally in milk (a ratio of about 1:1.5, w/w), it is not possible to use 1% (w/w) of micellar casein to stabilize an emulsion containing 20% oil (45). On the other hand, 1% casein in a molecular form (sodium caseinate) is quite sufficient to form a finely dispersed, stable emulsion with 20% oil. Therefore, unless the micelles are expected to confer some specific advantage on the functional properties of the emulsion, or unless there are specific legislative reasons, it is generally more effective to use caseinate than casein micelles to stabilize an emulsion. With egg granules, the situation is different, inasmuch as the individual proteins from the egg granules are not readily available in a purified form analogous to caseinate. In this case, the choice between particulate and molecular forms of the protein does not arise.

## VII. FORMATION AND CHANGES OF THE INTERFACIAL LAYER

In many food emulsions, more than one surfactant is present, so that mixtures of proteins, small-molecule surfactants (oil soluble and water soluble), and lecithins may be present. The result of this is that the interfacial layer will contain more than one type of molecule. The properties of the emulsion (the sizes of the droplets and the functionality) will, in turn, depend on which of the molecules in the formulation is actually on the interface.

It has been shown that, in mixtures of proteins in emulsions, formed at neutral pH and moderate temperatures, there is generally no selectivity for the interface. For example, there is no preferential adsorption between the proteins when a mixture of $\alpha$-lactalbumin and $\beta$-lactoglobulin is homogenized with oil; the amounts of protein which are adsorbed are strictly in proportion to their concentrations (104). The same is true when a mixture of sodium caseinate and whey protein is used as the surfactant in an emulsion (55). The only case where preferential adsorption has been truly observed is when $\beta$-casein is used to displace adsorbed $\alpha_{s1}$-casein, and vice versa, so that there is a possibility that these two proteins adsorb according to thermodynamic equilibrium (105). Even this observation is complicated, however, because it may apply only to mixtures of highly purified caseins; the displacement reactions with commercial sodium caseinate (where a similar result is to be expected) give much less clear results (106, 107). It is sometimes possible to form an emulsion using one protein and then to attempt to displace that protein from the interface with another (Fig. 8a); however, usually the protein which is first on the interface resists displacement (108). Because of its high surface activity and flexibility, $\beta$-casein appears to be the best displacing agent, of the proteins tested to date, but it is by no means always capable of displacing an already adsorbed protein. It can displace $\alpha_{s1}$-casein and $\alpha$-lactalbumin from an interface (105, 109), but the process is more complex with adsorbed $\beta$-lactoglobulin (110), especially if the emulsion containing the $\beta$-lactoglobulin has been allowed to age before the $\beta$-casein is added.

This behavior is not surprising because proteins are adsorbed to the interface by many independent points of contact. For all of these to become desorbed at once is extremely unlikely, and so the spontaneous desorption of a protein molecule is very rare; this is why it is very difficult to simply wash proteins from the oil–water interface (97). Replacement of an adsorbed protein molecule by one from solution must presumably require a concerted movement of the two molecules; as parts of one are displaced, they are replaced by parts of the other, until finally one of the two proteins is liberated into the bulk solution. Even this process, although more likely than spontaneous desorption, is by no means certain to succeed, especially if the adsorbed protein has been on the interface for some time and has been able to form bonds with neighboring molecules. Moreover, given the very high concentration of

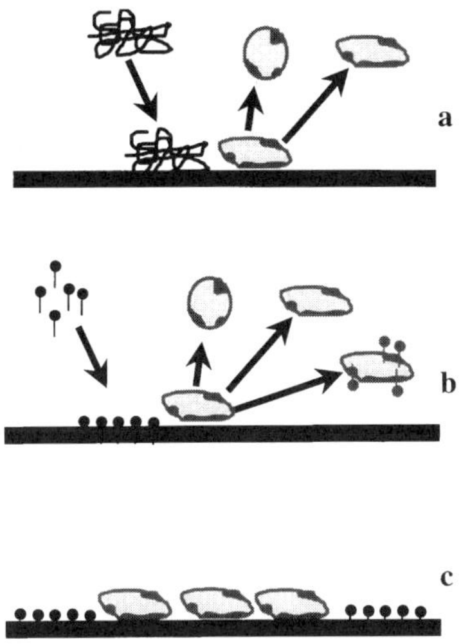


**Figure 8** Schematic of some possible displacement reactions of proteins from an interface. In (a) one type of protein is displaced by a second; the displaced protein may be either denatured or in a form close to native; (b) displacement by small-molecule surfactant – the displaced protein may adopt one of several forms, including ones in which a complex is formed with the surfactant; (c) as some protein is displaced, there is phase separation on the interface between adsorbed protein and the displacing surfactant.

protein in the adsorbed layer (see earlier text), it may even be difficult for a second type of protein to penetrate the adsorbed layer to initiate the displacement process. Therefore, although thermodynamic considerations may favor one protein over another, kinetic factors militate against rapid exchange. Nonetheless, there do seem to be factors which influence the competition between proteins. As has been suggested above, flexibility may be an important criterion (because $\beta$-casein is in many cases an effective displacing agent). Thus, whatever increases flexibility may lead to increasing competitiveness. The most obvious example of such a change is $\alpha$-lactalbumin; in its native state, this protein has a globular structure which is partly maintained by the presence of one ion of calcium within the molecule (111).

Removal of $Ca^{2+}$ leads to the protein adopting a "molten globule" state, whose tertiary structure is altered (112), and this leads to increased flexibility and competitiveness at the interface (113). The removal of $Ca^{2+}$ can be achieved by chelation with complexing agents or by reducing the pH, and under these conditions $\alpha$-lactalbumin outcompetes $\beta$-lactoglobulin for adsorption to the interface (47, 94, 113). To some extent, the competition can be reversed by reneutralizing, so that in this case there is dynamic competition between the proteins for the interface, not simply preferential adsorption during emulsion formation (94).

The above description of the only moderate exchange between adsorbed and free proteins refers to results which have been obtained at room temperature. However, it appears that the behavior may be rather dependent on the temperature at which the studies are made. It has recently been demonstrated that whey proteins (especially $\beta$-lactoglobulin) can displace $\alpha_{s1}$- and $\beta$-caseins from an oil–water interface during heating (this does not occur apparently at room temperature) (114). If whey protein isolate is added to an emulsion prepared from oil and sodium caseinate and the mixture is heated to a temperature in excess of about 40°C, the whey proteins rapidly become adsorbed, and, as they do so, the caseins are desorbed, so the surface coverage by protein remains approximately constant (115). It appears that only the major caseins are desorbed, however, since the $\alpha_{s2}$- and $\kappa$-caseins remain on the emulsion droplets. This behavior has been insufficiently studied to be fully understood; it is not clear why the whey proteins should begin to displace the caseins at what is a relatively low temperature, much below the denaturation temperature of the proteins. Nor is it known why the minor caseins resist displacement; the obvious possibility, that of the formation of disulfide bonds between these caseins and the whey proteins, does not seem to occur. It seems evident, however, that exchange of proteins may occur more readily than was previously thought, especially in food preparations where an emulsion is added to a solution containing other proteins and the mixture undergoes heat treatment.

Competition between adsorbed and free species can be considerably enhanced by the use of small surfactant molecules as well as proteins (116). Here, there is competition between the proteins and small molecules as well as between the proteins themselves. However, in such a case, instead of the desorption of a protein requiring the inefficient process of simultaneous detachment at all points, or the slow creeping displacement of one protein molecule by another, it is possible for a number of small molecules to displace a protein by separately replacing the individual points of attachment. It is known that small-molecule surfactants are capable of efficiently displacing adsorbed proteins (Fig. 8b), although the details of the reactions depend on the type of surfactant and whether it is oil or water soluble (117–120). Water-soluble surfactants are capable of removing all of the adsorbed protein from the oil–water interface, although they may require a molecular ratio of about 30:1 surfactant:protein (116). At lower ratios, some, but not all, of the protein is displaced (Fig. 9). Oil-soluble surfactants (low HLB numbers) are, in general, less efficient at completely displacing protein, or rather of preventing protein adsorption (107, 120, 121). For solubility reasons, these surfactants cannot be added to the emulsion once it has been formed, but must be incorporated at the time the emulsion is produced in the homogenizer. In addition to competing with proteins for adsorption to the oil–water interface, both during formation of the emulsion and its subsequent storage, some small-molecule surfactants also facilitate the exchange reactions of the proteins themselves. For example, although α-lactalbumin and β-lactoglobulin do not compete well with each other under normal circumstances at neutral pH (104), the presence of Tween causes the adsorption of α-lactalbumin to be favored over β-lactoglobulin (116). Presumably, the presence of surfactant enables a more thermodynamic equilibrium to be established, rather than the extremely slow kinetically determined exchange which normally occurs (if it occurs at all) between the two proteins. Alternatively, if the surfactant actually binds to the protein, its conformation may change so that it becomes more surface active, rather as was shown for the "molten globule" conformation of α-lactalbumin.

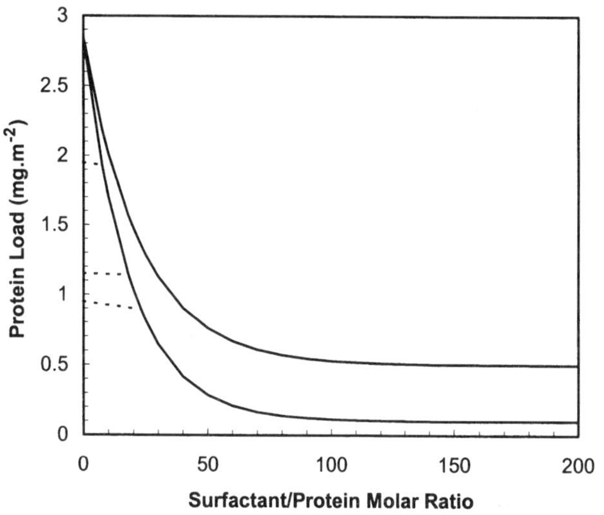


**Figure 9** Surface-protein load on oil droplets in caseinate/soya oil emulsions in the presence of hydrophilic (Tween, lower curve) or hydrophobic (Span, upper curve) surfactants. The tie lines are between the compositions of the surface of the emulsions in presence of no surfactant and those where surfactant is present. Note that for the hydrophobic surfactant it is often found that complete displacement of the protein is impossible.

Obviously, emulsions made using proteins have different properties from those made using small-molecule emulsifiers because their surfaces are very different. It is interesting to speculate on the manner in which the protein is replaced by increasing amounts of surfactant. It appears likely that the protein and surfactant tend to group on the surface (Fig. 8c), so that areas rich in protein and areas containing only surfactant are present (122). This phase separation on the interface is equivalent to creating "hot spots" on the surface of the emulsion droplet, which may lead to a form of directed reactivity of the particle.

It is apparent that "real" food emulsions are likely to behave in a more complex way than are simple model systems studied in the laboratory. This may be especially important when lecithins are present in the formulation. Although these molecules are indeed surfactants, they do not behave like other small-molecule emulsifiers. For example, they do not appear to displace proteins efficiently from the interface, even though the lecithins may themselves become adsorbed (123). They certainly have the capability to alter the conformation of adsorbed layers of caseins, although the way in which they do this is not fully clear; it is possibly because they can "fill in" gaps between adsorbed protein molecules (124). In actual food emulsions, the lecithins in many cases contain impurities, and the role of these (which may also be surfactants) may confuse the way that lecithin acts (125). It is possible also for the phospholipids to interact with the protein present to form vesicles composed of protein and lecithin, independently of the oil droplets in the emulsion. The existence of such vesicles has been demonstrated (126), but their functional properties await elucidation.

Therefore, in a real food emulsion, the composition of the interface may be exceedingly complex. Probably all of the types of surfactant present will be adsorbed to some extent, but it is at present impossible to do more than broadly predict what the composition of the interfacial layer will be, especially when the emulsion may be subjected to a variety of environmental changes (e.g., changes in pH and various sterilization procedures). Likewise, the prediction of stability or otherwise, and other functional properties of the emulsion, which depend on the composition and structure of the adsorbed layer, will become extremely complex.

## VIII. STABILITY OF FOOD EMULSIONS

Food emulsions need to be stable because many of them are designed to have a long shelf-life; for example, mayonnaise and concentrated homogenized milks may be required to have shelf-lives of several months. The aim of the food technologist is to maintain the structure of the emulsion droplets and to prevent their coagulation, creaming, and especially coalescence, which leads to irreversible phase separation, or, on the other hand, coagulation and gel formation. Simple creaming, being gravity driven, can be controlled in a number of physical ways, such as producing smaller droplets or increasing the viscosity of the continuous phase, but whether or not aggregation or coalescence occurs depends on the composition and properties of the adsorbed surface layers on the oil droplets.

Different food emulsions appear to be stabilized by a variety of mechanisms, but because the structures of the interfacial layers of the emulsion droplets can be very complex, it is difficult to define exactly why an emulsion may be stable or unstable. One need only consider the problem of age gelation and instability in heated milks (127) to understand the difficulties which attend the determinations of the causes of instability in stored-food emulsions. Of the two major mechanisms generally described for the stability of colloidal systems (DLVO and steric stabilization), it is not clear to what extent stabilization by a DLVO type of mechanism (3) is relevant to food emulsions, especially those stabilized by proteins. This may be because the diffuse surface layers of the particles will begin to overlap before appreciable electrostatic repulsion will be experienced between them (128). Because of the extended nature of adsorbed layers of protein, the actual thickness of the adsorbed layer may be longer than the Debye length, and so the full repulsive potential between the surfaces cannot be developed. Perhaps an alternative way of looking at this is that the primary minimum of energy between the approaching particles is not deep and that temporary aggregates will fall apart rapidly.

Classical DLVO theory shows that an increase in ionic strength should lead to instability, that is, coagulation of emulsion droplets. For emulsions stabilized by proteins, this is not necessarily true. Some evidence is available for casein-stabilized particles, whereby ionic strengths of > 1 M NaCl do indeed cause the emulsion to coagulate (64), but this applies under only rather specific conditions. The addition of $Ca^{2+}$ in quite low concentrations (< 10 mM) can destabilize casein-stabilized emulsions (129–131), but it is probable that this is a specific ion effect rather than a DLVO type of mechanism, since an equivalent amount of ionic strength in the form of NaCl does not cause coagulation (129). It is known that $Ca^{2+}$ binds to caseins, which may lead to the formation of calcium bridges between the emulsion droplets rather than to their coagulation as a result of general ionic strength effects. Binding of $Ca^{2+}$ to adsorbed caseins should neutralize some of the negative charge on the emulsion droplets and thus permit close approach and flocculation or aggregation, and measurement of the $\zeta$-potential of latices and emulsions shows a decrease in the absolute magnitude of the $\zeta$-potential as $Ca^{2+}$ is added (131, 132). However, the binding of $Ca^{2+}$ does not completely neutralize the $\zeta$-potential, so it may not be sufficient to destabilize the emulsion (Fig. 10).

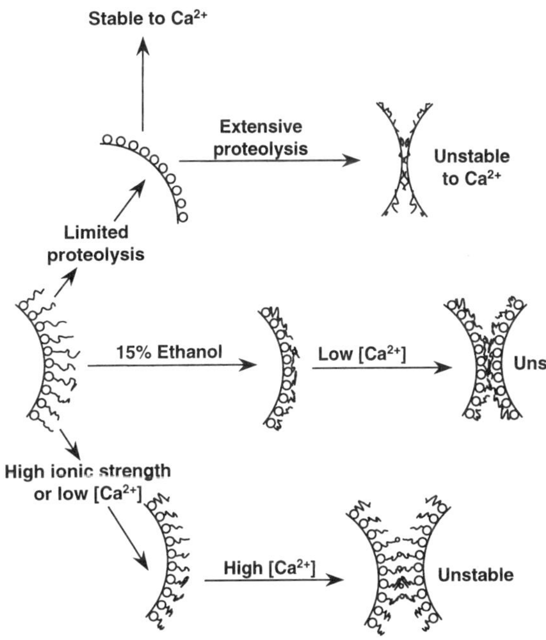


**Figure 10** Instability phenomena in caseinate-stabilized emulsions. The original emulsion is stabilized by the extended layer of adsorbed casein which can be partly destroyed by limited proteolysis, to increase the stability of the emulsion to added $Ca^{2+}$. Further proteolytic degradation of the casein leads to instability. Addition of ethanol causes the adsorbed layer to collapse, decreasing steric stabilization and making the emulsion increasingly sensitive to the presence of $Ca^{2+}$. High ionic strength, or even the presence of small concentrations of $Ca^{2+}$ cause a decrease in the thickness of the adsorbed layer, reducing steric stabilization, and high concentrations of $Ca^{2+}$ give an unstable emulsion, possibly by additionally bridging between emulsion droplets.

If stabilization is to be achieved by a mechanism other than DLVO, an obvious possibility is steric stabilization (although definitions of this phenomenon are rather variable). For casein-coated emulsion droplets, the extended layer of protein can almost certainly provide this kind of stabilization. As two emulsion droplets approach closely, the adsorbed layers may try to interpenetrate or to distort one another. These interactions lead to increases in enthalpy and decreases in entropy, consequences which reduce the possibility of spontaneous aggregation. In addition, the simple act of close approach causes the formation of a high concentration of macromolecules between the particles, with consequent increase in osmotic pressure, so that water will flow in and tend to push apart the two droplets (3). Although casein is the most obvious example because it extends far from the oil–water interface, and the interactions between emulsion droplets coated with caseins have been described (93, 133), it is possible that all proteins when adsorbed protrude from the interface sufficiently far (2–3 nm) to allow these mechanisms to be active. Even when adsorbed casein and $\beta$-lactoglobulin have been digested by proteolytic enzymes, the emulsions are not necessarily destabilized (54), so even some of the peptides remaining on the interface seem to have the capacity to maintain stability, even though they protrude much less into the solution than do the original proteins. It has not been established, however, how thick a layer of protein or peptide is necessary to provide adequate steric stabilization of emulsion droplets (Fig. 10).

The observation that proteins need not be intact to stabilize emulsions is reinforced by the ability of peptides, rather than proteins, to act as emulsifiers. In some cases, the emulsifying ability is even enhanced by partial proteolysis of the proteins (134–136). However, since individual amino acids are not effective surfactants, there must obviously be a length below which peptides become ineffective in forming emulsions. This length must depend on the composition, sequence, and structure of the peptide, so that any estimate of optimal length must be a generalization. For proteolyzed whey proteins, a moderate degree of hydrolysis gives the best emulsifying capacity for the mixture of peptides created by the proteolysis (137). Isolation of specific peptides with high emulsifying ability is a process which in general is too expensive to be used for food ingredients.

Both charge-dependent and steric stabilization mechanisms prevent the close approach of particles, so that flocculation is prevented, and if flocculation (or aggregation) cannot occur, then creaming will be slow, provided that the emulsion droplets are small to start with. If aggregation and creaming are slow, then coalescence is unlikely under normal storage conditions. However, the presence of protein in the adsorbed layers will also help to prevent coalescence. Although the precise properties are still a subject of dispute, it is probable that the viscoelastic properties of an interfacial layer help to define whether the membranes on adjacent oil droplets will rupture to allow the flow of oil from one to another. Some proteins form very viscous interfacial films (109, 138), and these may well be important when coalescence is to be controlled; also, the lack of mobility of adsorbed proteins, compared with the rapid movement of small surfactants, must lead to greater stability toward coalescence.

An additional aid to the stability of food emulsions may be the incorporation of polysaccharide stabilizers. With few exceptions, these act simply to increase the viscosity of the continuous phase of the emulsion, so that they slow down the kinetics of flocculation of the droplets and also slow creaming (139). They do not themselves necessarily participate in any reaction with the emulsion droplets themselves. Thus, they may prolong the life of an unstable emulsion, but do not provide absolute stability. An alternative class of polysaccharides form structures in the continuous phase, by forming a weak gel which is not only viscous but also possesses a yield stress. These additives have a more pronounced effect on stability, because the yield stress prevents the movement of droplets which have insufficient energy to overcome the stress (140). Thus, creaming, for example, may be prevented because the gravitational effect is too weak to overcome the yield stress of the continuous phase. Not all polysaccharides form gels; specifically, xanthan and carrageenans form significant gels in the presence of divalent ions (139). However, these polymers may also destabilize the emulsions by a mechanism of depletion flocculation (141, 142).

A few polysaccharides do appear to interact with emulsion droplets. Of these, the best known example is carrageenan. It is established from studies in milk and with isolated components from milk that $\kappa$-carrageenan can interact with $\kappa$-casein (143). Presumably, therefore, emulsions which contain carrageenans and caseins should show this interaction, and because the casein is likely to be found on the droplet interface, it is likely that the carrageenan will be found there as well. This is likely to result in a highly stable particle (144, 145), which will have very strong steric stabilization because the carrageenan molecules may protrude far into the solution from the emulsion interface (146).

Although polysaccharide molecules often act as stabilizers for the emulsions, it is also possible for them to act as destabilizers, by the mechanism of depletion flocculation. In this, the polysaccharide, once it has diffused from between emulsion droplets, acts to hold the droplets together. The result of this is an apparent flocculation, which can be reversed by dilution of the flocculating agent in the continuous phase. It is, however, necessary that both the emulsion and the macromolecule be present in sufficient amounts, otherwise the separation of phases does not occur. As well as extended hydrocolloids, it is possible that sodium caseinate may also cause depletion flocculation of an emulsion, if its concentration is high enough. Although the caseinate acts primarily as an emulsifier, there is a limit to which it can adsorb to the oil–water interface, as was shown earlier. When a sufficiently high concentration of the protein remains in the continuous phase, it can aggregate to form particles of protein sufficiently large to exert a destabilizing effect (147–149).

Heating may be an important factor in the destabilization of emulsions, especially if the emulsions contain proteins which can be thermally denatured. There is an extensive literature, for example, on the destabilization of homogenized milks (especially when concentrated) by heating (150). Heat will generally facilitate the interactions between emulsion droplets and promote the aggregation process, but as it alters the conformations of the proteins involved, they may also interact to form new structures, such as gels. In particular, when whey proteins are used to form the emulsions, it is possible to form gels (151,152), which, in effect, are protein gels with filler particles (i.e., the emulsion droplets). It seems that the gels are formed by the interaction of protein in solution with protein on the oil–water interface, because they are not formed when the amount of soluble protein is low (139). However, it is possible to make gels from fine emulsions containing 20% oil and as little as 2.5% whey protein (153); normally, gelation of whey protein alone only occurs at concentrations in the region of 8% and greater. Therefore, the presence of the interactive filler particles of the emulsion droplets considerably enhances the gelling capability of the protein (154).

The effects of heat on emulsions depend primarily on the type of protein which is present. However, other factors are important; for example, the pH and the presence in solution of specific minerals. Thus, an emulsion which is stable to heating at one pH value may not be stable at another (153). It is, for example, difficult to produce heat-stable emulsions at pH values of about 4, using milk proteins as emulsifying agents; other emulsifiers have to be used to offset the tendency of the proteins to aggregate when heated in this pH region. However, added emulsifiers themselves have an effect on the gel properties of the destabilized emulsions (155, 156).

## IX. KINETICS OF THE DESTABILIZATION OF EMULSIONS

Generally, emulsions are considered to be thermodynamically unstable, because the presence of large areas of oil–water interface provides positive free energy; adsorption of surfactants can only decrease the interfacial free energy, but cannot make it negative. Thus, it is the kinetics of destabilization which determine whether emulsions have long enough lives to be useful. Some of the factors governing the kinetics have already been described in the preceding section, but it may be appropriate to consider more mechanistic descriptions of the process.

If destabilizing interactions between emulsion droplets are possible, despite the stabilizing action of the adsorbed layers, the process of destabilization will consist of aggregation, followed, possibly, by coalescence. These two processes must be sequential, because coalescence is likely only to result from the prolonged proximity of droplets brought about either by creaming or aggregation (including depletion flocculation) followed by creaming. For an unstirred emulsion, the aggregation kinetics should follow Smoluchowski kinetics for perikinetic flocculation, in the presence of an energy barrier (73, 157). However, this does not appear to be generally true. Although Smoluchowski kinetics have been shown (158) to apply to the destabilization of dilute suspension of casein micelles (which we may take as an example of a protein-stabilized colloid similar to an emulsion), there is little evidence to show that the simple Smoluchowski formulation is appropriate for the destabilization of emulsions. What should be observed if all particles are equally reactive is a linear growth of the volume-average particle size; what tends to be observed is a lag time where growth of molecular weight is slow, followed by a steadily increasing rate of reaction (129, 130). Most of the relevant experiments have been conducted in dilute solution (to allow light scattering to be measured), but even in whole emulsions, the growth of particle size is not that which is predicted by Smoluchowski kinetics (55).

To explain this type of behavior, it is necessary to suggest either that there is a first aggregation reaction which is slow and rate determining, followed by one which is rapid, or that the interaction of particles depends on their size, being more efficient as the size increases. The first of these is difficult to explain, as the individual emulsion droplets are expected to be isotropic. The second can be explained at least partly by invoking the concept of fractality (159, 160); if the aggregates have a fractal form, because of random aggregation, then the space taken up by an aggregate particle will be dependent on the size of the monomer and the fractal dimension. In effect, the collision diameter of the aggregate will be larger than expected, so that its effective concentration will be increased and its reactions will be faster than if the emulsion droplets in the aggregate coalesced. The apparent lag phase of the aggregation reaction can, therefore, be seen to arise simply from the fractal structures of the aggregate particles. Fractality is not in itself a driving force; it is simply the result of the random aggregation of particles. However, it can be seen that fractality can exercise control over the reaction.

Such a mechanism will properly only be calculable for large aggregates, where the concept of self-similarity of structures can be used. Small aggregates might be expected to form by a more general Smoluchowski mechanism, and only when they reach a critical size can the concept of fractality be truly applied. This mechanism seems to be the best available at the moment, if we are to regard the emulsion droplets as possessing isotropic surfaces, which do not have specific "hot spots" through which reaction may take place (161). If reactive sites exist, it is possible in principle to use a polyfunctional approach to describe the kinetics; essentially, this leads to the same kind of kinetics as the fractal description, with a slow lag phase being followed by explosive growth of the particle size, because large particles react faster than do small ones, having more reactive centers (73, 129, 159, 162).

The possibility of the formation of nonisotropic surfaces on some types of emulsion droplets has been recently demonstrated. On interfaces of protein which have been treated with small-molecule emulsifiers, the protein is displaced. However, when insufficient emulsifier is added to cause desorption of all of the protein, there is a tendency for the different surfactants to form regions (i.e., to phase separate on the interface) (122). Clearly, such an interface offers the opportunity for directed aggregation because of the anisotropy of the surface. However, it depends on the presence of at least two surfactants.

The kinetics of orthokinetic flocculation of emulsions (i.e., when the unstable emulsion is being stirred or agitated) can perhaps be more easily explained because theory predicts that larger particles, whether fractal or not, will aggregate faster than small ones. Slow early aggregation of the emulsion becomes faster as time goes on because of this (157). The fractal concept, as described earlier, can also be used to describe the reactions because it describes how the effective particle size of the aggregates is, in fact, larger than would be expected from simple consideration of the degree of polymerization. In a stirred system, the shearing of the solution enhances particle growth, but it can also limit the size of the aggregates which can form, because the effect of shear is also to break up the largest particles (159), unless they rapidly coalesce, in which case only rehomogenization can reduce the particle size. So, precipitation can be avoided (at least while shearing is applied), even through the emulsion suspension remains unstable. However, it has been shown that, at least in some cases, the orthokinetic reaction appears to continue even when the stirring has been stopped (163).

The preceding discussion applies to emulsions which are inherently unstable, so the source of the instability is present from the time that the emulsion is formed, or arises from the addition of some other destabilizing material [e.g., $Ca^{2+}$ (129) or small surfactant (164, 165)] to the emulsion. It is possible, however, to propose another mechanism of destabilization, where the stability of the emulsion is decreased with time by the action of some chemical reaction. The breakdown of a layer of adsorbed protein by proteolytic enzymes present in the food product is a possible cause of instability, especially in milk-based products (166). Such proteolytic enzymes may arise from microbiological contamination at an early stage in manufacture; often a heat treatment will kill the microorganisms but will leave some of the proteases in an active state. Mechanisms of this type have been proposed for the age gelation of homogenized milks. In the course of storage or processing it is also possible to cause chemical modification of proteins which have been adsorbed so that their functional behavior is modified. In heated emulsions containing milk, for example, there is a possibility of a Maillard reaction (the reaction between reducing sugars and the amino groups of proteins). Depending on the severity of this reaction, it may be possible to crosslink the proteins adsorbed to the interface, or even to cause gelation of an emulsion subjected to strong heating. Studies of this type of reaction are few, and more information is needed on the potential of such a reaction to alter the behavior of the emulsions. A second type of modification occurs in emulsions containing oil which is capable of being oxidized; the formation of proteins modified by the enal products of lipid oxidation is a distinct possibility, but again too little is known of the details of this reaction, although the modified proteins bind very strongly to the interface, and presumably, therefore, cannot take part in exchange reactions so readily (167).

## X. CONTROLLED INSTABILITY OF EMULSION–PARTIAL COALESCENCE

Even though emulsions generally need to be as stable as possible, there are several food products where the opposite effect is desired, so a shelf-stable emulsion can be controllably destabilized when required. Such emulsions are the basis for products such as whipped toppings and ice cream and generally depend for their effect on the processes of partial coalescence (168–170) and destabilization by whipping air bubbles into the mixture, at which time the interfacial layer of the emulsions may be mechanically broken and liquid oil spread around the air–solution interface.

Because of the importance of ice cream as a product, much has been written on its structure and formation (171), and the process can only be summarized here. In toppings and ice cream (and indeed simply in whipping cream), it is first necessary to produce a stable emulsion. Ice-cream mix is a complex mixture, but the initial emulsion is basically homogenized milk, containing an admixture of small-molecule surfactants as well; in whipped toppings, the emulsion is made with oil and a surfactant mixture, which may or may not contain protein; and in cream, the natural membrane of phospholipid and protein surrounds the milk fat. In all of these, it is necessary to have some small-molecule emulsifiers so as to exchange with, and weaken the rigidity of, the adsorbed layer of protein (118). The second essential is that the fat or oil in the formulation is partly crystalline; neither completely liquid nor completely solid oil will perform optimally. If the oil is partly crystalline, then the emulsion droplets may not be truly spherical but may have protrusions of crystals on their surfaces.

When such an emulsion is subjected to shearing forces, the emulsion droplets will be subject to partial coalescence, where the protruding fat crystals on different droplets catch together (4). This is followed by breakage of the interfacial layer, which allows the linking of the emulsion droplets together. Although liquid fat will flow from one droplet to another under these circumstances, complete coalescence cannot be achieved because of the presence of the crystalline oil. The end result is that the emulsion droplets form a network, linked by crystalline oil. This effect is enhanced when the emulsion is not simply sheared but whipped so as to incorporate air into the product. Air interfaces are highly disruptive of the interfaces of emulsion droplets, especially if the interfacial layer is not too strong (which is why small-molecule surfactants are used in the formulations), and the result is that the partial coalescence of the droplets occurs preferentially around the air bubbles. The end product is a foam, in which partially coalesced oil droplets form a framework around the air bubbles to give a stable product. Although this is the basic process, the formulations of the original emulsions may be refined by the addition of stabilizers and flavor compounds, and of course in ice cream, the crystals of ice which are formed during the cooling of the final foam are also important texturizers. It is the emulsion, however, which defines the basic structure of the ice cream, although there must be as many detailed formulations as there are companies making the product.

## XI. MULTIPLE EMULSIONS

The emulsions so far described have been mainly of the simple O/W type. However, because of their utility in other fields (e.g., cosmetology), an interest is developing in food applications of multiple emulsions, i.e., water-in-oil-in-water emulsions, since they modify the behavior of the fat and also offer the potential to carry, in their interior water droplets, materials of nutritional interest (172, 173). However, the formulation and control of such preparations is much more difficult than for simple emulsions (174). The basic principles of such emulsion formulation are well known; the water droplets within the oil droplet need to be stabilized using a mixture of lipophilic emulsifiers, whereas the stabilization of the oil droplets requires rather a hydrophilic surfactant. Evidently, the preparation of such emulsions cannot be preformed in a single stage, but requires the preparation of a W/O emulsion first, and then dispersion of this emulsion into an aqueous medium.

Unfortunately, it is not sufficient to simply use any pairs of hydrophilic and lipophilic emulsifiers. In practice, the choice is very limited, because the ingredients of foods must be approved by legislation, and therefore many potentially useful surfactants cannot be used. However, the use of natural polymeric emulsifiers (proteins) is making the formulation of such food emulsions more possible (175). The major problem of these types of emulsions is their tendency to be unstable during their incorporation into a food. Not only are they subjected to high shear stresses during mixing, which may distort or destroy the rather fragile structures of the emulsion droplets, but they may also be suspended in media which have different osmotic pressures from those of the water contained in the oil droplets. Finally, there are the effects of temperature; very few multiple emulsions, even if they stable at room temperature, can be subjected to pasteurization or higher heat treatments without causing them to break and revert to a simple emulsion or even a two-phase oil–water system. It is possible to stabilize the multiple emulsions by using gelling agents either to solidify the internal water droplets or to rigidify their interfaces, but in general the droplets with fluid water and fluid interfaces are not readily stabilized at elevated temperatures. The consequence of this is that, for food applications, multiple emulsions would have to be made in sterile form at low temperature and then added to the food without the application of heat, thereby increasing the complexity of manufacture.

For these reasons, there is at present less application of these types of emulsions in foods that might perhaps be anticipated or desired. Furthermore, the emulsions are more expensive to produce than simple emulsions, and therefore tend not to be widely employed, since processed foods in general tend to rely heavily on minimum-cost formulations.

## XII. CONCLUDING REMARKS

The object of making food emulsions is to provide a stable and controllable source of food, whose texture, taste, and nutritional and storage properties are acceptable to the consumer. Although the number of possible ingredients is limited by the constraints of healthy nutrition, it is nevertheless evident that within the available range there is ample opportunity for variation in the properties of the emulsions, for instance, the particle size and the composition of the stabilizing layer of the interface, which, in turn, influence the stability and functional behavior of the emulsion. Nonetheless, many emulsions used in foods have their roots in established formulations, and an understanding of why certain emulsions behave as they do is still not established in a number of cases. It has been pointed out that not all of the aspects of the stability of emulsions are known, in terms of the mechanisms which may be operating to maintain the stability of the systems. Also, in many real food emulsions, the path followed during their production is critical, emphasizing once again that emulsions are not equilibrium systems and that the same overall composition may not necessarily lead to the same product. On this point, our knowledge is insufficient and needs to be extended. For example, the heat treatment of ingredient proteins either before or after the formation of an emulsion may critically affect the behavior of the emulsion. As the emulsions contain more ingredients, the level of complexity required to understand the details of their formation and properties is increased. The challenge for the future is to be able to describe and control some of the most complex emulsions so as to allow greater functional stability for these food systems.

A further aspect, which is becoming of ever-increasing importance in the public mind, is that of the nutritional function of food emulsions. Much interest is centered on the nutritional properties of fats and oils, and the way that they are perceived to affect health. Reduced-fat formulations are demanded, which nevertheless are required to possess textural and organoleptic properties as close as possible to those of the traditional types of food emulsion. This in itself provides a challenge to the emulsion technologist – how can I make the same amount of fat do twice the work? In addition, the demand for the incorporation of nutritionally beneficial lipid materials produces a challenge; the incorporation of these materials into foods, complete with antioxidants and other necessary ingredients, also requires increased ingenuity on the part of the technologist. In addition, there is the question of targeting the materials contained in a food. No longer is it sufficient simply to provide nutrition; ideally, it is necessary to define in which portion of the digestive tract the components of the emulsion are to be liberated. Already there are encapsulated materials available which can be targeted by the selection of the coating material. With emulsions of specific oils being part of the "functional food" system, we may expect to see increased demand for emulsions which can be controlled beyond the points of manufacture and consumption. This represents a real challenge for the emulsion technologist of the future.

## REFERENCES


1. E Pilman, K Larsson, E Tornberg. J Dispers Sci Technol 1: 267–282, 1980.
2. N Krog. In: K Larsson, SE Friberg, eds. Food Emulsions. 3rd ed. New York: Marcel Dekker, 1997, pp 141–188.
3. E Dickinson, G Stainsby. Colloids in Food. Barking, UK: Applied Science Publishers, 1982.
4. DF Darling, RJ Birkett. In: E Dickinson, ed. Food Emulsions and Foams. Cambridge: Royal Society of Chemistry, 1987, pp 1–29.
5. N Krog. In: M El-Nokaly, D Cornell, eds. Microemulsions and Emulsions in Foods. ACS Symposium Series 448. Washington, DC: American Chemical Society, 1991, pp 139–145.
6. E Li-Chan, S Nakai. In: JE Kinsella, WG Soucie, eds. Food Proteins. Champaign, IL: American Oil Chemists' Society, 1989, pp 232–251.
7. CC Bigelow. J Theor Biol 16: 187–211, 1967.
8. E Dickinson, SR Euston, CM Woskett. Progr Colloids Polym Sci 82: 65–75, 1990.
9. E Dickinson, DJ Pogson, EW Robson, G Stainsby. Colloids Surfaces 14: 135–141, 1985.
10. A Kato, N Tsutsui, N Matsudomi, K Kobayashi, S Nakai. Agric Biol Chem 45: 2755–2760, 1981.
11. E Keshavarz, S Nakai. Biochim Biophys Acta 576: 269–279, 1979.
12. S Nakai. J Agric Food Chem 31: 676–683, 1983.
13. A Kato, S Nakai. Biochim Biophys Acta 624: 13–20, 1980.
14. E Dickinson, JA Hunt, DG Dalgleish. Food Hydrocolloids 4: 403–414, 1991.
15. A Kato, S Miyazaki, A Kawamoto, K Kobayashi. Agric Biol Chem 51: 2989–2994, 1987.
16. E Dickinson, VJ Pinfield, DS Horne. J Colloid Interface Sci 187: 539–541, 1997.
17. S Damodaran, S. Xu. J Colloid Interface Sci 178:426–435, 1996.
18. BM Byrne, AD van het Schip, JAM van de Klundert, AC Arnberg, M Gruber, G Ab. Biochemistry 23: 4275–4279, 1984.
19. F MacRitchie, NF Owens. J Colloid Interface Sci 29: 66–71, 1969.
20. CA Haynes, W Norde. Colloids Surfaces B 2:517–566, 1994.
21. W Norde, F MacRitchie, G Nowicka, J Lyklema. J Colloid Interface Sci 112: 447–456, 1986.
22. RJ Green, I Hopkinson, RAL Jones. In: E Dickinson, JM Rodriguez Patino, eds. Food Emulsions and Foams. Cambridge: Royal Society of Chemistry, 1999, pp 285–295.
23. Y Fang, DG Dalgleish. J Colloid Interface Sci 196: 292–298, 1998.
24. Y Fang, DG Dalgleish. Food Hydrocolloids 12: 121–126, 1998.
25. W Norde, JP Favier. Colloids Surfaces 64: 87–93, 1992.
26. AL de Roos, P Walstra. Colloids Surfaces B 6: 201–208, 1996.
27. MCL Maste, EHW Pape, A van Hoek, W Norde, AJWG Visser. J Colloid Interface Sci 180: 632–638, 1996.
28. E Dickinson, Y Matsumura. Int J Biol Macromol 13: 26–30, 1991.
29. E. A. Foegeding. In: JE Kinsella, WG Soucie, eds. Food Proteins. Champaign, IL: American Oil Chemists' Society, 1989, pp 185–194.
30. KD Caldwell, J Li, JT Li, DG Dalgleish. J Chromatogr 604: 63–71, 1992.
31. CA Haynes, E Swilinsky, W Norde. J Colloid Interface Sci 164: 394–409, 1994.
32. M Corredig, DG Dalgleish. Colloids Surfaces B 4: 411–422, 1995.
33. C Holt, L Sawyer. Protein Eng 2: 251–259, 1988.
34. HE Swaisgood. In: P. F. Fox, ed. Advanced Dairy Chemistry: 1. Proteins. Barking, UK: Elsevier Science Publishers, 1992, pp 63–110.
35. DE Graham, MC Phillips. J Colloid Interface Sci 70: 415–426, 1979.
36. E Tornberg, Y Granfeldt, C Hakansson. J Sci Food Agric 33: 904–917, 1982.
37. MC Phillips. Chem Ind 170–176, 1977.
38. DE Graham, MC Phillips. J Colloid Interface Sci 70: 427–439, 1979.
39. DG Dalgleish, J Leaver. In: E Dickinson, ed. Food Polymers. Gels and Colloids. Cambridge: Royal Society of Chemistry, 1991, pp 113–122.
40. SG Hambling, AS McAlpine, L Sawyer. In: PF Fox, ed. Advanced Dairy Chemistry: 1. Proteins. Barking, UK: Elsevier Science Publishers, 1992, pp 141–190.
41. RD Bee, J. Hoogland, and RH Ottewill. In: E Dickinson, P Walstra, eds. Food Colloids and Polymers: Stability and Mechanical Properties. Cambridge: Royal Society of Chemistry, 1993, pp 341–353.
42. C Holt. Adv Protein Chem. 43: 63–151, 1992.
43. H Oortwijn, P Walstra, H Mulder. Neth Milk Dairy J 31: 134–147, 1977.
44. HS Rollema. In: PF Fox, ed. Advanced Dairy Chemistry: 1. Proteins. Barking, UK: Elsevier Science Publishers, 1992, pp 111–140.
45. JA Hunt and DG Dalgleish, University of Guelph, 1993, unpublished results.
46. H Pessen, JM Purcell, HM Farrell. Biochim Biophys Acta 828: 1–12, 1985.
47. J Hunt, DG Dalgleish. J Agric Food Chem 42: 2131–2135, 1994.
48. SL Turgeon, C Sanchez, SF Gauthier, P Paquin. Int Dairy J 6: 645–658, 1996.
49. Y Kamata, K Ochiai, F Yamauchi. Agric Biol Chem 48: 1147–1152, 1984.

50. P Plietz, G Damaschun, JJ Müller, KD Schwenke. Eur J Biochem 130: 315–321, 1983.
51. D Causeret, E Matringe, D Lorient. J Food Sci 56: 1532–1540, 1991.
52. LD Ford, R Borwankar, RW Martin Jr, N Holcomb. In: K Larsson, SE Friberg, eds. Food Emulsions. 3rd ed. New York: Marcel Dekker, 1997, pp 361–412.
53. KN Pearce, JE Kinsella. J Agric Food Chem 26: 716–723, 1978.
54. Y Fang, DG Dalgleish. J Colloid Interface Sci 156: 329–334, 1993.
55. J Hunt, DG Dalgleish. Food Hydrocolloids 8: 175–187, 1994.
56. P Walstra. In: JMV Blanshard, P Lillford, eds. Food Structure and Behavior. London: Academic Press, 1987, pp 87–106.
57. P. Walstra. In: P Becher, ed. Encyclopedia of Emulsion Technology. New York: Marcel Dekker, 1983, pp 57–128.
58. O Robin, N Remillard, P Paquin. Colloids Surfaces A 80: 211–222, 1993.
59. E Dickinson, FO Flint, JA Hunt. Food Hydrocolloids. 3: 389–397, 1989.
60. LV Ogden, P Walstra, HA Morris. J Dairy Sci 59: 1727–1737, 1976.
61. W Buchheim, P Dejmek. In: K Larsson, SE Friberg, eds. Food Emulsions. 3rd ed. New York: Marcel Dekker, 1987, pp 235–278.
62. T Gershanik, S Benzeno, S Benita. Pharm Res 15: 863–869, 1998.
63. E Dickinson, RH Whyman, DG Dalgleish. In: E Dickinson, ed. Food Emulsions and Foams. Cambridge: Royal Society of Chemistry, 1987, pp 40–51.
64. P Walstra, H Oortwijn. J Colloid Interface Sci 29: 424–431, 1969.
65. P Walstra. J Colloid Interface Sci 27: 494–500, 1968.
66. K Strawbridge, FR Hallet. Macromolecules 27: 2283–2290, 1994.
67. K Strawbridge, J Watton. Can J. Spectrosc 36: 53–60, 1991.
68. K Strawbridge, FR Hallet. Can J Phys 70: 401–406, 1992.
69. DG Dalgleish, FR Hallett. Food Res Int 28: 181–193, 1995.
70. FR Hallett, T Craig, J Marsh, B Nickel. Can J Spectrosc 34: 63–70, 1989.
71. KB Strawbridge, E Ray, FR Hallett, SM Tosh, DG Dalgleish. J Colloid Interface Sci 171: 392–398, 1995.
72. DS Horne, C Davidson. Colloids Surfaces A 77. 1–8, 1993.
73. A Lips, T Westbury, PM Hart, ID Evans, IJ Campbell. In: E Dickinson, P Walstra, eds. Food Colloids and Polymers: Stability and Mechanical Properties. Cambridge: Royal Society of Chemistry, 1993, pp 31–44.
74. DJ McClements. Langmuir 12: 3454–3461, 1996.
75. JN Coupland, DJ McClements. J Am Oil Chem Soc 74: 1559–1564, 1997.
76. AS Dukhin, PJ Goetz. Colloids Surfaces A 144: 49–58, 1998.
77. T Wade, JK Beattie. Colloids Surfaces B 10: 73–85, 1997.
78. B Bergenståhl, PM Claesson. In: K Larsson, SE Friberg, eds. Food Emulsions. 3rd ed. New York: Marcel Dekker, 1997, pp 57–110.
79. JV Boyd, C Parkinson, P Sherman. J Colloid Interface Sci 41: 359–370, 1972.
80. JMM Westerbeek, A Prins. In: M El-Nokaly, D Cornell, eds. Microemulsions and Emulsions in Foods. ACS Symposium Series 448. Washington, DC: American Chemical Society, 1991, pp 146–160.
81. F MacRitchie. Colloids Surfaces A 76: 159–166, 1993.
82. J Leaver, DG Dalgleish. Biochim Biophys Acta 1041: 217–222, 1990.
83. J Leaver, DG Dalgleish. J Colloid Interface Sci 149: 49–55, 1992.
84. DG Dalgleish. Colloids Surfaces 46: 141–155, 1990.
85. AR Mackie, J Mingins, AN North. J Chem Soc Faraday Trans 87: 3043–3049, 1991.
86. E Dickinson, DS Horne, JS Phipps, RM Richardson. Langmuir 9: 242–248, 1993.
87. DG Dalgleish, J Leaver. J Colloid Interface Sci, 141: 288–294, 1991.
88. DG Dalgleish. Colloids Surfaces B 1: 1–8, 1993.
89. M Shimizu, A Ametani, S Kaminogawa, K Yamauchi. Biochim Biophys Acta 869: 259–264, 1986.
90. AR Mackie, J Mingins, R Dann. In: E Dickinson, ed. Food Polymers, Gels and Colloids. Cambridge: Royal Society of Chemistry, 1991, pp 96–111.
91. FAM Leermakers, PJ Atkinson, E. Dickinson, DS Horne. J Colloid Interface Sci 178: 681–693, 1996.
92. E Dickinson, DS Horne, VJ Pinfield, FAM Leermakers. J Chem Soc Faraday Trans 93: 425–432, 1997.
93. E Dickinson, VJ Pinfield, DS Horne, FAM Leermakers. J Chem Soc Faraday Trans 93: 1785–1790, 1997.
94. J Hunt, DG Dalgleish. Food Hydrocolloids 10: 159–165, 1996.
95. M Britten, HJ Giroux. J Agric Food Chem 41: 1187–1191, 1993.
96. M Rosenberg, SL Lee. Food Struct 12: 267–274, 1993.
97. F MacRitchie. J Colloid Interface Sci 105: 119–123, 1985.
98. DG Dalgleish, EW Robson. J Dairy Res 52: 539–546, 1985.
99. H Oortwijn, P Walstra. Neth Milk Dairy J 33: 134–154, 1979.
100. P Walstra, H Oortwijn. Neth Milk Dairy J, 36: 103–113, 1982.

101. DG Dalgleish, S Tosh, S West. Neth Milk Dairy J 50: 135–148, 1996.
102. AV McPherson, BJ Kitchen. J Dairy Res 50: 107–133, 1983.
103. MA Tung, LJ Jones. Scanning Electron Microsc 3: 523–530, 1981.
104. E Dickinson, SE Rolfe, DG Dalgleish. Food Hydrocolloids 3: 193–203, 1989.
105. E Dickinson, SE Rolfe, DG Dalgleish. Food Hydrocolloids 2: 397–405, 1988.
106. EW Robson, DG Dalgleish. J Food Sci 52: 1694–1698, 1987.
107. SE Euston, H Singh, P Munro, DG Dalgleish. J Food Sci 60: 1124–1131, 1995.
108. E Dickinson. J Food Eng 22: 59–74, 1994.
109. E Dickinson, SE Rolfe, DG Dalgleish. Int J Biol Macromol 12: 189–194, 1990.
110. DG Dalgleish, SE Euston, JA Hunt, E Dickinson. In: E Dickinson, ed. Food Polymers, Gels and Colloids. Cambridge: Royal Society of Chemistry, 1991, pp 485–489.
111. M Ikeguchi, K Kuwajima, S Sugai. J Biochem 99: 1191–1201, 1986.
112. OB Ptitsyn. In: TE Creighton, ed. Protein Folding. New York: W H Freeman, 1992, p 243.
113. E Dickinson, Y Matsumura. Colloids Surfaces B 3: 1–17, 1994.
114. JM Brun, DG Dalgleish. Int Dairy J 9: 323–327, 1999.
115. DG Dalgleish. In: E Dickinson, JM Rodriguez Patino, eds. Cambridge: Royal Society of Chemistry, 1999, pp 1–16.
116. JL Courthaudon, E Dickinson, Y Matsumura, A Williams. Food Struct 10: 109–115, 1991.
117. JL Courthaudon, E Dickinson, DG Dalgleish. J Colloid Interface Sci 145: 390–395, 1991.
118. JL Courthaudon, E Dickinson, Y Matsumura, DC Clark. Colloids Surfaces 56: 293–300, 1991.
119. JA de Feijter, J Benjamins, M Tamboer. Colloids Surfaces 27: 243–266, 1987.
120. E Dickinson, S Tanai. J Agric Food Chem 40: 179–183, 1992.
121. E Dickinson, RK Owusu, S Tan, A Williams. J Food Sci 58: 295–298, 1993.
122. AR Mackie, AP Gunning, PJ Wilde, VJ Morris. J Colloid Interface Sci 210: 157–166, 1999.
123. JL Courthaudon, E Dickinson, WW Christie. J Agric Food Chem 39: 1365–1368, 1991.
124. Y Fang, D Dalgleish. Colloids Surfaces B 1: 357–364, 1993.
125. CH McCrae, DD Muir. J Dairy Res 59: 177–185, 1992.
126. Y Fang, DG Dalgleish. Langmuir 11: 75–79, 1995.
127. VR Harwalkar. In: PF Fox, ed. Advanced Dairy Chemistry: 1. Proteins. Barking, UK: Elsevier Science Publishers, 1992, pp 691–734.
128. TAJ Payens. J Dairy Res 46: 291–306, 1979.
129. S Agboola, DG Dalgleish. J Food Sci 60: 399–404, 1995.
130. J Chen, E Dickinson, G Iveson. Food Struct 12: 135–146, 1993.
131. E Dickinson, JA Hunt, DS Horne. Food Hydrocolloids 6: 359–370, 1992.
132. DG Dalgleish, E Dickinson, RH Whyman. J Colloid Interface Sci 108: 174–179, 1985.
133. H Casanova, E Dickinson. J Agric Food Chem 46: 72–76, 1998.
134. SO Agboola, DG Dalgleish. J Agric Food Chem 4: 3631–3636, 1996.
135. SL Turgeon, SF Gauthier, D Mollé, J Léonil. J Agric Food Chem 40: 669–675, 1992.
136. XL Huang, GL Catignani, HE Swaisgood. J Agric Food Chem 44: 3437–3443, 1996.
137. AM Singh, DG Dalgleish. J Dairy Sci 81: 918–924, 1998.
138. E Dickinson, BS Murray, G Stainsby. J Chem Soc Faraday Trans, 84: 871–883, 1988.
139. EM Ozu, IC Baianu, L-S Wei. In: IC Balanu, H Pessen, TF Kumosinski, eds. Physical Chemistry of Food Processes. Vol 2. New York: 1993, Van Nostrand Reinhold, pp 487–517.
140. P Walstra. In: E Dickinson, ed. Food Emulsions and Foams. Cambridge: Royal Society of Chemistry, 1987, pp 242–257.
141. H Luyten, M Jonkman, W Kloek, T van Vliet. In: E Dickinson, P Walstra, eds. Food Colloids and Polymers: Stability and Mechanical Properties. Cambridge: Royal Society of Chemistry, 1993, pp 224–234.
142. E Dickinson. ACS Symposium Series 650: 197–207, 1996.
143. K Osawa, R Niki, S Arima. Agric Biol Chem 48: 627–632, 1984.
144. DG Dalgleish, AL Hollocou. In: E Dickinson, B Bergenståhl, eds. Food Colloids – Proteins, Lipids, Polysaccharides. Cambridge: Royal Society of Chemistry, 1997, pp 69–76.
145. E Dickinson, K Pawlowsky. Food Hydrocolloids 12: 417–423, 1998.
146. DG Dalgleish, ER Morris. Food Hydrocolloids 2: 311–320, 1988.
147. E Dickinson, M Golding. Food Hydrocolloids 11: 13–18, 1997.
148. E Dickinson, M Golding, MJW Povey. J Colloid Interface Sci 185: 515–529, 1997.
149. E Dickinson, M Golding. Colloids Surfaces A 144: 167–177, 1998.
150. AWM Sweetsur, DD Muir, J Soc Dairy Technol 35: 120–125, 1982.
151. G Masson, R Jost. Colloid Polym Sci 264: 631–638, 1986.
152. R Jost, R Baechler, G Masson. J Food Sci 51: 440–449, 1986.

153. JA Hunt, DG Dalgleish. J Food Sci 60: 1120–1123, 1995.
154. E Dickinson, J Chen. J Dispers Sci Technol 20: 197–213, 1999.
155. E Dickinson, Y Yamamoto. J Food Sci 61: 811–816, 1996.
156. E Dickinson, ST Hong, Y Yamamoto. Neth Milk Dairy J 50: 199–207, 1996.
157. M von Smoluchowski. Z Phys Chem 92: 129–168, 1917.
158. DG Dalgleish. Biophys Chem 11: 147–155, 1980.
159. E Dickinson, A Williams. Colloids Surfaces A 88: 317–326, 1994.
160. P Walstra. T van Vliet, LGB Bremer. In: E. Dickinson, ed. Food Polymers, Gels and Colloids. Cambridge: Royal Society of Chemistry, 1991, pp 369–382.
161. PJ Flory. Faraday Discuss Chem Soc 57: 7–18, 1974.
162. E Dickinson, RK Owusu, A Williams. J Chem Soc Faraday Trans 89: 865–866, 1993.
163. EP Schokker, DG Dalgleish. Colloids Surfaces A 145: 61–69, 1998.
164. HD Goff, WK Jordan. J Dairy Sci 72: 18–29, 1989.
165. NM Barfod, N Krog, G Larsen, W Buchheim. Fat Sci Technol 93: 24–35, 1991.
166. TAJ Payens. Neth Milk Dairy J 32: 170–183, 1978.
167. J Leaver, AJR Law, EY Brechany. J Colloid Interface Sci 210: 207–214, 1999.
168. MAJS van Boekel, P Walstra. Colloids Surfaces 3: 109–118, 1981.
169. DF Darling. J Dairy Res 49: 695–712, 1982.
170. K Boode, P Walstra. In: E Dickinson, P Walstra, eds. Food Colloids and Polymers: Stability and Mechanical Properties. Cambridge: Royal Society of Chemistry, 1993, pp 23–30.
171. KG Berger. In: K Larsson, SE Friberg, eds. Food Emulsions. 3rd ed. New York: Marcel Dekker, 1997, pp 413–490.
172. RK Owusu, Q Zhu, E Dickinson Food Hydrocolloids 6: 443–453, 1992.
173. E Dickinson, J Evison, JW Gramshaw, D Schwope. Food Hydrocolloids 8: 63–67, 1994.
174. N Garti, C Bisperink. Curr Opinion Colloid Interface Sci 3: 657–667, 1998.
175. N Garti, A Aserin. Adv Colloid Interface Sci 65: 37–69, 1996.

# 10

# Ultrasonic Characterization of Food Emulsions

**John N. Coupland**
*Pennsylvania State University, University Park, Pennsylvania*

**D. Julian McClements**
*University of Massachusetts, Amherst, Massachusetts*

## I. INTRODUCTION

### A. Food Emulsions

Emulsions are defined as a fine dispersion of one liquid in a second, largely immiscible, liquid. In food science this definition is often extended to include large particles, solid phases, and mixtures including other colloidal species (1,2). By this wider definition most foods contain emulsions as part of their structure or have done so at some stage of their processing. Some examples of food emulsions include beverages, ice cream, infant formulations, sauces, mayonnaise, salad dressings, soups, butter, margarine, and even chocolate. For reasons of clarity and simplicity we restrict this discussion to liquid or solid oil droplets in a liquid aqueous continuous phase, but many of the principles discussed here could be similarly applied to other colloidal food materials. Extensive reviews of the colloid science pertinent to food materials are available in the literature (1–3).

Food scientists are interested in the colloidal properties of emulsions because of their influence on the overall quality and physicochemical properties (texture, stability, appearance, and taste) of products. The emulsion properties most important in determining the bulk properties of the foods containing them are oil concentration, particle size distribution, particle–particle association, particle crystalinity, and the spatial distribution of particles. In some circumstances a stable emulsion is desirable in foods (e.g., a cloud emulsion in a beverage forms an ugly ring around the bottle neck if it creams) but in other cases controlled instability is required (e.g., breaking of a milk emulsion during churning to form butter). Hence, knowledge of the kinetic changes in colloidal properties that occur during manufacture, storage, and usage are also important to the food scientist.

Quantification of the colloidal properties of food emulsions has been difficult using traditional methods (e.g., microscopy, light scattering, and electrical-pulse counting) because of the complexity of their composition and microstructure. Consequently, the development of novel methods of characterizing food emulsions has gained impetus in recent years. The ideal analytical technique would be easy to use, versatile, rapid, reproducible, and reasonably priced. It would also be valuable to be able to effect measurements on emulsions during processing, so the technique should meet hygienic processing standards and be robust enough to survive in a factory environment. As we shall see in this chapter, ultrasonics meets many of these requirements.

## B. Ultrasonic Propagation in Food Emulsions

Ultrasound is high-frequency mechanical vibrations qualitatively similar to the sound we hear and create but at much higher frequencies. (As an illustration, the musical note "A" is 440 Hz; ultrasound is conventionally defined as starting above 20 kHz.) Sound waves are transmitted through materials as deformations in their physical structure and so measurements of the capacity to transmit sound are related to physicochemical properties. There are two distinct types of ultrasonic waves. The most commonly used in fluids are *longitudinal waves* where the deformations occur in the direciton of transmission of the wave. In the second case, *shear waves*, the wave passes through the material with a shearing action, causing deformations normal to the movement of the wave front. Combinations of shearing and longitudinal propagation are also possible. Shear waves are very strongly attenuated in fluids and so are very rarely used in food emulsions which are by definition liquid. Longitudinal ultrasonic waves with frequencies between about 0.1 and 100 MHz are most commonly used in the ultrasonic analysis of fluids. The power levels used in ultrasonic analysis are so low that the deformations caused in a material are extremely small and reversible, which means that the technique is nondestructive.* The ultrasonic techniques described here should not be confused with high-powered ultrasonic techniques that are used for cleaning, welding, homogenization (4,5).

The primary measurable ultrasonic parameters of a material are velocity, $c$, and attenuation coefficient, $\alpha$, defined as follows:

$$c = \frac{d}{t} = \lambda f \tag{1}$$

$$\frac{A}{A_0} = e^{-\alpha d} \tag{2}$$

where $d$ is the distance traveled by the wave in time $t$, $\lambda$ is the wavelength, $f$ is the frequency, and $A$ and $A_0$ are the initial and final amplitudes. Both ultrasonic parameters can be concisely expressed as a complex wavenumber, $k = \omega/c + i\alpha$, where $\omega$ is the angular frequency ($= 2\pi f$) and $i$ is $\sqrt{-1}$.

*This may not be strictly true in the case of very fragile structures (e.g., vesicles) or certain physically or chemically reacting systems (e.g., during crystallization).

In many cases, it is possible to establish empirical relationships between the ultrasonic properties of an emulsion and its composition, and then use these relationships to measure the properties of an unknown sample. This approach may work well for a specific application but, as there is no real understanding of what properties of the system the ultrasound is sensitive to, it may be misleading. Where possible it is preferable to develop a mechanistic understanding of the interactions between an ultrasonic wave and an emulsion.

The propagation of ultrasound in a fluid is given by the following equation:

$$\left(\frac{k}{\omega}\right)^2 = \frac{\rho}{K + \frac{4}{3}G} \tag{3}$$

where $\rho$ is the density, $K$ is the adiabatic bulk modulus, and $G$ is the adiabatic shear modulus. For most fluids $G \ll K$, so the shear component can be neglected and Eq. (3) is reduced to:

$$\left(\frac{k}{\omega}\right) = \sqrt{\kappa\rho} \tag{4}$$

Where $\kappa$ is the adiabatic compressibility ($= 1/K$). All of the parameters in Eq. (3) and (4) are complex and frequency dependent, and while the imaginary part is often neglected in simple and polymeric solutions it is essential in understanding the properties of emulsions. Equation (2) may be useful for predicting the ultrasonic velocity in simple solutions by using volume-weighted averages of density and compressibility; however, for emulsions a more complex theory is needed to calculate their frequency-dependent values. We take a two-stage approach to this problem, first calculating the scattering from a single particle (6), and second, accounting for the interactions between scattered waves from neighboring droplets (7).

A plane wave of ultrasound passing across an emulsion droplet is partially scattered in different directions. Usually the particles (~ µm) are much smaller than the wavelength of the ultrasound (~mm) and analytical expressions for the effects of scattering can be developed using the relatively simple Rayleigh approximation to Mie theory. [Small particles at high frequencies may enter the intermediate wavelength region in which case a more complex theory is required (8).] For Rayleigh scattering, there are two principle modes:

- *Thermal losses* (Fig. 1a). The droplets and surrounding liquid expand and contract to differ-

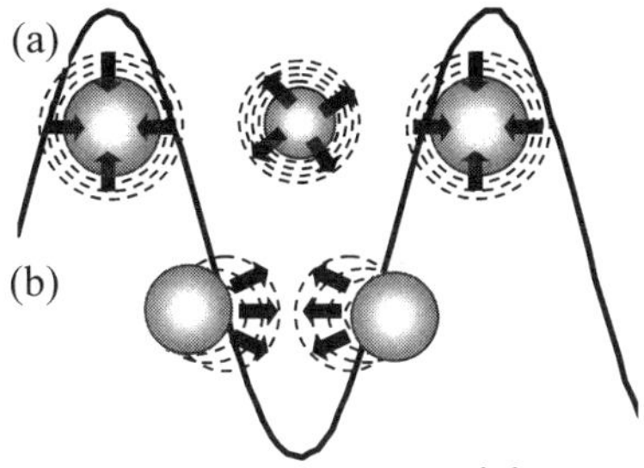


**Figure 1** Diagram (not to scale) illustrating modes of ultrasonic scattering from an emulsion droplet. In each case the pressure–temperature waves (dotted lines and arrows) scattered from the particle (shaded circle) are shown at three time intervals as the pressure changes (sinusoidal curve) due to the passing ultrasonic wave. (a) *Thermal (monopolar) scattering*: the droplet is compressed by the passing acoustic wave; pressure causes the fluid to heat up and thermal energy flows in and out of the droplet. (b) *Viscous (dipolar) scattering*: the droplet has a different inertia to the surrounding fluid and so moves in response to the changing pressure gradient; energy is lost due to relative movement of the continuous and dispersed phases.

ent extents in the presence of the pressure fluctuations associated with an ultrasonic wave. Consequently, the droplet pulsates and generates a monopolar pressure wave (thermal scattering losses) that propagates into the surrounding liquid. In addition, because of pressure–temperature coupling, there is a fluctuating temperature gradient between the droplet and surrounding liquid that causes heat to flow across the interface. This process is not usually completely reversible because the heat flow into the droplet is not equal to the heat flow outwards, consequently there is an associated energy loss (thermal-absorption losses).

- *Viscoinertial losses* (Fig. 1b). As an ultrasonic wave passes through an emulsion it causes the droplets to oscillate backwards and forwards because of the density difference between them and the surrounding liquid. The movement of the droplets leads to the generation of a dipolar pressure wave; the energy of the new wave is not detected and hence contributes to measured attenuation. In addition, the oscillation is damped because of the viscosity of the surrounding liquid, and so some of the ultrasonic energy is lost as heat.

It is possible to derive expressions for the effective density ($\rho_{eff}$) and compressibility ($\kappa_{eff}$) of an emulsion that take into account the above mechanisms by using multiple scattering theory (9):

$$\rho_{eff} = \rho_1\left(1 - \frac{3i\phi A_0}{k_1^3 r^3}\right) \quad (5)$$

$$\kappa_{eff} = \kappa_1\left(1 - \frac{9i\phi A_1}{k_1^3 r^3}\right) \quad (6)$$

where $\phi$ is the volume fraction, $r$ is the particle radius, and $\kappa_1$ and $\rho_1$ refer to the properties of the continuous phase; $A_0$ and $A_1$ are the scattering coefficients of the individual droplets that account for thermal and viscoinertial losses, respectively. The scattering coefficients are complex functions of the emulsion particle size distribution, temperature, and thermophysical properties of the component phases (i.e., ultrasonic velocity, attenuation coefficient, density, specific heat capacity, thermal conductivity, density, and coefficient of volume expansion). Analytical expressions for $A_0$ and $A_1$ are available in the long-wavelength limit (9). Equations (5) and (6) can then be substituted into Eq. (4) to calculate the complex wavenumber of the emulsion. The ultrasonic velocity of the emulsion is given by $\omega/Re(k)$ and the attenuation coefficient by $Im(K)$.

To reduce the number of unknowns in $A_0$ and $A_1$ it is necessary to input the physicochemical properties of the component phases. Some data are tabulated in the literature (10), but in many cases it will be more reliable for the experimenter to carry out measurements on pure aqueous and lipid phases. Using published data and scattering theory it is therefore possible to calculate the frequency-dependent velocity and attenuation of ultrasound in an emulsion as a function of the size and concentration of the particles present. This firm analytical base is an important reason for the success of ultrasound as a characterization tool for food emulsions.

## C. Ultrasonic Measurements

Ultrasonic measurement techniques applicable to food emulsions are reviewed extensively elsewhere (11–15) and only a brief introduction is presented here. Most ultrasonic measurement systems require an electrical signal generator that is used to excite an ultrasonic transducer to produce an acoustic wave which, after passing through the emulsion, is detected by a second transducer (or the first after a reflection). Finally, a display system, usually a computer or oscilloscope, is used to record and store the electrical signals. These components are arranged in two main groups of methods.

### 1. Pulse Methods

A pulse ultrasound is measured after transmission through a known pathlength of emulsion. The two major variations of this device are: *pulse–echo* – the sound is reflected from a reflector plate and detected at the original transducer (Fig. 2a), and *through transmission* – the sound from one transducer is detected by a second (Fig. 2b). If required, the frequency dependence can be calculated by using pulses that contain a number of cycles at a single frequency at single-frequency sound (i.e., tone burst). Alternatively, an electrical spike can be used to excite a broad-band transducer, which generates a narrow pulse of sound containing a range of frequency components. Using a fast Fourier transformation to compare the frequency content of the signal before and after transmission through the material, it is possible to measure a region of the spectrum around the center frequency of the transducer from a single pulse (11).

### 2. Resonator Methods

There are two groups of ultrasonic resonators, fixed pathlength and fixed wavelength. In a fixed pathlength resonator, a continuous wave containing a single ultrasonic frequency is transmitted across the measurement cell. The frequency is slowly increased and the change in amplitude of the resulting signal is measured. When the cell pathlength is an integer number of whole wavelengths, constructive interference occurs and there is a maximum in the detected energy (14,15). The shape and position of these resonance peaks can be used to calculate the velocity and attenuation of the liquid in the cell to very high precision. In a variable-pathlength resonator, a continuous wave containing a single ultrasonic frequency is again transmitted across the measurement cell, and the amplitude of the resulting signal is measured. However, in this case, the change in amplitude is measured as the pathlength of the cell is varied while keeping the frequency constant. The velocity and attenuation of the sample are determined by analyzing the shape and position of the resonance peaks (16).

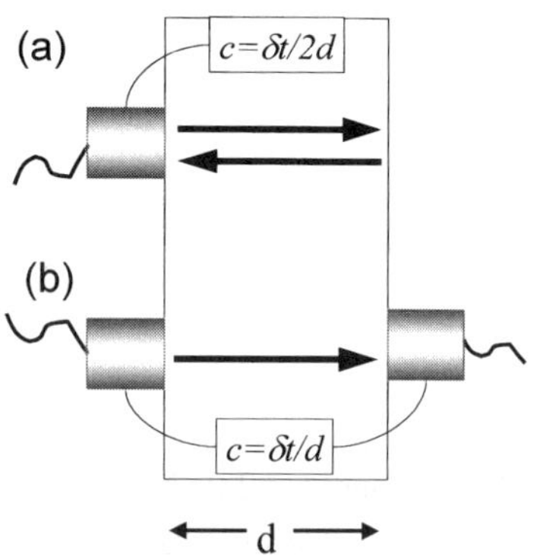


**Figure 2** Pulsed measurement techniques. (a) *Pulse–echo measurement*: the time taken and energy lost for sound to travel through the emulsion, reflect at a boundary, and return to the transducer is used to calculate velocity and attenuation, respectively. (b) *Through-transmission measurement*: a second transducer is used to detect the sound pulse after travelling a known distance through the emulsion.

Whichever method is selected, certain practical considerations are essential to making good measurements in food emulsions. The ultrasonic properties of the components in food emulsions are particularly sensitive to temperature and therefore it is usually important to control the measurement temperature carefully (i.e., $\pm 0.2\,\mathrm{m\,s^{-1}}$ or better). At approximately 18°C, $c_{oil} = c_{water}$ and so the ultrasonic technique becomes relatively insensitive to oil concentration (10). At higher and lower temperatures, ultrasonic measurements become increasingly independent on concentration because $|c_{oil}-c_{water}|$ increases. It may, therefore, be possible to enhance the sensitivity of an ultrasonic analysis by carefully selecting the temperature at which the measurements are carried out. The attenuation coefficient is typically less dependent on temperature than is the ultrasonic velocity and may be a more reliable parameter to measure in situations when thermal control is poor (e.g., on-line measurements).

Probably the single most common reason for poor quality ultrasonic measurements is the presence of small air cells trapped in viscous liquids. The large impedance mismatch between air cells and the surrounding emulsion leads to extensive scattering of ultrasound that can obscure the effects of the emulsion itself. In many cases, air cells can be eliminated by judicial use of gentle centrifugation or degassing before measurement.

## II. APPLICATIONS

Ultrasound, like any experimental technique, should not be considered as a panacea for problems of emulsion characterization. It is almost always better to use a battery of techniques to extract complementary information about the system under investigation. Having said that, ultrasound is perhaps uniquely sensitive to the important properties of a food emulsion (17).

## A. Crystallization

The large difference in the physical properties of liquids and solids, most importantly $c_{solid} \gg c_{liquid}$, means ultrasound is very sensitive to the melting and crystallization of a dispersed phase. As an example, the effect of heating and cooling on the ultrasonic velocity and attenuation of an n-hexadecane-in-water emulsion is shown in Fig. 3.

When an emulsion containing solid fat droplets is heated, there is an abrupt decrease in measured velocity at approximately the melting point of the bulk fat. However, when the liquid oil emulsion is cooled, the dispersed phase shows a large degree of supercooling and typically the velocity does not return to the starting solid-fat line until several degrees below its thermodynamic freezing point. Emulsion droplets supercool to such a large extent because each droplet is effectively pure oil containing no heterogenous nucleation sites. The fraction of the fat present as solid at any temperature can then be calculated as:

$$\text{fractional fat content} = \frac{\dfrac{1}{c^2} - \dfrac{1}{c_l^2}}{\dfrac{1}{c_s^2} - \dfrac{1}{c_l^2}} \tag{7}$$

where $c$ is the measured ultrasonic velocity, and $c_s$ and $c_l$ are, respectively, the velocities in solid fat and liquid

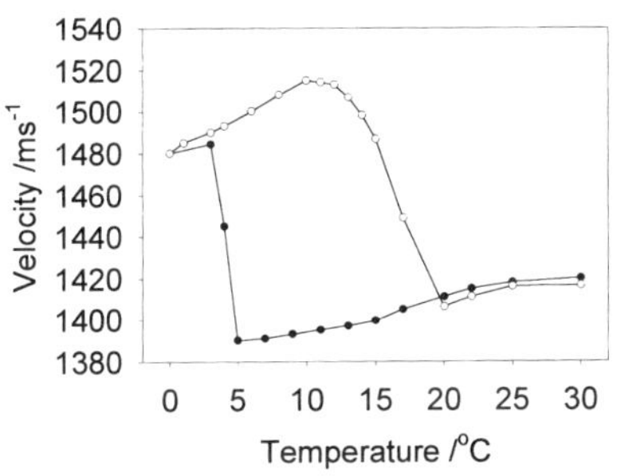


**Figure 3** Effect of heating (open points)–cooling (filled points) cycle on ultrasonic velocity in a 20% hexadecane-in-water emulsion (adapted from Ref. 23). The speed of sound in the emulsion decreases with temperature and there is an abrupt change corresponding to the phase transition in the droplet oil. Supercooling of the liquid oil is responsible for the hysteresis loop observed.

oil extrapolated to that temperature. This equation was developed from Eq. (1) by assuming solid fats and liquid oils have similar densities and behave ideally as a mixture (18). Ultrasonic measurements of solid fat in emulsions are more precise than the more commonly used pulsed NMR method for low oil contents (19).

A similar hysteresis loop occurs in the attenuation coefficient of an emulsion during the melting and freezing of the droplets. However, in the experiment reported in Fig. 3 there was also a large attenuation peak and extensive velocity dispersion at the melting point of the dispersed phase (not shown). This occurs because pressure–temperature fluctuations associated with the acoustic wave perturb the solid–liquid equilibrium and some of the ultrasonic energy is lost as heat. The magnitude of the attenuation peak is dependent on the frequency of the ultrasound. At low frequencies, the temperature/pressure gradients associated with the ultrasonic wave are small and so the equilibrium is maintained and the attenuation is low. At high frequencies, the changes in temperature and pressure occur so rapidly that there is no time for the system to respond and the ultrasonic absorption is also low. However, at intermediate frequencies, the temperature/pressure gradients are sufficiently large and persist for a sufficiently long time that an appreciable amount of material is able to undergo an ultrasonically induced phase transition. The energy required to crystallize and melt rapidly a proportion of the fat leads to significant absorption of the ultrasonic wave observed. The excess of absorption (and velocity dispersion) is related to the kinetics of the phase transition (20) and could be exploited to probe the molecular dynamics of the system. [The excess of attenuation was less during solidification because the phase transition was not reversible: once the droplets crystallized the ultrasonic wave was not capable of causing them to melt again because of the high degree of supercooling (21).]

Real food oils are a complex mixture of chemicals and therefore show a much broader melting profile than that seen in Fig. 3 because of the mutual solubility of the oils and because of the range of melting points of the individual components (22). Nevertheless, ultrasonic methods have been successfully used to measure the supercooling of the emulsified triacyl glycerols (23), margarine and butter (24), and milk fat (18).

## B. Droplet Concentration

The effect of droplet concentration on the ultrasonic velocity and attenuation of a model food emulsion is shown in Fig. 4. The monotonic change in measured

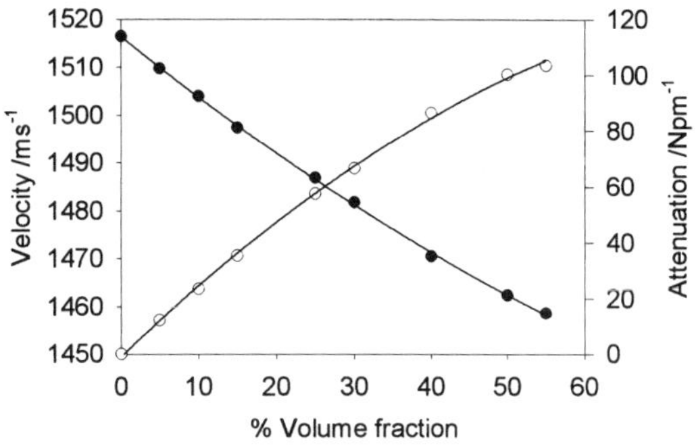


**Figure 4** Effect of dispersed phase-volume fraction on velocity (filled symbols) and attenuation (open symbols) of a corn oil-in-water emulsion. Quadratic regressions are shown with the data.

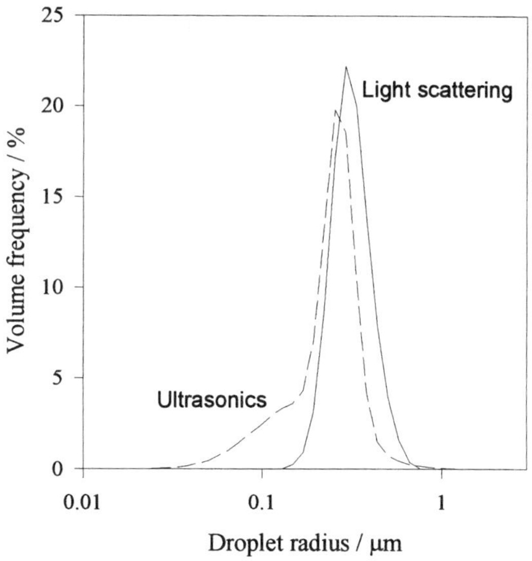


**Figure 5** Particle size for a 10% corn oil-in-water emulsion measured by ultrasonic attenuation spectroscopy (Malvern Ultrasizer, Malvern, UK) and a static light-scattering measurement (LA-900, Horiba Instruments, Irvine, CA) of a dilution ($\phi < 0.05\%$) of the same emulsion. Both techniques give comparable results, but ultrasound is reliable in more realistic, concentrated emulsions.

parameters with concentration can easily be used to measure the concentration of an unknown emulsion by appropriate calibration or use of a theroetical model. For relatively dilute emulsions ($\phi < 0.3$) the multiple-scattering theory set out above can adequately predict the effects of concentration on the ultrasonic properties of an emulsion. At higher concentrations the thermal waves evanescent from one droplet overlap with their neighbors and this simple theory becomes progressively unreliable. Concentrated emulsions can be more adequately described using a shell-model approach (25). Ultrasonic measurements have been used to measure the oil volume fraction of emulsified vegetable oil (26), milk (27), and salad cream (28).

## C. Particle Size

Ultrasonic spectroscopy is an increasingly popular method of measuring the droplet size in concentrated food emulsions and is one of its key unique applications. In this case it is essential to develop a theory to give good predictions of the ultrasonic properties of an emulsion as a function of particle size over the range of conditions used. The size parameters in the model can then be iteratively adjusted to give the best fit to an experimentally measured velocity/attenuation spectrum of an unknown emulsion and hence the size distribution. Figure 5 compares the particle size of a 10% corn oil-in-water emulsion measured by ultrasonic attentuation spectroscopy with the results of a light-scattering study on a dilution of the same emulsion. Ultrasonic spectroscopy is the only practical method for measuring the size distribution of concentrated dispersions. The ultrasonic technique has been used to determine particle size distributions in a number of food products, including emulsified vegetable oils (29), salad cream (28), and milk fat globules (27) to good agreement with other methods.

Faster and more robust size determinations may be achieved by reducing the number of unknown parameters in the iterative model-fitting routine. It is often desirable to make some simplifying assumption about the shape of the particle size distribution (often log normal), thus reducing the full distribution to one or two variables. Another unknown can be eliminated by independently measuring the dispersed phase-volume fraction.

## D. Surface Charge

Surface charge is important for the electrostatic stabilization of emulsions (1) and in certain cases protecting the oil from oxidization catalysts (30). A dynamic accumulation of oppositely charged ions surrounds a charged droplet surface. When the droplet oscillates in an ultrasonic wave, the charge distribution is perturbed. If the droplet is moving slowly (low frequency, large particle inertia) the distribution is never out of phase with the droplet, and when the droplet is moving quickly the ions never have time to reorient themselves (high frequency, small particle inertia). At intermediate frequencies, the relative movement of oppositely charged species (ionic friction) converts some of the ultrasonic energy into a measurable alternating current. The voltage, phase, and frequency dependence of the excess attentuation/current is sensitive to the size, concentration, and composition of the particles as well as their surface charge.

Instrumentation can be based on this principle by either (1) measuring the current generated by an oscillating charged droplet in the presence of an ultrasonic field; or (2) measuring the ultrasonic wave generated by an oscillating charged droplet in the presence of an alternating electrical field. The latter method is believed to be the most practical as it is directly proportional to the electrophoretic mobility of the particles and easier to measure in high-conductivity fluids. Electroacoustic techniques have been used to measure the surface charge in concentrated model suspensions (31), casein micelles (32), and flocculated and polydisperse colloids (33).

Surface charge is otherwise determined electrokinetically by measuring the velocity that a particle moves in an electrical field by light scattering or microscopy, but these methods often suffer from the need to dilute the emulsion. More problematically, some closely bound ions move with the particle, and the charge calculated is that at the plane of shear ($\zeta$-potential) between the moving particle and the surrounding continuous phase. Reducing the complex and dynamic ionic atmosphere to a single variable is an experimental necessity but requires a sacrifice of much important information. A full analysis of the frequency dependence of ultrasonic losses is, in principle, capable of revealing far more detail.

## E. Flocculation

Many of the most important quality attributes of food emulsions (e.g., creaming stability, rheology, and appearance) are strongly influenced by the degree of association of the droplets. Flocculation is desirable in the formation of some gelled and viscous food products, but undesirable in other products as it reduces their shelf-life. Consequently, it is important to have analytical techniques to be able to measure the degree of droplet flocculation in food emulsions. If it is possible to identify flocculation before changes in bulk properties are detected (e.g., formation of a cream layer, oiling off, or thickening) it would be possible to reject a defective product without the need for lengthy storage trials.

A number of experimental studies have shown that droplet flocculation causes an alteration in the ultrasonic properties of emulsions (34–36). Ultrasonic attenuation spectroscopy has been used to study droplet aggregation in protein-stabilized emulsions in which flocculation was induced by decreasing the electrostatic repulsion between droplets (37), or by adding a nonabsorbing biopolymer to the continuous phase (38). These studies have shown that ultrasound is sensitive to the spatial distribution of the droplets within an emulsion. Recently, a theory has been developed to describe the ultrasonic properties of flocculated emulsions (39). This theory assumes that a flocculated emulsion can be treated as a two-phase system, which consists of spherical "particles" (the flocs) dispersed in a continuous phase. The flocs are treated as an "effective medium" whose properties depend on the size, concentration, and packing of the droplets within them. Calculation of the ultrasonic properties of a flocculated emulsion involves two stages: (1) determination of the thermophysical and ultrasonic properties of the effective medium within the flocs; and (2) determination of the ultrasonic properties of a suspension of these flocs dispersed in a continuous phase, using ultrasonic scattering theory. This theory has been shown to give good agreement with experimental measurements of flocculated oil-in-water emulsions (40). Sample theoretical predictions of the attentuation spectra of flocculated emulsions are presented in Fig. 6. These were calculated by assuming varying proportions of the droplets (diameter 1 μm) were present in flocs (diameter 10 μm).

The attenuation coefficient in the flocculated emulsion is lower at low frequencies and higher at high frequencies than that of the nonflocculated emulsions. The decrease in attenuation at low frequencies on flocculation as a result of the thermal overlap effects mentioned earlier, whereas the increase at high frequencies results from increased scattering of ultrasound by the flocs. The same ultrasonic spectroscopy technique has been used to study the disruption of flocs in a shear field (38). As the emulsions are exposed to higher shear rates the flocs become disrupted and their attenuation spectra become closer to that of nonflocculated droplets.

## F. Creaming

The separation of oil-in-water emulsions under gravity leads to the formation of a droplet-rich "cream layer" at the top and a droplet-depleted "serum layer" at the

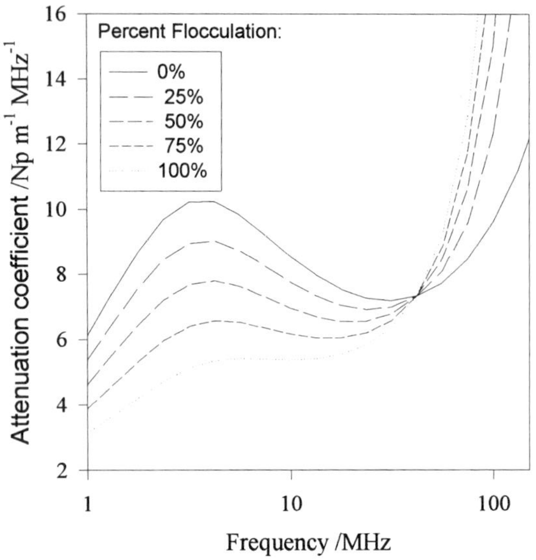


**Figure 6** Theoretical prediction of attenuation of a fine (1 μm diameter particles) corn oil-in-water emulsion with varying proportions of the droplet present in larger (10 μm diameter) flocs.

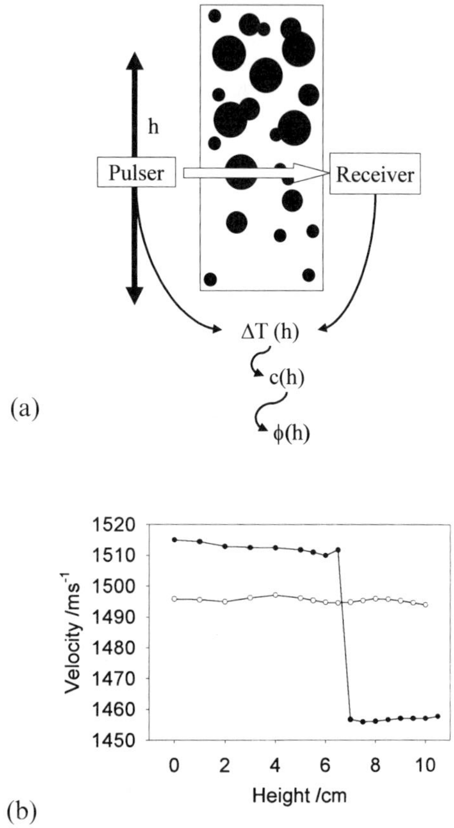


**Figure 7** Creaming of emulsions. (a) Diagram (not to scale) illustrating the operation of an ultrasonic method of detecting creaming in emulsions. The principle of operation is similar to the through-transmission measurement method (cf. Fig. 2b), but the transducers are mounted on stepper motors to make measurements relative to position as a function of height, $h$. Ultrasonic time of flight, $\Delta T(h)$, is converted in to speed of sound, $c(h)$, and hence to dispersed phase-volume fraction, $\phi(h)$. (b) Creaming profiles for corn oil-in-water emulsions containing either 0% (filled symbols) or 0.02% (open symbols) xanthan gum. Ultrasonic velocity was measured as a function of height by a pulse–echo method. Creaming induced in the gum-containing sample by a depletion interaction is clearly seen as a discontinuity in the measured velocity corresponding to the boundary between an oil-rich cream layer and an oil-poor serum layer.

bottom. Creaming is undesirable in most food emulsions because it frequently leads to a decrease in the acceptability of the product. Ultrasonic velocity (or attenuation) can be used to measure the concentration of oil in an emulsion (see above) as a function of height by using a simple one-dimensional imaging apparatus (Fig. 7a). A stepper motor moves an ultrasonic transducer along the vertical axis of the emulsion, and a series of ultrasonic measurements are made at different heights and times. Alternatively, a two-dimensional image can be generated by moving the transducer across the face of the sample (41), allowing several emulsions to be investigated simultaneously. The ultrasonic measurements can then be converted into oil concentrations using a calibration curve or ultrasonic scattering theory. A typical creaming profile for emulsified corn oil is shown as Fig. 7b.

While it is possible to control the position of the transducer to within 1 mm, the face width of the transducer is larger (frequently over 1 cm). The signal recorded is therefore an average of the oil content over that distance, and high-resolution measurements are frequently impossible. This situation is compounded when the transducer is positioned at the cream–serum transition where there may be a splitting of the returning detected signal.

Using a theoretical model of creaming behavior in concentrated emulsions, Pinfield et al. (42) predicted a segregation of different sized particles within the cream layer, leading to an accumulation of small particles at the top. The effect of particle size on ultrasonic velocity would, therefore, give misleading results in concentration measurements if appropriate scattering theory were not used (43–45). If an ultrasonic spectrum rather than a single frequency were recorded at each point it would, in principle, be possible to measure size and concentration simultaneously as a function of time and position. However, the volume fractions in cream layers are large, often approaching close packing, and scattering theory predictions have so far proved inadequate for the deconvolution of such data.

The precise time–concentration–position data generated ultrasonically are powerful tools for measuring creaming and distinguishing between different types of creaming. For example, xanthan gum can cause an emulsion to cream due to depletion flocculation (34), but the type of cream layer formed depends on other ingredients present. Ultrasonic imaging of the process was also able to detect an influence of added salt on the structure of the cream layer induced in the emulsion by xanthan gum (46). Creaming measurements have also been used to investigate the effect of nonadsorbed caseinate (47) on food emulsion stability.

Creaming measurements were used in an indirect approach to particle sizing taken by workers at the Institute of Food Research, Norwich (48). Droplets were assumed to cream according to Stokes' law retarded to account for the finite volume fraction. The ultrasonically measured change in concentration profiles with time was then used in the Stokes equation to calculate particle size distributions, which were in good agreement with light-scattering data. This approach was extended to measure the effective floc size and percentage flocculation in an aggregated emulsion.

## III. CONCLUSIONS

The colloidal properties of emulsions are responsible for the quality of many foods. Ultrasound is sensitive to most of the properties of interest and can be used as both a research and a process-control tool by food scientists. As a research tool, ultrasonic measurements are particularly powerful as they can be used to generate information not readily available by other methods – importantly, physical state, particle size, concentration, and flocculation in concentrated and optically opaque emulsions. In a process environment, ultrasonic measurements can be effected noninvasively in process lines and are therefore compatible with the stringent hygiene and cleaning requirements of food production.

In the past, ultrasound has not been widely used in studies of food emulsions. This is probably because the scientist would have to build his/her own measurement cell, specify and assemble the required transducers and electronics, transfer the data to a computer for analysis, and implement the quite complex scattering theory required for data analysis. These start-up requirements have meant that scientists more interested in ultrasonics than food emulsions have performed much of the research. Currently, several companies are marketing commercial devices to make simple single-frequency measurements, record precise spectra, and even determine the particle size and concentration. As ultrasonic measurements become accessible to more food scientists the number of practical applications and theoretical insights will surely increase.

## REFERENCES


1. E Dickinson. An Introduction to Food Colloids. Oxford: Oxford University Press, 1992.
2. E Dickinson, DJ McClements. Advances in Food Colloids. Glasgow: Blackie Academic and Professional, 1995.
3. P Walstra. In: OR Fennema, ed. Food Chemistry. New York: Marcel Dekker, 1996, pp 95–156.
4. TJ Mason. In: MJW Povey, TJ Mason, eds. Ultrasound in Food Processing. London: Blackie Academic and Professional, 1998, pp 105–126.
5. TG Leighton. In: MJW Povey, TJ Mason, eds. Ultrasound in Food Processing. London: Blackie Academic and Professional, 1998, pp 151–182.
6. JA Allegra, SA Hawley. J Acoust Soc Am 51: 1545–1564, 1971.
7. PC Waterman, R. Truel. J Math Phys 2: 512–537, 1961.
8. DJ McClements. Langmuir 12: 3454–3461, 1996.
9. DJ McClements, JN Coupland. Colloids Surfaces A 117: 161–170, 1996.
10. JN Coupland, DJ McClements. J Am Oil Chem Soc 74: 1559–1564, 1997.
11. DJ McClements, P Fairley. Ultrasonics 29: 58–62, 1991.
12. EP Papadakis. In: RN Thurston, AD Pierce, eds. Ultrasonic Measurement Methods. San Diego, CA: Academic Press, 1990, pp 81–106.
13. EP Papadakis. In: RN Thurston, AD Pierce, eds. Ultrasonic Measurement Methods. San Diego, CA: Academic Press, 1990, pp 107–155.
14. AP Sarvazyan. Ultrasonics 20: 151–154, 1982.
15. AP Sarvazyan, TV Chalikian. Ultrasonics 29: 119–124, 1991.
16. J Blitz. Ultrasonics: Methods and Applications. London: Butterworths, 1971.
17. DJ McClements. Food Emulsions: Principles, Practice, and Techniques. Boca Raton, FL: CRC Press, 1999.
18. CA Miles, GAJ Fursey, RCD Jones. J Sci Food Agric 36: 218–228, 1985.

19. DJ McClements, MJW Povey. Int J Food Sci Technol 23: 159–170, 1988.
20. AJ Matheson. Molecular Acoustics. London: John Wiley, 1971.
21. DJ McClements, MJW Povey, E Dickinson. Ultrasonics 31: 433–437, 1993.
22. PJ Lawler, PS Dimick. In: CC Akoh, DB Min, eds. Food Lipids: Chemistry, Nutrition, and Biotechnology. New York: Marcel Dekker, pp 229–250, 1998.
23. JN Coupland, E Dickinson, DJ McClements, MJW Povey, C de Rancourt de Mimmerand. In: E Dickinson, P Walstra, eds. Food Colloids and Polymers: Stability and Mechanical Properties. London: Royal Society of Chemistry, 1993.
24. DJ McClements. The Use of Ultrasonics for Characterizing Fats and Emulsions. PhD dissertation. Leeds: Leeds University, 1988.
25. N Herrmann. Application de Techniques Ultrasonores a L'Etude de Dispersions, PhD dissertation, Strasbourg: Universite Louis Pasteur, 1997.
26. DJ McClements, MJW Povey. J Phys D: Appl Phys 22: 38–47, 1989.
27. CA Miles, D Shore, KR Langley. Ultrasonics 28: 245–256, 1990.
28. DJ McClements, MJW Povey, M Jury, E Betsanis. Ultrasonics 28: 266–272, 1990.
29. JN Coupland, DJ McClements. J Food Eng, in press, 2000.
30. L Mei, DJ McClements, J Wu, EA Decker. Food Chem 61: 307–312, 1997.
31. RW O'Brien, BR Midmore, A Lamb, RJ Hunter. Faraday Discuss Chem Soc 90: 301–312, 1990.
32. T Wade, JK Beattie, WN Rowlands, MA Augustin. J Dairy Res 63: 387–404, 1996.
33. M James, RJ Hunter, RW O'Brien. Langmuir 8: 420–423, 1992.
34. DJ McClements. Colloids Surfaces A 90: 25–35, 1994.
35. D Hibberd, A Holmes, M Garrood, A Fillery-Travis, M Robbins, R Challis. J Colloid Int Sci 193: 77–87, 1997.
36. Y Hermar, N Hermann, P Lemarechal, R Hocquart, F Lequeux. J Physique II 7: 637–647, 1997.
37. K Demetriades, DJ McClements. Colloids Surfaces A 150:45–54, 1999.
38. R Chanamai, N Herrmann, DJ McClements. J Colloid Int Sci 204: 268–276, 1998.
39. DJ McClements, N Herrmann, Y Hemar. J Phys D 31: 2950–2955, 1998.
40. R Chanamai, N Herrmann, DJ McClements. J Phys D 31: 2956–2963.
41. TK Basaran, K Demetriades, DJ McClements. Colloids Surfaces A 136: 169–181, 1998.
42. VJ Pinfield, E Dickinson, MJW Povey. J Colloid Int Sci 186: 80–89, 1997.
43. VJ Pinfield, E Dickinson, MJW Povey. J Colloid Int Sci 166: 363–374, 1994.
44. VJ Pinfield, MJW Povey, E Dickinson. Ultrasonics 33: 243–251, 1995.
45. VJ Pinfield, MJW Povey, E Dickinson. Ultrasonics 34: 695–698, 1996.
46. E Dickinson, J Ma, MJW Povey. Food Hydrocolloids 8: 481–497, 1994.
47. E Dickinson, M Golding, MJW Povey. J Colloid Int Sci 185: 515–529, 1997.
48. MM Robins. In: MJW Povey, TJ Mason, eds. Ultrasound in Food Processing. London: Blackie Academic and Professional, 1998, pp 219–234.

# 11

# The Structure, Mechanics, and Rheology of Concentrated Emulsions and Fluid Foams

**H. M. Princen***
*Mobil Technology Company, Paulsboro, New Jersey*

## I. BACKGROUND

Whether enjoying the luxury of a bubble bath or enduring the drudgery of washing dishes, one is likely to be struck by the beauty and intricate structure of foams, froths, or "suds". Keen observers may even notice the unusual elastic and yield properties, not seen in the constituent aqueous and gaseous phases. Scientifically, the interest in, and the study of, foams have been truly multidisciplinary and have not been confined to chemists, engineers, and physicists. Foams have traditionally inspired mathematicians for their geometric properties and as equilibrium structures in which the surface area is minimized (1). Metallurgists (2) have realized the similarity between foams and polycrystalline metals, both in their structure and coarsening behavior (grain growth). Similarly, botanists and life scientists in general have noticed strong structural parallels between foams and living tissues (3).

Gas–liquid foams are abundant in nature and their technological applications are numerous. They are used to advantage in fire fighting, enhanced oil recovery, foods (e.g., whipped cream), cosmetics (e.g., shaving cream), and in many other ways. The "world of foams" may be considerably expanded by the realization that concentrated liquid/liquid emulsions, although generally characterized by a much smaller mean size of the dispersed units, are structurally identical to gas/liquid foams, which is readily revealed under the microscope. Macroscopically they behave like viscoelastic gels, mayonnaise being a good example. Such emulsions have been variously referred to as high-internal-phase-ratio emulsions (HIPREs), biliquid foams, "aphrons", or, simply, highly concentrated emulsions. Although they lack the compressibility of gas/liquid foams, they behave similarly in all other respects. Detailed study of such emulsion systems started relatively recently and may perhaps be traced to the attempts of Lissant (4–6) and Beerbower and coworkers (7–10) to design safer aviation and rocket fuels, in which fuel droplets are tightly packed inside a continuous aqueous phase. Reverse, i.e., concentrated water-in-oil systems can be readily prepared as well. They find application in the high-explosives area, but have particular appeal in the foods and cosmetics industries. What entrepreneur's mouth would not water at the prospect of being able to sell a product that is at least 90% water and yet is luxuriously rich and creamy? Lissant in particular patented numerous potential applications in these areas (e.g., 11). In yet other applications, the oil phase, either external or internal, can consist of a polymerizable monomer. Subsequent polymerization by heat or radiation can lead to interesting polymers or structurally unique materials (e.g., 6, 12–16).

**Current affiliation*: Consultant, Flemington, New Jersey.

Because of all these scientific and technological aspects, a thorough understanding of foams and concentrated emulsions is highly desirable. In response to this need, there has lately been a clear upsurge in interest, again from a variety of disciplines, and considerable progress has been and is being made. Several comprehensive textbooks on emulsions and foams have recently been published (17–20). We believe that the overlap with this review is minimal.

## II. INTRODUCTION

In general, when a fluid phase (liquid or gas) is dispersed in an immiscible liquid to form drops or bubbles, there is a tendency for the phases to separate again to reduce the augmented surface free energy. With pure phases, this proceeds by rapid coalescence of approaching dispersed entities, as there is no barrier against rupture of the intervening liquid film. Stability or, more correctly, metastability, can be conferred by adsorption of surfactants, polymers, or finely divided solid particles at the interface. By this expedient, coalescence can often be suppressed completely. However, this will not prevent ultimate phase separation, as there is another mechanism for reducing the surface area, namely, Ostwald ripening. By this mechanism, large bubbles or drops grow at the expense of small ones by dissolution and diffusion of the dispersed phase in response to the higher Laplace pressure in the latter ones. Because gases tend to have greater solubility and diffusivity in a given continuous liquid than do other liquids, this process is generally much more rapid in foams than in emulsions. Indeed, while most foams will not survive for more than a few hours – even in the absence of coalescence – it is relatively easy to prepare concentrated emulsions whose drop size distribution does not change perceptibly for months or years. They are kinetically or operationally (although not thermodynamically) stable. For this and many other reasons, emulsions may be better characterized, and their properties more reliably investigated experimentally, than is possible with foams. Thus, to learn about *foam* behavior through experiments, we recommend that one look at concentrated emulsions instead. In the same vein, we may use the terms "bubble" and "drop" interchangeably.

In this review, we shall only consider stable dispersions, in which coalescence has been totally suppressed. We further restrict ourselves to highly concentrated dispersions, in which the volume fraction of the dispersed phase, $\phi$, exceeds a critical value $\phi_0$ where the properties start to change drastically. This critical volume fraction corresponds to that of a system of close-packed spheres having the same drop volume distribution as the dispersion. The term "close packed" is somewhat ambiguous and the corresponding volume fraction is not always clearly defined and/or established. Although monodisperse spheres can in principle be packed to a maximum density of $\phi_0 = 0.7405$, this value is rarely achieved. In practice, one is more likely to achieve only *random* close packing, which is considerably less dense ($\phi_0 \approx 0.64$) due to the voids created by "arching". There is a persistent myth that the packing density of a *polydisperse* system is characterized by $\phi_0 > 0.7405$. It is true that the voids in a close-packed system of spheres can be filled sequentially with ever smaller spheres of very specific sizes until $\phi_0 \approx 1$. However, this would require a unique multimodal size distribution, as well as a unique spatial distribution, neither of which are likely to be ever encountered in practice. It is our experience with typical, unimodal polydisperse emulsions that the spherical droplets arrange themselves at a packing density that, though considerably larger than the 0.64 expected for the random close-packed monodisperse case, is close to, but slightly *smaller* than, 0.74. Although the actual value must depend somewhat on the details of the size distribution, we estimate that $0.70 < \phi_0 < 0.74$ in most practical cases (21, 22).

There are reasons why the *effective* value of $\phi$, including that of $\phi_0$, may deviate from the apparent value. If the thickness, $h$, of the stabilized film of continuous phase, separating the dispersed drops or bubbles, is not insignificant compared to the drop or bubble radius, $R$, then the effective volume of each drop must be augmented by that of a surrounding sheath of thickness $h/2$. This leads to a somewhat larger effective volume fraction, $\phi_e$, which is given (21) by

$$\phi_e = \phi\left[1 - \left(\frac{\phi}{\phi_0}\right)^{1/3} \frac{h}{2R}\right]^{-3} \approx \phi(1 + 3.15h/2R) \qquad (1)$$

The latter form is a good approximation for any $\phi > \phi_0$ and $h/R \ll 1$. In most foams, the effect is expected to be minimal, as the bubbles tend to be relatively large. For emulsions of small drop size, however, the effect may be considerable and the peculiar properties resulting from extreme crowding may commence at an apparent volume fraction that is considerably smaller than one would expect for zero film thickness. For example, in an emulsion with droplets of $2R = 1$ μm and $h = 50$ nm, the effective volume fraction already reaches a value of 0.74 at an apparent volume fraction of only about 0.64! The finite film thickness may, for example, result from electrostatic double-layer forces (23) or adsorbed polymers. In what follows, we shall assume zero film thickness, with the understanding that Eq. (1) is to be invoked whenever $h/R \neq 0$.

Another complication arises when strong attractive forces operate between the drops or bubbles. This may lead to a finite contact angle, $\theta$, between the intervening film (of reduced tension) and the adjacent bulk interfaces (21, 24–26). Under those conditions, droplets will spontaneously deform into truncated spheres upon contact and can thus pack to much higher densities. For monodisperse drops, the ideal close-packed density, consistent with minimization of the system's surface free energy, is given (21) by

$$\phi_0(\theta) = 0.7405\left(-\frac{5}{\cos^3\theta} + \frac{9}{\cos^2\theta} - 3\right) \tag{2}$$

which is valid up to $\theta = 30^\circ$, where $\phi_0 = 0.964$. For $\theta = 0$, we recover $\phi_0 = 0.7405$, while $\phi_0$ is expected to reach unity when $\theta$ exceeds 35.26° (21, 26). In the latter limit, all of the continuous phase (except that in the intervening films) should, in principle, be squeezed out spontaneously. In practice, however, one tends to find just the opposite, i.e., when $\theta$ is large, the droplets spontaneously flocculate into a rather open structure in which $\phi_0 < 0.7405$. The situation is similar to that of a flocculated solid dispersion whose sediment volume is generally greater than that of a stable dispersion. Apparently, the strong attractive forces prevent the droplets from sliding into their energetically most favorable positions, leaving large voids in the otherwise dense structure. Nevertheless, the structure may be irreversibly densified to approach the condition prescribed by Eq. (2) by centrifugation and subsequent relaxation (21, 25). Foams and emulsions in which $\theta \neq 0$ have only been studied occasionally and will rarely be touched upon in this review.

## III. STRUCTURAL ELEMENTS

As discussed above, the nature and properties of fluid/fluid dispersions start to change drastically when the volume fraction approaches or exceeds $\phi_0$. A certain rigidity sets in, because the drops or bubbles can no longer move freely past each other.

As the volume fraction is raised beyond $\phi_0$, the drops lose their sphericity and are increasingly deformed while remaining separated by thin stable films of continuous phase. At sufficiently high $\phi$, the drops become distinctly polyhedral, albeit with rounded edges and corners. At this stage the continuous phase is confined to two structural elements: *linear* Plateau borders with essentially constant cross-section over some finite length, and *tetrahedral* Plateau borders where four linear borders converge (Fig. 1a).

Each linear border is generally curvilinear and fills the gap between the rounded edges of three adjoining polyhedral drops. In cross-section, its sides are formed by three arcs, each pair of which meet tangentially to form the thin film separating the corresponding droplet

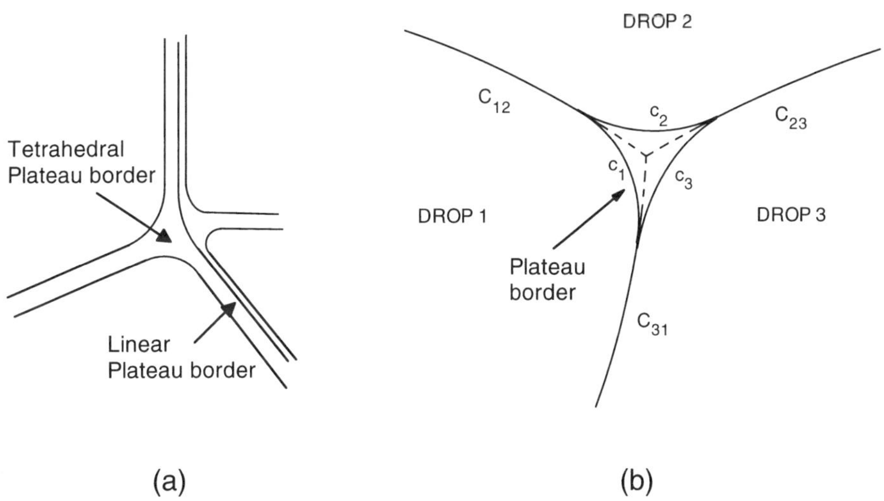


**Figure 1** (a) Four linear Plateau borders meeting in a tetrahedral Plateau border; (b) cross-section through a linear Plateau border and its three associated films and drops.

pair (Fig. 1b). The pressures in the drops are related to the mean curvatures of the intervening films through

$$\begin{aligned} p_1 - p_2 &= 2\sigma C_{12} \\ p_2 - p_3 &= 2\sigma C_{23} \\ p_3 - p_1 &= 2\sigma C_{31} \end{aligned} \tag{3}$$

where $\sigma$ is the interfacial tension between the continuous and dispersed phases, and the sign of each film curvature $C$ is taken as positive (negative) if the pressure in the drop indicated by the first index is the higher (lower) one. Adding Eqs (3) leads to the following relationship between the three mean film curvatures:

$$C_{12} + C_{23} + C_{31} = 0 \tag{4}$$

The pressure inside the linear Plateau border, $p_b$ is given by

$$p_b = p_1 - \sigma c_1 = p_2 - \sigma c_2 = p_3 - \sigma c_3 \tag{5}$$

where $c_1$, $c_2$, and $c_3$ are the curvatures of the border walls and are all counted as positive. Since all Plateau borders are connected, they are in hydrostatic equilibrium.

Normally, an ambient gaseous atmosphere of pressure $P$ surrounds the dispersion. Relative to this ambient pressure, $p_b$ is lower and given (27) by

$$p_b = P - \sigma_c |C_t| \tag{6}$$

where $|C_t|$ is the absolute value of the curvature of the free continuous-phase surface at the dispersion/atmosphere boundary (i.e., *between* the exposed bubbles), and $\sigma_c$ is the *surface tension* of the continuous phase [$\sigma_c = \sigma$ for foams, but $\sigma_c \neq \sigma$ for emulsions (27) unless the "ambient atmosphere" consists of the bulk dispersed liquid].

The excess pressures in the drops, relative to that in the interstitial continuous phase, $p_b$ are often referred to as their *capillary pressures*, $p_c$. For example,

$$(p_c)_1 = p_1 - p_b = \sigma c_1 \tag{7}$$

It is clear that, in general, the capillary pressure varies from drop to drop.

When Eqs (3) are combined with Eqs (5), the following relationships between the curvatures of the films and those of the Plateau border walls are obtained:

$$\begin{aligned} 2C_{12} &= c_1 - c_2 \\ 2C_{23} &= c_2 - c_3 \\ 2C_{31} &= c_3 - c_1 \end{aligned} \tag{8}$$

For each film to be stable, it must be able to develop an internal, repulsive *disjoining pressure* $\Pi_d$ to counteract the capillary suction acting at the film/Plateau border junction. At equilibrium, it can be readily shown from the above that

$$\begin{aligned} (\Pi_d)_{12} &= \sigma(c_1 + c_2)/2 \\ (\Pi_d)(\Pi_d)_{23} &= \sigma(c_2 + c_3)/2 \\ (\Pi_d)_{31} &= \sigma(c_3 + c_1)/2 \end{aligned} \tag{9}$$

Thus, the disjoining pressures in three confluent films are, in general, unequal. It turns out that the difference in the disjoining pressures in two of the films is defined by the curvature of the third film. For example, from Eqs (9) and (8):

$$(\Pi_d)_{31} - (\Pi_d)_{23} = \sigma(c_1 - c_2)/2 = \sigma C_{12} \tag{10}$$

The inequality of the disjoining pressures implies that the films may have slightly different equilibrium thicknesses and tensions. In extreme cases (28) this may lead to sensible deviations from Plateau's first law of foam structure, stated below.]

As the volume fraction approaches unity, the linear Plateau border shrinks into a line. In this "dry-foam" limit, mechanical equilibrium demands that the three films – of presumed equal tensions – meet pairwise at angles of 120° along this line (Plateau's first law of foam structure). However, even when the Plateau border is finite and the films do not really intersect, the principle may well hold when applied to the virtual line of intersection that is obtained when the films, while maintaining their curvatures, are extrapolated into the border (dashed lines in Fig. 1b). A rigorous proof has been published by Bolton and Weaire (29) for two-dimensional (2-D) foams, in which the Plateau borders are rectilinear. To our knowledge, no proof has yet been presented for the more general case of curvilinear borders in three-dimensional (3-D) space. In fact, since the Plateau border can be viewed as a line with a line tension (30), this broader statement of Plateau's first law may *not* strictly apply when the border has some finite longitudinal curvature.

A *tetrahedral* Plateau border is formed by the confluence of four linear Plateau borders (Fig. 1a). It fills the gap between the rounded corners of four adjoining polyhedral drops. The pressure in the tetrahedral border is, of course, equal to that in each of the outgoing linear borders, which sets the curvature of each of its four bounding walls. In the dry-foam limit ($\phi \to 1$), the tetrahedral border reduces to a point ("vertex" or "node"), where the four linear borders meet pairwise at the angle of $\cos^{-1}(-1/3) = 109.47°$ (Plateau's second law of foam structure). The principle probably remains valid for finite borders, when applied to the point where the four virtual lines of film-intersection (see above) meet upon extension into the tetrahedral border.

## IV. OVERALL STRUCTURE AND OSMOTIC PRESSURE

Having described the structural elements of foams approaching the dry-foam limit ($\phi \to 1$), it is still a daunting task to describe the structure and properties of the system as a whole. The task is even more difficult for systems in which $\phi_0$ is exceeded, but the polyhedral regime has not yet been reached. In this case, the drops have exceedingly complex shapes, and linear and tetrahedral Plateau borders, as defined above, are not present. Much can be learned about the qualitative behavior by considering *2-D model systems*, in which the drops do not start out as spheres but as parallel circular cylinders, and tetrahedral Plateau borders do not arise. We shall first consider the particularly simple monodisperse case, with a subsequent gradual increase in complexity.

[Lest the reader think that 2-D foams are just figments of the imagination, it must be pointed out that they can be generated – or at least closely approximated – by squeezing a 3-D foam between two narrowly spaced, wetted, transparent plates (2, 31–35). Structurally even closer realizations may be obtained in phase-coexistence regions of insoluble monolayers of surface-active molecules at the air–water interface (36), where the role of surface tension is taken over by the line tension at the phase boundaries.]

### A. Monodisperse, 2-D Systems

Such a system has been discussed in detail in Ref. 37. In the absence of gravity, the circular cylinders of radius $R$ arrange in hexagonal packing (Fig. 2a) at a volume fraction $\phi_0 = \pi/2\sqrt{3} = 0.9069$. In cross-section, each circular drop can be thought to be contained within a regular hexagon of side length $a_0 = 2R/\sqrt{3}$. As the volume fraction is increased, the drop is flattened against its six neighbors to form a hexagon of side length $a (< a_0)$ but with rounded corners described by circular arcs of radius $r$ (Fig. 2b). At constant drop volume, one finds

$$\frac{r}{R} = \left(\frac{\phi_0}{1-\phi_0}\right)^{1/2}\left(\frac{1-\phi}{\phi}\right)^{1/2} \tag{11}$$

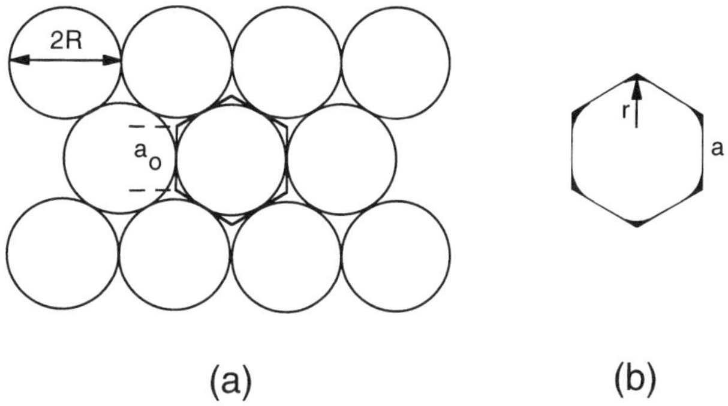


**Figure 2** (a) Uncompressed cylindrical drops in hexagonal close packing ($\phi = \phi_0 = 0.9069$); (b) compressed drop ($0.9069 < \phi < 1$).

The capillary pressure in each drop is given by $p_c = \sigma/r$ or, when scaled by the initial capillary pressure $(p_c)_0 = \sigma/R$, by

$$\tilde{p}_c = p_c/(p_c)_0 = R/r \tag{12}$$

In the above process, the surface area of each drop, per unit of length, increases from $S_0 = 2\pi R$ to $S = 6(a - 2r/\sqrt{3}) + 2\pi r$ which, at constant drop volume, can be shown to lead to

$$\frac{S}{S_0} = \frac{1}{(\phi_0\phi)^{1/2}}[1 - (1-\phi_0)^{1/2}(1-\phi)^{1/2}] \tag{13}$$

This function has been plotted in Fig. 3. In the limit of $\phi = 1$, the scaled surface area reaches a maximum that is given by

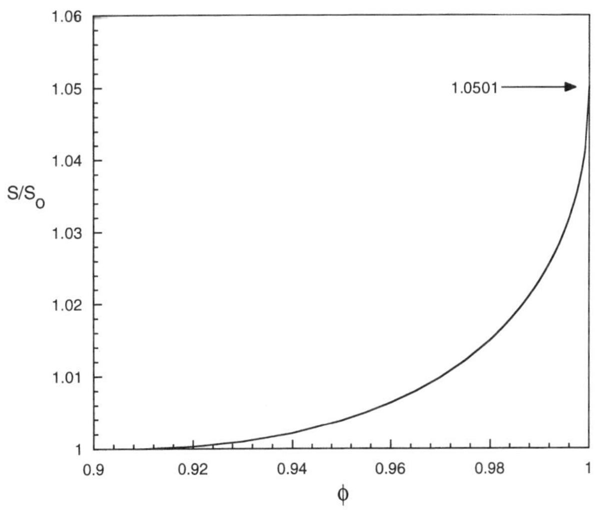


**Figure 3** Scaled surface area, $S/S_0$, for monodisperse 2-D drops as a function of volume fraction.

$$\frac{S_1}{S_0} = \frac{1}{\phi_0^{1/2}} = 1.0501 \tag{14}$$

The scaled surface area and its variation with $\phi$ are of crucial importance in the definition and evaluation of the *"osmotic pressure"*, $\Pi$, of a foam or emulsion. We introduced the concept in Ref. 37, where it was referred to as the "compressive pressure", $P$. It has turned out to be an extremely fruitful concept (22, 27, 38). The term "osmotic" was chosen, with some hesitation, because of the operational similarity with the more familiar usage in solutions. In foams and emulsions, the role of the solute molecules is played by the drops or bubbles; that of the solvent by the continuous phase, although it must be remembered that the nature of the interactions is entirely different. Thus, the osmotic pressure is defined as the pressure that needs to be applied to a semipermeable, freely movable membrane, separating a fluid/fluid dispersion from its continuous phase, to prevent the latter from entering the former and to reduce thereby the augmented surface free energy (Fig. 4). The membrane is permeable to all the components of the continuous phase but not to the drops or bubbles. As we wish to postpone discussion of compressibility effects in foams until latter, we assume that the total volume (and therefore the volume of the dispersed phase) is held constant.

As long as the membrane is located high up in the box in Fig. 4, the emulsion or foam may be characterized by $\phi < \phi_0$ and $\Pi = 0$. As the membrane moves down, a point is reached where $\phi = \phi_0$. Any further downward movement requires work against a finite pressure $\Pi$, reflecting the increase in the total surface area as the drops are deformed, i.e.,

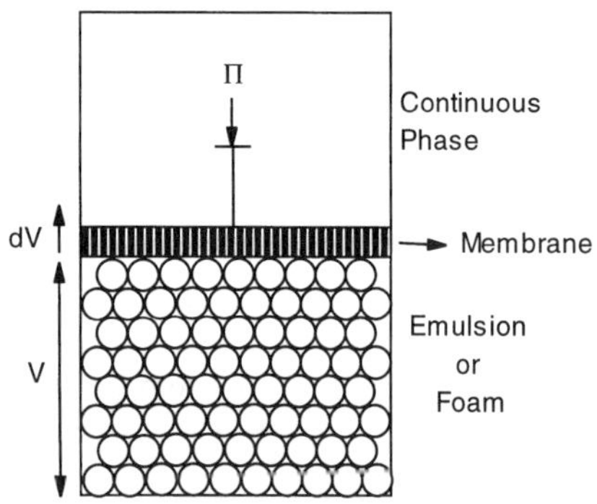


**Figure 4** Semipermeable membrane separating dispersion from continuous phase; pressure to prevent additional continuous phase from entering the dispersion is the "osmotic" pressure, $\Pi$. [From Ref. 38. Copyright (1986) American Chemical Society.]

$$-\Pi \mathrm{d}V = -\Pi \mathrm{d}V_2 = \sigma \mathrm{d}S \qquad (\text{constant } V_1) \tag{15}$$

where $V$ is the dispersion volume, $V_1$ is the volume of the dispersed phase, $V_2$ is the volume of the continuous phase in the dispersion, and $\sigma$ is assumed to be constant. As $V = V_1 + V_2$ and $\phi = V_1/V$, Eq. (15) leads to the completely general expression:

$$\Pi = \sigma\phi^2 \frac{\mathrm{d}(S/V_1)}{\mathrm{d}\phi} = \sigma\phi^2 \frac{S_0}{V_1}\frac{\mathrm{d}(S/S_0)}{\mathrm{d}\phi} \tag{16}$$

where $S/V_1$ is the surface area per unit volume of the dispersed phase. Alternatively, as shown in Ref. 27, $\Pi$ may be equated to the pressure difference between an ambient atmosphere and the continuous phase in the dispersion, or from Eq. (6):

$$\Pi = P - p_\mathrm{b} = \sigma_\mathrm{c}|C_t| \tag{17}$$

For yet a third useful way to express $\Pi$, see Refs 22, 27 and 38.

For the special case of a monodisperse, 2-D system:

$$\frac{S_0}{V_1} = \frac{2}{R} \tag{18}$$

which, when combined with Eqs (16) and (13), results in

$$\Pi = \frac{\sigma}{R}\left(\frac{\phi}{\phi_0}\right)^{1/2}\left[\left(\frac{1-\phi_0}{1-\phi}\right)^{1/2} - 1\right] \tag{19}$$

or, in reduced form:

$$\tilde{\Pi} \equiv \frac{\Pi}{\sigma/R} = \frac{\Pi}{(p_\mathrm{c})_0} = \left(\frac{\phi}{\phi_0}\right)^{1/2}\left[\left(\frac{1-\phi_0}{1-\phi}\right)^{1/2} - 1\right] \tag{20}$$

where $\phi_0 = 0.9069$. Figure 5 shows the dependence of $\tilde{\Pi}$ on $\phi$.

The suggestion has been made (D.R. Exerowa, personal communication, 1990), since withdrawn (20, 39), that $\Pi$ and $p_\mathrm{c}$ are really identical. It is clear from the above that this is not so. In fact, examination of Eqs (20), (11), and (12) shows that, at least for this simple model system:

$$\tilde{p}_\mathrm{c} - \tilde{\Pi} = \left(\frac{\phi}{\phi_0}\right)^{1/2} \tag{21}$$

At $\phi = \phi_0$ :

$$\tilde{p}_\mathrm{c} - \tilde{\Pi} = 1 \tag{22}$$

At the upper limit of $\phi = 1$:

$$\tilde{p}_\mathrm{c} - \tilde{\Pi} = \phi_0^{-1/2} = S_1/S_0 = 1.0501 \tag{23}$$

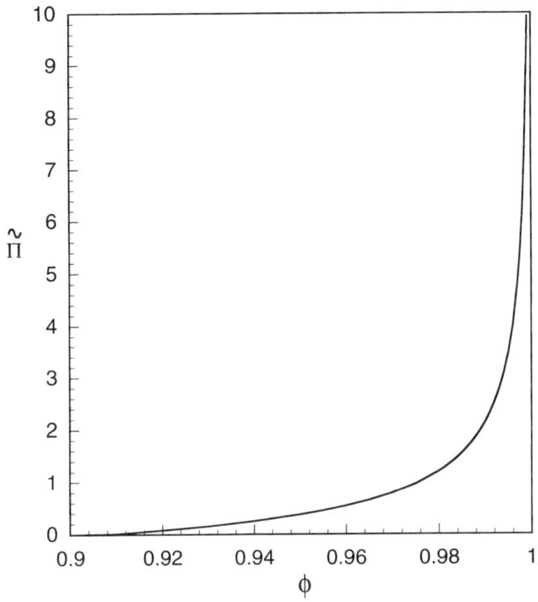


**Figure 5** Reduced osmotic pressure as a function of $\phi$ for perfectly ordered 2-D system.

Both $\tilde{p}_c$ and $\tilde{\Pi}$ tend to infinity in this limit, but the *relative* difference between them tends to zero. This is the regime of concern in much of the interesting work of Exerowa *et al.* (e.g., 40, 41), where the difference between the capillary and osmotic pressures may, therefore, indeed be safely ignored (39). However, this is not so in general and we shall demonstrate below that $\Pi$ is a much more useful and informative parameter than $p_c$.

Before leaving this topic, it should be mentioned that modifications of most of the above expressions have been derived to take account of finite film thickness, finite contact angle at the film/Plateau border junction, or both (37). Finally, it must be realized that a monodisperse, 2-D system does not necessarily pack in the highly ordered, hexagonal state depicted in Fig. 2. Herdtle *et al.* (personal communication, 1993) have constructed highly disordered, yet monodisperse, 2-D dry foams with periodic boundaries (Fig. 6), in which all films meet at angles of 120° and all film curvatures satisfy Eq. (4). These are equilibrium structures, whose surface energy, though at a *local* minimum, must be higher than that of the perfectly ordered hexagonal system. Because the bubble pressures are not the same, such a system is bound to coarsen, thereby reducing its total surface energy. In practice, disorder of this type may be imposed by the finiteness of any system with bounding walls. If the walls are wetted by the continuous phase, then the outer films must be directed normal to the walls, which is generally incompatible with a perfectly ordered internal structure. As we shall see, this complication arises in 3-D foams as well.

## B. Polydisperse, 2-D Systems

In the last decade or so, much progress has been made toward a more complete understanding of these disordered structures. Most work relies on the computer generation of disordered, polydisperse structures with periodic boundary conditions, in which the film angles and curvatures obey the rules set forth above. For a recent review, see Ref. 31. An example, taken from Ref. 42, is shown in Fig. 7. The structure contains many bubbles that are not hexagons, but it is readily proven that the *average* number of sides is still six (42). Simpler and very special types of polydispersity and disorder have been considered by Khan and Armstrong (43) and Kraynik *et al.* (44). In these cases, illustrated in Fig. 8, all bubbles are still hexagons and all films remain flat; the bubbles, therefore, do not coarsen with time. The first system (Fig. 8a) is simply bimodal and is obtained by increasing or decreasing the height of all bubbles in a given row. The second system (Fig. 8b) is much more disordered and can be generated from the monodisperse system by randomly increasing (or decreasing) each bubble area, as illustrated in Fig. 9, with the limitation that no vertices ever touch or cross over, lest Plateau's first law be violated and resulting (so-called T1) rearrangements lead to a much more complex structure. The total surface area is not affected by such transformations, so that, as in the monodisperse case:

$$S_1/S_0 = 1.0501 \tag{24}$$

This is not necessarily true for the more general structures such as that in Fig. 7. Unfortunately, although presumably available as a result of the numerical simulations, the value of $S_1/S_0$ and how it varies with the details of the size distribution, appears not to have been reported for these cases.

Starting from a dry-foam system as in Fig. 7, the volume fraction can be lowered by "decorating" each vertex with a Plateau border, whose wall curvatures obey the rules set forth above (29). As the volume fraction is lowered by increasing the size of the Plateau borders, a point is soon reached where adjacent Plateau borders "touch" and subsequently merge into single four-sided borders. Bolton and Weaire (45) have followed this process down to the volume fraction

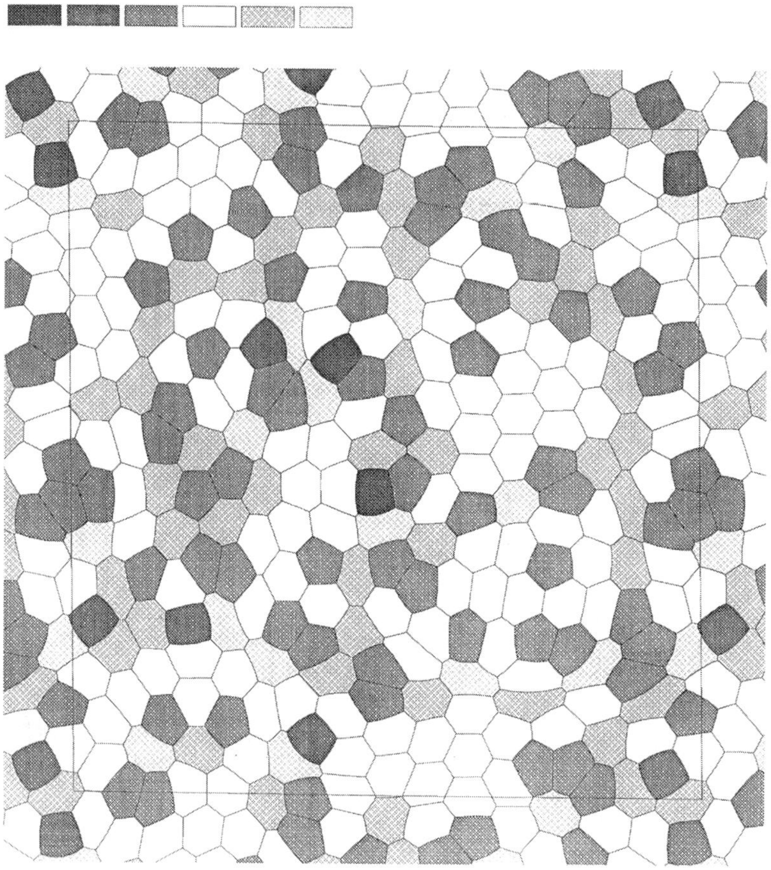

Figure 6 Disordered, monodisperse 2-D system ($\phi = 1$) with periodic boundaries; each shade corresponds to drops with a certain number of sides, e.g., the unshaded drops all have 6 sides. (Courtesy of T. Herdtle and A.M. Kraynik.)

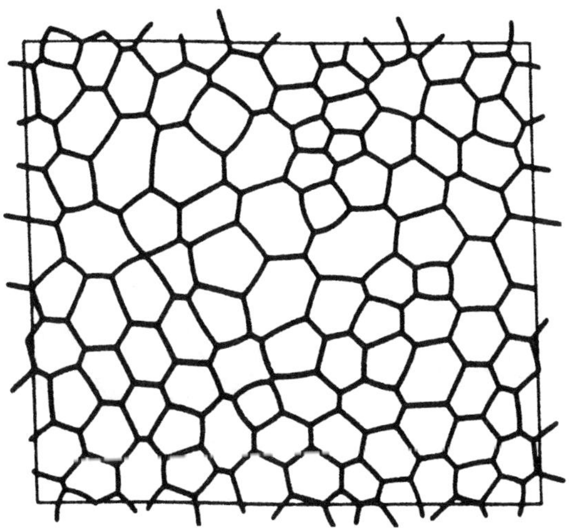

Figure 7 Computer-generated polydisperse 2D system ($\phi = 1$) with periodic boundaries. (From Ref. 42, with permission from Taylor & Francis Ltd.)

$\phi_c$, where all bubbles are spherical and structural rigidity is lost. This is perhaps the most satisfactory definition of $\phi_0$. Their finding suggests that, for that particular system, $\phi_c$ equaled 0.84 (not 0.9069), which happens to be close to the random packing density of (monodisperse) circular disks. Using similar computer simulations, Hutzler and Weaire (46) calculated the osmotic pressure and found it to obey Eq. (19) closely in the "drier" regime. It started to deviate at lower volume fraction and did not reach zero until $\phi$ dropped to about 0.82, which is close to the above rigidity loss transition.

## C. Monodisperse, 3-D Systems

Ideally, uniform spheres arrange in "hexagonal close" packing, which is face-centered cubic (fcc), at $\phi_0 = \pi\sqrt{2}/6 = 0.7405$. The role of the circumscribing hexagon in monodisperse 2-D systems is taken over by the rhombic dodecahedron (Fig. 10). As the volume

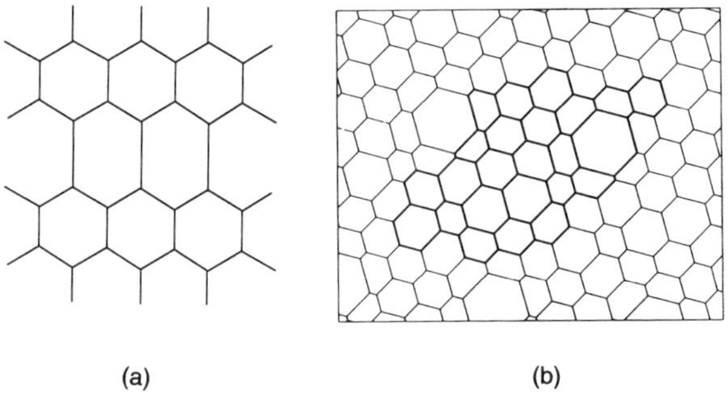


**Figure 8** (a) Simplest case of bimodal 2-D system; (b) more highly disordered, polydisperse hexagonal 2-D system ($\phi = 1$). The cluster of darkly outlined drops forms the repeating unit. (Courtesy of A.M. Kraynik. Similar structures appear in Ref. 44.)

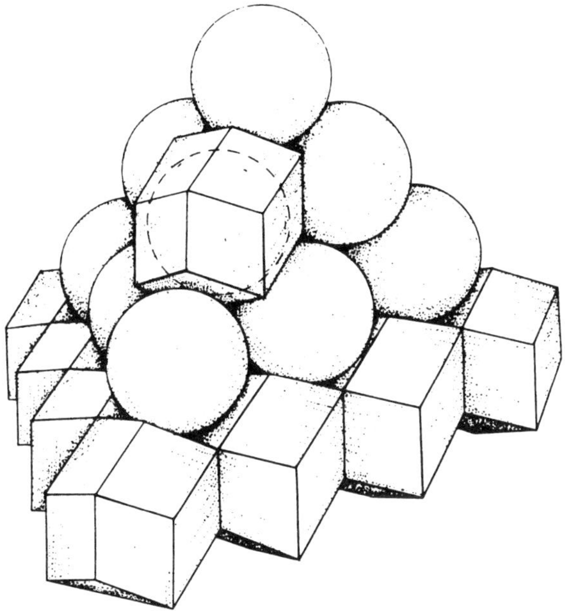

**Figure 10** Spheres in hexagonal close packing (fcc), each occuping a rhombic dodecahedron. (From Ref. 4, with permission from Academic Press.)

fraction is raised, each drop flattens against its 12 neighbors. This process has been described by Lissant (4, 5), who considered the drop to be transformed into a truncated sphere and each film to be circular, at least until it reaches the sides of the diamond faces (Fig. 11). This is incompatible with a zero contact angle at the film edge. Moreover, at constant drop volume, this model would imply *decreasing* capillary (and osmotic) pressure with increasing $\phi$, which is clearly inconsistent. In reality, the problem is much more complicated; the drop cannot remain spherical and the films must be noncircular. Using Brakke's now-famous "Surface Evolver" computer software (47), Kraynik and Reinelt (48), and Lacasse *et al.* (49) have correctly and accurately solved this problem for this and other structures (see below).

As suggested already by Lissant (4, 5), the packing is likely to change above some critical value of $\phi$. It is clear that, if the dodecahedral packing were to persist up to $\phi = 1$, Plateau's second law would be violated

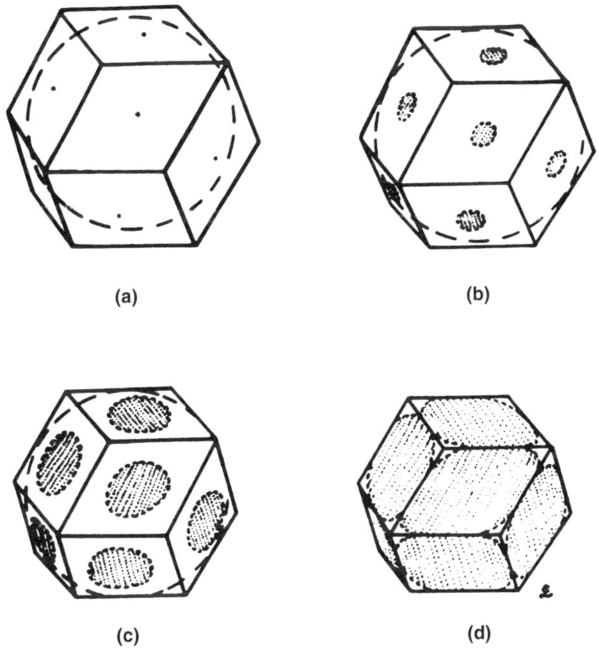


**Figure 11** Each drop flattens against its neighbors as the volume fraction increases; a stable thin film of continuous phase separates neighboring drops. (From Ref. 4, with permission from Academic Press.)

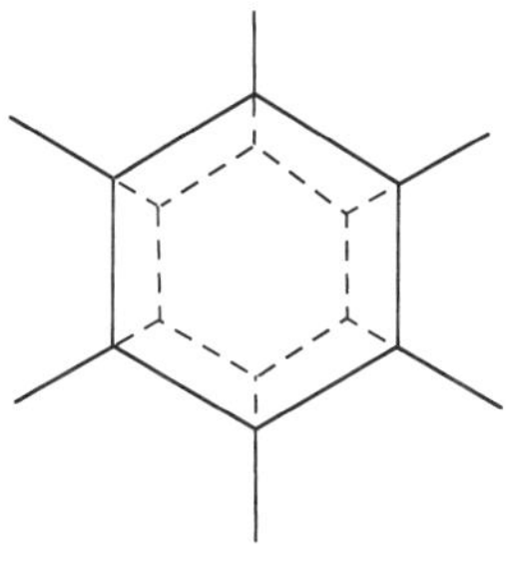

**Figure 9** Recipe for creating polydisperse hexagonal system from perfectly ordered 2-D system; total surface energy remains unchanged.

at six of the 14 corners of the polyhedron, since eight linear borders would converge there, rather than the mandatory four. Lissant proposed that the structure changes to a body-centered cubic (bcc) packing of planar tetrakaidecahedra (truncated octahedra; see Fig. 12a). However, such a structure satisfies neither of Plateau's laws. In this dry-foam regime, Kelvin's "minimal tetrakaidecahedron" (Fig. 12b), which is obtained by slight distortion of its planar counterpart, solves this problem and has long been considered as the most satisfactory candidate for the drop shape. It has six planar quadrilateral faces, eight nonplanar hexagonal faces of zero mean curvature, and 36 identical curved edges. In a space-filling ensemble of such polyhedra, Plateau's first and second laws are fully satisfied. Kelvin derived approximate expressions for the shape of the hexagons and the sides (50–52). Based on that model, Princen and Levinson (53) calculated the length of the sides, and the surface areas of the quadrilateral and hexagonal faces, relative to those of the parent planar tetrakaidecahedron of the same volume. They arrived at the following result for the increase in surface area as a spherical drop transforms into a Kelvin tetrakaidecahedron of the same volume:

$$S_1/S_0 = 1.0970 \qquad (25)$$

(This compares to values of 1.0990 for the planar tetrakaidecahedron; 1.1053 for the rhombic dodecahedron; and 1.0984 for the regular pentagonal dodecahedron. The latter – though often considered as a unit cell in foam modeling – is not really a viable candidate either, as it not only violates Plateau's laws but is also not space filling.)

More recently, Reinelt and Kraynik (54) have carried out more exact numerical calculations on the Kelvin cell, leading to the slightly higher value of

$$S_1/S_0 = 1.0972 \qquad (26)$$

Kelvin's polyhedron would indeed represent the ideal drop shape in the dry-foam limit by effecting, in Kelvin's own words, "a division of space with minimum partitional area," if he had added the proviso that this division is to be accomplished with identical cells. It has been proven by Weaire and Phelan (55) that at least one structure of even lower energy exists, if this restriction is lifted. The Weaire–Phelan structure (Fig. 13), whose surface area is about 0.34% lower than that of Kelvin's (i.e., $S_1/S_0 = 1.0936$), has repeating units that contain eight equal-volume cells: two identical pentagonal dodecahedra and six identical tetrakaidecahedra that each have 12 pentagonal and two hexagonal faces. The pressure in the dodecahedra is slightly higher than that in the tetrakaidecahedra. Perhaps surprisingly, neither the Kelvin nor the Weaire–Phelan structure is rarely, if ever, encountered in actual, monodisperse foams (3). The reason for this may lie in small deviations from monodispersity or, more likely, in the disturbing effects of the container walls, as alluded to already in connection with 2-D foams. Alternatively, as the continuous phase is removed from between the initially spherical drops in

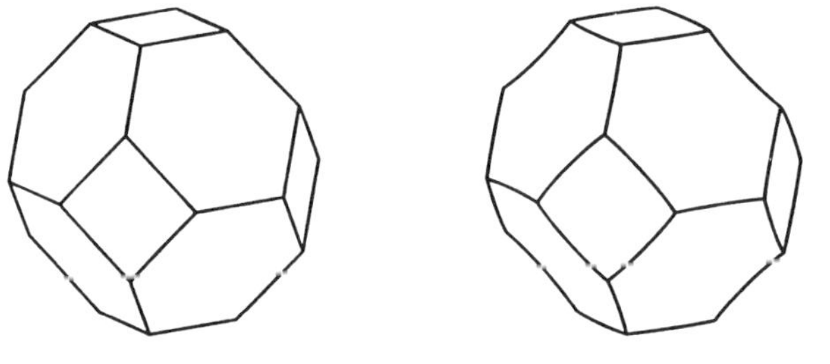

(a) (b)

**Figure 12** (a) Planar tetrakaidecahedron (or truncated octahedron); (b) Kelvin's minimal tetrakaidecahedron (bcc).

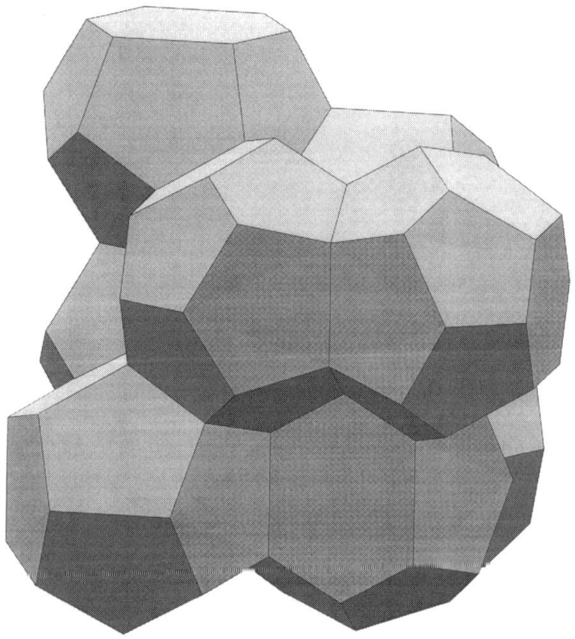

**Figure 13** Unit cell in Weaire–Phelan structure, containing two pentagonal dodecahedra and six tetrakaidecahedra, each having 12 pentagonal and two hexagonal faces. (Courtesy of A.M. Kraynik.)

fcc packing, slight irregularities in this drainage process may force the system to get trapped in a less-ordered structure that may be at a local surface area minimum but is separated from the lower-energy Kelvin and Weaire–Phelan structures by a significant barrier [cf. the difficulty one encounters in trying to build a 15-bubble cluster that has a Kelvin polyhedron at its center (56)].

Kraynik and Reinelt (48) and Lacasse *et al.* (49) have accurately computed the changes in surface area as a drop transforms from a sphere into a regular dodecahedron (fcc) or a Kelvin cell (bcc) with increasing volume fraction, while maintaining zero contact angle. Expressed in terms of $S/S_0$, the results are shown in Fig. 14. The Kelvin structure is internally unstable below $\phi \approx 0.87$. The results further indicate that the Kelvin cell becomes the more stable structure above $\phi \approx 0.93$. Also indicated is the limiting law for $\phi \to 1$ for the dodecahedron. In that regime, linear Plateau borders of constant cross-section run along the edges of the polyhedron. Their volumes and surface areas can be evaluated as a function of $\phi$, while the volumes and surface areas of the tetrahedral borders become negligible. For the rhombic dodecahedron (22) this leads to

$$\frac{S}{S_0} = 0.0686[1 - 1.892(1-\phi)^{1/2}]^3 + 1.0367(\phi \to 1)$$

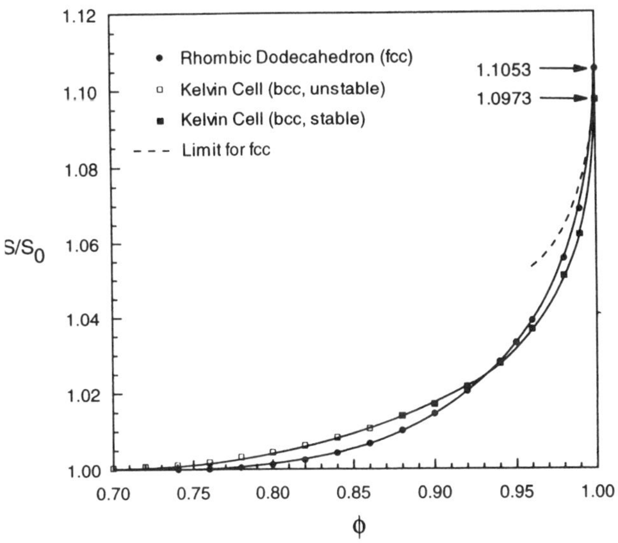


**Figure 14** Scaled surface areas as a function of volume fraction for the rhombic dodecahedral (fcc) and Kelvin structures (bcc). (From data kindly provided by A.M. Kraynik and D.A. Reinelt.)

Kraynik and Reinelt (48) also evaluated the all-important osmotic pressure $\Pi(\phi)$, which, for 3-D structures, is given by [cf. Eq. (16)]:

$$\Pi = \sigma\phi^2 \frac{S_0}{V_1}\frac{d(S/S_0)}{d\phi} = \frac{3\sigma\phi^2}{R}\frac{d(S/S_0)}{d\phi}$$

where $R$ is the radius of the initially spherical drops, or

$$\tilde{\Pi} \equiv \frac{\Pi}{\sigma R} = 3\phi^2 \frac{d(S/S_0)}{d\phi}. \tag{27}$$

For the dodecahedron, the appropriate limiting law for $\phi \to 1$ is given (22) by

$$\tilde{\Pi} = \frac{\Pi}{\sigma/R} = 0.5842\phi^{1/3}\frac{[1-1.892(1-\phi)^{1/2}]^2}{(1-\phi)^{1/2}}(\phi \to 1) \tag{28}$$

Figure 15 shows $\Pi(\phi)$ for the dodecahedron and Kelvin cell.

Detailed numerical calculations have been carried out by Bohlen *et al.* (57) for the transition of monodisperse spheres in simple *cubic* packing ($\phi_0 = 0.5236$) to cubes ($\phi = 1$), for both zero and finite contact angles. Unfortunately, although the results are interesting, this kind of packing is not realistic for foams and emulsions, and will not be discussed further.

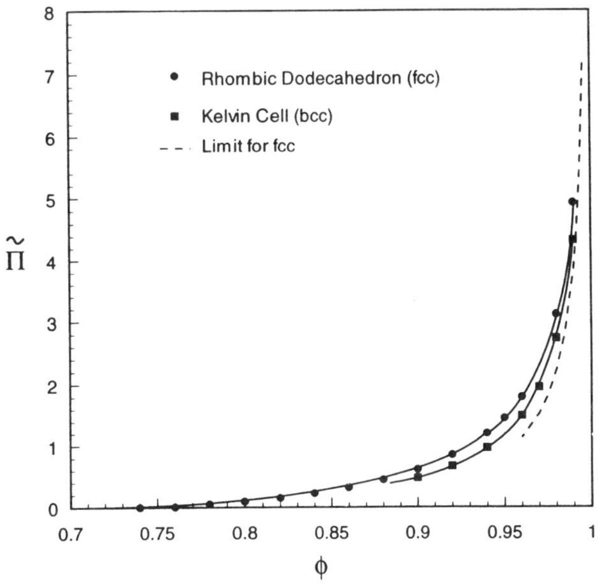


**Figure 15** Reduced osmotic pressure as a function of volume fraction for the rhombic dodecahedral and Kelvin structures. (From data kindly provided by A.M. Kraynik and D.A. Reinelt.)

## D. Polydisperse, 3-D Systems

This is, of course, the system of greatest interest from a practical point of view. The detailed structure is exceedingly complex. As mentioned above, even the value of $\phi_0$ is not precisely defined and is expected to depend somewhat on the details of the size distribution. Nevertheless, there is clear experimental evidence (21, 22) that $\phi_0$ is close to – or slightly smaller than – 0.7405 for "typical", polydisperse, unimodal emulsions.

In the dry-foam limit, each polyhedral drop must satisfy Euler's formula, i.e.,

$$v - e + f = 2 \tag{29}$$

where $v$ is the number of vertices, $e$ is the number of edges, and $f$ is the number of faces. For an infinite number of space-filling polyhedra that are subject to Plateau's rules, a number of statistical relationships can be derived from Eq. (29) (58–60). Perhaps the most interesting of these is

$$\langle f \rangle = \frac{12}{6 - \langle e \rangle} \tag{30}$$

where $\langle f \rangle$ is the average number of faces per cell, and $\langle e \rangle$ is the average number of edges per face. Equation (30) is consistent with what is expected for a monodisperse "Kelvin foam", where $\langle f \rangle = f = 14$ and $\langle e \rangle = (6 \times 4 + 8 \times 6)/14 = 5.143$, or a Weaire–Phelan structure, where $\langle f \rangle = (2 \times 12 + 6 \times 14)/8 = 13.5$ and $\langle e \rangle = [2 \times 12 \times 5 + 6 \times (12 \times 5 + 2 \times 6)]/108 = 5.111$. As mentioned before, Matzke (3) found that, in a real, supposedly monodisperse foam, Kelvin's polyhedra did not occur and that pentagonal faces were predominant. He found that $\langle f \rangle = 13.70$ and $\langle e \rangle = 5.124$ which is again consistent with Eq. (30). For a real polydisperse dry foam, Monnereau and Vignes-Adler (61) found $\langle f \rangle = 13.39 \pm 0.05$ and $\langle e \rangle = 5.11$, again in close agreement with Eq. (30). These authors did not encounter any Kelvin cell (or Weaire–Phelan structure) either.

For $\phi_0 < \phi < 1$, the drops go through a complex transition from spheres to pure polyhedra. In this most general system, the osmotic pressure is given by

$$\Pi(\phi) = \sigma\phi^2 \frac{S_0}{V_1} \frac{\mathrm{d}(S/S_0)}{\mathrm{d}\phi} = \frac{3\sigma\phi^2}{R_{32}} \frac{\mathrm{d}(S/S_0)}{\mathrm{d}\phi} \tag{31}$$

where $R_{32}$ is the surface/volume or Sauter mean radius of the initially spherical drops:

$$R_{32} \equiv \frac{\sum n_i R_i^3}{\sum n_i R_i^2} = \frac{3V_1}{S_0} \tag{32}$$

Although $R_{32}$ can be readily measured for any practical system, the complex geometry does not allow evaluation of $S(\phi)/S_0$ and $\Pi(\phi)$ from first principles. Instead, in the next section we shall show how these and other important functions can be derived from experiment.

# V. UTILITY AND EXPERIMENTAL EVALUATION OF OSMOTIC PRESSURE

We have repeatedly emphasized the importance and utility of the osmotic pressure $\Pi$ of foams and concentrated emulsions. Once known as a function of $\phi$, it may be used quantitatively to link and predict a large number of other important properties. Some of these are listed below. In addition, these considerations lead to a convenient method for evaluating $\Pi(\phi)$ experimentally (see subsection D below).

## A. Motion of Continuous Phase Between Different Systems in Contact

Let two concentrated dispersions with the same type of continuous phase (e.g., an aqueous foam and an O/W emulsion, or two different O/W emulsions) be brought into contact, either directly or via a freely movable semipermeable membrane. If the osmotic pressures are unequal (e.g., as a result of differences in the volume fractions, mean drop size, interfacial tension, or combinations thereof), it is obvious that the (common) continuous phase will flow from the dispersion with the lower osmotic pressure into that with the higher osmotic pressure until the two pressures are equalized. The final volumes and volume fractions of the two dispersions may be predicted in a straightforward manner, once $\tilde{\Pi}(\phi)$ is known. It is important to point out that equality of the (mean) *capillary pressures* does not necessarily rule out flow, nor does their inequality imply it.

## B. Vapor Pressures of Continuous and Dispersed Phases

It can be shown (27) that the vapor pressure, $p_v^c$, of the continuous phase is reduced to below that of the bulk continuous phase, $(p_v^c)_0$, according to

$$p_v^c = (p_v^c)_0 \exp(-\Pi \overline{V}_2 / \Re T) \tag{33}$$

where $\overline{V}_2$ is the partial molar volume of the solvent, $\Re$ is the gas constant, and $T$ is the absolute temperature.

Similarly, the vapor pressure of the dispersed phase, $p_v^d$, in a concentrated emulsion can be related to that of the bulk dispersed phase, $(p_v^d)_0$, by

$$p_v^d \approx (p_v^d)_0 \exp\left(\frac{2\sigma}{R_{32}} \cdot \frac{\overline{V}_1}{\Re T} \cdot \frac{S}{S_0}\right) \tag{34}$$

where $\sigma$ is the interfacial tension, $R_{32}$ is the Sauter mean drop radius, $\overline{V}_1$ is the molar volume of the dispersed liquid, and $S/S_0$ is the relative increase in surface area at the volume fraction $\phi$. For $\phi < \phi_0$, where $S/S_0 = 1$, we recover a variant of Kelvin's equation; for $\phi > \phi_0$, the increased vapor pressure is augmented further by the appearance of the factor $S/S_0$ in the exponent, with $S/S_0$ being related to $\Pi(\phi)$ through Eq. (31).

## C. Gradient in $\phi$ in Gravitational Field

So far, we have assumed that gravity is absent or negligible, so that the volume fraction is uniform throughout the system. In gravity, however, a sufficiently tall column will develop a significant gradient in $\phi$ (22). Even if each individual drop is small enough to be essentially unaffected by the field, i.e., when the Bond number is very small, the combined buoyant force of the underlying drops causes increasing drop deformation (and volume fraction) in the higher regions (Fig. 16). At the boundary between the dispersion and the bulk continuous phase, where $z = 0$, we have $\phi = \phi_0$, and the drops are purely spherical. At higher $z$, they increasingly deform until, as $z \to \infty$, they acquire a purely polyhedral shape and $\phi \to 1$. It is clear that, at any level, the combined buoyant force of all underlying drops per unit area must equal the local osmotic pressure:

$$\Pi(\phi) = \Delta\rho \cdot g \int_0^z \phi \mathrm{d}z \tag{35}$$

or

$$\tilde{\Pi}(\phi) = \frac{\Delta\rho \cdot gR_{32}}{\sigma} \int_0^z \phi(z)\mathrm{d}z = \int_0^{\tilde{z}} \phi(\tilde{z})\mathrm{d}\tilde{z} \tag{36}$$

where $\Delta\rho$ is the density difference between the phases, $g$ is the acceleration due to gravity, and $\tilde{\Pi}(\phi)$ is the reduced osmotic pressure:

$$\tilde{\Pi}(\phi) \equiv \frac{\Pi(\Phi)}{\sigma/R_{32}} \tag{37}$$

and $\tilde{z}$ is the reduced height:

$$\tilde{z} \equiv \frac{R_{32}z}{a_c^2} \tag{38}$$

where $a_c = [\sigma/(\Delta\rho.g)]^{1/2}$ is the capillary length.

In all the above it is assumed that there is no gravitational segregation by drop size, i.e., the drop size distribution does not vary with height.

Thus, once $\tilde{\Pi}(\phi)$ is known, $\phi(\tilde{z})$ can be evaluated from Eq. (36) in the form:

$$\tilde{z}(\phi) = \int_0^{\tilde{\Pi}} \frac{d\tilde{\Pi}(\phi)}{\phi} = \int_{\phi_0}^{\phi} \frac{1}{\phi}\left(\frac{\mathrm{d}\tilde{\Pi}(\phi)}{\mathrm{d}\phi}\right)\mathrm{d}\phi \tag{39}$$

As mentioned earlier, the only system for which $\tilde{\Pi}(\phi)$ is known exactly is the monodisperse 2-D system [cf. Eq. (16)]. When Eq. (39) is applied to this case, we find

$$\tilde{z}(\phi) = \frac{1}{(\phi_0\phi)^{1/2}}\left[1 + \left(\frac{1-\phi_0}{1-\phi}\right)^{1/2}(2\phi - 1)\right] - 2 \tag{40}$$

where $\phi_0 = 0.9069$. This result has been obtained also by Pacetti (62). The volume fraction profile is shown in Fig. 17.

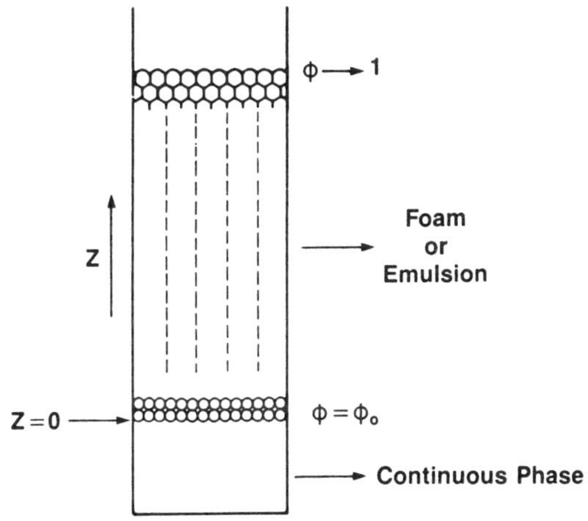


**Figure 16** Transition from spherical to polyhedral drops in vertical column. [From Ref. 38. Copyright (1986) American Chemical Society.]

## D. Experimental Determination of $\tilde{\Pi}(\phi)$ for Real Systems

From the above, it is clear that $\tilde{\Pi}(\phi)$ may be evaluated experimentally from Eq. (36) by determining the volume fraction as a function of height in an equili-

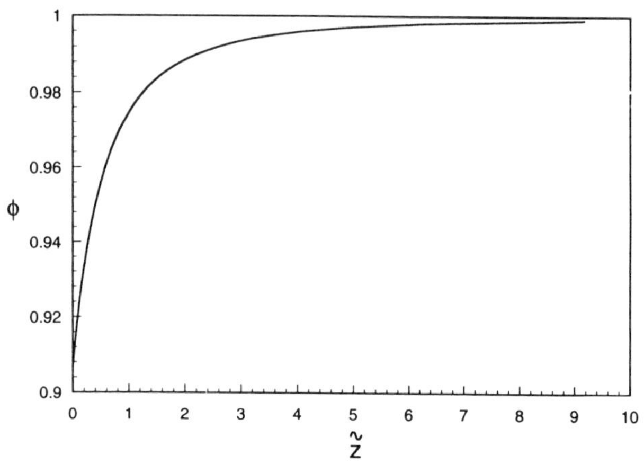


**Figure 17** Volume fraction vs. reduced height for perfectly ordered 2-D case.

brated, i.e., completely drained, dispersion column. This has been done very carefully for a typical, well-characterized polydisperse emulsion of paraffin oil in water (22). The emulsion had a Sauter mean drop radius of $R_{32} = 44.7$ μm, an interfacial tension of 7.33 mN/m, and a density difference of 0.144 g/cm$^3$. The experimental profile $\phi(\tilde{z})$ is given in Fig. 18 and may be compared with that in Fig. 17 for the monodisperse 2-D system. It could be numerically fitted to the following equations, covering three different ranges of $\phi$.

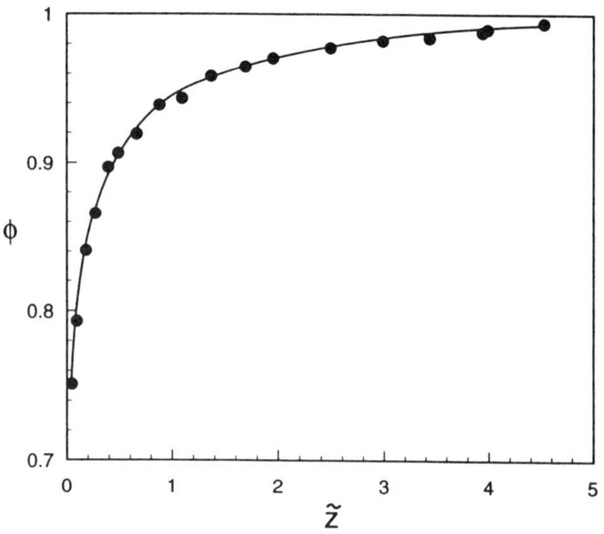


**Figure 18** Experimental profile of volume fraction vs. reduced height for typical polydisperse emulsion. (From Ref. 22. Copyright (1987) American Chemical Society.)

*"Low" volume fraction* ($0.715 < \phi < 0.90$ or $0 < \tilde{z} < 0.5$):

$$\tilde{z} = 0.237\left(\frac{\phi - 0.715}{1 - \phi}\right) \tag{41}$$

or

$$\phi = \frac{\tilde{z} + 0.169}{\tilde{z} + 0.237} \tag{42}$$

This leads to

$$\tilde{\Pi}(\tilde{z}) = \tilde{z} - 0.068\ln(\tilde{z} + 0.237) - 0.098 \tag{43}$$

which, upon substitution for $\tilde{z}$ according to Eq. (41), leads to $\tilde{\Pi}(\phi)$. Equation (41) shows that $\phi = \phi_0 = 0.715$ at $\tilde{z} = 0$. This is one of our reasons for concluding that typical polydisperse systems pack slightly less tightly than ideally close-packed monodisperse systems, where $\phi_0 = 0.7405$.

*Intermediate volume fraction* ($0.90 < \phi < 0.99$ or $0.5 < \tilde{z} < 4.0$):

$$\phi = 1.037\left[1 - (117.6\tilde{z} + 4.0)^{-1/2}\right] \tag{44}$$

and

$$\tilde{\Pi} = \frac{0.00819\phi^2}{(1 - 0.9639\phi)^2} \tag{45}$$

*High volume fraction* ($0.99 < \phi < 1$ or $\tilde{z} > 4.0$:

$$\tilde{z} \approx \tilde{\Pi} = 0.5842\frac{\left[1 - 1.892(1 - \phi)^{1/2}\right]^2}{(1 - \phi)^{1/2}} \tag{46}$$

which is the appropriate limiting solution for the polyhedral system.

Equations (42), (43), (45), and (46) describe the dependence of $\tilde{\Pi}$ on $\phi$, as shown in Fig. 19. It may be compared with that for the monodisperse 2-D and 3-D systems in Figs 5 and 15, respectively. Close examination shows that the experimental osmotic pressure is consistently lower than those for the idealized structures in Fig. 15.

Even though these relationships were derived for one particular emulsion, its size distribution was "typical", so that we believe that they can be applied with reasonable confidence in most practical situations. Nevertheless, more work remains to be done to elucidate the effect of the details of the size distribution. There is a particular need for the equivalent expressions for the *monodisperse* system, which would serve as a benchmark. Bibette's (63) novel way of preparing emulsions of low polydispersity (±10% in radius) has opened up experimentation along these lines. Unfortunately, the technique appears to be capable

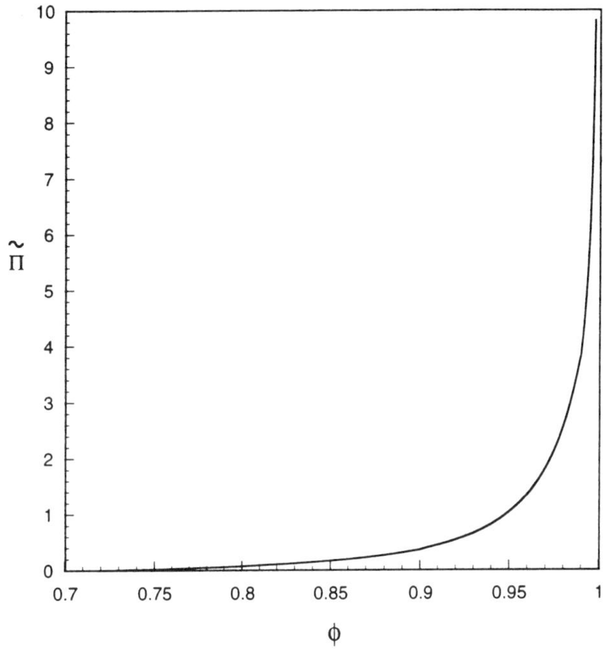


**Figure 19** Reduced osmotic pressure as a function of volume fraction for typical polydisperse emulsion. [From Ref. 22. Copyright (1987) American Chemical Society.]

only of generating emulsions of extremely small drop size ($R < 1$ μm), which complicates matters in several ways. First, estimates of the effective volume fractions [cf. Eq. (1)] become questionable, unless detailed quantitative information is available on the equilibrium film thickness as a function of the apparent volume fraction (or capillary pressures). This is usually not the case, potentially leading to significant errors. Secondly, droplets of such small size are Brownian, which may lead to an entropic contribution to the osmotic pressure, in addition to the energetic contribution considered so far. These and other factors may be responsible for some of the differences between the above results and those of Mason *et al.* (64), who measured $\Pi(\phi)$ for an oil-in-water "Bibette emulsion" of $R = 0.48$ μm. To cover the whole range of $\phi$, they used three different ways to generate the osmotic pressure: gravitational compaction, centrifugation, and dialysis of the emulsion against the continuous phase containing various levels of dextran, a polymer to which the dialysis membrane is impermeable. The osmotic pressure was found to rise at an estimated effective $\phi$ of $(\phi_0)_e \approx 0.60$ (rather than 0.715). This is close to 0.64, the value for random close packing of uniform spheres. Up to $\phi_e = 0.80$, the data could be fitted well to

$$\tilde{\Pi} \propto \phi^2(\phi - 0.60) \qquad (\phi < 0.80)$$

For $\phi > 0.80$, the results of the two studies appear to be quite consistent, in spite of the disparity in the degree of polydispersity of the emulsions employed. The apparent discrepancy at the lower volume fractions may be entirely the result of the large difference in mean drop size, for the reasons cited above.

## E. Gravitational Syneresis or Creaming

In the absence of gravity (or with fluids of matched densities), a perfectly stable emulsion or foam with $\phi > \phi_0$ will remain uniform and not "phase separate", i.e., it will not exude a bottom layer of continuous phase. In a gravitational (or centrifugal) field such syneresis may occur, however, as a result of compaction in the upper region (assuming that we are dealing with a foam or O/W emulsion; continuous phase would separate at the *top* in W/O emulsions). In a consumer product, such behavior could be detrimental, as it might suggest instability, breakdown, and limited shelf-life, even though simple shaking would restore (temporary) uniformity. With the knowledge contained in the previous subsection, it is possible to predict exactly when such syneresis will in fact occur (65). For a container of constant cross-section, the parameters of importance are the overall volume fraction, $\bar{\phi}$, and the reduced height of the sample, $\tilde{H}$, defined by

$$\tilde{H} \equiv \frac{HR_{32}}{a_c^2} = HR_{32}\Delta\rho \cdot g/\sigma \tag{47}$$

where $H$ is the actual height of the sample. It is clear that, for any $\bar{\phi}$, there must be a critical reduced sample height, $\tilde{H}_{cr}$, above which syneresis will occur and below which it will not. From a material balance and Eq. (36), it is readily shown that $\tilde{H}_{cr}$ must obey the condition:

$$\tilde{H}_{cr} = \frac{\tilde{\Pi}(\tilde{H}_{cr})}{\bar{\phi}} \tag{48}$$

Figure 20 shows how the resulting $\tilde{H}$–$\bar{\phi}$ diagram is bisected by $\tilde{H}_{cr}(\bar{\phi})$. Reference 65 provides procedures for determining the height of the separated layer of continuous phase, if any, as well as the precise variation of $\phi$ with height in the sample. The method may be extended to containers with varying cross-section (65). The following general conclusions may be drawn: (1) everything else being equal, syneresis is less likely the higher the overall concentration of the dispersed phase, $\bar{\phi}$; of course, when $\bar{\phi} < \phi_0$, syneresis will *always* occur;

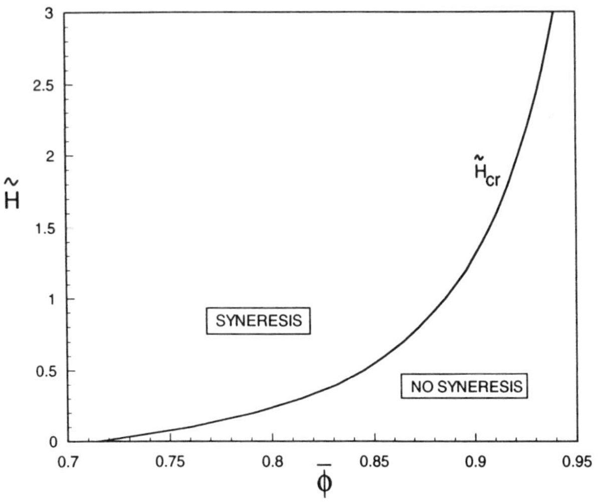


**Figure 20** Critical sample height for occurrence of syneresis as a function of overall volume fraction.

(2) for given $\bar{\phi}(> \phi_0)$, the tendency toward syneresis is less pronounced the smaller $\tilde{H}$, i.e., for small drop size, high interfacial tension, small density difference, and small sample height [cf. Eq. (47)]; and (3) for a foam or typical O/W emulsion, the tendency toward syneresis is reduced if the container is shaped with its widest part at the bottom. The reverse is true for typical W/O emulsions.

## F. Increase in Specific Surface Area with $\phi$

We have seen that the osmotic pressure is directly linked to the scaled specific surface area, $S/S_0$, as $\phi$ increases from $\phi_0$ through Eq. (31). For the monodisperse 2-D system, $S/S_0$ is given by Eq. (13) and is plotted in Fig. 3.

To the extent that the real emulsion studied in Ref. 22 is representative of typical polydisperse, 3-D systems, one can derive $S/S_0$ from the expressions for $\tilde{\Pi}(\phi)$ in Section V.D. The results (22) are

For $0.715 < \phi < 0.90$:

$$\frac{S}{S_0} = 1 + \frac{1}{3}\left[\frac{0.084}{\phi} - \frac{0.068}{\phi}\ln(1-\phi) - 0.237\right] \quad (49)$$

For $0.90 < \phi < 0.99$:

$$\frac{S}{S_0} = \frac{0.00283}{1 - 0.9639\phi} + 0.989 \quad (50)$$

For $0.99 < \phi < 1$:

$$\frac{S}{S_0} = 1.014 + 0.0686[1 - 1.892(1-\phi)^{1/2}]^3 \quad (51)$$

The combined results are shown in Fig. 21, where it is seen that the transition from spheres to completely developed polyhedra is accompanied by an increase in surface area of 8.3%. As mentioned above, for the *monodisperse* case one predicts an increase in surface area of 9.7% on the basis of Kelvin's polyhedron as the ultimate drop shape, or 9.4% for the Weaire–Phelan structure. Polydispersity appears to give rise to an even somewhat *smaller* overall change in surface area. Recent computer simulations of various monodisperse and polydisperse structures by Kraynik *et al.* (66) confirm this result almost quantitatively.

## G. Surface Area in Films versus Total Surface Area

At any given volume fraction $\phi$, a fraction $S_f/S$ of the total surface area forms part of the films separating the droplets, while the remainder is still "free" in the Plateau borders ($S_f/S = 0$ at $\phi = \phi_0$; $S_f/S = 1$ at $\phi \to 1$). This parameter may play an important role in problems relating to the stability of, and mass transfer in, such systems. We have shown (27) that

$$\frac{S_f}{S} = \frac{S_1/S_0}{S/S_0} \cdot \frac{f(\phi)}{\phi^{2/3}} \approx \frac{1.083}{\phi^{2/3}} \cdot \frac{f(\phi)}{S/S_0} \quad (52)$$

where $S/S_0$ is given by Fig. 21, and $f\phi)$ is the fraction of a confining wall that is "contacted" by the flattened

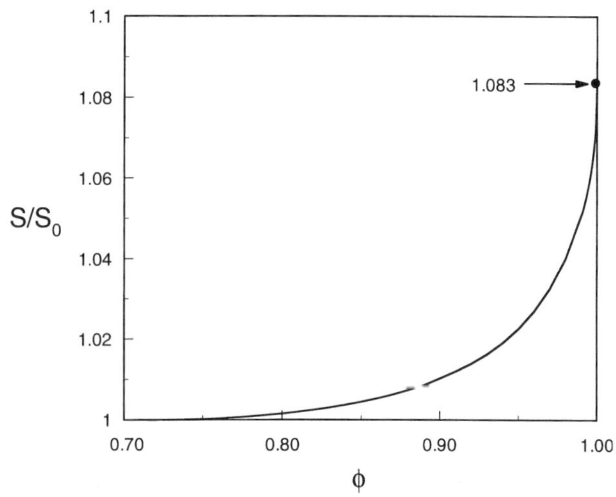


**Figure 21** Scaled specific surface area as a function of volume fraction for typical polydisperse emulsion. [From Ref. 22. Copyright (1987) American Chemical Society.]

parts of the drops pushing against it, under the assumption that the wall is perfectly wetted by the continuous phase. This fraction, which varies from $f = 0$ at $\phi_0$ to $f = 1$ at $\phi = 1$, can be measured experimentally (67) and was found empirically to be given by

$$f(\phi) = 1 - \frac{3.20}{\left(\frac{\phi}{1-\phi} + 7.70\right)^{1/2}} \tag{53}$$

for $\phi_0 < \phi < 0.975$. (By solving for $\phi$ at $f = 0$, we again obtain evidence that $\phi_0 \approx 0.72$ for real, polydisperse systems.) For $\phi > 0.975$, we expect that $f(\phi)$ is given, to a good approximation (38), by

$$f(\phi) = [1 - 1.892(1-\phi)^{1/2}]^2 \tag{54}$$

Combining Eqs (53) and (54) with Eq. (52) leads to the approximate dependence of $S_f/S$ on $\phi$ as shown in Fig. 22.

These are just some of the examples of where and how the osmotic pressure, or its related properties, can be used to define the overall equilibrium behavior of these complex fluids, even though their detailed microscopic structure may not be fully known. Other examples are to be found in the following section, where we describe the only properties that are unique to foams as a result of the compressibility of their dispersed phase.

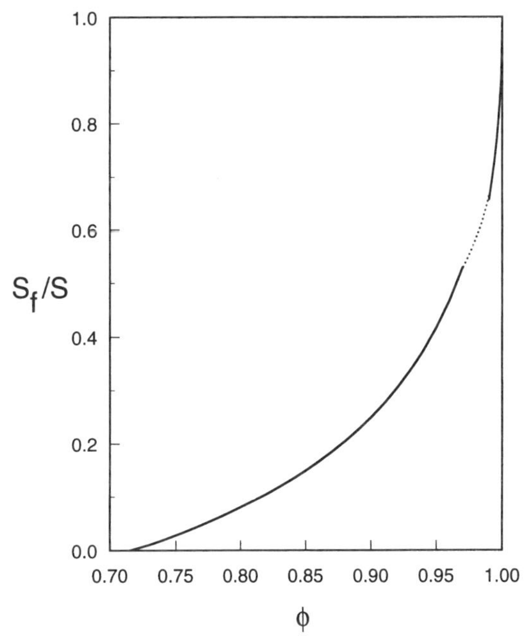


**Figure 22** Fraction of total surface area contained in films as a function of volume fraction for typical polydisperse emulsion. Solid curve at right is limiting solution for fcc; the dashed curve connects it to the lower experimental region. [From Ref. 27. Copyright (1988) American Chemical Society.]

## VI. FOAMS: INTERNAL PRESSURE, EQUATION OF STATE, AND COMPRESSIBILITY

Up to this point we have emphasized the common structural and other properties of concentrated emulsions and foams. However, because of their gaseous dispersed phase, foams are compressible and, just as gases themselves, can be characterized by an equation of state that relates their volume, external pressure, and temperature.

### A. Dry-Foam Limit ($\phi = 1$)

For a polydisperse dry foam one can define an average *internal pressure* $\bar{p}_i$ that is given by

$$\bar{p}_i = \frac{\sum p_i v_i}{\sum v_i} = \frac{\sum p_i v_i}{V} \tag{55}$$

where $p_i$ and $v_i$ are the pressure and volume of bubble $i$, and $V$ is the total foam volume. Derjaguin (68) has shown that

$$\bar{p}_i = P + \frac{2\sigma S_1}{3V} \qquad (\phi = 1) \tag{56}$$

where $P$ is the external pressure and $S_1/V$ is the specific surface area of the foam. Assuming ideality of the gas phase, this leads to the *equation of state*:

$$\left(P + \frac{2\sigma S_1}{3V}\right) V = n\Re T \qquad (\phi = 1) \tag{57}$$

where $n$ is the number of moles of gas in the foam. The same results were later obtained by Ross (69).

Morrison and Ross (70) have indicated that, while Eqs (56–57) are undoubtedly correct for monodisperse foams, a rigorous proof of their validity for *polydisperse* systems was lacking. Such proof has since been provided by Hollinger (71), Crowley (72), and Crowley and Hall (73).

Derjaguin further showed (68) that the compression modulus $K$ is given by

$$K \equiv -V\frac{dP}{dV} = \frac{(P + 2\bar{p}_i)}{3} = P + \frac{4\sigma}{9}\frac{S_1}{V} \qquad (\phi = 1) \tag{58}$$

which compares to $K = P$ for a simple ideal gas.

The specific surface area in Eqs (56–58) may be replaced by

$$\frac{S_1}{V} = \frac{S_1}{S_0} \cdot \frac{S_0}{V} = \frac{3}{R_{32}} \cdot \frac{S_1}{S_0} \tag{59}$$

where, as before, $R_{32}$ is the Sauter mean bubble radius, and $S_1/S_0 \approx 1.083$ is the increase in surface area associated with the transition from spherical to polyhedral bubbles at equal volume.

### B. Foams with Finite Liquid Content ($\phi < 1$)

We have shown (27) that, for this general case, Eqs (56–58) are to be modified as follows:

$$\bar{p}_i = P + \frac{1-\phi}{\phi}\Pi + \frac{2\sigma S}{3V_1} \tag{60}$$

$$\left(P + \frac{1-\phi}{\phi}\Pi + \frac{2\sigma S}{3V_1}\right)\phi V = n\Re T \tag{61}$$

$$K = \frac{1}{\phi}\left[P + \frac{1-\phi}{3\phi}\Pi + (1-\phi)^2\frac{d\Pi}{d\phi} + \frac{4\sigma}{9}\frac{S}{V_1}\right] \tag{62}$$

where $\Pi$ is the osmotic pressure, $V_1$ is the volume of the dispersed gas phase, and $V$ is the total foam volume ($V_1 = \phi V$). For $\phi = 1$, Eqs (56–58) are recovered.

Equations (60–62) may be written in the form:

$$\bar{p}_i - P = \frac{\sigma}{R_{32}}\left(\frac{1-\phi}{\phi}\tilde{\Pi} + 2\frac{S}{S_0}\right) \tag{63}$$

$$\left[P + \frac{\sigma}{R_{32}}\left(\frac{1-\phi}{\phi}\tilde{\Pi} + 2\frac{S}{S_0}\right)\right]\phi V = n\Re T \tag{64}$$

$$K = \frac{1}{\phi}\left[P + \frac{\sigma}{R_{32}}\left(\frac{1-\phi}{3\phi}\tilde{\Pi} + (1-\phi)^2\frac{d\tilde{\Pi}}{d\phi} + \frac{4}{3}\frac{S}{S_0}\right)\right] \tag{65}$$

where $\tilde{\Pi}$ is the reduced osmotic pressure. The terms within the round brackets depend on $\phi$ only and can be evaluated from the data presented above. It may be shown (27) that the "osmotic" terms, while significant, provide only a rather small correction (<6%) to the dominant "Derjaguin terms" in $S/S_0$. Of perhaps trivial but greater significance is the correction for the volume fraction outside the brackets of Eqs (61), (62), (64), and (65).

## VII. MECHANICAL AND RHEOLOGICAL PROPERTIES

It has long been realized that the crowding of deformable drops and bubbles in concentrated emulsions and foams gives rise to interesting mechanical and rheological properties, not shown by the separate constituent fluid phases. When subjected quasistatically to a small stress, these systems respond as purely elastic solids, characterized by a static elastic modulus, $G$. Under dynamic conditions, the modulus has a real, elastic component (the storage modulus, $G'$) and a complex, viscous component (the loss modulus, $G''$). Once a critical or yield stress is exceeded, the systems flow and behave as viscoelastic fluids, whose effective viscosity decreases from infinity (at the yield stress) with increasing shear rate. Thus, in rheological terms, they are *plastic* fluids with viscoelastic *solid* behavior below, and viscoelastic *fluid* behavior above, the yield stress.

A number of early experimental studies have provided qualitative evidence for some or all of these behavioral aspects (e.g., 4, 74–80), but the techniques employed were usually crude and/or the systems were poorly characterized, if at all. This makes it impossible to use these early expermental data to draw conclusions as to the quantitative relationships between the rheological properties on the one hand, and important system variables, such as volume fraction, interfacial tension, mean drop size (and size distribution), fluid viscosities, shear rate, etc., on the other. In the last decade or so, interest in this area has intensified and much progress has been and is being made along several fronts: theoretical modeling, computer simulation, and careful experimentation. For other recent, though by now somewhat outdated, reviews, see Refs 81–84.

### A. Theoretical Modeling and Computer Simulation

In view of the exceedingly complex structure of 3-D systems – even when monodisperse – initial efforts were confined almost exclusively to their 2-D analogs. Although unrealistic in some ways, these models provide important kinematic insights and their behavior may be extrapolated, with caution and limitations, to real systems. At first, for the sake of mathematical tractability, the complexity was reduced even further by considering perfectly ordered, *monodisperse* 2-D systems. Gradually, the degree of complexity has been increased by allowing disorder. It is only very recently that some intrepid investigators have begun to tackle the 3-D problem in earnest.

### 1. Elastic and Yield Properties: Shear Modulus and Yield Stress

*a. Two-dimensional Systems*

For the perfectly ordered case, the unstrained equilibrium structure has been discussed above. The (cylindrical) drops are arranged on a perfectly ordered hexagonal lattice, decorated at its vertices with Plateau borders, whose wall curvatures are determined by the drop size and volume fraction according to Eq. (11). The system can be thought to be confined between two parallel plates, with rows of drops being forced to align with the plates. As one of the plates is now moved within its own plane to induce shear, all drops respond by being deformed identically. In the process the surface area increases. With the assumption of constant interfacial tension, this results in a force (stress) versus deformation (strain) behavior that has been analyzed in detail, using straightforward geometrical arguments, by Princen (85) for any value of $\phi \geq \phi_0$. The simplest, dry-foam case of $\phi = 1$ has been considered independently by Prud'homme also (86).

The sequence of events in the dry-foam limit is illustrated in Fig. 23 for a single unit cell, i.e., the parallelogram formed by the centers of four adjacent drops. As the cell is strained at constant volume, the angle between the films must remain at 120°, which causes the central film to shorten until its length shrinks to zero. At that point, four films meet in a line. The resulting instability resolves itself by a rapid so-called T1 rearrangement or "neighbor switching". In the process, new film is generated from the center to restore the original, unstrained configuration. A different, perhaps clearer, view of the system as it moves through such a cycle is shown in Fig. 24. At any stage, the stress per unit cell is given by the horizontal component of the tension of the originally vertical films, i.e.,

$$F = 2\sigma \cos \psi \tag{66}$$

where $\psi$ is the angle between these films and the horizontal shear direction. The resulting stress–strain curve per unit cell is given by curve #8 in Fig. 25, where $\tilde{F}$ is the dimensionless stress per unit cell.

$$\tilde{F} = F/2\sigma = \cos \psi \tag{67}$$

Khan and Armstrong (43, 87, 88), using a slightly different analysis, arrived at the following simple analytical result for curve #8:

$$\tilde{F} = \frac{\gamma}{(\gamma^2 + 4)^{1/2}} \tag{68}$$

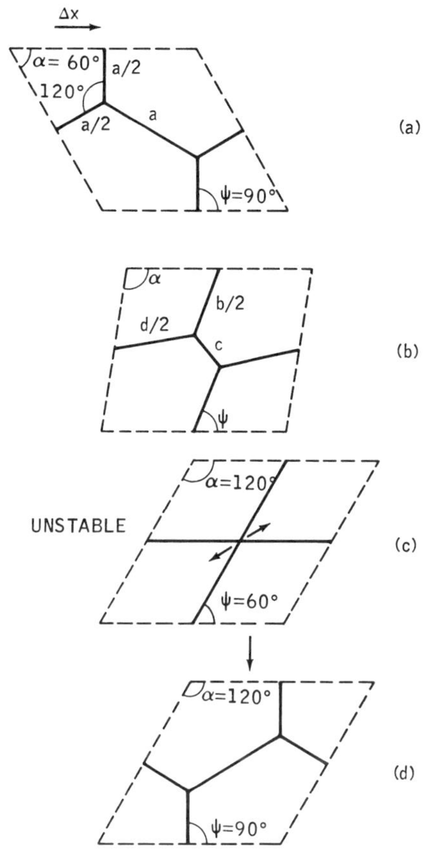


**Figure 23** Shear deformation of unit cell of perfectly ordered 2-D system in dry-foam limit ($\phi = 1$); the transition from (c) to (d) is rapid and is often referred to as a T1 rearrangement or neighbor switching. (From Ref. 85, with permission from Academic Press.)

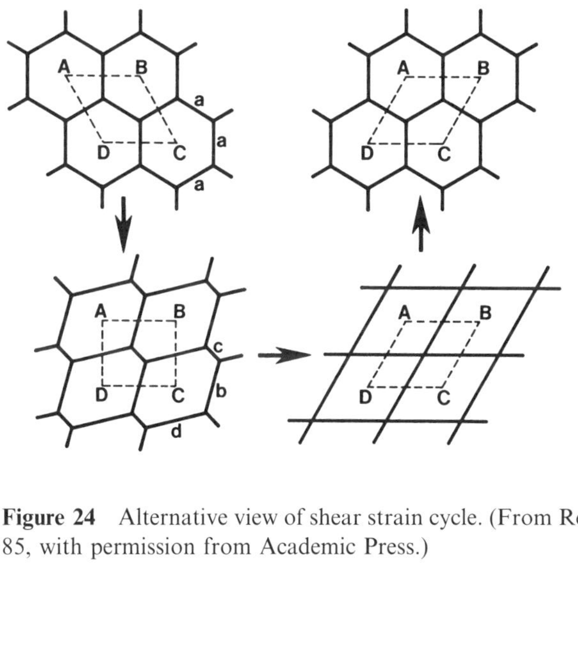


**Figure 24** Alternative view of shear strain cycle. (From Ref. 85, with permission from Academic Press.)

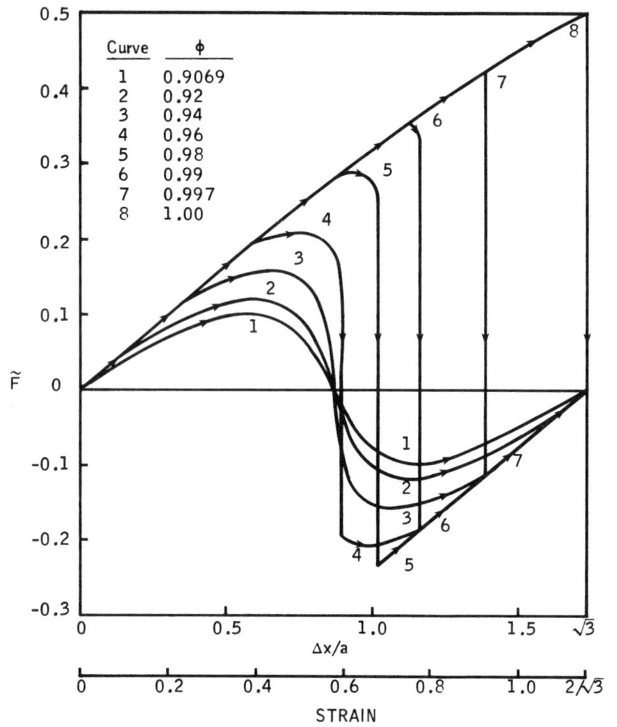


**Figure 25** Shear stress per unit cell vs. shear strain for perfectly ordered 2-D system at different volume fractions. (From Ref. 85, with permission from Academic Press.)

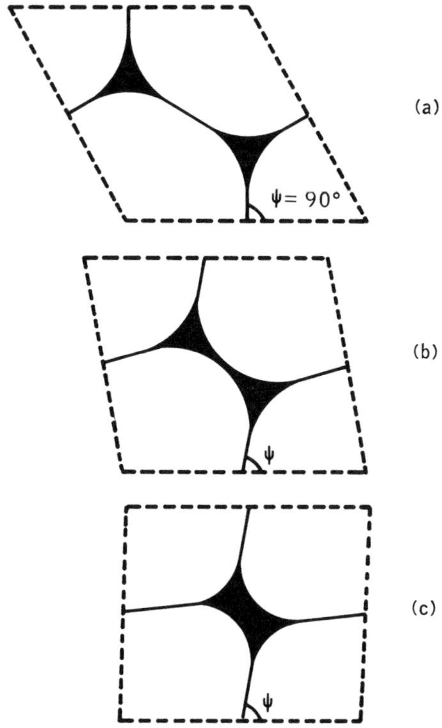


**Figure 26** Increasing strain for systems with $0.9069 < \phi < 1$. Between (a) and (b), system is in Mode I; between (b) and (c), system is in Mode II. (From Ref 85, with permission from Academic Press.)

where $\gamma$ is the imposed strain, which varies from zero to $2/\sqrt{3}$ at the point of instability. The cycle then repeats itself.

When $\phi < 1$, the situation is considerably more complicated (Fig. 26). As long as the two Plateau borders within the unit cell remain separated ("Mode I"), the stress/unit cell is unaffected. However, beyond a given strain, which depends on $\phi$, the Plateau borders merge to form a single, four-sided border. In this "Mode II" regime, the films no longer meet at 120°, and the stress/strain curve deviates from that for the dry-foam limit. It passes through a (lower) maximum and ultimately reverses sign, either continuously or via a T1 rearrangement (85). The resulting curves are collected in Fig. 25. In each case the maximum $\tilde{F}_{max}$ corresponds to the static *yield stress*/unit cell. It is plotted in Fig. 27 as a function of $\phi$, together with the corresponding yield strain. Realizing that there are $1/a\sqrt{3}$ unit cells per unit of length in the shear direction and that $a$ may be expressed in terms of the more practical drop radius $R$ and volume fraction $\phi$, one finds for the stress ($\tau$)/strain ($\gamma$) relationship:

$$\tau = 1.050\frac{\sigma}{R}\phi^{1/2}\tilde{F}(\gamma, \phi) \tag{69}$$

while the *yield stress*, $\tau_0$, is given by

$$\tau_0 = 1.050\frac{\sigma}{R}\phi^{1/2}\tilde{F}_{max}(\phi) \tag{70}$$

where $\tilde{F}_{max}(\phi)$ may be read from Fig. 27. It is expected to start deviating from zero when adjacent layers of close-packed drops or bubbles can freely slide past each other, i.e., at $\phi = \pi/4 = 0.7854$.

The small-strain, static *shear modulus*, $G$, is defined as

$$G \equiv \left(\frac{d\tau}{d\gamma}\right)_{\gamma=0} \tag{71}$$

and can be obtained from Eqs (69) and (68):

$$G = 1.050\frac{\sigma}{R}\phi^{1/2}\left(\frac{d\tilde{F}}{d\gamma}\right)_{\gamma=0} = 0.525\frac{\sigma}{R}\phi^{1/2} \quad (\phi > \phi_0) \tag{72}$$

The model predicts zero shear modulus for $\phi < \phi_0$.

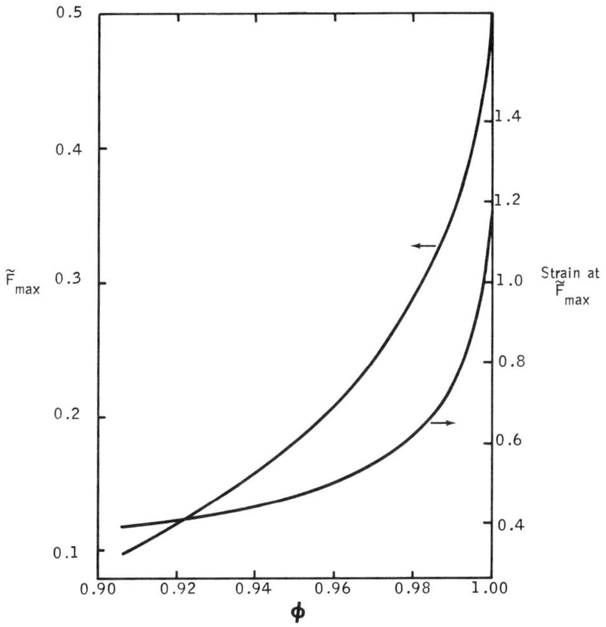


**Figure 27** Static yield stress per unit cell and yield strain as a function of volume fraction for perfectly ordered 2-D system. (From Ref. 85, with permission from Academic Press.)

Both the yield stress and the shear modulus scale with $\sigma/R$ but, while the yield stress increases strongly with volume fraction, the shear modulus is affected only very weakly through $\phi^{1/2}$. In the dry limit of $\phi = 1$, both reach identical limiting values of

$$\tau_0 = G = 0.525\frac{\sigma}{R} \qquad (\phi = 1) \tag{73}$$

The analysis may be extended to systems, in which the film thickness, $h$, or the contact angle, $\theta$ between the films and the Plateau border walls are finite (85). The effect of a finite film thickness is to increase the effective volume fraction [cf. Eq. (1)], which raises the yield stress and shear modulus in a predictable fashion. The effect of a finite contact angle on the shear modulus is to simply reduce it by a factor of cos $\theta$. The effect on the yield stress is more complex. In most but not all cases the yield stress is increased. Furthermore, a finite contact angle can give rise to interesting new instability modes and to hysteretic behavior. The reader is referred to Ref. 85 for further details.

Subsequently, Khan and Armstrong (87, 88) and Kraynik and Hansen (89) considered the effect of the orientation of the unit cell, relative to the shear direction, for the dry-foam case. They found that the shear modulus is unaffected, but that the yield stress is sensitive to the orientation. In addition, they considered planar extension, as well as shear.

The sudden jump of the shear modulus from zero to a finite value at $\phi_0$ and its subsequent weak sensitivity to $\phi$ for $\phi > \phi_0$ are rather peculiar and appear to be associated with the perfect order of the model. The pure cyclical character of the stress/strain curves is – by itself – a symptom of "perfection pathology". As discussed below, real systems do not exhibit these particular features, since they are invariably disordered, which causes T1 rearrangements to occur even at very small strains, as well as randomly throughout the system, rather than simultaneously at all vertices.

The shear modulus of polydisperse hexagonal systems of the type depicted in Fig. 8b is still given by Eq. (72) when $R$ is replaced by $R_{av} = (\sum R_i^2/n)^{1/2}$, a characteristic drop radius that is based on the average drop area (44). However, as expected, the "elastic limit", i.e., the stress and strain where the first T1 rearrangement occurs, is reduced relative to that of the monodisperse case of the same volume fraction.

The elastic and yield properties of 2-D systems with the most general type of disorder (cf. Fig. 7) have been simulated by Hutzler *et al.* (90) for both dry and wet systems. Indeed, as the number of polydisperse drops in the simulation is increased, the jumps in stress associated with individual or cooperative T1 rearrangements become less and less noticeable. Instead, the stress increases smoothly with increasing strain until it reaches a plateau that may be identified with the yield stress. The yield stress was found to increase sharply with increasing volume fraction, very much as in the monodisperse case. Furthermore, the shear modulus for the dry system ($\phi = 1$) was essentially identical to that for the monodisperse case, as given by Eq. (73) with $R_{av}$, as defined above, replacing $R$. Its dependence on $\phi$ was very different from that in Eq. (72), however. When expressed in our terms, their results for $1 > \phi > 0.88$ could be fitted to

$$\frac{G}{\sigma/R_{av}} = 0.51 - 21(1-\phi)^2 \tag{74}$$

Assuming that this relationship continues to hold for $\phi < 0.88$ (where their simulations ran into difficulties because of the large number of T1 processes the program had to deal with), the authors concluded that $G$ reaches zero at $\phi = \phi_0 \approx 0.84$. As mentioned earlier, this "rigidity-loss transition" can be identified as the random close packing of hard disks. The drop in $G$ with decreasing $\phi$ could further be correlated with the average number of sides of the Plateau borders, which gradually increased from three close to $\phi = 1$ to about four at $\phi = 0.84$. Although these simulations involved a rather small number of drops and leave some questions unanswered, they do indicate a type of elastic behavior that – as we shall see later – much more closely reflects that of real systems. Clearly, disorder plays a critical role.

*b. Three-dimensional Systems*

The first expression for the shear modulus of random dry foams (and emulsions) was derived by Derjaguin (91). It is based on the assumption that the foam is a collection of randomly oriented films of constant tension $2\sigma$ and negligible thickness, and that each film responds affinely to the applied shear strain, as would an imaginary surface element in a continuum. Evaluating the contribution to the shear stress of a film of given orientation and averaging over all orientations then leads to

$$G = \frac{4}{15}\sigma\frac{S_1}{V} \qquad (\phi \approx 1) \tag{75}$$

where $S_1/V$ is the surface area per unit volume. Since $S_1/V \approx 1.083 S_0/V = 3.25/R_{32}$, this may be written as

$$G \approx \frac{13}{15}\frac{\sigma}{R_{32}} \approx 0.87\frac{\sigma}{R_{32}} \qquad (\phi \approx 1) \tag{76}$$

Much later, Stamenović and Wilson (92) rediscovered Eq. (75), using similar arguments but pointing out at the same time that it probably represents an overestimate. Indeed, using 2-D arguments, Princen and Kiss (93) concluded that the affine motion of the individual films violates Kelvin's laws and leads to an overestimate of $G$ by a factor of two, at least in 2-D. (Kraynik, in a private communication, pointed out an internal inconsistency in Ref. 93 and concluded that $G$ was overestimated by a factor of only 3/2.) Furthermore, Derjaguin's model does not allow for T1 rearrangements; it does not predict a yield stress, nor does it have anything to say about the effect of $\phi$ in "wet" systems. On the other hand, the model correctly predicts that $G$ scales with $\sigma/R$.

Stamenović (94) analyzed the deformation of an idealized single foam vertex, where four Plateau borders meet and concluded that

$$G = \frac{1}{6}\sigma\frac{S_1}{V} \approx 0.54\frac{\sigma}{R_{32}} \qquad (\phi \approx 1) \tag{77}$$

As pointed out by Reinelt and Kraynik (54), however, the idealized vertex does not adequately represent an equilibrium structure. Similar reservations apply to the work of Budiansky and Kimmel (95), who considered the behavior of an isolated foam cell in the form of a rectangular pentagonal dodecahedron and obtained a shear modulus between the two above values.

Using Brakke's surface evolver (47), Reinelt and coworkers (54, 66, 96–100) have explored in detail the elastic response of monodisperse, perfectly ordered structures, both "dry" and "wet", to extensional and shear strain. Structures considered included the rhombic dodecahedron, the regular ("planar") tetrakaidecahedron, the Kelvin cell, and the Weaire–Phelan structure. Some degree of disorder was introduced by considering bidisperse Weaire–Phelan systems (101), in which the relative volumes of the dodecahedra and tetrakaidecahedra were varied, as well as random, though monodisperse, systems (66). As in the 2-D case, the stress/strain behavior depends on the cell orientation relative to the strain direction. Because of the multitude of edges and faces of each cell, a variety of T1 transitions may occur at increasing strain, leading to very complex behavior. Some of their results for the shear moduli of dry systems ($\phi = 1$) are listed in Table 1.

The ordered structures are all anisotropic, have cubic symmetry, and can be characterized by two shear moduli, $G_1$ and $G_2$. To simulate orientation disorder, the authors introduced an "effective isotropic shear modulus", $G_{av} = \frac{2}{5}G_1 + \frac{3}{5}G_2$, which is obtained by averaging over all orientations. The first three columns of Table 1 give the moduli in units of $\sigma V^{-1/3}$, where $V$ is the cell volume; the last column in units of $\sigma/R$, where $R = (3V/4\pi)^{1/3}$. The orientation-averaged results are surprisingly close to the 2-D prediction of

**Table 1** Shear Moduli of Dry Systems

| | $G_1/\sigma V^{-1/3}$ | $G_2/\sigma V^{-1/3}$ | $G_{av}/\sigma V^{-1/3}$ | $G_{av}/\sigma R^{-1}$ |
|---|---|---|---|---|
| Reg. Tetrakaidecahedron | 0.5525 | 0.9696 | 0.8028 | 0.4980 |
| Kelvin | 0.5706 | 0.9646 | 0.8070 | 0.5006 |
| Weaire-Phelan | 0.8902 | 0.8538 | 0.8684 | 0.5387 |
| Random (monodisperse) | | | $0.78 \pm 0.08$ | $0.48 \pm 0.05$ |

$G/\sigma R^{-1} = 0.525$ [cf. Eq. (73)], to Stamenović's prediction of $G/\sigma R^{-1} = 0.54$ [cf. Eq. (77)], and to the extrapolated experimental result of Princen and Kiss (93) for polydisperse emulsions, which indicated that $G/\sigma R_{32}^{-1} = 0.509$ (see below). The small influence of polydispersity is also suggested by the finding that $G_{av}$ varies less than 0.5% when the volume ratio of the two types of cells in bidisperse Weaire–Phelan structures is varied between 0.039 and 2.392 (101).

Simulations of this type can pinpoint an "elastic limit" where the first (or subsequent) T1 transition(s) take(s) place. It depends extremely strongly on orientation, as does the "dynamic yield stress", i.e., the stress integrated over a complete strain cycle. The relevance to the yield stress of real disordered systems is, therefore, quite limited (98). As in 2-D simulations, simulations on more highly disordered systems will undoubtedly bring increased insight.

Simulations on "wet" rhombic dodecahedra and Kelvin cells have been carried out by Kraynik and coworkers (66, 100). The effective isotropic shear moduli were found to depend slightly on the volume fraction, but did not show the linear dependence on $\phi - \phi_0$ found experimentally for disordered systems (93). Again, simulations on highly disordered wet systems should improve our understanding.

Buzza and Cates (102) also addressed the question whether disorder or the increased dimensionality from two to three dimensions is responsible for the observed experimental behavior of the shear modulus. In particular, they explored the lack of the sudden jump in $G$ from zero to a finite value at $\phi = \phi_0$ that is predicted by the perfectly ordered 2-D model. We have seen above that disorder appears to remove that abrupt jump in two dimensions (90). For drops on a simple cubic lattice, Buzza and Cates analyzed the drop deformation in uniaxial strain close to $\phi = \phi_0$, first using the model of "truncated spheres". (For reasons given above, we believe this to be a very poor model.) They showed that this model did not eliminate the discontinuous jump in $G$. An exact model, based on a theory by Morse and Witten (103) for weakly deformed drops, led to $G \propto 1/\ln(\phi - \phi_0)$, which eliminates the discontinuity, but still shows an unrealistically sharp rise at $\phi = \phi_0$ and is qualitatively very different from the experimentally observed linear dependence of $G$ on $(\phi - \phi_0)$. Similar conclusions were reached by Lacasse and coworkers (49, 104). A simulation of a disordered 3-D model (104) indicated that the droplet coordination number increased from 6 at $\phi_0$ to 10 at $\phi = 0.84$, qualitatively similar to what is seen in disordered 2-D systems (90). Combined with a suitable (anharmonic) interdroplet force potential, the results of the simulation were in close agreement with experimental shear modulus and osmotic pressure data. It therefore appears again that disorder is responsible for many of the features of real systems.

## 2. Shear Viscosity

Compared to the quasistatic elastic and yield behavior of concentrated emulsions and foams, the rate-dependent viscous properties are even more complex and relatively unexplored. Formally, the shear stress, $\tau$, may be expressed as a function of the shear rate, $\dot{\gamma}$, as

$$\tau(\dot{\gamma}) = \tau_0 + \tau_s(\dot{\gamma}) \tag{78}$$

where $\tau_0$ is the (elastic) yield stress, and $\tau_s(\dot{\gamma})$ is the contribution from any rate-dependent dissipative processes; or, in terms of the effective shear viscosity, $\mu_e$,

$$\mu_e \equiv \tau(\dot{\gamma})/\dot{\gamma} = \frac{\tau_0}{\dot{\gamma}} + \frac{\tau_s(\dot{\gamma})}{\dot{\gamma}} \tag{79}$$

The first term is, to a large extent, responsible for the shear-thinning behavior of these systems. As is clear from the previous discussion, $\tau_0$ is determined primarily by $\sigma$, $R$, and $\phi$, while the size distribution may play a secondary role. The dynamic stress, $\tau_s$, is expected to depend on these and other variables, e.g., the shear rate, the viscosities of the continuous and dispersed phases, and surface-rheological parameters. So far, the predictive quality of theoretical and modeling efforts has been very restricted because of the complexity of the problem.

Buzza *et al.* (105) have presented a qualitative discussion of the various dissipative mechanisms that may be involved in the small-strain linear response to oscillatory shear. These include viscous flow in the films, Plateau borders, and dispersed-phase droplets (in the case of emulsions); the intrinsic viscosity of the surfactant monolayers, and diffusion resistance. Marangoni-type and "marginal regeneration" mechanisms were considered for surfactant transport. They predict that the zero-shear viscosity is usually dominated by the intrinsic dilatational viscosity of the surfactant monolayers. As in most other studies, the discussion is limited to small-strain oscillations, and the rapid events associated with T1 processes in steady shear are not considered, even though these may be extremely important.

It is now generally recognized that surfactants are indeed crucial, not only in conferring (meta)stability to the emulsion or foam, but also in controlling the rate-dependent rheology of the film surfaces and that of the system as a whole. Several early, spatially periodic 2-D models neglected this aspect and made other simplifying assumptions. Khan and Armstrong (43, 87, 88) and Kraynik and Hansen (106) assumed that all the continuous phase resides in the films (i.e., there were no Plateau borders) and that there is no exchange of fluid *between* the films. The film surfaces were assumed to be completely mobile (no surfactant!). When such a system is strained globally, the uniform films respond with simple planar extension (or compression) at constant volume. This mechanism predicts significant structural changes, but leads to viscous terms in Eqs (78) and (79) that are insignificant compared with the elastic terms up to extremely high shear rates that are unlikely to be encountered in practice. Experimentally, one finds a much more significant contribution (see below).

A more complete 2-D analysis of simple shear is that of Li *et al.* (107). It solves the detailed hydrodynamics in the drops, films, and Plateau borders for the case of equal viscosities of the continuous and dispersed phases. Again, large structural changes are predicted. However, surfactants (and surface tension gradients) are assumed to be absent, which severely limits the practical implications of the analysis. An interesting conclusion is that, under certain conditions, shear flow can stabilize concentrated emulsions, even in the total absence of surfactants.

An approach that is almost diametrically opposed to the earlier models of Khan and Armstrong, and Kraynik and Hansen, was advanced by Schwartz and Princen (108). In this model, the films are negligibly thin, so that all the continuous phase is contained in the Plateau borders, and the surfactant turns the film surfaces immobile as a result of surface-tension gradients. Hydrodynamic interaction between the films and the Plateau borders is considered to be crucial. This model, believed to be more realistic for common surfactant-stabilized emulsions and foams, draws on the work of Mysels *et al.* (109) on the dynamics of a planar, vertical soap film being pulled out of, or pushed into, a bulk solution via an intervening Plateau border. An important result of their analysis is commonly referred to as Frankel's law, which relates the film thickness, $2h_\infty$, to the pulling velocity, $U$, and may be written in the form:

$$\frac{h_\infty}{r} = 0.643(3Ca^*)^{2/3} \tag{80}$$

where $Ca^* = \mu U/\sigma (\ll 1)$ is the film-level capillary number; $\mu$ and $\sigma$ are, respectively, the viscosity and surface tension of the liquid (the "continuous phase"); $r$ is the radius of curvature of the Plateau border where it meets the film and is given by capillary hydrostatics, $r = (\sigma/2\rho g)^{1/2}$, where $\rho$ is the density of the liquid; and $g$ is the gravitational acceleration.

Frankel's law has its close analogs in a number of related problems (110–112) and has been verified experimentally (113, 114) in the regime where the drawn-out film thickness, $2h_\infty$, is sufficiently large for disjoining-pressure effects to be negligible. Below some critical speed, the thickness of the drawn-out film equals the finite equilibrium thickness, $2h_{eq}$, which is set by a balance of the disjoining pressure, $\Pi_d(h)$, and the capillary pressure, $\sigma/r$, associated with the Plateau border. Thus, Frankel's law, and the following analysis, apply only as long as $1 \gg Ca^{2/3} \gg h_{eq}/r$. It is expected to break down as the capillary number approaches zero. Disjoining pressure effects may, in principle, be included (e.g., 115) but at the expense of simplicity and generality of the model.

The interesting hydrodynamics and the associated viscous-energy dissipation are confined to a transition region between the emerging, rigidly moving film and the macroscopic Plateau border. The lubrication version of the Stokes equation may be used in this region, as the relative slope of the interfaces remains small there.

It is reasonable to assume that the same basic process operates in moving emulsions and foams. Lucassen (116) has pointed out that, for such systems to be stable to deformations such as shear, the dilational modulus of the thin films must be much greater than that of the surfaces in the Plateau border. However, this is equivalent to the assumption of inextensible film surfaces that underlies Frankel's law. Therefore, it may well be that, by implication, emulsions and foams that are stable to shear (and we are interested in such systems only) have the appropriate surface rheology for Frankel's law to apply. Of course, in emulsions and foams, each Plateau border of radius $r$ (set by drop size and volume fraction) is now shared by three films. At any given moment, one or two of the films will be drawn out of the border, while the other(s) is/are pushed into it, at respective quasisteady velocities $U(t)$ that are dictated by the macroscopic motion of the system (Fig. 28). Using a perfectly ordered 2-D system, Schwartz and Princen (108) considered a periodic uniaxial, extensional strain motion of small frequency and amplitude, so that inertial effects are negligible, and complications due to merger of adjacent Plateau borders and associated rapid T1 processes are avoided. They proceeded by calculating the instantaneous rate

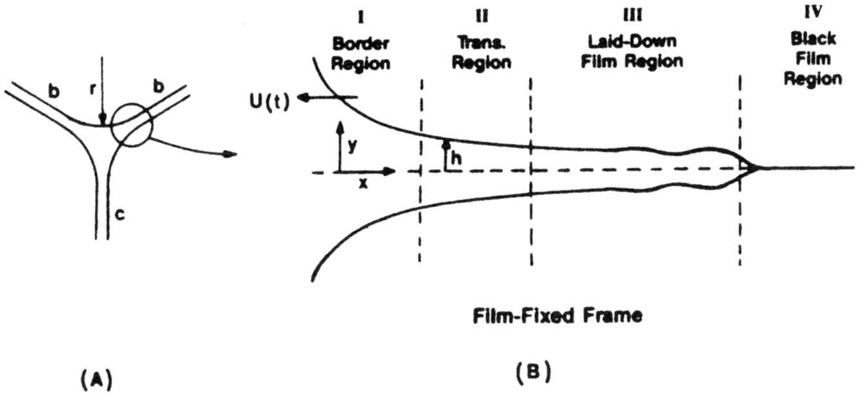


**Figure 28** Film being pulled out of a Plateau border with velocity $U(t)$; all viscous dissipation occurs in the transition region (II). (From Ref. 108, with permission from Academic Press.)

of energy dissipation in the transition region of each of the three films associated with a Plateau border, and integrated the results over a complete cycle. When the effective strain rate is related to the frequency of the imposed motion, the result can be expressed as an effective viscosity that is given by*

$$\mu_e = 5.3\mu Ca^{-1/3} \tag{81}$$

where the macroscopic capillary number $Ca \equiv \mu a\dot{\gamma}/\sigma$, $a$ is the length of the hexagon that circumscribes a drop or bubble, and $\mu$ is the viscosity of the continuous phase. Because of the small amplitude of the imposed motion, the result does not depend on the volume fraction. It was further argued that, in the case of emulsions, the effect of the dispersed-phase viscosity, $\mu_d$, is relatively insignificant. Reinelt and Kraynik (117) later estimated that this is a good approximation as long as

$$\mu_d/\mu \ll Ca^{-1/3} \tag{82}$$

Apart from a change in the numerical coefficient, Eq. (81) is expected to apply also to a periodic, small-amplitude *shearing* motion. However, in *steady shear*, rapid film motions associated with the T1 processes, whose effect has so far not been analyzed, periodically interrupt the above process. Further, as the strain at the instability depends on the volume fraction (Fig. 27), the viscous term may become $\phi$ dependent. Provided that the effect of the T1 jumps may be neglected, or that the associated viscous contribution also scales with $\mu Ca^{-1/3}$, this model would then predict for the shear viscosity:

$$\mu_e = \frac{\tau_0}{\dot{\gamma}} + C(\phi)\mu Ca^{-1/3} = \frac{\tau_0}{\dot{\gamma}} + C'(\phi)\frac{\mu^{2/3}\sigma^{1/3}}{R^{1/3}}\dot{\gamma}^{-1/3} \tag{83}$$

or, for the shear *stress*:

$$\tau = \tau_0 + C(\phi)\frac{\sigma}{a}Ca^{2/3} = \tau_0 + C'(\phi)\frac{\mu^{2/3}\sigma^{1/3}}{R^{1/3}}\dot{\gamma}^{2/3} \tag{84}$$

where $C(\phi)$ and $C'(\phi)$ are of order unity, and the yield stress $\tau_0$ is given by Eq. (70). Equations (83) and (84) describe a particular type of "Herschel–Bulkley" behavior, characterized in general by $\tau = \tau_0 + K\dot{\gamma}^n$ and $\mu_e = \tau_0/\dot{\gamma} + K\dot{\gamma}^{n-1}$. The special case of $n = 1$ is referred to as "Bingham plastic" behavior. Occasionally, foams and concentrated emulsions are claimed to behave as Bingham fluids. As we shall see, this is not so. (In fact, it is extremely unlikely that *any* fluid, when examined carefully, can be described as such.)

Reinelt and Kraynik (118) improved on the above model by including structural changes that result from the fact that the film tensions deviate from the equilibrium value of $2\sigma$ as they are being pulled out of or pushed into the Plateau border. These changes are of order $(Ca^*)^{2/3}$, as already pointed out by Mysels *et al.* (109). As the values and signs of $Ca^*$ at any instant are different for the three films emanating from a Plateau border, their tensions are generally unequal and the

*In the original paper (108) the numerical coefficient was given as 6.7. This and a few other minor numerical errors were pointed out by Reinelt and Kraynik [118 and personal communication].

angles between them deviate from 120°, while the Plateau border radius, $r$ is also affected. However, these refinements do not alter the qualitative conclusion of the original model, as embodied in Eq. (81), for either planar-extensional or shear deformations. Applying this approach to uniform dilatation of a foam, Reinelt and Kraynik (118) also derived an expression for the dilatational viscosity, which again scales with $\mu Ca^{-1/3}$. Using a different surface-rheological description, Edwards and coworkers (119–121) arrived at alternative expressions for the dilatational viscosity of wet and dry foams.

In yet another extension, Reinelt and Kraynik (117) applied the approach to *steady shearing* and *planar-extensional* flow of perfectly ordered 2-D systems for $0.9069 < \phi < 0.9466$. This is the range of "very wet" systems, for which the shear stress varies continuously with strain over a complete strain cycle (cf. Fig. 25), so that rapid film events associated with T1 processes are avoided. They also investigated the effect of orientation, while structural effects due to changes in film tension were again included. As before, the effective viscosity was found to be proportional to $\mu Ca^{-1/3}$. Interestingly, the model indicates that the effective viscosity increases with increasing volume fraction, which parallels practical experience.

Okuzuno and Kawasaki (122) simulated the shear rheology of dry, random 2-D systems, using their "vertex model" in which the films are uncurved and do not generally meet at 120° angles. Although Plateau's condition is therefore violated, the model offers the advantage of being computationally more efficient than other, more realistic models. By solving the "equations of motion" for all the vertices, while taking account of T1 rearrangements and using the energy-dissipation approach of Schwartz and Princen (108), these authors tentatively concluded that the system behaves like a Bingham plastic fluid. However, since the number of simulations were quite limited, they did not rule out Herschel–Bulkley behavior with $n \neq 1$ (see above). In a later study, the same investigators (123) observed violent flows like that of an avalanche in their simulations in the large strain regime at small shear rate. Similar avalanche-like flows were observed in simulations by Jiang *et al.* (124).

This review is not exhaustive by any means. Other studies have been and are being published regularly, as the topic continues to enjoy considerable interest. It appears, however, that theoretical analyses and computer simulations can only go so far. There is a need for careful experimental work in order to establish the *actual* behavior of real systems. As has been the case in the past, further progress will be optimal when the two approaches go hand in hand.

## B. Experimental Approaches and Results

The rheological parameters of primary scientific and practical concern are the static and dynamic shear modulus, the yield stress, and the shear rate-dependent viscosity. The aim is to understand and predict how these depend on the system parameters. In order to accomplish this with any hope of success, there are two areas that need to be emphasized. First, the systems studied must be characterized as accurately as possible in terms of the volume fraction of the dispersed phase, the mean drop size and drop size distribution, the interfacial tension, and the two bulk-phase viscosities. Second, the rheological evaluation must be carried out as reliably as possible.

### 1. System Characterization

The bulk phases are generally Newtonian and their *viscosities* can be measured with great accuracy with any standard method available.

The *nominal volume fraction* of the dispersed phase can be obtained very accurately from the relative volumes (or weights) of the phases used in the preparation of a highly concentrated emulsion (67). A series of emulsions, differing only in volume fraction, may be conveniently prepared by dilution of a mother emulsion with varying known amounts of the continuous phase (67). Alternatively, if the phases differ greatly in volatility, the volume fraction may be obtained, albeit destructively, from the weight loss associated with evaporation of the more volatile phase, usually water (125). Another destructive method is to destroy the emulsion by high-speed centrifugation in a precision glass tube, followed by accurate measurement of the relative heights of the separated liquid columns (22). To arrive at the *effective volume fraction*, the nominal volume fraction may need to be corrected for a finite film thickness according to Eq. (1). Since all rheological parameters depend more or less strongly on the volume fraction, it is important that the vertical gradient in volume fraction due to gravity be kept to a minimum, if reliable rheological evaluations are to be expected. The gradient in volume fraction may be predicted quantitatively (65). Since the drop size and the density difference between the phases are generally much larger in foams than in emulsions, the gradient in $\phi$ is usually much more pronounced in the former than in the latter.

The rheologies of both types of systems being governed by identical laws, it is preferable – for this and many other reasons (see below) – to use *emulsions*, rather than foams, to learn about *foam* rheology.

The *mean drop size* and *drop size distribution* can be measured to within a few per cent accuracy with a number of techniques, such as the "Coulter Counter" (67, 93, 126) and dynamic light scattering. The Coulter Counter is eminently suitable for oil-in-water emulsions but has a lower practical limit of about 1 μm. Various light-scattering techniques are equally suitable for oil-in-water and water-in-oil emulsions and afford a larger dynamic range. In either case, the concentrated emulsion must be diluted with the continuous phase to a level where coincidence counting or multiple scattering, respectively, is avoided. One popular method that should perhaps be avoided is optical microscopy, which is not only tedious but also relatively inaccurate when applied to polydisperse systems because of depth-of-focus limitations and wall effects. At any rate, a practical lower limit for accurate, quantitative optical microscopy is well in excess of 1 μm. Whatever method is used, it is desirable that complete size distributions be reported. At the very least, when only a mean drop size is reported, the *type* of mean should be specified. Finally, it appears that size determinations are a lot easier to obtain in emulsions than in foams. Moreover, while it is easy to prepare emulsions whose drop size distribution changes imperceptibly over a period of months, the bubble size distribution in foams changes very rapidly as a result of Ostwald ripening. It is, therefore, almost impossible to have accurate knowledge of the bubble size distribution at the moment a rheological measurement is being made. These are yet additional reasons for using emulsions in order to investigate foams.

The *interfacial tension* may be determined to within about 1% accuracy with the spinning-drop method (127, 128). It is an absolute and static method that requires only small samples and, in contrast to most other methods, does not depend on the wettability of a probe, such as a ring or Wilhelmy plate. The stabilizing surfactant is commonly used at concentrations in the bulk continuous phase that are far above the critical micelle concentration (cmc). This ensures that the concentration remains above the cmc after adsorption on to the vastly extended interface has taken place, which is clearly needed to maintain emulsion stability. It is tempting, therefore, to assume that the interfacial tension in the finished emulsion equals that between the unemulsified bulk phases and that it remains constant when a "mother emulsion" is diluted with continuous phase in order to create a series of emulsions in which only $\phi$ is varied (67). This may be a reasonable assumption when a pure surfactant is used, but there is evidence that this may not be so when impure commercial surfactants or surfactant mixtures are employed (93, 126).

## 2. Rheological Evaluation

Most studies have used standard rheological techniques, such as rotational viscometers of various types and geometries, such as concentric-cylinder, cone-and-plate, and parallel-plate rheometers, each of which may be operated in various modes (constant stress, constant strain, steady shear, or dynamic, i.e., oscillatory shear). The relative advantages and/or limitations of these and other techniques may be found in any standard textbook on practical rheometry [e.g., (129)]. When applied to highly concentrated emulsions and foams – or suspensions in general, for that matter – these techniques are fraught with many difficulties and pitfalls that are often overlooked, leading to results of questionable validity. Some of these difficulties are the following.

### *a. Wall-induced Instability*

Princen (67) has reported that, otherwise very stable, oil-in-water emulsions showed extremely erratic behavior when sheared in a commercial concentric-cylinder viscometer with stainless-steel parts. The problem could be traced to "coalescence" of the dispersed oil droplets with the steel walls and the formation of a thick oil layer. Apparently, the thin films of continuous phase separating the walls from the first layer of individual droplets were unstable and ruptured. Coating all relevants parts with a thin film of silica, which assured adequate film stability and complete wetting of the steel by the continuous phase, solved the problem (67). Later, an even more satisfactory solution consisted of replacing the steel inner and outer cylinders with glass parts, combined with other improvements in design (93, 126, 130). Some of the glass cylinders were highly polished; others were roughened and equipped with vertical grooves to eliminate or reduce wall slip (see below). Wall-induced instability may or may not be a problem, depending on the wall material, the emulsion (W/O or O/W), and surfactant type.

### *b. End and Edge Effects*

In the analysis of raw data obtained with any type of rotational viscometer, it is assumed that the flow field is known and simple. For example, in the conventional concentric-cylinder viscometer, it is assumed that the fluid moves in concentric cylindrical layers that extend unchanged from the precise top to the precise bottom of the inner cylinder. This is true only when the cylinders are infinitely long. For cylinders of finite length, complications at the top are usually minor and can often be neglected. In the lower region of the viscometer, however, the flow is seriously disturbed. In addition, the bottom of the inner cylinder may contribute a substantial fraction of the total measured torque. This can lead to serious errors. Various suggestions have been made to deal with the problem (129) but their practical value is questionable. In addition to making other improvements, including the use of a *hollow* inner cylinder, Princen (93, 126, 130) effectively isolated the bottom region by filling it with a layer of mercury. That way, the sample of interest is strictly confined to the space between the cylinders. As long as its effective viscosity is much greater than that of mercury, flow between the cylinders is undisturbed and the torque on the bottom of the inner cylinder is negligible. The arrangement is shown schematically in Fig. 29.

In the cone-and-plate viscometer, there are similar, though perhaps somewhat less severe, problems associated with the outer edge (129).

*c. Wall Slip*

Along with wall-induced instability, the occurrence of slip between the sample and the viscometer walls is one of the most serious and prevalent, though often neglected, problems one encounters in assessing the rheology of dispersed systems in general, and concentrated emulsions in particular. Since concentrated emulsions have a yield stress, wall slip – if present – can be readily demonstrated by painting a thin line of dye on top of the sample in a wide-gap rotating-cylinder viscometer (67). As long as the yield stress is not exceeded at the inner cylinder wall, the sample is not sheared at all but is seen to move around in the gap as an elastically strained solid! In this regime, shear is confined to the thin films of continuous phase separating the wall from the adjacent droplets. For a sufficiently smooth wall, it is possible to estimate the thickness of these films from the measured wall stress and angular velocity (67).

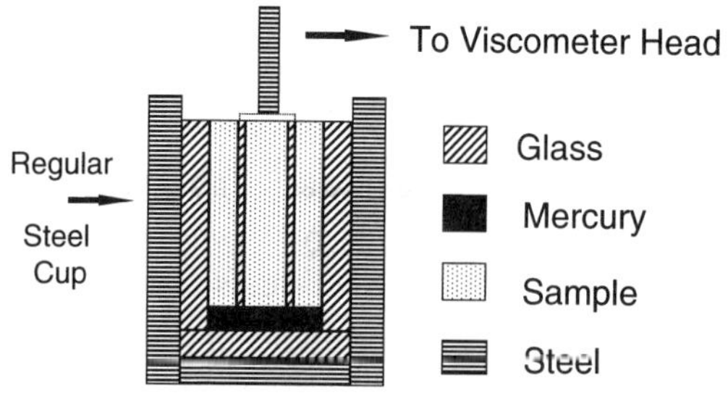


**Figure 29** Modified concentric-cylinder viscometer with glass outer cylinder, hollow glass inner cylinder, and pool of mercury to confine sample to gap and thus to minimize end effect.

It is obvious that neglect of wall slip may lead to meaningless conclusions as to the system's rheology. There are two different approaches to dealing with this particular problem. First, one can try to eliminate slip by roughening the viscometer surfaces. Princen and Kiss (93) successfully used roughened and grooved glass cylinders to determine the static shear modulus of concentrated emulsions. This worked well in the low-stress, linear elastic regime, although even here some wall creep did occur (which could be readily corrected for). However, massive wall slip was noted to commence at shear stresses exceeding only about one-half of the bulk yield stress. Thus, even though the roughness was commensurate with the drop size and served the intended purpose, the arrangement would have been inadequate for determining the yield stress and shear viscosity. Therefore, the question remains how rough a surface must be to eliminate slip up to the maximum shear stress considered. As an extreme case, large radial vanes have been recommended, at least for yield stress measurments (131). Although undoubtedly effective in preventing slip, the vanes do lead to some uncertainty in the strain field.

Many published rheological studies declare that wall slip was checked for and found to be absent. Unless solid evidence is provided, it behooves the reader to approach such assertions with a healthy dose of skepticism.

A second approach is to permit slip and to correct for it. This usually involves running the sample in two or more viscometer geometries, e.g., at different gap widths (129, 132, 133). Doubts have been expressed as to the validity of this approach (134). At any rate, the procedure is rather tedious and may not be very accurate. In an alternative method, Princen and Kiss (126), using their improved design with polished glass cylinders, established empirically that the torque versus angular velocity data for concentrated emulsions may be linearized over most of the all-slip/no-flow regime. The stress at which the data deviated from this linear behavior was identified as the yield stress. Under the further, reasonable assumption that the linearized slip behavior persists above the yield stress, where flow commences, the angular velocity could be corrected for wall slip. Following standard rheological procedures for yield-stress fluids in a wide-gap concentric-cylinder viscometer, the dependence of the effective viscosity on shear rate could then be determined.

It is clear from the above that extreme care must be exercised in the characterization and rheological evaluation of concentrated emulsions. Few, if any, commercial viscometers are designed to give reliable results for nonNewtonian fluids. Not only are modifications of the hardware often called for, but also the software of automated instruments is generally incapable of dealing with yield-stress fluids, end effects, and wall slip. For example, to correct for end effects, it will not do to use a calibration or "instrument factor" for any but Newtonian fluids. Unfortunately, there are no shortcuts in this field!

### 3. Experimental Results

For reasons indicated above, accurate physical characterization and rheological evaluation of *foams* is extremely difficult. Indeed, although there is much published material on foams that is qualitatively consistent with what one would expect (and much that is not), we are not aware of any such studies that can stand close quantitative scrutiny. Therefore, we shall restrict ourselves to what has been learned from *highly concentrated emulsions*, whose rheology is, in any case, expected to be identical to that of foams in most respects. However, even in the emulsion area, the number of carefully executed studies is severely limited. Admittedly not without some prejudice, we shall concentrate on the systematic experimental work by two groups that were active at different times at the Corporate Research Laboratory of Exxon Research and Engineering Co., i.e., Princen and Kiss (67, 93, 126) and Mason and coworkers (64, 125, 135, 136). Both groups used oil-in-water emulsions but, while Princen and Kiss used "typical" polydisperse emulsions with a mean radius of 5 to 10 μm, Mason and coworkers opted for sub-micrometer, monodisperse "Bibette emulsions". The term "monodisperse" is relative; there remained some polydispersity in drop radius of about 10%, and the emulsions were structurally disordered on a macroscopic scale. The mean drop size in Princen's emulsions was at least an order of magnitude greater, which may account for some of the differences in the results (see below). Princen and Kiss used their customized concentric-cylinder viscometer exclusively, either in steady shear with wall slip (to give the yield stress and viscosity) or as a constant-strain device without wall slip (to give the static shear modulus). Mason and coworkers were more eclectic in choosing their techniques (concentric-cylinder and cone-and-plate geometries in steady-shear and dynamic modes, as well as optical techniques).

#### a. *Shear Modulus*

Princen and Kiss (93) used a series of well-characterized, polydisperse oil-in-water emulsions of essentially identical Sauter mean drop size, $R_{32}$, and drop size distribution, but varying dispersed-phase volume fraction, $\phi$. Their modified Couette viscometer was purposely equipped with ground and grooved glass cylinders to eliminate wall slip*, and the emulsion was strained by turning the outer cylinder over a small, precisely measured angle in the linear elastic regime. From the measured stress at the inner cylinder, the static shear modulus, $G$, could be obtained in a straightforward manner. The results in Fig. 30 show that, over the range considered ($0.75 < \phi < 0.98$), $GR_{32}/\sigma\phi^{1/3}$ varies linearly with $\phi$, and we may write

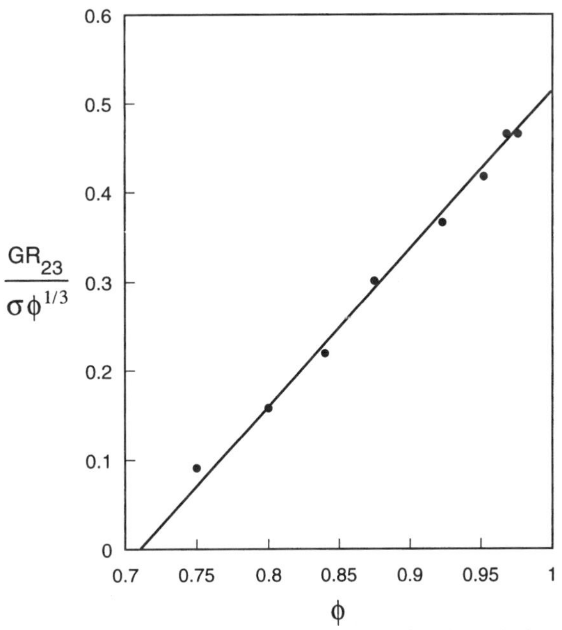


**Figure 30** Scaled static shear modulus, $GR_{32}/\sigma\phi^{1/3}$, vs. $\phi$ for typical polydisperse emulsions. Solid points are experimental data; solid line is drawn according to Eq. (85). (From Ref. 93, with permission from Academic Press.)

*This fact was unfortunately misrepresented in Ref. 64.

$$G = 1.77\frac{\sigma}{R_{32}}\phi^{1/3}(\phi - 0.712) = 1.77\frac{\sigma}{R_{32}}\phi^{1/3}(\phi - \phi_0) \quad (85)$$

where $\phi_0 = 0.712$ may be identified as the "rigidity-loss transition" for the particular size distribution in these emulsions. This is surprisingly close to that for ideal close packing of monodisperse spheres ($\phi_0 = 0.7405$) but clearly in excess of that for random close packing of monodisperse spheres ($\phi_0 \approx 0.64$). The exact value of $\phi_0$ is expected to depend somewhat on the details of the drop size distribution.

In the "dry-foam" limit ($\phi = 1$), Eq. (85) reduces to

$$G_{\phi=1} = 0.509\frac{\sigma}{R_{32}} \quad (86)$$

As indicated above, this is in close agreement with various theoretical estimates.

It may be argued which mean drop size is most appropriate for describing the rheology of polydisperse systems. The selection of $R_{32}$ is based on limited evidence (67) and some other mean might ultimately turn out to be preferable.

A simple extension of the perfectly ordered 2-D model to a 3-D model would have suggested that $G = 0$ for $\phi < \phi_0 = 0.74$, with a sudden jump to an almost constant, finite value of $G \propto \sigma\phi^{1/3}/R$ for $\phi > 0.74$ [cf. Eq. (72)]. As discussed above, it is now generally agreed that the absence of the discontinuity and the essentially linear dependence on $\phi$ above $\phi_0$, found experimentally, is as a result of structural disorder.

Mason *et al.* (64) used small-amplitude, dynamic, oscillatory methods (both in cone-and-plate and concentric-cylinder geometries) to probe the viscoelastic properties, i.e., the storage (elastic) and loss (viscous) moduli, $G'$ and $G''$, as a function of frequency, $\omega$. No mention is made of wall-induced instability, or end and edge effects. Having roughened the viscometer walls, the authors claim that wall slip was nonexistent. At low frequencies, $G'$ reached a plateau that may be equated with the static shear modulus, $G$. Plots of the scaled modulus, $GR/\sigma$, versus the effective volume fraction, $\phi_e$, for four emulsions of different drop size essentially overlapped, as expected. The drops were so small that significant corrections had to be made to the nominal volume fractions to account for the finite (estimated) film thickness, $h$, according to Eq. (1). In the dry-foam limit ($\phi_e = 1$), the scaled modulus approached a value of about 0.6, which is reasonably close to Princen's value of 0.51, but even for $\phi < 1$, the data of the two groups are remarkably similar. For example, for $\phi_e = 0.85$ and 0.75, Mason *et al.* show values for $GR/\sigma$ of about 0.30 and 0.10, respectively, while Eq. (85) yields 0.23 and 0.061 for $GR_{32}/\sigma$. The differences are roughly commensurate with the scatter in Mason's data. At any rate, the difference in polydispersity in the two sets of emulsions, or some experimental factor in either study (end/edge effects?), may well explain these minor systematic discrepancies.

Overall, Mason *et al.* found that their data may be described by

$$G \approx 1.7\frac{\sigma}{R}\phi(\phi - 0.64) = 1.7\frac{\sigma}{R}\phi(\phi - \phi_0) \quad (87)$$

where $\phi_0 \approx 0.64$ is the value for random close packing of monodisperse spheres. Except for the difference in $\phi_0$, this is very similar to Eq. (85).

Because of the limited sensitivity of their viscometer, and the increased potential effect of a gradient in $\phi$ due to gravity, Princen *et al.* (93) did not explore the range of $\phi < 0.75$ and reasonably assumed that the linear behavior in Fig. 30 continues down to $G = 0$ at $\phi = \phi_0 \approx 0.71$. It is unclear what significance, if any, must be attached to the apparent difference in $\phi_0$ found in the two studies. Had it been possible to explore that regime properly, Princen's data might have shown some curvature for $\phi < 0.75$ and a similar smooth decline in $G$ toward zero at $\phi_0 \approx 0.64$. More likely, the difference is real and simply attributable to the differences in polydispersity and associated random-packing density. Another factor of potential significance is the large difference in mean drop size. The drops in Mason's emulsions were submicrometer and, therefore, Brownian, which may contribute an entropic (thermal) component to the modulus, as well as affect the packing density.

Direct support for Eq. (85) has been reported by, among others, Taylor (137), Jager-Lézer *et al.* (138), Pal (139), and Coughlin *et al.* (140). Indirect support has been obtained by Langenfeld *et al.* (141) who compared the specific surface areas of a number of water-in-oil emulsions as determined by two independent methods; (1) from the measured shear modulus – which yields $R_{32}$ from Eq. (85), and thus the specific surface area from $3\phi/R_{32}$ – and (2) from small-angle neutron scattering. The agreement was very satisfactory.

### b. *Yield Stress and Shear Viscosity*

Using their modified concentric-cylinder viscometer – equipped in this case with polished glass inner and outer cylinders to allow unimpeded wall slip, and a mercury pool to eliminate the lower end effect – Princen and Kiss (126) determined the yield stresses,

$\tau_0$, and effective viscosities, $\mu_e$ ($\dot{\gamma}$), of a series of well-characterized, polydisperse oil-in-water emulsions. They empirically established that in all cases the all-slip/no-flow regime at slow steady shear was characterized by a linear dependence of $\tau_1$ on $\omega/\tau_1$ (where $\tau_1$ is the stress on the inner cylinder, and $\omega$ is the angular velocity of the outer cylinder). The stress at which the data deviated from this linearity was identified as the yield stress. At higher angular velocity, it was reasonably assumed that the same linear slip behavior continued to operate, which permitted a straightforward slip correction. Using conventional rheometric analyses, the stress and viscosity were finally obtained as a function of shear rate.

The yield stress data could be expressed in the form:

$$\tau_0 = \frac{\sigma}{R_{32}} \phi^{1/3} \mathrm{Y}(\phi) \tag{88}$$

The experimental values of $Y(\phi)$ are shown in Fig. 31 and may be empirically fit to

$$Y(\phi) = -0.080 - 0.114 \log(1 - \phi) \tag{89}$$

Equation (89) should be used only within the range considered, i.e., $0.83 < \phi < 0.98$.

Data from Pal (139) support Eqs (88) and (89), once the volume fraction is corrected for a finite film thickness of 90 nm. Earlier data from Princen (67) are consistently somewhat higher, probably because of significant end effects in the original, unmodified viscometer.

Figures 32 and 33 show the fully corrected plots of shear stress versus shear rate. Taking account of small differences in the measured interfacial tensions, all data could be accurately represented by

$$\begin{aligned} \tau &= \tau_0 + 32.0(\phi - 0.73) \frac{\sigma}{R_{32}} Ca^{1/2} \\ &= \tau_0 + 32.0(\phi - 0.73) \left( \frac{\sigma \mu \dot{\gamma}}{R_{32}} \right)^{1/2} \end{aligned} \tag{90}$$

where $\mu$ is the viscosity of the continuous phase, and $Ca$ is the capillary number:

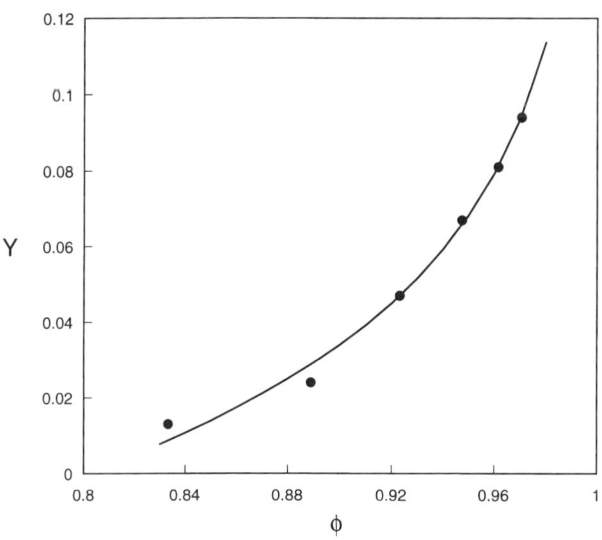


**Figure 31** Yield stress function $Y(\phi) = \tau_0 R_{32}/\sigma\phi^{1/3}$ vs. $\phi$ for typical polydisperse emulsions. Solid points are experimental data; curve is drawn according to Eq. (89). (From Ref. 126, with permission from Academic Press.)

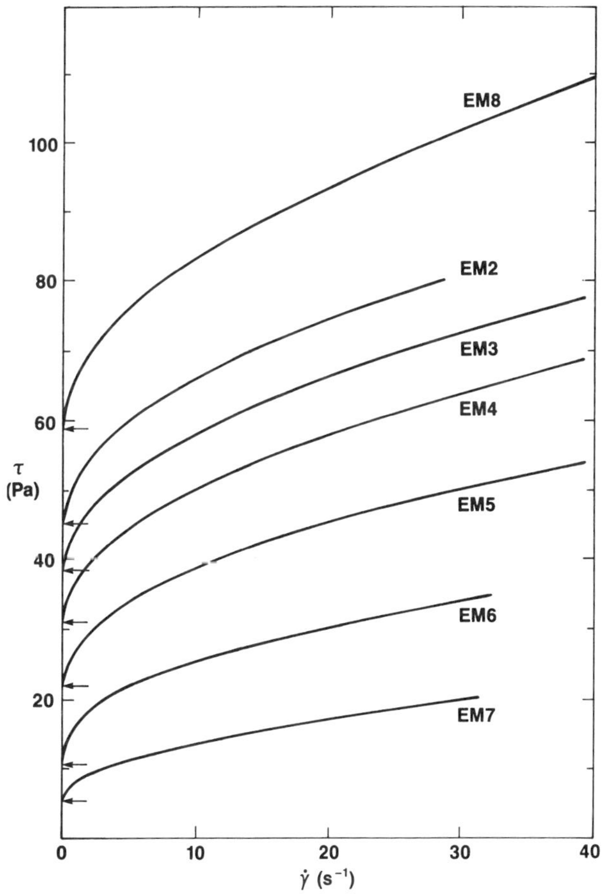


**Figure 32** Fully corrected plots of shear stress vs. shear rate for series of typical polydisperse emulsions. Arrows indicate the yield stress, $\tau_0$. Emulsions EM 2–7 have the same drop size ($R_{32} = 10.1 \pm 0.1$ μm) but different volume fractions ($\phi = 0.9706, 0.9615, 0.9474, 0.9231, 0.8889$, and 0.8333, respectively). For EM8, $R_{32} = 5.73$ μm and $\phi = 0.9474$. (From Ref. 126, with permission from Academic Press.)

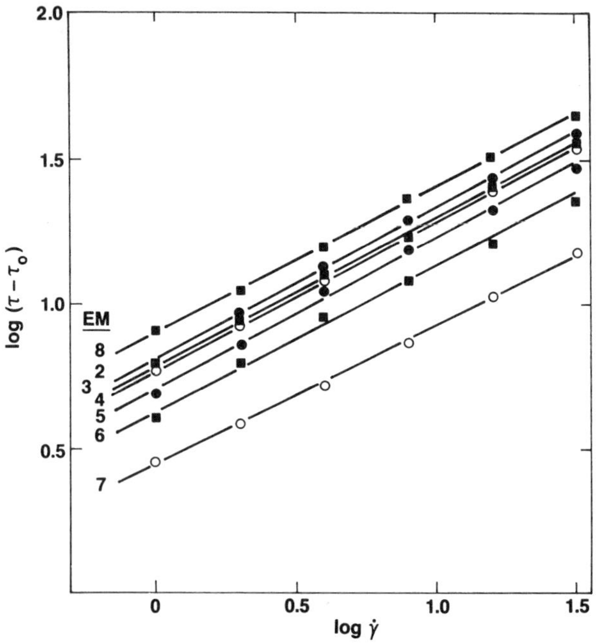


**Figure 33** Plots of log $(\tau - \tau_0)$ vs. shear rate for same emulsions as in Fig. 32. In all cases, the slope is very close to ½. (From Ref. 126, with permission from Academic Press.)

$$Ca \equiv \frac{\mu R_{32}\dot{\gamma}}{\sigma} \tag{91}$$

which did not exceed a value of $10^{-4}$ in any of the experiments.

For the effective viscosity, this leads to

$$\begin{aligned}\mu_e \equiv \frac{\tau}{\dot{\gamma}} &= \frac{\tau_0}{\dot{\gamma}} + 32.0(\phi - 0.73)\mu Ca^{-1/2} \\ &= \frac{\tau_0}{\dot{\gamma}} + 32.0(\phi - 0.73)\left(\frac{\sigma\mu}{R_{32}\dot{\gamma}}\right)^{1/2}\end{aligned} \tag{92}$$

where $\tau_0$ is given by Eqs (88) and (89). Again, Eq. (92) should not be used outside the range considered. It is interesting to point out that, as with so many other properties, the viscous term tends to zero at $\phi = \phi_0 \approx 0.73$.

It is encouraging that Eqs (90) and (92) have the same form as Eqs (84) and (83), respectively, except for the exponent of the capillary number. Several reasons for this difference have been advanced (126), including the neglect of T1 rearrangements and disjoining pressure effects in the original model. At any rate, considering that this is the first and only systematic study of its kind, it is not yet clear how generally applicable Eqs (90) and (92) will turn out to be. Although some other qualitative experimental support exists (142–144), there is a great need for additional, careful studies to explore this area further. It may be significant in this context that Liu *et al.* (145), using diffusing-wave spectroscopy [a light-scattering technique (146)] have found a contribution to the dynamic shear modulus that is proportional to $\omega^{1/2}$ (or $Ca^{1/2}$) and increases roughly linearly with volume fraction. Mason *et al.* (136) investigated the steady shear behavior of some monodisperse emulsions in the low-$\phi$ range. They found that the viscous stress contribution varies as $\dot{\gamma}^{2/3}$ for $\phi = 0.58$ and as $\dot{\gamma}^{1/2}$ for $\phi = 0.63$. For $\phi > 0.65$, no clear power-law behavior was observed. These authors claim that meaningful steady-shear measurements cannot be made on emulsions of higher volume fractions because of the occurrence of "inhomogeneous" strain rates. They presumably refer to the fact that, e.g., in a concentric-cylinder viscometer, only part of the emulsion (i.e., within a given radius) is being sheared, while the outer part is not. However, this situation, common to all yield-stress fluids, has been well recognized and analyzed in the rheology literature, and can be handled in a quite straightforward manner (126).

Mason *et al.* (136) determined the yield stresses and yield strains of a series of monodisperse emulsions, using either a cone-and-plate or double-wall Couette geometry in oscillatory mode. Wall-induced coalescence and wall slip were claimed to be absent, but no mention is made of attempts to reduce end or edge effects. Estimated film thicknesses were used to arrive at the effective volume fractions. Their data for the yield stress could be fit to

$$\tau_0 = 0.51\frac{\sigma}{R}(\phi - 0.62)^2 \tag{93}$$

and, for high $\phi$, are claimed to be "about an order of magnitude greater than those measured for polydisperse emulsions," as given by Eqs (88) and (89). This appears to be a misrepresentation. It is readily demonstrated that the two sets of data are, in fact, quite comparable. For example, for $\phi = 0.85$ and $\phi = 0.95$, the values of the scaled yield stress, $\tau_0 R/\sigma$, are 0.027 and 0.055 according to Eq. (93), and 0.013 and 0.067 according to Eq. (88). In fact, as $\phi \to 1$, Mason *et al.* predict that the scaled yield stress reaches a limiting value of 0.074, whereas extrapolation of Princen and Kiss's data in Fig. 31 suggest a value that is well in excess of 0.1 and perhaps as high as 0.15 (the yield stress must remain finite in this limit and use of Eq. (89) is unwarranted in this regime). Mason *et al.* further assert that, at high $\phi$, the yield *strain* of their monodisperse emulsions is also over an order of magnitude greater than that of the polydisperse emulsions of Princen and Kiss. This conclusion appears to be equally unfounded. In fact, the rheological behavior of concentrated emulsions appears to be remarkably unaffected by polydispersity.

We are not aware of any other systematic experimental studies that meet the criteria set out above and there remains a great need for additional careful work in this fascinating area.

## VIII. ADDITIONAL AREAS OF INTEREST

Although this review covers many aspects of highly concentrated emulsions and foams, it does not deal with a number of issues that are of considerable interest. Foremost is the issue of emulsion and foam *stability*. A great deal of information can be gleaned from recent books on foams and conventional emulsions (17–20). The stability of highly concentrated emulsions is a rather more delicate and specialized problem. The reader may consult a number of publications that specifically deal with this subject (147–152).

One of the main driving forces for the recent upsurge in interest in foams – and one that has been responsible for the entrance of so many physicists into the field – has been their presumed usefulness in modeling grain growth in metals. The coarsening of foam through gas diffusion (a special form of Ostwald ripening) is thought to follow similar laws. This, among other things, inspired the first computer simulations of foams by Weaire and coworkers and remains an active area of research (31).

As indicated above, highly concentrated emulsions provide attractive starting materials for the synthesis of novel materials, e.g., polymers and membranes. Ruckenstein has been particularly active in this area. In addition to the references cited earlier (6, 12–16), the reader may wish to consult a recent comprehensive review of this area (153).

## ACKNOWLEDGMENTS

Special thanks are due to A. M. Kraynik for the many stimulating discussions we have had over the years, for keeping me informed on recent developments, and for kindly providing some of the unpublished results and illustrations. I have also benefited from illuminating discussions with P.-G. de Gennes and D. Weaire. My interest in these fascinating systems goes back to my years at Unilever Research, where E. D. Goddard and M. P. Aronson provided invaluable and much appreciated support and collaboration.

## NOMENCLATURE

### Latin Symbols

| | |
|---|---|
| $a$ | side of hexagon circumscribing compressed 2-D drops in perfect order |
| $a_o$ | side of hexagon circumscribing uncompressed (circular) 2-D drops in perfect order |
| $a_c$ | capillary length $= [\sigma/(\Delta\rho.g)]^{1/2}$ |
| $c_i$ | mean curvature of surface between Plateau border and drop $i$ |
| $C_{ij}$ | mean curvature of film between drops $i$ and $j$ |
| $C_t$ | mean curvature of free surface of continuous phase at dispersion/atmosphere boundary |
| $Ca$ | macroscopic capillary number $= \mu a\dot{\gamma}/\sigma$ or $\mu R_{32}\dot{\gamma}/\sigma$ |
| $Ca^*$ | film-level capillary number $= \mu U/\sigma$ |
| $e$ | number of edges of a polyhedral drop |
| $f$ | number of faces of a polyhedral drop |
| $f(\phi)$ | fraction of surface of confining wall "in contact" with dispersed drops |
| $F$ | stress per unit cell |
| $F_{max}$ | maximum or yield stress per unit cell |
| $g$ | acceration due to gravity |
| $G$ | static shear modulus |
| $G'$ | storage modulus |
| $G''$ | loss modulus |
| $h$ | film thickness |
| $h_{eq}$ | equilibrium film thickness |
| $h_\infty$ | half the film thickness pulled out of Plateau border |
| $H$ | sample height |
| $H_{cr}$ | critical sample height for separation of continuous phase |
| $K$ | compression modulus |
| $p_b$ | pressure in Plateau border |
| $p_c$ | capillary pressure |
| $p_i$ | pressure in drop $i$ |
| $p_v^c$ | vapor pressure of continuous phase in dispersion |
| $(p_v^c)_0$ | vapor pressure of bulk continuous phase |
| $p_v^d$ | vapor pressure of dispersed phase in dispersion |
| $(p_v^d)_0$ | vapor pressure of bulk dispersed phase |
| $P$ | external pressure |
| $r$ | radius of Plateau border surfaces in 2-D close-packed drops |
| $R$ | radius of spherical or circular drop |

$R_{av}$ average drop radius
$R_{32}$ surface-volume or Sauter mean drop radius
$\mathfrak{R}$ gas constant
$S$ surface area of compressed drops
$S_0$ surface area of uncompressed (spherical or circular) drops
$S_f$ surface area contained in films
$T$ absolute temperature
$U$ film velocity
$v$ number of vertices of a polyhedral drop
$V$ dispersion volume
$V_1$ volume of the dispersed phase
$V_2$ volume of the continuous phase in the dispersion
$\overline{V}_1, \overline{V}_2$ partial molar volume of phases 1 and 2, respectively
$Y(\phi)$ yield stress function
$z$ vertical height in dispersion column

## Greek Symbols

$\gamma$ strain
$\dot{\gamma}$ rate of strain
$\Delta\rho$ density difference
$\theta$ contact angle at film/Plateau border junction
$\mu$ viscosity of continuous phase
$\mu_d$ viscosity of dispersed phase
$\mu_e$ effective viscosity of dispersion
$\Pi$ osmotic pressure
$\Pi_d$ disjoining pressure
$\rho$ density
$\sigma$ surface or interfacial tension
$\tau$ stress
$\tau_0$ yield stress
$\tau_s$ stress due to dissipative processes
$\phi$ volume fraction of dispersed phase in emulsion or foam
$\phi_0$ volume fraction of close-packed spherical drops
$\phi_e$ effective volume fraction, after correction for finite film thickness
$\psi$ angle between films and shear direction
$\omega$ frequency or angular velocity

## REFERENCES

1. FJ Almgren, JE Taylor. Sci. Am 235: 82, 1976.
2. CS Smith. Metall Rev 9: 1 (1964).
3. EB Matzke. Am J Bot 33: 58, 130, 1946.
4. KJ Lissant. J Colloid Interface Sci 22: 462, 1966.
5. KJ Lissant. J Soc Cosmet Chem 21: 141, 1970.
6. KJ Lissant, KG Mayhan. J Colloid Interface Sci 42: 201, 1973.
7. A Beerbower, J Nixon, TJ Wallace. J Aircraft 5: 367, 1968.
8. A Beerbower, J Nixon, W Philippoff, TJ Wallace. SAE Trans, Sec. 2, 76: 1446, 1968.
9. J Nixon, A Beerbower, TJ Wallace. Mech Eng 90: 26, 1968.
10. J Nixon, A Beerbower. Preprints, Div Petrol Chem, Am Chem Soc 14: 49, 1969.
11. KJ Lissant. US Patent 3 892 881.
12. KJ Lissant, BW Peace, SH Wu, KG Mayhan. J Colloid Interface Sci 47: 416, 1974.
13. JM Williams. Langmuir 4: 44, 1988.
14. E Ruckenstein, K Kim. J Appl Polym Sci 36: 907, 1988.
15. E Ruckenstein. Colloid Polym Sci 267: 792, 1989.
16. E Ruckenstein, JS Park. Chem Mater 1: 343, 1989.
17. J Sjöblom, ed. Emulsions and Emulsion Stability. Surfactant Science Series, Vol 61. New York: Marcel Dekker, 1996.
18. RK Prud'homme, SA Khan, eds. Foams: Theory, Measurements, and Applications. Surfactant Science Series, Vol 57. New York: Marcel Dekker, 1996.
19. LL Schramm, ed. Foams: Fundamentals and Applications in the Petroleum Industry. Advances in Chemistry Series #242. Washington, DC: American Chemical Society, 1994.
20. D Exerowa, PM Kruglyakov, eds. Foam and Foam Films: Theory, Experiment, Application. Studies in Interface Science, Vol 5. Amsterdam: Elsevier, 1998.
21. HM Princen, MP Aronson, JC Moser. J Colloid Interface Sci 75: 246, 1980.
22. HM Princen, AD Kiss. Langmuir 3: 36, 1987.
23. DMA Buzza, ME Cates. Langmuir 9: 2264, 1993.
24. MP Aronson, HM Princen. Nature 286: 370, 1980.
25. MP Aronson, HM Princen. Colloids Surfaces 4: 173, 1982.
26. HM Princen. Colloids Surfaces 9: 47, 1984.
27. HM Princen. Langmuir 4: 164, 1988.
28. AV Neimark, M Vignes-Adler. Phys Rev E 51: 788, 1995.
29. F Bolton, D Weaire. Phil Mag B 63: 795, 1991.
30. VV Krotov, AI Rusanov. Mendeleev Commun 177, 1998.
31. JA Glazier, D Weaire. J Phys: Condens Matter 4: 1867, 1992.
32. JA Glazier, J Stavans. Phys Rev A 40: 7398, 1989.
33. JA Glazier, SP Gross, J Stavans. Phys Rev A 36: 306, 1987.
34. J Stavans, JA Glazier. Phys Rev Lett 62: 1318, 1989.
35. J Stavans, Phys Rev A 42: 5049, 1990.
36. J Lucassen, S Akamatsu, F Rondelez. J Colloid Interface Sci 144: 434, 1991.
37. HM Princen. J Colloid Interface Sci 71: 55, 1979.
38. HM Princen. Langmuir 2: 519, 1986.

39. PM Kruglyakov, DR Exerowa, KI Khrystov. Langmuir 7: 1846, 1991.
40. K Khrystov, P Kruglyakov, D Exerowa. Colloid Polym Sci 257: 506, 1979.
41. K Khrystov, D Exerowa, PM Kruglyakov. Colloid J (Engl. Transl.) 50: 765, 1988.
42. D Weaire, JP Kermode, Phil Mag B 50: 379, 1984.
43. SA Khan, RC Armstrong. J Non-Newtonian Fluid Mech 25: 61, 1987.
44. AM Kraynik, DA Reinelt, HM Princen. J Rheol 35: 1235, 1991.
45. F Bolton, D Weaire, Phil Mag B 65: 473, 1992.
46. S Hutzler, D Weaire. J Phys: Condens. Matter 7: L657, 1995.
47. KA Brakke. Exp Math 1: 141, 1992.
48. AM Kraynik, DA Reinelt (private communication; to be published).
49. M-D Lacasse, GS Grest, D Levine. Phys Rev E 54: 5436, 1996.
50. W Thomson (Lord Kelvin). Phil Mag. 24: 503, 1887.
51. W Thomson (Lord Kelvin). Acta Math 11: 121, 1887–1888.
52. W Thomson (Lord Kelvin). Mathematical and Physical Papers. Vol V. London/New York: Cambridge University Press, 1911, p 297.
53. HM Princen, P Levinson. J. Colloid Interface Sci. 120: 172, 1987.
54. DA Reinelt, AM Kraynik. J Colloid Interface Sci 159: 460, 1993.
55. D Weaire, R Phelan. Phil Mag Lett 69: 107, 1994.
56. S Ross, HF Prest. Colloids Surfaces 21: 179, 1986.
57. DS Bohlen, HT Davis, LE Scriven. Langmuir 8:892, 1992.
58. HW Schwartz. Rec Trav Chim 84: 771, 1964.
59. H Aref, T Herdtle. In: H Moffat, A Tsinober, eds. Topological Fluid Mechanics. Cambridge University Press, 1990, p 745.
60. T Herdtle. PhD thesis. University of California, San Diego, 1991.
61. C Monnereau, M Vignes-Adler. Phys Rev Lett 80: 5228, 1998.
62. SD Pacetti. Master thesis. University of Houston, 1985.
63. J. Bibette. J Colloid Interface Sci 147: 474, 1991.
64. TG Mason, M-D Lacasse, GS Grest, D Levine, J Bibette, DA Weitz. Phys Rev E 56: 3150, 1997.
65. HM Princen, J Colloid Interface Sci 134: 188, 1990.
66. AM Kraynik, MK Neilsen, DA Reinelt, WE Warren. In: JF Sadoc, N Rivier, eds. Foams and Emulsions. Dordrecht: Kluwer Academic, 1999, pp 259–286.
67. HM Princen. J Colloid Interface Sci 105: 150, 1985.
68. BV Derjaguin. Kolloid Z 64: 1, 1933.
69. S Ross. Ind Eng Chem 61: 48, 1969.
70. ID Morrison, S Ross. J Colloid Interface Sci 95: 97, 1983.
71. HB Hollinger. J Colloid Interface Sci 143: 278, 1991.
72. TL Crowley. Langmuir 7: 430, 1991.
73. TL Crowley, DG Hall. Langmuir 9: 101, 1993.
74. M Blackman. Trans Faraday Soc 44: 205, 1948.
75. JO Sibree. Trans Faraday Soc 30: 325, 1934.
76. A David, SS Marsden. SPE paper #2544, 44th Annual Meeting, Society of Petroleum Engineers, AIME, Denver, 1969.
77. V Sanghani, CU Ikoku. Trans ASME 105: 362, 1983.
78. BJ Mitchell. Oil Gas J 96, 1971.
79. SH Raza, SS Marsden. Soc Petrol Eng J 7: 359, 1967.
80. RJ Mannheimer. J Colloid Interface Sci 40: 370, 1972.
81. JP Heller, MS Kuntamukkula. Ind Eng Chem 26: 318, 1987.
82. JH Aubert, AM Kraynik, PB Rand. Sci. Am. 254: 58, 1986.
83. AM Kraynik. Annu Rev Fluid Mech 20: 325, 1988.
84. D Weaire, MA Fortes. Adv Phys 43: 685, 1994.
85. HM Princen. J Colloid Interface Sci 91: 160, 1983.
86. RK Prud'homme. Presented at Annual Meeting of the Society of Rheology, Louisville, KY, 1981.
87. SA Khan. PhD thesis. Massachusetts Institute of Technology, 1985.
88. SA Khan, RC Armstrong. J Non-Newtonian Fluid Mech 22: 1, 1986.
89. AM Kraynik, MG Hansen. J Rheol 30: 409, 1986.
90. S Hutzler, D Weaire, F Bolton. Phil Mag B 71: 277, 1995.
91. B Derjaguin. Kolloid Z 64: 1, 1933.
92. D Stamenović, TA Wilson. J Appl Mech 51: 229, 1984.
93. HM Princen, AD Kiss. J Colloid Interface Sci 112: 427, 1986.
94. D Stamenović. J Colloid Interface Sci 145: 255, 1991.
95. B Budiansky, E Kimmel. J Appl Mech 58: 289, 1991.
96. DA Reinelt. J Rheol 37: 1117, 1993.
97. AM Kraynik, DA Reinelt. Forma 11: 255, 1996.
98. DA Reinelt, AM Kraynik. J Fluid Mech 311: 327, 1996.
99. AM Kraynik, DA Reinelt. J Colloid Interface Sci 181: 511, 1996.
100. AM Kraynik, DA Reinelt, Proceedings of the XIIth International Congress on Rheology, Quebec City, Canada, 1996, p 625.
101. AM Kraynik, DA Reinelt. Chem Eng Comm 148/150: 409, 1996.
102. DMA Buzza, ME Cates. Langmuir 10: 4502, 1994.
103. DC Morse, TA Witten. Europhys Lett 22: 549, 1993.
104. M-D Lacasse, GS Grest, D Levine, TG Mason, DA Weitz. Phys Rev Lett 76: 3448, 1996.
105. DMA Buzza, C-YD Lu, ME Cates. J Phys II (France) 5:37, 1995.
106. AM Kraynik, MG Hansen. J Rheol 31: 175, 1987.
107. X Li, H Zhu, C Pozrikidis. J Fluid Mech 286: 379, 1995.
108. LW Schwartz, HM Princen. J Colloid Interface Sci 118: 201, 1987.

109. KJ Mysels, K Shinoda, S Frankel. Soap Films: Studies of Their Thinning and a Bibliography. New York: Pergamon Press, 1959.
110. L Landau, B Levich. Acta Physicochim URSS 17: 42, 1942.
111. FP Bretherton. J Fluid Mech 10: 166, 1961.
112. LW Schwartz, HM Princen, AD Kiss. J Fluid Mech 172: 259, 1986.
113. KJ Mysels, MC Cox. J Colloid Interface Sci, 17: 136, 1962.
114. J Lyklema, PC Scholten, KJ Mysels. J Phys Chem 69: 116, 1965.
115. GF Teletzke, HT Davis, LE Scriven. Rev Phys Appl 23: 989, 1988.
116. J Lucassen. In: EH Lucassen-Reijnders, ed. Anionic Surfactants: Physical Chemistry of Surfactant Action. Surfactant Science Series, Vol 11. New York: Marcel Dekker, 1981, Ch. 6, p. 217.
117. DA Reinelt, AM Kraynik. J Fluid Mech 215: 431, 1990.
118. DA Reinelt, AM Kraynik. J Colloid Interface Sci 132: 491, 1989.
119. DA Edwards, H Brenner, DT Wasan. J Colloid Interface Sci 130: 266, 1989.
120. DA Edwards, DT Wasan. J Colloid Interface Sci 139: 479, 1990.
121. DA Edwards, DT Wasan. In: RK Prud'homme, SA Khan, eds. Foams: Theory, Measurements, and Applications. Surfactant Science Series, Vol 57. New York: Marcel Dekker, 1996, Ch. 3, p189.
122. T Okuzuno, K Kawasaki. J Rheol 37: 571, 1993.
123. T Okuzono, K Kawasaki. Phys Rev E 51: 1246, 1995.
124. Y Jiang, PJ Swart, A Saxena, M Asipauskas, JA Glazier. Phys Rev E 59: 5819, 1999.
125. TG Mason, J Bibette, DA Weitz. Phys Rev Lett 75: 2051, 1995.
126. HM Princen, AD Kiss. J Colloid Interface Sci 128: 176, 1989.
127. HM Princen, IYZ Zia, SG Mason. J Colloid Interface Sci 23: 99, 1967.
128. JL Cayias, RS Schechter, WH Wade. In: KL Mittal, ed. Adsorption at Interfaces. AC Symposium Series No. 8. Washington, DC: American Chemical Society, 1975, p 234.
129. RW Whorlow. Rheological Techniques. Chichester: Ellis Horwood, 1980.
130. HM Princen. J Rheol 30: 271, 1986.
131. PV Liddell, DV Boger. J Non-Newtonian Fluids 63: 235, 1996.
132. AS Yoshimura, RK Prud'homme. Soc Petrol Eng 735, 1988.
133. AS Yoshimura, RK Prud'homme, J Rheol 32: 53, 1988.
134. P Brunn, M Müller, S Bschorer. Rheol Acta 35: 242, 1996.
135. TG Mason, DA Weitz. Phys Rev Lett 74: 1250, 1995.
136. TG Mason, J Bibette, DA Weitz. J Colloid Interface Sci 179: 439, 1996.
137. P Taylor. Colloid Polym Sci 274: 1061, 1996.
138. N Jager-Lézer, J-F Tranchant, V Alard, C Vu, PC Tchoreloff, JL Grossiord. Rheol Acta 37: 129, 1998.
139. R Pal. Colloid Polym Sci 277: 583, 1999.
140. MF Coughlin, EP Ingenito, D Stamenović. J Colloid Interface Sci 181:661, 1996.
141. A Langenfeld, F Lequeux, M-J Stébé, V Schmitt. Langmuir 14: 6030, 1998.
142. F van Dieren. In: P Moldenaers, R Keunings, eds. Theoretical and Applied Rheology. Proceedings of the XIth International Congress on Rheology, Brussels: Elsevier, 1992, p 690.
143. Y Otsubo, RK Prud'homme. Soc Rheol 20: 125, 1992.
144. Y Otsubo, RK Prud'homme. Rheol Acta 33: 303, 1994.
145. AJ Liu, S Ramaswamy, TG Mason, H Gang, DA Weitz. Phys Rev Lett 76: 3017, 1996.
146. DJ Pine, DA Weitz, PM Chaikin, E Herbolzheimer. Phys Rev Lett 60: 1134, 1988.
147. E Ruckensten, G Ebert, G Platz. J Colloid Interface Sci 133: 432, 1989.
148. HH Chen, E Ruckenstein. J Colloid Interface Sci 138: 473, 1990.
149. HH Chen, E Ruckensten. J Colloid Interface Sci 145: 260, 1991.
150. MP Aronson, K Ananthapadmanabhan, MF Petko, DJ Palatini. Colloids Surfaces A 85: 199, 1994.
151. MP Aronson, MF Petko. J Colloid Interface Sci 159: 134, 1993.
152. J Bibette, DC Morse, TA Witten, DA Weitz. Phys Rev Lett 69: 2439, 1992.
153. E Ruckenstein. Adv Polym Sci 127: 1, 1997.

*Note*: Since this manuscript went to press, at least two additional books have appeared on the subject of foams and emulsions.

D Weaire, S Hutzler. The Physics of Foams. Oxford: Clarendon Press, 1999.
JF Sadoc, N Rivier, eds. Foams and Emulsions. Dordrecht: Kluwer Academic, 1999.

# 12

# Emulsions—the NMR Perspective

**Balin Balinov**
*Nycomed Imaging AS, Oslo, Norway*

**Olle Söderman**
*University of Lund, Lund, Sweden*

## I. INTRODUCTION

Very early in the development of colloid science, emulsions received considerable attention. This is due to the fact that emulsions are of great fundamental as well as technical importance. They occur in a multitude of situations ranging from biological systems, such as in the digestion of fats, to the extraction of crude oil. Therefore, it is not surprising that the practical knowledge about emulsions is quite extensive. People engaged in the production generally know how to produce emulsions with desired properties such as droplet size distributions and shelf-life. The same level of empirical knowledge is at hand for the opposite process of breaking an emulsion. When it comes to the basic scientific understanding of emulsions, we are a little worse off. It is quite clear that several fundamental properties of emulsions, such as what factors determine the stability of emulsions, what is the importance of the properties of the continuous phase, and so on, are not fully understood. To some extent, we also lack or have not yet applied suitable techniques in the study of emulsions.

The majority of emulsions are stabilized by surfactants, and Bancroft, one of the pioneers in emulsion science, realized that the stability of an emulsion was related to the properties of the surfactant film. This insight has become important during latter years when attempts have been made to draw on the rather detailed and profound understanding that today exists about bulk surfactant systems in the description of emulsions (1). This high level of understanding about bulk surfactant systems stems not least from the application of modern physicochemical techniques such as scattering methods and nuclear magnetic resonance (NMR). It seems reasonable to expect that the application of these techniques to emulsion systems would lead to an increased basic understanding of such systems.

In this contribution we will attempt to show how NMR can be used to study various emulsion systems and how NMR may be used in emulsion applications. We first discuss the NMR technique as such, with special emphasis on features important for the study of emulsions. Subsequently, we treat a few important examples of how NMR can be used to obtain important information about central aspects of emulsions.

## II. THE NMR TECHNIQUE

### A. Fundamentals

NMR is a spectroscopic technique and as such it is very suited for investigations of molecule aspects such as molecular arrangements and molecular dynamics. NMR has gone through a dramatic development over the last decades. For this reason there are numerous monographs treating many different aspects of the method (2–4). Here, we will attempt to introduce the technique from the point of view of emulsion science.

NMR spectroscopy is based on the fact that some nuclei possess a permanent nuclear magnetic moment. When placed in an external magnetic field, they take a certain well-defined state which correspond to distinct energy levels. Transitions between neighboring energy levels take place due to adsorption of electromagnetic radiation of characteristic wavelengths at radio frequencies. In NMR, the signal is obtained by a simultaneous excitation of all transitions with multifrequency pulses followed by detection of the free-induction decay (FID) of the irradiation emitted as the system returns to the equilibrium state. The recorded FID may be used to study the system of interest but Fourier transformation of the time-resolved signal is usually performed to obtain the NMR spectrum.

Modern NMR spectrometers, operating at 500–800 MHz for protons, have a high resolution that allows one to identify up to hundreds of lines in a complex NMR spectra. In most cases only a limited number of lines are used to obtain the information needed. In some applications the spectral resolution is not a necessary step of the data analyses and the information may be obtained from the time dependence of the amplitude of the FID. The FID reflects all the nuclei of a given type, for example, protons, and is used to obtain information on the average relaxation phenomena of the nuclei interest. For this type of signal detection, a 10–20 MHz NMR spectrometer for protons is an option suitable for many industrial applications.

NMR is an extremely versatile spectroscopic technique for three reasons: (1) It is not a destructive technique. Thus, the system may be studied without any perturbation that influences the outcome of the measurements. The system can be characterized repeatedly with no time-consuming sample preparation in between runs. (2) There is a large number of spectroscopic parameters that may be determined by NMR relating to both static and dynamic aspects of a wide variety of systems. (3) A large number of atomic nuclei carry nuclear spins, and for essentially any element in the periodic table it is possible to find at least one suitable nucleus. This has the important consequence that for systems showing local segregation (such as surfactant and emulsion systems) it is possible to investigate different domains of the microheterogeneous system by studying different nuclei.

The most commonly used nucleus in NMR is $^{1}H$, but $^{13}C$, $^{19}F$, $^{31}P$, and $^{129}Xe$ are also of importance. These nuclei are naturally occurring isotopes, and need not be inserted chemically. Another very useful nucleus is $^{2}H$; however, the natural abundance of this species is in general too low to allow for reasonable measuring times. Therefore, this nucleus is usually inserted by chemical labeling. This can in fact be used to advantage since the $^{2}H$ nucleus can be directed to a particular part of the molecule, hence making it possible to investigate different parts of a given molecule.

It is convenient to divide the parameters obtained the NMR spectra into dynamic and static parameters.

### B. Static Parameters

The static parameters are obtained from the observed resonance frequencies. The general frequency range where a particular nucleus shows a spectroscopic line is determined by the magnetogyric ratio, which is a nuclear property without chemical interest. However, the precise value of the resonance frequency is determined by molecular properties. For isotropic systems the two most important parameters determining the resonance frequency is the chemical shift and the scalar spin–spin coupling.

The chemical shift is determined by the screening due to electrons in the vicinity of the investigated nucleus. This in turn is determined mainly by the primary chemical structure of the molecule, but other factors such as hydrogen bonding, conformation, and the polarity of the environment also influence the chemical shift. In systems with local segregation, such as emulsions, this has the consequence that one may observe two separate signals from the same molecule if it resides in two different environments. This occurs if the exchange of the molecule between the two environments is slow on the relevant time scale, which is given by the inverse of the shift difference between the two environments. This then gives us the possibility to investigate the distribution of molecules and exchange rates in microheterogeneous systems.

The scalar coupling is exclusively of intramolecular origin and thus of little importance in emulsion systems.

Another static parameter of importance is the integrated area under a given peak. This quantity is proportional to the number of spins contributing to a given peak. Thus, the area is a quantity by which (relative) concentrations can be determined. This, as will be shown below, has some important applications in emulsion science.

### C. Dynamic Parameters

Dynamic processes on the molecular level influence the nuclear spin system by rendering the spin Hamiltonian time dependent. Depending on the relation between the characteristic times of the molecular motions, $\tau_c$, and the strength of the modulated interaction, $\omega_I$, one can identify different regimes.

For slow motions, when $\tau_c\omega_I \gg 1$, the system is in the solid regime and $\omega_I$ contributes to the resonance frequencies observed. This is the situation encountered in anisotropic liquid crystals where the NMR spectra are usually dominated by static effects.

In the other extreme, when $\tau_c\ \omega_I \ll 1$, the phenomenon of motional narrowing occurs. Here, the spin relaxation is characterized by a limited number of time constants, the most readily observed being the longitudinal, $T_1$, and transverse, $T_2$, relaxation times. The first of these characterizes the decay of the $M_z$ magnetization to equilibrium and the second the return of the $M_{x,y}$ magnetization to equilibrium. The measurement of the relaxation times requires that the sample is perturbed by various pulse sequences. The one most often used for measuring $T_1$ is the inversion recovery sequence (or variations of it), while different echo sequences are usually used for measuring $T_2$.

For the case when $\tau_c\ \omega \approx 1$, the shape of the NMR spectral line is strongly affected by the characteristic features of the molecular motions.

We shall make two remarks concerning two features which are peculiar to the topic of NMR and emulsions.

The first deals with the fact that an emulsion system may actually contain molecules that fulfill more than one of the motional regimes described above. Consider for instance molecules in the continuous phase. For them the condition $\tau_c\ \omega_I \ll 1$ holds. For molecules residing at the interface between the two liquids of the emulsion, the situation may be different, and depending on the size of the emulsion droplet, they may experience any of the three time domains described above.

Our second remark deals with the effect of diamagnetic susceptibility variations. In heterogeneous structures, the magnetic field is perturbed in the vicinity of regions of differing susceptibilities. As a consequence the random motions of the molecules in this spatially varying magnetic field induces relaxation effects. $T_1$ relaxation requires fluctuations at the characteristic nucleus frequency, called Larmor frequency, while $T_2$ relaxation is also sensitive to slower fluctuations. Since the fluctuations brought about by diffusion in the locally varying field typically occur with rates which are slower than the Larmor frequencies, $T_2 \ll T_1$ for this effect. Peculiar to systems with spherical structures, such as reasonably dilute emulsions, is that the gradient of the magnetic field *inside* the droplet is not affected. Thus, the effect described above only operates for the continuous phase, which fact can be used to identify whether an emulsion is of the O/W or W/O type in a straightforward manner (5).

### D. Measurement of Self-diffusion

A very important dynamic molecular process is that of transport of molecules due to thermal motion. This can conveniently be followed by the NMR pulsed field gradient (PFG) method. Since this approach has been of particular importance in the field of microheterogeneous surfactant systems in general and in emulsion systems in particular we will spend some time introducing this application of NMR. The technique has recently been described in a number of review articles [cf. (2, 6–8)], so here we will merely state that the technique requires no isotopic labeling (avoiding possible disturbances due to addition of probes) and that it gives component-resolved self-diffusion coefficients with great precision in a minimum of measuring time. The main nucleus studied is the proton, but other nuclei, such as those of Li, F, Cs, and P are also of interest.

The method monitors transport over macroscopic distances (typically in the micrometer regime). Therefore, when the method is applied to the field of surface and colloid chemistry, the determined diffusion coefficients reflect aggregate sizes and obstruction effects for colloidal particles. This is the origin of the success the method has had in the study of microstructures of surfactant solutions and also forms the basis of its applications to emulsion systems. We expect that the PFG method will also be increasingly important in the study of emulsion systems and therefore we will discuss the method in some detail, with particular focus on its application to emulsions.

The fact that the information is obtained without the need to invoke compliated models, as is the case for the NMR relaxation approach, is particularly important. In this context it should be stressed that the PFG approach measures the self-diffusion rather than the collective diffusion coefficient, which is measured by, for instance, light-scattering methods.

In its simplest version, the method consists of two equal and rectangular gradient pulses of magnitude **g** and length $\delta$, sandwiched on either side of the 180° radio-frequency (r.f.) pulse in a simple Hahn echo experiment. For molecules undergoing free (Gaussian) diffusion characterized by a single diffusion coefficient of magnitude $D$, the echo attenuation due to diffusion is given by (9, 10):

$$E(\Delta, \delta, g) = E_0 \exp\left[-\gamma^2 g^2 \delta^2 \left(\Delta - \frac{\delta}{3}\right) D\right] \tag{1}$$

where $\Delta$ represents the distance between the leading edges of the two gradient pulses, $\gamma$ is the magnetogyric ratio of the monitored spin, and $E_0$ denotes the echo intensity in the absence of any field gradient. By varying either $g$, $\delta$, or $\Delta$ (while at the same time keeping the distance between the two r.f. pulses constant), $D$ can be removed by fitting Eq. (1) to the observed intensities.

As mentioned above, the key feature of PFG diffusion experiments is the fact that the transport of molecules is measured over a time $\Delta$ which we are free to choose at our own will in the range of from a few milliseconds to several seconds. This means that the length scale over which we are measuring the molecular transport is in the micrometer regime for low molecular weight liquids. When the molecules experience some sort of boundary with regard to their diffusion during the time $\Delta$, the molecular displacement is lowered as compared to free diffusion, and the outcome of the experiment becomes drastically changed (2, 11, 12).

This situation applies to the case of restricted motion inside an emulsion droplet. In this case the molecular displacements cannot exceed the droplet size, which indeed often is in the micrometer regime. Until recently, no analytical expressions that describe the echo delay in restricted geometries for arbitrary gradient pulses have been available. However, Callaghan and coworkers have published two approaches that work for arbitrary gradient pulses (13, 14). This is a very important step, and will undoubtedly lead to an increased applicability of the method. Since the application of these approaches are somewhat numerically cumbersome, and since PFG work performed up until now rely on one of two approximative schemes, we will describe these schemes below.

In one of these, one considers gradient pulses which are so narrow that no transport during the pulse takes place. This has been termed the short gradient pulse (SGP) (or narrow gradient pulse, NGP) limit. This case leads to a very useful formalism whereby the echo attenuation can be written as:

$$E(\delta, \Delta, g) = \int\int \rho(\mathbf{r}_0) P(\mathbf{r}_0|\mathbf{r}, \Delta) \exp[i\gamma\mathbf{g}\delta \cdot (\mathbf{r} - \mathbf{r}_0)] d\mathbf{r} d\mathbf{r}_0 \tag{2}$$

where $\rho(\mathbf{r}_0)$ is the normalized spin density and $P(\mathbf{r}_0|\mathbf{r}, \Delta)$ is the propagator which gives the probability of finding a spin at position $\mathbf{r}$ after a time $\Delta$ if it was originally at position $\mathbf{r}_0$.

As discussed by Kärger and Heink (15) and Callaghan (2), Eq. (2) can be used to obtain the displacement profile, which is the probability for a molecule to be displaced d$z$ during $\Delta$ in the direction of the field gradient irrespective of its starting position. Modern NMR spectrometers are capable of producing gradient pulses with a duration less than 1 ms and with strengths around 10 T/m. Under these conditions the SGP limit is often valid. The challenge is to perform such an $E(\delta, \Delta, g)$ experiment that is sensitive to a particular motion in a system of interest. The problem varies from studying free diffusion in the continuous phase in an emulsion, restricted diffusion inside emulsion droplets, molecular exchange between emulsion droplets in highly concentrated emulsions, or drug release from emulsion droplets. Some of those examples will be considered below.

For free diffusion, $P(\mathbf{r}_0|\mathbf{r}, \Delta)$ is a Gaussian function and if this form is inserted into Eq. (2), then Eq. (1) with the term $(\Delta - \delta/3)$ replaced by $\Delta$ is obtained, which is the SGP result for free diffusion. For cases other than free diffusion, alternative expressions for $P(\mathbf{r}_0|\mathbf{r}, \Delta)$ have to be used. Tanner and Stejskal (16) solved the problem of reflecting planar boundaries, while the case of interest to us in the context of emulsion droplets, i.e., that of molecules confined to a spherical cavity of radius $R$, was presented by Balinov et al. (17). The result is:

$$E(\delta, \Delta, g = \frac{9[\gamma g\delta R\cos(\gamma g\delta R) - \sin(\gamma g\delta R)]^2}{(\gamma g\delta R)^2} + 6(\gamma g\delta R)^2 \times \sum_{n=0}^{\infty}[j_n'(\gamma g\delta R)]^2 \sum_m \frac{(2n+1)\alpha_{nm}^2}{\alpha_{nm}^2 - n^2 - n} \exp\left(-\frac{\alpha_{nm}^2 D\Delta}{R^2}\right) \frac{1}{[\alpha_{nm}^2 - (\gamma g\delta R)^2]^2} \tag{3}$$

where $j_n(x)$ is the spherical Bessel function of the first kind and $\alpha_{nm}$ is the $m$th root of the equation $j_n'(\alpha) = 0$; $D$ is the bulk diffusion of the entrapped liquid, and the remaining quantities are defined above. The main point to notice about Eq. (3) is that the echo decay does indeed depend on the radius, and thus the droplet radii can be obtained from the echo decay for molecules confined to the sphere, provided that the conditions underlying the SGP approximation are met.

The second approximation used is the so-called Gaussian phase distribution. Originally introduced by Douglass and McCall (18), the approach rests on the approximation that the phases accumulated by the spins on account of the action of the field gradients are Gaussian distributed. Within this approximation and for the case of a steady-gradient, Neuman (19) derived the echo attenuation for molecules confined within a sphere, within a cylinder, and between planes. For spherical geometry, Murday and Cotts (20) derived the equation for pulsed field gradients in the Hahn echo experiment described above. The result is:

$$\ln[E(\delta, \Delta, g)] = -\frac{2\gamma^2 g^2}{D}\sum_{m=1}^{\infty}\frac{\alpha_m^{-4}}{\alpha_m^2 R^2 - 2} \times \left\{ 2\delta - \frac{\begin{array}{c} 2+\exp[-\alpha_m^2 D(\Delta-\delta)] - 2\exp(-\alpha_m^2 D\delta) \\ -2\exp(-\alpha_m^2 D\Delta) + \exp\left[-\alpha_m^2 D(\Delta+\delta)\right] \end{array}}{\alpha_m^2 D} \right\} \tag{4}$$

where $\alpha_m$ is the $m$th root of the Bessel equation $\frac{1}{(\alpha R)} J_{3/2}(\alpha R) = J_{5/2}(\alpha R)$. Again, $D$ is the bulk diffusion coefficient of the trapped liquid.

Thus, we have at our disposal two equations with which to interpret PFG data from emulsions in terms of droplet radii, neither of which are exact for all values of experimental and system parameters. As the conditions of the SGP regime are technically demanding to achieve, Eq. (4) (or limiting forms of it) have been used in most cases to determine the droplet radii. A key question is then under what conditions Eq. (4) is valid. That it reduces to the exact result in the limit of $R \to \infty$ is easy to show and also obvious from the fact that we are then approaching the case of free diffusion, in which the Gaussian phase approximation becomes exact.

Balinov *et al.* (17) performed accurate computer simulations aimed at further testing its applicability over a wide range of parameter values. An example is shown in Fig. 1. The conclusion reached in Ref. 17 was that Eq. (4) near deviates by more than 5% in predicting the echo attenuation for typically used experimental parameters. Thus, it is a useful approximation and we shall use it in the next section of this paper.

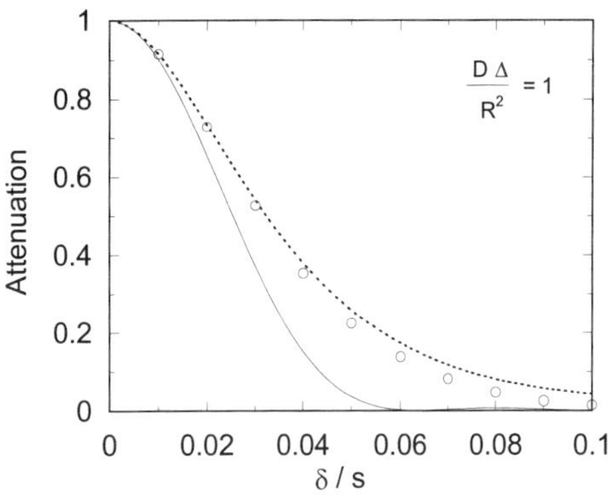


**Figure 1** Results of a simulation of the diffusion of water molecules inside an emulsion droplet of radius $R$, given as the echo amplitude vs. the duration $\delta$ of the field gradient pulse. The ratio $D\Delta/R^2$ is 1. The dotted line is the prediction of the Gaussian phase approximation [Eq. (4)], whereas the solid line is the prediction of the short gradient pulse [Eq. (3)]. (Adapted from Ref. 17.)

As pointed out above, the NMR echo signal $E$ depends on the droplets radius, which can be estimated by measuring $E$ at different durations $\delta$ of the pulse gradient. A typical echo attenuation, generated by using Eq. (4), is presented in Fig. 2, which demonstrates the sensitivity of the NMR self-diffusion experiment to resolve micrometer droplet sizes.

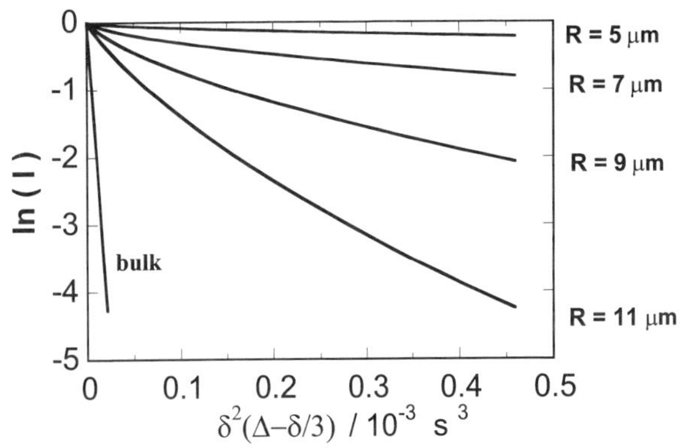


**Figure 2** Echo attenuation as a function of $\delta^2(\Delta - \delta/3)$ for different radii of emulsion droplets [according to Eq. (4)] with $\Delta = 0.100$ s, $\gamma g = 107$ rad $m^{-1}s^{-1}$, and $D = 2 \times 10^{-9}$ $m^2\ s^{-1}$.

We end this section by noting that the discussion above applies to the situation when the dispersed phase cannot exchange between droplets or between the droplets and the continuous medium (at least on the relevant time scale). This is the case for most emulsions. However, for some concentrated emulsions this is not necessarily the case and the molecules of the dispersed phase may actually exchange between droplets during the characteristic time of the measurement. This leads to special effects which will be discussed below.

## III. DETERMINATION OF EMULSION DROPLET RADII BY MEANS OF THE NMR PFG METHOD

As pointed out in Sec. II the echo attenuation curve for the PFG experiment, when applied to molecules entrapped in an emulsion droplet, is a signature of the size of the emulsion droplet, is a signature of the size of the emulsion droplet (cf. Fig. 2). As a consequence, droplet sizes can be determined by means of the PFG experiment.

The NMR sizing method, which was apparently first suggested by Tanner 10, has been applied to a number of different emulsions ranging from cheese to crude oil emulsions (21–27).

When applied to a real emulsion one has to consider the fact that the emulsion droplets in most cases are polydisperse in size. This effect can be accounted for if the molecules confined to the droplets are in a slow exchange situation, meaning that their lifetime in the droplet must be longer than $\Delta$. For such a case, the echo attenuation is given by:

$$E_{\text{poly}} = \frac{\int_0^\infty R^3 P(R) E(R) \mathrm{d}R}{\int_0^\infty R^3 P(R) \mathrm{d}R} \tag{5}$$

where $P(R)$ represents the droplet size distribution function and $E(R)$ the echo attenuation according to Eq. (4) [or, within the SGP approximation, Eq. (3)] for a given value of $R$.

The principle goal is to extract the size distribution function $P(R)$ from the experimentally observed $E(\delta)$. Two approaches have been suggested. In the first, one assumes the validity of a model size distribution with a given analytical form (26, 27), while in the second, the size distribution which best describes the observed experimental $E(\delta)$ dependence is obtained without any assumption on the form of the size distribution (28, 29).

A frequently used form for the size distribution (26, 27) is the lognormal function as defined in Eq. (6) as it appears to be a reasonable description of the droplet size distribution of many emulsions. In addition, it has only two parameters which makes it convenient for modeling purposes.

$$P(R) = \frac{1}{2R}\sigma\sqrt{2\pi}\exp\left[-\frac{(\ln 2R - \ln d_0)^2}{2\sigma^2}\right] \tag{6}$$

In Eq. (6), $d_0$ represents the diameter median, and $\sigma$ is a measure of the width of the size distribution.

To illustrate the method and discuss its accuracy we will use as an example some results for margarines (low-calorie spreads) (21). This system highlights some of the definite advantages of using the NMR method to determine emulsion droplet sizes, since other nonperturbing methods hardly exist for these systems.

Given in Fig. 3 is the echo decay for the water signal of a low-calorie spread containing 60% fat. These systems are W/O emulsions and as can be seen the water molecules do experience restricted diffusion (in the representation of Fig. 3, the echo decay for free diffusion would be given by a Gaussian function). Also given in Fig. 3 is the result of fitting Eqs (4)–(6) to the data. As is evident, the fit is quite satisfactory and the parameters of the distribution function obtained are given in the figure caption. However, one might wonder how well determined these parameters are, given the fact that the equations describing

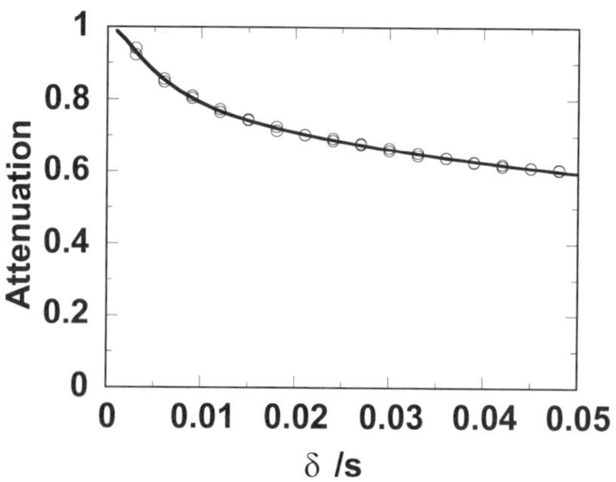


**Figure 3** Echo intensity for the entrapped water in droplets formed in a low-calorie spread containing 60% fat vs. $\delta$. The solid line corresponds to the predictions of Eqs (4), (6), and (7). The results from the fit are $d_0 = 0.82$ µm and $\sigma_0 = 0.72$. (Adapted from Ref. 21.)

the echo attenuation are quite complicated. To test this matter further, Monte Carlo error investigations were performed in Ref. 21. Thus, random errors were added to the echo attenuation and a least-squares minimization was repeated 100 times as described previously (30). A typical result of such a procedure is given in Fig. 4. As can be seen in Fig. 4 the parameters are reasonably well determined, with an uncertainty in $R$ (and $\sigma$, data not shown) of about ±15%.

In the second approach, Ambrosone *et al.* (28, 29) have developed a numerical procedure based on a solution of the Fredholm integral equation to resolve the distribution function $P(R)$ without prior assumptions of its analytical type. The method involves the selection of a generating function for the numerical solution which may not be trivial in some cases. The method was successfully tested by computer simulation of $E(\delta)$ for a hypothetical emulsion with bimodal distribution (31). Figure 5 shows the reconstruction of the true droplet size distribution used to test the method by calculating the size distribution from the corresponding synthesized $E(\delta)$ relationship.

Creaming or sedimentation of emulsions with droplet sizes above 1 μm causes some experimental difficulty because of the change in the total amount of spins in the NMR-active volume of the sample tube during the experiment. This can be accounted for by extra reference measurements with no gradient applied before and after each NMR scan at a particular value of $\delta$. In addition, such reference measurements may provide information on the creaming rate which is a useful characteristic of emulsions. Creaming or sedimentation is not a problem in the study of most food emulsions (such as low-calorie spreads), highly concentrated emulsions, or viscous water-in-crude oil emulsions (22).

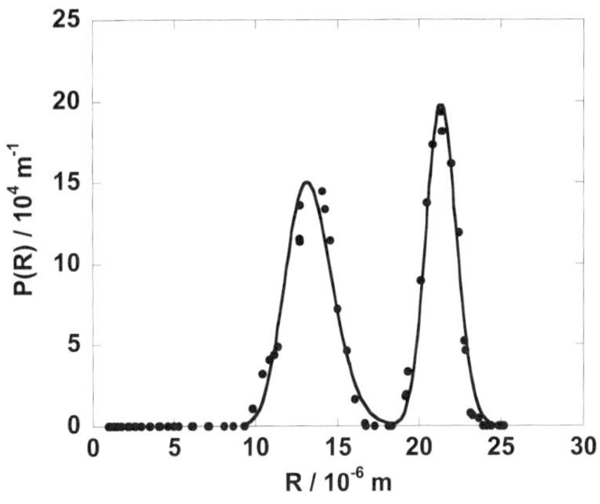


**Figure 5** Determination of the size distribution function in the case of a bimodal distribution. The solid line represents the "true" volume fraction distribution function. The dots represents its values evaluated from the generated NMR data. (Adapted from Ref. 31.)

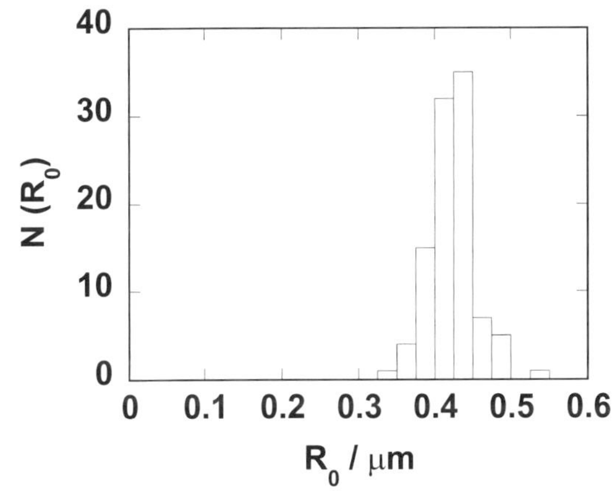


**Figure 4** Monte Carlo error analysis of the data in Fig. 3. The value of the parameter $d_0$ in Eq. (6) is $d_0 = 0.82 \pm 0.044$ μm (note that $R_0 = d_0/2$ is plotted). (Adapted from Ref. 21.)

Emulsion droplet sizes in the range from 1 to 50 μm can be measured with rather modest gradient strengths of about 1 T/m. Note that the size determination rests on measuring the molecular motion of the dispersed phase, so the method cannot be applied to dispersed phases with low molecular mobility. In practice, oils with self-diffusion coefficients above $10^{-12} m^2\ s^{-1}$ is required for sizing of O/W emulsions. Of course, W/O emulsions with most conceivable continuous media can be sized.

Emulsion droplets below 1 μm can often be characterized by the Brownian motion of the droplet as such (exceptions are concentrated emulsions or other emulsions where the droplets do not diffuse). This is the approach taken in the study of microemulsion droplets, where the diffusion behavior of the solubilized phase is characterized by the droplets' (Gaussian) diffusion.

In conclusion, we summarize the main advantages of the NMR diffusion method as applied to emulsion droplet sizing. It is nonperturbing, requiring no sample manipulation (such as dilution with the continuous phase) and nondestructive, which means that the same sample may be investigated many times, which is important if one wants to study long-term stability or the effect of certain additives on the droplet size. It requires small amounts of sample (typically of the order of a few hundred milligrams). Moreover, the total NMR signal from the dispersed phase in emulsions is usually quite intense because of the large amount of spins. This fact allows for rapid measurements with a single scan per $\delta$ point.

## IV. PFG STUDIES OF CONCENTRATED EMULSIONS

The discussion carried out in Sec. III applies to the case where the molecules are confined to the droplets on the time scale of the experiment. This is a reasonable assumption for many emulsions, and it can in fact be tested by the NMR diffusion method by varying $\Delta$. However, there are some interesting emulsion systems where this is not always the case. These are the so-called highly concentrated emulsions (often termed high internal phase emulsions) (32, 33), which may contain up to (and in some cases even more than) 99% dispersed phase. Here, the droplets are separated by a liquid film which may be very thin (of the order of 100 Å), and which may in some instances be permeable to the dispersed phase.

The case when the lifetime of the dispersed phase in the droplet is of the same order of magnitude as $\Delta$ is particularly interesting. Under these conditions, one may in some cases obtain one (or several) peak(s) in the plot of the echo amplitudes. This is a surprising result at first sight, as we are accustomed to observe a monotonic decrease in echo amplitude with $q$, $\left(q = \frac{\gamma g \delta}{2\pi}\right)$, but it is actually a manifestation of the fact that the diffusion is no longer Gaussian. Such peaks can be rationalized within a formalism related to the one used to treat diffraction effects in scattering methods (34), and the analysis of the data may yield important information regarding not only the size of the droplets, but also the permeability of the dispersed phase through the thin films as well as the long-term diffusion behavior of the dispersed phase. We show in Fig. 6 some preliminary results which display such a diffraction-like effect in a concentrated emulsion system. The particular example pertains to a concentrated three-component emulsion based on a nonionic dodecyltetraoxyethylene glycol ether, $C_{12}E_4$, with the composition $C_{12}E_4/C_{10}H_{22}/H_2O$ (1 wt% NaCl), (3/7/90) wt %. The concentrated emulsion was prepared according to a protocol described in Ref. 35.

The data in Fig. 6 are presented with the value of the quantity $q$ on the abscissa. This quantity has the dimension of inverse length and it is in fact related to the scattering vector used in describing scattering

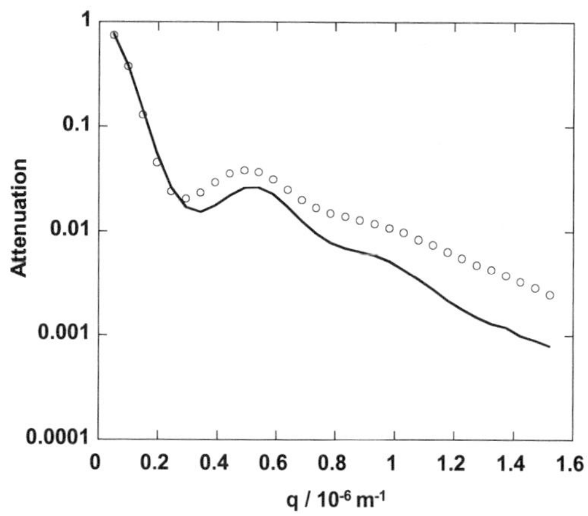


**Figure 6** Diffraction-like effects in a concentrated emulsion system based on a nonionic surfactant with the composition $C_{12}E_4/C_{10}H_{22}/H_2O$ (1 wt % NaCl) (3/7/90 wt %).

experiments. In fact, the inverse of the value of the position of the peak can be related to the center-to-center distance of the droplets. In the example given in Fig. 6 this value is about 1.8 μm which is in rough agreement with twice the droplet radii as judged from microscope pictures taken of the emulsion.

In order to analyze the data in more detail one needs access to a theory for diffusion in these interconnected systems. One such theory was developed by Callaghan *et al.* (12). It assumes that the SGP limit described above is valid and is based on a number of underlying assumptions of which pore equilibration is perhaps the most serious one. This latter assumption implies that an individual molecule in a droplet samples all the positions in the interior of the droplet fully before it migrates to a neighboring droplet. The echo decay for such a case is given by the product of a structure factor for the single pore and a function that depends on the motion of the molecules between the pores. The pore-hopping formalism takes as input the radius of the sphere, the long-term diffusion coefficient, the center-to-center distance between the droplets, and a spread in the center-to-center distance (to account for polydispersity of the droplets). The prediction of the pore-hopping theory for the data in Fig. 6 is included as a solid line. The agreement is not quantitative (the difference most likely owing to problems in defining a relevant structure factor for our system of polydisperse droplets), but the main features of the experimental data are certainly reproduced. The results are: for the pore-to-pore distance 1.2 μm, with a spread of 0.2 μm and, finally, for the long-term diffusion we obtain $D = 9 \times 10^{-11}$ $m^2s^{-1}$. From the last value one can estimate a value for the lifetime of a water molecule in a droplet and the value obtained is 3 ms.

A different starting point in the analysis of the data, such as the one in Fig. 6, is to make use of Brownian simulations (36). These are essentially exact within the specified model, although they do suffer from statistical uncertainties. For the present case, one allows a particle to perform a random walk in a sphere with a semipermeable boundary. With a given probability the particle is allowed to leave the droplet after which it starts to perform a random walk in a neighboring droplet. We are in the process of applying this model to data, of which those presented in Fig. 6 are a subset. That the approach yields peaks in the echo-decay curves can be seen in Fig. 7, where such simulations have been performed under some different conditions (36). The simulation scheme yields essentially the same kind of information as the pore-hopping theory. Thus, one obtains the droplet size and the lifetime of a molecule in the droplet (or quantities related to this, such as the permeability of the film separating the droplet).

Clearly data such as those presented in Fig. 6 can be used to study many aspects of concentrated emulsisons of which a few are exemplified above. It can also be used to study the evolution of the droplet size with time (recall that the method is nonperturbing) and also as a function of changes in external parameters such as temperature, which is an important variable for the properties of nonionic surfactant films. A full

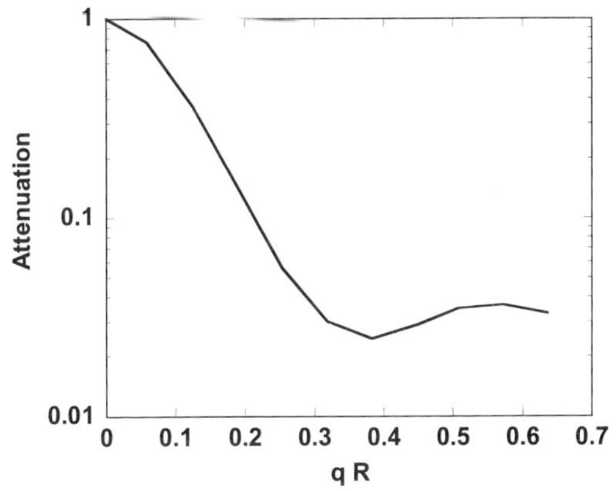


**Figure 7** Brownian dynamic simulations of the echo–decay at various $qR$ at various $q = \gamma g\delta/2\pi$ [$R = 4$ μm, $P_{wall} = 0.032$ (probability of a molecule penetrating the film), $\Delta = 100$ ms]. The simulation scheme yields essentially the same kind of information as the pore-hopping theory. (Adapted from Ref. 36.)

account of the concentrated emulsion work presented above is under preparation.

Finally, we note that concentrated emulsions are excellent model systems in the development of PFG methods as applied to the general class of porous systems, where the method has a great potential in providing relevant and important information.

## V. PFG STUDIES OF MULTIPLE EMULSIONS

Multiple emulsions usually refer to series of complex two-phase systems that result from dispersing an emulsion into its dispersed phase. Such systems are often referred to as water-in-oil-in-water (W/O/W) or oil-in-water-in-oil (O/W/O) emulsions, depending on the type of internal, intermediate, and continuous phase. Multiple emulsions were early recognized as promising systems for many industrial applications, such as in the process of immobilization of proteins in the inner aqueous phase (37) and as liquid membrane systems in extraction processes (38). W/O/W emulsions have been discussed in a number of technical applications, e.g., as prolonged drug-delivery systems (39–44), in the context of controlled-release formulations (45), and in pharmaceutical, cosmetic, and food (46) applications.

Multiple emulsions have a complex morphology and various important parameters for their preparation and characterization have been described (39, 47). Examples are the characteristics of the W/O globules in W/O/W systems, such as their size and volume fraction, W/O ratio inside the W/O globules, and average number and size of water droplets inside the W/O globules. The time dependence of those parameters are closely related to the stability of multiple emulsions and their morphology. Other important features are transport properties of substances encapsulated into discrete droplets and the permeability of the layer separating the internal from the external continuous phase. As was shown above, the NMR PFG method is a sensitive tool to study structure and complex dynamic phenomena, and therefore it is a promising technique in the study of multiple emulsions.

An example of possible use of the method is demonstrated in Fig. 8 where the echo signal from the water in a W/O emulsion (panel a), and from the resulting double emulsion obtained when the original W/O emulsions is emulsified in water (panel b), are displayed (5). Also given are the resulting size distributions (panel c). The state of the water in multiple W/O/W emulsions was examined by the water proton

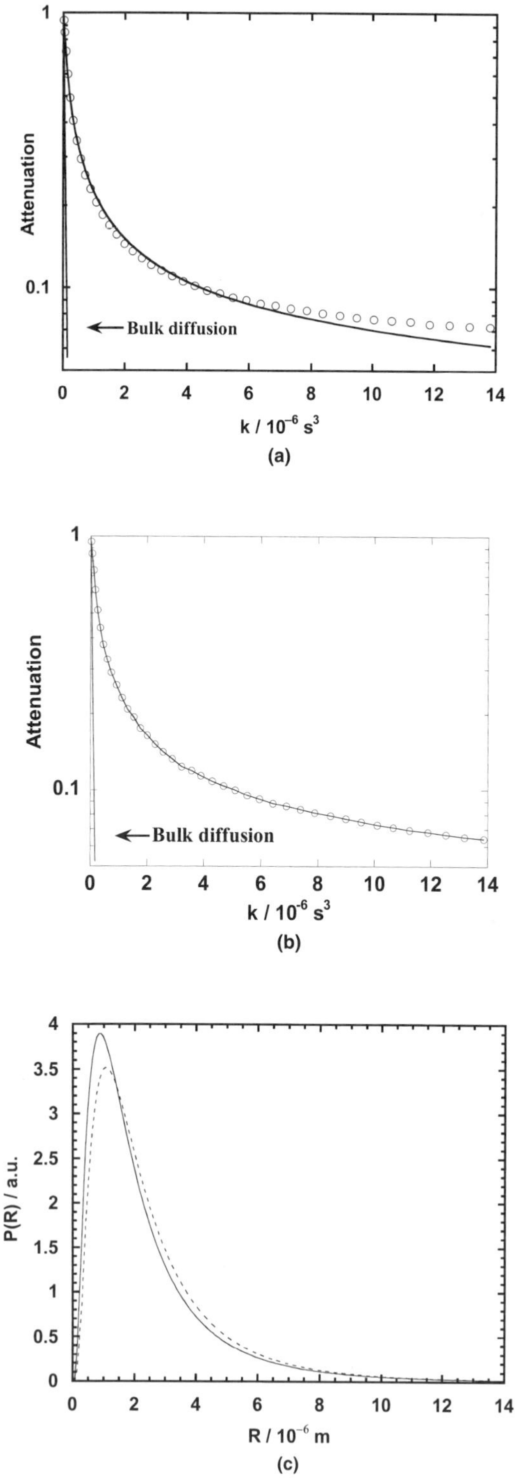


**Figure 8** Echo signal from the water in a W/O emulsion (a), and from the resulting double emulsion obtained when the original W/O emulsions is emulsified in water (b). Also given is the resulting size distributions (c), — is W/O emulsion and - - - is W/O/W emulsion. (Adapted from Ref. 5.)

spectra (48). A narrow signal indicated water in a simple emulsion, whereas a broad signal indicated a multiple emulsion containing dispersed water.

For studies of these complex systems NMR is very well suited, as few other methods exist that can determine such basic properties as the state of water and the size distribution of internal emulsion droplets. In addition, the PFG NMR method is sensitive to the molecular transport from emulsion droplets, a quantity which is relevant in the context of release mechanisms from such emulsion carriers.

## VI. TRANSPORT FROM EMULSION DROPLETS

### A. Drug Delivery

We have shown that the NMR self-diffusion method is sensitive to the mean displacement of a molecule of interest on the time scale of the NMR experiment ($\Delta$). This fact allows us to measure molecular transport inside the emulsion droplets, as in the case of determination of droplet sizes, and the exchange between the emulsion droplets, as in the case of highly concentrated emulsions. In more complex systems the NMR self-diffusion method is sensitive to the molecular exchange between the emulsion droplets and the continuous phase, as in the case of multiple emulsions. Many emulsion systems are currently used as carriers for drugs or other bioactive substances, such as pesticides. The selective measurement of the diffusivity of the individual components within the emulsion system is therefore of theoretical and practical relevance. The NMR self-diffusion technique is an appropriate tool to study the drug release from emulsion droplets. This useful information may be obtained in a rapid and nondestructive way.

Similar theoretical and experimental topics were considered in Refs 49 and 50 where the water diffusions inside a discrete were distinguished from the free diffusion in the surrounding continuous medium. For example, in Ref. 49, the water-diffusion permeability of human erythrocytes was measured by the PFG NMR technique. The measurement of exchange rates was based on restricted diffusion of water molecules within red blood cells, and the average residence time of water (17 ms) inside human erythrocytes was estimated. A study of the transport properties of a model drug by the PFG NMR self-diffusion method has been reported (51). The poorly water-soluble drug, clomethiazole, was dissolved in a pharmaceutical O/W emulsion that is used as a potential drug-delivery system. The drug transport was characterized in terms of slow diffusion within the submicrometer emulsion droplets and fast diffusion of the drug in the continuous phase. The drug exchange between the discrete droplets and the continuous phase was investigated and it was concluded that about 15% of the drug remained in the same emulsion droplet during the measuring time of 140 ms. We stress that a detailed picture of the release mechanism from emulsion droplets may be obtained by NMR self-diffusion methods.

## B. Characteristics of the Displacement Profile

The methodological background to obtaining the transport properties from discrete compartments is the formalism used by Cory and Garroway (50) to obtain the displacement profile (15) of molecules in a dispersed system. Detailed information on the molecular motion may be obtained by measuring the $\Delta$ dependence of the apparent diffusion coefficient caused by a possible obstruction of the spin motion. The stimulated echo sequence, Fig. 9a, is usually used to probe various diffusion times, $\Delta$. As is seen from the figure, time in which $T_1$ relaxation of spins takes place is also varied. To avoid the influence of $T_1$ behavior of the spins a simple sequence (Fig. 9b) is suggested in Ref. 52. This sequence allows one to vary solely the diffusion time $\Delta$ at a constant period of $T_1$ relaxation and to obtain information about the spin displacements.

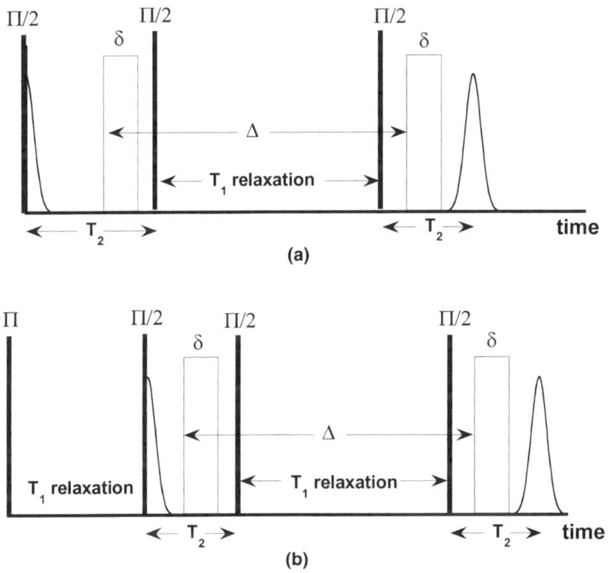


**Figure 9** (a) Stimulated echo pulse sequence with length of gradient pulse $\delta$ and diffusion time $\triangle$; longitudinal and transverse relaxation takes place within the time intervals as indicated. (b) Stimulated echo sequence with longitudinal prerelaxation.

The temporal development of the displacement profile reflects the presence of restricted motion, which may be studied in detail. Examples are motion inside compartments, between compartments, and motion of the compartment itself. We may anticipate progress in the use of PFG NMR methods to study release mechanisms and kinetics from emulsion carriers.

# VII. DETERMINATION OF THE EMULSION COMPOSITION

Many industrial emulsion systems, such as cosmetic or pharmaceutical formulations, have well-defined compositions, while for other systems the composition and nature of the ingredients may be unknown. An example of the latter is water/crude-oil emulsions. The high sensitivity of the NMR experiment and its ability to identify substances by their characteristic spectra may be used to quantify the emulsion composition. This method relies on the fact that the NMR spectra appears as a set of separated signals corresponding to the nuclei of interest and where the separation results from the varying electronic environment within the molecule. The intensity of each signal, determined from the area under the signal, is proportional to the number of equivalent protons. Many emulsion systems are based on water and hydrocarbon oils which have well-resolved lines even in quite primitive NMR spectrometers. This fact allows one to quantify the emulsion ingredients of interest (such as oil, water, surfactant, or additives) without the need to separate the dispersed phase from the continuous phase. This quantitative characterization of emulsion systems may be particularly valuable for quality or process control where an accurate and rapid analysis of the emulsion composition is a major requirement.

The NMR characterization of emulsion composition may be quite valuable also in fundamental studies of emulsion systems and their applications. Various emulsion processes, such as creaming, solvent evaporation, and extraction of substances by emulsions, often require a quantitative analyse of the emulsion components, which can be performed with high precision by NMR techniques.

## VIII. ESTIMATING THE CREAMING OR SEDIMENTATION RATE

It is difficult to measure an absolute value for the creaming (or sedimentation) rate of an emulsion, as there is often a broad size distribution of the emulsion droplets. For an isolated droplet in a continuous medium, the creaming rate is dependent on the difference in density between the droplet and the continuous medium, the size of the droplet, and the viscosity of the continuous medium. It is obvious that different droplet sizes will give different creaming rates (assuming that there is a density difference between the droplet and continuous medium). We note two NMR methods that may be principally valuable in obtaining information on emulsion creaming or sedimentation. The first one is based on the quantitative analysis of the amount of dispersed phase. Emulsion containing large droplets gradually redistribute in a test-tube, and the creaming of the emulsion can be studied by determining the amount of droplet phase suspended in the emulsion at a fixed position as a function of time. This could be done directly in the NMR tube or by analyzing the amount of dispersed phase in a sample withdrawn from a fixed position of the test-tube at distinct time intervals (53). Additional centrifugation of the emulsion, followed by NMR comparison of the composition of the lower and upper fractions is a preferred method for more stable emulsions. The second method for estimating the sedimentation rate is based on flow measurements by PFG NMR. As in the self-diffusion measurements, the method is sensitive to flow rates in the micrometer per second range along the direction of the gradient of the magnetic field.

In many cases the creaming or sedimentation occurs simultaneously with coalescence and is related to emulsion stability. In the next section, we will briefly consider the assessment of emulsion shelf-life by NMR.

## IX. DETERMINATION OF THE EMULSION SHELF-LIFE AND EMULSION STABILITY

Traditionally, emulsion stability is characterized by evaluating the droplet size distribution as a function of time and relating the results to various formulation parameters. On the basis of such studies, the thermodynamic instability of conventional emulsions is understood and well documented (54–56). As discussed above, PFG NMR is able to yield the evolution of the droplet size distribution of the same sample as a function of time without any destruction of the sample. The change in size distribution may then be interpreted in terms of the change in the average droplet size or the total droplet area. There are two mechanisms for the decrease in the total droplet area. The first is the coalescence of two droplets involving the rupture of the film formed in the contact region of two neighboring droplets. The second process is Ostwald ripening involving the exchange of the molecules of the dispersed phase through the continuous phase. For concentrated O/W emulsions [at least the ones we have investigated (57)], the permeability across the thin liquid film between the droplets is so slow that the stability is given by the film-rupture mechanism. In Fig. 10 the total droplet area is depicted as a function of time for a concentrated emulsion consisting of 98 wt % heptane, water, and cetyltrimethylammonium bromide (CTAB). As can be seen, there is a rapid initial decrease in the area, which levels out after 12 h. From the initial part of the curve, the film-rupture rate may be obtained, and for the data in Fig. 10 the value is $4 \times 10^{-5} s^{-1}$. At longer times, the emulsion becomes remarkably stable, and there is little or no further decrease in the total droplet area during a period of one year.

In some concentrated emulsions, molecular exchange between the emulsion droplets occurs on the time scale $\Delta$. This situation is at hand if the dispersed phase crosses the film by some mechanism, the

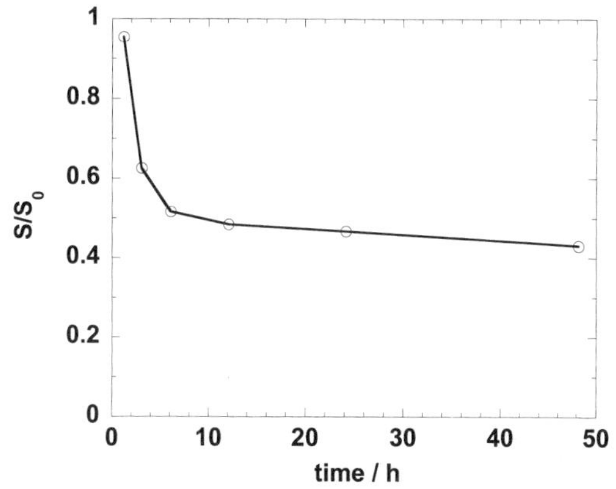


**Figure 10** Decrease of the relative droplet surface area ($S/So$) with time for a highly concentrated O/W emulsion containing 98 wt % heptane and 0.4 wt % CTAB as emulsifier; $S/So$ is calculated from the droplet size distribution as obtained by the NMR self-diffusion technique. The initial slope corresponds to a film-rupture rate of $J = 4 \times 10^{-5}$ $s^{-1}$. (Adapted from Ref. 57.)

detailed nature of which need not concern us here. We are then dealing with a system with permeable barriers (on the relevant time scale), and the system can now be regarded as belonging to the general class of porous systems. As was shown above, interpretation of the PFG NMR signal in this case also provides information on the droplet size. In Fig. 11 the droplet size is determined at two moments (58), demonstrating that the variation in stability of the emulsion system with time can be conveniently followed by this method.

## X. STUDY OF THE DISPERSED AND CONTINUOUS PHASES

### A. Identification of the Dispersed Phase in Emulsion

Emulsions are formed by mixing two liquids, a process which creates discrete droplets in a continuous phase. During emulsification,by mechanical agitation for example, both liquids tend to form droplets resulting in a complex mixture of O/W and W/O emulsions. Which of the components forms the continuous phase depends on the emulsifier used since one of the types of droplet is unstable and coalesces. Therefore, there is a need to identify the continuous phase in emulsion systems not only in the final emulsion system, but also at short times after emulsion formation or even during the emulsification process. The NMR self-diffusion method may easily distinguish the continuous and dispersed phases based on the transport properties of the component molecules. For example, molecules confined in the discrete droplets have an apparent diffusion coefficient which is much lower than the corresponding value in the bulk phase. Molecules in the continuous phase, on the other hand, have an apparent diffusion coefficient similar to the value in the corresponding bulk phase and this fact can be used to identify the type of continuous phase. This is particularly relevant in the case of emulsification by the phase-inversion technique (59) where a single surfactant may form either O/W or W/O emulsions, depending on the formulation conditions, for example, the temperature. In many cases the inversion of the emulsion from O/W to the W/O type is a required and important step of emulsion formation. Due to its sensitivity to the transport properties of the dispersed phase, the NMR self-diffusiosn method is a useful tool for studying the phase-inversion process.

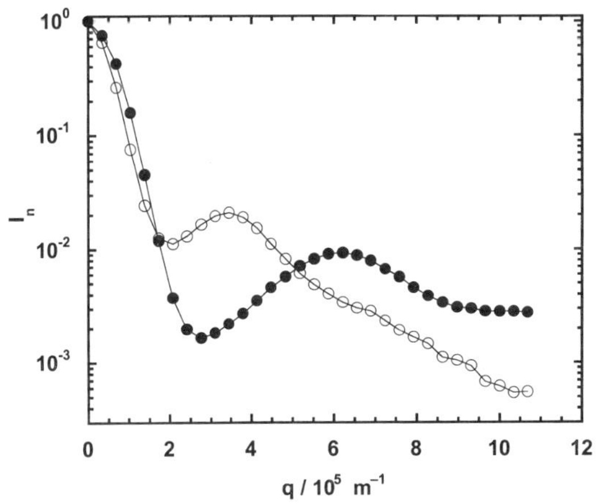


**Figure 11** Normalized intensities vs. $q(q = \gamma g\delta/2\pi)$ for one diffusion time ($\Delta = 50$ ms) at 2 h (full circles) and 8.5 h (open circles) after emulsion preparation. Parameters used in the experiment were $\delta = 3$ ms and a maximum gradient strength of 8.37 T $m^{-1}$. (Adapted from Ref. 58.)

### B. Study of Properties of the Continuous Phase

Intensive work has been carried out in order to establish a relationship between emulsion properties and the properties of surfactant systems. The classical HLB (hydrophile–lipophile balance) concept is widely used in emulsion science to describe the balance of the hydrophilic and lipophilic properties of a surfactant at oil/water interfaces. The HLB value determines the emulsion inversion point (EIP) at which an emulsion changes from W/O to O/W type. This was of particular importance for nonionic surfactants that change their properties with changes in temperature (59). Various NMR techniques have provided significant contributions to this basic understanding of surfactant systems and some of those were reviewed in Ref. 7. The usefulness of NMR techniques in studying surfactant solutions lies in the direct information they provide about the microstructure of microheterogeneous systems (8, 60–64). It is beyond the scope of this chapter to summarize the use of NMR techniques in the study of surfactant systems, but we will present some representative examples related to emulsions.

In order to study the influence of the microstructure of the continuous phase on the stability of emulsions, the present authors investigated the system sodium dodecyl sulfate (SDS)/glycerolmono(2-ethylhexyl)ether/decane/brine (3 wt % NaCl) (65). In this system, emulsions of the O/W or W/O type can be made, depending on the ratio between surfactant and cosurfactant. The total amount of surfactant and cosurfactant is kept constant at 5 wt %. The samples are made with equal weights of brine and decane and with a varying ratio between surfactant and cosurfactant. The emulsions in this system are made in two phase areas of the phase diagram which for the O/W emulsions consists of an oil-rich phase and a phase of normal micelles. For the W/O emulsions, it consists of a water-rich phase and a micellar phase of reversed micelles. The micellar phase is the continuous medium for both types of emulsions. In order to determine the structure of this continuous medium, we let the emulsion samples cream (or sediment) and separated the clear continuous medium from each sample. These solutions were then characterized by the NMR self-diffusion method and the diffusion coefficients of both the oil and the water were determined. The result is shown in Fig. 12, where the reduced diffusion coefficients ($D/D_0$, where $D$ is the actual diffusion coefficient and $D_0$ is the diffusion coefficient of the neat liquid at the same temperature) for the oil and the water are plotted versus the ratio between cosurfactant and surfactant. For the O/W emulsion where SDS is the only surfactant, one finds that the continuous medium consists of small spherical micelles, that the water diffusion is fast [slightly lowered relative to neat water due to obstruction effects (66)], and that the oil diffusion is low and corresponds to a hydrodynamic radius of the oil-swollen micelle of about 50 Å (according to Stokes law).

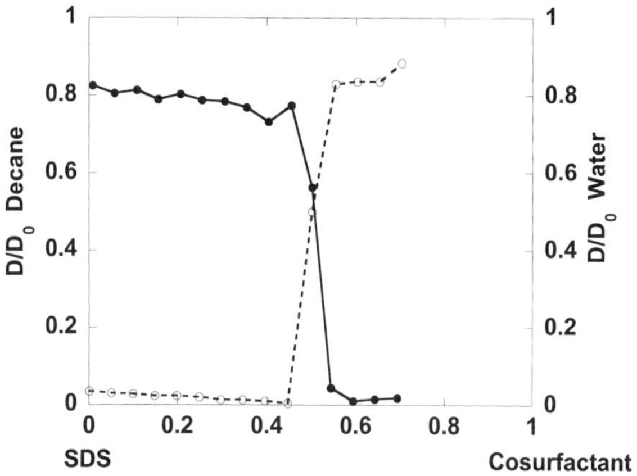


**Figure 12** Microemulsion structure in the continuous phase studied by the diffusion coefficients ($D$) divided by the diffusion coefficients ($D_0$) for the neat liquid vs. the relative amount of SDS in the SDS – surfactant mixture. (Adapted from Ref. 127.)

When the cosurfactant is introduced and its amount is increased, the size of the micelle is increased, as can be inferred from the diagram by the lowered value of the reduced diffusion coefficient of the oil. Close to the three-phase area, the continuous medium is bicontinuous as the value of the reduced diffusion coefficient is almost the same for both the oil and the water and is equal to approximately 50% of the value for the bulk liquids. When the amount of cosurfactant is increased further, one passes over to the W/O emulsion region where the continuous medium is bicontinuous near the three-phase area and then changes to closed reversed micellar aggregates as can be seen from the reverse in order of the magnitudes of the values of oil and water diffusion coefficients. The hydrodynamic radius of the inverse micelles is about 70 Å.

In another study (67), NMR self-diffusion measurements of the continuous oil phase show that a stable, highly concentrated W/O emulsion is formed when the continuous phase is a reverse micellar solution, while an extremely unstable emulsion is formed when the continuous phase is a bicontinuous microemulsions.

## C. $^{31}$P-NMR of Emulsion Components

The linewidths of $^{31}$P-NMR can be used to characterize the motional properties of phospholipids. In emulsions, the linewidths are affected by the aqueous phase pH, size of dispersion states of particles, and methods of emulsification. The hydrophilic head-group motions of emulsified egg-yolk PC and lyso PC are examined by $^{31}$P-NMR to evaluate their phospholipid states and stability. The results suggest that the head-group motions of phospholipids are related to emulsion stability (68, 69).

Many emulsion systems are stabilized by phospholipids that form various self-organized structures in the continuous phases. Examples are fat emulsions containing soy triacylglycerols and phospholipids that are used for intravenous feeding. Studies have shown that these emulsions contain emulsion droplets and excess of phospholipids aggregated as vesicles (liposomes), which remain in the continuous phase upon separation of the emulsion droplets by ultracentrifugation. The lamellar structure of the vesicles in the supernatant was characterized by $^{31}$P-NMR (70), which distinguished lipids in the outer and inner lamellas. $^{31}$P-NMR was used (71) to confirm that the resulting structures in lipid emulsions are emulsion droplets rather than lipid bilayers. The composition of the fat particles of parenteral emulsions was obtained by $^{31}$P-NMR (72). Analysis of the data identified the ratio of phospholipid/triacylglycerol in various fractions of emulsion droplets separated by centrifugation. $^{31}$P-NMR showed (73) that approximately 48 mol % of the phospholipid emulsifier in model intravenous emulsion forms particles smaller than 100 nm in diameter.

$^{31}$P-NMR and $^{13}$C-NMR may be used to study the emulsifier properties at the O/W interface. The analysis of $T_1$ relaxation times of selected $^{13}$C and $^{31}$P nuclei of $\beta$-casein in oil/water emulsions indicates (74) that the conformation and dynamics of the N-terminal part of $\beta$-casein are not strongly altered at the oil/water interface. A large part of the protein was found in a random-coil conformation with restricted motion and a relatively long interal correlation time.

We can conclude that similar NMR techniques are powerful in the study of both the emulsion systems and the surfactant systems from which the emulsions are formed.

## XI. DEGREE OF SOLIDIFICATION OF THE DISPERSED PHASE

Many emulsion-based formulations also contain solid particles. Typical examples are food products, such as margarine and salad dressing or petroleum products, such as crude oil emulsions or bitumen emulsions.

In the food industry the determination of the amount of solid fat is an essential part of the process control. An example is the monitoring of the fat hardening in margarine after its formation as a W/O emulsion. Close control of the solid fat content is needed to give the margarine its characteristic properties. The method of determination of the solid fat content will be briefly described as well as its application to the study of emulsion stability.

### A. Determination of the Solid Fat Content

The determination of the solid fat content by pulsed NMR is based on the fact that the transverse magnetization of solid fat decays much faster than that of oil. The spin–spin relaxation time ($T_2$) of solid fat is about 10 μs, and that of oil is about 100 ms. The NMR signal, derived from the amplitude of the FID, of partially crystallized fat after a 90° r.f. pulse is schematically shown in Fig. 13. The magnetization of the solid fat decays very fast. As a consequence, its contribution to the signal is far less than 0.1% of the initial value after about 70 μs. The decrease in the liquid-oil signal

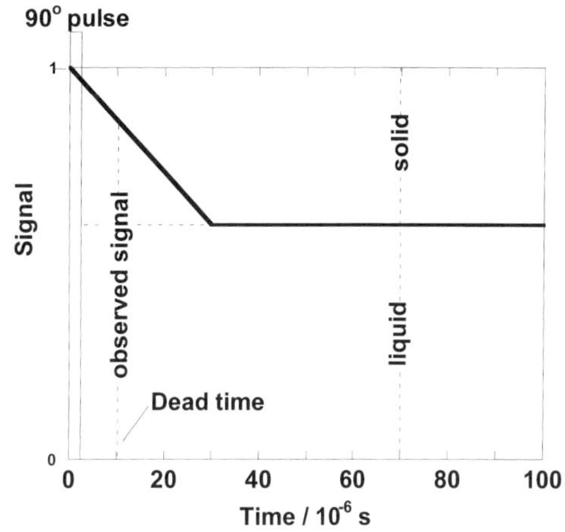


**Figure 13** Signal of partially crystallized fat after a 90° r.f. pulse.

at this moment (70 μs) is less than 1% and, therefore, the signal intensity will be directly proportional to the number of protons in the liquid.

The solid fat content may be determined by a direct method from the signals corresponding to the total amount of solid and liquid fat (measured at time $t \approx 0$) and the amount of liquid fat (measured at a long enough time). Another method, known as the indirect method, determines the solid fat content $S_{\text{ind}}$ by comparing the signal from the liquid fraction $I_l$ with the signal $I_m$ from completely melted fat. We have:

$$S_{\text{ind}} = \frac{cI_m - I_l}{cI_m} \cdot 100\% \tag{7}$$

where the factor $c$ corrects the signal for the temperature dependence of both the equilibrium magnetization and the $Q$-factor of the receiver coil. The correction factor $c = \frac{I_0 t}{I_{0m}}$ is obtained by measuring the signals $I_{0t}$ and $I_{0m}$ of a reference liquid sample at both the measuring temperature and the melting temperature, respectively. The indirect method is mainly used as a reference for the direct method when the signals from both the liquid and the solid fat are processed. The indirect method is similar to the previously used continuous-wave (wide-line) technique that has some disadvantages compared to pulsed NMR. For example, saturation conditions are needed to obtain a sufficiently large signal-to-noise ratio. Even with a well-chosen reference sample the systematic error is in the range 1–2% (75). The wide-line analyzer gives only a narrow peak from the liquid fat which should be compared to the signal for melted fat. This requires a second measurement to be performed after at least an hour, and automation of the measurement is hardly possible. Pulsed NMR solves these problems and measures the solid fat content, based on the processing of both the liquid and the solid signal (76). The measurement procedure may be fully automatic and the percentage of solid is displayed immediately after the measurements. The measuring time is a few seconds and the solid fat content is determined with a standard deviation of about 0.4%. The signal can be obtained directly from the magnetization decay of the solid fat protons and is equal to the signal immediately after the 90° pulse (Fig. 13). Due to the dead time of the receiver it is not possible to measure the initial signal height of solid and liquid ($s + l$), but only a signal ($s' + 1$) after a certain time of about 10 μs after the 90° pulse. To determine the solid fat content the "true" NMR signal from the solid fat, $s$, may be obtained from the measured signal, $s'$, from the solid fraction multiplied with the correction factor $f$, which depends on the $T_2$ of the solid fat protons. The solid fat fraction $S$ can be expressed by $s'$ (equal to the difference between the observed and liquid signal according to Fig. 13):

$$S = \frac{fs'}{l + fs'} 100\% \tag{8}$$

The correction factor $f = s/s$ can be determined from the measured $s'$ and the solid fraction $s$ obtained by the indirect method [Eq. (7)] using a reference sample. The calibration should be performed once for a series of similar samples. The preparation of the sample is simple and takes about 15 s, resulting in an instrument capacity of about 200 samples per hour.

Commercial spectrometers are available that converts the NMR signal into the percentage of solid fat content in fats and margarine. Examples of NMR spectrometers that are suitable for characterization of solid fat content are PC100, NMS100, the Minispec from Brucker, Qp20+ from Oxford Instruments, and Maran Ultra from Resonance Instruments. Determination of the solid fat content by NMR is a recognized international ISO standard (77).

## B. Studies of Emulsion Stability

It was demonstrated (78, 79) that pulsed NMR may be used to measure the extent of oil solidification during cooling of O/W emulsions. Pulsed proton-NMR can distinguish between the oil and aqueous phases because of the large differences between the relaxation times of oil and water protons. On cooling, the NMR signal obtained for the aqueous phase is relatively small compared to that for the oil phase during the entire cooling cycle. For example, the water signal was approximately 10 times lower than the water signal for an O/W emulsion containing approximately 50% hydrogenated oil (80). When the lipid phase containing the lipophilic emulsifier is cooled the NMR signal decreases as the sample solidifies. The total NMR signal from all the protons of the emulsion system is approximately additive (the sum of the aqueous and oil phases). In practice, the signal obtained for the supercooled emulsion is always larger than expected from a mixture with a solid fat and water, indicating that emulsification has had an inhibiting effect on fat solidification. The dispersion of fat into small droplets suppresses the rate of solidification under supercooling due to a nucleation phenomenon in confined emulsion droplets (81–83). The extent of solidification at supercooling is high for very large emulsion droplets and correlates with a low emulsion stability. Unstable emulsions show little supercooling, but those that are relatively stable to creaming and phase separation are resistant to oil solidification. The greater the degree of dispersion, the slower the rate of phase separation. A correlation was made between the emulsion stability and the NMR signal from the emulsion when compared to the NMR signal from the fat and water phases alone (80). A parameter called "percent interaction" was derived from the NMR signal (79) that correlated well with actual resistance of the emulsion to creaming and phase separation during storage. For example, the NMR signals from an emulsion and its corresponding fat phases were determined for an emulsion containing 48% hydrogenated oil, 1% acetylated monoglyceride (a 49% total fat phase), 1% Tween 20, and 50% water (a 51% aqueous phase) (80) as follows:

Fat phase signal, $71.5 \times 0.49 = 35.0$
Aqueous phase signal, $7.51 \times 0.51 = 3.8$.
Expected emulsion signal, 38.8
Observed emulsion signal, 41.5.

An "interaction percentage" (Int%) of 7% was calculated by the equation:

$$Int\% = \left( \frac{s_1 + l + w}{s + w} - 1 \right) \cdot 100\% \tag{9}$$

The observed emulsion signal ($s_1 + l + w$) corresponding to the signal from the solidified fat fraction, $s_1$ and unsolidified fat, $l$, and water, $w$, is compared to the expected emulsion signal ($s + w$) corresponding to solid fat and water.

An emulsion is considered to be relatively stable if the observed NMR signal is more than 30% larger than the sum of the NMR signals from the corresponding bulk water and oil (80). Pulsed-NMR cooling-curve measurements on emulsions offer an improved method for prediction of emulsion stability. By using a flow-through cell in the NMR magnet, the rate and extent of phase separation was measured accurately (79). The method was useful in the selection of optimum types and levels of emulsifiers for each system (78). NMR measurements can also assist in optimizing surfactant blends to obtain a stable emulsion. At the optimum emulsion formulation for a particular oil, the "interaction percentage" will be at a maximum, which indicates higher emulsion stability.

## XII. NMR STUDY OF FLUOROCARBON EMULSIONS

The intensive study of fluorocarbon emulsions is mainly driven by potential biomedical applications such as their use as blood substitutes and in medical imaging. There has been considerable interest in developing a blood substitute capable of transporting, and delivering, oxygen to the tissues. The classical experiment of Clark and Gollan showed that perfluorocarbon (PFC) may transport oxygen, and laboratory animals could survive while totally immersed in PFC liquid (84). An O/W emulsion of inert perfluorochemicals has been proposed as a possible substitute for blood because of the high solubility of gases such as oxygen and carbon dioxide in perfluorocarbons. This fact allows oxygen to be dissolved in the emulsion droplets, i.e., inserted between the fluorocarbon molecules without any specific bonding site. The research on this approach for oxygen transport culminated in a production of the first PFC-based blood substitute, Fluorosol-DA© (Green Cross Corp., Japan). This preparation, a 20% emulsion of a mixture of perfluorodecalin and perfluoropropylamine, in a balanced electrolyte solution, was first used in human volunteers. Fluosol-DA© stimulated the basic research and the considerable interest in potential applications of PFCs in many areas of clinical medicine, such as in eliminating gaseous microemboli (85), as a therapeutic treatment after myocardial infarction (86–91), and in cancer therapy (89, 92–100).

NMR provides useful methods for studying perfluorocarbon emulsions. Fluorine is of special interest for biomedical applications. The $^{19}F$ nucleus is magnetically active and the gyromagnetic ratio is only slightly less than that of protons, so that a high sensitivity is obtained. In addition, there are only traces of naturally occurring $^{19}F$, so $^{19}F$ may be used to label substances, the distribution and the local properties of which are expected to reflect pathological states of the organism.

### A. Measuring Oxygen Concentration in Fluorocarbon Emulsions

For PFC compounds pertinent to blood substitutes, the oxygen-dissolving capacities range from 40 to 50 vol %, and those for carbon dioxide from 140 to 230 vol % (101). The individual resonances of PFC are sensitive to oxygen tension (*p*O2) (102–105). For example, perfluorotributylamine shows four peaks in the $^{19}F$-NMR spectrum obtained with a 75.38-MHz spectrometer (103). An example of such a $^{19}F$ spectrum from a perfluorotributylamine emulsion is given in Fig. 14 (106). When the $^{19}F$ longitudinal relaxation rates of perfluorotributylamine were plotted against the partial pressure of oxygen, they showed straight lines with different slopes for the fluorine atoms at four different positions. Longitudinal-relaxation rates of each fluorine nucleus in emulsions of perfluorotributylamine in water with Pluronic F-68 (BASF Corporation) as emulsifier also shows a linear relationship with respect to the partial pressure of oxygen (Fig. 15).

The reason for this dependence is the difference in the $^{19}F$ longitudinal relaxation rate between the oxygen-free fluorocarbon and the ones with oxygen in

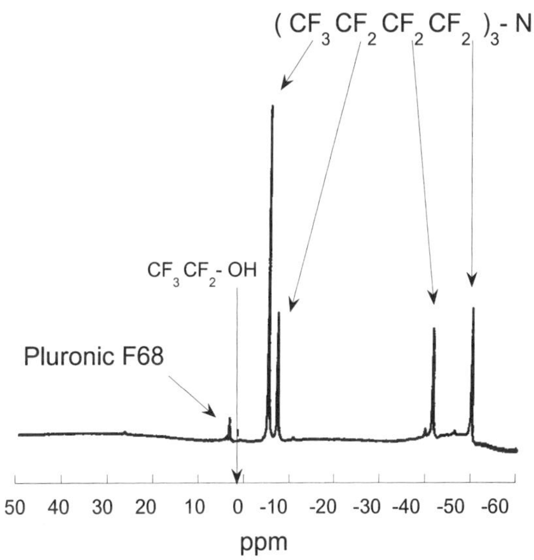


**Figure 14** $^{19}F$ spectrum of perfluorocarbon emulsion FC-43 containing 20(w/v)% perfluorotributylamine and Pluronic F-68 as emulsifier; the position of trifluoroethanol line is indicated. (Adapted from Ref. 106.)

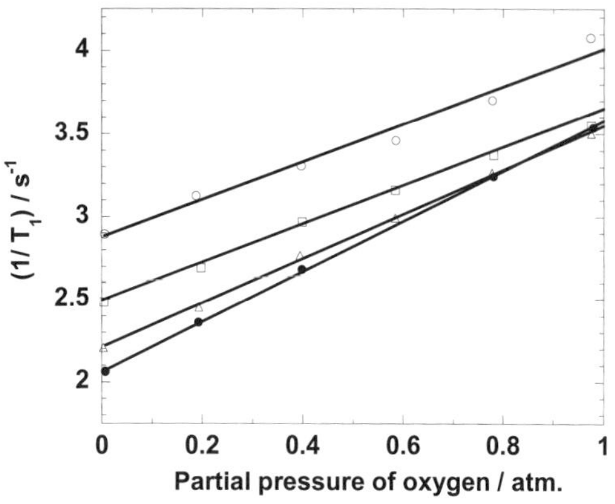


**Figure 15** Dependence of the $^{19}$F longitudinal relaxation rates on the oxygen tension for a 10% emulsion of perfluorotributylamine in water (4% Pluronic F-68) at 75.38 MHz and 298 K; △–$\alpha$ fluorine, –$\beta$ fluorine, o–$\gamma$ fluorine, •–$\delta$ fluorine. (Adapted from Ref. 103.)

their immediate vicinity. Since the oxygen molecules rapidly diffuse in the perfluorocarbon solvent, the observed relaxation rate for each type of fluorine atom is a weighted average:

$$\frac{1}{T_1} = (1-x)\frac{1}{T_{1d}} + x\left(\frac{1}{T_{1d}} + \frac{1}{T_{1p}}\right) = \frac{1}{T_{1d}} + \frac{x}{T_{1p}} \tag{10}$$

where $x$ is the mole fraction of oxygen, $1/T_{1d}$ is the longitudinal relaxation rate of the oxygen-free fluorine nucleus, and $1/T_{1p}$ is the paramagnetic relaxation rate due to the presence of oxygen.

Since the solubility of oxygen in the PFCs is proportional to the partial pressure $p_{O_2}$ of oxygen (107), Eq. (10) becomes:

$$\frac{1}{T_1} = \frac{1}{T_{1d}} + \frac{p_{O_2}}{kT_{1p}} \tag{11}$$

where $k$ is the Henry's law constant. Thus, the linear relationship between $1/T_1$ and $p_{O_2}$ gives an opportunity to use $^{19}$F NMR for determination of the amount of oxygen dissolved in a PFC emulsion. It is particularly valuable to be able to determine the amount of oxygen dissolved in body fluids and distinguish it from the oxygen dissolved in the emulsion droplets.

## B. Probing the Molecular Conformation

The measurement of $T_{1p}$ in Eq. (11) allows one to obtain information on the preferred location of oxygen on the PFC molecule (103) due to the differences in the paramagnetic relaxation rates. The relaxation rate of a nuclear spin system due to the interaction with paramagnetic species is (108):

$$\frac{1}{T_{1p}} = \frac{2S(S+1)\gamma^2 g^2 \beta^2}{15r^6}\left[\frac{3\tau_c}{1+\omega_I^2\tau_c^2} + \frac{3\tau_c}{1+\omega_0^2\tau_c^2}\right] \tag{12}$$

where $S$ is the total electron spin of the paramagnetic species ($S = 1$ for $O_2$), $\gamma$ is the nuclear gyromagnetic ratio, $\gamma\beta$ is the electron magnetic moment, $r$ is the distance between the paramagnetic center and the nucleus of interest, $\omega_0$ is the angular frequency of the electron resonance, $\omega_I$ is the angular frequency of the nuclear resonance, and $\tau_c$ is the correlation time. The strong dependence of $1/T_{1p}$ on $r$ makes it possible to "probe" the average distance between the oxygen molecule and various fluorine atoms in PFC emulsion. The slope of $1/T_1$ versus $pO_2$ would be larger for fluorine nuclei that are close to the oxygen molecule. This also allows one to interpret effects of the PFC conformation, such as obtained for *cis*- and *trans*-perfluorodecalin.

## C. Simultaneous Measurement of Oxygen Tension and Temperature

There are several reports on the estimation of the oxygen tension in vivo (105, 109, 110). The practical implementation of the method is limited due to the temperature dependence of $T_1$. Mason et al. (111) determined the relationship of $R_1 = 1/T_1$ with $pO_2$ and temperature for each $^{19}$F NMR resonance of Oxyphenol-ET (an emulsion of perfluorotributylamine). He also demonstrated how to measure $pO_2$ and temperature simultaneously. A relationship between the $R_1$ relaxation rate and the temperature $T$ (in °C) and the oxygen tension (in % atm; 100 % atm = 760 Torr = 101 kPa) was suggested:

$$R_1 = a + bT + cP + dTP \tag{13}$$

and the coefficients a,b, c, and d were determined empirically for the $CF_3$ resonance of perfluorocarbon emulsion. The result in the temperature range between 27°C and 36.6°C were summarized (112) as

$$R_1 = 0.442 - 0.0050T + 0.02442P - 0.000215TP \tag{14}$$

## XIII. MAGNETIC RESONANCE IMAGING WITH FLUOROCARBON EMULSIONS

Perfluorocarbon emulsions have been studied as contrast agents for X-ray, ultrasound, and magnetic resonance imaging (MRI) and as agents for direct $^{19}F$ MRI. Perfluoro-octyl bromide (PFOB) was first developed as a gastrointestinal MRI contrast agent, with the trade name Imagent MR (Alliance). PFOB contains no protons and as such appears as a dark void in MR images. Rather more attractive is the possibility to use the advantages of the $^{19}F$ nucleus which is highly sensitive as a result of the high gyromagnetic ratio and lack of natural background signal. This potential has been recognized in several studies reported in the literature. An example is the determination of oxygenation by perfluorocarbon emulsions (104), considered also as a labeling agent for studies of capillary blood flow and other tissue properties. This is a promising field of biomedical research and even diagnosis may develop on this basis (113). The basic principles of MRI that are relevant for use with emulsions are presented below.

### A. Basic Principles of MRI

Magnetic resonance found an important application in medical imaging. In this case the magnetic resonance signal from the nucleus has to contain information on its position. This is possible owing to the linear dependence of the Larmor frequency on the strength of the magnetic field. If the main magnetic field is uniform across the sample, all the nuclei in the sample will have the same frequency. However, one can vary the frequency of the observed signal by changing the magnetic field linearly across the sample. The field gradient varies with the position along the main $(x, y, z)$ axes and creates a range of resonance frequencies. Since the Fourier transformation can convert the NMR signal from the time domain into the frequency domain we can obtain a frequency spectrum which represents the spatial spin concentration. With the help of gradients, one can select the spatial region being excited, normally a slice within the sample. The selective excitation by various encoding techniques allows one to choose the spins from which the signal is obtained and basic pulse techniques can be used to manipulate the signal itself. The form of the magnetic resonance signal is determined by a large number of factors including the spin density, $T_1$, $T_2$, flow, and diffusion, and images reflecting those parameters may be obtained.

### B. Emulsion-based $^{19}F$ MRI Contrast Agents

Oil-in-water emulsions in combination with paramagnetic ions are used as oral MRI contrast agent (114). Typical paramagnetic substances include $Mn^{2+}$, $Fe^{2+}$, and $Gd^{3+}$, as well as molecular oxygen and free radicals. Paramagnetic contrast agents mainly shorten the $T_1$ relaxation. An emulsion in such formulations is usually used as an inert medium that delivers the contrast agent and improves the taste and biodistribution (114). Early attempts to use $^{19}F$ imaging were successful but not clinically useful (113, 115, 116) because of the fact that most PFCs have multiple signals causing misregistration of signals and dilution of the amount of $^{19}F$ signal per molecule. Confounding this is the short $T_2$ time of most PFCs (< 6 ms). Thus, the signal is not only weak but also short-lived.

A solution to the problem of multiple signals was suggested (117) by using perfluoro-15-crown-ether in which all 22 of the fluorine atoms are identical and form a single peak. In addition, the $T_2$ relaxation time is 200 ms compared to 6 ms for PFOB. Excellent liver, spleen, tumor, and vascular images have been obtained at doses of 3 ml $kg^{-1}$ PFC in animals (118).

### C. Simultaneous $^{19}$ and $^1H$ Images

The starting point for the $^{19}F$ MRI is often a $^1H$ image as a standard to identify the structures of interest. Taking both a $^1H$ and a $^{19}F$ image often requires a change of the probehead to tune the tomograph to $^{19}F$ resonance. That prevents the localization of the fluorine image relative to the proton one. A hardware modification which permits the record of $^{19}F$ images with a $^1H$ resonator is described (106) that allows one to obtain both $^{19}F$ and $^1H$ images without removing the object. Such simultaneous images from $^{19}F$ and $^1H$ have been obtained by a phantom with a porcine kidney perfused with the PFC emulsions (Fig. 16). A fluorine-based contrast was obtained as well as relaxation curves from resolved fluorine lines in a selected sample volume (106). The fluorinated material used in this study was a 10% aqueous solution of trifluoroethanol and the perfluoroemulsion FC-43 (Green Cross Corp., Osaka, Japan) consisting mainly of perfluorotributylamine (20 w(v%) and emulsifier Pluronic F-68 (2.56 w/v %). The $^{19}F$ image in Fig. 16 is based on the $^{19}F$ lines between 0 and $-10$ ppm (Fig. 14). This study showed that the localization of

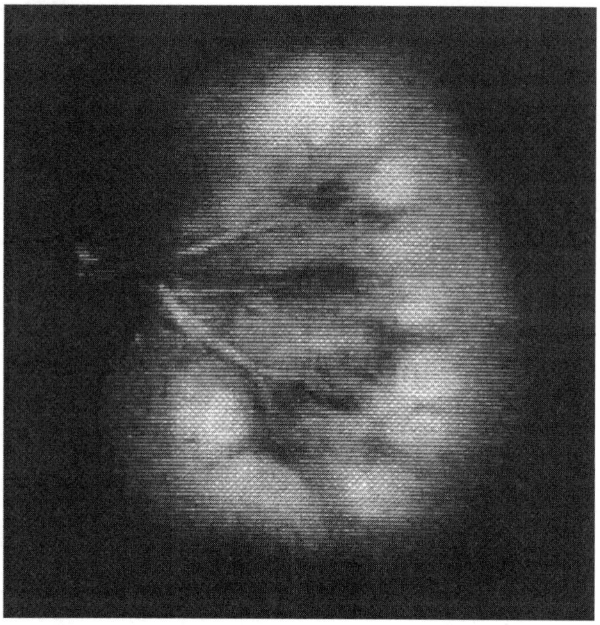

Figure 16 $^{19}F$ image of porcine kidney perfused with PFC emulsion. The fluorine image is based on the lines between 0 and −10 ppm from Fig. 14. (Adapted from Ref. 106.)

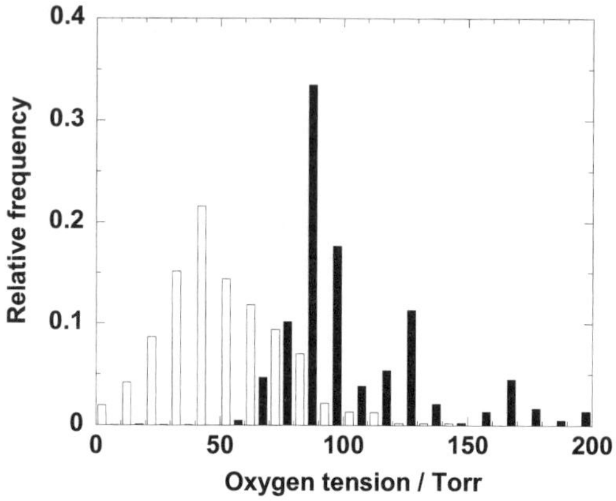


Figure 17 Oxygen tension of $pO_2$, histogram obtained from 2-D $^{19}F$ images of mouse liver and spleen when the mouse is breathing air and carbogen, respectively (empty bar: air; filled bar: carbogen). Average $pO_2$ for air breathing was 43 Torr and that for carbogen breathing was 92 Torr. (Adapted from Ref. 118.)

the informative fluorine images may be performed on the basis of the sensitive proton images.

## D. Oxygen Mapping with $^{19}F$

Clark *et al.* discovered that the $T_1$ relaxation times of PFCs change with the oxygen tension (119). They first postulated that the dependence of $T_1$ an oxygen tension could be mapped by PFC, and the concept was used in animal studies where part of the animal blood was replaced with perfluorotripropylamine, and oxygen maps of the brain were obtained (109). The oxygen maps of the liver, spleen, and tumors in mice were obtained (118) by using perfluoro-15-crown-ether emulsion when the differences in oxygen tension were compared while breathing air compared to "carbogen" (a mixture of 95% $O_2$ and 5% $CO_2$). A histogram of the oxygen tension obtained from $^{19}F$ images of mouse liver and spleen when the mouse was breathing air and carbogen, respectively, is presented in Fig. 17. The average $pO_2$ for air breathing was 43 Torr and that for carbogen breathing was 92 Torr. The method is a sensitive map of the oxygen tension as demonstrated in Fig. 18 (109). The $^{19}F$ image of a cat brain in Fig. 18a shows the intravascular PFC in veins and arteries while the image in Fig. 18b is the calculated $pO_2$ map. Bright shades correspond to high values of $pO_2$ as in arteries and dark shades correspond to lower $pO_2$ as in veins.

Other applications of MRI imaging of emulsions are related to the study of the biodistribution of PFC emulsions (120).

# XIV. XENON NMR WITH EMULSION CARRIERS

## A. The $^{129}Xe$ NMR

The rare gas xenon contains two NMR-sensitive isotopes in high natural abundance: $^{129}Xe$ has a spin of 1/2 and $^{131}Xe$ is a quadrupolar nucleus with a spin of 3/2. The complementary NMR characteristics of these nuclei provide a unique opportunity for probing their environment. The method is widely applicable because xenon interacts with a useful range of condensed phases including pure liquids, protein solutions, and suspensions of lipid and biological membranes. It was found that the range of chemical shifts of $^{129}Xe$ dissolved in common solvents is $\approx$ 200 ppm, which is 30 times larger than that found for $^{13}C$ in methane dissolved in various solvents (121). The high solubility of xenon in PFCs (122) and the long $^{129}Xe$ $T_1$ relaxation time of dissolved gas (123) suggest that PFC is suitable for delivery of xenon.

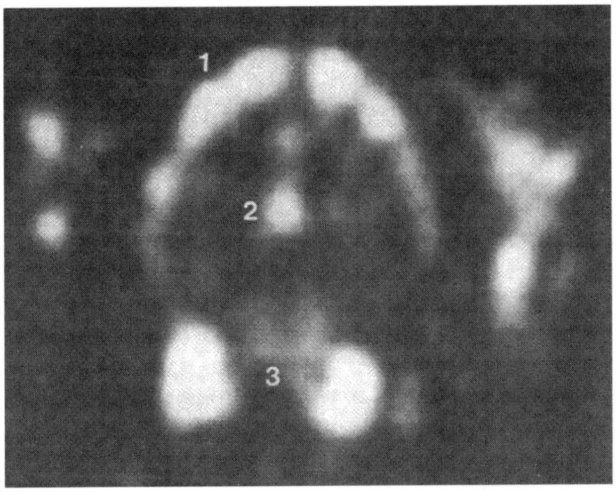


(a)

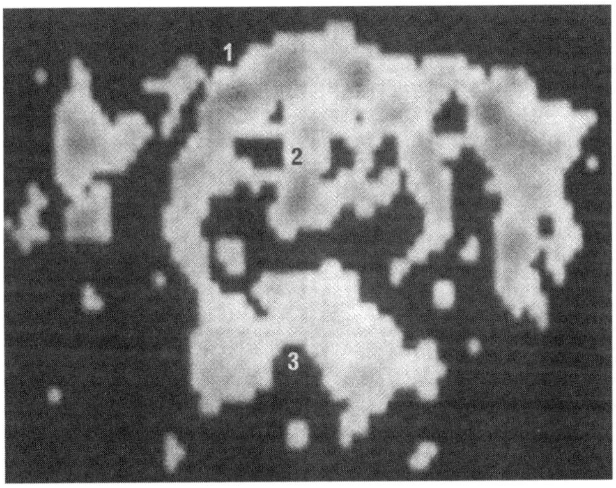


(b)

**Figure 18** (a) $^{19}$F image of cat brain showing intravascular PFC in (1) veins, and (2) arteries. (b) A calculated $pO_2$ map of cat brain. Bright shades correspond to high values of $pO_2$ as in arteries (2); dark shades correspond to lower $pO_2$ as in veins. (Adapted from Ref. 109.)

## B. Emulsion as Carriers for Hyperpolarized Gases

The interest in efficient carriers for xenon is increased due to the finding that the polarization of $^{129}$Xe and $^{3}$He nuclei can be enhanced up to five orders of magnitude by using optical pumping methods (124). A dramatically enhanced signal of such hyperpolarized gases in NMR has been utilized. A fluorocarbon emulsion, Fluosol (20% w/v), has been suggested as a possible carrier for laser-polarized xenon (123). A recent study shows that PFOB emulsions have potential as carriers for hyperpolarized $^{129}$Xe (125). The $^{129}$Xe NMR spectra of xenon dissolved in pure PFOB and in pure water show peaks at 106 and 196 ppm, respectively, compared with the NMR spectra of xenon gas (Fig. 19). The $^{129}$Xe NMR spectra from emulsions were found to correlate strongly with the emulsion droplet size distribution due to exchange of xenon with the aqueous environment (Fig. 20). Hydrocarbon emulsions were also considered as carriers of hyporpolarized gases and the commercially available Intralipid 30% (Pharmacia, Clayton, NC), a 30% (w/w) soybean O/W emulsion, was chosen (126). Figure 21 shows the $^{129}$Xe NMR spectrum of hyperpolarized xenon gas compared to that of xenon dissolved in Intralipid 30%. The chemical shift is relative to the xenon gas resonance with the plus sign referring to higher frequencies. A high spin–lattice relaxation time $T_1$ of approximately 25 s was observed in the emulsion. Images from $^{129}$Xe in Intralipid emulsion were obtained in animal models, and additional information on the blood flow velocity was obtained.

The fact that the peak location of xenon in emulsion is approximately 100 ppm different from either gaseous xenon or xenon dissolved in blood or in tissue is a unique advantage for medical MRI. This could be exploited, for instance, for studying tumor vascularity

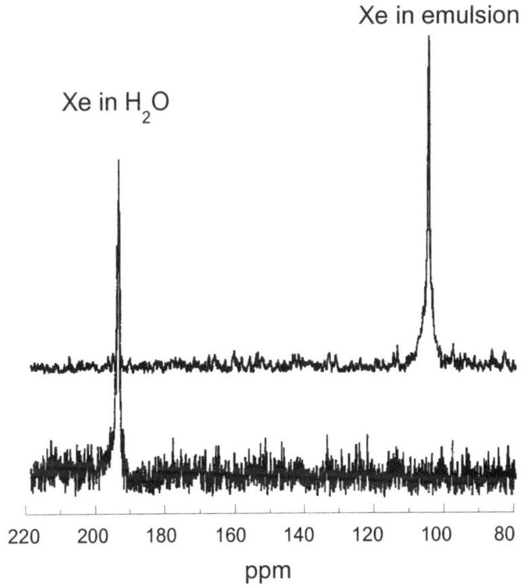


**Figure 19** $^{129}$Xe NMR spectra of hyperpolarized Xe dissolved in perfluoro-octyl bromide (PFOB) emulsions and water, respectively; chemical shifts (in ppm) are expressed relative to Xe gas. (Adapted from Ref. 125.)

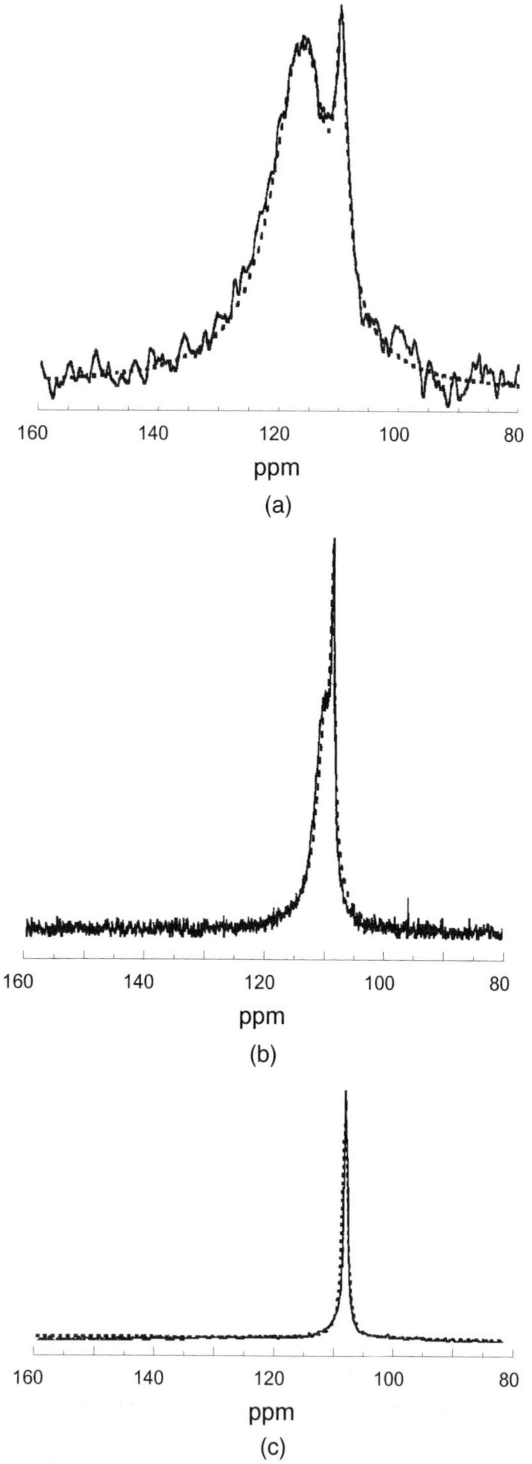


**Figure 20** Spectra of PFOB emulsions with three different size ranges: (a) small droplets, (b) intermediate droplet size, (c) large droplets; solid lines are measured $^{129}$Xe spectra, and dotted lines are simulated spectra based on Xe exchange between emulsion droplets and water. (Adapted from Ref. 125.)

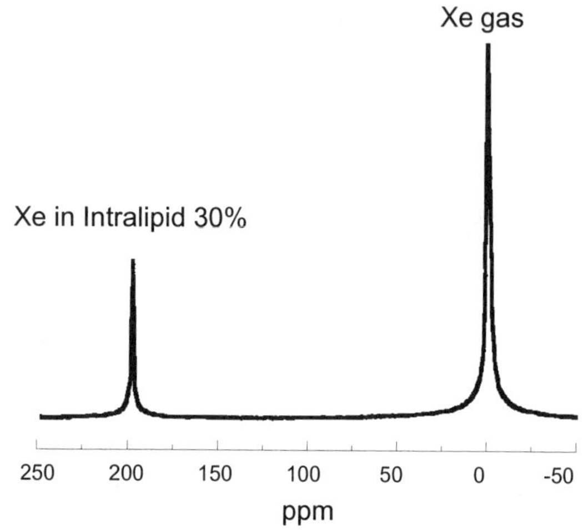


**Figure 21** $^{129}$Xe NMR spectrum of hyperpolarized Xe gas (2 atm) and Xe dissolved in Intralipid 30%; the chemical shift is relative to the Xe gas resonance with the plus sign referring to higher frequencies. (Adapted from Ref. 126.)

by using hyperpolarized $^{129}$Xe in emulsions, as blood-pool contrast agent.

## XV. CONCLUSIONS

Above we have presented various applications of the NMR technique in the study of emulsions. NMR is a versatile spectroscopic technique. This is also reflected in the span of questions pertaining to various aspects of emulsions that can be addressed with the NMR technique. The topics covered above include the determination of droplet size distributions, aspects of emulsion stability, crystallization of fat, and medical imaging.

It seems quite clear that emulsions will become increasingly important in more specialized applications in the future. In such applications the demands for accurate design of properties of the emulsion systems, process, and quality control, will be a major challenge. As should be clear from the above, NMR will become increasingly important in this regard. Controlled delivery of drugs as well as the use in the administration of pesticides are two examples of emulsion technology that will require accurate and reproducible characterization. An example that serves to illustrate this point is furnished by the use of emulsions in the delivery of drugs, where the release kinetics of the active compound from the carrier is of importance in the design. As discussed above, this information can be obtained from PFG NMR.

## ACKNOWLEDGMENT

The work was financially supported by the Swedish Board for Industrial and Technical Development (NUTEC).

## REFERENCES


1. A. Kabalnov, H Wennerström. Langmuir 12: 276–292, 1996.
2. PT Callaghan. Principles of Nuclear Magnetic Resonance Microscopy. Oxford: Clarendon Press, 1991.
3. RR Ernst, G Bodenhausen, A Wokaun. Principles of Nuclear Magnetic Resonance in One and Two Dimensions. Oxford: Oxford University Press, 1987.
4. D Canet. Nuclear Magnetic Resonance. Concepts and Methods. Chichester: John Wiley, 1996.
5. I Lönnqvist, B Håkansson, B Balinov, O Söderman. J Colloid Interface Sci 192: 66–73, 1996.
6. PT Callaghan, CM Trotter, KW Jolley. J Magn Reson 37: 247–259, 1980.
7. P Stilbs. Prog Nucl Magn Reson Spect 19: 1–45, 1987.
8. O Söderman, P Stilbs. Prog Nucl Magn Reson Spect 26: 445–482, 1994.
9. EO Stejskal, JE Tanner. J Chem Phys 42: 288–292, 1965.
10. JE Tanner. Use of a Pulsed Magnetic-field Gradient for Measurements of Self-diffusion by Spin–Echo Nuclear Magnetic Resonance with Applications to Restricted Diffusion in Several Tissues and Emulsions. PhD dissertation, University of Wisconsin, 1966.
11. PT Callaghan, A Coy. In: P Tycko, ed. NMR Probes of Molecular Dynamics. Dordrecht: Kluwer Academic 1993.
12. PT Callaghan, A Coy, TPJ Halpin, D MacGowan, JK Packer, FO Zelaya. J Chem Phys 97: 651–662, 1992.
13. AV Barzykin. Phys Rev B 58: 14171–14174, 1998.
14. SL Codd, PT Callaghan. J Magn Reson 137: 358–372, 1999.
15. J Kärger, W Heink. J Magn Reson 51: 1–7, 1983.
16. JE Tanner, EO Stejskal. J Chem Phys 49: 1768–1777, 1968.
17. B Balinov, B Jönsson, P Linse, O Söderman. J Magn Reson, A 104: 17–25, 1993.
18. DC Douglass, DW McCall. J Phys Chem 62: 1102, 1958.
19. CH Neuman. J Chem Phys 60: 4508–4511, 1974.
20. JS Murday, RM Cotts. J Chem Phys 48: 4938–4945, 1968.
21. B Balinov, O Söderman, T Wärnheim. J Am Oil Chem Soc 71: 513–518, 1994.
22. B Balinov, O Urdahl, O Söderman, J Sjöblom. Colloids Surfaces 82: 173–181, 1994.
23. PT Callaghan, KW Jolley, R Humphrey. J Colloid Interface Sci 93: 521–529, 1983.
24. X Li, JC Cox, RW Flumerfelt. AIChE J 38: 1671–1674, 1992.
25. I Lönnqvist, A Khan, O Söderman. J Colloid Interface Sci 144: 401–411, 1991.
26. KJ Packer, C Rees. J Colloid Interface Sci 40: 206–218, 1972.
27. JC Van den Enden, D Waddington, H Van Aalst, CG Van Kralingen, KJ Packer. J Colloid Interface Sci 140: 105–113, 1990.
28. L Ambrosone, A Ceglie, G Colafemmina, G Palazzo. J Chem Phys 110: 797–804, 1999.
29. L Ambrosone, A Ceglie, G Colafemmina, G Palazzo. J Chem Phys 107: 10756–10763, 1997.
30. P Stilbs, M Moseley. J Magn Reson 31: 55–61, 1978.
31. L Ambrosone, G Colafemmina, M Giustini, G Palazzo, A Ceglie. Prog Colloid Polym Sci 112: 86–88, 1999.
32. KJ Lissant. J Colloid Interface Sci 22: 462–468, 1966.
33. HM Princen. J Colloid Interface Sci 91: 160–176, 1983.
34. PT Callaghan, A Coy, D MacGowan, KJ Packer, FO Zelaya. Nature (London) 351: 467–469, 1991.
35. R Pons, I Carrera, P Erra, H Kunieda, C Solans. Colloids Surfaces A 91: 259–266, 1994.
36. B Balinov, P Linse, O Söderman. J Colloid Interface Sci 182: 539–548, 1996.
37. RH Engel, SJ Riggi, MJ Fahrenbach. Nature 219: 856–857, 1968.
38. NN Li, AL Shrier. In: NN Li, ed. Recent Developments in Separation Science. Cleveland: Chemical Rubber Co., 1972.
39. S Matsumoto. In: M Schick, d. Nonionic Surfactants. New York: Marcel Dekker, 1987, pp 549–600.
40. PJ Taylor, CL Miller, TM Pollack, FT Perkins, MA Westwood. J Hyg (London) 67:485–890, 1969.
41. LA Elson, BC Mitchlev, AJ Collings, R Schneider. Rev Eur Etud Clin Biol 15: 87–90, 1970.
42. CJ Benoy, LA Elson, R Schneider. Br J Pharmacol4 5: 135–136, 1972.
43. AF Brodin, DR Kavaliunas, SG Frank. Acta Pharm Suec 15: 1–12, 1978.
44. D Whitehill. Chem Drug 213: 130–135, 1980.
45. RK Owusu, Z Qinhong, E Dickinson. Food Hydrocolloids 6: 443–453, 1992.
46. E Dickinson, J Evison, RK Owusu. Food Hydrocolloids 5:481–485, 1991.
47. M Frenkel, R Shwartz, N Garti. J Colloid Interface Sci 94: 174–178, 1983.
48. A Rabaron, PA Rocha-Filho, C Vaution, M Seiller. Study and demonstration of multiple w/o/w emulsions by nuclear magnetic resonanace. Fifth International Congress on the Technology of Pharmaceuticals, Chatenay Malabry, France, 1989.

49. J Andrasko. Biochim Biophys Acta 428: 304–311, 1976.
50. DG Cory, AN Garroway. Magn Reson Med 14: 435–444, 1990.
51. S Lundquist, M Malmsten, T Petersson Norted, B Siekmann. Methods to study the physico-chemical properties of a drug containing phospholipid stabilized emulsion for intravenous administration. Twelfth International Symposium on Surfactants in Solution, 6–7 November, Stockholm, 1998, p 123.
52. W Heink, J Kärger, H Pfeifer. Z Phys Chem (Munich) 170: 199–206, 1991.
53. M Gangoda, BM Fung, EA O'Rear. J Colloid Interface Sci 116: 230–236, 1987.
54. C Washington. Int J Pharm 66: 1–21, 1990.
55. J Boyd, C Parkinson, P Sherman. J Colloid Interface Sci 41: 359–370, 1972.
56. N Weiner. Drug Dev Ind Pharm 12:933–951, 1986.
57. O Söderman, B Balinov. In: J Sjöblom, ed. Emulsion and Emulsion Stability. New York: Marcel Dekker, 1996, pp 369–392.
58. B Håkansson, R Pons, O Söderman. Langmuir 15: 988–991, 1999.
59. L Marszall. In: M Schick, ed. Nonionic Surfactants. New York: Marcel Dekker, 1987, pp 493–547.
60. B Lindman, O Söderman, H Wennerström. In: R Zana, ed. Novel Techniques to Investigate Surfactant Solutions. New York: Marcel Dekker, 1987, pp 295–357.
61. B Balinov, U Olsson, O Söderman. J Phys Chem 95: 5931–5936, 1991.
62. B Lindman, P Stilbs. In: HL Rosano, M Clausse, eds. Microemulsion Systems. New York: Marcel Dekker, 1987, pp 129–144.
63. B Lindman, K Shinoda, U Olsson, D Andersen, G Karlström, H Wennerström. Colloids Surfaces 38: 205–224, 1989.
64. A Khan. In: GA Webb, ed. Specialist Periodical Reports, Nuclear Magnetic Resonance. Cambridge: The Royal Society of Chemistry, 1993, pp 498–540.
65. K Shinoda, H Kunieda, T Arai, H Saijo. J Phys Chem 88: 5126–5129, 1984.
66. B Jönsson, H Wennerström, P Nilsson, P Linse. Colloid Polymer Sci 264: 77, 1986.
67. C Solans, R Pons, S Zhu, HT Davis, DF Evans, K Nakamura, H Kunieda. Langmuir 9: 1479–1482, 1993.
68. K Chiba, M Tada. Nippon Nogei Kagaku Kaishi 62: 859–865, 1988.
69. K Chiba, M Tada. Agric Biol Chem 53: 995–1001, 1989.
70. M Rotenberg, M Rubin, A Bor, D Meyuhas, Y Talmon, D Lichtenberg. Biochim Biophys Acta 1086: 265–272, 1991.
71. J Drew, A Liodakis, R Chan, H Du, M Sadek, R Brownlee, WH Sawyer. Biochem Int 22: 983–992, 1990.
72. J Férézou, TL Nguyen, C Leray, T Hajri, A Frey, Y Cabaret, J Courtieur, C Lutton, AC Bach. Biochim Biophys Acta 1213: 149–158, 1994.
73. K Westesen, T Wehler. J Pharm Sci 82: 1237–1244, 1993.
74. LC ter Beek, M Ketelaars, DC McCain, PE Smulders, P Walstra, MA Hemminga. Biophys J 70: 2396–2402, 1996.
75. AJ Haighton, K van Putte, LF Vermaas. J Am Oil Chem Soc 49: 153–156, 1972.
76. K van Putte, J van den Enden. J Phys E: J Sci Instrum 6: 910–912, 1973.
77. ISO 8292: 1991. Animal and vegetable fats and oils – Determination of solid fat content–Pulsed nuclear magnetic resonance method.
78. J Trumbetas, JA Fioriti, RJ Sims. J Am Oil Chem Soc 53: 722–726, 1976.
79. J Trumbetas, JA Fioriti, RJ Sims. J Am Oil Chem Soc 55: 248–251, 1978.
80. JL Cavallo, DL Chang. Chem Eng Prog 86: 54–59, 1990.
81. D Turubull. J Chem Phys 20: 411, 1952.
82. JPC Cordiez, G Grange, B Mutaftschiev. J Colloid Interface Sci 85: 431–441, 1982.
83. D Clausse. In: P Becher, ed. Encyclopedia of Emulsion Technology. New York: Marcel Dekker, 1985, pp 77–157.
84. LCJ Clark, F Gollan. Science 152: 1755–1756, 1966.
85. BD Spiess, RJ McCarthy, KJ Tuman, AW Woronowicz, KA Tool, AD Ivankovich. Undersea Biomed Res 15: 31–37, 1988.
86. M Cleman, CC Jaffe, D Wohlgelernter. Circulation 74: 555–562, 1986.
87. CC Jaffe, D Wohlgelernter, H Cabin, L Bowman, L Deckelbaum, M Remetz, M Cleman. Am Heart J 115: 1156–1164, 1988.
88. DH Glogar, RA Kloner, J Muller, LW DeBoer, E Braunwald, LCJ Clark. Science 211: 1439–1441, 1981.
89. GR Nunn, G Dance, J Peters, LH Cohn. Am J Cardiol 52: 203–205, 1983.
90. AK Bajaj, MA Cobb, R viramani, JC Gay, RT Light, MB Forman. Circulation 79: 645–656, 1989.
91. TC Wall, RM Califf, J Blankenship, JD Talley, M Tannenbaum, M Schwaiger, G Gacioch, MD Cohen, M Sanz, JD Leimberger. Circulation 90: 114–120, 1994.
92. CW Song, I Lee, T Hasegawa, JG Rhee, SH Levitt. Cancer Res 47: 442–446, 1987.
93. T Mate, S Rockwell. Perfluorochemical emulsions do not affect bone marrow radiosensitivity. American Society of Therapy Radiation Oncologists Meeting, Washington, DC, 1984.
94. C Rose, R Lustig, N McIntosh, B Teicher. Int J Radiat Oncol Biol Phys 12: 1325–1327, 1986.

95. R Lustig, N McIntosh-Lowe, C Rose, J Haas, S Krasnow, M Spaulding, L Prosnitz. Int J Radiat Oncol Biol Phys 16: 1587–1593, 1989.
96. R Lustig, N Lowe, L Prosnitz, M Spaulding, M Cohen, J Stitt, R Brannon. Int J Radiat Oncol Biol Phys 19: 97–102, 1990.
97. R Evans, B Kimler, R Morantz, S Batnitzky. Int J Radiat Oncol Biol Phys 26: 649–652, 1993.
98. RG Evans, BF Kimler, RA morantz, TS Vats, LS Gemer, V Liston, N Lowe. Int J Radiat Oncol Biol Phys 19: 415–420, 1990.
99. BA Teicher, SA holden, G Ara, CS Ha, TS Herman, D Northey. J Cancer Res Clin Oncol 118: 509–514, 1992.
100. FK Schweighardt, D Woo. US Patent 4 781 676, 1988.
101. JG Riess, M Le Blank. Pure Appl Chem 54: 2383–2406, 1982.
102. RP Mason, RL Nunnally, PP Antich. Magn Reson Med 18: 71–79, 1991.
103. P Parhami, BM Fung. J Phys Chem 87: 1928–1931, 1983.
104. RS Reid, CJ Koch, ME Castro, JA Lunt, EO Treiber, DJ Boisvert, PS Allen. Phys Med Biol 30: 677–686, 1985.
105. SR Thomas. In: L Partain, R Price, J Patton, M Kulkavni, E James, eds. Magnetic Resonance Imaging. Philadelphia: WB Saunders, 1988, p 1536.
106. G Schnur, R Kimmich, R Lietzenmayer. Magn Reson Med 13: 478–489, 1990.
107. J Riess, M Le Blanc. Angew Chem Int Ed engl 17: 621–634, 1978.
108. I Solomon. Phys Rev 99: 559, 1955.
109. D Eidelberg, G Johnson, D Barnes, PS Tofts, D Delpy, D Plummer, WI McDonald. Magn Reson Med 6: 344–352, 1988.
110. JE Fishman, PM Joseph, MJ Carvlin, M Saadi-Elmandjra, B Mukherji, HA Sloviter. Invest Radiol 24: 65–71, 1989.
111. RP Mason, H Shukla, PP Antich. Magn Reson Med 29: 296–302, 1993.
112. BR Barker, RP Mason, N Bansal, RM Peshock. J Magn Reson Imag 4: 595–602, 1994.
113. PM Joseph, Y Yuasa, HL Kundel, B Mukherji, HA Sloviter. Invest Radiol 20: 504–509, 1985.
114. KC Li, PG Ang, RP Tart, BL Storm, R Rolfes, PC Ho-Tai. Magn Reson Imag 8: 589–598, 1990.
115. E McFarland, JA Koutcher, BR Rosen, B Teicher, TJ Brady. J Comp Assisted Tomogr 9: 8–15, 1985.
116. HE Longmaid, DF Adams, RD Neirinckx, CG Harrison, P Brunner, SE Seltzer, MA Davis, L Neuringer, RP Geyer. Invest Radiol 20: 141–145, 1986.
117. FK Schweighardt. US Patent 4 838 274, 1989.
118. BJ Dardzinski. CH Sotak. Magn Reson Med 32: 88–97, 1994.
119. LCJ Clark, JL Ackerman, SR Thomas, RW Millard, RE Hoffman, RG pratt, H Ragle-Cole, RA Kinsey, R Janakiraman. Adv Exp Med Biol 180: 835–845, 1984.
120. LJ Jäger, U Nöth, A Haase, J Lutz. Adv Exp Med Biol 361: 129–134, 1994.
121. K Miller, N Reo, A Schoot Uiterkamp, D Stengle, T Stengle, K Williamson. proc Natl Acad Sci USA 78: 4946–4949, 1981.
122. GL Pollack, RP Kennan, GT Holm. Biomater Artif Cells Immobil Biotech 20: 1101–1104, 1992.
123. A Bifone, YQ Song, R Seydoux, RE Taylor, BM Goodson, T Pietrass, TF Budinger, G Navon, A Pines. Proc Natl Acad Sci USA 93: 12932–12936, 1996.
124. W Happer, E Miron, S Schaefer, D Schreiber, W van Wijngaarden, X Zeng. Phys Rev A 29: 3092–3110, 1984.
125. J Wolber, IJ Rowland, MO Leach, A Bifone. Magn Reson Med 41: 442–449, 1999.
126. HE Möller, MS Chawla, XJ Chen, B Driehuys, LW Hedlund, CT Wheeler, GA Johnson. Magn Reson Med 41: 1058–1064, 1999.
127. O Söderman, I Lönnqvist, B Balinov. In: J Sjöblom, ed. Emulsions – A Fundamental and Practical Approach. Dordrecht: Kluwer Academic, 1992, pp 239–258.

# 13

# Surface Forces and Emulsion Stability

**Per M. Claesson, Eva Blomberg, and Evgeni Poptoshev**
*Royal Institute of Technology and Institute for Surface Chemistry, Stockholm, Sweden*

## I. INTRODUCTION

Emulsions are dispersions of one liquid in another liquid, most commonly water-in-oil or oil-in-water. The total interfacial area in an emulsion is very large, and since the interfacial area is associated with a positive free energy (the interfacial tension), the emulsion system is thermodynamically unstable. Nevertheless, it is possible to make emulsions with an excellent long-term stability. This requires the use of emulsifiers that accumulate at the oil/water interface and create an energy barrier towards flocculation and coalescence. The emulsifiers can be ionic, zwitterionic, or nonionic surfactants, proteins, amphiphilic polymers, or combinations of polymers and surfactants. The structure of the adsorbed layer at the water/oil interface may be rather complex, involving several species adsorbed directly to the interface as well as other species adsorbing on top of the first layer.

The first question one may ask is if an oil-in-water emulsion or a water-in-oil emulsion is formed then the two solvents are dispersed into each other with the use of a given emulsifier. There are several empirical roles addressing this problem. The first is due to Bancroft (1) who stated that if the emulsifier is most soluble in the water phase, then an oil-in-water emulsion will be formed. A water-in-oil emulsion will be obtained when the reverse is true. The HLB (hydrophilic–lipophilic balance) concept is used for describing the nature of the surfactant. It was first introduced by Griffin (2) and later extended by Davies (3). Rather hydrophobic emulsifiers having a low HLB number, say below 6, are predicted to be suitable for forming water-in-oil emulsions whereas more hydrophilic emulsifiers with high HLB values, above about 10, are suggested to be suitable for forming oil-in-water emulsions. The HLB value can easily be calculated from the structure of the emulsifier (3). An HLB value has also been assigned for most common oils. It is defined as the HLB number of the emulsifier in a homologous series that produces the most stable oil-in-water emulsion. A nonpolar oil is found to have a lower HLB number than a polar oil. Hence, the choice of emulsifier has to be adjusted to the type of oil to be emulsified.

The use of an HLB value for nonionic emulsifiers of the oligo (ethylene oxide) type has its drawbacks since their properties are strongly temperature dependent. This is clearly seen in three-component oil–water–surfactant phase diagrams. At low temperatures, microemulsions of oil droplets in water (Winsor I) are formed. In a small temperature interval, bicontinuous microemulsions (Winsor III) are stable, followed at higher temperatures by a microemulsion consisting of water droplets in oil (Winsor II). These transitions are due to a change in the spontaneous monolayer curvature from positive at low temperatures to negative at high temperatures. This behavior is closely mimicked by the thermodynamically unstable (macro)emulsions, and it is common to describe these emulsions in terms of the phase-inversion temperature (PIT). Below the PIT the emulsion is of the oil-in-water type whereas above the PIT it is of the water-in-oil type. Very close to the PIT no stable (macro)emulsions can be formed. It has been argued that this change in behavior, as for the microemulsions, is due to the change in spontaneous curvature of such surfactant films at the oil–water interface, particularly the ease with which hole formation leading to coalescence occurs (4). Note that the PIT does not only depend on the nature of the emulsifier but also on the type of oil used, which often can be explained by the degree of oil penetration into the emulsifier film.

When two droplets approach each other they will interact with hydrodynamic forces and with surface forces of molecular origin. Finally, when the droplets are close enough they may coalesce and form one larger droplet. An emulsion will have a long-term stability if the droplets are prevented from coming close to each other by strong repulsive forces and if they are prevented from coalescing even when they are close to each other. However, in this case also a slow destabilization due to Ostwald ripening will occur.

## II. INTERACTIONS AND HOLE FORMATION

In this section we will give a short overview of hydrodynamic and surface forces as well as hole formation leading to coalescence. References will be provided for the reader who wants to penetrate further into these subjects.

### A. Hydrodynamic Interactions

When liquid drains from the gap between two approaching spherical emulsion droplets of equal size a hydrodynamic force is produced resulting from viscous dissipation. As long as the surfaces do not deform (i.e., small forces) and the liquid next to the surface is stationary (no slip condition, see below) the hydrodynamic force is given by (5):

$$F = -\frac{3\pi\mu R^2}{2D}\frac{\mathrm{d}D}{\mathrm{d}t} \tag{1}$$

where $R$ is the radius of the spheres, $\mu$ is the viscosity of the draining liquid, $D$ is the separation between the spheres, and $t$ is the time. This equation describes the hydrodynamic interaction when the droplets are far apart and do not interact with each other very strongly. However, as soon as the interaction between the surfaces is sufficiently large, the emulsion droplets will deform and Eq. (1) is no longer valid.

In concentrated emulsions we meet another extreme case. A thin planar liquid film now separates the emulsion droplets, and they may change their shape from spherical to polyhedral (6). In this case, the liquid drains out of the flat part of the film owing to the capillary suction pressure. The outflow of liquid between rigid parallel disks was considered by Reynolds and others (7, 8) who found that the pressure varied with the radial distance from the center of the disk as:

$$P = P_0 + \frac{3\mu}{D^3}(r_0^2 - r^2)V_\mathrm{R} \tag{2}$$

where $P$ is the pressure at a distance $r$ from the center; $r_0$ is the radius of the plate; $P_0$ is the hydrostatic pressure which equals the total pressure at the edge of the contact, i.e., at $r = r_0$; and $V_\mathrm{R}$ is the rate of approach, i.e., $-\mathrm{d}D/\mathrm{d}t$.

The repulsive hydrodynamic force acting on the plates is obtained by integrating over the plate area and subtracting the hydrostatic pressure contribution:

$$F = 2\pi\int_0^{r_0}(P - P_0)r\mathrm{d}r = 2\pi\int_0^{r_0}\left[\frac{3\mu}{D^3}(r_0^2 - r^2)V_R\right]r\mathrm{d}r = \frac{3\pi r_0^4\mu V_R}{2D^3} \tag{3}$$

The average excess pressure (which equals the capillary pressure), between circular plates, can be expressed as:

$$\overline{P} = \frac{F}{\pi r_0^2} \tag{4}$$

Hence, we obtain the well-known Reynolds equation:

$$V_R = -\frac{\mathrm{d}D}{\mathrm{d}t} = \frac{2D^3\overline{P}}{3\mu r_0^2} \tag{5}$$

We immediately see that the film-thinning rate is reduced, and thus the emulsion stability increased, by an increase in bulk viscosity. In the case where the liquid film is so thin that surface forces no longer can be neglected, the capillary pressure term in the Reynolds equation should be replaced by the total driving force ($\Delta P$) for the thinning. This is equal to the difference between the capillary pressure and the disjoining pressure ($\Pi$) due to the surface forces acting between the emulsion droplet surfaces, $\Delta P = (\overline{P} - \Pi)$. Clearly, a positive disjoining pressure, i.e., a repulsive force, reduces the driving force for film thinning and thus the drainage rate.

Experimentally determined rates of thinning do not always agree with the predictions of the Reynolds model. For foam films stabilized by an anionic surfactant, sodium dodecyl sulfate (9, 10) it has been shown that typical thinning rates exhibited a much weaker dependence on the film radius ($r^{-0.8-0.9}$) than the predicted $r^{-2}$ dependence. To obtain an understanding for why the Reynolds theory of thinning does not always agree with experimental results it is worthwhile to consider two assumptions made when arriving at Eq. (5). First, the result is valid only under "no slip" conditions, i.e., the velocity of the liquid at the film interface is assumed to be zero. This is the case when the drainage takes place between solid hydrophilic surfaces. In contrast, only the adsorbed emulsifier layer provides the surface rigidity in foam and emulsion films, and it is not obvious that the no-slip condition is fulfilled. The drainage rate would be larger than predicted by Eq. (5) if this condition was not valid. Jeelani and Hartland (11), who calculated the liquid velocity at the interfaces of emulsion films for numerous systems studied experimentally, addressed this point. They showed that even at low surfactant concentration the liquid mobility at the interface is dramatically reduced by the adsorbed surfactant. Hence, it is plausible that when the adsorption density of the emulsifier is large (nearly saturated monolayers) the surface viscosity is high enough to validate the no-slip condition. It has been pointed out that a nonzero liquid viscosity at the interface is not expected to have an influence on the functional dependence of the drainage rates upon the film radius (9). Hence, the deviations found experimentally have to have another origin.

A second assumption made when arriving at Eq. (5) is that the drainage takes place between parallel surfaces. Experimental studies on liquid films (9, 10) have shown that during the thinning process it is common that nonuniform films are formed which have a thicker region, a dimple, in the center. For larger films even more complicated, multidimpled profiles have been found. To calculate the drainage rate for interfaces with such a complex shape is far from easy. However, recently Manev et al. (9) proposed a model for the drainage between nonparallel, immobile surfaces. The following expression has been proposed for the rate of thinning:

$$V = \frac{8}{3\mu} \sqrt[5]{\frac{4D^{12}\Delta P^{8}}{\alpha_1^{12}\sigma^{3}r^{4}}} \tag{6}$$

Here, $\alpha_1$ is the first root of the first-order Bessel function of the first kind, and $\sigma$ is the surface tension. Note that in the above equation the rate of thinning is inversely proportional to $r^{4/5}$. This is in good agreement with some experimental observations.

At sufficiently small droplet separations, say below 100 nm, surface forces have to be considered. These forces affect the drainage rate as well as the equilibrium interactions, particularly if flocculation occurs. The most commonly encountered forces are briefly described below. For a general reference to surface force, see the book by Israelachvili (12).

## B. Van der Waals Forces

Van der Waals forces originate mainly from the motion of negatively charged electrons around the positively charged atomic nucleus. For condensed materials (liquids or solids) this electron motion gives rise to a fluctuating electromagnetic field that extends beyond the surface of the material. Thus when, e.g., two particles or emulsion droplets are close together the fluctuating fields associated with them will interact with each other. The energy of interaction per unit area ($W_{vdw}$) between two equal spheres with radius $R$ a distance $D$ apart is given by:

$$W_{vdw} = -\frac{A}{6}\left[\frac{2R^2}{D^2+4RD} + \frac{2R^2}{D^2+4RD+R^2} + \ln\left(\frac{D^2+4RD}{D^2+4RD+R^2}\right)\right] \tag{7}$$

where $A$ is the nonretarded Hamaker constant. When the particle radius is much larger than the separation of the particles, Eq. (7) is reduced to:

$$W_{vdw} = -\frac{AR}{12D} \tag{8}$$

The Hamaker constant depends on the dielectric properties of the two interacting particles and the intervening medium. When these properties are known one can calculate the Hamaker constant. An approximate equation for two identical particles (subscript 1) interacting across a medum (subscript 2) is:

$$A = \frac{3kT}{4}\left(\frac{\varepsilon_1 - \varepsilon_2}{\varepsilon_1 + \varepsilon_2}\right)^2 + \frac{3h\nu}{16\sqrt{2}}\left(\frac{(n_1^2 - n_2^2)^2}{(n_1^2 + n_2^2)^{3/2}}\right) \tag{9}$$

where $k$ is the Boltzmann constant, $T$ is the absolute temperature, $\nu$ is the main adsorption frequency in the UV region (often about $3 \times 10^{15}$ Hz), $h$ is Planck's constant, $\varepsilon$ is the static dielectric constant, and $n$ is the refractive index in visible light.

From Eqs (8) and (9) it is clear that the van der Waals interaction between two identical particles or emulsion droplets is always attractive. One may also note that the Hamaker constant for two oil droplets interacting across water is identical to the Hamaker constant for two water droplets interacting across oil.

## C. Electrostatic Double-layer Forces

Electrostatic double-layer forces are always present between charged particles or emulsion droplets in electrolyte solutions. Counterions to the emulsion droplet (ions with opposite charges to that of the drop) are attracted to the surfaces and coions are repelled. Hence, outside the charged emulsion droplet, in the so-called diffuse layer, the concentration of ions will be different to that in bulk solution, and the charge in the diffuse layer balances the surface charge.

An electrostatic double-layer interaction arises when two charged droplets are so close together that their diffuse layers overlap. The electrostatic double-layer interaction, $W_{dl}$, for two identical charged drops with a small electrostatic surface potential and a radius large compared to their separation is approximately given by:

$$W_{dl} = 2\pi R \varepsilon \varepsilon_0 \Psi_0^2 \exp(-\kappa D) \tag{10}$$

where $\varepsilon_0$ is the permittivity of vacuum, $\varepsilon$ is the static dielectric constant of the medium, $\psi^0$ is the surface potential, and $\kappa^{-1}$ is the Debye screening length given by:

$$\kappa^{-1} = \sqrt{\frac{\varepsilon_0 \varepsilon kT}{1000 N_A e^2 \sum_i c_i z_i^2}} \tag{11}$$

where $e$ is the elementary charge, $N_A$ is Avogadro's number, $c_i$ is the concentration of ion $i$ expressed as mol/dm$^3$, and $z_i$ is the valency of ion $i$.

The double-layer interaction is repulsive and it decays exponentially with surface separation with a decay length equal to the Debye length. Further, the Debye length and consequently the range of the double-layer force decreases with increasing salt concentration and the valency of the ions present. The famous DLVO theory for colloidal stability (13, 14) takes into account double-layer forces and van der Waals forces.

## D. Hydration and Steric-protrusion Forces

Hydration and steric-protrusion forces are repulsive forces that have been found to be present at rather short separations between hydrophilic surfaces such as surfactant head-groups. At least two molecular reasons for these forces have been identified. First, when two polar surfaces are brought close together the polar groups will be partly dehydrated, which gives rise to a repulsive force (15). Second, as two surfaces are brought close together the molecules at the interface will have a decreased mobility perpendicular to the surface, which decreases the entropy of the system and this gives rise to a steric type of repulsion (16). Empirically it has been found that the hydration/steric repulsion between surfactant and lipid head-groups decays roughly exponentially with distance:

$$W_{hyd} = W_{hyd}^0 \exp(-D/\lambda) \tag{12}$$

where $\lambda$ is the decay length of the force, typically 0.2–0.3 nm.

## E. Polymer-induced Forces

The presence of polymers on surfaces gives rise to additional forces that can be repulsive or attractive. Under conditions when the polymer is firmly anchored to the surface and the surface coverage is large a steric repulsion is expected. As the surfaces are brought together the segment density between them increases, which results in an increased number of segment–segment contacts and a loss of conformational entropy of the polymer chains. The conformational entropy loss always results in a repulsive force contribution that dominates at small separations. The increased number of segment–segment contacts may give rise to an attractive or a repulsive force contribution. This is often discussed in terms of the chi-parameter ($\chi$-parameter) or in terms of solvent quality. Under sufficiently poor solvent conditions ($\chi > 1/2$), when the segment–segment interaction is sufficiently favorable compared to the segment–solvent interaction, the long-range interaction is attractive. Otherwise it is repulsive. The steric force can be calculated by using lattice mean field theory (17) or scaling theory (18). The actual force encountered is highly dependent on the adsorption density, the surface affinity, the polymer architecture, and the solvency condition. Hence, no simple equation can describe all situations. However, a high-density polymer layer, a "brush" layer, in a good solvent, provides good steric stabilization. The scaling approach provides us with a simple formula that often describes the measured interactions under such conditions rather well (19). It states that the pressure $P(D)$ between two flat polymer-coated surfaces is given by:

$$P(D) \approx \frac{kT}{s^3}\left[\left(\frac{D^*}{D}\right)^{9/4} - \left(\frac{D}{D^*}\right)^{(3/4}\right] \tag{13}$$

where Eq. (13) is valid provided that the separation, $D$, is less than $D^*$ (where $D^*$ is twice the length of the polymer tail), and $s$ is the linear distance between the anchored chains on the surface. For the interactions between two spheres with a radius significantly larger than their separation this relation is modified to:

$$\frac{F(D)}{R} \approx -\pi \int_{D^*}^{D} P(D')\mathrm{d}D' \tag{14}$$

The parameters needed in order to calculate the force are the length of the extended polymer chain and the separation between the polymer chains on the surface. The latter parameter can be estimated from the adsorbed amount whereas the length of the polymer chains enters as a fitting parameter. The formula predicts a repulsion that increases monotonically with decreasing separation.

## F. Coalescence and Hole Formation

When studying drainage and equilibrium interactions in single foam films above the critical micellar concentration (cmc) of the surfactant, it is often found that the film thickness undergoes sudden changes (20, 21). This phenomenon is known as stratification. Below the cmc one sudden change from a water-rich common black film to a very thin Newton black film may occur. This transition does not occur uniformly over the whole film area but initially in some small regions. The thinner regions are often called black spots since they appear darker than the rest of the film when viewed in reflected light. Once formed, the size of a black spot grows as the liquid drains out from the foam lamellae. Bergeron and coworkers noted that the viscous resistance to the flow in the thin film is large, and that this leads to an increase in the local film thickness next to the black spot (22, 23). The suggested shape of the thin liquid layer, which is supported by experimental observations and theoretical calculations (22, 23) is illustrated in Fig. 1. In many cases no or unstable Newton black films are formed. In these cases the films rupture due to formation of a hole that rapidly grows as a result of surface-tension forces. Emulsion coalescence occurs in a similar manner.

The mechanism of black spot formation and rupture has been extensively studied (24). It is generally recognized that the liquid film is unstable in regions of the disjoining pressure (Π) isotherm (force curve) where the derivative with respect to film thickness ($D$) is larger than zero, i.e., $\mathrm{d}\Pi/\mathrm{d}D > 0$. Hence, close

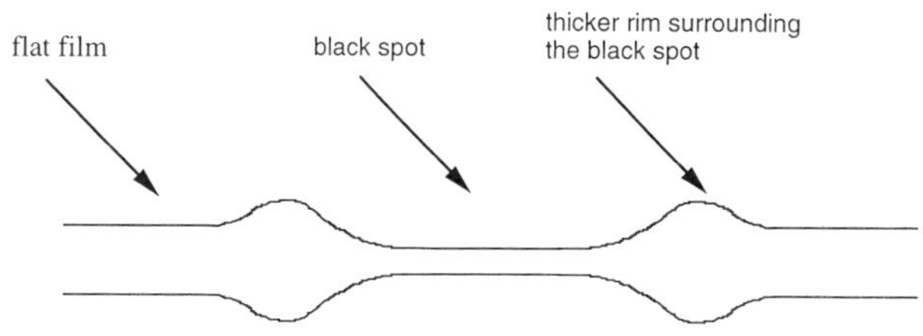


**Figure 1** Illustration of shape of the thin liquid film around the position of a newly formed black spot.

to a maximum in the disjoining pressure isotherm (see Fig. 2) a small disturbance causing a change in film thickness and/or capillary pressure may spontaneously grow and lead to a significant change in film thickness, e.g., Newton black film formation or rupture.

The stability of foams and emulsions depends critically on whether formation of a stable Newton black film or a hole leading to coalescence is favored. Kabalnov and Wennerström (4) addressed this question by developing a temperature-induced hole nucleation model applicable to emulsions. They point on that the coalescene energy barrier is strongly affected by the spontaneous monolayer curvature. The authors consider a flat emulsion film, covered by a saturated surfactant monolayer, in thermodynamic equilibrium with

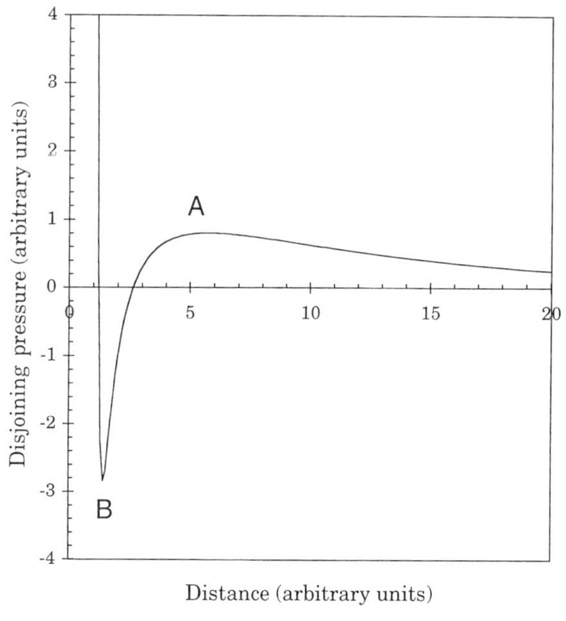


**Figure 2** Typical disjoining pressure isotherm showing one maximum (A) and one minimum (B); the film is unstable between points A and B.

a micellar bulk solution. The emulsion breaks if an induced hole grows along the film having a thickness $h = 2b$ (Fig. 3). The change in free energy occurring when a hole is formed is given as the difference in the interfacial tension integrals over the interface for a film with a hole compared to that for a planar film:

$$W = \underset{\text{(film with hole)}}{\int \sigma(x, y, z)\mathrm{d}A} - \underset{\text{(flat film)}}{\int \sigma(x, y, z)\mathrm{d}A} \quad (15)$$

The driving force for formation of a hole is the reduction in free energy owing to a decrease in surface area of the planar part of the film, whereas it is counteracted by the increased free energy due to the surface area created around the hole. In general, the change in free energy goes through a maximum as the hole radius increases. One new feature of the Kabalnov–Wennerström model is that the surface tension at the hole edge is considered to be different to that at the planar film surface. The reason for this is that the curvature of the interface is different, leading to a difference in surfactant monolayer bending energy. This can be expressed as (4):

$$\sigma^{\text{curved}} = \sigma^{\text{planar}} + 2\kappa(H - H_0)^2 - 2\kappa H_0^2 \quad (16)$$

Here, $H$ and $H_0$ are the mean and the spontaneous curvatures and $\kappa$ is the bending modulus.

Clearly, the surface tension has a minimum when the spontaneous curvature of the surfactant film equals the mean curvature of the interface. The mean curvature for a flat interface is zero, larger than zero for an interface curving towards the oil (oil-in-water emulsions), and smaller than zero for a water-in-oil emulsion. Hence, a large positive spontaneous monolayer curvature, as for a strongly hydrophilic surfactant, favors oil-in-water emulsions and vice versa. The Kabalnov–Wennerström model also allows the thickness of the film to vary in order to minimize the free energy of hole formation, i.e., the mean curvature of the film close to the hole can approach the spontaneous monolayer curvature. The Kabalnov–Wennerström model has to be solved numerically in order to calculate the coalescence activation energy. However, a "big hole approach" where $a \gg b$ (see Fig. 3) gives surprisingly good results. In this model the energy for creating a hole with radius $a$ is given as:

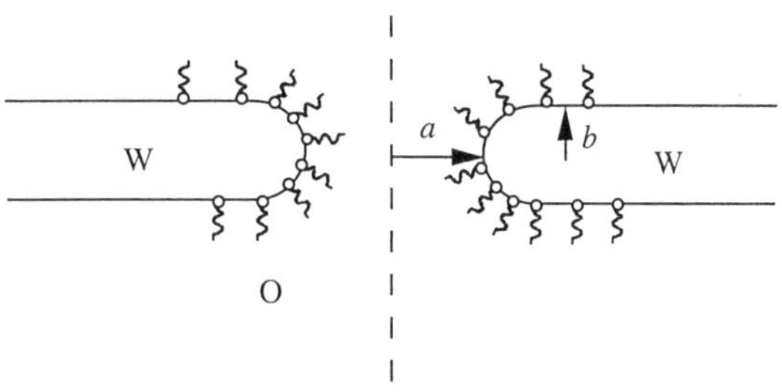


**Figure 3** Geometry of the thin film just after a hole has been created. (Redrawn from Ref. 4, with permission.)

$$W = 2\pi a\gamma - 2\pi a^2\sigma \quad (17)$$

where $2\pi a$ is the circumfrence of the hole, $\gamma$ is the line tensions, $\pi a^2$ is the area of the hole at each interface, and $\sigma$ is the surface tension. The second term is the free energy gain obtained by reducing the flat area of the film, and the first term is the energy penalty of creating the inside of the hole. The value of the line tension can be calculated when the spontaneous monolayer curvature and the monolayer bending modulus is known (4). The activation energy of coalescence ($W^*$) is obtained by finding the point where $\mathrm{d}W/\mathrm{d}a = 0$, which gives the final expression:

$$W^* = \frac{\pi\gamma^2}{2\sigma} \quad (18)$$

The particular feature with ethylene oxide based surfactants is that their interaction with water is less favorable at higher temperatures. This leads to a decrease in the spontaneous monolayer curvature with temperature, explaining the transition from oil-in-water emulsions below the PIT to water-in-oil emulsion above the PIT. In the vicinity of the phase inversion temperature the energy barrier against coalescence ($W^*$) varies very strongly with temperature. For the system n-octane–$C_{12}E_5$–water the following approximate relation was obtained in terms of $\Delta T = T - T_d$, where $T_d$ is the PIT (4):

$$W^*/kT = 0.43 + 30.9|\Delta T| \quad (19)$$

The predicted very steep increase in the coalescence barrier away from the PIT is qualitatively in good agreement with the experimentally observed macroemulsion behavior (25).

## III. SURFACE FORCE TECHNIQUES

There are several methods available for measuring forces between two solid surfaces, two particles, or liquid interfaces (26). In this section we briefly mention some of the features of the techniques that have been used in order to produce the results described in the later part of this contribution. The forces acting between two solid surfaces were measured either with the interferometric surface force apparatus (SFA) or with the MASIF (measurements and analysis of surface and interfacial forces). Interactions between fluid interfaces were determined using various versions of the thin film balance (TFB).

## A. The Interferometric Surface Force Apparatus (SFA)

The forces acting between two molecularly smooth surfaces, normally mica or modified mica, can be measured as a function of their *absolute* separation with the interferometric SFA (Fig. 4) (27). This provides a convenient way to measure not only long-range forces but also the thickness of adsorbed layers. The absolute separation is determined interferometrically to within 0.1–0.2 nm by using fringes of equal chromatic order. The surfaces are glued on to optically polished silica disks and mounted in the SFA in a crossed cylinder configuration. The surface separation is controlled either by adjusting the voltage applied to a piezoelectric crystal rigidly attached to one of the surfaces, or by a synchronous motor linked by a cantilever spring to the other surface. The deflection of the force measuring spring is also determined interferometrically, and the force is calculated from Hooke's law. For further details, see Ref. 27.

When an attractive force component is present the gradient of the force with respect to the surface separation, $\partial F/\partial D$, may at some distance become larger than the spring constant, $k$. The mechanical system then enters an unstable region causing the surfaces to jump to the next stable point (compare instabilities in free liquid films that occur when $d\Pi/dD > 0$). The adhesion force, $F(0)$, normalized by the local mean radius of curvature, $R$, is determined by separating the surfaces from adhesive contact. The force is calculated from:

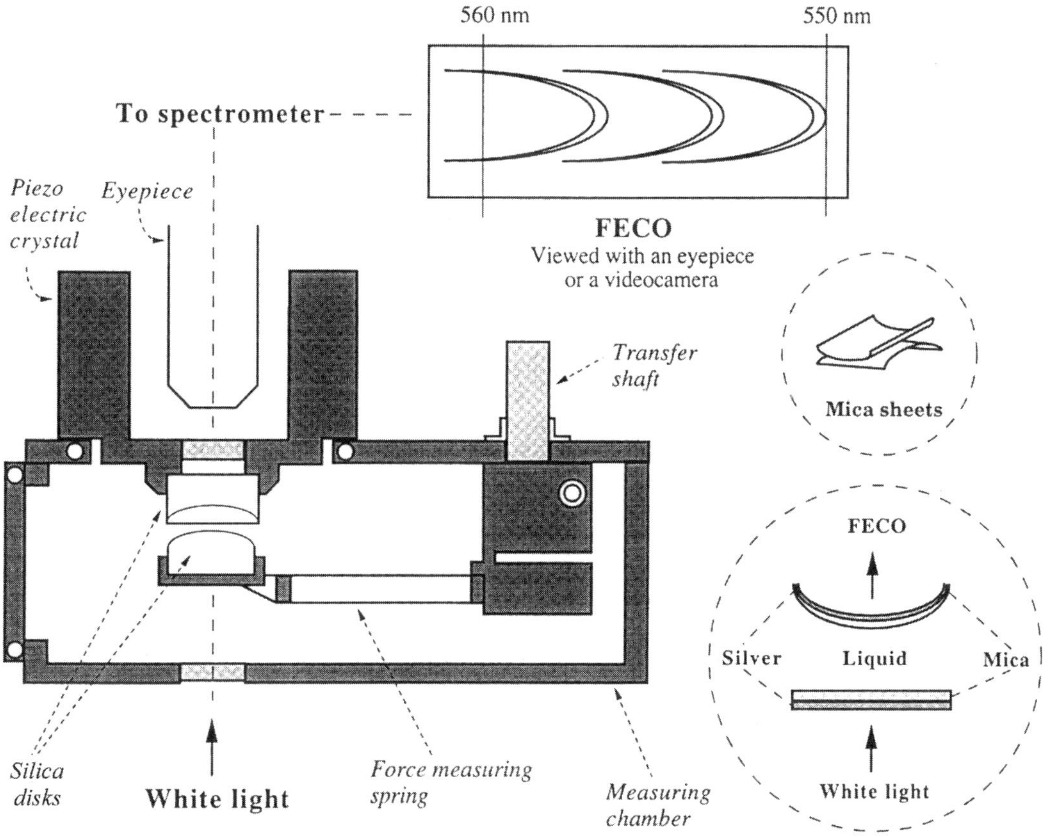


**Figure 4** Schematic picture of a surface-force apparatus (SFA). The measuring chamber is made from stainless steel. One of the surfaces is mounted on a piezoelectric tube that is used to change the surface separation; the other surface is mounted on the force measuring spring. (From Ref. 26, with permission.)

$$\frac{F(0)}{R} = \frac{kDj + F(Dj)}{R} \tag{20}$$

where $F(D_j)$ is the force at the distance $(D_j)$ to which the surfaces jumped on separation, and $R$ is the mean radius of the surfaces.

## B. The Bimorph Surface Force Apparatus (MASIF)

The force as a function of surface separation between glass substrate surfaces was measured with a MASIF instrument [28]. This apparatus is based on a bimorph force sensor to which one of the surfaces is mounted (Fig. 5). The other surface is mounted on a piezoelectric tube. The bimorph (enclosed in a Teflon sheath) is mounted inside a small measuring chamber, which is clamped to a translation stage that serves to control the coarse position of the piezoelectric tube and the upper surface.

The voltage across the piezoelectric tube is varied continuously and the surfaces are first pushed together and then separated. The bimorph will deflect when a force is experienced and this generates a charge in proportion to the deflection. From the deflection and the spring constant the force follows simply from Hooke's law. The motion of the piezo is measured during each force run with a linear displacement sensor. This signal together with the signal from the bimorph charge amplifier, the voltage applied to the piezoelectric tube, and the time are recorded by a computer. The speed of approach, the number of data points, and other experimental variables can easily be controlled with the computer software.

When the surfaces are in contact the motion of the piezoelectric tube is transmitted directly to the force sensor. This results in a linear increase of the force sensor signal with the expansion of the piezoelectric tube. The sensitivity of the force sensor can be calibrated from this straight line, and this measuring procedure allows determination of forces as a function of separation from a hard wall contact with a high precision (within 1–2 Å in distance resolution). Note, however, that the assumption of a "hard wall" contact is not always correct (29).

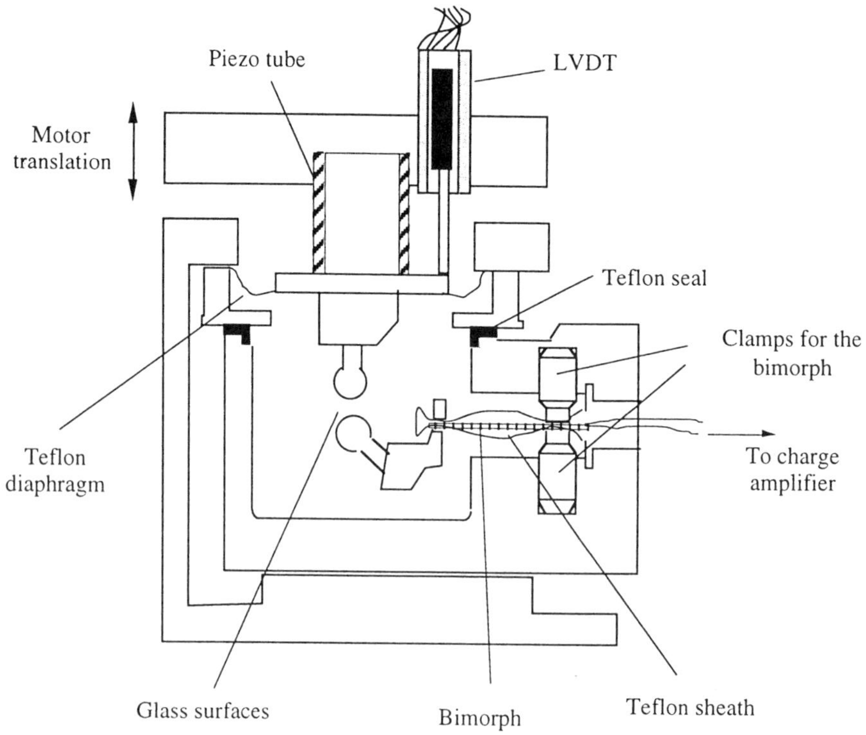


**Figure 5** Schematic illustration of the MASIF (measurement and analysis of surface interaction forces) SFA. The upper surface is mounted on a piezo ceramic actuator that is used for changing the surface separation; the hysteresis of the piezo expansion/contraction cycle can be accounted for by using a linear variable displacement transducer (LVDT). The lower surface is mounted on a bimorph force sensor. (From Ref. 26, with permission.)

The MASIF instrument does not use interferometry to determine surface separations which leads to the drawback that the layer thickness cannot be determined, but to the advantage that the instrument can be used with any type of hard, smooth surfaces. In most cases spherical glass surfaces are used. They are prepared by melting a 2 mm diameter glass rod until a molten droplet with a radius of 2 mm is formed.

## C. Derjaguin Approximation

The force measured between crossed cylinders ($F_c$), as in the SFA, and between spheres ($F_s$), as in the MASIF, a distance $D$ apart is normalized by the local geometric mean radius ($R$). This quantity is related to the free energy of interaction per unit area between flat surfaces ($W$) according to the Derjaguin approximation (30):

$$\frac{F_c}{2\pi R} = \frac{F_s}{\pi R} = W \tag{21}$$

This approximation is valid when the radius (about 2 cm in the SFA; 2 mm in the MASIF) is much larger than the surface separation (typically $10^{-5}$ cm or less), a requirement fulfilled in these experiments. With the SFA the local radius is determined from the shape of the standing wave pattern, whereas in the MASIF we have used the assumption that the local radius is equal to the macroscopic radius, determined using a micrometer. The radius used in Eq. (21) is that of the undeformed surfaces. However, under the action of strongly repulsive or attractive forces the surfaces will deform and flatten (31, 32). This changes the local radius and invalidates Eq. (21) since $R$ becomes a function of $D$.

## D. The Thin-film Balance

Accurate information about the rate of thinning, the critical thickness of rupture, and the forces acting between two air–water interfaces, betwen two oil–water interfaces, and between one air–water and one oil–water interface can be gained by using thin-film balance techniques. The thickness of the separating water film is determined by measuring the intensity of reflected white light from a small flat portion of the film (33). Due to interference of the light reflected from the upper and lower film surfaces, characteristics interference colors are observed during the thinning. These colors correspond to a shift in the wavelengths undergoing constructive and destructive interference. When such a process is recorded (normally as intensity of a given light wavelength versus time) a sequence of intensity minima and maxima appear. The equivalent water film thickness can be calculated from the following equation (33):

$$h_{eq} = \left(\frac{\lambda}{2\pi n_1}\right) \arcsin \sqrt{\frac{\Delta}{1 + \left[\frac{4R(1-\Delta)}{(1-R)^2}\right]}}$$
$$R = \left(\frac{n_1 - n_2}{n_1 + n_2}\right)^2 \tag{22}$$
$$\Delta = \left(\frac{I - I_{min}}{I_{max} - I_{min}}\right)$$

where $\lambda$ is the wavelength, and $n_1$ and $n_2$ are the bulk refractive indices of the continuous and the disperse phases respectively (in the case of foam films $n_2 = 1$); $I_{max}$ and $I_{min}$ are the intensity values of the interference maximum and minimum, and $I$ is the instantaneous value of the light intensity. The above equation gives the equivalent film thickness, $h_{eq}$, i.e., the film thickness plus the thickness of the adsorbed layers calculated by assuming a constant value of the refractive index equal to $n_1$. A better approximation to the true film thickness can be obtained by correcting for the difference in refractive index between the bulk film and the adsorbed layer. The corrected film thickness is (23):

$$h = h_{eq} - 2h_{hc}\left(\frac{n_{hc}^2 - n_1^2}{n_1^2 - n_2^2}\right) - 2h_{pg}\left(\frac{n_{pg}^2 - n_1^2}{n_1^2 - n_2^2}\right) \tag{23}$$

In the above equation $h_{hc}$ and $h_{pg}$ are the thickness of the region occupied by the surfactant hydrocarbon chain and polar group, respectively. Similarly, $n_{hc}$ and $n_{pg}$ are the corresponding refractive indices. The thickness values needed in order to use Eq. (23) can be estimated from the volume of the two parts of the molecule together with values of the area per molecule at the interface obtained from adsorption data, e.g., the surface-tension isotherm. Finally, the thickness of the core layer (water in case of foam films) can be calculated as:

$$h_{core} = h - 2(h_{hc} + h_{pg}) \tag{24}$$

The apparatus used for studying thin liquid films is schematically depicted in Fig. 6. This device, commonly known as a thin-film balance, allows drainage patterns of single foam, emulsion, or wetting films to be recorded. The film is formed in a specially constructed cell that is placed on the state of an inverted microscope. The reflected light from the film is split into two parts, one directed to a CCD camera and another to a fiber-optic probe tip located in the microscope eyepiece. The radius of the tip is only about 20

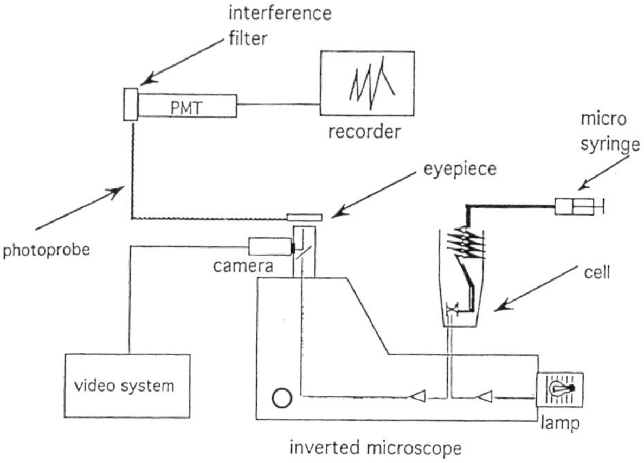


**Figure 6** Schematic representation of the main components of a typical thin-film balance

μm which allows light from a small portion of the film to be registered. The light signal is then passed through a monochromatic filter and finally directed on to a high-sensitive photomultiplier. The output of the photomultiplier is connected to a chart recorder and the data are collected in the form of intensity (as a photocurrent) versus time. This graph is called an interferogram.

An essential part of the thin-film balance is the cell holding the thin film. The cell can be constructed in several ways depending on the type of measurement to be done and the systems under investigation. For emulsion films a type of cell proposed by Scheludko (33) is often used. The cell is illustrated on Fig. 7. The film is formed between the tips of the menisci of a biconcave drop held in a horizontal tube with radius 1.5–2 mm. The tube and the spiral capillary are filled with the aqueous phase and immersed in a cuvette (the lower part of the cell) containing the oil phase. A small suction pressure applied through the capillary controls the film radius. Recently, a cell that is similar to that of Scheludko, but miniaturized about 10-fold was used by Velev *et al.* (34). This allowed film sizes and capillary pressures found in real emulsion systems to be studied. Bergeron and Radke (35) used a cell with a porous frit holder as suggested by Mysels and Jones (36) for measuring equilibrium forces across foam and pseudoemulsion films. Their construction is shown in Fig. 8. The main advantage of this so-called porous-plate technique is that one can vary the pressure in the film by simply altering the gas pressure in the cell, and thus the stable part of the equilibrium disjoining pressure isotherm (where $d\Pi/dD < 0$) can be obtained.

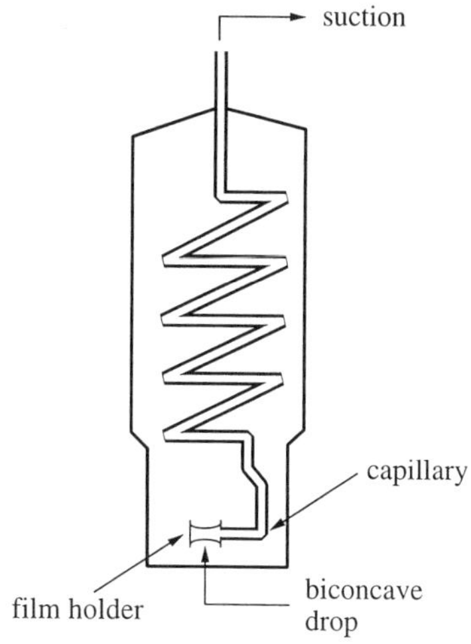


**Figure 7** Illustration of the Scheludko cell used for investigation of single, horizontal foam and emulsion films.

## IV. RESULTS AND DISCUSSION

### A. Ionic Surfactants on Hydrophobic Surfaces

Many oil-in-water emulsions are stabilized by an adsorbed layer of surfactants. One example is asphalt oil-in-water emulsions that often are stabilized by cationic surfactants (37). The surfactants fill two pur-

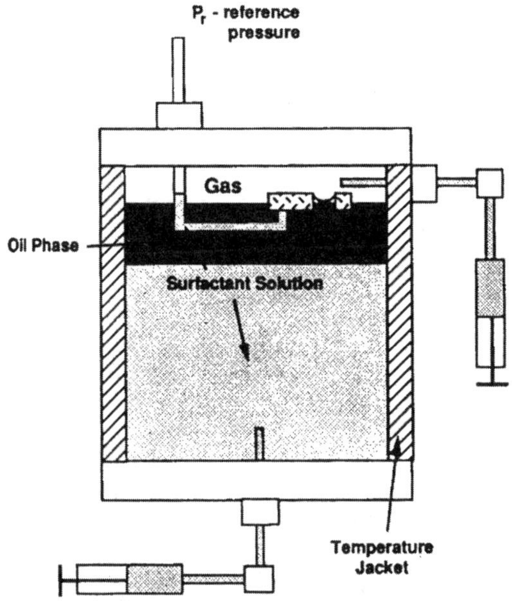


**Figure 8** Modified porous-plate cell for investigation of pseudoemulsion films. (From Ref. 35, with permission.)

poses. First, they generate long-range repulsive forces that prevent the emulsion droplets from coming close to each other. Second, the surfactant layer acts as a barrier against coalescence if the emulsion droplets by chance come close to each other despite the long-range repulsive forces. The coalescence is hindered by a high spontaneous monolayer curvature, monolayer cohesive energy, surface elasticity, and surface viscosity, which increase the activation energy for hole formation and slow down the depletion of surfactants from the contact region. The importance of the cohesive energy for foam films stabilized by a homologous series of cationc surfactants was particularly clearly demonstrated by Bergeron (38). We note that an increased cohesive energy in the monolayer increases the bending modulus and thus the free energy cost for the surfactant film is to have a curvature different to the spontaneous curvature.

Surface-force measurements using a hydrophobic solid surface as a model for a fluid hydrocarbon/water interface provide a good picture of the long-range forces acting between emulsion droplets. However, the coalescence behavior of emulsions will not be accurately described from such measurements. One reason is that the fluid interface is much more prone to deformation than the solid surface (facilitating hole formation), and the surfactant chains can readily penetrate into the fluid oil phase, but not into the solid hydrocarbon surface. Further, the mobility of the surfactants on a solid hydrophobic surface will be different from the mobility at a fluid interface.

The forces acting between two hydrophobic surfaces across dodecylammonium chloride surfactant solutions are illustrated in Fig. 9 (39). The long-range repulsion is due to the presence of an electrostatic double-layer force. This force is generated by the cationic surfactants that adsorb to the hydrophobic surface whereby giving them a surface-charge density. The range of the double-layer force decreases with increasing surfactant concentration, which is simply as a result of the increased ionic strength of the aqueous media. On the other hand, the magnitude of the double-layer force at short separations increases with increasing surfactant concentrations. This is a consequence of the increased adsorption of the ionic surfactant that results in an increase in surface-charge density and surface potential. The surfactant concentration will influence the long-range interactions between oil-in-water emulsions in the same way as observed for the model solid hydrophobic surfaces, i.e., the range of the double-layer force will decrease and the magnitude of the force at short

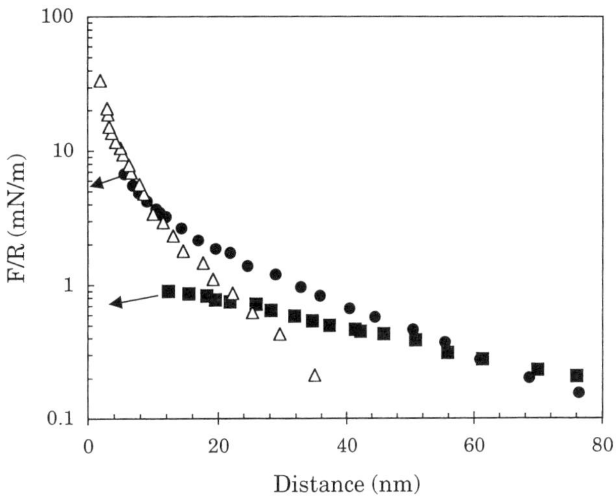


**Figure 9** Force normalized by radius measured between two hydrophobized mica surfaces in crossed cylinder geometry across aqueous solutions of dodecylammonium chloride; the surfactant concentration was 0.01 mM (■), 0.1 mM (●), and 1 mM (△), respectively. The arrows indicate inward jumps occurring when the force barrier has been overcome. (Redrawn from Ref. 39, with permission.)

separations will increase. However, the adsorbed amount at a given surfactant concentration may not be the same on the emulsion surface as on the solid hydrophobic surface.

At low surfactant concentrations it is observed that an attraction dominates at short separations. The attraction becomes important at separations below about 12 nm when the surfactant concentration is 0.01 mM, and below about 6 nm when the concentration is increased to 0.1 mM. Once the force barrier has been overcome the surfaces are pulled into direct contact between the hydrophobic surfaces at $D = 0$, demonstrating that the surfactants leave the gap between the surfaces. The solid surfaces have been flocculated. However, at higher surfactant concentrations (1 mM) the surfactants remain on the surfaces even when the separation between the surfaces is small. The force is now purely repulsive and the surfaces are prevented from flocculating. Emulsion droplets interacting in the same way would coalesce at low surfactant concentrations once they have come close enough to overcome the repulsive barrier, but remain stable at higher surfactant concentrations. Note, however, that the surfactant concentration needed to prevent coalescence of emulsion droplets cannot be accurately determined from surface-force measurements using solid surfaces.

For application purposes it is often found that asphalt emulsions stabilized by cationic surfactants function better than such emulsions stabilized by anionic surfactants. One main reason is that the interaction between the emulsion droplet and the road material differs depending on the emulsifier used (37). When the asphalt emulsion is spread on thc road surface it should rapidly break and form a homogeneous layer. The stones on the road surface are often negatively charged and there will be an electrostatic attraction between cationic emulsion droplets and the stones. This attraction facilitates the attachment and spreading of the emulsion. On the other hand, when the emulsion droplet is negatively charged there will be an electrostatic repulsion between the stones and the emulsion droplets.

## B. Nonionic Surfactants on Hydrophobic Surfaces

Nonionic ethylene oxide based surfactants are commonly used as emulsifiers. Since these surfactants are uncharged they are not able to generate stabilizing long-range electrostatic forces. Instead, they generate short-range hydration/protrusion forces that prevent the emulsion droplets from coming into direct contact with each other. The short-range forces acting between hydrophobic solid surfaces as a function of temperature are illustrated in Fig. 10 (40). The zero distance in the diagram is defined as the position of the hard wall (at a value of $F/R \approx 100$ mN/m). The force present at distances above 4 nm is a weak double-layer force. It originates from remaining chages on the hydrophobic substrate surface. The force observed at smaller separations has a pronounced temperature dependence. It becomes less repulsive with increasing temperature. At the same time the adsorbed layer thickness increases, demonstrating that the repulsion betwen the adsorbed molecules within one layer is also reduced at higher temperatures, facilitating an increased adsorption. The increase in layer thickness is not seen in Fig. 10 due to our definition of zero separation.

A decreasing inter- and intra-layer repulsion with increasing temperature is common for all surfactants and polymers containing oligo(ethylene oxide) groups. This shows that the interaction between ethylene oxide groups and water becomes less favorable at higher temperatures, i.e., the ethylene oxide chain becomes

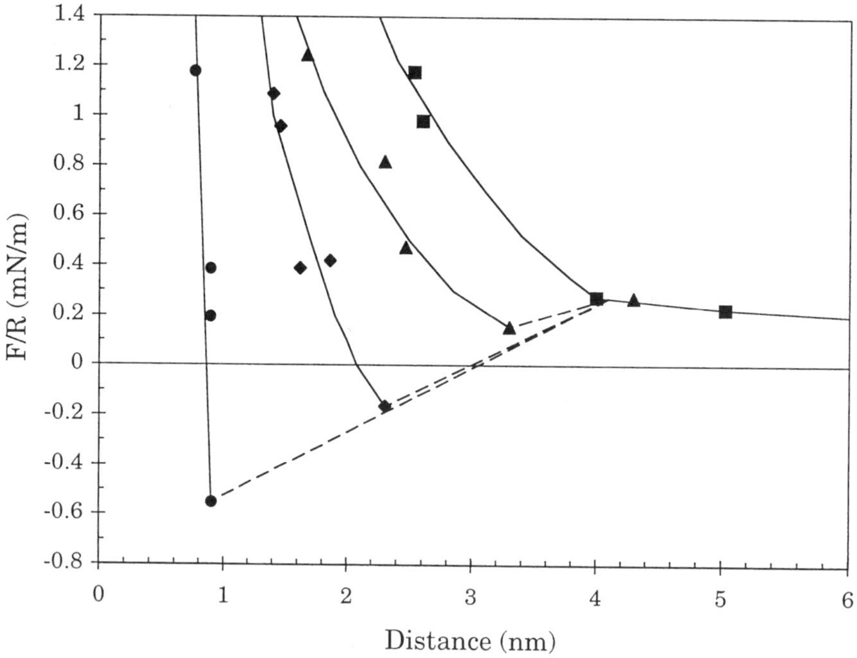


**Figure 10** Force normalized by radius measured between hydrophobized mica surfaces in crossed cylinder geometry across a $6 \times 10^{-5}$ M aqueous solution of penta(oxyethylene) dodecyl ether. The temperature was 15°C (■), 20°C (▲), 30°C (◆), and 37°C (●). The lines are guides for the eye. (Redrawn from Ref. 40, with permission.)

more hydrophobic. There are several theoretical attempts to explain this phenomenon, but it is outside the scope of this chapter to discuss them and the reader is referred to the original literature (41–48). The temperature dependence of the interaction between oligo(ethylene oxide) chains and water has several important consequences. The micellar size increases with temperature (49), and the micellar solution has a lower consolute temperature, i.e., a phase separation occurs on heating (50). The cloud point for a range of micellar alkyl ethoxylate solutions are provided in Fig. 11 (51). The cloud point increases with the number of ethylene oxide units. The reason is that a longer ethylene oxide chain gives rise to a more long-range intermicellar repulsion and a larger optimal area per headgroup (favoring smaller micelles). On the other hand, the cloud point decreases with the length of the hydrocarbon chain. By considering the geometry of the surfactant as described by the packing parameter (52), one realizes that the micellar size is expected to increase with the hydrocarbon chain length. It is also found that surfaces coated with (ethylene oxide containing) polymers often have good protein-repellent properties at low temperatures whereas proteins adsorb more readily to such surfaces at higher temperatures (53).

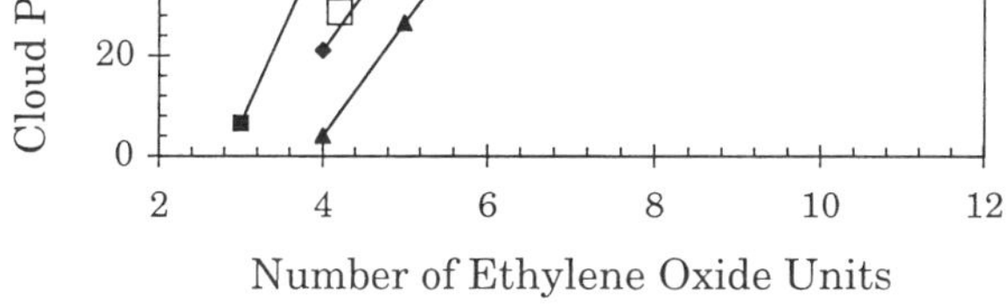


**Figure 11** Cloud-point temperature of micellar solutions as a function of the ethylene oxide chain length: the hydrophobic part is an alkyl chain with 8 (■), 10 (◆), 12 (▲), or 16 (●) carbon atoms. Data from Ref. 51. The symbols (□) represent the phase-inversion temperature for a 1:1 cyclohexane–water emulsion containing 5% of commercial ethylene oxide based emulsifiers having dodecylalkyl chains as a hydrophobic group. (Data from Ref. 54.)

For emulsions the most important aspect may be that the optimal area per head-group in an adsorbed layer decreases with increasing temperature, which reduces the spontaneous monolayer curvature (4). This is the reason why emulsions stabilized by ethylene oxide based surfactants may change from oil-in-water to water-in-oil when the temperature is increased. The temperature when this occurs is known as the phase inversion temperature (PIT). The PIT depends on the length of the hydrocarbon chain and the ethylene oxide chain in a similar way as the cloud point (54), see Fig. 11. However, the PIT also depends on the type of oil used (55), which is partly due to differences in solubility of the ethylene oxide surfactants in the different oils and to differences in oil penetration in the surfactant layer. We also note that if the emulsifier concentration is high enough a liquid crystalline phase may accumulate at the oil–water interface. In such cases emulsions which are very stable towards coalescence may be formed (56). This is as a result of the decreased probability of hole formation (4). In this case the type of oil used has a dramatic effect on the emulsion stability, which can be understood by considering the three-component phase diagram.

## C. Nonionic Polymers on Hydrophobic Surfaces

We discussed above how the length of the oligo(ethylene oxide) chain influences the properties of emulsions stabilized by alkyl ethoxylates. When the ethylene oxide chain becomes sufficiently long one normally refers to the substance as a diblock copolymer rather than as a surfactant. Of course, there is no clear distinction but the properties vary in a continuous fashion with increasing ethylene oxide chain length. It is of interest to follow how the forces acting between two surfaces carrying adsorbed diblock copolymeres vary with the length of the adsorbing (anchor) block, and the nonadsorbing (buoy) block. A nice experimental work addressing this question is that of Belder *et al.* (57). Fleer *et al.* give a thorough theoretical treatment in their book (17), where it is suggested that the most efficient steric stabilization is obtained when the anchor block has a size that is 10–20% of that of the buoy block. The reason for this optimum is that when the anchor block is too small the driving force for adsorption is weak and the adsorbed amount will be low. On the other hand, when the anchor block is too large the area per adsorbed molecule will be large. As a consequence the buoy block layer will be dilute and it will not extend very far out into the solution, leading to a not very pronounced steric force.

The forces acting between solid hydrophobic surfaces coated with different ethylene oxide based diblock polymers are illustrated in Fig. 12. The forces acting between surfaces coated with penta(oxyethylene) dodecyl ether, $C_{12}E_5$, becomes significantly repulsive at distances below about 2 nm, calculated from the hard wall contact at $D = 0$. Note that the surfactant layer remains between the surfaces and the range of the force given is thus relative to the position of direct contact between the compressed adsorbed surfactant layers. The forces between hydrophobic surfaces coated with a diblock copolymer containing eight butylene oxide units and 41 ethylene oxide units, $B_8E_{41}$, are significantly more long ranged. The interaction at distances above 4 nm from the "hard wall' is dominated by a weak electrostatic double-layer force originating from remaining charges on the silanated glass surface. However, at shorter distances a steric force predominates. Hence, the molecules with the longer ethylene oxide chains give rise to a more long-range force. Note that this is true even when the range is calculated from the position of the hard wall, i.e., without considering the difference in compressed layer thickness. A much more long-range force is observed in the case of $B_{15}E_{200}$, where the steric force extends to more than 10 nm away from the hard-wall contact.

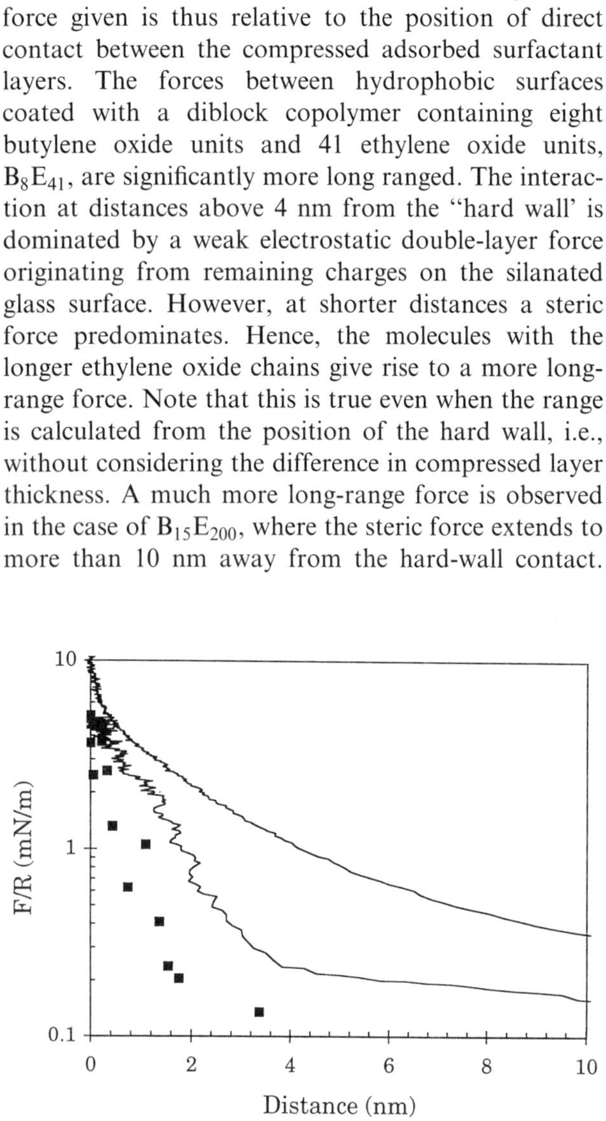


**Figure 12** Force normalized by radius between hydrophobized mica or glass surfaces coated by penta(oxyethylene) dodecyl ether at 20°C (■), and copolymers of butylene oxide (B) and ethylene oxide (E) with composition $B_8E_{41}$ (lower line) and $B_{15}E_{200}$ (upper line). All data have been recalculated to spherical geometry.

From the above it is clear that rather large diblock copolymers are efficient in generating long-range repulsive steric forces, which is beneficial for increasing the stability of dispersed particles and emulsion droplets. Even higher stability may be obtained if a mixture of diblock copolymers and charged surfactants are used, thus providing both steric and electrostatic stabilisation.

## D. Polyelectrolytes on Surfaces

Both steric and electrostatic stabilization was utilized by Fäldt et al. (58) when making soybean oil emulsions. They first made the emulsion using a mixture of phosphatidylcholine and glycoholic acid (bile salt) with a p*K*a of 4.4. The emulsion droplets obtained a net negative surface charge due to dissociation of the glycoholic acid. To improve the stability of the emulsion a weak cationic polyelectrolyte, chitosan, with a p*K*a of 6.3–7 was added. The polyelectrolyte adsorbs to the negatively charged emulsion droplet surface, which becomes positively charged at low pH. It was found that the emulsion was stable at high and low pH but not at pH values around 7, where irreversible aggregation was observed. This clearly shows that the forces acting between the emulsion droplets change with pH. To shed light on this behavior the forces acting between negatively charged solid surfaces coated by chitosan were measured as a function of pH (Fig. 13).

A repulsive double-layer force dominated the long-range interaction at pH values, below 5. However, at distances below about 5 nm, the measured repulsive force is stronger than expected from DLVO theory due to the predominance of a steric force contribution. The layer thickness obtained under a high compressive force was 1 nm per surface. Hence, it is clear that positively charged chitosan adsorbs in a very flat conformation on strongly oppositely charged surfaces such as mica with only short loops and tails. When the pH is increased to 6.2, the mica–chitosan system becomes uncharged, because the charge density of the chitosan molecules has decreased. The decrease in charge density of the chitosan also results in a decrease in segment–segment repulsion and therefore an even more compact adsorbed layer. At this pH value there is an attraction between the layers at a surface separation of about 2 nm. The steric repulsion is in this case very short range (< 2 nm) and steep. A further increase in pH to 9.1 results in a recharging due to the fact that the charges on the polyelectrolyte no longer can compensate for all of the mica surface

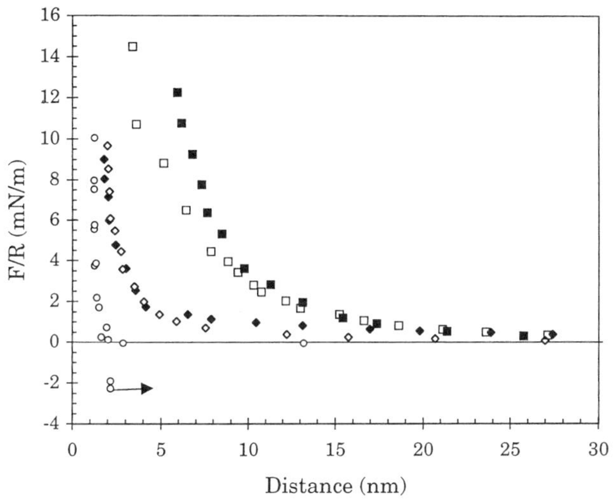


**Figure 13** Force normalized by radius between negatively charged mica surfaces in crossed cylinder geometry precoated with a layer of chitosan, a cationic polyelectrolyte. The forces were recorded at pH 3.8 (◆), 4.9 (◇), 6.2 (○), 7.6 (□), and 9.1 (■); the arrow indicates an outward jump. (From Ref. 58, with permission.)

charges. Further, as the charge density of the polyelectrolyte is reduced the range of the steric force increases again due to the lower affinity of the polyelectrolytes for the surface. Clearly, the mica–chitosan system is positively charged at low pH (i.e., the charges on the polyelectrolyte overcompensate for the charges on the mica surface), uncharged at pH 6.2, and negatively charged at high pH due to an undercompensation of the mica surface charge by the polyelectrolyte charges.

The flocculation behavior of the soybean emulsion can now be better understood. A stable emulsion is formed at low pH owing to the electrostatic repulsion generated by the excess charges from the adsorbed chitosan. At intermediate pH values the soybean emulsion is uncharged and the adsorbed chitosan layer is very flat. Hence, no long-range electrostatic force or any long-range steric force is present that can stabilize the emulsion. At high pH, the charges due to ionization of the glycoholic acid are no longer compensated for by the, at high pH nearly uncharged, chitosan. Thus, stabilizing electrostatic forces are once again present. Further, the range of the stabilizing steric force is most likely also increased.

## E. Proteins on Hydrophobic Surfaces

Amphiphilic proteins have properties similar to those of block copolymers and surfactants in the sense that they have clearly separated hydrophilic and hydrophobic domains that allow formation of monodisperse aggregates or micelles in solution. For $\beta$-casein the association process starts at a protein concentration of around 0.5 mg/ml at room temperature (59). Amphiphilic proteins adsorb strongly to nonpolar surfaces in contact with aqueous solutions, and they may generate stabilizing steric and electrostatic forces. In fact, caseins isolated from milk are widely used in different technical products ranging from food to paint and glue. One reason for this is that the caseins have excellent properties as emulsifiers and foaming agents, and emulsions stabilized by proteins constitute the most important class of food colloids. The caseins protect the oil droplets from coalescing and also provide long-term stability during storage and subsequent processing (60). $\beta$-Casein is more hydrophobic compared to the other caseins and the charged domain is clearly separated from the hydrophobic part, which makes the $\beta$-casein molecule as whole distinctly amphiphilic (61). At pH 7, the isolated $\beta$-casein molecule carries a net charge of about −12 (61).

Nylander and coworkers investigated the interactions between adsorbed layers of $\beta$-casein in order to clarify the mechanism responsible for the ability of $\beta$-casein to act as a protective colloid (62, 63). The force as a function of surface separation between hydrophobic surfaces across a solution containing 0.1 mg $ml^{-1}$ $\beta$-casein and 1 mM NaCl (pH = 7) is illustrated in Fig. 14. At separations down to about 25 nm an electrostatic double-layer force dominates the interacation. The hydrophobic substrate surface was uncharged so the charges responsible for this force had to come from the adsorbed protein. When the surfaces are compressed closer together the repulsive force is overcome by an attraction at a separation of about 25 nm, and the protein-coated surfaces are sliding into contact about 8 nm out from the hydrophobized mica surface (Fig. 14, inset). This observation, as well as the adhesive force found on separation, was interpreted as being due to bridging of the hydrophilic tails that extend out into solution. Further compression does not significantly change the surface separation. The results indicate that the adsorbed $\beta$-casein layer consists of an inner compact part and a dilute outer region.

This conclusion compares favorably with what is known from studies of the adsorption of $\beta$-casein on to air/liquid, liquid/liquid, and solid/liquid interfaces using a range of other techniques. It has generally been found that the adsorbed amount of $\beta$-casein on hydrophobic surfaces is between 2 and 3 mg $m^{-2}$ over

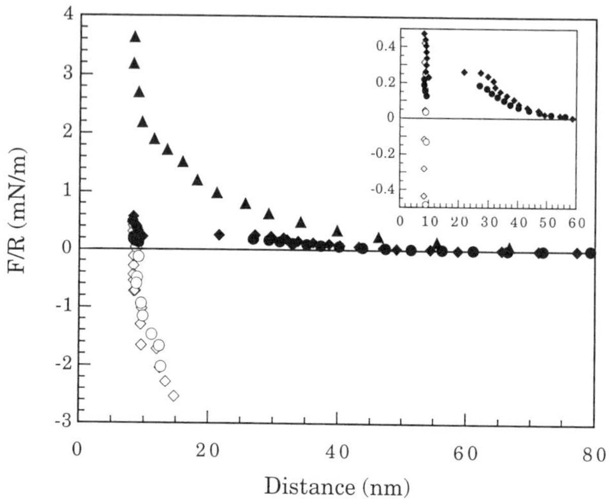


**Figure 14** Normalized force measured between hydrophobized mica surfaces in crossed cylinder geometry coated with $\beta$-casein in a solution containing 0.1 mg $\beta$-casein/ml (pH = 7; 1 mM NaCl) (◆,◇) and after dilution with 1 mM NaCl (●,○). Filled and unfilled symbols represent the force measured on compression and decompression, respectively; ▲ represent the force measured between hydrophobized mica surfaces across a 0.1 mM NaCl solution at pH 5.6 containing 0.2 mg proteoheparan sulfate/ml. The inset shows the measured forces between adsorbed layers of $\beta$-casein before and after dilution with 1 mM NaCl on an expanded scale. (From Ref. 26, with permission.)

a wide range of bulk concentrations. This is the case for planar air/water and planar oil/water interfaces (59), for hydrocarbon oil/water interfaces in emulsions (64), and for interfaces between water and polystyrene latex particles (65–67) and hydrophobized silica (68). At the triglyceride/water interface, however, the adsorbed amount is somewhat lower (69).

Information about the adsorbed layer structure of $\beta$-casein at the hydrophobic surface can be obtained by employing neutron reflectivity, small-angle X-ray scattering (SAXS), and dynamic light scattering. It was found that the layer of $\beta$-casein adsorbed to a hydrocarbon oil/water interface or an air/water interface (70, 71) consisted of a dense inner part, 2 nm thick, and a protein volume fraction of 0.96, immediately adjacent to the interface. Beyond that a second dilute region with a protein volume fraction of 0.15 extended into the aqueous phase. A similar structure of $\beta$-casein adsorbed on to polystyrene latex particles was observed with SAXS (65). The electron-density profile calculated from the SAXS data indicated that most of the protein resided within 2 nm from the surface. The profile also showed a small amount of protein extending some 10 nm into the aqueous phase. Further, the hydrodynamic layer thickness estimated from the diffusion coefficient determined by dynamic light scattering of latex particles (67, 72, 73) and emulsion droplets (69) coated with $\beta$-casein was found to be 10–15 nm. The fact that different experimental techniques give a different value of the layer thickness is simply because they have a different sensitivity to the extending tails.

This type of layer structure, with a compact inner region and a dilute outer region, was also predicted by self-consistent field theory and by computer simulations. For instance, Monte Carlo simulations show that a dense layer (1–2 nm thick) is present close to the planar interface (74). This layer contained about 70% of the segments. Further out a region of much lower density was found to extent about 10 nm into the aqueous phase. Similar results were obtained by self-consistent field calculations (75), which also showed that the most hydrophilic segments reside predominantly in the outer layer.

The properties of adsorbed $\beta$-casein layers can be changed by the action of enzymes. Leaver and Dalgleish (69) have observed that the N-terminal end is more accessible to trypsinolysis than the rest of the adsorbed molecule, and that loss of the tail leads to a reduced layer thickness. A similar change was observed by Nylander and Wahlgren (68) who found that addition of endoproteinase ASP-N to an adsorbed layer of $\beta$-casein reduced the adsorbed amount by approximately 20%. The removal of the extending tails will clearly reduce the range of the stabilizing steric force and thus reduce the emulsion stability against flocculation. We note that the forces generated by adsorbed $\beta$-casein are not very strongly repulsive (Fig. 14). Hence, the excellent stability of emulsion droplets coated by $\beta$-casein is most likely because the hole nucleation energy barrier is high.

Proteoheparan sulfate is an amphiphilic membrane glycoproten. It has, like $\beta$-casein, one large hydrophobic region. Proteoheparan sulfate has 3–4 hydrophilic and strongly charged side-chains whereas $\beta$-casein has only one less charged tail. Protoheparan sulfate is not used for stabilizing emulsions. However, it is nevertheless of interest to compare the forces generated by this protein with those generated by $\beta$-casein. The interaction between hydrophobic surfaces across a 1 mM NaCl solution containing 0.2 mg proteoheparan sulfate/ml is shown in Fig. 14 (62, 76). The long-range interaction is dominated by a repulsive double-layer force, considerably stronger than that observed for $\beta$-casein. This is simply because proteoheparan sulfate is more strongly charged than $\beta$-casein. A steric force dominates the short-range interaction for both proteins.

## F. Phospholipids on Polar Surfaces in Oil

We have seen above that surface-force measurements provide important information about interactions between solid hydrophobic surfaces coated with surfactants and polymers, and that some of the information obtained is directly relevant for oil-in-water emulsions. However, the details of the interaction profiles are expected to be different for liquid hydrocarbon droplets coated with the same molecules as the model solid surfaces. In particular, the coalescence behavior of the emulsion droplets cannot be modelled. It is even more difficult to make a solid model surface that mimicks the behavior of water-in-oil emulsions. At present, the best one can do is to use a polar surface that attracts the polar part of the emulsifier. In this way the orientation of the emulsifier on the model surface and at the water-in-oil emulsion surface will be the same. This will allow us to draw some conclusions about how polar solid surfaces coated with emulsifiers interact across oil, but care should be taken when using such results to draw conclusions about water-in-oil emulsions.

The forces between polar mica surfaces interacting across trilolein containing 200 ppm of soybean phosphatidylethanolamine (PE) have been studied (77). Some results obtained at two different water activities are illustrated in Fig. 15. When the water activity is 0.47 a monolayer of PE is adsorbed on each surface. The orientation is such that the polar group is attached to the mica surface with the nonpolar part directed towards the oil phase. Thus, adsorption of the phospholipid renders the mica surface nonpolar. No force is observed until the surfaces are about 6 nm apart. At this point a very steep repulsion is experienced which is due to compression of the adsorbed layers. A weak attraction is measured on separation. The forces change significantly when the triolein is saturated with water (water activity = 1). The adsorbed layer becomes significantly thinner, and now only a rather weak compressive force is needed in order to merge the two adsorbed layers into one. The reason is that water molecules adsorb next to the polar mica surface and in the region of the zwitteronic lipid head-group. This increases the mobility of the adsorbed phospholipid and decreases the force needed to merge the two adsorbed layers. Interestingly, it is not possible to remove the last adsorbed layer even by employing a very high compressive force.

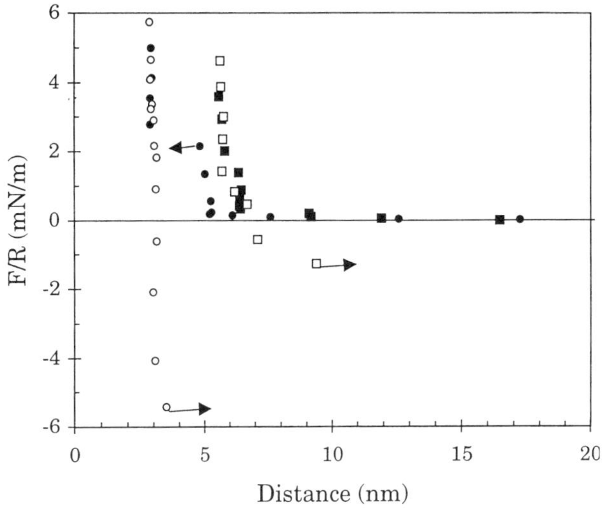


**Figure 15** Force normalized by radius between mica surfaces in crossed cylinder geometry interacting across a triolein solution containing 200 ppm OPPE. The forces were measured at water activities of 0.47 on approach (■) and separation (□), as well as at a water activity of 1 on approach (●) and separation (○); the arrows indicate inward and outward jumps. (From Ref. 77, with permission.)

From these observations we can draw some conclusions that are relevant for water-in-oil emulsions. First, no long-range electrostatic forces are present in the nonpolar media. This is because the dissociation of surface groups is very unfavorable in low-polarity media. Hence, generally it is very difficult to utilize electrostatic forces for generating long-range stabilizing forces in oil. Surfactants like phospholipids or alkyl ethoxylates adsorbed in monolayers will only generate short-range repulsive forces due to compression of the hydrocarbon chains penetrating into the oil medium. These substances will be efficient in preventing coalescence of water-in-oil emulsions only when the adsorbed amount is high enough and the spontaneous monolayer curvature is sufficiently negative.

## G. Polymers on Polar Surfaces in Oil

We saw above that surfactants adsorbed in monolayers only give rise to rather short-range forces in oil media. The range of the forces can be increased considerably if liquid crystalline phases are accumulated at the interface, or if amphiphilic oil-soluble polymerse are used instead of low molecular weight surfactants. An example of such a polymer is PGPR (polyglycerol polyricinoleate), which is a powerful water-in-oil emulsifier used in the food industry (78). PGPR has a complex branched structure as indicated in Fig. 16. This polymer was used for studying interactions between polar mica surfaces in triolein(79). The forces obtained at a polymer concentration of 200 ppm are shown in Fig. 17. In this system, repulsive steric forces are observed at distances below 15 nm. The magnitude of the force increases rather slowly with decreasing surface separation until the surfaces are about 5 nm apart. A further compression of the layers results in a steep increase of the steric force. The force profile indicates that the adsorbed layer consists of an inner dense region and an outer dilute region with some extended tails and loops. When dense polymer layers that generate long-range steric forces and have a high surface elasticity and viscosity are adsorbed at the interface of water-in-oil emulsions one can expect that the emulsion stability against flocculation and coalescence will be good.

## H. Forces Between Surfaces Across Emulsions

Emulsion droplets cannot only break by coalescing with each other, but they may also break by attaching to a solid surface. Depending on the application this may be wanted or unwanted. In order to study emulsion–surface interactions a model oil-in-water emulsion was prepared from purified soybean oil (20 wt %) using fractionated egg phosphatides (1.2 wt %) as emulsifier. The major components of the emulsifier were phosphatidylcholines and PEs. The mean diameter ($D_z$ average) of the emulsion was 320 nm, as determined with photon-correlation spectroscopy. A small amount of negatively charged lipids was also present, giving the emulsion droplets a net negative zeta-potential of about −40 mV (80). This emulsion was then placed inside a surface force apparatus.

The forces acting between two glass surfaces across the 20% oil-in-water emulsion, measured by using the MASIF are illustrated in Fig. 18 (81). A repulsive force dominates the interactions at separations below 200 nm. The force increases strongly with decreasing distance. This illustrates that large aggregates, with a diameter of at least 100 nm, are associated with each surface and the repulsion between 40 and 200 nm is due to deformation and eventual breaking of these aggregates. The range of the repulsive force is consistent with the layer thickness obtained in a previous ellipsometric study by Malmsten *et al.* (80). They found that the thickness of a layer adsorbed from the emulsion on to a negatively charged silica surface was around 100 nm, independent of surface coverage.

In some force curves one or two distinct steps are present. Figure 18 illustrates one such force curve where a clear step is seen at a separation of about 40 nm. At a separation of about 10 nm another step, but less pronounced, is seen. These steps are interpreted as being due to coalescence of adsorbed emulsion droplets and/or due to materials that collectively leave the zone between the surfaces. On subsequent approaches of the surfaces on the same position the

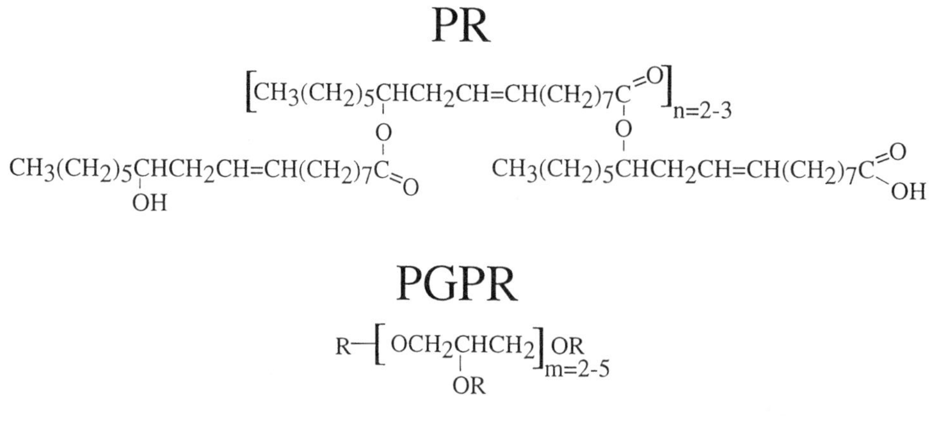


**Figure 16** Illustration of the structural elements of PGPR. The upper structure is that of the polyricinoleate moiety; the lower structure shows the polyglycerol backbone. The R in the structure can be either hydrogen, a fatty acid residue, or a polyricinoleate residue. In PGPR at least one of the side chains is polyricinoleate.

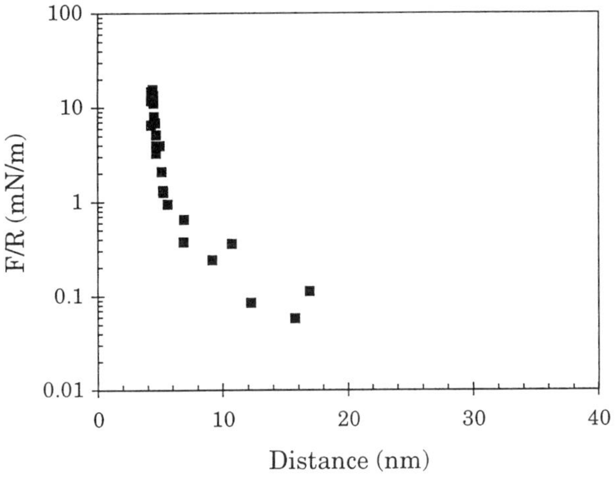


**Figure 17** Force normalized by radius between mica surfaces interacting across a triolein solution containing 200 ppm of PGPR measured on approach.

range of the repulsion remains at about 200 nm. However, the steps in the force profile become less pronounced or disappear completely, indicating a change in the adsorbed layer when exposed to a high compressive force.

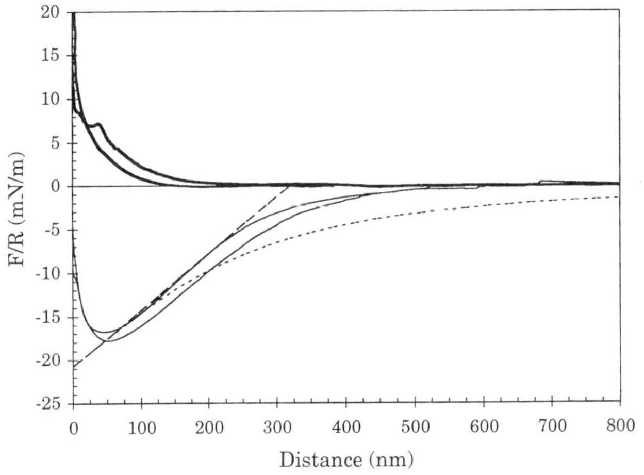


**Figure 18** Force normalized by local geometric mean radius as a function of surface separation between glass across a concentrated emulsion solution (20 wt % oil and 1.2 wt % phospholipid). The thinner lines correspond to the force measured on separation; the dashed line represents the calculated force between two spherical surfaces connected by a capillary condensate in the full equilibrium case [Eq. (25)] and the dotted line represents the force between two spherical surfaces connected by a capillary condensate in the non-equilibrium case [Eq. (26)]. (From Ref. 81, with permission.)

A strong and long-range force is observed when the surfaces are separated. It is plausible that this attraction is due to the formation of a capillary condensate of oil between the surfaces (Fig. 19). This capillary condensate originates from the emulsion droplets that have been destroyed when the surfaces are brought together. The forces between two spherical surfaces connected by a capillary condensate are in the full equilibrium case given by (82):

$$\frac{F}{R} = 2\pi(\sigma_{sc} - \sigma_{sb})\left(1 - \frac{D}{R_K}\right) \tag{25}$$

where $\sigma$ is the interfacial tension and the subscript s, c, and b stand for surface, capillary condensate, and bulk, respectively; $R_K$ is the Kelvin radius of the capillary condensate. In cases when the surfaces are separated too rapidly to allow the volume of the capillary to change with separation one instead obtains (82):

$$\frac{F}{R} = 2\pi(\sigma_{sc} - \sigma_{sb})\left[1 - \frac{1}{\sqrt{(1 + R_K^2/D^2)}}\right] \tag{26}$$

Two theoretical force curves calculated by using Eqs (25) and (26) are shown in Fig. 18. In these calculations we used a Kelvin radius of 320 nm and an interfacial tension difference of 3.3 mN/m. The measured force curves fall in between the extreme cases of full equilibrium, where the volume of the condensate is changing with distance to minimize the free energy, and the case of no change in condensate volume with separation. Long-range forces due to capillary condensation have been observed previously by Petrov *et al.* who found that water condensed between two surfaces immersed in a microemulsion. (83). Capillary condensation of sparingly soluble surfactants between surfaces close to each other in surfactant solutions has also been reported (84).

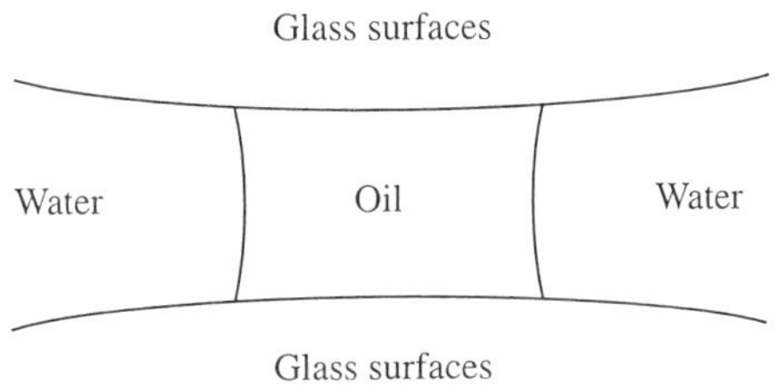


**Figure 19** Schematic illustration of the capillary condensate formed between glass surfaces due to breakdown of adsorbed emulsion droplets. The figure is not according to scale. (From Ref. 81, with permission.)

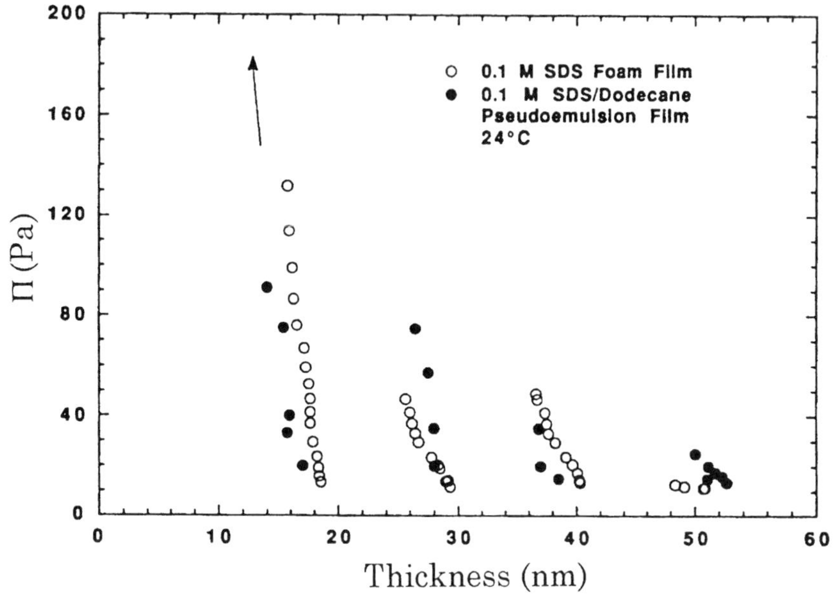


**Figure 20** Low-pressure region of the disjoining pressure isotherm across a 0.1 M SDS solution in a single foam lamella, and across a 0.1 M SDS solution separating a dodecane/solution interface from an air/solution interface. (Reproduced from Ref. 35, with permission.)

It is worth pointing out that the functional form of the measured attraction shows that the volume of the capillary condensate decreases with increasing separation. However, not fast enough compared to the speed of the measurements (the attractive part took about 30 s to measure) to allow full equilibrium to be established. Also, the range of the measured repulsion on approach does not increase with the number of times the surfaces are brought into contact but rather the reverse. Both these observations point to the fact that the material present in the capillary condensate is spontaneously re-emulsified when the surfaces are separated.

In order to obtain information about whether a monolayer, a bilayer, or a multilayer was firmly attached to the surfaces, we employed the interferometric SFA and mica surfaces rather than glass surfaces (81). In these measurements a drop of the emulsion was placed between the surfaces. The emulsion was very opaque and no interference fringes could be seen until the surfaces were close to contact. The force measured between mica surfaces across a concentrated emulsion were repulsive and long range (several hundred nanometers), which was in agreement with the results obtained using glass surfaces. Since the forces were so highly repulsive no attempt was made to measure them accurately, but under a high compressive force the surfaces come to a separation 8.5 nm. This corresponded to a bilayer of phospholipid on each surface.

### I. Forces Due to Stratification in Foam and Pseudoemulsion Films

The thinning of thin liquid films in micellar solution is found to occur in a stepwise fashion, known as stratification. Bergeron and Radke (35) set out to study the forces responsible for this phenomenon using the porous frit version of the thin-film balance. They found that the equilibrium disjoining pressure curve (force curve) showed an oscillatory behavior both for foam and pseudoemulsion (i.e., asymmetric oil/water/gas) films stabilized by the anionic surfactant sodium dodecyl sulfate above the cmc (Fig. 20). The reason for this oscillatory force profile was the layering of micelles in the confined space in the thin aqueous film separating the two interfaces. The periodicity of the oscillations was the same for foam films and for pseudoemulsion films. The main difference between the two systems was found in the high-pressure region of the disjoining pressure isotherm. The pseudoemulsion films ruptured at much lower imposed pressures than the foam films. This was attributed to the action of the oil phase as a foam destabilizer.

## V. SUMMARY

Several techniques are available for studying long-range interactions between solid surfaces and fluid interfaces. The forces generated by surfactants, polymers, and proteins have been determined. For oil-in-water emulsions both steric and electrostatic stabilizing forces are of importance whereas only steric forces are operative for the case of water-in-oil emulsions. These forces are well understood theoretically. The experimental techniques employed give very detailed information on the long-range forces, and in this respect the results obtained for the model systems can be useful for understanding interactions in emulsion systems. However, the surface-force techniques employed are not suitable for modeling the molecular events leading to coalescence of emulsion droplets once they have been brought in close proximity to each other. Some data illustrating the breakdown and reemulsification of emulsion droplets in the gap between two macroscopic solid surfaces were also presented. This is a new research topic and very little is known about how the surface properties and the type of emulsifier influence the stability of emulsion droplets at surfaces in narrow gaps between surfaces.

## ACKNOWLEDGMENT

This work was partly sponsored by the Competence Centre for Surfactants Based on Natural Products (SNAP).

## REFERENCES

1. WD Bancroft. J Phys Chem 19: 275–309, 1915.
2. WC Griffin. J Soc Cosmet Chem 1: 311–326, 1949.
3. JT Davies. Proc 2nd Int Congr Surf Activity 1: 426–438, 1957.
4. A Kabalnov, H Wennerström. Langmuir 12: 276–292, 1996.
5. DYC Chan, RG Horn. J Chem Phys 83: 5311–5324, 1985.
6. J Kizling, B Kronberg. Colloids Surfaces 50: 131–140, 1990.
7. O Reynolds. Phil Trans R Soc London 177: 157–212, 1886.
8. FW Cain, JC Lee. J Colloid Interface Sci 106: 70–85, 1985.
9. ED Manev, R Tsekov, BP Radoev. J Disp Sci Technol 18: 769–788, 1997.
10. BP Radoev, AD Scheludko, ED Manev. J Colloid Interface Sci 95: 254–265, 1983.
11. SAK Jeelani, S Hartland. J. Colloid Interface Sci 164: 296–308, 1994.
12. JN Israelachvili. Intermolecular and Surface Forces. London: Academic Press, 1991.
13. EJN Verwey, JTG Overbeek. Theory of Stability of Lyophobic Colloids. Amsterdam: Elsevier, 1948.
14. BV Derjaguin, L Landau. Acta Physicochim USSR 14: 633–662, 1941.
15. VA Parsegian, N Fuller, RP Rand. Proc Natl Acad Sci USA 76: 2750–2754, 1979.
16. JN Israelachvili, H Wennerström. J Phys Chem 96: 520–531, 1992.
17. GJ Fleer, MA Cohen Stuart, JMHM Scheutjens, T Cosgrove, B Vincent. Polymers at Interfaces. London: Chapman & Hall, 1993.
18. P-G de Gennes. Scaling Concepts in Polymer Physics. Ithaca, NY: Cornell University Press, 1979.
19. P-G de Gennes. Adv Colloid Interface Sci 27: 189–209, 1987.
20. AD Nikolov, DT Wasan. J. Colloid Interface Sci 133: 1–12, 1989.
21. V Bergeron, CJ Radke. Langmuir 8: 3020–3026, 1992.
22. V Bergeron, AJ Jiménez-Laguna, CJ Radke. Langmuir 8: 3027–3032, 1992.
23. V Bergeron. Forces and Structure in Surfactant-laden Thin-liquid films. PhD thesis, University of California, Berkeley, 1993.
24. AJ de Vries. Recl Trav Chim Pays-Bas 77: 383–389, 1958.
25. K Shinoda, H Saito. J Colloid Interface Sci 30: 258–263, 1969.
26. PM Claesson, T Ederth, V Bergeron, MW Rutland. Adv Colloid Interface Sci 67: 119–183, 1996.
27. JN Israelachvili, GE Adams. J Chem Soc Faraday Trans I 74: 975–1001, 1978.
28. J Parker. Prog Surf Sci 47: 205–271, 1994.
29. K Schillén, PM Claesson, M Malmsten, P Linse, C Booth. J Phys Chem 101: 4238–4252, 1997.
30. B Derjaguin. Kolloid Zeits 69: 155–164, 1934.
31. RG Horn, JN Israelachvili, F Pribac. J Colloid Interface Sci 115: 480–492, 1987.
32. P Attard, JL Parker. Phys Rev A 46: 7959–7971, 1992.
33. A Scheludko. Adv. Colloid Interface Sci 1: 391–464, 1967.
34. OD Velev, GN Constantinides, DG Avraam, AC Paytakes, RP Borwankar. J Colloid Interface Sci 175: 68–76, 1995.
35. V Bergeron, CJ Radke. Colloid Polym Sci 273: 165–174, 1995.
36. KJ Mysels, MN Jones. Disc Faraday Soc 42: 42–50, 1966.
37. RL Ferm. In: KJ Lissant, ed. Emulsion and Emulsion Technology. Vol 1. New York: Marcel Dekker, 1974, pp 387.

38. V Bergeron. Langmuir 13: 3474–3482, 1997.
39. PC Herder. J Colloid Interface Sci 134: 336–345, 1990.
40. PM Claesson, R Kjellander, P Stenius, HK Christenson. J Chem Soc Faraday Trans I 82: 2735–2746, 1986.
41. R Kjellander. J Chem Soc Faraday Trans II 78: 2025–2042, 1982.
42. RE Goldstein. J Chem Phys 80: 5340–5341, 1984.
43. G Karlström. J Phys Chem 89: 4962–4964, 1985.
44. A Matsuyama, F Tanaka. Phys Rev Lett 65: 341–344, 1990.
45. P-G de Gennes. CR Acad Sci Paris 313: 1117–1122, 1991.
46. S Bekiranov, R Bruinsma, P Pincus. Europhys Lett 24: 183–188, 1993.
47. P Linse, B Björling. Macromolecules 24: 6700–6711, 1991.
48. M Björling. Macromolecules 25: 3956–3970, 1992.
49. W Brown, R Johnsen, P Stilbs, B Lindman. J Phys Chem 87: 4548–4553, 1983.
50. GJT Tiddy. Phys Rep 57: 3–46, 1980.
51. DJ Mitchell, GJT Tiddy, L Waring, T Bostock, MP McDonald. J Chem Soc Faraday Trans I 79: 975–1000, 1983.
52. JN Israelachvili, DJ Mitchell, BW Ninham. J Chem Soc Faraday Trans II 72: 1525–1568, 1976.
53. SI Jeon, JD Andrade. J Colloid Interface Sci 142: 159–166, 1991.
54. K Shinoda, H Takeda. J Colloid Interface Sci 32: 642–646, 1970.
55. BA Bergenståhl, PM Claesson. In: SE Friberg, K Larsson, eds. Food Emulsions. New York: Marcel Dekker, 1997.
56. SE Friberg, C Solans. Langmuir 2: 121–126, 1986.
57. GF Belder, G ten Brinke, G Hadziioannou. Langmuir 13: 4102–4105, 1997.
58. P Fäldt, B Bergenståhl, PM Claesson. Colloidds Surfaces A 71: 187–195, 1993.
59. DG Schmidt, TAJ Payens. J Colloid Interface Sci 39: 655–662, 1972.
60. E Dickinson. J Dairy Sci 80: 2607–2619, 1997.
61. HE Swaisgood. Development in Dairy Chemistry – 1. London: Applied Science Publishers, 1982.
62. PM Claesson, E Blomberg, JC Fröberg, T Nylander, T Arnebrandt. Adv Colloid Interface Sci 57: 161–227, 1995.
63. T Nylander, NM Wahlgren. Langmuir 13: 6219–6225, 1997.
64. J-L Courthaudon, E Dickinson, DG Dalgleish. J Colloid Interface Sci 145: 390–395, 1991.
65. AR Mackie, J Mingins, AN North. J Chem Soc Faraday Trans 87: 3043–3049, 1991.
66. JR Hunter, PK Kilpatrick, RG Carbonell. J Colloid Interface Sci 142: 429–447, 1991.
67. DV Brooksbank, CM Davidson, DS Horne, J Leaver. J Chem Soc Faraday Trans 89: 3419–3425, 1993.
68. T Nylander, NM Wahlgren. J Colloid Interface Sci 162: 151–162, 1994.
69. J Leaver, DG Dalgleish. J Colloid Interface Sci 149: 49–55, 1992.
70. E Dickinson. J Chem Soc Faraday Trans 88: 2973–2983, 1992.
71. E Dickinson, DS Horne, JS Phipps, RM Richardson Langmuir 9: 242–248, 1993.
72. DG Dalgleish. Colloids Surfaces 46: 141–155, 1990.
73. DG Dalgleish, J. Leaver. J Colloid Interface Sci 141: 288–294, 1991.
74. E Dickinson, SR Euston. Adv Colloid Interface Sci 42: 89–148, 1992.
75. FAM Leermakers, PJ Atkinson, E Dickinsin, DS Horne. J Colloid Interface Sci 178: 681–693, 1996.
76. M Malmsten, PM Claesson, G Siegel. Langmuir 10: 1274–1280, 1994.
77. A Dedinaite, PM Claesson, B Campbell, H Mays. Langmuir 14: 5546–5554, 1998.
78. R Wilson, BJ van Schie, D Howes. Food Chem Toxicol 36: 711–718, 1998.
79. A Dedinaite, B Campbell. Langmuir 16: 2248–2253, 2000.
80. M Malmsten, A-L Lindström, T Wärnheim. J Colloid Interface Sci, 173: 297–303, 1995.
81. E Blomberg, PM Claesson, T Wärnheim. Colloids Surf. A, 159: 149–157, 1999.
82. DF Evans, H Wennerström. The Colloidal Domain. 2nd ed. New York: VCH, 1998.
83. P Petrov, U Olsson, H Wennerström. Langmuir 13: 3331–3337, 1997.
84. Å Waltermo, PM Claesson, I Johansson. J Colloid Interface Sci 183: 506–514, 1996.

# 14

# Microcalorimetry

**Christine S. H. Dalmazzone**
*Institut Français du Pétrole, Rueil-Malmaison, France*

**Danièle Clausse**
*Université de Technologie de Compiègne, Compiègne, France*

## I. INTRODUCTION

Owing to their complex formulation, some emulsions are difficult to characterize by classical methods, which generally need disturbing dilution before analysis. This is the case for water-in-oil (W/O) emulsions in the petroleum industry, for example. These emulsions may be encountered at all stages in the petroleum recovery and processing industry: drilling fluids, production, process plant, transportation, etc. (1). They may be desirable or not, or desirable in one part of the process and undesirable at the next stage. They may contain not just oil and water, but also solids and even gas. Among all the types of petroleum emulsions, concentrated W/O emulsions are the most difficult to characterize. Another example is found in the field of cosmetics (2). The emulsions formulated for that purpose are opaque and very concentrated as well.

Emulsion stability can be assessed by a great number of techniques which are generally based on the analysis of the droplet size distribution. Most of these techniques are easily applicable to diluted oil-in-water (O/W) emulsions (e.g. light scattering, Coulter counter, and microscopy), but they are unfortunately not adaptable to opaque W/O emulsions. Even optical microscopy, which is considered as a "universal" technique in the field of emulsions, is very complicated to use with systems containing different types of solids (3). A rapid way to assess the stability of opaque and concentrated emulsions is still the "bottle test", which consists in monitoring the extent of phase separation with time. This test provides a significant amount of information relating to both the emulsion stability and the clarity of the separated water, but it is very empirical (4).

In the field of emulsions' characterization, it is well known that dilution may create perturbation on the surface properties of the droplets and on interactions between the droplets. To give an example, matter transfers resulting from osmotic shocks may occur causing polydispersity changes as has been shown when such events are required (5,6). In fact, very few techniques avoid dilution, namely, dielectric or hertzian spectroscopy (7–9), rheology (2), conductimetry (6,7), and more recent ones based on acoustical methods (10), focussed beam reflectance (11,12), or microwave attenuation (13). All these techniques are complementary and new techniques are always welcome.

In this article, a technique based on the properties of solidification and melting of the droplets and the medium wherein they are dispersed is described. The proposed technique is microcalorimetry, which allows one to detect such transitions through the energies involved, as previous studies have shown (14–18). The aim of this chapter is to point out how it is possible, from a single test performed on an emulsion sample without dilution, to obtain information about:

- The emulsion kinds: simple (W/O or O/W) or multiple (W/O/W or O/W/O).
- The amount of water and its state: bound, dispersed, or in bulk.
- The compositions of the dispersed and bulk phases.
- The mean diameter of the droplets and its evolution versus time due to coalescence, and Ostwald and composition ripening.
- The mass transfers between droplets due to their differences in composition.

## II. METHOD

### A. Principle

The principle of the method is based on the relationships between the most probable temperature of solidification $T^*$ of a sample and its volume $V$ on the one hand and between the amount of melting energy $\Delta H$ and its mass or volume on the other. Referring to emulsions, various volumes have to be taken into account, namely:

- The volume of the continuous phase that can be considered as roughly equal to the volume of the cell wherein the emulsion sample has been placed. This approximation could not be acceptable if the emulsion is highly concentrated. Nevertheless, the boundaries of the continuous phase are the ones of the cell.
- The volume of each droplet.
- The total volume of the droplets representing the dispersed phase of the emulsion.

Therefore, the knowledge of both relationships allows one to obtain information about the structure of the emulsions at time $t$ and versus $t$ if experiments are performed periodically on samples taken from the emulsions.

To understand how these relationships are obtained, it is necessary to have some information about the ways by which solidification and melting occur in a given sample submitted to regular cooling and heating. Many studies have been devoted to the description of these phenomena, especially for water (19–25), as this compound is found in a great variety of materials taken in different manners: clouds, biological cells, ice creams, emulsions in the field of petroleum products, etc. (26–31), and full descriptions can be found in these papers. Let us give in this chapter only the main points.

First, solidification needs the formation of what is called a germ of the solid phase to occur. This germ is very tiny and capillary effects have to be taken into account. Assuming a spherical germ, the pressure inside the germ of radius $R$ is higher than it is outside by the amount:

$$\Delta P = \frac{2\gamma}{R} \tag{1}$$

where $\gamma$ is the interfacial tension between the solid germ and the surrounding liquid, which is necessarily undercooled as is shown thereafter.

The consequence is that the solid–liquid equilibrium temperature $T(R)$ is different from that of the bulk phase, which can be referred to as being one of a sphere of infinite radius, $T(\infty)$. The relationship between these two temperatures is:

$$\ln\frac{T(R)}{T(\infty)} = -\frac{2\gamma V^{\mathrm{s}}}{RL_{\mathrm{m}}} \tag{2}$$

where $V^{\mathrm{s}}$ is the solid molar volume, and $L_{\mathrm{m}}$ the molar melting enthalpy.

From Eq. (2) it is seen that the liquid has to be cooled under the bulk temperature in order for it to solidify. Furthermore, the germ formation is the result of local density fluctuations, and kinetics aspects have to be considered. This is done by considering the nucleation rate $J$ that gives the number of germs formed by time unit and volume unit.

The stationary value of $J$ is given by:

$$J = A\exp\left(-\frac{\Delta\Phi}{kT}\right) \tag{3}$$

where $\Delta\Phi$ is the formation energy, and $A$ is a quantity depending on the viscosity medium among other factors; $\Delta\Phi$ is given by:

$$\Delta\Phi = \frac{4}{3}\pi\gamma R^2 \tag{4}$$

It appears that the lower the temperature the higher the nucleation rate and consequently it is very difficult to get the solidification of a pure liquid near its bulk solid–liquid equilibrium temperature. Another less obvious consequence is that the temperature at which a given sample will solidify is not unique. Therefore, only a most probable temperature can be given and this temperature $T^*$ appears to be volume dependent and of course it also depends on the sample composition as it is shown thereafter.

The determination of $T^*$ requires the study of the solidification of a sufficient number $N_0$ of identical samples. From these preliminary experiments it is possible to define the probability $P$ of solidification of one sample at any temperature during the regular cooling as:

$$P = \frac{N(T)}{N_0} \tag{5}$$

$N(T)$ being the number of samples having been found solid at the same temperature. Assuming that only one germ is needed to obtain a solid sample, it is possible to express the number of samples that are expected to become solid during a further cooling between $T$ and $T + \mathrm{d}T$, namely:

$$\mathrm{d}N = -J(N_0 - N)V\left(\frac{1}{\dot{T}}\right)\mathrm{d}T \tag{6}$$

where $\dot{T}$ is the cooling rate. It appears that the proportion of solidified samples, $\mathrm{d}N/N_0$, goes through a maximum as the number of liquid samples $(N_0 - N)$ vanishes and the nucleation rate increases exponentially with decreasing temperature as was stated before [Eq. (3)]. It is possible to deduce from this analysis schematic drawings of the histograms that give the proportion of solid samples versus temperature in the temperature interval $\mathrm{d}T$ (Fig. 1a) and the percentage of solid samples at any temperature $T$ (Fig. 1b).

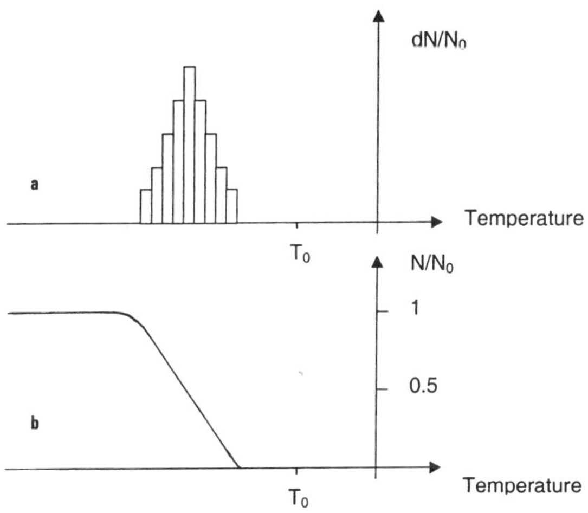


**Figure 1** Schematic histograms showing the transformation of water droplets into ice: (a) proportion of solid samples vs. temperature in temperature interval $\mathrm{d}T$; (b) percentage of solid samples at any temperature $T$.

Integration of Eq. (6) between $T(\infty)$ and $T$ gives:

$$\ln(1 - P) = \frac{V}{\dot{T}}\int_{T(\infty)}^{T} J(T)\mathrm{d}T \tag{7}$$

From Eq. (7), it can be seen that the smaller the sample volume $V$, the lower its solidification probability $P$ at any temperature $T$ in the range of undercooling of the material considered.

Referring to an emulsion submitted to a regular cooling, it is now possible to predict the ways by which the solidification of the various phases, occupying various volumes as has been described before, will occur.

### 1. Continuous Phase

Only one germ could initiate the solidification, and the temperature is expected to vary from one emulsion sample to another. Nevertheless, from preliminary experiments performed on a sufficient number of samples of equal volume and prepared from the continuous-phase material, it is possible to determine the range wherein this temperature is included and the most probable temperature of solidification of this sample as well. Referring to an O/W emulsion placed in a cell the volume of which is a few cubic millimeters, this temperature is expected to be around −20°C according to the large number of studies available about the formation of ice (32–37). Should the cell volume be larger this temperature will be higher.

### 2. Droplet of the Dispersed Phase

Generally the volume is around a few cubic micrometers which is quite different from the continuous phase. Referring to the same material, water, the temperature of solidification will be lower. From previous studies on water (32–37), this temperature is not unique either and it is expected to be between −35° and −45°C.

### 3. Whole Droplets of the Dispersed Phase

The number of droplets is very high and undoubtedly sufficient to allow a statistical treatment of the various solidification temperatures of the whole droplets. Therefore, Eq. (6) can be applied directly and the schematic drawings (Fig. 1) of the distributions of the solidification temperatures is a reliable view of the solidification. Of course, the extent of the range of solidification of the whole droplets and the most probable temperature given by the apex are a characteristic of the material dispersed. For water droplets the volume of which is around a few cubic micrometers, the most probable solidification temperature has been found to be around −39°C and the melting temperatures, equal for the whole droplets, occur at 0°C as there is no noticeable discrepancies between the inside and outside droplet pressures (19,20). For droplets, the volumes of which are much smaller, the solidification and melting temperatures have been found to be directly radius dependent (26). From the theory of thermoporometry described in Ref. 38, it is possible to evaluate the melting point of water, $T_m$, or $T(\infty)$, versus the radius $R$ of the ice particle assumed to be spherical (Table 1).

It is found that the smaller the radius, the higher the difference $P_s - P_l$ between the inside and outside pressures of the ice particle according to Eq. (1) and the lower the melting temperature of ice. Therefore, studying the melting temperatures of the solid particles obtained from the solidification of the droplets dispersed in the emulsion is a means to indicate very tiny droplets the radii of which are difficult to determine.

## B. Differential Scanning Calorimetry (DSC)

An actual way for denoting the solidification and the melting of a sample is to perform the cooling and the heating of the sample in a calorimeter. Furthermore, for the emulsion characterization test proposed, it is better to use a differential scanning calorimeter, the phase transitions being detected through the energies involved. A schematic drawing of the sensitive part of the DSC apparatus is given Fig. 2.

The following quantities:

- $dh/dt$, the heat generated by solidification (> 0) or absorbed by melting (< 0) of the sample per unit time;

**Table 1** Melting Point of Water $T_m$ vs. Radius $R$ of Ice Particle; $P_s - P_l$ is the Difference in Pressure Between Inside and Outside of Ice Particle

| Radius of ice particle, $R$ (nm) | Melting point of water, $T_m$ (°C) | Difference in pressure, $P_s - P_l$ (bars) |
|---|---|---|
| 10 | −6.8 | 76.5 |
| 50 | −1.3 | 16.2 |
| 100 | −0.6 | 8.1 |
| 1000 | −0.06 | 0.8 |

*Source*: Ref. 38.

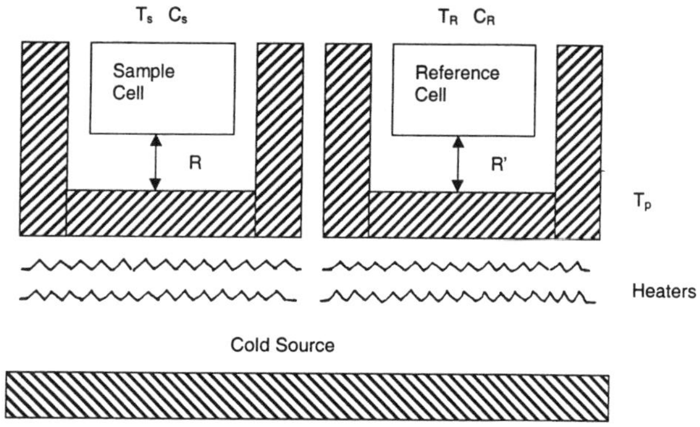


**Figure 2** Schematic drawing of the sensitive part of the DSC apparatus. $T_S$ and $T_R$: temperatures of the sample and the reference; $C_S$: heat capacity of the cell including the sample; $C_R$: heat capacity of the reference cell (generally empty); $R$ and $R'$: thermal resistance between the cell and the heater; $T_P$: programmed temperature of the cell holder.

- $dq/dt$, the difference in energies exchanged between the sample and the reference cells and the cell holders in order to compensate for the temperature difference between the sample and the reference;
- $\dot{T}_P = \frac{dT_P}{dt}$, the scanning rate, a positive constant during heating and a negative one during cooling;

are related by the following expression (39):

$$\frac{dh}{dt} = -\frac{dq}{dt} + (C_S - C_R)\dot{T}_P - RC_s\frac{d^2q}{dt^2} \tag{8}$$

$dh/dt$ is the result of the sum of three terms: the first being given by the signal measured from the zero line, the second being the baseline displacement due to the difference in heat capacity between the sample and the reference, and the third being the slope of the recorded signal multiplied by the product $RC_s$, $R$ being the thermal resistance between the cell and the heater.

As no energy is involved, $dh/dt$ is zero before and after the transition. The calorimeter then yields $dq/dt$, which is expressed by:

$$\frac{dq}{dt} = (C_S^s - C_R)\dot{T}_p \qquad \text{(solid sample)} \tag{9}$$

$$\frac{dq}{dt} = (C_S^l - C_R)\dot{T}_p \qquad \text{(liquid sample)} \tag{10}$$

for the melting of a sample. On the thermogram, the parts dealing with the heating of the solid sample and the same liquid sample will be indicated by straight baselines that must be out of line due to the differences between the solid and liquid heat capacities, $C_S^s$ and $C_S^l$, respectively.

During the thermal transition, $dh/dt$ is not zero and neither is $dq/dt$: a signal is therefore recorded. It is very important to examine the shape of the signal as it is dependent on the characteristics of the sample under study. It is actually this point that makes the technique interesting for obtaining information about emulsions, which can be considered as a gathering of samples with their own specificity. In the previous section, it has been shown that solidification of the continuous and the dispersed phases show different characteristics and, therefore, this must be noticeable through the signal shapes as will be described in the next section.

Another problem is to delimit correctly the signal in order to calculate the energy involved. For that purpose, this can be reliably done only if some information is available on the transition. Nevertheless, the simple drawing of a straight line between the noticeable beginning and end of the phase transition gives acceptable results if great precision is not required. This being done, the signal area $S$ is directly related to the energy involved, especially for the lowest scanning rates. The correlation factor is determined from a study of the melting of pure compounds the melting heats of which are known.

From a knowledge of $S$ and by way of the energy $\Delta H$ involved during the melting, it is possible to deduce the mass concerned in the transition and, therefore, information about the emulsion. Referring to a W/O emulsion, the possible amount of water $M$ can be deduced from the relation:

$$M = \frac{\Delta H}{L_m} \tag{11}$$

where $L_m$ is the energy involved by a mass unit and which is 80 cal/g for pure water.

## C. Test

The test is very simple as no special treatment of the emulsion is needed. It is recommended to fill at the same time several cells with the emulsion under study. Generally, the cell volume is small (a few cubic millimeters) in order to assure temperature homogeneity within the sample and to allow use of scanning rates as high as 2.5 K/min. Higher scanning rates need corrections of the data given by the calorimeter, namely, $dq/dt$ and $T_p$ (cell-holder temperature), which may be quite different from the required data, namely, $dh/dt$ and $T_s$ (sample temperature), and furthermore temperature homogeneity is no longer guaranteed (16). Nevertheless, absolute data are not required because the principle of the test proposed is based on a comparison between the thermograms. Because of the small volume it is better to select samples from the emulsion core in order to avoid heterogeneity.

After insertion of one emulsion-filled cell into the calorimeter, and thermal equilibrium has been reached, the calorimeter is programmed to be cooled down and heated between two limits of temperature. As the test is especially suitable for W/O emulsions, focus is placed on the freezing of water and the melting of ice. Therefore, the scanning is performed from +20°C down to −80°C at least, as late solidification, due to undercooling and dissolved solute in water, may occur. The other cells left at room temperature are studied in the same way.

It is also necessary to scan the thermograms of the materials of the continuous and dispersed phases separately in order to obtain reference values, especially for their melting, as the same behavior is expected for the emulsion.

To derive information from the thermograms obtained, it is necessary to compare them to schematic ones deduced from theoretical considerations of the solidification and melting of samples. This can be done from the analysis made in Sec. III and from knowledge of the technique as has been outlined in this section. For that purpose, these schematic and typical thermograms will be described.

# III. SCHEMATIC DRAWINGS OF THERMOGRAMS AND INTERPRETATION

Referring to emulsions for which the test is particularly appropriate, as was mentioned in Sec. I, it is obvious that pure compounds are not found either in the continuous phase or in the dispersed phase. Nevertheless, it is necessary to know the behavior of pure compounds as references. Therefore, the thermograms dealing with pure compounds will be described first and their possible distribution in an emulsion, in the bulk, dispersed, or bound phase, will be considered. Afterwards, the example of solutions will be examined and finally the response of emulsions to the DSC test will be analyzed. All the thermograms are depicted in Fig. 3 for direct comparison.

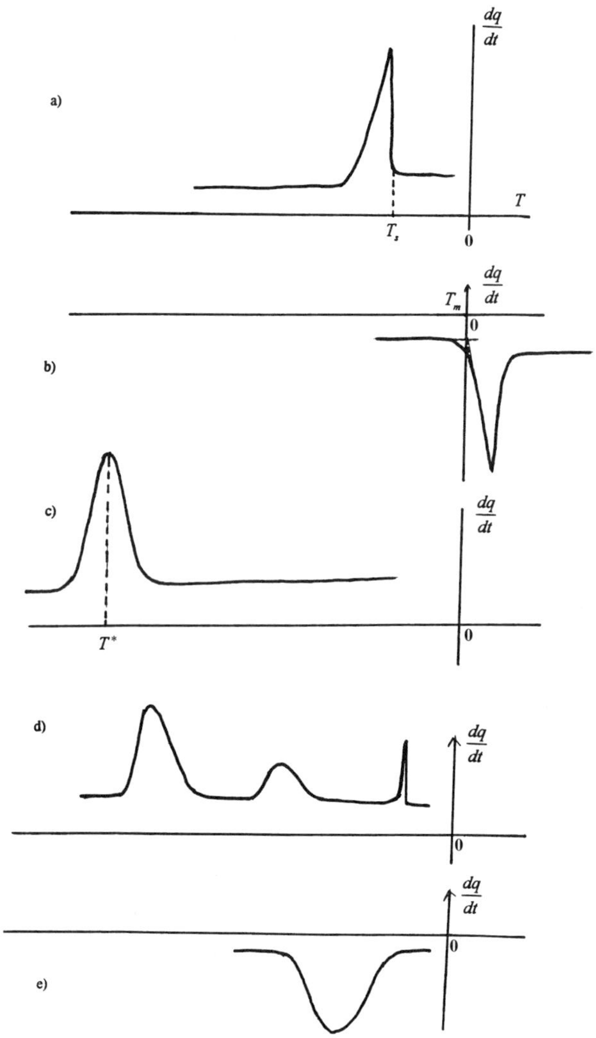


**Figure 3** Schematic drawings of thermograms: (a) solidification of pure compound in bulk; (b) melting of a pure compound; (c) solidification of a monodispersed pure compound; (d) solidification of a polydispersed pure compound; (e) melting of nanosized droplets (Gaussian function of size distribution). $dq/dt$: energy registered by the calorimeter; $T_s$: temperature of solidification; $T_m$: temperature of melting; $T^*$: most probable temperature of solidification.

## A. Pure Compound

### 1. Bulk

According to Sec. II.A, the solidification will occur by undercooling breakdown at a temperature that is less than that of the solid–liquid equilibrium or of melting. Therefore, total solidification is expected instantaneously and the amount of solidification energy released is instantaneous as well. The consequence of this is an almost perpendicular deviation of the baseline as is shown in Fig. 3a. The solidification temperature $T_s$ is given by the one corresponding to the beginning of the deviation. After the transition, a straight baseline is reached the sooner, the lower the time lag of the calorimeter used and the higher the scanning rate. The drawings of the baselines before and after the signal are not on line due to the differences in the solid and liquid heat capacities of the sample (see Sec. II.B). The melting thermogram, described thoroughly elsewhere (40) is a classical one as the melting is expected at $T_m$ without delay (Fig. 3b).

### 2. Dispersed

In this section emphasis is put on the response to the DSC test of the dispersed phase of an emulsion alone. It has been shown in Sec. II.A that the solidification temperatures are scattered with a maximum of solidification events at the temperature corresponding to a proportion of solidified droplets around 50% (Fig. 1). A correlation between the energy released by the time unit $dh/dt$ and the proportion of solidified droplets has already been carried out (14). It has been found that $dh/dt$ is close to $dq/dt$ for the conditions of realization of the test described above. In these conditions, $dh/dt$ can be related to $dN/dt$ through the expression:

$$\frac{dh}{dt} = V\rho L_s \frac{dN}{dt} \qquad (12)$$

assuming that all the droplets have the same volume $V$; $L_s$ is the heat of solidification per mass unit, and $\rho$ is the mass per unit volume. In these conditions, the solidification peak is expected to have a Gaussian shape, the temperature maximum giving the most probable solidification temperature $T^*$ (Fig. 3c) that will be lower than the one obtained for the same material in bulk as the volume is smaller. As $T^*$ is dependent on the mean droplet diameter, the higher is found $T^*$, the larger the droplets. It is therefore possible from the test to compare the further evolution of two emulsions (c) and (d), the thermograms of which being those of Fig. 3c and 3d. Emulsion (d) exhibits larger droplets and therefore more instability than emulsion (c).

The thermogram obtained during the melting will be the same as the one obtained for the bulk (Fig. 3b) as the droplets are not too small. For smaller droplets showing pressure gradients between the inside and outside, the melting temperature is the lower, the smaller the droplets (see Table 1). This can be noticeable on the thermogram (Fig. 3e) through a Gaussian shape, if the distribution size is described by a Gaussian function.

### 3. Bound

Bound water is generally the term employed for referring to water that is not found to freeze during cooling. One has to be very cautious with this term as the experimental conditions used may play a role in obtaining the transition. Furthermore, solute may create vitreous states that prevent solidification for impure compounds. If this is not the case, the fact that no signal is observed may be attributed to the possible presence of bound material.

## B. Solutions

The study by DSC of the solidification and the melting of bulk and dispersed solutions within emulsions is a very suitable means for detecting how these transitions occur versus the composition of the solutions. Therefore, the DSC test will give information about the phase compositions of the emulsions under study and about the way they possibly evolve. In this section, only binary solutions prepared from water and a salt (e.g., NaCl, $MgSO_4$, $BaCl_2$, or urea) will be considered as these are the ones that are mostly found in the emulsions described in Sec. 1.

To understand what is going on when such a solution is submitted to a regular cooling and heating cycle, it is necessary to know the solid–liquid phases diagram of the binary system. For that purpose, a schematic drawing of such a diagram is given in Fig. 4. The compositions of the solutions are given by the molar fraction of salt:

$$x = \frac{n_s}{n_0} \tag{13}$$

where $n_s$ is the number of moles of salt, and $n_0$ is the total number of moles (water plus salt); $x$ is supposed to vary between 0 and 1, theoretically. Nevertheless, it is well known that very often the whole domain is not attainable for physicochemical reasons: sublimation, decomposition, etc.

Three typical solutions, the compositions of which are given by the abscissa of the points $M_1$, $M_2$, and $M_3$, have been chosen to illustrate the general behavior. These solutions maintained at the temperature $T_0$ are totally liquid. Let us now suppose that they are placed at different temperatures, namely, $T_1$, $T_2$, and $T_3$, respectively. From the phases diagram, it can be deduced that:

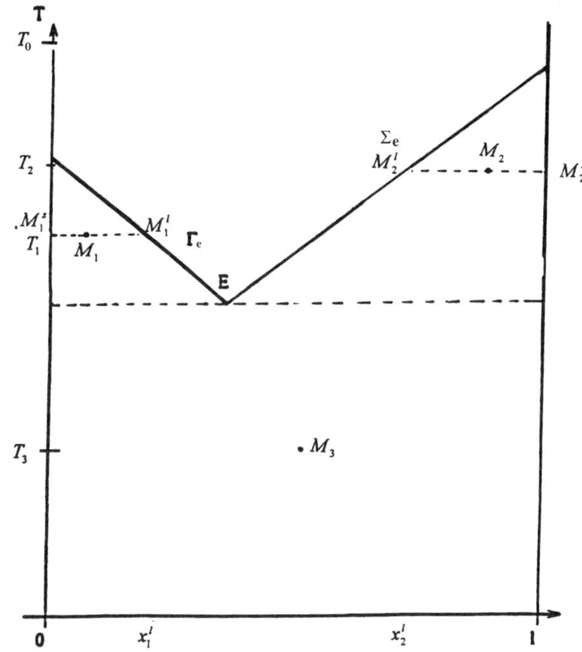


**Figure 4** Schematic drawing of a solid–liquid phase diagram of a binary system. $x$: salt molar fraction; $T$: temperature; $M_1$, $M_2$, $M_3$: points representing three typical solutions at temperatures $T_1$, $T_2$, $T_3$, respectively; $x_1^l$, $x_2^l$: composition of the liquid phase at points $M_1$ and $M_2$; $\Gamma_e$: freezing curve; $\Sigma_e$: solubility curve; $E$: eutectic point.

- At temperature $T_1$, a solid–liquid equilibrium is expected, pure ice being the solid (composition $x_1^s = 0$, point $M_1^s$) and a more concentrated solution being the liquid (composition $x_1^l$, point $M_1^l$). The proportion of ice in the system is given by the ratio of the lengths of the segments $M_1M_1^l$ and $M_1^sM_1^l$.
- At temperature $T_2$, a solid–liquid equilibrium is expected as well but different from the preceding one, pure salt being the solid (composition $x_2^s = 1$, point $M_2^s$) and a less concentrated solution being the liquid (composition $x_2^l$, point $M_2^l$). The proportion of salt is given by the ratio of the lengths of the segments $M_2M_2^l$ and $M_2^lM_2^s$.
- At temperature $T_3$, a totally solid system is expected composed of ice and salt. This picture is correct for any temperature less than the eutectic one given by point $E$ temperature.

The picture is quite different when a sample is regularly cooled in so far as solid–liquid equilibrium is not reached at any stage of the cooling. Kinetic aspects have to be taken into account as has been done for the pure material in Sec. II.A, and huge delays may be observed. Nevertheless, for the test proposed this fact appears as an advantage: the results are volume dependent, and the analysis of the behaviors of the solutions present in the emulsions and pointed out by DSC allow us to obtain considerable information about the emulsions themselves as will be described later on. Let us first describe the expected thermograms when a sample is cooled and heated regularly.

### 1. Bulk

The thermograms expected when a bulk solution is cooled regularly from $T_0$ are drawn in Fig. 5a. One or two signals are obtained, revealing differences in the solidification processes.

Two signals are obtained for the most diluted solutions in which ice is formed first at $T_1$ and total solidification (ice plus solid salt) is obtained at a lower temperature $T_1'$. At $T_1$ the sample is changing from a liquid to a mixture of ice and a supersaturated solution with respect to the solute. Therefore, the composition of this solution is given by the abscissa of point $M_1^l$ belonging to the extension $\Gamma_e'$ of the equilibrium curve $\Gamma_e$. The amount of ice formed is given by the ratio of the lengths of segments $M_1 M_1^l$ and $M_1^s M_1^l$, and the energy released in a short time is quite important. So, the corresponding signal shows an abrupt slope. During further cooling, more ice is formed and the remaining solution becomes increasingly concentrated. At any temperature the composition of the solution is given by the abscissa of point $M_1^l(x, T)$ belonging to $\Gamma_e'$. When the conditions of salt germination are reached, salt is formed in the remaining solution, which is diluted so much that water is transformed instantaneously into ice. That gives rise to a release of energy, but this is generally less important than the first signal, and the corresponding signal observed at this temperature $T_1'$ is less abrupt.

One signal is obtained when ice is formed in such a way that the remaining solution is so concentrated that salt germination instantaneously occurs and total solidification of the sample is obtained. That is the case for the sample the composition of which is given by point $M_2$ in Fig. 5a.

The signals expected during the melting of totally solid samples are easier to draw and interpret. There is no delay and the signals obtained reveal the eutectic melting and progressive ice melting (Fig. 5b). More interesting is the case of achievement of partial solidification during the cooling. In that case, only one sig-

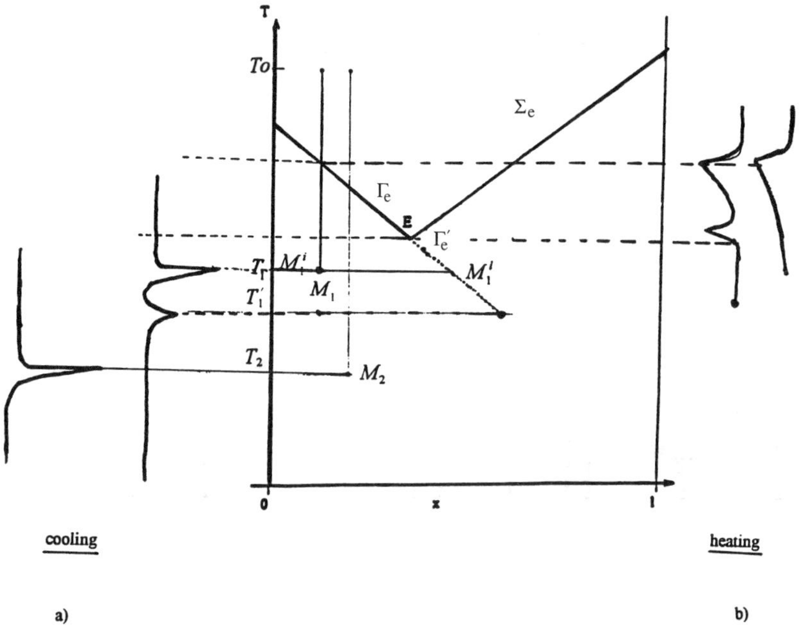


**Figure 5** Schematic thermograms expected for a bulk solution, related to the corresponding binary diagram: (a) cooling; (b) heating. $x$: salt molar fraction; $T$: temperature; $M_1$ and $M_2$: points representing two solutions of different concentrations; $\Gamma_e$ : freezing curve; $\Gamma_e'$: extension of the freezing curve; $\Sigma_e$: solubility curve; $E$: eutectic point.

nal, corresponding to the progressive melting of ice, will be observed (Fig. 5b). There may be different reasons for the cause of such an event. The obvious one is that the cooling was not sufficient and that the temperature $T_I'$ was not reached. The test has to be done again by cooling the sample down to a lower temperature. If by doing so, total solidification is not achieved, the formation of a vitreous state of the remaining solution may be suspected. More will be said about this situation in the section dealing with droplets for which the probability of encountering such a phenomenon is more probable. Furthermore, such an observation will be used to obtain information about the evolution of the emulsion as the volume is playing a role.

### 2. Dispersed

To illustrate the general behavior, dispersed solutions prepared from water and a dissolved salt will be considered. The differences from the case previously studied are first, the volume of the sample that is around a few cubic micrometers, and second, the number of samples which is very large, each droplet being a sample. A statistical response to the DSC test will, therefore, be obtained. However, even if the behavior for each droplet is similar to the one given for a bulk sample in Sec. III.B.1, the solidification signals will show differences in the shape and in the temepraturе ranges wherein they are observed, as has already been described for pure compounds in Secs. III.A.1 and 2.

As for bulk solutions, and referring to the solid–liquid phases diagrams as well, one or two signals will be obtained according to the composition, provided that there is no problem in obtaining totally solidified droplets during the cooling. The expected thermograms are depicted in Fig. 6. The solid–liquid phase diagram is drawn in the middle and the thermograms are found on its left side for the cooling and on its right side for the heating. Circles that represent the droplets are drawn in the peak surfaces. They are shared in order to indicate the phases composing the droplets. "I" means ice, "L", liquid solution, and "S", solid salt.

Four solutions, the compositions of which are referred to by the numbers 1, 2, 3, and 4, have been chosen to denote different typical behaviors. During

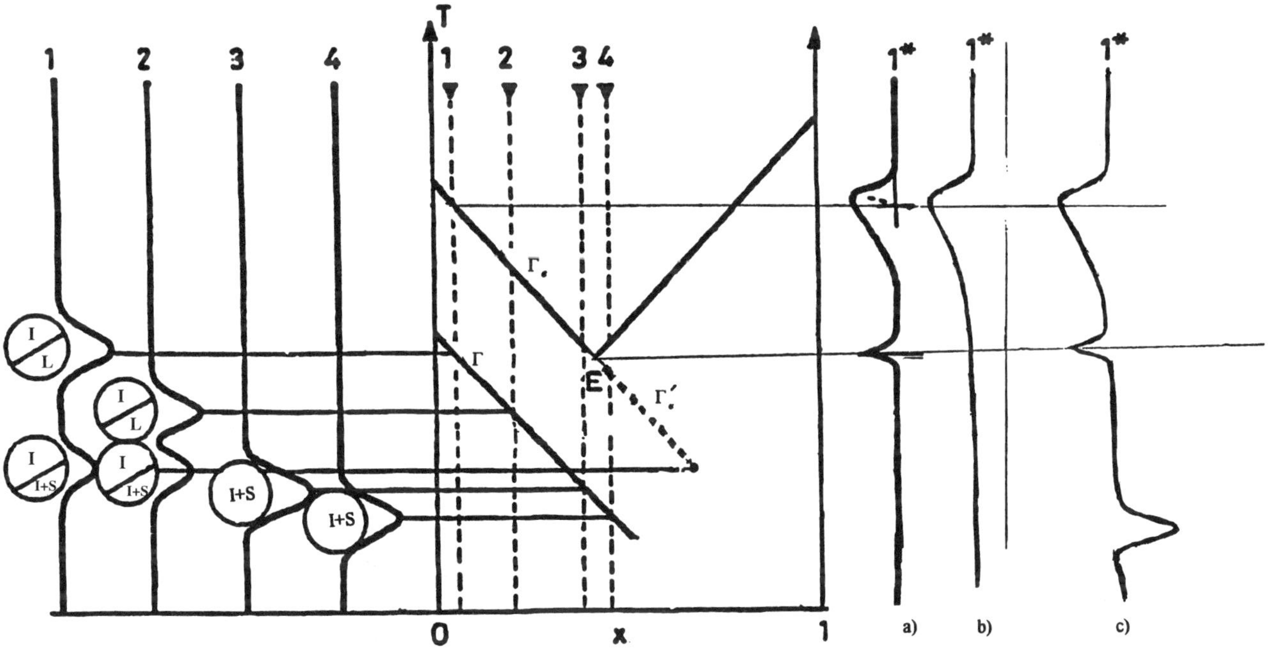


**Figure 6** 1, 2, 3, and 4: schematic thermograms expected during cooling of four dispersed solutions with the corresponding binary diagram. I: ice; L: liquid solution; S: solid salt. 1*: schematic thermograms expected during heating of dispersed solution 1 after (a) complete solidification; (b) partial solidification; (c) partial solidification during cooling and complete solidification during heating. $x$: salt molar fraction; $T$: temperature; $\Gamma_e$: freezing curve; $\Gamma_e'$: extension of the freezing curve; $\Gamma$: curve giving the variation of the most probable ice formation temperature vs. the droplets' composition; $E$: eutectic point.

the cooling of the most diluted solution 1, two signals revealing energy release due to solidification are observed. The first peak indicates ice formation inside the droplets and the second one the total solidification of the droplets. Although the amount of ice formed in each droplet is relatively larger than in a bulk in so far as the solidification temperature is lower, the volume of each droplet is much smaller and the energy released by the partial solidification of each droplet is low. Nevertheless, the number of droplets is large enough to allow detection of the droplets' solidification even if the solidification is not total and there is a scattering of the solidification temperatures as for a pure compound dispersed into droplets (see Sec. III.A.1). The second signal shows the formation of salt in the remaining solution, which instantaneously induces total solidification. The same behavior is noted for the more concentrated solution 2. The first signal is expected at a lower temperature as the concentration of solution 2 is higher than that of solution 1. It is noteworthy that the second signal is expected at the same temperature as for solution 1. This can be explained by the fact that the droplets have reached equal states of composition after ice formation, the remaining solution having its composition varying according to the same law given by the $\Gamma'_e$ curve during the further cooling. Instantaneous and total solidification is indicated by a single signal for dispersed solutions 3 and 4, the lower the temperature, the higher the concentration. From knowledge of the most probable temperature $T^*$ of ice formation in the liquid droplets, it is possible to draw a curve $\Gamma$ that gives the variation of this temperature versus the composition of the droplets. Conversely, from knowledge of this curve it is possible to detect by the DSC test performed on an emulsion sample, any composition change in the droplets due to composition ripening, as will be shown in the next sections.

Studying the thermograms obtained during the heating is a way to evaluate what has been solidified during the cooling. Should the solidification be complete and the thermogram is the one referred to as 1*(a), the solution under study is 1. No difference is noticeable compared to the one corresponding to a bulk solution (Fig. 5b) so long as the droplets are not too small (see Sec. II.A). If only ice is formed within the droplets during cooling, the thermogram obtained during heating may be different, depending essentially on the droplet diameters. The expected thermograms are those referred to as 1*(b) and (c) in Fig. 6. The thermogram 1*(b) is expected when no further solidification is occurring even if the lowest temperature during cooling has been reached. The unique signal observed is as a result of the progressive melting of the ice formed during cooling; that melting begins as soon as the heating is started. The thermogram 1*(c) shows a eutectic melting followed by progressive melting of the remaining ice. In that case, total solidification has not been achieved during cooling but it has occurred during heating, as the solidification signal is found before the eutectic one shows. This behavior may be attributed to the formation of a vitreous state of the remaining solution that can be very concentrated after the ice formation. The lower the ice-formation temperature, the higher the concentration given by curve $\Gamma'_e$ (Fig. 5, point $M_1^l$). As the smaller the droplets, the lower the ice-formation temperature, it can be deduced that an evolution with time of a given emulsion should be noticeable by the thermograms in the way that 1*(b), 1*(c), and 1*(a) show, the droplets becoming increasingly larger.

## IV. RESULTS

In this section, actual results, showing what information can be deduced from the DSC test performed on different emulsions, are given.

First, W/O emulsions will be considered, emphasis being put on complex emulsions found in the petroleum industry for which there is a great need of specific techniques as was already mentioned in Sec. I. Numerous studies have been carried out on less complex W/O emulsions, in order to establish the conditions for solidification of dispersed water, with a view to obtaining information about the behavior of materials that contain dispersed water, e.g., clouds, biological cells, and food emulsions (24). These experiments have been performed by using DSC, and the results obtained have been the basis of the setting up of the test proposed in this chapter. The reader can obtain more fundamental information from the literature already cited.

Next, what has been called mixed emulsions and multiple emulsions will be examined. Actually, these kinds of emulsion can be found during the simple W/O emulsion process of fabrication, during the transportation of complex fluids, in crude oil production, and also during the evolution of simple emulsions. Therefore, it appears to be very important to detect them and the DSC test is very efficient as it will be seen in the following.

## A. W/O Emulsions

### 1. Chocolate Mousses

Chocolate mousses are water-in-crude oil emulsions resulting from oil weathering after an oil spill. These concentrated and viscous emulsions are known to complicate drastically the operations of recovery and clean-up (41). It is now generally recognized that emulsion formation at sea is the result of surfactant-like behavior of polar compounds and asphaltenes naturally present in the crude oil (42,43). In the initial crude oil, aromatic compounds are solvents of the heavier compounds. After an oil spill, when the aromatics evaporate, asphaltenes and resins precipitate and stabilize water droplets in the oil volume, forming a very resistant interfacial film that avoids the coalescence of droplets (44). The emulsification process strongly affects the physicochemical properties of spilt oil. Stable emulsions can contain between 50 and 80% of seawater, and the resulting volume of spilt oil is therefore rapidly expanded from two to five times the initial volume. The viscosity of the material changes from a few Pa s to about a 1000 Pa s. Chocolate mousses are often heavy materials, hard to recover mechanically, treat, or burn. It is therefore essential to assess their stability in order to optimize their treatments.

Different types of synthetic chocolate mousses were studied by DSC (30,31). The objective was to assess the water droplet size distribution by the analysis of the solidification signal, in order to establish a correlation between the form of the solidification signal and the stability of the emulsion. Water-in-crude oil emulsions were prepared from two different types of asphaltenic crude oils: an Arabian Light crude oil topped at 150°C (BAL 150) and a Safaniya crude oil. They were both made at water contents of 50 and 75% (v/v). The synthetic seawater was prepared with 33 g/l of aquarium salt in distilled water. Very stable emulsions were made with an Ultra-Turrax apparatus: seawater was poured drop by drop into the crude oil during mixing; the resulting emulsion was fine and monodispersed (mean diameter: about 2 μm). Unstable emulsions were prepared with a rotating-flask apparatus (45), which was developed in order to be representative of the natural conditions of formation of chocolate mousses at sea. Emulsions prepared with this apparatus were polydispersed. At first, seawater and crude oils were studied separately. It was shown that the solidification or melting signals obtained with crude oils under investigation did not interfere with signals due to water in the temperature range −150° to 20°C. The synthetic emulsions of BAL 150 and Safaniya were studied in the same way at a constant scanning rate of 5°C/min between 20° and −150°C, during both cooling and warming. In all cases, the thermograms of melting were similar, because melting occurs at the thermodynamic equilibrium temperature, whatever the form or volume of the sample (Fig. 7; see Sec. II.B). The solidification thermograms give more information about the emulsified water. Figure 8 shows the solidification peaks obtained with both types of BAL 150 emulsion prepared with 50% (v/v) of seawater: the straight line represents the solidification peak of an emulsion made with the rotating-flask apparatus, and the dotted line corresponds to the signal recorded with the Ultra-Turrax emulsion. These thermograms are typically similar to all the thermograms observed in the case of the other types of chocolate mousses. The solidification signal obtained with the stable emulsion (Ultra-Turrax) is a single bell-shaped peak. The temperature at the top, which corresponds to the most probable temperature of solidification of seawater droplets, is around −45°C. With the rotating-flask emulsions, a single peak can sometimes also be observed, but the most probable temperature of solidification is always higher than in the previous case. Most of the time, the thermogram exhibits a series of peaks between −20° and −40°C, as shown in Fig. 8. The calculation of the ice-melting enthalpies from the warming DSC thermogram allows an accurate determination of the water content in the emulsion [see Eq. (11)]. It is then quite easy to measure the water content of a given emulsion. Results from water contents calculated from DSC measurements are presented in Table 2. The samples were taken at the same time from the top of the test-tube containing the synthetic emulsion. It is clear that the calculated water contents of the rotating-flask emulsions are significantly lower than the theoretical

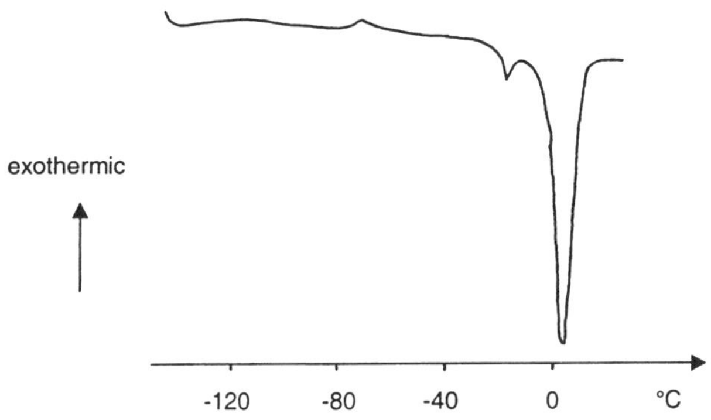


**Figure 7** Melting thermogram of "chocolate mousse" emulsions. (From Ref. 31.)

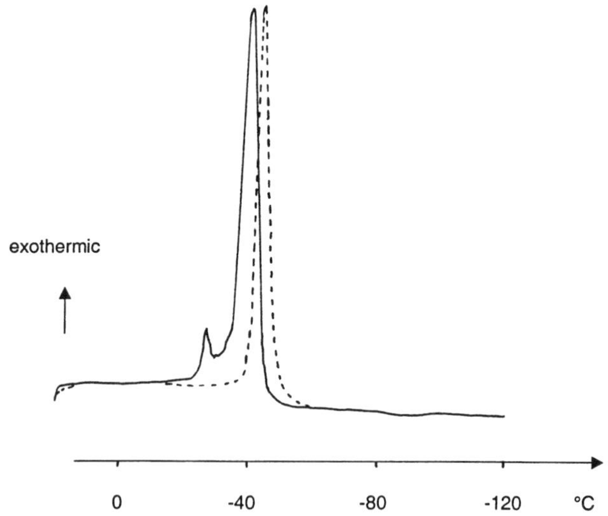


**Figure 8** Cooling thermograms obtained with BAL150 emulsions prepared with 50% (v/v) seawater; straight line: "rotating-flasks" emulsion; dotted line: "Ultra-Turrax" emulsion. (From Ref. 31)

values. It can be explained by the relative instability of these polydispersed emulsions. On the other hand, the calculated water contents determined with Ultra-Turrax emulsions exactly correspond to the theoretical values.

## 2. Water-in-Crude Oil Emulsions Obtained During Production

During crude oil production, all conditions for forming an emulsion are gathered:

- The presence of two immiscible liquids: produced water and crude oil.
- Zones of strong agitation to disperse one liquid into small droplets: this agitation is generally due to turbulence or shear forces encountered in the different parts of the process facilities.
- The presence of surfactants or emulsifiers that stabilize the dispersed droplets.

**Table 2** Experimental Determination of Water Content in the Case of Seawater-in-Safaniya Crude Oil Emulsions

| | "Rotating-flasks" emulsions | | "Ultra-Turrax" emulsions | |
|---|---|---|---|---|
| Theoretical water content (%) | 50 | 75 | 50 | 75 |
| Experimental water content (%) | 15 | 53 | 51 | 73 |

*Source*: Ref. 31.

Very stable emulsions can form at the wellhead because of the sudden pressure drops that can occur in the choke valve or in the pumps (46).

In the oilfield, water-in-crude oil emulsions are generally called regular, while O/W emulsions are called inverse or reverse (47). It is noteworthy that more than 95% of the crude oil emulsions formed in the field are of the W/O type. Nevertheless, multiple or complex emulsions (O/W/O or W/O/W) can also be encountered.

In regular oilfield emulsions, the dispersed aqueous phase is usually called "sediment and water" (S&W) and the continuous phase is crude oil. The dispersed S&W phase is essentially saline water, but different types of solids such as sand, mud, scale, corrosion residues, or precipitates are often present and can participate in the mechanisms of stabilization of emulsions. Petroleum emulsions vary from one field to another because crude oils differ by their geological age, chemical composition, and associated impurities, and furthermore, the water exhibits physical and chemical properties that are also specific to each reservoir. Nevertheless, all fields have in common the fact that a great number of emulsifying agents are present in the fluids produced (48):

- Indigenous surface-active compounds such as asphaltenes and resins, which can play the role of high molecular weight surfactants.
- Finely divided solids such as clay, sand, shale, silt, gilsonite, drilling muds, workover fluids, corrosion products, crystallized paraffins or waxes, and precipitated asphaltenes and resins.
- Chemical products used during production such as corrosion inhibitors, paraffin dispersants, biocides, cleaners and surfactants, wetting agents, etc.

As in the case of chocolate mousses, W/O oilfield emulsions are often difficult to characterize because of their opacity, their high water concentration, the presence of various solids in both phases, and the organic nature of the continuous phase. It is of prime importanhce to be able to assess the droplet size distribution in oilfield emulsions, because of the strong dependence between the rate of separation and the size of the droplets. A knowledge of the drop size distribution is clearly an important factor in the design of separation equipment (46). Furthermore, the phase continuity of the mixture is also a parameter which must be known since the separation rate of oil droplets from water is different from that for water droplets from oil in the same system, even if the droplet size distribution is the same.

Few studies can be found in the literature about the influence of equipment in the production facilities on the evolution of the dispersions or emulsions (46,49,50). According to the technical difficulties related to the characterization of these emulsions, experimental studies in the laboratory are performed with very diluted dispersions, especially of the O/W type, in order to use classical techniques for granulometry, such as light scattering (46,49). In the other cases, separation efficiency is deduced from "bottle tests" results.

At the IFP (Institut Français du Pétrole), we have used DSC to study the influence of different agitation conditions on the drop size distributions of actual samples of oilfield emulsions. The use of optical microscopy was very difficult because of the presence of finely divided solids in the emulsion. The DSC techniques allowed the study of the emulsion without dilution or any other perturbation. After separation of the organic and aqueous phases by centrifugation, each phase was studied separately during cooling and warming, in order to check that the oil did not crystallize in the given interval of temperatures. Samples of emulsion were then submitted to various mechanical perturbations:

- gentle shaking;
- homogenization with an ultra-turrax apparatus (17,500 and 24,000 rpm).

The different samples were finally studied in the same way during cooling and warming, between 20° and −80°C, at a constant rate of 5°C/min. Figure 9 shows the typical cooling thermogram of a sample of emulsion after homogenization by gentle shaking. Three main peaks can be identified:

- A small bell-shaped peak between −43° and −39°C, corresponding to the solidification of very fine droplets (a few micrometers) (see Fig. 3c).
- A large peak with a vertical part at −15°C, characteristic of the solidification of free water (or very large droplets) (see Fig. 3a).
- A wide peak at −30°C (temperature at the top of the peak), which is overlapped by a succession of badly defined peaks between −25° and −18°C.

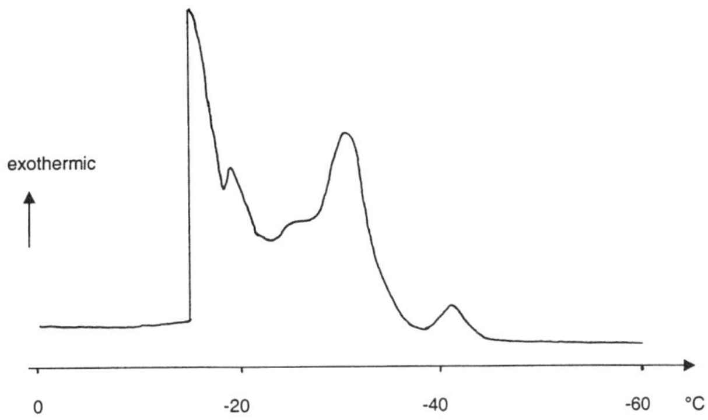


**Figure 9** Thermogram of cooling of an oilfield emulsion after homogenization by gentle shaking.

This emulsion is, therefore, strongly polydispersed and contains a great number of medium and large droplets. Figure 10 shows the superposition of the solidification thermograms obtained with the emulsion submitted to different levels of agitation: handshaking, and Ultra-Turrax at position 3.5 (17,500 rpm) and position 6 (24,000 rpm). The effect of a vigorous agitation is clearly demonstrated. The first level of agitation with the ultra-turrax system involves an enlargement of the medium-size droplets' population while the very large droplets (free water) disappear. The second level of agitation shows an increase in the smaller droplets' population: the bell-shaped peak becomes wider and the temperature of the apex is shifted to a lower temperature (−45°C).

In that specific case, DSC measurements allowed us to study quite easily the influence of agitation on the emulsion polydispersity while other classical techni-

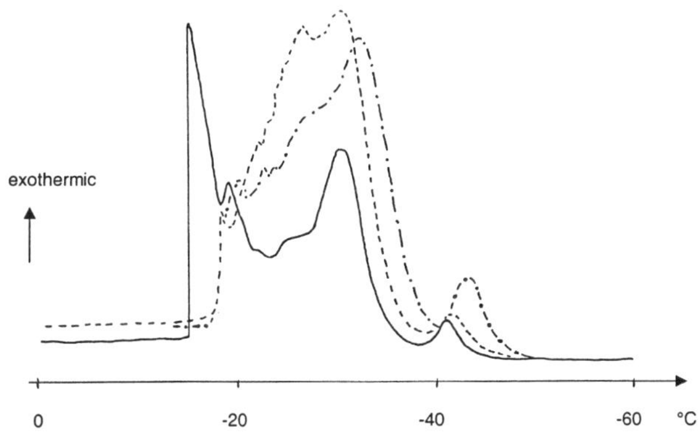


**Figure 10** Influence of agitation on the thermogram of cooling of an oilfield emulsion: — gentle shaking; - - - 17,500 rpm; –·– 24,000 rpm.

ques were not applicable, due to the opacity of oilfield emulsions and the coexistence of suspended solids and dispersed liquid droplets.

### 3. Drilling Fluids

The success of any well-drilling operation depends on many factors, one of the more important being the drilling fluid. The fluid performs a variety of functions that influence the drilling rate, the efficiency, the safety, and, of course, the cost of the operation (51). Drilling fluids are generally composed of liquids (water or oils) and suspended, finely divided solids of different nature. They are classified as to the nature of the continuous phase: gas, water, or oil. Within each broad classification are divisions based on composition or chemistry of the fluid or the dispersed phase. For many years, oil-based muds have proved to be the best-performance and cost-effective fluids in difficult drilling situations. Typical muds are W/O emulsions with an aqueous phase (saline water) varying from 5 to 40%. These reverse emulsions contain three main types of compounds:

- Emulsifiers for improving the emulsion stability.
- Organophilic clays for controlling rheological properties, especially thixotropy.
- Weighting agents, such as hematite or barite, to adjust the fluid density.

Considering the specific conditions for use of oil-based muds (high pressure/high temperature), the characterization of their stability, especially with temperature, is essential. However, the complicated nature of these drilling fluids makes these studies very difficult to perform and explain. Stability is, therefore, generally assessed from simple bottle-test experiments and empirical standardized tests, such as electrical-stability measurements. Classical techniques for the determination of water-droplet size distribution cannot be applied because of the great amount of solids that compound drilling muds. With bottle tests, the only obvious indication of destabilization is given by the kinetics of clarification of the supernatant phase. Considering that all solids, as well as water droplets, tend to deposit by sedimentation processes, it is then quite impossible to observe the coalescence of liquid droplets. It is therefore difficult to determine if destabilization is simply due to flocculation and sedimentation of particles or if both liquid phases (oil and water) are completely separated (total breakage of the emulsion).

At the IFP, we tried to apply DSC to various oil-based muds before and after thermal treatments (16 h at 180°C) in order to characterize their stability. The organic and aqueous phases were at first studied separately during cooling and warming, in order to check that the oil did not crystallize in the given interval of temperatures. Each sample of mud was then studied between 20° and −120°C, during both cooling and warming, at a constant scanning rate of 5°C/min. Figure 11 shows typical thermograms of solidification and melting of two fresh muds (muds 1 and 2) which differ from their emulsifying system: their respective behaviors were completely similar before thermal ageing. The dispersed aqueous phase was saline water (200 g/l $CaCl_2$). In that case, only solidification of ice was observed, which was confirmed by the drawing of the melting thermogram (see Sec. III.B.2 and Fig. 6). This single exothermic bell-shaped peak was obtained at about −90°C.

After thermal aging of both muds, the DSC thermograms were quite different (Figs. 12 and 13). For mud 1 (Fig. 12a), two exothermic peaks were found at about −55° and −75°C during cooling. Compared to the fresh mud, the peaks were shifted to higher temperatures, which was characteristic of an enlargement of the droplet size (coalescence). In the case of mud 2, a slightly different behavior could be observed during cooling (Fig. 13a): a unique exothermic peak was found at about −55°C, but no other signal was

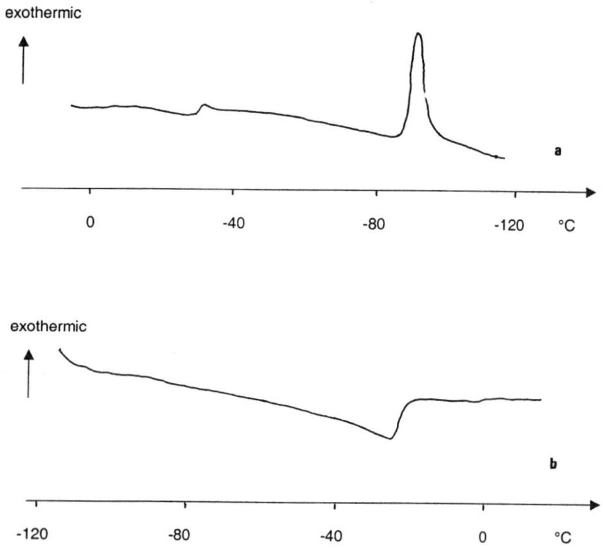


**Figure 11** Typical thermograms of (a) cooling and (b) heating of fresh muds 1 and 2.

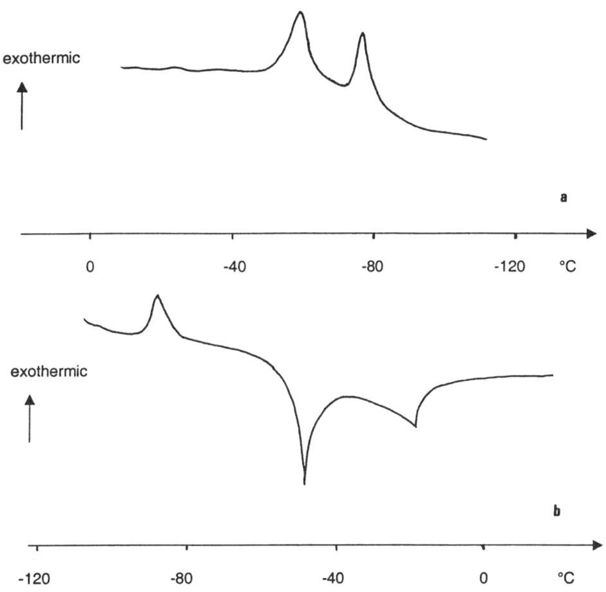


**Figure 12** (a) Cooling and (b) heating thermograms of mud 1 after thermal aging.

observed at −75°C. Mud 2 seems, therefore, to be less stable after thermal aging than mud 1 (see Fig. 3d; Sec. III.A). This behavior was further confirmed from filtration tests. It is noteworthy that melting thermograms were similar for both muds, but differed from those obtained for the fresh muds (Figs. 12b and 13b).

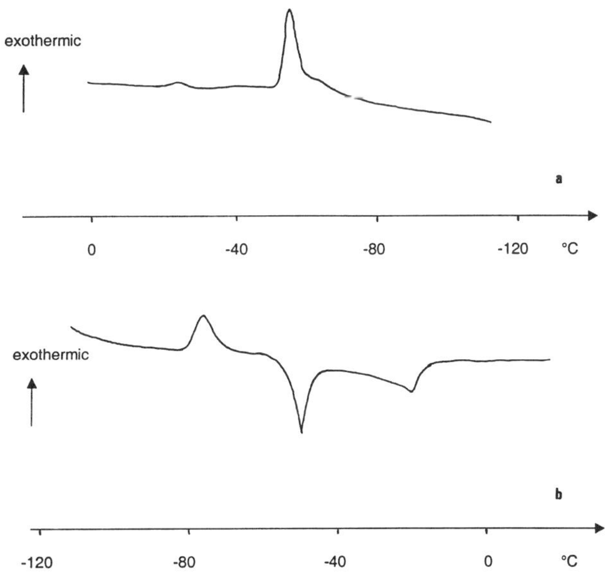


**Figure 13** (a) Cooling and (b) heating thermograms of mud 2 after thermal aging.

First, solidification of concentrated saline solutions, which was not observed during cooling, occurred during heating (exothermic peak at −80°C). This phenonenon could be easily explained: during cooling, nucleation did certainly occur in the remaining solution, but the high viscosity of the concentrated saline solution, in equilibrium with ice, prevented the growth of crystal germs. During warming, the viscosity decreased and the growth of small crystals was allowed. After that, the dispersed aqueous phase was completely solid. Further warming allowed the eutectic melting of the solid at −52°C (first endothermic peak which corresponded to the eutectic temperature for the system $H_2O/CaCl_2$), and the progressive melting of ice was then observed (second endothermic peak) (see Sec. III.B.2 and Fig. 6).

It is noteworthy that microcalorimetry could, therefore, be easily used to compare the thermal stability of oil-based muds while even bottle tests did not allow us to observe the coalescence of droplets.

## B. Mixed Emulsions

Mixed emulsions are so called because they are obtained from the mixing of two simple emulsions that differ by the compositions of the dispersed phases. The mixing is done gently in order to avoid coalescence at the very maximum. The resulting emulsion has the particularity of containing droplets that are different in composition and close together (Fig. 14). Should the medium wherein they are dispersed be permeable to their components then mass transfers between them

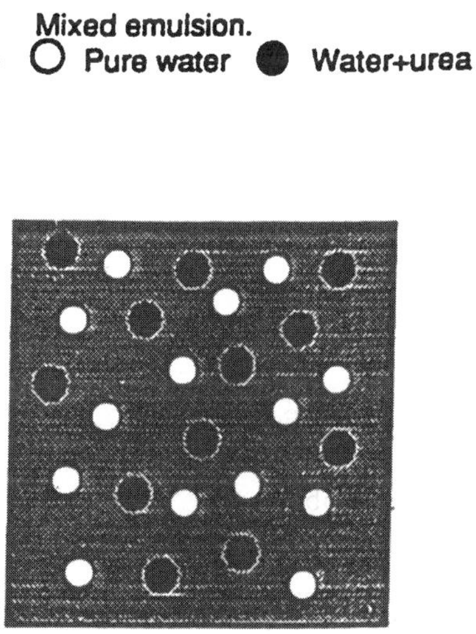


**Figure 14** Schematic view of a mixed emulsion.

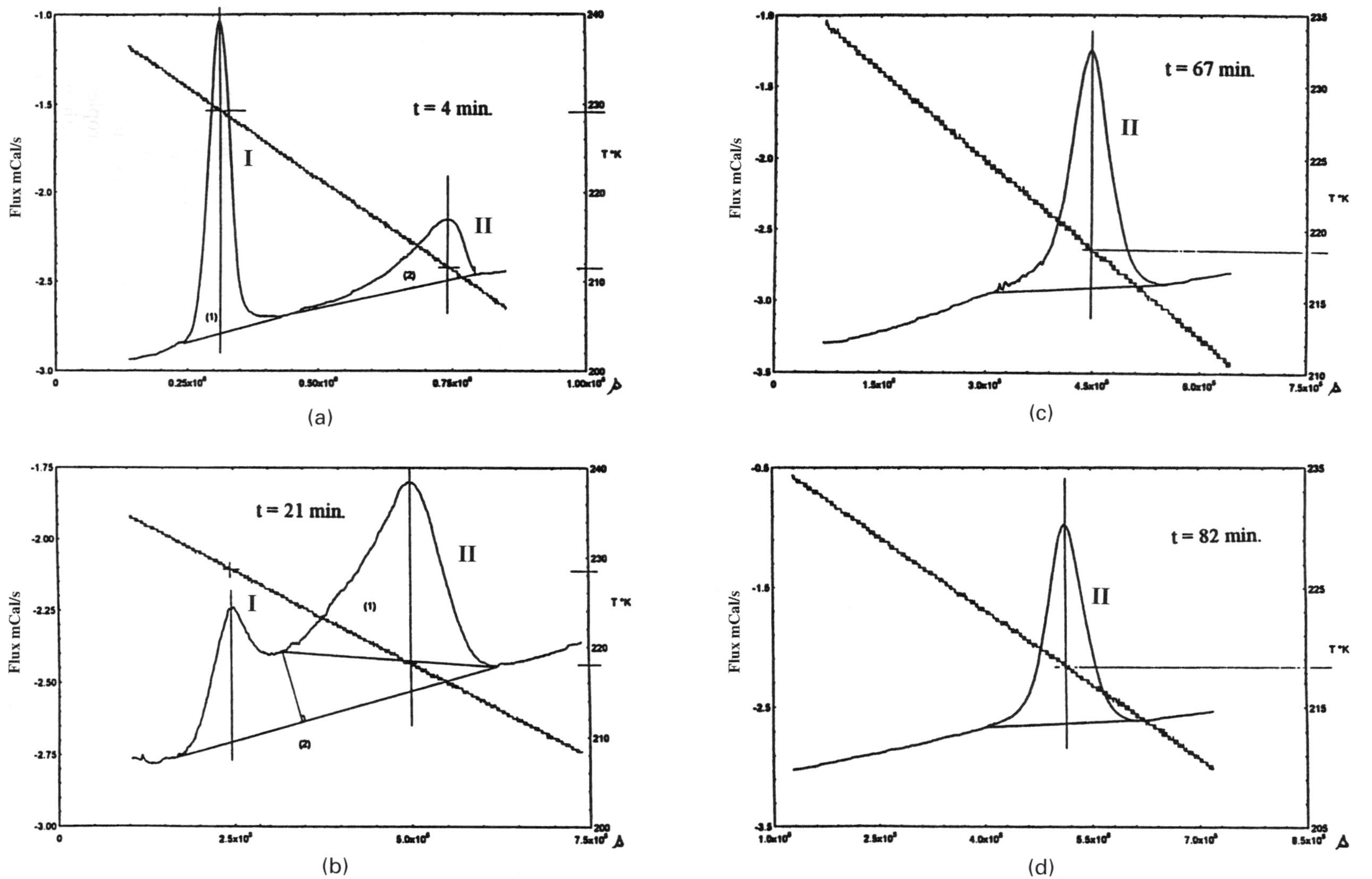


**Figure 15** Cooling thermogram of a mixed emulsion: (a) $t = 4$ min; (b) $t = 21$ min (c) $t = 67$ min; (d) $t = 82$ min. I: solidification of pure water; II: solidification of water plus urea droplets. (Courtesy of A. Gauthier, UTC, France.)

may occur. This phenomenon is called composition ripening as it leads to a homogeneity of the droplet compositions (52–54). It has been proposed as a means to obtain size-controlled droplets thanks to a fitting formulation of the father–mother emulsions (52).

Let us see what has been deduced from DSC tests performed on mixed emulsions prepared from a W/O emulsion (emulsion I) and a (water plus urea)-in-oil emulsion (emulsion II), the compositions of which are:

*Emulsion I*

1. Dispersed phase (20% w/w):
   a. deionized water (conductivity 4 μS);
2. Continuous phase (80% w/w):
   a. paraffin oil (Prolabo) : 47.5%;
   b. pure vaseline paste (Prolabo) : 23.8%;
   c. lanolin (Prolabo) : 8.7% (lipophilic emulsifier).

Emulsion II differs from emulsion I only in the dispersed phase, the composition of which is 30% of urea dissolved in water. Equal amounts of emulsions I and II are mixed manually and the resulting emulsion is kept at ambient temperature. From time to time an emulsion sample is submitted to cooling and heating in a differential scanning calorimeter.

From the cooling thermogram depicted in Fig. 15, it can be deduced that the scattering of the solidification temperatures of the dispersed droplets is varying versus time. For a preservation time $t = 4$ min, two distinct solidification signals, I and II, are observed and at the end of the study corresponding to a preservation time of 82 min, only one well-defined signal is obtained. Therefore, according to the analysis performed in Sec. III.B.2, an evolution of the droplets' composition is indicated in this way. To describe this evolution, it is necessary to know the change in the most probable solidification temperature $T^*$ versus the composition of the droplets. For that purpose, several emulsions, the dispersed phases of which being of various solution compositions, have been submitted to the DSC test. The results obtained are presented in Fig. 16. The change $T^*$ versus the composition of the dispersed phase is given by curve $\Gamma$. As the eutectic point $E$ is characterized by a relatively high value of the temperature $-11°C$, and the amount of dissolved urea is 30%, the extension curve $\Gamma_e'$ shows a low slope. Therefore, it is expected that total solidification occurs instantaneously, giving rise to only one signal (see Sec. III.B.2). Actually, this is what has been observed; Fig. 17 illustrates this behavior: only one signal during cooling, and the eutectic

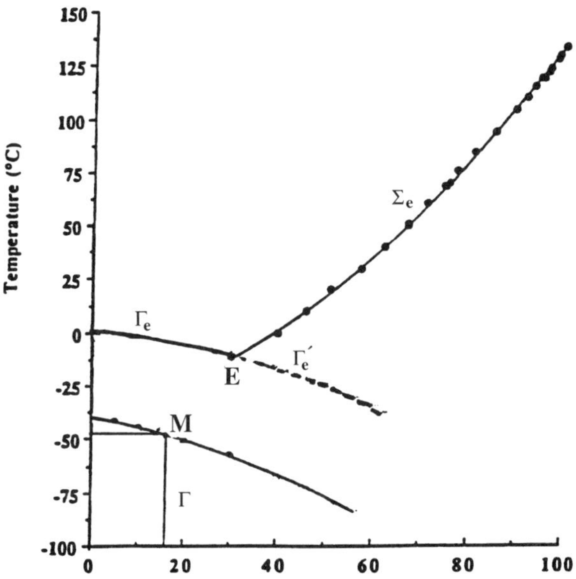


**Figure 16** Solid–liquid phases diagram of the binary (water plus urea) with the curve $\Gamma$, giving the most probable solidification temperature vs. composition of the dispersed phase. $x$: salt molar fraction; $T$: temperature; $\Gamma_e$: freezing curve; $\Gamma_e'$: extension of the freezing curve; $\Sigma_e$: solubility curve; E: eutectic point. (Adapted from Ref. 55.)

melting signal followed by the progressive ice-melting signal during heating. Returning to the mixed emulsions, the two signals observed during the cooling have been attributed to the solidification of pure water (signal I) and to the solidification of the water plus urea droplets (signal II). At the beginning, these signals are well separated, but as time goes on the boundary is more difficult to draw. At the end of the evolution, only one well-defined signal is observed, showing a homogenization of the droplets' composition. This phenomenon, showing a composition ripening, has been thoroughly studied elsewhere (55) and has been attributed to a water transfer between the pure water droplets and the water plus urea droplets. From curve $\Gamma$ and the value of $T^*$ obtained at the end of the evolution, namely $-49°C$, it has been checked that the composition of 16.1% corresponding to this temperature (point $M$, Fig. 16) is in good agreement with the value of 15.8% deduced from the formulation and assuming a total transfer of water from the pure water droplets towards the water plus urea droplets. This kind of evolution has also been seen with mixed emulsions prepared from pure water droplets and water plus sodium chloride droplets (56).

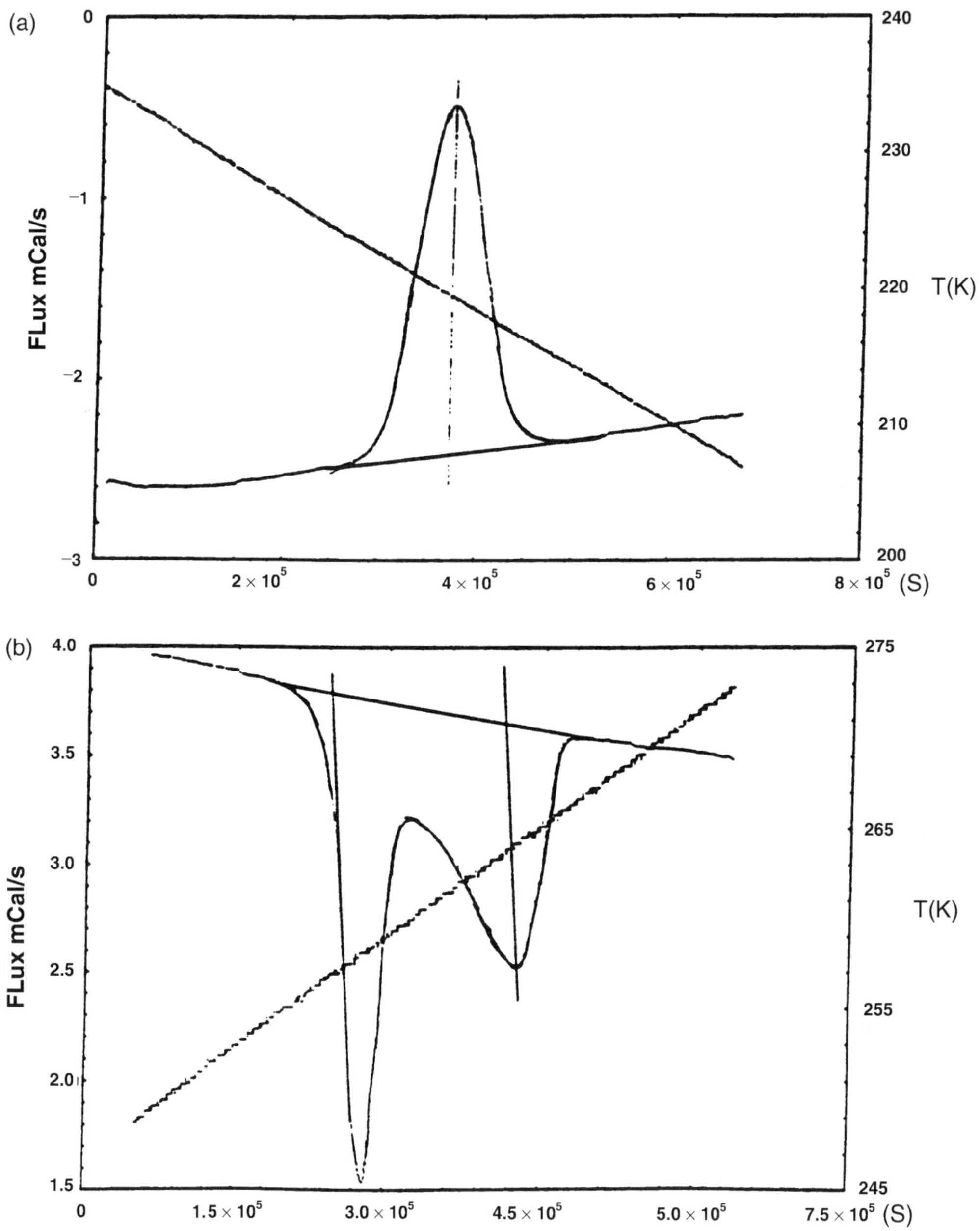


**Figure 17** (a) Cooling and (b) heating thermograms of a (water plus urea)-in-oil emulsion. (Courtesy of A. Gauthier, UTC, France.)

## C. Multiple Emulsions

Two kinds of multiple emulsions exist. Water-in-oil-in-water (W/O/W) emulsions are made of oil globules containing water droplets and are dispersed in water. Oil-in-water-in-oil (O/W/O) emulsions are made of water globules containing oil droplets and are dispersed in oil. These emulsions have been formulated to trap specific substances and to effect their release at will (5,6,57–61), but they may be also encountered spontaneously when for instance a simple emulsion is changing from a W/O emulsion into an O/W emulsion, the multiple emulsion being an intermediate state. Figure 18 illustrates the thermograms obtained by submitting a multiple emulsion to a DSC test. Let us see what is possible to deduce from them. Signals I and

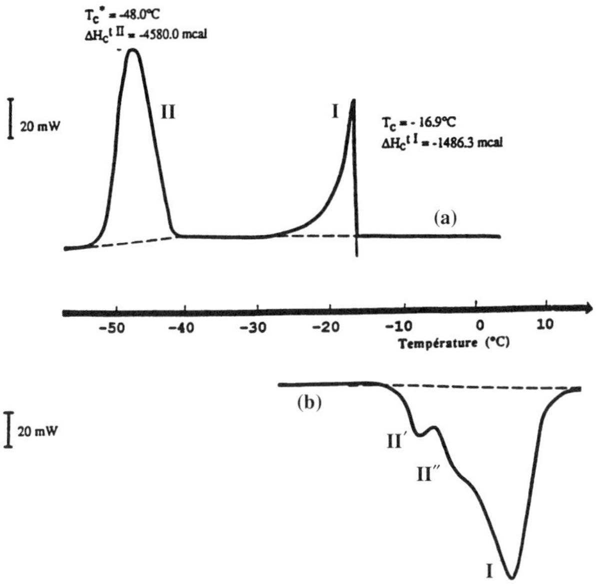


**Figure 18** (a) Cooling and (b) heating thermograms of a multiple emulsion (W/O/W). I(a): solidification of outer aqueous phase; II(a): solidification of dispersed water; I(b): melting of outer aqueous phase; II′(b): eutectic melting; II″(b): progressive ice melting. (Adapted from Ref. 57.)

II show a release of energy and therefore solidification of the emulsion by stages. Signal I has the characteristic shape of the solidification of a bulk material (see Fig. 3a). It could be either the outer aqueous phase of a W/O/W emulsion or the outer oil phase of an O/W/O emulsion. The melting signal I found around 0°C is in favor of a W/O/W emulsion. The second one is characteristic of the solidification of a dispersed phase (see Fig. 3c). As the kind of emulsion has been identified, the only possibility left is the solidification of the water droplets trapped inside the oil globules. Pure water would have given a $T^*$ value around −39°C. The lower temperature found, −48°C, indicates that some solute is present inside the droplets. This statement is confirmed by observation of the heating thermograms that show melting at two times II′ and II″. These signals let us suppose that eutectic melting followed by progressive ice melting has taken place as has been found for dispersed solutions. For the present case, urea was the solute, and the data provided in the preceding section allow us to obtain further information. $T^*$ is found to be the temperature corresponding to a 10% solution. Should this temperature be found to vary versus time, a change in the composition due to a mass transfer may be assumed. The DSC test allows the possibility to quantify this transfer. This kind of analysis has been performed and more information about this may be found in Ref. 6.

## V. CONCLUSION

In this chapter a very simple test to characterize complex emulsions, i.e., water-in-crude oil emulsions, mixed emulsions, and multiple emulsions, has been proposed. The test is based on the correlation between the conditions of solidification and melting of the various phases encountered within the emulsions and their characteristics: bulk, dispersed (microsize or nanosize), bound, pure materials or solutions, and their respective amounts. Differential scanning calorimetry is a technique that permits such an investigation. The test, which is easy to set up, consists in submitting an emulsion sample, that does not need to be diluted, to a regular cooling and melting cycle. It is very important to note that two thermograms are needed to give a reliable interpretation. Schematic thermograms have been drawn to help the user to characterize the emulsions under study. Practical examples show how to use the thermograms and how it is possible to obtain information about the changes occurring within emulsions versus time. Other results obtained from the DSC test performed on emulsions can be found elsewhere, e.g., formation of hydrates (62), and influence of partial solidification on the stability (63). More can be also found elsewhere on theoretical studies dealing with the solidification of undercooled droplets and from experimental results deduced from the test (64,65).

## REFERENCES

1. LL Schramm, ed. Emulsions – Fundamentals and Applications in the Petroleum Industry. Washington, DC: Advances in Chemistry Series 231, American Chemical Society, 1992, pp 1–49.
2. S Torandell. Extrapolation de conditions opératoires de fabrication d'émulsions cosmétiques lors du passage de l'échelle laboratoire aux échelles pilote et industrielle [Extrapolation of operating conditions for cosmetic emulsion making from laboratory scale to pilot and industrial scales]. PhD dissertation, Institut National Polytechnique de Lorraine, Nancy, France, 1999.
3. RJ Mikula. In: LL Schramm, ed. Emulsions – Fundamentals and Applications in the Petroleum Industry. Washington DC: Advances in Chemistry Series 231, American Chemical Society, 1992, pp 79–129.

4. KJ Lissant. Demulsification – Industrial Applications. New York: Marcel Dekker, 1983, pp 105–132.
5. S Raynal, JL Grossiord, M Seiller, D Clausse. J Controlled Release 26: 129–140, 1993.
6. S Raynal, I Pezron, L Potier, D Clausse, JL Grossiord, M Seiller. Colloids Surfaces A: Physicochem Eng Aspects 91: 191–205, 1994.
7. M Clausse. In: P Becher, ed. Encyclopedia of Emulsion Technology, Vol 1. New York: Marcel Dekker, 1983, pp 481–715.
8. T Jackobsen, J Sjöblom, P Ruoff. Colloid Surfaces A 112: 73–84, 1996.
9. J Sjöblom, H Fordedal, T Skodvin. In: J Sjöblom, ed. Emulsions and Emulsion Stability. New York: Marcel Dekker, 1996, pp 393–435.
10. K-E Froysa, O Nesse. In: J Sjöblom, ed. Emulsions and Emulsion Stability. New York: Marcel Dekker, 1996, pp 437–468.
11. AM El-Hamouz, AC Steward. On-line drop size distribution measurement of oil–water dispersion using a Par-Tec M300 laser backscatter instrument. Proceedings of SPE Annual Technical Conference and Exhibition, Denver, CO, 1996, pp 785–796.
12. P Fawel Mraci, W Richmong Mraci, L Jones, M Collison Graci. Chem Australia 64: 4–6, 1997.
13. C Thomas, JP Perl, DT Wasan. J Colloid Interface Sci 139: 1–13, 1990.
14. JP Dumas, D. Clausse, F Broto. Thermochim Acta 13: 267–275, 1975.
15. DH Rasmussen, CR Loper. Acta Metall 24: 117, 1976.
16. JP Dumas, M Krichi, M Strub, Y Zeraouli. Int J Heat Mass Transfer 37: 737, 1993.
17. RW Michelmore, F Franks. Cryobiology 19: 163, 1982.
18. C Jolivet-Dalmazzone, P Guigon, J-F Large, D Clausse. Ind Eng Chem Res 36: 874–880, 1997.
19. F Broto, D Clausse. J Phys C/Solid State Phys 9: 4251–4257, 1976.
20. D Clausse, L Babin, F Broto, M Aguerd, M Clausse. J Phys Chem 87: 4030-4034, 1983.
21. D Clausse, JP Dumas, PHE Meijer, F Broto. J Dispersion Sci Technol 8: 1–28, 1987.
22. D Clausse, I Sifrini, JP Dumas. Thermochim Acta 122: 123–133, 1987.
23. D Clausse. J Thermal Anal 51: 191–201, 1998.
24. D Clausse. In: P Becher, ed. Encyclopedia of Emulsion Technology. Vol 2: Applications. New York and Basel: Marcel Dekker, 1985, pp 77–157.
25. CA Angell, J Donnella. J Chem Phys 64: 4560, 1977.
26. L Dufour, R Defay. Thermodynamics of Clouds. New York: Academic Press, 1963.
27. H Pruppacher, D Klett. Microphysics of Clouds and Precipitation. Dordrecht: D Reidel, 1980, pp 162–180.
28. G Vali. In: RE Lee, GJ Warren, LV Gusta, eds. Biological Ice Nucleation and its Applications. St Paul, MN: The American Phytopathological Society, 1995, pp 1–28.
29. NM Barford. In: AG Gaonkar, ed. Characterization of Food – Emerging Methods. New York: Elsevier, 1995, pp. 59–91.
30. C Dalmazzone. Use of the DSC technique to characterize water-in-crude oil emulsions stability. Proceedings of the Second World Congress on Emulsion, Bordeaux, France, 23–26 Sept. 1997, Vol. 2, pp 2-1-069/01-05.
31. C Dalmazzone, H Seris. Rev Inst Franç Pétrole 53: 463–471, 1998.
32. EK Bigg. Proc Phys Soc London 66B: 688–694, 1953.
33. BJ Mason. Sci Progr GB 44: 479–499, 1956.
34. GM Pound, IA Madonna, SL Leake. J Colloid Sci 8: 187–193, 1953.
35. HR Pruppacher. J Chem Phys 39: 1586, 1963.
36. JR Heverly. Trans Am Geophys UN. 30: 205, 1949.
37. C Lafargue. CR Acad Sci 230: 2022-2025, 1950.
38. M Brun, A Lallemand, J-F Quinson, C Eyraud. Thermochim Acta 21: 59–88, 1977.
39. AP Gray. In: RJ Porter, JF Johnson, eds. Analytical Calorimetry, Vol 1. New York: Plenum Press, 1968, p 209.
40. CM Guttman, JH Flynn. Analyt Chem 45: 408, 1973.
41. AL Bridié, TH Wanders, W Zegveld, HB Van der Heijde. Mar Pollut Bull 11: 343–348, 1980.
42. M Bobra. Water-in-oil emulsification: a physicochemical study. Proceedings of International Oil Spill Conference, San Diego, CA, 1991, pp 483–488.
43. M Desmaison, C Piekarski, S Piekarski, JP Desmarquest. Rev Inst Franç Pétrole 39: 603–615, 1984.
44. TJ Jones, EL Neustadter, KP Whittingham. J Can Petrol Technol April–June: 100–108, 1978.
45. C Dalmazzone, C Bocard, D Ballerini. Spill Sci Technol Bull 2: 143–150, 1995.
46. GA Davies, FP Nilsen, PE Gramme. The formation of stable dispersions of crude oil and produced water: the influence of oil type, wax and asphaltene content. Proceedings of SPE Annual Technical Conference and Exhibition, Denver, CO, 1996, pp 163–171.
47. G Leopold. In: LL Schramm, ed. Emulsions – Fundamentals and Applications in the Petroleum Industry. Washington, DC: American Chemical Society, 1992, pp 341–383.
48. JA Svetgoff. Petrol Eng Int 61: 28–35, 1989.
49. MJ van der Zande, WMGT van den Broek. Break-up of oil droplets in the production system. Proceedings of the ASME Energy Sources Technology Conference & Exhibition, Houston, TX, 1998, ETCE98-4744.
50. HP Ronningsen, O Urdahl. A North Sea crude oil and its water-in-crude oil emulsions. Comparison between small scale laboratory experiments and more realistic

conditions. Proceedings of Seventh BHR Group Ltd et al. Multiphase Production International Conference, Cannes, France, 1995, pp 33–49.
51. J-P Nguyen. Drilling. Paris: Technip Editions, 1993, pp 115–138.
52. BP Binks, JH Clint, PDI Fletcher, S Rippon, SD Lubetkin, PJ Mulqueen. Langmuir 15: 4495–4501, 1999.
53. L Taisne, P Walstra, B Cabane. J Colloid Interface Sci B 184: 378–390, 1996.
54. D McClements, S Dungan. J Phys Chem 97: 7304–7308, 1993.
55. D Clausse, I Pezron, A Gauthier. Fluid Phase Equilibria 110: 137–150, 1995.
56. D Clausse. J Dispersion Sci Technol 20: 315–316, 1999.
57. L Potier, S Raynal, M Seiller, JL Grossiord, D Clausse. Thermochim Acta 204: 145–155, 1992.
58. D Clausse, I Pezron, S Raynal. Cryo-Letters 16: 219–230, 1995.
59. S Matsumoto. In: DO Shah, ed. ACS Symposium Series 272. Washington DC: American Chemical Society, 1987, pp 415–436.
60. N Garti, S Magdassi. J Colloid Interface Sci 104: 587, 1985.
61. JL Grossiord, M Seiller, F Puisieux. Rheol Acta 32: 168–180, 1993.
62. B Fouconnier, V Legrand, L Komunjer, D Clausse, L Bergflodt, J Sjöblom. Progr Colloid Polym Sci 112: 105–108, 1999.
63. D Clausse, I Pezron, L Komunjer. Colloids Surfaces A: Physicochem Eng Aspects 152: 23–29, 1999.
64. PHE Meijer, D Clausse. Physica B 119: 243–247, 1983.
65. D Kashchiev, D Clausse, C Jolivet-Dalmazzone. J Colloid Interface Sci 165: 148–153, 1994.

# 15

# Video-enhanced Microscopy Investigation of Emulsion Droplets and Size Distributions

**Øystein Sæther**
*Norwegian University of Science and Technology, Trondheim, Norway*

## I. INTRODUCTION

Video-enhanced microscopy (VEM, or video microscopy, VM) is a technique that combines the magnification power of a microscope with the image-acquisition capability of a video camera. The resulting data matrix, from which information about the sample can be extracted, is an image or a series of images. This intimately relates VM to image-analysis techniques, now frequently with the assistance of a computer. Current image-analysis software provides a wide range of analytical features, in addition to image enhancement (the improvement of image quality prior to analysis), which is only briefly treated here. It is obvious that image analysis is not restricted to VM, but finds application within any technique where the data take the form of an image, e.g., electron microscopy, and other video or photographic techniques.

Typical information that can be found in images is sample state, geometry, dispersity, etc. For emulsions, this generally means droplet size and concentration, which are important properties of any emulsion. Figure 1 shows a coarse and a fine emulsion, the behavior of which can be expected to differ strongly due to droplet size and concentration. Further, the state of flocculation will indicate droplet/droplet interactions. Series of images or continuous video provide information about droplet interactions and the kinetics of important processes within the emulsion, like flocculation and coalescence. All the above parameters are central to the understanding of emulsion behavior and emulsion stability.

Microscopy (1–5), photomicrography (6–28), and VM (29–49) have combined a long history in the determination of particle and droplet size. A number of studies have been performed comparing the microscopy methods to alternative methods, such as those of light scattering (10, 32, 50), Coulter counting (2, 5, 24, 32, 50), turbidimetry (3, 9, 27), NMR (33, 45, 46), and others (8, 15, 28). Generally, the comparison is favorable and objections often relate to the labor-intensity of the derived methods. Amongst many applications reported in the literature are the study of vesicles (size and shape) (51, 52), particle trajectories (53) and emulsion (suspension) kinetics (26, 38, 41, 48, 54–61), measurement of pair potentials (62), film studies and interfacial tension measurements (63–68), and emulsions in electric fields (3, 13, 69, 70) (Fig. 2), to name but a few, to illustrate the versatility of such techniques.

## II. CHARACTERISTICS OF THE TECHNIQUE

Figure 3 shows a schematic of a typical VEM set-up. A video camera (digital or analog) is attached to the phototube of a microscope. The image is transferred

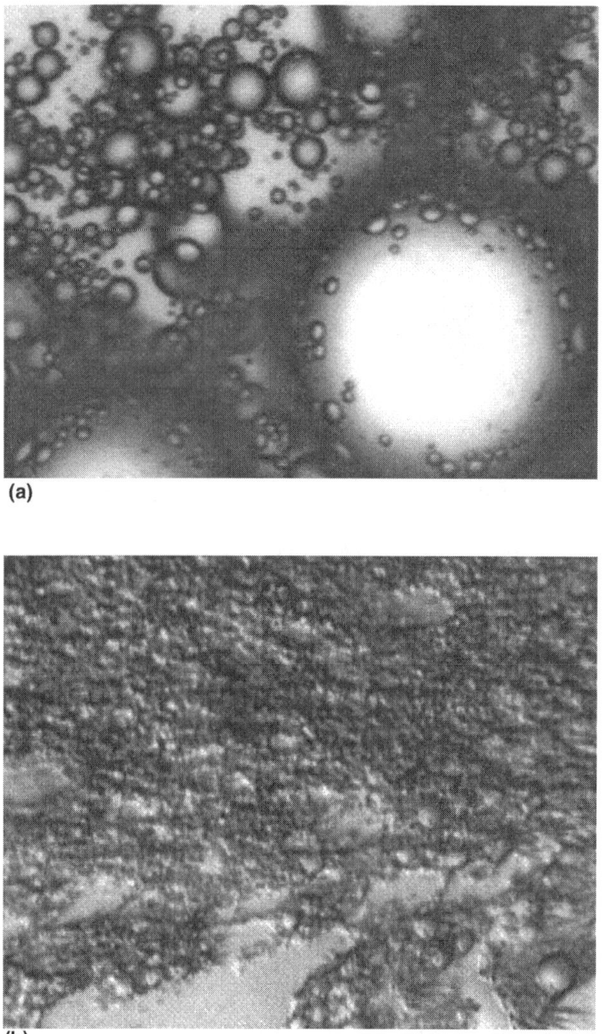

(a) (b)

**Figure 1** (a) Emulsion mixed with a simple rotor paddle; the uneven distribution of mechanical energy on the liquids has caused a broad DSD. (b) Ultrasonically prepared emulsion; the dispersed volume is to a high degree present as very small droplets in a narrowly distributed fraction of the population.

to the image captureboard installed on to the computer motherboard (digitization). Image enhancement and analysis is accomplished with image-analysis software.

Careful adjustment of the microscope is essential for the achievement of reliable and reproducible results. Light-source intensity, the focussing of optics, and the adjustment of field and condenser diaphragms must be carefully controlled. As the droplets are defined by their circumference gray tone (or color) levels, deviations in the above from image to image may cause differing measurements of equally sized dro-

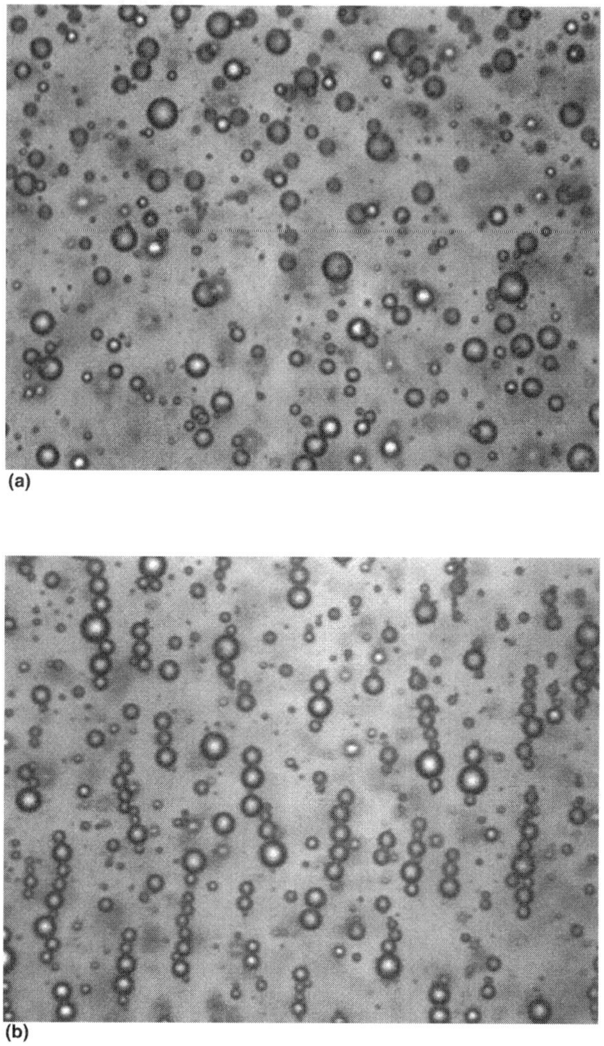

(a) (b)

**Figure 2** Aqueous droplets dispersed in crude oil and subjected to an electric field: (a) no field; (b) 5 s, 1 kV/cm – droplet orientation in chains along the direction of the field. The droplets become small net dipoles in the dielectric oil continuum and are attracted to each other, forming chains in the direction of the field. High field strengths will cause interdroplet membrane rupture and coalescence. The principle has been utilized for measuring emulsion stability (i.e., resistance to electrically forced breakdown) in the high voltage-time domain spectroscopy (HiV-TDS) (71,72) and conductivity techniques (73).

plets. Also, it is important to realize the operational characteristics of the components in the system. For example, the video camera is a vital link in the chain, and different models will handle the relay of the microscope image differently. For CCD cameras, an important property is the pixel geometry – some cameras have square, others rectangular, pixels, influencing the data matrix forwarded to the captureboard.

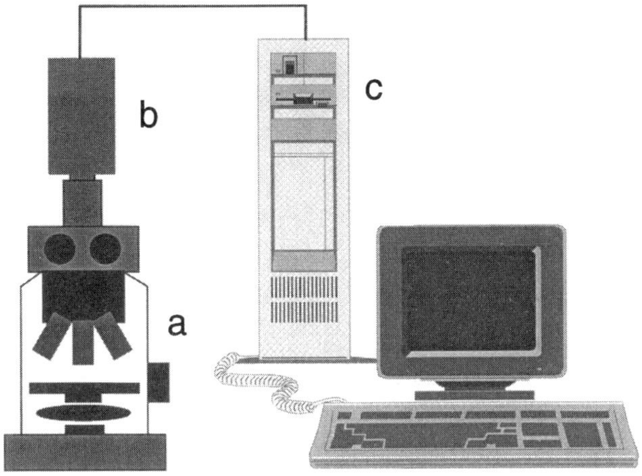


**Figure 3** Main components of the VEM experimental setup: (a) the microscope, including optics; (b) the videocamera; (c) the computer with captureboard and image-analysis software.

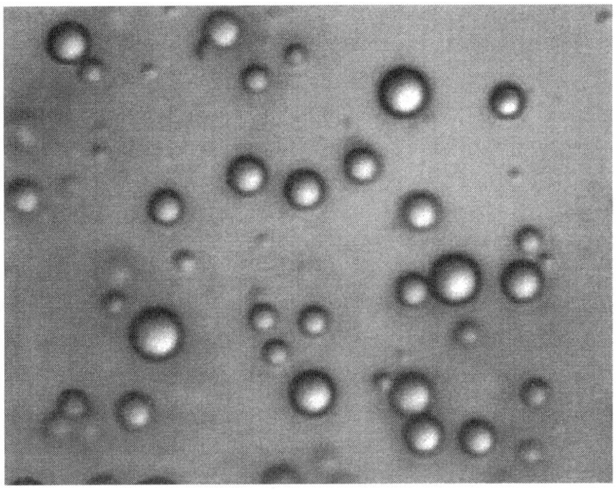

**Figure 4** DIC image of O/W emulsion. The droplets appear in relief; droplets beyond the in-focus section appear blurred.

Tremendous advances in the development of microscope optical components (filters, objectives) have gradually increased the range of applications for microscopy, especially by improving contrast between the objects of interest and the background. Much favored examples are differential (Nomarski) interference contrast (DIC) (74) and phase contrast (PC) optics. Often, there is little or no color or transmission contrast between objects and background. There are, however, differences in refraction index, which give rise to a change in the optical path through the object, along with a change in the phase of the light passing through the object relative to that of the light passing through the surrounding medium. These phase differences can then be translated into visible intensity differences between the object and the background.

For the study of emulsions and the measurement of droplet size, high sophistication of the optics is not generally necessary. However, when droplets flocculate into complex structures, ordinary optics may not be able to provide a satisfactory clear image of the structure; DIC can much improve this. Figure 4 shows the three-dimensional (3-D) representation of the sample which DIC can provide.

It is natural at this point to define the factors limiting the applicability of VM. First, the sample must have certain optical properties, since the technique relies on the reflection, refraction, scattering, and absorption of radiation, for instance, visible light, as is the case for optical microscopy. For emulsions, this means that the sample must be transparent and that the continuous liquid and the droplets must have different refractive indices or different colors, i.e., properties which make them optically distinguishable. Second, the *resolution limit*, and hence the operational size domain, is governed by the wavelength of the illumination. This feature is known as the Rayleigh limit (75) [Eq. (1)] and results in a physical limit of about 0.2 μm (half the illumination wavelength) (76). The practical limit tends to be slightly higher, because of rapidly increasing measurement error with decreasing object dimensions (p. 47 in Ref. 77). This is caused by diffraction; the image of an object is actually a diffraction pattern, and the overlapping patterns of closely spaced objects result in image blurring. Regarding *magnification*, there is no theoretical upper limit. Still, increasing the magnification only renders larger, blurred images of the objects. Innovations in optics have, however, proven the diffraction-imposed barrier not to be absolute (78).

The Rayleigh criterion (75):

$$R = \frac{0.61\lambda}{N.A.} \tag{1}$$

where $\lambda$ is the illumination wavelength, and $N.A.$ is the numerical aperture.

VEM, and in particular when coupled with PC and DIC optics, permits some bending of the Rayleigh criterion. Jokela et al. (32) experienced a VEM resolution limit that was about half that stated by the criterion (0.1 μm). As described above, the absence of a magnification limit allows observation of objects smaller than the resolution limit, but the images will appear blurred with lack of detail. However, the contrast-enhancing ability of VEM, PC, and DIC can help clarify minute features normally lost owing to the blurring diffraction patterns; e.g., Allen (79) observed the behavior of individual 25-nm diameter microtubules.

For VM, there is in addition the discretization of the image resulting from the digitization. The image is transformed into a matrix of colored or gray-scale dots, the pixels. The spatial dimensions of the pixels set the actual resolution of the image.

A further restriction relates to the three-dimensionality of the sample. The image is necessarily two-dimensional (2-D), although an increased depth of field (in the direction normal to the image plane) can provide information about a thicker optical section of the sample. For emulsions, this feature is clearly rather important, as these are very much 3-D and structurally dynamic as well. The choice of where to place the focal plane can strongly influence the data. For examining droplets and sampling a population for determination of the droplet size distribution (DSD), the simplest way will be to accumulate the droplets along a narrow focal plane. If the method of sample preparation (i.e., the container) provides a sample with a volume large enough in three dimensions for the droplets to move by gravity (e.g., microslides, see Sec. III), the droplets will ultimately accumulate along either the upper (ceiling) or the lower (floor) wall of the cell. Thermal movement may cause a size distribution function within the cream or sediment (55), as smaller droplets will diffuse more strongly than larger ones in a direction normal to the sediment plane. However, this need not be a factor, given large enough droplets or depth of field. It is more likely that small droplets will avoid measurement by not sedimenting into the focal plane within the time of measurement (Fig. 5). This exemplifies one of the clearer shortcomings of the technique – the 2-D representation of 3-D data.

Frequently, another restriction arises from the nature of the emulsion sample, namely that of disperse concentration. High droplet concentrations can cause droplet overlap, and small droplets tend to be obscured in strongly flocculating systems. Often, some degree of dilution is required, the chemical system allowing. As this may lead to changes in emulsion stability and consequently droplet size, it is important to apply dilution methods that do not influence the sample adversely in an uncontrolled manner. Basically, in surfactant-stabilized emulsions where the parameter of interest is droplet size, it may prove useful to dilute with the liquid of the continuous phase containing the surfactant at corresponding concentrations. This is, however, system dependent. For example, it usually proves sufficient to dilute crude oil-based W/O emulsions by adding

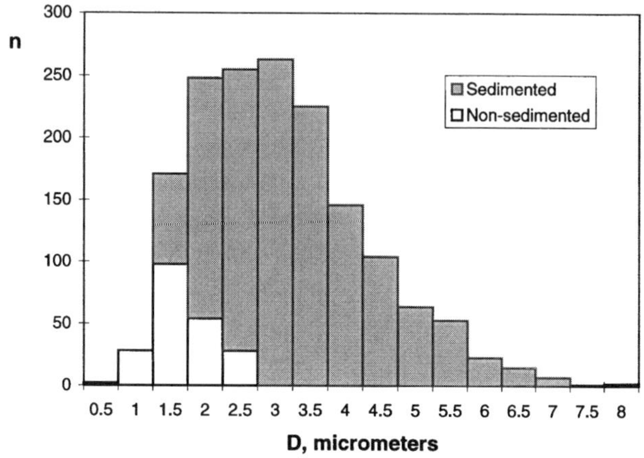


**Figure 5** Effect of Brownian motion on the measurement of the DSD in a sediment. Small droplets take part in chaotic thermal motion in a direction normal to the sediment plane. The histogram shows that a significant part of the smaller droplets are withdrawn from the DSD as measured in the sediment; 67% of the droplets measured in the 1-μm class were not found within the sediment.

the original crude oil. In any case, it is vital to control any changes in component concentrations which involve the crossing of phase boundaries and the distribution of components between phases. In creaming/sedimenting emulsions the thickness of the cell will influence the degree of dilution necessary – a "thick" sample (long viewpath) represents a larger pool from which the droplet concentration at the chosen focal plane can increase.

## III. SAMPLE PREPARATION

Sample preparation is a science in itself, due to the diversity of the systems studied with microscopy. Emulsions can be prepared in several ways, according to which parameters one seeks to observe and measure. The most common way of studying a sample droplet deployed between an object slide and a cover slide is prone to pollution and distortion (evaporation, shear). Often, some form of sample cell may be used with advantage. Hollow, flat microcapillaries are one example (Fig. 6). Within such a cell, the sample can remain protected against the surrounding working environment, which makes them ideal for long-term observation.

A microslide is a flat, rectangular glass tube with plane-parallel cross-section. A liquid sample can be introduced simply by letting capillary forces pull the liquid into the slide. The prepared sample can then be

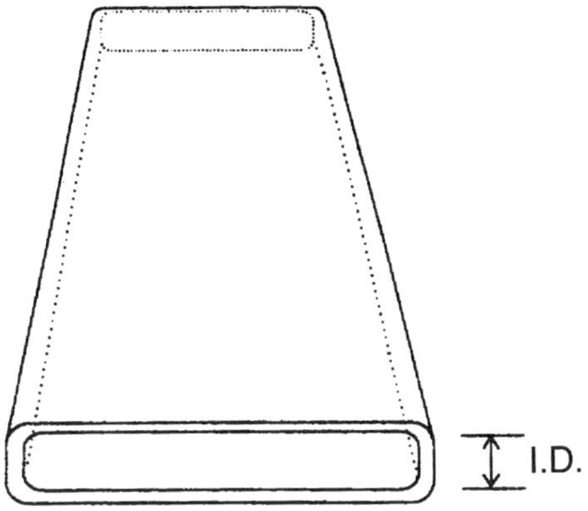


**Figure 6** The microslide – a thin, flat rectangular microcapillary of glass, useful for preparation of liquid samples vulnerable to evaporation or shear.

secluded from the surroundings by covering the tube ends, e.g., with some inert wax or grease. This way, the risk of evaporation and contamination altering the sample can be reduced significantly. However, the slide is made from glass and is therefore susceptible to influence from the sample. The surface is not perfectly smooth. Further, glass is slightly negatively charged, and will interact with other charged species in the sample, e.g., charged droplet surfaces. Adsorption at the interface by an anionic surfactant would expectedly reduce droplet/glass attraction, extending the lifetime of the sample and allowing observations of droplet kinetics over time.

Well-stabilized emulsions are not so vulnerable as to change through immediate coalescence. This fact can be utilized by letting droplets cream or sediment to the upper or lower cell wall, forming a slightly concentrated layer within a narrow focal zone. This simplifies the accumulation of data. However, if the droplets tend to coalesce rapidly upon collision, or it is desirable to retain the 3-D structure of the sample, one may increase the viscosity of the continuous phase (e.g., glycerin) or solidify the sample altogether (freezing). For kinetic studies, a cell preparation technique must be used (29, 30, 38, 41, 48, 54, 55).

Jokela et al. (32) developed a flow-cell system for VEM-assisted DSD measurements. Images of nonsedimented droplets were analyzed, and the method performed favorably compared to light-scattering and Coulter-counting methods. It follows that such a technique would work better with less-stabilized droplets than would the microslide technique, as droplet contact (with each other or the cell walls) could be reduced. The central features of VEM-assisted DSD determination are discussed in Ref. 32.

Further, the properties of glass surfaces can be changed to suit the current experiment. A common procedure in, for example, chromatography, is silylation, which allows alteration of the surface hydrophilicity by introducing a less polar substituent at the silanol groups. An example of a silylating agent is HMDS (hexamethyldisilazane), which reacts with the glass by the following reaction (80):

$$(CH_3)_3Si - N - Si(CH_3)_3 + 2 - Si - OH$$
$$\rightarrow 2 - Si - O - Si(CH_3)_3 + NH_3$$

The less polar and geometrically restricting $-Si(CH_3)_3$ now extends outwards from the surface, efficiently reducing surface polarity and also functioning as a steric repulsor. Other reagents can provide a range of properties, e.g., through different alkyl-substitutent chain lengths.

## IV. IMAGE ENHANCEMENT

Image enhancement signifies any process, which when applied to the image, improves its quality, hereunder clarifying the features of interest for the subsequent analysis and measurements. Before the arrival of the digital age, simple but valuable enhancement operations were performed with the aid of specialized equipment. Now, image-enhancement software permits the same and more to be done digitally, increasing method versatility tremendously. The different processes vary greatly in complexity and, hence, the computational power required. However, currently available computers provide this in affluence at minimal cost, leaving the main issue to be the flexibility of the software (often three times as expensive as the machinery on which it is run).

The first category of enhancement processes work on every pixel, disregarding its immediate neighborhood. Typically, this encompasses enhancement of contrast and adjustment of brightness. These functions have been available since before the introduction of the computer into the microscopy set-up, through analog image enhancers (made obsolete through digital treatment of the image). Seemingly trivial, a little effort here can greatly contribute to the quality of the data to be extracted at a later stage, as well as making the task easier. The second class of procedures works on each pixel relative to the adjacent pixels. Typically, a matrix assigning new values to the central pixel and its neighbors, altering their relative intensities, is run across the image. This is used for enhancing edges, filtering out noise, etc., and represents a very powerful way in which to improve image quality. Figure 7 shows how blur and noise can be removed, enhancing detail and preparing the image for analysis.

## V. IMAGE ANALYSIS AND MEASUREMENT

An image contains a lot of information that in different ways can be useful when attempting to describe the sample. However, when the task at hand is that of determining droplet size and the size distribution, the procedure of measurement uses only a small amount of this information. In its purest sense, the procedure seeks to distinguish the droplets from their surroundings (the background) and from each other, and then to perform the measurement on each of the defined droplets. The most primitive way is, of course, when the operator performs both the defining and measurements manually, a course which does not really need the assistance of a computer (although this may somewhat ease the tedious work). For complex systems this may be the only way to go, because the decision process of defining separate droplets is too complicated for a practical and reliable use of automated procedures that may be found within the software. For simpler systems, such procedures may tremendously simplify the generation of statistically sufficient amounts of reliable data, making the method a competitive alternative. Readily analyzable samples are typically dilute and nonflocculated, with a rather narrow distribution of droplet sizes. Figure 8 shows an image which does not permit automated measurement.

The general procedure is simple: first, define the parameters distinguishing the droplets from the background. This is typically accomplished by performing a thresholding on the basis of gray scale (or color) pixel values characteristic to the droplets. The second step is based on shape criteria; a droplet has a monotonic curvature, and a break in the monotony represents a droplet–droplet contact. The images in Fig. 9 show the analysis and measurement procedure. The resulting histogram is shown in Fig. 9e.

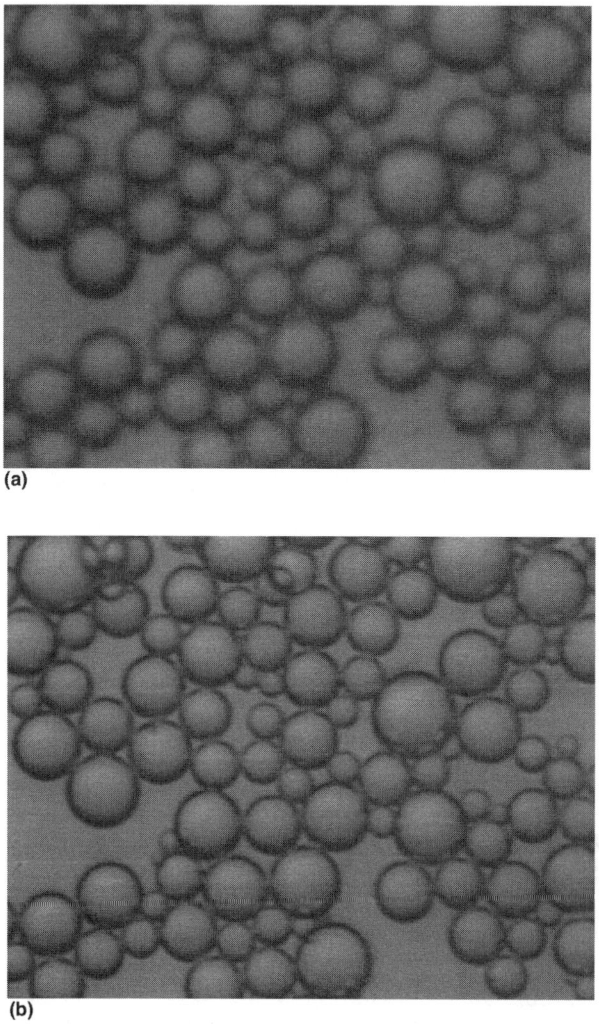

**Figure 7** Sharpening of image features: before (a) and after (b) spatial filtering.

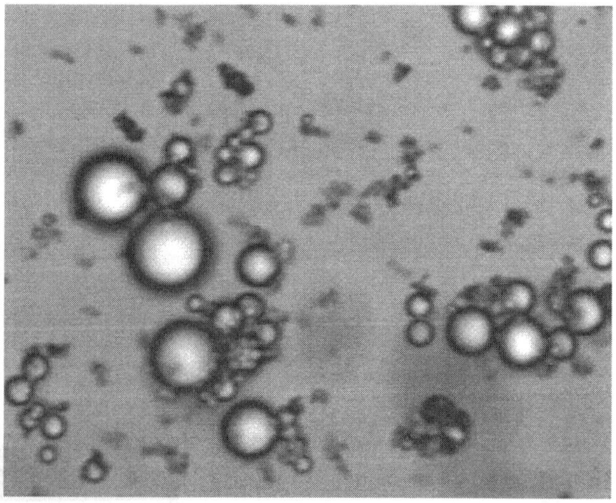

**Figure 8** Analytically demanding emulsion: droplets are largely coagulated into 3-D flocs; particles are present in the droplet size range. Such an image is extremely hard to analyze by automated software routines, and consequently demands strong participation by the operator. On the other hand, this procedure remains the only true alternative for handling such systems, as other techniques will not be able to resolve flocs or even discriminate between particles and droplets.

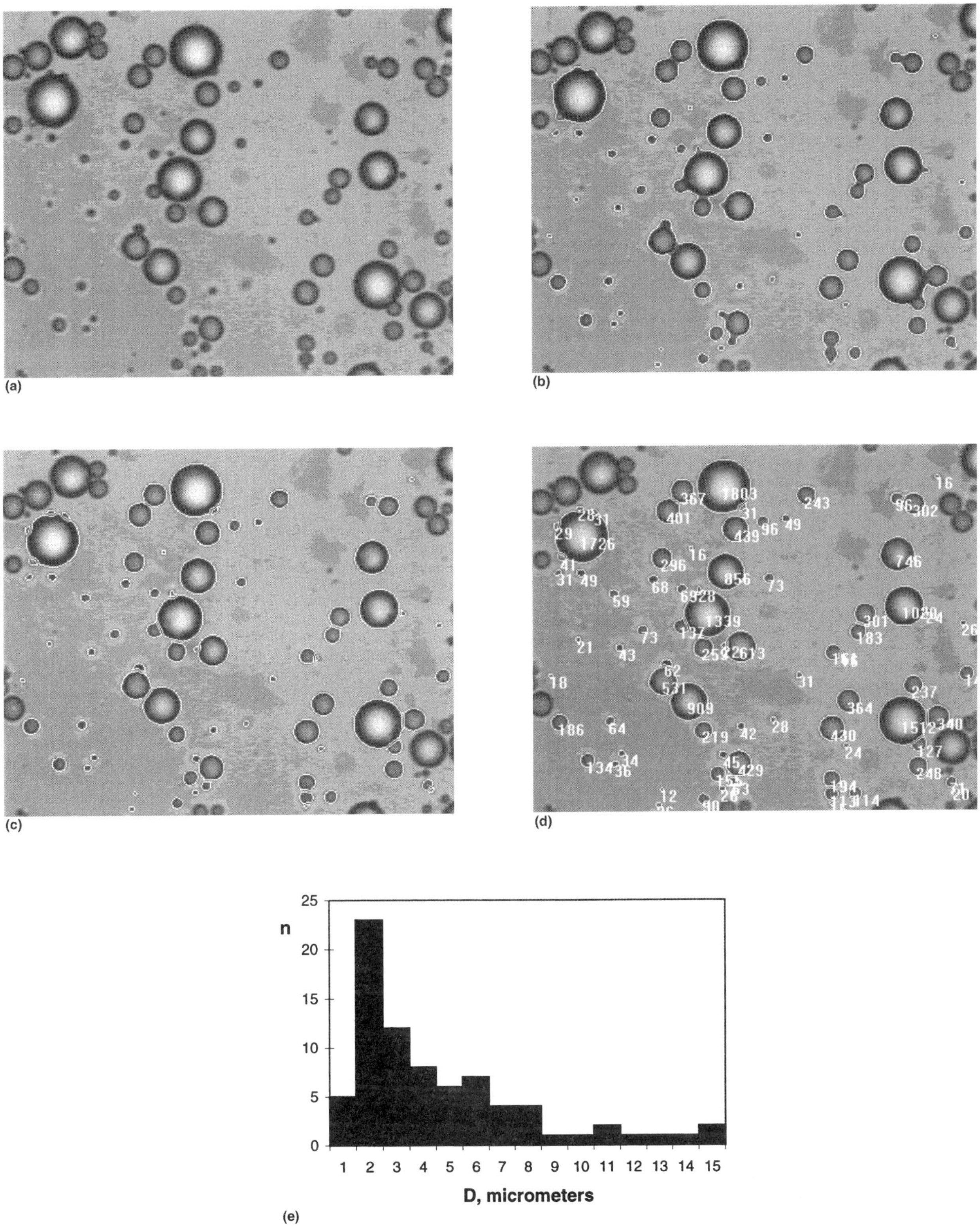


**Figure 9** A readily analyzable image; droplets are clearly distinguished from the background by characteristic gray-tone values, and flocs are 2-D: (a) the raw image; (b) thresholding – defining droplet pixel values; (c) separating droplets within flocs; (d) measurement of resolved image; (e) DSD of image.

When measuring the droplet size manually, the diameter comes out directly. The automated procedure will attempt to calculate the diameter from the droplet outline resulting from the thresholding. If, for some reason, the droplet is distorted and has an ellipsoidal shape, the return value might be the maximum, the minimum, or an average value, according to the set preferences. It may instead prove useful to measure the area within the outline and calculate the diameter on the basis of this. In any event, it is crucial to be aware of the criteria on which the program founds its return values.

Figure 10 serves as an example of calculation of the droplet diameter from the area of the pixel disk that was defined as belonging to a droplet through thresholding (Waddel disk diameter, $D_{WD}$). The pixel dimension is 0.23 µm, giving a pixel area of 0.0529 µm$^2$. The number of pixels in the 1-µm diameter disk is 16, while 61 pixels give 2 µm, and 132 pixels give 3 µm. Equation (2) yields the $D_{WD}$:

$$D_{WD} = \left(\frac{A_p n_p}{\pi}\right)^{\frac{1}{2}} \cdot 2 \quad (2)$$

where $A_p$ is the area of each pixel, and $n_p$ is the number of pixels in the disk. From the Fig. 10 it is clear that the potential error accompanying the $D_{WD}$ increases rapidly with decreasing droplet (disk) size.

## VI. TREATMENT OF VM SIZE DATA

The digital image consists of a regular matrix of pixels, which means that the number of different values that can be measured is limited. In reality, of course, the distribution is continuous; the digitization imposes discretization. In any case, since this is a direct method and the number of droplets counted and measured is finite, the resulting distribution will take the shape of a histogram. The histogram shows the frequency distribution for the assigned number

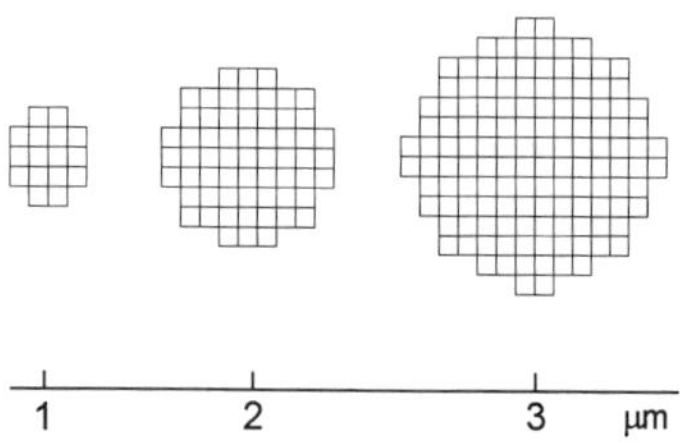


**Figure 10** Influence of digital resolution on the exactness of measurements.

and width of classes. A higher number will, when applied to a sufficiently high number of measurements, give a better representation of the overall shape of the real population distribution.

It is suggested (81, 82) that the class width should be chosen according to the distribution width; arithmetic progression (width constant, independent of droplet size) is sufficient for narrow distributions, while broad distributions should be presented with progressively increasing class width with increasing droplet size (equal differences between logarithms of the diameters, geometric progression). The reason for this is the fact that emulsion droplet diameters tend to be lognormally distributed (81, 83, 84).

The frequency distribution of diameters is the most widely used way of presenting population size data. It contains useful information which aids the prediction of emulsion kinetic behavior; e.g., sedimentation and diffusion are functions of droplet size. Also, one can follow the evolution of the DSD as a function of time, the shift towards fewer/larger droplets being evidence of droplet-depletion mechanisms, such as coalescence and Ostwald ripening. From the distribution, the kinetic coefficients can be calculated, allowing prediction of how the DSD will develop (e.g., 48, 55). This is described in detail by Dukhin et al., Chapter 4, this volume. Figure 11 shows how the addition of a demulsifier can destabilize an emulsion and bring about emulsion resolution. The example is a water-in-crude oil emulsion, the demulsifier a phenolic resin alkoxylate.

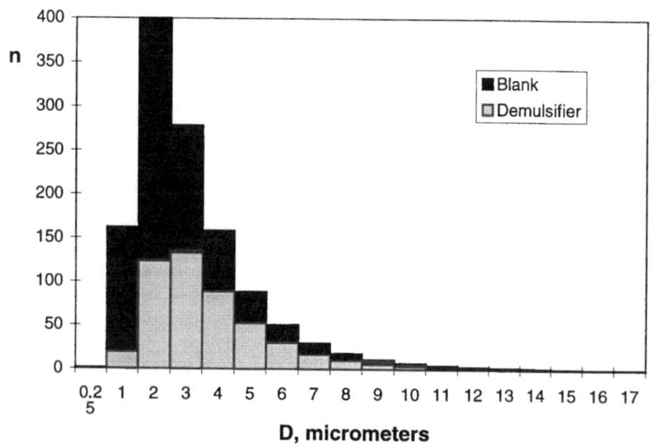


**Figure 11** Effect of a demulsifier; a phenolic resin alkoxylate commercial demulsifier accelerates water/crude oil resolution. The demulsifier was added to 50 ppm to a 40% (v/v) water/oil [7.8 wt% asphaltene in 30/70 (v/v) toluene/decane]. The histogram shows the DSDs of the emulsion with demulsifier and the reference emulsion after about 2 h.

It is often more useful to apply cumulative distributions to illuminate characteristic features of and difference between datasets. The cumulative distribution adds the contents of the next class to the sum of all previous classes, yielding the well-known S-curve for normal or lognormal distributions. Figure 12 shows the normalized cumulative volumes of four emulsions with varying contents of surface-active matter. Asphaltenes and resins are classes of large, surface-active molecules found in crude oil. These are expected to be responsible for the stabilization of water droplets mixed into the oil, and the relative amounts and properties of asphaltenes and resins present in the oil will affect their state and emulsifying behavior. Figure 12 shows how the introduction of a resinous fraction may improve the emulsifying power of an asphaltene surfactant fraction. All emulsions were prepared as a 40% (v/v) solution of 3.5% NaCl in a 70/30 (v/v) decane/toluene oil with asphaltene/resin. The asphaltene fraction was the pentane-insoluble part of a crude oil, while the resin fraction was adsorbed from the pentane eluate on to silica, then washed with benzene and desorbed with a methanol/dichlorodecane mixture. As can be seen in Fig. 12, a "high" asphaltene content (1.5 wt% of the dispersed phase) can stabilize a larger interfacial area than a "low" concentration (0.5 wt%), putting a greater part of the dispersed volume in small droplets. The "low" asphaltene content emulsion also shows that a great part of the dispersed volume is found within a few large droplets, reducing the accuracy of the DSD determination. Introducing a resinous fraction, a "low" concentration (6.31% the mass of asphaltene) has little effect on the relative distribution of droplet sizes within the two emulsions, while a "high" concentration (25%) further increases the emulsifying power of the asphaltene fraction, resulting in a still higher part of the dispersed volume to be found amongst small droplets. It has been suggested (85) that resins may assist the inclusion of asphaltene aggregates into the water/oil interface, thus improving its stabilization. A 70/30 (v/v) decane/toluene oil contains both monomeric and aggregated asphaltene; aromatic toluene is a good asphaltene solvent, while decane is not.

Measurement of small droplets is more prone to error than measurements of large (though undistorted, see below) droplets. However, given a 1-μm class width and 10% error in the diameter of a 2-μm droplet [D ∈ (1.8–2.2 μm)], the measurement will still be recognized in the histogram as an element in the 2-μm class. In a common VEM set-up with a 60× magnification microscope objective a typical digital resolution is of the order 0.1 μm/pixel. A 10% measurement error for a 1-μm droplet is then the equivalent of 1 pixel, 2 pixels for a 2-μm droplet, and so on. As a consequence of this, it is clear that the relative error of a measurement decreases with increasing droplet size. Figure 13 shows the problem of approaching the digital resolution limit.

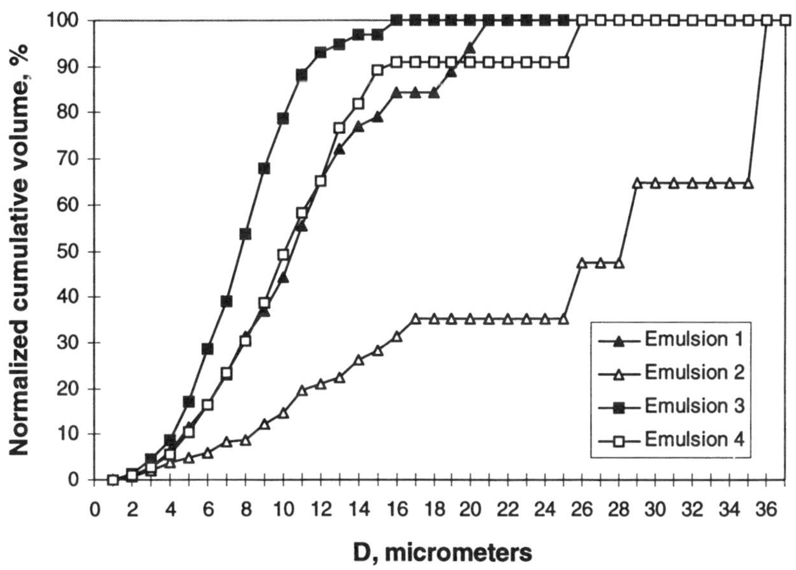


**Figure 12** Water droplets in model oils with asphaltene and resin fractions extracted from a crude oil; normalized cumulative volume showing the effect of asphaltene and resin content. Key: emulsion 1 – high asphaltene, no resin; 2 – low asph., no resin; 3 – high asph., high resin; 4 – high asph., low resin.

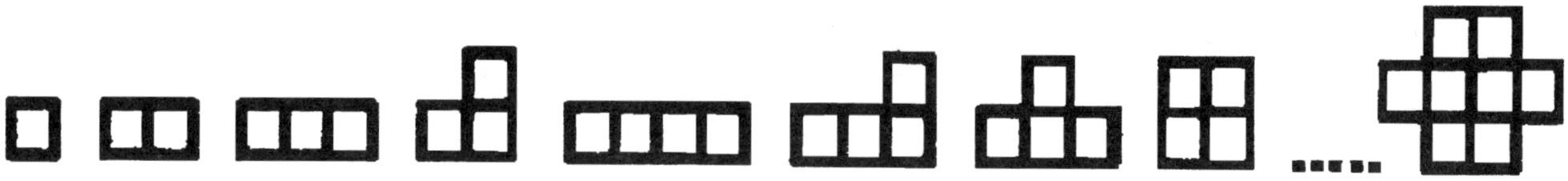

Figure 13 Example of decreasing reliability of measurement of small droplets when approaching the digital resolution of the image (example: 0.2326 μm/pixel): 1 pixel renders a Waddle disk diameter of 0.25 μm; 4 pixels – 0.5 μm; 8 pixels – 0.75 μm. All objects will register within the 0.5-μm class (given 0.5-μm width.)

As droplets grow larger, gravity can affect their shape. This is also a function of interfacial tension – higher interfacial tension acts to uphold the spherical shape. This means that there is an upper limit above which measurements will be increasingly marred by error. When observing droplets residing on the cell wall in a direction parallel to the direction of gravity, which is normally the case, the droplets will appear to be circular, but the diameter of the oblate will be larger than that of the volume-equivalent sphere, as illustrated in Fig. 14.

Often the brightness levels characterizing the droplet outline may vary with droplet size. This can influence the thresholding process, where the outer boundaries of the droplets may be wrongly set, causing underestimation of the size. Again, this underlines the importance of proper image preprocessing prior to analysis.

Due to the directness of the VM method, the exactness of the single measurements should be high. However, to describe the true shape of the population profile a high number of measurements is needed for statistical reliability. To achieve an error of 5% at the 95% confidence level, 740 droplets must be counted (86).

Different representations of the data will be differently influenced by missing data, which is, typically, an undercounting of droplets within the extreme size classes of the population. When using a mass- or volume-based distribution, an underestimation of droplets at the upper end of the size scale may severely reduce the correctness of the distribution, as one 20-μm droplet has the mass and volume of 1000 2-μm droplets. It is easily realizable that if 1000 droplets are counted from, say, three or four images, the number of large droplets counted may not be truly representative of the sample. This will cause a shift in the distribution. However, it is to some extent possible to perform a mathematical fit of the data set (when key features of the distribution function are known), thus reducing the adverse influence of missing data. Such a fit is shown in Fig. 15.

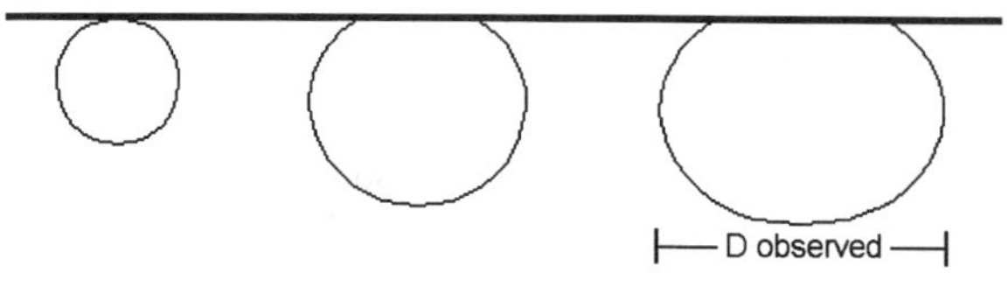


Figure 14 Gravity-induced deformation of droplets.

## VII. SUMMARY

The use of microscopy for emulsion studies is well established and the technique is generally regarded as a reliable way of generating, for example, DSDs. Traditionally, a labor demanding and tedious task, emerging image-analysis software technology provides opportunities for automated or partially automated droplet counting and measurement. However, alternative methods are often favored owing to a higher degree of simplicity of operation (which in parallel makes them quicker). Still, microscopy remains the only technique that offers direct observation of the sample, a feature that allows a greater control of sample state – e.g., degree of flocculation – which is unrivalled by any other technique. Another important advantage is the ability to follow the behavior of single droplets or a set of droplets, which creates opportunities for studies of droplet–droplet interactions. The examples included in the chapter attempt to underline the sensitivity and versatility of derived methods. Through developments in the fields of optics and image enhancement and analysis, VM will find continuingly expanding applicability within very diverse research disciplines.

## ACKNOWLEDGMENTS

The technology program Flucha, financed by The Norwegian Research Council (NFR) and the oil

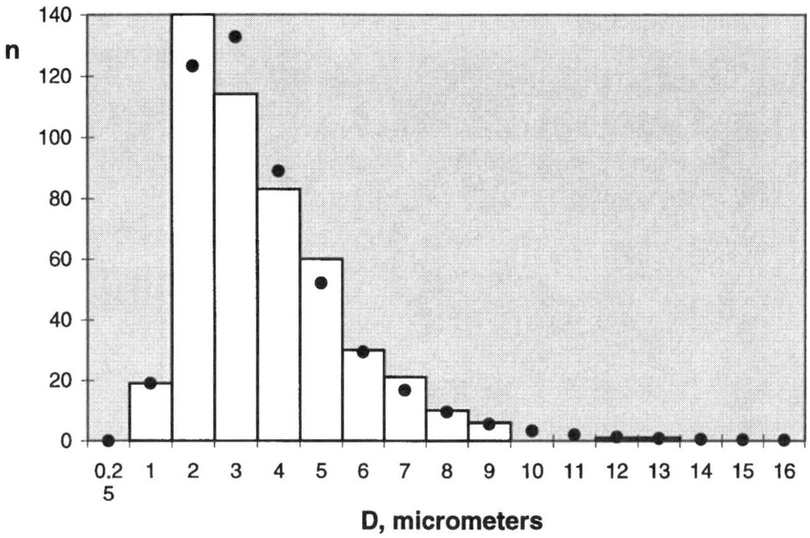


**Figure 15** Curve fitting of data to a lognormal distribution function.

industry, is acknowledged for a PhD grant and financial support.

## REFERENCES

1. S Bradbury. Microsc Anal (May): 7–12, 1990.
2. P Walstra, H. Oortwijn. J Colloid Interface Sci 29: 424–431, 1969.
3. A Takamura, S Noro, S Ando, M Koishi. Chem Pharm Bull 25: 2644–2649, 1977.
4. J Drelich, G Bryll, J Kapcynski, J Hupka, JD Miller, FV Hanson. Fuel Process Technol 31: 105–113, 1992.
5. K Eberth, J Merry. Int J Pharm 14: 349–353, 1983.
6. O Flint. Microsc Anal (March): 19–23, 1991.
7. RA Mohammed, AI Bailey, PF Luckham, SE Taylor. Colloids Surfaces 83: 261–271, 1994.
8. C Orr. In: P Becher, ed. Encyclopedia of Emulsion Technology. Vol 3. New York: Marcel Dekker, 137–169, 1988.
9. A Takamura, S Noro, S Ando, M Koishi. Chem Pharm Bull 25: 2617–2623, 1977.
10. RH Müller, S Heinemann. Clin Nutr 11: 223–236, 1992.
11. T Kubo, S Tsukiyama, A Takamura, I Takashime. Yakugaku Zasshi (Japan) 91: 518–521, 1971.
12. SK Mason, K May, S Hartland. Colloids Surfaces 96: 85–92, 1995.
13. C-J Lee, S-S Wang, C-C Chan. J Chin Inst Chem Eng 26: 263–275, 1995.
14. A Bhardwaj, S Hartland. J Disp Sci Technol 15: 133–146, 1994.
15. EE Isaacs, H Huang, AJ Babchin, RS Chow. Colloids Surfaces 46: 177–192, 1990.
16. M Sato. IEEE Trans Ind Applic 27: 316–322, 1991.
17. M Deitel, KL Friedman, S Cunnane, PJ Lea, A Chaiet, J Chong, B. Almeida. J Am Coll Nutr 11: 5–10, 1992.
18. FD Rumscheidt, SG Mason. J Colloid Sci 16: 238–261, 1965.
19. RD Hamill, RV Petersen. J Pharm Sci 55: 1268–1274, 1966.
20. RD Steele, JE Halligan. Sep Sci 9: 299, 1974, 299–311, 1974.
21. DD Eley, MJ Hey, JD Symonds. Colloids Surfaces 32: 87–101, 1988.
22. Y Otsubo, RK Prud'homme. Rheol Acta 33: 303–306, 1994.
23. M Zerfa, BW Brooks. Chem Eng Sci 51: 3591–3611, 1996.
24. N Garti. Colloids Surfaces 123/124: 233–246, 1997.
25. R Pal. AIChE J 42: 3181–3190, 1996.
26. PJ Halling. CRC Crit Rev Food Sci Nutr 15: 155–203, 1981.
27. M Britten, HJ Giroux. J Food Sci 56: 792–795, 1991.
28. GH Hanna, KM Larson. Ind Eng Chem Prod Res Dev 24: 269–274, 1985.
29. Ø Sæther, SS Dukhin, J Sjöblom, Ø Holt. Colloid J 57: 793–799, 1995.
30. Ø Holt, Ø Sæther, J Sjöblom, SS Dukhin, NA Mishchuk. Colloids Surfaces 141: 269–278, 1998.
31. B Balinov, O Urdahl, O Söderman, J Sjöblom. Colloids Surfaces 82: 173–181, 1994.
32. P Jokela, PDI Fletcher, R Aveyard, JR Lu. J Colloid Interface Sci 134: 417–426, 1990.

33. I Fourel, JP Guillement, D Lebotlan. J Colloid Interface Sci 169: 119–124, 1995.
34. C Solans, R Pons, S Zhu, HT Davis, DF Evans, K Nakamura, H Kunieda. Langmuir 9: 1479–1482, 1993.
35. AW Pacek, IPT Moore, RV Calabrese, AW Nienow. Trans Inst Chem Eng A: Chem Eng Res Des 71: 340–341, 1993.
36. AW Pacek, AW Nienow, IPT Moore. Chem Eng Sci 49: 3485–3498, 1994.
37. AW Pacek, AW Nienow. Trans Inst Chem Eng A: Chem Eng Res Des 73: 512–518, 1995.
38. NA Mishchuk, SV Verbich, SS Dukhin, Ø Holt, J Sjöblom. J Disp Sci Technol 18: 517–537, 1997.
39. TC Scott, WG Sisson. Sep Sci Technol 23: 1541–1550, 1988.
40. AW Pacek, IPT Moore, AW Nienow, RV Calabrese. AIChE J 40: 1940–1949, 1994.
41. Ø Holt, Ø Sæther, J Sjöblom, SS Dukhin, NA Mishchuk. Colloids Surfaces 123/124: 195–207, 1997.
42. SI Pather, SH Neau, S Pather. J Pharm Biomed Anal 13: 1283–1289, 1995.
43. AH Kamel, SA Akashah, FA Leeri, MA Fahim. Comput Chem Eng 11: 435–439, 1987.
44. RD Hazlett, RS Schechter, JK Aggarwal. Ind Eng Chem Fundam 24: 101–105, 1985.
45. B Balinov, O Söderman, T Wärnheim. J Am Oil Chem Soc 71: 513–518, 1994.
46. X Li, JC Cox, RW Flumerfelt. AIChE J 38: 1671–1674, 1992.
47. Y Wang, S Bian, D Wu. Pestic Sci 44: 201–203, 1995.
48. Ø Sæther, J Sjöblom, SV Verbich, NA Mishchuk, SS Dukhin. Colloids Surfaces 142: 189–200, 1998.
49. UT Lashmar, JP Richardson, A Erbod. Int J Pharm 125: 315–325, 1995.
50. WL Lammers, HJ van der Stege, P Walstra. Neth Milk Dairy J 41: 147–160, 1987.
51. B Kachar, DF Evans, BW Ninham. J Colloid Interface Sci 100: 287–301, 1984.
52. K Florine-Casteel. Biophys J 57: 1199–1215, 1990.
53. JC Crocker, DG Grier. J Colloid Interface Sci 179: 298–310, 1996.
54. Ø Holt, Ø Sæther, J Sjöblom, SS Dukhin, NA Mishchuk. Colloids Surfaces 141: 269–278, 1998.
55. Ø Sæther, J Sjöblom, SV Verbich, SS Dukhin. J Disp Sci Technol 20: 295–314, 1999.
56. SV Verbich, SS Dukhin, A Tarovski, Ø Holt, Ø Sæther, J Sjöblom. Colloids Surfaces 123/124: 209–223, 1997.
57. H Matsumura, K Watanabe, K Furusawa. Colloids Surfaces 98: 175–184, 1995.
58. PT Spicer, W Keller, SE Pratsinis. J Colloid Interface Sci 184: 112–122, 1996.
59. SR Deshiikan, KD Papadopoulos. J Colloid Interface Sci 174: 302–312, 1995.
60. W Bartok, SG Mason. J Colloid Sci 14: 13–26, 1959.
61. J Bongers, H Manteufel, K Vondermassen, H Versmold. Colloids Surfaces 142: 381–385, 1998.
62. JC Crocker, DG Grier. Phys Rev Lett 73: 352–355, 1994.
63. A Bhardwaj, S Hartland. Ind Eng Chem Res 33: 1271–1279, 1994.
64. KL Alexander, D Li. Colloids Surfaces 106: 191–202, 1996.
65. P Cheng, D Li, L. Boruvka, Y Rotenberg, AW Neumann. Colloids Surfaces 43: 151–167, 1990.
66. D Li. Colloids Surfaces 116: 1–23, 1996.
67. SS Susnar, HA Hamza, AW Neumann. Colloids Surfaces 89: 169–180, 1994.
68. DY Kwok, P Chiefalo, B Khorshiddoust, S Lahooti. ACS Symp Ser 615: 374–386, 1995.
69. SE Taylor. Colloids Surfaces 29: 29–51, 1988.
70. R. Isherwood, BR Jennings, M Stankiewicz. Chem Eng Sci 42: 913–914, 1987.
71. H Førdedal, E Nodland, J Sjöblom, OM Kvalheim. J Colloid Interface Sci 173: 396–405, 1995.
72. H Førdedal, Y Schildberg, J Sjöblom, J-L Volle. Colloids Surfaces 106: 33–47, 1996.
73. H Kallevik, OM Kvalheim, J Sjöblom. J Colloid Interface Sci 225: 494–504, 2000.
74. RD Allen, GB David, G Nomarski. Z Wiss Mikrosk 69: 193, 1969.
75. M Spencer. Fundamentals of Light Microscopy. New York: Cambridge University Press, 1982.
76. EM Chamot. Elementary Chemical Microscopy. New York: John Wiley, 1915.
77. DJ Shaw. Introduction to Colloid and Surface Chemistry. 4th ed. Oxford: Butterworth-Heinemann, 1991.
78. PM Cooke. Analyt Chem 64: R219–R243, 1992.
79. RD Allen. Annu Rev Biophys Chem 14: 265, 1985.
80. T Takei, A Yamazaki, T Watanabe, M Chikazawa. J Colloid Interface Sci 188: 409–414, 1997.
81. C Orr. In: P Becher, ed. Encyclopedia of Emulsion Technology. Vol 1. New York: Marcel Dekker, 369–404, 1983.
82. T Allen. Particle Size Measurement. 2nd ed. London: Chapman & Hall, 1974.
83. I Bajsic, B Blagojevic. Strojniski Vestnik [Mech Eng J] 36: E1–E8, 1990.
84. HHG Jellinek. J Soc Chem Ind 69: 225, 1950.
85. JD MacLean, PK Kilpatrick. J Colloid Interface Sci 196: 23–34, 1997.
86. WJ Dixon, FJ Massey Jr. Introduction to Statistical Analysis. 3rd ed. New York: McGraw-Hill, 1969, p 550.

# 16

# Lignosulfonates and Kraft Lignins as O/W Emulsion Stabilizers Studied by Means of Electrical Conductivity

**Stig Are Gundersen**
*University of Bergen, Bergen, Norway*

**Johan Sjöblom**
*Statoil A/S, Trondheim, Norway*

## I. INTRODUCTION

### A. Creaming of Emulsions

Emulsions are thermodynamically unstable systems and will, as a function of time, separate to minimize the interfacial area between the oil phase and the water phase. If a density difference exists between the dispersed and continuous phases, dispersed droplets experience a vertical force in a gravitational field. The gravitational force is opposed by the fractional drag force and the buoyancy force. The resulting creaming rate $v_0$ of a single droplet is given by Stokes law:

$$v_0 = \frac{2r^2(\rho_1 - \rho_2)}{9\eta} g \tag{1}$$

in which $r$ is the hydrodynamic radius of the droplet, $\rho_1$ and $\rho_2$ are the densities of the dispersed and the continuous phases, respectively, $\eta$ is the macroscopic shear viscosity of the continuous phase, and $g$ is the gravitational constant. Stokes law has several limitations and is strictly applicable only for noninteracting spherical droplets at low concentration with a monodisperse droplet size distribution. Prediction or calculation of creaming rates in concentrated emulsions where other than hydrodynamic factors come to account is therefore complicated. Conductivity measurements have appeared to be a suitable method for determination of emulsion stability in such systems. The conductivity of emulsifons is sensitive to small changes in the volume frame of the dispersed phase, and the technique has previously been applied to predict the inversion points of emulsions (1, 2). Also, studies of sedimentation processes, creaming stability, and phase separation, utilizing this technique have been reported (3–5).

In this chapter two methods based on conductivity measurements were used to determine creaming profiles and creaming rates of water-continuous emulsions stabilized with lignosulfonates (LSs) and Kraft lignins. In the first part the influence of LS and Kraft lignin concentration on emulsion stability was studied. These emulsions had a narrow dispersion band when creaming, which made it possible to read the amount of water separated from the emulsions directly from the creaming profiles. This was achieved by designing the electrode in such a way that the conductivity was measured in a stepwise manner as the creaming progressed. In the second part the influence of electrolyte concentration on emulsion stability was studied. In this part, two pairs of electrodes were used to measure the conductivity progressively in the time interval under study, and the stability was reported in terms of initial creaming rates and as the change in specific conductivity as a function of time. The two methods of measuring emulsion stability are discussed in Secs II and III, respectively.

### B. Lignosulfonates and Kraft Lignins

Lignosulfonates and Kraft lignins are isolated from spent liquors used in the sulfite pulping and Kraft pulping process, respectively (6). Lignosulfonates are cross-linked polydisperse polyelectrolytes in which the molecules are compact spheres in aqueous solutions (7). The molecule contains sulfonate groups as well as carboyxlic, phenolic, and methoxyl groups, and the basic repeating building unit in the molecule is a phenylpropane derivative (8). The structure may be nonuniform with regard both to number and distribution of anionic groups, and also in the structure of the hydrocarbon backbone. Owing to the ionic groups in the interior of the molecule, LSs show typical polyelectrolyte expansion, where the LS molecule swells or shrinks as the concentration of the counterions varies from low to high, respectively (9, 10). The charged sulfonate groups near the surface of the molecule matrix makes LSs readily soluble in water (11). Purified LSs have found widespread practical applications because of their dispersing, stabilizing, binding, and complexing properties (12–14). Like LSs, Kraft lignins are cross-linked polydisperse polyelectrolytes, and sulfonated Kraft lignins find similar uses as LSs.

### C. Lignosulfonates and Kraft Lignins as Emulsion Stabilizers

Several authors have studied the colloidal properties of particulate dispersions stabilized with Kraft lignins and LSs (15–18). From a commercial point of view the dispersing, stabilizing, binding, and complexing properties of these polyelectrolytes have made lignosulfonates useful in a range of practical applications (19–21). Several studies have also revealed that LSs are exceptional oil-in-water (O/W) (22, 23) emulsion stabilizers. Lignosulfonates have also been utilized in technical applications such as viscosity controllers in oil-well drilling fluids (24) and in improvement of the oil recovery efficacy (25–30).

Although LS does not form micelles, the molecule has both hydrophilic and lipophilic moieties. However, those two parts are not separated in a way that promotes high surface activity. The nonsolubility of LS in aliphatic and aromatic hydrocarbons, and the lack of a pronounced surface and interfacial tension-lowering properties, indicates that LS adsorbs at the oil–water interface rather than in the interface (31). A solution pressure has to be developed by the LS to force it into the interface. This means that a relatively high amount of LS must be added to give stable emulsions. The semirigid film gives rise to mechanical, steric, and electrostatic stabilization (12, 15). The solubility of LS is reduced in electrolyte-contaminated solutions, and this will force more LS to the oil–water interface. A reduction in emulsion stability, due to a reduction in the zeta potential on the oil droplets, will be compensated for by an increase in the condensed-layer adsorption. This leads to stable emulsions even in saturated salt solutions (12).

Lignosulfonates are of a highly polydisperse nature. There are indications that the low molecular weight fractions associate in solution and that high molecular weight fractions have a higher degree of molecular branching than the lower fractions (32, 33). Several investigations have shown that the LSs can function as dispersants or flocculants, depending on their molecular weight (13, 17, 18). Low molecular weight LSs act as stabilizers, while higher molecular weight LSs are of sufficient molecule length to enable them to be adsorbed on to the surface of adjacent particles. This bridging will promote flocculation and destabilize the dispersion. Other authors have reported better dispersant properties for high molecular weight LSs in concentrated kaolin suspensions (17). The influence of molecular weight on dispersant properties indicates that this molecular property will also affect the stabilities of water-continuous emulsions. Adsorption of LS on to negatively charged polystyrene latex surfaces indicates that the adsorption process is not only governed by electrostatic interactions, but also by a sort of hydrophobic interaction. It has been proposed that the hydrophobic phenlypropane ring may be of significance in this respect (17).

## II. CONDUCTIVITY MEASUREMENTS AND INFLUENCE OF KRAFT LIGNIN AND LIGNOSULFONATE CONCENTRATION ON EMULSION STABILITY

### A. Experimental

#### 1. Materials and Sample Preparation

The sodium LSs (UP364, UP365, UP366, UP407, and UP411) and the Kraft lignin (Diwatex UP329) referred to in this section were delivered by Borregaard Lignotech. The properties of the lignin samples are described in a previous work (22) and here summarized in Tables 1, 2, and 3. Details of the emulsion-prepara-

**Table 1** Properties of the Sodium Lignosulfonate Fractions

| Sample description | UP364 | UP365 | UP366 | UP407 | UP411 |
|---|---|---|---|---|---|
| pH (10% solution) | 9.4 | 9.6 | 8.8 | | |
| Solids (%) | 92 | 96.5 | 97.2 | | |
| Calcium (%) | 0.11 | 0.02 | 0.06 | | |
| Sodium (%) | 5.94 | 6.03 | 7.17 | | |
| Total S (%) | 6.2 | 6.1 | 7.1 | 4.7 | 6.4 |
| Inorg. S (%) | 0.01 | 0.07 | 0.28 | 0.2 | 0.1 |
| Org. S (%) | 6.1 | 6.0 | 6.8 | 4.5 | 6.3 |
| Methoxyl (%) | 11.9 | 12.1 | 9.2 | 12.7 | 10.8 |
| Carboyxl (%) | 4.5 | 5.4 | 5.9 | | |
| Phenolic OH (%) | 2.4 | 2.4 | 2.4 | | |
| Reducing Sugars (%) | 0.6 | 0.4 | 4.3 | | |
| Mol. weight (HPLC) | | | | | |
| $M_w$ | 62,700 | 76,400 | 9300 | | |
| $M_n$ | 13,900 | 16,300 | 5100 | | |

tion procedures are described in the same paper, although it is worth mentioning that the oil/water ratio was held constant [40/60 (w/w)], and that the LS and Kraft lignin concentrations are reported on the basis of the internal phase weight.

### 2. Conductivity Measurements and Electrode Design

Immediately after production, a cylindrical glass container ($d = 25$ mm; $h = 185$ mm) containing the electrode was filled with 90 ml of the emulsion. The electrode was connected to a Hewlett-Packard 4263B LCR Meter operating at a frequency of 1 kHz, and the conductivity was measured every 60 s under constant-temperature conditions ($25.0^\circ \pm 0.1^\circ$C). The electrode was coated with gold to eliminate polarization effects and corrosion of the electrode material.

The electrode was made up of two 210-mm iron rods of 1 mm diameter. The lower 30 mm of the electrode was divided into insulated and noninsulated areas in the following way: 0–7, 10–17, and 20–27 mm and from 30 mm up to the connection points were insulated areas, while the areas between were noninsulated. Teflon was used as insulating material. Figure 1 demonstrates how the electrode was placed in the measuring cell.

## B. Results and Discussion

### 1. Stepwise Conductivity Measurements

The creaming process in the emulsions studied resulted in a distinct boundary between the water-rich and oil-rich phases. A freshly prepared emulsion had a white appearance. After a few minutes a dark-brown zone containing LS and water could be seen in the lower part of the cell. Samples from the water-rich phases were studied using VEM (video-enhanced microscopy)

**Table 2** Molecular Weight of Desulfonated and High Sulfonated Sodium Lignosulfonates

| Molecular weight (g/mol) | UP407 (%) | UP411 (%) |
|---|---|---|
| > 59,000 | 14.9 | 2.8 |
| < 59,000 | 85.1 | 97.2 |
| < 38,000 | 72.2 | 89.2 |
| < 16,000 | 47.9 | 70.9 |
| < 6600 | 26.2 | 48.8 |
| < 2755 | 12.2 | 26.0 |
| < 1740 | 8.3 | 16.6 |

**Table 3** Properties of the Kraft Lignin Fraction

| Sample description | Diwatex UP329 | Molecular weight (g/mol) | (%) |
|---|---|---|---|
| pH (10% solution) | 10.6 | > 12,500 | 30.8 |
| Solids (%) | 97.9 | < 12,500 | 69.2 |
| Total S (%) | 5.5 | < 10,000 | 58.7 |
| Inorg. S (%) | 0.6 | < 7000 | 45.5 |
| Org. S (%) | 4.8 | < 5000 | 33.1 |
| Methoxyl (%) | 10.2 | < 3000 | 19.2 |
| Carboxyl (%) | 8.1 | < 2000 | 10.3 |
| Phenolic OH (%) | 1.9 | < 1000 | 1.5 |

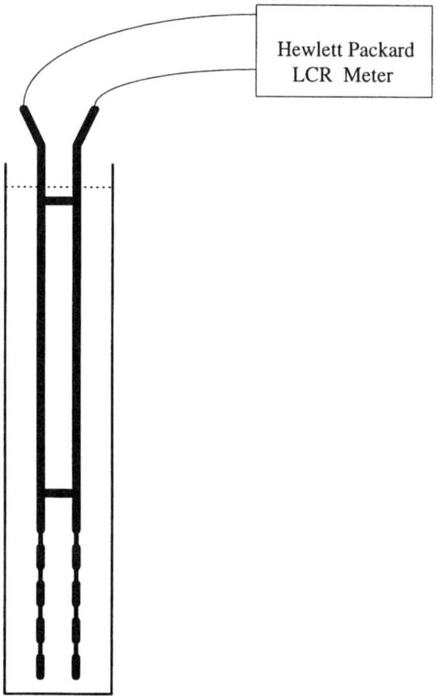


**Figure 1** Schematic figure of the electrode where the isolated and nonisolated areas are shown.

(34, 35), and it was shown that the water phases contained only traces of oil. The electrode was designed to utilize the narrow extension of the zone separating the water-rich and oil-rich phases, and the dividing of the electrode into alternating insulated and noninsulated parts made it possible to follow this zone as it propagates in the measuring cell as a function of time. Figure 2 demonstrates how the conductivity in a 40/60 (w/w) oil/water emulsion stabilized with 2.0% UP366 increases as the creaming process progressed. The trend in the measured conductivity of this sample is representative for all the emulsions studied. The data are given as the normalized conductivity, $G$:

$$G_{\mathrm{nt}} = \frac{(g_{\mathrm{nt}} - g_{\mathrm{initial}})}{(g_{\mathrm{end}} - g_{\mathrm{initial}})} \tag{2}$$

in which $G_{\mathrm{nt}}$ is the normalized conductivity at a given time $t$, $g_{\mathrm{nt}}$ is the measured conductivity at the same time $t$, $g_{\mathrm{initial}}$ is the measured conductivity at time $t = 0$ and $g_{\mathrm{end}}$ is the last value measured. When the measurements start at time $t = 0$, it is reasonable to assume that each of these levels has an equal contribution to the total conductivity in the system. This initial situation is rapidly changed as a result of the creaming process. The first plateau in Fig. 2 reveals the frontier moving along the first insulated area on the electrode. Only small changes in the sample conductivity are measured in this area. When the frontier reaches the first noninsulated area a dramatic increase in the con-

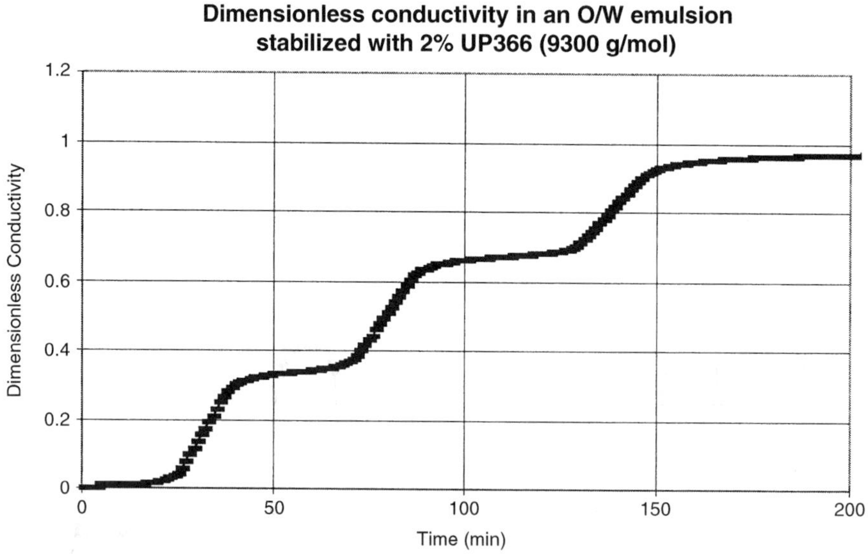


**Figure 2** Dimensionless conductivity in an O/W emulsion [40/60 w/w)] stabilized with 2.0% lignosulfonate (UP366) as a function of time.

ductivity is observed. The total conductivity of the sample will at this time be dominated by the changes that occur in this area. The conductivity profile flattens out when the frontier reaches the next insulated area, and so on.

The derivatives of the conductivity plot in Fig. 2 shows how fast the conductivity changes as a function of time. Figure 3 shows that there is a fast growth and a fast reduction when the water-rich phase reaches the noninsulated and the insulated area, respectively. Highest on each top there is a horizontal plateau where the derivative shows only small alteration. The plateau represents the linear increase in the conductance that can be seen in Fig. 2, and the midpoint of each plateau represents the phase boundary located in the middle of each noninsualted level.

Generally, the conductivity is directly proportional to $\Phi_{water}$ (36); VEM revealed that $\Phi_{water}$ in the water-rich phases was very close to 100%. This makes it possible to calculate the amount of water separated from the emulsions directly from the conductivity profiles when the fixed positions of the noninsulated areas (7–10, 17–20, and 27–30 mm) are known.

**Table 4** Separation of Water from O/W Emulsion Stabilized with Different Amounts of Lignosulfonate and Kraft Lignin

| Sodium lignosulfonate and Kraft lignin samples | 10% Separated water (min) | 20% Separated water (min) | 30% Separated water (min) |
|---|---|---|---|
| UP364 (0.5%) | 35 | 75 | 125 |
| UP364 (1.0%) | 35 | 80 | 140 |
| UP365 (2.0%) | 50 | 120 | 190 |
| UP365 (0.5%) | 30 | 70 | 115 |
| UP365 (1.0%) | 40 | 95 | 155 |
| UP365 (2.0%) | 50 | 115 | 180 |
| UP366 (1.0%) | 40 | 95 | 150 |
| UP366 (2.0%) | 35 | 85 | 145 |
| UP407 (0.5%) | 30 | 65 | 100 |
| UP407 (1.0%) | 35 | 75 | 125 |
| UP407 (2.0%) | 50 | 115 | 195 |
| UP411 (0.5%) | 25 | 60 | 100 |
| UP411 (1.0%) | 30 | 70 | 125 |
| UP411 (2.0%) | 45 | 105 | 185 |
| Diwatex UP329 (1.0%) | 40 | 90 | 150 |
| Diwatex UP329 (2.0%) | 70 | 150 | 240 |

## 2. Effect of LS and Kraft Lignin Concentration on Creaming Stability

The amount of water separated from different LS and Kraft lignin samples was calculated at three points in time. The results are listed in Table 4 and are given in percentages of the total water-cut of the emulsions. Although all the LSs and Kraft lignins studies are efficient O/W emulsion stabilizers, the fast creaming rates observed reflect a relatively high mean droplet size as a result of the moderate homogenization energy used.

The high molecular weight (UP364 and UP365) and the desulfonated and high sulfonated LSs (UP407 and

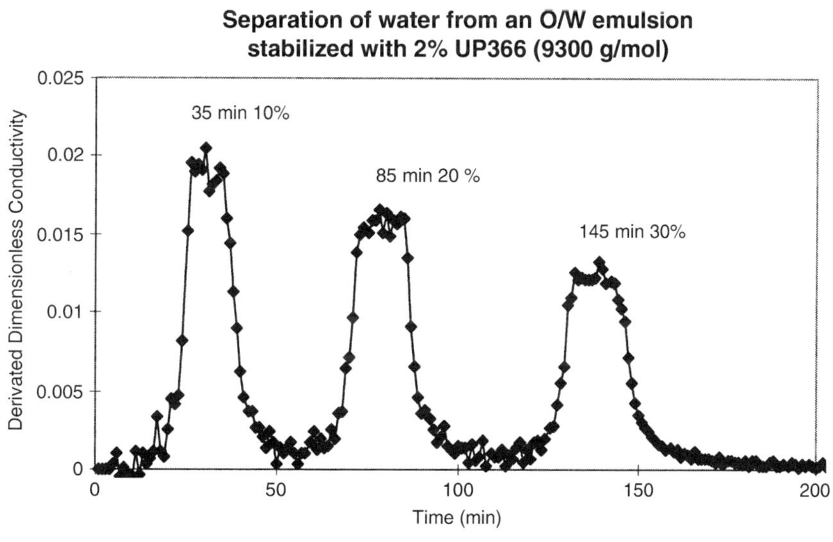


**Figure 3** Derivatives of the conductance plot shown in Fig. 2.

UP411, respectively) give rise to approximately identical emulsion stabilities. The stability is significantly improved when increasing the LS concentration, which indicates an increased adsorption on to the droplet surface when the LS concentration is high. For the low molecular weight LS (UP366) the picture is somewhat different. The creaming process is slightly more rapid, and the loss in stability caused by increasing the amount of LS indicates effective adsorption on to the droplet surface at low LS concentrations. The increased emulsion stability observed using high molecular weight LS could be due to multilayer adsorption, giving both steric and electrostatic stabilization (12, 13). Also, for the Kraft lignin studied (Diwatex UP 329) the creaming rate declines when increasing the polymer concentration. The 2% sample has the slowest creaming rate of all the emulsions studied. It has been shown that the adsorption of LS on to polystyrene latex particles increases when the degree of sulfonation is low (17). The level of emulsion stability obtained, and the low degree of sulfonation of the Kraft lignin fraction indicate effective adsorption on to the oil droplets. The structural differences that are known to exist between LSs and Kraft lignins could also be important in this respect (37).

VEM was used to study diluted samples from the oil-rich phases. Figures 4 and 5 show images of O/W emulsions stabilized with low and high molecular weight LSs, respectively. The images reveal a distinct difference in the oil droplet interaction in the two samples. The high molecular weight LS gives rise to a strongly flocculated system, indicating that the molecule is of sufficient length to be adsorbed on to more than one droplet, thereby promoting flocculation (13, 18, 33). The low molecular weight LS, on the other hand, gives rise to discrete droplets with a low degree of flocculation. This is also the case for the Kraft lignin, indicating that these molecules are of insufficient length to be adsorbed on to more than one droplet. For high molecular weight LS a stretched configuration has been reported at low concentrations, and a coiled configuration at higher concentrations due to the suppression of dissociation of the ionic groups (33). The creaming rates measured indicate a reduction in the bridging ability of the LS molecules in the more concentrated emulsions.

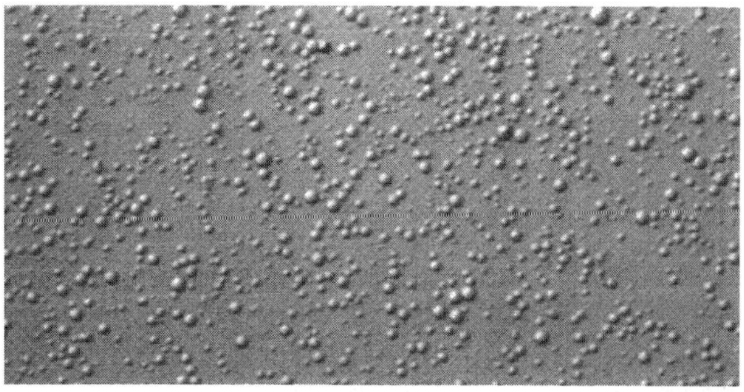

**Figure 4** Oil droplets stabilized with low molecular weight lignosulfonate (UP366); 1 cm on the image correspond to a distance of 15.5 μm.

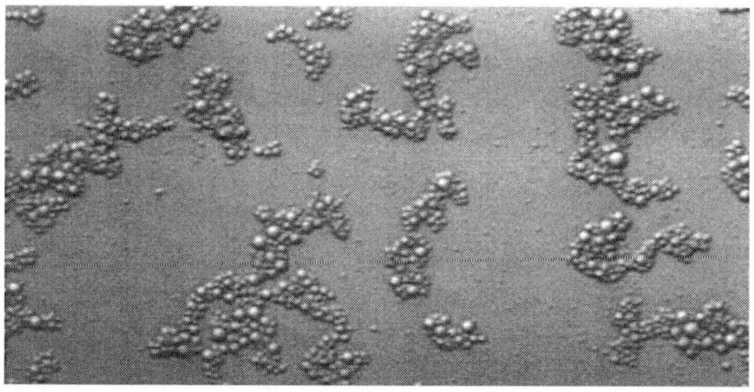

**Figure 5** Oil droplets stabilized with high molecular weight lignosulfonate (UP365); 1 cm on the image correspond to a distance of 15.5 μm.

### 3. Accuracy of Conductivity Measurements

The validity or accuracy of the conductivity measurement technique was verified by comparing results based on conductivity data with results based on visual inspections. The creaming process was studied in emulsions stabilized with LS and LS/hexadecylpyridinum chloride (CPC) complexes. The oil/water ratio and the sample-preparation procedure are the same as outlined in Sec. II.A.1. The homogenization time was 2 min, and the high molecular weight LS (UP364) was used as polyelectrolyte. The amounts of emulsifiers used are listed in Table 5, and are given in percentages of the internal phase weight.

Figure 6 shows how the dimensionless conductivity evolves as a function of time in the emulsions studied. In Fig. 7 the creaming profile for equivalent emulsions is given on the basis of visual observations. For each system the percentage of water separated from the emulsions was calculated at different stages in the creaming process. From the values listed in Table 5 it can be seen that there is a good correlation between data found by the two techniques. This is the case both when the creaming rate is fast, as in emulsions stabilized with LS, and in emulsions stabilized with LS/CPC complexes, where the creaming rate is slower. The deivation in the results is most likely due to uncertainty in

**Table 5** Separation of Water from O/W Emulsions Stabilized with Lignosulfonate (UP364) and Lignosulfonate/CPC Complexes Calculated from Conductivity Measurements and Visual Observations

| Sample | Time (min) | Separated water (%): conductivity measurements | Separated water (%): visual observations |
|---|---|---|---|
| UP364 | 35 | 17 | 22 |
| | 110 | 34 | 34 |
| | 185 | 43 | 43 |
| LS (1%) + CPC (0.14%) | 50 | 9 | 12 |
| | 230 | 27 | 27 |
| | 445 | 38 | 42 |
| LS (1%) + CPC (3.45%) | 80 | 10 | 12 |
| | 360 | 25 | 27 |
| | 730 | 37 | 42 |

the visual readings. To some extent there is also uncertainty connected with reproducing the emulsions.

## III. CONDUCTIVITY MEASUREMENTS AND INFLUENCE OF SALT ON EMULSION STABILITY

### A. Experimental

#### 1. Materials and Sample Preparation

In this section a further investigation of the LS fractions UP365 and UP366 described in Sec. II.A.1 was carried out. Additionally, a Kraft lignin fraction (Diwatex XP9), also delivered by Borregaard Lingotech, was studied. The properties of this fraction are summarized in Table 6. The LS and Kraft lignin concentration was 2% on the basis of the external phase weight. The polymer/salt mixture was dissolved in the aqueous phase before emulsification, and the emulsions had an internal phase weight of 30%.

#### 2. VEM and Droplet Size Measurements

The O/W emulsions were diluted with a continuous phase to a dispersed phase concentration of about

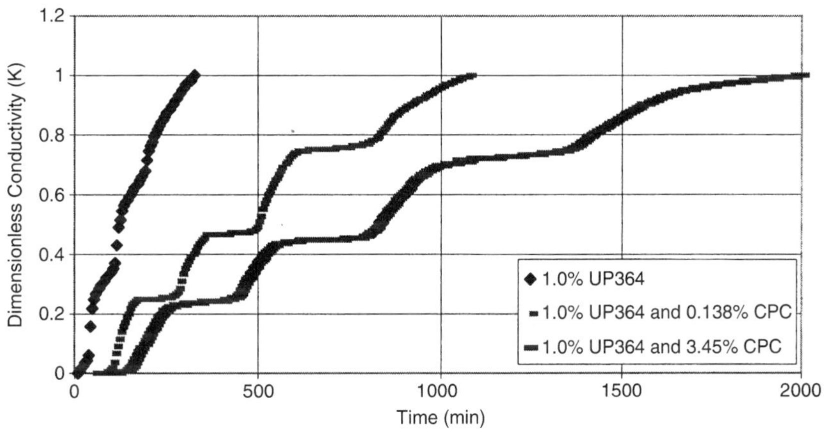


**Figure 6** Dimensionless conductivity in O/W emulsions [40/60 (w/w)] stabilized with lignosulfonate (UP364) and lignosulfonate/CPC complexes as a function of time.

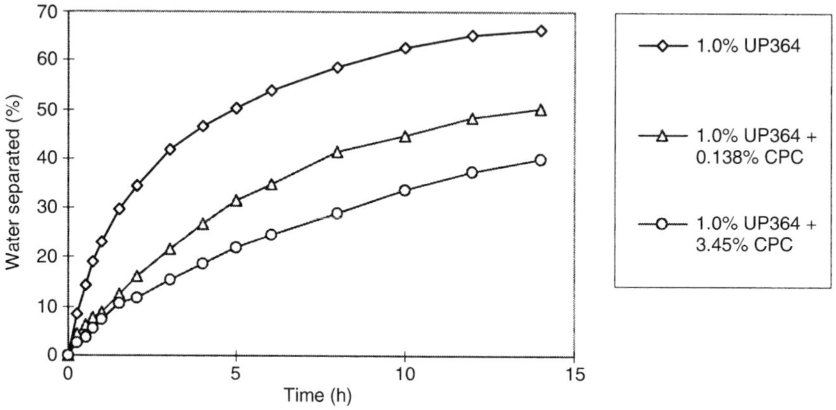


**Figure 7** Water separated from O/W emulsions [40/60 (w/w)] stabilized with lignosulfonate (UP364) and lignosulfonate/CPC complexes as a function of time.

1% (v/v) to allow a microscopy study of the droplet population. The diluted emulsion was turned repeatedly to ensure homogeneous redistribution of droplets and droplet aggregates throughout the entire volume. A microslide preparative technique was chosen to allow a long-term study of the sample with minimum risk of sample distortion through evaporation or contamination. In this technique, a small volume of the diluted emulsion was introduced into a flat, open-ended glass capillary of rectangular cross-section. When the ends were sealed, the prepared sample was protected and could be studied over an extended period, in this case with a Nikon Optiphot-2 microscope. For determination of droplet size, images of the dilute emulsion were digitized with a black/white CCD camera (Hitachi KP-160) and a framegrabber captureboard (Integral Technologies Flashpoint PCI) into a computer for image analysis. Due to the complexity of the images, featuring in some instances severe three-dimensional flocculation and reduced transparency as well as broad size distributions (prohibiting full automation of measurement), the analysis required an investment of manual labor.

**Table 6** Properties of the Kraft Lignin Fraction

| Sample description | Diwatex XP9 | Molecular weight | |
|---|---|---|---|
| | | (g/mol) | (%) |
| pH (10% solution) | 10.6 | > 12,500 | 43.5 |
| Solids (%) | 91.7 | < 12,500 | 56.5 |
| Total S (%) | 4.8 | < 10,000 | 46.1 |
| Inorg. S (%) | 1.3 | < 7000 | 35.1 |
| Org. S (%) | 3.5 | < 5000 | 26.9 |
| Methoxyl (%) | 8.4 | < 3000 | 16.8 |
| Carboxyl (%) | 9.5 | < 2000 | 9.6 |
| Phenolic OH (%) | 2.1 | < 1000 | 1.7 |

### 3. Conductivity Measurements and Electrode Design

The conductivity measurements were performed using the same instrumentation as described in Sec. II.A.2. Figure 8 shows how the electrodes were placed in the sample cell containing the emulsion. The two electrode pairs were designed to measure the conductivity in the lower and the upper part of the emulsion, respectively. This was achieved by insulating parts of the electrodes with Teflon tubes. By using this design the conductivity was measured in the lower 35 mm and the upper 35 mm of the sample container.

## B. Results and Discussion

### 1. Progressive Conductivity Measurements

Initially, in a freshly prepared emulsion, the oil droplets were homogeneously distributed in the entire sample volume. This situation was rapidly changed as a result of the creaming process. As the creaming process progressed, the conductivity in the lower part of the emulsion increased because of the decreased volume fraction of oil in this region. At the same time the conductivity in the upper part declined. The difference in the measured conductivities was cal-

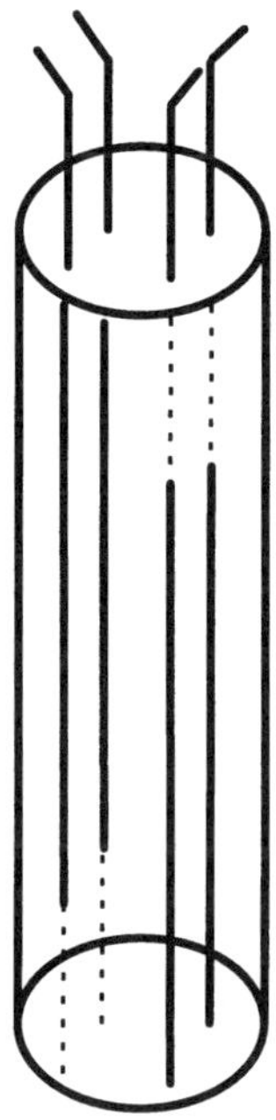

**Figure 8** Schematic views of the electrodes immersed in the measuring cell containing the emulsion.

culated, in terms of percentage, from the following equation:

$$\Delta C = (C_{\text{lower}} - C_{\text{upper}})/C_{\text{lower}} \times 100 \qquad (3)$$

The creaming profiles are presented as the deviation in specific conductivity as a function of time. Creaming rates were quantitatively determined by calculating the slope of the first linear part of the creaming profiles.

The validity and accuracy of the conductivity-measurement method were verified by comparing the signals from the electrodes at the initial stage. At this point an homogeneous emulsion should give equivalent conductivity readings in the whole sample volume. However, as a general trend, the initial conductivity was slightly higher in the upper part of the measuring cell. This could be due to formation of air bubbles when pouring the emulsion into the measuring cell. The measured values leveled after a few minutes and then $\Delta C$ progressively increased. Experimental data used for calculation of creaming rates and creaming profiles were therefore obtained after an initial stabilization period of 10 min. The reproducibility of the method was evaluated, and it was found, as shown in Fig. 9, that the deviation between creaming profiles of equivalent emulsions was negligible.

### 2. Salt Effect on Creaming Stability

Kraft lignin- and LS-stabilized emulsions were studied, using different electrolytes at different concentrations: no salt, $10^{-2}$ M NaCl, and $10^{-4}$ and $10^{-2}$ M $AlCl_3$, $FeCl_3$, and $Cr(NO_3)_3$. Figure 10 shows the typical creaming profiles obtained, and illustrates how these electrolyte affect the emulsion stability. Tables 7 and 8

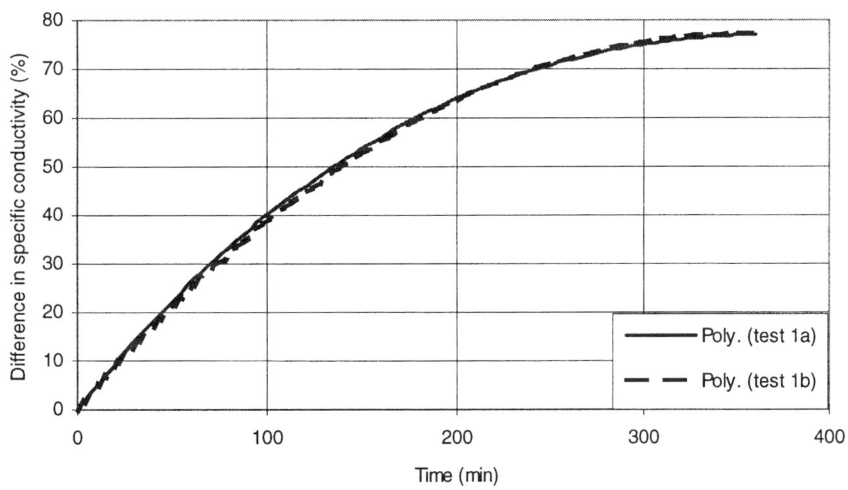


**Figure 9** Reproducibility of the conductivity measurements of lignosulfonate-stabilized emulsions measured at 25°C.

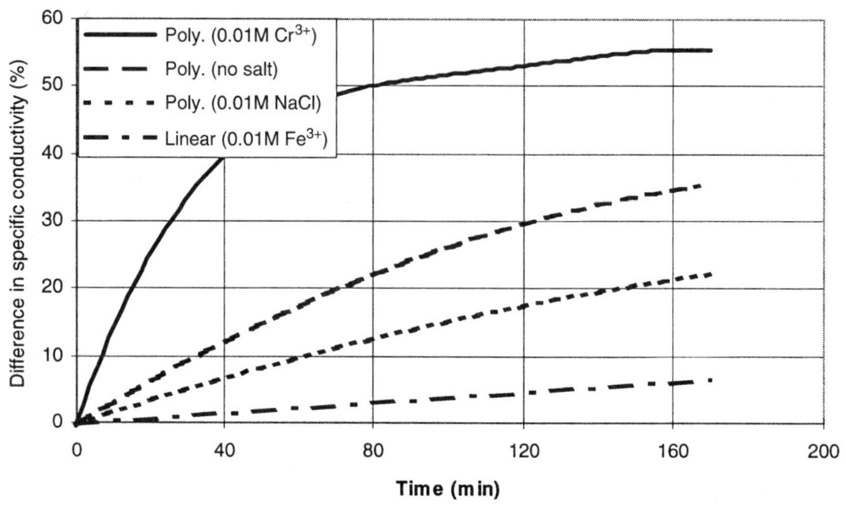


**Figure 10** Creaming profiles of emulsions stabilized with low molecular weight lignosulfonate (UP366) measured at 25°C.

summarize initial creaming rates and differences in specific conductivities ($\Delta C$) at two points in time for the LS- and the Kraft lignin-stabilized emulsions, respectively.

The creaming rates presented in Table 7 show that in the emulsions with no added salt, the low molecular weight LS fraction gives rise to creaming rates slower than those of the high molecular weight fraction. This observation reflects the better flocculating properties of UP365. Adding NaCl to the UP365 emulsion increased further the flocculation due to shielding of the electrostatic repulsion between the droplets. VEM showed single droplets and high numbers of dimers and trimers in the emulsion without salt, while flocs containing less than five droplets were negligible in the NaCl sample. VEM showed flocculation also in the UP366 samples but to a lower extent. Most likely this is because of a high fraction of molecules with insufficient molecular length to be adsorbed on to more than one droplet. It has been shown that both the size of the LS molecules and the adsorbed layer thickness decrease in electrolyte

**Table 8** Initial Creaming Rates and Changes in Specific Conductivities in Kraft Lignin-Stabilized Emulsions

| Concentration of electrolyte | Kraft lignin (XP9) | |
|---|---|---|
| | Slope | $\Delta C$ (160 min) |
| No salt | 0.002 | 0.4% |
| $10^{-2}$ M NaCl | 0.007 | 1.2% |
| $10^{-2}$ M $Al^{3+}$ | 0.01 | 1.7% |
| $10^{-2}$ M $Fe^{3+}$ | 0.01 | 1.6% |
| $10^{-2}$ M $Cr^{3+}$ | 0.005 | 0.8% |

**Table 7** Initial Creaming Rates and Changes in Specific Conductivities in Low and High Molecular Weight-Stabilized Emulsions

| Concentration of electrolyte | Low MW lignosulfonate (UP366) | | | High MW lignosulfonate (UP365) | | |
|---|---|---|---|---|---|---|
| | Slope | $\Delta C$ (80 min) | $\Delta C$ (160 min) | Slope | $\Delta C$ (80 min) | $\Delta C$ (160 min) |
| No salt | 0.34 | 23% | 35% | 0.63 | 34% | 43% |
| $10^{-2}$ M NaCl | 0.19 | 13% | 22% | 0.80 | 44% | 55% |
| $10^{-2}$ M $Al^{3+}$ | 2.90 | 55% | 65% | 0.02 | 1.5% | 3% |
| $10^{-2}$ M $Fe^{3+}$ | 0.04 | 3% | 6% | 0.02 | 2% | 4% |
| $10^{-2}$ M $Cr^{3+}$ | 1.63 | 50% | 56% | 0.01 | 1% | 2% |
| $10^{-4}$ M $Al^{3+}$ | 0.36 | 23% | 36% | 0.37 | 24% | 43% |
| $10^{-4}$ M $Fe^{3+}$ | 0.33 | 25% | 38% | 0.74 | 34% | 45% |
| $10^{-4}$ M $Cr^{3+}$ | 0.39 | 21% | 34% | 0.74 | 34% | 43% |

solutions (10, 17). The increased creaming stability observed when adding NaCl to the UP366 emulsion could be due to the formation of a condensed surface layer with a further reduction of the already moderate bridging ability of UP366.

Regarding coalescence, the two LS fractions also gave different stabilities. The high molecular weight LS gave stable emulsions both with and without NaCl added. while the separation of clear oil using the low molecular weight fraction was significant. The addition of salt increased the separation process considerably, and the amount of clear oil phase after 4 weeks was 60 and 10% (of the total oil cut) in the emulsions with and without NaCl, respectively. In the very dense top layer of the creamed emulsions the maintenance of the droplet stability requires an effective mechanical barrier to suppress coalescence. The instability of the emulsions using UP366 shows that the mechanical strength of the adsorbed LS layer is weak and that the electrostatic contribution to the stabilization is crucial to prevent coalescence. It is well known that adsorption preference occurs with respect to the molecular weight and that, at equilibrium, high molecular weight polymers adsorb preferentially over lower molecular weight ones (38, 39). The stability differences observed could imply that the 2% LS concentration is sufficient to give multilayer adsorption in the UP365 emulsions but not in the UP366 emulsions. The coalescence rate could also be affected by molecular diffusion: the Ostwald ripening effect (40–45). Ostwald ripening is observed when the oil from the emulsion droplets exhibits minimal solubility in the continuous phase. The chemical potential of the oil in the larger droplets is lower than in the smaller droplets, which leads to a diffusion transport of oil from the smaller toward the larger droplets. The importance of Ostwald ripening in the systems studied is not clear, but most likely it is of minor importance because of the very low solubility of Exxol D80 in the water phase. The diffusion across the interface may also be sterically impeded by the adsorbed LS layer (46). If the separation of clear oil was significantly affected by Ostwald ripening one would expect to observed an increase in the population of large droplets, reducing the fraction of the smaller droplets. However, measured droplet size distributions of the emulsion containing no salt revealed identical droplet populations in the freshly prepared and the 4-week samples, with droplet diameters of less than 2.5 μm. Droplets of diameter greater than 4,5 μm were not observed in the fresh emulsion, while approximately 10% of the dispersed volume was found in droplets of diameters greater than 4.5 μm in the aged emulsion. These results strongly indicate that the coalescence observed in the UP366 samples is governed by gravitation.

Metal salts at high concentrations had a pronounced effect on the creaming rates of the high molecular weight LS samples. As can be seen from Table 7, the $\Delta C$(160 min) values fall from about 44% in the emulsion containing no salt to about 2–4% in the samples containing metal salt. The initial creaming rates were also very efficiently reduced. When adding metal salt to the aqueous phases the pH may be lowered due to hydrolysis of the $[M(H_2O)_6]^{3+}$ ions to hydroxy species (47). The pH values of the aqueous phases are listed in Table 9. As the pH is lowered from about 9 in the pure LS solutions to about 3–5 in the solutions containing $10^{-2}$ M salt, the ionization of the carboxyl and sulfonate groups will be suppressed, although the sulfonate group will be appreciably dissociated also at low pH ($pK_a \approx 2$) (15). The suppression of dissociated groups increases the tendency of hydrogen-bond formation between LS molecules, and reduces the swelling of the molecules (15). The effect of pH variations on the emulsion stability was studied by adjusting the water phase containing UP365 to pH 4 with hydrochloloric acid, and it was found that the creaming stability at this pH was very close to that of the sample at pH 9.5. Also, the fact that the pH ranges from 3 to 5 in the LS/metal salt samples and that these emulsions gave rise to approximately identical creaming rates indicates that the LS-electrolyte interaction directs the emulsion stability. How this interaction alters the properties of the LS and the LS adsorption at the oil/water interface is of vital importance in this respect.

The increased emulsion stability indicates that the addition of metal ions modifies the high molecular weight LS fraction so that the flocculating properties are reduced. This was confirmed by studying VEM images. While the UP 365 sample without electrolyte

**Table 9** pH of the Aqueous Phase of the Lignosulfonate- and Kraft Lignin-Stabilized Emulsions

| Concentration of electrolyte | LS (UP 366): pH | LS (UP365): pH | KL (XP9): pH |
|---|---|---|---|
| No salt | 8.5 | 9.5 | 9.9 |
| $10^{-2}$ M NaCl | 9.0 | 9.7 | 9.8 |
| $10^{-2}$ M $Al^{3+}$ | 4.1 | 4.3 | 5.5 |
| $10^{-2}$ M $Fe^{3+}$ | 2.7 | 3.3 | 5.0 |
| $10^{-2}$ M $Cr^{3+}$ | 4.4 | 4.9 | 5.9 |

added was characterized by flocs existing of three to more than 50 droplets, the aluminum and chromium samples were characterized by discrete droplets. The iron sample was also much less flocculated than the pure LS emulsion, but flocks containing up to 10 droplets were observed. Also, a displacement of the DSDs (droplet size distributions) to smaller droplet diameters in the metal salt-containing emulsions would reduce the creaming rates. According to Stokes law the creaming rate of a single emulsion droplet is proportional to the square root of the droplet radius. Normalized DSDs, however, revealed very similar distributions for all the emulsions under study, with the main droplet population in the region 0.5–3.0 μm. Figure 11 shows the DSD of UP365-stabilized emulsions containing NaCl and different metal salts. The DSDs were based on approximately 1000 measured droplets for each sample. According to work done by Orr this brings the error to well below 5% at the 95% confidence level (48).

VEM was also used to study the colloidal stability of LS in the different solutions. Particulate LS was not observed in the samples containing $10^{-2}$ M electrolyte, although it was shown that coagulation occurred with further addition of metal salt and that UP365 was more sensitive than UP366 to coagulation. At a concentration of $2 \times 10^{-2}$ M Al the UP365 sample became turbid, and a rigid macroscopic network of coagulated LS was observed in the microscope. The iron and chromium concentrations had to be $6 \times 10^{-2}$ M to obtain the same effect. In comparison UP366 was not coagulated even at metal salt concentrations of $8 \times 10^{-2}$ M. This is in agreement with the findings of Nyman and Rose, showing that high molecular weight LS fractions are more sensitive to coagulation than are fractions at lower molecular weights (37). The strong coagulation quality of $Al^{3+}$ indicates that the UP365 emulsion may be stabilized by particulate aluminum/LS complexes. Although particles were not observed at the emulsification concentration, the principles described by Chamot (49) limit the lower resolving power of light microscopy to approximately 0.2 μm, which implies that particles of lower dimensions would not be observed. The critical coagulation concentrations of LS, on the other hand, are reported to be sharp and well defined (37). This, together with almost identical creaming rates obtained when using high salt concentrations, indicates that Al, Fe, and Cr modify UP365 in the same manner. The most important effect on the emulsion characteristic seems to be the reduced flocculation properties. At lower concentrations small differences in the effects of the salts were observed. The initial creaming rate of the aluminum sample was only half of what was found in the chromium and the iron sample. Compared to the pure LS sample $10^{-4}$ M Al had a stabilizing effect while the effect of chromium and iron was slightly destabilizing. The strong interction between aluminum and UP365 indicates that the flocculating properties of the LS are partially reduced also at low concentrations. The iron and chromium concentrations seem to be too low to alter the LS properties, but sufficient to reduce the electrostatic repulsion between the oil droplets. The

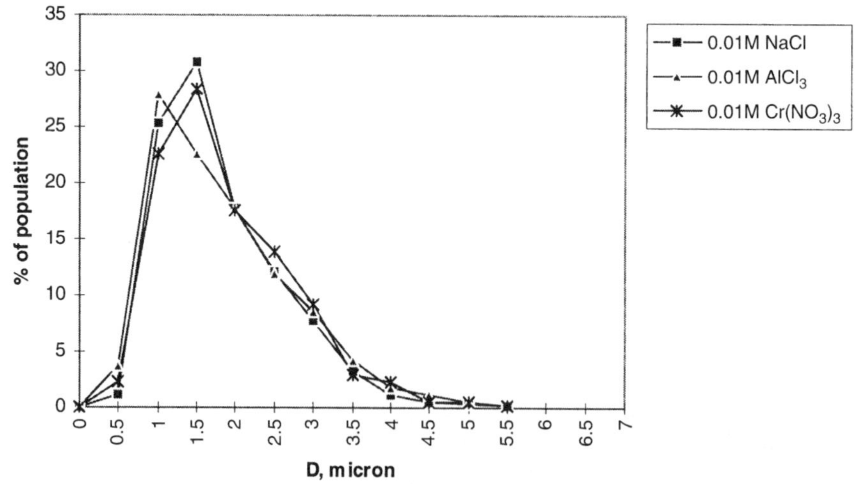


**Figure 11** Droplet size distributions of UP365-stabilized emulsions containing electrolyte.

slightly increased creaming rates indicate that a small growth in droplet flocculation takes place. The pH in the $10^{-4}$ M LS–electrolyte solutions was only slightly less than in the pure LS solutions.

Metal salts strongly affected the creaming rates also in the low molecular weight LS-stabilized emulsions. The $10^{-2}$ M Fe sample became stable and had a initial creaming rate and $\Delta C$ values close to what was found in the corresponding UP365 sample. The aluminum and chromium, on the other hand, had a very destablizing effect, being most pronounced for aluminum. Obviously the effects of the salts depend on the molecular weight of the LS. Figure 12 shows the droplet size distributions of UP365 and UP366 stabilized emulsions containing $10^{-2}$ M Al. The distributions are similar and do not reflect the very different creaming rates. VEM pictures showed that aluminum effectively reduced the droplet flocculation in both emulsions. The differences in the creaming rates then have to be addressed to differences in the electrostatic interaction of the droplets. Aluminum obviously reduces the charge of UP366 more effectively than of UP365. This is because of the smaller molecules in the UP366 fraction. The reduced electrostatic repulsion between the emulsion droplets implies that aluminum acts as a destabilizer in the UP366 emulsion. This is in agreement with findings by Askvik (50), which have shown that the charge of LS/aluminum complexes, in addition to the pH and concentration ratio, vary with the molecular weight of the LS. The less-pronounced increase in the creaming rate of the chromium sample indicates a more moderate decrease in the charge of the droplets. The sample was also more flocculated than the aluminum sample. Iron strongly reduced the creaming rates independently of the molecular weight of the LS. It has been shown that both high and low molecular weight LSs contain their negative charge under the prevailing conditions in this work (50). This is due to hydrolysis of the aquo ions of iron(III) yielding $Fe(H_2O)_4(OH)_2^{2+}$ which does not reduce the LS charge as effectively as does $Al(H_2O)_6^{3+}$ (51). This, together with the reduction of the flocculation properties of both LS fractions, leads to Fe acting as an emulsion stabilizer in both samples. The emulsions containing $10^{-2}$ M metal salt all became stable against coalescence. These factors indicate formation of a condensed surface layer of enhanced mechanical strength on the droplets. Metal salts at lower concentrations had little effect on the emulsion characteristic.

The creaming rates of the Kraft lignin-stabilized emulsions were all very low with a slightly destabilizing effect observed when adding electrolytes. Kraft lignins in electrolyte solutions, and the coagulating effects of hydrolzyed metal ions on Kraft lignin sols have been reported by Lindström (52, 53). The critical coagulation concentrations (CCCs) of trivalent metal ions were reported to be sharp and well defined. It has also been shown that Kraft lignins are very resistant to coagulation by NaCl although critical coagulation concentrations have been reported at high concentrations. In this work coagulated Kraft lignin was observed in the samples containing $10^{-2}$ M metal

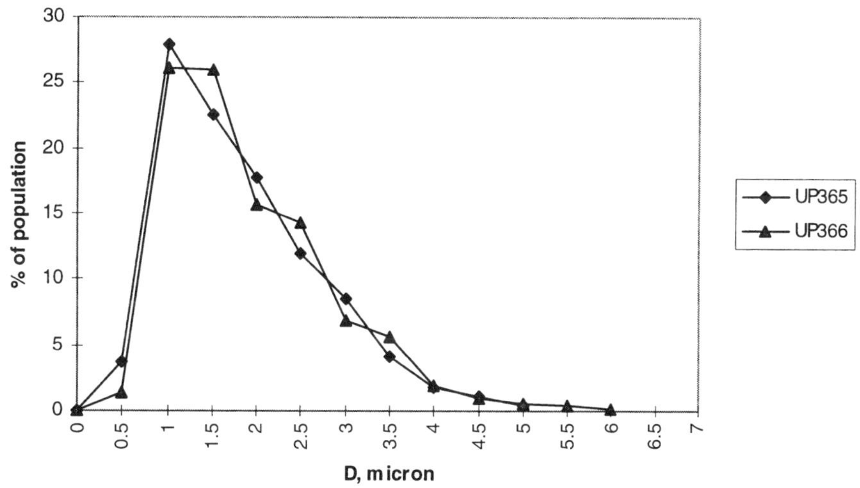


**Figure 12** Droplet size distributions of UP365- and UP366-stabilized emulsions containing $10^{-2}$ M Al.

salts, but not in the NaCl sample, which is in agreement with earlier findings that state that Kraft lignins are more easily coagulated than lignosulfonates (37). As a consequence the metal salt-containing emulsions were stabilized by particles, while the pure Kraft lignin XP9 and the NaCl-containing emulsions were characteized by polymer stabilization. In the latter case, flocculation was observed, while the metal salt-containing samples consisted of discrete nonflocculated droplets. Figure 13 shows DSDs of Kraft lignin- and LS-stabilized emulsions. As can be seen from the figure the DSDs are almost identical. When adding metal salts to the Kraft lignin samples a small dislocation of the droplet distribution towards higher diameters was observed. This, and a possible lower net charge of the emulsion droplets, may explain the increased creaming rates of these emulsions despite the fact that they are nonflocculated.

The creaming rate of the Kraft lignin-stabilized emulsion without added metal salts was low compared to that of the corresponding LS samples. Although both Kraft lignins and LSs are viewed as spherical polyelectrolytic microgels, the polyelectrolytic behavior is reported to be less marked for the Kraft lignins (52). Kraft lignins have also been reported to be less hydrophilic than LSs (37). It is probable that this, and the type and amount of anionic groups in the lignin derivates, will influence their behavior as dispersants (18). The much higher amount of carboxyl groups in the Kraft lignin fraction compared to that in the LS fraction could be of importance in this respect. The low creaming rate observed indicates an adsorbed layer of high charge density on the emulsion droplets. Most likely this and the moderate flocculation properties of the Kraft lignin fraction give rise to the high stability. All the Kraft lignin-stabilized emulsions were also stable against coalescence.

## ACKNOWLEDGMENTS

Financial support from Borregaard Lignotech, the Norwegian Research Council, and the technology program Flucha is gratefully acknowledged. Doctor Scient Øystein Sæther is thanked for assistance with the microscopy images.

## REFERENCES

1. KH Lim, DH Smith. J Disp Sci Technol 11: 529–545, 1990.
2. BP Brinks, J Dong. Colloids Surfaces A: Physicochemical Eng Aspects 132: 289–301, 1998.
3. M Bury, J Gerhards, W Erni. Int J Pharm 76: 207–216, 1991.
4. M Bury, J Gerhards, W Erni, A Stamm. Int J Pharm 124: 183–194, 1995.
5. F Kiekens, A Vermeire, N Samyn, J Demeester, JP Remon. Int J Pharm 146: 239–245, 1997.
6. E Sjöström. Wood Chemistry, Fundamentals and Applications. 2nd ed. New York: Academic Press 1993, pp 114–164.

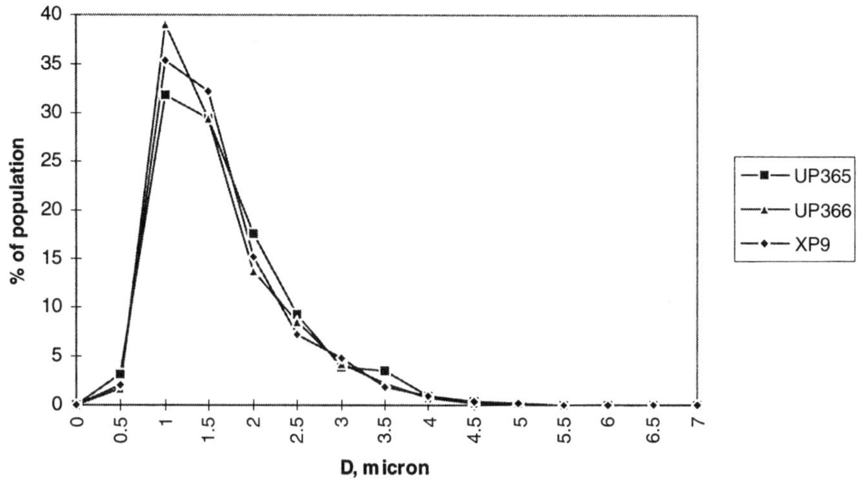


**Figure 13** Droplet size distributions of Kraft lignin- and lignosulfonate-stabilized emulsions.

7. AK Kontturi, K Kontturi. J Colloid Interface Sci 120: 256–262, 1987.
8. K Forss, KE Fremer. Papper och Trä 8: 443–454, 1965.
9. AK Konturri. J Chem Soc Faraday Trans I 84: 4033–4041, 1988.
10. PR Gupta, JL McCarthy. Macromolecules 1: 236–244, 1968.
11. A Rezanowich, DAI Goring. J Colloid Sci 15: 452–471, 1960.
12. WC Browning. Appl Polym Symp 28: 109–124, 1975.
13. SLH Chan, CGJ Baker, JM Beeckmans. Powder Technol 13: 223–230, 1976.
14. JM Hachey, VT Bui. J Appl Polym Sci; Appl Plym Symp 51: 171–182, 1992.
15. JC Le Bell. The Influence of Lingnosulphonate on the Colloidal Stability of Particulate Dispersions. PhD dissertation, Åbo Akademi, ÅBo, 1983.
16. CA Herb, S Ross. Colloids Surfaces 1: 57–77, 1980.
17. TF Tadros. Colloid Polym Sci 258: 439–446, 1980.
18. A Rezanowich, FJ Jaworzyn, DAI Goring. Pulp Paper Mag Can 62: 172–181, 1961.
19. JR Salvesen, WC Browning. Rts Com Chem Ind Week 61: 232–234, 1947.
20. T Lindström, C Södermark, L Westman. J Appl Polym Sci 21: 2873–2876, 1980.
21. GG Allan, DD Halabisky. Pulp Paper Mag Can 17: 64–70, 1970.
22. SA Gundersen, J Sjöblom. Colloid Polym Sci 277: 462–468, 1999.
23. WC Browning. J Petrol Technol 7: 9–15, 1955.
24. RV Lauzon, JS Short. The colloidal interaction of ferrochrome lignosulfonate with montmorillonite in drilling fluid applications. The 54th Annual Fall Technical Conference and Exhibition of the Petroleum Engineers of AIME, Las Vegas, NV, 1979, SPE 8225.
25. CI Chiwetelu, V Hornof, GH Neale, AE George. Can J Chem Eng 72: 534–540, 1994.
26. V Hornof, R Hombek. J Appl Polym Sci 41: 2391–2398, 1990.
27. V Hornof. Cellulose Chem Technol 24: 407–415, 1990.
28. C Chiwetelu, V Hornof, GH Neale. Trans IChemE 60: 177–182, 1982.
29. JE Son, GH Neale, V Hornof. J Can Petrol Technol (July–Aug.): 42–48, 1982.
30. C Chiwetelu, G Neale, V Hornof. J Can Petrol Technol (July–Sept): 91–99, 1980.
31. WC Browning. Surface Active Properties of Lignosulfonates. Proceedings IV International Congress on Surface Active Substances. Vol 1, Sect A, Brussels, 1964, pp 141–155.
32. PG Shotton, PC Hewlett, AN James. Tappi 55: 407–415, 1972.
33. JL Gardon, SG Mason. Can J Chem 33: 1491–1501, 1955.
34. B Kachar, DF Evans, BW Ninham. J Colloid Interface Sci 99: 593–596, 1984.
35. B Kachar, DF Evans, BW Ninham. J Colloid Interface Sci 100: 287–301, 1984.
36. DAG Bruggerman. Ann Phys 24: 636–664, 1935.
37. V Nyman, G Rose. Colloids Surfaces 21: 125–147, 1986.
38. JM Kolthoff, RG Gutmacher. J Phys Chem 56: 740–745, 1952.
39. RE Felter, LN Ray. J Colloid Interface Sci 32: 349–460, 1970.
40. C Wagner. Z Elektrochem 65: 581–591, 1961.
41. M Kalhweit. Adv Colloid Interface Sci 5: 1–35, 1975.
42. M Kalhweit. Faraday Discuss Chem Soc 61: 48–52, 1976.
43. K Parbhakar, J Lewandowski, LH Dao. J Colloid Interface Sci 174: 142–147.
44. AS Kabalnov, AV Pertzov, ED Shchukin. Colloids Surf 24: 19–32, 1987.
45. AS Kabalnov, ED Shchukin. Adv Colloid Interface Sci 38: 69–97, 1992.
46. B Vincent. In: TF Tadros, ed. Surfactants. London: Academic Press, 1984, pp 183–184.
47. TL Brown, HE LeMay Jr, BE Bursten. Chemistry: The Central Science. 5th ed. London: Prentice Hall, 1991, pp 894.
48. C Orr. Determination of particle size. In: P Becher, ed. Encyclopedia of Emulsion Technology, Basic Theory, Measurement Applications. New York: Marcel Dekker, 1988, pp 137–169.
49. EM Chamot. Elementary Chemical Microscopy. 1st Ed. New York: John Wiley, 1915, pp. 15.
50. KM Askvik, J Sjöblom, P Stenius. Colloids Polymer Sci, submitted.
51. JW Akitt. Progr NMR Spectrosc 21: 1–149, 1989.
52. T Lindström. Colloid Polym Sci 257: 277–285, 1979.
53. T Lindström. Colloid Polym Sci 258: 168–173, 1980.

# 17

# Double Emulsions for Controlled-release Applications—Progress and Trends

**Nissim Garti and Axel Benichou**
*Casali Institute of Applied Chemistry, The Hebrew University of Jerusalem, Jerusalem, Israel*

## I. INTRODUCTION

Double emulsions are complex liquid dispersion systems known also as "emulsions of emulsions," in which the droplets of one dispersed liquid are further dispersed in another liquid. The inner dispersed globule/droplet in the double emulsion are separated (compartmentalized) from the outer liquid phase by a layer of another phase (1–9).

Several types of double emulsions have been documented. Some consist of a single, internal compartment while others have many internal droplets and are known as "multiple-compartment emulsions." A schematic presentation of some double emulsions is shown in Fig. 1. The most common double emulsions are of W/O/W, but in some specific applications O/W/O emulsions can also be prepared. The term "multiple emulsion" was "coined" historically because microscopically it appeared that a number (multiple) of phases were dispersed one into the others. In most cases it was proven that in practice most systems are composed of double (or duplex) emulsions. A more suitable and more accurate term for such systems should be, therefore, "emulsified emulsions."

Potential applications for double emulsions are well documented and many of these applications have been patented (10–14). The important applications are in agriculture, pharmaceuticals, cosmetics, and foods. In most cases, double emulsions are aimed for slow and sustained release of active matter from an internal reservoir into the continuous phase (mostly water). In some applications the double emulsions can serve also as an internal reservoir to entrap matter from the outer diluted continuous phase into the inner confined space. These applications are aimed to remove toxic matter. In other applications, double emulsions are reservoirs for improved dissolution or solubilization of insoluble materials. The materials will dissolve in part in the inner phase, in part at the internal interface, and occasionally at the external interface. Applications related to protection of sensitive and active molecules from the external phase (antioxidation) have been recently mentioned (15, 16). Many more applications are expected to emerge in the near future. Special attention must be paid to the most promising new application of double emulsions as intermediate systems in preparation of solid or semisolid microcapsules (17–19).

## II. THE EMULSIFIERS

Double emulsions consist of two different interfaces which require two sets of different types of emulsifiers. In O/W/O double emulsions the first set of emulsifiers, for the internal interface, must be hydrophilic, while the second set of emulsifiers, for the external interface, must be hydrophobic (Table 1). For W/O/W double emulsions the order of the emul-

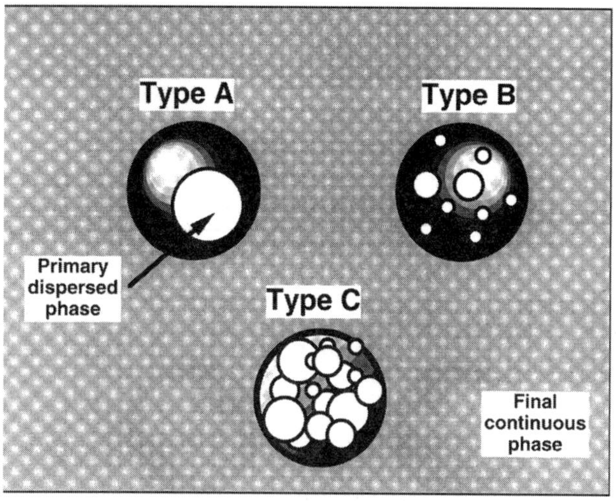


**Figure 1** Schematic presentation of the three common types of multiple-emulsion droplets.

sifiers is the opposite; the inner emulsifiers are hydrophobic while the outer ones are hydrophilic. In many cases a blend of two or more emulsifiers in each set is recommended for better stabilization results. This review will discuss mostly W/O/W double emulsions since most of the important applications require such emulsions. Some O/W/O emulsions applications are also given.

In early reports on the formation of double emulsions only one set of emulsifiers and an inversion process were used. Such preparations were done in one step, but the stability was in most cases questionable. It was difficult to control the distribution of the emulsifiers within the two interfaces. There was fast migration of the emulsifiers between the phases that destabilized the emulsions. In most recent emulsion formulations the emulsions are prepared in two steps. At first, a high-shear homogenization is applied to the water that is added to the solution of the oil and the hydrophobic emulsifiers, to obtain a stable W/O emulsion. In the second step the W/O emulsion is gently added with stirring (not homogenization) to the water and hydrophilic emulsifiers solution (Fig. 2). The droplet size distribution of a typical classical double emulsion ranges from 10 to 50 μm.

Some more sophisticated preparation methods have been reported in the literature out of which two are interesting and worth being mentioned. the "lamellar

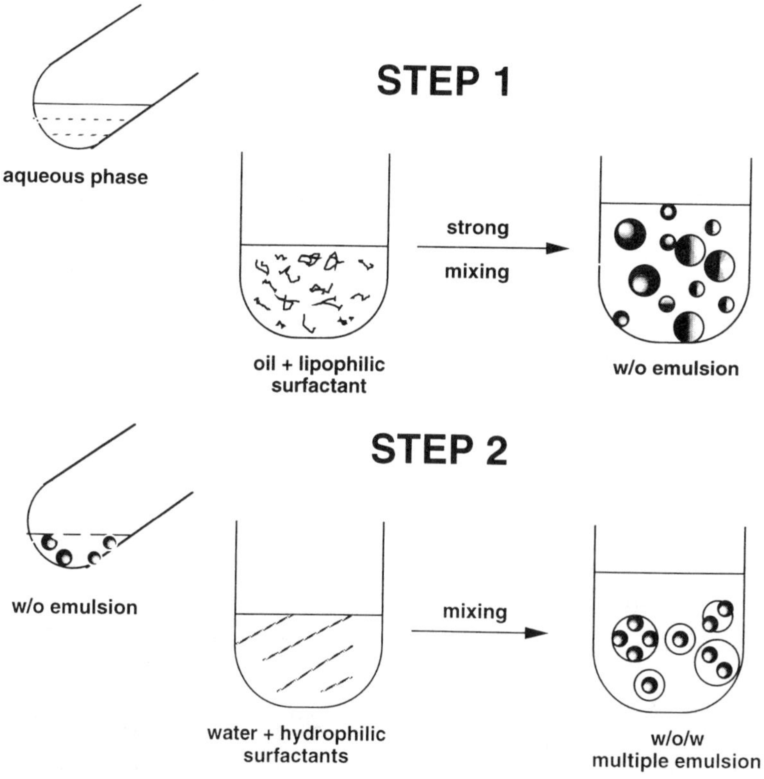


**Figure 2** Schematic illustration of a two-step process in formation of a double emulsion.

phase dispersion process" was reported by Vigie (20) (Fig. 3). The procedure is derived from the process employed to obtain liposome-like vesicles with nonionic emulsifiers. This process can be used only when the constituents form a lamellar phase by mixing with water in definite proportions. This procedure offers advantage since it requires only a simple emulsification step. The mesophase formed by an ideal ratio of lipophilic emulsifiers in water is thermodynamically stable and can be obtained rapidly and easily. The method's main limitation is derived from the fact that most emulsifiers do not form lamellar phases. When the lamellar mesophase exists, the HLB of the mix of emulsifiers is often too high, which is disadvantageous for the stability of a multiple emulsion. In addition, the quantity of oil incorporated into the lamellar phase is always low, rarely higher than 10 wt%. Another drawback of this process is the weak control of the rate of encapsulation of the active substances.

Grossiord et al. (21) discuss an additional method termed, by them, the "oily isotropic dispersion process." We prefer a more accurate terminology of "emulsified microemulsions." The idea is to disperse an oil phase within water by surfactant and to form an $L_2$ phase. This phase is further emulsified with water to form a double emulsion (Fig. 4). The problem is that there is no evidence that the formation of the microstructures by this method leads indeed to multiple emulsions. Moreover, there is no good evidence that the internal phase, an $L_2$ phase of submicrometer droplets, remains after the second emulsification process. It seems that in some cases the process is a well-characterized two-step emulsification that leads to relatively large double-emulsion droplets. This process is worth further investigation and should be more carefully evaluated. If one can prove that the internal compartmentalization is of a stable microemulsion it might bring a breakthrough to this field since the sizes of the external droplets could be reduced to values below 1 μm. Such formulations will allow formation of indictable double emulsions.

Higashi et al. (22), described a new method of producing W/O/W multiple emulsions by a "membrane emulsification technique." This method permits the formation of monodispersed liquid microdroplets con-

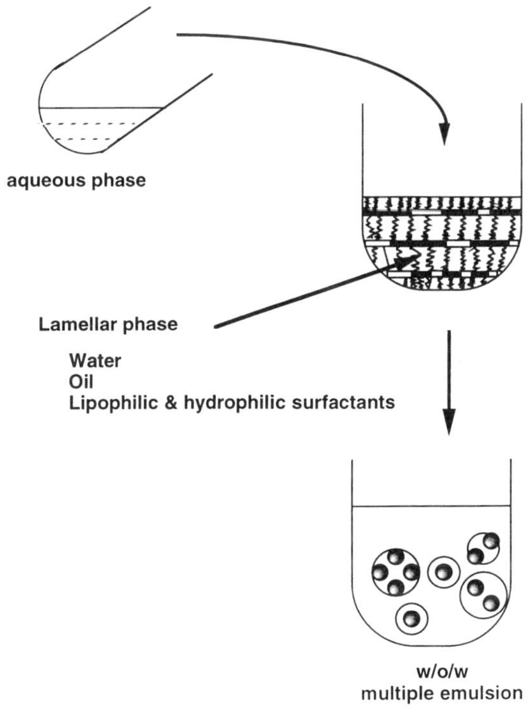


**Figure 3** Preparation of W/O/W multiple emulsion by lamellar phase dispersion. (From Ref. 21.)

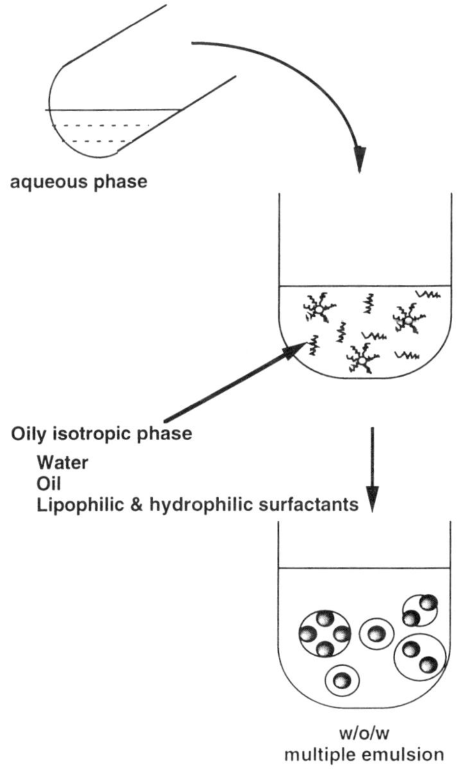


**Figure 4** Preparation of W/O/W multiple emulsion by "oily isotropic dispersion" (emulsified microemulsion). (From Ref. 21.)

**Table 1** W/O/W Double Emulsions (with 20 wt % W/O emulsion) Prepared with (A) 25 wt % Water, 1 wt % PGPR, 0.5 wt % Tristearin (TS), and 73.5 wt % Soybean Oil; and (B) 25 wt % Water, 1 wt % PGPR, 1 wt % TS and 73 wt % Soybean Oil[a]

| | Tw-80 (wt %) | Sp-80 (wt %) | Size (W/O/W), $t = 0$ (μm) | Size (W/O/W), $t = 14$ days (μm) | Conductivity, $t = 0$ (mS)$^{-1}$ | Conductivity, $t = 10$ days (mS)$^{-1}$ |
|---|---|---|---|---|---|---|
| *Set A* | 0.5 | — | 8 | 20 | 0.4 | 1.0 |
| (1% PGPR, | 1.0 | — | 7 | 22 | 0.8 | 1.4 |
| 0.5% TS) | 2.0 | — | 6.5 | 25 | 1.2 | 1.4 |
| $\varphi_1^{w/o} = 0.25$[b] | 3.0 | 2.0 | 14 | 14 | 1.0 | 1.1 |
| $\varphi_2^{w/o} = 0.2$[c] | 6.0 | 4.0 | 18 | 18 | 1.8 | 2.2 |
| *Set B* | 0.5 | — | 11 | 11 | 1.1 | 0.8 |
| (1% PGPR, | 1.0 | — | 10 | 9 | 1.0 | 1.1 |
| 1% TS) | 2.0 | — | 13 | 10 | 0.7 | 1.2 |
| $\varphi_1^{w/o} = 0.25$ | 3.0 | 2.0 | 12 | | | |
| $\varphi_2^{w/o} = 0.2$ | 6.0 | 4.0 | 13 | | | |

[a]The external phase contained various compositions of Tween 80 (Tw-80) and Span 80 (Sp-80). W/O/W, water-in-oil-in-water; W/O, water-in-oil.
[b]$\varphi_1^{w/o} = 0.25$, weight fraction of oil phase for W/O emulsion consisting of 25 wt % oil.
[c]$\varphi_2^{w/o} = 0.2$, weight fraction of W/O emulsion in the double W/O/W emulsion.

taining aqueous microdroplets to form a W/O/W system. In this method, the aqueous internal phase is mixed with an oil phase containing lipophilic emulsifier. The mixture is sonicated to form a W/O emulsion, and the upper chamber of a special apparatus (Fig. 5) is filled with the emulsion. The external aqueous phase containing the hydrophilic emulsifier is continuously injected into the lower chamber to create a continuous flow. Nitrogen gas fed into the upper chamber initiates permeation of the W/O droplets through the controlled-pore glass membrane into the emulsifying chamber, forming a W/O/W multiple emulsion. The emulsion is progressively removed from the apparatus. This process is claimed to be, at present, used on the industrial scale.

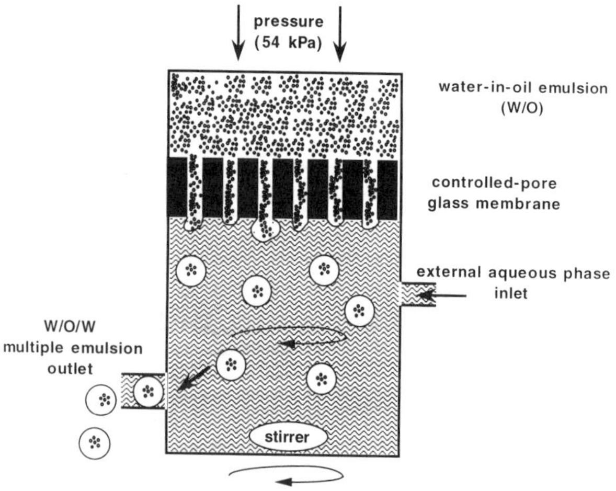


**Figure 5** Preparation of W/O/W multiple emulsion by a "membrane emulsification technique." (From Ref. 22.)

It should be noted that low-molecular-weight emulsifiers migrate from the W/O interface to the oil phases and alter the required hydrophilic/hydrophobic balance of each of the phases. Most of the studies, in the years 1970 to 1985, searched for a proper monomeric emulsifier's blend or combination (hydrophilic and hydrophobic) to be used at the two interfaces and the proper ratios between the two. Matsumoto and coworkers (23–29) established a "magic" weight ratio of minimum 10 of the internal hydrophobic to the external hydrophilic emulsifiers (Fig. 6). Garti and coworkers (30–34), proved that the free exchange between the internal and the external emulsifiers required a calculation of an "effective HLB [hydrophilic–lipophilic balance] value" of emulsifiers to optimize the stabilization of the emulsion. Complete parametrization work was performed on almost every possible variation in the ingredients and compositions (1–9). In most cases the internal emulsifiers are used in great excess to the external emulsifiers. The nature of the emulsifiers also dictates the number of compartments and the internal volume that the inner phase occupies.

Many of the more recent studies explore various more sophisticated emulsifiers such as sphingomyelins (35), modified or purified phospholipids (36), cholates,

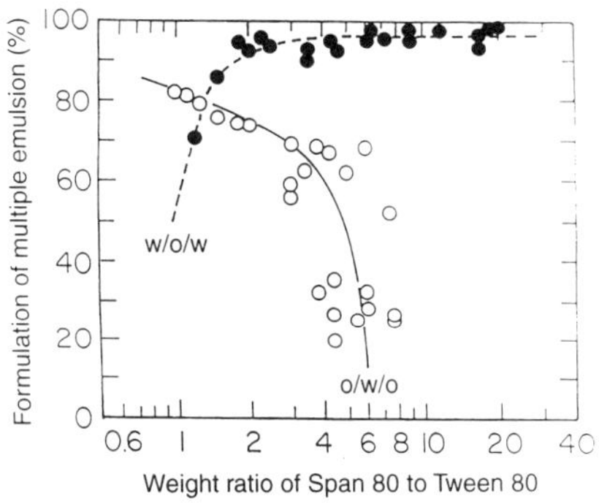


**Figure 6** Yield of formation of double emulsions of W/O/W and O/W/O as a function of the w/w ratio of the internal hydrophobic emulsifier, Span 80, and the external hydrophilic emulsifier, Tween 80. (From Ref. 29.)

etc. The principals for selecting the proper emulsifiers are similar to those known for classical emulsions. Some of the emulsions might have better stability than others, but the general trend remains unchanged. It should be also stressed that one must adjust the emulsifiers to the application in mind and must substitute one emulsifier from the other, depending on the total composition of the system.

It must be also recognized that 'empty' double emulsions will behave differently from those containing active matter (electrolytes, biologically active materials, proteins, sugars, drugs, etc.) owing to osmotic pressure gradients (caused by the additives) between the outer and the inner phases. In addition, many of the active ingredients have some hydrophobicity and surface properties. Such molecules (pep tides, drugs, pesticides) will migrate from the inner bulk and will adsorb on to the interface, changing the delicate emulsifiers' HLB. The emulsifiers around the water or the oil droplets will not fully cover the droplets, and the stability will be reduced. These phenomena are very frequently neglected or overlooked by many of us. The conclusions that are frequently derived from these oversimplified experiments and incorrectly structured emulsions can be misleading and incorrect. We will discuss further some of these aspects in this review.

## III. THE OIL PHASE

In many food and pharmaceutical applications only a limited number of different "oil phases" (water-immiscible liquids) have been suggested and tried throughout years of research. In most applications the oil phase is based on vegetable or animal unsaturated triglycerides such as soya oil, cotton, canola, sunflower, and others. It was also suggested to replace the long-chain triglycerides (LCTs), which are "oxygen and hydrolysis sensitive," by medium-chain triglycerides (MCTs) that are fully saturated and thus oxidation resistant. The MCT is easier to emulsify and requires less shear.

In cosmetic applications the freedom in using different oils is greater. Long-chain fatty acids, fatty alcohols, and simple esters such as isopropyl myristate (IPM), jojoba oil, and essential oils are only a few of the examples. In addition, various waxes, sterols, and paraffin oils have been tried.

In agricultural applications, various organic solvents have been used. Solvents such as toluene and its derivatives, as well as chlorinated hydrocarbons, have been suggested.

Hydrocarbons, aromatic compounds, silicone oils, and fluorinated or halogenated hydrocarbons are only few of the main oils in use in industrial applications.

Several scientists have tried to correlate the nature of the oil phase and its volume fraction to the stability of the double emulsions. No unusual or surprising findings were observed. Double-emulsion interfaces behave very much like simple emulsions except for the severe limitations on sizes of the droplets and the internal distribution of the emulsifiers.

## IV. STABILITY CONSIDERATIONS

Double emulsions consisting of low-molecular-weight emulsifiers (the so-called monomeric emulsifiers) are mostly unstable thermodynamically, mainly because in the second stage of the emulsification severe homogenization or shear takes place and, as a result, large droplets are obtained. During years of research attempts have been made to find proper and more suitable combinations of emulsifiers to reduce droplet sizes and to improve the emulsion stability. Aggregation, flocculation, and coalescence (occurring in the inner phase and between the double-emulsion droplets) are major factors affecting the instability of the emulsions, resulting in rupture of droplets and separation of the phases.

Double emulsions are usually not empty. Water-soluble active materials are entrapped during the emulsification in the inner aqueous phase. It is well documented that, because of the difference in osmotic pressure through a diffusion-controlled mechanism, the active matter tends to diffuse and migrate from the internal phase to the external interface, mostly through a mechanism known as "reverse micellar transport" (Fig. 7a). The dilemma that researchers were faced with was how to control the diffusion of water molecules as well as the emulsifier molecules, and mostly the active matter from the internal phase to the outer phase. It seemed almost impossible to retain the active material within the water phase upon prolonged storage. Attempts to increase the HLB of the external emulsifier or to increase its concentration in order to improve the stability of the emulsion worsened the situation and ended in faster release of the drug or electrolytes.

Much work was devoted to establish the effects of osmotic-pressure differences between the internal and the external phases on the stability of the emulsions, and on the release rates of the markers from the internal phase, and the engulfment of the internal droplets by the flow of water from the outer continuous phase to the inner droplets. Mono- and multiple-compartment emulsions were prepared and evaluated in view of the

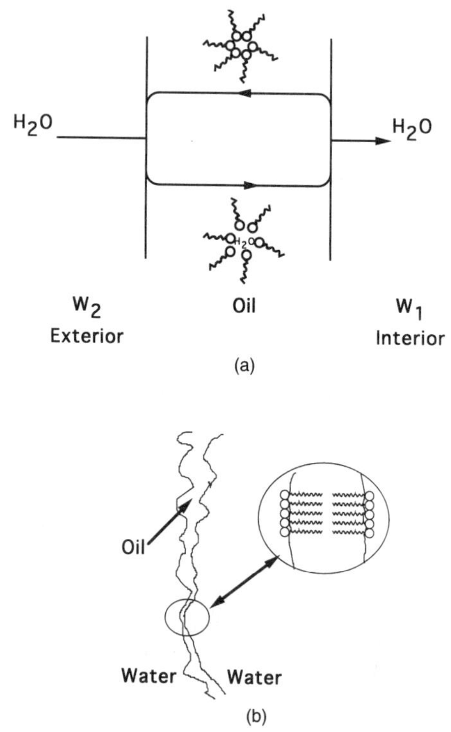


**Figure 7** Schematic illustration of the two possible transport mechanisms: (a) reverse micellar; and (b) lamellar thinning transport of marker from the inner aqueous phase to the continuous aqueous phase.

enormous potential that these low-viscosity liquid systems have in slow delivery of water-soluble drugs.

Additional instability mechanisms and release pathways have been demonstrated and discussed in detail by various authors. These mechanisms include "transport through thinned lamella" (Fig. 7b), transport of adducts or complexes that are formed in the oil phase, and other variations of these mechanisms. It seems, however, that the main instability and release mechanisms are the parallel or simultaneously occurring phenomena of "reverse micellar transport" and coalescence.

All the above mechanisms have been well established, but it seems that the stability and the release patterns of these complex double-emulsion systems depend on various parameters that simultaneously interplay and that a simplified or unique mechanism cannot explain all the in-parallel pathways that take place in the double emulsions.

In a recent study, Ficheux et al (37), identified two types of thermodynamic instability that are responsible for the evolution of double emulsions. Both mechanisms are within good agreement with the old Bancroft rule, but stress different aspects of the previously mentioned mechanisms. The mechanisms elucidated by the authors result in different behavior of the entrapped matter in the double emulsion. The first is a "coalescence of the small inner droplets with the outer droplets interface" which is due to the rupture of the thin nonaqueous film that forms between the external continuous phase and the inner small water droplets (Fig. 8). This instability irreversibly transforms a double droplet into a simple direct emulsion. Such a mechanism is suitable for delivery of water-soluble substances. The second mechanism is a "coalescence between the smaller inner droplets within the oil globule." The first type of instability leads to a complete delivering of the small inner droplets toward the external phase whereas the second one does not. The second mechanism leads to an increase in the average diameter of the internal droplets and a decrease in their number. The authors worked both with anionic (SDS, sodium dodecyl sulfate) and cationic (TTAB, tetradecyltrimethylamonium bromide). It was demonstrated that the kinetics associated with the release of the small inner droplets, resulting from the former instability, is clearly related to the hydrophilic surfactant concentration in the external phase (Fig. 9). Depending on the value of this concentration, double emulsions may be destabilised on a time scale ranging from several months to a few minutes.

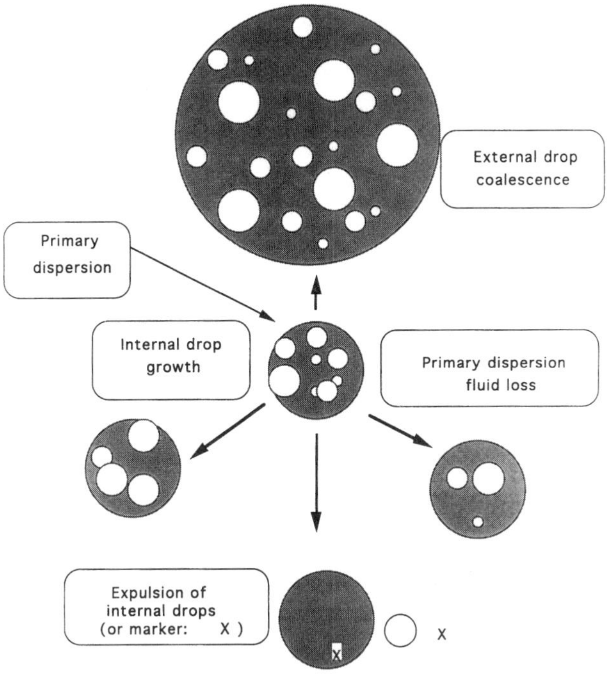


**Figure 8** Schematic representation of the possible pathways for breakdown in multiple emulsions.

Rosano et al. (38) explored the influence of "ripening and interfacial interactions" on the stability of the W/O/W double emulsions. The oil-insoluble solute was shown to stabilize both the first W/O emulsion (of the inner water droplets) and the external O/W interface. The authors used a theoretical model and experimental results (video-microscope observations). It was shown that the presence of an electrolyte in the inner water phase is necessary for the stability of a multiple emulsion. The stability is achieved from the osmotic-pressure equalization derived from the differences (excess) in the Laplace pressure. This effect stabilizes the inner W/O emulsion. It is possible also to determine the right salt concentration necessary to balance the osmotic pressure between the two water phases. In a set of experiments (Tables 2 and 3) a total of 15 formulations are shown in which both the oil phase and the $W_2$ phase are constant, and the only parameter varied is the NaCl concentration in the $W_1$ phase. The first six formulations were prepared with betaine, whereas the others were prepared with SDS. In the case of betaine, without any salt in the inner phase, an unstable multiple emulsions is observed, and both Ostwald ripening and release of the $W_1$ phase into the $W_2$ phase occur. As the concentration of the salt is increased in the $W_1$ phase, the systems do not separate and the structure remains multiple, but a large increase in viscosity is observed (9500 cP for 0.07 wt % to 27,600 cP for 0.4 wt % NaCl). The authors conclude that one must consider three possible factors that influence the stability: the Laplace and osmotic pressure effects between the two aqueous phases, the interaction between the low and high HLB emulsifiers at the outer O/W interface, and the polymeric thickener–hydrophilic emulsifier interactions in the outer phase ($W_2$).

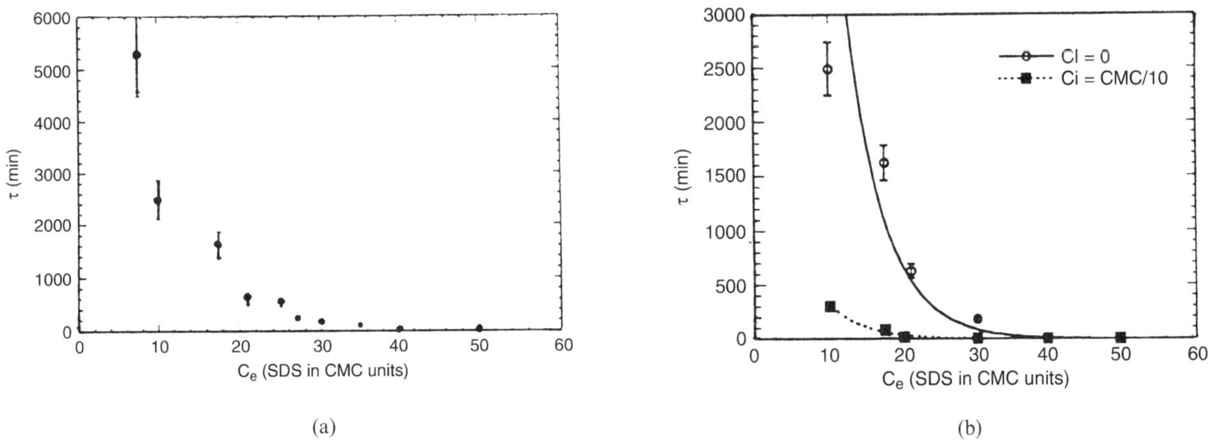


**Figure 9** (a) Plot, at 20°C, of the life time $\tau$, of internal droplets entrapped in the oil globules as a function of the external phase surfactant concentration $C_e$. The double emulsions are composed of 90% external phase and 10% double droplets. There is 10% water within the large double globules; 2% Span 80 was used within the oil, and SDS in the external water phase. (b) Influence of the internal surfactant concentration $Ci$ on the $\tau = f(Ce)$ curve, at 20°C. System: Span 80/SDS as in Figure 9a. The dashed and solid lines are only guides to the eyes (From Ref. 37.)

Table 2 Role of the Concentration of Salt in the Inner $W_1$ Phase for Formulations Prepared with Cocamidopropylbetaine

| | Composition (wt %) | | | | | |
|---|---|---|---|---|---|---|
| Formulation number | 1 | 2 | 3 | 4 | 5 | 6 |
| $W_1H_2O$ | 27.5 | 27.43 | 27.35 | 27.2 | 27.1 | 27.2 |
| NaCl | 0 | 0.07 | 0.15 | 0.3 | 0.4 | 0.3 |
| Light paraffin | 21.5 | 21.5 | 21.5 | 21.5 | 21.5 | 21.5 |
| Abil EM 90 | 1 | 1 | 1 | 1 | 1 | 1 |
| $W_2H_2O$ | 48.9 | 48.9 | 48.9 | 48.9 | 48.9 | 48.9 |
| Betaine | 0.3 | 0.3 | 0.3 | 0.3 | 0.3 | 0.3 |
| Carbopol 1342 | 0.1 | 0.1 | 0.1 | 0.1 | 0.1 | 0.05 |
| Separation (%) | 50 (20)[a] | 0 (80) | 0 (80) | 0 (80) | 0 (80) | 0 (110) |
| Structure | Simple | Multiple | Multiple | Multiple | Multiple | Multiple |

[a]Number of days.

Table 3 Role of the Concentration of Salt in the Inner $W_1$ Water Phase for Formulation Prepared with SDS

| | Composition (wt %) | | | | | | | | |
|---|---|---|---|---|---|---|---|---|---|
| Formulation number | 7 | 8 | 9 | 10 | 11 | 12 | 13 | 14 | 15 |
| $W_1H_2O$ | 27.43 | 27.4 | 27.36 | 27.2 | 27.1 | 25.95 | 26.9 | 26.6 | 27.2 |
| NaCl | 0.07 | 0.1 | 0.14 | 0.3 | 0.4 | 0.45 | 0.6 | 0.9 | 0.3 |
| Light paraffin | 21.5 | 21.5 | 21.5 | 21.5 | 21.5 | 21.5 | 21.5 | 21.5 | 21.5 |
| Abil EM 90 | 1 | 1 | 1 | 1 | 1 | 1 | 1 | 1 | 1 |
| $W_2H_2O$ | 48.9 | 48.9 | 48.9 | 48.9 | 48.9 | 50 | 48.9 | 48.9 | 48.9 |
| SDS | 1 | 1 | 1 | 1 | 1 | 1 | 1 | 1 | 1 |
| Carbopol 1342 | 0.1 | 0.1 | 0.1 | 0.1 | 0.1 | 0.1 | 0.1 | 0.1 | 0.2 |
| Separation (%) | 4 (40)[a] | 3 (40) | 17 (40) | 38 (40) | 9 (40) | 16 (40) | 52 (90) | 40 (40) | 0 (20) |
| Structure | Multiple | Multiple | Multiple | Multiple | Multiple | Multiple | Multiple | Multiple | Multiple |

[a]Number of days.

The variations between the different suggested mechanisms are not dramatic. Some authors tend to stress certain mechanistic aspects and to neglect others while other authors stress, in very specific formulations, the more relevant pathways. It seems that most suggested mechanisms are basically very similar.

## V. STABILIZATION BY POLYMERIZABLE EMULSIFIERS

Florence and coworkers (39–41) were the first to recognize the need to improve the stability of double emulsions by improving the interfacial coverage. They suggested strengthening the "seal" of the external interface by nonconventional methods, using tailor-made monomeric emulsifiers. They made some attempts to use monomeric amphiphilic molecules with reactive functional group (serving as a ballast). The so-called "polymerizable emulsifiers" were adsorbed on to the external interface and were polymerized in situ. A cross-linked thick film of polymeric surfactant was formed at the interface. The procedure is tedious and very expensive since no commercial polymerizable and emulsifiers are easy to find. The results were somewhat disappointing. It is not clear if the relatively poor stability performance was derived from the uncontrolled or insufficient polymerization process or from the poor alignment of the polymerizable emulsifiers. No further attempts were made by other authors to explore this idea. We believe that the lack of further attempts by other investigators is mainly because the polymerizable surfactants are difficult to synthesize and to orient at the interface, and because there is only a slim chance that the polymerizable surfactants will be approved by any health authorities.

## VI. STABILIZATION BY BIOPOLYMERS

Macromolecules adsorb on to interfaces and facilitate more (or better) coverage than monomeric emulsifiers. The amphiphilic macromolecules form, in most cases, thick and flexible films that are strongly anchored in the dispersed and the dispersion phases. The adsorbed polymers are known to enhance steric stabilization mechanisms and have proven to be good emulsifiers in food colloids, mainly, in some food-grade O/W emulsions. The use of macromolecular amphiphiles and stabilizers such as proteins and polysaccharides has long been adopted by scientists exploring the stability of double emulsions. Gelatin (42, 43), whey proteins (44), bovine serum albumin (BSA) (45–48), human serum albumin (HSA), caseins, and other proteins were mentioned and evaluated. The proteins were used usually in combination with other monomeric emulsifiers (46). A significant improvement in the stability of the emulsions was shown when these macromolecules were encapsulated on to the external interface. In most cases the macromolecule was used in low concentrations (maximum 0.2 wt %) and in combination with a large excess of nonionic monomeric emulsifiers. In some cases, casein alone served as the external emulsifier (46). Furthermore, from the release curves it seems that the marker transport is more controlled (Figs. 10–12). Dickinson and coworkers (47–50) concluded that proteins or other macromolecular stabilizers are unlikely to replace completely lipophilic monomeric emulsifiers in double emulsions. However, proteins in combination with stabilizers do have the capacity to confer some enhanced degree of stability on a multiple-emulsion system and, therefore, the lipophilic emulsifier concentration is substantially reduced.

The authors of this review (46) have used BSA along with monomeric emulsifiers, both in the inner and the outer interfaces (in low concentrations of up to 0.2 wt %), and found significant improvement both in the stability and in the release of markers as compared to the use of the protein in the external phase only (Fig. 13). It was postulated that while the BSA has no stability effect at the inner phase it has strong effect on the release of the markers (mechanical film barrier). On the other hand, BSA together with small amounts of monomeric emulsifiers (or hydrocolloids) serve as good steric stabilizers, improve stability and shelf-life, and slow down the release of the markers. The BSA plays, therefore, a double role in the emulsions: as film former and

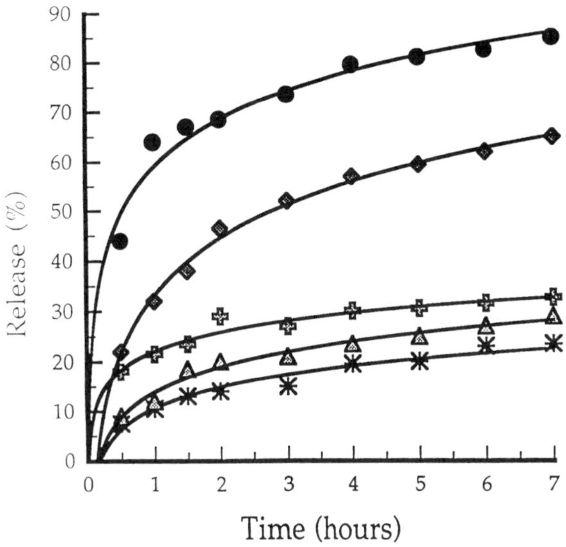


**Figure 10** Release of methotrexate (MTX) from multiple W/O/W emulsions. The emulsions contained 2.5 wt % Span 80 and 0.2 wt% BSA as primary emulsifiers, with MTX (1 mg/ml) in the internal phase and the following oil phases: * – octane; Δ –dodecane; ✣ – hexadecane; ◇ – octadecane; and ● – isopropyl myristate (IPM). (From Ref. 40.)

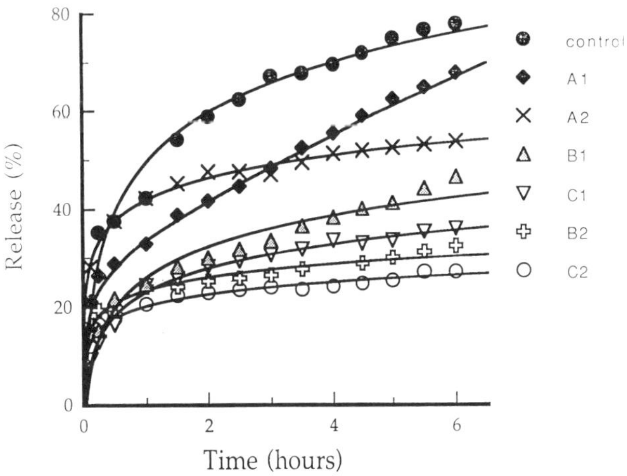


**Figure 11** Profile of chloroquine phosphate release from W/O/W multiple emulsions. A1: freshly prepared PVP emulsions; A2: PVP emulsions stored for 2 weeks; B1: freshly prepared gelatin emulsions; B2: gelatin emulsions stored for 2 weeks; C1: emulsion prepared with gum acacia; C2: acacia emulsion stored for 2 weeks. (From Ref. 41.)

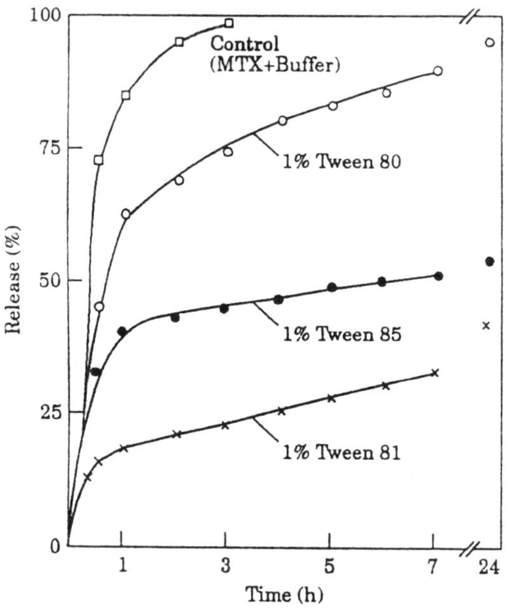


**Figure 12** Effect of the secondary emulsifier (E2) on the relase of MTX from double emulsions prepared with 2.5% Span 80 and 0.2% BSA as primary emulsifiers with IPM (isopropyl myristate) as the oil phase. (From Ref. 3.)

barrier to the release of small molecules at the internal interface, and as steric stabilizer at the external interface. The release mechanisms involving reverse micellar transport were also established (Fig. 14).

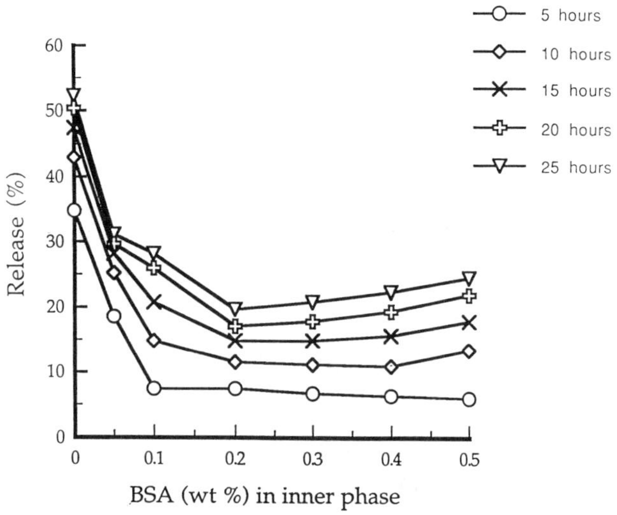


**Figure 13** Percentage release of NaCl with time, from double emulsion prepared with 10 wt % Span 80 and various BSA concentrations in the inner phase and 5 wt % Span 80 + Tween 80 (1:9) in the outer aqueous phase.

In more recent studies (51, 52) the biopolymer chitosan was used as an emulsifier in food double emulsions. Chitosan has surface activity and seems to stabilize W/O/W emulsions. Chitosan reacts with anionic emulsifiers such as sodium dodecyl sulfate at certain ratios to form a water-insoluble complex that has strong emulsification capabilities. Chitosan solution was used to form double emulsions of O/W/O as intermediates from which by a simple procedure of stripping the water the authors formed interesting porous spherical particles of chitosan (52).

Cyclodextrins ($\alpha$, $\beta$, and $\gamma$) were shown to be potential emulsifiers for O/W/O emulsions (53). The advantages of the cyclodextrins as emulsifiers are their ability to complex with the oil components at the oil/water interface which obviates the need for additional surfactant. It appears that the emulsifier efficacy depends on the nature of the oil and the type of cyclodextrin ($\alpha > \beta > \gamma$). The presence of any active matter in the inner phase (such as benzophenone) destabilized the emulsion. Only the $\alpha$ cyclodextrin yielded stable emulsions. The reason is an interfacial interaction between the components that are present at the interface, which changes the HLB and causes a destabilization effect. This elegant thought of interfacial complexation between the "oil components" and the surfactant cannot be a universal solution. The idea suffers from a very severe intrinsic disadvantages since once different additive is included in the emulsion. For every additive at any concentration an adjustment must be made and the given cyclodextrin or the complexing agent might be totally suited.

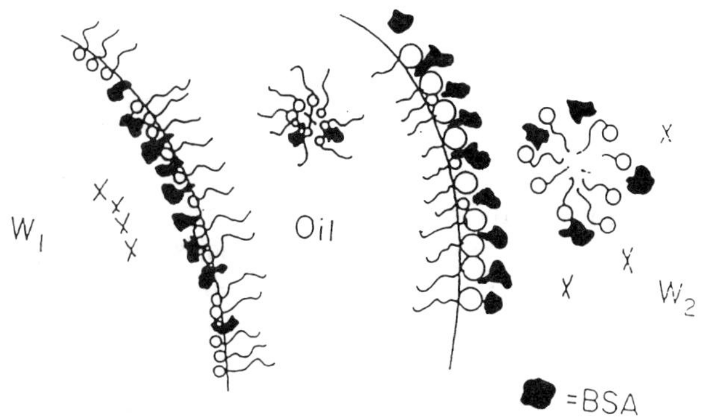


**Figure 14** Schematic illustration of possible organization and stabilization mechanism of BSA and monomeric emulsifiers (Span 80) at the two interfaces of double emulsion.

## VII. STABILIZATION BY SOLID PARTICLES

Stabilization of margarine and other similar food emulsions is achieved by partial adsorption of solid fat particles ($\beta'$-polymorphs) on to the water–oil interface, bridged by monomeric hydrophobic emulsifiers. The complex stabilization is achieved by "wetting" the oil phase by solid fat particles and emulsifiers (lecithins and monoglycerides of fatty acids). The concept was re-examined and reconsidered by Bergenstahl and coworkers (54, 55) and the mechanism was somewhat better elucidated.

The concept of stabilizing emulsions by solid particles (mechanical stabilization) was partially demonstrated, in an old and incomplete study (56), showing that colloidal microcrystalline cellulose (CMCC) particles can adsorb in a "solid form" on to oil droplets at the interface of a W/O emulsion and thus improve their stability by mechanical action. Oza and Frank (56) tried the mechanical stabilization concept on double emulsions. Their report shows some promise in improving stability and in slowing down the release of drug. This study was, for several years, the only example of applying the concept of mechanical stability to double emulsions. In a recent paper, Khopade and Jain (57) repeated the use of a similar process and managed to stabilize W/O/W emulsions by using MCC (microcrystalline colloidal cellulose) particles at both interfaces. The droplets were small and the yield of the multiple emulsion was fairly good. The increasing concentration of MCC in either the internal or external phase increased droplet sizes. These systems showed promise in tuberculosis therapy.

Recently, Garti et al. (58) carried out some experiments with micronized particles of the $\alpha$- and $\beta'$-polymorphs of tristearin fat together with polyglycerol polyricinoleate (PGPR) as the internal emulsifiers in double emulsions. Solid fat particles did not sufficiently stabilize the W/O emulsion and similarly the PGPR (at the concentrations used in the formulation) did not provide good stability. It was, however, shown that the use of the blend of the two components composed of solid submicrometer fat particles of $\alpha$- and $\beta'$-polymorphs (which are more hydrophilic than the $\beta$-form and thus wet better the oil–water interface) "precipitate" on to the water droplets and "cover" them. The fat particles bridge between the water droplets and sinter them only if a lipophilic surfactant (PGPR) was coadsorbed on to the water–oil interface (the W/O emulsion) (Fig. 15). The authors interpretation of the results is that "the fat particles adsorb on to the hydrophobic emulsifier film and both, the solid particles and the emulsifier, wet the water and spread at the interface." The double emulsions prepared by this technique were more stable than those prepared with monomeric emulsifiers (Fig. 16). The release patterns were also examined.

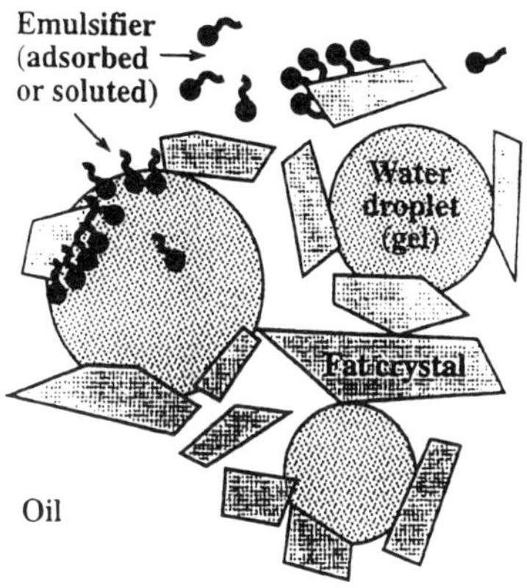


**Figure 15** Schematic illustration of colloidal margarine structure demonstrating the role of emulsifiers and fat crystals in stabilizing W/O emulsion of margarine by colloidal fat crystals.

Organophilic montmorillonite is an interesting clay that gained some interest in emulsion technology. Stable O/W/O emulsions with components consisting of hydrophilic nonionic surfactant (hydrogenated, ethoxylated castor oil, HCO-60), organophilic montmorillonite and commercial nonionic surfactant (DIS-14) were made (59). The montmorillonite was added in the second step at the outer W/O/W interface. The droplet sizes decreased with increase in the HCO-60 (0.1–3 wt %) concentration. The viscosity of the double emulsion increased as the concentration of the montomorillonite and DIS-14 increased, indicating that the excess amount of inner oil phase is adsorbed by the outer oil phase. The results indicate that the weight fraction of the inner oil phase should not exceed 0.3 wt % for a stable O/W/O emulsion since the viscosity of the double emulsion is so high that the formulation becomes semisolid.

## VIII. STABILIZATION BY INCREASED VISCOSITY

It is obvious that restricting the mobility of the active matter in the different compartments of the double emulsion will slow down coalescence and creaming, as well as decrease the transport rates of the drug

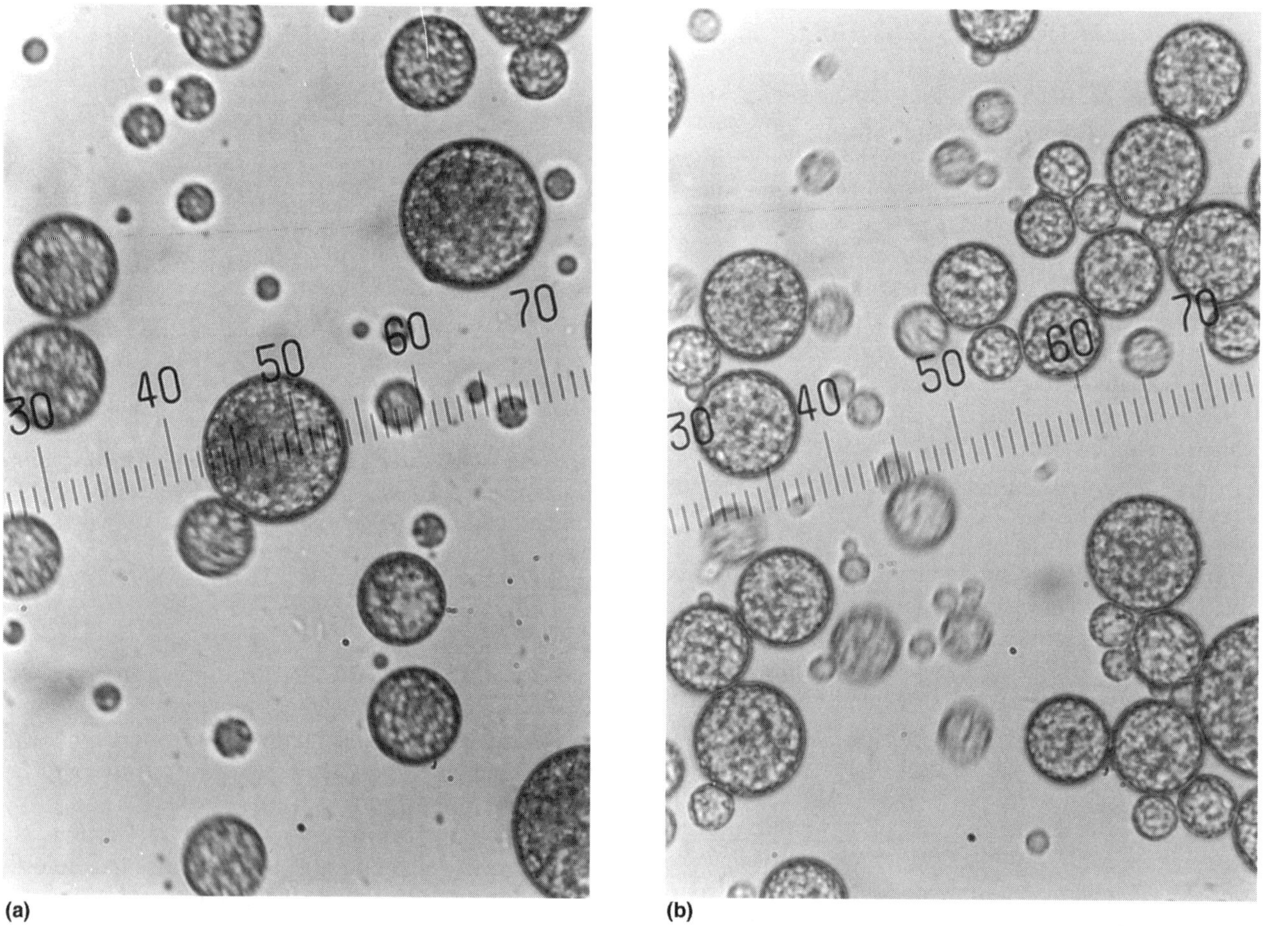


**Figure 16** Photomicrographs of double emulsions (W/O/W) stabilized with fat crystals and PGPR (polyglycerol polyricinoleate) at the inner interface and Tween 80 at the outer interface: (a) after 24 h; and (b) after 3 weeks.

or the marker from the water phase through the oil membrane. Attempts were made: (1) to increase the viscosity of the internal aqueous phase by adding gums/hydrocolloids to the inner water phase. Such a thickener may affect also the external continuous phase since the entrapment is not quantitative and the yields of entrapment are limited and emulsifier-dependent; (2) increase the viscosity of the oil phase (fatty acids salts); and (3) thickening or gelling the external water by gums is limited only to cosmetic or similar applications in which semisolid emulsions are directly applied (Tables 4 and 5). Some of these examples are topical skin-care products, creams, and body lotions (60, 61).

Double emulsions that were solidified after preparation may suffer from destabilization effects. This phenomenon is scarcely considered but in practice it can occur very often. The solidification occurs because of temperature changes [temperatures can fluctuate from subzero of (ca −20°C) to ca 40°C] during transport or storage. Clausse and coworkers (62, 63) studied the phenomenon in W/O/W emulsions by microcalorimetric (DSC) techniques. It was concluded that, out of thermodynamic equilibrium, double emulsions may suffer from water transfer during the solidification. This phenomenon occurs even if partial solidification takes place. In addition, a change in the size distribution of emulsion droplets is observed. The mean diameter of the droplets in the W/O emulsion may shift toward O/W emulsion and the double emulsion can invert. Therefore, it is not always obvious that increasing the viscosity, gelation, or partial solidification improves emulsion stability.

**Table 4** Example of a W/O/W Multiple Emulsion Stabilized by a Calcium Alginate Gel Layer at the Internal Aqueous/Oil Interface

| | | |
|---|---|---|
| Oil Phase | Hexaglycerol mixed ester | 1% |
| | CSL (calcium stearyl-2-lactylate) | 0.75% |
| | Soybean oil | 23.25% |
| Internal aqueous phase | 2% Solution of SAlv (sodium alginate low viscosity) | 25% |
| Outer aqueous phase | Polysorbate 20 | 0.36% |
| | Water | 9.14% |
| | Vinegar | 12% |
| | Sucrose | 6% |
| | NaCl | 2.5% |
| | Aqueous phase + 1% xanthan gum | 20% |

**Table 5** Composition for Low-fat Spread Using a W/O/W Formulation

| | | |
|---|---|---|
| External aqueous phase | Salt | 1% |
| | Gelatin | 3% (270 bloom) |
| | Maltodextrin DE = 7[a] | 10% |
| | Water | 85% |
| | Sodium caseinate | 1% |
| W/O emulsion | Water | 57.9% |
| | Rapeseed oil | 40% |
| | PGPR | 2% |
| | Flavor | 0.1% |

[a]De: dextrose equivalent.

## IX. MICROCAPSULES OR MICROSPHERES IN THE INTERNAL PHASE

Entrapping the active matter in solid or semisolid particles will dramatically decrease their release rates. Such double emulsions can be stored, before use, for prolonged periods without transporting the active matter to the outer interface. Upon use the double emulsion will be heated or sheared and the solid internal matrix will be ruptured and the active matter should be released. The major problem in practising such technology is the difficulties arising in dispersing (and in keeping them stable) the micro- or nano-particles in the continuous water phase. Using solid-encapsulation techniques, microspheres and nanoparticles, were tested as a replacement for the classical W/O emulsion. A few experiments were carried out showing that release can be slowed down, but the stability of these systems is very limited (64–78). Such methods are applicable only for emulsions that can be freshly prepared prior to their use. It is our hope that more efforts will be made in this direction.

## X. VESICLES-IN-WATER-IN-OIL (V/W/O) EMULSIONS

There are some potential applications in which the external phase is nonaqueous. Florence and co-workers (79, 80) and Albert et al. (81) described systems in which the aqueous suspension of vesicles are dispersed in the continuous oil phase. The technique was exercised both for the study of its use in drug-delivery systems and as immunological adjuvants, as well as an intrinsically interesting colloidal system. The system is an emulsion prepared from a dispersion of niosomes in water, re-emulsified in an oil using a surfactant mixture of low HLB to achieve a stable W/O emulsion.

The product, which was termed by Florence and coworkers a V/W/O emulsion, is a close analog to an O/W/O emulsion. The authors have studied parameters such as the type of surfactant on the in-vivo release of a drug and have found that the degree of hydrophobicity of the Span surfactant had a significant influence on the release rate. Spans 60 and 40 released the drug at the lowest rate, while the more hydrophilic surfactants increased the release rates. The authors have interpreted the results in terms of gelation of the oil phase in the presence of the more hydrophobic surfactants (Fig. 17). Other factors such as the nature of the oil phase (octane, hexadecane, and ispropyl myristate), and the effect of temperature were also studied. The results were encouraging, but no dramatic improvement was made on the release and transport phenomena.

## XI. EMULSIFIED MICROEMULSIONS ($L_2$/W)

In an attempt to reduce the sizes of the internal droplets it was suggested, by Pilman et al. (82), to replace the internal emulsion by a microemulsion or with thermodynamically swollen stable reverse micelles (an $L_2$ phase). In a short report they conclude that such an option is feasible. One can emulsify $L_2$ reverse micelles

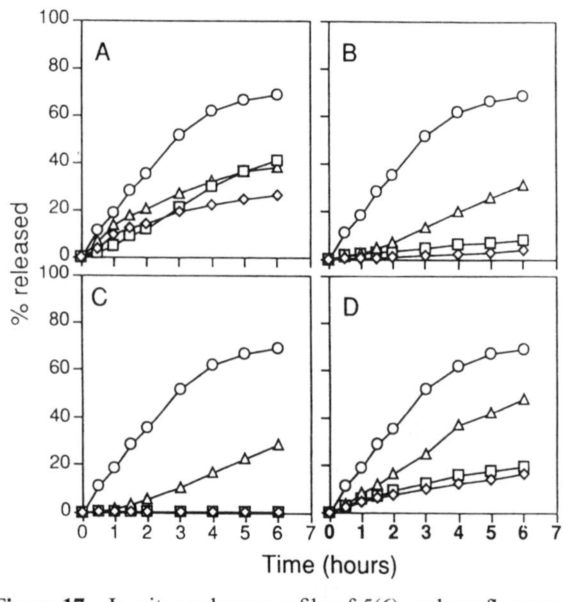


**Figure 17** In-vitro release profile of 5(6)-carboxyfluorescein (CF) from various formulations prepared with (A) Span 20, (B) Span 40, (C) Span 60, and (D) Span 80; ○: CF solution; △: vesicles, □: w/o emulsion; ◇: V/O/W emulsions. (From Ref. 80.)

or microemulsions to form very small, external droplets (dictated by the size of the internal droplets). The paper does not give sufficient experimental details nor does it have much stability and structural data, but it serves as seeded idea to retry the concept. We (83) have prepared reverse microemulsions with Aerosol OT (AOT), in which small amounts of marker (2 wt % NaCl) were entrapped. The reverse micelles are thermodynamically stable. The $L_2$ phase was further added to a water solution containing various hydrophilic emulsifiers. The "emulsified microemulsion" has a stability similar to that of the classical monomeric emulsion, with better (slower) release rates. The problem is that since the internal droplets are so small it is impossible to detect them by classical optical methods, and it requires the use of advanced techniques such as cryo-TEM, SAXS (small angle x-ray scattering), and SANS (small angle neutron scattering) linked to self-diffusion NMR methods in order to evaluate the internal microstructure. The work is still in progress.

## XII. STABILIZATION BY POLYMERIC SYNTHETIC AMPHIPHILES

Stable double emulsions, based on various block copolymers of polyethylene oxides and polypropylene oxides known as Pluronics, have been used. In a recent example, Cole and Whateley (84) have used complexes of Pluronic F127:PAA (polyacrylic acid) in the internal aqueous phase. In the oil phase, Span 80 and Pluronic L101 (5 wt %) were used. The outer interface was stabilized by xanthan gum (0.25 wt %) and Tween 80 (1 wt %). Theophylline and $^{125}$I-insulin (iodinated insulin) were incorporated in the internal aqueous phase of the stabilized multiple emulsion, and the release rates were studied. The release rates were found to be related to the droplet sizes of the emulsion which were dependent on the particle size of the pluronic F127:PAA complex in the internal aqueous phase and the type of the lipophilic surfactant in the oil phase. The authors have used the complex between the poloxamer surfactant and PAA that occurred at pH 2 and at low molar ratio as a barrier for the release of active matter from the inner to the outer phase.

The fact that poloxamers and proteins were excellent steric stabilizers for double emulsions encouraged scientists to try and design an "optimal synthetic polymeric emulsifier" to be adsorbed both at the internal and the external interfaces. However, only few commercial polymeric surfactants are available, many of which are designed for other applications. It was, therefore, a challenging task for Garti et al. (85–87) to design and prepare a comb-like graft copolymer based on a modification made on the commercial polymer known as PHMS–PEG, which is a polyhydrogensiloxane grafted with polythene glycol side chains. The modification added a flexible hydrophobic side chain (termed "spacer") to the backbone of the polymer on which the hydrophilic groups were grafted. A family of more than 16 products were synthesized and characterized. The synthesis opened an option of preparing several new families of emulsifiers with lipophilic and hydrophilic characteristics: (1) polymers with a wide range of grafting with various degrees of substitution (various DgS′); (2) polymers with different side ethylene glycol chain lengths [various EO repeating units $(EO)_n$]; and (3) various lengths of the backbone (various degrees of polymerization, DPs′). The surface properties of all the amphiphilic polymers (four hydrophobic and 12 hydrophilic polymers) were measured. It was demonstrated that most of them could reduce the interfacial tensions of paraffin oils to very low values (below 10 mN/m). Emulsification was performed with various oils at various fractions. The emulsification did not require severe homogenization. The W/O droplets were stable and small and so also were the O/W emulsion droplets. In a two-stage emulsification process the water was added to the solution of the lipophilic emulsifiers and oil, and the hydrophilic emulsifiers were added to the external aqueous phase. The hydrophilic emulsifiers adsorbed on to the external interface. Excellent double emulsions were obtained, upon adding the W/O emulsion to the water phase without homogenization (Fig. 18), with small droplets and a narrow droplet size distribution. Excellent stability was achieved and the ability to keep the emulsions stable upon dilution was also remarkable. Microscopic observations indicated that the polymeric emulsifier had coated the interfaces with thick multilayer films.

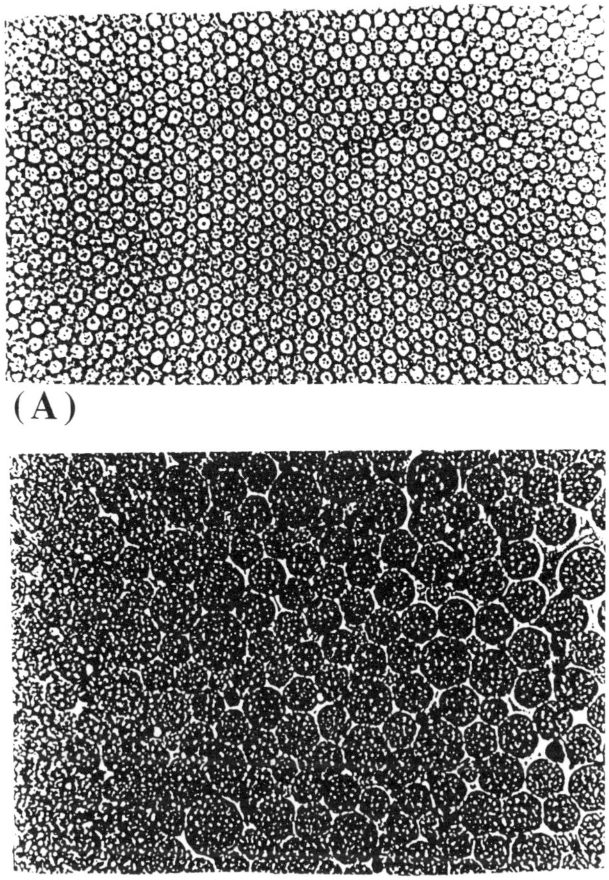

(A)

(B)

**Figure 18** Photomicrograph of typical W/O/W double emulsion stabilized with tailor-made grafted PEG side chains on to poly(dimethylsiloxane) backbone (see text) after homogenization and centrifugation: (A) magnification ×200; and (B) magnification ×500

The 'marker' (NaCl) was released extremely slowly (Fig. 19). When polymeric emulsifiers were used at both interfaces the release rates were extremely low. The authors also prepared emulsions with blends of the new polymeric emulsifiers and hydrophobic low-molecular-weight emulsifiers such as sorbitan esters (Span 80). It was found that the low-molecular-weight emulsifiers increased the release rates. The migration kinetics were strongly dependent on emulsifier concentration. The release rates were enhanced also when other emulsifiers, such as EDT–PGPR, were used in the internal interface. It was concluded that, while the polymeric emulsifiers, even if aggregating in the oil phase, are restricted in their solubilizing capacity as hydrophilic compounds, the hydrophobic monomeric emulsifiers (such as Span 80) can easily form, in the oil layer, reverse micelles that carry out the markers. Strong evidence for such a mechanism was found. Such a combination of monomeric and polymeric emulsifiers can serve as an efficient tool for controlling the release of markers from double emulsions.

However, the most surprising phenomenon was the fact that the double emulsions were easily homogenized without "transporting" the markers into the aqueous solution during the emulsification process. The homogenization process yielded almost monodispersed double-emulsion droplets with very narrow size distribution ranging from 3 to 7 μm (Fig. 18).

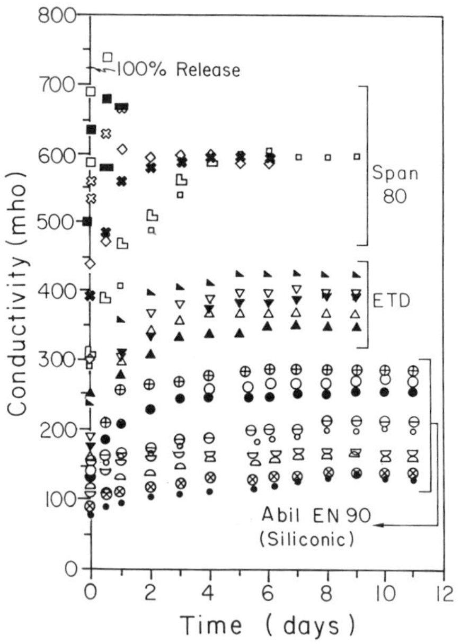


**Figure 19** Plot of conductivity of the outer aqueous phase (reflecting the concentration of NaCl in the outer phase (% release) vs. time (days) in three sets of double emulsions. (1) All circles indicate use of Abil EM-90 as hydrophobic Emulsifier I. The lower curves (●) indicate the most hydrophilic PHMS–PDMS–UPEG Emulsifier II, and the upper curves (○) in each set indicate the most hydrophobic PHMS–PDMS–UPEG with 52% substitution and 45 EO units. Each circle represents a different polymeric emulsifier. (2) All triangles represent use of polyglycerol polyricinoleate (ETD) as Emulsifier I, and the curves are arranged again with increasing hydrophobicity of Emulsifier II. The lower curve (▲) represents the most hydrophilic emulsifier, and the upper curve in the set represents the most hydrophobic one (▶). (3) All squares represent the use of Span 80 as Emulsifier I, and the curves are arranged with increasing hydrophobicity of Emulsifier II.

A close examination of the release profiles of markers (NaCl or drug) from these double emulsions revealed that the release occurs in three stages. A long lag time (Stage A) in which almost no release takes place is observed immediately after preparation (Fig. 20, slope A). The lag time depends on the nature of the internal lipophilic polymeric emulsifier and its concentration ratio to the monomeric lipophilic emulsifier. In the second stage (Fig. 20, slope B) the release slope is gradual and diffusion controlled, and complies with the Higuchi mechanism of release from a "slab into a sink" by a reverse micellar mechanism. In the third stage (Fig. 20, slope C) the release is very slow and levels off. The fact that the release mechanism is via reverse micelles has a significant advantage since it allows us to control the transport through the liquid membrane film by controlling the number of micelles present in the oil membrane. The polymeric emulsifiers that dissolve or aggregate in the oil phase do not have the characteristics of reverse spherical micelles. It is assumed that the polymers aggregate in rather "open" (unfolded) random coils. Therefore, such structures cannot solubilize much water and cannot transport the water molecules or the water-soluble marker. Emulsions prepared with polymeric emulsifiers alone practically do not release the marker upon storage. Very long lag times were detected and only minor quantities of marker were released at very slow rates. On the other hand, the use of variable concentrations of Span 80 (the lipophilic monomeric emulsifier) allowed one to control the lag time, the released amount of matter, and the rates. As one increases the amounts of Span 80 more marker is released. The lag time is shortened and the rates increase with the increase of added Span 80 (Fig. 21). One can add the monomeric lipophilic emulsifier either during emulsification (with given and calculated release kinetics) or, better, by adding it dropwise and/or "at need" after the emulsification or before using the double emulsion. Such "delayed addition" will allow one to store the emulsion for prolonged periods without transporting the active matter and to trigger its release at another release rate upon application.

The relevance of the control of the reverse micelles and dependence of the release kinetics on the reverse-micelles' concentration and size was demonstrated. The amount of solubilized water in the oil phase, as a function of Span concentration, was measured (after centrifugation). It was found that water is present only in very minor quantities when the siliconic emulsifier is employed by itself. The water concentration increased as the amounts of Span 80 were raised (87). These findings are also in good correlation with the release rates and the lag time.

It was, therefore, concluded that the internal polymeric emulsifier controls the release rates both by improving the film formation on the interface and by restricting the formation of reverse micelles in the oil phase. It is assumed that the presence of the two emulsifiers (Span 80 and silicone lipophilic emulsifier) form "reverse hemimicelles." These structures are capable of solubilizing less water and, therefore, less marker, a fact that leads to slower release and the ability to control better the release rates. Emulsions prepared from polymeric emulsifiers and without the lipophilic mono-

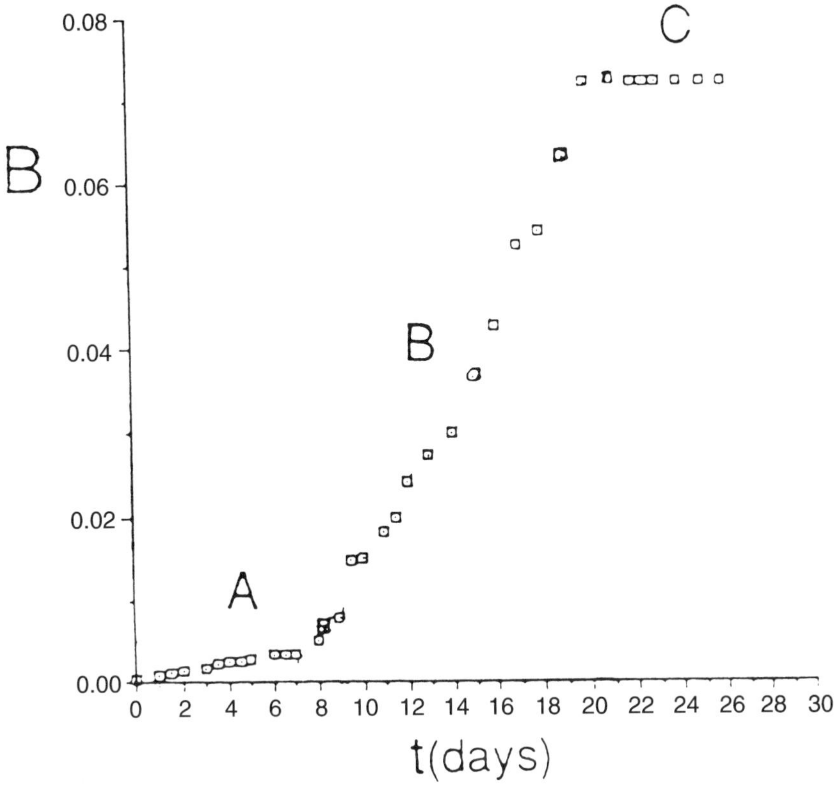


**Figure 20** Plot of factor $B$ (proportional to the release rate) as a function of time ($t$) for double emulsions stabilized with 5 wt % Abil EM-90 as the inner emulsifier and PHMS–PDMS–52.5% UPEG-45 (see Ref. 87) as the external emulsifier.

meric emulsifier remained stable on the shelf for over 6 months and did not show any leaking of the marker upon storage. Once the monomeric emulsifier was added dropwise, at various concentration levels, the release started and was completely controllable by

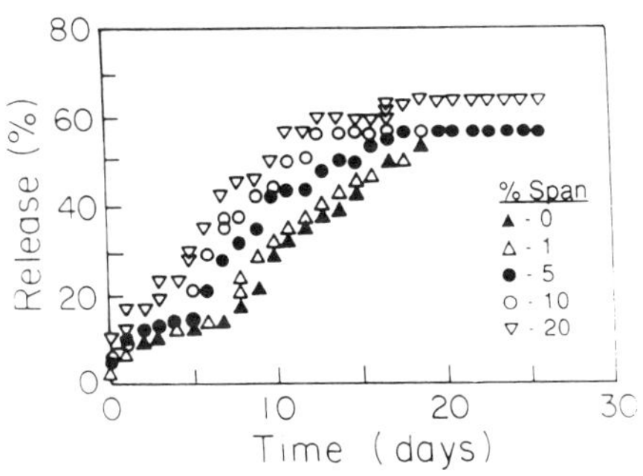


**Figure 21** Release profiles of double emulsions stabilized with polymeric hydrophobic emulsifier at the internal interface (Abil-EM-90) and a combination of hydrophilic siliconic polymeric emulsifier (PHMS–PDMS–52% UPEG-45EO) with various concentrations of Span 80.

both the rate of addition of the monomeric emulsifier and by its concentration.

An interesting modification of using polymeric materials for stabilizing double emulsions was exercised by Khopade and Jain (88). "Stealth" multiple emulsions of small size, consisting of 6-mercaptopurine, were prepared by incorporation of sphingomyelins (SMs) and monosialogangliosides [GM(1)] in the oil phase and by coating it with lipid–graft polyethylene glycol (PEG-PC). The three types of "stealth" double emulsions were characterized for size distribution, viscosity, encapsulation efficacy, drug release, and stability. The results suggest that PEG-PC-coated multiple emulsions are superior as prolonged release and extended blood-circulation carriers compared with double emulsions bearing either SM or GM(1). The authors (35) have used other drugs as well and included also a sonication step. The reported results on the anticancer activities of drugs entrapped in the double emulsions by this technique are quite encouraging. Similarly, the authors used also cocavalin-A and carbodiimide as ingredients for the formation of "lectin-functionalized double emulsions" (89). The anticancer activity was, similarly, superior to that of the noncoated double emulsions (Fig. 22).

In summary, all the recent improvements in the double-emulsion compositions (emulsifiers and oils), the improved evaluation related to the "weight" given to any of the possible instability mechanisms, and the better understanding of the instability factors that were achieved in the last 15 years of research work were supposed to solve most of the scientific problems of this technology. Yet, only very limited improvements in the stability of the emulsions and in extending their shelf-life have been recorded. There is practically very limited control of the release of the additives or the active matter.

## XIII. DOUBLE EMULSIONS AS INTERMEDIATES FOR NANOPARTICULATION

Advanced double-emulsion formulations are no longer prepared for the purpose of simple delivery and release of active matter, but have changed application directions. Three main new directions can be clearly seen: (1) double emulsions as intermediates for the preparation of solid microspheres or microcapsules; (2) O/W/O double emulsions for improved solubilization and chemical protection of water-insoluble active matter; and (3) double emulsions for selective adsorption of certain compounds for extraction and purification purposes.

Some major examples among the classical delivery applications in pharmaceuticals and food technology will be described that will emphasize the new emerging applications.

### A. Controlled-delivery Applications

The intrinsic instability of the double emulsion caused difficulties to formulators and many of the investigators have decided to abandon this technology (8, 9). However, one should bear in mind that the potential applications for double emulsions appears to be enormous, mainly in the areas of food, cosmetics, medicine, and pharmaceutics. Throughout the years potential applications have been demonstrated in: (1) improved biological availability (parenteral nutrition, anticancer drugs); (2) delivery of drugs (sustained release of narcotic antagonistic drugs, prolonged release of corticosteroids, slow release of Bleomycin, and targeted release of anticancer drugs); and (3) adsorption of toxic compounds. The technology has much promise also in non-

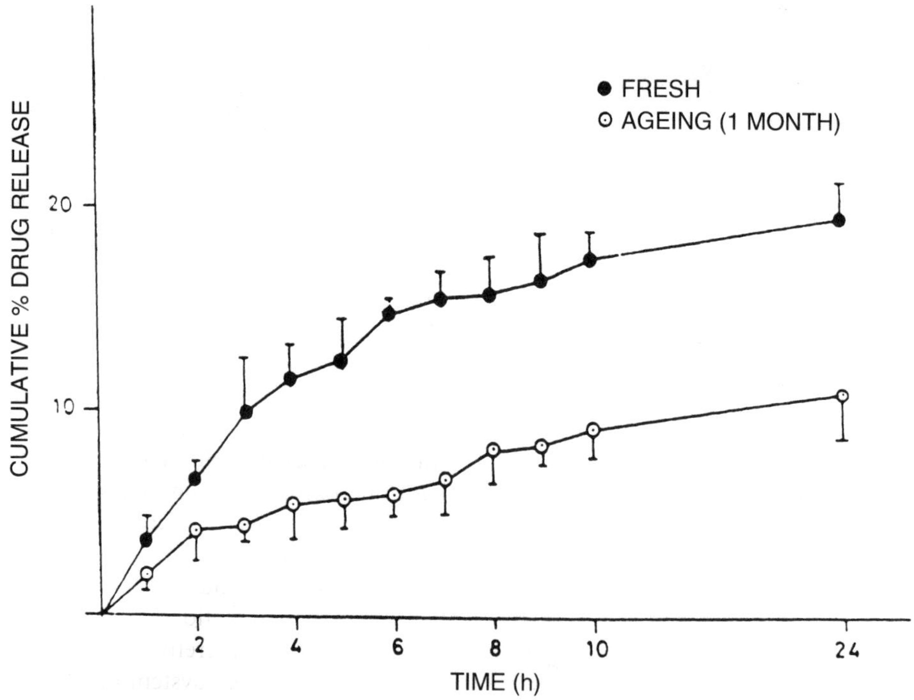


**Figure 22** Drug-release profile of isoniazid from multiple emulsion based on uni/oligo/droplet internal phase. (From Ref. 57.)

pharmaceutical areas of slow and controlled release of materials such as fertilizers and pesticides for agricultural formulations as well as for controlled release for cosmetic, industrial, and food applications (8, 9).

A preparation of Ketamine [2-(Chlorophenyl)-2-(methylamino) cyclohexanone, $C_{13}H_{16}ClNO$, an anesthetic agent] in an O/W/O multiple emulsion for prolonged drug release was formulated and evaluated. Ketamine is used as a short-acting anesthetic in humans and in some animal species (90). Ketamine is poorly bound to plasma proteins and has a half-life of approximately 4 h following intravenous injection.

Ketamine leaves the blood very rapidly to be distributed into the tissues. The recommended dosage of intravenous Ketamine is 2.5–20 mg/kg. The objective of the study was to test the concept that a multiple emulsion could be formulated which has high porosity and lower viscosity at 37°C consistent with its intended use for sustained drug release and to prolong the half-life of the anesthesia. The results showed that 8.2% of the Ketamine was released (100 mg/ml in the inner phase) after 10 min, 67.0% at 30 min, and 95.5% at 60 min from the Ketamine/O/W multiple emulsion in a well-controlled manner (Figs. 23 and 24).

In another example, long-circulating multiple-emulsion systems for improved delivery of an anticancer agent of small droplet sizes, containing 6-mercaptopurine (88), were prepared by incorporation of sphingomyelins (SMs) and monosialogangliosides (GMs) in the oil phase and by coating with lipid-grafted polyethylene glycol (PEG-PC). Three types of "stealth" multiple emulsions were characterized for size distribution, zeta potential, viscosity, encapsulation efficiency, drug release, and stability. Drug-disposition studies were performed with formulated multiple emulsions to assess "stealth" behavior. The results suggest that PEG-PC-coated multiple emulsions are superior as prolonged-release and extended blood circulating carriers compared to multiple emulsions bearing either SM or GM.

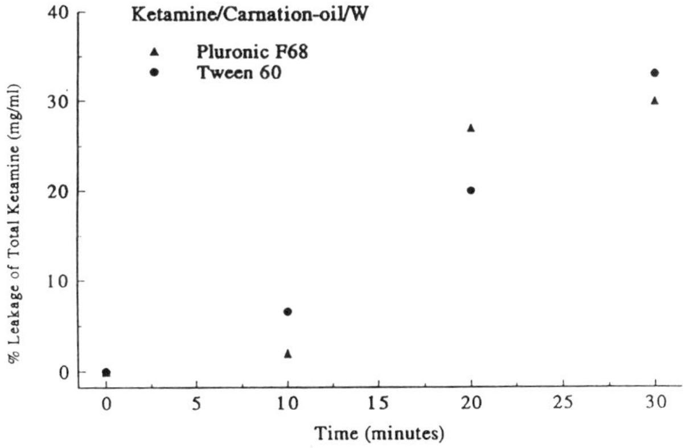


**Figure 23** Effect of surfactant on Ketamine release from the multiple emulsion with shearing. (From Ref. 90.)

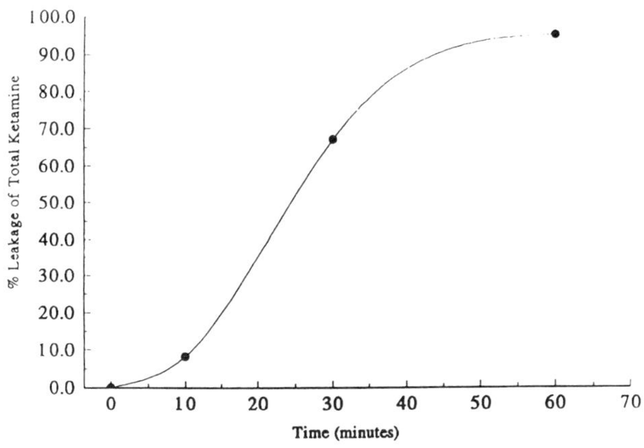


**Figure 24** Ketamine release from the multiple emulsion with shearing. (From Ref. 90.)

The insulin efficacy in double emulsions is an additional interesting example. The purpose of that study was to evaluate the hypoglycemic effects of W/O/W insulin emulsions containing lipoidal absorption after enteral administration to rats (91). The hypoglycemic effects of insulin were examined during an in-situ loop method in rats. The insulin emulsions prepared with soybean oil, triolein, or trilinolein slightly but significantly decreased the serum glucose levels compared to the insulin solution. By addition of 3 wt % limonene or 3 wt % menthol to the triolein emulsion, the hypoglycemic effect of insulin was promoted in the ileum but not in the colon. Strong hypoglycemic effects were observed with the triolein emulsion containing 2 wt % fatty acids such as oleic, linoleic, and linolenic acids. The remarkable enhancing effects occurred more predominantly in the colon than in the ileum. The effect of degree of unsaturation of the fatty acids was not observed. No tissue damage was noted by light-microscopic examination of both regions treated with triolein emulsion, or triolein emulsion containing menthol or oleic acid. W/O/W emulsions containing unsaturated fatty acids are able to enhance the ideal and colonic adsorption of insulin without tissue damage and may, therefore, be useful in dosage form in an enteral delivery system for insulin.

A stable multiple emulsion containing rifampicin (92) in the internal aqueous phase was prepared by the incorporation of additives in both the aqueous and oily phases. The formulation and process variables were optimized for primary and secondary emulsification. Drug-release studies were performed on selected multiple emulsions to observe the effect of dilution. The release data were fitted to the Higuchi equation for slab and spherical geometry, and effective diffusion coefficients were calculated. Stability studies undertaken for three months revealed good stability of the multiple emulsions with respect to creaming, phase separation, viscosity change, drug leakage, change in droplet size upon storage, and exposure to osmotic and shear stress. The in-vivo studies performed with stable multiple emulsions administered orally in human volunteers showed prolonged plasma drug levels. The multiple emulsions were found suitable for improved tuberculosis therapy.

Double emulsions may offer some advantages for food applications mainly with relation to their capability to encapsulate (or entrap) in the internal compartments some water-soluble substances that are then slowly released. The double emulsion can also be used in the food industry where an external water phase is more acceptable in terms of palatability than an oil one (93, 94). On this basis several new products have been patented in the form of W/O/W emulsions, as salted creams (encapsulation of salt), aromatic mayonnaise, etc. (95–98). Further food applications are related to the double-emulsion dielectric properties; for example: one can prepare a W/O/W system having the same volume fraction of the dispersed phase and the same texture as a simple emulsion, but with a lower oil content (due to the existence of the aqueous compartments in the food globules), i.e., low-calorie mayonnaise (93).

## B. Double-emulsion–Solvent-evaporation Techniques for Preparation of Microspheres

Drugs, cosmetic ingredients, and food additives are microencapsulated for a variety of reasons, which include reducing local side-effects, controlled release, site-specific (drug) delivery, and drug targeting. A tremendous amount of research work has been done in a search for suitable methods to achieve an effective encapsulation of water-soluble active matter. The physical characteristics of the microspheres produced largely determine their suitability for use for different objectives. Microspheres are prepared from both natural and synthetic polymers.

Among microencapsulation techniques, the double emulsion–solvent-evaporation method, also known as "the in-water drying method," is one of the most useful methods for entrapping water-soluble compounds (64–78). Figure 25 shows schematically the preparation technique. Over 100 papers and many patents have been published in the course of the last five years on thee use of this technique. It is behind the scope of this review to screen or to categorize them. We have selected only some examples that are of a more significant value.

It is known that the preparation of microspheres by using O/W emulsions is not an efficient method for the entrapment of water-soluble drugs as the compounds rapidly dissolve in the aqueous continuous phase and are lost. It has been widely accepted that the problem of inefficient encapsulation of water-soluble drugs can be overcome by using the double-emulsion solvent-evaporation technique. Water-insoluble drugs are usually satisfactorily encapsulated by the O/W emulsion technique (64, 65).

The W/O/W emulsions are generally used for encapsulating proteins or peptides. These highly water-soluble molecules are quantitatively introduced into the internal aqueous phase of the multiple emulsions, which results in an increased encapsulation efficiency for the microcapsules in comparison with particles produced by the single-emulsion–solvent-evaporation method. The particular location of the proteins induces a stabilizing effect on the two emulsions which, in turn, contributes to a successful stabilization of the double emulsion and loading.

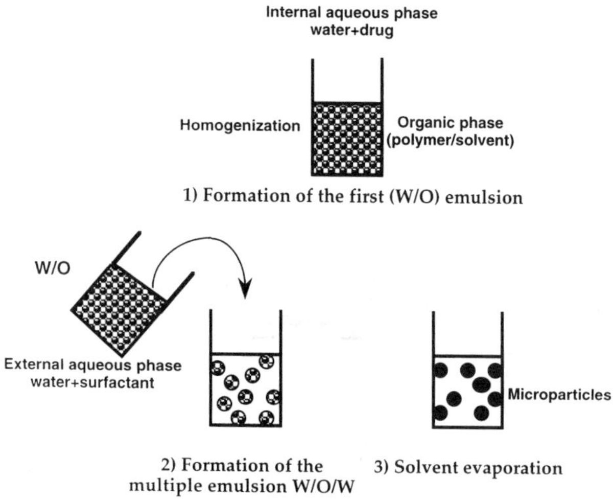


**Figure 25** Preparation of microsphere-containing peptides by the multiple-emulsion solvent-evaporation technique.

The double-emulsion–solvent-evaporation technique is commonly used to prepare biodegradable hydrophobic microspheres containing hydrophilic pharmaceuticals, proteins, and polypeptides for sustained-release applications (66, 70–78). In most cases the microspheres are in the range size 10–100 μm. However, recently, Blanco-Prieto et al. (71) managed to reduce the microcapsule sizes to less than 5 μm.

The microspheres, consisting of poly(L)lactide, poly(DL)lactide, poly(DL)lactide-*co*-glycolide, polyhydroxybutyrates of various molecular weights, and poly(hydroxybutyrate-*co*-valerate), were characterized for their structure, size distribution, drug loading, release kinetics, surface morphology, and hydrophobicity (69, 78) (Figs. 26–29). The influence of these properties on the dynamics of the immune response, following topical administration, was studied. The hydrophobic nature of polyhydroxybutyrate microspheres, compared with those formed using polylactides, was confirmed and the generation of a significant immune response was delayed using these preparation. Also, the time course of immune responses generated using a range of polyhydroxybutyrate molecular weights was examined.

Couvreur et al. (70) reviewed the preparation and characterization of many of the different types of solvent-evaporation microspheres and mostly discussed small poly(lactic-*co*-glycolic acid) microspheres (mean size lower than 10 μm) containing small peptides (Fig. 30). Three main evaporation strategies have been utilized in order to increase the encapsulation capacity: an interrupted process, a continuous process, and the rotary evaporation procedure.

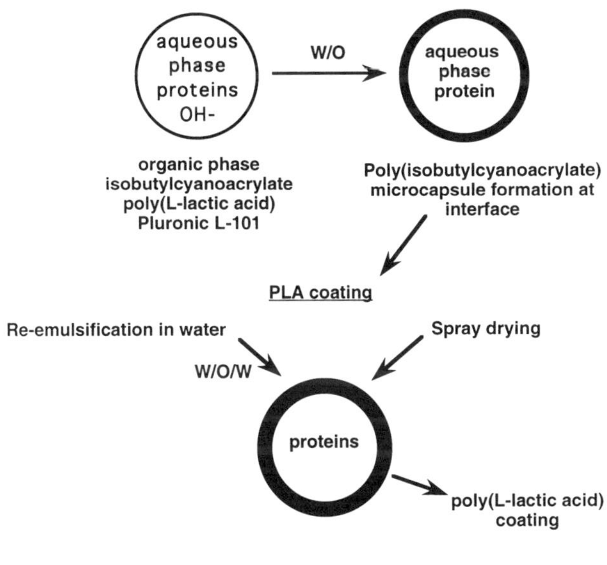


**Figure 26** Schematic description of preparing PLA-coated PIBCA (polyisobutylcyanoacrylate) microcapsules that contain protein molecules. (From Ref. 78.)

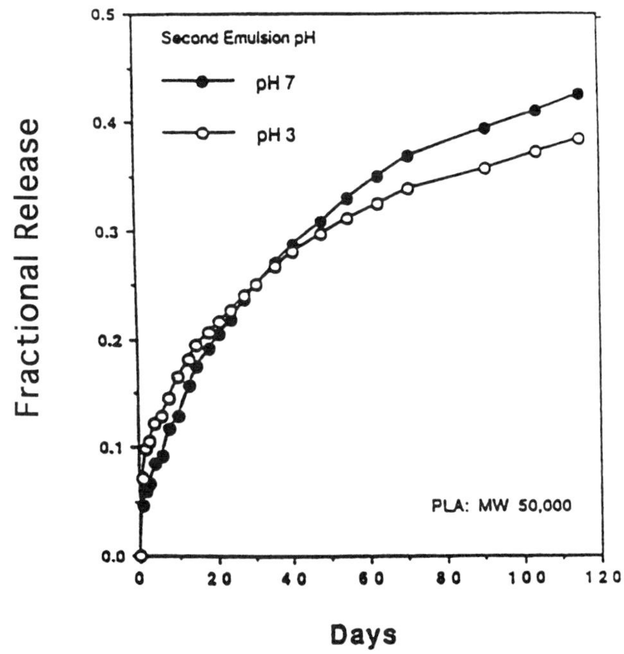


**Figure 27** BSA release profiles from the PLA MW 50,000 coated PIBCA microcapsules prepared by the re-emulsification method at two different pH values. (From Ref. 78.)

Much work has been devoted in recent years to preparing microparticles of narrow size distribution, with different biodegradable polymers. The sizes of common microcapsules are 40 to 50 μm (70, 71, 76–78).

Liquid–liquid emulsification is a critical step in the double-emulsion microencapsulation process (W/O/W or O/W/O). It was found that the size of these droplets decreases with increasing homogenization intensity and duration. The emulsion droplet size depends as expected on viscosity, total volume size, and volume ratio of the continuous phase to the dispersed phase; the rotor/stator design was also investigated. All these physical parameters influence the structure of the microspheres obtained by this technique.

An interesting example is the incorporation of a protein-based drug in microspheres prepared from a hydrophobic polymer via double liquid–liquid emulsification (W/O/W) or by dispersing a powdered protein in a polymer solution followed by liquid–liquid emulsification (S/O/W) (72). Bovine serum albumin (BSA) was used as the model protein and poly(methyl methacrylate) (PMMA) was used as the model polymer. The

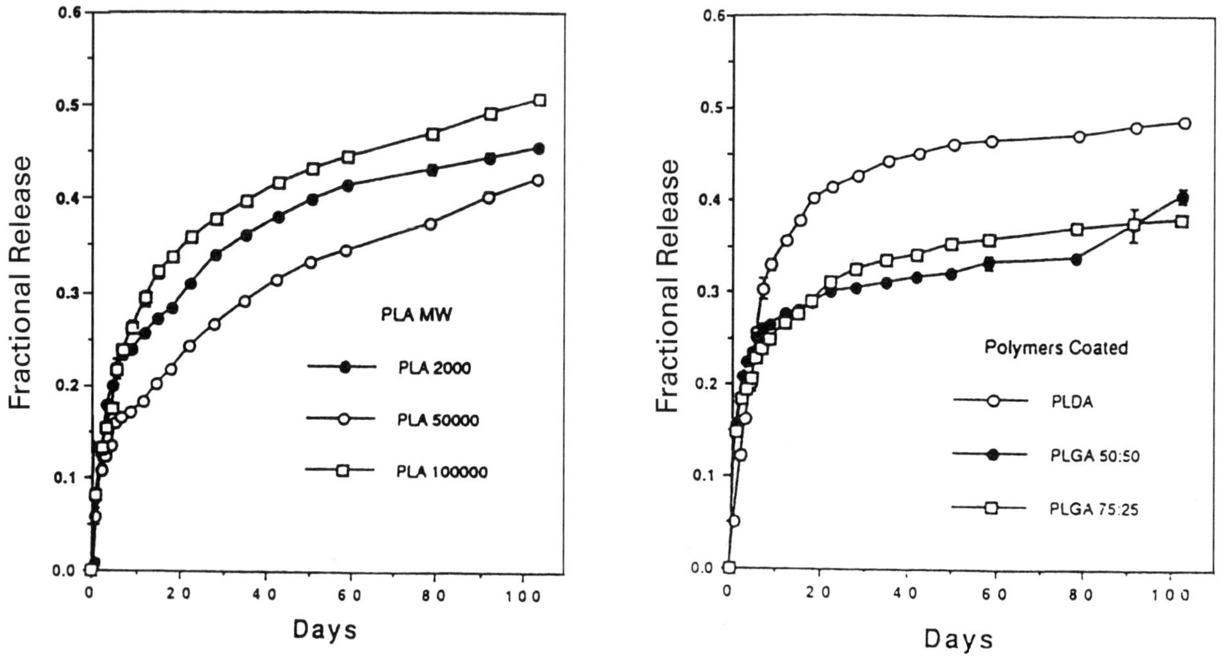


**Figure 28** (A) BSA-release profiles from the PLA-coated PIBCA microcapsules prepared by spray-drying method; (B) BSA-release profiles from poly(DL-lactic acid) and poly(lactic-*co*-glycolic acid)-coated PIBCA microcapsules. (From Ref. 78.)

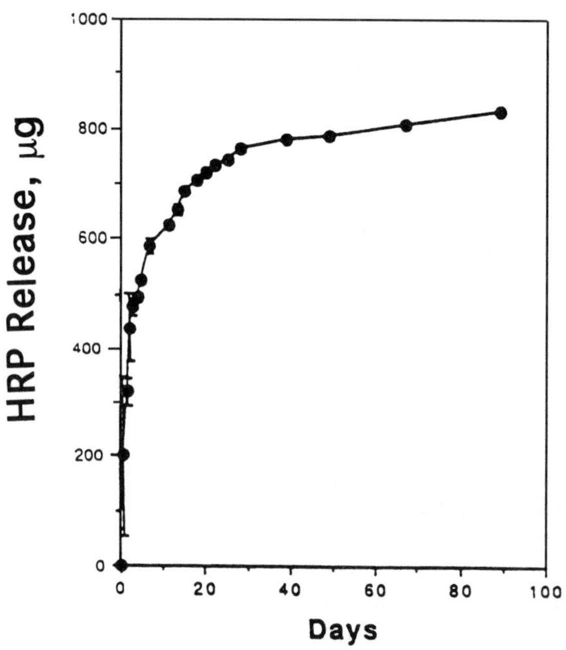


**Figure 29** Horseradish peroxidase (HRP) release profiles from PLA (MW 2000) coated PIBCA microcapsules. (From Ref. 78.)

droplet sizes of the W/O emulsion were controlled using rotor/stator homogenization. The S/O emulsion was characterized on the basis of protein powder size and shape, in both emulsification processes. The size of the micospheres thus prepared was found to increase with increasing size of the protein powder in the S/O/W system, but increased with decreasing size of the liquid emulsion droplets in the W/O/W system. Empirical correlation could accurately predict the size of the microspheres. Protein loading in the microspheres decreased with respect to increase in W/O emulsion droplet size or in protein powder size. These phenomena were attributed to two possible mechanisms: fragmentation along the weak routes in the W/O/W system and particle redistribution as the result of terminal velocity in the S/O/W system. The role of protein powder shape was not significant until the protein powder size exceeded 5 µm.

In a recent paper (77) it was demonstrated that a milk model protein, $\beta$-lactoglobulin (BLG), was encapsulated into microspheres prepared by the solvent-evaporation technique. The effect of the pH of the outer aqueous phase on the protein encapsulation and release as well as on the microsphere morphology has been investigated. It was demonstrated that as the amount of the BLG increased the stability of the inner emulsion decreased and the entrapment was

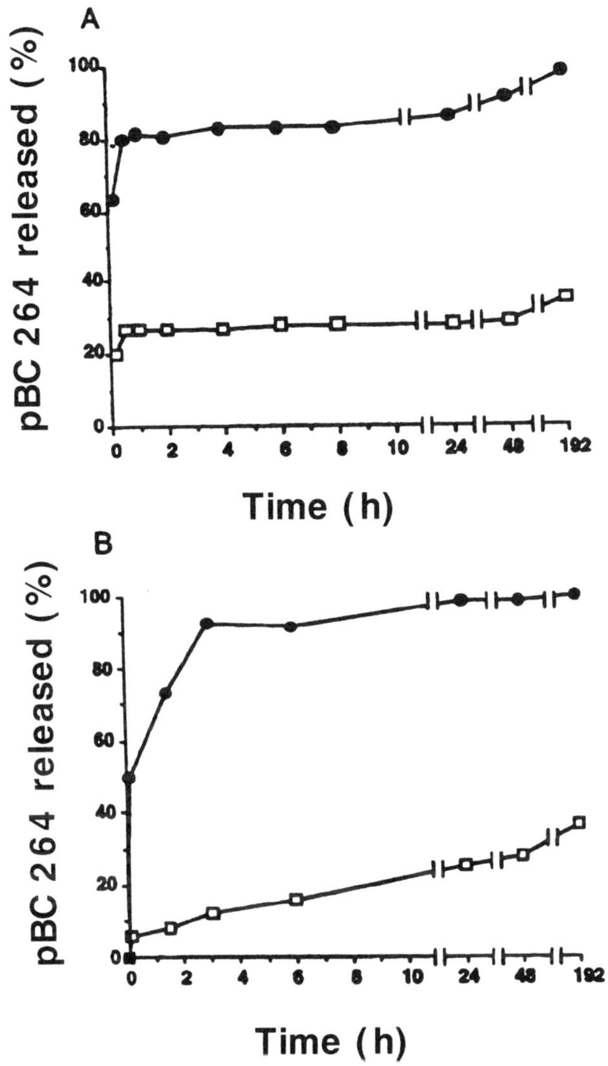


**Figure 30** Release kinetic of pBC 264 in phosphate-buffered saline (pH 7.4) (A) and in vivo, in brain tissue (B) from PLGA microspheres prepared with either ovalbumin (●) or Pluronic F-68 (□) (From Ref. 70.)

less efficient. Therefore, adjusting the combined effect of pH and the stability of the inner emulsion may lead to better entrapment. As to the release, it was demonstrated that the "burst effect," attributed to a morphology change in the microcapsules characterized by the presence of pores or channels able to accelerate the release of the BLG, was the most signficant release factor. These pores were attributed to the presence of a large amount of BLG on the surface, which aggregates during microsphere formation at pH 5.2. It was concluded that it is beneficial to lower the solubility of the protein in the outer phase in order to improve the encapsulation efficiency, although this benefit is provided by strong adsorption of the protein on the microsphere surface.

Maa and Hsu (75) reported the formation of nanoparticles by the double-emulsion method (W/O/W), using methylene chloride as an organic solvent and poly(vinyl alcohol) (PVA) or human serum albumin (HSA) as a surfactant. Experimental parameters such as the preparation temperature, the solvent-evaporation method, the internal aqueous phase volume, the surfactant concentration, and the polymer molecular weight were investigated for particle size, the zeta potential, the residual surfactant percentage, and the polydispersity index. Preparation parameters leading to particles with well-defined characteristics such as an average size around 200 nm and a polydispersity index lower than 0.1 were identified.

Some more interesting papers on protein entrapments for various oral and other intakes can be found in the literature (72–74).

Porous spherical particles of chitosan were prepared from O/W/O double emulsions stabilized with chitosan aqueous solution. The particulation was obtained by a simple evaporation technique (52).

## XIV. O/W/O DOUBLE EMULSIONS

Oil-in-water double emulsions were considered to have less potential applications and therefore were less extensively studied. However, in more recent years several new applications have been reported for O/W/O double emulsions, which sound interesting and are worth being mentioned.

The modulated release of triterpenic compounds from an O/W/O multiple emulsion formulated with dimethicones, studied with infrared spectrophotometric and differential calorimetric approaches, is one of these examples (99). The authors explored the advantages in the release of triterpenic compounds from O/W/O emulsions. They found two principal advantages: (1) the use of low molecular weight silicones decreased the oily touch of the final preparation; and (2) owing to the large range of viscosity, these excipients influenced the skin distribution of the active matter after the topical application. The effects of different dimethicones incorporated within multiple emulsions were studied, through in-vitro penetration results. The residual film on the skin was also evaluated. Correlations were established between the silicones structure and the distribution of drugs at different skin levels or between the silicone structure and the percutaneous penetration. The incorporation of silicones within O/W/O multiple emulsions seems to be an efficient means of modulating the penetration and the distribution of drugs in the skin.

In another study the stability of retinol (vitamin A alcohol) was compared in three different emulsions: O/W, W/O, and O/W/O (15). The stability in the O/W/O emulsion was the highest among the three types of emulsions. The remaining percentages, at 50°C after 4 weeks, were of 56.9, 45.7, and 32.3, in the O/W/O, W/O, and O/W emulsions, respectively. However, it was also reported that with increasing peroxide value of O/W and W/O emulsifiers, the remaining percentage of vitamin A palmitate and retinol in the emulsions increased significantly, indicating that peroxides in the formulas accelerate the decomposition of vitamin A. Organophilic clay mineral tan-oil gelling agent and a W/O emulsifier also affected the stability of retinol. The stability of retinol in the O/W/O emulsion increased with increasing inner oil phase ratio, whereas in O/W it was unaffected by the oil fraction. The encapsulation percentage of retinol in the O/W/O emulsion, and the ratio of retinol in the inner oil phase to the total amount in the emulsion, increased with increasing the oil fraction. The remaining percentage of retinol in the O/W/O emulsion was in excellent agreement with the encapsulation percentage, suggesting that retinol in the inner oil phase is more stable than that in the outer oil phase. Addition of antioxidants (*tert*-butylhydroxytoluene, sodium ascorbate, and EDTA) to the O/W/O emulsion improve the stability of retinol up to 77.1% at 50°C after 4 weeks. The authors concluded that the O/W/O emulsion is a useful formula to stabilize vitamin A.

## XV. MULTIPLE EMULSIONS RHEOLOGY

The understanding of the rheological behavior of double emulsions is quite important in the formulation, handling, mixing, processing, storage, and pipeline transformation of such systems. Furthermore, rheological studies can provide useful information on the stability and internal microstructure of the double emulsions. Some attention has been given to this subject in recent years and the results are significant since they help to clarify certain aspects of stability and release properties of the double emulsions (100). Few publications deal with W/O/W double emulsions and only one deals with the O/W/O emulsion. Most of the work on rheology is old and does not contribute to new understanding. Therefore, we will discuss only recent relevant work.

An interesting characterization of the mechanical properties of the oil membrane in W/O/W emulsions was carried out by an aspiration technique (101). It was adapted from techniques related to the evaluation of globule or cell deformability. The deformability of an individual globule during total or partial flow into a cylindrical glass tube, calibrated under well-controlled conditions of aspiration, was determined. An analysis was performed on the behavior of the multiple emulsion by migration of the lipophilic surfactant to the interface between the oily and the external aqueous phases. It was shown that the elastic shear modulus and the interfacial tension of the oily membrane increased with the lipophilic surfactant concentration. This study also provides an explanation of the mechanism related to the swelling–breakdown process from physical and mechanical considerations.

Grossiord and coworkers (100, 102) applied linear shear flow to W/O/W double emulsions, which contained active matter, and from the rheological patterns they learned of the bursting effect of the droplets with the release of entrapped substances, and of the composition of the system. The authors (102) described a set of two types of experiment: oscillatory dynamic tests and steady-state analysis. They measured the stress and strain of the emulsions by applying sinusoidal shear. These parameters (shear or complex modulus $G^*$, the lag phase between stress and strain $\delta$; the storage modulus $G'$, and the loss modulus $G''$) provide a quantitative characterization of the balance between the viscous and the elastic properties of the multiple emulsions. At the lag phase $\delta = 0°$ and when equal to 90° the system is viscoelastic. The shear sweep and the temperature sweep characterize the multiple emulsion at rest. Figure 31 describes a transition between elastic and viscous behavior, which occurs at critical stress values. The change in these parameters indicate a pronounced structural breakdown. The authors considered the influence of the formulation parameters on the swelling and release kinetics by using the rheological properties. Parameters such as the nature of the oil, the width of the oil membrane, and the lipophilic and hydrophilic nature of the surfactant have been evaluated. The two main parameters that were identified to affect the swelling/break-up kinetics were the difference in concentration in water-soluble molecules between the internal and the external aqueous phase (Fig. 32), and the lipophilic surfactant concentration. It was also observed that the maximum viscosity values increase with the

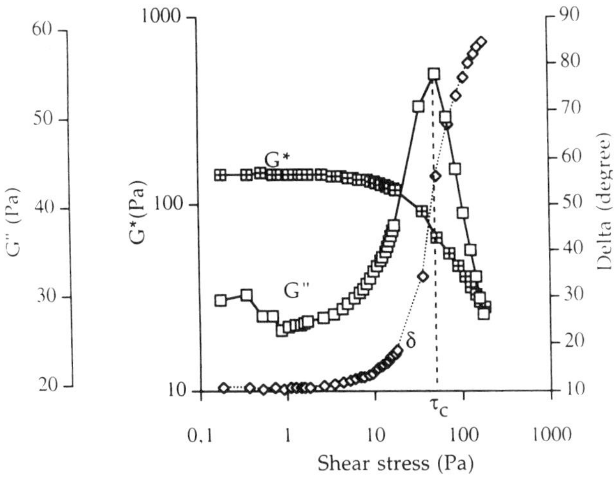


**Figure 31** Changes in $G^*$, $G''$, and $\delta$ for increasing stress at fixed frequency in double emulsions. The double-emulsion composition of the W/O emulsion is 24% oil, various lipophilic surfactant concentrations, and 0.7% $MgSO_4$. The O/W/O emulsion contains 80% water in oil, 2% hydrophilic surfactant, and demineralized water. (From Ref. 102.)

surfactant ratio (under hypo-osmotic conditions, Fig. 33). The same trend was observed when the release of water-soluble materials, $B(t)$, was followed (Fig. 34). The authors concluded that a progressive migration of the excess of lipophilic surfactant in the oil phase toward the primary or the secondary interface occurs.

Stroeve and Varanasi (103) examined also the break up of the multiple-emulsion globules in a simple shear flow and concluded from the critical Weber number $[(We)_{cr}]$ (Figs. 35 and 36) that the multiple emulsion exhibits behavior that is similar to that of simple emulsions. From Fig. 35 one can see at least qualitatively, from the evolution of $(We)_{cr}$ as a function of $p$ (the viscosity ratio between the drop and the continuous phase) for a simple and multiple emulsion, that there are some differences between the two emulsions. The double emulsion has more heterogeneous characteristics. From Fig. 36 it is possible to obtain the minimum shear rate value that is able to produce the break up of

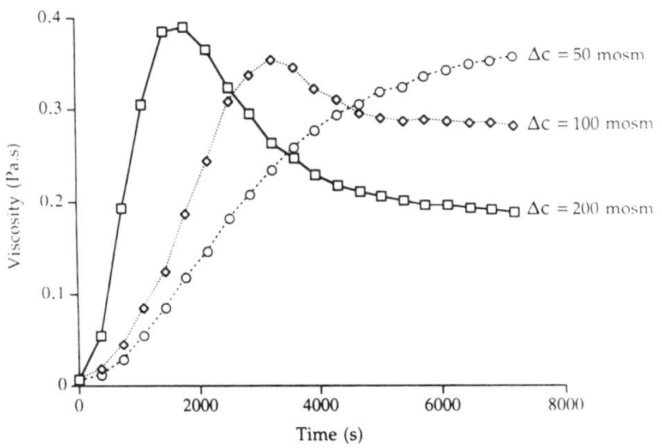


**Figure 32** Change in viscosity vs. time for different concentration gradients in the inner phase. (From Ref. 102.)

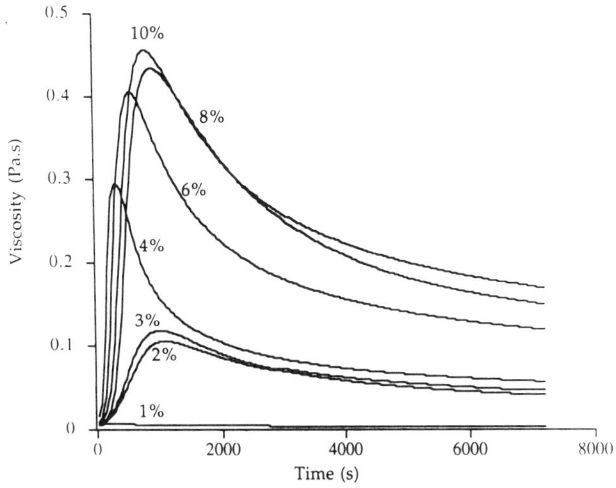


**Figure 33** Change in viscosity vs. time under hypo-osmotic conditions for different lipophilic surfactant concentrations. (From Ref. 102.)

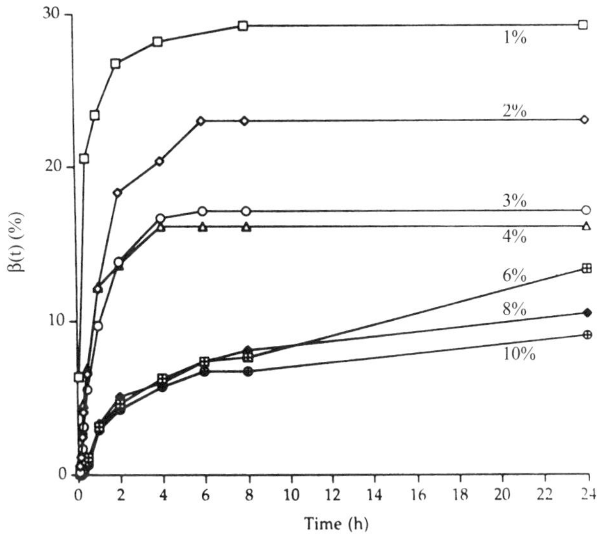


**Figure 34** Change in released fraction vs. time under hypo-osmotic conditions for different lipophilic surfactant concentrations. (From Ref. 102.)

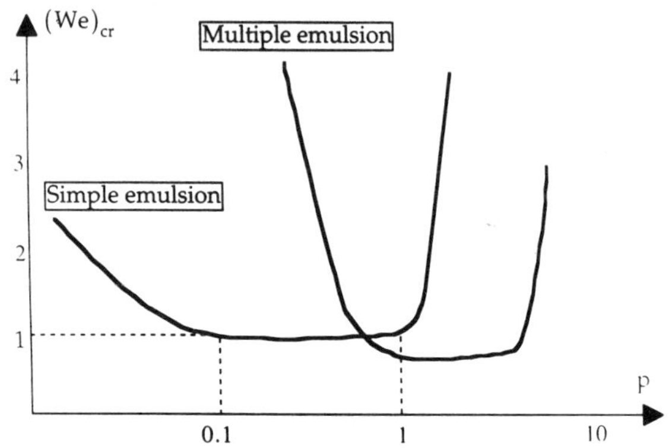


**Figure 35** The critical Weber number for simple and multiple emulsions' disruption as a function of the viscosity ratio dispersed to continuous phase. (From Ref. 103.)

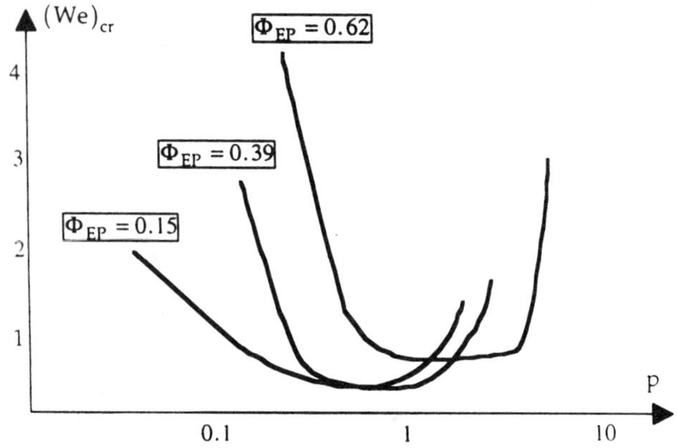


**Figure 36** The critical Weber number for simple and multiple emulsions' disruption as a function of the viscosity ratio dispersed to continuous phase, for different primary emulsion volume fractions. (From Ref. 103.)

the oil globules and the release of water-soluble encapsulated molecules.

The studies showed also that the mechanisms taking place during the break up were complex and did not always lead to total release of the entrapped electrolyte. Some phenomena such as a partial leakage of the internal aqueous compartment or the expulsion of the aqueous microglobules covered by a residual liophilic film were able to restrict the release.

De Cindio et al. (94) prepared food double emulsions and studied their rheological behavior by steady shear and oscillatory measurements. They concluded that the W/O/W appeared to have rheological properties similar to those of a simple O/W emulsions having the same fraction of dispersed phase but lower oil content (Fig. 37). It was also demonstrated that the plot of both storage modulus $G'$ and $G''$ versus oscil-

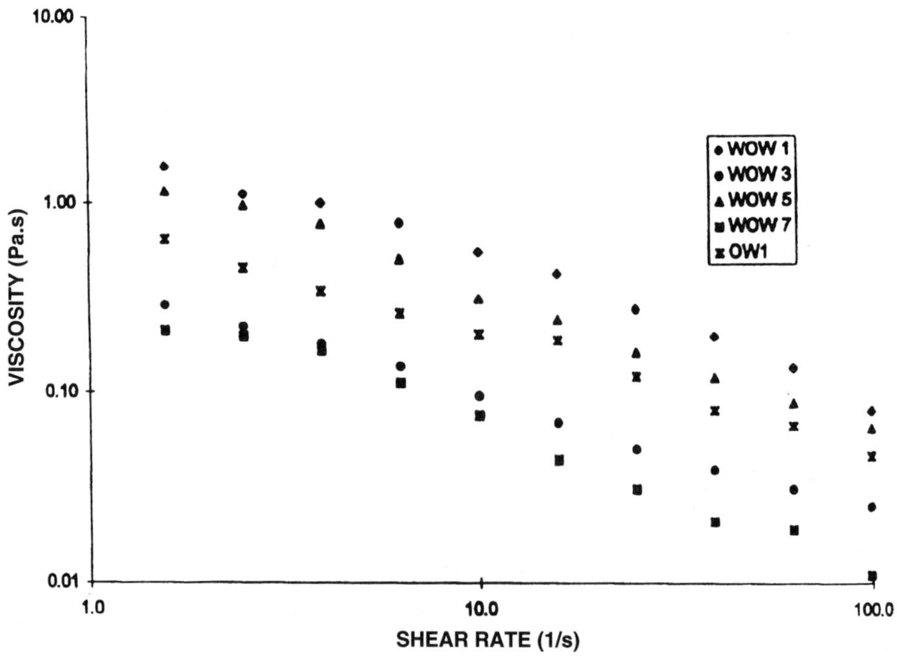


**Figure 37** Apparent viscosity vs. shear rate for both W/O/W and O/W systems at a volume fraction of disperse phase = 0.3. (From Ref. 94.)

lation frequency, $W$, are similar in all eight prepared emulsions, the loss tangent is about 1, and both elastic and viscous contributions to viscoelastic behavior of double emulsions are of similar magnitude. The similarity of texture between simple and double emulsions is very encouraging, leading to some interesting conclusions and new perspectives. The influence of a mixture of emulsifiers on the double-emulsion stability was studied by an oscillatory ring-surface rheometer from which the interfacial elasticity at the oil–aqueous interface could be evaluated.

Pal (104) studied the rheology of O/W/O double emulsions. The simple O/W emulsions were found to be Newtonian up to a dispersed-phase concentration of 45% by volume and nonNewtonian above this volume fraction. All the double O/W/O emulsions are highly nonNewtonian. The degree of shear thinning increases with the increase in primary O/W emulsion concentration (Fig. 38). The oscillatory measurements indicate that the multiple emulsions are predominantly viscous in that the loss modulus falls above the storage modulus over the entire frequency range investigated (Fig. 39). Upon aging, the storage and loss moduli of the double emulsions show a significant increase. However, the increase in viscosity with aging is only marginal.

The rheological behavior of W/O/W emulsion studied with a cone-and-plate viscometer has shown a negative thixotropic flow pattern, mostly under low shear rate. Upon raising the shear rate or the shear time an increase in the shear stress was observed, which induced phase inversion to a W/O of a semisolid type emulsion. The hydrodynamic parameters (dissipated energy, kinetic energy, and impulse

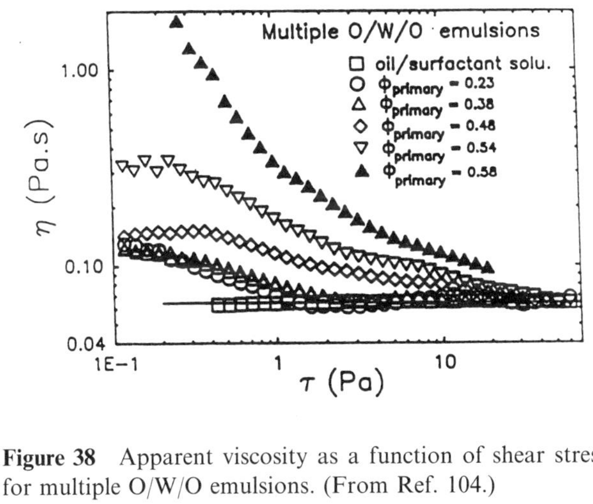


**Figure 38** Apparent viscosity as a function of shear stress for multiple O/W/O emulsions. (From Ref. 104.)

applied to the emulsion by the rotating cone) causing the phase inversion were determined and a mechanism for such inversion was suggested. Figure 40 shows a proposed mechanism for phase inversion under shear rate (105). Induced shear, causing phase inversion, can serve as a possible technique for the release of drugs.

## XVI. CONCLUDING REMARKS

Double emulsions have been known for over three decades and were extensively studied in the last 15 years. The internal phase is an excellent reservoir for active matter that needs protection and can be released at a controlled rate. However, the sizes of the droplets and the thermodynamic instability was a significant drawback of this technology.

The use of conventional low-molecular-weight emulsifiers did not solve these problems. However, much progress has been made with the introduction of amphiphilic macromolecules as emulsifiers. These multianchoring flexible macromolecules can improve the steric stabilization by forming thick multilayered coating on the droplets. A variety of hybrids, complexes, adducts between the amphiphiles, and coemulsifiers as cosolvents have been studied. These molecules improved the stability significantly and slowed the release rates.

Physical methods of separation, filtration, and extraction also had a positive effect on the release patterns of any drug or active matter. Progress was also made in the characterization of the parameters and mechanisms that are involved in the coalcscence, aggregation, and rupture of double-emulsion droplets, and effective control of the rheological parameters was achieved by a better understanding of their effect on the static and shear-induced stability.

It seems that double-emulsion technology can now be applied in various areas, mainly in food, cosmetics, and pharmaceuticals (for nonintravenous applications).

The main goals remaining are: to obtain submicrometer double-emulsion droplets with long-term stability (possibly with emulsified microemulsions) and to trigger and control the release at will.

Compatible blends of biopolymers (hydrocolloids and proteins) are excellent future amphiphilic candidates that under certain combinations will serve both as 'release controllers' and 'stability enhancers' for the future preparations of double emulsions.

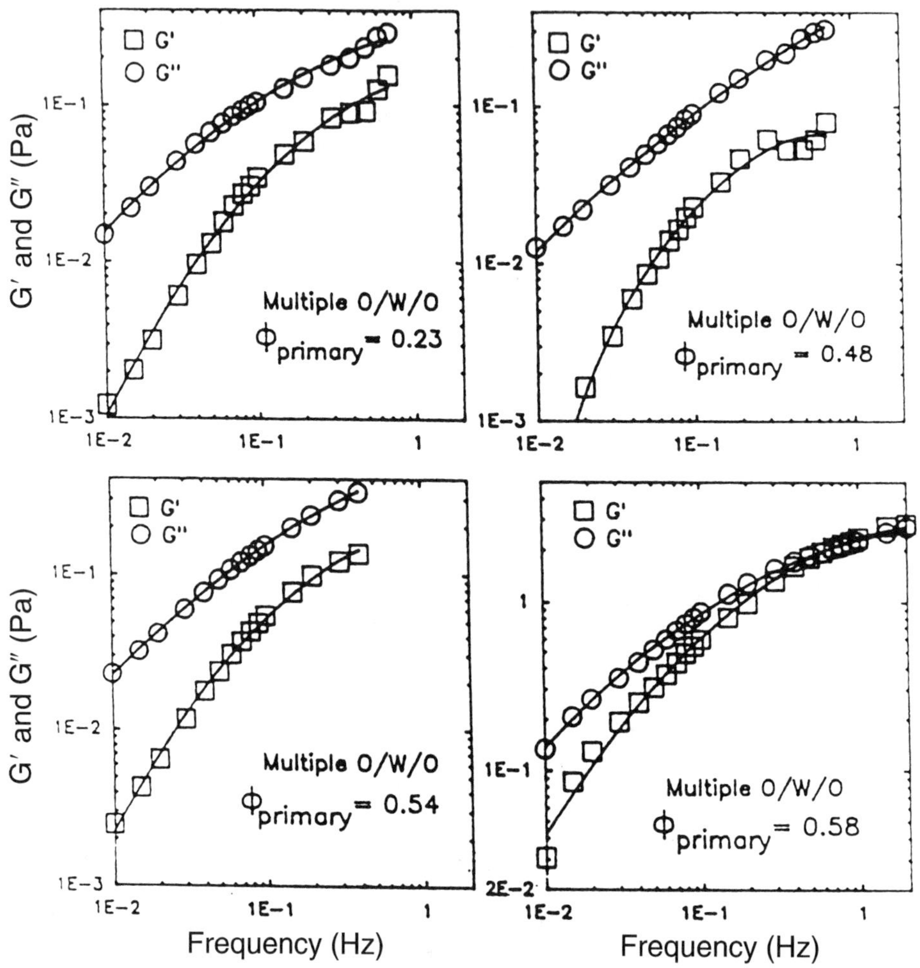


Figure 39 Storage and loss moduli for multiple O/W/O emulsions as functions of frequency. (From Ref. 104.)

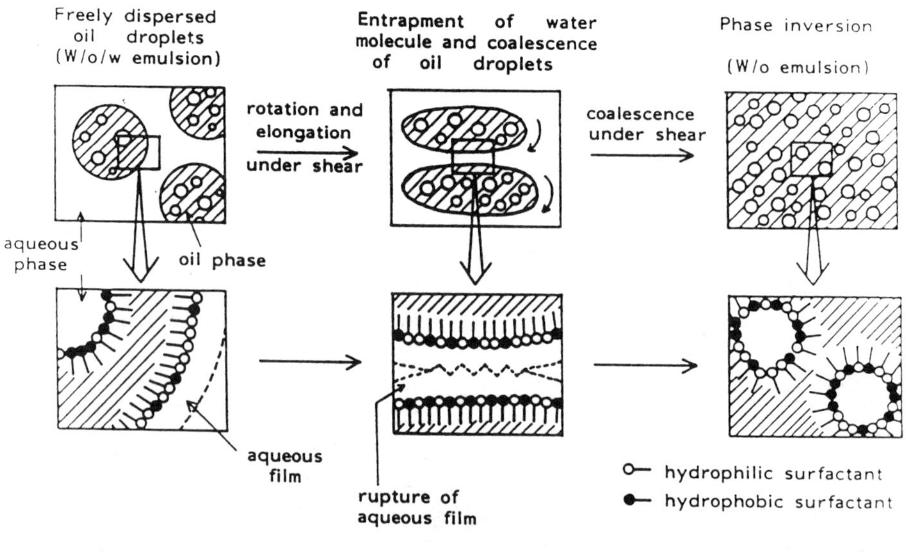


Figure 40 Proposed mechanism of phase inversion under shear. (From Ref. 105.)

## ACKNOWLEDGMENTS

The authors are grateful to Dr Abraham Aserin for his valuable help and critical reading of the present manuscript.

## REFERENCES

1. S Matsumuto, WW Kang. J Dispers Sci Technol 10: 455–482, 1989.
2. C Fox. Cosmet Toilet 10: 101–106; 109–112, 1986.
3. S Matsumoto. In: M Schick, ed. Non-ionic Surfactants. Vol 23, Surfactant Science Series. New York: Marcel Dekker, 1987, pp 549–600.
4. N Garti. Colloids Surfaces 123: 233–246, 1997.
5. N Garti, A Aserin. In: AK Chattopadhyay, KL Mittal, eds. Surfactants in Solution. Vol 64, Surfactant Science Series. New York: Marcel Dekker, 1996, pp 297–332.
6. N Garti. In: Y Barenholtz, DD Lasic, eds. Handbook of Non Medical Applications of Liposomes. Vol 3. New York: CRC Press, 1996, pp 143–199.
7. N Garti. In: M Seiller, JL Grossiord, eds. Multiple Emulsions: Structure, Properties and Applications. Paris: Editions de Santé, 1998, pp 81–116.
8. N Garti. Food Sci Technol 30: 222–235, 1997.
9. N Garti, C Bisperink. Curr Opinion Colloid Interface Sci 3: 657–667, 1998.
10. L Thill-Francis. Stable double emulsions containing finely divided particles. US Patent 93/00007, 1993.
11. AG Gaonkar. Method for preparing a multiple emulsion. US Patent 5322704, 1994.
12. Y Takahashi, T Yoshida, T Takashi. Process for producing a W/O/W type multiple emulsion for medicines, cosmetics, etc. US Patent 94/4985173, 1994.
13. P Strauel, G Friour. Process for preparing photographic emulsions having a low fog level. US Patent 94/420106, 1994.
14. CA Herb. Rinse-off water-in-oil-in-water compositions. Eur Patent 0717978A2, 1996.
15. K Yoshida, T Sekine, F Matsuzaki, T Yanaki, M Yamaguchi. J Am Oil Chem Soc 76: 195–200, 1999.
16. SY Kim, YM Lee. J Ind Eng Chem 5: 306–313, 1999.
17. BBC Youan, J Gillard, B Rollmann. STP Pharm Sci: 9:175–181, 1999.
18. JL Grossiord, M Seiller, I Pezron, D Clausse. Oléagineux Corps Gras Lipides 3: 158–162, 1996.
19. T Uchida, K Yoshida, S Goto. J Microencapsulation 13: 219–228, 1996.
20. L Vigie. Emulsions multiples h/l/h: nouveaux procédés de fabrication et de caractérisation. Graduate thesis, University Paris XI, 1992.
21. JL Grossiord, M Seiller, A Silva-Cunha. In: M Seiller, JL Grossiord, eds. Multiple Emulsions: Structure, Properties and Applications. Paris: Editions de Santé, 1998, pp 57–80.
22. S Higashi, M Shimizu, T Nakashima, K Iwata, F Uchiyama, S Tateno, S Tamura, T Setoguchi. Cancer 75: 1245–1254, 1995.
23. S Matsumoto, Y Kida, D Yonezawa. J Colloid Interface Sci 57: 353–361, 1976.
24. S Matsumoto, M Koda, S Murata. J Colloid Interface Sci 62: 149–157, 1977.
25. S Matsumoto, Y Ueda, Y Kita, D Yonezawa. Agric Biol Chem 42: 739–743, 1978.
26. S Matsumoto. J Colloid Interface Sci 94: 362–368, 1983.
27. S Matsumoto, T Inoue, M Koda, T Ota. J Colloid Interface Sci 77: 564–565, 1980.
28. S Matsumoto, Y Koh, A Michiura. J Dispers Sci Technol 6: 507–521, 1985.
29. S Matsumoto, WW Kang. J Dispers Sci Technol 10: 455–482, 1989.
30. M Frenkel, R Schwartz, N Garti. J Colloid Interface Sci 94: 174–178, 1983.
31. N Garti, M Frenkel, R Schwartz. J Dispers Sci Technol 4: 237–252, 1983.
32. S Magdassi, M Frenkel, N Garti. J Dispers Sci Technol 5: 49–59, 1984.
33. S Magdassi, M Frenkel, N Garti. Drug Ind Pharm 11: 791–798, 1985.
34. S Magdassi, N Garti. J Controlled Release 3: 273–277, 1986.
35. AJ Khopade, NK Jain. Drug Delivery 6: 107–110, 1999.
36. MPY Piemi, M De Luca, JL Grossiord, M Seiller, JP Marty. Int J Pharm 171: 207–215, 1998.
37. MF Ficheux, L Bonakdar, F Leal-Calderon, J Bibette. Langmuir 14: 2702–2706, 1998.
38. HL Rosano, FG Gandolfo, JDP Hidiot. Colloids Surfaces A 138: 109–121, 1998.
39. AT Florence, TK Law, TL Whateley. J Colloid Interface Sci 107: 584–588, 1985.
40. JA Omotosho, TL Whateley, TK Law, AT Florence. J Pharm Pharmacol 38: 865–870, 1986.
41. JA Omotosho, TL Whateley, AT Florence. J Microencapsulation 6: 183–192, 1989.
42. A Vaziri, B Warburton. J Microencapsulation 11: 649–656, 1994.
43. W Zhang, T Miyakawa, T Uchida, S Goto. J Pharm Soc Jpn 112: 73–80, 1992.
44. M Cornec, PJ Wilde, PA Gunning, AR Mackie, FA Husband, ML Parker, DC Clark. J Food Sci 63: 39–43, 1998.
45. E Fredrokumbaradzi, A Simov. Pharmazie 47: 388–389, 1992.
46. N Garti, A Aserin, Y Cohen. J Controlled Release 29: 41–51, 1994.

47. E Dickinson, J Evison, RK Owusu, A Williams. In: GO Phillips, PA Williams, DJ Wedlock, eds. Gums and Stabiliziers for the Food Industry. Vol. 7. Oxford: IRL University Press, 1994, pp 91–101.
48. E Dickinson, J Evison, RK Owusu. Food Hydrocolloids 5: 481–485, 1991.
49. J Evison, E Dickinson, RK Owusu Apenten, A Williams. In: E Dickinson, D Lorient, eds. Food Macromolecules and Colloids. Cambridge: Royal Society of Chemistry, 1995, pp 235–243.
50. A Koberstein Hajda, E Dickinson. Food Hydrocolloids 10: 251–254, 1996.
51. PC Schulz, MS Rodriguez, LF Del Blanco. Colloid Polym Sci 276: 1159–1165, 1998.
52. T Adachi, J Ida, M Wakita. Polym J 31: 319–323, 1999.
53. SC Yu, A Bochot, M Cheron, M Seiller, JL Grossiord, G Le Bas, D Duchene. STP Pharm Sci 9: 273–277, 1999.
54. D Johansson, B Bergenstahl, E Lundgren. J Am Oil Chem Soc 72: 921–932, 1995.
55. D Johansson, B Bergenstahl. J Am Oil Chem Soc 72: 933–938, 1995.
56. KP Oza, SG Frank. J Dispers Sci Technol 7: 543–561, 1986.
57. AJ Khopade, NK Jain. Drug Dev Ind Pharm 24: 289–293, 1998.
58. N Garti, A Aserin, I Tiunova. J Am Oil Chem Soc 76: 383–389, 1999.
59. T Sekine, K Yoshida, F Matsuzaki. J Surfactants Deterg 2: 309–315, 1999.
60. A Vaziri, B Warburton. J Microencapsulation 12: 1–5, 1995.
61. S Susuki, JK Lim. J Microencapsulation 11: 197–203, 1994.
62. D Clausse. J Therm Anal Calorim 51: 191–201, 1998.
63. D Clausse, I Pezron, L Komunjer. Colloids Surfaces A 152: 23–29, 1999.
64. Y Ogawa, YM Yamamoto, H Okada, HT Yashiki, T Shimamoto. Chem Pharm Bull 36: 1095–1103, 1988.
65. MJ Blanco-Prieto, F Delie, A Fattal, A Tartar, F Puisieux, A Gulik, P Couvreur. Int J Pharm 111: 137–145, 1994.
66. CH Schugens, N Laruelle, N Nihant, C Grandfils, R Jerome, P Teyssie. J Controlled Release 32: 161–176, 1994.
67. N Nihant, C Schugens, C Grandfils, R Jerome, P Teyssie. J Colloid Interface Sci 173: 55–65, 1995.
68. MJ Blanco-Prieto, E Leo, F Delie, A Gulik, P Couvreur, E Fattal. Pharm Res 13: 1137–1139, 1996.
69. R Cortesi, F Bortolotti, E Menegatti, C Nastruzzi, E Esposito. Int J Pharm 129: 263–273, 1996.
70. P Couvreur, MJ Blanco-Prieto, F Puisieux, B Roques, E Fattal. Adv Drug Delivery Rev 28: 85–96, 1997.
71. MJ Blanco-Prieto, E Fattal, A Gulik, B Dedieu, P Couvreur. J Controlled Release 43: 81–87, 1997.
72. Y Maa, C Hsu. J Controlled Release 38: 219–228, 1996.
73. H Rafati, A Coombes, J Adler, J Holland, SS Davis. J Controlled Release 43: 89–102, 1997.
74. N Ubrich, J Ngondi, C Rivat, M Pfister, C Vigneron, P Maincent. J Biomed Mater Res 37: 155–160, 1997.
75. YF Maa, CC Hsu. J Microencapsulation 14: 225–241, 1997.
76. E Leo, S Pecuet, J Rojas, P Couvreur. J Microencapsulation 15: 421–430, 1998.
77. MF Zambaux, F Bonneaux, R Gref, P Maincent, M Dellacherie, J Alonso, P Labrude, C Vigneron. J Controlled Release 50: 31–40, 1998.
78. Park TG, MJ Alonso, R Langer. J Appl Polym Sci 52: 1797–1807, 1994.
79. AT Florence, JA Omotosho, TL Whateley. In: M Rossof, ed. Controlled Release of Drug: Polymers and Aggregated Systems. New York: VCH, 1989, pp 163–184.
80. AT Florence, T Yoshioka. In: Y Barenholtz, DD Lasic, eds. Handbook of Nonmedical Applications of Liposomes. Vol 3. New York: CRC Press, 1996, pp 199–223.
81. EC Albert, DFH Wallach, R Mathur. Lipid vesicle-containing water-in-oil emulsions. WO-92/17179, 1992.
82. E Pilman, K Larsson, E Torenberg. J Dispers Sci Technol 1: 267–281, 1980.
83. N Garti, H Houminer. Emulsified Microemulsions for Sustained Release of Drugs. PhD internal report, The Hebrew University of Jerusalem, 1996.
84. ML Cole, TL Whateley. J Controlled Release 49: 51–58, 1997.
85. Y Sela, S Magdassi, N Garti. Colloid Polym Sci 272: 684–691, 1994.
86. Y Sela, S Magdassi, N Garti. Colloids Surfaces A 83: 143–150, 1994.
87. Y Sela, S Magdassi, N Garti. J Controlled Release 33: 1–12, 1995.
88. AJ Khopade, NK Jain. Drug Delivery 6: 181–185, 1999.
89. AJ Khopade, KS Nandakumar, NK Jain. J Drug Target 6: 285–292, 1998.
90. S Zheng, RL Beissinger, LR Sehgal, D Wasan. J Dispers Sci Technol 20: 235–245, 1999.
91. M Morishita, A Matsuzawa, K Takayama, K Isowa, T Nagai. Int J Pharm 172: 189–198, 1998.
92. AJ Khopade, NK Jain. Pharmazie 54: 915–919, 1999.
93. B De Cindio, D Cacace. Int J Food Sci Technol 30: 505–514, 1995.
94. B De Cindio, G Grasso, D Cacace. Food Hydrocolloids 4: 339–353, 1991.
95. DK Asahi. Double emulsified oil-fat composition comprises outer phase of oil and fat inner phase of oil-in-water emulsified fats, and oil and fat phase containing mustard oils. Jap Patent 87016619, 1987.

96. KK Yoshihara Seiyu. Oil and fat cooking product containing double emulsion of liquid oil and fat of above 30°C melting points. Jap Patent 60091939, 1985.
97. K Suzuki, K Uehara, H Omura. Oil-in-water-in-oil double emulsified fat or oil composition for cream, spread confectionery or baking use. Eur Patent 425958, 1991.
98. S Tsukishima. Double emulsion type filling cream obtained by adding fat or oil to aqueous phase containing saccharides, proteins and/or emulsifiers and mixing with plastic fat or oil. Jap Patent 3290152, 1991.
99. C Laugel, P Rafidison, G Potard, L Aguadisch, A Baillet. J Controlled Release 63: 7–17, 2000
100. V Muguet, M Seiller, G Barratt, D Clausse, JP Marty, JL Grossiord. J Colloid Interface Sci 218: 335–337, 1999.
101. S Geiger, N Jager-Lezer, S Tokgoz, M Seiller, JL Grossiord. Colloids Surfaces A 157: 325–332, 1999.
102. JL Grossiord, M Seiller. In: JL Grossiord, M Seiller, eds. Multiple Emulsions: Structure, Properties and Applications. Paris: Editions de Santé, 1998, pp 169–192.
103. P Stroeve, PP Varanasi. J Colloid Interface Sci 99: 360–373, 1984.
104. R Pal. Langmuir 12: 2220–2225, 1996.
105. Y Kawashima, T Hino, H Takeushi, T Niwa, K Horibe. Int J Pharm 72: 65–77, 1991.

# 18

# Environmental Emulsions

**Merv Fingas and Benjamin G. Fieldhouse**
*Environment Canada, Ottawa, Ontario, Canada*

**Joseph V. Mullin**
*U.S. Minerals Management Service, Department of the Interior, Herndon, Virginia*

## I. INTRODUCTION

Emulsification is the process whereby water-in-oil emulsions are formed. These emulsions are often called "chocolate mousse" or "mousse" by oil-spill workers. Emulsions change the properties and characteristics of oil spills to a very large degree. Stable emulsions contain between 60 and 80% water, thus expanding the volume of spilled material from two to five times the original volume. The density of the resulting emulsion can be as great as 1.03 g/mL compared to a starting density as low as 0.80 g/mL. Most significantly, the viscosity of the oil typically changes from a few hundred mPa.s to about 100 Pa.s, a typical increase of 1000. Thus, a liquid product is changed to a heavy, semisolid material.

Many feel that emulsification is the second most important behavioral characteristic of oil after evaporation. Emulsification has a very great effect on the behavior of oil spills at sea. As a result of emulsification, evaporation of oil spills slows by orders of magnitude, spreading slows by similar rates, and the oil rides lower in the water column, showing different drag with respect to the wind. Emulsification also significantly affects other aspects of a spill. Spill countermeasures are quite different for emulsions as they are hard to recover mechanically, to treat, or to burn.

In terms of its understanding of emulsions and emulsification, the oil-spill industry has not kept pace with the petroleum production industry and colloid science generally. Workers in the spill industry often revert to old papers published in oil-spill literature, which is frequently incorrect and reflects very old knowledge. A basic understanding of the formation, stability, and processes of emulsions is now evident in literature in both the colloid science and oil-spill fields, although some new papers still appear with references to 15-year-old literature and no newer literature.

A very important part of emulsion study is the availability of methodologies to study emulsions. In the past ten years, both dielectric methods (1) and rheological methods (2) have been exploited to study formation mechanisms and the stability of emulsions formed from many different types of oils. Standard techniques, including NMR, chemical analysis techniques, microscopy, interfacial pressure, and interfacial tension, are also being applied to emulsions. These techniques have largely confirmed findings noted in the dielectric and rheological mechanisms.

## II. BACKGROUND

The mechanism and dynamics of emulsification were poorly understood until the 1990s. It was not recognized until recently that the basics of water-in-oil emulsification were understood in the surfactant industry, but not in the oil-spill industry. In the later 1960s, Berridge et al. were the first to describe emulsification in detail and measured several physical properties (3). Berridge described the emulsions as forming because of the asphaltene and resin content. Workers in the 1970s concluded that emulsification occurred primarily due to increased turbulence or mixing energy (4, 5). The oil's composition was not felt to be a major factor. Some workers speculated that particulate matter in the oil may be a factor and others suggested it was viscosity. Evidence could be found for and against all these hypotheses.

Twardus studied emulsions in 1980 and found that emulsion formation might be correlated with oil composition (6). It was suggested that asphaltenes and metal porphyrins contributed to emulsion stability. Bridie et al. (7) studied emulsions in the same year and proposed that the asphaltenes and waxes stabilized water-in-oil emulsions. The wax and asphaltene content of two test oils correlated with the formation of emulsions in a laboratory test (7). Mackay and Zagorski hypothesized that emulsion stability was due to the formation of a film in oil that resisted water-droplet coalescence (8–10). The nature of these thin films was not described, but it was proposed that they were caused by the accumulation of certain types of compounds. later work led to the conclusion that the compounds were asphaltenes and waxes. A standard procedure was devised by making emulsions and measuring stability. This work formed the basis of much of the emulsion theory and emulsion formation in the oil-spill literature over the past two decades.

In 1983, Thingstad and Pengeurd conducted photo-oxidation experiments and found that photo-oxidized oil formed emulsions (11). Nesterova et al. studied emulsion formation and concluded that it was strongly correlated with both the asphaltene and tar content of oil and also the salinity of the water with which it was formed (12). Mackay and Nowak studied emulsions and found that stable emulsions had low conductivity and therefore a continuous phase of oil (13, 14). Stability was discussed and proposed to be a function of oil composition, particularly waxes as asphaltenes. It was proposed that a water droplet could be stabilized by waxes, asphaltenes, or a combination of both. The viscosity of the resulting emulsions was correlated with water content. Later work by the same group reported examination of Russian hypotheses that emulsions are stabilized by colloidal particles which gather at the oil–water interface and may combine to form a near-solid barrier that resists deformation and thus water–water coalescence (15). It was speculated that these particles could be mineral, wax crystals, aggregates of tar and asphaltenes, or mixtures of these. Asphaltenes were felt to be the most important of these particles and controlled the formation of all particles. A formation equation relating the asphaltenes, paraffin, aromatic, and silica gel (resin) content was proposed, but it was later shown to be a poor predictor of oil-emulsion tendencies.

Desmaison et al. conducted studies on Arabian crudes and noted the emulsion formation was correlated with two factors – photo-oxidation exposure and amount of asphaltenes (16). The photo-oxidation was found to occur on the aromatic fractions of the oil. Asphaltenes were found to become structured with time and this was associated with emulsion formation. Miyahara reported that the stability of emulsions was primarily controlled by the composition of the oil, specifically that which resided in the hexane-insoluble fraction of the oil, but he did not define what this content was (17). Miyahara also reported that salt and freshwater emulsions showed relatively similar stabilities, although in one case the salt-water emulsion appeared to be more stable. Payne and Phillips reviewed the subject in detail and reported on their on experiments of emulsification with Alaskan crudes in the presence and absence of ice (18). Their studies showed that emulsion formation could occur in an ice field, thus indicating that there was sufficient energy in this environment and that the process could occur at relatively low temperatures.

Because of the many differing theories in the literature, many oil-spill workers were confused as to the stability, source of stability, and properties of water-in-oil emulsions. Furthermore, until about 1995, neither advanced rheological techniques nor other techniques such as dielectric studies were applied to emulsions.

## III. CURRENT STUDIES AND RESULTS

### A. Field Studies

In the oil-spill trade, much of the information on emulsions has been obtained by practical studies in the laboratory or the field. In 1991, Jenkins et al. studied emulsions formed in the laboratory and concluded that the formation did not correlate with previously established codes of properties, nor with pour point, asphaltene, and wax contents of the fresh oils (19). Jenkins et al. suggested that, in the absence of any correlation, characterization of every oil should be made using a standardized procedure in the laboratory.

Other examples of empirical studies include a two-year study conducted on emulsions by Walker at Warren Spring Laboratory in Britain in which approximately 40 North Sea crude oils were prepared and characterized in the laboratory (20). Some of these oils were subsequently spilled at sea and some of their properties measured. Walker concluded that the laboratory procedures did not result in emulsions similar to those found at sea, but also noted that there was a marked lack of characterization techniques to study emulsions. The same group participated in another field trial conducted in 1994 (21). The correlation between parallel experiments, physical properties, and emulsion characteristics was poor. It was concluded that delays in sampling and analysis were partially responsible for the poor results as well as the lack of standard measurement and characterization techniques. It was also noted that slight differences in release conditions resulted in major differences in slick behavior. The energy required to form emulsions was found to be high, and oil must be weathered to a degree before release. Stability could not be characterized, but appeared to be a continuum through the process.

## B. General Reviews and General Influences on Emulsions

In 1992, Schramm reviewed the basics of emulsions and provided the oil industry with the basis for much subsequent understanding of water-in-oil emulsions (22). In 1993, Becher reviewed emulsion stability in mathematical terms (23).

## C. Stability

Sjöblom and coworkers surveyed several oils from the Norwegian continental shelf. After the interfacially active fraction was removed from the oils, none would form water-in-oil emulsions (24–26). Model emulsions could be made from the extracted interfacially active fractions. Stability was gaged by measuring the separation of water with time. Destabilization studies showed that the rigidity of the interfacial film or reaction with the film components are the principle methods of emulsion breakdown. Medium-chain alcohols and amines destabilized emulsions the most.

In 1992, Friberg and Yang reviewed the stability of emulsions, noting that a primary measure of stability is the separation into two phases (27). Friberg noted the focus on two factors: the rheology of the continuous phase and the barrier between the dispersed droplets. It was demonstrated that an increase in viscosity of the continuous phase of the emulsion is not a viable alternative to increasing the half-life of the emulsion. Friberg noted that the continuous phase must show a small yield value to demonstrate stability.

In 1994, Tambe and Sharma proposed a model for the stability of colloid-stabilized emulsions (28). They noted that colloidal particles stabilize emulsions both by providing steric hindrance to drop–drop coalescence and by modifying the rheological properties of the interfacial region. Tambe and Sharme noted that the effectiveness of colloidal particles in stabilizing emulsions depends in part on the ability of these particles to reside in a state of equilibrium at the oil–water interface and showed that the adsorption of particles at the oil–water interface also affects the rheological properties of the interfacial region. If the concentration of the particles is high, the colloid-laden interface will exhibit viscoelastic behavior. Viscoelastic interfaces, in turn, affect emulsion stability by retarding the rate of film drainage between coalescing emulsion droplets and by increasing the energy required to displace particles from the contact region between water droplets or, in other words, by increasing the magnitude of the steric hindrance.

## D. Source of Stability

### 1. Asphaltenes

Over 30 years ago, asphaltenes were found to be a major factor in emulsion stability (3). Specific roles of emulsions have not been defined until recently. The Sjöblom group in Norway defined the interfacial properties of asphaltenes in several local offshore crudes (29). Asphaltenes were separated from the oils using consecutive separations involving absorption to silica. Molecular weights ranged from 950 to 1450 Da. Elemental analysis revealed that 99 mole % of the asphaltenes was carbon and hydrogen, while up to 1% was nitrogen, oxygen, and/or sulfur. The films form monomolecular layers at the air/water interface. Aromatic solvents such as benzyl alcohol have a strong influence on the asphaltenes and will destabilize water-in-oil emulsions. Asphaltenes were shown to be the agent responsible for the stabilization of the Norwegian crudes tested.

Workers in the same group separated resins and asphaltenes and studied the Fourier-transform infrared (FT-IR) spectrum and the emulsions formed by each fraction (30). The asphaltenes were separated by pentane precipitation and the resins by desorption from silica gel using mixtures of benzene and methanol. The fractions were tested in model systems for their emulsion-forming tendencies. Model emulsions were stabilized by both asphaltene and resin fractions, but the asphaltene fractions were much more stable.

Acevedo et al. studied the interfacial behavior of a Cerro Negro crude by planar rheology (31). Distilled water and salt water were used with a 30% and a 3.2% xylene-diluted crude. The elasticity and viscosity were obtained from creep compliance measurements. The high values of viscoelastic and elastic moduli were attributed to the flocculation of asphaltene:resin micelles at the interface. The high moduli were associated with the elastic interface. In the absence of resins, asphaltenes were not dispersed and did not form stable interface layers and then, by implication, stable emulsions. Mohammed et al. studied surface pressure, as measured in a Langmuir film balance, of crude oils and solutions of asphaltenes and resins (32). They found that the pseudodilational modulus has high values for low resin-to-asphaltene ratios and low values for high resin-to-asphaltene ratios. They suggest that low resin-to-asphaltene ratios lead to more stable emulsions and vice versa.

Chaala et al. studied the flocculation and the colloidal stability of crude fractions (33). Stability was defined as the differential in spectral absorption between the bottom and top of a test vessel. The effects of temperature and of additions of waxes and aromatics on stability were noted. Increasing both waxes and aromatics generally decreased stability. Temperature increased stability up to 60°C and then stability decreased. In another study, the resins and asphaltenes were extracted from four crude oils by various means (34). It was found that different extraction methods resulted in different characteristics as measured by FT-IR spectroscopy as well as different stabilities when the asphaltenes and resins were used as stabilizers in model systems. It was concluded that the interfacially active components in crude oil were interacting and were difficult to distinguish. Both the resins and asphaltenes appeared to be involved in interfacial processes.

Urdahl and Sjöblom studied stabilization and destabilization of water-in-crude oil emulsion (35). It was concluded that indigenous interfacially active components in the crude oils are responsible for stabilization. These fractions would be the asphaltenes and resins. Model systems stabilized by extracted interfacially active components had stability properties similar to those of the crude oil emulsions. The same group studied the aging of the interfacial components (36). Resins and asphaltenes were extracted from North Sea crudes and exposed to aging under normal atmospheric and ultraviolet conditions. The FT-IR spectra showed that the carbonyl peak grew significantly as indicated by the C=O mode. Spectra also showed that condensation was occurring. The interfacial activity increased in all fractions as the aging process proceeded. In the case of two crude oils, the aging was accompanied by an increase in the water/oil emulsion stability.

McLean and Kilpatrick studied asphaltene aggregation in model emulsions made from heptane and toluene (37). The resins and asphaltenes were extracted from four different crude oils – two from Saudia Arabia, Alaskan North Slope, and San Joaquin Valley crudes. The asphaltenes were extracted by using heptane, and the resins, by using open silica columns. Asphaltenes dissolved in heptol, consisting of only about 0.5% asphaltenes, generated emulsions that were more stable than those generated by the originating crude oils. Although some emulsions could be generated using resins, they were much less stable than those generated by asphaltenes. The model emulsions showed that the aromaticity of the crude medium was a prime factor. This was adjusted by varying the heptane:toluene ratio. It was also found that the concentration of asphaltenes and the availability of solvating resins were important. The model emulsions were most stable when the crude medium was between 30 and 40% toluene and had low resin:asphaltene ratios. McLean and Kilpatrick put forward the thesis that asphaltenes were the most effective in stabilizing emulsions when they are near the point of incipient precipitation (38). It was noted that there are specific resin–asphaltene interactions, as differing combinations yielded different results in the model emulsions. The resins and asphaltenes were characterized by elemental and neutron-activation analyses. The most effective emulsion stabilizers of the resins and asphaltenes were the most polar and the most condensed. McLean and Kilpatrick concluded that the most significant factor in emulsion stability is the solubility state of the asphaltenes.

In 1998, Mouraille et al. studied the stability of emulsions by using separation/sedimentation tests and high-voltage destabilization (39). It was found that the most important factor was the stabilization state of the asphaltenes. The wax content did not appear to affect the stability except that a high wax content displayed a high temperature dependence. Resins affected the solubilization of the asphaltenes and thus indirectly the stability. In the same year,

McLean et al. reviewed emulsions and concluded that the asphaltene content is the single most important factor in the formation of emulsions (40). Even in the absence of any other synergistic compounds (i.e., resins, waxes, and aromatics), asphaltenes were found to be capable of forming rigid, cross-linked, elastic films which are the primary agents in stabilizing water-in-crude oil emulsions. It was noted that the exact conformations in which asphaltenes organize at oil-water interfaces and the corresponding intermolecular interactions have not been elucidated. McLean and colleagues suggest that the intermolecular interactions must be either $\pi$-bonds between fused aromatic sheets, H-bonds mediated by carboxyl, pyrrolic, and sulfoxide functional groups or electron donor–acceptor interactions mediated by porphyrin rings, heavy metals, or heteroatomic functional groups. It is suggested that specific experimental designs to test these concepts are needed to understand the phenomenon on a molecular level. Such knowledge would aid in the design of chemical demulsifiers. The oleic medium plays a large role in the surface activity of asphaltenic aggregates and in the resulting emulsion stability. It is noted that the precise role of waxes and inorganic solids in either stabilizing or destabilizing emulsions is not known. Emulsions are primarily stabilized by rigid, elastic asphaltenic films.

Recently, Singh et al. studied the effect of fused-ring solvents, including naphthalene, phenanthrene, and phenanthridine, in destabilizing emulsions (41). They note that the primary mechanism for emulsion formation is the stability of asphaltene films at the oil–water interface. They suggest that the mechanism is one in which planar, disk-like asphaltene molecules aggregate through lateral intermolecular forces to form aggregates. The aggregates form a viscoelastic network after absorption at the oil–water interface. The network is sometimes called a film or skin, and the strength of this film correlates with emulsion stability. The film strength can be gaged by shear and elastic moduli. Singh et al. probed the film-bonding interactions by studying the destabilization by aromatic solvents. It was found that fused-ring solvents, in particular, were effective in destabilizing asphaltene-stabilized emulsions. It is suggested that both $\pi$-bonds between fused aromatic sheets and H-bonds play significant roles in the formation of the asphaltene films.

Sjöblom et al. used dielectric spectroscopy to study emulsions over a period of years (42). It is concluded that the stabilizing fraction in water-in-oil emulsions is the asphaltenes and not the resins. However, it is noted that some resins must be present to give rise to stability. It is suggested that the greater mobility of the resins is needed to stabilize the emulsions until the asphaltenes, which migrate slowly, can align at the interface and stabilize the emulsions.

### 2. Resins

In 1981, Neuman and Paczynka-Lahme studied the stability of petroleum o/w emulsions and found that they are stabilized by "thick films" which appeared to be largely composed of petroleum resins (43). These thick films demonstrate elasticity and thus increase stability. Temperature increases showed increasing structure formation of the films. Isolated petroleum resins showed structure formation as well.

Rønningsen et al. studied the aging of crude oils and its effect on the stability of emulsions (44). The oil was exposed to air and light and it was found that the interfacial tension of the oil towards formation of water decreased as a result of the aging. This was caused by the formation of various oxidation products, mainly carbonyl compounds. In general, the emulsions became more stable. In some cases, however, the opposite was observed, namely, that although the interfacial tension was high, the emulsion stability was lower, presumably because new compounds with less beneficial film properties are formed. Presumably, the compounds that were formed could be loosely classified as resins.

### 3. Waxes

Johansen et al. studied water-in-crude emulsions from the Norwegian continental shelf (45). The crudes contained a varying amount of waxes (2–15%) and a few asphaltenes (0–1.5% by weight). Emulsion stability was characterized by visual inspection as well as by ultracentrifugation at 650 to 30,000 *g*. Mean water droplet sizes of 10 to 100 μm were measured in the emulsions. It was found that higher mixing rates reduced the droplet size and a longer mixing time yielded a narrow distribution of droplet size. The emulsion stability correlated with the emulsion viscosity, the crude oil viscosity, and the wax content.

McMahon studied the effect of waxes on emulsion stability as monitored by the separation of water over time (46). The size of the wax crystals showed an effect in some emulsions but not in others. Interfacial viscosity indicated that the wax crystals form a barrier at the water/oil interface which retards the coalescence of colliding water droplets. Studies with octacosane, a model crude oil wax, show that a limited wax/asphaltene/resin interaction occurs. A wax layer, even with absorbed asphaltenes and resins, does not by itself stabilize an emulsion. McMahon concludes that the effect of wax on emulsion stability does not appear to be through action at the interface. Instead, the wax may act in the bulk oil phase by inhibiting film thinning between approaching droplets or by a scavenging demulsifier. The asphaltenes and resins were found to affect stability via interfacial action and are primarily responsible for the emulsion formation.

Puskas et al. extracted paraffinic deposits from oil wells and pipelines (47). This hydrophobic paraffin derivative had a high molar mass and melting point and contained polar end groups (carbonyls). This paraffinic derivative stabilized water-in-oil emulsions at concentrations of 1 to 2%.

## E. Methodologies for Studying Emulsions

### 1. Dielectric

One of the methods used to study emulsions has been the use of dielectric spectroscopy. The permittivity of the emulsion can be used to characterize an emulsion and assign a stability (1, 42, 48–54). The Sjöblom group has measured the dielectric spectra using time-domain spectroscopy (TDS) technique. A sample is placed at the end of a coaxial line to measure total reflection. Reflected pulses are observed in time windows of 20 ns, Fourier transformed in the frequency range from 50 MHz to 2 GHz, and the complex permittivity calculated. Water or air can be used as reference sample. The total complex permittivity at a frequency ($\omega$) is given by:

$$\varepsilon^*(\omega) = \frac{\varepsilon_s - \varepsilon_\phi}{1 + i\omega\tau} \qquad (1)$$

where $\varepsilon_s$ is the static permittivity, $\varepsilon_\phi$ is the permittivity at high frequencies, $\omega$ is the angular frequency, and $\tau$ is the relaxation time.

The measuring system used by the Sjöblom group includes a digital sampling oscilloscope and a pulse generator. A computer is connected to the oscilloscope and controls the measurement timing as well as performing the calculations.

The data are used to give an indication of stability and the geometry of the droplets. Flocculation of the emulsion can be detected. In tests of flowing and static emulsions, it was shown that the flowing emulsions have lower static permittivities (52). This was interpreted as indicating flocculation in the static emulsions.

Skodvin and Sjöblom used dielectric spectroscopy in conjunction with rheology to study a series of emulsions (54). A close connection was found between the viscosity and dielectric properties of the emulsions. The large effects of shear on both the static permittivity and the dielectric relaxation time for the emulsion was ascribed, at least in part, to the degree of flocculation in the emulsion system. At high shear rates, at which emulsions are expected to have a low degree of flocculation and high stability, the dielectric properties still varied from those expected from a theoretical model for spherical emulsion droplets.

Førdedal and Sjöblom used dielectric spectroscopy to study several real crude oil emulsions and model systems stabilized with either separated asphaltenes and resins from crude oil or by commercial surfactants (55). Emulsions could be stabilized by the asphaltene fraction alone, but not by the resin fraction alone. A study of a combination of mixtures shows an important interaction between emulsifying components. Førdedal et al. used dielectric spectroscopy to study model emulsions stabilized by asphaltenes extracted from crude oils (56). Analysis showed that the choice of organic solvent and the amount of asphaltenes, as well as the interaction between these variables, were the most significant parameters for determining the stability of the emulsions.

Ese et al. found similar results on model emulsions stabilized with resins and asphaltenes extracted from North Sea oil (57). The dielectric spectroscopy results showed that the stability of model emulsions could be characterized. Stability was found to depend mainly on the amount of asphaltenes, the degree of aging of asphaltenes and resins, and the ratio between asphaltenes and resins.

### 2. Rheology

In 1983, Steinborn and Flock studied the rheology of crude oils and water-in-oil emulsions (58). Emulsions with high proportions of water exhibited pseudoplastic behavior and were only slightly time dependent at higher shear rates. Omar et al. also measured the rheological characteristics of Saudi crude oil emulsions (59). NonNewtonian emulsions exhibit pseudoplastic behavior and followed a power-law model. Mohammed et al. studied crude oil emulsions using a biconical bob rheometer suspended at the interface (60). More stable emulsions displayed viscoelastic behavior and a solid-like interface. Demulsifiers changed the solid-like interface into a liquid one.

Tadros summarized the fundamental principles of emulsion rheology (61). Emulsions stabilized by surfactant films (such as resins and asphaltenes) behave like hard sphere dispersions. These dispersions display viscoelastic behavior. Water-in-oil emulsions show a transition from predominantly viscous to predominantly elastic response as the frequency of oscillation exceeds a critical value. Thus, a relaxation time can be determined for the system which increases with the volume fraction of the discontinuous phase. At the critical value, the system shows a transition from predominantly viscous to predominantly elastic response. This reflects the increasing steric interaction with increases in volume of the discontinuous phase.

In 1996, Pal studied the effect of droplet size and found it had a dramatic influence on emulsion rheology (62). Fine emulsions have much higher viscosity and storage moduli than the corresponding coarse emulsions. The shear thinning effect is much stronger in the case of fine emulsions. Water-in-oil emulsions age much more rapidly than oil-in-water emulsions. More recently, Lee et al. (63) and Aomari et al. (64) examined model emulsions and found that a maximum shear strain existed which occurred around $100\,s^{-1}$.

### 3. Nuclear Magnetic Resonance (NMR)

Urdahl et al. studied crude oils and silica-absorbed compounds (asphaltenes and resins) using $^{13}$C-NMR techniques (65). It was found that the asphaltenes and resins were enriched in condensed aromatics compared to the whole crude oils. There were strong indications of a long straight-chain aliphatic compound containing a heteroatom substituent which is abundant in paraffinic oils. There was also reason to believe that this compound was active in the formation of stable water-in-crude oil emulsions. The range from 130 to 210 ppm in the $^{13}$C NMR spectra was particularly of interest. This region represents quaternary aromatic and heteroatom-bonded carbons.

Balinov et al. studied the use of $^{13}$C NMR to characterize emulsions using the NMR self-diffusion technique (66). The technique uses the phase differences in consecutive pulses to measure the diffusion length of the target molecules. As such, the technique indicates the relative particle size in an emulsion. The NMR technique was compared to optical microscopy and showed good correlation over several experiments involving aging and breaking of the emulsions.

LaTorraca et al. used proton NMR to study oils and emulsions (67). The amount of hydrogen was found to be inversely proportional to the viscosity. The amount of water could be determined in an emulsion because of the separation downfield of the proton on water and on hydrocarbons. The viscosity and water content of emulsions could be simultaneously determined.

### 4. Interfacial Properties

Sjöblom et al. studied model emulsions stabilized by interfacially active fractions from crude oil (68). A good correlation was found between interfacial pressure of the fractions, as measured in a Langmuir trough, and the stability of emulsions as measured by the amount of water separated with time. Surface tension, as measured by the drop-volume method, did not show a surfactant-like behavior for the asphaltenes and resins. Børve et al. studied the pressure–area isotherms and relaxation behavior in a Langmuir trough (69). In one study, model polymers, styrene and allyl alcohol (PSAA, molecular weight 150 $g\,mol^{-1}$), and mixtures of PSAA with 4-hexadecylaniline or eicosylamine, were used as comparative polymers to the natural surfactants in oils. The mixtures of PSAA with the amines reproduce the $\pi$-$A$ isotherms and relationship properties shown by the interfacially active fractions of crude oils. The main difference was found to be a lower maximum compressibility and a higher surface activity. The crude oil fractions are well modeled by a relatively low-molecular-weight aromatic, weakly polar, water-insoluble hydrocarbon polymer to which has been added long-chain amines.

In a similar study, Ebeltoft et al. mixed surfactants (sodium dodecyl sulfate, cetyltrimethylammonium bromide, or cetylpyridinium chloride) with PSAA and studied the pressure–area isotherms (70). All the surfactants appeared to interact with the PSAA and in similar ways. It was concluded that PSAA monolayers are good models for emulsion-stabilizing monolayers in Norwegian crude oils. Monolayers of both PSAA and crude oil interfacially active fractions responded similarly to the presence of ionic surfactants, indicating analogous dissolution mechanisms.

Hartland and Jeelani performed a theoretical study on the effect of interfacial-tension gradients on emulsion stability (71). Dispersion stability and instability were explained in terms of a surface mobility which is proportional to the surface velocity. When the interfacial tension gradient is negative, the surface mobility is negative under some conditions, which greatly reduces the drainage so that the dispersion is stable. This is a normal situation as surfactant is present at the interface. Demulsifier molecules penetrate the interface within the film, thereby lowering the interfacial tension sufficiently and causing a positive interfacial-tension gradient. Drainage is subsequently increased and the emulsion becomes unstable.

Ese et al. studied the film-forming properties of asphaltenes and resins by using a Langmuir trough (72). Asphaltenes and resins were separated from different crude oils. It was found that the asphaltenes appear to pack closer at the water surface and form a more rigid surface than the resins. The size of asphaltene aggregates appears to increase when the spreading solvent becomes more aliphatic and with increasing asphaltene bulk concentration. Resin films show high compressibility, which indicates a collapse of the monomolecular film. A comparison between asphaltenes and resins shows that resins are more polar and do not aggregate to the same extent as the asphaltenes. Resins also show a high degree of sensitivity to oxidation. When resins and asphaltenes are mixed, resins begin to dominate the film properties when the former exceed about 40 wt %.

#### 5. Optical Methods

Miller and Böhm studied the coalescence of water-in-oil emulsions, using a specially designed optical instrument dependent on light scattering of the emulsions (73). The instrument could be used in either a transmission or backscatter mode.

### F. Physical Studies

#### 1. Structure and Droplet Sizes

Eley et al. studied the size of water droplets in emulsions, using optical microscopy and found that the droplet sizes followed a lognormal distribution (74). The number mean diameters of the droplets varied from about 1.4 to 5.6 μm. Paczynska-Lahme studied several mesophases in petroleum, using optical microscopy (75). Petroleum resins showed highly organized laminar structures and water-in-oil emulsions were generally unstructured, but sometimes hexagonal. Toulhoat et al. studied alphaltenes extracted form crude oils of various origins, using atomic-force microscopy (76). The asphaltenes were dried on to freshly cleaved mica and in some cases were present in diskoids of dimensions of approximately 2 nm × 30 nm. It was noted that these dimensions were similar to those measured using neutron scattering of asphaltenes in solvents. An increase in diskoid size was observed with increasing sulfur content of the asphaltenes, but no correlation in size was observed with increasing asphaltene molecular weight. Absorption of asphaltenes from unfiltered solutions revealed fractal-like asphaltene clusters with lengths of a few micrometers, width 1 μm and a thickness of 10 to 20 nm.

Balinov et al. studied the use of $^{13}C$ NMR to characterize emulsions, using the NMR self-diffusion technique, and compared this to optical microscopy (66). The optical microscopy showed an average droplet size of about 4 μm with a mean volume of approximately 8 $\mu m^3$ (estimated by the present author).

#### 2. Dynamics and Thermodynamics

Eley et al. studied the formation of emulsions and found that the rate was first order with respect to stirring time (74). As the asphaltene content increased, the rate constant decreased.

## IV. LABORATORY STUDIES ON EMULSION STABILITY AND SEPARATION OF STABILITY CLASSES

### A. Studies on Stability Classes

The most important characteristic of a water-in-oil emulsion is its "stability". The reason for this importance is that one must first characterize an emulsion as stable (or unstable) before one can characterize the properties. Properties change very significantly for each type of emulsion. Until recently, emulsion stability has not been defined (77). Therefore, studies were difficult because the end points of analysis were not defined. This section of the chapter summarizes studies to measure the stability of water-in-oil emulsions and to define characteristics of different stability classes. Four "states" in which water can exist in oil will be described. These include: stable emulsions, mesostable emulsions, unstable emulsions (or simply water and oil), and entrained water. These four "states" are differentiated by visual appearance as well as by rheological measures.

Studies in the past three years have shown that a class of very "stable" emulsions exists, characterized by their persistence over several months. The viscosity of these stable emulsions actually increases over time. These emulsions have been monitored for as long as three years in the laboratory. "Unstable" emulsions do not show this increase in viscosity and their viscosity is less than about 20 times greater than the starting oil. The viscosity increase for stable emulsions is at least three orders of magnitude greater than the starting oil. The present authors have studied emulsions for many years (77–82). The last two of these references as well as Ref. 77 describe studies to define stability. The findings of these earlier studies are summarized here. It was concluded both on the basis of the literature and experimental evidence above, that certain emulsions can be classed as stable. Some (if not all or many) stable emulsions increase in apparent viscosity with time, i.e., their elasticity increases. It is suspected that the stability derives from the strong viscoelastic interface caused by asphaltenes, perhaps along with resins. Increasing viscosity may be caused by increasing alignment of asphaltenes at the oil–water interface.

Mesostable emulsions are emulsions that have properties between stable and unstable emulsions (really oil/water mixtures) (77). It is suspected that mesostable emulsions either lack sufficient asphaltenes to render them completely stable or still contain too many destabilizing materials, perhaps some aromatics and aliphatics. The viscosity of the oil may be high enough to stabilize some water droplets for a period of time. Mesostable emulsions may degrade to form layers of oil and stable emulsions. Mesostable emulsions can be red or black in appearance and are probably the most commonly formed emulsions in the field.

Unstable emulsions are those that rapidly decompose (largely) to water and oil after mixing, generally within a few hours. Some water (usually less than about 10%) may be retained by the oil, especially if the oil is viscous.

The most important measurements for characterizing emulsions are forced oscillation rheometry studies. The presence of significant elasticity clearly defines whether or not a stable emulsion has been formed. The viscosity by itself can be an indicator of the stability of the emulsion, although it is not necessarily conclusive, unless one is fully certain of the viscosity of the starting oil. Color is an indicator, but may not be definitive. This laboratory's experience is that all table emulsions were reddish. Some mesostable emulsions were also reddish and unstable emulsions were always the color of the starting oil. Water content is not an indicator of stability and is error prone because of the "excess" of water that may be present. It should be noted, however, that stable emulsions have water contents greater than 70% and that unstable emulsions or entrained water-in-oil generally have water contents less than 50%. Water content after a period of about one week is found to be more reliable than immediate water content because separation will occur in those emulsions that are less stable.

### 1. Experimental Summary

Detailed experimental procedures are given in the literature (2). Water-in-oil emulsions were prepared in a rotary agitator and then the rheometric characteristics of these emulsions were studied over time. Eighty-two oils were used, taken from the storage facilities at the Emergencies Science Division. The properties of these oils are given in standard references (83).

Emulsions were prepared in an end-over-end rotary mixer (Associated Design). The apparatus was located in a temperature-controlled cold room at a constant 15°C. The mixing vessels were 2.2-liter FLPE wide-mouthed bottles. The mixing vessels were approximately one-quarter full, with 600 mL of salt water (3.3% w/v NaCl) and 30 mL of the sample crude oil or petroleum product. The vessels were mounted in the rotary mixer and allowed to stand for 3 h to equilibrate thermally. The vessels were then rotated for a period of 12 h at a rate of 55 rpm. The vessels were approximately 20 cm in height, providing a radius of rotation of about 10 cm. The resulting emulsions were then collected into jars, covered, and stored in the same 15°C cold room. Analysis was performed on the day of collection, and again one week later.

The following apparatuses were used for rheological analysis: Haake RS100 RheoStress rheometer, IBM-compatible PC with RheoStress RS Ver.2.10 P software, 35- and 60-mm parallel plates with corresponding base plates, clean air supply at 40 psi, and a circulating bath maintained at 15°C. Analysis was performed on a sample spread on to the base plate and raised to 2.00 mm from the measuring plate, with the excess removed using a Teflon spatula. This was left for 15 min to equilibrate thermally at 15°C. A stress sweep at a frequency of 1 $s^{-1}$ was performed first to determine the linear viscoelastic range (stress-independent region) for frequency analysis. This also provided values for the complex modulus, the elasticity and viscosity moduli, the low shear dynamic viscosity, and the tan ($\delta$) value. A frequency sweep was then performed at a stress value within the linear viscoelastic range, ranging from 0.04 to 40 Hz. This provided the data for analysis to determine the constants of the Ostwald-de Waele equation for the emulsion. The apparent dynamic viscosity was determined on the plate–plate apparatus as well, and corrected using the Weissenberg equation. A shear rate of 1 $s^{-1}$ was applied for 1 min, without ramping.

A Metrohm 701 KF titrino Karl-Fischer volumetric titrator and Metrohm 703 Ti Stand were used to measure water content. The reagent was Aquastar Comp 5 and the solvent 1:1:2 methanol:chloroform:toluene.

## 2. Results and Discussion

The rheological data are given in Table 1. The second column of the table is the evaporation state of the oil in mass percentage lost. The third column is the assessment of the stability of the emulsion based on both visual appearance and rheological properties. The power law constants, $k$ and $n$, are given next. These are parameters from the Ostwald–de Waele equation which describes the Newtonian (or nonNewtonian) characteristics of the material. The viscosity of the emulsion is next and in column 7, the complex modulus which is the vector sum of the viscosity and elasticity. Column 8 lists the elasticity modulus and column 9, the viscosity modulus. In column 10, the isolated, low-shear viscosity is given. This is the viscosity of emulsion at very low shear rate. In column 9, the tan $\delta$, the ratio of the viscosity to the elasticity component, is given. Finally, the water content of the emulsion is presented.

Observations were made on the appearance of the emulsions and were used to classify the emulsions. All of the stable emulsions appeared to be stable and remained intact over 7 days in the laboratory. All of the mesostable emulsions broke after a few days in water, free oil, and emulsion. The time for these emulsions to break down varied from about 1 to 3 days. The emulsion portion of these break-down emulsions appears to be somewhat stable, although separate studies on this portion have not been performed because of the difficulty in separating these portions from the oil and water. All entrained water appeared to have larger suspended water droplets. The appearance of nonstable water in oil was just that; the oil appeared to be unchanged and a water layer was clearly visible.

Table 2 provides the data on the oil properties as well as a new parameter called "stability" which is the complex modulus divided by the viscosity of the starting oil. It is noted from this table that this parameter correlates quite well with the assigned behavior of the oils. High-stability parameters imply stable emulsions and low ones imply unstable emulsions. Stability has nominally the units of mPa/mPa.s or $s^{-1}$; however, it can be converted into a unitless parameter by multiplying by "s" which is 1 in these cases.

The "stability" parameter was used to study the correlations between the properties of the oil and the stability of the resulting emulsion. The correlations are summarized in Table 3 which shows the regression coefficient ($r^2$) correlations of stability with the starting-oil properties. The regression coefficients were calculated using the program TableCurve (Jandel Scientific, San Rafael, CA). The regression coefficients are taken from the highest consistent value of the simple curves fit to a given set of data. Table 3 shows that the correlations vary with each type of emulsion. For all the emulsions and oil-in-water situations, none of the parameters correlate well with stability, except for the final water content of the emulsion. This is because the less stable emulsions have little water content. It should also be noted that this is not a starting-oil property. For stable emulsions, there is only a slight correlation with density and saturates. For mesostable emulsions, there is a relatively good correlation with density, viscosity, resins, saturates, and aromatics. This may indicate that these emulsions are temporarily stabilized by a combination of viscous forces and resin stabilization. Entrained water stability correlates best with density, aromatics, viscosity, and resin content. This indicates that these may be dominated most by viscous forces. Finally, in the case of unstable emulsions, no parameter correlates well. This appears to confirm the findings that none of the stabilization forces noted is operative.

It is important to recognize that there may be a strong interaction between parameters. To check for this, the program TableCurve 3D (Jandel Scientific, San Rafael, California) was used to correlate three parameters simultaneously. Results of this are shown in Table 4. Again, only the consistently highest regression coefficient ($r^2$) was taken. This table shows that several two-way interactions exist. For all water-in-oil forms, there is no significant correlation between the parameters tested. For stable emulsions, there is a strong correlation between stability and viscosity and asphaltenes. This did not show on a two-way parameter correlation, presumably because of the interaction between parameters. The best correlation for mesostable emulsions is that of stability with resins and viscosity, followed very closely by correlations of stability, viscosity with asphaltenes, aromatics, and density. The correlation of stability with viscosity and resins for water entrained in oil shows that the stability of entrained water correlates best with aromatics and density. Similarly, unstable emulsion "stability" correlates highly with aromatics and density along with the viscosity.

In all four correlations, sharply defined regions of stability are noted. it is also noted that different forces are evident on the basis of these correlations. For stable emulsions, there appears to be a region where viscosity, asphaltenes, and resins interact to form a stable emulsion. This is similar in mesostable emulsions, except that the importance of asphaltenes and resins are reversed. There is a region defined for entrained water on the basis of aromatic content and density. Similarly for unstable emulsions this is also defined by aromatics and density. This confirms previous findings that stable emulsions are the result of stabilization by asphaltenes and to a secondary extent by viscous retention. Resins are only partially responsible and, in fact, if the resin/asphaltene ratio rises, the result appears to be a mesostable emulsion. Mesostable emulsions are largely the result of resin and viscosity stabilization. Asphaltenes play a secondary role. It is interesting to note that there is a viscosity "window" for both stable and mesostable emulsions. Very high-viscosity oils do not appear to make either stable or mesostable emulsions.

It is noted that a stability index calculated by dividing the complex modulus of the emulsion (or remains) after one week, divided by the starting-oil viscosity, correlates very well to the assignment of the stability class. The stability parameter was correlated with other parameters that might be used as indicators of stability. The results are shown in Table 5. It can be seen from this table that the correlation varies with the type of emulsion or water-in-oil state considered. For all types, there is only a moderate correlation with the water content. It should be noted that this would be expected since the correlation is with the 7-day-old sample, and all but stable emulsions have lost a significant amount of their water. There is also a moderate correlation of the Ostwald–de Waele equation parameters, which indicates nonNewtonian behavior in the case of both stable and mesostable emulsions and Newtonian behavior in the case of the entrained water. The entrained water class shows a high correlation with the low-shear viscosity, indicating that these are largely viscosity stabilized. given the high correlation of the "stability" index with the assigned properties of the emulsion, there does not appear to be other rheological parameters that can discriminate to the same extent. This is largely a result of the fact that the other parameters are generally relevant to specific water-in-oil classes and none other than the stability covers all four classes.

Ternary diagrams of the aromatic, resin and asphaltene components of the four classes of water-in-oil states discussed here are shown in Figs 1–4. These diagrams show that there is overlapping regions for composition of all four types. Unfortunately, a simple compositional analysis will not discriminate between those oils that form emulsions and those that do not. This again indicates that there is a complex interaction between components and the viscosity (and perhaps density) of the oil.

The differences in the starting-oil properties and the final states after one week are shown in Table 6. This shows that the factor defined as emulsion stability is capable of discriminating among the various states of water-in-oil studied here. Although there are overlapping ranges, the differences are generally sufficient to act as a single-value discriminator. It is noted that there are different viscosity ranges for the different states. This shows that viscous forces are responsible for part of the stability, but that after viscosity of the starting oil rises to a given point, about 20 mPa.s, mesostable or stable emulsions are no longer produced. This may also explain two outstanding mysteries, that of why Bunker C generally does not form emulsions and why stable emulsions are not commonly seen in actual spills. The viscosity of Bunker C, especially after a short period of weathering, is too great to form either stable or mesostable emulsions. Further, if the viscosity of formation is too great, perhaps the weathering of an emulsion will increase its viscosity past a certain point and then destabilization may occur.

Table 6 also shows that the deviation from Newtonian behavior (as shown by the power-law constants) is greatest for the stable emulsions and secondly for mesostable emulsions with almost no deviation noted for the entrained and unstable cases. This is the result of a high elastic component to the viscosity, as evidenced by the elastic modulus and tan $\delta$ for the stable emulsions and slightly elastic for the mesostable emulsions. As would be expected, the water content correlates very highly with the state after one week. This is accentuated by the fact that mesostable emulsions and entrained water-in-oil have separated to a significant degree after this time.

Table 7 shows the properties of the water-in-oil studied here. This shows that the starting-oil properties differ somewhat for oils that form the various states. The oil properties for stable and mesostable emulsions are similar. These are oils of medium viscosity that contain a significant amount of resins and asphaltenes. Mesostable emulsions may form oils that have higher or lower viscosities than those that might form stable emulsions. Stable emulsions are more likely to form from those oils having more asphaltenes than resins. Entrained water is likely to form from more viscous oils with relatively high densities. Oils of very high or very low viscosities (and

**Table 1** Rheological Data on the Emulsions Produced from the Oils

| | | | Day of formation | | | | | | | | | One week after formation | | | | | | | | |
|---|---|---|---|---|---|---|---|---|---|---|---|---|---|---|---|---|---|---|---|---|
| Oil | Percentage evap. | Visual stability | Power law constants ($K$) | Power law constants ($n$) | Viscosity (mPa.s) | Complex modulus (mPa) | Elasticity modulus (mPa) | Viscosity modulus (mPa) | Isolated low-shear viscosity (mPa.s) | tan $\delta$ ($V/E$) | Water content (%w/w) | Power law constants ($K$) | Power law constants ($n$) | Viscosity (mPa.s) | Complex modulus (mPa) | Elasticity modulus (mPa) | Viscosity modulus (mPa) | Isolated low-shear viscosity (mPa.s) | tan $\delta$ ($V/E$) | Water content (%w/w) |
| Arabian Light | 0 | Stable | 4.044E+04 | 0.0613 | 2.3E+04 | 4.7E+05 | 4.7E+05 | 5.0E+04 | 8.0E+04 | 0.11 | 87.42 | 3.537E+04 | 0.0860 | 2.3E+04 | 4.6E+05 | 4.6E+05 | 4.0E+04 | 6.3E+04 | 0.14 | 86.93 |
| Arabian Light | 12.04 | Stable | 6.031E+04 | 0.0450 | 4.6E+04 | 4.0E+05 | 4.0E+05 | 3.7E+04 | 6.0E+04 | 0.10 | 88.86 | 3.596E+04 | 0.096 | 3.1E+04 | 2.0E+05 | 2.0E+05 | 2.7E+04 | 4.2E+04 | 0.13 | 85.82 |
| Arabian Light | 24.20 | Stable | 6.560E+04 | 0.0785 | 4.8E+04 | 5.1E+05 | 5.0E+05 | 5.7E+04 | 8.6E+04 | 0.11 | 84.71 | 4.091E+04 | 0.123 | 4.3E+04 | 3.6E+05 | 3.5E+05 | 7.3E+04 | 1.1E+04 | 0.20 | 83.62 |
| Arabian Medium | 0 | Stable | 8.987E+04 | 0.0764 | 5.3E+04 | 5.5E+05 | 5.5E+05 | 5.2E+04 | 8.2E+04 | 0.09 | 84.68 | 8.828E+04 | 0.944 | 5.5E+04 | 6.8E+05 | 7.7E+05 | 7.9E+04 | 1.2E+04 | 0.11 | 84.36 |
| Arabian Medium | 13.15 | Stable | 1.870E+04 | 0.5689 | 2.2E+04 | 1.5E+05 | 1.1E+05 | 1.0E+05 | 1.6E+04 | 0.90 | 76.52 | 2.663E+04 | 0.4521 | 2.8E+04 | 2.1E+05 | 1.8E+05 | 1.2E+05 | 1.9E+04 | 0.64 | 77.06 |
| Arabian Medium | 20.77 | Stable | 9.727E+03 | 0.8349 | 2.2E+04 | 7.4E+04 | 2.4E+04 | 7.2E+04 | 1.1E+04 | 2.9 | 73.10 | 1.117E+04 | 0.8129 | 2.3E+04 | 1.0E+05 | 5.0E+04 | 9.2E+04 | 1.5E+04 | 1.8 | 72.86 |
| Arabian Medium | 30.93 | Stable | 2.565E+04 | 0.6982 | 5.0E+04 | 1.9E+05 | 1.0E+05 | 1.7E+05 | 2.6E+04 | 1.7 | 64.92 | 5.757E+04 | 0.442 | 5.4E+04 | 4.0E+05 | 3.2E+05 | 2.2E+05 | 3.6E+04 | 0.70 | 65.60 |
| Belridge Heavy | 0 | Entrained | 2.452E+04 | 0.7025 | 4.5E+04 | 1.4E+05 | 7.8E+00 | 1.1E+05 | 2.0E+04 | 1.5 | 54.23 | 2.638E+04 | 0.6938 | 5.4E+04 | 1.6E+05 | 8.2E+04 | 1.2E+05 | 2.2E+04 | 1.6 | 44.39 |
| Belridge Heavy | 2.74 | Entrained | 2.658E+04 | 0.6235 | 5.2E+04 | 2.0E+05 | 1.1E+05 | 1.6E+05 | 2.5E+04 | 1.4 | 59.55 | 2.779E+04 | 0.6374 | 6.2E+04 | 1.8E+05 | 1.0E+05 | 1.4E+05 | 2.4E+04 | 1.4 | 45.19 |
| Bunker C (Anchorage) | 0 | Entrained | 1.824E+04 | 0.8035 | 2.8E+04 | 1.3E+05 | 4.6E+04 | 1.2E+05 | 1.8E+04 | 2.8 | 34.74 | 1.903E+04 | 0.8567 | 1.5E+05 | 1.4E+05 | 3.2E+04 | 1.3E+05 | 2.1E+04 | 4.0 | 30.96 |
| Bunker C (Anchorage) | 8.41 | No | | | | | | | | | 5.83 | | | | | | | | | |
| Bunker C (1987) | 0 | Entrained | 2.350E+05 | 0.4204 | 1.1E+05 | 7.2E+05 | 4.8E+05 | 6.6E+05 | 8.8E+04 | 1.3 | 26.44 | 9.818E+04 | 0.834 | 1.4E+05 | 6.5E+05 | 6.2E+05 | 5.6E+05 | 8.4E+04 | 1.0 | 24.02 |
| Carpenteria | 0 | No | | | | | | | | | 8.73 | | | | | | | | | |
| Carpenteria | 10.31 | Meso | 1.077E+04 | 0.6232 | 2.3E+04 | 7.3E+04 | 4.7E+04 | 5.6E+04 | 9.0E+03 | 1.2 | 71.80 | 4.815E+03 | 0.9765 | 2.3E+04 | 3.6E+04 | 3.2E+03 | 3.6E+04 | 5.8E+03 | 12 | 28.19 |
| Carpenteria | 14.87 | Meso | 6.434E+04 | 0.4340 | 3.4E+04 | 1.3E+05 | 8.2E+04 | 9.9E+04 | 1.6E+04 | 1.2 | 54.26 | 9.734E+03 | 0.8845 | 2.1E+04 | 2.9E+05 | 2.6E+05 | 1.5E+05 | 2.3E+04 | 0.57 | 30.87 |
| Coal Oil Point Seep Sample | 0 | Same[a] | 1.828E+05 | 0.7172 | 2.8E+05 | 1.2E+06 | 5.3E+05 | 1.1E+06 | 1.7E+04 | 1.8 | 32.15 | 2.187E+05 | 0.7201 | 4.0E+05 | 1.5E+06 | 6.7E+05 | 1.4E+06 | 6.7E+05 | 2.0 | 39.24 |
| Cook Inlet – Granite Point | 0 | No | | | | | | | | | < 1 | | | | | | | | | 1.89 |
| Cook Inlet – Granite Point | 45.32 | Meso | 1.683E+01 | 0.1983 | 9.6E+03 | 3.4E+05 | 3.3E+05 | 1.1E+05 | 1.8E+04 | 0.30 | 83.05 | 7.725E+04 | 0.1401 | 8.5E+03 | 2.6E+05 | 2.6E+05 | 7.4E+04 | 1.2E+04 | 0.30 | 57.58 |
| Cook Inlet – Swanson River | 0 | Meso | 9.878E+02 | 0.6880 | 2.4E+03 | 1.0E+04 | 6.9E+03 | 7.7E+03 | 1.2E+03 | 0.80 | 75.95 | 4.376E+02 | 0.4120 | 1.9E+03 | 2.7E+04 | 2.2E+04 | 1.7E+04 | 2.6E+03 | 0.80 | 57.50 |
| Cook Inlet – Swanson River | 39.69 | Stable | 5.914E+04 | 0.1624 | 1.6E+04 | 2.9E+05 | 2.8E+05 | 8.8E+04 | 1.4E+04 | 0.30 | 81.48 | 1.393E+05 | 0.271 | 1.8E+05 | 8.2E+05 | 7.4E+05 | 3.3E+05 | 5.2E+04 | 0.43 | 80.76 |
| Cook Inlet – Trading Bay | 0 | Meso | | | | | | | | | 80.63 | | | | | | | | | 1.89 |
| Cook Inlet – Trading Bay | 33.30 | Meso | 9.508E+04 | 0.1852 | 3.8E+04 | 4.5E+05 | 4.3E+05 | 1.4E+05 | 2.3E+04 | 0.34 | 76.22 | 1.361E+05 | 0.1861 | 4.3E+04 | 5.3E+05 | 4.9E+05 | 1.5E+05 | 2.6E+04 | 0.27 | 61.43 |
| Diesel (Anchorage) | 0 | No | | | | | | | | | < 1 | | | | | | | | | |
| Diesel (Anchorage) | 37.44 | No | | | | | | | | | < 1 | | | | | | | | | |
| Diesel (Mobile Burn #3) | 0 | No | | | | | | | | | < 1 | | | | | | | | | |
| Diesel (Mobile Burn #3) | 8.21 | No | | | | | | | | | < 1 | | | | | | | | | |
| Diesel (Mobile Burn #3) | 16.32 | No | | | | | | | | | < 1 | | | | | | | | | |
| Dos Cuadras | 0 | No | | | | | | | | | 12.57 | | | | | | | | | |
| Dos Cuadras | 11.17 | Meso | 6.570E+02 | 1.0000 | 8.0E+02 | 3.4E+03 | 1.3E+03 | 3.3E+03 | 5.0E+02 | 2.5 | 47.60 | 5.203E+02 | 1.0000 | 5.3E+02 | 4.6E+03 | 1.6E+03 | 4.2E+03 | 7.0E+02 | 3.0 | 29.49 |
| Dos Cuadras | 20.30 | Meso | 6.638E+03 | 0.5776 | 9.8E+03 | 3.3E+04 | 2.3E+04 | 2.2E+04 | 3.6E+03 | 1.0 | 68.55 | 2.145E+03 | 1.0000 | 1.9E+03 | 1.7E+04 | 3.0E+03 | 1.3E+04 | 2.1E+03 | 4.2 | 28.72 |
| Hondo | 0 | Stable | 1.612E+05 | 0.1797 | 1.1E+05 | 9.2E+05 | 9.1E+05 | 2.2E+05 | 3.6E+04 | 0.24 | 80.93 | 1.665E+05 | 0.1825 | 1.7E+05 | 8.8E+05 | 8.3E+05 | 2.5E+05 | 4.2E+04 | 0.32 | 79.96 |
| Hondo | 16.67 | Stable | 2.44E+05 | 0.2870 | 1.9E+05 | 1.3E+06 | 1.1E+06 | 5.3E+05 | 8.4E+04 | 0.45 | 66.20 | 1.537E+05 | 0.5753 | 2.8E+05 | 8.4E+05 | 6.2E+05 | 6.1E+05 | 9.0E+04 | 0.96 | 64.23 |
| Hondo | 32.29 | No | | | | | | | | | 5.24 | | | | | | | | | |
| Jet fuel (Anchorage) | 0 | No | | | | | | | | | < 1 | | | | | | | | | |
| Jet fuel (Anchorage) | 52.72 | No | | | | | | | | | < 1 | | | | | | | | | |
| North Slope (middle pipeline) | 0 | No | | | | | | | | | < 1 | | | | | | | | | |
| North Slope (middle pipeline) | 30.54 | Meso | 5.975E+03 | 0.7398 | 6.8E+03 | 1.2E+05 | 1.0E+05 | 4.8E+04 | 7.6E+03 | 0.52 | 61.92 | 1.411E+03 | 0.9844 | 1.8E+03 | 1.1E+04 | 1.3E+03 | 1.1E+04 | 1.7E+03 | 8.4 | 21.76 |
| North Slope (middle pipeline) | 0 | No | | | | | | | | | 1.00 | | | | | | | | | |
| North Slope (northern pipeline) | 31.14 | Meso | 3.757E+03 | 0.9195 | 2.6E+03 | 1.1E+05 | 96.E+04 | 4.7E+04 | 7.4E+03 | 0.50 | 69.82 | 1.603E+03 | 0.8623 | 1.7E+03 | 9.8E+03 | 2.2E+03 | 9.6E+03 | 1.5E+03 | 4.2 | 15.00 |
| North Slope (southern pipeline) | 0 | No | | | | | | | | | < 1 | | | | | | | | | |

| | | | | | | | | | | | | | | | | | | | | |
|---|---|---|---|---|---|---|---|---|---|---|---|---|---|---|---|---|---|---|---|---|
| North Slope (southern pipeline) | 29.62 | Meso | 6.061E+04 | 0.1207 | 8.7E+03 | 1.9E+05 | 1.7E+05 | 8.1E+04 | 1.3E+04 | 0.46 | 53.47 | 2.586E+03 | 1.001 | 2.2E+03 | 2.0E+04 | 7.7E+03 | 1.8E+04 | 2.8E+03 | 2.2 | 21.14 |
| Pitas Point | 0 | No | | | | | | | | | < 1 | | | | | | | | | |
| Pitas Point | 23.56 | No | | | | | | | | | < 1 | | | | | | | | | |
| Platform Irene | 0 | Entrained | 2.201E+05 | 0.6272 | 3.9E+05 | 1.4E+06 | 7.6E+05 | 1.2E+06 | 1.9E+05 | 1.52 | 62.22 | 2.744E+05 | 0.6255 | 5.4E+05 | 3.3E+06 | 2.1E+06 | 2.5E+06 | 4.0E+05 | 1.20 | 34.94 |
| Point Arguello Comingled | 0 | Stable | 1.798E+05 | 0.1822 | 1.8E+05 | 7.8E+05 | 7.2E+05 | 3.2E+05 | 5.0E+04 | 0.43 | 82.31 | 1.493E+05 | 0.2883 | 1.8E+05 | 1.1E+06 | 1.0E+06 | 4.1E+05 | 6.4E+04 | 0.36 | 82.19 |
| Point Arguello Comingled | 9.05 | Stable | 1.838E+05 | 0.2176 | 1.5E+05 | 8.5E+05 | 7.9E+05 | 3.0E+05 | 4.7E+04 | 0.38 | 67.92 | 1.651E+05 | 0.3100 | 1.5E+05 | 6.2E+05 | 4.7E+05 | 4.3E+05 | 7.0E+04 | 0.95 | 69.41 |
| Point Arguello Comingled | 15.19 | Entrained | 8.871E+04 | 0.7296 | 1.4E+05 | 6.1E+05 | 2.7E+05 | 5.5E+05 | 8.7E+04 | 2.0 | 30.15 | 9.211E+04 | 0.762 | 1.6E+05 | 8.0E+05 | 3.5E+05 | 7.0E+05 | 1.1E+05 | 2.10 | 28.42 |
| Point Arguello Comingled | 22.12 | No | | | | | | | | | 4.96 | | | | | | | | | |
| Point Arguello Heavy | 0 | Stable | 7.554E+04 | 0.4375 | 1.5E+05 | 4.9E+05 | 4.0E+05 | 2.8E+05 | 4.5E+04 | 0.71 | 72.95 | 8.767E+04 | 0.4322 | 1.8E+05 | 7.2E+05 | 5.8E+05 | 3.9E+05 | 6.6E+04 | 0.72 | 74.97 |
| Point Arguello Heavy | 17.78 | No | | | | | | | | | < 1 | | | | | | | | | |
| Point Arguello Light | 0 | Stable | 8.088E+04 | 0.0679 | 6.7E+04 | 6.5E+05 | 6.4E+05 | 6.8E+04 | 1.0E+04 | 0.11 | 93.14 | 9.671E+04 | 0.0581 | 9.6E+04 | 8.3E+05 | 8.2E+05 | 1.4E+05 | 2.2E+04 | 0.17 | 93.79 |
| Point Arguello Light | 10.19 | Stable | 4.891E+05 | 0.0281 | 2.8E+05 | 3.3E+06 | 3.3E+06 | 1.7E+05 | 2.6E+04 | 0.05 | 88.84 | 4.167E+05 | 0.0372 | 2.5E+05 | 2.8E+06 | 2.8E+06 | 3.3E+05 | 52.E+04 | 0.12 | 87.78 |
| Point Arguello Light | 19.04 | Stable | 5.026E+05 | 0.0492 | 2.7E+05 | 3.4E+06 | 3.4E+06 | 2.8E+05 | 4.3E+04 | 0.08 | 85.50 | 5.054E+05 | 0.0539 | 3.1E+05 | 3.5E+06 | 3.6E+06 | 4.2E+05 | 7.2E+04 | 0.13 | 85.63 |
| Point Arguello Light | 28.33 | Stable | 1.808E+05 | 0.1835 | 1.4E+05 | 9.8E+05 | 9.6E+05 | 2.5E+05 | 4.0E+04 | 0.26 | 79.80 | 1.920E+05 | 0.1723 | 1.6E+05 | 1.5E+06 | 1.4E+06 | 4.2E+05 | 6.6E+04 | 0.28 | 75.89 |
| Port Hueneme | 0 | Entrained | 9.389E+03 | 0.7889 | 1.6E+04 | 6.4E+04 | 2.4E+04 | 6.1E+04 | 8.8E+03 | 2.6 | 37.97 | 7.253E+03 | 0.9633 | 8.8E+03 | 4.3E+04 | 4.8E+03 | 4.1E+04 | 6.9E+03 | 9.4 | 20.06 |
| Port Hueneme | 3.14 | Entrained | 2.464E+04 | 0.6939 | 4.6E+04 | 1.7E+05 | 8.2E+04 | 1.5E+05 | 2.3E+04 | 1.7 | 45.33 | 1.918E+04 | 0.8713 | 2.7E+04 | 1.3E+05 | 2.1E+04 | 1.2E+05 | 2.0E+04 | 5.7 | 29.20 |
| Port Hueneme | 6.37 | Entrained | 3.995E+04 | 0.6873 | 7.1E+04 | 2.7E+05 | 1.4E+05 | 2.3E+05 | 3.6E+04 | 1.7 | 43.37 | 3.438E+04 | 0.8070 | 5.4E+04 | 2.6E+05 | 6.4E+04 | 2.4E+05 | 4.0E+04 | 3.9 | 26.42 |
| Santa Clara | 0 | Meso | 7.315E+03 | 0.9272 | 2.1E+04 | 1.8E+04 | 2.2E+03 | 1.8E+04 | 2.8E+03 | 6.5 | 60.63 | 2.125E+03 | 0.9800 | 2.6E+03 | 2.9E+04 | 1.7E+04 | 2.4E+04 | 3.9E+03 | 1.4 | 12.75 |
| Santa Clara | 11.40 | Meso | 9.441E+04 | 0.1924 | 2.0E+04 | 7.0E+05 | 6.9E+05 | 1.0E+05 | 1.5E+0.4 | 0.15 | 50.43 | 1.087E+05 | 0.169 | 1.5E+04 | 7.5E+04 | 2.0E+04 | 7.3E+04 | 1.1E+04 | 3.5 | 38.72 |
| Santa Clara | 21.63 | Meso | 5.440E+04 | 0.6664 | 1.0E+05 | 3.6E+05 | 2.0E+05 | 2.9E+05 | 4.7E+04 | 1.4 | 38.99 | 5.794E+04 | 0.675 | 9.2E+04 | 3.7E+05 | 2.0E+05 | 3.3E+05 | 5.0E+04 | 1.65 | 40.15 |
| Sockeye | 0 | Stable | 1.075E+06 | 0.0462 | 6.9E+05 | 6.4E+06 | 6.3E+06 | 4.0E+05 | 6.2E+04 | 0.06 | 87.12 | 8.596E+05 | 0.0544 | 6.9E+05 | 4.3E+06 | 1.0E+06 | 3.1E+05 | 8.8E+04 | 0.11 | 86.87 |
| Sockeye | 12.50 | Stable | 1.605E+05 | 0.1873 | 1.7E+05 | 9.3E+05 | 8.9E+05 | 2.4E+05 | 3.9E+04 | 0.27 | 80.05 | 1.806E+05 | 0.1871 | 2.2E+05 | 1.1E+06 | 4.3E+06 | 4.9E+05 | 5.0E+04 | 0.31 | 74.35 |
| Sockeye | 22.10 | Stable | 2.369E+05 | 0.1980 | 2.6E+05 | 1.4E+06 | 1.3E+05 | 3.9E+05 | 9.8E+04 | 0.30 | 77.94 | 2.396E+05 | 0.1948 | 3.1E+05 | 1.5E+06 | 1.4E+06 | 5.0E+05 | 7.8E+04 | 0.35 | 70.39 |
| Sumatran Heavy | 0 | No | | | | | | | | | 20.97 | | | | | | | | | 12.73 |
| Sumatran Heavy | 5.26 | No | | | | | | | | | 2.35 | | | | | | | | | |
| Sumatran Light | 0 | No | | | | | | | | | 12.54 | | | | | | | | | 11.20 |
| Taching | 0 | No | | | | | | | | | 3.53 | | | | | | | | | |
| Takula | 0 | Stable | 2.573E+05 | 0.1063 | 8.9E+04 | 1.1E+06 | 1.1E+06 | 1.6E+05 | 2.5E+04 | 0.14 | 86.40 | 1.609E+05 | 0.137 | 1.2E+05 | 1.2E+06 | 1.2E+06 | 2.1E+05 | 3.1E+04 | 0.15 | 84.18 |
| Takula | 8.31 | Stable | 1.398E+05 | 0.1655 | 1.1E+05 | 1.1E+06 | 1.0E+06 | 2.1E+05 | 3.0E+04 | 0.21 | 82.21 | 1.528E+05 | 0.1744 | 1.3E+05 | 1.2E+06 | 1.1E+06 | 2.3E+05 | 3.6E+04 | 0.20 | 81.08 |
| Takula | 15.88 | Stable | 1.402E+05 | 0.1952 | 1.5E+05 | 1.1E+06 | 1.1E+06 | 2.5E+05 | 4.0E+04 | 0.23 | 75.56 | 1.877E+05 | 0.1791 | 1.7E+05 | 1.2E+06 | 1.2E+06 | 2.8E+05 | 4.4E+04 | 0.24 | 73.83 |
| Tapis | 0 | No | | | | | | | | | 15.87 | | | | | | | | | 9.06 |
| Tapis | 13.90 | No | | | | | | | | | 22.68 | | | | | | | | | 20.03 |
| Tapis | 28.62 | No | | | | | | | | | 9.02 | | | | | | | | | |
| Tapis | 43.43 | No | | | | | | | | | 7.75 | | | | | | | | | |
| Udang | 0 | Entrained | 1.921E+04 | 0.7288 | 3.2E+04 | 1.3E+05 | 5.6E+04 | 1.1E+05 | 1.8E+04 | 2.0 | 37.05 | | | | | | | | | 19.65 |
| Waxy Light Heavy Blend | 0 | No | | | | | | | | | 4.11 | | | | | | | | | |
| Waxy Light Heavy Blend | 12.00 | Entrained | 4.028E+04 | 0.7738 | 6.0E+03 | 3.1E+04 | 1.0E+04 | 3.0E+04 | 4.6E+03 | 2.8 | 33.97 | 2.763E+03 | 0.919 | 3.7E+03 | 1.5E+04 | 2.4E+03 | 1.5E+04 | 2.4E+03 | 6.2 | 12.21 |
| Waxy Light Heavy Blend | 19.60 | Meso | 2.754E+04 | 0.6687 | 4.5E+04 | 1.9E+05 | 9.6E+04 | 1.6E+05 | 2.7E+04 | 1.7 | 34.65 | 1.884E+04 | 0.709 | 3.7E+04 | 1.1E+05 | 5.3E+04 | 1.0E+05 | 1.6E+04 | 2.1 | 33.58 |

[a]Starting sample is an emulsion itself.

**Table 2** Oil Properties and Comparison with the One-week Emulsion Properties

| | Starting-oil properties | | | | | | | | | | Emulsion rheological properties – one week after formation | | | | | | | | |
|---|---|---|---|---|---|---|---|---|---|---|---|---|---|---|---|---|---|---|---|
| Oil | Percentage evap. | Visual stability | Density | Viscosity | Saturates | Aromatics | Resins | Asphaltenes | Waxes | Emulsion stability[a] | Power law (*K*) | constants (*n*) | Viscosity (mPa.s) | Complex modulus (mPa) | Elasticity modulus (mPa) | Viscosity modulus (mPa) | Isolated low-shear viscosity (mPa.s) | tan δ (V/E) | Water content (%w/w) |
| Sockeye | 0 | Stable | 0.897 | 45 | 48 | 31 | 13 | 8 | 6 | 95555.56 | 8.596E+05 | 0.0544 | 6.9E+05 | 4.3E+06 | 1.0E+06 | 3.1E+05 | 8.8E+04 | 0.11 | 86.87 |
| Point Arguello Light | 0 | Stable | 0.874 | 22 | 57 | 27 | 9 | 6 | 6 | 37727.27 | 9.671E+04 | 0.0581 | 9.6E+04 | 8.3£+05 | 8.2E+05 | 1.4E+05 | 2.2E+04 | 0.17 | 93.79 |
| Point Arguello Light | 10.19 | Stable | 0.898 | 76 | 54 | 30 | 9 | 8 | 6 | 36842.11 | 4.167E+05 | 0.0372 | 2.5E+05 | 2.8E+06 | 2.8E+06 | 3.3E+05 | 5.2E+04 | 0.12 | 87.78 |
| Arabian Light | 0 | Stable | 0.866 | 14 | 51 | 39 | 6 | 3 | 5 | 32857.14 | 3.537E+04 | 0.0860 | 2.3E+04 | 4.6E+05 | 4.6E+05 | 4.0E+04 | 6.3E+04 | 0.14 | 86.93 |
| Arabian Medium | 0 | Stable | 0.878 | 29 | 54 | 32 | 7 | 6 | 6 | 23448.28 | 8.828E+04 | 0.0944 | 5.5E+04 | 6.8E+05 | 7.7E+05 | 7.9E+04 | 1.2E+04 | 0.11 | 84.35 |
| Point Arguello Light | 19.04 | Stable | 0.913 | 183 | 48 | 31 | 12 | 9 | 7 | 19125.68 | 5.054E+05 | 0.0539 | 3.1E+06 | 3.5E+06 | 3.6E+06 | 4.2E+05 | 7.2E+04 | 0.13 | 85.63 |
| Takula | 0 | Stable | 0.864 | 110 | 65 | 22 | 8 | 5 | 8 | 10909.09 | 1.609E+05 | 0.1371 | 1.2E+05 | 1.2E+06 | 1.2E+06 | 2.1E+05 | 3.1E+04 | 0.15 | 84.18 |
| Sockeye | 12.50 | Stable | 0.917 | 163 | 44 | 32 | 15 | 9 | 5 | 6748.47 | 1.806E+05 | 0.1871 | 2.2E+05 | 1.1E+06 | 4.3E+06 | 4.9E+05 | 5.0E+04 | 0.31 | 74.35 |
| Arabian Light | 12.04 | Stable | 0.892 | 33 | 49 | 37 | 8 | 5 | 5 | 6060.61 | 3.596E+04 | 0.0957 | 3.1E+04 | 2.0E+05 | 2.0E+05 | 2.7E+04 | 4.2E+04 | 0.13 | 85.82 |
| Cook Inlet – Swanson River | 39.69 | Stable | 0.914 | 152 | 56 | 29 | 7 | 7 | | 5394.74 | | | | 8.2E+05 | 7.4E+05 | 3.3E+05 | 5.2E+04 | 0.43 | 80.75 |
| Arabian Light | 24.20 | Stable | 0.911 | 94 | 46 | 39 | 10 | 6 | 5 | 3829.79 | 4.091E+04 | 0.1226 | 4.3£+05 | 3.6E+05 | 3.5E+05 | 7.3E+04 | 1.1E+04 | 0.20 | 83.62 |
| Sockeye | 22.10 | Stable | 0.926 | 628 | 39 | 34 | 15 | 15 | 5 | 2388.54 | 2.396E+05 | 0.1948 | 3.1E+05 | 1.5E+06 | 1.4E+06 | 5.0E+05 | 7.8E+04 | 0.35 | 70.39 |
| Arabian Medium | 13.15 | Stable | 0.91 | 91 | 42 | 44 | 7 | 7 | 5 | 2307.69 | 2.663E+04 | 0.4521 | 2.8E+04 | 2.1E+05 | 1.8E+05 | 1.2E+05 | 1.9E+04 | 0.64 | 77.06 |
| Point Arguello Light | 28.22 | Stable | 0.929 | 671 | 45 | 32 | 12 | 11 | 8 | 2235.47 | 1.920E+05 | 0.1723 | 1.6E+05 | 1.5E+06 | 1.4E+06 | 4.2E+05 | 6.6E+04 | 0.28 | 75.89 |
| Point Arguello Comingled | 0 | Stable | 0.925 | 533 | 36 | 25 | 23 | 16 | 8 | 2063.79 | 1.493E+05 | 0.2883 | 1.8E+05 | 1.1E+06 | 1.0E+06 | 4.1E+05 | 6.4E+04 | 0.36 | 82.19 |
| Takula | 8.31 | Stable | 0.886 | 844 | 62 | 24 | 10 | 4 | 8 | 1421.80 | 1.528E+05 | 0.1744 | 1.3E+05 | 1.2E+06 | 1.1E+06 | 2.3E+05 | 3.6E+04 | 0.20 | 77.18 |
| Hondo | 0 | Stable | 0.936 | 735 | 33 | 31 | 24 | 12 | 4 | 1197.28 | 1.665E+05 | 0.1825 | 1.7E+05 | 8.8E+05 | 8.3E+05 | 2.5E+05 | 4.2E+04 | 0.32 | 79.96 |
| Arabian Medium | 20.77 | Stable | 0.926 | 275 | 40 | 46 | 8 | 7 | 5 | 363.64 | 1.117E+04 | 0.8129 | 2.3E+04 | 1.0E+05 | 5.0E+04 | 9.2E+04 | 1.5E+04 | 1.8 | 72.85 |
| Point Arguello Heavy | 0 | Stable | 0.945 | 3250 | 32 | 32 | 17 | 19 | 6 | 221.54 | 8.767E+04 | 0.4322 | 1.8E+05 | 7.2E+05 | 5.8E+05 | 3.9E+05 | 6.6E+04 | 0.72 | 74.97 |
| Arabian Medium | 30.93 | Stable | 0.95 | 2155 | 33 | 54 | 9 | 7 | 5 | 185.61 | 5.757E+04 | 0.4417 | 5.4E+04 | 4.0E+05 | 3.2E+05 | 2.2E+05 | 3.6E+04 | 0.70 | 65.60 |
| Point Arguello Comingled | 9.05 | Stable | 0.953 | 4988 | 31 | 33 | 19 | 17 | 4 | 124.30 | 1.651E+05 | 0.3100 | 1.5E+05 | 6.2E+05 | 4.7E+05 | 4.3E+05 | 7.0E+04 | 0.95 | 69.41 |
| Hondo | 16.67 | Stable | 0.967 | 9583 | 27 | 33 | 29 | 12 | 4 | 87.66 | 1.537E+05 | 0.5753 | 2.8E+05 | 8.4E+05 | 6.2E+05 | 6.1E+05 | 9.0E+04 | 0.96 | 64.23 |
| North Slope (southern pipeline) | 29.62 | No | 0.943 | 961 | 42 | 39 | 13 | 7 | | 197.71 | N/M[b] | N/M | 1.5E+03 | 1.9E+05 | 1.7E+05 | 7.4E+04 | 1.2E+04 | 0.42 | 24.48 |
| North Slope (northern pipeline) | 31.14 | No | 0.94 | 748 | 44 | 37 | 12 | 7 | | 147.06 | N/M | N/M | 1.2E+03 | 1.1E+05 | 9.4E+04 | 4.7E+04 | 7.6E+03 | 0.50 | 12.60 |
| North Slope (middle pipeline) | 30.54 | No | 0.942 | 900 | 42 | 38 | 12 | 7 | | 122.22 | N/M | N/M | 1.4E+03 | 1.1E+05 | 1.0E+05 | 4.8E+04 | 7.4E+03 | 0.49 | 19.67 |
| Santa Clara | 0 | No | 0.92 | 304 | 36 | 22 | 29 | 13 | 6 | 95.39 | 2.125E+03 | 0.9800 | 2.6E+03 | 2.9E+04 | 1.7E+04 | 2.4E+04 | 3.9E+03 | 1.4 | 12.75 |
| Bunker C (Anchorage) | 8.41 | No | 1.005 | 280,000 | 23 | 42 | 20 | 15 | 2 | 1 | | | | 280000 | | | | | |
| Carpenteria | 0 | No | 0.916 | 164 | 44 | 30 | 17 | 9 | 7 | 1 | | | | 164 | | | | | |
| Hondo | 32.29 | No | 0.988 | 449,700 | 27 | 28 | 32 | 13 | 4 | 1 | | | | 449700 | | | | | |
| Jet fuel (Anchorage) | 0 | No | 0.811 | 2 | 81 | 19 | 0 | 0 | 0 | 1 | | | | 2 | | | | | |
| Jet fuel (Anchorage) | 52.72 | No | 0.835 | 3 | 80 | 19 | 0 | 0 | 0 | 1 | | | | 3 | | | | | |
| North Slope (Middle pipeline) | 0 | No | 0.876 | 16 | 52 | 35 | 9 | 5 | | 1 | | | | 16 | | | | | |
| North Slope (northern pipeline) | 0 | No | 0.872 | 14 | 51 | 34 | 9 | 5 | | 1 | | | | 14 | | | | | |
| North Slope (southern pipeline) | 0 | No | 0.877 | 18 | 54 | 32 | 8 | 6 | | 1 | | | | 18 | | | | | |
| Pitas Point | 0 | No | 0.834 | 2 | 80 | 18 | 3 | 0 | 0 | 1 | | | | 2 | | | | | |
| Pitas Point | 23.56 | No | 0.854 | 2 | 62 | 35 | 0 | 0 | 0 | 1 | | | | 2 | | | | | |
| Point Arguello Comingled | 22.12 | No | 0.985 | 2E+06 | 24 | 33 | 21 | 22 | 5 | 1 | | | | 2266000 | | | | | |
| Point Arguello Heavy | 17.78 | No | 0.991 | 5E+06 | 25 | 34 | 21 | 22 | 4 | 1 | | | | 4953000 | | | | | |
| Sumatran Heavy | 0 | No | 0.931 | 13,300 | 46 | 30 | 13 | 8 | 5 | 1 | | | | 13300 | | | | | 12.73 |
| Sumatran Heavy | 5.26 | No | 0.937 | 17,000 | 45 | 32 | 16 | 8 | 4 | 1 | | | | 17000 | | | | | |
| Sumatran Light | 0 | No | 0.86 | 41,480 | 70 | 15 | 6 | 7 | 24 | 1 | | | | 41480 | | | | | 11.20 |
| Taching | 0 | No | 0.87 | 5E+06 | 74 | 12 | 9 | 4 | 23 | 1 | | | | 5138000 | | | | | |
| Waxy Light Heavy Blend | 0 | No | 0.931 | 184 | 39 | 35 | 21 | 5 | 4 | 1 | | | | 184 | | | | | |

| | | | | | | | | | | | | | | | | | | | |
|---|---|---|---|---|---|---|---|---|---|---|---|---|---|---|---|---|---|---|---|
| Cook Inlet – Granite Point | 0 | No | 0.829 | 4 | 72 | 22 | 5 | 1 | | 1 | | | | 4 | | | | | |
| Diesel (Anchorage) | 0 | No | 0.83 | 3 | 74 | 24 | 1 | 0 | 0 | 1 | | | | 3 | | | | | |
| Diesel (Anchorage) | 37.44 | No | 0.862 | 5 | 75 | 23 | 1 | 0 | 0 | 1 | | | | 5 | | | | | |
| Diesel (Mobile Burn #3) | 0 | No | 0.839 | 4 | 76 | 22 | 2 | 0 | 0 | 1 | | | | 4 | | | | | |
| Diesel (Mobile Burn #3) | 8.21 | No | 0.843 | 5 | 78 | 20 | 2 | 0 | 0 | 1 | | | | 5 | | | | | |
| Diesel (Mobile Burn #3) | 16.32 | No | 0.845 | 6 | 78 | 20 | 2 | 0 | 0 | 1 | | | | 6 | | | | | |
| Dos Cuadras | 0 | No | 0.9 | 51 | 48 | 30 | 17 | 6 | 6 | 1 | | | | 51 | | | | | |
| Cook Inlet – Swanson River | 0 | Meso | 0.842 | 6 | 65 | 25 | 6 | 5 | | 4500 | 4.376E+02 | 0.4120 | 1.9E+03 | 2.7E+04 | 2.2E+04 | 1.7E+04 | 2.6E+03 | 0.80 | 57.50 |
| Cook Inlet – Trading Bay | 33.30 | Meso | 0.924 | 278 | 51 | 32 | 9 | 8 | | 1906.47 | 1.361E+05 | 0.1861 | 4.3E+04 | 5.3E+05 | 4.9E+05 | 1.5E+05 | 2.6E+04 | 0.27 | 61.43 |
| Takula | 15.88 | Meso | 0.896 | 3148 | 60 | 25 | 11 | 4 | 8 | 381.19 | 1.877E+05 | 0.1791 | 1.7E+05 | 1.2E+06 | 1.2E+06 | 2.8E+05 | 4.4E+04 | 0.24 | 60.22 |
| Carpenteria | 14.87 | Meso | 0.948 | 3426 | 31 | 36 | 22 | 11 | 4 | 84.65 | 9.734E+03 | 0.8845 | 2.1E+04 | 2.9E+05 | 2.6E+05 | 1.5E+05 | 2.3E+04 | 0.57 | 16.04 |
| Cook Inlet – Granite Point | 45.32 | Meso | 0.903 | 4119 | 62 | 28 | 7 | 3 | | 63.12 | 7.725E+04 | 0.1401 | 85.E+03 | 2.6E+05 | 2.6E+05 | 7.4E+04 | 1.2E+04 | 0.30 | 57.58 |
| Carpenteria | 10.31 | Meso | 0.93 | 755 | 40 | 30 | 19 | 11 | 5 | 47.68 | 4.815E+03 | 0.9765 | 2.3E+04 | 3.6E+04 | 3.2E+03 | 3.6E+04 | 5.8E+03 | 12 | 13.82 |
| Santa Clara | 11.40 | Meso | 0.948 | 1859 | 32 | 28 | 27 | 13 | 4 | 40.34 | 1.087E+05 | 0.1688 | 1.5E+04 | 7.5E+04 | 2.0E+04 | 7.3E+04 | 1.1E+04 | 3.5 | 38.72 |
| Dos Cuadras | 11.17 | Meso | 0.927 | 187 | 42 | 31 | 20 | 7 | 4 | 24.6 | N/M | N/M | 5.3E+02 | 4.6E+03 | 1.6E+03 | 4.2E+03 | 7.0E+02 | 3.0 | 29.49 |
| Dos Cuadras | 20.30 | Meso | 0.947 | 741 | 41 | 31 | 19 | 9 | 6 | 22.94 | N/M | N/M | 1.9E+03 | 1.7E+04 | 3.0E+03 | 1.3E+04 | 2.1E+03 | 4.2 | 28.72 |
| Santa Clara | 21.63 | Meso | 0.967 | 22,760 | 28 | 32 | 23 | 17 | 5 | 16.26 | 5.794E+04 | 0.6751 | 9.2E+04 | 3.7E+05 | 2.0E+05 | 3.3E+05 | 5.0E+04 | 1.65 | 40.15 |
| Belridge Heavy | 0 | Meso | 0.975 | 12,610 | 28 | 39 | 30 | 3 | 1 | 12.69 | 2.638E+04 | 0.6938 | 5.4E+04 | 1.6E+05 | 8.2E+04 | 1.2E+05 | 2.2E+04 | 1.6 | 27.16 |
| Belridge Heavy | 2.74 | Meso | 0.977 | 17,105 | 29 | 38 | 30 | 4 | 1 | 10.52 | 2.779E+04 | 0.6374 | 6.2E+04 | 1.8E+05 | 1.0E+05 | 1.4E+05 | 2.4E+04 | 1.4 | 29.47 |
| Waxy Light Heavy Blend | 19.60 | Meso | 0.975 | 17,280 | 30 | 35 | 28 | 6 | 1 | 6.37 | 1.884E+04 | 0.7088 | 3.7E+04 | 1.1E+05 | 5.3E+04 | 1.0E+05 | 1.6E+04 | 2.1 | 33.58 |
| Cook Inlet – Trading Bay | 0 | Meso | 0.86 | 10 | 62 | 26 | 7 | 5 | | 1 | | | | 10 | | | | | 1.89 |
| Platform Irene | 0 | Entrained | 0.991 | 7600 | 26 | 29 | 22 | 22 | 2 | 434.21 | 2.744E+05 | 0.6255 | 5.4E+05 | 3.3E+06 | | | 4.0E+05 | 1.20 | 34.94 |
| Point Arguello Comingled | 15.19 | Entrained | 0.969 | 41,860 | 27 | 33 | 21 | 19 | 4 | 19.11 | 9.211E+04 | 0.7623 | 1.6E+05 | 8.0E+05 | 3.5E+05 | 7.0E+05 | 1.1E+05 | 2.10 | 28.42 |
| Port Hueneme | 3.14 | Entrained | 0.975 | 7833 | 23 | 41 | 21 | 14 | 3 | 16.6 | 1.918E+04 | 0.8713 | 2.7E+04 | 1.3E+05 | 2.1E+04 | 1.2E+05 | 2.0E+04 | 5.7 | 29.20 |
| Bunker C (Anchorage) | 0 | Entrained | 0.989 | 8706 | 25 | 47 | 17 | 11 | 2 | 16.08 | 1.903E+04 | 0.8567 | 1.5E+05 | 1.4E+05 | 3.2E+04 | 1.3E+05 | 2.1E+04 | 4.0 | 30.96 |
| Bunker C (1987) | 0 | Entrained | 0.983 | 45,030 | 24 | 55 | 15 | 7 | 12 | 14.43 | 9.818E+04 | 0.8336 | 1.4E+05 | 6.5E+05 | 6.2E+05 | 5.6E+05 | 8.4E+04 | 1.0 | 24.02 |
| Port Hueneme | 6.37 | Entrained | 0.979 | 20,990 | 23 | 37 | 28 | 13 | 3 | 12.39 | 3.438E+04 | 0.8070 | 5.4E+04 | 2.6E+05 | 6.4E+04 | 2.4E+05 | 4.0E+04 | 3.9 | 26.42 |
| Port Hueneme | 0 | Entrained | 0.966 | 4131 | 24 | 43 | 20 | 12 | 5 | 10.41 | 7.253E+03 | 0.9633 | 8.8E+03 | 4.3E+04 | 4.8E+03 | 4.1E+04 | 6.9E+03 | 9.4 | 20.06 |
| Waxy Light Heavy Blend | 12.00 | Entrained | 0.958 | 2002 | 32 | 38 | 24 | 6 | 1 | 7.49 | 2.763E+03 | 0.9194 | 3.7E+03 | 1.5E+04 | 2.4E+03 | 1.5E+04 | 2.4E+03 | 6.2 | 12.21 |
| Beta | 0 | Entrained | 0.974 | 13,380 | 21 | 39 | 31 | 7 | 3 | 1 | | | | 13380 | | | | | 20.74 |
| California API 11.0 | 0 | Entrained | 0.988 | 34,000 | | | | 16 | 1 | 1 | | | | 34000 | | | | | 29.72 |
| California API 15.0 | 0 | Entrained | 0.977 | 6400 | 19 | 35 | 23 | 22 | 1 | 1 | | | | 6400 | | | | | 34.62 |
| Point Arguello Heavy | 8.88 | Entrained | 0.971 | 59,390 | 26 | 35 | 18 | 20 | 4 | 1 | | | | 59390 | | | | | |
| Udang | 0 | Entrained | 0.97 | 10.700 | 32 | 41 | 24 | 3 | 1 | 1 | | | | 10700 | | | | | 19.65 |

[a]Emulsion stability is the complex viscosity of the emulsion divided by the starting-oil viscosity.

[b]Not measurable.

**Table 3** Correlation of Stability with Oil Parameters [a]

| Oil parameter | All water in oil | Stable emulsions | Mesostable emulsions | Entrained water | Unstable |
|---|---|---|---|---|---|
| Density | 0.05 | 0.23 | 0.73 | 0.99 | 0.14 |
| Viscosity | 0.05 | 0.40 | 0.84 | 0.36 | 0.16 |
| Saturates | 0.04 | 0.16 | 0.83 | 0.02 | 0.11 |
| Aromatics | 0 | 0.04 | 0.38 | 0.99 | 0.2 |
| Resins | 0.04 | 0.06 | 0.80 | 0.69 | 0.06 |
| Asphaltenes | 0.02 | 0.09 | 0.04 | 0.43 | 0.08 |
| Waxes | 0.05 | 0.09 | 0.09 | 0.25 | 0.04 |
| Water in emulsion | 0.49 | | | | |

[a]Results presented as regression coefficients ($r^2$).

**Table 4** Two-way Correlation of Stability with Oil Parameters[a]

| Oil parameters | All water in oil | Stable emulsions | Mesostable emulsions | Entrained water | Unstable |
|---|---|---|---|---|---|
| Viscosity – asphaltenes | 0.19 | 0.83 | 0.85 | 0.54 | 0.45 |
| Viscosity – resins | 0.19 | 0.66 | 0.87 | 0.08 | 0.29 |
| Viscosity – aromatics | 0 | 0.40 | 0.85 | 0.99 | 0.98 |
| Viscosity – density | 0.19 | 0.40 | 0.85 | 0.98 | 0.97 |
| Viscosity – asphaltenes and resins | 0.19 | 0.72 | 0.86 | | |

[a]Results presented as regression coefficients ($r^2$).

**Table 5** Correlation of Stability with Emulsion Parameters[a]

| Oil parameter | All water in oil | Stable emulsions | Mesostable emulsions | Entrained water | Unstable Water |
|---|---|---|---|---|---|
| $k$ (Ostwald–de Waele equation) | 0.60 | 0.64 | 0.17 | 0.89 | nr |
| $n$ (Ostwald–de Waele equation) | 0.64 | 0.65 | 0.23 | 0.82 | nr |
| Emulsion viscosity | 0.66 | 0.61 | 0.06 | 0.94 | Few data |
| Complex modulus | 0.21 | 0.50 | 0.16 | 0.96 | 0.01 |
| Elasticity modulus | 0.14 | 0.14 | 0.21 | 0.17 | Few data |
| Viscosity modulus | 0.02 | 0.23 | 0.08 | 0.52 | Few data |
| Low-shear viscosity | 0.05 | 0.17 | 0.09 | 0.95 | Few data |
| $\tan\delta\,(V/E)$ | 0.64 | 0.56 | 0.11 | 0.42 | Few data |
| Water content | 0.74 | 0.76 | 0.51 | 0.40 | 0.70 |

[a]Results presented as regression coefficients ($r^2$).

densities) are unlikely to uptake water in any form. These oils typically have no asphaltenes or resins (associated with low viscosity and density) or very high amounts of these.

Table 7 also shows that the differences between the four water-in-oil states is readily discernible by appearance and rheological properties. The reddish or brown appearance on formation indicates either a stable or mesostable emulsion; however, stable emulsions always have a more solid appearance. The increase in apparent viscosity (from the starting oil) on formation averages about 1100 for a stable emulsion, 45 for a mesostable emulsion, 13 for entrained water-in-oil, and an unstable emulsion shows little or no increase. This difference increases after one week. The increase in apparent viscosity after one week averages about

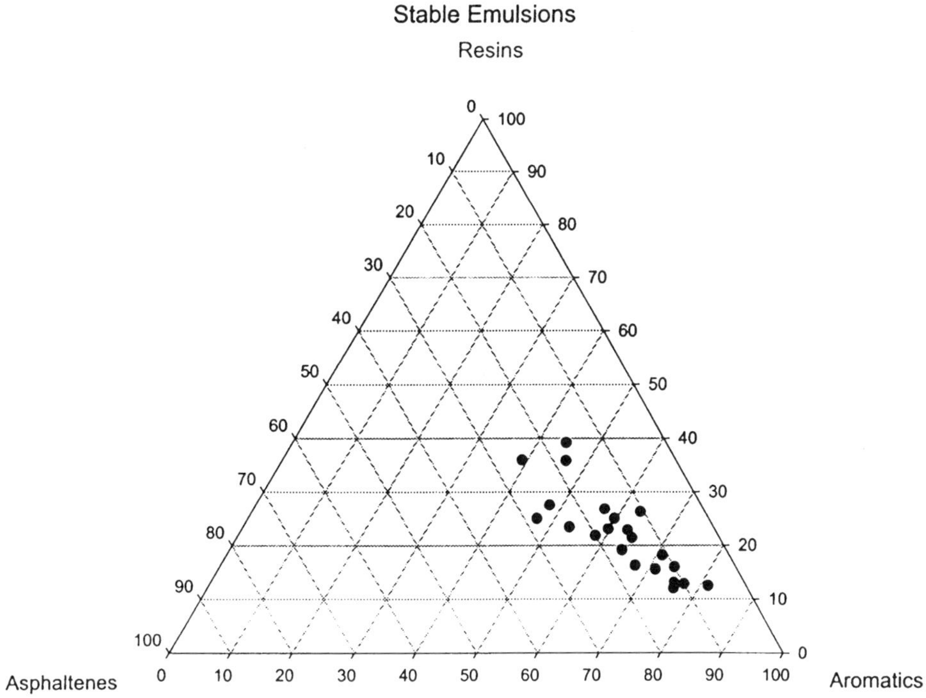


**Figure 1** Ternary diagram of group contents – stable emulsions.

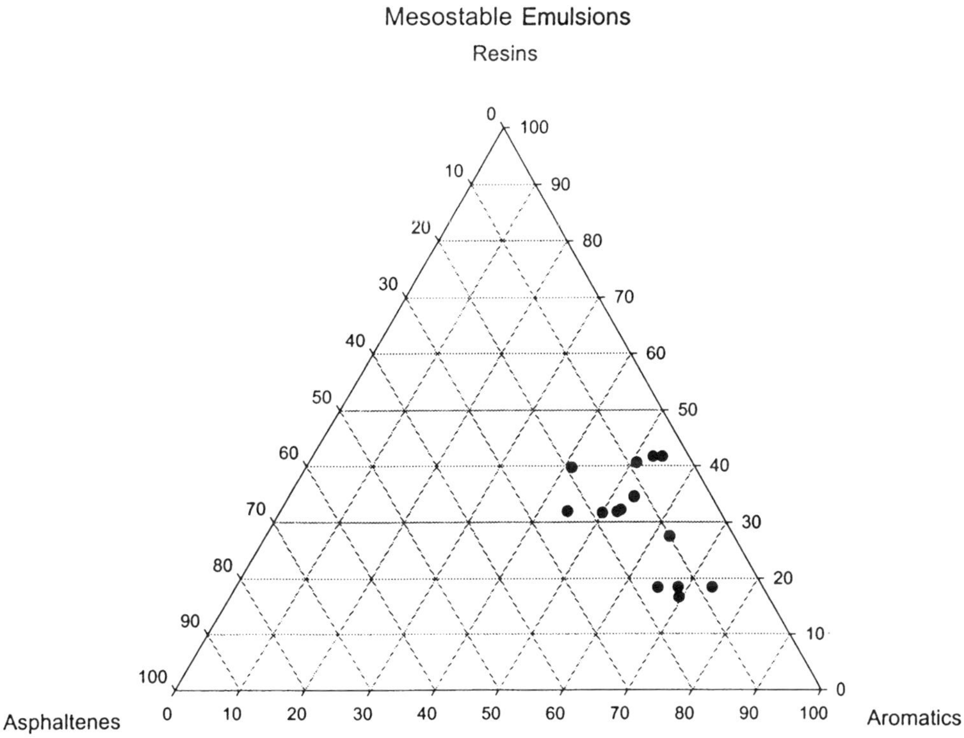


**Figure 2** Ternary diagram of group contents – mesostable emulsions.

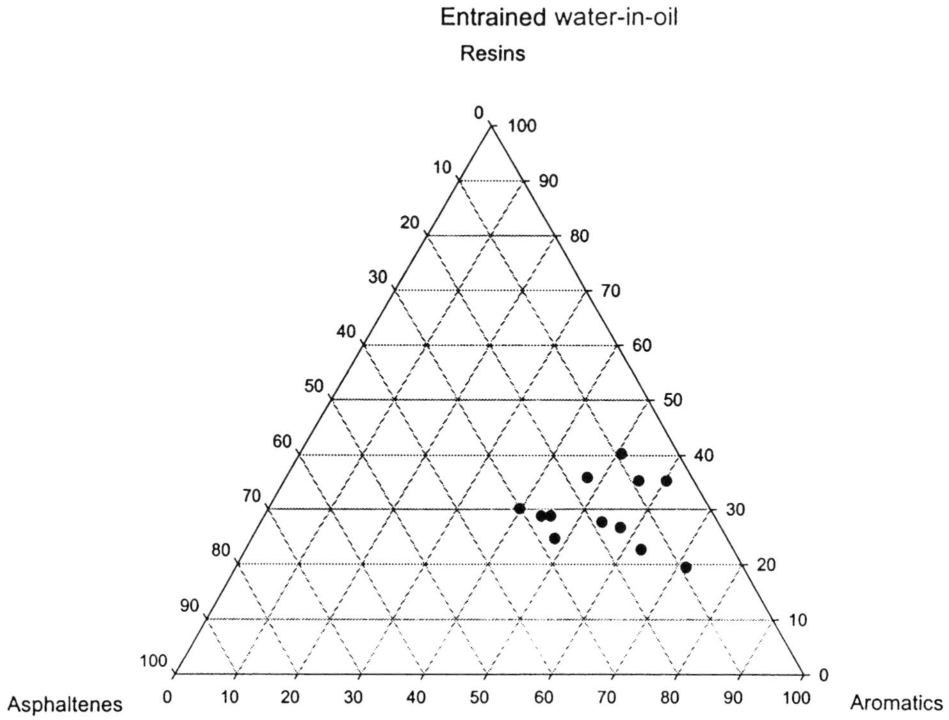


**Figure 3** Ternary diagram of group contents – entrained water-in-oil.

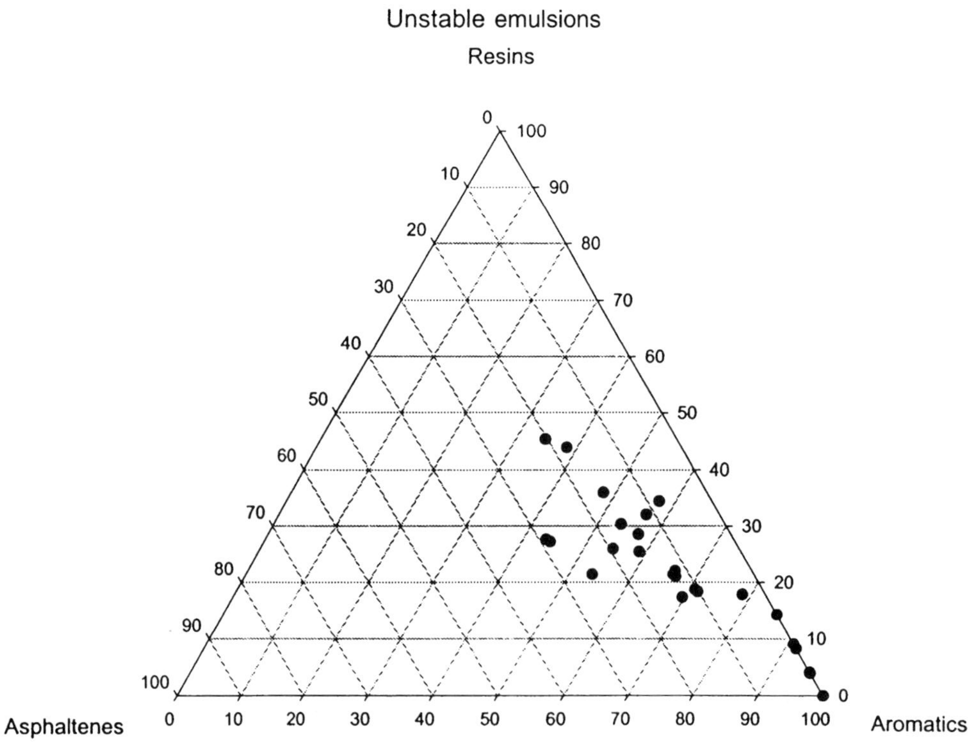


**Figure 4** Ternary diagram of group contents – unstable emulsions.

**Table 6** Ranges of Properties for the Various Emulsion Stabilities

| | | Starting-oil properties | | | | | | | | Emulsion rheological properties – one week after formation | | | | | | | | | |
|---|---|---|---|---|---|---|---|---|---|---|---|---|---|---|---|---|---|---|---|
| Emulsion stability | Property range | Density (g/mL) | Viscosity (mPa.s) | Saturates (%) | Aromatics (%) | Resins (%) | Asphaltenes (%) | Waxes (%) | Emulsion stability | Power law constants (*K*) | Power law constants (*n*) | Viscosity (mPa.s) | Complex modulus (mPa) | Elasticity modulus (mPa) | Viscosity modulus (mPa) | Isolated low-shear viscosity (mPa.s) | tan δ (*V*/*E*) | Water content (%w/w) |
| Stable | High | 0.9674 | 9580 | 65 | 54 | 29 | 19 | 8 | 95600 | 8.596E+05 | 0.8129 | 6.9E+05 | 4.3E+06 | 4.3E+06 | 6.1E+05 | 9.0E+04 | 1.8 | 93.79 |
| | Low | 0.8637 | 14 | 27 | 22 | 6 | 3 | 4 | 88 | 1.117E+04 | 0.0372 | 2.3E+04 | 1.0E+05 | 5.0E+04 | 2.7E+04 | 1.1E+04 | 0.11 | 64.23 |
| *High difference from unstable* | | −0.0376 | −5, 128, 000 | −16 | 12 | −3 | −3 | −16 | 95400 | 857,475 | −0.1671 | 687,000 | −838, 000 | 4,130,000 | 536,000 | 78,000 | 0.4 | 69.31 |
| *Low difference from unstable* | | 0.0527 | 12 | 4 | 10 | 6 | 3 | 4 | 87 | 11,170 | 0.0372 | 23,000 | 99, 998 | 50,000 | 27,000 | 11,000 | 0.11 | 64.23 |
| *High difference from mesostable* | | −0.0096 | −13, 200 | 0 | 15 | −1 | 2 | 0 | 93700 | 671,900 | −0.1636 | 520,000 | 3, 100, 000 | 3,100,000 | 280,000 | 40,000 | −10.2 | 32.36 |
| *Low difference from mesostable* | | 0.0217 | 8 | −1 | −3 | 0 | 0 | 3 | 87 | 107,324 | −0.1029 | 22,470 | 99, 990 | 48,400 | 22,800 | 10,300 | −0.13 | 62.34 |
| Meso | High | 0.977 | 22,760 | 65 | 39 | 30 | 17 | 8 | 1900 | 1.877E+05 | 0.9765 | 1.7E+05 | 1.2E+06 | 1.2E+06 | 3.3E+05 | 5.0E+04 | 12 | 61.43 |
| | Low | 0.842 | 6 | 28 | 25 | 6 | 3 | 1 | 1 | 4.376E+02 | 0.1401 | 5.3E+02 | 10 | 1.6E+03 | 4.2E+03 | 7.0E+02 | 0.24 | 1.89 |
| *High difference from unstable* | | 0.028 | −5, 120, 000 | −16 | −3 | −2 | −5 | −16 | 1700 | 185,575 | −0.0035 | 167,400 | −3, 938, 000 | 1,030,000 | 256,000 | 38,000 | 10.6 | 36.95 |
| *Low difference from unstable* | | 0.031 | 4 | 5 | 13 | 6 | 3 | 1 | 0 | 437.6 | 0.1401 | 530 | 8 | 1600 | 4200 | 700 | 0.24 | 1.89 |
| Entrained | High | 0.9907 | 59,390 | 32 | 55 | 31 | 22 | 12 | 430 | 2.744E+05 | 0.9633 | 5.4E+05 | 3.3E+06 | 6.2E+05 | 7.0E+05 | 4.0E+05 | 9.4 | 34.94 |
| | Low | 0.9688 | 2002 | 19 | 29 | 15 | 3 | 1 | 1 | 2.763E+03 | 0.6255 | 3.7E+03 | 6400 | 2.4E+03 | 1.5E+04 | 2.4E+03 | 1.0 | 12.21 |
| *High difference from unstable* | | −0.0143 | −5, 080, 000 | −49 | 13 | −1 | 0 | −12 | 230 | 272,275 | −0.0167 | 537,400 | −1, 838, 000 | 450,000 | 626,000 | 388,000 | 8 | 10.46 |
| *Low difference from unstable* | | 0.1578 | 2000 | −4 | 17 | 15 | 3 | 1 | 0 | 2763 | 0.6255 | 3700 | 6398 | 2400 | 15,000 | 2400 | 1 | 12.21 |
| Unstable | High | 1.005 | 5E+06 | 81 | 42 | 32 | 22 | 24 | 198 | 2.125E+03 | 0.9800 | 2.6E+03 | 5, 138, 000 | 1.7E+05 | 7.4E+04 | 1.2E+04 | 1.4 | 24.48 |
| | Low | 0.811 | 2 | 23 | 12 | 0 | 0 | 0 | 1 | 0.000E+00 | 0 | 0.0E+00 | 2 | 0.0E+00 | 0.0E+00 | 0.0E+00 | 0.0 | 0.0 |

[a] Emulsion stability is the complex viscosity of the emulsion divided by the starting-oil viscosity.

**Table 7** Summary Properties for the Water-in-oil States

| Starting oil (property) | | Stable | | Meso | | Entrained | | Unstable | |
|---|---|---|---|---|---|---|---|---|---|
| | | High | Low | High | Low | High | Low | High | Low |
| Density | (g/mL) | 0.9674 | 0.8637 | 0.977 | 0.842 | 0.9907 | 0.9688 | 1.005 | 0.811 |
| Viscosity | (mPa.s) | 9580 | 14 | 22,800 | 6 | 59,400 | 2002 | 5,140,000 | 2 |
| Saturates | (%) | 65 | 27 | 65 | 28 | 32 | 19 | 81 | 23 |
| Aromatics | (%) | 54 | 22 | 39 | 25 | 55 | 29 | 42 | 12 |
| Resins | (%) | 29 | 6 | 30 | 6 | 31 | 15 | 32 | 0 |
| Asphaltenes | (%) | 19 | 3 | 17 | 3 | 22 | 3 | 22 | 0 |
| Waxes | (%) | 8 | 4 | 8 | 1 | 12 | 1 | 24 | 0 |
| Asphaltene–resin ratio | | 1.12 | 0.4 | 0.89 | 0.1 | 1.11 | 0.13 | 1.17 | 0 |
| *Properties on day of formation* | | | | | | | | | |
| Appearance | | Brown solid | | Brown viscous liquid | | Black with large droplets | | Like oil | |
| Average ratio of viscosity increase | | 1100 | | 45 | | 13 | | 1 | |
| Range | | 15,000 | 14 | 250 | 2 | 70 | 1 | 8 | 1 |
| Average water content | | 80 | | 62 | | 42 | | 5 | |
| Range | | 93 | 65 | 83 | 35 | 62 | 26 | 23 | 1 |
| Stability[a] | | 15,000 | 20 | 400 | 1 | 50 | 1 | 60 | 1 |
| *Properties after one week* | | | | | | | | | |
| Appearance | | Brown solid | | Broken, 2 or 3 phases | | Separated oil and water | | Like oil | |
| Average ratio of viscosity increase | | 1500 | | 30 | | 2 | | 1 | |
| Range | | 15,000 | 20 | 150 | 1 | 3 | 1 | 2 | 1 |
| Average water content | | 79 | | 38 | | 15 | | 2 | |
| Range | | 94 | 64 | 61 | 2 | 35 | 12 | 24 | 0 |
| Stability[a] | | 95,000 | 88 | 1900 | 1 | 434 | 1 | 198 | 1 |
| Power law constants | ($K$) | 8.596E+05 | 1.117E+04 | 1.877E+05 | 4.376E+02 | 2.744E+05 | 2.763E+03 | 2.125E+03 | 0.000E+00 |
| | ($n$) | 0.8129 | 0.0372 | 0.9765 | 0.1401 | 0.9633 | 0.6255 | 0.9800 | 0 |
| Viscosity | (mPa.s) | 6.9E+05 | 2.3E+04 | 1.7E+05 | 5.3E+02 | 5.4E+05 | 3.7E+03 | 2.6E+03 | 0.0E+00 |
| Complex modulus | (mPa) | 4.3E+06 | 1.0E+05 | 1.2E+06 | 10 | 3.3E+06 | 6400 | 5138000 | 2 |
| Elasticity modulus | (mPa) | 4.3E+06 | 5.0E+04 | 1.2E+06 | 1.6E+03 | 6.2E+05 | 1.4E+03 | 1.7E+05 | 0.0E+00 |
| Modulus | (mPa) | 6.1E+05 | 2.7E+04 | 3.3E+05 | 4.2E+03 | 7.0E+05 | 1.5E+04 | 7.4E+04 | 0.0E+00 |
| Shear viscosity | (mPa.s) | 9.0E+04 | 1.1E+04 | 5.0E+04 | 7.0E+02 | 4.0E+05 | 2.4E+03 | 1.2E+04 | 0.0E+00 |
| $\delta$ | (V/E) | 1.8 | 0.11 | 12 | 0.24 | 9.4 | 1.0 | 1.4 | 0.00 |
| Water content | (%w/w) | 93.79 | 64.23 | 61.43 | 1.89 | 34.94 | 12.21 | 24.48 | 0.00 |

[a]Complex modulus/viscosity of starting oil.

1500 for a stable emulsion, 30 for a mesostable emulsion, 3 for entrained water-in-oil, and an unstable emulsion shows little or no increase. It is noted that apparent viscosity does not decrease after one week for stable emulsions only.

There are several other features noted in the summary data presented in Table 7. An examination of the wax content shows that it has no relation to the formation of any of these states. While there may be some correlation with viscosity, the specific wax content is not associated with the formation of any state. It is noted that density is associated with the viscosity and somewhat with the state. It is also noted that the water content correlates somewhat with the state. The average water content of stable emulsions is 80% on the day of formation, of mesostable is 62%, of entrained is 42%, and is 5% for unstable emulsions. One must be cautious in using this as a sole discriminator, however, because of overlapping ranges. As would be expected, the water content after one week correlates very highly with the state. As was noted above, this is accentuated by the fact that mesostable emulsions and entrained water-in-oil have separated to a significant degree.

These data indicate that there are "windows" of composition and viscosity which result in the formation of each of the types of water-in-oil states. The important oil composition factors are the asphaltene and resin contents. While asphaltenes are responsible for the formation of stable emulsions, a high asphaltene content can also result in a high viscosity, one that is above the region where stable emulsions form. The asphaltene/resin ratio is generally higher for stable emulsions. In a previous work by the present authors, it was shown that the migration rate of asphaltenes in emulsions is very slow (81). This indicates that, in very viscous oils, the migration of asphaltenes may be too slow to allow for the stabilization of emulsions.

## 3. Conclusions on the Stability Studies

Four, clearly defined states of water-in-oil have been shown to be defined by a number of measurements and by their visual appearance, both on the day of formation and one week later. The differences between these states and the oils that form them are summarized in Table 8.

The results of this study indicate that the formation of both stable and mesostable emulsions is due to the combination of surface-active forces from resins and asphaltenes and from viscous forces. Each type of water-in-oil state exists in a range of compositions and viscosities. the difference in composition between stable and mesostable emulsions is small. Stable emulsions have more asphaltenes and less resins and have a narrow viscosity window. Instability results when the oil has a high viscosity (over about 50 Pa.s) or a very

**Table 8** Typical Properties for the Water-in-oil states

| | | Stable | Meso | Entrained | Unstable |
|---|---|---|---|---|---|
| Day of formation appearance | | Brown solid | Brown viscous liquid | Black with large droplets | Like oil |
| Water content on first day | (%) | 80 | 62 | 42 | 5 |
| Appearance after one week | | Brown solid | Broken, 2 or 3 phases | separated oil and water | Like oil |
| Water content after week | (%) | 79 | 38 | 15 | 2 |
| Stable time | (days) | > 30 | < 3 | < 0.5 | not |
| *Starting Oil* | | | | | |
| Density | (g/mL) | 0.85–0.97 | 0.84–0.98 | 0.97–0.99 | 0.8–1.03 |
| Viscosity | (mPa.s) | 15–10000 | 6–23,000 | 200000–60,000 | 2–5.1 × $10^6$ |
| Saturates | (%) | 25–65 | 25–65 | 19–32 | 23–80 |
| Aromatics | (%) | 20–55 | 25–40 | 30–55 | 5–12 |
| Resins | (%) | 5–30 | 6–30 | 15–30 | 0–32 |
| Asphaltenes | (%) | 3–20 | 3–17 | 3–22 | 0–32 |
| Asphaltenes/resins | | 0.74 | 0.47 | 0.62 | 0.45 |
| *Properties on day of formation* | | | | | |
| Average ratio of viscosity increase | | 1100 | 45 | 13 | 1 |
| *Properties after one week* | | | | | |
| Average ratio of viscosity increase | | 1500 | 30 | 2 | 1 |

low viscosity (under about 6 mPa.s) and when the resins and asphaltenes are less than about 3%. Water entrainment occurs rather than emulsion formation when the viscosity is between about 2 and 50 Pa.s. The formation of stable or mesostable emulsions may not occur in highly viscous oils because the migration of asphaltenes (and resins) is too slow to permit droplet stabilization.

The role of other components is still unclear at this time. Aromatics dissolve asphaltenes and there is a small correlation observed with the stabilities. Waxes appear to have no role in emulsion formation. The density of the starting oil is highly correlated with viscosity and thus shows a correlation with stability.

The state of the final water-in-oil can be correlated with the single parameter of the complex modulus divided by the starting-oil viscosity. This stability parameter also correlates somewhat with the nonNewtonian behavior of the resulting water-in-oil mixture, with the elasticity of the emulsion, and also the water content. These properties are more decisive in defining the state one week after formation because all states have largely separated into oil and water except for stable emulsions. The water content retained one week after the formation process is a very clear discriminator of state.

## B. Studies on Energy Threshold of Formation

An important aspect of emulsions that has not been studied extensively to date is the kinetics of emulsion formation and the energy levels associated with their formation. Such information is needed to understand the emulsification process and to model the process. This section reports on initial experiments to examine the kinetics and the formation energy of emulsions. It is important to note that turbulent energy is thought to be the most important form of energy related to emulsion formation. Turbulent energy could not be measured in this experiment, so the total energy was used as an estimate of the energy available for emulsion formation.

### 1. Experimental Summary

Details of the experimental work are given in Ref. 84. Water-in-oil emulsions were made in a rotary agitator and then the rheometric characteristics of these emulsions were studied over time. Oils were taken from the storage facilities at the Emergencies Science Division. Properties of these oils are given in standard references (83). The energy threshold measurements were conducted by varying the rotational rate, and hence the energy of the apparatus used to make the emulsions. Analysis of the emulsions was conducted using rheological measurements as described herein and by standard visual observations.

Emulsions were made in an end-over-rotary mixer as noted in Sec. IV.A.1. After temperature equilibration, the vessels were rotated for 12 h at a rate between 1 and 55 rpm. The resulting emulsions were then collected into jars, covered, and stored in a 15°C cold room. Analysis was performed on the day of collection a short time after formation.

The rheology and water content were measured in the same manner as noted in Sec. IV.A.1.

Energy calculation was related to the total kinetic energy exerted on the oil/water in the device. The total kinetic energy in each bottle is given by:

$$KE = \tfrac{1}{2}MV^2 \tag{2}$$

where $KE$ is the energy in ergs, $M$ is the mass in grams, here approximately 620 g of water and oil, and $V$ is the velocity in cm/s which is $2\pi r$ – which is rpm/60 × 7.5 cm.

Kinetic energy by this formula is then 196 × rpm$^2$ ergs. Ergs were used in this study because they are a much more convenient unit than the SI unit joules at these low energy levels. This simple formulation was used to assign an energy level to each rotational velocity. Again, it is important to note that the energy estimated here is the total energy input to the system, and not turbulent energy which is the prime factor in emulsion formation.

### 2. Results and Discussion

The rheological data associated with the energy threshold experiments are given in Table 9. The second column of Table 9 is the rotational rate of the formation vessel. The third column is the calculated kinetic energy applied to the system in ergs. The fourth column is the complex modulus which is the vector sum of the viscosity and elasticity. The fifth column shows stability of the emulsion which is the complex modulus divided by the starting-oil viscosity. The sixth column gives the water content of the emulsion. the seventh column gives the assessment of the stability of the emulsion based on both visual appearance and rheological properties. The eighth and ninth columns give the viscosity of the emulsions. The eighth column is the viscosity as given by the RV-20 instrument and the

**Table 9** Experimental Parameters and Results

| Oil | rpm | Energy (ergs)[a] | Complex modulus (mPa) | Stability ($s^{-1}$) | $H_2O$ (w/w) | Visual stability | Viscosity | | Oil viscosity (mPa.s) |
|---|---|---|---|---|---|---|---|---|---|
| | | | | | | | RV20 | RS100 | |
| Arabian Light | 1.0 | 196 | | 1 | | Unstable | | | 14 |
| Arabian Light | 1.3 | 331 | | 1 | | Unstable | | | 14 |
| Arabian Light | 2.8 | 1537 | | 1 | | Unstable | | | 14 |
| Arabian Light | 3.1 | 1884 | | 1 | | Unstable | | | 14 |
| Arabian Light | 5.1 | 5098 | 4.0E+04 | 2860 | 84.73 | Stable | 4.9E+03 | 5.5E+03 | 14 |
| Arabian Light | 5.3 | 5506 | 8.2E+03 | 590 | 81.82 | Stable | | 1.5E+03 | 14 |
| Arabian Light | 10.3 | 20,794 | 1.5E+04 | 1070 | 85.60 | Stable | 7.0E+03 | 4.2E+03 | 14 |
| Arabian Light | 10.5 | 21,609 | 2.4E+05 | 17,140 | 90.31 | Stable | | 1.1E+03 | 14 |
| Arabian Light | 20.4 | 81,567 | 2.5E+05 | 17,500 | 93.22 | Stable | 5.9E+04 | | 14 |
| Arabian Light | 21.0 | 86,436 | 8.3E+04 | 5930 | 85.06 | Stable | | 1.2E+04 | 14 |
| Arabian Light | 29.8 | 174,056 | 3.1E+05 | 22,140 | 92.30 | Stable | | 4.0E+04 | 14 |
| Arabian Light | 29.8 | 174,056 | 3.3E+05 | 23,570 | 92.56 | Stable | 6.0E+04 | 5.0E+04 | 14 |
| Arabian Light | 39.8 | 310,472 | 7.9E+05 | 56,430 | 91.58 | Stable | 9.1E+04 | | 14 |
| Arabian Light | 39.8 | 310,472 | 6.6E+05 | 47,140 | 90.73 | Stable | 1.0E+05 | | 14 |
| Arabian Light | 40.3 | 318,322 | 3.0E+05 | 21,070 | 92.06 | Stable | 5.9E+04 | 7.0E+04 | 14 |
| Arabian Light | 41.8 | 342,459 | 8.2E+05 | 58,570 | 90.40 | Stable | | 3.0E+04 | 14 |
| Arabian Light | 55 | 592,000 | 9.4E+05 | 67,140 | 88.71 | Stable | 1.4E+05 | 5.0E+04 | 14 |
| Bunker C | 1.0 | 196 | 2.2E+05 | 1 | 46.83 | Entrained | | 6.5E+04 | 45,030 |
| Bunker C | 1.3 | 331 | 2.4E+05 | 10 | 49.91 | Entrained | 9.7E+04 | 9.0E+04 | 45,030 |
| Bunker C | 2.8 | 1537 | 3.6E+05 | 10 | 58.22 | Entrained | | 1.2E+05 | 45,030 |
| Bunker C | 3.1 | 1884 | 3.6E+05 | 10 | 61.72 | Entrained | 1.5E+05 | 1.3E+05 | 45,030 |
| Bunker C | 5.1 | 5098 | 4.0E+05 | 10 | 60.32 | Entrained | 1.7E+05 | 1.3E+05 | 45,030 |
| Bunker C | 5.3 | 5506 | 4.4E+05 | 10 | 62.96 | Entrained | | 1.3E+05 | 45,030 |
| Bunker C | 10.3 | 20,794 | 2.9E+05 | 10 | 49.91 | Entrained | 1.2E+05 | 1.0E+05 | 45,030 |
| Bunker C | 10.5 | 21,609 | 3.6E+05 | 10 | 54.71 | Entrained | | 1.3E+05 | 45,030 |
| Bunker C | 20.4 | 81,567 | 5.1E+05 | 10 | 64.80 | Entrained | 1.7E+05 | | 45,030 |
| Bunker C | 21.0 | 86,436 | 4.5E+05 | 10 | 54.16 | Entrained | | 1.2E+05 | 45,030 |
| Bunker C | 29.8 | 174,056 | 3.4E+05 | 10 | 48.89 | Entrained | | 9.5E+04 | 45,030 |
| Bunker C | 29.8 | 174,056 | 3.5E+05 | 10 | 45.82 | Entrained | 9.1E+04 | 8.5E+04 | 45,030 |
| Bunker C | 40.3 | 318,322 | 3.8E+05 | 10 | 49.18 | Entrained | 1.1E+05 | 9.5E+04 | 45,030 |
| Bunker C | 41.8 | 342,459 | 3.9E+05 | 10 | 47.46 | Entrained | | 1.1E+05 | 45,030 |
| Bunker C | 55 | 592,000 | 3.9E+05 | 10 | 49.27 | Entrained | 1.1E+05 | 1.1E+05 | 45,030 |
| Prudhoe Bay | 1.0 | 196 | | 1 | | Unstable | | | 900 |
| Prudhoe Bay | 1.3 | 331 | | 1 | | Unstable | | | 900 |
| Prudhoe Bay | 2.8 | 1537 | | 1 | | Unstable | | | 900 |
| Prudhoe Bay | 3.1 | 1884 | | 1 | | Unstable | | | 900 |
| Prudhoe Bay | 5.1 | 5098 | | 1 | | Unstable | | | 900 |
| Prudhoe Bay | 5.3 | 5506 | | 1 | | Unstable | | | 900 |
| Prudhoe Bay | 10.3 | 20,794 | | 1 | | Unstable | | | 900 |
| Prudhoe Bay | 10.5 | 21,609 | | 1 | | Unstable | | | 900 |
| Prudhoe Bay | 20.4 | 81,567 | | 1 | | Unstable | | | 900 |
| Prudhoe Bay | 21.0 | 86,436 | | 1 | | Unstable | | | 900 |
| Prudhoe Bay | 29.8 | 174,056 | | 1 | | Unstable | | | 900 |
| Prudhoe Bay | 29.8 | 174,056 | | 1 | | Unstable | | | 900 |
| Prudhoe Bay | 40.3 | 318,322 | 8.8E+03 | 10 | 28.49 | Meso | 5.1E+03 | | 900 |
| Prudhoe Bay | 41.8 | 342,459 | 8.2E+03 | 10 | 72.52 | Meso | | 1.3E+03 | 900 |
| Prudhoe Bay | 55 | 592,000 | 4.2E+03 | 5 | 55.81 | Meso | 6.7E+03 | 4.0E+02 | 900 |

*(Continued)*

**Table 9** Experimental Parameters and Results (*contd.*)

| Oil | rpm | Energy (ergs)[a] | Complex modulus (mPa) | Stability ($s^{-1}$) | $H_2O$ (w/w) | Visual stability | Viscosity RV20 | Viscosity RS100 | Oil viscosity (mPa.s) |
|---|---|---|---|---|---|---|---|---|---|
| Sockeye | 1.0 | 196 | | 1 | | Unstable | | | 45 |
| Sockeye | 1.3 | 331 | 1.4E+04 | 310 | 82.39 | Stable | 7.6E+03 | 3.0E+04 | 45 |
| Sockeye | 2.8 | 1537 | 1.7E+05 | 3780 | 89.45 | Stable | | 2.5E+05 | 45 |
| Sockeye | 3.1 | 1884 | 2.1E+05 | 4670 | 90.92 | Stable | 2.7E+04 | 2.0E+04 | 45 |
| Sockeye | 5.1 | 5098 | 3.9E+05 | 8670 | 92.60 | Stable | 5.4E+04 | 7.0E+04 | 45 |
| Sockeye | 5.3 | 5506 | 3.6E+05 | 8000 | 92.50 | Stable | | 5.0E+04 | 45 |
| Sockeye | 10.3 | 20,794 | 1.5E+06 | 33,330 | 90.57 | Stable | 4.0E+05 | 3.0E+05 | 45 |
| Sockeye | 10.5 | 21,609 | 1.8E+06 | 40,000 | 90.56 | Stable | | 2.8E+05 | 45 |
| Sockeye | 20.4 | 81,567 | 8.3E+06 | 18,330 | 92.03 | Stable | 2.3E+05 | | 45 |
| Sockeye | 20.7 | 83,984 | 1.2E+06 | 26.670 | 90.51 | Stable | 2.8E+05 | | 45 |
| Sockeye | 20.7 | 83,984 | 7.6E+05 | 16,890 | 91.23 | Stable | 2.1E+05 | | 45 |
| Sockeye | 21.0 | 86,436 | 8.7E+05 | 19,330 | 91.31 | Stable | | 1.6E+05 | 45 |
| Sockeye | 29.8 | 174,056 | 2.8E+06 | 62,220 | 89.12 | Stable | | 3.5E+05 | 45 |
| Sockeye | 29.8 | 174,056 | 3.0E+06 | 66,670 | 88.17 | Stable | 6.6E+05 | 5.0E+05 | 45 |
| Sockeye | 40.3 | 318,322 | 3.6E+06 | 80,000 | 88.95 | Stable | 9.2E+05 | 3.0E+05 | 45 |
| Sockeye | 41.8 | 342,459 | 3.4E+06 | 75,560 | 87.95 | Stable | | 4.0E+05 | 45 |
| Sockeye | 55 | 592,000 | 4.1E+06 | 90,000 | 86.44 | Stable | 7.6E+05 | 5.8E+05 | 45 |

[a]One erg = $10^{-7}$ Joules.

ninth column gives the viscosity derived from the RS 100 instrument. The eleventh column gives the viscosity of the starting oil as measured by the RV-20 instrument. Differences between the viscosity determined by these two instruments are a result of the differences between the two instruments as well as normal measurement variances.

Observations were made on the appearance of the emulsions and were used to classify the emulsions. All of the stable emulsions appeared to be stable and remained intact over 7 days in the laboratory. All of the mesostable emulsions broke after a few days in water, free oil, and emulsion. The time for these emulsions to break down varies from bout 1 to 3 days. The emulsion portion of these break-down emulsions appears to be somewhat stable, although separate studies on this portion have not been performed because of the difficulty in separating these portions from the oil and water. All entrained water appeared to have larger suspended water droplets. The appearance of non-stable water in oil was just that; the oil appeared to be unchanged and a water layer was clearly visible.

The stability and energy of formation are plotted for the four emulsions in Fig. 5. The stability in these figures is the complex modulus divided by the starting-oil viscosity. In summary, the "stability," as here defined, was found to be the only single parameter that could be used to describe the emulsions mathematically. Furthermore, stability was found to correlate very highly with other indices related to the formation of emulsions. Figure 5 shows that the onset of stability for Arabian Light oil is rapid and increases somewhat after onset. Stability is generally taken as the point at which the stability is approximately 1000 and this is achieved at a very low energy level corresponding to a rotational rate of about 5 rpm. Figure 5 shows the uptake of water by one sample of Bunker C. The Bunker C takes up water very rapidly between 200 and 300 ergs (1–3 rpm). After the rapid initial uptake, the stability of the water–oil mixtures remains the same and is typical of entrained water in oil. Figure 5 also shows the relationship of stability Prudhoe Bay with increasing energy. Water uptake is again rapid as in Bunker C, but at a higher energy threshold and a mesostable emulsion is formed. The uptake of water for Sockeye is very rapid at first, between the energy levels of 300 and 1500 ergs (1.3–2.8 rpm) and after this point stability increases slowly with increasing energy. All four oils show several similar features: initial water uptake occurs very rapidly over a short energy range; the energy threshold for initial water uptake is very low, typically around 300 ergs, except for that of Prudhoe Bay which is about 250,000 ergs; there is no stability increase for the Bunker C in which the water is entrained and for the Prudhoe Bay which forms a mesostable emulsion; and there is a slow increase in

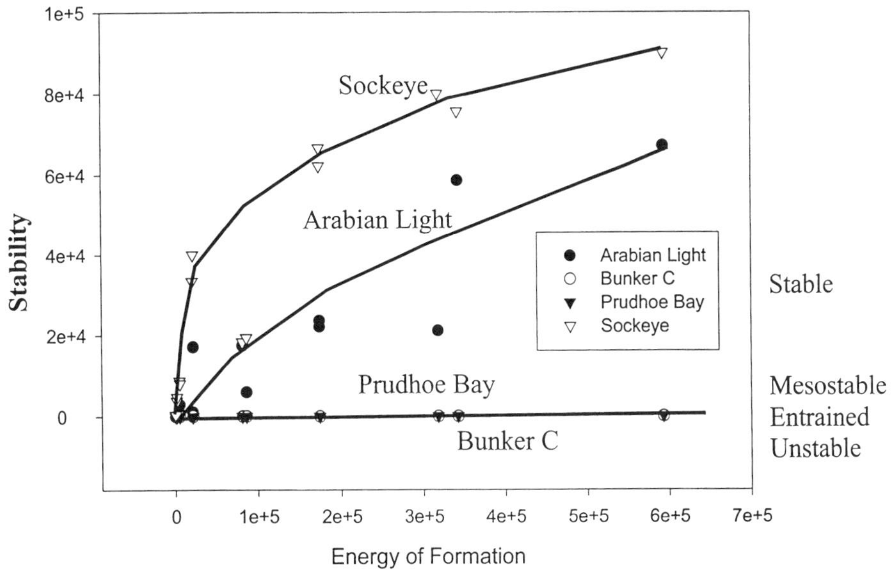


**Figure 5** Relationship between stability and energy of formation for four oil emulsions.

stability with increasing energy for the oils, Sockeye and Arabian Light, which form stable emulsions.

### 3. Conclusions on Energy Threshold

The energy threshold to the onset of the two states known as stable and entrained water, is very low, 300 to about 1500 ergs, corresponding to a rotational rate in the formation apparatus of about 1 to 3 rpm. The total energy applied to the system was used as an indicator value. Turbulent energy could not be measured. This study also shows that for the one oil type, Bunker C, which forms an entrained water state, there is no increase in stability with increasing energy input after the initial formation point. the oil that forms a mesostable emulsion, Prudhoe Bay, shows a similar tendency in that after the energy onset, which occurs at a high level of about 25,000 ergs, there is no apparent increase in stability. Both oils that form stable emulsions, Arabian Light and Sockeye, show an increasing stability with increasing energy, although the rate of increase is gradual with increasing energy.

## C. Studies on Asphaltene and Resin Migration

A series of studies were conducted to indicate the rate of asphaltene and resin migration in emulsions. Basically the technique was to measure the asphaltene and resins content of the starting oil, then in the bulk emulsion, and then at the oil–water interface in the emulsion.

### 1. Experimental Summary

Emulsions were formed using the specified crude oil according to selected standard emulsion-formation procedures outlined above. The emulsion was then placed in a large beaker and allowed to stand in a 10°C cold room for one week. The oil layer on top was removed using a syringe with a large-gage needle. The oil was collected as close to the surface as possible, with care taken to avoid the emulsion below. This sample was called the "free oil." If the emulsion was semisolid, the beaker was tipped to concentrate the oil at one end. Remaining oil on top of the emulsion was collected later and discarded. The emulsion remaining after the free oil was removed constituted the emulsion layer. The emulsion was broken using freeze/thaw cycles from −36°C to room temperature. The thawed emulsion was then centrifuged at >25000 rpm for 30 min to separate as much water as possible. After several cycles, the water content was minimal. The method of analysis of the oil for asphaltenes, saturates, aromatics, and resins has been shown to be able to tolerate a small quantity of water without significantly affecting the results. Therefore, this method was deemed to be acceptable for the given application. The asphaltene content of the oil sample was determined by asphaltene precipitation according to

ASTM Standard Method D 2007. The eluted maltenes were blown dry using compressed air. The maltene components of the oil were then determined according to the methods described in Ref. 83. Only the nonvolatile portions of the oil were analyzed.

For the long-term experiment, a 1 liter volume of emulsion was then placed in a large beaker and allowed to stand in a 10°C cold room for three months.

The oil layer was removed using a syringe with a large-gage needle. The oil was collected as close to the surface as possible, with care taken to avoid the emulsion below. If the emulsion was semisolid, the beaker was tipped to concentrate the oil at one end. Remaining oil on top of the emulsion was collected after and discarded. An emulsion that has survived three months was found to have elasticity, giving the emulsion some rigidity. This allowed the collection of the top 20% of the emulsion using a spatula. The top layer of the emulsion was scooped up in small quantities covering the surface of the emulsion, and placed in a graduated cylinder until 200 mL had been collected. The middle layer of emulsion between the top and bottom 20% was removed in the same manner as the top portion. It was not possible to collect a full 600 mL as coalesced water on the bottom distorted the proportion. An estimation was made to leave approximately 200 mL of emulsion, which was collected for extraction.

The extraction procedure was used on both of the emulsion layers from the experiment, as well as the oil layer. The sampling procedure collected approximately 10 mL of oil. The sample was homogenized by simple mixing/stirring, and an estimated amount of emulsion sampled to yield 10 to 15 mL of oil. In the case of the oil layer, 10 mL of mixed oil was sampled for extraction using a 10-mL disposable plastic syringe. The sample was placed into a 500-mL glass separatory funnel, then 100 mL of dichloromethane (DCM) and 50 mL of salt water (3.3% NaCl) were added to the sample. The separatory funnel was shaken for 1 min and the contents were allowed to settle until most of the water and DCM had separated. The DCM layer was drained off to the turbid layer between the water and DCM phases, and collected into a 500-mL beaker. Care was taken to ensure that there were no water droplets in the DCM layer, as the dark color would make it difficult to determine the presence of water. A 70/30 mixture of DCM and pentane, respectively, was added to the separatory funnel. This was again shaken for 1 min and the contents allowed to settle until most of the water and DCM phases had separated. Again, the DCM layer was drained off into the 500-mL beaker. The rinsing of the sample with 50-mL aliquots of DCM/pentane was continued until the DCM layer was clear, usually between four and six rinse cycles, depending on the oil. When the DCM layer was clear and most of the DCM/pentane removed, 50 mL of benzene was added. The separatory funnel was shaken for 1 min and the contents again allowed to settle. The water layer was then drained off, down to the turbid layer, into a separate beaker to be discarded. Two rinses of deionized water (100 mL each rinse) were performed, discarding the water from each rinse. The remaining benzene layer and the turbid layer containing water were both collected in the 500-mL beaker containing the rest of the effluent. The contents of the 500 mL collection beaker were evaporated down in a 100-mL boiling flask until the oil sample was obtained. The oil sample was then placed under a blow-down apparatus and blown with compressed air until remaining solvent was driven off.

The asphaltene content of the oil sample was determined by asphaltene precipitation according to ASTM Standard Method D 2007. The maltenes were blown dry using compressed air. Weight difference was used in both instances to determine quantities. Only the nonvolatile portions of the oil were analyzed.

Centrifuging was used to extract oil for analysis for some experiments. Salt water (2 mL 3.3% NaCl) was poured into a 15-mL disposable centrifuge tube. Oil (10 mL) was injected over the water from a 10-mL disposable plastic syringe. A total of six tubes were filled in this manner. The centrifuge tubes were then placed in a centrifuge and spun at 3300 rpm for 2.5 h. The tubes were not moved from their places in the centrifuge as 2 mL of oil was removed from each tube by a syringe with a large-gage needle, keeping the tip as close to the surface as possible. The oil was collected for later analysis. Next, 6 mL of oil was removed from the centrifuge tube using the needle-tipped syringe, again without moving the tube, from the top of the remaining oil. The oil was sucked up slowly to reduce turbulence in the oil remaining in the tube. After all six tubes were sampled, the water under the remaining 2 mL of oil was removed by a needle-tipped syringe and discarded. The oil and small layer of water were then rinsed with two 5-mL volumes of deionized water. At this point, the contents of two centrifuge tubes were combined in a 25-mL beaker by washing with DCM. The oil sample was blown down with compressed air until all the solvent was driven off.

One series of experiments consisted of placing oil and emulsion side-by-side to measure the migration of asphaltenes and resins without the influence of gravity. Emulsion (120 mL) was placed in a 125-mL wide-mouthed bottle. Teflon tape was wound around the threads of the bottle and upper rim. The mouth of the bottle was covered with a 10 × 10 cm square of 105-μm nylon mesh. A 60-mm ID Teflon collar was forcefully inserted over the mouth of the bottle, such that a firm seal was made between the mesh and the rim of the bottle, aided by Teflon tape. An aliquot of 120 mL of the crude oil was placed in another 125-mL, wide-mouthed bottle and used to form the emulsion. Teflon tape was wound around the threads and rim of the bottle, and covered with a 10 × 10 cm square of 105-μm nylon mesh. The first bottle was placed over the second and inserted into the Teflon collar, using the necessary force to complete the union. The bottles in the collar were laid on their sides, and clamped into place with a C-clamp. Neoprene spacers were used to protect the bottles from the contact pressure of the C-clamp. The bottles remained horizontal for a period of one week in a 10°C cold room.

The extraction procedure was used on both the emulsion side of the experiment, as well as the source oil side. The procedure varied, depending on the quantity of water expected to be contained in the emulsion. If 25 mL or less of oil was expected in the emulsion, the entire sample was extracted. If there was more oil present, then the sample was homogenized by simple mixing/stirring, and an estimated amount of emulsion was sampled to yield 10 to 15 mL of oil. In the case of the oil layer, 10 mL of mixed oil was sampled for extraction using a 10-mL disposable plastic syringe. The liquid extraction procedure was described above was used.

## 2. Results and Discussion

Table 10 shows the results of all four series of experiments. Table 11 provides the summary results. The experiments entitled "one-week standing" were designed to determine if there was a separation of asphaltenes between the top oil layer and the lower emulsion layer. Table 10 shows that there is a concentration of both asphaltenes and resins in the emulsion layer. One particular experiment shows low concentration (−0.04%); however, this result is felt to be anomalous. It is interesting to note that both the resins and asphaltenes are concentrated in the emulsion layer. In terms of relative percentage, asphaltenes are concentrated an average of 18% and resins an average of 10%. When the emulsions are formed in the blender, perhaps leading to a more stable emulsion, asphaltenes are concentrated an average of 32% and resins an average of 1%. This appears to indicate that asphaltenes move downward to the emulsion layer, whereas a much lesser amount of resins migrate. Because the emulsion layer is underneath the oil layer in this case, at least part of this migration may be due to separation of the heavier asphaltenes by gravity.

The second set of experiments involved the testing of an emulsion that had been standing for 3 months. Three layers were sampled, a free oil layer from the top, the top 20% (by height measurement) of the emulsion, and the lower 20% of the emulsion. As shown in Table 11, the oil layer is depleted 0.23% in asphaltene content in absolute terms or 6% in relative terms. The top 20% is enriched by 15% in asphaltenes (relative percentage) and the bottom by 74%. This indicates a strong partitioning of asphaltenes to the lower part of the emulsion system. Again, part of this may be as a result of gravitational settling of the asphaltenes.

A third experiment measured the asphaltene content of a salt water–emulsion–oil system in a centrifuge tube. This experiment was designed to measure whether asphaltenes would migrate to the oil–water interface. Gravity might be a factor, because the centrifugal force should move the heavier asphaltenes to the bottom. In fact, the results as illustrated in Table 11, show that there is a greater concentration of asphaltenes at the oil–water interface (47% relative). This shows that the asphaltenes will move to the oil–water interface and will be influenced by gravity.

A fourth series of experiments was conducted to examine how asphaltenes would migrate in the absence of a strong gravity effect. Two vessels were placed side by side, one with oil and the other with emulsion. Only a mesh separated the two materials. Sampling after one week showed an increase of 38% (relative) in asphaltenes in the emulsion formed from Arabian light crude and an increase on 17% in the emulsion formed from Transmountain blend oil.

These experiments show that asphaltenes migrate to the oil–water interface from the oil. This shows why an emulsion that sits for a period of time may become more viscous and more stable as time progresses. During this time, asphaltenes are still migrating to the oil–water interface, thus rendering the emulsion more stable. The experiments show that migration still occurs after one or more weeks of contact. Furthermore, these experiments provide evidence that asphaltenes are the primary hydrocarbon group responsible for emulsion stability. Further work is necessary to determine if the resins will act in the same manner.

**Table 10** Asphaltene–Resin Migration Experiment Results

*Hydrocarbon group analysis following one week standing*

| | Maltenes | Asphaltenes Total | Asphaltenes % | Saturates | Saturates % | Aromatics | Aromatics % | Resins 1 | Resins 2 | Resins Total | Resins % |
|---|---|---|---|---|---|---|---|---|---|---|---|
| Fresh Arabian Light crude oil | | | | | | | | | | | |
| Source Oil | | | | | | | | | | | |
| 1 | 1.4259 | 0.0587 | 3.95 | 0.1769 | 46.59 | 0.1706 | 44.93 | 0.0278 | 0.0044 | 0.0322 | 8.48 |
| 2 | 1.3610 | 0.0530 | 3.75 | 0.2119 | 47.33 | 0.1994 | 44.54 | 0.0291 | 0.0073 | 0.0364 | 8.13 |
| 3 | 1.4326 | 0.0571 | 3.83 | 0.2209 | 47.34 | 0.2107 | 45.16 | 0.0284 | 0.0066 | 0.0350 | 7.50 |
| Average | | | 3.85 | | 47.09 | | 44.88 | | | | 8.04 |
| Std. dev. | | | 0.10 | | 0.43 | | 0.31 | | | | 0.50 |
| Arabian Light oil emulsion formed on the rotary agitator | | | | | | | | | | | |
| Free oil layer | | | | | | | | | | | |
| 1 | 1.5345 | 0.0668 | 4.17 | 0.2241 | 47.11 | 0.2131 | 44.80 | 0.0289 | 0.0096 | 0.0385 | 8.09 |
| 2 | 1.5902 | 0.0635 | 3.84 | 0.2162 | 46.45 | 0.2104 | 45.21 | 0.0303 | 0.0085 | 0.0388 | 8.34 |
| 3 | 1.5530 | 0.0651 | 4.02 | 0.2178 | 47.20 | 0.2040 | 44.21 | 0.0295 | 0.0101 | 0.0396 | 8.58 |
| Average | | | 4.01 | | 46.92 | | 44.74 | | | | 8.34 |
| Std. dev. | | | 0.17 | | 0.41 | | 0.50 | | | | 0.24 |
| Free oil layer | | | | | | | | | | | |
| 1 | 1.7085 | 0.0702 | 3.95 | 0.2277 | 46.95 | 0.2141 | 44.14 | 0.0344 | 0.0088 | 0.0432 | 8.91 |
| 2 | 1.7575 | 0.0732 | 4.00 | 0.2257 | 46.92 | 0.2144 | 44.57 | 0.0336 | 0.0073 | 0.0409 | 8.50 |
| 3 | 1.8284 | 0.0709 | 3.73 | 0.2245 | 46.79 | 0.2174 | 45.31 | 0.0299 | 0.0080 | 0.0379 | 7.90 |
| 4 | 1.5519 | 0.0645 | 3.99 | 0.2285 | 46.22 | 0.2261 | 45.73 | 0.0308 | 0.0090 | 0.0398 | 8.05 |
| Average | | | 3.92 | | 46.72 | | 44.94 | | | | 8.34 |
| Std. dev. | | | 0.12 | | 0.34 | | 0.71 | | | | 0.46 |
| Emulsion layer | | | | | | | | | | | |
| 1 | 1.1458 | 0.0453 | 3.80 | 0.2018 | 46.86 | 0.1874 | 43.52 | 0.0294 | 0.0120 | 0.0414 | 9.61 |
| 2 | 1.2642 | 0.0502 | 3.82 | 0.2320 | 48.78 | 0.1994 | 41.93 | 0.0329 | 0.0113 | 0.0442 | 9.29 |
| 3 | | | | 0.2040 | 47.46 | 0.1879 | 43.72 | 0.0295 | 0.0084 | 0.0379 | 8.82 |
| 4 | | | | 0.2198 | 48.04 | 0.2009 | 43.91 | 0.0296 | 0.0072 | 0.0368 | 8.04 |
| Average | | | 3.81 | | 47.79 | | 43.27 | | | | 8.94 |
| Std. dev. | | | 0.01 | | 0.82 | | 0.91 | | | | 0.68 |
| Emulsion | | | | | | | | | | | |
| 1 | 1.6908 | 0.0917 | 5.14 | 0.2242 | 46.41 | 0.2169 | 44.90 | 0.0307 | 0.0113 | 0.0420 | 8.69 |
| 2 | 1.6979 | 0.0923 | 5.16 | 0.2207 | 45.70 | 0.2206 | 45.68 | 0.0316 | 0.0100 | 0.0416 | 8.61 |
| 3 | 0.9414 | 0.0514 | 5.18 | 0.2204 | 46.29 | 0.2118 | 44.49 | 0.0349 | 0.0090 | 0.0439 | 9.22 |
| 4 | | | | 0.2209 | 45.91 | 0.2191 | 45.53 | 0.0312 | 0.0100 | 0.0412 | 8.56 |
| Average | | | 5.16 | | 46.08 | | 45.15 | | | | 8.77 |
| Std. dev. | | | 0.02 | | 0.33 | | 0.56 | | | | 0.30 |
| Arabian Light oil emulsion formed in the blender[a] | | | | | | | | | | | |
| Emulsion | | | | | | | | | | | |
| 1 | 1.5282 | 0.0789 | 4.91 | 0.2273 | 47.37 | 0.2126 | 44.31 | 0.0323 | 0.0076 | 0.0399 | 8.32 |
| 2 | 1.0931 | 0.0610 | 5.29 | 0.2020 | 47.44 | 0.1870 | 43.92 | 0.0300 | 0.0068 | 0.0368 | 8.64 |
| 3 | | | | 0.2197 | 47.67 | 0.2080 | 45.13 | 0.0253 | 0.0079 | 0.0322 | 7.20 |
| 4 | | | | 0.20.76 | 46.87 | 0.1979 | 44.68 | 0.0318 | 0.0056 | 0.0374 | 8.44 |
| Average | | | 5.10 | | 47.34 | | 44.51 | | | | 8.15 |
| Std. dev. | | | 0.27 | | 0.34 | | 0.52 | | | | 0.65 |

**Table 10** Asphaltene–Resin Migration Experiment Results (*contd.*)

| | | | | Asphaltene content | |
|---|---|---|---|---|---|
| *Long-term standing (3 months)* | | | | | |
| Free oil layer | | | | | |
| 1 | 1.7294 | 0.0644 | 3.59 | Average | Std. dev. |
| 2 | 1.7857 | 0.0677 | 3.65 | 3.62 | 0.04 |
| Top 20% of emulsion | | | | | |
| 1 | 1.7116 | 0.0803 | 4.48 | Average | Std. dev. |
| 2 | 1.7604 | 0.08 | 4.34 | 4.41 | 0.10 |
| Bottom 20% of emulsion | | | | | |
| 1 | 1.8953 | 0.14 | 6.83 | Average | Std. dev. |
| 2 | 1.9559 | 0.14 | 0.60 | 6.71 | 0.16 |
| *Centrifuging fresh oil over salt water* | | | | | |
| Arabian Light crude oil | | | | | |
| Top 20% of oil layer | | | | | |
| 1 | 1.9181 | 0.0915 | 4.55 | Average | Std. dev. |
| 2 | 1.9521 | 0.0934 | 4.57 | 4.56 | 0.01 |
| Bottom 20% of oil layer | | | | | |
| 1 | 1.7630 | 0.1066 | 5.70 | Average | Std. dev. |
| 2 | 2.3609 | 0.1443 | 5.76 | 5.65 | 0.14 |
| 3 | 2.2993 | 0.1335 | 5.49 | | |
| *Side-by-side standing for 1 week* | | | | | |
| Arabian Light crude oil | | | | | |
| Emulsion side | | | | | |
| 1 | 1.9373 | 0.1053 | 5.16 | Average | Std. dev. |
| 2 | 1.9038 | 0.1109 | 5.50 | 5.33 | 0.25 |
| Oil side | | | | | |
| 1 | 2.0258 | 0.0952 | 4.49 | Average | Std. dev. |
| 2 | 1.8806 | 0.0910 | 4.62 | 4.55 | 0.09 |
| Transmountain blend | | | | | |
| Emulsion side | | | | | |
| 1 | 1.8456 | 0.1082 | 5.54 | Average | Std. dev. |
| 2 | 1.8152 | 0.1008 | 5.26 | 5.40 | 0.20 |
| Oil side | | | | | |
| 1 | 1.5350 | 0.0807 | 4.99 | Average | Std. dev. |
| 2 | 1.7556 | 0.0972 | 5.25 | 5.12 | 0.18 |

[a]Free oil layer contaminated after sampling.

### 3. Conclusions on Asphaltene and Resin Migration

Asphaltenes are the primary agents responsible for the formation and stability of water-in-oil emulsions. These large compounds are interfacially active and behave like surfactants. Like surfactants, they can stabilize droplets of oil and water within each other, in this case water-in-oil. The role of resins may be important; however, the experimental results in this paper did not encompass resins to the same extent as asphaltenes.

Asphaltenes migrate to the oil–water interface from solution in oil. This process can continue over weeks. Experiments conducted as long as 3 months after emulsion formation indicate that the asphaltene migration may still continue. This migration may explain the observation that many emulsions increase in stability and viscoelasticity after sitting for periods of time. Furthermore, bonding may be occurring in the film, thus raising its elasticity and strength.

## V. SUMMARY AND CONCLUSIONS

Four clearly defined states of water-in-oil have been shown to exist. These are established by their stability over time, their appearance, and by rheological mea-

**Table 11** Summary of Asphaltene and Resin-partitioning Studies

| | | |
|---|---|---|
| Arabian Light crude and emulsion | | |
| Separation following one-week standing | | |
| Location | | |
| Rotary shaker | *Absolute change from starting oil (%)* | |
| | Asphaltenes | Resins |
| Free oil layer | 0.16 | 0.3 |
| Free oil layer | 0.07 | 0.3 |
| Emulsion layer | −0.04 | 0.9 |
| Emulsion layer | 1.31 | 0.73 |
| Blender formation | | |
| Emulsion layer | 1.25 | 0.11 |
| Long-term standing (3 months) | | |
| Free oil layer | −0.23 | |
| Top 20% of emulsion | 0.56 | |
| Bottom 20% of emulsion | 2.86 | |
| Centrifuging oil over salt water | | |
| Top 20% of oil layer | 0.71 | |
| Bottom 20% of oil layer | 1.8 | |
| Side-by-side experiment (1 week) | | |
| Oil side | 0.7 | |
| Emulsion side | 1.48 | |
| Transmountain blend oil | | |
| Oil side | 0 | |
| Emulsion side | 0.85 | |
| Separation following one-week standing | | |
| Location | | |
| Rotary shaker | *Relative change from starting oil (%)* | |
| | Asphaltenes | Resins |
| Free oil layer | 4.2 | 3.7 |
| Free oil layer | 1.8 | 3.7 |
| Emulsion layer | −1 | 11.2 |
| Emulsion layer | 34 | 9.1 |
| Blender formation | | |
| Emulsion layer | 32.5 | 1.4 |
| Long-term standing (3 months) | | |
| Free oil layer | −6 | |
| Top 20% of emulsion | 14.5 | |
| Bottom 20% of emulsion | 74.3 | |
| Centrifuging oil over salt water | | |
| Top 20% of oil layer | 18.4 | |
| Bottom 20% of oil layer | 46.8 | |
| Side-by-side experiment (1 week) | | |
| Oil side | 18.4 | |
| Emulsion side | 38.4 | |
| Transmountain blend oil | | |
| Oil side | 0 | |
| Emulsion side | 18.7 | |

surements. The states are stable water-in-oil emulsions, mesostable water-in-oil emulsions, entrained water, and unstable water-in-oil.

Stable emulsions are brown solid materials with an average water content of about 80% on the first day of formation and about the same one week later. Stable emulsions remain stable for at least four weeks under laboratory conditions. The properties of the starting oil are as follows: density 0.85 to 0.97 g/mL, viscosity 15 to 10,000 mPa.s, resin content 5 to 30%, asphaltene content 3 to 20%, asphaltene-to-resin ratio 0.74, and average increase in viscosity 1100 at day of formation and 1500 one week later.

Mesostable water-in-oil emulsions are brown or black viscous liquids with an average water content of 62% on the first day of formation and about 38% one week later. Mesostable emulsions remain so less than 3 days under laboratory conditions. The properties of the starting oil are as follows: density 0.84 to 0.98 g/mL, viscosity 6 to 23,000 mPa.s, resin content 6 to 30%, asphaltene content 3 to 17%, asphaltene-to-resin ratio 0.47, and average increase in viscosity 45 at day of formation and 3 one week later. The largest differences between the starting oils for stable and mesostable emulsions is the asphaltene-to-resin ratio (stable – 0.74; mesostable – 0.47) and the ratio of viscosity increase (stable 1100 first day and 1500 in one week; mesostable 45, first day and 30 in one week).

Entrained water-in-oil are black liquids with an average water content of 42% on the first day of formtion and about 15% one week later. Entrained water-in-oil remain so less than 1 day under laboratory conditions. The average properties of the starting oil are as follows: density 0.97 to 0.99 g/mL, viscosity 2 to 60 Pa.s, resin content 15 to 30%, asphaltene content 3 to 22%, asphaltene-to-resin ratio 0.62, and average increase in viscosity 45 at day of formation and 30 one week later. The largest differences between the starting oils for entrained water-in-oil compared to stable and mesostable emulsions is the narrow density range (entrained = 0.97 to 99; stable = 0.85 to 0.99; mesostable = about the same as stable) and the ratio of viscosity increase (entrained = 13 first day and 2 in one week; stable 1100 first day and 1500 in one week; mesostable 45 first day and 30 in one week). Furthermore, the starting-oil viscosity is 2 to 60 Pa.s compared to 15 to 10,000 mPa.s for stable emulsions and 6 to 22,000 mPa.s for mesostable emulsions.

Unstable water-in-oil is characterized by the fact that the oil does not hold significant amounts of water and when it does so, only for a short time. All starting-oil properties are of a much broader range than for the other three water-in-oil states. For example, viscosities are very low or very high. Included in this group are light fuels such as diesel fuel and very heavy, viscous oil products.

The stability of emulsions is due to the formation of asphaltene and resin films at the oil and water interface. Asphaltenes form strong, elastic films which are largely responsible for the stability of emulsions. There is clear evidence of interaction between resins and asphaltenes in forming emulsions; however, asphaltenes can form emulsions without resins, but the most stable emulsions are formed when the asphaltene/resin ratio is about 0.75. The migration experiments show that asphaltenes migrate to the interface very slowly. There is evidence that the migration can continue for longer than one month. This leads to the possibility that the resins migrate very quickly and temporarily stabilize water droplets before stronger asphaltene films form and displace the weaker resin films.

Asphaltene films have been found to be a highly viscoelastic barrier to the coalescence of water droplets. The films may be strengthened by H- or $\pi$-bonding between individual asphaltene molecules.

Oil viscosity alone may be a partial barrier to recoalescence of water droplets. This mechanism is propsed as the primary stabilizer for entrained water and partially for mesostable emulsions. This may also explain why waxes are seen as important in certain circumstances. They may increase viscosity sufficiently to allow the formation of entrained-water states. Waxes are not a factor in the formation of either stable or mesostable emulsions.

Weathering of oil is a factor in the stability of emulsions. First, the elimination of saturates and smaller aromatic compounds leads to the formation of emulsions. Second, the viscosity increases as oil weathers, inhibiting the recoalescence of water droplets. Finally, oxidation and photo-oxidation create more polar compounds, some of which may be regarded as resins.

The energy required to form emulsions is quite low in most cases. Further study is required on a wide variety of emulsions to determine if there is a relationship to oil properties or to emulsion types.

Emulsion properties and stability can be measured by rheological studies and dielectric spectroscopy. Rheological studies include forced oscillation experiments. The formation of stable emulsions is marked by a sharp increase in the elastic modulus. Water content is not a good indicator of emulsion characteristics other than that low water contents ($<50\%$) indicate that an emulsion has not been formed and that the product is entrained water-in-oil. Interfacial measurements are useful for measuring the film strength of asphaltene and resin components.

## REFERENCES


1. J Sjöblom, T Skovin, T Jakobsen, SS Dukhin. Dispers Sci Technol 15: 401–421, 1994.
2. MF Fingas, B Fieldhouse, JV Mullin. Studies of water-in-oil emulsions: stability and oil properties. Proceedings of the Twenty-First Arctic and Marine Oil Spill Program Technical Seminar, Ottawa, ON, 1998, pp 1–25.
3. SA Berridge, RA Dean, RG Fallows, A Fish. J Inst. Petrol 54: 300–309, 1968.
4. T Haegh, T Ellingsen. The Effect of Breaking Waves on Oil Spills; 1. Emulsification of Crude Oil at Sea. Trondheim, Norway: SINTEF, 1977, p 23.
5. H Wang, CP Huang. The effect of turbulence on oil emulsification. The Workshop on the Physical Behavior of Oil in The Marine Environment. Princeton, NJ, 1979, pp 81–100.
6. EM Twardus. A Study to Evalaute the Combustibility and Other Physical and Chemical Properties of Aged Oils and Emulsions. Ottawa, ON: Environment Canada, 1980, p 159.
7. AL Bridie, TH Wanders, W. Zegveld, HB Van Der Heijde. Mar Poll Bull 11: 343–348, 1980.
8. D Mackay, W Zagorski. Studies of the formation of water-in-oil emulsions. Proceedings of the Fourth Annual Arctic Marine Oilspill Program Technical Seminar. Ottawa, ON, 1981, pp 75–86.
9. D Mackay, W Zagorski. Studies of Water-in-Oil Emulsions. Ottawa, ON: Environment Canada, 1982, 93 pp.
10. D Mackay, W Zagorski. Water-in-oil emulsions: a stability hypothesis. Proceedings of the Fifth Annual Arctic Marine Oilspill Program Technical Seminar, Ottawa, ON, 1982, pp 61–74.
11. T Thingstad, B Pengerud. Mar Poll Bull 14: 214–216, 1983.
12. MP Nesterova, OS Mochalova, AB Mamayev. Oceanology 23: 734–737, 1983.
13. D Mackay, M Nowak. Water-in-oil emulsions: some recent advances. Proceedings of the Seventh Annual Arctic Marine Oilspill Program Technical Seminar. Ottawa, ON, 1984, pp 37–46.
14. D Mackay. The Fate and Behaviour of Oil in Cold Climates. Ottawa, ON: Environment Canada, 1984 (unpublished).
15. W Stiver, M Novak, WY Shiu, D Mackay. Recent University of Toronto studies on the fate and effects of oil in the marine environment. Proceedings of the Sixth Annual Arctic Marine Oilspill Prgram Technical Seminar. Ottawa, ON, 1983, pp 24–29.
16. M Desmaison, C Pierkarski, S. Pierkarski, JP Desmarquest. Rev Inst Franç Petrole 39, No. 5: 603–615, 1984.
17. S Miyahara. On the formation of water-in-oil emulsions. Proceedings of Oil in Freshwater: Chemistry, Biology, Countermeasure Technology, New York, 1985.
18. JR Payne, CR Phillips. Petroleum Spills in the Marine Environment – The Chemistry and Formation of Water-in-Oil Emulsions and Tar Balls. Chelsea, MI: Lewis Publishers, 1985.
19. RH Jenkins, SJW Grigson, J McDougall. The formation of emulsions at marine oil spills and the implications for response strategies. Proceedings of the First International Conference on Health, Safety and Environment in Oil and Gas Exploration and Production. Vol 2, Richardson, TX, 1991, pp 437–443.
20. M Walker. Crude oil emulsification: a comparison of laboratory and sea trials data. Formation and Breaking of Water-in-Oil Emulsions Workshop. Washington, DC, 1993, pp 163–178.
21. MI Walker, T Lunel, PJ Brandvik, A Lewis. Emulsification processes at sea – forties crude oil. Proceedings of the Eighteenth Arctic Marine Oilspill Program Technical Seminar. Ottawa, ON, 1995, pp 471–491.
22. LL Schramm. Advances in Chemistry Series. Vol 231. Washington, DC: American Chemical Society, 1992, pp 1–49.
23. P Becher. Why are W/O emulsions stable? Formation and Breaking of Water-in-Oil Emulsions Workship, Technical Report Series 93–018, Washington, DC, 1993, pp 81–88.
24. J Sjöblom, O Urdahl, H Høiland, AA Christy, EJ Johansen. Progr Colloid Polym Sci 82: 131–139, 1990.
25. J Sjöblom, H Söderlund, S Lindblad, EJ Johansen, IM Skjärvo. Colloid Polym Sci 268: 389–398, 1990.
26. J Sjöblom, O Urdahl, KGN Borve, L Mingyuan, JO Saeten, AA Christy, T Gu. Adv Colloids Interface Sci 41: 241–271, 1992.
27. SE Friberg, J Yang. In: J Sjöblom, ed. Emulsions and Emulsion Stability. New York: Marcel Dekker, 1996, pp 1–40.
28. DE Tambe, MM Sharma. Adv Colloid Interface Sci 52: 1–63, 1994.
29. KG Nordli, J Sjöblom, P Stenius. Colloids Surfaces 57: 83–98, 1991.
30. L Mingyuan, AA Christy, J Sjöblom. In: J Sjöblom, ed. Emulsions: A Fundamental and Practical Approach. Dordrecht, The Netherlands: Kluwer Academic, 1992, pp 157–172.
31. S Acevedo, G Escobar, LB Gutierrez, H Rivas. Colloids Surfaces A 71: 5–71, 1993.

32. RA Mohammed, AI Bailey, PF Luckham, SE Taylor. Colloids Surfaces A 80: 237–242, 1993.
33. A Chaala, B Benallal, S Hachelef. Can J Chem Eng 72: 1036–1041, 1994.
34. Y Schildberg, J Sjöblom, AA Christy, J-L Volle, O Rambeau. J Dispers Sci Technol 16: 575–605, 1995.
35. O Urdahl, J Sjöblom. J Dispers Sci Technol 16: 557–574, 1995.
36. J Sjöblom, L Mingyuan, AA Christy, HP Ronningsen. Colloids Surfaces A 96: 261–272, 1995.
37. JD McLean, PK Kilpatrick. J Colloid Interface Sci 196: 23–34, 1997.
38. JD McLean, PK Kilpatrick. J Colloid Interface Sci 189: 242–253, 1997.
39. O Mouraille, T Skodvin, J Sjöblom, J-L Peytavy. J Dispers Sci Technol 19: 339–367, 1998.
40. JD McLean, PM Spiecker, AP Sullivan, PK Kilpatrick. In: OY Mullins and EY Sheu, eds. Structure and Dynamics of Asphaltenes. New York: Plenum Press, 1998, pp 377–422.
41. S Singh, JD McLean, PK Kilpatrick. J Dispers Sci Technol 20: 279–293, 1999.
42. J Sjöblom, H Førdedal, T Skodvin, B, Gestblom. J Dispers Sci Technol 30: 921–943, 1999.
43. HJ Neumann, B Paczynska-Lahme. Progr Colloid Polym Sci 101: 101–104, 1996.
44. HP Rønningsen, J Sjöblom, L Mingyuan. Colloids Surfaces A 97: 119–128, 1995.
45. EJ Johansen, IM Skjärvo, T Lund, J Sjöblom, H Söderlund, G Boström. Colloids Surfaces 34: 353–370, 1988/89.
46. AJ McMahon. In: J Sjöblom, ed. Emulsions: A Fundamental and Practical Approach. Dordrecht, The Netherlands: Kluwer Academic, 1992, pp 135–156.
47. S Puskas, J Balazs, A Farkas, I Regdon, O Berkesi, I Dekany. Colloids Surfaces 113: 279–293, 1996.
48. H Førdedal, Ø Midttun, J Sjöblom, OM Kvalheim, Y Schildberg, J-L Volle. J Colloid Interface Sci 182: 117–125, 1996.
49. B Gestblom, H Førdedal, J Sjöblom. J Dispers Sci Technol 15: 450–464, 1994.
50. J Sjöblom, H Førdedal. In: J Sjöblom, ed. Emulsions and Emulsion Stability. New York: Marcel Dekker, 1996, pp 393–435.
51. J Sjöblom, T Skodvin, O Holt, FP Nilsen. Colloids Surfaces A 123/124: 593–607, 1997.
52. T Skodvin, T Jakobsen, J Sjöblom. J Dispers Sci Technol 15: 423–448, 1994.
53. T Skodvin, J Sjöblom, JO Saeten, O Urdahl, B Gestblom. J Colloid Interface Sci 166: 43–50, 1994.
54. T Skodvin, J Sjöblom. J Colloid Interface Sci 181: 190–198 1996.
55. H Førdedal, J Sjöblom. J Colloid Interface Sci 181: 589–594, 1996.
56. H Førdedal, Y Schildberg, J Sjöblom, J-L Volle. Colloids Surfaces 106: 33–47, 1996.
57. M-H Ese, J Sjöblom, H Førdedal, O Urdahl, HP Ronningsen. Colloids Surfaces A 123/124: 225–232, 1997.
58. R Steinborn, DL Flock. J Can Petrol Technol 22: 38–52, 1983.
59. AE Omar, SM Desouky, B Karama. J Petrol Sci Eng 6: 149–160, 1991.
60. RA Mohammed, AI Bailey, PF Luckham, SE Taylor. Colloids Surfaces A 80: 223–235, 1993.
61. TF Tadros. Colloids Surfaces A 91: 39–55, 1994.
62. R Pal. AIChE J 42: 3181–3190, 1996.
63. HM Lee, JW Lee, OO Park. J Colloid Interface Sci 185: 297–305, 1997.
64. N Aomari, R Gaudu, F Cabioc'h, A Omari. Colloids Surfaces A 139: 13–20, 1998.
65. O Urdahl, T Brekke, J Sjöblom. Fuel 71: 739–746, 1992.
66. B Balinov, O Urdahl, O Soderman, J Sjöblom. Colloids Surfaces A 82: 173–181, 1994.
67. GA LaTorraca, KJ Dunn, PR Webber, RM Carlson. Magn Reson Imaging 16: 659–662, 1998.
68. J Sjöblom, L Mingyuan, AA Christy, T Gu. Colloids Surfaces 66: 55–62, 1992.
69. KGN Børve, J Sjöblom, P Stenius. Colloids Surfaces 63: 241–251, 1992.
70. H Ebeltoft, KGN Børve, J Sjöblom, P Stenius. Progr Colloid Polym Sci 88: 131–139, 1992.
71. S Hartland, SAK Jeelani. Colloids Surfaces A 88: 289–302, 1994.
72. M Ese, X Yang, J Sjöblom. Colloids Polym Sci 276: 800–809, 1998.
73. DJ Miller, R Böhm. J Petrol Sci Eng 9: 1–8, 1993.
74. DD Eley, MJ Hey, JD Symonds. Colloids Surfaces 32: 87–101, 1988.
75. B Paczynska-Lahme. Progr Colloid Polym Sci 83: 196–199, 1990.
76. H Toulhoat, C Prayer, G Rouquet. Colloids Surfaces A 91: 267–283, 1994.
77. MF Fingas, B Fieldhouse, L Gamble, JV Mullin. Studies of water-in-oil emulsions: stability classes and measurement. Proceedings of the Eighteenth Arctic and Marine Oil Spill Program Technical Seminar, Ottawa, ON, 1995, pp 21–42.
78. M Bobra, M Fingas, E Tennyson. Chemtech (April): 214–236, 1992.
79. M Fingas, B Fieldhouse, M Bobra, E Tennyson. The physics and chemistry of emulsions. Proceedings of the Workshop on Emulsions, Washington, DC, 1993, 7 pp.
80. MF Fingas, B Fieldhouse. Studies of water-in-oil emulsions and techniques to measure emulsion treating agents. Proceedings of the Seventeenth Arctic and Marine Oil Spill Program Technical Seminar, Ottawa, ON, 1994, pp 213–244.
81. MF Fingas, B Fieldhouse, JV Mullin. Studies of water-in-oil emulsions: the role of asphaltenes and

resins. Proceedings of the Nineteenth Arctic and Marine Oil Spill Program Technical Seminar, Ottawa, ON, 1996, pp 73–88.
82. MF Fingas, B Fieldhouse, JV Mullin. Studies of water-in-oil emulsions: stability studies. Proceedings of the Twentieth Arctic and Marine Oil Spill Program Technical Seminar, Ottawa, ON, 1997, pp 21–42.
83. P Jokuty, S Whiticar, Z Wang, M Fingas, P Lambert, B Fieldhouse, J Mullin. A Catalogue of Crude Oil and Oil Product Properties. Ottawa, ON: Environment Canada, 1996.
84. MF Fingas, B Fieldhouse, JV Mullin. Studies of water-in-oil emulsions: energy threshold of emulsion formation. Proceedings of the Twenty-Second Arctic and Marine Oil Spill Program Technical Seminar, Ottawa, ON, 1999, pp 57–68.

# 19

# Towards the Atomic-level Simulation of Water-in-Crude Oil Membranes

**Bjørn Kvamme and Tatyana Kuznetsova***
*University of Bergen, Bergen, Norway*

## I. INTRODUCTION

Water-in-crude oil (W/O) emulsions form a ubiquitous part of reality for the majority of oil-production sites. The water component of W/O emulsions comes from either formation water or reinjected water mixed up with the crude oil produced, with the rate of water injection increasing as the deposits are worked out. When this mixture progresses through pipes, valves, and chokes, and the energy required to produce a colloidal system is dissipated, formation of emulsion will follow almost inevitably (though the latter could break down in a matter of seconds if certain demulsifiers are present or the content of natural stabilizers is insufficient) (1–4). Indeed, as a massive molecular-dynamics simulation (5) has proved, when placed in a two-phase water/oil environment even a system composed by very crudely defined hydrophilic headgroups and lipophilic tails will undergo self-assembly into a surfactant layer and micelles.

The stability of the resulting emulsion is known to depend heavily on surfactant stabilizers, both naturally occurring in crude oil and artificial ones (usually commercial). Natural surfactant stabilizers of W/O include asphaltenes and resins. Asphaltenes are often defined as pentane-insoluble and benzene-soluble portions of crude oil (6); on the other hand, the authors of Ref. 4 argue that the distinction between asphaltenes and resins is not so well defined and given the closeness in molecular weights, asphaltenes could be called heavy resins, an argument supported by showing the similarities in the infrared diffuse-reflectance spectra of asphaltenes and resins (7). According to Table 1 of Ref. 4, the molecular weight of the two surfactant components found in the North Sea crude ranges from 1400–1300 (asphaltenes) to 1200–900 (resins).

Commercial stabilizers include among others, ionic sodium nonylphenol polyoxyethylene-25 sulfate (SNP-25S) (8), as well as nonionic Tween 20™ [sorbitan mono-9-octadecenoate poly(oxy-1,1-ethanediyl)] and Tween 80 (8), tetraoxylene nonylphenolether ($C_9PhE_4$, NP-4), octaoxyethylene nonylphenolether ($C_9Ph_8$, NP-8), sorbitan monolaureate (Span 20), and sorbitan mono-oleate (Span 80) (4).

### A. W/O Emulsion: Macroscopic Factors Affecting Stability

Colloidal systems containing both natural and commercial surfactants display a wide range of behavior, for instance, 90% of the continuous phase decants after 12 h when NP-4 is added, while the figure is only 5% for Span 80 (4). Under the same mixing conditions, Span 80 produces small, monodisperse

*On leave from Institute of Physics, St. Petersburg University, St. Petersburg, Russia.

droplets, while span 20 and NP-4 give substantially larger and more polydisperse droplets. The addition of NP-8 (a compound with eight ethoxy units as opposed to four units in NP-4) causes the system to invert from a W/O to an O/W emulsion as a result of interfacial conditions changing with the hydrophilic–lipophilic balance.

The composition of the oil phase can also influence the stability of W/O emulsions insofar as it affects the interfacially active components. It was found (2) that emulsion stability underwent dramatic alterations as the oil phase changed from aliphatic to aromatic; an initial increase in stability, peaking at a 4:1 decane-to-toluene ratio, was followed by monotonic decline towards total instability (separation within seconds of emulsification). Reference 4 explains this decline by monomerization and dissolution of the asphaltenes in toluene and their resulting removal from the interface.

Understanding the factors deciding either stability or instability of W/O emulsions means deciphering the stabilization mechanism. This last hinges on insights into the chemistry, physics, and dynamics of their interfacially active constituents. Experimental data show that interfacial activity alone could not be considered a measure of stabilizing capacity. Resins, though good surfactants, fail as stabilizers, whereas asphaltenes are able to stabilize emulsions by themselves if their fraction is high enough (2%), but not 1%). On the other hand, when the 1% aslphaltene fraction is combined with 1% of resin the resulting emulsion proves to be much more stable than one with only 1% of asphaltenes.

While biological-scale permeation by water and ions apparently does not destroy lipid bilayers, we believe that evidence of any heavy mutual incursion of water and hydrocarbon components will indicate an imminent breakdown of the surfactant-stabilized interface and open the way to flocculation in the given W/O system. Consider Ref. 1, which states that the ability of alcohols to dissolve into the different regions of the emulsified system is an important parameter of their efficiency as destabilizers. Medium-chain alcohols readily dissolve into all three pseudophases (interface, and aqueous and oil phases). Kravczyk (9) concluded that the interfacial region becomes less rigid and structured as a result of considerable interfacial fluctuations occurring in the presence of medium-chain alcohols.

There is a considerable negative surface charge present on the interface between pure oil and water phases and even on water–oil interfaces containing nonionic surfactants (8). This relatively high surface potential ($-50$ to $-70\,\text{mV}$) was found to contribute to an appreciably longer lifetime of emulsion drops. Conversely, a surfactant offsetting the surface potential either by displacing the surface charges or by creating a surface potential with opposite charge to that of the bare water–oil interface will undoubtedly destabilize the W/O emulsion.

Under hydrate-formation conditions, the destruction of the colloidal system will bring available water into contact with hydrocarbon hydrate formers, allowing thermodyanmically favored growth of the solid phase (hydrate crystals); these crystals can plug a pipe in a matter of minutes.

## II. LIPID BILAYERS: CELLULAR VERSIONS OF SURFACTANT-STABILIZED INTERFACES

Cellular membranes are complex multicomponent structures assembled from various lipids and proteins. It could be said that the majority of processes essential for the functioning of living cells involve the interface between water and those biomembranes. These include unassisted and mediated transport ions and nutrients, transmission of neural signals, mediation of immune response, and membrane fusion (10). The structural features of membranes are mainly determined by a bilayer arrangement of their basic amphiphilic components, lipids, with the polar headgroups facing the aqueous exterior, and the hydrocarbon tails extended towards the membrane interior. Lipid membrane formers include dipalmitoylphosphatidylcholine (DPPC), dilauroylphosphatidylcholine, palmitoyloleoylphosphatidylcholine, and phosphatidylserine.

As one can see from Figs 1 and 2, membranes in lipid bilayers and W/O emulsions have in common the general sequence of water/aqueous solution – polar/ionic heads – hydrocarbon tail regions. The only bilayer section that does not find a direct parallel in W/O membranes is the region of low tail (and overall) density; its W/O counterpart does exist but its overall density will not be all that low due to the hydrocarbon phase filling the free volume. However, since the time scale available to state-of-the-art simulations does not allow for spontaneous penetration of this area (as opposed to deliberate placing of molecules for calculation purposes), the bulk of the insights gained from bilayer simulations could probably be transferred to a basic understand-

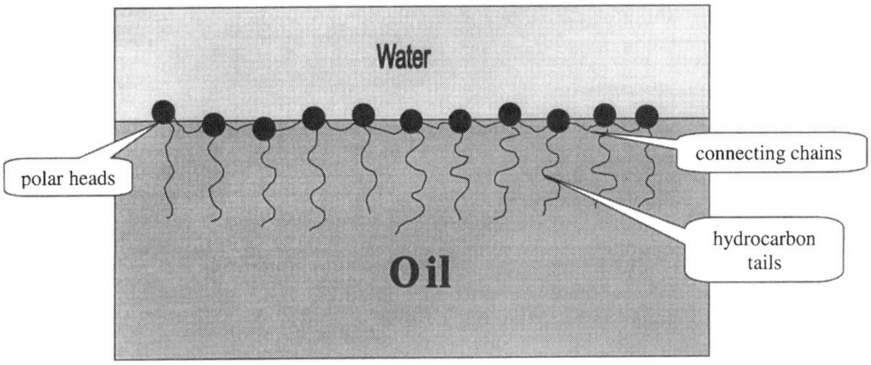


**Figure 1** Schematic representation of W/O membrane: water; headgroups, shown as spheres, are connected by chains; hydrocarbons tails protrude into the oil phase.

ing of the factors determining stability/instability in W/O emulsions. Another structural difference between the membranes involves the fact that asphaltenes must be polymeric to be able to act as stabilizers. This implies headgroups connected by hydrocarbon chains lying near the interface surface and imposing certain restrictions on the headgroup positions and movements. Yet another difference is the lipid bilayers' lack of "backwash" of apolar hydrocarbon components, occurring to a certain extent in real-life W/O membranes. On the other hand, water solubility in hydrocarbons is surprisingly high, much higher than hydrocarbon solubility in water, so the excursions of hydrocarbons into the water section could be treated as additive disturbances of the main process: penetration of the surfactant membrane by water molecules.

### A. State of the Art in Molecular Modeling of Lipid Bilayers

Once the computer resources advanced far enough to allow their atomic-level simulation, few interfacial phenomena have attracted more attention from researches in the field of molecular modeling as the transport of small molecules in lipid bilayers. Simulation of unassisted transport of water (11) and ions (12); study of energetic and structural effects that a net charge (13), an inclusion of cholesterol (14), or an amphiphilic peptide (15) has on the membrane; and general behavior of anesthetics and their interaction with the water–bilayer interface (16–18), to name only a tiny fraction of papers in this steadily growing field.

Existing models for permeation of small molecules through lipid membranes apply, for the most part,

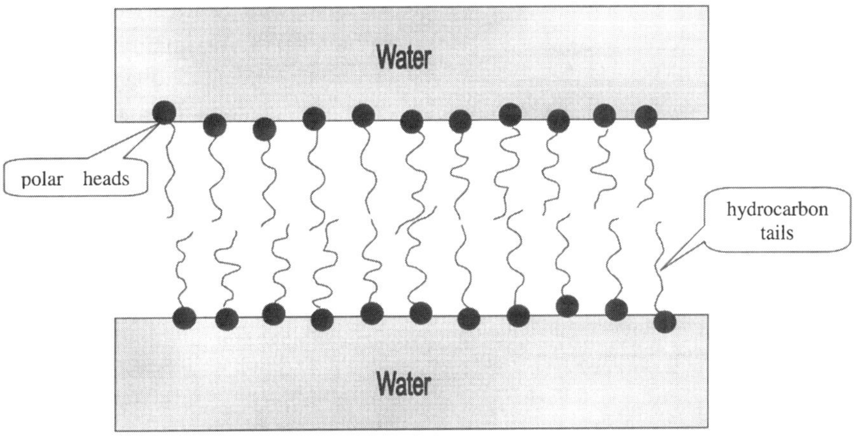


**Figure 2** Schematic representation of the bilayer membrane: water; two lipid membranes with headgroups, shown as spheres; and hydrocarbon tails protruding into membrane interior.

basic ideas developed for diffusion across polymer membranes, therefore treating lipid bilayers as soft polymer membranes with sharp boundaries. This approach emphasizes the significance of free volume and its fluctuations (11, 16). Small diffusing molecules are thought to spend most of the time confined to a cavity formed from the immediate neighboring molecules. These cavities exist in all amorphous polymers whether penetrants are present or not and they typically have the size to hold a gas molecule or a small solvent molecule. They fluctuate in size and shape, but do not migrate or disappear on the nanosecond time scale. Displacement of a molecule contained in such a "cage" is enabled by transient channels of free volume which are constantly being opened and closed owing to thermal fluctuations. Thus, solute diffusion across membranes should be facilitated by the dynamics of free volume pockets in the membrane, a process which could be studied by means of molecular simulations. However, as we shall see, free volume is only one (and often not the deciding) factor governing transport across the membrane.

## B. Simulation of Water Transport Across a Lipid Membrane

Work was undertaken on an atomic-level study of water transport through a phospholipid/water bilayer system (11). The simulation cell contained a bilayer composed of 64-DPPC molecules as well as 736 water molecules modeled as SPC (simple point charge). Periodic boundary conditions in all three dimensions have been applied and a GROMOS (19, 20) force field was used. The system was weakly coupled with constant-temperature and constant-pressure baths (21) (at 350 K and 1 atm, respectively). Molecular-dynamics simulation showed that four drastically different regions could clearly be discerned in the system, namely, (1) a low headgroup density zone with comparable density of water and headgroups; (2) a zone of high headgroup density with water density under 1%; (3) a region of high tail density; and (4) a low-density membrane interior. All of the regions except for the interior have their counterparts in W/O membranes. The W/O system's counterpart of zone 4 would be a region of low tail velocity blending into the bulk hydrocarbon fluid. Judging by the temperature-dependent data (11), neither of the subregions will present any great resistance to water transport [diffusion coefficient of water in hexadecane is estimated as $12 \times 10^{-5}\,cm^2/s$ at 350 K; self-diffusion in bulk SPC is $7.5 \times 10^{-5}\,cm^2/s$ (11)]. The relative wdith of zone 1 indicates that the interface between the dipolar groups and water is significantly diffused, with distributions of water and headgroups demonstrating a wide overlap. These findings are confirmed by experimental data (11). No evidence for single-dispersed molecules of water has been found in the course of simulations. It appears that water molecules penetrate the membrane "in a school crocodile," trying to keep at least one hydrogen bond to a neighboring water molecule. Radial distribution functions exhibited well-defined first and second hydration shells for choline-methyl groups, but phosphate-oxygen and carbonyl-oxygen groups were found to have only the first shell.

The free-energy profiles were roughly trapezoidal, with a maximum reached in the high tail density region. This is in contrast to the step function assumed in soft-polymer models with sharp boundaries. As expected, the four different membrane regions have widely varying diffusion coefficients. The highest diffusion rate corresponds to the membrane interior, the lowest one to the start of the high tail density region. Diffusion on the local scale appears essentially anisotropic everywhere except for the high tail density zone where the alignment of tails appears to favor the diffusion along the membrane normal in comparison with the lateral one. On the other hand, this difference might be something of an artifact resulting from the particular estimation technique. The low diffusion rate in zone 2 does not necessarily mean high resistance to permeation, since this is mostly determined by the free-energy barrier, which is almost non-existent there.

## C. Charged Biological Membrane

The importance of net charge on the lipid membrane has been highlighted in the molecular-dynamics study of dipalmitoylphosphatidylserine (DPPS) lipid membrane (13). Experimental data suggest that at neutral pH aqueous dispersions of DPPS assemble into lamellas with a net negative charge. The membrane model consisted of 64 $DPPS^-$, 64 $Na^+$ and 732 $H_2O$ molecules. Weak coupling (21) with the constant-temperature bath (350 K) and reference pressure bath (1 atm) was used to maintain temperature and pressure. The simulated system of Ref. 13 has been able to reproduce seemingly illogical experimental findings that the per-lipid surface area in charged DPPS is smaller than in the case of neutral phospholipid DPPC. It appears that the ammonium group displays a strong intermolecular coordination, preferring to be bound to serine carbonyl oxygen rather than to phospatidyl or carbonyl tail oxygen, which could lead to an increase in the per-phospholipid surface area. Thus, the charge interactions between phospholipids are able to offset the electrostatic repulsion.

The model system has also showed much lesser hydration of phosphate groups, a feature probably contributing to easier penetration by water, since the water molecules do not have to remain bound within the hydration shells. The atom charge distribution across the membrane showed that phospholipid headgroups provide an electrostatic environment conductive to penetration of the headgroup zone by water molecules and sodium ions. This conclusion was also borne out by the fact that water diffused faster in the interface region of the DPPS membrane than in the DPPC membrane.

## D. A Cholesterol-containing Bilayer

It is a well-known experimental fact that incorporation of cholesterol in lipid bilayers affects the mechanical and transport properties of membranes. This includes their increased bending elasticity (22) and reduced passive permeability to small molecules (23, 24). Despite a great deal of theoretical and experimental research, no definitive microscopic understanding of phospholipid–cholesterol interactions has been proposed yet.

Incidentially, this modification of bilayers by cholesterol could be considered a rough parallel to increased rigidity and stability of heterogeneous W/O membranes resulting from addition of resins to asphaltene-containing crude.

Constant-temperature and constant-pressure molecular-dynamics simulations of a cholesterol-containing DPPC bilayer were reported in Ref. 14 where eight well-separated DPPC molecules were replaced in a configuration representing a fully hydrated liquid crystal phase bilayer at 50°C. The analysis of changes brought about by addition of cholesterol proves that the impact plays itself on the microscopic scale so far attainable only through atomistic molecular simulations. For example, though cholesterol is known to have a "condensing" effect on the bilayer (a definite macroscopic-scale event), the examination of the free-volume fractions suggests that cholesterol-induced reduction of bilayer passive permeabilities (by a factor of 3.5 for acetic acid upon addition of 20% cholesterol in DPPC at 50% does not result exclusively from a "condensation" or repacking of chains.

Cholesterol impact translates into a roughly one-third decrease in diffusion motion of both DPPC and cholesterol molecules, with different sections of lipid bilayer being affected to a very different degree: the bilayer interior shows only a slight influence, manifesting as a small decrease in empty free volume. It is the water/bilayer interface that undergoes significant changes. The bilayer/water interface is more pronounced, with the DPPC and water densities decreasing more abruptly at the edge of the membrane, in the cholesterol-containing bilayer. The peaks of constituent electron densities have been shifted slightly towards the bilayer center, consistent with the decreased bilayer thickness. The choline density, roughly symmetrical in the pure DPPC bilayer, has been skewed significantly towards the center and its peak is now coincident with the phosphate peak. Thus, the presence of the cholesterol causes the choline group to move inward and lie nearly flat to the bilayer plane (P–N vector's average inclination is 6° compared to 17° for pure DPPC). The authors of Ref. 14 suggest that this effect is caused by cholesterol lying low and leaving holes in the bilayer surface that are generally filled by choline ammonium groups from neighboring DPPC molecules. Cholesterol hydroxyl groups were found to lack strong preferences for interacting with specific DPPC moieties; statistical analysis revealed that the hydroxyls interact exclusively with water about half of the time, with the other half equally split between the phosphate and carbonyl groups. As shown by both electrostatic measurements and molecular-dynamics simulations, water molecules in the DPPC/cholesterol bilayer vicinity are "orientationally polarized" to a greater extent than in pure DPPC bilayer. The high percentage of time spent in hydrogen bonding with water could explain the lower permeability of the cholesterol-containing DPPC bilayer.

Another factor contributing to a higher free-energy barrier for a passively penetrating solute is the cholesterol-induced narrowing of the interface. A significant influence of cholesterol on the subnanosecond lipid dynamics, namely, freezing of the center of mass and large-amplitude chain motions, could translate into increased microscopic viscosity for the unassisted transport of solute.

# III. POSSIBLE SIMULATION STRATEGY FOR MOLECULAR DYNAMICS INVESTIGATION OF WATER/OIL LIQUID MEMBRANE SYSTEMS

## A. Grand Canonical Ensemble: Fixing the Chemical Potential

While able to simulate constant temperature and pressure/temperature ensembles in atomistic detail, all the membrane simulations reviewed so far have one feature in common – the number of molecules remains a constant. If surfactants under study fail to stabilize W/O membrane and it is breached by water and oil molecules at an appreciable rate, an essentially nonequilibrium process (flux) would develop, leaving open the possibility of water density deviating from its bulk value event at positions furthest from the water–surfactant interface. With this in mind we believe it would be instructive to review approaches and set-ups used to model systems at constant temperature, constant pressure/volume, and, most important, constant chemical potential gradient, even though the membrane modeled were fixed.

## B. Dual-control Volume Approach for Gradient-driven Diffusion: Monte Carlo versus Molecular Dynamics

Earlier simulations of flux in slit pores and diffusion of gas through micro- and nano-pore membranes (25–28) essentially simulated particle flow between a source and sink regions with a pore or a fixed membrane in the middle. Particles were removed once they reached the sink area. The source region density has been kept constant by means of particle insertion. However, since it is the chemical potential gradient that is the driving force of diffusion, simulation in the grand canonical ensemble (μVT) would be the one most true to real-life diffusion situations. The authors of Refs 28 and 29 proposed using the recipes of grand canonical Monte Carlo (GCMC) to control the chemical potential in two control volumes placed at a distance of a half-cell length from each other. Periodic boundary conditions could then be applied in all three dimensions. The hybrid GCMC–molecular dynamics (GCMC/MD) scheme employed two types of moves, stochastic MC moves which aimed at adjusting the density in control volumes to match the driving chemical potential, and dynamic ones providing mass transport across the system. This dual-control volume approach has been implemented by means of a massively parallel algorithm LADERA (30) and has yielded correct values of transport properties for a simple Lennard–Jones system undergoing color diffusion (31), "uphill diffusion" in a bulk ternary Weeks-Chamber-Andersen potential (WCA) Lennard–Jones system (32), and gradient-driven diffusion through polymers (33).

Typical hybrid MC/MD approaches suffer from two major drawbacks. First, they call for insertion of a fully-fledged molecule, and the acceptance probability of such a step will be quite low for dense fluids (as witnessed by the poor sampling of Widom particle insertion, which makes it fail in the case of water) and/or systems containing molecules widely differing in size. Indeed, according to Ref. 34, "because the accompanying change in energy of the insertion or deletion of large molecules is so great as to make the probability of accepting such a move prohibitively small, the range of applicability of the DCV-GCMD method as well as all other grand canonical simulation methods is limited." This limitation forced the authors of Ref. 34 to abandon constant-chemical potential treatment for the larger species in favor of the constant number of particle technique. However, the above statement is strictly true only for Monte Carlo-based grand canonical simulations. Pure GC/MD pioneered by Pettitt and coworkers (35–38) entails insertion/deletion of virtually ghost-like particles, whose interaction with all the other particles, both full and fractional, is attenuated by a factor of 0.01–0.02. This feature has permitted successful introduction of water molecules at real-life densities in two different integration schemes (36, 39) and should, in principle, be perfectly feasible for a constant–μ (constant-chemical potential) simulation involving large molecules.

Lynch and Pettitt (36) employed a pure Nosé thermostat and RATTLE-like algorithm for bond constraint, as well as a single fraction particle. Our approach (40) used quaternions (41) to allow separate Nosé–Hoover thermostats for translational and rotational modes. We found that the system's tendency to freeze in metastable states could be overcome by introduction of multiple fractional particles (four were used). Curve (*b*) of Fig. 3 is an extreme example of runaway insertions at high chemical potential as compared to curve (*a*) where the actual chemical potential of water resulted in a stable density virtually undistinguishable from the experimental one.

The second, and no less important disadvantage of hybrid MC/MD approaches lies in the fact that if the insertion step is accepted, the newly created particle is assigned a velocity drawn from Maxwell's distribution. In one study (27) the particle was also assigned a streaming velocity calculated from the averaged previous flux, which raises the problem of self-consistency. Given the relatively small size of control volumes and the possibility of several particles being created or deleted within just a few time steps, there are no guarantees against the GCMC/MD procedure completely destroying the dynamics of the system and thus defeating the purpose of the exercise – simulating mass transport driven by the gradient in chemical potential. In the case of the pure GC/MD the newly created

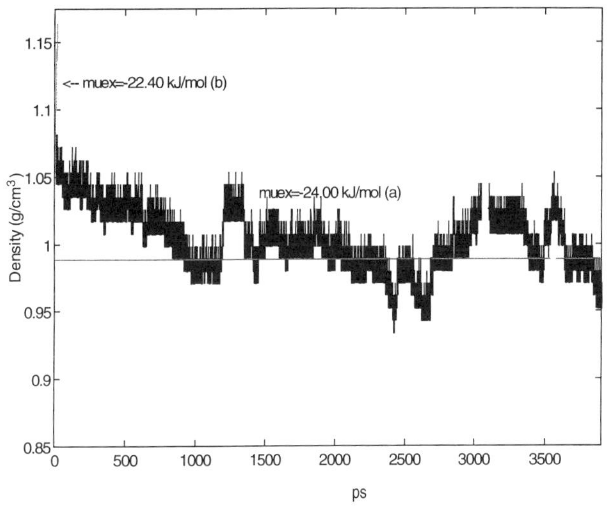


**Figure 3** Time evolution of TIP4P water system density in a grand canonical ensemble driven by two different chemical potentials; solid horizontal line corresponds to experimental water density at 298 K (0.9982 g/cm$^3$).

almost-zero particle is assigned zero velocity, the system is undisturbed, and the fledgling particle is free to probe its surrounding and gain appropriate velocity. Playing the devil's advocate, we must point out that the advantage of the MC scheme lies in the fact that, while any straightforward application of GC/MD relies on the ensemble's intrinsic responses to correct for the possible density deviation in the control volumes, the hybrid GCMC/MD technique allows one to tailor the ratio between stochastic and dynamic steps to match the system's dynamics.

Application of the relationship between density and chemical potential allows one to use both GCMC/MD and GC/MD to calculate chemical potential in model single-component systems of interest (e.g., 35, 36). The application of the schemes to interface systems could follow the general set-up suggested in Refs 26 and 31. One possible way to overcome the problem of slow response to pure GC/MD mentioned earlier will be to take a leaf out of the hybrid technique book and periodically freeze the entire system outside the control volumes (the distance between them insures that they do not affect each other). While the GC/MD steps are executed in two control volumes, the rest of the system is treated as static background. Full and fractional particles alike are confined to control volumes. In between the grand canonical steps it is the number dynamics that will be put on hold, with the fractional particles bouncing off the control volume walls in a direction normal to the interface.

## C. Nosé–Hoover Thermostat, Parameter Optimization, and the Issue of Ergodicity

Ever since its introduction, the Nosé–Hoover thermostate (42, 43) became a technique of choice for implementation of molecular-dynamics simulations in various constant-temperature ensembles, especially the canonical one (NVT). Since the use of molecular dynamics is mainly prompted by a desire to obtain the systems' dynamic characteristics, any method used to control the temperature must yield canonical distribution not only in coordinate space but also in velocity space. This in its turn raises the problem of the ergodic property of the system behavior. While it is universally accepted that even the most straightforward of constant-temperature schemes such as velocity scaling, will yield canonical distribution in configurational space (and yield, for instance, very good estimates of potential energy), canonical distribution of dynamic properties is a much more difficult problem. Curve (*a*) of Fig. 4 demonstrates the typical behavior of translational modes in the NVE ensemble. We should draw attention to the highly irregular shape of the waveform, indicating a substantial coupling between translational and rotational modes reported previously by DiCola and Deriu (44) and contributing to the ensemble's ergodicity. Curve (*b*) of Fig. 4 corresponds to a

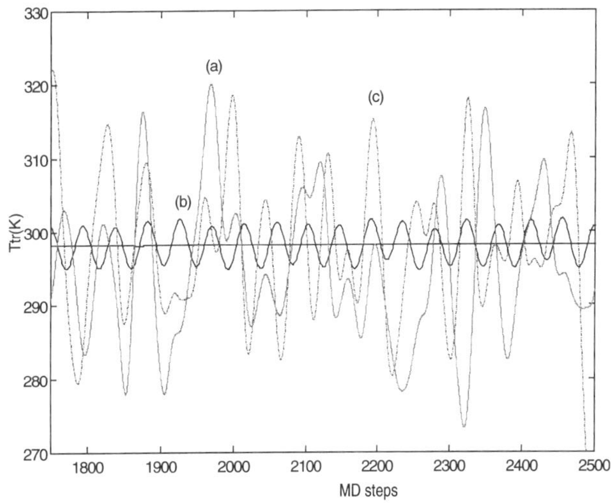


**Figure 4** Variations of translational kinetic energy: (a) microcanonical ensemble; (b) tight thermostating (frequency about four times higher than the natural one); (c) optimum thermostating (response time about half again of the resonance value).

case of short thermostat response times; it shows clearly that chaoticity of motion, ergodicity's prerequisite, is clearly violated because of too tight restrictions imposed on the individual degrees of freedom. It is also obvious from the curve that low thermostat masses will severely inhibit mode mixing as well, since the modes will be driven by their own thermostats suppressing any possible interactions. We should also note that temperature fluctuations in this case are much smaller than those corresponding to intermediate and large thermostat response times.

It has been argued (42) that since the thermsotat's purpose is to provide the transfer of energy between the system and the heat "bath" the most efficient coupling would be insured when the thermostat frequency is at resonance with the natural frequencies of the system. Our analysis shows that the resonance conditions might be far from ideal where the ergodicity considerations are concerned. Substantial coupling between the modes, present in the NVE ensemble, appears to be absent at resonance, with the complex shape of the translational NVE mode substituted by a regular sine wave though with almost identical frequency. Curve (*c*) corresponds to thermostat response times, which are about half again of the resonance values; it is obvious that the mode behavior shows a striking resemblance to the natural one. This seems to indicate that thermostat response frequencies should be set somewhat below the resonance values to allow for the proper coupling of translational and rotation modes.

## D. Simulating Water Penetration into NAM Membrane: Combining Single Control Volume with Pure Molecular Dynamics in a Grand Canonical Ensemble

*N*-Acetylmorpholine (NAM, see Fig. 5) is a solvent used in a mixture with *N*-formylmorpholine (NF) for the removal of acidic gas compoounds ($CO_2$, $H_2S$) from subquality natural gases by Morphysorb® technology from Krupp Uhde GmbH, Germany. The key economic advantages of using the process in a hydrocarbon-transporting pipeline stems from the very low solubility of $C_1$–$C_6$ hydrocarbons in NAM/NFM mixtures and the high capacity for acidic gases. This is why we believed it would be highly instructive to study the processes of an NAM membrane being penetrated by polar and apolar compounds. To this end we have used the suitably modified methodology of gradient-driven diffusion (30, 32–34). One modification consisted of using just one control volume (for water only). Another one involved application of Pettitt's purely molecular dynamical version of the grand canonical ensemble, instead of Monte Carlo, to insure a constant chemical potential (35–38).

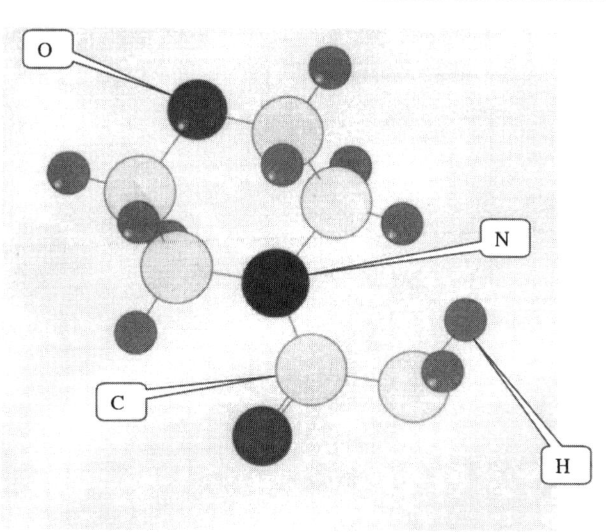


**Figure 5** NAM molecule.

We constructed our simulation system by adding two identical slabs of 512 TIP4P water molecules on both sides of a cubic pre-equilibrated section of 64 NAM molecules (subsequently kept fixed throughout the simulation run). Water was chosen as a quintessential example of a polar solute [with a proven ability to reproduce the correct density at an experimental chemical potential as well (36, 39). The resulting simulation cell measured 52.25 × 22.8 × 22.8 Å, with the four fractional particles confined to an 11 Å-wide control volume located in the middle of the water region. Periodic boundary conditions were applied in all three directions. The cut-off radius was set at 10 Å for all interactions. The electrostatic part of the intermolecular forces was handled by means of Ewald summation (45). Separate Nosé–Hoover thermostats with optimum parameters established in our previous research constrained the kinetic energy of translational and rotational modes.

## E. Simulation Details, Results, and Discussion

Our simulation program was a modification of the MCMOLDYN package (46); the alterations involved changing the integration scheme to allow for the fractional particle dynamics with a Nosé–Hoover thermostat and parallelizing it by means of a shared-memory approach. The parallel version was run on the 128-processor Cray Origin 2000 at Parallab

(Høgteknologi Center in Bergen); the sequential one on a DEC Workstation 400.

Some results obtained over the 0.8 ns run are presented in Figs 6–11. The number of water molecules in the chemical potential-control system first fell to 505 and stayed at this number for 0.3 ns, then the system corrected its own density, with the number of water molecules growing to 512, and subsequently starting to rise in response to gradual penetration of water into the NAM membrane. The number of full particles grew to 516–520 and then to 533 molecules (current number corresponding to the configuration shown in Figs 8 and 11). Consider the patterns of water penetration obtained in two systems, starting from identical initial configurations, but one run with the constant chemical potential within the control volume, and another under constant particle conditions. As can be seen by comparing Figs 7 and 8, and 10 and 11, water penetration proceeded significantly more vigorously under the constant chemical-potential regime, since the fractional particles that scan the control volume have been able to "feel" the decrease in density resulting from water molecules moving into the NAM membrane, and compensate for it. No such compensation mechanism is available in constant-number schemes, causing them to underestimate the permeability of a membrane by a given species.

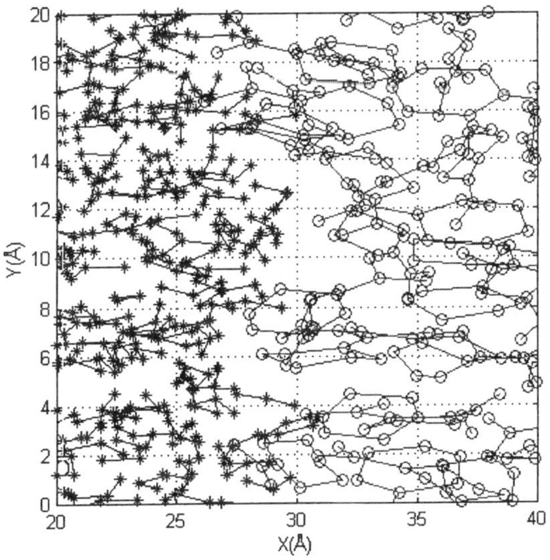


**Figure 6** Starting configuration of the composite system in *XY* projection. Atoms in NAM molecules are connected by dotted lines. Section shown includes one of two water–NAM interfaces.

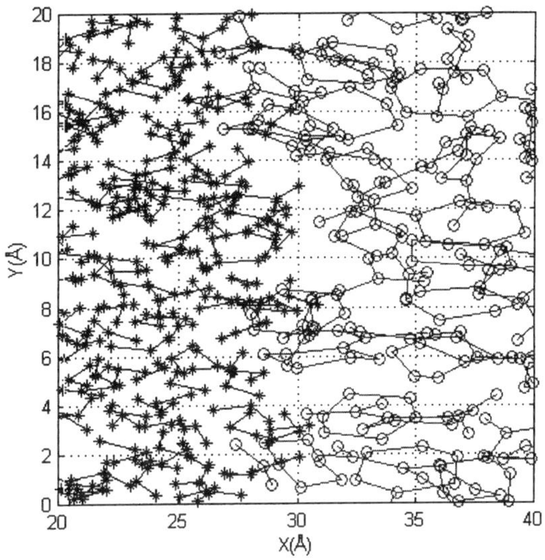


**Figure 7** Same projection and area after 0.8 ns. No chemical potential control has been imposed in the control volume.

## F. Free Energy and Chemical Potential: the Stumbling Block of Conventional Canonical Ensemble Calculations

A number of important physical properties, such as internal energy, heat capacity, diffusion coefficient, and pressure, could be expressed as NVE and NVT

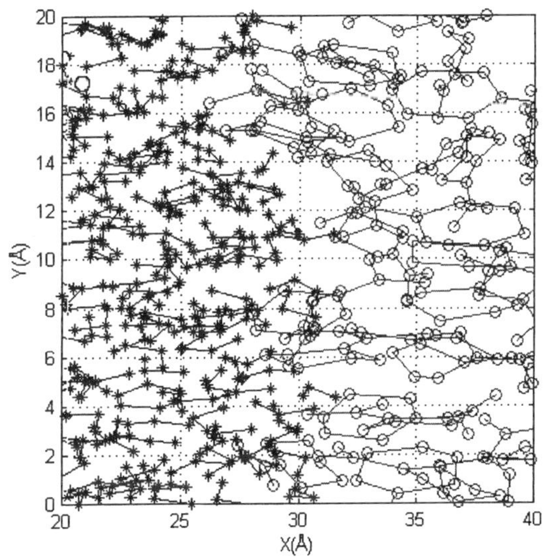


**Figure 8** Projection, area, and time elapsed are the same as in Fig. 7. Chemical potential control was switched on from the start.

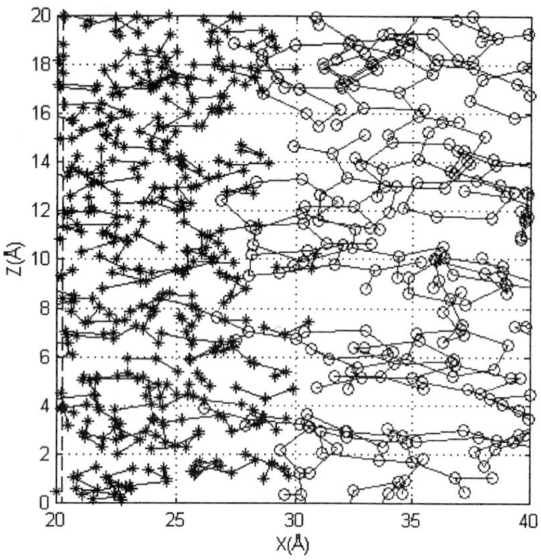


**Figure 9** *XZ* projection of the configuration in Fig. 6.

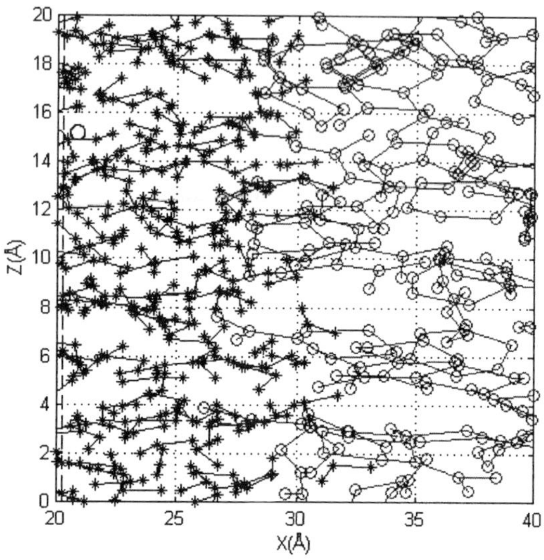


**Figure 11** *XZ* projection of the configuration in Fig. 10.

ensemble averages over the phase-space trajectory. And as such they could be evaluated in the course of a single simulation run with constant thermodynamic parameters. This is possible because these "mechanical" properties are related to the ratio of two high-dimensional integrals rather than integrals themselves. Entropy and entropy-related functions prove to be a very important exception. By its very definition entropy depends on the volume of phase space available to the system, which makes it very difficult to

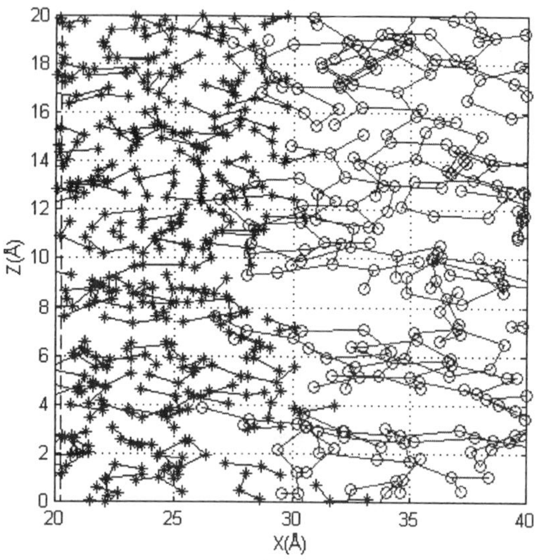


**Figure 10** *XZ* projection of the configuration in Fig. 7.

determine within a canonical ensemble. On the other hand, knowledge of chemical potential and free energy of different phases and components is essential for determination of phase equilibria in chemical reactions and multicomponent systems, while it is the spatial distribution of free energy that determines the barriers to unassisted transport across various membranes, as discussed earlier. This is why considerable effort has been expended by numerous researchers to formulate techniques able to estimate these thermodynamic functions from computer simulations. Thermodynamic integration, the simplest and perhaps the most reliable of these techniques, requires a series of simulations. Straightforward application of Widom's particle insertion method (47, 48) is known to fail at high densities because of poor sampling. Various versions of the cavity-biased insertion technique were used by several authors to overcome the sampling problem for water and other dense liquids (17, 49).

Extending the system under investigation into a discrete set of balanced subensembles differing in such parameters as temperature, number of particles, or "ghostliness" of chains gives rise to the method of expanded ensembles (50, 51). All information necessary for determination of free energy or chemical potential could be extracted in the course of a single simulation run, since the system's evolution takes it from one subensemble to another through a Monte Carlo routine. This approach appears to work even for such an exotic system as the quantum Heisenberg model (52). Yet another line of attack could be implemented within the framework of the grand canonical ensemble (36–38) where chemical potential is treated as an input parameter and the number of particles as a dynamic variable. Following the determination of density for several input values, chemical potential at the density of interest could be estimated by interpolation.

### G. Hybrid Method for Chemical Potential Estimation

"Hybrid," or modified real particle method was proposed by Kumar (53) as a method suitable for calculation of chemical potential at high densities. This technique combines Widom's test particle and the so-called real particle methods and calls for fictious removal and subsequent reinsertion of particles already present in the ystem. The hybrid technique can be classified as a "nondestructive" one since it does not affect the proper time evolution of the system. It was suggested that this method would be particularly advantageous for simulation of macromolecules when a removal of a whole polymer chain is likely to create substantial free space and thus facilitate the reinsertion. This technique was tested in the original paper on Lennard–Jones particles and proven to yield good results at densities up to 1.1 and $T^*$ down to 0.7.

The hybrid technique was applied by us (40) to the bulk TIP4P water system at 273 and 298 K. The "instant" chemical potentials provided by the method proved to be heavily dependent on the number of insertions, an expected feature shared with the test particle method, and on the particular molecule chosen for removal and reinsertion. The hybrid method also yielded chemical potentials lying below both experimental and model-specific values, and our simulations indicate that it may give rise to a "premature" convergence of results. We have come to the conclusion that this technique could be treated as a sort of convenient mathematical trick enabling one, once the "correct" density of reinsertion points has been determined for a bulk system at "reference" temperature, to estimate the chemical potential at different temperatures and system sizes. When applied to the grand canonical ensemble, Kumar's method (53) proved to be totally unsuitable, thus indicating the need for a reliable technique capable of calculating the spatial distribution of free energy and chemical potential for both solute and solvents.

## ACKNOWLEDGMENT

Support of the Norwegian Supercomputing Project through the grant of CPU time quota is gratefully acknowledged.

## REFERENCES

1. O Udahl, J Sjöblom. J Dispers Sci Technol 16: 557–574, 1995.
2. J Sjöblom, H Førdedal, T Jakobsen, T Skovdin. In: KS Bird, ed. Handbook of Surface and Colloid Chemistry. Boca Raton: CRC Press, 217–237, 1997.
3. H Førdedal, Y Schildberg, J Sjöblom, JL Volle. Colloids Surfaces A 106: 33–47, 1996.
4. J Sjöblom, T Skovdin, Ø Holt, FP Nilsen. Colloids Surfaces A 123/124: 593–607, 1997.
5. B Smit, PAJ Hilbers, K Esselink, LAM Rupert, NM van Os, AG Schlijper. Nature 348: 624–625, 1990.
6. AA Christy, B Dahl, OM Kvalheim. Fuel 68: 430–435, 1989.
7. DJ McClements. Adv Colloid Interface Sci 37: 33–72, 1991.
8. IB Ivanov, PA Kralchevsky. Colloids Surfaces A 128: 155–175, 1997.
9. M Kravczyk. PhD thesis, Illinois Institute of Technology, Chicago, 1990.
10. RB Gennis. Biomembranes: Molecular Structure and Function. New York: Springer-Verlag, 1989.
11. SJ Marrink, HJC Berendsen. J Phys Chem 98: 4155–4168, 1994.
12. MA Wilson, A Pohorille. J Am Chem Soc 118: 6580–6587, 1996.
13. JJ Lòpez Cascales, J Garcia de la Torre, SJ Marrink, HJC Berendsen. J Phys Chem 105: 2713–2720, 1996.
14. K Tu, M Tarek, ML Klein, DJ Tobias. Biophys J 75: 2147–2156, 1998.
15. K Belohorcová, JH Davis, TB Woolf, B Roux. Biophys J 73: 3039–3055, 1997.
16. K Tu, M Tarek, ML Klein, D Scharf. Biophys J 75: 2123–2134, 1998.
17. A Pohorille, MA Wilson. J Chem Phys 104: 3760–3773, 1996.
18. JJ Lòpez Cascales, JG Hernández Cifre, J Garcia de la Torre. J Phys Chem 102: 625–631, 1998.
19. WF van Gundsteren, HJC Berendsen. GROningen Molecular Simulation is a Software Package. Groningen, The Netherlands: Biomos, 1997.
20. WF van Gundsteren, HJC Berendsen. Angew Chem Int Ed Engl 29: 992–1023, 1990.
21. HJC Berendsen, JPM Postma, WF van Gundsteren, A DiNola, JR Haak. J Chem Phys 8: 3684, 1984.

22. PC Méléard, C Gerbeaud, T Pott, L Fernandez-Puente, I Bivias, MD Mitov, J Dufourcq, P Bothorel. Biophys J 72: 2616, 1997.
23. A Caruthers, DL Malchior. Biochemistry 22:5797–6010, 1983.
24. TX Xiang, BD Anderson. Biophys J 72: 223, 1997.
25. S Furukawa, T Shigeta, T Nitta. J Chem Eng Jpn 29: 725–728, 1996.
26. PI Pohl, GS Heffelfinger, DM Smith. Molec Phys 89: 1725–1731, 1996.
27. RF Cracknell, D Nickolson, N Quirke. Phys Rev Lett 74: 2463–2466, 1995.
28. GS Heffelfinger, F van Swol. J Chem Phys 100: 7548–7552, 1994.
29. JMD MacElroy. J Chem Phys 101: 5274–5280, 1994.
30. DM Ford, GS Heffelfinger. Molec Phys 94: 659–671, 1998.
31. GS Heffelfinger, F van Swol. J Chem Phys 100: 7548–7552, 1994.
32. AP Thompson, DM Ford, GS Heffelfinger. J Chem Phys 109: 6406–6414, 1998.
33. DM Ford, GS Heffelfinger. Molec Phys 94: 673–683, 1998.
34. AP Thompson, GS Heffelfinger. J Chem Phys 110: 10693–10705, 1999.
35. S Weerasinghe, BM Pettitt. Molec Phys 82: 897–912, 1994.
36. CG Lynch, BM Pettitt. J Chem Phys 107: 8594–8610, 1997.
37. T Çagin, BM Pettitt. Molec Simul 6: 5–26, 1991.
38. J Ji, T Çagin, BM Pettitt. J Chem Phys 96: 1333–1342, 1992.
39. T Kuznetsova, B Kvamme. Molec Phys 97: 423–431, 1999.
40. T Kuznetsova, B Kvamme. Molec Simul 21: 205–225, 1999.
41. D Fincham. Molec Simul 8: 165–178, 1992.
42. S Nosé. Progr Theor Phys 103: 1–46, 1991.
43. WG Hoover. Phys Rev A 31: 1695, 1985.
44. D Di Cola, A Deriu. J Chem Phys 104: 4223–4232, 1996.
45. MP Allen, DJ Tildesley. Computer Simulation of Liquids. Oxford: Clarendon Press, 1990.
46. A Laaksonen. MCMOLDYN. Daresbury Laboratory, UK, 1995.
47. B Widom. J Chem Phys 39: 2808–2812, 1963.
48. B Widom. J Chem Phys 86: 869–872, 1982.
49. Y Tamai, H Tanaka, K Nakahashi. Fluid Phase Equil 104: 363–374, 1995.
50. AP Lyubartsev, AA Martsinovski, SV Shevkunov, PN Vorontsov-Velyaminov. J Chem Phys 96: 1776–1783, 1992.
51. FA Escobedo, JJ De Pablo. J Chem Phys 103: 2703–2710, 1995.
52. TV Kuznetsova, PN Vorontsov-Velyaminov. J Phys: Cond Matt 5: 717–724, 1993.
53. S Kumar. J Chem Phys 97: 3550–3556, 1992.

# 20

# Heavy Hydrocarbon Emulsions
## Making Use of the State of the Art in Formulation Engineering

**Jean-Louis Salager, María Isabel Briceño,* and Carlos Luis Bracho**
*Universidad de Los Andes, Mérida, Venezuela*

## I. INTRODUCTION

### A. Heavy Crude Oil Reserves of the Planet

Surface tar deposits have been known since ancient times, particularly in what is now the Middle East. In the Genesis, God advised Noah to caulk the wood shell of his vessel with bitumen before the Deluge. Again in Genesis it is said that the Babel tower was made with bricks and asphaltic concrete. Asphalt was collected floating on the high-density waters of the Dead Sea. It was used in Egyptian embalming procedures, particularly in the wrapping of mummies, a word that comes from "mum", tar in Iranian (1). Asphalt comes from a Greek root, bitumen from Latin. Both terms refer to the heavy fraction of this oil that comes from the earth (*erdöl* in German) or from the stones (petroleum in other Western languages).

Before the 1973 oil embargo, most oil production came from light to moderately heavy crude oil fields. Little interest was shown in tapping the so-called extraheavy crudes, bitumens, or other low hydrogen-to-carbon ratio substances found in tar sands or shales. With the 10-fold increase in petroleum price in the 1970s, energy-resource alternatives started to bloom, from sugar cane fermentation to devices able to harness energy from winds or tides.

**Previous affiliation*: INTEVEP, Los Teques, Venezuela

Some ventures headed for known hydrocarbon deposits. The 70% or more of oil that remained trapped in the reservoirs after secondary recovery could be retrieved by injecting surfactant solutions or supercritical carbon dioxide. On the other hand, known for 50 years but previously discarded, very heavy hydrocarbon deposits were given a second look. Thus, it may be said that extraheavy crude oils were not really discovered but uncovered in the late 1970s and early 1980s.

The usually accepted terminology (2, 3) distinguishes between "conventional crude oils", "heavy", and "extraheavy" crudes, the last being more dense than water. The words "bitumen and tar" do not signify any real difference from extraheavy crude oil. They have been used for convenience, particularly in Venezuela, to point out that these substances are not to be classified as petroleum, a suitable semantic difference for stating that these hydrocarbons do not compete directly on the conventional petroleum markets. Separation of tar-sand bitumens from extraheavy crude oils on the basis of their viscosity in reservoir conditions has been proposed, with the dividing split at 5 or 10 Pa.s. According to this, extraheavy crudes flow under reservoir conditions, while bitumens do not. However, the real difference is that tar-sand deposits cannot be recovered by conventional oilfield pumping but require surface mining techniques, and this may be as a result of external factors like the prevalent climate or reservoir depth.

According to Smith (2), a classification may be based on the aromatic and resins content of the fraction with a boiling point above 370°C. This makes more sense because it has to do with the structure and hence the history of the field (3). Most of the extraheavy crude oils and bitumens were trapped at low depth in their migration to the surface, and were slowly biodegraded. These huge deposits are thus the leftover of a microorganism banquet that lasted several million years.

Extraheavy crude oil and bitumen world reserves are estimated to be about 100 to 150 billion tons, depending upon the recovery factor used. In any case, and in spite of the well known subjectivity in estimating reserves, this amount is comparable with the current estimate of the world conventional oil reserves, including the undiscovered ones, as seen in Fig. 1.

Two countries, Canada and Venezuela, hold over 40% each of the total extraheavy hydrocarbon reserves. Alberta tar-sands fields in western Canada, particularly the mammoth Athabasca deposit, contain over 250 billions tons of bitumen which has to be produced by surface mining because it is not fluid under reservoir conditions, owing to the cold climate and the low depth (100–150 m). Because of relatively adverse conditions, "only" 40 billion tons can be recovered by current technology. The Orinoco Oil Belt in eastern Venezuela contains a staggering 200 billion tons of extraheavy crude oils, from which 40 billion ton reserves can be tapped by widespread oilfield techniques. Other reserves are found in countries of the former USSR, USA, and China. These figures are to be compared with Saudia Arabia's 36 billion tons of conventional crude oil reserves and South Africa's coal reserves which are equivalent to 37 billion tons of petroleum.

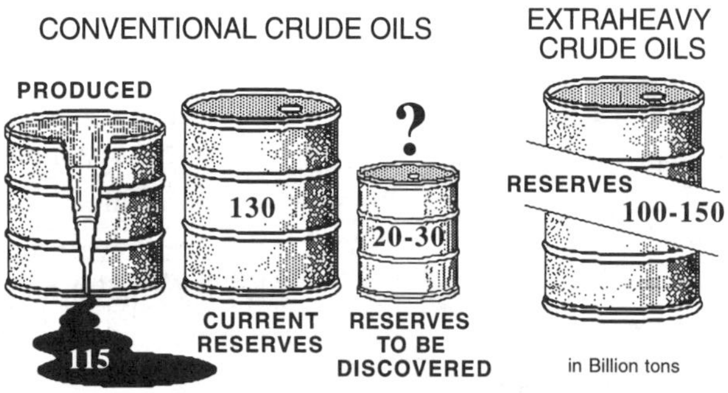


**Figure 1** Estimates of world conventional and extraheavy oil reserves (numbers as $10^9$ tons).

Since the current petroleum prices have almost returned to pre-embargo level (in constant dollars), it could be thought that extraheavy crudes are no longer an attractive source of energy. This is not exactly the case, and it seems that extraheavy crudes have a niche for cheap electrical energy generation, as well as for the manufacturing of synthetic light crudes, particularly if the conventional crude oil production starts declining at the end of this decade as forecast by experts (4, 5).

Electrical energy generation will increase with world economic growth, and the current trend and distribution forecast according to the International Energy Agency are indicated in Fig. 2. Coal is probably going to stay at a 40% level of all resources because it is cheap and abundant. Hydraulic resources would not grow more than the overall trend, while the share of nuclear energy would probably decline because of public rejection for such technology in most developed countries but France.

Natural gas use will be pushed to increase, particularly if prices stay low, a scenario which is not warranted, however. On the other hand the petroleum share that crested in 1970 at about one billion tons and decreased considerably after the 1973 and 1979 oil price hikes, will still go down, particularly because the current world oil production (3.6 billion tons per year) is probably near its all-time maximum and will decline by the end of this decade (6).

Under these circumstances, extraheavy oils may be an advantageous coal substitute if they can be produced at competitive prices.

## B. Orinoco Oil Belt Extraheavy Crude Deposit

The Orinoco Oil Belt is located in eastern Venezuela, along the Orinoco river's north shore. It extends 700 km from east to west and about 100 km from north to south, and contains numerous extraheavy crude deposits, among them the Cerro Negro and Hamaca fields, which are currently in production (7). Table 1 indicates some typical characteristic values (8–11). It is worth remarking that these low API gravity extraheavy crudes are not actually very viscous under reservoir conditions.

Overall conditions are quite favorable for oil exploitation. The climate is of a subtropical savanna type with an average day temperature around 30°C, which is of course quite higher than in Canada, and with little seasonal variations. Reservoirs are not deep (typically

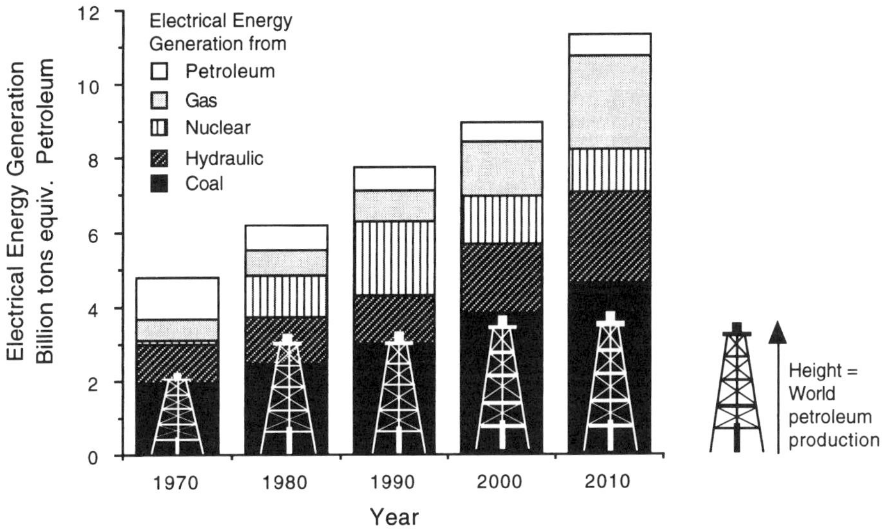


**Figure 2** Share of different fuels in world electrical generation.

less than 3000 ft), but still warm enough (typically 55°C) to maintain a reasonable viscosity (often less than 1 Pa.s).

Unconsolidated sand reservoirs exhibit a high porosity (30%) and a very high permeability (5–10 Darcys) that facilitates the recovery. As far as the production has gone, there is no evidence of significant compaction or soil subsidence.

Steam injection might not be necessary because of the existence of an unusual cold production mechanism called foamy solution gas drive, which results in a porous medium flow rate up to 10 times higher than predicted by Darcy's law, and an anomalously high final recovery, e.g., 15%. These features were found in heavy crude oilfields both in Canada and Venezuela, and appear to be highly favorable behavior of those deposits, though the main reasons for them are still unclear (12–18). As depletion takes place, very tiny gas bubbles start forming and do not coalesce, contrary to what happens usually in most oil production situations. Indeed, most apolar liquid foams are unstable because there is no polarity switch at the gas–liquid interface, and thus no possible surfactant action. In the case of extra heavy oils, there exists a very efficient stabilization mechanism, which is likely due to asphaltene deposition on their surface, resulting in some kind of bubble encapsulation. In spite of a low gas-to-oil ratio, the oil becomes a foam similar to a "chocolate mousse", and is able to flow, often with suspended sand particles. The actual foam viscosity is low, which is a rather paradoxical result, since the presence of dispersed phase fragments normally tends to increase viscosity. Some authors (15) have suggested that asphaltene deposition at the bubble surface results in a considerable asphaltene abatement in the liquid, so that the most important consequence is the associated decrease in viscosity of the continuous phase. Others suggested that the moving suspended sand particles produce tiny tsunami waves that push the foam. None of these explanations is fully satisfactory and a lot of research is still required to elucidate this bizarre and controversial phenomenology.

**Table 1** Characteristics of Crudes from Orinoco Oil Belt

| |
|---|
| Typical gravity 8–12°API |
| Viscosity 500–2500 mPa.s at reservoir $T$, $P$ |
| Down hole pressure ($P$) = 700–1300 psi at $T$ = 50–70°C |
| Gas-to-oil ratio 100–150 $ft^3$/barrel |
| Saturated 10–20%, Aromatics 40–50%, Resins 30–40%, Asphaltenes 10–20% |

*Source*: Refs 8–11.

Whatever the reason for this behavior, field and laboratory data indicate that the foamy solution gas drive regime is attained only at high drawdown pressure. The current state of the art indicates that a high depletion rate is required to trigger and maintain this mechanism, and that it should be applied early in the production history. The confirmation of these trends would probably compel producers to apply shorter well-spacing patterns.

In Venezuela, extraheavy crude oils are lifted by injecting a light diluent down-hole, a technique suitable for carrying out a complete desalting prior to emulsion manufacturing. Dilution can make long-distance transport easier as well, and the economy of the diluent recycling process seems to be favorable in some cases. Incidentally, dilution is used as well in Canada for pipelining the produced bitumen (19).

Research is likely to improve upon the currently highly favorable situation. Nevertheless, it can be said today that extraheavy crude oils are easy to produce, in spite of their high viscosity. Operators have not released their actual production cost, which is likely to go under 2 US dollars per barrel for large-scale projects, at least with mild outside temperatures as in Venezuela. This is an extremely important driving force behind the development of extraheavy oil ventures in Venezuela. Comparatively, the current surface-mining techniques for extracting tar sands in Canada results in a bitumen production cost of around 4–5 US dollars per barrel (3, 19).

### C. Alternate Markets for Extraheavy Crude Oils

The first natural market for extraheavy oils is electricity generation, in which the competition would be mostly with coal and marginally with residual fuels coming from oil refining. Since the cost structure is highly favorable, the limiting factors are essentially technological ones. As far as pollution abatement is concerned, extraheavy oils would not fare worse than coal. Thus, a main technological advantage over solid coal could be attained by conditioning these extraheavy oils into an easy to handle liquid fuel such as an emulsion.

Because of their low production cost and huge availability, extraheavy oils could be a suitable prime material for deep conversion processes tending to improve the hydrogen/carbon ratio to make a synthetic crude. Several technical alternatives are at hand and the decision is essentially a matter of cost and market opportunities that are expected to shift in favor of heavy crudes at the end of this decade or sometime later when world oil production will irremediably decline.

Since hydrogenation processes are likely to be ruled out because they require the operation of extremely high-pressure reactors, atmospheric pressure processes, such as delayed coking, are likely to prevail in the production of synthetic light crudes at a low price whenever a coke by-product can be used. This is, by the way, the conversion technology which is currently used to manufacture over 400,000 barrels/day of synthetic crude in Canada (20).

More technologically demanding processes like aquaconversion (21, 22), i.e., a catalytic modern version of the nineteenth century water–gas reaction, could lead to the production of cheap synthetic crudes with little pollution concern. It is not known whether one day it may be applied to the conversion of bitumen.

Nevertheless, all these chemical-transformation processes are ventures that require large-scale investments. For this reason they were not the first to be developed by the national oil company, Petróleos de Venezuela (PDV), which rather decided in favor of less capital intensive, but technologically more hazardous, emulsified fuel manufacture. On the other hand, more than 60% of the current Canada bitumen production (25 million tons per year) is upgraded into light synthetic crude (19).

## II. BITUMEN EMULSION IN VENEZUELA – ORIMULSION®

### A. History of Extraheavy Crude Oils Parallels Development of Orimulsion®

One of the first countries to undertake a serious program of exploitation of heavy and extraheavy oil reserves, in the late 1970s, was Venezuela. The motivation behind this enterprise was almost independent of the prevalent oil prices at that moment. On the one hand, Venezuela's heavy and extraheavy oil reserves were assessed at that time as being 10-fold the amount of conventional oil reserves. On the other hand, extraheavy crude oils were classified as bitumens or nonoil products according to OPEC standards, and would not be considered in the production quota. The last two conditions were the impelling forces that produced intensive exploration in the Orinoco Oil Belt, a huge area of 50,000 $km^2$, north of the Orinoco river.

One of the first problems to be addressed was how to transport the viscous crude oil from the field to the coast in which deep-conversion refineries or shipping terminals had been planned. It was clear that an innovative solution was required and a decade long research program was initiated, carried out mainly by INTEVEP, a research and development subsidiary of PDV. Joint ventures with the British Petroleum research center were undertaken along with the sponsoring of oriented basic research in national universities such as the Universidad de Los Andes and the Universidad Central de Venezuela.

## B. Transportation Alternatives

The problem of oil transportation was not a small undertaking. Typically, an extraheavy crude oil exhibits a viscosity from $10^3$ to $10^5$ mPa.s at the pumping temperature (8–11), as shown in Fig. 3. In this sense, any transportation scheme had to reduce the viscosity as an unavoidable requirement (23). On a first approach, conventional schemes were first considered. Reasonably enough, heating was the obvious primary alternative (8, 9, 24, 25). In such a scheme, interspersed pumping and heating stations would be required in order to maintain a low viscosity. However, it was foreseen that prolonged pumping interruptions, which were bound to occur, would end up with a cold crude oil, which would result in pipe plugging. Insulating hundred of kilometers of pipe was considered exceedingly expensive. On top of all this, heating the crude oil would consume a considerable amount of fuel, and would alter the economics.

The second alternative was the dilution with lighter crude oils or refinery cuts such as gas oil or kerosene (8, 26). Dilution would answer not only the viscosity problem but also the dehydration and desalting of these, very often, heavier-than-water crude oils, facilitating the process and enabling conventional units to be used for the purpose. This method, attractive as it was, encountered a major drawback. The large amount of diluent required was not readily available in the production area and, in any case, the diluent would have to be recovered at the pipe end and pumped back to the field hundreds of miles away. In most cases more valuable hydrocarbon fractions or light crude oils would be sacrificed in the process.

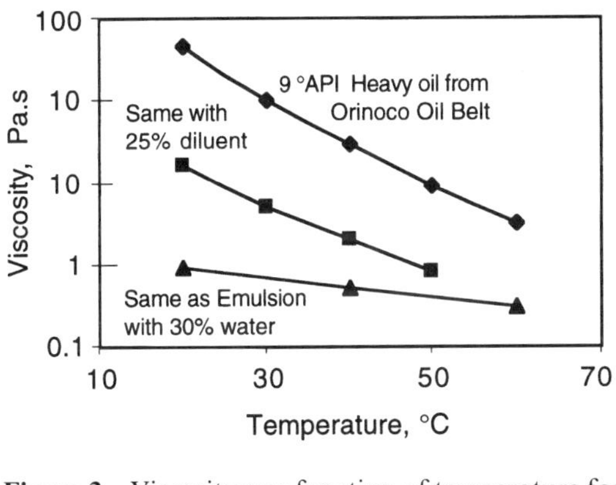


**Figure 3** Viscosity as a function of temperature for a 9° API extra heavy crude oil, its mixture with 25% diluent, and its emulsion with 30% water content.

The third alternative and, seemingly, the most cost-effective scheme was the transportation of a crude oil-in-water emulsion (10, 11). This idea was very attractive since only 30% of water, a low-cost commodity, can generate a much better result than dilution, as shown in Fig. 3. In contrast with the previous alternatives, there were few antecedents of emulsified transportation. Hence, it was clear that further research and development was necessary.

The background research in the field of multiphase transport started more than 40 years ago with the study of two-phase flow (27–31), mostly gas and liquid, that exhibits different regimes from stratified (or separated) to annular flow, passing through bubbly and slug flow.

When the fragmented-matter particle size is relatively large, as in an unfluidized bed or when the density difference is large, as in a bubbling column, heterogeneous patterns occur such as slugging, chaneling, spouting, or wave formation (27). In the case of fine droplet emulsions (which are neither large compared with the pipe diameter, nor exhibit a large density difference between fluids), these concerns were forgotten, maybe too quickly, since there is now evidence that some kind of segregation process could take place near the wall, as will be discussed later.

Generally, emulsion or suspension flow was treated as in the case of a homogeneous fluid, and the only issue addressed was to estimate the rheological characteristics of the dispersion, either Newtonian or nonNewtonian.

The purpose of all transportation processes was to reduce the pressure drop, by replacing a viscous or quasisolid hydrocarbon by a less viscous O/W emulsion either for pipeline transportation purposes of waxy (32, 33) or heavy crude (34–42), or by conventional crude in very cold conditions (43), or petroleum production by downhole emulsification (44–47).

Some systematic studies indicated that the previously known dependence of emulsion rheology on internal phase content and drop size characteristics could be applied to petroleum-in-water emulsions (48). However, most early studies were not concerned with the presence of stabilizing substances like surfactants (49–51), so that they were missing an extremely important issue in practice.

## C. Development of Emulsified Transportation – Orimulsion®

### 1. Early Stages

In the late 1970s, the aim of PDV (7) was to transport the extraheavy crude oil to refineries located at least 200 km away, where it would be converted into lighter fractions. At that time, oil prices had reached unprecedented high levels and the whole scheme was considered feasible. On a first approach, laboratory and pilot plant tests were conducted (52–55), aimed at testing the feasibility of transporting a sufficiently stable emulsion, with a typical crude oil content of about 65%. It has to be noted that with very few exceptions (56–58) no attention was paid to the physicochemical formulation. This is understandable since little information was available at that time of the effect of the many variables involved in the formulation, mixing, and rheology of emulsions.

The approach was quite simple. Trial and error tests were conducted in pilot plants, eventually leading to the selection of a convenient surfactant or surfactant package that produced an emulsion sufficiently stable to withstand shearing and storage from the field to the refinery. The emulsions were prepared in situ, by means of down-hole emulsification. A nonionic surfactant was the obvious choice since the formation water could be quite salty (53, 54, 59).

Crude reality put an end to this strategem (59). During the mid-1980s, oil prices declined considerably, turning deep conversion into a commercially unattractive business. Moreover, the investment capacity of PDV became limited by government policy. However, it was unthinkable in Venezuela not to do something to tap the enormous wealth of the Orinoco Oil Belt. It was clear that a different strategy had to be considered. An interesting idea then came up, which, in retrospect, was rather adventurous in the first place. Why not use the bitumen emulsion directly as a feedstock for power generation? Nevertheless, it did not seem easy to introduce a new fuel in an already saturated market.

In 1985, the transportation and the combustion research teams at INTEVEP combined their efforts and a preliminary round of combustion tests was carried out (10, 11). By 1986, it was established that the bitumen emulsion could be used as an efficient fuel (60, 61).

It was recognized that the combustion gases could be cleaner, on average, than that of coal burning (61). Besides, the amount of particulate emissions was drastically lower (61). An evaluation of the market showed that there were favorable conditions for a coal substitute to be introduced (61, 62). Additionally, the manufacturing of bitumen emulsions required four times less investment than deep conversion into synthetic crude.

This is how Orimulsion® was created (10, 63). Owing to the special characteristics of the product, the national oil company PDV created a new subsidiary, Bitúmenes del Orinoco (BITOR), which has been in charge of the manufacturing, quality control, and marketing of the product.

### 2. Emulsion Specifications

A decade of research and development gave birth to a commercial product whose typical specifications (10, 11, 61, 64) are shown in Table 2. Some of these specifications are rather fortuitous and depend on the bitumen characteristics. Properties such as the pour and flash point are rather debatable since they were established for homogeneous hydrocarbon products and not for heterogeneous, high water content fluids such as Orimulsion®. It seems that these two last properties were added in order to comply with prevalent fuel specifications even though they seem neither to make sense nor be really useful.

Other properties, such as water content and droplet size, are related to important emulsion properties such as stability, viscosity, and calorific content. Regarding water content, a reduced amount would be more convenient in order to increase the calorific value. Droplet

**Table 2** Typical Specifications of Orimulsion®

| Physical property | Approximate value |
|---|---|
| Water content (w/w%) | 27–30 |
| Density (g/cm$^3$) | 1.009–1.013 |
| Mean particle size (μm) | 8–15 |
| Apparent viscosity (mPa.s, at 30°C, 100 s$^{-1}$) | 350–570 |
| Pour point (°C) | 3–6 |
| Flash point (°C) | 65 |
| Gross calorific value (kcal/kg) | 7200 |
| Net calorific value (kcal/kg) | 6650 |
| Ash content (w/w%) | 0.13–0.25 |
| Sulfur content (w/w%) | 2.6–2.9 |
| Vanadium (mg/L) | 285–340 |
| Sodium (mg/L) | 14–65 |
| Magnesium (mg/L) | 350 |

*Source*: Refs 10, 11, and 63.

size should also be as small as possible to improve stability and enhance combustion. In effect, smaller droplets burn more efficiently, and the impingement of the combustion chamber is reduced. However, less water and smaller droplets would yield a much more viscous emulsion. Therefore, it was found during the development of the product that a trade-off between calorific value, stability, viscosity, and combustion efficiency was necessary in order to attain an optimum compromise. Finally, it is worth remarking that the magnesium content is not indigenous to the heavy crude oil, but is added to the aqueous phase as a corrosion inhibitor during the combustion process (10, 11).

### 3. Combustion Characteristics

As mentioned before, the historical merging of the transportation and combustion teams led to the development of an all-encompassing technological scheme called Imulsion™ (59, 60, 65). Once the idea was considered feasible, pilot tests were undertaken at INTEVEP pilot plants as well as at the Nagasaki Technical Institute in Japan (60). Several tests were also conducted at commercial facilities belonging to Combustion Engineering and New Brunswick Power in Canada and Babcock Power in the United Kingdom.

All of these tests were carried out in conventional boilers and similar conclusions were reached (61). The amount of ashes and flue-gas particles produced were many times inferior compared to coal, whereas $CO_2$ emissions were 20% lower, on the same energy output basis. $NO_X$ emissions also proved to be inferior, because of the lower flame temperature brought about by the presence of water, despite the higher levels of nitrogen existing in the heavy oil (61, 63). Regarding combustion efficiency, the heavy-oil emulsion performed much better than coal and similarly to conventional heavy fuel oil.

However, there were two unfavorable aspects. First, the amount of $SO_x$ emissions was significantly larger than that of commercial heavy fuel oil and low sulfur coal. Second, the high content of heavy metals such as vanadium could produce severe corrosion in the boiler chamber (61, 63). Fortunately, both problems could be corrected by means of already existing technologies, such as flue-gas desulfurization, and by the addition of small amounts of magnesium salt. This last forms a noncorrosive solid precipitate with vanadium that can be easily removed.

Herein after, Orimulsion® has been tested in various commercial systems such as boilers, diesel engines, steam turbines, and cement kilns. New firing technologies have also been evaluated, namely, gasification and reburning processes. The latter process has been proven to deliver an advantageous $NO_x$ emission control. Orimulsion® was also found to perform better than coal and, regarding combustion efficiency, better than conventional heavy fuel oil, for diesel engines and cement kilns (61, 63, 64).

Orimulsion® produces an amount of particulate emission larger than that of heavy fuel oil, chiefly due to the addition of magnesium salt. However, conventional electrostatic precipitators could be used to achieve over 90% removal efficiency (60, 61).

### 4. Environmental Aspects

In addition to the indispensable technology for producing clean flue gases, the control of eventual sea spills has been a matter of concern and serious evaluation (66). Orimulsion® is transported in double-hulled tankers (11), a feature which is thought to curtail drastically the probability of a spill.

Nonetheless, PDV has evaluated the characteristics of a sea spill, both in the laboratory and in field tests (64, 66). It has been found that, as soon as the product contacts the water, it disperses to form a 2 to 3 m deep floating plume located a few centimeters below the surface. Consequently, the extension of the spill is smaller than the one produced by conventional oils, and prevailing wind drift has almost no effect on emulsion spill in contrast to oil spill.

A special recovery process has been designed, which consists in retaining the spill by means of an extended skirted boom, and pumping the affected water column to a ship. Once in the ship, the fluid is conveyed to a flotation tank in which the coalescence of the oil droplets is promoted through a coagulation process called the forced adhesion and flotation system (66).

## D. Marketing of Orimulsion®

BITOR marketing strategy was, from the very beginning, to present the product as a convenient substitute for coal for thermoelectrical plants. The commercial conditions were quite favorable. Environmental regulations were becoming increasingly more demanding, forcing coal-firing plants to envisage expensive flue-gas treatments. Technical and economical evaluations proved that the conversion of coal plants to the burning of Orimulsion® could result in lower investment costs than those necessary for a cleaner coal combustion (61, 63).

Moreover, lower stable prices could be guaranteed since the Orimulsion® is classified as a "nonpetroleum" product according to OPEC standards, and is not included in the production quota. Consequently, Orimulsion® has been, slowly but surely, penetrating the energy market. Exports of the product, which started in 1990, have reached in 1998 more than 4 million tons a year to a variety of countries such as Canada, Japan, China, Denmark, Italy, and Lithuania (63, 64).

### E. Further Benefits Acquired as a Result of the Know-how

The applied research carried out to develop Orimulsion® was extensive and had quite an impact on several other technological issues such as:

- The possibility of transporting many viscous crude oils, from high pour-point paraffinic crude to not so heavy crude in a cold climate, as O/W emulsions.
- The understanding of the rheological properties of concentrated emulsions and the outstanding advances in formulation engineering allow one to use emulsions in other situations, e.g., for plugging high-permeability reservoir zones with the purpose of improving water or steam injection, as well as in many other applications found outside the petroleum production business.
- Extensive work on combustion of water–oil mixtures showed enhanced performance features, particularly for pollution abatement. Not only can the O/W conditioning be considered for other cases of viscous products such as refining residues, but also the introduction of water as a W/O emulsion can be beneficial for other (nonviscous) hydrocarbons. In all these cases the lower temperature favors the displacement of the water–gas shift reaction and the reduction of nitrogen oxide production, which are particularly annoying in high-altitude cities like Mexico, Bogota, or La Paz.
- The emulsified oil-in-water conditioning allows for a large interfacial area of contact between the two fluids, which is a favorable situation for the biotreatment of heavy fractions for removal of sulfur or metals, or for a two-phase chemical desulfurization.

Although most of the development work that led to the current commercial product is still confidential, many basic concepts and unveiled technological breakthroughs have contributed to the current state-of-the-art in emulsion-formulation engineering.

## III. EMULSION-FORMULATION ENGINEERING – STATE OF THE ART

Emulsions have been prepared for quite a while, and the basic "know how" was described several decades ago in books by Becher (67) and Sherman (68). At that time, emulsion manufacturing was an art, and many formulation and manufacturing procedures were jealously kept as whimsical recipes to be passed on from father to son. Since then, many investigations have broadened the phenomenological knowledge in emulsion science, which is reported in reviews (69–73), but there is still a lot of know how that comes from experience, particularly as far as the relationship between formulation and properties is concerned. The following sections are dedicated to rationalizing this know how, which will be used next to carry out the formulation engineering of heavy crude-oil emulsions.

### A. Emulsion Properties

Emulsions can be found as two basic types, i.e., O/W and W/O, but in some particular cases, multiple or double emulsions labeled $W_1/O/W_2$ and $O_1/W/O_2$ also occur. The emulsion type may be determined by different methods. In most applications the aqueous phase contains one or various electrolytes, and thus conducts electricity somehow, whereas the oil or organic phase does not. Consequently, the measurement of electrolytic conductivity is a handy way to ascertain the emulsion type. Moreover, the continuous monitoring of the electrolytic conductivity allows the determination of the change in emulsion type which is referred to as emulsion inversion.

Figure 4 (left) indicates (from left to right) the variation of conductivity of an O/W emulsion when increasing amounts of oil phase are added. The starting high value $\kappa_w$ corresponds to the aqueous phase conductivity. Then, as oil phase is added (under constant stirring to keep the system emulsified), the conductivity of the emulsion decreases, according to an almost linear variation with respect to the external water phase fraction $f_w$, which may be expressed as

$$\kappa_{em} = \kappa_w (f_w)^m \qquad (1)$$

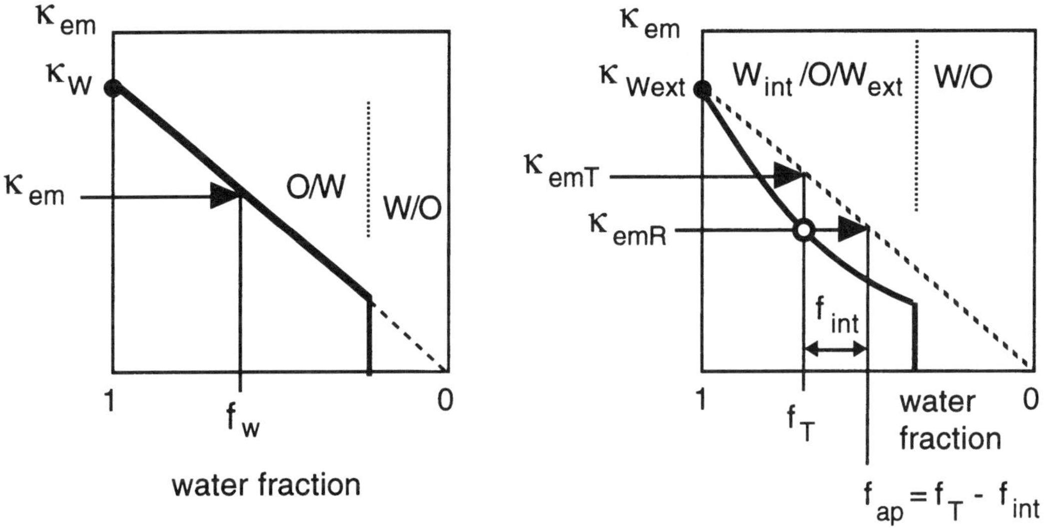


**Figure 4** Emulsion electrolytic conductivity variation with the water content. Simple (left) and multiple (right) emulsion cases.

where exponent $m$ is close to unity, eventually slightly higher (74–77). For instance, Bruggeman's relationship (74), applied to the case of a nonconductive internal oil phase, leads to a good experimental fit for O/W emulsions when $m = 1.5$. Nevertheless, it is worth noting that the difference from the linear approximation ($m = 1$) is not very significant. An exhaustive review (77) on the subject is available if more details are required. In most cases it may be assumed as a first approximation that $m$ is unity, so that the variation essentially follows a straight line pointing toward zero conductivity at pure oil content ($f_w = 0$) (78). This trend is interrupted at some point (inversion) at which the conductivity drops to zero. Actually the value is not zero but is equal to the oil phase conductivity, which is 100-or 1000-fold lower than the aqueous phase conductivity. As long as the conductivity follows the straight-line variation, the emulsion is O/W, while it is W/O when the conductivity becomes essentially nil. Deviations from this pattern are found, as shown in Fig. 4 (right) when multiple emulsions are formed, as in the present case $W_{int}/O/W_{ext}$ in which $W_{int}$ is the phase which is dispersed as droplets inside the oil drops, and $W_{ext}$ is the external aqueous phase.

In such a case, the conductivity of a multiple $W_{int}/O/W_{ext}$ emulsion decreases more rapidly than expected from the straight-line extrapolation between the conductivity of the external phase ($\kappa_{Wext}$) and zero. At the white circle, the real conductivity of the emulsion is $\kappa_{emR}$ while it is expected to be $\kappa_{emT}$ according to the total $f_T$ proportion of water in the emulsion. The conductivity value $\kappa_{emR}$ corresponds to an O/W emulsion containing a lower proportion of water which is called apparent water proportion $f_{ap}$ (Fig. 4 right). The difference, i.e., $f_{int}$, is the amount of water in the form of $W_{int}$ droplets inside the oil drops. Since this method depends on the accuracy of the conductivity measurement, which is often no better than 2 or 5%, it is used to detect the presence of multiple emulsion, rather than to measure the internal phase proportion $f_{int}$. In any case, use of the Bruggeman formula would be preferred for the accurate detection of a multiple emulsion by conductivity measurement.

Once the emulsion type is determined as one of the two simple cases, i.e., O/W or W/O, the emulsion structure is characterized by its drop size, or more exactly by its drop size statistical distribution. As a matter of fact, this is quite logical since most emulsions are made by a stirring process that often involves turbulence and thus random effects. Moreover, the emulsion is the result of opposite phenomena, i.e., drop breaking and coalescence, that cannot be described but in some statistical fashion. It may be said that the drop size distribution is the fingerprint of the emulsion.

The purpose of this section is not to analyze drop size statistics. Therefore, only basic features are discussed here. Depending on the way the emulsion is manufactured, particularly the fluid mechanical conditions in which the shear or turbulence have produced the droplets, the emulsion should contain drops of similar or very different sizes, with the associated variety in statistical distribution. Figure 5 illustrates the aspect of typical drop size distribution which is labeled

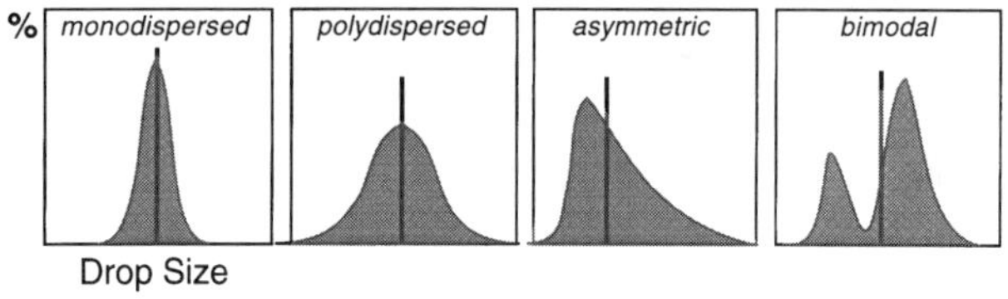


**Figure 5** Different shapes of drop size distribution.

as monodispersed or polydispersed, symmetrical or asymmetrical, unimodal or polymodal. These adjectives will be used later in the discussion since some of the emulsion properties depend on the qualitative and quantitative features of the emulsion drop size distribution.

An emulsion is usually characterized by its average diameter, a property that should be representative of all drops. If the emulsion is monodispersed such a characterization is warranted. It is not the case, however, when the distribution is flattened and when it is not symmetrical, because one part of the distribution could play a role if some particular property is concerned, while another part could alter another property. This is particularly critical when the distribution exhibits two or more modes, i.e., when the emulsion is a mixture of emulsions. For instance, it is seen that there are almost no drops with a diameter corresponding to the average diameter (position of the mean is indicated as a black line) of the bimodal emulsion in Fig. 5.

Maybe the paramount property of an emulsion is its viscosity. As a conditioning vehicle, emulsions are often required to be viscous, as in mayonnaise or paints, or contrariwise, to be the least viscous as possible as in heavy hydrocarbon fuel emulsions. Emulsion viscosity depends on many variables (79–83).

It may be said that the effects of the physical variables are well documented, though not completely elucidated for high internal phase ratio emulsions, which are very viscous and nonNewtonian. On the contrary, the basic formulation effects have been uncovered in the past 20 years, but many complex formulation-related phenomena still remain unclear.

Emulsion viscosity is set to be proportional to its external phase viscosity, and this assumption is obviously correct at low internal phase content, say up to 20–30% and often at higher content as long as Newtonian behavior is exhibited. The second most important factor related to emulsion viscosity is the internal phase content, i.e., the volumetric proportion of the drops. As increasing numbers of drops crowd the emulsion external phase, the interdrop interactions produce increased friction that results in escalating viscosity. Although the rise of the emulsion viscosity with internal phase ratio cannot be described by a simple equation because it also depends on other factors, it is known to increase in some exponential fashion above 50–60% of the internal phase (79, 80).

Many empirical formulas have been proposed to render the effect of internal phase ratio on the emulsion viscosity, but they are only valid in specific cases. Pal and Rhodes (84, 85) proposed and used a semi-empirical equation, that makes use of experimental data $\phi_{100}$ as the internal phase fraction $\phi$ at which the relative viscosity $\eta_r = 100$. This experimental value must be attained in the same formulation and emulsification conditions, particularly stirring characteristics, which is maybe why it significantly embodies the overall effects of all remaining variables:

$$\eta_r = \left[1 + \frac{\phi/\phi_{100}}{1.187 - \phi/\phi_{100}}\right]^{2.49} \tag{2}$$

Note that the occurrence of nonNewtonian behavior means that the viscosity concept is no longer valid and that it has to be replaced by the concept of apparent viscosity. It is found that the rheological behavior of many concentrated emulsions may be rendered, at least approximately, by a power law variation in which the shear stress $\tau$ is proportional to the $n$th power of the shear rate $\dot{\gamma}$:

$$\tau = k\dot{\gamma}^n \quad \text{and} \quad \frac{\tau}{\dot{\gamma}} = \eta_{ap} = \mathrm{k}\dot{\gamma}^{n-1} \tag{3}$$

where $k$ is the consistency index, which is the viscosity for Newtonian fluids ($n = 1$) and the apparent viscosity $\eta_{ap}$ at 1 $s^{-1}$ shear rate for nonNewtonian fluids.

In concentrated emulsions the fluidity index $n$ is less than unity, which is an indication of shear thinning or pseudoplastic behavior. Figure 6 indicates the shear stress–shear rate rheogram for O/W emulsions with different oil contents and similar drop size. The straight line variation is indicative of the compliance with the power-law model. The slope, i.e., the fluidity index $n$ in this plot, is essentially unity up to 50% internal phase, then it tends to decrease as the internal phase content increases (black lines with no data points in Fig. 6).

The friction is somehow due to the contact between drops, and it is thus related to the surface area of the drops, and consequently to their size. As a general rule of thumb the smaller the drop size, the higher the viscosity. The shape of the drop size distribution is important as well. The more polydispersed the distribution, the less viscous the emulsion, with all other characteristic parameters being equal. Bimodal emulsions are

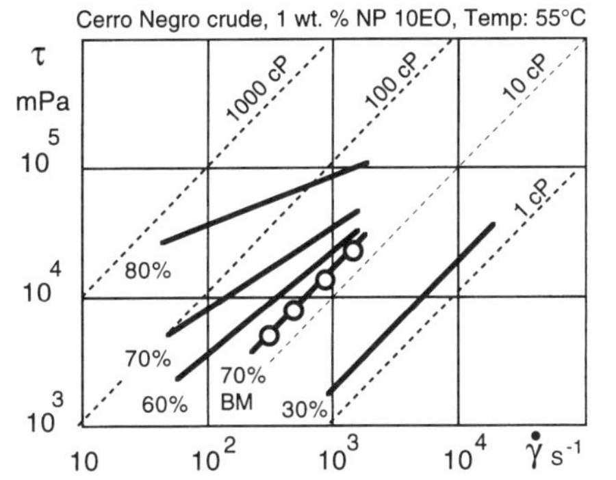


**Figure 6** Rheological diagrams (shear stress vs. shear rate) for heavy crude oil-in-water emulsions with different oil contents indicated in %. Data points correspond to a bimodal emulsion with 70% oil content.

special type of highly polydispersed emulsions whose viscosity is much lower than expected from their internal phase content. In Fig. 6 a bimodal emulsion prepared with a 70% oil content (line with white circle data points labeled 70% BM) exhibits Newtonian behavior ($n = 1$) and a much lower viscosity than its monomodal counterpart, a feature that will be used in a later discussion.

The concept of viscosity has been introduced and developed with homogeneous fluids, Newtonian or not. In case of an emulsion some phenomena can take place at the scale of a drop size, and the homogeneity assumption is no longer valid. For instance, it is known that in most cases of fluid transport in the presence of surfactant, there is some slipping velocity resulting from the electrical double-layer effect called the streaming potential (86).

As a consequence, the boundary condition at the wall is likely to be different from the vanishing velocity, as usually taken. The presence of different drop sizes can also result in drop segregation because of the so-called depletion interaction (87, 88), a complex mechanism of an entropic nature, which has been used to manufacture monodispersed emulsions (89). Another type of segregation can result from the variation in shear from the wall to the bulk of the liquid, and results in a drop depleted layer near the wall and a lubrication effect (90). These phenomena are not mastered yet but may prove to be either favorable or annoying in applications, and a lot of research work will have to be dedicated to them before they can be tackled.

Last but not least, the influence of physicochemical formulation on viscosity or apparent viscosity has been shown to be a determinant in some situations as will be discussed later on.

Another important property is the emulsion stability, i.e., its resistance to any change, which could be either a slow decay through drop coalescence and thus drop size increase, or a quick instability as in emulsion inversion (70, 71). When using the word stability, it is necessary to specify stability against which kind of perturbation. It is not the same to keep an emulsion at constant temperature at rest on a shelf, as to handle it through a process involving temperature cycles, high shear pumping, ocean navigation, or other different manipulations.

Emulsion break up is an ineluctable consequence of thermodynamic laws, but it can sometimes be delayed by kinetic means for a considerable length of time. The emulsion persistence depends on the occurrence and on the rate of the different mechanisms and steps that are involved in the decay (70–72, 91–93).

First, the drops' long distance approach, i.e., at several times their size, takes place according to different driving forces ranging from gravity pull, which produces sedimentation and creaming (94), to Brownian motion and drop collisions (95, 96).

As sedimentation proceeds, a cream can form and drop aggregation and disproportionation (Otswald ripening) can take place (97). The sedimentation step terminating the approach of neighboring drops ends up in a complex dampening process, which may leave a thin film between the drops (98). Numerous factors, from external phase fluid properties to interfacial properties and dynamic effects, can influence the thin film drainage. This is a very active area of research and many of the complexities have been understood (71, 72, 98, 99). However, these phenomena are too complex and intricate to be amenable to straightforward formulator handbook recipes.

In the past, emulsion properties were seen from a physical point of view to involve forces and hydrodynamic motion in the interdrop thin film. Thanks to the physicochemical concepts developed for the enhanced oil-recovery processes, a newer physicochemical approach, based on a molecular description of the interaction between the interfacially adsorbed surfactant and the oil and water phases, is now available and will be emphasized here.

## B. Physicochemical Formulation Basics

The physicochemical formulation of even the simplest surfactant-oil-water (SOW) systems at equilibrium involves many different variables. The first to be described in some qualitative manner is the nature of the different components.

If the oil phase is an n-alkane, it is characterized by its number of carbon atoms (ACN = alkane carbon number). For other oils or oil mixtures it turns out to be more complex. Even for a simple oil phase containing three or four components, which could be viewed as a rather simplistic description of a crude oil, the number of variables required to describe the nature of the oil phase could be quite high. However, an approximated single variable estimate can be found, as discussed later.

The nature of the water phase has to do mostly with the type and concentration of the dissolved substances it contains. This may also require several degrees of freedom, at least two of which are the type and concentration of a single electrolyte.

At least two parameters are needed to describe the nature of the surfactant. Usually these parameters are some characterisics of the polar and apolar groups, for instance, the number of carbon atoms in the alkyl or alkylbenzene group, and the number of ethylene oxide groups (EON) per nonionic surfactant molecule. Many real-world formulations contain other additives as well, e.g., alcohols.

Finally, physicochemical effects are also known to depend upon temperature and pressure. However, and in most cases, pressure effects could be disregarded, unless there is a large amount of dissolved gas is any liquid phase, which could obviously be the case with live crude oils.

Consequently, it can be said that there is an overwhelming number of variables, so many that thousands of experiments could be necessary to make an even simplistic screening.

However, the huge research and development effort, which was targeted at enhanced oil-recovery techniques in the 1970s, has shown that the handling of this large number of degrees of freedom can be simplified by referring to the elementary situations related to the phase behavior of SOW systems at equilibrium. It corroborated something which was known (intuitively) in the past decades, i.e., the formulation directly influences the phase behavior and in some more indirect way the emulsion properties (which also depend on other factors).

The development of formulation concepts is a long story (100) with theoretical and empirical episodes. Almost a century ago, Bancroft stated that the emulsion type (and some of its properties) was related to the fact that the surfactant was more soluble in one phase than in the other.

Fifty year later, Griffin (101) introduced the HLB number in order to quantify this concept of hydrophilicity (water solubility) and vice versa. The introduction of the HLB method was quite a breakthrough at the time, and it has been used and overused ever since because of its extreme simplicity and its approximate correlation with many surfactant properties (102, 103). To appreciate the overwhelming proliferation of scientific literature on the topic, it may be remembered that Becher (104) dedicated more than 100 pages of his Encyclopedia to the mere listing of HLB-related papers.

Nevertheless, it is now understood that HLB essentially depends on the surfactant, while the phase behavior and emulsion properties are also related to the water and oil phase nature, as well as to the temperature (100). The temperature was the preferred variable in the case of nonionic surfactants which are very sensitive to it, and an experimentally based concept was first introduced by Shinoda to quantify the formulation, i.e., the phase inversion temperature (PIT) (105, 106). It is known that the hydrophilicity of a nonionic surfactant decreses when temperature decreases. In water solution there exists a temperature at which the surfactant is no longer soluble and thus produces a separate phase. This so-called cloud point occurrence is related to the Shinoda PIT, which is essentially the same phenomenon, but in the presence of an oil phase whose nature could facilitate this separation and make it happen at a lower temperature. Although the PIT is limited to the liquid water temperature range of nonionic surfactants, its introduction was an important milestone because it was related not only to the surfactant, but also to the whole physicochemical environment (107), a feature that was shown to be essential by Winsor.

Winsor's pioneering theoretical work (108) showed that the formulation concept could be described through a single parameter that gathered all effects. This parameter was the ratio of the interaction of energy of the surfactant with the oil phase, to the interaction energy of the surfactant with the aqueous phase. The original definition of Winsor $R$ ratio was then slightly modified to accommodate the net interaction per unit interfacial area, a change that makes reasoning easier to carry out, but does not alter the essential framework:

$$R = \frac{A\text{co} - 1/2A\text{oo}}{A\text{cw} - 1/2A\text{ww}} \tag{4}$$

$A$co is the interaction energy between the surfactant (c) and the oil phase (o) molecules, $A$cw is the interaction energy between the surfactant and water, $A$oo is the interaction energy between two oil molecules, and so forth. All interaction energies are expressed per unit interfacial area. Since the $R$ ratio handling is qualitative the 1/2 coefficients are not necessary. They are introduced here for a better fit with the definition of net interaction energies. $A$co and $A$cw involve one surfactant molecule and both an oil molecule and a water molecule, while the interactions when the components are separated, i.e., $A$oo and $A$ww, involve two molecules of oil, and two of water, hence the 1/2 coefficient.

Winsor's research showed that the phase behavior at low surfactant concentration, say from 0.1 to 5%, as usually used in applications, is directly linked with the $R$ value (108, 109).

For $R < 1$ a Winsor I phase behavior is exhibited in which a surfactant-rich aqueous phase is in equilibrium with an essentially surfactant-free oil phase. In this case, the interaction of the surfactant with the aqueous phase exceeds the interaction of the surfactant with the oil phase. In Winsor II phase behavior, that is attained for $R > 1$, the opposite occurs, i.e., the surfactant-rich phase is the oil phase, which is in equilibrium with an aqueous phase that contains essentially no surfactant, and the dominant interaction of the surfactant is with the oil phase.

When $R = 1$, the surfactant interactions with the oil and water phases are equal and the system splits into three phases in equilibrium: a microemulsion that contains most of the surfactant and that cosolubilizes large amounts of oil and water, and two excess phases that contain essentially pure oil and pure water (108, 109). This is a complex but not uncommon phase behavior situation, which is found as well in systems that do not involve surfactants (110–112).

Since $R$ varies with one or more changes in interaction energy, those factors that affect the $A$s are likely to change the $R$, and consequently the phase behavior. For instance, an increase in surfactant lipophilic group length tends to increase $A$co and thus increases $R$. If the starting $R$ value is less than one, and the final value is greater than one, or vice versa, then a complete phase behavior transition is exhibited. Similar transitions may be attained by changing any of the formulation variable that can affect any of the $A$s, i.e., essentially for all formulation variables plus temperature. Sometimes, the deduction is not straightforward. For instance, if the oil phase molecule is lengthened, the $A$co term increases, but the $A$oo term also increases to a greater extent, and thus the numerator of $R$ decreases.

The Winsor approach is extremely pedagogical and quite helpful in qualitatively determining expected trends, but it is not amenable to the numerical estimation of a formulation yardstick because the interaction energies cannot be calculated accurately.

Thanks to the enhanced oil-recovery research drive of the 1970s, a complex but accurate description of the physicochemical formulation is now available through the so-called surfactant affinity difference (SAD), which translates into numbers the Winsor conceptual approach. According to early thermodynamics texts, the affinity is the negative of the chemical potential. The surfactant affinity difference is defined as

$$\mathrm{SAD} = -\mu^\circ_{\mathrm{o}} - (-\mu^\circ_{\mathrm{w}}) = \mu^\circ_{\mathrm{w}} - \mu^\circ_{\mathrm{o}} \tag{5}$$

where $\mu^\circ_{\mathrm{w}}$ and $\mu^\circ_{\mathrm{o}}$ are the standard chemical potentials of the surfactant in the water and oil phase, respectively. It was first suggested that this difference could be a way of quantify the formulation effects as the deviation from the optimum formulation which was empirically established (113, 114).

This difference is the Gibbs free-energy variation that occurs when a surfactant molecule is transferred from the oil phase to the water phase, and it is thus a measurement of the relative affinity of the surfactant for both phases, just like the Winsor $R$. SAD can be linked to the partitioning coefficient of the surfactant, according to an early suggestion made by Davies on the basis of coalescence kinetics (115–118).

It was finally related to the empirical relationships found to describe the occurrence of an optimum formulation for three-phase behavior as a function of the formulation variables. A complete updated definition of SAD can be found in recent reviews (119, 120).

For ionic surfactant systems the definition of SAD as a function of the formulation variables is

$$\frac{\mathrm{SAD}}{RT} = \ln S - K\,\mathrm{ACN} - f(A) + \sigma - a_{\mathrm{T}}\,\Delta T \tag{6}$$

For ethoxylated nonionic surfactant systems:

$$\frac{\mathrm{SAD}}{RT} = \alpha - \mathrm{EON} + bS - k\,\mathrm{ACN} - \phi(A) + c_T\,T \tag{7}$$

where $S$ is the salinity, expressed in wt% NaCl with respect to the aqueous phase, ACN is a characteristic parameter of the oil phase, $f(A)$ and $\phi(A)$ are functions of the alcohol type and concentration, $\sigma$ and $\alpha$ are parameters characteristic of the surfactant structure, and EON is the average number of ethylene oxide groups per molecule of nonionic surfactant; $\Delta T$ is the temperature deviation measured from a certain reference (25°C), $b$, $k$, $K$, $a_T$, and $c_T$ are empirical constants that depend on the type of system.

The values of the coefficients are reported in the literature, in the publications related to the attainment of an optimum formulation (109), i.e., SAD = 0, with anionic (121, 122), nonionic (123, 124), and cationic (125, 126) surfactants, as well as complementary information to handle more complex cases of mixtures of surfactants (127–130), and the effect of alcohols (131), electrolytes (132, 133), pressure (134), and temperature (122, 123, 135).

If the oil phase is not an alkane, but behaves similarly to an alkane, it is characterized by its equivalent alkane carbon number or EACN (136). It has been shown that on an "apolarity" scale, cyclohexane EACN is 3, alkylcyclohexane EACN is equal to its alkyl group ACN plus 3, while benzene EACN is 0, and alkylbenzene EACN is equal to its alkyl group ACN. As a matter of fact, the more polar the oil the lower its EACN. For instance the ethyl oleate EACN is about 6. Since it contains a $C_{18}$ chain, this means that the ester group accounts for a 12-unit reduction in the EACN. Complex hydrocarbon mixtures can be assigned an EACN too, according to a simple mixing rule which is more or less followed (136, 137).

The EACNs of crude oils are often in the 8–14 range, a value that might look quite low in view of the molecular weight of their components, but which is because of the aromatic character of the heavy fractions (137). Crude oil EACN can be measured experimentally (137, 138).

It is worth noting that the SAD linear expression for all formulation variables indicates that their contributions are independent from one another. When SAD = 0 the affinity of the surfactant for the oil phase exactly equals its affinity for the aqueous phase, which is equivalent to the Winsor III (WIII) case. When the SAD is negative or positive, the phase behavior is Winsor I (WI) or Winsor II (WII), respectively.

Thus, SAD is conceptually equivalent to $R$, but this time, the relative contribution of each variable, and the possible compensating trade off between variable effects, is fully attainable. It is worth noting that the SAD relationship is a generalization of the HLB concept which appears under the symbols $\sigma$, EON, and $\alpha$ (139), and the PIT that is included as the temperature at which optimum formulation occurs (140).

The numerical quantification of this generalized formulation concept is important for the formulator because it has been directly linked with the emulsion properties, as will be discussed next.

### C. Formulation Affects Emulsion Properties Through Basic Phenomena

The following phenomenology is valid for the cases in which the physicochemical formulation is the main driving factor concerning the emulsion properties. The exact conditions of occurrence of this case will be discussed in detail later on. For now it is enough to state that this simple case takes place at least when the emulsion is made with low-viscosity oil and water phases, and when it contains similar amounts of both fluids and a few per cent of surfactant.

The relationship between physicochemical formulation and emulsion properties is evidenced through a formulation scan. This study technique is carried out by preparing a series of SOW systems with identical composition and formulation, with the exception of one formulation variable which is selected as the scanned variable. In many cases the scanned variable is the salinity of the aqueous phase for ionic surfactant systems, and the average EON for nonionic ones. Oil ACN or temperature may been selected as well, depending of the purpose of the study. The change in formulation variable should be such that the whole W I ↔ WIII ↔ WII phase behavior transistion takes place in the center of the scanned region. It is advisable that the whole scan encompasses a wide zone on both sides of the optimum formulation, e.g., at least from $SAD/RT = -2$ to $+2$, preferably wider, whatever the variable used to produce a departure from SAD = 0. When the series of systems are equilibrated, in most cases after less than a week, the systems are emulsified according to a standard procedure (stirring equipment and duration). Typical emulsion properties are then measured, i.e., electrolytic conductivity, viscosity, stability, and drop size.

When the formulation is scanned from WI to WII phase behavior, i.e., when the surfactant affinity toward its physicochemical environment switches from hydrophilic to lipophilic, several transitions are known to take place at the so-called optimum formulation as illustrated in Fig. 7, which gathers a large number of experimental results from different research groups (113, 121–125, 133–154).

The interfacial tension undergoes a minimum, often a very deep one, e.g., with ultralow values (0.001 nM/m or less) at optimum formulation SAD = 0. In the three-phase region, there are two interfacial tensions involving a surfactant-rich phase, one between the microemulsion and the oil phase, and the other between the microemulsion and the water phase. At low surfactant concentration, or when no

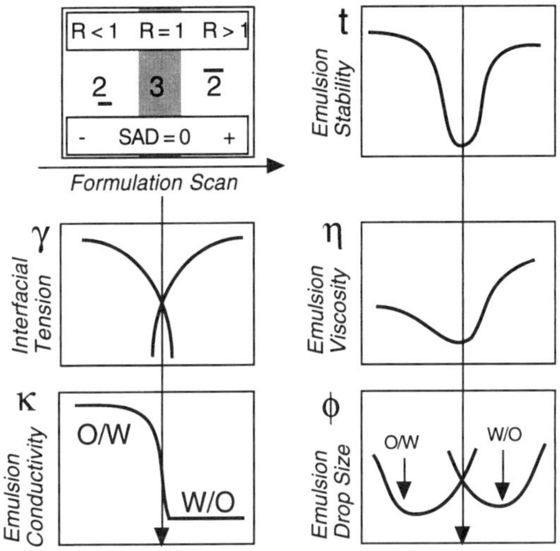


**Figure 7** Property transition as formulation is scanned through SAD = 0. (From Ref. 182.)

microemulsion can be formed, no three-phase behavior is observed at optimum formulation and the minimum interfacial tension is a good alternative to pinpoint SAD = 0 occurrence.

As seen in the typical variation indicated in Fig. 7 the electrolytic conductivity changes drastically inside the three-phase region, indicating that emulsion inversion takes place (142, 153, 155). According to the Bancroft rule, the wedge theory, and more modern curvature conceptualization (156), SAD < 0 is associated with O/W emulsions and SAD > 0 with W/O emulsions.

The exact formulation at which the inversion is located depends upon other variables, particularly the water-to-oil ratio, but it may be remembered for now that it is somewhere near optimum formulation. It is worth noting that the fact that the conductivity switch takes place over a very narrow range of formulation around SAD = 0 is quite general. However, it should be also remarked that the change is continuous as deduced from the existence of intermediate values, though difficult to catch in practice. This means that the emulsion inversion could involve some continuity in this case (155).

Emulsion stability undergoes a very deep minimum in the vicinity of optimum formulation, which was hinted at in an early study (142), and confirmed in more recent publications (56, 139, 157–159). Sometimes two maxima are observed as well on both sides at some SAD/$RT$ distance of optimum formulation, which may be ±3 units for instance, or sometimes less (159).

However, the existence and position of these maxima have been found to depend upon other system characteristics as well, and for now, there is no available prediction. It is worth noting that the stability minimum corresponds to emulsions that coalesce extremely rapidly. Indeed, it seems that in three-phase emulsions that are displayed near SAD = 0, the only delay is the sedimentation process, and that drops coalesce immediately upon contact. Several explanations have been advanced for that (160–162).

All of them are compatible with the experimental facts that the emulsion stability changes slowly with formulation far away from optimum formulation, while it changes suddenly several orders of magnitude away when the three–phase behavior region is attained from both sides. The quasioptimum systems are so unstable that the measurement of other properties can be in jeopardy, and in any case would require extremely rapid techniques.

Aside from the subject main stream, it is worth mentioning that this extremely low-stability feature is probably the key to the successful dehydration of crude oils (163).

As indicated in Fig. 7, the emulsion viscosity passes through a minimum at optimum formulation (164). The value of this minimum is quite low, unusually low as it is known that the ultralow interfacial tension tends to produce small droplets. Actually, this is not necessarily true as will be discussed later, because the droplets formed can coalesce at once. It seems that the low emulsion viscosity is due to the ease of deformation of the droplets along the streaming lines, a phenomenon similar to the drag reduction by polymers (165).

In any case the emulsion stability is so low that an homogeneous fluid viscosity could be measured only under vigorous stirring. The best data are attained with a mixer cell or by use of flow-through porous media.

Incidentally, it is the combination of the low viscosity of the multiphasic systems and the quick coalescence that is the key to enhanced oil-recovery processes using low-tension surfactant flooding (58, 166). It means that the low tension is necessary but not sufficient, and that the coalescence of oil droplets is imperative to produce mobilization.

For instance, alkaline flooding (167–170) produces low tension as well, but in some transient fashion, and not necessarily at optimum formulation. As a consequence the emulsions produced could be very stable and thus viscous enough to plug the porous medium.

The drop size undergoes a minimum at some distance (in the formulation scale) from optimum, on both sides, as a conseqence of antagonist effects (171). When approaching optimum formulation, from any side, the first effect is the tension lowering that enhances drop breaking up, with the resulting decrease in drop size. A little bit further in the direction of optimum formulation, the stability decreases very rapidly, with the corresponding drastic increase in coalescence rate. The breakup–coalescence dynamic equilibrium is displaced in favor of the coalescence, and the drop size rises again. In three-phase emulsions, it is not really possible to measure the drop size because of the extreme instability of the emulsion that instantly leads to large drops.

Before passing to the second-level description, it is worth noting the extreme importance of optimum formulation in the described phenomenology. It is worth remarking as well that the formulation scan can be carried out by changing any formulation variable according to $R$ and SAD concepts. However, the extention of the three-phase region and the values of the properties at optimum formulation and far from it, depend upon other effects that are not entirely understood. In most cases it is necessary to carry out experimental work to obtain these "scaling" characteristics, though the reported phenomenology is quite general.

## D. Formulation–Composition Interacting Influences

The previous description is valid for emulsions with similar oil and water proportions, and containing low-viscosity fluids. For extreme values of the water/oil ratio, the volume constraints tend to dictate more or less the emulsion type, depending upon the emulsifying device.

The study of the previous phenomenology (along a single generalized formulation scan) can be now extended to a bidimensional map that takes into account the water/oil ratio as well.

There is in practice at least another composition variable, i.e., the surfactant concentration, but strangely enough it has less effect than the water/oil ratio. This is of course related to the fact that the surfactant concentration is not allowed to change much in most practical cases for cost reasons, from say a minimum of 0.2% to a maximum of 5%. However, since the surfactant concentration does play a role, it will have to be taken into account in another way.

Figure 8 shows a formulation–composition map (172) in which the formulation is indicated as SAD and composition as the water content in the water–oil mixture. Since the surfactant content is always low, this is essentially the fraction of water in the system. It is worth noting that since temperature is a formulation variable, then this is equivalent to a temperature–composition map, as proposed over 30 years ago (173) and confirmed more recently (174) not only for the phase behavior, but also for emulsion inversion.

SOW systems are prepared at certain formulation–composition set points. They are, for instance, placed on the map at regular grid nodes. Each system is then left to equilibrate at constant temperature and its phase behavior is noted. Each system is then emulsified according to a standard procedure (equipment, energy, duration) and the emulsion

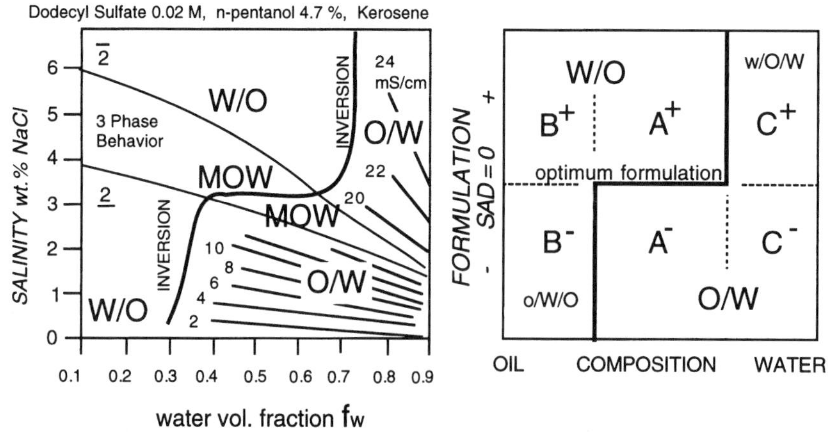


**Figure 8** Formulation–composition bidimensional map indicating isoconductivity lines (left) and region labeling (right); the bold line is the emulsion inversion locus. (From Ref. 172.)

properties are measured (conductivity, stability, viscosity, drop size).

The emulsion type is determined after its conductivity. The regions of the map that exhibit high conductivity (O/W emulsions) are separated from the regions of low conductivity (W/O emulsions) by what will be called the standard inversion line. Figure 8 (172) shows an actual experimental map and the corresponding schematic map. It is seen that the standard inversion (bold) line is composed of three branches. First there is a "horizontal" branch, located at optimum formulation (SAD = 0) in the central part of the map, i.e., when the relative amounts of oil and water are similar. This region is labeled A (172), with a + or − superscript depending on the sign of SAD. In A region that typically goes from 30 to 70% water, the emulsion type depends on formulation only, as mentioned in the previous paragraph.

The other two branches of the standard inversion line are essentially vertical, and are located typically at 30% water on the negative SAD side of optimum formulation, and at 70% water on the positive side. When the water content is low, the emulsion is always W/O, regardless of the formulation. Similarly, when the oil content is low, an O/W can be expected, whatever the formulation. In these extreme WOR regions, the phase which is present in larger volume becomes the external phase of the emulsion. It may be said that the composition dominates. However, a closer look at the conductivity value indicates the presence of multiple emulsions in the $B^-$ and $C^+$ zones, i.e., where the composition effects dominate over the normal formulation trend. These $B^-$ and $C^+$ regions have been called abnormal in opposition to the other ones which are labeled normal because they follow the Bancroft rule and the wedge theory (172).

The different regions are associated with some properties. From previously reported phenomenology, along a "vertical" cut located at 50% water, the formulation scan would result in emulsion inversion, minimum stability, and minimum viscosity at the crossing of optimum formulation. Figure 9, which summarizes many experimental data (73, 78, 155, 164, 171–182), shows the mapping of property general trends on the formulation–composition bidimensional chart.

Normal A regions and adjacent $B^+$ and $C^-$ normal regions are associated with stable emulsions. In many cases the maximum stability (of both O/W and W/O emulsions) is attained in the corresponding A zone near the inversion-line vertical branch and at some distance from optimum formulation, say 3–4 SAD units (shaded in map). In effect, far away from optimum formulation stability typically drops. The emulsion stability decreases as well when the internal-phase ratio decreases, because sedimentation and creaming are easier, sometimes with larger drops that settle quicker. On the other hand, the strip near optimum formulation, say SAD = 0 ± 1 unit, exhibits very unstable emulsions.

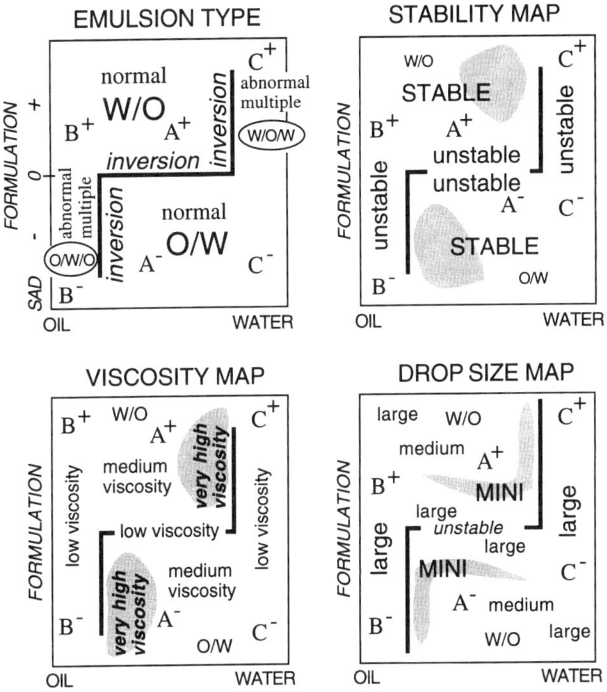


**Figure 9** General mapping of emulsion properties. (From Ref. 182.)

Unstable emulsions are also found in abnormal $B^-$ and $C^+$ regions. However, it is worth noting that multiple emulsions are often found in these regions, and that the low stability refers to the most external emulsion, e.g., the $O/W_2$ emulsion in a $W_1/O/W_2$ multiple emulsion located in the $C^+$ region. In other words, in the $C^+$ case the "large" O drops would coalesce quickly, but not the $W_1$ droplets that are inside them. Hence, in multiple emulsions located in abnormal regions the most stable emulsion is always the most internal one, which is stable according to the formulation influence or the Bancroft rule. The multiple-emulsion decay thus ends up in a macroscopic two-phase separation, but one of the phases contains fine droplets of the other one.

Viscosity increases in the normal region in the direction of higher internal-phase ratio (at constant formulation), so that the viscosity maximum is just near the inversion-line vertical branches. On the other hand, the viscosity decreases when the formulation tends to SAD = 0 at constant composition.

In most cases, the abnormal emulsions are not very viscous because their internal-phase content is low. There is, however, an exception to this trend, when multiple emulsions occur with a very high droplet content inside the drop. Such very high internal-phase content could happen at once in an emulsification process or slowly, e.g., a few days after emulsification and as a consequence of diffusional migration from the most external phase, e.g., $W_2$, to the most internal one, e.g. $W_1$ in the case of a multiple $W_1/O/W_2$ emulsion. Such migration can be driven by any chemical potential gradient such as in an osmotic pressure difference between two brines with different electrolytic concentrations.

The drop size distribution is the result of a dynamic equilibrium between opposite breaking and coalescence rates during emulsification (92). Since many different factors are susceptible to influence these processes, the overall result could be difficult to interpret, as well as the relationships between causes and effects. Consequently, drop size maps sometimes exhibit whimsical features that are the result of overlapping phenomena and trends, which should be recognized first of all.

At constant stirring, the physicochemical formulation affects both drop breaking (through interfacial tension) and coalescence (through surfactant adsorption) as discussed before. For a given SOW system, the nearer the formulation from optimum, the lower the tension, the easier the drop breaking, and thus the smaller the drop size that actually forms. However, the nearer the formulation to optimum, the quicker the coalescence, an effect which is just opposite to the previous one. It seems that the coalescence effect dominates in the three-phase region, while the tension lowering dominates a little bit further from optimum formulation. Consequently, there is a minimum drop size strip at some distance from SAD = 0, e.g., typically $SAD/RT = \pm 1$, in the shaded strips on both sides of optimum formulation.

At constant formulation and stirring, the water/oil ratio does affect the breaking–coalescence equilibrium. It is found that the observed variation depends upon the oil viscosity (171). For instance, in the case of O/W emulsions, and when the oil viscosity is 5–10 mPa.s or higher, the drop size always decreases as increasing amounts of internal phase are added, although the same stirring protocol is applied. With extremely low-viscosity oils, the drop size starts increasing as the oil content increases from zero to say 40 or 50%, then it decrease as in the previous case of more viscous oils (183). In both cases, and when the internal oil phase content reaches 70 or 80%, i.e., near the vertical branch of the standard inversion line, the emulsion is extremely viscous and highly nonNewtonian. It cannot be produced by violent stirring but by low-shear slow motion stirring. The low-energy stirring of high internal-phase ratio emulsions produces extremely fine droplets and is very energy efficient. It is a choice situation when making high-viscosity O/W emulsions like mayonnaise, paints, and extraheavy crude emulsions. However, it is worth remembering that this zone is near the inversion line, and that precaution is required so as not to overstep the limit inadvertently.

The previous section depicted the general patterns of change on a formulation composition map. Actual values of the properties, particularly of minima and maxima, as well as formulation and composition scale characteristics, may be changed by other factors.

For instance, it is well known that the emulsion viscosity depends not only on formulation and internal-phase content, but also on external-phase viscosity, and drop size average and distribution (79, 80, 83, 84). In concentrated emulsions, flocculation takes place and results in time-dependent properties like thixotropy.

The bimodal features were studied first on solid particle dispersions (184–186) then on emulsions (187–189). Whenever the distributions of the two-base emulsions do not ovelap too much and if the ratio of the coarse and fine emulsions' average diameter exceed three, then a considerable decrease in viscosity is attained by mixing the emulsions.

Figure 10 (190) indicates the viscosity of mixtures of two emulsions, a fine one and a coarse one, with a drop size ratio exceeding 3. It is seen that the mixture of small drops with large drops results in a considerable decrease in emulsion viscosity, particularly in the range from 20 to 70% of coarse emulsion. An even larger viscosity reduction can be attained by tailoring the drop size distribution, so that the coarse emulsion is rather monodispersed, whereas the fine one is quite polydispersed.

Stability depends upon so many things that it is easy to alter its value. However, in most cases the general phenomenology versus formulation and composition is valid. The presence of alcohol, particularly an intermediate-solubility alcohol, such as *sec*-butanol or *ter*-pentanol, or a mixture of propanol and butanol, tends to reduce the interfacial adsorption of the surfactant, thus reducing all associated effects, in particular the repulsion that contributes to stabilization. It is worth

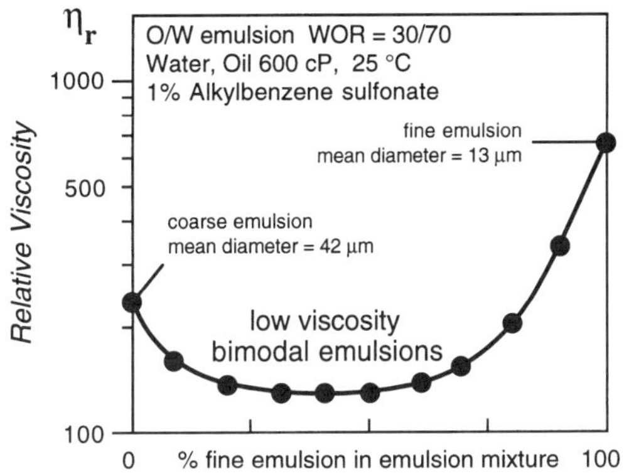


**Figure 10** Viscosity of emulsion mixtures. (Courtesy of M. Ramirez, Ref. 190.)

noting that the use of mixed surfactant systems, which is often advised in emulsion making manuals, can be detrimental in some cases in which a selective partitioning of surfactant species takes place (191, 192), and little surfactant is left at the interface.

On the contrary, some effects would enhance stability (71, 99, 193). Retardation of film drainage is probably the best way to increase stability. It can be done in different ways. First, the external-phase fluid can be made more viscous by adding thickening agents so that film drainage and Brownian motion effects are lessened. The film drainage can then be inhibited by any kind of repulsion between approaching interfaces. This is generally the role of the adsorbed surfactant or of deposited colloid particles. Finally, film drainage can be slowed down by dynamic effects, such as streaming potential or interfacial viscosity (194).

## E. Displacement of the Standard Inversion Line – Concentration and Stirring Energy Effects

The bidimensional mapping does not take into account several secondary factors that are, however, known to influence the emulsion type and properties. Recent research has shown that they may be accounted for as a modification of the bidimensional map characteristics. First, it was found that an increase in surfactant concentration tends to widen the A zone (195), i.e., the region in which the emulsion type is determined by the formulation.

On the other hand, an increase in oil-phase viscosity produces a displacement of the $A^+/C^+$ branch of the standard inversion line toward higher oil content, while it does not affect the other branches (172, 177).

Increased stirring energy was found to reduce the width of the A zone (196) contrary to a previous incorrect statement on this effect (182). As far as O/W emulsions are concerned it means that an increase in stirring energy would reduce the highest oil content attainable, while a decrease in stirring would do the opposite. This rule of thumb is to be remembered when manufacturing high internal-phase ratio O/W emulsions.

Nevertheless, there is another way to avoid the emulsion inversion at high internal-phase content, which is the dynamic emulsification technique which is discussed next.

## F. Programming Formulation or Composition

Until now, the emulsion was made by stirring a preequilibrated SOW system whose representative point was laid somewhere in the formulation–composition bidimensional map. The line that separated the two types of emulsion on the map was called the standard inversion line.

A dynamic process of emulsification takes place when an emulsion is modified under constant stirring, for instance, by adding some amount of one of the phases, or by changing the temperature. A dynamic process generally initiates with an emulsion made in the conventional way at some representative point (called initial). The formulation or composition (or both of them at the same time) is then changed, so that the representative point of the emulsion is shifted through the map. The change can be lumpwise, e.g., by adding a certain aliquot of one of the phases, or almost continuous as in a drop by drop addition or in a slow temperature variation. In all cases the stirring is kept at the same level to maintain the system fully emulsified during the process. As the formulation–composition conditions are changed, the position of the representative point of the emulsion follows a path on the map, which is the trace of the dynamic process.

Current knowledge may be found elsewhere (179, 197, 198), and only a summarization will be presented here by answering the two following questions. The first one is what happens if the representative point is displaced without trespassing on the inversion line, but wandering from a location on the map to another one in which the emulsion properties are different from those of the original one? How are the emulsion properties expected to change during such a process?

The second question deals with what happens when the representative point attains and crosses the standard inversion line? Does the emulsion invert at once or is there some delay? What become the properties of the emulsion after the inversion takes place?

### 1. Wandering on the Same Side of the Inversion Line

In this section it is assumed that the representative point of an emulsion is displaced on the bidimensional map, typically by modifying either the formulation or the composition, without crossing the standard inversion line so that such changes do not affect the emulsion type. Since each zone of the map is associated with typical emulsion properties, e.g., stability or viscosity, these properties could possibly change during the process.

For instance, the dilution of an O/W emulsion located in the $A^-$ zone by adding more water phase produces a shift to the right, which is associated with a decrease in viscosity. As far as the drop size is concerned, it is generally not changed by the dilution, unless the stability gets worse and the drops start coalescing, which is not the case if the formulation is located far enough from SAD = 0. It is worth noting, however, that adding pure water, instead of an aqueous phase consisting of the proper surfactant and electrolyte characteristics, can additionally produce a formulation change and the concomitant shift of the point toward or far away from SAD = 0.

As a general rule of thumb, it can be said that a change of location of the representative point of the emulsion on the same side of the inversion line would not change the drop size (unless a highly unstable emulsion region is attained) but would affect both its viscosity and stability. Thus, the original drop size characteristics will be conserved according to the original position, whereas the viscosity and stability will match the new position characteristics. Nevertheless, in most cases it will be necessary to make the appropriate corrections to take into account the secondary effect of the "memorized" drop size on both viscosity and stability.

For instance, it has been previously discussed that a very small drop size can be attained in two areas in the $A^-$ region. Along a formulation scan, the smallest drop size is attained at some distance from SAD = 0 that corresponds to the best compromise between the low tension and the low stability (171). Such a situation is located at the black circle origin of arrow (1) in Fig. 11. The corresponding emulsion would exhibit a very small drop size but would be unstable because of the near-zero value of SAD. A quick change of the formulation along the arrow allows one to reach a high-stability region in the $A^-$ zone (white circle), and the emulsion would stabilize before the drops start coalescing. This change can be accomplished at essentially constant composition by adding a small amount of a concentrated solution of hydrophilic surfactant, or by quickly lowering the temperature, say by 20°C in less than 1 min, if the surfactant is of the nonionic type. This procedure ends up in an emulsion which exhibits a drop size much smaller than the one attained by stirring a pre-equilibrated system prepared at the same (white circle) point. It may be said that the drop size value has been quenched in this process.

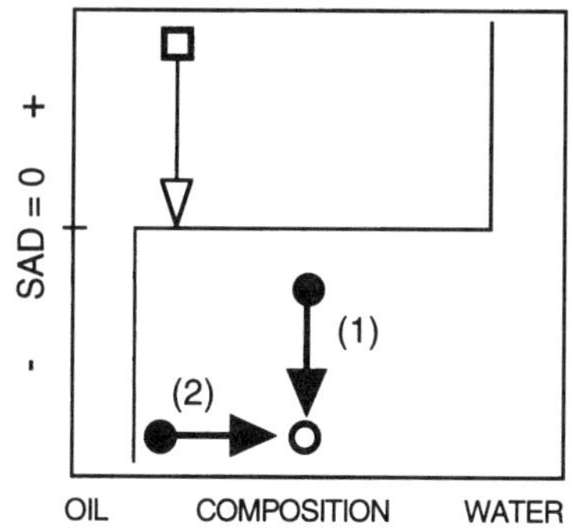


**Figure 11** Emulsion changes without trespassing on the inversion line.

Another way to attain a fine drop emulsion at this (white circle) position is to start with a much higher internal-phase content O/W emulsion, e.g., where the (black circle) origin of arrow (2) is located. In such a position, the emulsification is carried out at low shear in a very efficient way, to produce a viscous fine emulsion (183). This emulsion is then diluted with an aqueous solution of hydrophilic surfactant (to maintain the SAD constant) along the arrow (2) path until the final (white circle) location is reached. Also in this case, the final emulsion attained will exhibit a drop size much smaller than the emulsion that could be attained from a pre-equilibrated system directly in these (white circle) conditions.

As a consequence of their smaller drop size, the emulsions made through both dynamic processes will certainly exhibit a higher viscosity and probably a higher stability. Nevertheless, these two dynamically prepared emulsions will not be necessarily identical, since the actual drop size distribution depends upon the dynamic process characteristics, particularly the stirring efficiency.

These are two examples on how to capitalize on the variation of the properties in the map. Another example would be the case of crude-oil dehydration (163), in which the original W/O emulsion is stabilized by natural surfactants, such as resins, which are quite lipophilic. The emulsion representative point is thus typically in the $B^+$ region in a position indicated by a square in Fig. 11. It is known that the dehydrant chemicals contain highly hydrophilic surfactant species, so that the overall surfactant mixture (natural plus dehydrant) attains the SAD = 0 neighborhood, where the emulsion stability becomes extremely low.

It is worth noting that the type and amount of dehydrant chemicals are determinant because the end of the arrow must exactly reach the formulation that corresponds to SAD = 0. If it falls short, the W/O emulsion would still be stable, and if it overshoots it, an inverse O/W emulsion could result. This is, by the way, why the use of dehydration chemicals is such a fine-tuning business.

### 2. Crossing the Inversion Line

The shift in the representative point of an emulsion on the formulation–composition map is now allowed to trespass on the standard inversion line and to move well inside the other region.

The two different ways of crossing the inversion line are associated with quite different behaviors. The first one, which is known as transitional inversion, is produced by changing formulation at a constant water-to-oil ratio, i.e., along a vertical path in the bidimensional map. Such a crossing takes place in the A region in the central zone of the map. The experimental evidence indicates that, in this kind of dynamic process, the inversion takes place at the very moment the standard inversion line is crossed, i.e., essentially at SAD = 0, whatever the direction of change [from $A^-$ to $A^+$ or vice versa as indicated with white arrows in Fig. 12 (left)]. The horizontal branches of the standard and dynamic inversion lines are thus identical. The term transitional inversion was proposed to indicate that the change from an O/W to a W/O emulsion (or vice versa) occurs smoothly with an intermediate triphasic emulsion MOW which is extremely unstable. When the formulation moves away from optimum, the microemulsion M phase solubilizes water and rejects oil to become the W phase on the $A^-$ side, whereas it solubilizes oil and rejects water to turn into the O phase on the $A^+$ side. It is worth nothing that in such a process, the rejected phase separates as extremely fine droplets that could end up in a miniemulsion (199, 200), if they are protected from coalescence by some quenching mechanism.

The crossing of the vertical branches of the inversion line results in a completely different phenomenon called catastrophic inversion (172, 197) because it can be modeled as a cusp catastrophe transition as pointed out by Dickinson (201) and further discussed by others (195–203).

Actually there are two cusps which can interact in some cases (204) to form a higher order catastrophe called "butterfly" catastrophe because of the fancy shape of its bifurcation.

Figure 12 (left and center) indicates the position of the dynamic inversion line when the composition is changed left or right at constant formulation. The black arrows show the direction of change, and the tip of the arrows indicate the point where the inversion takes place. It is seen that the inversion does not happen at the same composition whether the shift is from left to right or opposite. This means that there are some areas [shaded in Fig. 12 (right)] in which the two types of emulsion can be found, depending on the direction of change in the dynamic process, a feature that was called hysteresis by Becher (205) no less than 40 years ago.

These hysteresis areas are found at the limit of the normal and abnormal regions, and exhibit a typical wedge shape that vanishes as SAD = 0. From the practical point of view of emulsion making these features

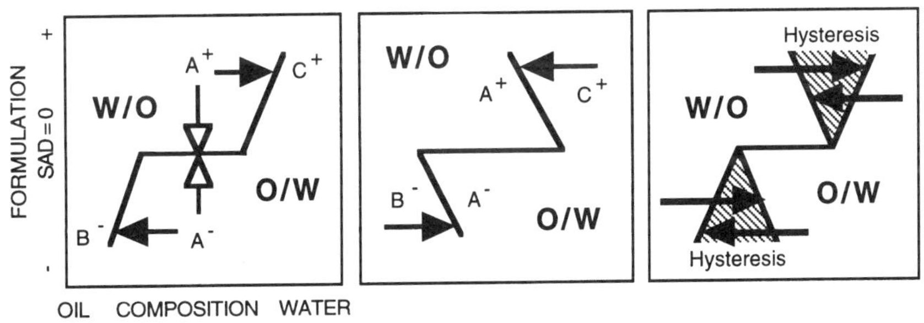


**Figure 12** Emulsion formulation or composition evolution up to dynamic inversion occurrence.

are quite useful. For instance, by starting in a normal A region and increasing the internal-phase ratio [path of black arrows in Fig. 12 (left)], the dynamic inversion is delayed much further than the typical 70% location of the standard inversion line. It is seen also that the more different from zero the value of SAD, either negative or positive, the higher the reachable internal-phase ratio prior to dynamic inversion. This indicates that there is a complex trade-off between formulation and composition influences.

The whole phenomenology of phase behavior and emulsion inversion was interpreted with a butterfly catastrophe model with amazing qualitative matching between theory and experiment. The phase behavior model used the Maxwell convention which allows the system to split into several states, i.e., phases at equilibrium. On the other hand, the emulsion-type model allows for only one state (emulsion type) at the time, with eventually catastrophic transition and hysteresis, according to the perfect delay convention. The fact that the same model potential permits the interpretation of the phase behavior and of the emulsion inversion (204, 206) is a symptomatic hint that both phenomenologies are linked, probably through formulation and water/oil composition which are two of the four manipulable parameters in the butterfly catastrophe potential.

The butterfly catastrophe model explains why the transitional inversion is not really an inversion but a surfactant transfer from one phase to the other, while the catastrophic inversion is a nonreversible hysteresis-type instability. This approach, which is out of the scope of this chapter, is well documented elsewhere (197).

Some experimental facts should be stressed anyway for the present purpose. The delayed inversion of a normal emulsion along a change in composition toward a higher internal-phase ratio, which may be called the memory feature of the dynamic inversion, can be harnessed to attain extreme values of internal-phase content when the formulation is quite far away from SAD = 0. It means that the representative point of an emulsion can be shifted somehow quite further than the standard inversion line, and that the dynamic inversion line is now "pushed" away like a curtain rather than crossed. This increases in practice the extension of the region in which the formulation–composition change takes place on the same side of the inversion line, for instance the $A^-$ region in the case of an O/W emulsion. This widening is illustrated in Fig. 13 by the modification from the left map (standard inversion) to the center map (dynamic inversion).

Recent studies have shown that a rise in surfactant concentration tends to increase the normal A regions' extension, including the extra width provided by the wedge-shaped hysteresis zone (195), whereas an augmentation in stirring energy does just the opposite (196), with some shrinking of the hysteresis zones altogether. Consequently, an even wider zone of O/W emulsion occurrence can be attained by using at the same time the hysteresis feature of the dynamic inversion (modification from left map to center map) and additional conditions like low-shear stirring and high surfactant concentration (change from center to right plot).

As a matter of fact, such a far reaching recipe has been used for years by cooks when preparing a homemade mayonnaise (a very high internal-phase content O/W emulsion) by using a drop by drop addition of oil (dynamic process), spoon stirring (low shear), and a dash of mustard (extra surfactant) on top of the egg yolk (water plus surfactant).

When the dynamic process is pushed too far, it finally results in inversion, and the emulsion type

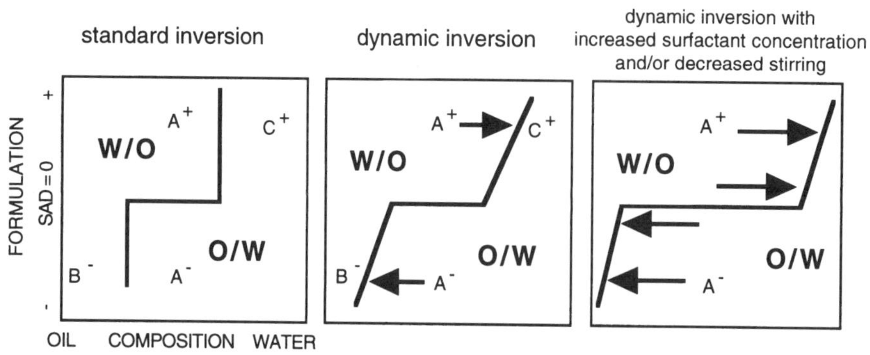


**Figure 13** Inversion line depends upon process conditions.

changes, often with the production of a multiple emulsion as an intermediate situation (207, 208).

The newly inverted emulsion properties often match the properties of standard emulsions made in the same (new) location, sometimes with extra features such as extremely small drop size formed during the inversion. In fact, industrial plants that produce extremely fine emulsions use a dynamic inversion process even more complex than the one described here, which promotes the surfactant mass transfer from one of the phases to the other in order to trigger a spontaneous emulsification (209, 210).

### G. Combined Influence of Formulation, Composition, and Stirring

It is known that drop size can be reduced either by longer or more energetic stirring, as well as by changing a formulation parameter, which alters the tension, or changing the composition, which produces a change in shear transfer through the emulsion. Although these changes involve very different physical and physicochemical effects, they can be said to be equivalent if they produce the same drop size reduction. Starting with these premises a new state of the art of the compensated effects of formulation, composition, and stirring is slowly emerging (183, 211–213) that will allow one to compare emulsification situations with a quantitative yardstick for the stirring conditions, such as the same Reynolds number, the same capillary number, or the same stirring energy.

## IV. PRODUCT ENGINEERING

### A. Product Engineering Problem Statement

A lot of attention has been dedicated in the research literature to emulsion formulation, manufacturing, and properties such as viscosity and stability, but no overall product engineering approach seems to be available, at least in the systemic way that is intended here.

The first step in product engineering will be to specify the product properties which are preferred or required, as well as any other constraint concerning the manufacturing or handling. The next step will be the formulation engineering stage in which the current know-how is used to translate the specifications into formulation, composition, and protocol alternatives. In this second step, the formulator will often encounter contradictory formulation requirements. Some of them could be resolved by using a strategem such as the memory feature, but others would not be defeated by ruse and would require the attainment of a compromise between opposite effects. The third step will be then a not necessarily straightforward process design in which the formulation engineering requirements will be translated into equipment design and operational conditions.

As far as heavy crude oil-in-water emulsions are concerned, two cases are encountered. The first and most important one, which is commercially available, is the O/W emulsion to be handled and used as a fuel for thermoelectric plants and other energy generation purposes. The second one is the use of the O/W conditioning as a low-viscosity vehicle for the pipeline transportation of viscous oil over long distances. In this case the oil has to be separated from the water at the end of the pipeline. Thus, the difference from the previous case is that the emulsion should be stable in the pipe, but easy to break up at the end station.

The following will be dedicated to the emulsified fuel case, with a few words on how the arguments could be changed to accommodate them to the other situation.

### B. Surveying Current Know-how and Getting Feedback from It

The specifications for an extraheavy crude oil emulsion for fuel purposes are as follows:

- S1. The emulsion should be of the O/W type (because it is much less viscous than its oil phase).
- S2. It should have a high internal-phase ratio (because it is a fuel whose calorific content is loaded in the oil phase) of at least 70%.
- S3. A viscosity low enough to be handled like any other liquid fuel (chief advantage over coal).
- S4. Stable when stored at rest (one year or more) at different ambient temperatures (from tropical to freezing).
- S5. Stable against perturbations produced by, e.g., pumping, pipelining, sea navigation (which are likely to occur).
- S6. Small drop size for good combustion (although drop size, within a certain range, does not appear to be really critical for applications).
- S7. Low surfactant content (to reduce cost and pollution prospects).
- S8. No sodium ions (because they damage refractory equipment).

S9. Handling should satisfy environmental pollution abatement regulations.

S10. There should be an adequate control of accidental spills on land and water.

The first seven specifications S1–S7 can be analyzed with the help of the previously discussed know-how which has been summed up in Fig. 9. The other ones will be dealt with after. The first thing to do is to list the meaning of the specifications as far as the position in the map and other characteristics are concerned.

The S1 specification corresponds to the $A^-B^-$ region of the formulation–composition map, while S2 specification restricts it to the extreme left of the $A^-$ zone, near the inversion line. It should be noted that the minimum 70% oil composition could be dangerously near the standard inversion line.

The S3 specification of low viscosity corresponds either to the $B^-$ region or to the neighborhood of SAD = 0.

The S4 specification implies that the point is located in or near the maximum stability region (far from SAD = 0 but not too far away from it, and near the inversion line), including at high temperatures when this variable is taken as a formulation variable (particularly with nonionic surfactants). It also means that the maximum stability region should exhibit a stability value which satisfies the requirement, a feature not associated with the map characteristics.

The S5 specification is more difficult to translate into a feature of the formulation–composition map. The only known experimental data related to this specification are those dealing with the effect of stirring, i.e., the $A^-$ region extension widens when stirring decreases, thus reducing the risk of inversion during handling.

The specification S6 is not really a condition on the drop size to be achieved. It refers rather to finding the best location to produce small drops in the most efficient way, since this is an economic concern. Only two zones are known to satisfy this predicament, which are (1) a strip parallel to the $A^-/A^+$ inversion branch at some distance from SAD = 0; and (2) a strip in the $A^-$ region near and along the inversion line (see shaded zones in Fig. 9).

The S7 specification at low surfactant concentration tends to reduce the width of the $A^-$ region, with a resulting reduction in the maximum attainable internal-phase ratio. However, and for economic constraints, the surfactant concentration will be often set at the lowest efficient level, which is often in the 0.2–0.5 wt % range.

It is obvious that the consequences of some specifications are in direct contradiction with others. The conciliation or optimization of these opposite factors is the first part of the formulation engineering problem, which may be solved according to the following arguments, starting from the most demanding conditions.

The high internal-phase ratio (S2) generally produces an increase in emulsion viscosity (contrary to S3). Since the high oil content is a *sine qua non* requirement, it is kept to the minimum acceptable value, say 70%, and it is decided that the viscosity will be cut down some other way. It is known that viscosity can be reduced by approaching SAD = 0, but this is no answer to the dilemma because of the concomitant reduced stability which is strictly prohibited by S4. The viscosity can be curbed as well by increasing the drop size, although not too much, because this could probably result in stability degradation (contrary to S4) and incomplete combustion (contrary to S6). Thus, a compromise drop size is selected, say in the 10–30 μm range, between S3 and S4 opposite requirements, but satisfying S6.

Once the average drop size is set, the viscosity can be further curtailed by another effect, i.e., by manipulating the drop size distribution shape. In effect, it is known that a polydispersed emulsion, or even better, a bimodal emulsion, will exhibit a reduced viscosity.

The above reasoning leads to the following characteristics for the proper product: an O/W emulsion, located at 70% oil (for adequate caloric content), a few SAD units from optimum formulation (for stability), and with 10–30 μm drops with a bimodal drop size distribution (for low viscosity and satisfactory combustion).

The issue is now to manufacture an emulsion with these properties, and to meet the remaining specifications. It has been discussed in previous sections that a dynamic process often makes the manufacturing easier or cheaper, and is able to improve upon the properties. There are probably many alternative dynamic processes that could end up at the right place on the map (214). The second part of the formulation engineering problem is to find those that fulfill the remaining specifications and that present some advantages over the direct emulsification under the final conditions.

There are several different ways to manufacture such an emulsion. Three of them are presented here as paths 1–3 in the following paragraphs.

In path 1 the initial emulsion (black circle in Fig. 14) is prepared at the formulation that exhibits the minimum drop size and is located at some distance from

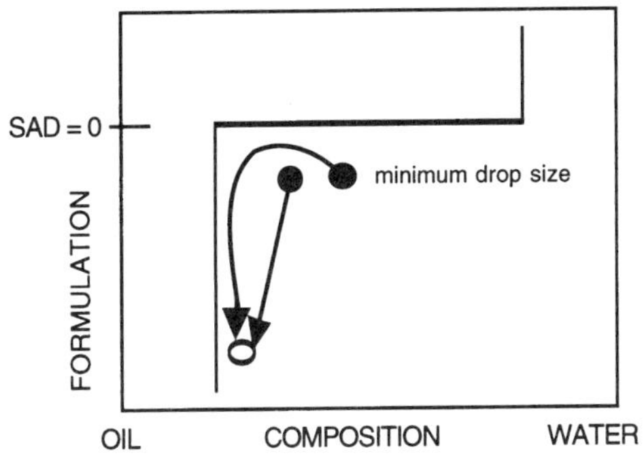


**Figure 14** Path 1.

optimum formulation on the negative SAD side, say at SAD/$RT = -1$. The original oil content is 50 or 60%, so that the initial emulsion is not too viscous, a concern since the drop size might be small. The representative point is then shifted in two directions: first toward a more negative SAD value to insure a better stability, and second toward a higher oil content to satisfy the final 70% specification. This can be done along different paths. In Fig. 14 the straight-line path is a combination of concomitant increases in hydrophilicity and oil content.

Along the curved path the extra oil is added first while the formulation stays at the value that corresponds to the minimum drop zone. The formulation is then made more hydrophilic on the final emulsion, while a low-shear stirring is applied. This path maintains the representative point of the emulsion always in the minimum diameter zone (shaded in Fig. 9), and will probably result in a smaller drop size. It is, however, difficult to operate because of the programming of subsequent changes in composition, formulation, and stirring.

If the system contains a nonionic surfactant, the proper increase in hydrophilicity may be attained by cooling, say by 20–30°C, or by adding a concentrated surfactant solution, or a combination of both. Such a method would take advantage of the lower viscosity of the oil phase when it is hot, an important factor in the case of extremely viscous oils.

The bimodal distribution feature may be attained in different ways, the most simple being the mixing of two streams with different stirring conditions. However, it is probably not the most economical, and may not be the safest one. In effect, since the final location of the emulsion is very near the inversion line, the extra stirring of one part of the emulsion might end up in inversion instead of in smaller drops. The curved path may be a more appropriate way to attain a bimodal emulsion, since the final condition at low-shear stirring might apply to only one part of the whole emulsion.

Path 2 makes use of the delay in the dynamic inversion when the oil content is increased. The initial formulation (black circle in Fig. 15) is taken far from optimum, and near the standard inversion line, say at 70–75% oil. The system is then stirred at low shear and oil is added to reach an 80% oil content or even more (black square).

It is worth remarking that with viscous oils it might not be necessary to increase the surfactant concentration in order to shift the inversion line so far to the left, whereas it is advisable to do so with low-viscosity oils. The emulsion can be stirred at this point under very low shear until the required drop size is attained. Hence, it is diluted to the final 70% oil content by adding water. Eventually a slight change in formulation (as indicated in Fig. 15) can be introduced by changing the temperature or the formulation of the dilution water.

The 80% oil emulsion is often very viscous, but the stirring efficiency is quite good and the drop size is reduced at a low energy expense. Such stirring of a very viscous emulsion often results in a bimodal emulsion of a lower viscosity than expected.

Figure 16 shows an example of such a stirring process, in which case the stirring device does not apply the same shear on all the emulsion volume. One part of the emulsion is submitted to a more intense or efficient shearing than the other part and this results in smaller drops that contribute to the emergence of a new peak on the small drop side of the distribution after a few minutes. The effect of the appearance of a bimodal distribution is a considerable viscosity abatement.

It is worth noting that a change in formulation or temperature, as seen in path 1, may be combined with the present path.

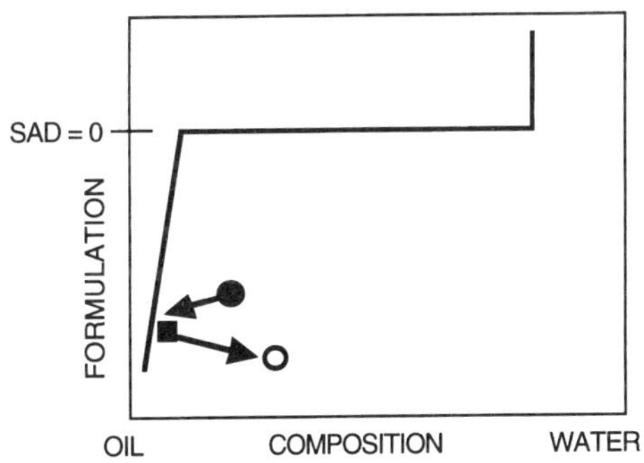


**Figure 15** Path 2.

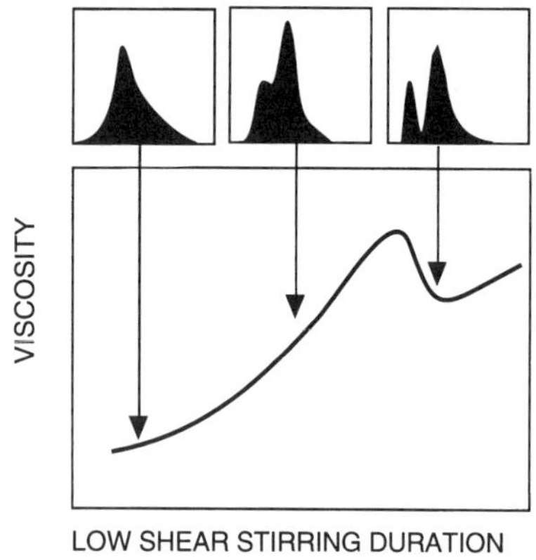


**Figure 16** Emulsion viscosity change as stirring duration increases in a nonhomogeneous blender.

In path 3 (Fig. 17) the initial system is set at SAD = 0 or at SAD slightly positive, and a formulation change shifts it to SAD = 0. This can happen by changing the temperature (emulsification by the PIT method) (215) or formulation, so that the surfactant passes from one phase to the other, often producing a spontaneous emulsification. Another way to trigger an easy emulsification is by adding an alkaline aqueous solution that reacts with carboxylic acids present in the oil phase and results in interfacial formation of surface-active substances (216).

In all these cases the tension temporarily becomes extremely low, in the μN/m range, and very small droplets are formed. Because of the unsteady state, the conditions for emulsion instability are not necessarily met at the same time and the drops might not coalesce at once. As a consequence, a fine emulsion may be produced, which is not the most stable anyway. In a second step (arrow) the formulation is changed to a more hydrophilic one and some oil is added. Because the efficiency of the stirring decreases in this second step, a bimodal emulsion can be readily made.

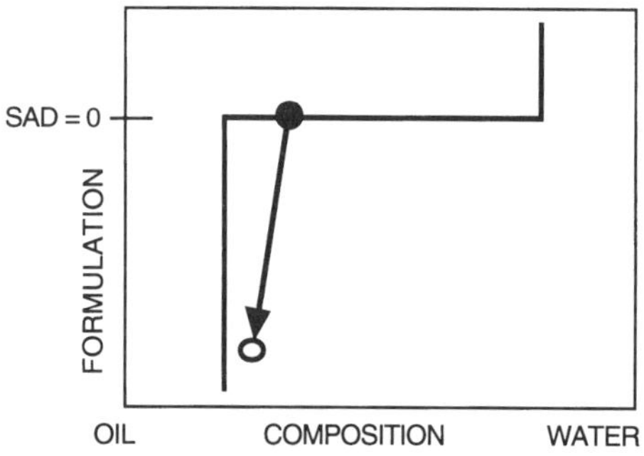


**Figure 17** Path 3.

There are still some specifications to be taken into account in making the final selections. The S4 specification requires a high stability at rest, and an efficient repulsion by the interfacially adsorbed surfactant. Anionic surfactants could do the job, but it must be remembered that sodium ions are prohibited (S8). Since divalent cations are likely to precipitate most anionic surfactants, organic ammonium derivative cations may be the answer. Cationic surfactants are likely to be ruled out for several reasons, among them their environment impact and hydrophobation properties. Nonionic surfactants may provide stabilization, mainly through steric repulsion.

The stability to pumping, pipelining, and sea transportation will have to be tested under simulation conditions. However, these hazards are essentially related to the eventual inversion produced by excessive shear, a very likely situation in high-speed rotatory pumps. Either the emulsion is quite robust, i.e., the inversion line is located far away from the representative point, or high-shear situations are to be avoided. Since the high oil content requirement is imperative, high-shear handling should be ruled out in most cases.

The S9 specifications on flue gas are probably easier to satisfy with an emulsion fuel than with coal, and do not depend significantly upon the surfactant formulation and emulsion manufacturing. The cooler flame results in less CO and $NO_x$ in the flue gas, while the $SO_2$ is removed as in coal-fired plants. In most cases, emulsified fuels create much less ash than coal does.

Emulsion spills are a concern in aquatic environments, and the surfactant should be nontoxic. Also, its compatibility with seawater should be appropriate.

## C. Practical Solutions for Orimulsion®

The following subsection contains a description of Orimulsion® production in terms of the formulation map and the path that is followed to obtain the final product. For reasons that are revealed later, the preparation of a product like this cannot be accomplished in a single step. However, the product-engineering approach may help to concoct the optimum procedure.

### 1. Description of Production Process

At the beginning of commercial production, the process consisted of the following steps (10, 11, 59, 189):

Down-hole emulsification resulted in a primary emulsion that contained about 60% oil. The representative point of this emulsion was located almost at the center of region $A^-$ in the formulation–composition map, where a stable O/W emulsion was attained. Since the emulsion would combine with the connate water, which was very salty, a nonionic surfactant of the ethoxylated nonylphenol type was used.

Destabilization of the primary emulsion by increasing the temperature was aimed at separating all water from crude, including dehydration and desalting of the oil. From the point of view of the formulation–composition map, this corresponded to moving upward along a vertical line (constant $f_w$) until the neighborhood of optimum formulation was reached. Once the emulsion was broken, the heavy crude oil was recovered in conventional (gravitational and electrostatic) separators. It is worth mentioning that the separation was facilitated by the fact that this oil was less dense than water at the treatment temperature, mostly because the water phase was a concentrated brine.

Preparation of a concentrated emulsion. This stage consisted in mixing the heavy crude oil and a fresh-water surfactant solution, first in static mixers, then in a dynamic in-line turbine-type blender. The oil/water proportion was quite high (about 85% of heavy crude oil). Consequently, mixing was carried out very near the inversion $A^-/B^-$ branch. Inversion to W/O was delayed by the use of a relatively high surfactant concentration (the new one from the surfactant solution plus the remaining one from the primary emulsion).

Finally the emulsion was diluted to 70% oil content by adding cold fresh water containing a corrosion inhibitor. The last step placed the emulsion formulation–composition representative point well within the $A^-$ region, where O/W emulsions are quite stable.

In essence the manufacturing of the first commercial emulsion was carried out following path 2, indicated in Fig. 15.

This process was, however, quickly abandoned owing to an unforeseen problem, which was first called "aging" (10), although it turned out to be different from what is usually called flocculation or coalescence aging (217–219).

It was found that the formation brine could not be completely eliminated from the heavy crude oil. Because of some limitations in the emulsion-breaking process, about 1% of this concentrated brine remained in the oil as a W/O emulsion. The brine droplets that were of very small size, typically less than 2 μm, did not join the water continuous phase during preparation of the commercial emulsion, but remained as droplets encapsulated inside the oil drops. It is also possible that the proximity of the $A^-/B^-$ transition region could have promoted the formation of a multiple brine/oil/fresh water emulsion. The result of this phenomenon was that the commercial emulsion was multiple and exhibited an osmotic pressure gradient from the external fresh water to the most internal brine droplets. Surprisingly enough, this osmotic gradient could draw toward the brine droplets about 20% of the more external water phase in less than 48 h. As a consequence, the emulsion apparent internal-phase ratio increased quickly and its viscosity rose to an unacceptable level that made it almost impossible to pump.

As mentioned in Sect. II.C.2, a small amount of electrolyte (as a water-soluble magnesium salt) was added to the emulsion as a corrosion inhibitor. This electrolyte could in theory balance the osmotic gradient and inhibit aging. However, the dehydration and desalting process was difficult to control and often the heavy crude oil ended up with larger amounts of brine that required larger amounts of electrolyte to offset the osmotic gradient.

An interesting anecdote may be told regarding the corrosion inhibitor. In the first place, magnesium sulfate was used. However, it was soon clear that the magnesium sulfate could be transformed into inopportune hydrogen sulfide by sulfate-reducing bacteria that resisted conventional biocide treatments. As a consequence, magnesium sulfate had to be changed for the more expensive magnesium nitrate (11).

This early version of the manufacturing process had another serious flaw. It was found that a considerable amount of the primary emulsion surfactant was not recovered with the separated oil, but rather was carried away by the water extracted during the dehydration and desalting (10). Together with the aging process, this further damaged the economics of the process.

The manufacturing process was thus modified into what is the current version (10, 11, 59, 64):

The extraheavy crude oil production is now carried out by down-hole injection of a light hydrocarbon solvent. Therefore, the blended oil exhibits a lower density and a greatly curtailed viscosity, which makes any further downstream treatment easier.

Dehydration and desalting are performed on the diluted crude oil in conventional (gravitational and electrostatic) separators to eliminate fully the formation water, so that the "aging" problem is avoided. After dehydration, the solvent is recovered by flash distillation and recycled to the production area.

The preparation of the concentrated emulsion is carried out as in the first process. Heavy crude oil is emulsified with an ethoxylated nonylphenol solution so that an 85% oil phase O/W emulsion is attained. Since the emulsification operation takes place near the $A^-/B^-$ branch of the inversion line, extra caution is taken to avoid the formation of a multiple emulsion. Very low mixing energy conditions are used to shift the position of the $A^-/B^-$ branch further left. To do so in practice, the concentrated emulsion is made by passing it through a sequence of low-shear blenders as indicated in Fig. 18: first a static mixer, then a custom-made dynamic mixer with a milder stirring than provided by the turbine blender used in the early process.

Dilution to 70% oil content is carried out by adding fresh water containing magnesium nitrate. In order to ensure a homogenous dilution and to avoid concentrated spots of magnesium nitrate solution which could promote coalescence, a static mixer battery is used as shown in Fig. 18 (10, 11, 59).

An overview of the current Orimulsion® manufacturing process is shown in Fig. 19 (64), from the heavy crude oil extraction to the shipping terminal. By 1998, over 4 million tons per year were produced (63).

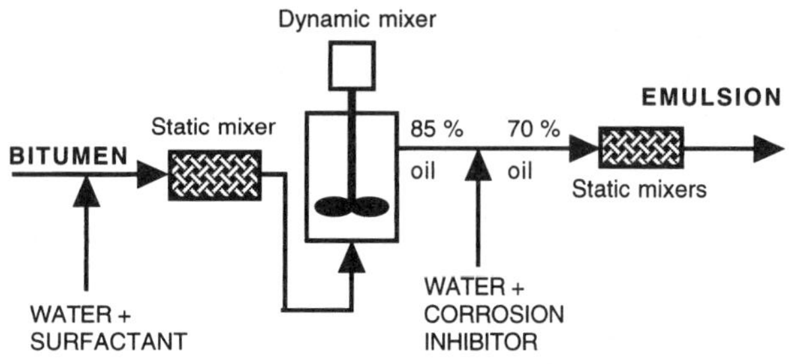


**Figure 18** Schematics of emulsification steps in the Orimulsion® current emulsification process.

One interesting question may be asked. Why not prepare the emulsion in just one step, since this would correspond to a point on the formulation–composition map, well within the $A^-$ region? The fact is that mixing a 70% oil content O/W emulsion requires a large amount of energy in order to attain a small drop size. This is associated with the high interfacial tension that is found distant from the optimum formulation. Mixing a higher oil content emulsion greatly improves the stirring efficiency for reasons that are still unclear. Nevertheless, it is a matter of fact that a shorter residence time and low-shear stirring are sufficient to attain the required drop size in these conditions. This process, which is usually referred to as the HIPR (high internal-phase ratio) emulsification method (220), has been noticed as well by cosmetic

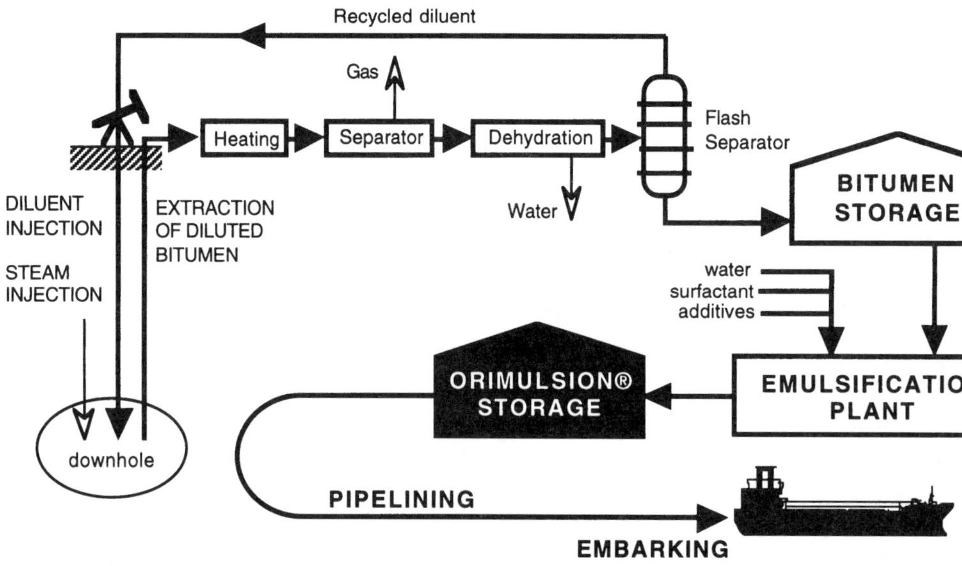


**Figure 19** Flow chart of Orimulsion® production.

formulators and others (221–226). Its high efficiency is probably linked with the high-momentum transfer insured by the viscous emulsion itself.

Scaling-up the emulsification process from the beaker to the production of several thousands tons per day was not an easy task. This was dealt with through fundamental and applied research. After understanding the basic formulation–composition conditions, many laboratory and pilot tests had to be carried out to find the appropriate mixing device and process parameters. Due to the proximity to the $A^-/B^-$ inversion branch, the risk of inversion to a W/O was high and it was necessary to "push away" the inversion line by applying low-shear mixing. Since adequate commercial in-line mixers were not available, it was necessary to design a mixing device, which was called Orimixer™, that would provide sufficiently low-shear mixing, together with a residence time long enough to allow the surfactant adsorption on to the interface.

If a sufficiently long residence time is essential to attain the required droplet size, it cannot be too long for two reasons. On the one hand, a long mixing time degrades the process economics. On the other, excessive recirculation of fluid through the impeller shear zone could occur with a risk of overmixing (227–229), a situation that would promote emulsion inversion.

An inversion produces an extremely viscous W/O emulsion almost instantaneously. Consequently, when inversion accidentally occurs, a rapid counteraction is imperative to protect the equipment gearbox.

Another problem that had to be dealt with was the preparation of the surfactant solution. In the first production scheme (down-hole emulsification), the water that was used to prepare the solution was warmed up so that surfactant dilution was facilitated. However, when switching to the second scheme (down-hole injection of diluent), the bitumen that came from the distillation tower was much hotter than in the previous process. The energy balance pointed to the fact that the water for the surfactant solution could not be warmed as much as before, otherwise the emulsification temperature would be too high, and thus the representative point on the formulation–composition diagram would be too near $SAD = 0$. This new requisite would not allow for the total, in-line dilution of the nonionic surfactant, which was likely to produce a gel when mixed with water that was not hot enough (230).

Therefore, it was necessary to study the dynamic and physicochemical aspects of the problem, in order to obtain the best combination of temperature and dynamic conditions (static mixer type, surfactant injection point in the mixer) that would accomplish the task (10, 11, 189).

## 2. Formulation Alternatives

The first choice was a nonionic surfactant because the primary emulsion contained a large amount of electrolytes. The selected surfactant was an ethoxylated nonylphenol with a mean content of 17.5 ethylene oxide groups per nonylphenol molecule (10). The surfactant was not modified when switching to the second production scheme, both because of convenience and for its excellent performance in making a stable O/W emulsion.

Nevertheless, the surfactant cost represents a significant part of the total emulsion production cost. Moreover, the phenol molecule has been suspected of exhibiting exoestrogenic effects that could affect marine wildlife in the case of a spill (11). The imperative is thus to swap this surfactant for an equally performing, less expensive, and less toxic alternative.

Anionic surfactants are usually less expensive and they perform in a similar way. However, this type of surfactant frequently contains a sulfur atom and a sodium cation, which are forbidden for the combustion application of the emulsion. At a first glance, cationic surfactants are also ruled out due to their cost, unless they can be used in very small proportions, which is not the convenient situation for a high internal-phase ratio emulsion, because this tends to shrink the A region width.

Nonetheless, there was still the option of activating the natural surfactants (231, 232) that are a part of the heavy crude oil composition (probably resins and asphaltenes). This is a well-known technology in enhanced oil recovery (233–236), in which a strong base, e.g., sodium hydroxide, is used to activate the carboxylic acids that are contained in the crude oil (237–240).

Since sodium ions are banned, extensive research was dedicated to organic bases, such as ethanolamines (10, 11), which were found to perform equally well, particularly in reducing interfacial tension to very low values (10). In fact, emulsification was made easier, probably as a result of the transient occurrence of ultralow interfacial tension, which enabled static mixers to produce a fine emulsion, maybe by spontaneous emulsification (216, 241, 242).

If the process is interpreted by means of the formulation–composition map (181), the system will be located near the inversion line, as in path 1, in which droplet size reaches a minimum. Stability is also near its minimum and, therefore, the system has to be displaced to well inside the A⁻ region in order to attain a stable emulsion. In the case of Orimulsion®, this could be done by adding a hydrophilic surfactant, such as a highly ethoxylated alcohol (10).

### 3. Future Developments in Emulsification Process

It has been mentioned that the presence of water in the heavy crude oil emulsion is detrimental to the calorific value, although the lower flame temperature produces less $NO_x$ emissions and lower corrosion. Nevertheless, an increase in oil content up to 80% would be beneficial to the combustion process.

Increasing the internal-phase content in very concentrated emulsions results in an exponential increase in viscosity and a growth in complexity of the rheological behavior (84, 243–245). Consequently, reducing only the amount of water is ruled out, and something else has to be done to compensate for the increase in viscosity due to the increase in oil content. A clever alternative is to use the viscosity-trimming feature of bimodal emulsions. A reduced-viscosity bimodal emulsion can be made in practice either by mixing two unequally sized unimodal emulsions (see Fig. 10), or by promoting the growth of a second mode through sophisticated mixing operations (see Fig. 16).

The first alternative has been already tested in the field (246). Current research is oriented to the scale-up of the formation of bimodal emulsions, which would allow increasing the heavy crude oil content up to 80% (11).

In summary, the current research efforts of PDV are focused at reducing costs (mainly by shifting to a lower-priced surfactant package) and by increasing the product value. This last may be achieved by increasing the oil content and, hence, the calorific value.

### 4. Rheological Behavior and Pipeline Transportation

The transportation of Orimulsion®, from the manufacturing plant in Morichal to the terminal port of Jose on the Caribbean Sea (Fig. 20), is accomplished in a 350 km long pipeline with a typical residence time of 3 to 4 days (10, 11). The flow regime is essentially laminar (Re < 1000) and rather low shear rates are reached in the pipe ($< 10\,s^{-1}$). At such a shear rate the heavy crude oil emulsion exhibits a viscosity near 2 Pa.s, sufficiently high to explain the low Reynolds

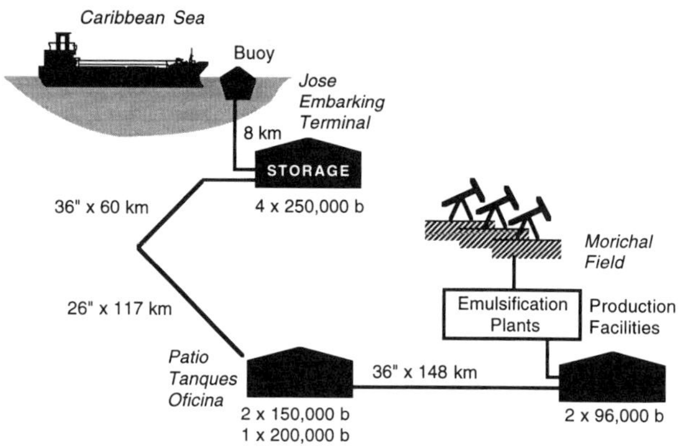


**Figure 20** Pipeline transportation of Orimulsion® from Morichal field and emulsion manufacturing plants to the seaport embarking facility. (From Refs 11 and 63.)

number, in spite of the flow rate that can reach an impressive 200,000 barrels/day (10, 11). In fact, the pipelining of Orimulsion® is probably a milestone in terms of the large-scale transportation of nonNewtonian fluids.

Predicting the pressure drop for pipeline transportation of such a fluid has not been an easy task. The rheological behavior of Orimulsion®, as measured in concentric cylinder rheometers of the Couette type, is shear thinning and only slightly viscoplastic and viscoelastic (63). At first, it was thought that these rheological data was reliable enough to predict the pressure drop in the pipeline. However, the field data have repeatedly showed that the actual pipeline pressure drop is systematically lower than the one predicted from the rheometric data (10).

Actual data for the commercial pipeline are still undivulged, but the basic characteristics of this unexpected and beneficial phenomenology can be ascertained from the published results (189) of one of the first field tests that was carried out in a pipe 70 km long and 24 inches in diameter.

As indicated in Fig. 21 (189), the flow rate was gradually increased for a period of 8 days, and the pumping pressure (or pressure drop) was found to remain constant and even to decrease. Although the effect is obvious, caution is advised when interpreting these data because steady-state conditions might not have been reached during the test. However, interesting comments may be advanced. During the first 5 days of testing, the pressure drop increased in the expected way; then, a four-fold increase in flow rate from the fifth to the seventh day barely resulted in a pressure-drop increase.

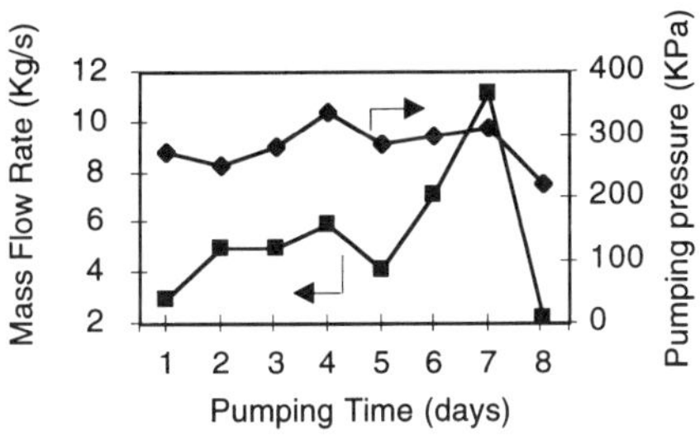


**Figure 21** Mass flow rate and pumping pressure as a function of pumping time for Orimulsion®, in a 24-inch diameter, 70 km long pipe. (Reproduced with permission from Revista Técnica INTEVEP, Vol. 10, No. 1, p. 13. Copyright INTEVEP S.A. 1990.)

At first, this phenomenon was attributed to the whimsical shear-thinning behavior of the emulsion. Further testing indicated that the lack of significant pressure increase was due to a very different cause and everything pointed to the occurrence of slip flow (247–249), already mentioned in Sect. III.A.

Further studies seemed to confirm this presumption. To that end, a once-through open-loop 7/8-inch pipe viscometer, which is described elsewhere (250), was used. The steady-state pressure drop readings could be made at two locations in the pipe: Leg 1 the nearest to the pump and Leg 2 the nearest to the discharge tank, i.e., further away in the flow sequence.

Figure 22 depicts the variation of the pressure drop as a function of mass flow rate for an 80% heavy crude oil O/W bimodal emulsion. It can be seen that the pressure drop readings are essentially the same for both legs up to a critical flow rate at which the pressure drop levels off in Leg 2. In this portion of the pipe, the mass flow rate increases two-fold while the pressure drop remains essentially unchanged (250). At a higher flow rate, the behavior becomes approximately Newtonian, as indicated by the straight-line variation.

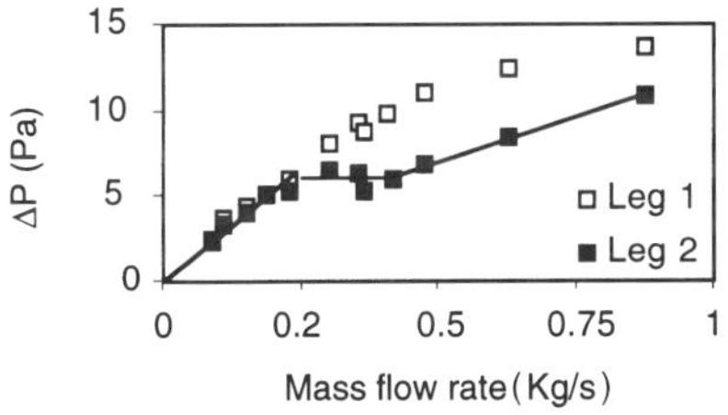


**Figure 22** Mass flow rate as a function of pressure drop for a bimodal, 80% heavy crude oil emulsion, in a once-through open-loop, 21.7 mm diameter pipe. (From Ref. 250.)

Pressure drop variations in the first leg (Leg 1) seem to indicate a pseudoplastic behavior, although a slight discontinuity appears at 0.35 kg/s.

It was speculated that hydrodynamic conditions could induce the migration of droplets away from the pipe walls, with a compensating counter-diffusion of the continuous phase toward the wall (250). This dynamic phenomenon seems to require not only a minimum speed to be triggered, but also a sufficiently long stretch of pipe for the lubricating layer to build up. The length over which the lubricating layer develops seems to be a function of the pipe diameter, or of the combined effect of the term $L/D$ ($L$ being pipe length). The larger the pipe diameter, the longer the stretch of pipe required to develop a lubricated regime. Therefore, a lubricating regime could evolve in a large-diameter pipe, provided that it is long enough to allow for the attainment of the aforementioned condition (250).

Although this clever diagnostic is sufficient to explain the unusual field data on pipe transportation of Orimulsion®, it is obvious that more research is essential to deepen the understanding of this phenomenon, which could be advantageously harnessed in this and other applications.

### D. Handling of Orimulsion®

As mentioned before, transportation is effected by means of a pipeline, up to the shipping terminal. The emulsion is stable in pipe flow, regardless the shear rate, as long as the size of the conduit is many times the size of the droplets (250), which is obviously the case in practical applications. Flow restrictions or contractions induce coalescence of droplets and, hence, an increase in mean drop size (10).

This is the case for many centrifugal pumps, especially high shear and multistage ones, in which the gap between the impellers and the casing is less than 1 mm. These types of pumps tend to damage the emulsion. In this sense, screw pumps have been strongly advised for Orimulsion® pumping (10, 11, 63).

Double-hulled tankers are used for oceanic transportation. The temperature is maintained around 30°C by hot water or glycol heating (11, 63, 66). It is worth nothing that no electric nor steam-heating systems are suitable since the heating elements could reach at their surface a temperature beyond the surfactant PIT, thus triggering the local inversion of the emulsion (189). In fact, both overheating beyond 80°C and freezing are to be avoided in order to ensure

Orimulsion® stability in storage tanks for long periods.

### E. Heavy Oil O/W Emulsified Transport

Emulsification as a low-viscosity vehicle for pipeline transportation shares some requirements with the O/W fuel emulsion, but differs in others. In effect, the transported emulsion should be stable during the pipeline pumping, but should be easy to break at the pipeline end.

The neighboring of SAD = 0 cannot be selected because of the extremely low stability. Consequently, a negative SAD value is selected, however, not too negative since the required stability does not warrant it. A not too robust emulsion can be made with a relatively high drop size, say 20–40 μm, and an oil content much lower than in the Orimulsion® case, say 50–60% oil. In such conditions the emulsion would probably be very fluid and would be moved by conventional techniques, even low-speed centrifugal pumps.

As far as the formulation variation is concerned, the best prospect seems to change the emulsion stability by means of a change in temperature with a nonionic-surfactant system. Figure 23 (left) indicates a typical path. The original emulsion can be made at some distance from optimum formulation on the negative side of SAD thanks to a combination of hydrophilic surfacant, e.g., nonylphenol with 10 EO groups and elevated temperature, say 70°C. The elevated temperature reduces the hydrophilicity of the surfactant to near SAD = 0 where the minimum drop size is attained and the oil-phase viscosity diminishes. This is the proper combination for easy emulsification. The emulsion is then rapidly cooled to ambient temperature, say 25°C, with a resulting increase in stability (black square box on left plot). At the end of the pipeline, the temperature is elevated to the one that corresponds to SAD = 0, e.g., 90°C, at which the emulsion breaks down easily. This was roughly the process used for making and transporting the primary emulsion in early Orimulsion® manufacturing (10, 189).

A slightly more complex scheme can be envisioned to curtail emulsification energy. As indicated in Fig. 23 (right) the emulsion can be prepared at the same favorable combination of hydrophilic surfactant and temperature, but this time with a higher internal-phase ratio, in order to make use of the favorable conditions for low-shear stirring. The emulsion is then cooled and diluted with cold water along any of the indicated paths in Fig. 23 (right). The destabilization is carried out by heating as in the first case.

If the oil is particularly acid, an alkaline water solution can be the answer to easy emulsification. However, the stability problem will have to be considered separately. In effect the activated natural surfactant would be able to produce an extremely stable emulsion.

It may be thought that emulsified transport does not require a completely dehydrated oil, since oil–water separation is to be carried out after pipelining. This is not the case, however, because of the possibility of osmotic swelling, which was discussed previously, that could become particularly annoying if the emulsion viscosity is already a concern.

## V. CONCLUSION

Although the transport of heavy crude oil as an O/W emulsion was first proposed in the early 1960s, it took until the 1970s to understand the importance of the

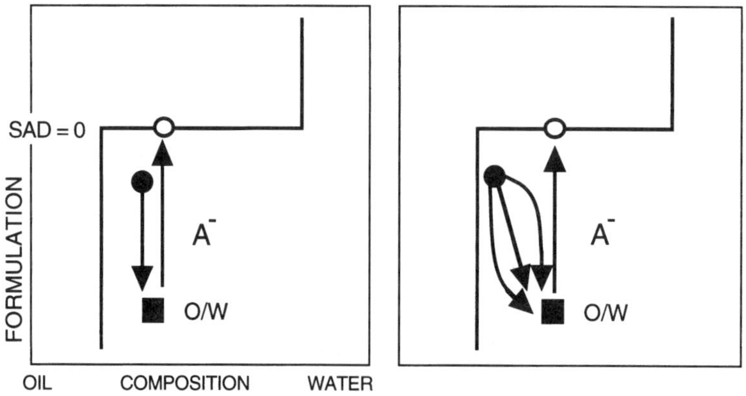


**Figure 23** Schematics for emulsion transport processes.

physicochemical formulation of SOW systems (thanks to the enhanced oil-recovery research drive), and the 1980s to apply these concepts to an actual case of emulsion making with all its scientific and engineering intricacies. The large-scale development of a commercial product such as Orimulsion® required 15 years of research and engineering effort and huge investments, in large part because of the extraordinarily rich variety of problems to be solved, both anticipated and unexpected, many of them to be elucidated from scratch (10, 59). This is a very serious lesson in humility for those who might think too quickly that since a considerable amount of know-how is at hand, emulsion making has become a straightforward business.

Even if it can be said that emulsion science has advanced some giant steps in the past two decades, and that extremely complex effects are now understandable and predictible, there is still an edge for a great deal of innovative research because of the large number of degrees of freedom in the formulation, composition, and fluid-mechanical conditions to be mastered during emulsification.

The pitfalls and drawbacks that paved the path of Orimulsion® development (59) clearly indicate that our knowledge is still too encapsulated in a protective cocoon, and that a lot of work has to be dedicated to real-world formulation engineering aspects such as the scale up of emulsification fluid-mechanical conditions, and the puzzling relationship between formulation, composition, and stirring effects.

One of the main know-how challenges is now to grasp the inherent nature of emulsion inversion (208) in order to push it away, to avoid it, or to harness it, depending on the application. Another one is to start treating emulsions as heterogeneous systems, probably a first step on the way to understand and take advantage of their complex rheological behavior near a solid boundary. This may have repercussions in many applications since fluids such as drilling foams, emulsified paints, and even blood are likely to exhibit such bizarre and extraordinary behavior.

The recently unveiled know-how indicates that it is possible to program changes in space and time to attain emulsion properties that might not exist in nature. Some day, this should allow us to improve upon the already enigmatic recipes used by cosmetic formulators and cooks alike.

As a final comment, it may be said that there is no doubt that often used random trial and error procedures are doomed to fail as research strategies, because of the enormous number of variables and degrees of freedom. One of the aims of this chapter was to try to convince the reader that a large amount of know-how does exist in a well-organized form, which could be extremely useful as a handy tool to carry out emulsion-formulation engineering tasks.

## ACKNOWLEDGMENTS

The authors would like to thank María L. Chirinos, Wladimiro Sarmiento, and Carlos Viloria (BITOR S. A.) for providing them with a written copy of their presentation material in the 1998 scientific meeting on Orimulsion® held in Valencia (Venezuela) and for calling their attention on specific issues. They also thank their colleague Professor Marta Ramirez (ULA) for providing Fig. 10 data, and Dr Arjan Kamp (INTEVEP S.A.) for his help in the literature search on foamy crudes. Finally, the authors would like to thank Ms Lylje Holmquist for her helpful comments.

The authors are indebted to the University of the Andes Research Council CDCHT, to the National Research Council CONICIT, particularly the "Agenda Petróleo" program, and to INTEVEP, R & D Subsidiary of Petróleos de Venezuela, for sponsoring Lab. FIRP research in Emulsion Science and Technology over the past 20 years.

## REFERENCES

1. R Sédillot. Histoire du Pétrole. Paris: Fayard, 1974.
2. JT Smith. A method for classifying tar sand oil versus heavy crude oil. Sixth UNITAR International Conference on Heavy Crude and Tar Sands. Houston, TX, 1995, Proceedings. Vol 1, p. 225.
3. The United Nations Institute for Training and Reseach (UNITAR) organized several International Conferences on Heavy Crudes and Tar Sands. The 1st was in Edmonton, Canada (1979), the 2nd in Caracas, Venezuela (1982), the 3rd in Long Beach, USA (1985), the 4th in Edmonton (1988), the 5th in Caracas (1991), the 6th in Houston, USA (1995), and the 7th in Beijing (1998). Although the proceedings of these symposia contain about 100 technical papers with an enormous amount of information on heavy crudes, tar sand oils, and bitumens, they carry rather limited information on extraheavy crude oil emulsions.
4. LF Ivanhoe. Updated Hubbert curves analyze World oil supply. World Oil 217: 91–94, 1996.
5. CJ Campbell. The coming Oil crisis. Brentwood, UK: Multi-Science Publishing and Petroleum Consultants, 1997.
6. CJ Campbell, JH Laherrère. The end of cheap oil. Sci Am (March): 60–65, 1998.

7. C Borregales. Como explotan el crudo de la Faja. Petról Int (Feb): 17–22, 1980.
8. ML Chirinos, J Gonzalez, IA Layrisse. Rheological properties of crude oils from the Orinoco oil belt and their mixtures with diluents. Rev Téc INTEVEP 3: 103–115, 1983.
9. C Guzmán, G Montero, MI Briceño, ML Chirinos, I Layrisse. Physical properties and characterization of Venezuelan heavy and extra heavy crudes and bitumens. Fuel Sci Technol Int 7: 571–598, 1989.
10. Seminario Técnico – 10 años de Orimulsion®, INTEVEP, Los Teques, Venezuela, 1985.
11. ML Chirinos. El negocio Orimulsion®. Seminario sobre Orimulsion®, Universidad de Carabobo, Valencia, Venezuela, 1998.
12. GE Smith, Fluid flow and sand production in heavy oil reservoir under solution gas drive. SPE Prod Eng (May): 169–177, 1988.
13. DJ Loughead, M Saltuklaroglu. Lloydminster heavy oil production. Why so unusual? Ninth Annual meeting of the Canadian Heavy Oil Association, Calgary, AB, 1992.
14. BB Maini, HK Sarma, AE George. Significance of foamy oil behavior in primary production of heavy oil. J Can Petrol Technol 32: 50–54, 1993.
15. EL Claridge, M Pratts. A proposed model mechanism for anomalous foamy heavy oil behavior. Paper SPE 29243. 1995 SPE International Heavy Oil Symposium, Calgary, Canada.
16. BB Maini. Foamy oil in heavy oil production. J Can Petrol Technol 35: 21–24, 1996.
17. M De Mirabal, R Gordillo, G Rojas, H Rodriguez, M Huerta. Impact of foamy oil mechanism on Hamaca oil reserves, Orinoco oil belt, Venezuela. Paper SPE 36140. 1996 Latin American Caribbean Petroleum Engineering Conference, Trinidad, pp 791–796.
18. M Huerta, C Otero, A Rico, I Jimenez, M de Mirabal, G Rojas. Understanding foamy oil mechanism for heavy oil reservoir during primary production. Paper SPE 36749. 1996 SPE Annual Technical Conference, Denver, CO, 1996.
19. AW Hyndmann. Regional development of Canada's oil sands. Sixth UNITAR International Conference on Heavy Crude and Tar Sands, Houston, TX, 1995. Proceedings Vol 1, p 85.
20. CT Purdy. Syncrude and the oil sands: meeting the need of North American refiners. Sixth Unitar International Conference on Heavy Crude and Tar Sands, Houston, TX, 1995. Proceedings, Vol 1, p 75.
21. P Pereira, R Marzin, MJ McGrath, H Feintuch, G Thompson, E Route. New residue process increases conversion, produces favorable residue in Curazao refinery. Oil Gas J 96:79–86, 1998.
22. P Pereira, R Marzin, L Zacarias, J Cordoba, J Carrazza, M Mariño. Steam conversion process and catalyst. US Patent 5 885 441, 1999.
23. TR Sifferman. Flow properties of difficult to handle waxy crude oils. J Petrol Technol (Aug.): 1047–1050, 1979.
24. RA Griffith. Getty-operated 20 inch heated crude pipeline. Paper SPE 3069. 45th Annual Fall Meeting SPE, Houston, TX, 1970.
25. RD Spade. Hot pipeline design and operating factors. Pipeline Gas J 200: 80–83, 1973.
26. H Cabrera. Production and transport of heavy oils by blending with lighter crudes. Second UNITAR International Conference on Heavy Crudes and Tar Sands. Caracas, Venezuela, 1982.
27. GB Wallis. One-dimensional two-phase flow. New York: McGraw-Hill, 1969.
28. TWF Russel, ME Charles. The effect of the less viscous liquid in the laminar flow of mixtures of two immiscible liquids. Can J Chem Eng 37: 18, 1959.
29. ME Charles, J Redberger. Reduction of pressure gradients in oil pipelines of stratified flow. Can J Chem Eng 40: 70, 1962.
30. G Achutaramayya, CA Sleicher. Analysis of stratified laminar flow of immiscible liquids in circular and non-circular pipes. Can J Chem Eng 47: 347, 1969.
31. D Hasson, U Mann. A Nir. Annular flow of two immiscible liquids. Can J Chem Eng 48: 514, 1970.
32. MJ Lamb, WC Simpson. Pipeline transportation of wax-laden crude oil as water suspension. Sixth World Petroleum Congress, Germany, 1963. Proceedings, Section VII, Paper 13, p 23.
33. MJ Lamb, WC Simpson. Waxy crude moves in water suspension. Oil Gas J 61: 140, 1963.
34. HJ Sommer, WC Simpson. Method of transporting liquid through pipelines. US Patent 3 006 354, 1961.
35. AV Kane. Method of moving viscous crude oil through a pipeline. US Patent 3 425 429, 1970.
36. R Simon, CD McAuliffe, WG Poynter, HY Jennings. Pipelining crude oil. US Patent 3 487 844, 1970.
37. R Simon, WG Poynter. Pipelining oil/water mixtures, US Patent 3 519 006, 1970.
38. JL Zakin, R Pinaire, ME Borgmeyer. Transportation of oils as oil-in-water emulsions. J Fluid Eng 101: 100, 1979.
39. GG McClaflin, CR Clark, TR Sifferman. The replacement of hydrocarbon diluent with surfactant and water for the production of heavy viscous crude oil. Paper SPE 10094. 56 Annual Fall Technical Conference SPE, San Antonio, TX, 1981.
40. GG McClaflin. Method of transporting viscous hydrocarbons. US Patent 4 246 920, 1981.
41. J Briant, DH Fruman, D Quemada, A Makria. Transport de pétroles bruts sous forme d'émulsions d'huile dans eau. Rev IFP 37: 809–821, 1982.
42. DH Fruman, J Briant. Investigation of the rheological characterization of heavy crude oil-in-water emulsions. International Conference on the Physical Modeling of Multiphase Flow, Coventry, UK, April 19–21, 1983, Paper J2.

43. SS Marsden, SC Rose. Cold emulsion line proposed in Arctic. Oil Gas J 69: 100–106, 1971.
44. CD McAuliffe, R Simon, CE Johnson. Method of pumping viscous crude. US Patent 3 380 531, 1968.
45. CD McAuliffe, R Simon, CE Johnson. Pumping viscous crude. US Patent 3 467 195, 1969.
46. R Simon, WG Poynter. Downhole emulsification for improving viscous crude production. JPT 20: 1349–1352, 1968.
47. AH Beyer, DE Osborn. Downhole emulsification for improving paraffinic crude production. Paper SPE 2767. 44th Annual Fall Meeting SPE, Denver, CO, 1969.
48. DA Alvarado, S Marsden. Flow of oil-in-water emulsions through tubes and porous media. SPE J 19: 369, 1979.
49. JA Cengel, AA Faruqui, JW Finnigan, CH Wright, JG Knudsen. Laminar and turbulent flow of unstable liquid–liquid emulsions. AIChE J 8: 335, 1962.
50. AA Faruqui, JG Knudsen. Velocity and temperature profiles of unstable liquid–liquid dispersions in vertical turbulent flow. Chem Eng Sci 17: 897, 1962.
51. JP Ward, JG Knudsen. Turbulent flow of unstable liquid–liquid dispersions: drop sizes and velocity distributions. AIChE J 13: 356, 1967.
52. JL Grosso, MI Briceño, J Paterno, I Layrisse. Influence of crude oil and surfactant concentration on the rheology and flowing properties of heavy crude oil in water emulsions. In: K Mittal, P Bothorel, eds. Surfactants in Solution. Vol 6. New York: Plenum Press, 1987, pp 1653–1673.
53. J Grosso, ML Chirinos, H Rivas, J Paternó, M Rivero, J González, I Layrisse. Transporte de Crudos Pesados Mediante Emulsiones. Bol Téc ARPEL 13: 225, 1984.
54. H Rivas, ML Chirinos, L Paz, I Layrisse, EL Murray, A Stockwell. Heavy and extra heavy crude oil in water emulsions for transportation: their formulation and characterization. Third UNITAR International Conference on Heavy Crude and Tar Sands. Long Beach, CA, 1985, p 1483.
55. J Grosso, I Layrisse, J González, JL Salager, J Villabona. La Influencia de los parámetros fluidomecánicos y fisico-químicos en la formación de emulsiones. Rev Téc INTEVEP 5: 3–8, 1985.
56. JL Salager, L Quintero, E Ramos, JM Andérez. Properties of surfactant-oil-water emulsified systems in the neighborhood of the three phase transition. J Colloid Interface Sci 77: 288, 1980.
57. FS Milos, DT Wasan. Emulsion stability of surfactant systems near the three phase region. Colloids Surfaces 4: 91, 1982.
58. JL Salager, JL Grosso, MA Eslava. Flow properties of emulsified surfactant-oil-water systems near optimum formulation. Rev Téc INTEVEP 2: 149–154, 1982.
59. LA Pacheco, J Alonso. Orimulsion® – The growing pains of an idea. Sixth UNITAR International Conference on Heavy Crude and Tar Sands. Houston, TX, 1995. Proceedings, Vol 1, p 203.
60. D Rodriguez, E Jimenez, J Izaguirre, J Salazar, R Carrizo, J Ancántara. La emulsion de bitumen Orinoco en agua (Orimulsion) como combustile excepcional en plantas termoeléctricas. Rev Téc INTEVEP 7: 17, 1987.
61. Vladimiro Sarmiento, Orimulsion® – sus usos, Seminario sobre Orimulsion®, Universidad de Carabobo, Valencia, Venezuela, 1998.
62. I Layrisse, H Rivas, M Rivero, E Guevara, M de Oliveira, L Matheus, C Kovach. Production, treatment and transportation of a new fuel Orimulsion, Twelfth International Conference on Slurry Technology, New Orleans, LA, 1987.
63. F Marruffo, W Sarmiento. Orimulsion®: an alternative source of energy. BITOR SA, Caracas, 1998.
64. Orimulsion®. Web site: http://www.pdv.com/orimulsion/english/index.html.
65. GA Nuñez, H Rivas. Imulsion® – A continuously evolving technology. Fifth UNITAR International Conference, Caracas, Venezuela, 1991, Vol 4, p 231.
66. C Viloria. Orimulsion® – aspectos ambientales. Seminario sobre Orimulsion®, Universidad de Carabobo, Valencia, Venezuela, 1998.
67. P Becher. Emulsions: Theory and Practice. Reprint. Huntington, NY: R Krieger, 1977.
68. P Sherman, ed. Emulsion Science. London: Academic Press, 1968.
69. KJ Lissant, ed. Emulsions and Emulsion Technology. In 2 parts, Vol 6, Surfactant Science Series. New York: Marcel Dekker, 1974.
70. P Becher, ed. Encyclopedia of Emulsion Technology. 4 Vols. New York: Marcel Dekker, 1985–1996.
71. J Sjöblom, ed. Emulsions and Emulsion Stability. Vol 61. Surfactant Science Series. New York: Marcel Dekker, 1996.
72. DT Wasan. Emulsion stability mechanisms. First World Congress on Emulsion, Paris, 1993. Proceedings, Vol 4, pp 93–112.
73. HT Davis, Colloid Surfaces A 91: 9, 1994.
74. DA Bruggeman. Ann Phys Lpz 24: 636, 1935.
75. T Hanai, N Koizumi, T Sugano, R Gotoh. Kolloid Z 171: 20, 1960.
76. T Hanai, N Koizumi, R Gotoh. In: P Sherman, ed. Rheology of Emulsions. London: Pergamon Press, 1963, pp 353.
77. M Clausse. Dielectric properties of emulsions and related systems. In: P Becher, ed. Encyclopedia of Emulsion Technology. Vol 1. New York: Marcel Dekker, 1983, p 481.
78. M Miñana, P Jarry, M Perez-Sanchez, M Ramirez-Gouveia, JL Salager. Surfactant-oil-water systems near the affinity inversion – Part V:

Properties of emulsions. J Dispers Sci Technol 7: 331, 1986.
79. P Sherman, ed. Rheology of Emulsions. New York: Pergamon Press, MacMillan, 1963.
80. P Sherman. Rheology of Emulsions. In: P Sherman, ed. Emulsion Science. London: Academic Press, 1968, pp 271–351.
81 VI Graifer, GA Lazarev, MI Leont'ev. Effects of various factors on the viscosity of crude oil and water emulsions. Int Chem Eng 15: 274–276, 1975.
82. P Sherman. Rheological properties of emulsion. In: P Becher, ed. Encyclopedia of Emulsion Technology. Vol 1. New York: Marcel Dekker, 1983, p 405.
83. AK Das, D Mukesh, V Swayambunathan, DD Kotkar, PK Ghosh. Concentrated emulsions. 3. Studies on the influence of continuous phase viscosity, volume fraction, droplet size and temperature on emulsion viscosity. Langmuir 8: 2427–2436, 1992.
84. R Pal, E Rhodes. J Rheol 33: 1021, 1989.
85. R Pal. Pipeline flow of unstable and surfactant-stabilized emulsions. AIChE J 39: 1754–1764, 1993.
86. JG Overbeek. Colloid and Surface Chemistry – A self Study Guide, Part 3. Electrokinetics and Membrane Phenomena. Cambridge: Massachusetts Institute of Technology, 1971.
87. S Asukara, F Oosawa. J Phys Chem 22: 1255, 1954.
88. A Vrij. J Pure Appl Chem 48: 471, 1976.
89. J Bibette. J Colloid Interface Sci 147: 474, 1991.
90. DD Joseph, YY Renardy. Fundamentals of Two Fluids Dynamics. Part II: Lubrication Transport, Drops and Miscible Fluids. Berlin: Springer, 1992.
91. GV Jeffreys, GA Davies. Coalescence of liquid droplets and liquid dispersions. In: C Hanson, ed. Recent Advances in Liquid–Liquid Extraction. Elmsford, NY: Pergamon Press, 1971, pp 495–584.
92. BJ Carroll. The stability of emulsions and mechanisms of emulsion breakdown. In: E Matijevic, ed. Surface and Colloid Science. Vol 9. New York: John Wiley, 1976, p 1.
93. L Lobo, I Ivanov, DT Wasan. Dispersion coalescence: Kinetic stability of creamed dispersions. AIChE J 39: 322, 1993.
94. TF Tadros, B Vincent. Emulsion stability. In: P Becher, ed. Encyclopedia of Emulsion Technology. Vol 1. New York: Marcel Dekker, 1983, p 129.
95. DH Melik, HS Fogler. Fundamentals of colloidal stability in quiescent media. In: P Becher, ed. Encyclopedia of Emulsion Technology. Vol 3. New York: Marcel Dekker, 1988, p 3.
96. H Wang, RH Davis. Droplet growth due to Brownian, gravitational or thermocapillary motion and coalescence in dilute dispersions. J Colloid Interface Sci 159: 108, 1993.
97. P Walstra. Emulsion stability. In: P Becher, ed. Encyclopedia of Emulsion Technology. Vol 4. New York: Marcel Dekker, 1996, p 1.
98. IB Ivanov, ed. Thin liquid films. New York: Marcel Dekker, 1988.
99. IB Ivanov, PA Kralchevsky. Stability of emulsions under equilibrium and dynamic conditions. Colloids Surfaces A: Physicochem Eng Aspects 128: 155–175, 1997.
100. JL Salager. Quantifying thc concept of physico-chemical formulation in surfactant-oil-water systems. Progr Colloid Polym Sci 100: 137–142, 1996.
101. WC Griffin. Classification of surface active agents by HLB. J Soc Cosmet Chem 1: 311, 1949; 5: 249, 1954.
102. P Becher. J Dispers Sci Technol 5: 81, 1984.
103. P Becher, WC Griffins. McCutcheon's Detergents and Emulsifiers, North American ed. Ridgewood, NJ: Allured Publishing, 1974, p 227.
104. P Becher. HLB: an updated bibliography. In: P Becher, ed. Encyclopedia of Emulsion Technology. Vols 2–4. New York: Marcel Dekker, 1983–1996, pp 425, 397, 337.
105. K Shinoda, H Arai. The correlation between PIT in emulsion and cloud point in solution of nonionic emulsifier. J Phys Chem 68: 3485, 1964.
106. K Shinoda. The comparison between the PIT system and the HLB-value system to emulsifier selection. Proceedings of the Fifth International Congress on Surface Activity. Barcelona, Vol 2, 1969, p 275.
107. K Shinoda, H Kunieda. Phase properties of emulsions: PIT and HLB. In: P Becher, ed. Encyclopedia of Emulsion Technology. Vol 1. New York: Marcel Dekker, 1983, p 337.
108. P Winsor. Solvent Properties of Amphiphilic Compounds. London: Butterworth, 1954.
109. M Bourrel, RS Schechter. Microemulsions and Related Systems. New York: Marcel Dekker, 1988.
110. JC Lang, B Widom. Equilibrium of three liquid phases and approach to the tricritical point in benezene-ethanol–water–ammonium sulfate mixtures. Physica 81 A: 190–213, 1975.
111. M Guerrero, JS Rowlinson, G Morrison. Model fluid mixture which exhibitss tricritical points. Part 1. J Chem Soc, Faraday Trans II 72: 1970–1979, 1976.
112. N Desrosiers, M Guerrero, JS Rowlinson, D Stubley. Model fluid mixture which exhibits tricritical points. Part 2. J Chem Soc, Faraday Trans II 73: 1632–1645, 1977.
113. WH Wade, J Morgan, J Jacobson, JL Salager, RS Schechter. Interfacial tension and phase behavior of surfactant systems. Soc Petrol Eng J 18: 242–252, 1978.
114. JL Salager. Fisico-química de los sistemas surfactante–agua–aceite: aplicaciones a la recuperación del petróleo. Rev Inst Mex Petrol 11: 59–71, 1979.
115. JT Davies. A quantitative kinetic theory of emulsion type. I. Physical chemistry of the emulsifying agent. Gas/liquid and liquid/liquid interfaces. Proceedings

Second International Congress on Surface Activity. Vol 1. London: Butterworths, 1957, p 426.
116. N Marquez, RE Antón, A Graciaa, J Lachaise, JL Salager. Partitioning of ethoxylated alkylphenol surfactants in microemulsion-oil-water systems. Colloids Surfaces A: Physicochemical Eng Aspects 100: 225–231, 1995.
117. N Marquez, RE Antón, A Graciaa, J Lachaise, JL Salager. Partitioning of ethoxylated alkylphenol surfactants in microemulsion-oil-water systems. Part II: Influence of hydrophobe branching. Colloids Surfaces A: Physicochemical Eng Aspects 131: 45–49, 1998.
118. H Schott. Hydrophilic–lipophilic balance, solubility parameter, and oil–water partitioning as universal parameters of nonionic surfactants. J Pharm Sci 84: 1215, 1995.
119. JL Salager. Microemulsions. In: G Broze, ed. Handbook of Detergents – Part A: Properties. Surfactant Science Series, Vol 82. New York: Marcel Dekker, 1999, pp 253–302.
120. JL Salager. Formulation concepts for the emulsion maker. In: F Nielloud, G Marti-Mestres, eds. Pharmaceutical Emulsions and Suspensions. New York: Marcel Dekker, 2000, Chap. 2, pp 19–72.
121. MC Puerto, WW Gale. Estimation of optimal salinity and solubilization parameter for alkylorthoxylene sulfonate mixtures. Soc Petrol Eng J 17: 193, 1977.
122. JL Salager, J Morgan, RS Schechter, WH Wade, E Vasquez. Optimum formulation of surfactant–oil–water systems for minimum tension and phase behavior. Soc Petrol Eng J 19: 107, 1979.
123. M Bourrel, JL Salager, RS Schechter, WH Wade. A correlation for phase behavior of nonionic surfactants. J Colloid Interface Sci 75: 451, 1980.
124. M Bourrel, C Koukounis, RS Schechter, WH Wade. Phase and interfacial tension behavior of nonionic surfactants. J Dispers Sci Technol 1: 13–35, 1980.
125. RE Antón, N Garcés, A Yajure. A correlation for three-phase behavior of cationic surfactant–oil-water systems. J Dispers Sci Technol 18: 539–555, 1997.
126. JL Salager, RE Antón. Ionic microemulsions. In: P Kumar, K Mittal, eds. Handbook of Microemulsions Science and Technology. Surfactant Science Series. New York: Marcel Dekker, 1999, pp 247–280.
127. JL Salager, M Bourrel, RS Schechter, WH Wade. Mixing rules for optimum phase behavior formulation of surfactant–oil–water systems. Soc Petrol Eng J 19: 271, 1979.
128. M Hayes, M El-Emary, RS Schechter, WH Wade. The relation between $EACN_{min}$ concept and surfactant HLB. J Colloid Interface Sci 68: 591, 1979.
129. RE Antón, JL Salager, A Graciaa, J Lachaise. Surfactant–oil–water systems near the affinity inversion – Part VIII: Optimum formulation and phase behavior of mixed anionic–nonionic systems versus temperature. J Dispers Sci Technol 13: 565–579, 1992.
130. RE Antón, D Gómez, A Graciaa, J Lachaise, JL Salager. Surfactant–oil–water systems near the affinity inversion – Part IX: Optimum formulation and phase behavior of mixed anionic–cationic systems. J Dispers Sci Technol 14: 401–416, 1993.
131. M Baviere, RS Schechter, WH Wade. The influence of alcohol on microemulsion composition. J Colloid Interface Sci 81: 266, 1981.
132. RE Antón, JL Salager. Effect of the electrolyte anion on the salinity contribution to optimum formulation of anionic surfactant microemulsions. J Colloid Interface Sci 140: 75, 1990.
133. M Puerto, R Reed. Surfactant selection with the three-parameter diagram. SPE Reservoir Eng (May): 198, 1990.
134. P Fotland, A Skange. Ultralow interfacial tension as a function of pressure. J Disper Sci Technol 7: 563–579, 1986.
135. A Skange, P Fotland. Effect of pressure and temperature on the phase behavior of microemulsions. SPE Reservoir Eng (Nov): 601, 1990.
136. R Cash, JL Cayias, G Fournier, D McAllister, T Shares, RS Schechter, WH Wade. The application of low interfacial tension scaling rules to binary hydrocarbon mixtures. J Colloid Interface Sci 59: 39, 1977.
137. JL Cayias, RS Schechter, WH Wade. Modeling crude oil for low interfacial tension. Soc Petrol Eng J 16: 351, 1976.
138. JL Salager, M Bourrel, RS Schechter, WH Wade. Physico-Chimie de la Récupération assistée du Pétrole par solutions micellaires: Récents développements. Bull Cent Rech Expl Prod 2: 399–417, 1978.
139. A Graciaa, Y Barakat, RS Schechter, WH Wade, S Yiv. Emulsion stability and phase behavior for ethoxylated nonyl phenol surfactants. J Colloid Interface Sci 89: 217, 1982.
140. H Kunieda, K Hanno, S Yamaguchi, K Shinoda. The three-phase behavior of a brine/ionic surfactant/nonionic surfactant/oil system: Evaluation of the hydrophile–lipophile balance (HLB) of ionic surfactant. J Colloid Interface Sci 107: 129, 1985.
141. H Saito, K Shinoda. The stability of W/O type emulsions as a function of temperature and of the hydrophilic chain length of the emulsifier. J Colloid Interface Sci 32: 617, 1970.
142. J Boyd, C Parkinson, P Sherman. Factors affecting emulsion stability and the HLB concept. J Colloid Int Sci 41: 359, 1972.
143. RN Healy, RL Reed. Physicochemical aspects of microemulsion flooding. Soc Petrol Eng J 14: 491, 1974.
144. P Doe, M El-Emary, JC Morgan, RS Schechter, WH Wade. Systematic studies of low interfacial tension parameters. Second ERDA Symposium on

Enhanced Oil and Gas Recovery, Tulsa, 1976, Paper B-6, Petroleum Publishing Co.
145. CA Miller, RN Hwan, W Benton, T Fort Jr. Ultralow interfacial tensions and their relation to phase separtion in micellar solutions. J Colloid Interface Sci 61: 554–568, 1977.
146. P Dunlap, PM Wilson, C Brandner. Aqueous surfactant solutions which exhibit ultralow tensions at the oil–water interface. J Colloid Int Sci 60: 473–479, 1977.
147. JL Cayias, RS Schechter, WH Wade. The utilization of petroleum sulfonates for producing low interfacial tensions between hydrocarbons and water. J Colloid Int Sci 59: 31–38, 1977.
148. WH Wade, JC Morgan, JK Jacobson, RS Schechter. Low interfacial tensions involving mixtures of surfactants. Soc Petrol Eng J 17: 122, 1977.
149. DO Shah, RS Schechter, eds. Improved Oil Recovery by Surfactant and Polymer Flooding. New York: Academic Press, 1977.
150. AM Bellocq, D Bourdon, B Lemanceau, G Fourche. Thermodynamic, interfacial and structural properties of polyphasic microemulsion systems. J Colloid Interface Sci 89: 427–440, 1982.
151. KE Bennet, CH Phelps, HT Davis, LE Scriven. Microemulsion phase behavior-observations, thermodynamic essentials, mathematical simulation. Soc Petrol Eng J21: 747–762, 1981.
152. K Shinoda, M Hanrin, H Kunieda, H Saito. Principles of attaining ultra-low interfacial tension: The role of hydrophile–lipophile balance of surfactant at oil/water interface. Colloids Surfaces 2: 301–314, 1981.
153. JC Noronha, DO Shah. Ultra-low interfacial tension, phase behavior and microstructure in oil/brine/surfactant/alcohol systems. AIChE Symp Ser, No. 212, 78: 42–57, 1982.
154. S Qutubuddin, CA Miller, T Fort. J Colloid Interface Sci 101: 46, 1984.
155. JL Salager, I Loaiza-Maldonado, M Miñana-Pérez, F Silva. Surfactant–oil–water systems near the affinity inversion – Part I: Relationship between equilibrium phase behavior and emulsion type and stability. J Disper Sci Technol 3: 279–292, 1982.
156. A Kalbanov, H Wennerström. Macroemulsion stability: the oriented wedge theory revisited. Langmuir 12: 276–292, 1996.
157. M Bourrel, A Graciaa, RS Schechter, WH Wade. The relation of emulsion stability to the phase behavior and interfacial tension of surfactant systems. J Colloid Interface Sci 72: 161, 1979.
158. JE Viniatieri. Correlation of emulsion stability with phase behavior in surfactant systems for tertiary oil recovery. Soc Petrol Eng J 20: 402, 1980.
159. LM Baldauf, RS Schechter, WH Wade, A Graciaa. The relationship between surfactant phase behavior and the creaming and coalescence of macroemulsions. J Colloid Interface Sci 85: 187–197, 1982.
160. RE Antón, JL Salager. Emulsion instability in the three-phase behavior region of surfactant–alcohol–oil–brine systems. J Colloid Interface Sci 111: 54–59, 1986.
161. R Hazzlett, RS Schechter. Colloids Surfaces 29: 53, 1988.
162. A Kalbanov, J Weers. Macroemulsion stability within the Winsor III region: Theory versus experiment. Langmuir 12: 1931–1935, 1996.
163. JL Salager. The fundamental basis for the action of a chemical dehydrant. Influence of the physical and chemical formulation on the stability of an emulsion. Int Chem Eng 30: 103–116, 1990.
164. JL Salager, M Miñana-Perez, J Andérez, J Grosso, CI Rojas, I Layrisse. Surfactant–oil–water systems near the affinity inversion – Part II: Viscosity of emulsified systems. J Dispers Sci Technol 4: 161, 1983.
165. ED Berger, WR Munk, HA Wahl. Flow increase in the trans-Alaska pipeline through use of a polymeric drag-reducing additive. J Petrol Technol (Feb): 377, 1982.
166. S Vijayan, C Ramachandran, H Doshi, DO Shah. Porous media rheology of emulsions in tertiary oil recovery. Symposium of Surface Phenomena in Enhanced Oil Recovery. Third International Conference on Surface and Colloid Science, Stockholm, 1979, p 327.
167. HY Jennings Jr. A study of caustic solution – crude oil interfacial tensions. Soc Petrol Eng J 15: 197, 1975.
168. VK Bansal, KS Chan, R McCallough, DO Shah. The effect of caustic concentration on interfacial charge, interfacial tension and droplet size: a simple test for optimum caustic concentration for crude oils. J Can Petrol Technol 17: 69, 1978.
169. PH Krumkine, JS Falcone, TC Campbell. Surfactant flooding. 1: The effect of alkaline additives on interfacial tension, surfactant adsorption, and recovery efficiency. Soc Petrol Eng J 22: 503, 1982.
170. TC Campbell. The role of alkaline chemicals in the recovery of low-gravity crude oils. J Petrol Technol 34: 2510, 1983.
171. JL Salager, M Pérez-Sánchez, Y Garcia. Physicochemical parameters influencing the emulsion drop size. Colloid Polym Sci 274: 81–84, 1996.
172. JL Salager, M Miñana-Perez, M Perez-Sanchez, M Ramirez-Gouveia, CI Rojas. Surfactant–oil–water systems near the affinity inversion – Part III: The two kinds of emulsion inversion. J Dispers Sci Technol 4: 313–329, 1983.
173. K Shinoda, H Arai. J Colloid Int Sci 25: 429–431, 1967.
174. RE Antón, P Castillo, JL Salager. Surfactant–oil–water systems near the affinity inversion – Part IV:

Emulsion inversion temperature. J Dispers Sci Technol 7: 319–329, 1986.
175. E Dickinson. Emulsions. Annual Reports C. London: The Royal Society of Chemistry, p 31.
176. P Jarry, M Miñana-Perez, JL Salager. Inversion of surfactant–oil–brine emulsified systems: generalized mapping and property transitions. In: K Mittal, P Bothorel, eds. Surfactants in Solution. Vol 6. New York: Plenum Press, 1987, pp 1689–1696.
177. JL Salager, G Lopez-Castellanos, M Miñana-Perez. Surfactant–oil–water systems near the affinity inversion – Part VI: Emulsions with viscous hydrocarbons. J Dispers Sci Technol 11: 397–407, 1990.
178. JL Salager, G Lopez-Castellanos, M Miñana-Perez, C Cucuphat-Lemercier, A Graciaa, J Lachaise. Surfactant–oil–water systems near the affinity inversion – Part VII: Phase behavior and emulsions with polar oils. J Dispers Sci Technol 12: 59–67, 1991.
179. BW Brooks, HN Richmond. Dynamics of liquid–liquid phase inversion using nonionic surfactants. Colloids Surfaces 58: 131, 1991.
180. RE Antón, H Rivas, JL Salager. Surfactant–oil–water systems near the affinity inversion – Part X: Emulsions made with anionic-nonionic surfactant mixtures. J Dispers Sci Technol 17: 553–566, 1996.
181. Z Mendez, RE Antón, JL Salager. Surfactant–oil–water systems near the affinity inversion – Part XI: pH sensitive emulsions containing carboxylic acids. J Dispers Sci Technol 20: 883–892, 1999.
182. JL Salager. Guidelines to handle the formulation, composition and stirring to attain emulsion properties on design (type, drop size, viscosity and stability). In: A Chattopadhyay, K Mittal, eds. Surfactants in Solution, Surfactant Science Series 64. New York: Marcel Dekker, 1996, pp 261–295.
183. JL Salager, M Perez-Sanchez, M Ramirez-Gouveia. JM Andérez, MI Briceño. Stirring–formulation coupling in emulsification. IXth European Congress on Mixing, Paris, 1997. Published in Récents Progrès en Génie des Procédés, Vol 11, No. 5: Multiphase Systems, 1997, pp 123–130.
184. C Parkinson, S Matsumoto, P Sherman. The influence of particle size distribution on the apparent viscosity of non-Newtonian dispersed systems. J Colloid Interface Sci 33: 150–160, 1970.
185. RJ Farris. Prediction of the viscosity of multimodal suspensions from unimodal viscosity data. Trans Soc Rheol 12: 281–301, 1968.
186. R Hoffman. Factors affecting the viscosity of unimodal and multimodal colloidal dispersions. J Rheol 36: 947–965, 1992.
187. JL Salager, M Ramirez-Gouveia, J Bullón. Properties of emulsion mixtures. Prog Colloid Polym Sci 98: 173–176, 1995.
188. MI Briceño, M Ramirez, J Bullón, JL Salager. Customizing drop size distribution to change emulsion viscosity. Second World Congress on Emulsion, Bordeaux, France, 1997, Proceedings Vol 2, paper 2-1-094-01/05.
189. MI Briceño, ML Chirinos, I Layrisse, G Martinez, G Nuñez, A Padron, L Quintero, H Rivas. Emulsion technology for the production and handling of extra-heavy crude oils and bitumens. Rev Téc INTEVEP 10: 5–14, 1990.
190. M Ramirez-Gouveia. Technical Report FIRP 9210. Universidad de Los Andes, Mérida, Venezuela, 1992.
191. A Graciaa, J Lachaise, JG Sayous, P Grenier, S Yiv, RS Schechter, WH Wade. J Colloid Interface Sci 93: 474, 1983.
192. A Graciaa, J Lachaise, M Bourrel, I Osborne-Lee, RS Schechter, WH Wade. SPE Reservoir Eng, 305, 1984.
193. TF Tadros, B Vincent. Emulsion stability. In: P Becher, ed. Encyclopedia of Emulsion Technology. Vol 1. New York: Marcel Dekker, 1983, p 129.
194. VG Levich. Physicochemical Hydrodynamics. Englewood Cliffs, NJ: Prentice Hall, 1962.
195. F Silva, A Peña, M Miñana-Perez, JL Salager. Dynamic inversion hysteresis of emulsions containing anionic surfactants. Colloids Surfaces A: Physicochemical Eng Aspects 132: 221–227, 1998.
196. A Peña, JL Salager. Effect of stirring energy upon the dynamic inversion hysteresis of emulsions. Colloids Surfaces, in press.
197. JL Salager. Phase transformation and emulsion inversion on the basis of catastrophe theory. In: P Becher, ed. Encyclopedia of Emulsion Technology. Vol 3. New York: Marcel Dekker, 1988, pp 79–134.
198. GEJ Vaessen. Predicting catastrophic phase inversion in emulsions. PhD dissertation, Eindhoven University of Technology, Netherlands, 1996.
199. BW Brooks, HN Richmond. Phase inversion in nonionic surfactant–oil–water systems. I. The effect of transitional inversion on emulsion drop size. Chem Eng Sci 49: 1953, 1994.
200. M Miñana-Perez, C Gutron, C Zundel, JM Andérez, JL Salager. Miniemulsion formation by transitional inversion. J Dispers Sci Technol 20: 893–905, 1999.
201. E Dickinson. Interpretation of emulsion phase inversion as a cusp catastrophe. J Colloid Interface Sci 84: 284, 1981.
202. DH Smith, KH Lim. Langmuir 6: 1071, 1990.
203. KH Lim, DH Smith. J Colloid Interface Sci 142: 278, 1991.
204. JL Salager. Applications of catastrophe theory to surfactant–oil–brine equilibrated and emulsified systems. In K Mittal, P Bothorel, eds. Surfactants in Solution. Vol 4. New York: Plenum Press, 1987, pp 439–448.
205. P Becher. The effect of the nature of the emulsifying agent on emulsion inversion. J Cosmet Chem 9: 141, 1958.

206. JL Salager. Phase behavior of amphiphile–oil–water systems related to the butterfly catastrophe. J Colloid Interface Sci 105: 21–26, 1985.
207. BW Brooks, H Richmond. Dynamics of liquid–liquid inversion using non-ionic surfactants. Colloids Surfaces 58: 131–148, 1991.
208. JL Salager, L Marquez, A Peña, M Rondon, J Silva, E Tyrode. Phenomenological know-how and modeling of emulsion inversion. Ind Eng Chem Res 39:2665, 2000.
209. DC England, JC Berg. The transfer of surface active agents across a liquid–liquid interface. AIChEJ 17: 313, 1971.
210. KJ Ruschak, CA Miller. Spontaneous emulsification in ternary systems with mass transfer. Ind Eng Chem Fundam 11: 534, 1972.
211. JL Salager. Influence of the physico-chemical formulation and stirring on emulsion properties – state of the art. International Symposium on Colloid Chemistry in Oil Production, ISCOP97, Rio de Janeiro, 1997.
212. JL Salager, M Pérez, M Ramirez, MI Briceño, Y Garcia. Combining formulation, composition and stirring to attain a required emulsion drop size. State of the art. Second World Congress on Emulsion, Bordeaux, France, 1997. Proceedings, Vol 2, paper 1-2-093-01/05.
213. JL Salager, RE Antón, CL Bracho, MI Briceño, A Peña, M Rondón, S Salager. Attainment of emulsion properties on design – A typical case of formulation engineering. Second European Congress in Chemical Engineering, Montpellier, France, 1999.
214. JL Salager, Emulsion properties and related know-how to attain them. In: F Nielloud, G Marti-Mestres, eds. Pharmaceutical Emulsions and Suspensions. New York: Marcel Dekker, 2000, Chap. 3, pp 73–125.
215. K Shinoda, H Saito. The stability of O/W type emulsions as a function of temperature and the HLB of emulsifiers: The emulsification by PIT-method. J Colloid Int Sci 30: 258–263, 1969.
216. DT Wasan, SM Shah, M Chan, K Sampath, R Shah. Spontaneous emulsification and the effect of interfacial fluid properties on coalescence and emulsion stability in caustic flooding. In: RT Johansen, RL Berg, eds. Chemistry of Oil Recovery. ACS Symposium Series 91, Washington, DC: American Chemical Society, 1979, p 115.
217. P Sherman. J Phys Chem 67: 2531, 1963.
218. P Sherman. J Colloid Sci 24: 97, 107, 1967.
219. P Sherman. Rheological change in emulsion aging. IV. O/W emulsions at intermediate and low rates of shear. J Colloid Interface Sci 27: 281, 1968.
220. A Stockwell, SE Taylor, EJ Taylor, EJ Murray, ML Chirinos. Viscous crude oil transportation: the preparation of bitumen, heavy and extraheavy crude oil in water emulsions. Third UNITAR International Conference On Heavy Crude and Tar Sands. Long Beach, CA, 1985. Proceedings, Vol. 4, p 1983.
221. TJ Lin, T Akabori, S Tanaka, K Shimura. Low energy emulsification. Part III: Emulsification in high alpha range. Cosmet Toilet 95: 33, 1980.
222. TJ Lin, T Akabori, S Tanaka, K Shimura. Low energy emulsification. Part IV: Effect of emulsification temperature. Cosmet Toilet 96: 31, 1981.
223. TJ Lin, T Akabori, S Tanaka, K Shimura. Low energy emulsification. Part V: Mechanism of enhanced emulsification, Cosmet Toilet 98: 67, 1983.
224. TJ Lin, YF Shen. Low energy emulsification. Part VI: applications in high-internal phase emulsions. J Soc Cosmet Chem 35: 357, 1984.
225. TG Masson, J Bibette. Shear rupturing of droplets in complex fluids. Langmuir 13: 4600, 1997.
226. TG Masson, J Bibette. Emulsification in viscoelastic media. Phys Rev Lett 77: 16, 1996.
227. RS Rajagopal. Proc Indian Acad Sci 49A: 333, 1959.
228. P Becher. The process of emulsification: a computer model. Langmuir 7: 1325, 1991.
229. J Lachaise, B Mendiboure, C Dicharry, G Marion, JL Salager. Simulation of the overemulsification phenomenon in turbulent stirring. Colloids Surfaces A: Physicochem Eng Aspects 110: 1–10, 1996.
230. R Laughlin. The aqueous phase behavior of surfactants. London: Academic Press, 1994.
231. O Kimbler, RL Reed, IH Silberberg. Physical characteristic of natural films formed at crude oil–water interfaces. Soc Petrol Eng J 6: 153, 1966.
232. JE Stassner. Effect of pH on interfacial films and stability of crude oil – water emulsions. J Petrol Technol 20: 303, 1968.
233. CE Cooke, RE Williams, PA Kolodzie. Oil Recovery by alkaline waterflooding. J Petrol Technol (Dec.): 1365, 1974.
234. N Mungan. Enhanced oil recovery using water as a driving fluid. Part 4 : Fundamentals of alkaline flooding. World Oil (June): 209, 1981.
235. CI Chiwetelu, V Hornof, GH Neale. A dynamic model for the interaction of caustic reagents with acidic oils. AIChE J 36: 233, 1990.
236. H Rivas, X Gutierrez, JL Zirrit, RE Antón, JL Salager. Microemulsion and optimum formulation occurrence in pH dependence systems as found in alkaline enhanced oil recovery. In: C Solans, H Kunieda, eds. Industrial Applications of Microemulsions. Surfactant Science Series 66. New York: Marcel Dekker, 1997, pp 305–329.
237. WK Seifert, WG Howells. Interfacially active acids in a California crude oil. Isolation of carboxylic acids and phenols. Analyt Chem 41: 554, 1969.
238. WK Seifert. Effect of phenols on the interfacial activity of crude oil (California) carboxylic acids and the identification of carbazoles and indoles. Analyt Chem 41: 562, 1969.

239. I Layrisse, H Rivas, S Acevedo. Isolation and characterization of natural surfactants present in extra heavy crude oils. J Dispers Sci Technol 5: 1, 1984.
240. I Layrisse, H Rivas, S Acevedo, R Medina, M Sanchez, M Utrera. Composicion y caracteristicas fisico-quimicas de crudos extrapesados. Rev Téc INTEVEP 4: 3–18, 1984.
241. MM Sharma, LK Jang, TF Yen. Transient interfacial tension behavior of crude-oil/caustic interfaces. SPE Reservoir Eng (May): 228, 1989.
242. J Rudin, DT Wasan. Interfacial turbulence and spontaneous emulsification in alkali/acidic oil systems. Chem Eng Sci 48: 2225, 1993.
243. HM Princen. Highly concentrated emulsions I. Real systems. The effect of film thickness and contact angle on the volume fraction of creamed emulsions. J Colloid Interface Sci 71: 246, 1980.
244. HM Princen. Rheology of foam and highly concentrated emulsions. I. Elastic properties and yield stress of a cylindrical model system. J Colloid Interface Sci 91: 160, 1983.
245. HM Princen, AD Kiss. Rheology of foam and highly concentrated emulsions. IV. Experimental study of the shear viscosity and yield stress of concentrated emulsions. J Colloid Interface Sci 128: 177, 1989.
246. H Rivas, G Nuñez, C Dalas. Emulsiones de viscosidad controlada. Vision Tecnol 1: 18, 1994.
247. R Pal, SN Bhattacharya, E Rhodes. Flow behavior of oil-in-water emulsions. Can J Chem Eng 64: 3–10, 1986.
248. J Zhang, Y Dafan. Rheological measurements of oil-in-water emulsions: concentric cylinder viscometer. Fourth UNITAR International Conference on Heavy Crude and Tar Sands. Edmonton, Canada, 1988.
249. ML Chirinos, J Colmenares, MI Briceño, G Nuñez. Slip flow in bitumen-in-water emulsions. Fifth European Congress of Rheology, Sevilla, Spain, 1994. Proceedings, Darmstadt: Editorial Steinkopff, pp. 182.
250. GA Nuñez, MI Briceño, C Mata, H Rivas. Flow characteristics of concentrated emulsions of very viscous oil in water. J Rheol 40: 405–423, 1996.

# 21

# Water-in-Oil Emulsions in Recovery of Hydrocarbons from Oil Sands

**Jan Czarnecki**
*Edmonton Research Centre, Syncrude Canada Ltd., Edmonton, Alberta, Canada*

## I. INTRODUCTION

Conventional oil reserves decline all over the world. The same is true for Canada, which is self-sufficient in oil production. Although some oil is imported to eastern Canada, an excess production from the west is sold, mainly to the USA. Overall, Canada's consumption of oil is balanced by its production. However, the current production markedly exceeds new oil discoveries. If this trend continues, in several years Canada would become a net oil importer. Fortunately for the country, huge oil sand deposits in northern Alberta constitute an alternative source of fuels. Oil sands are fluvial, estuarine, and marine deposits of sand saturated with bitumen, a form of heavy oil. Typically, the oil sand ore contains about 9–13 wt% bitumen, the rest being a mixture of silica sand and fine clays.

Figure 1 shows the location of major oil sand deposits. In the blown-up insert, which is a simplified map of Alberta, one of the western Canadian provinces, the locations of oil sand deposits are marked in black. In the Athabasca region, the ore is located close to the surface allowing for surface mining. Two major commercial plants operate here, some 40 km north of the city of Fort McMurray. Both operating companies, Syncrude Canada Ltd and Suncor Energy Inc., Oil Sands, use surface mining and water-extraction methods to recover the bitumen. Several new projects are on the drawing board, the most advanced is a new mine to be operated by Shell Canada. In the Cold Lake and Peace River areas, where the oil is covered with a thicker layer of overburden, in-situ enhanced oil-recovery methods are used for commercial bitumen recovery.

The bitumen extracted from oil sands is too heavy to be shipped directly to the market. Most of the bitumen produced is upgraded by a combination of standard refinery technologies. Both Syncrude and Suncor use coking as the main primary upgrading method. It is followed by hydrotreatment, eventually to produce light sweet blends that are sold at premium. A major pipeline links Fort McMurray area with the North American pipeline system. About 15% of Canadian oil production (or consumption) comes from the Athabasca Oil Sands. Producers from Cold Lake and Peace River use a diluent, mainly natural gas condensate, to pipeline the dilute bitumen produced to refineries where it is upgraded to make commercial products. The total contribution of oil sands hydrocarbons to Canadian oil production currently exceeds 25% and is projected to bypass the 50% mark in the first decade of the twenty-first century.

The total reserves of hydrocarbons contained in the Alberta Oil Sands are huge. As shown in Figure 2, the total amounts of hydrocarbons locked in the oil sands that are recoverable using surface mining followed by

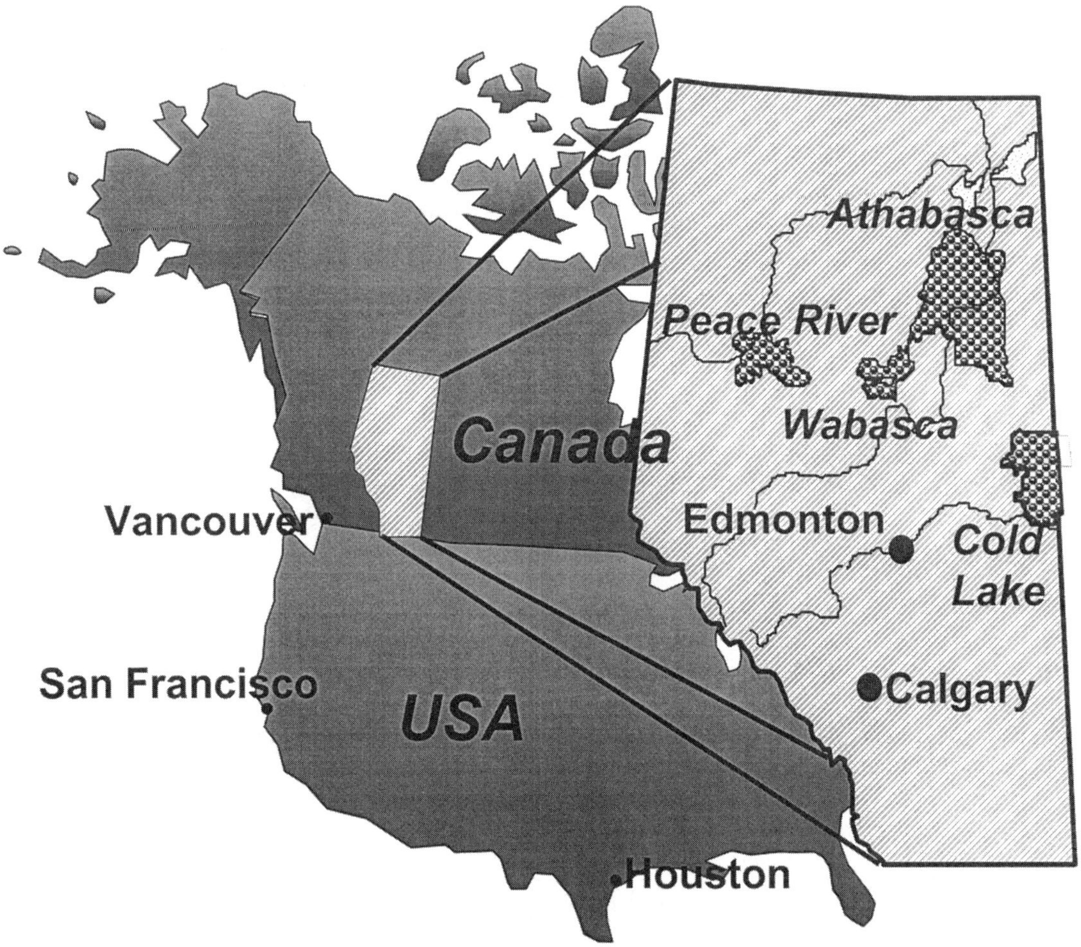


**Figure 1** Location of oil sand deposits in northern Alberta.

water-extraction technologies, are slightly higher than the total oil reserves of Saudi Arabia. However, the total hydrocarbon reserves contained in the oil sands located in northern Alberta are higher than the total known oil reserves of all OPEC countries combined.

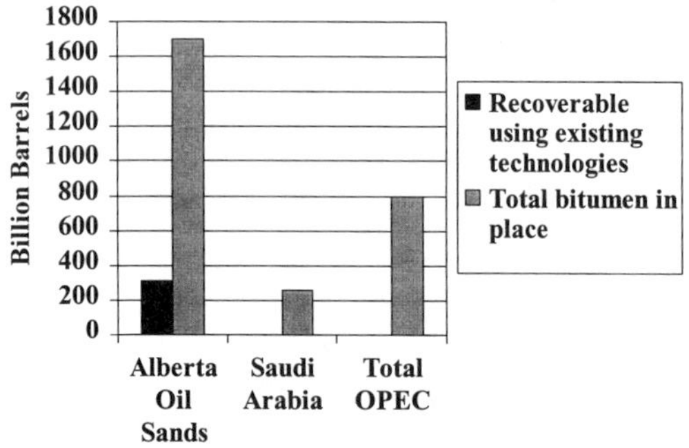


**Figure 2** Reserves of hydrocarbons locked in oil sands compared to other major oil sources.

The oil sand industry has succeeded in remarkable reduction in unit operation costs. In the first half of 1999, the costs of producing a barrel of upgraded synthetic sweet blend from oil sands by surface mining is hovering around Canadian $ 13, i.e., about US $ 8 (at the current exchange rate). Therefore, even at the low oil prices of late 1998 and early 1999, the industry is making a profit. New emerging technologies are likely to lower further the unit costs to about Canadian $ 10 or so within the next decade. It is thus evident that the oil sand industry is robust enough to become the main source of oil for Canada if not for North America for decades to come. This decline in the unit operating costs is mostly due to the introduction of new technologies in mining, extraction, and upgrading. New technologies were made possible because of extensive research efforts. The specific feature of the oil-sand industry is the need for knowledge, which is generated by joint efforts of industry, academia, and government research facilities. In the following paragraph, the technology for bitumen recovery, as used today, is briefly described.

The ore is primarily mined by a truck and shovel operation. Crushed ore is then mixed with water at the mine front and pumped to the extraction plant. This new hydrotransport technology has been already implemented by both operating plants and is to be utilized by all planned operations. In the pipeline the ore is digested, i.e., the existing lumps are ablated and the bitumen is liberated. The existing operations run the slurry pipelines at about 50°C. In the new planned operations coming on line in 2000, the pipeline temperature is going to be around 25°C, resulting in further reductions in energy use and in unit operating costs. Since the slurry preparation involves dumping the ore into water, screening, and pumping, some air is entrained in the slurry and dispersed into small air bubbles. Additional air is usually introduced to the pipeline. The bubbles collide with liberated bitumen droplets to form stable aggregates. All these processes make the slurry ready for consecutive bitumen separation. This takes place in large, relatively stagnant separation vessels, where the aerated bitumen rises to the surface forming a froth.

The froth contains typically about 60% bitumen, 30% water, and 10% solids. To remove solids and water, the froth is diluted with a diluent, in both existing operations this is locally produced naphtha. Bitumen density is about the same as that of water. Addition of a light solvent lowers the density of the oil phase and, at the same time, lowers its viscosity, making water and solids separation possible. In the froth-treatment operation, scroll and disk centrifuges as well as incline plate settlers (IPSs) are used for cleaning the froth diluted with naphtha. The product, which usually contains less than 2% water and less than 0.4% fine solids, is then fed to upgrading. Here, in the first operation, the diluent is recovered and recycled to froth treatment.

The problem is that the residual water mentioned above contains dissolved salts, mostly sodium chloride, which is then carried to the downstream upgrading operations, creating serious corrosion risks. The salt is coming from the ore, and because of the water recycle, accumulates in the process waters. Most of the water in diluted bitumen is present in the form of relatively large drops and lenses, which are relatively easy to remove in the froth-treatment operation. However, small quantities of water form a very stable emulsion, with a droplet size of about 2–5 $\mu$m. Removal of this water is of paramount importance, since excess of chloride salts, after treatment with hydrogen in hydrotreatment, is converted into hydrochloric acid. At the same time organic nitrogen is converted into ammonia. We thus have all the conditions to form volatile ammonium chloride than can migrate throughout the plant tending to precipitate in cooler and moister places, such as heat exchangers. In summary, the plant integrity depends on effective removal of water emulsified in diluted bitumen.

Pilot tests of conventional electrostatic desalters failed to improve water removal. In an attempt at improving dewatering efficiency, by making changes to the composition and amount of the diluent used for froth treatment, it was discovered that for paraffinic solvents, at sufficiently high diluent-to-bitumen ratio (D/B), a very clean and dry product can be made (1). This observation, first made on a bench scale, was then confirmed in extensive pilot studies. As the result, a new technology, the paraffinic froth-treatment process, was developed. In its future commercial operation, Shell Canada is planning to employ this technology by using natural gas condensate as the source of paraffinic diluent. The paraffinic process seems to have a number of advantages in a 'green field' situation. However, it is as always difficult to implement a new technology in an existing operating plant. Therefore, at Syncrude, several other coping strategies have been implemented ranging from the construction of a third-stage centrifuge plant to a joint research program with the demulsifier supplier, Champion Technologies, to optimize and customize the demulsifiers used in commercial operation. Despite considerable efforts and costs the problem is not yet satisfactorily solved.

There are two issues here – first, technological: how to lower the water content in the froth-treatment product, and second, scientific: what makes the water emulsion so stable. In this paper, I will focus on the latter and describe the results of collaborative studies performed by a group of researchers from the NSERC Research Chair in Oil Sands headed by Dr Jacob Masliyah from the Department of Chemical and Materials Engineering, University of Alberta in Edmonton,* CANMET Western Research Centre (Canadian Government research laboratory), and Syncrude's Edmonton Research Centre.

* Natural Sciences and Engineering Research Council (NSERC) support in the form of a grant to Dr Jacob Masliyah, the Chairholder of NSERC Research Chair in Oil Sands, is gratefully acknowledged.

One of the important characteristics of the oil sand industry is that it is unique to Alberta and, contrary to conventional oil which relies on worldwide research activities, there is very little known about the scientific needs of the oil sand industry outside the industry itself. Also, since the main competition to the oil sand industry is conventional oil, there is less need for extensive protection of intellectual property. Those two factors combined create an environment that strongly promotes collaboration in various areas, but especially in basic, precompetitive research. The material presented in this chapter summarizes the results of such collaborative efforts of the group mentioned above. Efforts were made to make flow of information as easy and boundary between various participating groups as fuzzy as possible.

## II. WASHING EXPERIMENTS

There are several hypotheses concerning the source of the water in bitumen emulsion stability. The most common says that the emulsion is stabilized by the asphalthene fraction of the bitumen. The following experiment (1) clearly shows that, although the stabilizers may come from the asphalthene fraction, it is not the whole asphalthene fraction that is responsible for the emulsion stability.

Figure 3 shows the schematic of the washing experiment. First, a solution of bitumen in toluene was prepared. Second, known aliquots of water were blended with the bitumen solution under standardized conditions. The emulsion formed (which we will call the first emulsion) was centrifuged under high rpm to create a clean supernatant. In the next step, 10% of water was blended into the supernatant to form the second emulsion. The stability of this secondary emulsion was studied as a function of the amount of water used to make the first emulsion. It is worth adding that the first emulsion was not broken even at high acceleration. Instead, the emulsion was concentrated at the bottom of the centrifuge cell in the form of a cake without any free water visible. Some of the results of washing experiments are shown in Figure 4.

As the amount of water used to make the first emulsion increases, the stability of the secondary emulsion decreases. This indicates that there is a possibility of exhausting the finite reserve of the material, which is responsible for the high emulsion stability, regardless what this material may be. It can be seen in the figure

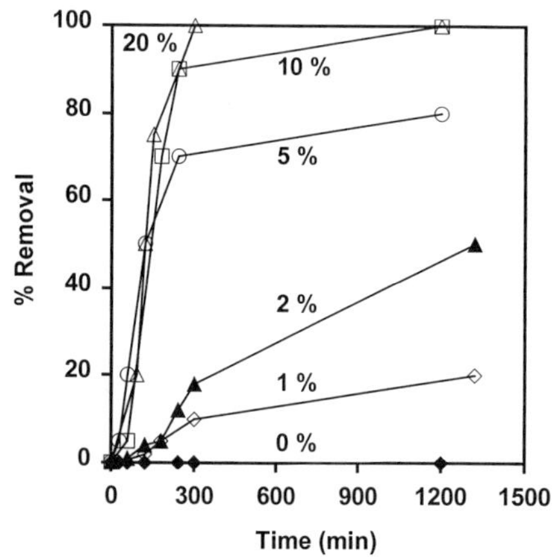


**Figure 4** Washing experiments results: percentage of water removed from the secondary emulsion for various amounts of water used to make the first emulsion, or to "wash" the oil.

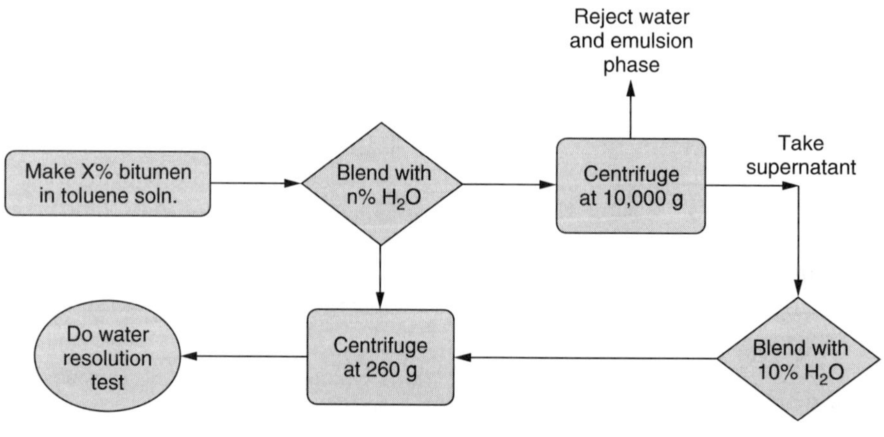


**Figure 3** Flow sheet of the washing experiment. Description in the text.

that, above the 5% water used to make the first, "washing" emulsion, the stability of the emulsion formed in the second step is already markedly reduced and is only slightly affected by increasing further the amount of washing water. In the first emulsion, most of the droplets are about 3 $\mu$m. We can calculate the total surface area of the water droplets in, say, 100 ml of 10% water emulsion. Then, assuming that the thickness of the layer that stabilizes the droplets via a steric mechanism is, say, 20 nm,* and that it has the same density as bitumen (i.e., 1 g/ml), we obtain the mass of the stabilizing layer as about 0.4 g. Since we had 25 g of bitumen in 10 ml toluene solution, this constitutes less than 2% of the bitumen. Now, the asphalthene content in Athabasca bitumen is about 18%. The conclusion is that, if asphalthenes are involved in stabilizing the emulsion, it can only be a small subfraction of the total asphalthene content of the oil.

## III. COLLOIDAL COLLIDER: COLLOIDAL FORCES BETWEEN EMULSIFIED WATER DROPLETS

Knowing the interaction forces between emulsified water droplets would help in understanding the reasons for the observed high stability of water in diluted bitumen emulsion. However, studies on liquid/liquid interactions are very limited, especially on forces between microscopic emulsion droplets. Recently, a new force-measuring technique, colloidal particle scattering (CPS), was developed to measure surface forces between micrometer-sized latex particles (2, 3). The apparatus, called a microcollider, is capable of determining forces of $10^{-14}$–$10^{-12}$ N, which is several orders of magnitude smaller than those detected by the surface-force apparatus (SFA) or atomic-force microscope (AFM). We performed experiments to measure the interaction forces between two 6.5-$\mu$m water droplets in a toluene-based solvent containing a small amount of bitumen.

The basic principles of CPS are described in detail in Ref. 2. In essence, the method is based on generating collisions between two micrometer-sized particles (droplets) under simple shear flow conditions and extracting the force–distance relationships by analyzing the asymmetry of collision trajectories before and after the collision. Figure 5 shows the schematic of the experiment set-up. The measuring cell consists of a fixed, upper wall made of optical glass and a movable bottom forming a gap of about 200 $\mu$m. The movements of the cell bottom are controlled by a computer and can be executed either by a joystick or by running a program resulting in the creation of a known shear flow within the cell. A collision is generated by bringing a random droplet in the emulsion, exhibiting Brownian motions, into contact with a previously found droplet attached to the upper cell wall. The collision is observed and recorded through the upper cell wall with a microscope equipped with a CCD camera and a VHS recorder. The image is then analyzed to find the position of the mobile droplet long before and after the collision. All three coordinates of the droplet can be extracted from the image. The $x,y$ coordinates can be deducted from the image directly. The third $z$ coordinate, normal to the plane of observation, can be calculated from the velocity of the droplet in the flow and the known shear rate, crated in the cell. Hydrodynamic considerations show that if there is no net attraction or repulsion between the droplets, the trajectory of the particle should be symmetrical with respect to the stationary droplet. However, if the droplets repel each other the mobile droplet will be pushed away from the stationary one

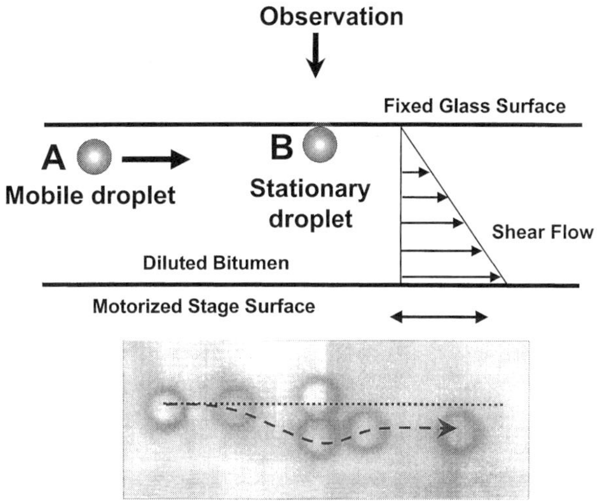


**Figure 5** Principle of the colloidal collider method. The shear flow is created between upper fixed plate and the mobile bottom of the cell. The insert shows several consecutive positions of the mobile droplet as it passes around the stationary one. The dotted line depicts a symmetrical trajectory expected when no surface forces exist. The dashed line shows the experimental trajectory indicating a repulsive interaction between the two droplets.

*In the following section, dealing with colloidal collider experiments, we present material indicating that the stability is as a result of a nonhomogeneous steric barrier with a thickness varying from 7 to 40 nm.

resulting in a shift of its final position (long after collision) versus the initial starting point. A detailed description of the method, including experimental procedure, theoretical basis for calculating the forces from the observed droplet trajectories, and error analysis are given in Refs 2–4.

To estimate colloidal forces acting between the droplets during the collision, $x$, $z$ coordinates of the initial $(x_i, z_i)$ and final $(x_f, z_f)$ positions of the mobile droplet before and after the collision are needed. When several pairs of $x$, $z$ coordinates are plotted on a graph a specific "scattering pattern" will appear. This pattern can be analyzed by comparing the experimental final positions with the ones calculated from a theory (2,3), which covers hydrodynamic interactions between the droplets and between the mobile droplet and the wall as well as all external forces acting on the mobile drop, e.g., those described by DLVO theory. Thus, in the calculations, we assume the existence of a certain force described by a certain function of the droplet–droplet separation. The final position of the droplet is then calculated and compared with the experimental results. The best match between experimental and theoretical final droplet positions yields the optimum set of parameters or the optimum force–distance profile.

One of the limitations of the collider method is the requirement of neutral buoyancy of the studied droplets in the medium. To match the density of the solution with that of the water, we mixed 71 wt% toluene with 29 wt% dibromobenzene (density 1.95 g/ml). The organic phase also contained 0.04 wt% bitumen extracted from Athabasca oil sand. Emulsion-stability tests showed that dibromobenzene had a negligible effect on the stability.

The experimental results are summarized in scattering diagrams (cf. Figs 6 and 7). The open circles represent initial positions of the mobile droplet $(x_i, z_i)$ and the solid circles represent final positions $(x_f, z_f)$. Owing to the poor light transmission in a dark bitumen solution, only seven collisions were analyzed.* The initial (open circles) and final (solid circles) droplet positions, numbered from 1 to 7, are shown in the figure. The scattering diagram shows that the final positions do not follow a "ring" pattern, normally observed in previous studies of polystyrene latex particle interactions (2, 4) where the force–distance relationship is consistent for all collisions.

It has been known that water droplets in a bitumen/toluene solution are stable. Since the van der Waals forces between two water droplets are always attractive, a repulsive force is required to stabilize the

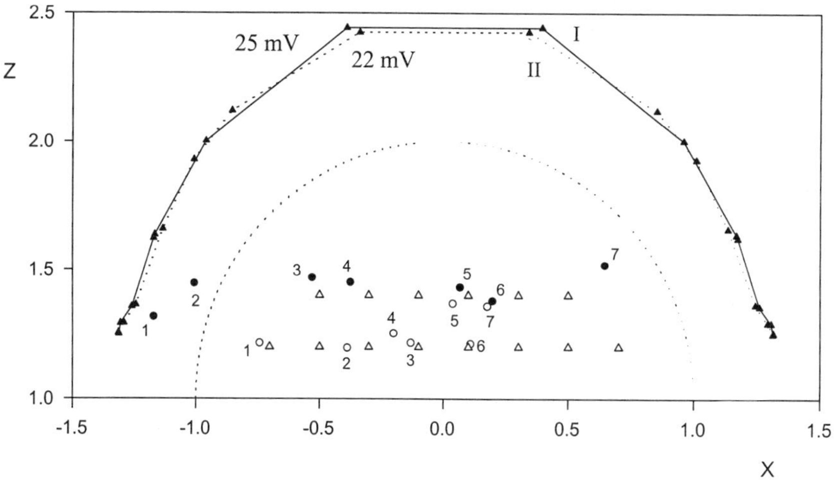


**Figure 6** Scattering diagram. Open circles: initial mobile droplet positions; solid circles: final mobile droplet positions. Open and solid triangles are initial and final positions taken for calculating scattering pattern if electrostatic forces were involved for surface potential 22 and 25 mV. Below 22 mV calculations suggested droplet coagulation.

* To enhance the contrast we added 0.01% of fluorescein isothiocyanate to the water. In the second instrument built we use reflected rather than transmitted light illumination and we use a polished stainless steel plate as the bottom of the cell that acts as a mirror. These modifications allowed us to work with solutions containing up to 10% bitumen without any dye in the aqueous phase.

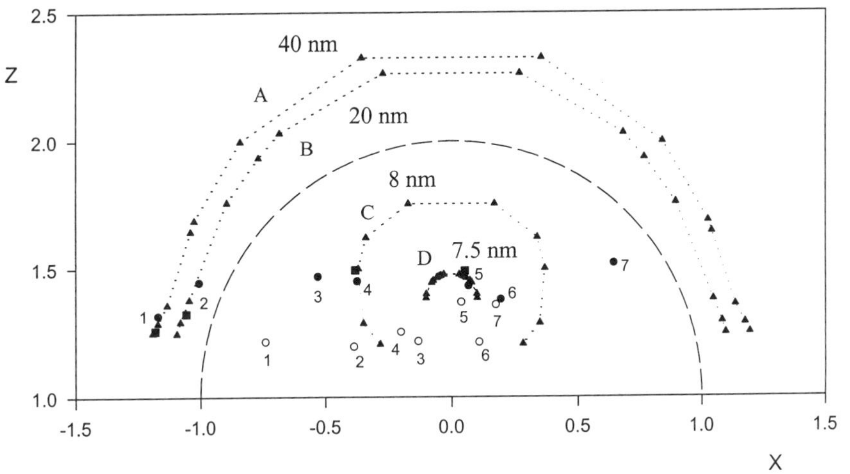


**Figure 7** Scattering diagram. Experimental initial and final positions are open and solid circles, respectively; the solid triangles depict final positions for a steric repulsion model with various thicknesses of the steric layer. The initial positions were the same as in Figure 6.

system. The most common repulsive forces are electrostatic and steric repulsion. The electrostatic interaction in a low dielectric constant organic medium is not fully understand. (We are currently working on a theory based on a cell model.) For the analysis reported herein we assumed that the electrostatic interactions in an organic medium are similar to that in an aqueous system of the same conductivity. The conductivity of our system is about 30 S/m. In an aqueous system, this would be the conductivity of $2 \times 10^{-6}$ M 1 : 1 electrolyte (e.g., NaCl). In an organic medium, however, ions are considerably smaller than their hydrated counterparts in water. This increases the equivalent conductivity of ions, so the argument above overestimates the actual ionic strength in an organic medium. Hence, we took a value of $1 \times 10^{-6}$ M as the ionic strength of our system. We also found that varying this value by one or two orders of magnitude does not change the final results qualitatively. This yields a $\kappa a$ value of 70 ($\kappa$ being the Debye parameter and $a$ the drop radius). The large $\kappa a$ value suggests that the double-layer interaction theory might still apply to our nonaqueous system. We used a modified Gouy–Chapman theory to calculate the electrostatic force.

Van der Waals forces were calculated on the basis of Hamaker's theory with a retardation function introduced by Schenkel and Kitchener (5). The Hamaker constant of water–toluene–water was taken from Ref. 6. Both the Hamaker constant and the retardation wavelength were kept constant.

Figure 6 gives two theoretical final positions "rings" (solid triangles) with surface potentials of 22 and 25 mV (signs unspecified). They correspond to arbitrarily chosen initial positions marked in the figure with open triangles. When the surface potential was below 22 mV, our calculations indicated that the mobile droplet should be captured permanently to form a doublet, contrary to our observations. It is obvious that the experimental final positions, which are mostly located close to the origin (0.0, 1.0), cannot be explained using the theoretical approach described above, no matter what is the value of the surface potential or ionic strength of the medium.

The steric repulsion mechanism is also difficult to model in our system. A bitumen/toluene solution itself is a very complex system containing high molecular weight asphaltenes, natural surfactants, and ultrafine particles. These components are very likely to be adsorbed on the water/toluene interface. Due to this complexity, it is hard to model the adsorption layer with a single elastic modulus, as was done for the analysis of poly(ethylene oxide) adsorption layers on latexes (7). However, all steric forces resemble hard-wall interactions. They can be approximately modeled by high-order polynomial functions. We used a simple expression $F_{steric} = c/h^7$, where $h$ is the separation distance between two droplets, and $c$ is a constant chosen in such a way that it makes $F_{steric}$ equal to the nonretarded van der Waals force at $h = 2L_s$ ($L_s$ being the adsorption layer thickness). The power of $h$ is arbitrary. The above formula describes a steric force that is zero slightly beyond $L_s$ and increases sharply as the distance $L_s$ is reached. It has been found that the final results are not sensitive to this power as long as it remains reasonably high. Based on this model, in Fig. 7 we plot theoretical final "rings" (solid triangles) using the same initial positions shown in Fig. 6 with open triangles. The scattering diagram shows that the experimental final position of collision 1 is located on ring A and the final position of collision 5 is on ring D. Other final positions located between rings A and D can also be predicted by varying $L_s$ between 7.5 and 40 nm. Therefore, we may state that our experimental results are consistent with the assumption of steric repulsion resulting from a heterogeneous steric layer. Another experimental finding by Yoon et al. (8) indicates that by using a SFA the interaction forces in an inverse system, i.e., bitumen/water/bitumen, resemble polymer/polymer interactions. If we assume that these interactions are caused by hydrophobic parts of the surfactants adsorbed on the bitumen/water interface, it is of no surprise to observe the steric interactions between hydrophobic parts of the same surfactants in a water/bitumen/water system. Error analysis indicated that the error is around 50 and 1% for the maximum and minimum value of $L_s$, respectively.

The most important message from the colloidal collider experiments is that our results will be consistent with any mechanism in which the repulsive force is decaying much faster than the attractive one (the van der Waals force). A steric force obviously meets this requirement. A well-screened double-layer force would also satisfy this condition at a certain separation $h$, and the resulting shallow minimum is called a secondary minimum in DLVO theory. However, this requires a high ionic strength ($\sim 0.01$ M) which is not possible in our system. If the actual ionic strength of our system is several orders of magnitude less than the assumed value ($10^{-6}$ M), the electric double layers cannot be fully developed and it is inappropriate to calculate the electrostatic force with a double-layer interaction equation. Although we do not know exactly what form the force equation will take, we know the expression of the Coulomb force, which is the extreme case of electrostatic interactions without screening. It does not decay faster with increasing $h$ than the van der Waals forces. For this reason, it seems that the electrostatic force is unlikely to be the main cause of stabilization in the system studied, although it may play a minor role in addition to the steric stabilization mechanism.

## IV. MICROPIPETTE STUDIES: THE INTERFACIAL PROPERTIES OF MICROMETER SIZED WATER DROPLETS IN DILUTE BITUMEN

The conclusion from the previous section was that the remarkable stability of water-in-diluted bitumen emulsions is due to the presence of a complex adsorbed layer at the surfaces of the dispersed droplets. Except for its role as a steric barrier, little is known about the properties of this interfacial layer. New insights were provided by direct, micrometer-scale measurements using the micropipet method, a technique borrowed from biology. It has long been noted that, in macroscale studies (involving sample sizes of millimeters or larger), structures appearing as "rigid skins" would often form at the crude oil–water interface (9–12). To make connections with emulsion systems, these skin-like structures are often likened to the adsorbed layer formed on the surfaces of micrometer-sized emulsion droplets. Such an extrapolation, however, may not necessarily be valid. For instance, interfacial tension at the emulsion drop surface (including water in crude oil cases) can markedly differ from interfacial tension measured with commonly used macroscale techniques, such as the spinning drop or Du Noüy ring. The difference results from the vastly different surface-to-volume ratios between the experimental set-ups (13). (It is likely that similar discrepancies exist between interfacial viscous and elastic properties as measured with micro or macro methods.)

The micropipet technique was developed to examine directly the interfacial properties of individual, micrometer-sized emulsion droplets, thus avoiding the extrapolative approach of necessity adopted by macroscale studies. The technique was originally developed as a tool for the biological and biophysical sciences (14–16). We have modified the technique for investigations of interfacial phenomena on individual, micrometer-sized emulsion droplets by using equally small suction pippets. A detailed description of our procedure can be found in Refs. 17 and 27.

### A. Emulsion Preparation

In the studies reported in this chapter, the hydrocarbon (continuous) phase of the emulsion was composed of coker feed bitumen, extracted from Athabasca Oil Sand deposit (18). To attain workable viscosities, bitumen is diluted in a 1 : 1 mixture, by volume, of n-heptane and toluene (both HPLC grade with no further purification). Such a solvent, which will be called "heptol," is chosen to simulate the aromatic/aliphatic ratio of the diluent used in commercial oil sand processing (18). In this work, bitumen contents of the oil phase, denoted by $C_o$, are expressed as volume percentages. Filtered, deionized water was used as the aqueous phase. An emulsion was prepared by adding 100 $\mu$l of water to 10 ml of diluted bitumen (1 vol.% water). The mixture was sonicated for several seconds, creating a macroemulsion of water droplets. Depending on the bitumen concentration and the duration of sonication, the average droplet size was between 5 and 30 $\mu$m. It was verified that the interfacial tension of the droplets (method of measurement discussed below) reaches an equilibrium value after the first minute of emulsification and remains unchanged for days. In this study, all emulsions were aged for 20 min.

## B. Interfacial Tension

As a first application, the micropipet was used to measure the interfacial tensions (IFTs) of individual emulsion drops. As shown in Figure 8, a single water droplet in an oil-continuous emulsion is held at the tip of a suction pipet (note at meniscus inside the capillary). From mechanical equilibrium, the critical pressure $P_{cr}$ needed to draw the water drop into an oil-filled pipet is related to the oil–water interfacial tension $\sigma$ by

$$\sigma = \frac{P_{cr} R_p}{2(1 - R_p/R_o)} \qquad (1)$$

where $R_p$ is the inner radius of the pipette, and $R_o$ is the radius of the droplet (19). Thus, the interfacial tension on a micrometer-sized emulsion droplet can be evaluated form simple measurements of pressure, and droplet and pipet sizes. This new method of tensiometry has been verified for emulsion systems consisting of pure liquids (17).

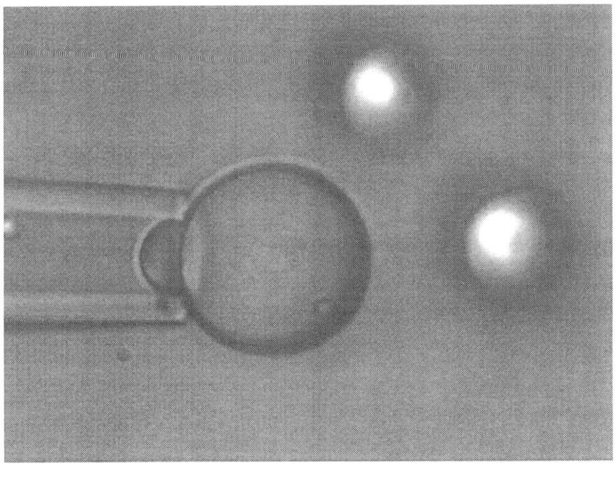


**Figure 8** A water droplet held with suction at the end of a micropipet. Note the curved meniscus inside the capillary. The pressure balance allows one to calculate the interfacial tension on an emulsion droplet in situ.

## C. Layers at Droplet Surfaces

Micropipet technique can also be used for studies of the adsorbed layer formed on emulsion droplets. Many natural components of crude oil are surface active and will tend to adsorb on the hydrocarbon–water interface (20). Surface excess of the adsorbed material can then be calculated from the Gibbs equation. In our case, we will use the concentration of bitumen in the solvent, since we do not know what is the chemical(s) responsible for droplet stabilization and what its concentration is. We will use the micropipet technique discussed above, to measure true IFT at emulsion drop surfaces.

Figure 9 shows changes in the equilibrium interfacial tension $\sigma$ as the bitumen content $C_o$ is varied over four orders of magnitude. Each data point is the average IFT of at least 12 droplets. The error bars extend over two standard deviations. The horizontal line in Fig. 9 represents the IFT between water and pure solvent. Note that $C_o$ is the bitumen content prior to emulsion preparation and it may generally be lower at equilibrium. Nevertheless, useful information can still be extracted from two limiting cases: the plateau regime at $C_o \leq 0.1\%$ and the linear regime at $C_o \geq 0.5\%$. They correspond to shortages and surpluses of surfactants relative to the available "interfacial sites," respectively.* It can be shown that, in both these limits, $C_o$ is proportional to the surfactant concentration in the oil phase (17). Assuming diluted bitumen to be an ideal, single-component surfactant solution (a simplistic picture which provides a lumped characterization of crude oil's surface active components), we can apply the Gibbs relation to the data shown in Fig. 9. It can thus be inferred that in the

* Such an observation remains valid despite the lack of precise control of the total interfacial area.

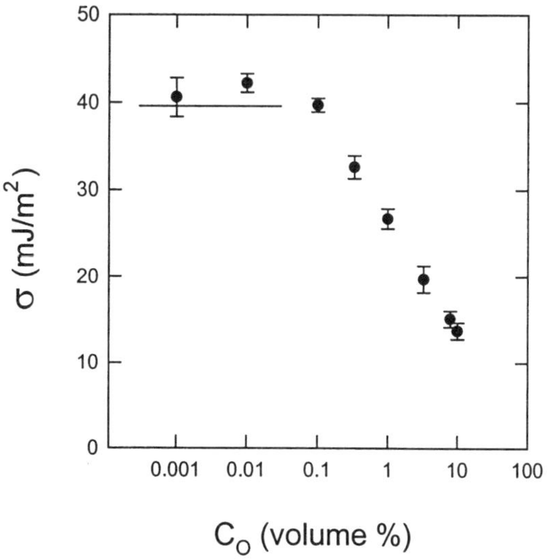


**Figure 9** Interfacial tension at water droplets in toluene containing increasing amounts of bitumen. The solid line on the figure represents toluene – water interfacial tension.

plateau regime ($C_o \leq 0.1\%$), $\Gamma \approx 0$, i.e., there should be little or no adsorbed material on the droplet surfaces. In the linear regime ($C_o \geq 0.5\%$), where the droplet surfaces are saturated with surfactants, the area per "site," given by the reciprocal of $\Gamma$, is 0.74 nm$^2$. Our estimate of the molecular cross-section at saturation is about an order of magnitude lower than most literature values (21, 22). The discrepancy may be due to factors such as the type of oil, solvent properties, and water chemistry. In addition, it remains an open question whether the crude oil–water interface in, say, a Du Noüy ring experiment (the most common technique), is structurally similar to the adsorbed layer at the surfaces of emulsified water drops (structures that are directly probed in this study).

## D. Mechanical Properties of the Adsorbed Layer

Macroscale studies in the past had pointed to the formation of a "rigid skin" at the crude oil–water interface (9–11). We observed similar structures using our micropipet technique but only under certain conditions. For this, a water-filled micropipet was first immersed in diluted bitumen. A droplet was then formed at the pipet tip by expelling a small amount of water into diluted bitumen, with $C_o$ ranging from 0.001 to 10%. Droplets thus formed were aged for 3 min. Figure 10a shows a photograph of a water droplet formed in a 0.1% bitumen. When the droplet was deflated and its area compressed, the surface crumpled abruptly (Fig. 10b), revealing a rigid cortical structure. This finding is similar to Langmuir's observations on protein films (23) and to reports on "skin" formation (9–11) at a water–crude oil interface.

The IFT isotherm in Fig. 9 suggests that, at 0.1% bitumen, there should be little or no adsorbed material on the interface. However, crumpling of the interface was observed for $C_o$ between 0.001 and 1%. A second anomaly is encountered when the deflation process is repeated for $C_o > 1\%$. As shown in Fig. 11, at high bitumen content, the interface loses its rigidity and remains spherical throughout deflation.* More interestingly, at some point during deflation, small surface protrusions begin to appear on the shrinking drop (Fig. 11b). As the interfacial area continues to decrease, these surface imperfections become more prominent and eventually detach as micrometer-sized droplets (Fig. 11c). Such a process, referred to as budding, is a new emulsification mechanism, which may have important implications for the petroleum industry (24). In an industrial setting, even at gentle agitation (which cannot be avoided), water droplets would undergo various deformations and their relaxation to spherical shape would be equivalent to area compression. New emulsion droplets can thus be formed through budding at much lower shear rates than

(a)
(b)

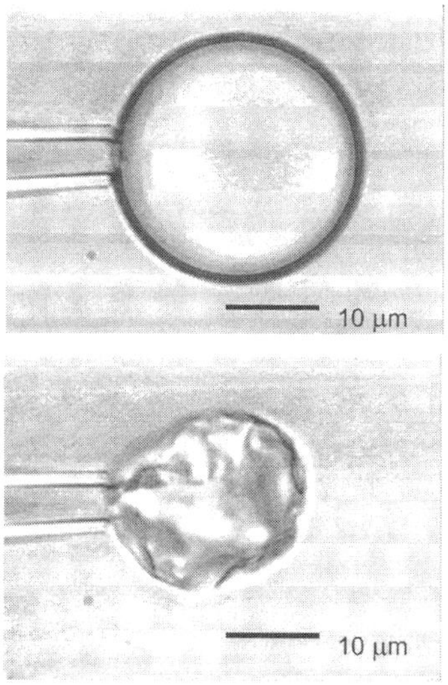


**Figure 10** Deflating a water droplet below 10% bitumen in toluene solution. The initially spherical droplet (a) crumbles like a paper bag (b). Note clear background of the image.

* Transformation of the adsorbed layer from a rigid to a fluid interface (at $C_o \approx 1\%$) does not appear abrupt.

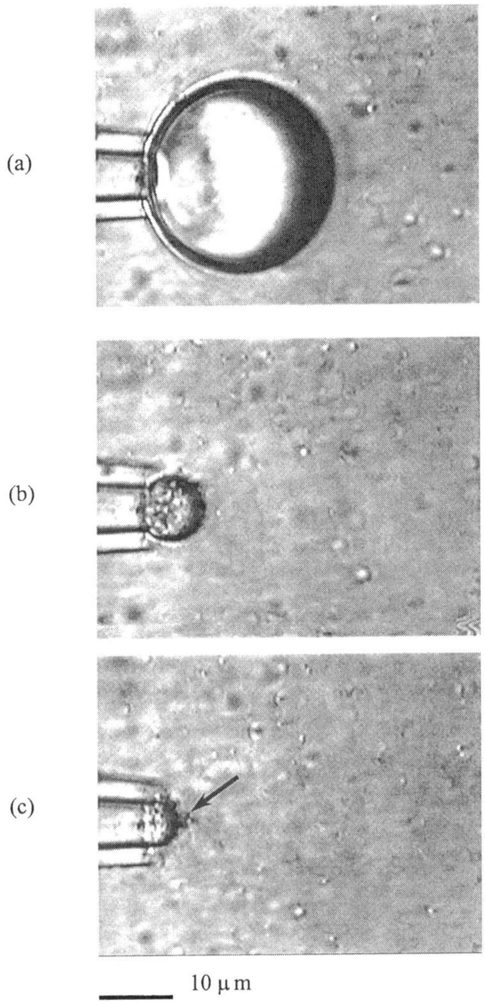


**Figure 11** Deflating a water droplet above 10% bitumen in toluene solution. The droplet retains its spherical shape (a–c). At high surface compression, undulations of the interface appear (b) and grow, eventually detaching as separate much smaller droplets (c, arrow). The background is filled with small Brownian water droplets.

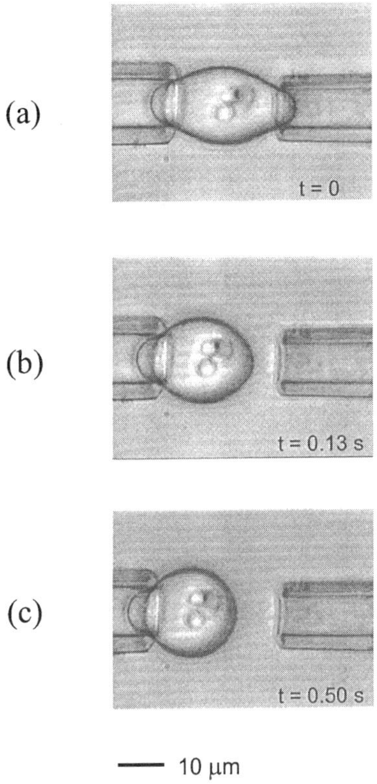


**Figure 12** Kinetics of the droplet shape recovery. Description in the text.

those required for a conventional droplet break-up mechanism. A detailed description of the budding mechanism has been published elsewhere (29). It is noted here that similar – although not identical – budding phenomena have been observed in biological and biophysical systems (19, 25, 26).

Adsorbed layers that crumple clearly possess resistance to surface deformations. Such resistance is manifested as surface viscosity. In a free suspension, micrometer-sized water droplets are normally spherical. Using two suction pipet as shown in Fig. 12a, such a droplet (here, formed in 0.1% diluted bitumen) is stretched and then released, thus allowing it to recover its spherical shape at constant volume (Fig. 12b,c). Typical observed times for the droplet-shape recovery are of the order of 1 s. The recovery is certainly driven by the interfacial tension $\sigma$. If the recovery process is limited by bulk viscosity, the recovery time would be of the order of

$$t_{rec} \sim \frac{\mu d}{\sigma} \tag{2}$$

where $\mu$ is the viscosity of either water or oil (assuming they have comparable values), and $d$ is the size of the droplet (28). Using typical values of $\mu \sim 10^{-3}\,\mathrm{N \cdot s/m^2}$, or $\sigma \sim 10\,\mathrm{mN/m}$ and $d \sim 10\,\mu\mathrm{m}$, the characteristic recovery time associated with bulk dissipation would be of the order of 1 $\mu$s. As the observed recovery rates are typically $10^6$ times slower, the dominant source of viscous dissipation must be due to the viscosity of the adsorbed layer at the droplet surface. As suspected, these viscous effects are only observed in the crumpling regime (roughly, for $C_o < 1\%$) where resistance to surface deformations are large. In the budding regime, the droplets recover their shape much faster. We could not monitor the process with a video speed of 30 frames per second.

A detailed analysis of the shape-recovery experiments can be found in Ref. 29, which provides, we believe, the first measurement of interfacial transport parameters on the micrometer scale. It is noted that, although different droplets in an emulsion have uniform IFT values, their rates of shape recovery – and hence surface viscosities – appear more varied. Further study on this variability will be required.

## E. Stability of Emulsion Droplets

Using the micropipet technique, two water droplets in diluted bitumen could be pressed together in an attempt to induce their coalescence (Fig. 13). At all bitumen concentrations, the emulsion drops remained stable to coalescence despite being pressed together for up to 5 min. It appeared that the droplets would remain stable for much longer had the experiments continued. The flattened contact regions had radii of several micrometers. As the two droplets were separated, no sign of droplet–droplet adhesion, as indicted by the drops' elongation, was observed at bitumen concentrations below 1%. At higher concentrations, however, these signs of adhesion became visible. As expected, in control experiments involving water droplets in pure solvent, the droplets coalesced immediately on contact.

Note that at low bitumen content, where there should be very little of adsorbed material on the interface (according to the IFT isotherm in Fig. 9), the droplets already display remarkable stability. It is also interesting that, with respect to the bitumen content $C_o$, the detection of droplet–droplet adhesion coincides with the disappearance of interfacial crumpling. The resistance to both coalescence and adhesion at low bitumen concentration ($C_o < 1\%$) is consistent with the steric stabilization mechanism concluded from our colloidal collider studies discussed above.

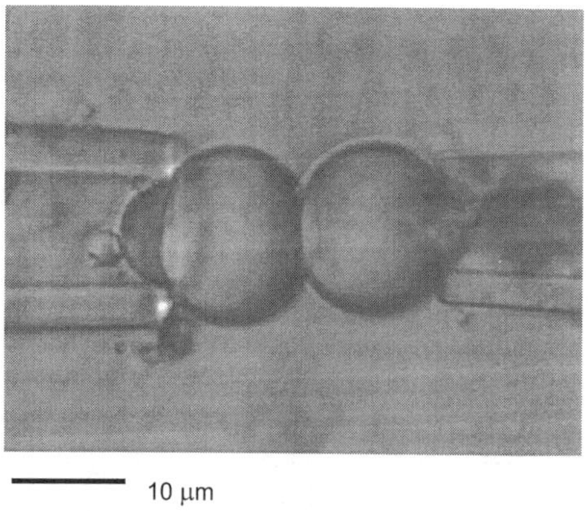


**Figure 13** Resistance to coalescence. Two water droplets are pressed together in 0.1% bitumen in toluene. The compressing force, calculated from droplet shape deformation, is equivalent to about 10,000 *g* acceleration.

Adhesive forces which begin to appear at $C_o > 1\%$ are just above our limit of detection, which is of the order of nanonewtons. For small shape elongation (relative to the drop size), a spherical drop behaves as an elastic spring with a stiffness that is of the same order as the IFT (30). At 1 to 10% bitumen, the IFT, and hence the effective elastic constants, is $\sim$ 10 mN/m (Fig. 9). With the smallest droplet elongation that can be observed, estimated to be $\sim 0.1\,\mu$m, the corresponding adhesive force is on the order of nanonewtons.

The force exerted to compress the droplets is of the order of $\Delta pr^2$, where $r$ is the radius of the contact region (about 1 $\mu$m, in our case) and $\Delta p$ is the pressure difference between the droplet interior and its surrounding (roughly $10^4\,N/m^2$) for a 10-$\mu$m drop. The resulting compressive force is about 10 nN. To create such forces by centrifugation, the required acceleration would have to be of the order of 10,000 *g*. This, as demonstrated here, would still be futile in causing coalescence in agreement with our observations. When we centrifuge our emulsions, even at 20,000 *g*, we can separate creamed emulsion in the form of a cake, without breaking the emulsion into separate water and oil phases.

The Marangoni effect, that is, the retardation of thin-film drainage by induced tension gradients, is believed to play an important role in emulsion stabilization (31). The most "severe" case of Marangoni's effect is, of course, one that involves immobile surfaces. The associated drainage time can be estimated from the Reynolds equation (31):

$$\frac{d}{dt}\left(\frac{1}{h^2}\right) = \frac{4}{3}\frac{\Delta p}{\mu r^2} \tag{3}$$

where $h$ is the film thickness, $\mu$ is the viscosity of diluted bitumen (roughly that of water), and $\Delta p$ and $r$ have the same meaning and values as stated above. Assuming that, as $h$ falls below 1 nm, film drainage is greatly accelerated by attractive colloidal forces, the corresponding film drainage time is of the order of 0.1 s. Yet, the water droplets in our studies remain stable for minutes or longer under the same drainage forces. It thus appears that the Marangoni effect is not the dominant stabilizing mechanism in water-in-diluted bitumen emulsions.

## V. THIN-FILM STUDIES: PROPERTIES OF THIN EMULSION FILM IN WATER-DILUTED BITUMEN–WATER SYSTEM

As the two water droplets of a W/O emulsion approach each other a thin oil film is formed between them (Fig. 14a). For the droplets to coalesce, this oil film has to break. Such a film can be created inside a specially designed measuring cell (32) of the thin liquid film–pressure balance technique (TLF-PBT) (Fig. 14b). Due to the curvature of the oil–water interface, a capillary pressure arises at the edge of the film, forcing the liquid to drain from the film. As the film becomes thinner, the interfaces which bind the film begin to interact through van der Waals, electrostatic, steric, or other surface forces. The overall effect of all these forces, which is known as disjoining pressure, determines whether the film will remain stable and thus directly determines the stability of the emulsion. The principle of the measuring technique involves balancing the capillary pressure with the film disjoining pressure. Although the technique has been extensively used for studies of foams, relatively little work has been done on emulsion films and even less on water–oil–water systems. The first attempt at applying this technique to study water-diluted bitumen–water system

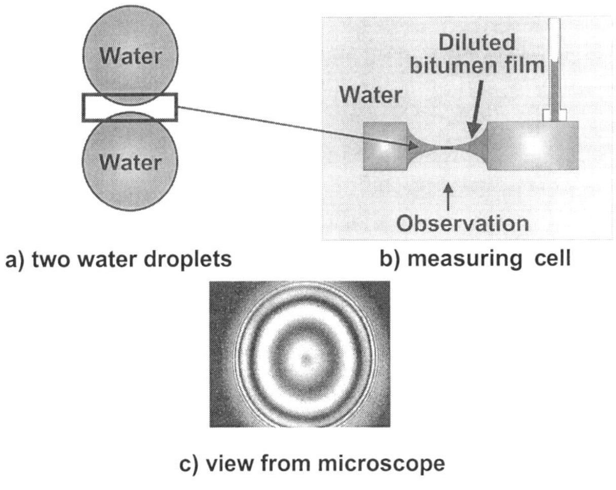


**Figure 14** Schematic of thin-film technique. The geometry of the oil film separating two water droplets (a) is recreated in a special holder in the thin-film measuring cell (b); (c) shows microscopic image at the beginning of the film-thinning process showing Newton interference rings.

was made in Wasan's laboratory in Chicago (33). Later on, a thin-film instrument was set up in our laboratory. A detailed description of our experimental set-up and procedures can be found in Ref. 34.

As in the previous sections, the objective of this study was to obtain insight into the mechanisms that stabilize water in dilute bitumen emulsion with particular attention to the relative importance of the resin, asphaltene, and solids fractions of the bitumen.

### A. Experimental

Our experimental set-up is described in detail in Ref. 34. The porous-plate measuring cell was placed inside a thermostated jacket on a Carl Zeiss Axiovert 100 inverted microscope. The film was viewed and recorded with a CCD video camera and a VCR. The capillary pressure was controlled by adjusting the height of the solution, using a manually operated micrometer syringe or by adjusting the air pressure inside the cell.

The film thickness was determined by a microinterferometric method (33–37) using heat-filtered light from a 100-W mercury-arc lamp and a monochromatic filter ($\lambda = 546\,\text{nm}$). The incident light was directed through a pinhole or iris diaphragm creating a $\sim 10\,\mu\text{m}$ spot focused on the center of the film. The intensity of reflected light passing through a second pinhole diaphragm was measured with a low-light, low-noise photodiode and recorded using a chart recorder. The equivalent thickness, $h$, was calculated following the procedure developed by Scheludko and Platikanov (38), assuming that the film was optically homogeneous with the refractive index of the film equal to the refractive index of the solvent used to prepare the bitumen solution studied.

Syncrude's coker feed bitumen, which was used for all experiments, had already been treated in commercial plant operations to remove coarse sand and water and was ready for upgrading. Deasphalted bitumen (i.e., bitumen with asphaltenes and solids fractions removed) was obtained by diluting coker feed bitumen with heptane to a volume ratio of 40 : 1 (heptane : bitumen), filtering the supernatant after 24 h and stripping the diluent by evaporation at 60°C to constant mass. The asphaltene fraction was recovered from the precipitate by dissolving it with toluene, centrifuging, and evaporating the solvent as before. The resin fractions (I and II) of bitumen were obtained using the SARA method (39) although only resin fraction I was used in the thin-film measurements. Solids-free bitumen was prepared by diluting coker feed bitumen with toluene to a volume ratio of 100 : 1 (toluene : bitumen), centrifuging at 20,000 g, filtering through a Millipore 22-$\mu$m filter, and stripping the solvent to a constant mass. HPLC-grade toluene and n-heptane were used to prepare all solutions. Three separate samples were prepared at each weight ratio and were used immediately after preparation to avoid any possible aging effects. All films were immersed into a solution of 0.014 M sodium chloride, 0.012 M sodium bicarbonate, and 0.04 M sodium sulfate (ph ~ 8.2) to imitate the composition of water used in Syncrude's commercial operation.

## B. Results

### 1. Toluene-diluted Bitumen Films

Figure 15a displays a series of images of the typical drainage pattern for a toluene-diluted bitumen film, from the stage of formation to the equilibrium gray film. A single center dimple appeared upon initial film formation; the liquid in the dimple would drain off through channels until a uniform white/yellow film was reached. The film would then continue to drain slowly via plane-parallel drainage to an equilibrium gray film. At high diluent/bitumen ratios (10 : 1 to 20 : 1), several "blurry" white dimples of trapped liquid approximately 3 to 5 $\mu$m in diameter often appeared in the dark gray films similar to those observed by Bergeron et al. in hydrocarbon foam films (40). The rate of film drainage was limited by the bulk viscosity of the bitumen solution.

Figure 16 (curve 2), depicts the experimentally obtained thickness for toluene-diluted bitumen emulsion films. At an industry-relevant diluent ratio of about 1 : 1 toluene to bitumen, the film thickness values scattered within a large range from about 50 to 60 nm. These films required long periods of time to reach equilibrium, and the film diameter would fluctuate resulting in a large variation in the measured film thickness. Such a film probably had a multilayer structure.

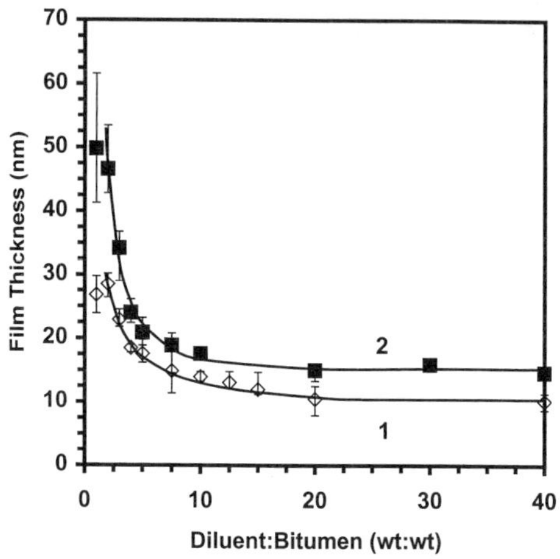


**Figure 16** Film thickness for bitumen in heptane (1) and bitumen in toluene (2) solutions for various diluent-to-bitumen ratios.

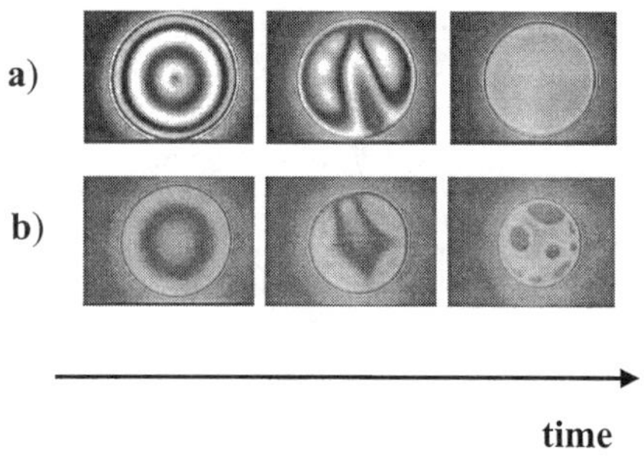


**Figure 15** Film-thinning process for bitumen in toluene (a) leads to a relatively thick gray film. For bitumen in heptane solution, thinning leads to formation of black spots, which eventually coalesce forming a thin black film (b).

### 2. Heptane-diluted Bitumen Films

With no asphaltene precipitation (i.e., at 1 : 1 heptane : bitumen ratio where no asphaltene precipitation occurs), the behaviour of a heptane-diluted bitumen film was very similar to that of toluene-diluted bitumen films described above. The film was formed with a single center dimple, followed by channel drainage to a uniform white/yellow film, which would continue to drain to an equilibrium gray film via plane-parallel drainage. When asphaltene precipitation began to occur (i.e., at heptane : bitumen ratios of 1.7 : 1 or higher) (41), black spots would appear within 5 to 10 s after film formation (Fig. 15b). The spots quickly coalesced in to a uniform black film with several small white spots of about 1 to 3 $\mu$m in diameter similar to the dimples observed by other researchers (40, 42, 43). A slight decrease in film thickness was observed once the black film had been formed. About half an hour after loading the cell, small aggregates of asphaltene precipitate began to appear near the oil/water interface. The aggregates were approximately 5 to 10 $\mu$m in diameter, which was close to the mean particle size of 7.0 ± 4.0 $\mu$m reported by Li and Wan (44). Over time (> 2 h), the number of asphaltene aggregates would continue to increase until any newly formed film would become completely clogged, preventing it from draining.

The stability of the heptane/bitumen emulsion films depended strongly on the diluent ratio. At a ratio of 1 : 1, the film remained stable for at least an hour while films of 2 : 1 to 3 : 1 dilution remained stable for at least 20 min. The stability of the film then dramatically decreased with increasing diluent ratio, causing the film lifetime to fall to less than 25 s for ratios of 20 : 1 or more (Fig. 17).

The thickness measurements for the heptane–diluted bitumen films are presented in Fig. 16 (curve 1). Below the onset of asphaltene precipitation at a heptane/bitumen ratio of about 1 : 1, the film drained to an equilibrium gray film of about 27 nm thickness. Above the precipitation onset at a heptane/bitumen ratio of 2 : 1, the black film reached a thickness of about 28 nm. The film thickness then decreased with increasing diluent : bitumen ratio to about 10 nm at a ratio of 20 : 1. The thickness then remained constant, indicating that a bilayer film was probably reached. At lower diluent ratios (< 20 : 1), the greater thickness of heptane/bitumen films may be caused by the presence of unprecipitated asphaltenes.

A comparison of the thickness of both heptane- and toluene-diluted films (Fig. 16, curves 1 and 2, respectively) indicated a consistently lower thickness for the former films. One would expect that the presence of the resins, asphaltenes, and solids fractions determined the thickness and behaviour of the emulsion films. The effect of each fraction can be isolated to determine which fraction(s) dominated the film behaviour in each solvent.

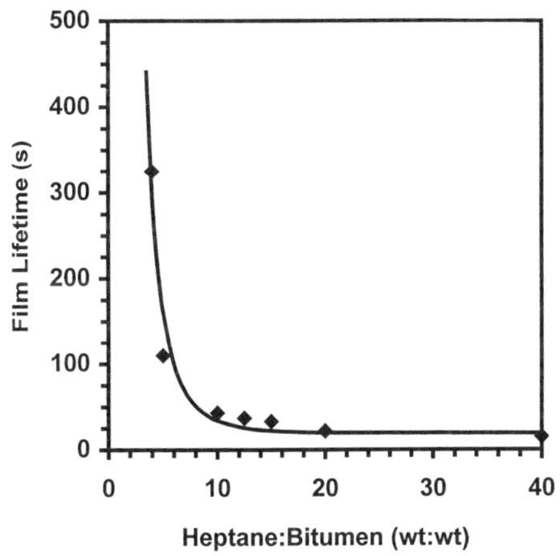


**Figure 17** Film lifetime for bitumen in heptane solutions.

### 3. Solids-free Bitumen

Fine solids are frequently mentioned as being responsible for W/O emulsions, especially in systems involving various crude oils. The solids fraction of bitumen consists of fine submicrometer clay particles that have been rendered "asphaltene-like" due to the adsorption of highly aromatic, polar material on the particle surfaces (45). To determine if this solid fraction played a role in the film stability, a "solids–free" bitumen was prepared where all solid material larger than 100 nm was removed from the sample. Films of toluene- and heptane-diluted solids-free bitumen showed little or no change in both the drainage patterns and the film thickness, indicating that the fine solids had little or no effect on the behaviour or stability of water/diluted bitumen/water films. This is consistent with the observation described in Sec. II, where removal of fine solids from diluted bitumen had no effect on subsequently formed water in diluted bitumen emulsion.

Solid particles were observed at or near the oil–water interface on several occasions in toluene-diluted bitumen films. The particles were easily pushed away from the film into the meniscus region when the films were first formed and never appeared in the film itself. For the solids fraction to play a role in emulsion stability, we thus expect that the fine clay particles probably build up in the Plateau borders around water droplets and clog the drainage routes through the emulsion, contributing to its additional, kinetic stability. Therefore, it may be expected that the fine solids may have a contribution to the overall W/O emulsion stability, but this is not the leading factor.

### 4. Resins

The effect of the resins was studied by observing a number of thin films of diluted deasphalted bitumen (i.e., free of asphaltene and solid fractions) and diluted resins I. Most of the experiments were conducted using deasphalted bitumen because of the high cost and long time needed to isolate a small amount of resin I through the SARA method (39). The properties of deasphalted bitumen and resin-I films in either toluene or heptane solutions were similar to the heptane-diluted bitumen films with a black film being formed via black spots. The film thickness was also independent of the solvent type, ranging from 14 nm at a dilution of 1 : 1 to 12 nm for a 10 : 1 diluent ratio. Experiments with resin I diluted to 5 : 1 and 10 : 1 in toluene displayed similar drainage patterns to those of the deasphalted bitumen films and a nearly identical film thickness.

These results indicate that the resin fraction determined the properties of heptane-diluted bitumen films at high diluent ratios. Since the thickness remained constant with further dilution, the film probably had a bilayer structure. The relative instability of heptane-diluted bitumen, deasphalted bitumen, and resin-I films agree well with the emulsion experiments of Yan (46) where deasphalted bitumen led to poorly stabilized water-in-bitumen emulsions.

### 5. Asphaltenes

Since asphaltenes have been identified as the most common stabilizers of water-in-bitumen emulsions (47), we expected that the asphaltenes could be responsible for the stability of the toluene-diluted bitumen films. A single experiment on a film of toluene-diluted asphaltenes at a weight ratio of 15 : 1 followed a similar drainage pattern to that of the toluene-diluted bitumen films with a uniform gray film being formed. However, the film unexpectedly ruptured after only a few minutes at a thickness of 36 nm. McLean and Kilpatrick (48) showed that stable water-in-oil emulsions could be obtained with pure asphaltenes. They also found that the combination of resin and asphaltene at a ratio of 1 : 3 (resin : asphaltene) resulted in the most stable emulsions. A thin film of resin I and asphaltene diluted in toluene (2 : 1 resin I : asphaltene by weight) displayed very similar behavior to the toluene-diluted bitumen films with slow drainage to a stable gray film. The equilibrium thickness of the resin I : asphaltene film was also very similar to the thickness of the toluene-diluted bitumen films. These results indicated that the combined interaction of resins and asphaltenes are important to the film stability. Again, it is worth noting that the conclusion from the washing experiments discussed above was that it is not the whole asphalthene fraction that is involved but only a small subfraction of the total asphalthene present in the oil. We are not sure what the chemical characteristics of this "bad actor" are.

While the asphaltene and resin fractions alone provide a partially stable film, the combination of resin and asphaltene produce extremely stable films. However, additional information is needed to confirm our assumptions that the films have a multilayer structure at lower diluent ratios and a bilayer structure at high diluent ratios. We hope that future measurements of the disjoining pressure–thickness isotherms will provide this confirmation as well as identify the surface forces that stabilize the water/diluted-bitumen/water films.

Bitumen is a complex mixture of hydrocarbons that can be divided into several material classes based on solubility. The classes include saturates, aromatics, resins, and asphaltenes. Asphaltenes are generally defined as the bitumen fraction that is soluble in toluene and insoluble in an aliphatic solvent. In the case of heptane, large asphaltene molecules begin to precipitate at a heptane : bitumen volume ratio of around 1.4 : 1 to 1.7 : 1 (48) with complete asphaltene precipitation occurring at volume ratios above 40 : 1. The asphaltene molecules are polyaromatic hydrocarbons that consist primarily of aromatic clusters and aliphatic chains along with a variety of functional groups (50, 51). Resins consist of similar chemical species except the molecules, on average, have a lower molar mass, fewer functional groups, and a higher H/C ratio. It is well known that asphaltenes play a significant role in the stability of the water-in-bitumen emulsions (48, 52–55). Resins are considered to be surface active (49, 52, 56) and, to a limited extent, are capable of stabilizing an emulsion. McLean and Kilpatrick have shown that the combination of resin and asphaltene at a ratio of around 1 : 3 (resin : asphaltene by weight) provide the most stable emulsions (48). The bitumen extracted from oil sands also contains very fine solids composed of clay particles coated with strongly bound toluene-insoluble organic material. These solids are also suspected of playing a role in water-in-oil emulsion stability (45, 46).

## VI. SUMMARY

Water-in-diluted bitumen emulsion is characterized by high stability, creating serious operational problems in commercial operations. Our studies indicate that this stability is due to an adsorbed layer at the water–oil interface responsible for a steric barrier to droplet coalescence. This steric repulsion was detected by using the colloidal collider technique, new tool for studying surface-to-surface interactions between particles or droplets of several micrometers size. The protective "skin" formed at the droplet surfaces was visualized by deflating water droplets formed at the tip of a micropipet. (The micropipet experimental technique was borrowed from biological studies, where it is commonly used to manipulate individual cells.) The deflating water droplets in diluted bitumen solutions in toluene (below 1 wt%) crumble like paper bags, indicating that the surface layer is rigid. At higher concentrations, the surface layer becomes highly flexible. As a result, the deflating droplets remain spherical, but instead of crumbling they spawn smaller droplets through a new spontaneous emulsification mechanism. This spontaneous emulsification may be responsible for formation of W/O emulsions in many oil industry related systems.

The transition from crumbling to spawning regime occurs at a characteristic bitumen in diluent concentration, which depends on the paraffinic/aromatic nature of the solvent. The higher solvent aromaticity, the higher is this transition concentration. Above this critical concentration, a very persistent emulsion is easily formed, no doubt partially through the spontaneous emulsification mechanism mentioned above. Below the critical concentration (or at higher diluent-to-bitumen ratios) a clean, dry oil phase can be produced using inclined plate settlers or centrifuges. For a paraffinic diluent, like heptane, or natural gas condensate, this transition takes place at bitumen-to-solvent ratio of about 2, allowing for development of a commercial paraffinic diluent technology.

Thin-film studies confirmed observations from other techniques. Among others, they revealed that the film separating two water droplets in heptane-diluted bitumen is about half the thickness when toluene is used as a solvent. Also, the film lifetime is considerably shorter in a paraffin-based system. Demulsifiers that are used in industry to lower the water content in the feed to refineries compete for the water–oil interface with a substance or substances that produce the protective steric layers.

It is still unknown what the mechanism is behind the transition from rigid to flexible character of the layer on the water–oil interface mentioned above. There is a strong possibility that a phase transition takes place there, most likely of a type known for microemulsion systems. An attempt at getting some insight into the phase equilibria in the system of bitumen, diluent, water, and perhaps added surfactants (demulsifiers) will be the subject of our future studies.

## ACKNOWLEDGMENTS

The research reported in this chapter was done in close collaboration between the NSERC Industrial Research Chair in Oil Sands at the University of Alberta, CANMET Western Research Centre, and Syncrude Research Centre in Edmonton. Natural Science and Engineering Research Council of Canada (NSERC) support to the above-mentioned Chair is greatly acknowledged. The author wishes to thank Syncrude Canada Ltd for the permission to publish this material and his colleagues and friends: Tad Dabros, Khristo Khristov, Jacob Masliyah, Kevin Moran, Shawn Taylor, Alex Wu , and Tony Yeung for providing the data, artwork, and numerous discussions, and for their help in preparation of the manuscript.

## REFERENCES

1. Y Xu, T Dabros, H Hamza, B Shelfantook. Petrol Sci Technol 17: 1051–1070, 1999.
2. TGM van de Ven, P Warszynski, X Wu, T Dabros. Langmuir 10: 3046–3056, 1994.
3. X Wu, TGM van de Ven. Langmuir 12: 3859–3865, 1996.
4. X Wu, TGM van de Ven, J Czarnecki. Colloids Surfaces A 149: 577–583, 1999.
5. JH Schenkel, JA Kitchener. Trans Faraday Soc 56: 161–173, 1960.
6. J Visser. Adv Colloid Interface Sci 3: 331–363, 1972.
7. X Wu, TGM van de Ven. J Colloid Interface Sci 183: 388–396, 1996.
8. RH Yoon, D Guzonas, BS Aksoy. Processing of hydrophobic minerals and fine coal. Proceedings of the First UBC-McGill Bi-Annual International Symposium, Vancouver, 1995, pp 277–289.
9. CG Dodd. J Phys Chem 64: 544–550, 1960.
10. FE Bartell, DO Niederhause. Research on Occurrence and Recovery of Petroleum. New York: American Petroleum Institute, 1949, pp 57–80.
11. JE Strassner. J Petrol Technol 20: 303–312, 1968.
12. OK Kimbler, RL Reed, IH Silberg. Soc Petrol Eng J 6: 153–165, 1966.
13. A Yeung, T Dabros, J Masliyah. J Colloid Interface Sci 208: 241–247, 1998.
14. EA Evans, R Skalak. Mechanics and Thermodynamics of Biomembranes. Boca Raton, FL: CRC Press, 1980.
15. E Evans, D Needham. J Phys Chem 91: 4219–4228, 1987.
16. A Yeung, E Evans. Biophys J 56: 139–149, 1989.
17. A Yeung, T Dabros, J Masliyah. J Colloid Interface Sci 208: 241–247, 1998.
18. RC Shaw, LL Schramm,. J Czarnecki. In: LL Schramm, ed. Suspensions: Fundamentals and Applications. Washington, DC: American Chemical Society, 1996, pp 639–675.
19. E Evans, W Rawicz. Phys Rev Lett 64: 2094–2097, 1990.
20. N Cyr, DD McIntyre, G Toth, OP Strausz. Fuel 66: 1709–1714, 1987.
21. A Bhardwaj, S Hartland. Ind Eng Chem Res 33: 1271-1279, 1994.
22. SE Taylor. Fuel 71: 1338–1339, 1992.

23. I Langmuir, DF Waugh. In: CG Suits, HE Way, eds. The Collected Works of Irving Langmuir. Vol 7. New York: Pergamon Press, 1961, pp 27–35.
24. T Dabros, A Yeung, J Masliyah, J Czarnecki. J Colloid Interface Sci 210: 222–224, 1999.
25. B Alberts, D Bray, J Lewis, M Raff, K Roberts, J Watson. Molecular Biology of the Cell. New York: Garland Publishing, 1989.
26. R Lipowsky. Nature 349: 475–481, 1991.
27. A Yeung, T Dabros, J Masliyah, J Czarnecki. Colloids Surfaces A 174: 169–181, 2000.
28. A Yeung, E Evans. J Phys II France 5: 1501–1523, 1995.
29. A Yeung, T Dabros, J Czarnecki, J Masliyah. Proc Royal Soc A, 445: 3709–3723, 1999.
30. E Evans, K Ritchie, R Merkel. Biophys J 68: 2580–2587, 1995.
31. DA Edwards, H Brenner, DT Wasan. Interfacial Transport Processes and Rheology. Boston: Butterworth–Heinemann, 1991.
32. D Exerowa, A Scheludko. Compt Rend Acad Bulg Sci 24: 47–50, 1971.
33. A Nikolov, D Wasan, J Czarnecki. Presented at 47th Canadian Chemical Engineering Conference, Edmonton, 1997.
34. K Khristov, S Taylor, J Czarnecki, J Masliyah. Colloids Surfaces A 174: 183–196, 2000.
35. D Exerowa, PM Kruglyakov. Foam and Foam Films. New York: Elsevier, 1998.
36. A Scheludko. Kolloid Z 155: 39–44, 1957.
37. A Scheludko. Adv Colloid Interface Sci 1: 391–464, 1967.
38. A Scheludko, D Platikanov. Kolloid Z 175: 150–158, 1961.
39. JT Bulmer, J Starr, eds. Syncrude Analytical Methods for Oil Sand and Bitumen Processing. Edmonton: Syncrude Canada Ltd, 1979, pp 121–124.
40. V Bergeron, JE Hanssen, FN Shoghl. Colloids Surfaces A 123/124: 609–622, 1997.
41. PF Clarke, BB Pruden. Petrol Sci Technol 16: 287–305, 1998.
42. JL Joye, CA Miller, GJ Hirasaki. Langmuir 8: 3083–3092, 1992.
43. Y Horiuchi, H Matsumura, K Furusawa. J Colloid Interface Sci 207: 41–45, 1998.
44. H Li, WK Wan. International Heavy Oil Symposium, Calgarry, 1995, pp 709–716.
45. LS Kotlyar, BD Sparks, JR Woods, S Raymond, Y LePage, W Shelfantook. Petrol Sci Technol 16: 1–19, 1998.
46. Z Yan. Interfacial behaviour of de-asphalted bitumen. MSc dissertation, University of Alberta, Edmonton, 1997.
47. HW Yarranton. Asphaltene solubility and asphaltene stabilized water-in-oil emulsions. PhD dissertation, University of Alberta, Edmonton, 1997.
48. JD McLean, PK Kilpatrick. J Colloid Interface Sci 196: 23–34, 1997.
49. PF Clarke, BB Pruden. Petrol Sci Technol 16: 287–305, 1998.
50. OP Strausz, TW Mojelsky, EM Lown. Fuel 71: 1355–1363, 1992.
51. DL Mitchell, JG Speight. Fuel 52: 149–152, 1973.
52. JD McLean, PK Kilpatrick. J Colloid Interface Sci 189: 242–252, 1997.
53. EJ Johansen, IM Skjärvö, T Lund, J Sjöblom, H Söderlund, G Boström. Colloids Surfaces 34: 353–370, 1988/89.
54. MH Ese, J Sjöblom, H Fφrdedal, O Urdahl, HP Rφnningsen. Colloids Surface A: 123/124: 225–232, 1997.
55. O Mouraille, T Skodvin, J Sjöblom, JL Peytany. J Dispers Sci Technol 19: 339–367, 1998.
56. EH Sheu, OC Mullins, eds. Asphaltenes: Fundamentals and Abdications. New York: Plenum Press, 1995.

# 22

# Interfacial Rheology of Crude Oil Emulsions

**Mingyuan Li, Bo Peng, Xiaoyu Zheng, and Zhaoliang Wu**
*University of Petroleum, Changping, Beijing, China*

## I. INTRODUCTION

It is well known that emulsions, on standing, may undergo a number of breakdown processes, namely, creaming or sedimentation, flocculation, Ostwald ripening, coalescence, and phase inversion (1–4). Most of these processes are determined by the interaction forces between the droplets, i.e., electrostatic repulsion, steric interaction, and van der Waals attraction. The balance of these forces and the properties of the interfacial film between the water and oil phases determine the stability of emulsions (4, 5). The degree of the stability of emulsions toward breakdown or coalescence is imparted by the adsorption of surface-active agents and/or the presence of macromolecules or fine particles at the interface (1). Although there have been some major breakthroughs in recognizing the factors that affect the process of coalescence, the phenomena are not yet completely understood. One of the main reasons is that the presence of surface-active agents and polymeric substances at the interface is known to display a variety of rheological properties (6, 7).

Based on an overview of more than 400 papers, Malhotra and Wasan (7) concluded that there is strong evidence that interfacial viscosity and elasticity play a significant role in determining the stability of dispersed systems. There appears to be a good correlation between coalescence rate and interfacial viscosity or elasticity, i.e., the higher the interfacial viscosity or elasticity, the lower the coalescence rate of drops and more stable the emulsion. They also concluded that the interfacial viscosity and elasticity had been demonstrated to be important properties in determining the drainage and stability of thin liquid films. High interfacial viscosity or elasticity lowers the rate of drainage of the film, which results in increased stability of the dispersed phases. High interfacial viscosity also provides resistance against rupture to thin films.

The stability of crude oil emulsions determines the effectiveness of enhanced oil recovery and the separation of water and oil in the oil production. Crude oils contains natural interfacially active fractions and particles, for example, resin, asphaltene, and wax particles. The interfacially active fractions and particles tend to present at the water/oil interface and form a tough film surrounding the dispersed droplets. The film can resist the coalescence of the droplets and stabilize the emulsion (8–11).

Rheology is the study of the deformation and flow of materials under the influence of an applied stress. The interfacial rheology of a surfactant film normally accounts for the interfacial viscosity and elasticity of the film. The interfacial viscosity can be classified with interfacial shear viscosity and interfacial dilational viscosity. Films are elastic if they resist deformation in the plane of the interface and if the surface tends to recover its natural shape when the deforming forces are removed. The interfacial elasticity can also be classified with interfacial shear elasticity and interfacial dilational elasticity (6, 7, 12). Malhotra and Wasan (7) gave a detail review of interfacial rheology of surfactant film in 1988. New technologies to study the interfacial viscosity, elasticity, and viscoelasticity of the interfacial film, and new theories about the interfacial rheology have been developed in recent years (13–16).

We have chosen in this paper only to update the information lacking in these reviews. We pay more attention to the correlation of the interfacial shear viscosity and interfacial primary yield value (IFPYV) of the interfacial film with the stability of a crude oil emulsion. This review will also deal with rheological phenomena of the interfacial film between oil and water phases, that have been investigated in our laboratory recently.

## II. INTERFACIAL FILM PROPERTIES OF CRUDE OIL EMULSION

### A. Interfacial Pressure

The presence of surfactants at the water/oil interface will lower the interfacial tension between the two phases. The interfacial pressure ($\pi$) is defined as

$$\pi = \gamma_{wo} - \gamma'_{wo} \tag{1}$$

where $\gamma_{wo}$ and $\gamma'_{wo}$ are the interfacial tensions between water and oil before and after addition of surfactant, respectively. When the concentration of the emulsifiers is sufficiently high, it may lead to the formation of a tough film surrounding the dispersed droplets. The film can resist the coalescence of the droplets and enhance the emulsion stability. Sjöblom et al. (17) found that there exists a good correlation between interfacial pressure and the stability of the emulsions stabilized by asphaltene fractions from North Sea crude oils, and the larger the interfacial pressure the more stable the emulsions. They also found that the addition of benzene to the pure decane phase reduced the interfacial pressure and the stability of the emulsions. The relationship between the interfacial pressure and the stability of the emulsions stabilized by the asphaltene fraction from Chinese crude oils is proven further by Li et al. (18). Their studies also show that when the interfacial pressure is large the interfacial shear viscosity and IFPYV of the interfacial film between water and oil phases are high (18).

### B. Interfacial Shear Viscosity

Interfacial viscosity plays an important role in determining the lifetime of the film formed between two approaching drops because it not only influences the rate of drainage but also tends to dampen the fluctuations that might lead to its rupture (7). Interfacial viscosity provides valuable information about the nature of intermolecular interactions, transport processes, and phase transitions in flowing monolayers. Malhotra and Wasan found that there was a decrease in the rate of coalescence with an increase in interfacial viscosity (7).

We have investigated the relationship between interfacial shear viscosity and the stability of a model emulsion in recent years. For preparing the model emulsions, the interfacially active fractions are separated from crude oils; the oil used is jet fuel, and the aqueous phase used is either distilled or synthetic formation water. The latter phase has the following composition:

| Ion | Concentration ($mg\,l^{-1}$) |
|---|---|
| $Na^+$ | 5473.92 |
| $Ca^{2+}$ | 32.06 |
| $Mg^{2+}$ | 2.43 |
| $Cl^-$ | 8484.46 |

The interfacial shear viscosity is measured with an SVR · S Interfacial Viscoelastic Meter (Kyowa Kagaku Co. Ltd, Japan). The schematic of the measuring part of the interfacial viscoelastic meter is shown in Fig. 1. The results show that the higher the interfacial shear viscosity of the interfacial film between jet fuel and water the more stable are the emulsions stabilized by the asphaltene or resin fractions from crude oils (18). We also found that the value of the interfacial shear viscosity is affected by the following factors (see below).

#### 1. Emulsifier Concentration

Figure 2 shows that, as the concentration of the interfacially active fractions from Daqing crude oil increased, the value of the interfacial shear viscosity of the interfacial film between jet fuel and the synthetic

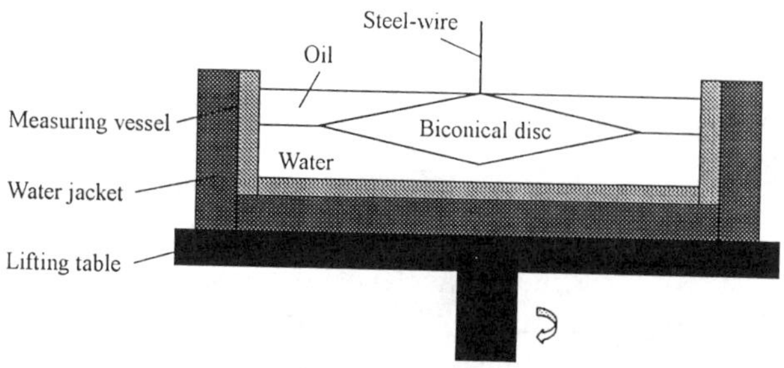


**Figure 1** Schematic of measuring part of SVR · S Interfacial Viscoelastic Meter.

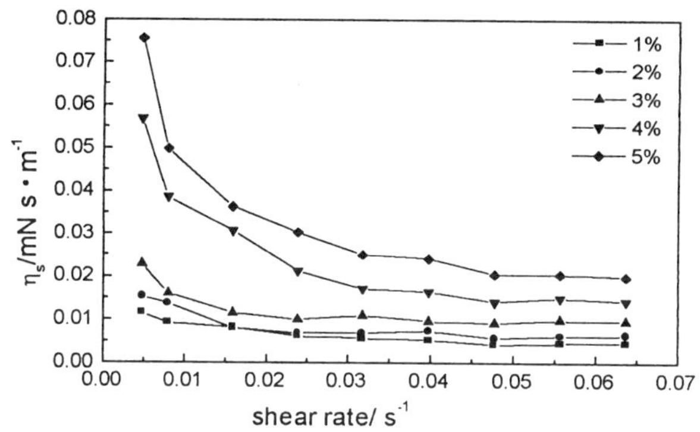


**Figure 2** Interfacial shear viscosity of the interfacial film between jet fuel and the synthetic formation water. The interfacially active fractions separated from Daqing crude oil; $T = 20^\circ C$.

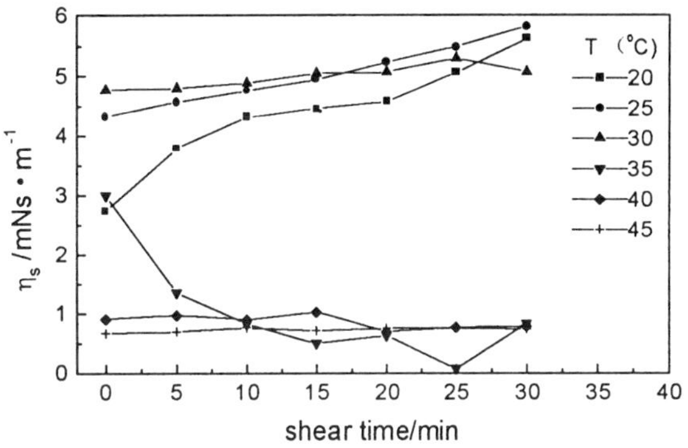


**Figure 3** Interfacial shear viscosity of the interfacial film between jet fuel and the synthetic formation water. The interfacially active fractions separated from Jilin crude oil. $C = 2\%$; shear rate: $0.0159\ s^{-1}$.

formation water also increased. These results show clearly that the higher the concentration of the interfacially active fractions the greater the value of the interfacial shear viscosity.

## 2. Shear Rate

From Fig. 2 it can also be seen that the interfacial shear viscosity of the interfacial film is reduced as the shear rate increases. This phenomenon shows a pseudoplastic behavior (shear thinning) as in a three-dimensional system (12). It seems that the structure of the interfacial film formed with the interfacially active fractions is broken as the shear rate increases. In this case, the decrease of the interfacial shear viscosity means the strength of the film is reduced. Such behaviour reflects the structural characteristics of the film that respond differently depending on the applied shear rate. It may be that the breaking of intermolecular bonds is required for molecules to flow. A low shear rate may then be less efficient for breaking these bonds, leading to apparent higher viscosity (6).

## 3. Shear Time

As shown in Fig. 3 when the interfacially active fractions from Jilin crude oil are used as emulsifier, the shear rate is fixed, and the temperature is below 35°C, the interfacial shear viscosity is raised as the shear time increases. This phenomenon indicates a negative thixotropic behavior as in a three-dimensional system (12). There might be solid particles at the interface between the oil and water phases at temperatures below 35°C. In this case, as the shear rate increases the particles are pushed more closely together in some regions. The effect is to reduce the free movement of the fluid and make the interface more resistant to shear (12). It was found that there were fine wax particles located at the interface at lower temperatures. A photograph of the wax particles at the interface (20°C) is shown in Fig. 4.

As seen in Fig. 5 the influence of wax particles on the interfacial shear viscosity was further verified. When the mixture of Daqing crude oil and the jet fuel is used as a model oil (the contents of the crude oil in jet fuel is 5 or 20%, respectively), the interfacial shear viscosity of the interfacial film between the model oil and distilled water increases markedly with shear time. The content of wax in Daqing crude oil is 18 wt%

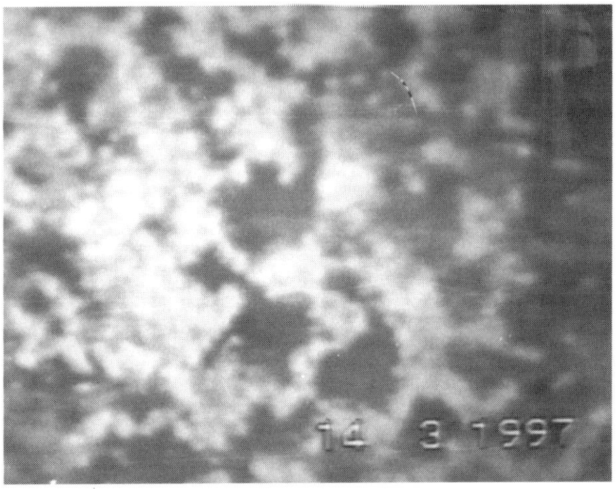


**Figure 4** Wax particles at the interface between jet fuel and the synthetic formation water. The white parts are wax particles, and the black parts are oil.

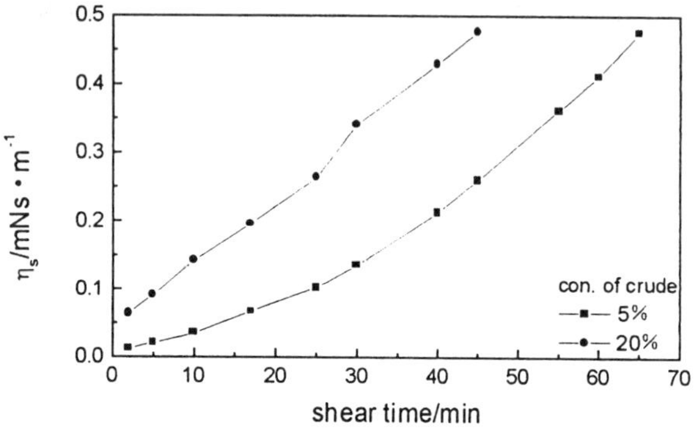


**Figure 5** Influence of wax particles on interfacial shear viscosity of the interfacial film between a model oil and distilled water. The model oil consisted of 5 or 20% Daqing crude oil in jet fuel. Shear rate: $0.0159\ s^{-1}$; $T = 45°C$.

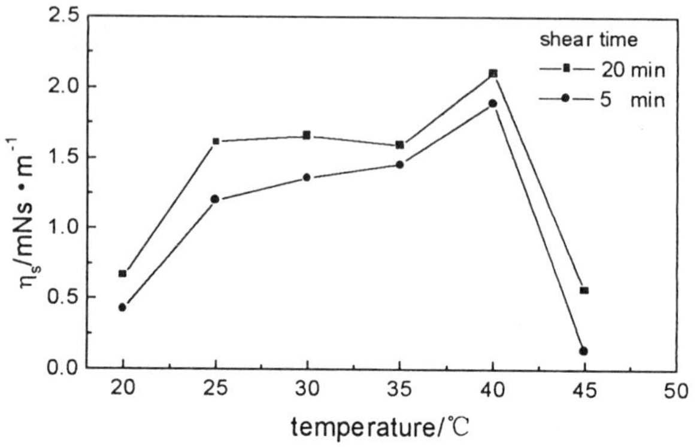


**Figure 7** Interfacial shear viscosity of the interfacial film between jet fuel and distilled water. The interfacially active fractions separated from Jilin crude oil. $C = 2\%$; shear rate: $0.0159\ s^{-1}$.

and the wax precipitation temperature in the crude oil is at 50°C (19).

## 4. Temperature

When the interfacially active fractions from Daqing crude oil are used as emulsifier, the interfacial shear viscosity of the interfacial film between jet fuel and synthetic formation water is decreased as the temperature is raised. The curve of the interfacial shear viscosity versus temperature is shown in Fig. 6.

As seen from Fig. 7 when the interfacially active fractions from Jilin crude oil are used as emulsifier and the temperature is between 20 and 40°C, the interfacial shear viscosity of the interfacial film between jet fuel and distilled water increases as the temperature rises. When the temperature is higher than 40°C the interfacial shear viscosity decreases dramatically. This phenomenon is also illustrated in Fig. 3. The difference in the temperature-dependence behavior of these systems is because the interfacially active fractions from Jilin crude oil contains wax. The precipitation temperature of the wax in Jilin crude oil is 43°C (19). When the temperature is below 40°C the wax particles at the interface expand as the temperature rises. Therefore, the wax particles occupy more space at the interface and make the interface more resistant to shear. When the temperature is higher than 40°C, the wax particles melt so that the interfacial shear viscosity is rapidly reduced.

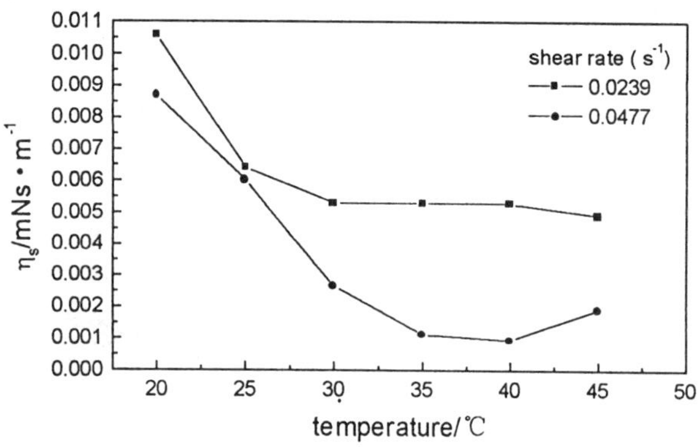


**Figure 6** Interfacial shear viscosity of the interfacial film between jet fuel and the synthetic formation water. The interfacially active fractions separated from Daqing crude oil; $C = 3\%$.

## 5. Water Properties

In comparing Fig. 7 with Fig. 8 it can be seen that, when the interfacially active fractions from Jilin crude oil are used as emulsifiers and the oil phase is jet fuel, the synthetic formation water gives an interfacial shear viscosity higher than that of distilled water when the temperature is below 30°C. However, it is reversed when the temperature is higher than 35°C. This result indicates that the interfacial shear viscosity is affected by the properties of water; the ions in the water may play an important role in the properties of the interfacial film.

## 6. Oil Properties

The stability of emulsions and the concentration of the interfacially active fractions at the interface between oil and water are strongly affected by the properties of the oil phase when the interfacially active fractions are oil soluble (6). Li et al. showed that the increase of the

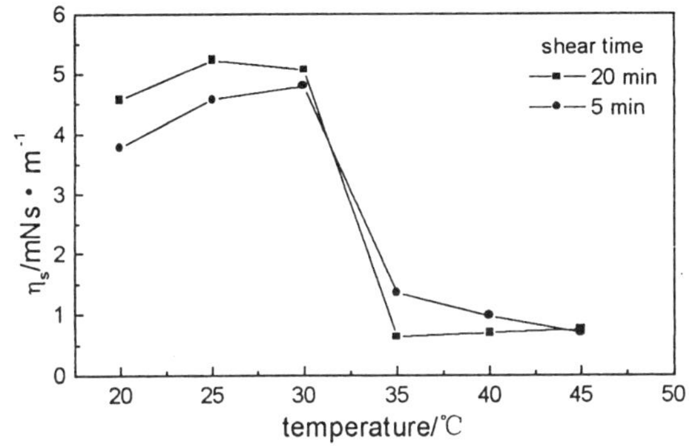


**Figure 8** Interfacial shear viscosity of the interfacial film between jet fuel and the synthetic formation water. The interfacially active fractions separated from Jilin crude oil. $C = 2\%$; shear rate: 0.0159 $s^{-1}$.

aromaticity of the oil phase can reduce the interfacial pressure and the stability of emulsions that are stabilized with asphaltene fractions from North Sea crude oils (11). As shown in Fig. 9 an increase in the aromaticity of the oil phase also reduces the interfacial shear viscosity of the film. When 1% of asphaltene fraction from Gaosheng heavy crude oil was used as emulsifier the interfacial shear viscosity of the interfacial film between the oil and distilled water decreased as the ratio of jet fuel/*p*-xylene(oil phase) changed from 100/0 to 60/40 (20).

## 7. Asphaltene and Resin Fractions

Generally, the asphaltene fractions from most crude oils give a high stability of emulsions and the resin fractions give a low emulsion stability (21). Yang (20) found that, for a fixed oil/water system, asphaltene gives a high interfacial shear viscosity and resin gives a low interfacial shear viscosity. When a mixture of the asphaltene and resin fractions are used as emulsifier the values for the interfacial shear viscosity vary. As seen in Fig. 10 it is clear that the addition of small amounts of resin fraction to the asphaltene decreases the interfacial shear viscosity dramatically.

## 8. Wax Particles

The above experimental results show that wax particles can affect the rheological properties of the interfacial film between the water and oil phases. In order to confirm these phenomena we added a synthetic wax to the jet fuel/synthetic formation–water system and took the interfacially active fractions from Daqing crude oil as emulsifier. The content of the synthetic wax in jet fuel was 5% and the melting temperature of the synthetic wax was 54–56°C. The results shown in Fig. 11 demonstrate that the interfacial shear viscosity increased as the temperature rose to the range between 20 and 30°C, and the interfacial shear viscosity decreased when the temperature was higher than 30°C. It is obvious that the presence of synthetic wax particles at the interface makes the properties of the interfacial film greatly different from those shown in Fig. 2.

Figure 12 shows that when the temperature is lower than 30°C, the interfacial shear viscosity increases as the shear rate increases. This phenomenon shows a dilatant behavior (shear thickening) as in a three-dimension system (12). These results further prove that wax particles can contribute to the rheological

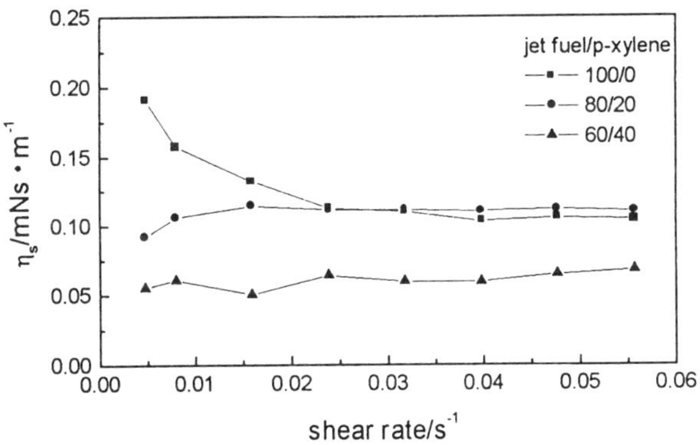


**Figure 9** Interfacial shear viscosity of the interfacial film between jet fuel/xylene and distilled water. The interfacially active fractions are asphaltenes separated from Gaosheng heavy crude oil. $C = 1\%$; $T = 25°C$.

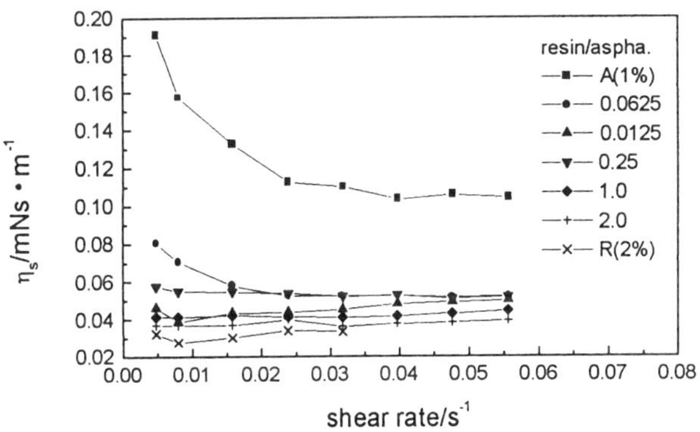


**Figure 10** Interfacial shear viscosity of the interfacial film between jet fuel and distilled water. The interfacially active fractions are a mixture of asphaltene(A) and resin(R) separated from Gaosheng heavy crude oil. $T = 25°C$.

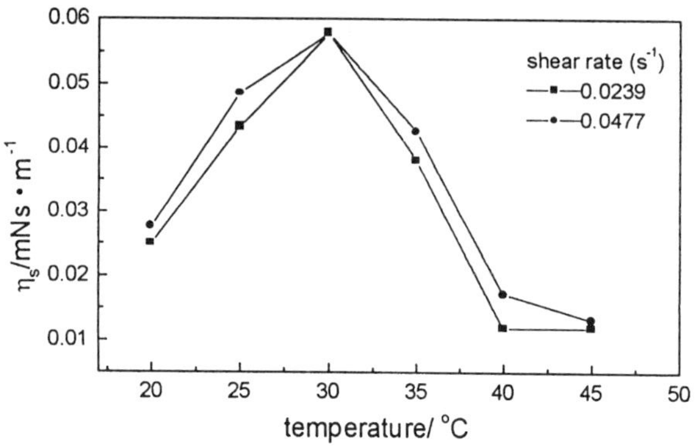


**Figure 11** Interfacial shear viscosity of the interfacial film between jet fuel and the synthetic formation water. The interfacially active fractions separated from Daqing crude oil; $C = 3\%$, and the content of synthetic wax in fuel is 5%.

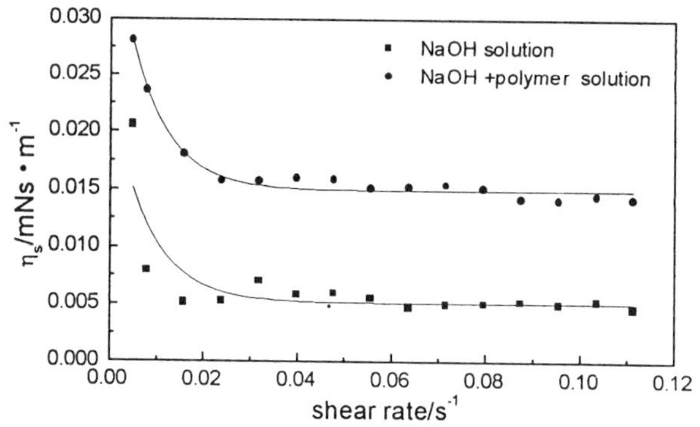


**Figure 13** Interfacial shear viscosity of the interfacial film between a model oil and NaOH solution or polymer solution. The model oil consists of 20% Daqing crude oil in jet fuel. The contents of NaOH, ORS41, and a biological surfactant in the NaOH solution are 1.2, 0.15, and 0.15%, respectively. The content of the polymer adds 150 mg/L; $T\% = 45°C$.

properties of the interfacial film between the oil and water phases and the stability of the emulsions (10, 22).

## 9. Polymers

The presence of polymers at the interface between oil and water makes for excellent stabilization of emulsions (1, 4). Figure 13 shows the interfacial shear viscosity of the interfacial film between a model oil and NaOH solution or polymer solution at 45°C. The model oil consisted of 20% Daqing crude oil in jet fuel. The contents of NaOH, ORS41, and a biological surfactant in the NaOH solution were 1.2, 0.5, and 0.15%, respectively. The concentration of polymer hydrolyzed polyacrylamide (HPAM) in the solution was 150 mg/L. It can be seen that the interfacial shear viscosity of the system with the polymer is three times higher than that of the system without the polymer. It is also shown in Fig. 14 that the higher the concentration of the polymer in the solution the higher the interfacial shear viscosity of the interfacial film. It should be noted that the interfacial film also presents shear thinning characteristics.

## 10. Aging

The aging of crude oils or the interfacially active fractions from crude oils are able to enhance the stability of emulsions (23, 24). Sjöblom et al. (23) found that the Fourier-transform infrared spectra reveal that the carbonyl peak grows markedly on account of the C=C

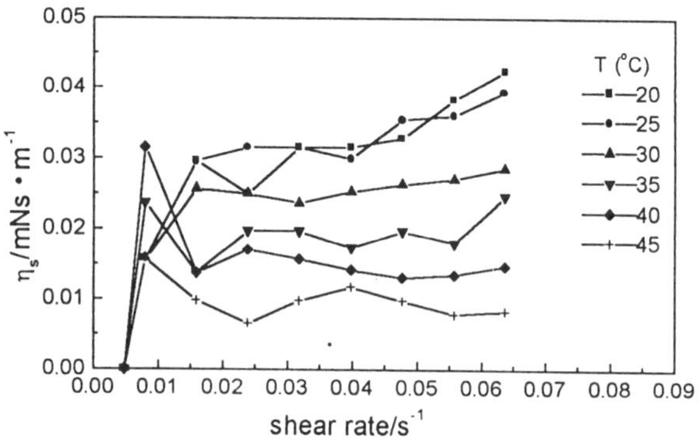


**Figure 12** Interfacial shear viscosity of the interfacial film between jet fuel and distilled water. The interfacially active fractions separated from Daqing crude oil; $C = 3\%$, and the content of the synthetic wax in fuel is 5%.

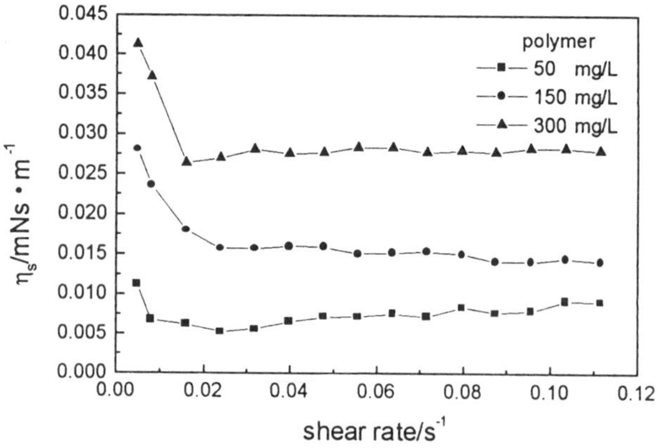


**Figure 14** Interfacial shear viscosity of the interfacial film between jet fuel and polymer solution. The contents of NaOH, ORS41, and a biological surfactant in the polymer solution are 1.2, 0.15, and 0.15%, respectively; $T = 45°C$.

mode. At the same time the spectral region between 900 and 700 $cm^{-1}$ reveals that a condensation process takes place upon aging. Yang (20) shows that aging can also raise the interfacial shear viscosity of the interfacial film between oil and water. As illustrated in Fig. 15 the interfacial shear viscosity of the interfacial film between jet fuel and distilled water increases significantly with aging time, and the aged film has obvious shear thinning characteristics.

### 11. Demulsifiers

Demulsifiers can reduce the stability of emulsions efficiently (3). We found that the emulsifiers which can destabilize the heavy crude oil emulsions from Liaohe Oil Field can also decrease the interfacial shear viscosity of the interfacial film of the emulsions markedly. As seen from Fig. 16 when 0.25% of the asphaltene from Gaosheng heavy oil is used as emulsifier, the interfacial shear viscosity of the interfacial film between jet fuel and distilled water reduced significantly at 25°C by adding 50 mg/L of S-9, S-10, or S-11 demulsifiers.

## C. Interfacial Primary Yield Value

The IFPYV is defined as the shear stress when the shear rate is zero (25). Taubman and Koretskii (26) found that the yield stress of the interfacial film between $CCl_4$ and an aqueous solution of $AlCl_3$ was related to the mechanical strength of the emulsifier film and the emulsion stability. Their study concluded that the lifetime of the emulsion, the yield stress, and the interfacial viscosity increase simultaneously. The

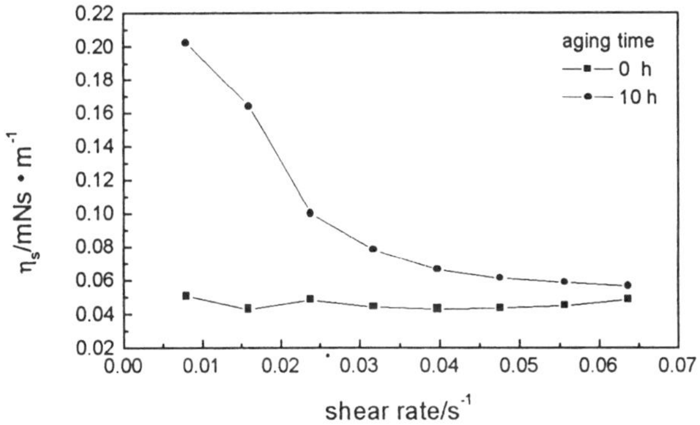


**Figure 15** Aging of the interfacial shear viscosity of the interfacial film between jet fuel and distilled water. The interfacially active fractions are a mixture of asphaltene and resin separated from Gaosheng heavy crude oil. The content of asphaltene in the jet fuel is 0.25% and the ratio of resin(R) to asphaltene(A) is 2; $T = 25$°C.

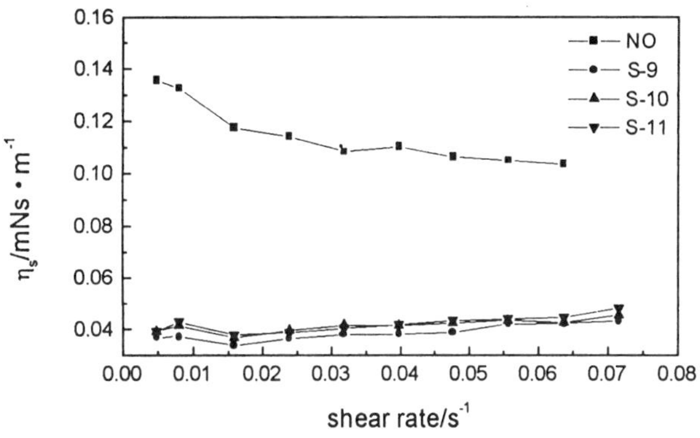


**Figure 16** Interfacial shear viscosity of the interfacial film between jet fuel and distilled water with demulsifiers. The interfacially active fraction is asphaltene separated from Gaosheng heavy oil; $C = 0.25\%$. The content of the demulsifiers in water is 50 mg/L; $T = 25$°C.

experiments in our laboratory show that the interfacial film formed from the interfacially active fractions from crude oils, such as asphaltenes and, resin fractions, has a markedly high IFPYV. When wax particles, asphaltene particles, or polymer exist at the interface, the IFPYV of the film is high. These results revealed that the interfacial film between the oil and water is strongly structured, and the degree of the IFPYV of the film is related to the structural strength of the film. It is obvious that the IFPYV of the film can be an important parameter for evaluating the strength of the film and the stability of the emulsions.

Generally, the factors that affect the interfacial shear viscosity can also affect the IFPYV of the interfacial film. In most cases, the influence of the factors on both the interfacial shear viscosity and IFPYV is identical. The following gives a brief description of the influence of the emulsifier concentration and temperature on the IFPYV of the interfacial film.

### 1. Emulsifier Concentration

Figure 17 shows that when the interfacially active fractions from Daqing crude oil are used as emulsifier, the IFPYV of the interfacial film between jet fuel and waters increases as the concentration of the emulsifier increases. Experimental results also reveal that the IFPYV of the interfacial film formed from 2% of the interfacially active fraction from Jilin crude oil is 0.015336 mN $m^{-1}$, which is about 18 times of the IFPYV of the interfacial film formed from 1% of the fraction (18). These results indicate that the interfacial film between the oil and waters is strongly structured,

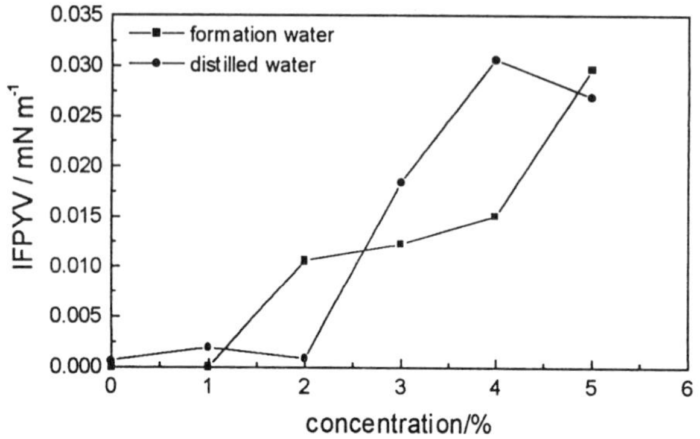


**Figure 17** Interfacial primary yield value of the interfacial film between jet fuel and waters. The interfacially active fraction separated from Daqing crude oil. Shear rate: 0.0048 $s^{-1}$; $T = 20°C$.

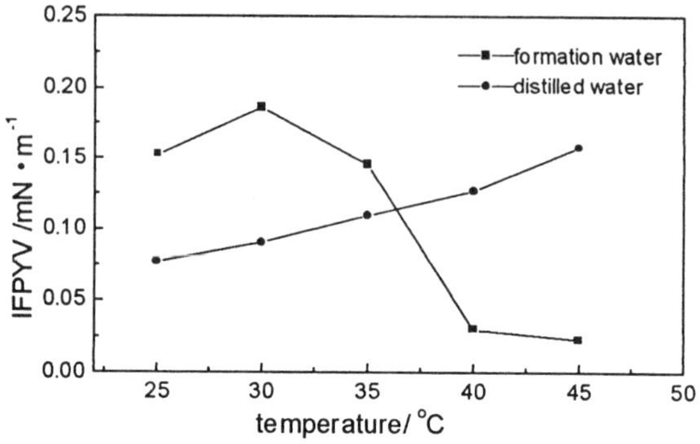


**Figure 19** Interfacial primary yield value of the interfacial film between jet fuel and waters. The interfacially active fraction separated from Jilin crude oil; $C = 2\%$. Shear rate: 0.0048 $s^{-1}$.

and the strength of the film structure depends on the concentration of the interfacially active fractions.

## 2. Temperature

As seen from Fig. 18 when the interfacially active fractions from Daqing crude oil are used as emulsifier, the IFPYV of the interfacial film between jet fuel and distilled water decreases rapidly when the temperature rises from 20 to 30°C, and the IFPYV of the film levels out when the temperature is higher than 30°C. When the interfacially active fractions from Jilin crude are used as emulsifier, the IFPYV of the interfacial film between jet fuel and distilled water increases as the temperature rises from 25 to 4 5°C (see Fig. 19). This phenomenon is similar to the temperature-dependence behavior of the interfacial shear viscosity of the system (see Fig. 7). When the synthetic formation water is used as the aqueous phase and the temperature is lower than 30°C, the IFPYV of the interfacial film increases as the temperature rises. This phenomenon is also similar to the temperature-dependence behavior of the interfacial shear viscosity of the system (see Fig. 8). These results demonstrate that the IFPYV of the interfacial film depends also on the temperature, wax particles, and the properties of the aqueous phase.

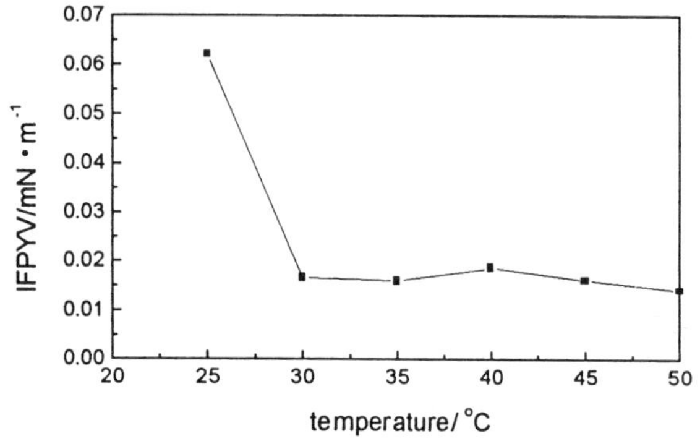


**Figure 18** Interfacial primary yield value of the interfacial film between jet fuel and distilled water. The interfacially active fraction separated from Daqing crude oil; $C = 3\%$. Shear rate: 0.0048 $s^{-1}$.

# III. CONCLUSIONS

This review shows that when the interfacially active fractions separated from crude oils, such as asphaltene, resin, and wax fractions, are used as emulsifiers, the rheological properties of the interfacial film between oil and water strongly affects the stability of crude oil emulsions. There exists a good correlation between interfacial pressure, interfacial shear viscosity, IFPYV of the interfacial film, and the stability of crude oil emulsions. In particular, the interfacial shear viscosity and the IFPYV of the interfacial film are more valuable for evaluating the strength of the film and the stability of the emulsions. The experimental results prove that the interfacial film between oil and water is strongly structured, and that the level of the interfacial shear viscosity and the IFPYV of the film are related to the strength of the film structure. The strength of the film structure is affected by the shear rate, shear time, temperature, aging, the properties of the oil, the water, the emulsifiers, and the demulsifiers. The interfacial shear viscosity and IFPYV of the film can be important parameters for evaluating the strength of the film and the stability of the emulsions.

## ACKNOWLEDGMENTS

China National Petroleum Cooperation (CNPC) and State Key Laboratory of Heavy Oil Processing are acknowledged for financial support. Our colleagues Shuling Ji, Zhengxin Tong, and Weidong Liu, and my students Peng Zhen and Tao Wang are acknowledged for the experimental work.

## REFERENCES

1. SE Friberg. In: J Sjöblom, ed. Emulsions – A Fundamental and Practical Approach. NATO ASI Series C363, Dordrecht: Kluwer, 1992, p 1–29.
2. TF Tadros, B Vincent. In: P Becker, ed. Encyclopedia of Emulsion Technology, Vol 1. New York: Marcel Dekker, 1985, p 1–29.
3. VB Menon, DT Wasan. In: P Becker, ed. Encyclopedia of Emulsion Technology. Vol. 2. New York: Marcel Dekker, 1985, p 1.
4. TF Tadros. In: J Slöblom, ed. Emulsions and Emulsion Stability. New York: Marcel Dekker, 1996, p 173.
5. S Dukhin, J Sjöblom. In J Sjöblom, ed. Emulsions and Emulsion Stability. New York: Marcel Dekker, 1996, p 41.
6. F MacRitchie. Chemistry at Interfaces. New York: Academic Press, 1990, p 81.
7. AK Malhotra, DT Wasan. In: IB Ivanov, ed. Surfactant Science Series. Vol 29. New York: Marcel Dekker, 1988, p 829.
8. J Sjöblom, O Urdahl, H Hoiland, AA Christy, EJ Johansen, Progr Colloid Polym Sci 82: 131–139, 1990.
9. J Sjöblom, H Soderlund, S Lindblad, EJ Johansen, IM Skjarvo. Colloid Polym Sci 268: 389, 1990.
10. M Li, J Sjöblom. J Dispers Sci Technol 12: 303–320, 1991.
11. M Li, AA Christy, J Sjöblom. In J Slöblom, ed. Emulsion – A Fundamental and Practical Approach. NATO ASI Series C363, Dordrecht: Kluwer, 1992, p 157–172.
12. RJ Hunter. Foundations of Colloid Science. Vol 2. Oxford: Clarendon Press, 1992, p 992.
13. J Benjamins, EL Reynders, and A Cagna. Proceedings of Second World Congress on Emulsions. Bordeaux, France, 1997, Vol 2, 2-2-086.
14. G Pratt, C Thoraval. Proceedings of Second World Congress on Emulsions. Bordeaux, France, 197, Vol 2, 2-2-125.
15. KD Danov, IB Ivanov, PA Kralchevsky. Proceedings of Second World Congress on Emulsions. Bordeaux, France, 1997, Vol 2, 2-2-152.
16. M Thoma, T Pfohl, H Mohwald. Langmuir 11: 2881, 1995.
17. J Sjöblom, M Li, AA Christy, T Gu. Colloids Surfaces 66: 55–62, 1992.
18. M Li, P Zhen, Z Wu, S Ji. Proceedings of Second World Congress on Emulsions. Bordeaux, France, 1997, Vol 2, 2-2-053.
19. M Li, J Su, Z Wu, Y Yang, S Ji. Colloids Surfaces A 123: 635–649, 1997.
20. X Yang. Study on Stabilization of Water-in-Crude Oil Emulsions – Film Properties of Asphaltenes and Resins. Doctoral thesis, Research Institute of Petroleum Processing, Beijing, 1998.
21. M. Li. Separation and Characterization of Indigenous Interfacially Active Fractions in North Sea Crude Oils. Correlation to Stabilization and Destabilization of Water-in-Crude Oil Emulsions. Doctoral thesis, University of Bergen, Norway, 1993.
22. AJ McMahon. In: J Sjöblom, ed. Emulsions – A Fundamental and Practical Approach. NATO ASI Series C363, Dordrecht: Kluwer, 1992, p 135–156.
23. J Sjöblom, M Li, AA Christy, HP Ronningsen. Colloids Surfaces A 96: 261–272, 1995.
24. HP Ronningsen, J Sjöblom, M Li. Colloids Surfaces A 97: 119, 1995.
25. RJ Hunter. Foundations of Colloid Science. Vol 2. Oxford: Clarendon Press, 1992, p 998.
26. AB Taubman, AF Koretskii. Kolloidn Zh 20: 676, 1985.

# 23

# Film Properties of Asphaltenes and Resins

**Xiaoli Yang and Wanzhen Lu**
*Research Institute of Petroleum Processing, Beijing, China*

**Marit-Helen Ese**
*University of Bergen, Bergen, Norway*

**Johan Sjöblom**
*Statoil A/S, Trondheim, Norway*

## I. INTRODUCTION

Since the wide application of new recovery technology much more crude oil has been produced with various amounts of free and emulsified water (1). Stable water-in-crude oil emulsions not only increase the cost of oil recovery and transportation, but also increase the cost of petroleum processing (2, 3). It has been well known that some of the components in crude are interfacially active in nature (4, 5), such as asphaltenes, resins, and naphthenic acids. Asphaltenes, which are believed to be the major materials involved in emulsion stabilization, can adsorb to and accumulate at water-in-crude oil interfaces to form a rigid film surrounding the water droplets to protect the interfacial film from rupturing during droplet–droplet collisions. Hence, interfacial rheological properties of the interfacial film should be closely related to the stability of the crude oil emulsions (6, 7). During recent years, the availability of much better analytical tools, instruments, and advanced computers (8–17) have increased interest in these subjects.

Asphaltenes are defined as being insoluble in n-pentane and n-heptane, but soluble in toluene (18, 19). Asphaltenes contain millions of different molecules of different polarity having the same solubility properties in oil or the precipitation solvent. Asphaltenes have a large number of polynuclear aromatic ring systems bearing alkyl side chains plus some heteroatoms (such as O, N, and S), essentially in the aromatic structure, which impart the amphiphilic characteristics of asphaltenes. In a real crude system, asphaltenes are believed to aggregate into micelles which are kept peptized or dispersed by resins (19). Oxygen-containing side chains are usually one cause for the interfacial activity of the asphaltenes. The aromatic moieties are mainly responsible for the aggregation of the asphaltic molecules (20–23).

Because asphaltenes contain highly complex macromolecules, only their averaged chemical structure is known. Yen and coworkers (11, 13) worked out a model structure, which explains most properties of asphaltenes. In this model, asphaltenes consist of flat sheets of condensed aromatic systems that may be interconnected by sulfide, ether, or aliphatic chains. An average of five of these sheets are stacked by Π–Π interactions. The hydrogen bonding and dipole interactions cause the asphaltenes to aggregate into "micelles" when the concentration is high enough. Studies of molecular weights of asphaltenes indicate that asphaltenes have a stronger propensity for self-aggregation in aliphatic organic solvents. They form molecular aggregates even in dilute solutions (15, 20–24). Many physical methods have been used to study these aggregates (15, 17, 25–34).

In 1991, Andersen and Birdi (25) first reported a critical micelle concentration (cmc) of asphaltenes in a mixture of n-alkane and toluene, using a calorimetric titration method. From their study, the aggregation process of asphaltic molecules in solutions was suggested as the following:

$$\text{monomers} \rightarrow (\text{particles} \rightarrow \text{micelles}) \rightarrow \text{aggregates}$$

The particles are formed by stacks of three to five aromatic disks as in Yen's model. Asphaltenes were assumed to be associated into different kinds of micelles in different solvents. Sheu et al. (27) verified the existence of a cmc when Ratawi asphaltenes were dissolved in pyridine solvent by measuring the surface tension. When the concentration of asphaltene is over the cmc, asphaltene molecules can be further aggregated. The same phenomenon was observed for east Texas asphaltenes in heptane/toluene by Krawczyk et al. (28).

Galtsev et al. (29) studied asphaltene and resin association in real crude oil by using electron nuclear double-resonance spectroscopy. They found that most of the asphaltene molecules are associated with each other from room temperature up to 90°C, and have a core of a stack of condensed aromatic sheets with a radius of 1 nm.

Bardon et al. (17) used scattering methods to study the structure of asphaltenes both in real systems and in organic solution. The lamellar structural model for asphaltenes was confirmed. From their experiment results, they concluded that asphaltene particle size decreased in a mixture of asphaltenes and resins.

The above results suggest that the behavior of asphaltenes in solutions is governed by some sort of aggregation equilibrium. Many factors such as the nature of solvent, the concentration of asphaltenes or resins, temperature, and so on, can influence the level of aggregation.

In order to understand the properties of the interfacially active components in crude oil systems, model systems with chemical properties identical or similar to those of the original crude oil were usually used. Fordedal and coworkers (35, 36) proposed model oils with asphaltenes from the Norwegian shelf dissolved in a series of decane/toluene mixtures as the oil phase. They studied the influence of the aromaticity of oil phase on the stability of water-in-crude oil emulsion. It was found that the stability of the water-in-oil emulsion is related to changes in the aromaticity of the oil phase. When the decane/toluene ratio is 80/20 (v/v), the asphaltene model oil emulsion is the most stable.

McLean and Kilpatrick (37) studied the effects of asphaltene aggregation in heptane/toluene mixtures on the stability of water-in-oil emulsions. The asphaltenes were separated from four different crudes with various heptane/toluene ratios, and various concentrations of asphaltenes and resin/asphaltene ($R/A$) ratios. The emulsions were most stable when the crude medium contained 30–40% toluene and a lower $R/A$ ratio, i.e., $R/A < 1$. These results also show the significance of the solubility of the asphaltenes in determining the emulsifying potential for these crude oils (38–42).

In crude oil and water systems asphaltenes are adsorbed at the water–oil interface and flocculate yielding a three-dimensional structured film (as Mesophase C) (43). Such structured films at the W/O interface were verified by Siffert et al. (44). They separated oil, water, and a sticky mass between sheets in asphaltene particles for this black mass. These regularly stacked lamellar structures have close similarity to surfactant liquid crystals.

Sheu and coworkers (15, 34) measured both dynamic and equilibrium surface tensions for two vacuum residue fractions (derived from the Ratawi and Neutral Zone vacuum residues) in solvents to investigate the self-association process. It was found that asphaltenes require several molecules of different structures to pack into a shape and a size that satisfies the minimum free-energy requirement under the given conditions. Sheu et al. (27) also studied the dynamic interfacial activities of asphaltene molecules in toluene against aqueous solutions containing sodium hydroxide of various concentrations. The interfacial tension was reduced considerably as compared with that of pure toluene/water. The equilibrium kinetics were evaluated. A reaction-like process, believed to be initiated by molecular packing, was observed as the system approached equilibrium. The results suggest that the adsorption/desorption kinetics were diffusion controlled initially, but become reaction controlled in the long term.

Hartland and coworkers (45–47) have extensively studied the dynamics of emulsification and demulsification of water-in-crude oil emulsion. They found that temperature, the concentration of emulsifier or demulsifier, and the nature of the medium (crude oil or brine) were very important parameters governing the adsorption of emulsifier at the interface.

Acevedo and coworkers (9, 10, 22, 48) studied the absorption of natural surfactants, i.e., asphaltenes, resins, and carboxylic acids at the W/O interface and their influence on $\gamma$ (surface tension)–pH and $\gamma$–time behavior. They noted that the changes in $\gamma$ could be accounted for in term of the basic (i.e., amine) and acidic (i.e., carboxylic acid) functional groups of these materials. The $\gamma$ versus time behavior depends on the diffusion-controlled adsorption of high molecular weight aggregates at the oil–water interface. They found that $\gamma$ for natural surfactants of Tia Juana crude oil reached equilibrium after 7 days in the acidic region. In this case the amphiphile should diffuse through the flocculated asphaltene–resin network, thus accounting for the extremely slow change in $\gamma(t)$.

Eley et al. (49) studied the rheological properties of asphaltene films adsorbed at the oil/water interface. Its elastic property is consistent with the formation of a network structure in the films, possibly arising from focculated asphaltene particles appearing at the water/oil interface. Both the dilatancy and the "stick-slip" flow could arise from thick films of asphaltene particles building up at the interface.

Mohammed et al. (50, 51) measured the interfacial rheological properties of asphaltene and resin at the W/O interface. It was found that the rheological properties of asphaltenes are time dependent and that the film needs at least 8 h to attain equilibrium. During this period, the surface elasticity and viscosity increase markedly with time (surface viscosity increasing from $3 \times 10^2$ mNs/m after 2 h of aging to $3 \times 10^3$ mNs/m after 8 h aging, and the surface elasticity increasing from zero to 2 mN/m). They also found that the strength of the asphaltene film is higher than that of the resin film.

Resins are the materials which remain oil soluble after asphaltenes are precipitated in n-pentane or n-heptane, but are adsorbed on to surface-active material such as silica gel. They are a comparatively little known fraction (52, 53). Their composition is very much dependent on the separation procedure. The resin molecule is structurally similar to, but smaller than the asphaltene molecule and appears to be a good solvent for asphaltenes (17, 19). Because of its dispersion function for asphaltenes, researchers (35, 36, 53, 54) started to note over recent years the influence of resins on the stability of water-in-crude emulsions.

As mentioned above, interfacial film properties play a very important role in the stabilization of water-in-crude oil emulsions. In order to obtain a better understanding of asphaltenes and resins affecting the stability of water-in-crude emulsions, it is of interest to investigate the film properties of these components.

## II. LANGMUIR FILMS OF ASPHALTENES AND RESINS

On the surface between air and water, amphiphilic molecules will orient with the hydrophilic part in the water phase, while the hydrophobic chains reject the water surface. The Langmuir technique is used to measure the surface pressure–area ($\Pi$–$A$) isotherms to give important information on amphiphilic molecules and their interaction in the air/water films. The interaction forces will change according to the free space between the molecules at the air/water surface. We may consider that the state of the films changes from gas-like, liquid-like, to solid-like, depending on the free space (55–57). In gas-like films, there is no interaction. A higher level of interaction gives liquid-like films; a close packing of the molecules at the surfaces forms solid-like films that can be quite rigid. In $\Pi$–$A$ isotherms, the slope reflects the compressibility of the film. When a water droplet is surrounded by a highly non-compressible and rigid film the droplet is free to coalesce.

The Langmuir–Blodgett technique also makes it possible to monitor the monolayer stability. This is done by compressing the film to a certain pressure that is held constant. The decrease in film area is measured as a function of time. An observed loss of film area may be as a result of rearrangements of the film molecules, dissolution of film molecules into a different state, and/or collapse by nucleation (58); subsequently, solid bulk fragments start to grow.

Langmuir films of interfacially active fractions of crude have been investigated by some groups (59–62). In 1989, Leblanc and Thyrion (60) tried to use the Langmuir technique to measure the average molecular weight and the size of asphaltene molecules. Monolayers of deasphalted oil containing $C_5$ and $C_7$ asphaltenes were studied using pressure–area ($\Pi$–$A$) isotherms. It was shown that the asphaltenes form stable monolayers at the air/water interface. The hydrophilic head group and hydrophobic tail are in balance.

Gonzalez et al. (61) studied air–water surface films of asphaltenes and resins from Australia. They used chloroform as the spreading solvent and obtained duplex resin films, but thermodynamically stable asphaltene monolayers were not obtained because of the poor spreading solvent.

Singh and Pandey (62) studied interfacially active fractions from Indian crudes by using the film pressure–area; they observed the influence of the nature of the water phase on the pressure–area isotherm and found that there was a direct relationship between film pressure and the stability of crude-oil emulsion.

In our work, the influence of the aromaticity of the spreading solvent and the concentration of asphaltenes, resins, or asphaltene and resin mixtures on the air/water film properties were investigated. The asphaltenes and resins were from "North Sea crude" and French Venezuela crude, respectively. The separation procedure for asphaltenes and resins is described in Ref. 63.

## A. Influence of Bulk Concentrations of Asphaltenes or Resins

Figures 1 and 2 depict the Π–*A* isotherms for asphaltenes and resins spread from pure toluene.

Figure 1 shows that the films of asphaltenes have a low compressibility. At high surface areas there is no variation in surface pressure with decreasing surface area, until a sharp increase is observed. Within this region the asphaltene molecules/aggregates on the surface start to interact, the film having been transferred from gaseous to liquid state. As the polarity of the asphaltenes is similar to that of the poorly spreading polymers (64), their Π–*A* isotherms are alike. When asphaltenes are spread from aromatic solvents, they probably are present as small association structures or even as monomers. However, larger aggregates or particles may be formed when the bulk concentration increases (65). Therefore, the quantity along the *X* axis in the Π–*A* isotherms does not give the correct size of the molecules. The trend is clearly noticed, i.e., the surface pressure starts to increase at higher surface areas when the bulk concentrations of asphaltenes are reduced. In order to obtain comparable quantitative results, the surface concentration of asphaltenes needed to create a surface pressure equal to 10 mN/m is calculated. The results given in Table 1 indicate that higher bulk concentrations of asphaltenes make the asphaltene structures formed more compact. Therefore, more asphaltenes are needed on the surface to entail the same surface pressure (10 mN/m). When the bulk concentration of asphaltenes in pure toluene is lower than 1 mg/ml, the Π–*A* isotherms are almost overlapping. This is a consequence of asphaltenes being dissolved as single molecules, owing to low concentrations and a good aromatic solvent.

Figure 2 shows the Π–*A* isotherms for resins with different bulk concentrations in pure toluene. Comparing with Fig. 1, evident differences can be observed. First, the bulk concentrations of resins have less effect on the Π–*A* isotherms. Second, the resin films show a high degree of compressibility. Like that of linear polymers (64), in the low-pressure region, the energy arises largely from entropic effects

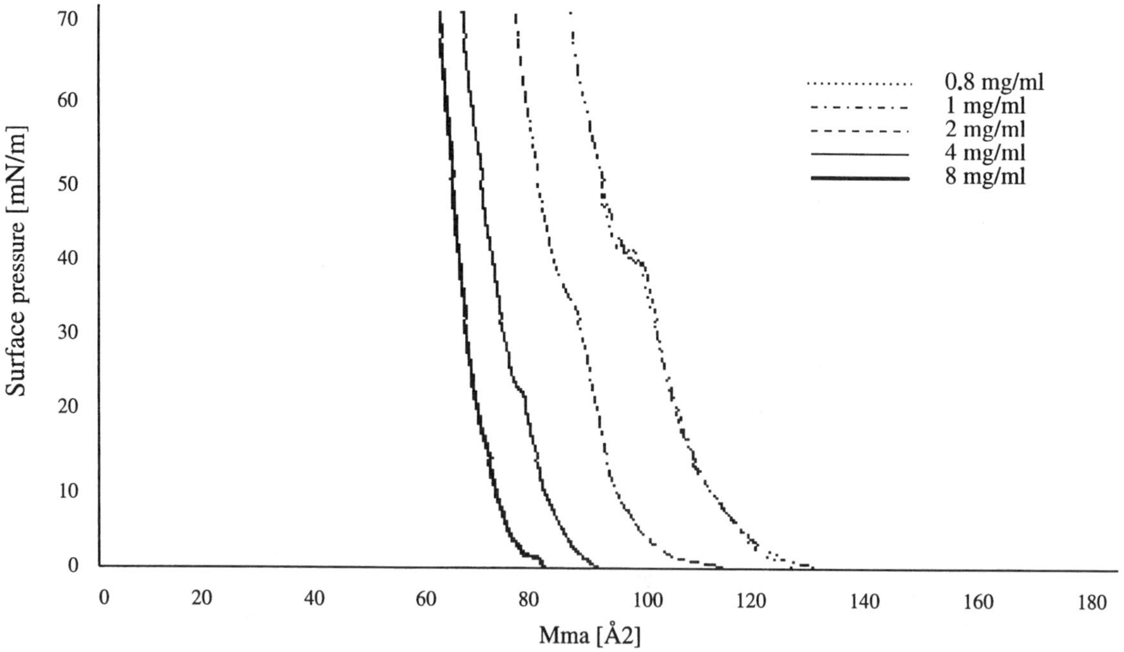


**Figure 1** Π–*A* isotherms for different bulk concentrations of asphaltenes spread from pure toluene on pure water.

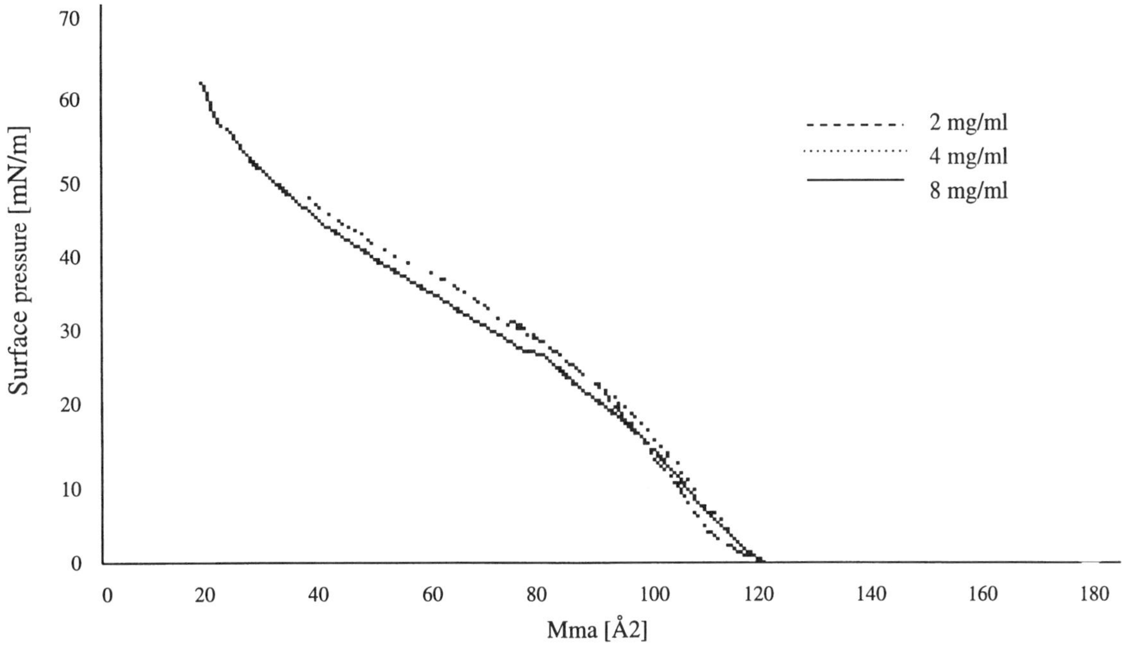


**Figure 2** Π–*A* isotherms for different bulk concentrations of resins spread from pure toluene on pure water.

associated with the arrangements formed by the flexible hydrocarbon chains. As the film is compressed, the equilibrium shifts in favor of the displaced segments of molecule chains which are pushed away from the surface and into the bulk phase to build up a multilayer.

## B. Influence of Aromaticity of Spreading Solvents

The Π–*A* isotherms for asphaltenes spread from solvents of different aromaticity (varying volume ratio of toluene/hexane) show that, when the spreading solvents contain $\leq 20$ or $\geq 40\%$ toluene, the film properties are distinctly different. However, when the amount of toluene is reduced from 100 to 40% in the spreading solvents, the film properties are almost the same. When the amount of toluene in the spreading solvent is $\leq 20\%$ a large part of the asphaltenes are present as particles, which is a result of poor dissolution of the heavy fraction in highly aliphatic solvents. With increased bulk concentrations of asphaltenes, the surface concentration needed to achieve the same surface pressure (10 mN/m) increases. Asphaltenes are near the point of critical precipitation when the aromaticity of the solvent is low.

Figure 3 summarizes the surface concentrations of asphaltenes needed to obtain a surface pressure equal to 10 mN/m, when spread from a series of solvents. In order to achieve the same pressure, the concentration of asphaltenes on the surface is increased both with increasing solvent aromaticity and with increasing bulk concentration of the film-forming material.

**Table 1** Surface Concentration of Asphaltenes and Resins Necessary to Obtain Surface Pressure Equal to 10 m/Nm

| Bulk concentration of asphaltenes or resins (mg, $ml^{-1}$) | Surface concentration of asphaltenes (mg, $m^{-2}$) | Surface concentration of resins (mg/$m^{-2}$) |
|---|---|---|
| 0.8 | 5.6 | — |
| 1.0 | 5.7 | — |
| 2.0 | 6.8 | 2.6 |
| 4.0 | 7.7 | 2.7 |
| 8.0 | 9.0 | 2.6 |

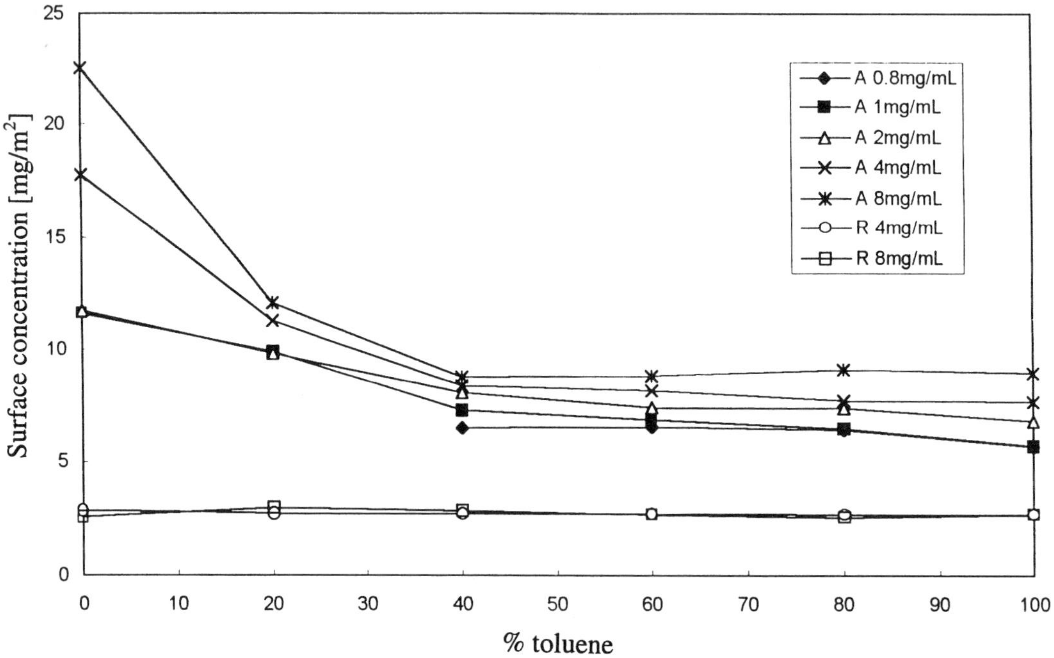


**Figure 3** Surface concentration of asphaltenes and resins needed to obtain $\Pi = 10$ mN/m vs. vol% toluene in the spreading solvent with varying bulk concentration.

However, only small changes in surface concentration of resins are observed with these kinds of variations in the system.

Figure 4 shows the results of the kinetic tests on asphaltene films, where the variation in surface pressure is measured against time while the area is kept constant. The $\Pi$–$t$ curves for asphaltenes spread from 20%/80% toluene/hexane are markedly different from the $\Pi$–$t$ curves for asphaltenes spread from pure toluene. Asphaltenes spread from solvents containing less than 20% toluene give rise to an increase in surface pressure with increasing bulk concentration of asphaltenes (Fig. 4). A multilayer structure may exist on the surface when the solvent contains less toluene ($\leq$ 20%). When the amount of toluene in the spreading solvent is high, even 8 mg/ml asphaltenes may be dissolved, and no change in the surface pressure is observed. This may be explained as a result of the asphaltene fraction being dissolved in the aromatic solvent, preventing formation of a multilayer.

The $\Pi$–$A$ isotherms for resins spread from solvents of different aromaticity show that the solvent has less influence on the film properties of resins. This is a consequence of low self-association of resins (Fig. 3).

Kinetic studies ($\Pi$–$t$ curves) of resin films spread from 20%/80% toluene/hexane, show a small increase in pressure during the first couple of hours. A stronger affinity of the film material toward the surface may be due to oxidation of the resin film, which results in a higher surface pressure.

## C. Isotherms for Asphaltene/Resin Mixtures

It is well known that resins are good dispersing agents for asphaltenes in crude oil. From previous results, the resin films are not strongly influenced by bulk concentration or the nature of the spreading solvent. Hence, it is of major interest to study mixed films of resins and asphaltenes. Figure 5 shows the surface concentration of asphaltene/resin mixtures necessary to achieve a surface pressure equal to 10 mN/m, when both solvent aromaticity and bulk concentration of resins are varied with the bulk concentration of asphaltenes is fixed (4 mg/ml).

When small amounts of resins are present in the bulk ($R/A = 0.125$), the effect of the solvent aromaticity on the surface concentration is reduced. With increased R/A ratio, a reduced surface concentration of asphaltene/resin mixtures is observed. When $R/A = 0.5$, the nature of the solvent has a minor effect on the surface concentration.

The $\Pi$–$A$ isotherms for asphaltene/resin mixtures in 20% toluene are presented in Fig. 6. The com-

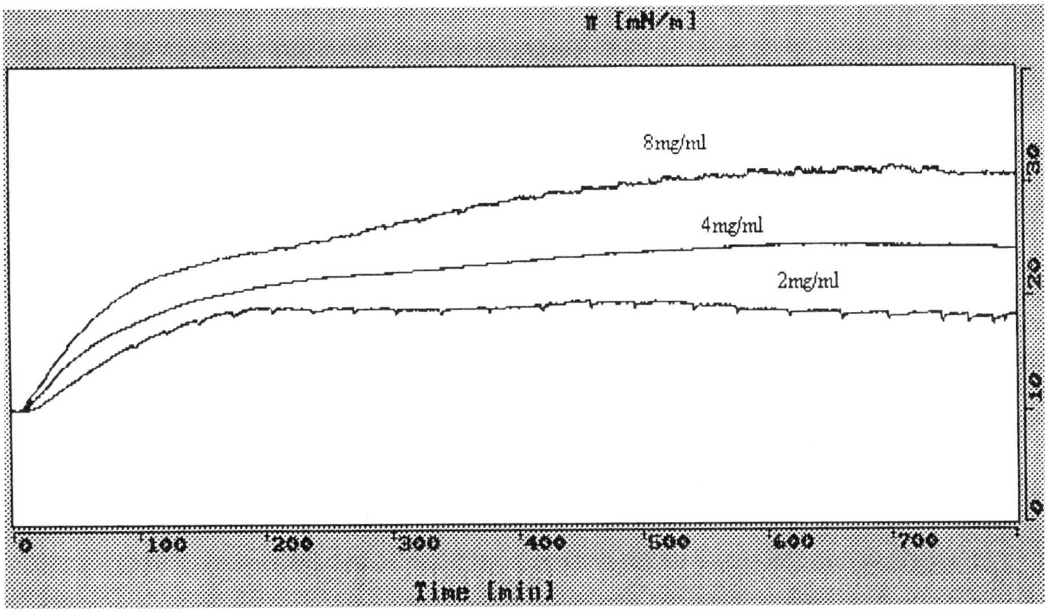


**Figure 4** Π–*t* curves for different bulk concentrations of asphaltenes spread from 20/80 toluene/hexane on pure water.

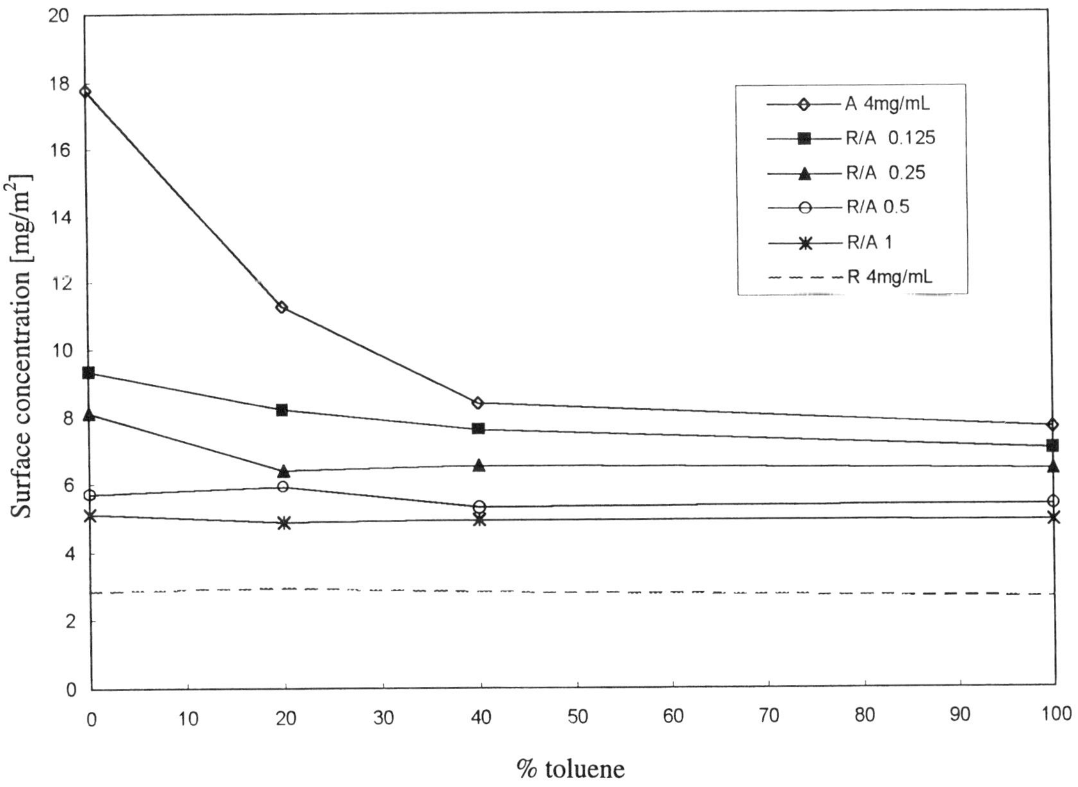


**Figure 5** Surface concentration of mixed film material needed to obtain $\Pi = 10$ mN/m vs. vol% toluene in the spreading solvent with varying bulk concentrations of resins (bulk concentration of asphaltenes = 4 mg/ml).

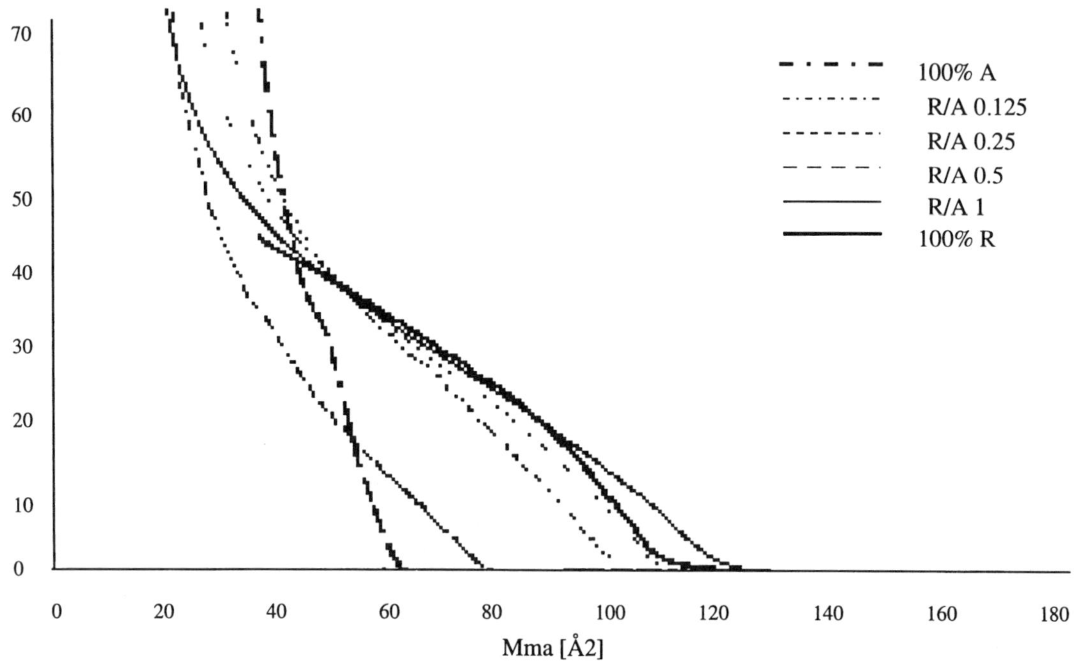


**Figure 6** Π–*A* isotherms for asphaltene/resin mixtures spread from pure 20/80 toluene/hexane on pure water (bulk concentration of asphaltenes = 4 mg/ml).

pressibility of the mixed films increases with increased resin content. When the *R*/*A* ratio is higher than 0.25, a distinct increase in compressibility at high surface pressure is observed. This phenomenon may be explained as an effect of the interactions between resins and asphaltenes; the resins disperse the asphaltenes and hence hamper the self-association of asphaltenes. The film properties are dominated by the resin fraction when the *R* /*A* ratio is increased.

Figure 7 shows the Π–*t* curves for mixed films spread from 20% toluene in hexane. The bulk concentration of asphaltenes is 4 mg/ml while the bulk concentration of resins is in the range 0.5–5 mg/ml. The film is less affected by aging when the *R*/*A* is high.

These results indicate that resins do interact with asphaltenes. Addition of resins clearly changes the Langmuir-film properties of the asphaltenes.

## III. INTERFACIAL FILM PROPERTIES OF ASPHALTENES AND RESINS

Interfacial film properties between water and model oils containing asphaltenes or resins were investigated, when the concentration of asphaltenes or resins and the aromaticity of the oil phase were varied. Interfacial tension (IFT) or interfacial pressure (IFP), interfacial viscosity (IFV), and interfacial primary yield value (IFPYV) were used to characterize interfacial film properties of asphaltenes and resins. The measuring temperatures were 20°C for interfacial tension and 25°C for IFV and IFPYV. IFT measurements were carried out by the drop-volume method, while IFV and IFPYV were measured with an SVR-S Interfacial Viscoelastic Meter (Kyowa Kagku Co., Japan) (66).

In order to understand the basic rules of flocculation behavior of asphaltenes, jet fuel/*p*-xylene was used as oil phase in the model system. The aromaticity of the oil phase was varied by varying the volume ratio of jet fuel and *p*-xylene. Double-distilled water was used as aqueous phase. Asphaltenes and resins were separated from two Chinese crudes – Gaosheng (GS) crude and Shuguang (SG) crude. The bulk concentrations of asphaltenes or resins in the model oil are given in weight per cent.

### A. Interfacial Tension and Interfacial Pressure

#### 1. Interfacial Tension

Several groups (9, 25, 33, 66) have investigated the interfacial tension between aqueous phases and organic phases containing asphaltenes. Several measurements

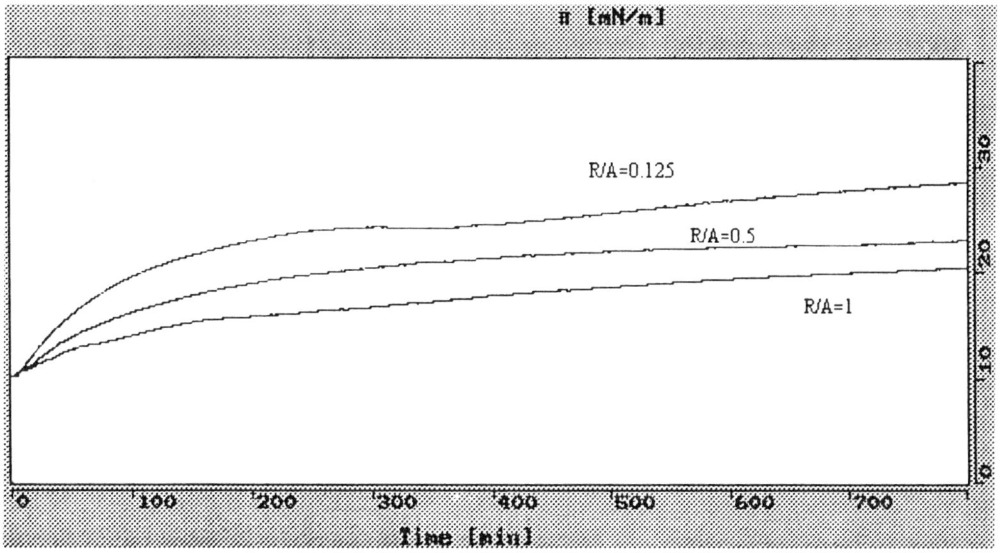


**Figure 7** Π–*t* curves for asphaltene/resin mixtures spread from 20/80 toluene/hexane on pure water, with varying bulk concentrations of resins (bulk concentration of asphaltenes = 4 mg/ml).

confirm that asphaltenes have low interfacial activity, the minimum interfacial tension lying between 25 and 35 mN/m (67). Table 2 shows the interfacial tension between water and asphaltene model oils.

For a certain W/O system at constant temperature and constant surfactant concentration, $\gamma_{w/o}$ is reduced when the surfactant's surface activity is increased. Comparing the interfacial tensions in Table 2 ([asphaltene] = 1%) and in Table 3 ([resin] = 1%), it is clear that resins are more interfacially active than are asphaltenes.

## 2. Interfacial Pressure

Interfacial pressure, Π, is defined as:

$$\Pi = \gamma_{wo} - \gamma_{w/o'} \quad (1)$$

where $\gamma_{w/o}$ and $\gamma_{w/o'}$ are the interfacial tensions between the water and oil before and after addition of interfacially active fractions, respectively. It has been found (62, 66, 68, 69) that there exists a good correlation between interfacial pressure and the stability of the emulsions.

Gibb's equation expresses the relationship of interfacial tension, concentration of surfactant and adsorbed amount of surfactant at the interface:

$$\Gamma = -\frac{c}{RT}\frac{\partial \gamma}{\partial c} \quad (2)$$

where Γ is the surface excess of the solute, $\partial\gamma/\partial c$ is the change of interfacial tension according to the change of surfactant concentration $c$, $R$ is the gas constant, and $T$ is the temperature. Based on Eq. (2), Γ is posi-

**Table 2** IFT (mN/m) Between Water and Model Oils with Different Concentrations of Asphaltenes

| Bulk concentration (wt%) | Oil phase composition: 100% Jet fuel | 80% Jet fuel/ 20% *p*-Xylene | 60% Jet fuel/ 40% *p*-Xylene |
|---|---|---|---|
| Blank | 48.2 | 43.7 | 41.5 |
| 0.25 | GS 39.7 | GS 37.9 | GS 37.3 |
| | SG 33.5 | SG 32.3 | SG 31.8 |
| 0.5 | GS 37.8 | GS 35.7 | GS 35.6 |
| | SG 32.3 | SG 27.9 | SG 27.9 |
| 1 | GS 35.1 | GS 31.0 | GS 30.9 |
| | SG 27.9 | SG 27.1 | SG 25.6 |

**Table 3** IFT (mN/m) Between Water and Model Oils with Different Concentrations of Resins

| Bulk concentration (wt%) | Oil phase composition | | |
|---|---|---|---|
| | 100% Jet fuel | 80% Jet fuel/ 20% *p*-Xylene | 60% Jet fuel/ 40% *p*-Xylene |
| Blank | 48.2 | 43.7 | 41.5 |
| 1 | GS 25.2 | GS 24.9 | GS 24.1 |
| | SG 26.7 | SG 24.8 | SG 24.7 |
| 2 | GS 23.8 | GS 23.2 | GS 22.7 |
| | SG 24.5 | SG 23.7 | SG 22.9 |

tive when $\partial\gamma/\partial c$ is negative, which means that $\gamma$ decreases when $c$ increases. Hence, the concentration of surfactant at the interface is higher than in the bulk. The opposite result is obtained if $\partial\gamma/\partial c$ is positive, then $\Gamma$ is negative, and the concentration of surfactant at the interface is lower than in the bulk.

In dilute solutions of surfactant, there is a linear relationship between $\gamma$ and $c$, and Eq. (1) can be expressed as Eq. (3), and Eq. (2) can be changed to Eq. (4):

$$\Pi = \gamma_{w/o} - \gamma_{w/o'} = Bc \tag{3}$$

$$\Gamma = \frac{Bc}{RT} \tag{4}$$

where $B$ is a constant. These modifications leads to the following relationship:

$$\Gamma = \frac{Bc}{RT} = \frac{\Pi}{RT} \tag{5}$$

Based on Eq. (5), the concentration of surfactant at the interface ($\Gamma$) and interfacial pressure ($\Pi$) can be correlated.

Table 4 shows that $\Pi$ increases as the bulk concentration of asphaltenes or resins increases; hence, the adsorbed amount of asphaltenes or resins at the interface increases. However, when the aromaticity of the oil phase increases (the jet fuel/*p*-xylene ratio decreases) the interfacial pressure decreases. So, higher aromaticity of the solvent prevents the interfacially active fraction from accumulating at the W/O interface; instead, the surfactants remain dissolved in the bulk.

## B. Interfacial Rheological Properties

Interfacial rheological properties are expressed as interfacial elasticity and interfacial viscosity. Based on the study of coalescence of crude oil droplets, Malhotra and Wasan (70, 71) concluded that there is a good correlation between coalescence time and inter-

**Table 4** ($\pi$,mN/m) Between Water and Model Oils with Different Concentrations of Asphaltenes or Resins

| Bulk concentration (wt%) | Oil phase composition | | |
|---|---|---|---|
| | 100% Jet fuel | 80% Jet fuel/ 20% *p*-Xylene | 60% Jet fuel/ 40% *p*-Xylene |
| [Asphaltene] = 0.25 | GS 8.5 | GS 5.8 | GS 4.2 |
| | SG 14.7 | SG 11.4 | SG 9.7 |
| [Asphaltene] = 0.5 | GS 10.4 | GS 8.0 | GS 5.9 |
| | SG 18.1 | SG 15.8 | SG 13.6 |
| [Asphaltene] = 1 | GS 13.1 | GS 12.7 | GS 10.6 |
| | SG 20.3 | SG 20.3 | SG 15.9 |
| [Resin] = 1 | GS 23.0 | GS 18.8 | GS 17.4 |
| | SG 21.5 | SG 18.9 | SG 16.8 |
| [Resin] = 2 | GS 24.4 | GS 20.5 | GS 18.8 |
| | SG 23.7 | SG 20.0 | SG 18.6 |

facial viscosity, i.e., the higher interfacial viscosity, the longer the time required for coalescence or more stable the emulsion system. Elastic films can resist deformation in the plane of the interface and also recover their natural shape when the deforming force is removed.

Figure 8 shows the interfacial viscosity of the interfacial films between the water and oil phases containing different amounts of asphaltenes in pure jet fuel. The shear rate is $0.03\,rad.s^{-1}$, and the temperature is 25°C. When resins are used instead of asphaltenes, only small variations are seen.

The IFPYV is defined as the shear stress at zero shear rate (72). The IFPYV of the film is related to its structural strength. From Fig. 9, it is revealed that the IFPYV or strength of the film increases with increased asphaltene concentration in the oil phase. Both IFPYV and interfacial viscosity of the films between water and asphaltene model oils increases with reduced aromaticity of the oil phase.

In aging experiments on the interfacial film, the films was aged from 0.5 to 40 h before measurement. Figure 10 show the effect of time on the interfacial viscosity between water and oil phases containing asphaltenes and resins. In a 0.25% asphaltene/jet fuel/water system, the interfacial viscosity and also the IFPYV dramatically increase (about 10 times) when the aging time is increased from 0.5 to 2 h.

These results confirm that asphaltene film properties are heavily influenced by the nature of the oil phase and the asphaltene concentration. The complex molecular structures and aggregation propensity of asphaltenes are the main effects which influence the interfacial properties of these components. Small concentrations of asphaltenes, present in a highly aromatic oil phase, represent conditions which makes it possible to dissolve asphaltenes as small association structures. Diffusion of the asphaltene molecules from the bulk toward the interface requires time in order to rearrange into structured interfacial films. Hence, the IFPYV and the interfacial viscosity increases markedly after aging, meaning increased strength and elasticity of the interfacial film.

The IFPYV of the interfacial films between water and resin model oils is zero. Hence, no structured film is formed on the interface between the water and the resin model oils. The interfacial viscosity is almost the same as that between water and blank oil, and much smaller than the values found for the interface between water and model oils containing asphaltenes. Neither the variation of resin concentration nor the nature of the oil phase have any effect on the interfacial viscosity. No aging phenomenon of the film between water and model oils containing 1% resins were observed, even when the aging time was extended to 40 h (Fig. 10).

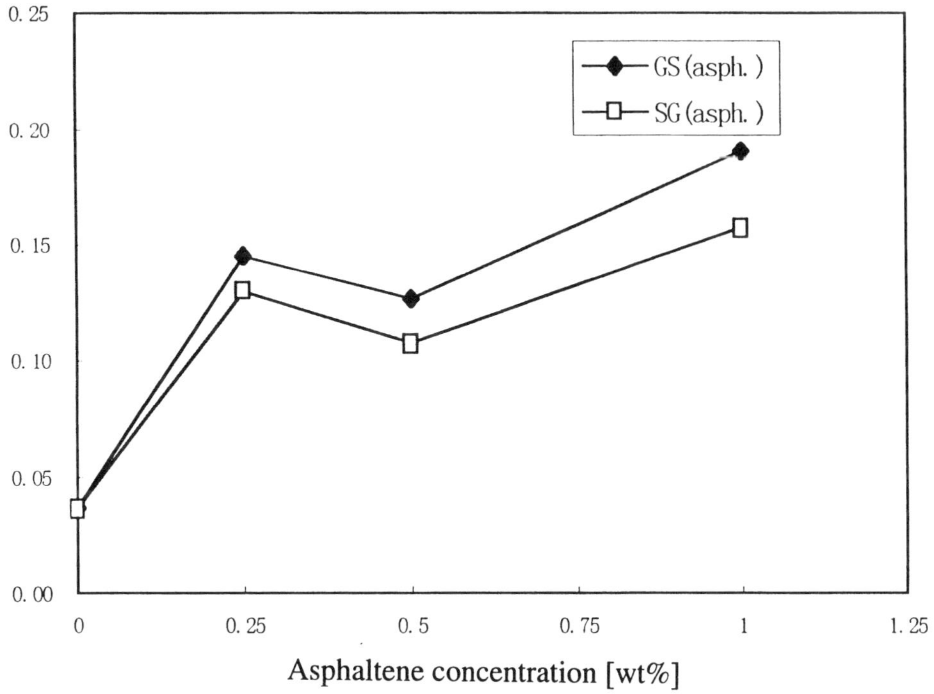


**Figure 8** Interfacial viscosity of interfacial films between water and jet fuel with different concentrations of asphaltenes.

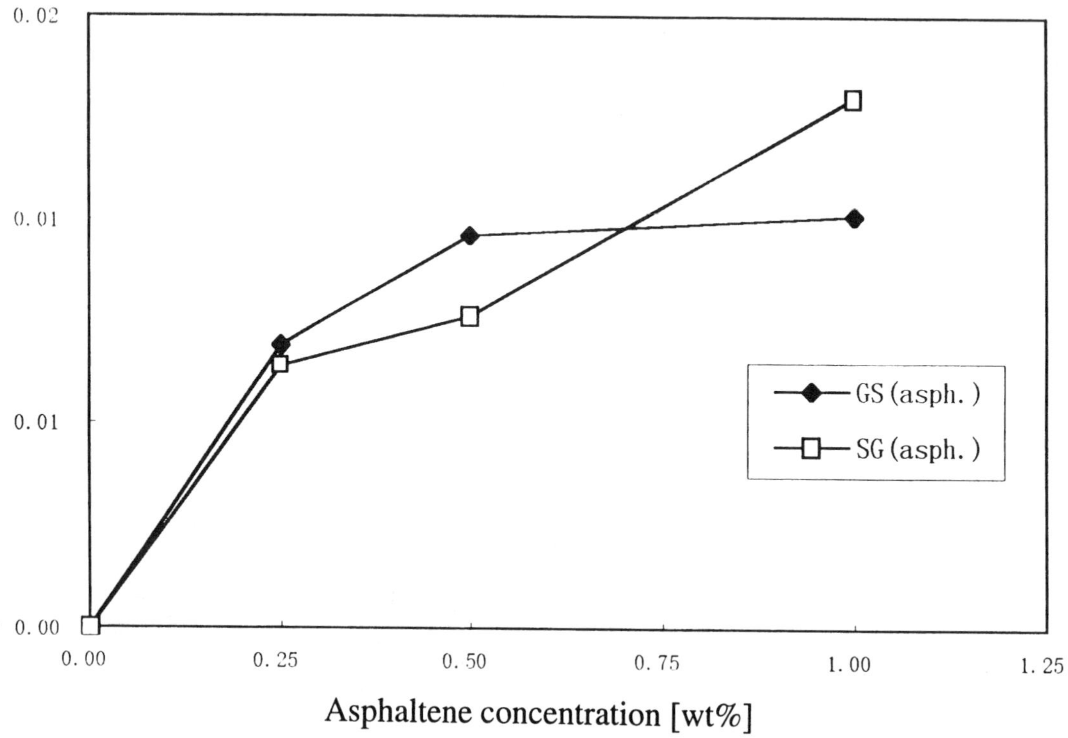


**Figure 9** Interfacial primary yield value of interfacial films between water and jet fuel with different concentrations of asphaltenes.

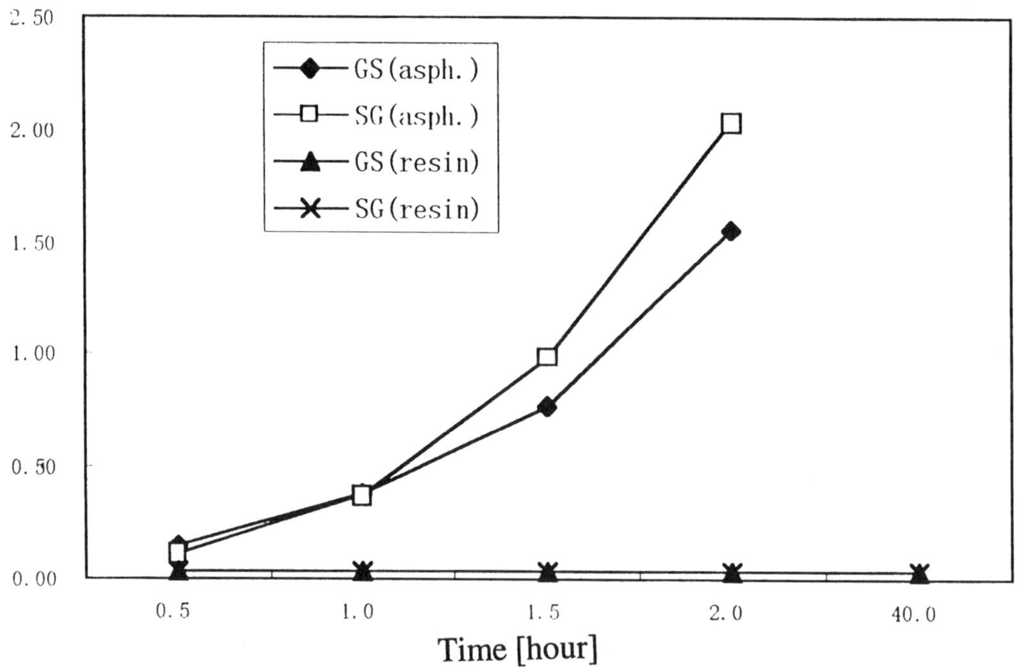


**Figure 10** Effect of time on interfacial viscosity of interfacial films between water and jet fuel with 0.25% asphaltenes or 1% resins.

These results may be a consequence of the fact that resin molecules are smaller, and do not aggregate to the same extent as asphaltenes.

### C. Interfacial Viscosity Between Water and Oil Phases Containing Both Asphaltenes and Resins

The interfacial properties of asphaltenes and resins have been shown to be quite different (73, 74). Hence, it is of significant importance to study the interfacial properties of the interfacial films between water and model oils containing both asphaltenes and resins.

Figures 11 and 12 show the interfacial viscosity of asphaltene/resin mixtures from two Chinese crudes. According to the results illustrated in Fig. 11, the interfacial viscosity of the films formed between water and Gaosheng asphaltene/resin mixtures decreases with increasing resin concentration (at any shear rate). However, as seen when comparing Figs 11 and 12, the interfacial viscosity changes differently, depending on which crude the fractions are extracted from. The IFPYVs for the two systems show trends similar to those of interfacial viscosity. These results suggest that resins from Gaosheng crude have a stronger effect on the interfacial properties between water and the asphaltene model oils. Infrared spectra of the fractions from the two crudes indicate different chemical structures.

Figure 13 shows the results from the aging experiments on the interfacial films formed by asphaltene/resin mixtures. The data in Fig. 13 also indicate that resins, which contains smaller molecules with a structure that resembles the asphaltenes, will disperse the asphaltenes in the bulk oil phase. More asphaltenes are dissolved in the oil phase, and the strength of the films formed by the asphaltenes becomes weaker when resins are present.

## IV. INFLUENCE OF AGGREGATION OF ASPHALTENES ON FILM PROPERTIES OF ASPHALTENES AND ON THE STABILIZATION OF CRUDE OIL EMULSIONS

As listed in Table 5, higher bulk concentration of asphaltenes or lower aromaticity of the oil phase can result in stabilization of water-in-crude oil emulsions. Our studies on the film properties (both surface films and interfacial films) show that the strength of films formed by asphaltenes are stronger when the conditions favor formation of aggregates, such as high

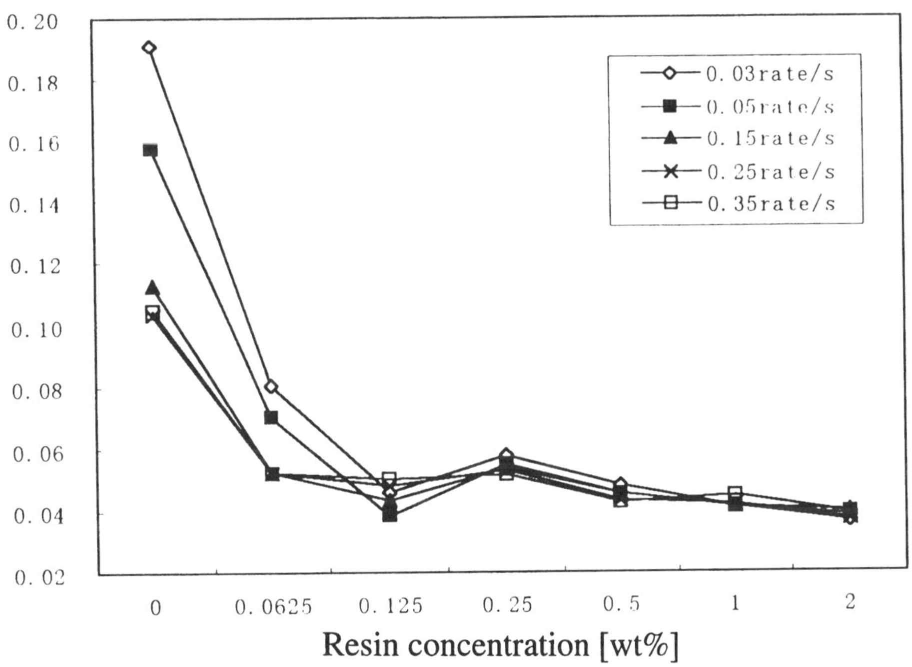


**Figure 11** Interfacial viscosity of interfacial films between water and model oils of Gaosheng asphaltene/resin mixtures with increasing bulk concentration of resins at different shear rates (the bulk concentration of asphaltene is 1%).

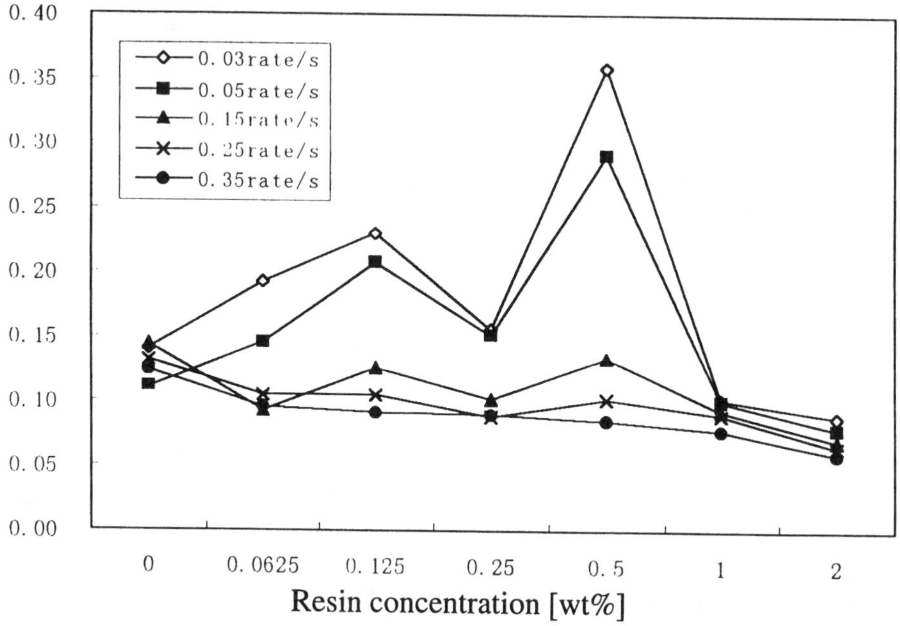


**Figure 12** Interfacial viscosity of interfacial films between water and model oils of Shuguang asphaltene/resin mixtures with increasing bulk concentration of resins at different shear rates (the bulk concentration of asphaltene is 1%).

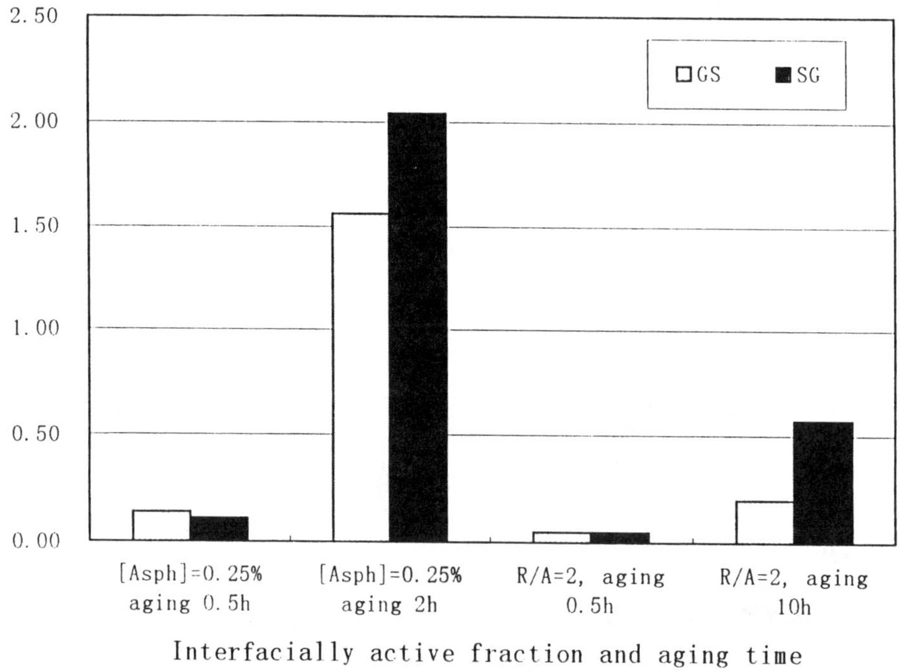


**Figure 13** Effect of time on interfacial viscosity of interfacial films between water and model oils of asphaltenes or asphaltene/resin mixtures.

**Table 5** Influence of Aggregation of Asphaltenes on Film Properties of Asphaltenes and on Stabilization of Water-in-Crude Oil Emulsions

| Effect factors | Stabilization of emulsion | Film properties measured: Surface film | Film properties measured: Interfacial film | Aggregation of asphaltenes |
|---|---|---|---|---|
| Bulk concentration ↑[a] | ↑ Known before 1980 | | IFP ↑ | Molecules → |
| Aromaticity of oil phase ↓[b] | ↑ Refs 37, 53 (1997) | Surface concentration ↑ | IFV ↑ | Aggregation → |
| | Refs 35, 36 (1996) | Low compressible | IFPYV ↑ | Particles |
| R/A ↑ | ↓ Refs 37, 53 (1997) | Surface concentration ↓ | IFV ↓ | Dispersing |
| | Ref. 54 (1998) | Compressible ↑ | IFPYV ↓ | aggregation |

[a] ↑ = Increase.
[b] ↓ = Decrease.

bulk concentrations and/or low aromaticity of the oil phase. Since resins are good dispersion agents for asphaltenes, and hence prevent self-association of asphaltenes, they are able to change the strength or compressibility of the analysed film. It is likely that any condition which may reduce the aggregation of asphaltenes will reduce the stability of water-in-crude oil emulsions.

## ACKNOWLEDGMENTS

Marit-Helen Ese acknowledges the technology program Flucha, financed by the Norwegian Research Council (NFR), and industry for a PhD grant. Elf Aquitaine, Norsk Hydro, and Statoil are thanked for providing the crudes. The Langmuir instrumentation was also financed by the NFR.

## REFERENCES

1. DU Bessler, GV Chilingarian. In: GV Chilingarian, ed. Surface Operation in Petroleum Production. I. Amsterdam: Elsevier, 1987.
2. SE Taylor. Chem Ind 19: 770–773, 1992.
3. JR Harris. Hydrocarb Process (8): 63–68, 1996.
4. J Reisberg, TM Doscher. Producers Monthly 20: 46, 1956.
5. O Urdahl, J Sjöblom. J Dispers Sci Technol 16: 557–574, 1995.
6. S Dukhin, J Sjöblom. In J Sjöblom, ed. Emulsions and Emulsion Stability. New York: Marcel Dekker, 1996, p 41.
7. DT Wasan. In J Sjöblom, ed. Emulsions – A Fundamental and Practical Approach. NATO ASI Series C363. Dordrecht: Kluwer, 1992, 283–295.
8. J Sjöblom, O Urdahl, H Hoiland, AA Christy, EJ Johansen. Progr Colloid Polym Sci 82: 131–139, 1990.
9. S Acevedo, G Escobar, L Gutierrez, H Rivas. Fuel 71: 619–624, 1992.
10. I Layrisse, H Rivas, S Acevedo. J Dispers Sci Technol 5: 1, 1984.
11. JP Dickie, TF Yen. Analyt Chem 39: 1847–1857, 1967.
12. TF Yen. Energy Sources 1: 447, 1974.
13. TF Fen, WH Wu, GV Chilinger. Energy Sources 7: 203, 1984.
14. JG Speight, DL Wernick, KA Gould, RE Overfield, BML Rao, DW Savage. Rev Inst Petrol 40: 51, 1985.
15. DA Storm, EY Sheu. Fuel 74: 1140–1145, 1995.
16. SE Moschopedis, JF Fryder, JG Speight. Fuel 55: 227–232, 1976.
17. C Bardon, D Espinat, V Guille, M Li, J Lambard, JC Ravey, E Rosenberg, T Zemb. Fuel Sci Technol Int 14: 203–242, 1996.
18. PR Waller, A Williams, KD Bartle. Fuel 68: 520–526, 1989.
19. JP Pfeiffet, RN Saal. J Phys Chem 44: 139, 1940.
20. SE Moschopedis, JG Speight. Fuel 55: 184, 334, 1976.
21. T Ignasiak, OP Strausz, DS Montgomery. Fuel 56: 359–365, 1977.
22. S Acevedo, B Mendez, A Rojas, I Layrisse, H Rivas. Fuel 64: 1741–1748, 1985.
23. DA Storm, EY Sheu. In TF Yen, GV Chilingarian, eds. Asphaltenes and Asphalts. New York: Elsevier, 1993.
24. JH Fendler. J Phys Chem 84: 1485–1491, 1980.
25. SI Andersen, KS Birdi. J Colloids Interface Sci 142: 497–502, 1991.
26. EY Sheu, DA Storm, MM Tar. J Non-Cryst Solids 131: 133 and 341, 1991.
27. EY Sheu, MM Tar, DA Storm. Fuel 71: 1277–1281, 1992.
28. MA Krawczyk, DT Wasan, CS Shetty. Ind Eng Chem Res 30: 367–375, 1991.
29. VE Galtsev, IM Ametov, OY Grinberg. Fuel 74: 610–673, 1995.

30. EY Sheu. J Phys: Condens Matter 8: A125–A141, 1996.
31. EY Sheu, MM Tar, DA Storm. Fuel 73: 45–50, 1994.
32. EY Sheu, DA Storm, MB Shields. Fuel 74: 1475–1479, 1995.
33. EY Sheu, MM Tar, DA Storm. In MK Sharma, TF Yen, eds. Asphaltene Particles in Fossil Fuel – Exploration, Recovery, Refining, and Production Processes. New York: Plenum Press, 1994, pp 115, 155.
34. EY Sheu, MM Tar, DA Storm, SJ DeCanio. Fuel 71: 299–303, 1992.
35. MY Li. Separation and Characterization of Indigenous Interfacially Active Fractions in North Sea Crude Oils. Correlation to Stabilization of Water-in-Crude Oil Emulsions. Doctoral thesis, University of Bergen, Norway, 1993.
36. H Fordedal, Y Schildberg, J Sjöblom, J-L Volle. Colloids Surfaces 106: 33, 1996.
37. JD McLean, PK Kilpatrick. J Colloid Interface Sci 189: 242, 1997.
38. J-R Lin, TF Yen. Energy Fuel 7: 111, 1993.
39. DL Mitchell, JG Speight. Fuel 52: 149–152, 1973.
40. JG Speight. In LL Schramm, ed. Suspensions: Fundamentals and Applications in the Petroleum Industry. Washington, DC: American Chemical Society, 1996.
41. JP Pfeiffer. The Properties of Asphaltic Bitumen. Amsterdam: Elsevier, 1950, p 285.
42. JG Speight. The Chemistry and Technology of Petroleum. New York: Marcel Dekker, 1991.
43. SE Friberg, L Mandell, M Larsson. J Colloids Interface Sci 29: 155–160, 1969.
44. B Siffert, C Bourgeois, E Papirer. Fuel 63: 834–837, 1984.
45. A Bhardwaj, S Hartland. Ind Eng Chem Res 33: 1271–1279, 1994.
46. PK Das, S Hartland. Chem Eng Commun 92: 169–181, 1990.
47. S Hartland, SAK Jeelani. Colloids Surfaces 88: 289–302, 1994.
48. S Acevedo, MA Ranaudo, LB Gutierrez, G Escobar. In: AK Chattopadhyay, KL Mittal, eds. Surfactants in Solution. Surfactant Science Series 64. New York: Marcel Dekker, 1996, p 221.
49. DD Eley, MJ Heyand, MA Lee. Colloids Surfaces 24: 173–182, 1987.
50. RA Mohammed, AI Bailey, PF Luckham, SE Taylor. Colloids Surfaces 91: 129–139, 1994.
51. RA Mohammed, AI Bailey, PF Luckham, SE Taylor. Colloids Surfaces 80: 223–237, 1993.
52. JA Koots, JG Speight. Fuel 54: 179–184, 1975.
53. JD McLean, PK Kilpatrick. J Colloids Interface Sci 196: 23–34, 1997.
54. XL Yang. Study on stabilization of Water-in-Crude Oil Emulsion – Film Properties of Asphaltenes and Resins. Doctoral thesis, Research Institute of Petroleum Processing, Beijing, 1998.
55. WD Harkins, D Boyd. J Phys Chem. 45: 20, 1941.
56. NK Adam, Physics and Chemistry of Surfaces. 3rd ed. London: Oxford University Press, 1941.
57. GL Gains Jr. Insoluble Monolayers at Liquid-Gas Interfaces, New York: John Wiley, 1966.
58. RH Doremus, BW Roberts, D Turnbull. Growth and Perfection of Crystals. New York: John Wiley, 1958.
59. KG Nordli, S Sjöblom, J Kizling, P Stenius. Colloids Surfaces, 57: 83, 1991.
60. RM Leblanc, FC Thyrion. Fuel 68: 260–262, 1989.
61. G Gonzalez, SM Saraiva, JF de Oliveira, F MacRitchie. In: AK Chattopadhyay, KL Mittal, eds. Surfactants in Solution. Surfactant Sciences Series. New York: Marcel Dekker, 1996, p 233.
62. BP Singh, BP Pandey. Indian J Technol. 29: 443, 1991.
63. Ø Midttun, J Sjöblom, OM Kvalheim. Progr Colloid Polym Sci, 1997.
64. DJ Crisp. In: F Daniell, KGA Pankhurst, AC Diddiford, eds. Surface Phenomena in Chemistry and Biology. Oxford: Pergamon Press, 1958, p 42.
65. F MacRitchie. Chemistry at Interfaces. San Diego, CA: Academic Press, 1990.
66. MY Li, P Zheng, ZL Wu, SL Ji. Acta Petrol Sin (Petrol Process Sec) 14: 1, 1998 (in Chinese).
67. J Sjöblom, H Soderlund, S Lindblad, EJ Johansen, IM Skjarvo. Colloid Polym Sci 268: 389–396, 1990.
68. J Sjöblom, MY Li, AA Christy, T Gu. Colloids and Surfaces 66: 55, 1992.
69. PC Hiemen. Particles of Colloid and Surface Chemistry. Published and translated by Beijing University Press, Beijing, 1986, p. 381.
70. AK Malhotra, DT Wasan. In: B Ivanov, ed. Thin Liquid Films. Surfactant Science Series 29. New York: Marcel Dekker, 1988, p. 829.
71. TF Tadros. In: J Sjöblom ed. Emulsions – A Fundamental and Practical Approach. NATO ASI Series C363. Dordrecht: Kluwer, 1992, p 173–188.
72. RJ Hunter. Foundations of Colloid Science II. Vol 2. Oxford: Clarendon Press, 1992, p 998.
73. J Sjöblom, MY Li, AA Christy, HP Ronningsen. Colloids and Surfaces, 96: 261, 1995.
74. HP Ronningsen, J Sjöblom, MY Li. Colloids and Surfaces 97: 119, 1995.

# 24

# Chemical Demulsification of Stable Crude Oil and Bitumen Emulsions in Petroleum Recovery—A Review

**Chandra W. Angle**
*Natural Resources Canada, Devon, Alberta, Canada*

## I. INTRODUCTION

The presence of emulsions in petroleum recovery operations is generally undesirable. Dehydration of the oil is demanded for various reasons. Among the foremost reasons are the high costs associated with transportation, corrosion, and heat demands, in addition to the problems caused by the presence of water/solids in the refining of crudes or in the upgrading of heavy oils and bitumen. This chapter reviews the chemical dehydration of crudes, heavy oils, and bitumen. First, we present a brief introduction on the extraction processes and the emulsions involved, followed by an outline of the scope of this review.

Emulsions formed in crude oil and bitumen during extraction operations are usually water-in-oil (W/O) macroemulsions (>0.1 to 100 $\mu$m in diameter). Macroemulsions are kinetically stable, unlike microemulsions, which are thermodynamically stable. In conventional oil recovery (high-energy process), the crude is often in contact with formation water or injection water, as in secondary recovery. In tertiary or enhanced oil recovery, surfactants are used purposely in water floods to make microemulsions for enhancing the flowability of the crude. Crude-oil macroemulsions are produced when two immiscible liquid phases such as oil and water are mixed via the input of mechanical or thermal energy into the processes. Conventional crudes held under high pressures and temperatures amidst porous rocks are recovered by drilling. Emulsions in these oils form mainly through contact with formation water. As crude oil is pumped through various pipes, valves, chokes, etc., under high pressure and/or high temperature, fine water droplets are formed, producing macroemulsions.

The recovery of heavy oil requires stimulation for flow. Flow is often achieved by reducing the viscosity of the oil by heating, as in steam-assisted gravity drainage (SAGD) and fire floods, or by the addition of viscosity-reducing agents. The recovery of water-wet bitumen from oil sands begins with a low-energy process of mining followed by conditioning with process water to release the bitumen. After release, the oil is separated in a series of process stages. Flotation of the bitumenous froth from the middlings and tailings is followed by removal of solids and excess water by dilution, demulsification, and centrifugation. The froth invariably contains 30–60% water before dropout. After dropout, the bitumen product contains 2–3% water which must be dehydrated. Bitumen production from a hot-or cold-water extraction process invariably entails a high degree of emulsification of water and air in the oil. The froth (oil-rich phase) must be treated to remove water, solids, and entrapped air.

Generally, in all recovery processes, if saline water is in contact with the oil, then the salts present in the W/O emulsion droplets must be removed. A washing process is used to remove the salts from the oil after pumping from the reservoir and before transportation. In all these processes W/O emulsion formation is prevalent.

Spills of crude oil on the sea quickly form a W/O emulsion known as "chocolate mousse" which contains approximately 80–90% water. Wave action supplies the mixing energy. These must be demulsified (1) as well.

The crude-oil market demands that water in crudes from all these processes must be removed to a level of less than 0.5% BS&W (bottoms, solids, and water) (2, 3). In order to remain competitive, emulsions must be resolved economically. The available treatment options are mechanical, thermal, via electrotreators (electrocoalescers), chemical, or a combination of physical and chemical methods. Chemical demulsification is one of the most economical means of dehydrating oil.

### A. Scope

The following review presents the chemical demulsification of W/O emulsions by first introducing crudes and bitumens in terms of the diagenetic diversity and chemistries of their components. Based on the premise that a full appreciation of demulsification must be preceded by an understanding of the basics of the field and laboratory emulsions, we have reviewed demulsification and some of the characteristics of light crude, heavy oil, and bitumen emulsions researched globally. Thus, in this work the composition and behavior of the natural emulsifiers present in the crudes and some factors responsible for emulsification in the field and laboratory are addressed first.

The second approach is understanding how the emulsions' natural stabilizers and their environment can be modified to augment destabilization. This is addressed by a description of the interfacial architecture of the emulsions in terms of how the structure may be first understood and then destabilized by probing the pseudostatic film behavior. This is followed by a description of dynamic properties of the interdroplet lamella, and thin-film behavior with and without demulsifiers. In all these systems the modification of the chemistries of the indigenous surfactants at the interfaces together with the dispersed water chemistry are suggested as tools that may be used toward destabilization. The effects of temperature and heat are addressed briefly.

Thirdly, a fourth section discusses chemical demulsification processes. Flocculation, creaming/sedimentation, and coalescence and the lamella drainage model are covered. The fifth section discusses the expected performance demanded of demulsifiers for various systems and processes.

This is followed by a description of the chemicals used as demulsifiers in practice and in research. The agricultural and petroleum sources of the basic chemical building blocks are indicated. The typical responses for selected types of chemicals are discussed, in terms of published research findings to date.

Lastly, the impact of demulsifier choices and chemistries on petroleum recovery operations are discussed. We conclude with identification of the need for cooperation between research providers, petroleum operations, and chemical suppliers geared towards an effort for full scientific understanding of demulsification. Examples of crude-oil properties and demulsification are drawn from over three decades of published results of researchers worldwide. Examples of demulsification of bitumen W/O emulsions are excerpted from work performed in the author's own laboratory at CANMET.

## II. PROPERTIES OF CRUDES, BITUMEN, AND WATER

### A. General Properties of Crudes and Bitumen

The world's fossil fuel resources consists of natural gas, liquids (oil sand bitumen and petroleum), and solids such as coal and oil shale. Petroleum represents associated gas, crude oil, and heavy oil (4). The appearance of crude oil ranges from watery white to black liquid. The thin nearly colorless liquid is mobile and flows easily, and the almost black liquid is viscous and thick. Light oils have a low boiling point and heavy oils a high boiling point. Petroleums are fairly balanced systems in terms of the interactions of their components in forming a smooth "solution" while occurring "in situ". Crude oil and bitumen are mixtures of organic compounds normally separated by fractionation through boiling-point differences in the components. Bitumens have a higher proportion of higher-boiling-point constituents than conventional crude oil.

Unlike coal, which is solid and whose surface properties and characteristics are reflections of a Cretaceous (5, 6) or Carboniferous (7, 8) geologic period corresponding with depth of burial, etc., crude oil does not bear such a clear correlation. The fluidity of crude oil made it highly mobile during diagenesis and maturation. The crude moved from its source fossil biomass location. In oil sands the heavy bitumens were integrated in loosely held fine rocks consisting of clays and silica. Crude-oil properties are more closely linked to the rocks through which they moved (9, 10). Thus, the composition of the oil varies, depending on the precursors, i.e., the nature of the biomass sediment, the underground environment, the temperatures and pressures experienced under the maturation conditions, and the naturally occurring migrations or separation processes involving inorganic catalysts (11).

Experience has shown that crudes from adjacent wells can be different in composition. The surface, physical, and chemical properties of associated emulsions can be expected to be as diverse and complex as the source crudes and water. The variances in the elemental C, H, and N composition of components such as asphaltenes of crudes from various geologic origins as compiled (10, 12) show no apparent pattern emerging from the data as was the case with coal (5, 7). Sharma et al. (13) have shown the use of bitumen asphaltenes as thermal maturation indicators.

## B. Chemistries of Crude Oils and Bitumen

### 1. Chemical Characteristics

In a series of published works (4, 9), Speight has shown the distribution of various organic structures in conventional crudes relative to heavy oil and bitumen. Heavy oils and bitumens have the largest proportions of polynuclear aromatics and polycycloparaffins, and the lowest proportions of normal and branched-chain paraffins and monocycloparaffins. Heavy oils and bitumens have boiling points higher than those of conventional crudes. The high degree of isomerization in organic chemicals found in all crudes and bitumens accounts for some of the complexity and variability of crudes of similar elemental analytical C, H, N, O, and S composition.

Crude oils have the same elemental make up globally. The carbon content of crude oils is relatively constant, but the varying quantities of hydrogen and heteroatoms are responsible for the major differences between the petroleums. The elemental composition ranges are carbon (83–87%), hydrogen (10–14%), nitrogen (0.1–2%), oxygen (0.05–1.5%), and sulfur (0.05–6.0%). Metals such as vanadium, nickel, and iron occur as metalloporphyrins, which add polar character to the oils. In petroleum refining the metals also poison catalysts. In refining, the nitrogen decreases the yield, and the presence of sulfur not only demands extra processing, but also indicates a lower quality product. These elements constitute a mixture of organic molecules classified as saturates, aromatics, resins, and asphaltenes (SARA), and waxes. The classification is empirically based on distillation fractions and solubility in alkanes (9).

Data published on crude-oil composition are often based on SARA components. This is illustrated in Table 1. Variations arise from the sources and the methods of extraction. Although the major components of crude oil are the same throughout the geologic

**Table 1** Properties of Crude Oils and Bitumen

| Crude oil property | Athabasca bitumen[a] | Heavy oil – Boscan[b] | California[b] | Ninian North sea[b] | Alberta light[b] | Kuwait[b] | Arabian light[b] | West Texas intermediate[b] | Udang[b] |
|---|---|---|---|---|---|---|---|---|---|
| API°gravity | 8.2–9.1 | 10.9 | 13.2 | NA | 36.8 | 31.1 | 31.8 | 36.4 | 14.3 |
| Viscosity at 0° and 15°C (cP) | 640,000 to > 72,000 at 39°C | 8,826,000<br>485,500 | 31,000<br>6400 | NA | 18<br>6 | 90<br>22 | 31<br>14 | 140<br>49 | 81,890<br>10,700 |
| *Hydrocarbon content* (wt %) | | | | | | | | | |
| Saturates | 16.9 | 25 | 19 | 61 | 78 | 28.5 | 51 | 60 | 32 |
| Aromatics | 18.3 | 35 | 35 | 30 | 18 | 56.2 | 39 | 26 | 41 |
| Resins | 44.8 | 22 | 23 | 8 | 3 | 15.3 | 6 | 6 | 24 |
| Asphaltenes | 17.2 | 18 | 22 | 2 | 1 | 2.7 | 3 | 1 | 3 |
| Waxes | | 4 | 3 | 6 | 7 | 6 | 4 | 4 | 1 |

[a] *Source*: Ref. 36.
[b] *Source*: Ref. 35.

origin, the proportions of the SARA components and their chemistries differ from source to source. Figure 1 shows a schematic of petroleum as represented by Pfeiffer and Saal (14).

Saturates describe paraffinics and cycloparaffinics or napthenics which are alkyl (methyl, ethyl, isopropyl) substituted cyclopentanes and cyclohexanes. The normal or branched alkanes and cyclic structures consist of one to five rings and various degrees of alkylation. According to Mackay (15), cyclic structures dominate in degraded oils. The saturates are the solvents for the higher-molecular-weight components.

If an oil is described as highly paraffinic and waxes are identified, these are usually "paraffinic waxes" which have basically straight- and branched-chained hydrocarbons ($C_{18}H_{38}$ to $C_{40}H_{82}$). Wax contents up to 50% (Altamount Utah) and as low as 1% in Louisiana can be found (16). Wax solubility in the crude oil is dependent on the chemical composition of the crude, as well as the pressure and the temperature the waxes experience (17, 18). Wax crystallizes out as an equilibrium temperature and pressure is reached around the cloud point of approximately 77°C. Leontaritis (19) describes the wax deposition envelope as the thermodynamic point in the pressure, temperature, phase composition diagram where crystallization occurs. Wax crystals are partially responsible for increased crude-oil viscosity and some of the changes in flow behavior from Newtonian to nonNewtonian. As temperature decreases crude becomes very viscous (pour-point range 16.5°–51.5°C). Thus, the viscosity of the waxy crude will depend on both the oil viscosity and the aggregation of wax crystals, whose sizes depend on the rates of cooling. Methane prevents wax crystal agglomeration, while butane decreases it by dilution (20). Wax is completely hydrophobic and is not surface active. However, the waxes are believed

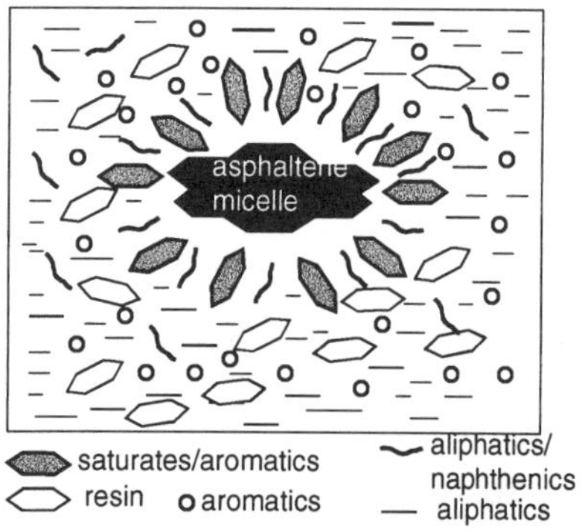


**Figure 1** Schematic of petroleums. (From Ref. 14.)

to attach themselves to the nonpolar ends of polar surface-active components of crudes (21). In this form they contribute to emulsion stability by participating in the interfacial film architecture. Their contribution to increased oil phase bulk viscosity assists in preventing coalescence by decreasing the mobility of the droplets (22, 23).

Aromatics in crude, on the other hand, are similar in structure to saturates, but contain many condensed aromatic rings instead. Low concentrations of oxygen and nitrogen are found in some polycyclic aromatics. Increased amounts of condensed rings of napthenics are attached to the aromatic rings. Molecular weight increases with increased condensation (4).

The polars describe mainly the resins, asphaltenes, and the poryphyrins, as well as the trace nitrogen found in bases, the nonbasic poryphyrins, the oxygen in the phenols, the napthenic acids and esters, and the sulfur in sulfide and disulfide bonds. The polars and the metalloporphyrins are indicated as emulsifier species involved in stabilizing the emulsions (24, 25). Resins, which will be discussed later with asphaltenes, contain O, N, and S in the form of carbazoles, fluorenones, fluorenols, carboxylic acids, and sulfoxides. These are attracted to water interfaces.

## 2. Crude Oil Acidity

The relative acidity of the crude is an indicator of the presence of polar acidic species such as phenols, napthenics, heterogeneous organic species, and anionic surfactants in the sample (26).

The acid numbers (mg KOH/g crude needed for neutralization) are indicative of the corrosivity of the crude and so suggest a value detrimental to crude (27). Jennings studied 164 crudes from 78 fields to determine an overall view of acidity (28). Worldwide crude oil acid numbers vary from less than 0.01 to as high as 3 (29). Relative to this, values for Canadian heavy oil (0.3–2.8) and bitumen (3.7) appear on the high side, closer to 3 (30, 31). There have been attempts at correlating acid numbers with the stability of emulsions, and using this as an empirical means of selecting demulsifiers for treating Venezuelan crudes (32). However, the correlations have not been tested on other crudes.

A polarity indicator was developed by Bruning, using inverse gas chromatography for 98 Brazilian and 5 foreign (Far East and Russia) crudes (33). The API° gravities [crude oils are classified by the American Petroleum Index (API°) gravity, which is a number derived empirically (141.5 divided by specific gravity at 15°C minus 131.5)] were correlated with these numbers. Low API° corresponded to high polarity. Crudes of extra high polarity (above 500) appeared to be highly degraded or to have high heteroatomic content and a propensity to form very stable emulsions that would require demulsification for more economic processing. However, according to Bruning, high numbers do not necessarily indicate degraded crude (34). Tests must be done to confirm such degradation or oxidation. In most cases the number reflected high asphaltenes and resin contents. Crudes with polarities between 300 and 400 have emulsion separation problems, while low-polarity crudes (from 200 to 300) appear to have no treatment difficulties. These characteristics of crude oils are indicative of the possibility of the formation of tight (very stable) or loose (not very stable) emulsions that require demulsification treatments. Demulsifiers are often selected on this basis as well.

### 3. Physical Characteristics

#### *a. Specific Gravities and Viscosities*

The specific gravities of crudes and bitumen vary from 0.75 (57 API°) to 0.95 (17 API°). API° gravity indicates the relative categorization of the crude as light (> 30), medium (15–30), and heavy (< 15). A high API° indicates lighter products that are saleable. Bitumen and heavy oil are of low API° with densities closer to water. This presents difficulty in gravity separation techniques. However, viscosity differences are large between crudes and water. Table 1 describes some physical and chemical properties of bitumen and crudes from various sources (35, 36).

The increase in API° has been correlated with increased reservoir depth (synonymous with increased temperature), which results in an increased fraction of compounds with less than 12 carbons (more paraffinic) in the crudes (9). High API° is normally associated with low asphaltene content and high sulfur content (9). The major physical differences between bitumens, heavy oils, and conventional oils reside in the specific gravity and the viscosity. These are very high in bitumen and heavy oils due to a preponderance of high-molecular-weight aromatic components such as resins and asphaltenes. Bitumen has the highest viscosity, followed by heavy oils, and then conventional crudes, at all temperatures. In a comparison of viscosity with increasing temperature for conventional crudes, heavy oils, and bitumen, there is an almost parallel asymptotic decrease in viscosity as temperature is increased. The highest viscosity values at each temperature were attributed to bitumen, followed by heavy oils and conventional crudes [4] (see Fig. 2).

Waxy crude oils, such as those found in the North Sea (20), would not produce a smooth decline in viscosity with increases in temperature. Wax as fine particles would contribute to a high viscosity and when heated between 50°C and 65°C would melt or solubilize in low-molecular-weight components, producing a more homogeneous fluid (17, 18). The properties of crude oils and bitumen are listed in Table 1.

#### *b. Molecular Weights and Micellar Sizes*

As determined by vapor-pressure osmometry (VPO), the average molecular weights (MWts) of Canada's heavy oil from Norman Wells, Countess, and Cold Lake were reported to be 197, 334, and 585 g/mol, respectively (37). Relative to other feedstocks the range appears as given in Table 2.

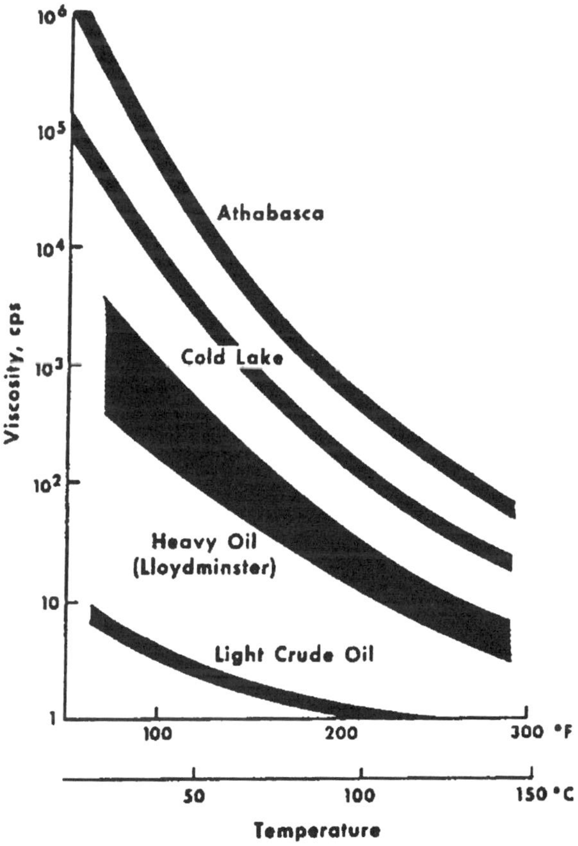


**Figure 2** Viscosity decreases with increasing temperature for conventional crudes compared with heavy oils and bitumens. (From Ref. 4.)

**Table 2** Range of Molecular Weights for Petroleum Feedstocks

| Property | Coal | Vacuum residue | Heavy oil bitumen | Crude oil | Gasoline |
|---|---|---|---|---|---|
| H/C atomic ratio | 0.5–0.8 | 1.3–1.35 | 1.4–1.5 | 1.6 | 2.0 |
| Molecular weight | 10,000 | 1000 | 500 | 300 | 100 |

*Source*: Ref. 4.

For additional data on crude oils with MWt around 300 g/mol the reader is referred to Speight (4). High asphaltene content is synonymous with higher MWt in oils. In VPO determinations all asphaltenes in heavy oil tend to have higher MWt than their resin counterparts (4, 38). The resins are generally of lower MWt than the asphaltenes. Ferworn et al. (39) have published data on the average MWt (300–500) of some of Alberta's crude oils and their properties and their distributions in various solvents (37). The agglomeration (40) of the asphaltene components and the bitumen behavior with paraffinic solvents are also discussed by Funk (41).

Extensive data on the chemical and physical properties of asphaltenes have been published in several texts. The MWt of asphaltenes in various solvents and at various temperatures are reported by Speight (42). The MWts range from 2000 to 7000 g/mol based on geologic period (Cretaceous, Carboniferous, Devonian) and depending on the methods of precipitation. For the pentane-precipitated asphaltenes, when solubilized in benzene, the MWt reported was 5120 g/mol; when solubilized in pyridine the MWt was 13,390 g/mol. Asphaltenes extracted by 3-pentanone showed higher MWts of 18,000 g/mol. For the heptane-precipitated asphaltenes, when dissolved in benzene, the reported value for MWt was 8500 g/mol, and when dissolved in nitrobenzene, the MWt was 2880 g/mol. The reported MWt of asphaltenes appeared to decrease with polarity of the solvent and increase with concentration in the solvent, suggesting that the association structures or micellar sizes are influenced by thermodynamic solubility-parameter differences between the asphaltenes and the solvent. Resins appear to have a MWt close to 700 for pentane solubles and 1050 for heptane solubles. The MWts of resin are less affected by solvents and are generally significantly lower than those of the corresponding asphaltenes (4).

Thus, asphaltenes (43) are high MWt pentane- or heptane-insoluble fractions of crudes and are readily peptized by resins which are the alkane-soluble fraction. Figure 3a–c illustrates a few of the various chemical models developed for asphaltenes from structural and chemical analysis. Resins make up the outer protective coating of asphaltene micelles or clusters (43–45). The resins are dissolved in the oil, are also surface active and polydisperse with a range of polarities and aromaticities. Figure 4 illustrates the chemistry for a typical resin molecule (46). Resins sometimes from reversible micelles. The resins are less polar and of lower MWt than asphaltenes and appear as dark sticky semisolid liquids in n-alkanes. Manek (23) characterized the asphaltenes and resins of three Canadian heavy oils, one Texas heavy oil, and one California heavy oil to show the relative surfactant-enhanced dispersibility of asphaltenes in resins.

The insolubility of asphaltenes causes deposits in pipes, wells, and valves, and in the formations. Dubbed as the colloids of crudes (47, 48), the asphaltene chemistry has received considerable attention. Asphaltenes are considered as major polar species with high aromaticity and are known as the major building blocks of the mechanical barriers or interfacial films formed at the W/O interface. Increased MWts is consistent with high aromaticity and greater numbers of incorporated heterocyclic structures containing the heteroatoms. Their structures have received considerable attention of late.

Even in dilute solutions they associate (49, 50). Published sizes of the micelles vary from 2 to 4 nm. Sophisticated analytical techniques such as small-angle X-ray diffraction (SAXS), small-angle neutron scattering (SANS), and NMR were used to study the asphaltene particle or "micelle" sizes (51). MacKay (15) reported that a MWt of 10,000 g/mol would correspond to a 2 to 4-nm cluster. This is very much smaller than a 1-$\mu$m water droplet, and considered to be 1/100 to 1/1000 the droplet diameter. This topic is worthy of a review on its own. However, the colloidal properties

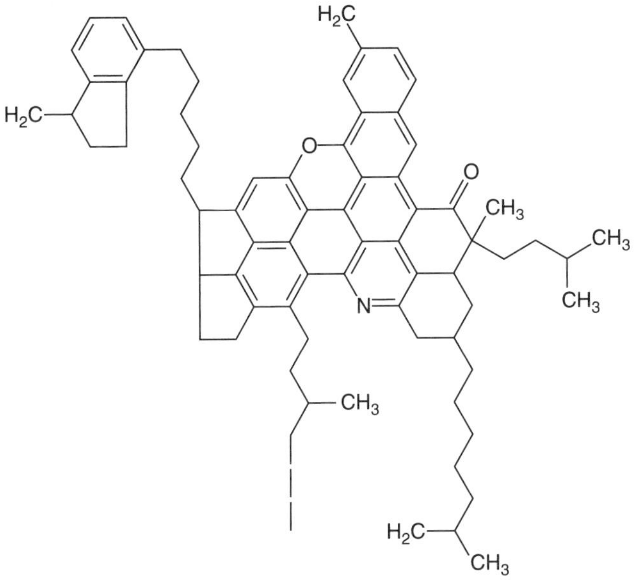


(a)

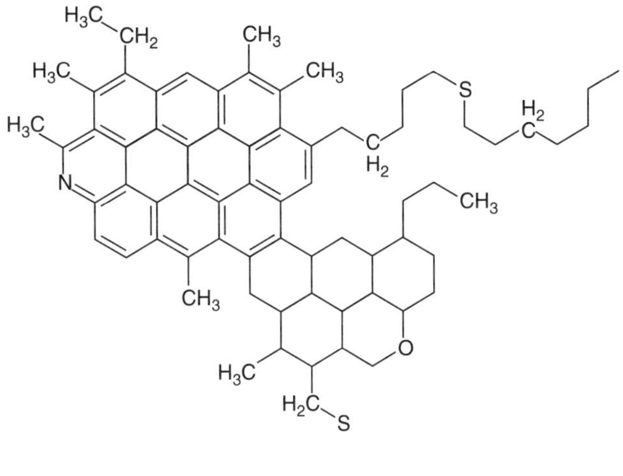


(c)

**Figure 3** Models of asphaltene chemical structures: (a) California; (b) Venezuela; (c) Athabasca. (From Refs 4,42, and 43.)

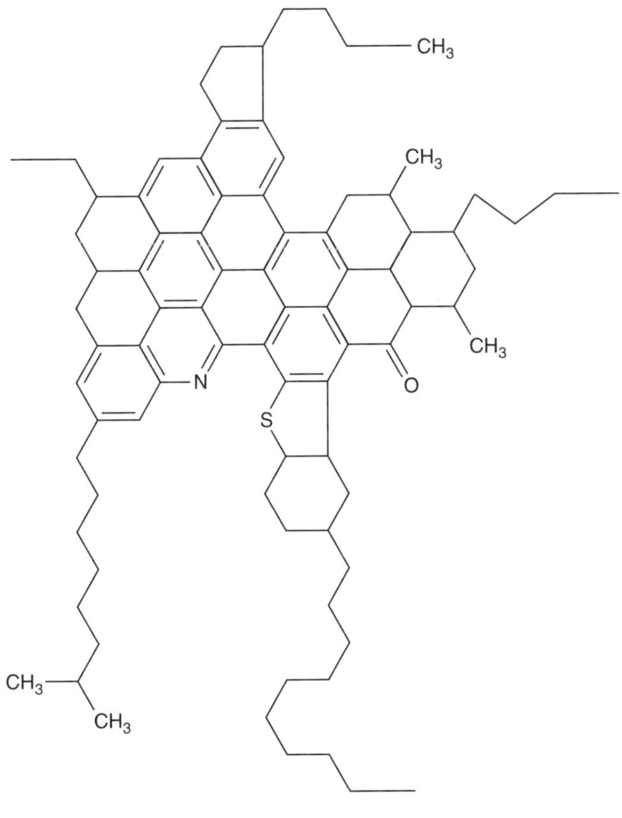


(b)

of asphaltenes, micelles, and solvent effects are compiled and reported by several authors (52, 53).

From data obtained by SANS analysis, Sheu and Storm (48) reported that the sizes of asphaltene micelles in a good solvent such as toluene/pyridine fall in the range 3.0–3.2 nm.

Although asphaltenes have been extensively studied in terms of colloidal (47, 54–56) and fractal properties (37, 57) they are not yet fully understood in terms of their interactions with other components of crudes, solubilities in various solvents and crudes, precipitation, and micellization (58). Asphaltenes appear to be the key component, together with resins, in forming a mechanical skin often described as structurally rigid, viscoelastic, or a deformable interfacial film – and the barrier to coalescence. Much of emulsion stability research which focussed on the film stability described the films in terms of mechanical or rheological behavior with varied characteristics of the system. Figure 5 shows a schematic of these fractions arranged about the interface of a water droplet. The natural surfactants are illustrated with the head group in the water phase and tails in the oil phase. As time progresses the organization of the surface-active species changes in the interfacial region. The changes, which lead to more concentration of species at the interface over time, will be discussed later in this chapter. Figure 6 describes

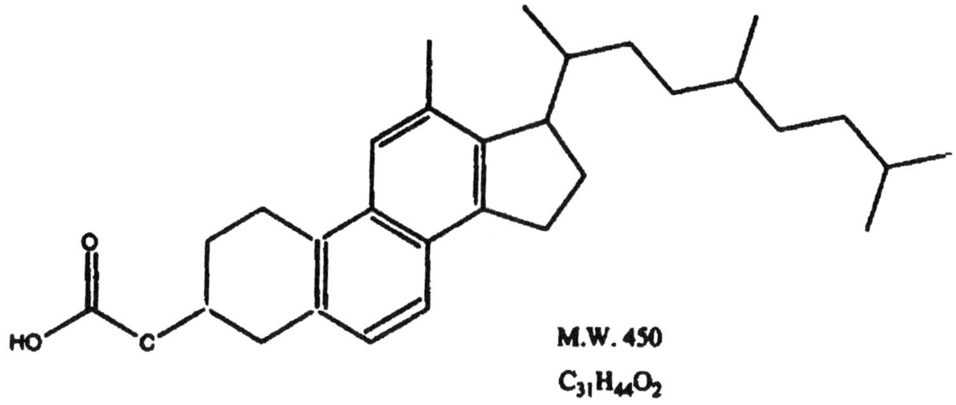


Figure 4 A model of resin chemical structure. (From Ref. 46.)

other components that may be present inside and outside the droplets.

## C. Water

In the petroleum industry, not all W/O emulsions are the same. The nature of emulsions formed in crude oil often depends on many factors: the geologic source and the engineering processes utilized in the crude oil recovery, the chemical and physical characteristics of the crudes and their thermal history, the type of mixing and energy introduced, and, most importantly, the formation and process water chemistry and composition. All these determine the stability and the sizes of the emulsion droplets.

All formation or extraction water differs in ionic composition and pH. Generally, formation water for the North Sea crudes is far more saline than that for the heavy oils and bitumen. Table 3 compares typical water compositions found in emulsions (59–61).

Thus, the W/O emulsions reflect the complexity of the above factors. Since the natural chemistries of crude oils or petroleum contain the stabilizers of the W/O emulsions, the chemical destabilization or demulsification requires knowledge of not only the emulsion interfaces, but also the physicochemical characteristics of the oil and water.

Parts of this complexity are addressed in published information on chemical destabilization of crude W/O emulsions. Examples include studies on California (62), Salem (63), Texas (64), Louisiana (65), North

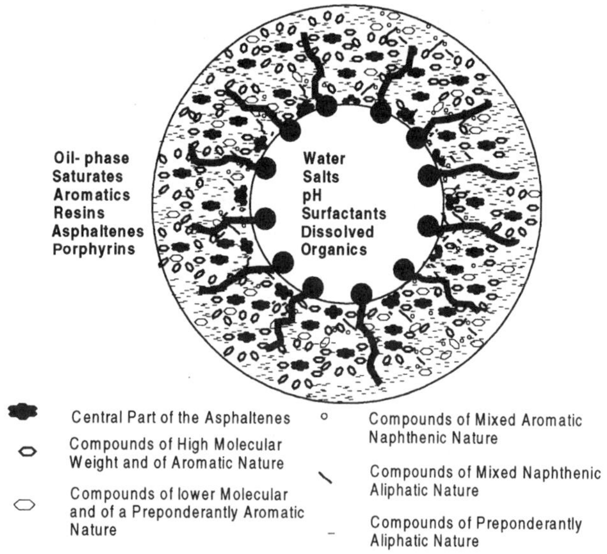


Figure 5 Schematic of droplet of W/O emulsion with pertroleum fractions arranged in the interfacial layer or skin around the droplet at early stage of formation. Solids and waxes are not indicated.

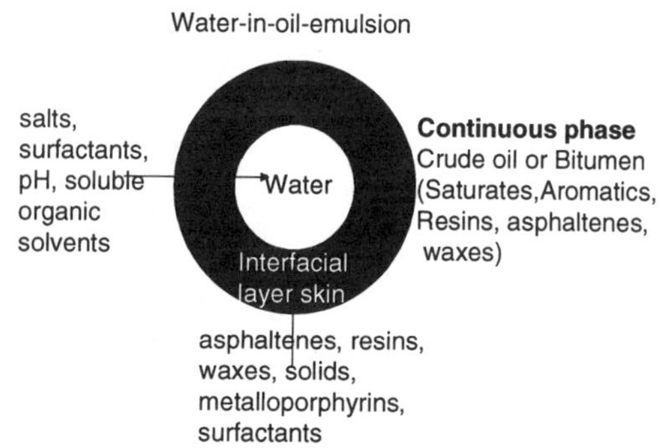


Figure 6 Schematic of components of crude oils and bitumen to be considered in an emulsion droplet and the interfacial layer.

**Table 3** Water Composition – North Sea and Alberta Oil Sands Pond Water

| Ions | Forties[a] formation (ppm) | Ninian[a] formation (ppm) | Magnus[a] formation (ppm) | North sea water[a] (ppm) | Athabasca Pond 2 water[b] (ppm) |
|---|---|---|---|---|---|
| $Na^+$ | 33,000 | 8300 | 6000 | 12,000 | 350 |
| $Cl^-$ | 58,500 | 12,500 | 9000 | 19,600 | 40 |
| $Ca^{2+}$ | 3950 | 12,500 | 80 | 400 | 6.3 |
| $Mg^{2+}$ | 500 | 190 | 15 | 1325 | 3.8 |
| $HCO^-{}_3$ | 450 | 30 | 1700 | 140 | 932 |
| $SO^{2-}{}_4$ | < 10 | 15 | 70 | 2750 | 9.9 |
| $Ba^{2+}$ | 210 | – | 100 | – | 0.16 |
| $Fe^{2+}$ | 170 | – | – | – | 0.4 |
| pH | 5.8 | 7.8 | 7.4- | 7.8 | 8.5 |
| Dissolved solids | 99,500 | 22,600 | 17,000 | 38,140 | n/a |

[a] *Source*: Ref. 59.
[b] *Source*: Refs 60 and 61.

Sea (66, 67), Kuwait (68), Assam (69), Indian (70, 71), Boscan and other Venezuelan (72, 73), Velden (74, 75), Bavarian (76), Hungarian (22), Egyptian (77, 78), Norwegian (79–81), and Canadian (82–88) heavy oils and bitumen. The crudes, heavy oils, and bitumens differ geographically, and hence in their diagenetic histories. Their physical and chemical properties reflect their diversity not only in the W/O emulsions formed in them, but also in the response of their emulsions to demulsifiers. Because of the water and electrolytes, petroleum W/O emulsions translate into high processing heat requirements, corrosion problems, and increased transportation costs. Demulsification is a necessity.

## III. FORMATION AND STABILITY OF CRUDE OIL AND BITUMEN EMULSIONS

### A. Production Emulsions and Stability

In the process of crude oil extraction and transportation the formation of emulsions is inevitable. As the immiscible fluid mixtures pass through piping valves, porous rocks, etc., and experience turbulence, especially at high pressure and/or high temperature, breakup and deformation occur. If the ionic composition or pH of the water is favorable, and surface-active agents are present, emulsion formation is enhanced (89–92). The degree of emulsification depends on several factors: the energy of the process, the amounts of surface-active components in the crude oil, the physicochemical properties of crudes, water, and surfactants, and the residence time, as emulsions age. Aged crude-oil interfacial films become more resilient over time (74, 93) and are therefore more difficult to demulsify (94, 95). Finer and more stable emulsions occur with prolonged high-energy input in these systems.

In a destabilization scheme, one first assesses the degree of stability of the emulsion. There is a need to know how much time and energy are crucial for the process, the phase compositions, and the emulsifier(s) chemistry. For the experienced engineer, this information then allows some deductions on the configuration of emulsifiers around the droplets, and the rheological properties of both the interfacial film on the droplet and the interphase lamella between droplets. The viscosities of the emulsions and continuous oil phase are also important for decisions on destabilization (96).

In production the quantities of water in emulsions vary from 30% W/O formed in oil sands extraction processes to 80 or 90% in the form of "chocolate mousse" during an oil spill at sea (97, 98). Real systems are more complex and heterogeneous than the ideal systems. The W/O droplets are fine, well dispersed, and very stable. Asphaltene content is around 17% in bitumenous emulsions and around 2% for North Sea crudes. Figure 7 shows a confocal photomicrograph of a bitumen froth emulsion freshly extracted at CANMET. The solids present are shown as white specks; the dark spots are emulsion droplets in a bitumen continuous phase. The composition is 41% bitumen, 44% water, and 15% solids.

Some emulsions exist for a few minutes; others can stay in suspension for years. The life/death cycles are dependent on the stability and the destabilizers. The sizes of the droplets and the rigidity of the surface film barrier determine the lifetime of the emulsion droplets. For convenience in treatment, the terms for definition of stable emulsions are determined by their

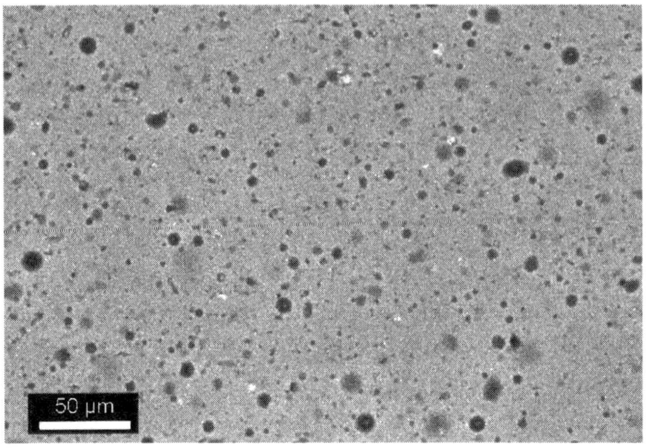


**Figure 7** Confocal photomicrograph of freshly extracted bitumen froth – Dark spots are water droplets, white specks are solids particles. Composition by weight is 41% bitumen, 44% water, 15% solids.

persistence in the process time scale, and thus are categorized kinetically only in operational terms. Planned studies on emulsions should consider operational conditions in the experimental design.

It has been found that the finer the emulsions the more stable they are in terms of resistance to coalescence, as the fine emulsions behave as hard nondeformable spheres (99). The increased ratios of the dispersed water phase to continuous oil phase determine the packing factor. Most often a packing density of 0.74 tends to be critical for destabilization (100) for ideal monodisperse spheres in a hexagonal arrangement in the systems (101). This packing factor increases considerably for polydisperse systems such as crude oil emulsions. With creaming there is increased packing and some deformation occurs in the more elastic systems; the emulsions may appear as foams, with phase ratios between 0.8 and 0.9. Any increase in the critical packing factor can cause instability of the emulsions. Thus, valuable information on stability or instability (102) can also be obtained by varying the volume fraction of the dispersed phase under flow conditions. Here, the emulsion bulk rheological responses are a function of imposed external stresses and constraints and can be signatures to the system stability in a process.

The period of time in which the droplets do not coalesce without mechanical, electrical, or chemical aids has become the kinetic guideline for stable emulsions. In the extraction of oil in the petroleum industry very stable emulsions are undesirable. The exceptions are "orimulsion" fuels (103), made in Venezuela, and/or emulsions especially tailored in formulation for transportation in pipelines (104, 105). These must be destabilized on delivery (106, 107).

Much of the early literature published on chemical demulsification involved emulsions of crudes from offshore operations and other conventional land-extracted light oils. Thus, much of the knowledgc acquired about crude-oil emulsion stability was based on paraffinic crudes of low asphaltene content. It is essential to be aware that results for crudes of the North Sea operations were based on a high wax content, and low asphaltenes and resins content (108) (see Table 1), when compared with results for highly aromatic and asphaltenic heavy oils and bitumen, which are low in wax, and higher in resins. The latter emulsions would be those of Canadian (Alberta) bitumen, heavy oils from Boscan (Venezuela), and Canada and California crudes, which are all highly aromatic and asphaltenic. These W/O emulsions are known to be very stable or "tight" (23).

Generally, oilfield emulsions are most often W/O with the surface-active emulsifiers residing in the crude-oil continuous phase. According to the Bancroft rule (109) the phase for which the emulsifiers are most soluble is the continuous phase. The emulsifiers possess some degree of polarity which attracts them to the water phase. Solid emulsifiers would be very fine particles in a state of incipient flocculation (110). The emulsifiers may be one or more of the following: solids which are partially hydrophobic with contact angle ($\theta \geq 90°$), polar asphaltenes and resins with some partial insolubility induced by solvents which dilute the crude oils, or metalloporphyrins integrated within the asphaltenes (24, 25).

A knowledge of solvents and solubilities is crucial for understanding the (in)stability. The solvent power is a function of the molecular structures. Aromatics have the greater dispersion forces and higher solution energies and thus superior solvent power. For Athabasca bitumen (111) the solvencies are as follows: paraffins < olefins < napthenes < cyclo-olefins < condensed napthenes < aromatics < condensed aromatics. This is also true for other crude and heavy oils (50) to various degrees. Thus, the choice of solvent can determine the degree of emulsion stability.

The structures of asphaltenes, as first deduced by Pfeiffer and Saal (112), showed them as fine aggregated particles solubilized in resins (14). Later, the solvent effects on their colloidal natures and their roles in the resiliency of the interfacial film formed at the boundary between the water droplet and the oil phase became important in the destabilization models (113, 114).

The role of waxes in emulsion stability has been studied in great detail for paraffinic crudes, especially for crudes that were spilt in the sea (115, 116). Waxes are believed to interact hydrophobically with the asphaltene and resin micelles present in the crude oils (18, 20, 21) and become part of the emulsion film. Waxes were hypothesized to adhere to the asphaltenes and resins, since waxes are oil soluble and not surface active. With such emulsions, the temperature prehistory and cooling rates have a greater impact on the emulsion stability. The rate of cooling influences the wax crystal sizes (17, 18). Thus, the type and history of emulsions must be considered before a demulsification treatment is selected.

Generally, the less soluble, surface-active materials tend to accumulate at the water/oil interface building the film structure. The thickness and concentration of these materials around the droplet periphery build over time until the layer becomes a structural barrier against coalescence with other droplets.

## B. Emulsions on a Bench Scale

Emulsification and demulsification of bench emulsions have been reviewed extensively in the past (89–92). The reader is referred to these publications for a more detailed fundamental background on theories and developments in laboratory measuring techniques for stable emulsions (117). To determine a destabilization program for emulsions it is often advisable to seek to understand first the interfacial properties in the emulsions (118). Most of this understanding is derived from bench studies.

Often, in bench studies aimed at understanding emulsion-stabilization mechanisms, a hypothesis is devised. Most often the components of crudes are first separated, and a model emulsion is prepared from various combinations of the components in a model oil and in water of quality similar to that of formation or process water. The stability or instability is traced either by water resolution or by observing the interfacial film properties under some form of externally applied stress over time. The stress may include temperature increases or solvent changes. The system may then be modified by the demulsifier and the changes in behavior are compared to that without the demulsifier. Deductions are then made about the film mechanics of the system in response to the variables.

When actual crude oil is studied, comparisons of the responses of the components in crude are usually made with those of the whole crude oil under similar conditions. Often the actual field system is used to test conclusions arrived at on model systems. In many cases the behaviors of several crude oils are compared under similar conditions for rigor. It is therefore important that representative crude oil samples from a process be obtained.

In the following discussions the published experimental findings are presented interrelatedly first in terms of internal oil chemistry at the interface and instabilities based on its composition, secondly in terms of effects of water chemistry, and thirdly in terms of demulsifier interaction. We include the activity of interfacial components involved in the structure of the protective skin, the behavior(s) of this structure with changes to water chemistry or solvency, or the effects of changes in film structure itself due to modification of relative proportions of interfacially active components. In some examples, developments in interfacial rheology, which is both a tool for understanding stable films and a means of rationalizing the effects of demulsifiers in demulsification, are discussed interrelatedly. Films may be sensitive to crude oil type, gas content, aqueous pH, salt content, temperature, age, and the presence of demulsifiers. Demulsifier performance is also influenced by many of these variables.

We distinguish first between the adsorbed skin at the W/O interface and the gap or lamella between approaching drops, which may be described as: W/I-interface/O/I-interface/W. Here, "I" is the boundary layer of the interface on the droplet surface, and O is crude oil or dilute bitumen between two water droplets, interfaces. Figure 8 is a photomicroph of two bitumen-stabilized water/oil droplets showing the interfacial skin and the interdroplet lamella. Figure 9 is a photomicrograph of several water droplets show-

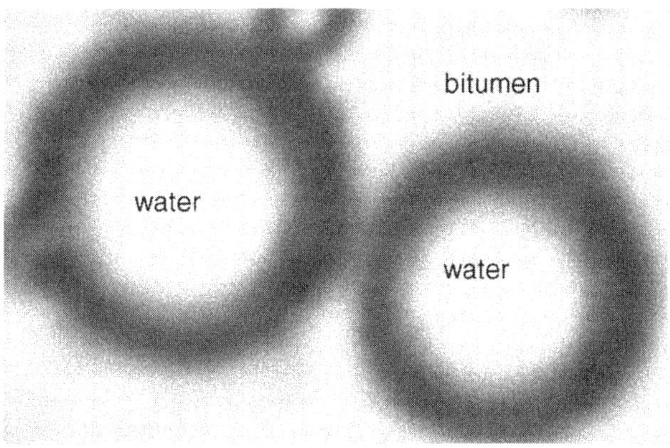


**Figure 8** Photomicrograph of two water droplets showing the interfacial skin and the interdroplet lamella.

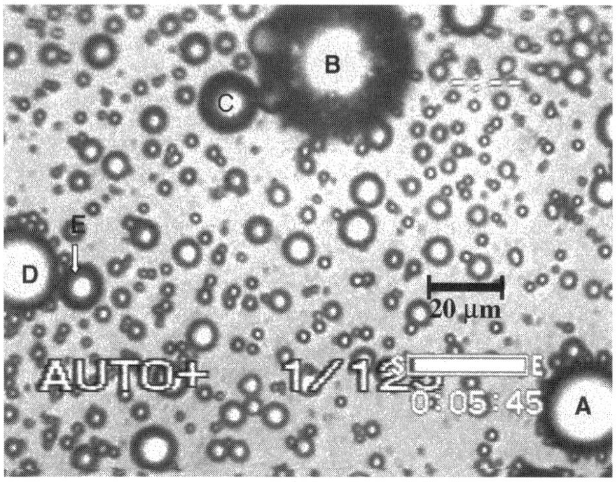


**Figure 9** Photomicrograph of water in bitumen emulsion distribution with interfacial layer shown as the dark ring around the white droplet.

ing thick bitumen interfacial skins as dark circles around the bright water droplet. The continuous phase is the diluted bitumen.

Emulsion formation mechanisms are not the reverse of demulsification mechanisms. Crude-oil emulsion formation may involve one or more mechanisms based on the process of immiscible phases interacting, in time, with energy. To date, the detailed mechanisms are not yet understood completely, especially for petroleum emulsions.

In the laboratory, very reproducible W/O emulsions of monodispersed size distributions can be prepared when all the variables for emulsification are controlled. The variables for a bench laboratory study are: emulsifier type and concentration, energy of mixing, time of mixing, method of mixing, volume fractions of oil and water phases, type and viscosity of oil, quality of water, and temperature. The mixture is blended in specific vessels, usually with rest intervals to control the rigidity of the film. The conditions are reproduced from batch to batch. In real production this is not often the case. The immiscible phases are subject to variable high shear for 2–8 min in offshore production and 40–50 min in the oil sands extraction process. Emulsion size distributions therefore vary with different systems.

Several bench-scale model emulsions were prepared from bitumen in toluene, by varying only the water quality to observe the basic differences in the emulsions.

Figure 10 (a,c,e) shows a series of photomicrographs of model bitumen emulsions freshly prepared with deionized, 1 mM NaCl and synthetic pond water. Figure 10 (b,d,f) shows that emulsions become larger with age, supporting the kinetic instability. After 24 h there are differences in the final appearance of the emulsion for each sample of water used. The growth of droplets is due to Oswald ripening. It is expected that the interfacial film would differ in strength. Note the increasingly finer emulsions initially formed as the water changed from deionized to NaCl to synthetic pond water under identical experimental conditions. After 24 h the same trend is still observed.

## C. Emulsions' Interfacial Film Structure and Instabilities

Undertaking first to achieve an understanding of the building blocks of the mechanical film could lead to the detection of surface weaknesses at which a demulsifier can be targeted. We illustrate this with examples from the literature.

In a review on formation and stability, Mackay (15) has suggested that the most stable emulsions have water particles in the 1 to 5 $\mu$m range, with a film thickness of 1/100 the diameter of the droplet. For bitumen W/O emulsions we have found in our laboratory that these are typical average droplet sizes (see Figs 7, 9 and 10). If we deduce the film structure from the components of crudes, then droplets are much larger than the components in the film. Mackay and Mason (119) have reported that asphaltene clusters can be about 2–4 nm in diameter and Neumann and Paczynka-Lahme (120) reported that asphaltenes were retained on filters of pores of 35 to 10 nm, while resins were retained on pores of 5 nm. Speight (4) reported that the asphaltene sheets are 10–15 Å by 6–15 Å. Bhardwaj and Hartland (74) reported that on an emulsion droplet the average cross-sectional interfacial area occupied by surface-active species from crude oil was experimentally shown to be 366 Å$^2$, while that for the asphaltenes in toluene was 253 Å$^2$. Larger water droplets ($>10$ $\mu$m), which will cream faster, tend to have a thinner film of between 1/100 and 1/1000 of the droplet diameter, on crude oil emulsions according to MacKay (15).

### 1. Observations of Destabilization of the Pseudostatic Film

The thick protective interfacial skins are considered as largely responsible for the stability of crude oil emul-

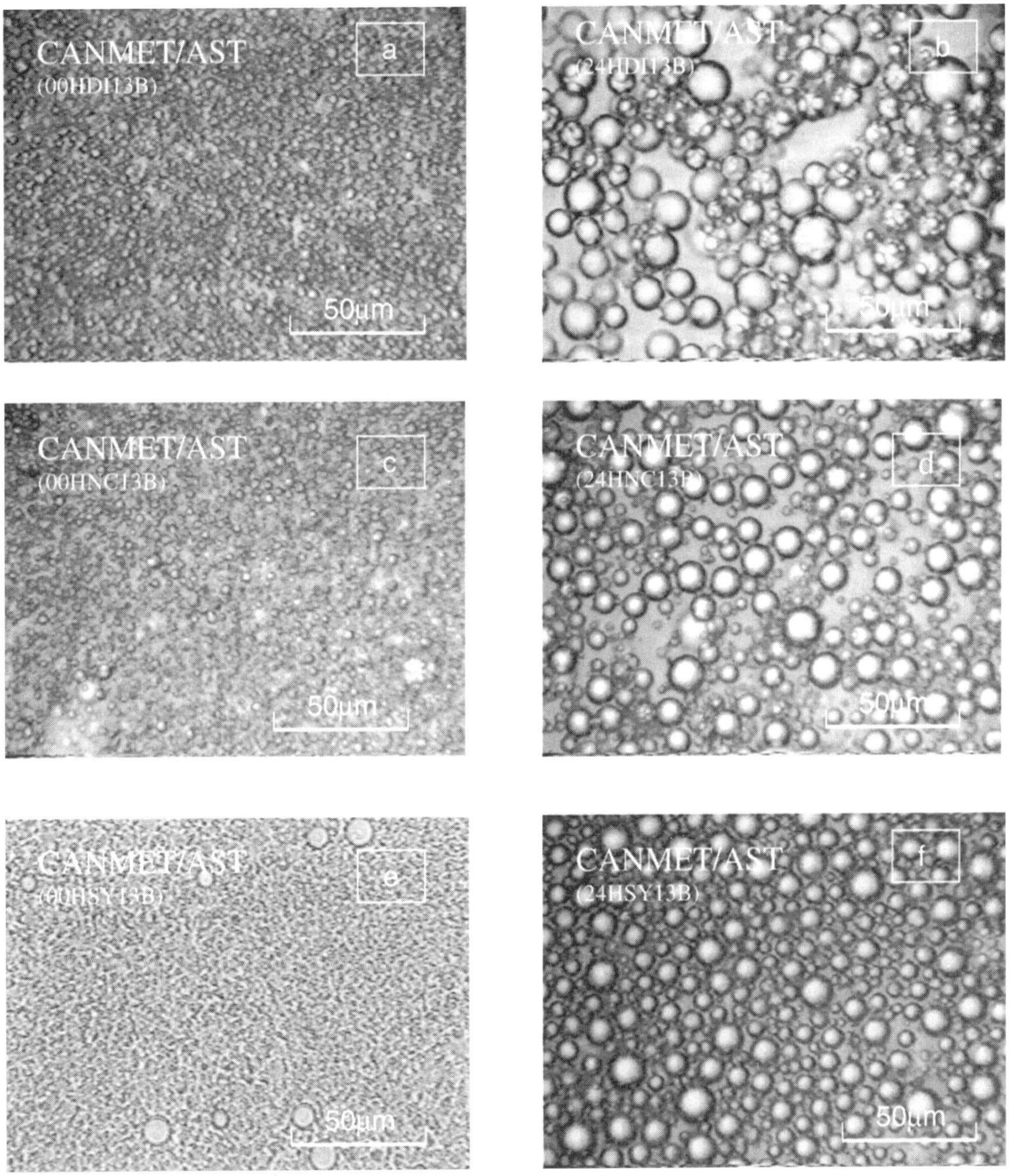


**Figure 10** Photomicrographs of model bitumen W/O emulsion made with 13% synthetic pond water in a 30% bitumen-in-toluene solution. Left – droplets immediately after preparation; right – droplets after 24 h of aging. Deionized water (a,b); 0.001M NaCl (c,d); synthetic water (e,f). (Courtesy of C.W. Angle, in house research, CANMET.)

sions. This has been established in the past half a century in early studies (121). Hunter (122) offered a description of various types of interfaces where he considered two bulk phases separated by a planar phase or skin containing a structure of adsorbed materials and a liquid-like film phase. Based on the complexity of composition and curvatures he categorized them as A,B,C, and D. The differences are based on the mass of adsorbed materials, the sizes of head groups, and the charged ions present. Crude oils mostly match category C which has no coulombic contributions. By using the retraction of sessile drops of water/oil/

water, Roberts, as early as 1932 (123), made the observation of a thick film of oil at the crude oil/water interface.

In 1948, Lawrence and Killner (124) indicated that both good and poor solvents affected the surface viscosity of the film. Poor solvents such as the aliphatics showed higher surface viscosity, while a good solvent such as toluene containing 1% asphaltene in hexane had lower surface viscosity with a plastic film behavior.

In 1954, Blakey and Lawrence indicated that asphaltic components were responsible for the stability of sea water in admiralty fuel oil (125). In 1956, Reisberg and Doscher (126) showed that crude oil films existed at temperatures close to 90°C, indicating the importance of the rigid oil/water interfacial films in stabilizing emulsions.

In 1960, Blair (94) published findings obtained on an elastic membrane, which when compressed, developed wrinkles and thick striations about the point of disturbance for several crude oil/water interfaces. He found that in time the striations disappeared by an annealing process, and the interface then finally assumed a more uniform appearance. After using a surface film balance to measure the spreading forces (from 16 to 31 dyne/cm) for several crude oils on water with and without demulsifiers, he concluded that stability arises from the formation of a condensed and viscous interfacial film of adsorbed soluble material from the petroleum phase. Specific demulsifiers have spreading pressures sufficient to displace the petroleum film, leaving a thin film with little resistance to coalescence.

Since then, many studies were undertaken to understand the nature, strength, and weaknesses of the mechanical barrier or surface film on a droplet relative to emulsion resolution. Often these studies used a combination of microscopic observation of droplets formed and resolved, and interfacial rheological studies.

Thus, the discussions that follow include several important factors. These are that: (1) the activity of interfacial components is involved in the structure of the protective skin; (2) the behavior of this structure changes with water chemistry or solvency due to mass transfer and interfacial dissipation effects; (3) the changes in structure may be due to modification of the relative proportions of components; and (4) for understanding stable films and as a means of measuring demulsification, one may adapt the new developments in interfacial rheology as tools. These are all factors considered in past studies and which are described in the following sections.

### 2. Measuring Instability of the Film by Pendant-drop Retraction

Pendant-drop retraction experiments on 10 crudes were conducted by Strassner (113, 127). A pendant drop is illustrated in the schematic of Fig. 11. Strassner studied asphaltene and resin films as the interfacial area of the drop was varied. He described the films as either mobile and/or as transition films. The mobile film was a semisolid skin which, under compression, momentarily distorted. The transition films showed no distortion under drop contraction. Pendant-drop retraction experiments were conducted on medium to light crude oils of varied asphaltene-to-resin ratios for which he categorized film rigidities. Since heavy Venezuelan crude oil of high viscosity could not be measured by this technique, he extracted the resin and asphaltenes from this crude. He then varied the asphaltene-to-resin ratio in both the emulsions and in the films. He showed that the resins contributed to mobile films and the asphaltenes to more rigid incompressible films. He concluded that asphaltene and resin interactions were mainly responsible for the film properties of crude oil. The rigid films had high interfacial viscosity and the mobile films low interfacial viscosity.

Further examination of the films by changes in pH of the brine phase led to the conclusions that the rigid films formed by asphaltenes were stronger in acidic pH, became intermediate at neutral pH, and were mobile at basic pH. This behavior was characteristic of the asphaltenes' amphotericity. In addition, stronger oil wetting of silica occurred at acidic pH. The mobile

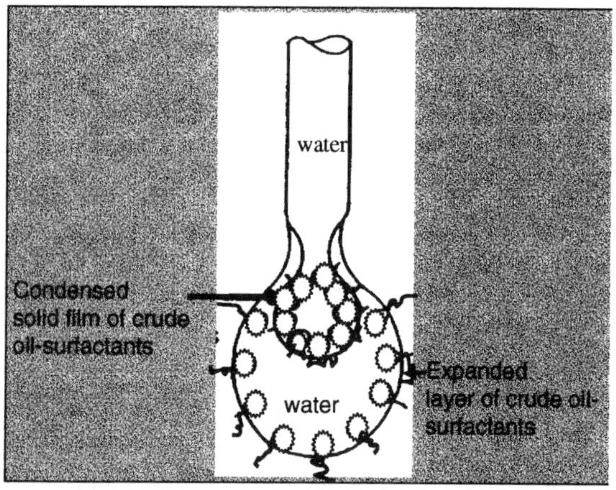


**Figure 11** Schematic of a pendant drop with a condensed film of surface-active material on retraction, and expanded film during expansion. (From Ref. 74.)

resin films were stronger in basic pH, and weaker in acidic pH. This was also an indication of the resins weakly acidic nature.

Strassner (127) examined a third component, "waxes". He showed that resin and waxes do not oil-wet silica. Waxes had no significant effect. He concluded that the waxes only contribute to increased viscosity of the oil phase. At low salt concentrations asphaltenes plus resins will oil-wet silica at acidic pH, but will water-wet silica at basic pH. He examined Venezuelan crude oil and distilled water, and found the most breakout of water occurred at pH 10. In this case there was a transition to a mobile weak film, and interfacial tension was still high. However, when the water was changed to a bicarbonate solution, this transition occurred at pH 6, where maximum water breakout was observed at the high interfacial tension. The role of interfacial tension is discussed later in this text.

MacLean and Kilpatrick (21) recently confirmed that the integrity of the films was sensitive to solvency parameters such as aromaticity and asphaltene-to-resin ratios, as well as polar functional groups. They confirmed Strassner's findings, using North Sea crudes (113). They suggested that modifying the state of dissolution of asphaltenes should decrease the ability to stabilize the emulsions, and that asphaltenes are solubilized with resins and can only stabilize in a state of incipient flocculation. They inferred that as a molecular solution asphaltenes do not stabilize emulsions. The state of dispersion of asphaltenes (molecular versus colloidal) is critical to the strength or rigidity of the interfacial films and hence the stability of petroleum emulsions (48, 52). The balance of acidic to basic groups in its structure, and with that of resins, determines the degree of association for film formation. Earlier, Fordedahl et al. (128) in studies of crude oil emulsions in high electric fields by dielectric spectroscopy showed the influence of interaction between indigenous surfactants. They claimed that resins alone are not stabilizers, and that asphaltenes are the stabilizers even though resins have high interfacial activity. A mixture of 1% asphaltenes and 1% resins gave the critical ratios required to give the necessary film rigidity for a stable emulsion. However, these emulsions were still less stable than the original crude from which the fractions of asphaltenes and resins were derived (129).

Cottingham et al. (130) destabilized and restabilized shale oil emulsions by adding proportions of pentane solubles (resins) to shale oil emulsions, thus quickly separating the phases. They stabilized the emulsions indefinitely by adding both solubles and insolubles to the emulsions. These factors among others can be considered in demulsification.

Recently, Puskas et al. (131), using FTIR, Raman, VPO, X-Ray, rheology, SAXS, and UV techniques, conducted an extensive characterization study of fractions that were deemed to be structural components of Hungarian paraffinic-based crude-oil emulsions. They attributed stability of the interfacial emulsion film to a solid hydrophobic paraffin derivative or organocolloid of lamellar structure containing polar end groups. Together with asphaltenes, resins and oleophilized solids as components of the film structure, they deduced the formation of a cohesive three-dimensional film structure. The interfacial film had mechanical stability and flexibility. The increased viscosity of the crude oil also increased the coherent structure of the paraffin particles or waxes. They found that toluene softened the structure containing the paraffin particles and caused an increase in viscosity.

Mackay and coworkers (15, 132) performed emulsification studies, using wax and asphaltene mixtures as stabilizers. They showed that waxes are not stabilizers on their own but, together with asphaltenes at a high ratio and proper blending, form stable emulsions. They tried to quantify stability by indexing it to the proportion of mass ratios of each crude oil component to total oil mass. The number was indicative of a strong or weak mechanical film barrier. They claimed that asphaltenes alone form a strong interfacial film, resins make the asphaltene barrier less rigid and more easily deformed, and wax aids in film rigidity. All these findings suggest that the internal chemistries of the crude oil components or solvents may be a few of the control parameters useful for selection of demulsifiers.

Bridie et al. (1) found that asphaltenes are two to five times more effective than waxes as stabilizing agents. In a stable emulsion containing 6.6% asphaltenes and 9.8% wax, the emulsion was still stable if 90% of asphaltenes were removed. Waxes as crude oil stabilizers for North Sea crudes have been investigated by Thompson et al. (133, 134) who showed that waxes melt as the temperature increases, and separation efficiency increases at 60°C, which is above the wax melt temperature.

The roles of interfacially active fractions of crude in emulsion stability were confirmed by Felian et al. (135) and Sjöblom et al. (136). According to Sjöblom et al., when discussing the impact of chemical destabilization, there must be a strong basis for understanding the stability mechanism. The interactions between the destabilizer and the active components of crude oil will dictate the sequence for destabilization in the system. They studied 10 Norwegian crudes of various SARA and wax contents. Some had high ratios of resins to asphaltenes. They found palmitic acid as an active stabilizer. When compared with heavy oil standards, asphaltene contents in North Sea crudes are extremely low, which suggests that other factors must enter into forming the stable interfacial films. Sjöblom et al. (136) indicated that palmitic acids, which they identified in their crudes, would adsorb very readily at the W/O interface and, because of their nonbulky nature, may tend to pack efficiently depending on the concentration. On the other hand, Acevedo et al. (52) attributed surface activity of the interfacial film partially to carboxylic acids, which they were able to extract from asphaltenes of Venezuelan heavy crudes. They concluded that carboxylic acids were integrated within the asphaltene micelles at the W/O interface. Other studies (137), conducted on Russian crudes, show the influence of the components on emulsion film strength (138, 139).

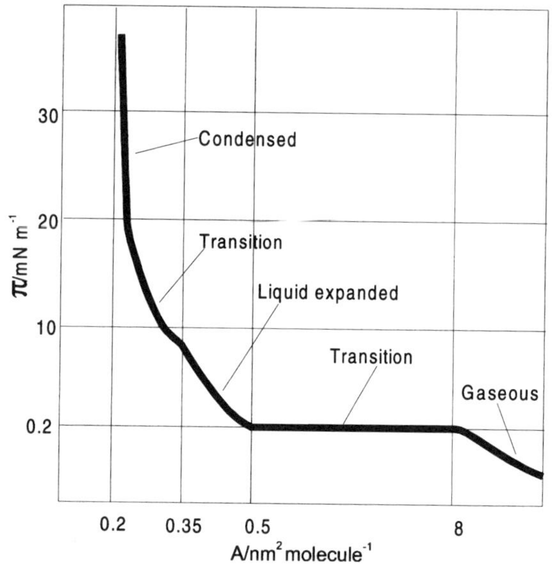


**Figure 12** Schematic showing transitions of a typical surface pressure/area($\Pi/A$) isotherm for a compressed monolayer of surfactant myristic acid spread on 0.1M HCl in a Langmuir-trough apparatus. (From Ref. 141.)

### 3. Surface Pressures Used as Measure of Film Instabilities with Demulsifiers

A Langmuir balance was adapted for studying the interfacial films of crude oil/water by Kimbler et al. (140). They plotted the surface pressures, Π, versus area as the film was compressed and observed the normal pattern of phase transitions such as gaseous (flat slightly rising), expanded (rise), and condensed liquid (change in slope of rise) phases. Figure 12 shows a schematic of the typical surface pressure/area isotherm for a simple surfactant, a monolayer of myristic acid measured in a Langmuir apparatus (141). Variations in the shapes of the curve indicate the state of the film. Collapsed films were identified by observation of the changes in the shapes of the isotherm.

Kimbler et al. (140) studied the addition of a destabilizer such as Triton X-100, which produced only expanded crude-oil films. It was noted that a drop in film pressure at a constant compressed area might be indicative of film dissolution or reorientation of structural elements. Changes in temperatures may change the adsorption rate and the buildup of interfacial film, and decrease the bulk viscosity. Temperature may change the rate of the relaxation process at constant compressed area (fall in film pressure), or change the rate of buildup of resistance to film compression (142).

Jones et al. (59) used a similar Langmuir-like trough for studying crude oil/water interfaces. They examined crudes such as Kuwait, Iranian, Ninian, Forties, and Magnus. Using the changes in the shapes of the Π–area curves and the fall in pressures (relaxing), they distinguished three types of film behavior: incompressible nonrelaxing (Iranian heavy and Ninian), incompressible relaxing (Forties and Kuwait), and compressible relaxing (Forties freshly added to water). On heating, the temperature increase caused the incompressible relaxing film to become a compressible relaxing film. The Magnus crude made a weak compressible film which became rigid and nonrelaxing and insensitive to temperature increases even in low bulk phase, when the light ends were evaporated. They suggested that high interfacial viscosities can be obtained by compressing the interfacial film, showing how the rate of thinning of these films can be greatly reduced based on their responses to compression. The kinetics of the relaxation process would determine the extent of the dynamic barrier to coalescence. Demulsifiers would change this.

By observing the responses of the films to two demulsifiers, they found that the ethoxylated phenol was a film displacer and inhibitor for aged films and that the carboxylic acids containing alkoxylated ester were film inhibiting if added to crude oil before contact with the water. They also found that $Ca^{2+}$ ions rendered the film incompressible and this resulted in a more stable emulsion. They also suggested that pendant-drop retraction could not be used to distinguish between relaxing compressible films and incompressible relaxing films. They suggested that viscous buildup should be shown; otherwise, interpretations based on the pendant-drop experiments can be misleading.

The Langmuir-balance technique was adapted for the study of Indian crudes by Singh and Pandey (69). They separated the crudes into anionic, cationic, and nonionic fractions and studied the effects of electrolytes and pH on their film properties. Maximum film pressure was observed for films made with the anionic fraction and minimum film pressure for the cationic fractions. Increased electrolyte concentrations caused increased viscosity and less resolution of emulsions. Film pressures were maximum at pH 12 and more stable emulsions resulted. The electrolyte also had an adverse effect on the demulsifier.

Singh (143) followed up this study to understand the performance of unidentified demulsifiers with a change in solvent properties. By noting the relative decrease in surface pressures resulting from added demulsifiers in various solvents, he found that benzene was the best in that it helped the demulsifier to lower the surface film pressures. The lowering of interfacial tension was measured at the same time and the results suggested that rapid adsorption occurred. It was concluded that structure, orientation, and film pressures were the most important factors in demulsifier performance.

Mohammed et al. (144, 145), in a series of studies, examined several aspects of emulsion films with and without demulsifiers as well as their chemistries. Using the Langmuir balance for studying the air/crude/water interface, they examined the surface pressure Π–area isotherm for monolayers of Buchan crude's asphaltenes and resins and their mixtures, spread on distilled water at pH 6.2 and 25°C. They found that the asphaltenes upon compression formed solid films, that could withstand pressures up to 45 mN $m^{-1}$ in contrast to the resin films at 7 mN $m^{-1}$ which thereafter collapsed. The asphaltenes formed highly stable emulsions in contrast to the resins alone, which formed the least stable emulsions. They found that film compressibility and emulsion stability decreased as resin content increased. Temperature increases caused no significant effects on asphaltene monolayer compressibility as was observed earlier by Reisberg and Doscher (126) for natural crude oil films.

Nordli et al. (146), using the Langmuir-balance technique, studied the monolayer properties of the interfacially active fractions extracted from six North Sea crude oils over a subphase of distilled water and simulated formation water. The pH and salinity were varied. They compared additives such as butanol, benzyl alcohol, and octylamine, added to the subphase, to note changes in film compressibilities. A typical phase-change pattern for Langmuir curves of surface pressure versus area was observed for all cases. The smallest specific area attributed to the liquid expanded states was reached by the lowest-molecular-weight species. Films, in this case, were more condensed as the temperature increased. At constant pressure, the changes in the surface pressure were monitored over time for each film. The time for relaxation was 60 min for the low-molecular-weight fraction, and 320 min for the high-molecular-weight fraction. Distilled water in the subphase showed decreased specific area for the same film, and a pH of 2.6 caused a shift to higher specific area. Salt had no significant effects.

According to Taylor and Mingins (142) a drop in film pressure at constant compressed area may be indicative of film dissolution or reorientation of structural elements. If there is no resistance to compression, an unstable emulsion results. Changes in temperature may change the adsorption rate and the buildup of the interfacial film. A decrease in the bulk viscosity may change the rate of the relaxation process (fall in film pressure) at constant compressed area and change the rate of buildup of resistance to film compression.

### D. Destabilization and Interfacial Film Rheology

Developments in measurement techniques for monitoring interfacial changes occurred for the purpose of understanding emulsion stability. However, the techniques were equally useful in measuring destabilization induced by changes to the internal natural chemistries of the system or with addition of chemical demulsifiers. The adaptation of the Langmuir film balance was only one tool for a pseudostatic system. However, as the fluid dynamics in the films and interphase (or lamella) of two approaching droplets or in a highly concentrated emulsion dispersion was discovered to be more complex and bore similarity to the polyhedral structures of foams, the need to study the emulsion interfacial films under stresses became apparent. Much was learned from foam studies.

A detailed treatise on the fundamentals and applications of thin films, i.e., lamella plus surface film, is found in Ivanov (147). The two major forces involved in the lamella behavior are thermodynamic (disjoining forces) and hydrodynamic. Further in-depth studies on the thermodynamics are presented in de Feijter (148) and Hirasaki (149) and on the hydrodynamics in Maldarelli and Jain (150). Because of the complexity of this topic, which is not within the scope of this chapter, the following discussions will be limited to the understanding that this research brought about concerning the nature of the films and the process of film thinning by chemical demulsification.

The interfacial rheology of films has been extensively discussed by Edwards et al. (151) and briefly by Tadros (152), and reviewed in relation to emulsions and foams by Malhotra and Wasan (153). A two-dimensional fluid interface can only be treated as an independent body when it is highly viscous or highly elastic. It cannot be treated separately from the bulk fluid which may provide viscous drag and influence the direction of flow. Simply, interfacial rheology describes the flow behaviour in the interfacial region between two immiscible fluid phases. The adsorbed surface-active components at the crude oil/water interface alter the hydrodynamic resistance to interfacial flow. As the molecular weight of the adsorbed species increases, the interface exhibits viscoelastic behavior. This is most likely in the case of bitumen W/O emulsions. The chemical demulsifier interaction alters not only the physical/chemical properties but the rheological behavior. Figure 13 is a schematic of the physicochemical and dynamic factors involved in droplet interactions.

### 1. Crude Oil Interfacial Rheology – Background

There are four rheological parameters which describe the response to imposed interfacial stresses or deformation. For a Newtonian interface, the significant rheological properties that determine interfacial motion are the interfacial shear viscosity, $\eta_s$, the interfacial dilational viscosity, $\eta_d$, and the interfacial tension gradient. The interfacial shear elasticity, $\varepsilon_s$ and viscosity, $\eta_s$, describe the resistance of the interface to changes in shape of the interface element. Here, the area is kept constant and the resistance is measured. On the other hand, the interfacial dilational elasticity, $\varepsilon_d$, and viscosity, $\eta_d$, describe surface resistance to changes in interfacial area. The interface is expanded without shear and the resistance is measured. Figure 14 shows schematically the shear and dilational forces on an interface in determining shear and dilational viscosity.

The need for the dilational data arose due to the manifestation of Gibbs–Marangoni effects or surface-tension gradients occurring during film thinning. The surface-tension gradients and hence the elastic behavior are influenced by changes in temperature, concentration of surface-active agents, and/or compression and expansion of the interface.

Detailed theoretical treatments of the dilational properties of liquid films can be found in several studies. Kristov et al. (154) performed a parametric study determining the compositional surface elasticity of model systems. Loglio et al. (155) reported on the dilational viscoelasticity of fluid interfaces modeled for transient processes, Edwards and Wasan (156) discussed foam dilational viscosity, and Lucassen-Reynders and Van den Temple (157) reported on the surface dilational modulus caused by variation of surface tension from a small-amplitude sinusoidal area variation. Recently, Yeung et al. (158) modeled the expanding pendant drop in their explanation of interfacial mass-dissipation effects.

### 2. Measuring Destabilization by Dynamic Interfacial Shear Viscometry

Dynamic shear measurements may be conducted by using several interfacial rheological instruments. A detailed reference list on these is found in a treatise by Malhotra and Wasan (153). Among those specially

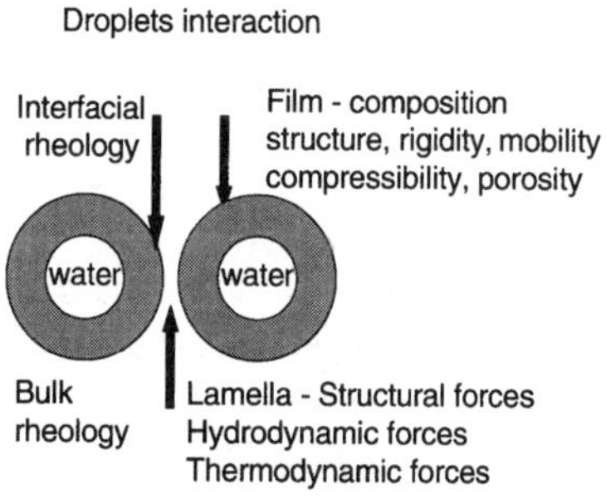


**Figure 13** Schematic on the physicochemical and dynamic factors involved in the droplet interaction.

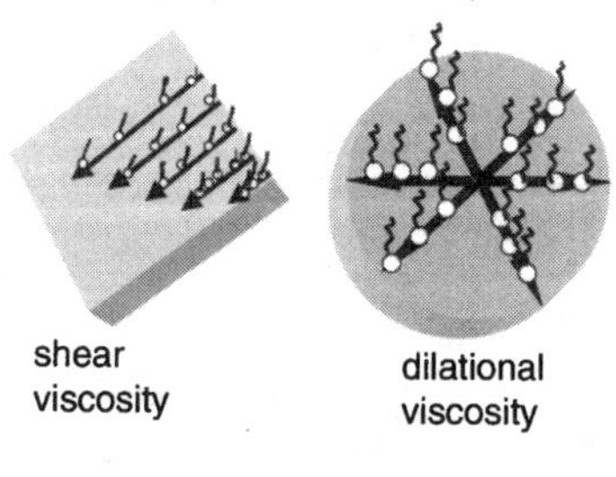


**Figure 14** Schematic of shear and dilational forces involved in measuring the corresponding viscosities at a W/O interface.

used for interfacial shear viscometry are a ring viscometer, a disk viscometer, a knife-edge viscometer, a deep-channel surface viscometer, and various modifications to these as described by Edwards et al. (151) and Boyd and Sherman (159). The majority of these are for interfacial shear measurements.

Cairns et al. (160) gave a detailed description of interfacial shear viscometry of crude oil/aqueous 1% NaCl systems. Kuwait crude had a low $\eta_s$ and produced unstable emulsions, while Iranian heavy had a high $\eta_s$ and stable emulsions. They showed that $\eta_s$ decreased with increases in pH and NaCl, and $\eta_s$ increased with aging. The aging effect invariably is a result of interfacial material build up over time, recently described by Neumann and Paczynska-Lahme (120) for crude W/O emulsions. Cairns et al. (160) concluded that the stability of emulsions was favorable when there was a rapid adsorption of surface-active agents (detected by a fall in interfacial tension), followed by a rapid rise in $\eta_s$, and that maximum emulsion stability occurred when pH and $\eta_s$ were highest. When comparing Zakum and Murban (0.08% asphaltenes) with Tia Juana (3.05% asphaltenes) in the pH range 2–11 a high asphaltene content did not correspond with high $\eta_s$. With $Ca^{2+}$, $\eta_s$ was high and interfacial tension fell rapidly, leading to intermediate emulsion stability; yet with $Na^+$, $\eta_s$ was low, interfacial tension fell slowly, and emulsions were unstable.

Grist et al. (161) provided a detailed critique of the biconical bob tension pendulum viscometer in interfacial shear viscosity measurements of Forties water/crude oil systems. They reported that a highly viscous interface reduces oil film drainage and is responsible for the high water content of "chocolate mousse" in oil spills as well as increased emulsion stability. Graham et al. (162) showed reduced $\eta_s$ with demulsifier and methanol, and that overdosing led to both increased $\eta_s$ and emulsion stability. Wasan et al. (163) in their study of Salem crude/brine, measured the $\eta_s$ responses with a deep-channel interfacial viscometer with and without a petroleum sulfonate demulsifier. The demulsifier activity was enhanced with a cosurfactant such as hexanol. They reported that as $\eta_s$ increased coalescence decreased. Later Pasquarelli and Wasan (164), using a viscous traction shear viscometer, showed that, with Salem crude and demulsifier TRS 10–80, the emulsion became more stable at high $\eta_s$. On examination of the $\eta_s$ of the fractionated crude (one high-asphaltenes fraction and one high-resin fraction) against brine and alkaline water, they found that $\eta_s$ was highest in the high-asphaltene fraction film against first brine and then water, giving 7.1 and 6.7 surface poise, respectively. Under similar conditions, against brine and water phase, respectively, the resins' $\eta_s$ showed 0.004 and 0.00074 surface poise, while the $\eta_s$ of the source crude oil were 0.25 and 0.00043 surface poise, respectively. Taylor (165) also indicated that the aging of the Romashkino, Ninian, and Kuwait crudes with an external added surfactant can retard the buildup of $\eta_s$. Neustadter et al. (166) used the conical bob torsion pendulum rheometer to study the interfacial rheology of Iranian heavy, Kuwait, and Forties crude oil–water films. They showed the same $\eta_s$ for all crudes, and suggested that one cannot use only interfacial shear viscosity to predict emulsion behaviour and that the compressibilities and elasticities are not reflected in this viscosity.

## 3. Destabilization and Dynamic Interfacial Viscoelasticity

To achieve both shear viscous and shear elastic properties, oscillatory measurements can be performed where one movable component of the rheometer is oscillated at given amplitudes and frequencies, and a sinusoidal wave is propagated. The complex modulus, $\varepsilon^*$, is then related to the elastic and viscous vector components, which are the real and imaginary coefficients of the frequency function.

Mukerjee and Kushnick (167) suggested that the interfacial dilational modulus can be obtained by a Fourier-transformed pulsed-drop technique similar to the method used by Clint et al. (168) in which the Langmuir trough was used in studying the interfacial tension ($\gamma$) variation with periodic variation in interfacial area. The frequency-dependent complex modulus, $\varepsilon^*(f)$, is equated to a real elastic modulus, $\varepsilon'(f)$ (dilational elastic modulus), plus an imaginary modulus $i\varepsilon''(f)$ (dilational viscosity modulus), and set equal to $(d\gamma/dA/A)$. The dilational elastic modulus is the interfacial tension gradient which is in phase with the area, $A$, change. The dilational viscosity modulus is 90° out of phase with the area change.

Mukerjee and Kushnick (167) showed that at low frequency the demulsifier behaves as a soluble monolayer, and at high frequency as an insoluble monolayer. Variation in interfacial tension from a local change in area is virtually instantaneous. This gradient is short circuited when the demulsifier molecule moves to and from the surface to bulk or is sufficiently soluble in the bulk phase.

Mukherjee and Kushnick's definitions were as follows: the interfacial tension increment, $d\gamma$, per unit fractional area change, $dA/A$, is eq'uated to the complex modulus, $\varepsilon^*(f)$;

$$\varepsilon^*(f) = \varepsilon'(f) + i\varepsilon''(f) = d\gamma / dA/A \quad (1)$$

$$= d_\gamma / d \ln A \quad (2)$$

Where $\varepsilon'(f)$ and $\varepsilon''(f)$ are the real and imaginary components at a frequency $f$.

For an air/liquid system a measure of the surface-tension variation resulting from the imposed periodic area variation in the Langmuir trough is performed. If both dilational viscous $\eta_d = \varepsilon''(f)$ and dilational elastic $\varepsilon_d = \varepsilon'(f)$ data are needed, and if a Langmuir-type trough is used, then one barrier can be oscillated and another barrier can be used to adjust the extent of the interfacial area. The calculation of the complex modulus, $\varepsilon^*$, requires complete scans at different frequencies.

There are many difficulties associated with this technique when applied to crude oil systems according to Mukherjee and Kushnick (167). They by-passed these difficulties by using the method developed by Neustadter and coworkers (168, 169) for crude oil systems by following the dynamic interfacial tension with step changes in interfacial area. A complete frequency spectrum is obtained by Fourier transform (FT) of the dynamic interfacial tension data:

$$\varepsilon^*(f) = 2\pi if / \Delta\ln A \int_0^\infty \Delta\gamma(f)e^{-2\pi ift}\, dt \quad (3)$$

$$\varepsilon'(f) = 2\pi\, if/\Delta\ln A\, [\text{Real FT}\ (\Delta\gamma)] \quad (4)$$

This is the dilational elastic modulus or the interfacial tension gradient which is in phase with the area change.

At low frequency they found that the demulsifier behaved as soluble monolayers and the tension was governed by the bulk concentration and did not change with change in area. At high frequency the demulsifiers behaved as insoluble monolayers and the change in interfacial tension resulting from area change was instantaneous:

$$\varepsilon''(f) = 2\pi\, if/\Delta\ln A\, [\text{Imaginary FT}\ (\Delta\gamma)] \quad (5)$$

This is the dilational viscosity modulus which occurs when the demulsifier is soluble in the bulk liquid. Area compression and expansion produce a tension gradient, which is diminished by the periodic transfer of demulsifier to the bulk from the surface and vice versa. The viscosity component amplitude is 90° out of phase with the area change.

This technique was used to show that demulsification effectiveness can be correlated with a low dynamic interfacial tension gradient, especially at low interfacial shear viscosity and this is evidenced by a sharp rise in interfacial tension at low frequencies for ineffective demulsifiers.

Alternatively, interfacial tension can also be measured continuously with area changes, with an expanding or contracting pendant-drop instrument adapted to oscillatory measurements as described by Bhardwaj and Hartland (74). Dilational data are obtained without introducing shear as the interface is expanded and contracted. This is also often achieved with an expanding drop volume or bubble-pressure tensiometer (65). Nikolov et al. (170) have recently developed this technique and an instrument to study oil/water systems.

Tambe et al. (171) used model systems of colloidal-laden interfaces of emulsions made up of 2 wt% graphite, 20 ml deionized water, and 30 ml decane. They modeled the system to understand factors that control the colloid-stabilized emulsions as was studied previously by Wasan and Menon (172, 173). They showed that, for viscoelastic films of finite dilational elasticity, interfacial rheology plays a dominant role in film-drainage rates. They suggested that the characteristic relaxation times are related to the ability of a demulsifier molecule to diffuse to the interface in response to a concentration gradient while minimizing the Marangoni flows, which retard film drainage. They suggested that the film-drainage rate is sensitive to the dilational elasticity of the interfaces. For viscous interfaces (zero dilational elasticity) the drainage rate is independent of interfacial dilational viscosity.

Neustadter et al. (166) earlier measured interfacial dilation elasticities, $\varepsilon_d$, and viscosity, $\eta_d$, for Iranian crude oil/water and deduced that the extent of the relaxation process was not a function of time due to the lack of change in viscosity, $\eta_d$, at fixed frequency. However, as frequency changed, $\eta_d$ decreased, indicating that the relaxation involved the interchange of bulk material to and from the interface. The increased elasticy, $\varepsilon_d$, with time suggested that there was irreversible adsorption of high-molecular-weight species.

Nordli et al. (146) later found larger relaxation time for the films with the higher-molecular-weight surface-active species derived from North Sea crude. For crude W/O emulsions, Neumann and Paczynska-Lahme (120) concurred that the emulsions are stabilized by thick films with Gibbs elasticity, and the more the interfacial activities of the components differ from each other the higher the elasticity modulus of the multicomponent films.

During compression and expansion, a molecular relaxation process occurs for that time scale. It is expected that there will be characteristic relaxation rates (times). It is expected that the molecular relaxation process will include: (1) increased packing by surface diffusion of low-molecular-weight species; (2) molecular conformational changes of irreversibly adsorbed high-molecular-weight species; (3) desorption; and (4) readsorption of low-molecular-weight reversibly adsorbed molecules. These may occur during the compression and expansion stages. The response can also be interpreted as solvent losses from the molecular films.

Neustadter et al. (166) indicated that, if the frequency of the longitudinal wave is varied, the bulk-to-interface interchange occurs during compression/expansion of the film. There may be a short-circuiting of the interfacial tension gradient, and then information on the relaxation rates and dilational properties may be obtained. In this interchange one may observe a low apparent compression modulus. A phase shift between the imposed strain (change in area) and resultant stress (change in interfacial tension) would occur. Different interfacial tensions with compression/expansion would be produced with molecular reorientation of irreversibly adsorbed species. With the increase in frequency of oscillation there is less bulk to interface interchange. Only for species of greater solubilities and shorter chain length would diffusional interchange be important. For the long-chain irreversibly bound components, which are almost insoluble, slower reorientation will occur. Neustadter et al. (166) emphasized that relaxation plays an important role in the behavior of the crude oil/water interface. With age, low dilation elasticities increase and no change in dilation viscosities occur.

## E. Further Studies on Shear and Dilational Viscosity of Crudes

In 1987, Eley et al. (72) used the pendant-drop retraction method in the study of film compressibilities of crude oil/water interfaces for three crudes. These varied in asphaltene content. Libya (Brega) had 0.46 g/L, Kuwait 3.7 g/L, and Tia Juana (Venezuela) 5.94 g/L. They added a dispersant containing a nonionic oil-soluble surfactant, and observed increased film compressibilities. The concentration of effective dispersant correlated with the asphaltene content. At the same time the decreased interfacial tension corresponded with increased film compressibilities and this was explained as a displacement of asphaltenes from the interface by the dispersant.

They adapted an interfacial shear rheometer (plate/rod) to measure the shear viscoelasticity of the system with and without dispersant. At an applied shear stress, creep curves for the system were monitored. There were no instantaneous elasticity and viscosity for the Kuwait and Tia Juana crudes with and without dispersant. They attributed this to a network structure of flocculated asphaltenes in the films. They found that there was some dilatancy in their crude oil films, described as a "stick/slip" flow in their flow curves. However, this flow was attributed to thick films of asphaltene particles building up at the interface. Using creep measurements, they examined a model system of asphaltenes/n-heptane/toluene. They found a retarded elastic deformation, which was different from the response of the crude oils. This suggested to them that there was a different type of interfacial structure formed with the model oil, and this may be attributed to the solvency of the medium and not to the lower asphaltenes content in the model system.

Further details of interfacial rheology of crude-oil emulsion films are discussed extensively by Menon et al. (174), Neustadter et al. (166), Mohammed et al. (175), Tambe et al. (176), and Mukherjee and Kushnick (167). They discussed the effects of demulsifiers on the interfacial properties governing the crude-oil demulsification.

Neustadter et al. (166) first discussed the crude-oil interfacial film in terms of pseudostatic film compressibility when the Langmuir balance was used for studying the liquid/liquid interface. This technique showed less variation in adsorption and showed that all but the most surface-active species were squeezed out during compression. The whole interphase region was then probed using a biconical bob torsion pendulum device suitable for high interfacial shear measurements. It responded to small changes in adsorption in the interphase region.

Highly viscous interfacial films retarded the rate of film drainage during coalescence of water droplets. They found that a highly viscous film at pH 6 led to maximum stability. They also suggested that to increase the interfacial viscosity of a low interfacial viscosity film, one should compress it. Interfacial shear viscosities were investigated with all three crudes, Iranian heavy, Kuwait, and Forties, which had similar interfacial shear viscosities.

Interfacial rheology should be used with care as a predictive tool for emulsion stability/instability. Graham et al. (162) showed that there was reduction in interfacial shear viscosity of Forties crude oil/water interface with demulsifier added in methanol. Cairns et al. (177) revealed for Zakum, Murban, and Tia Juana crudes/water systems, that interfacial shear viscosity, $\eta_s$, decreased with increased pH, and $\eta_s$ decreased with increase in electrolyte NaCl. Also, $\eta_s$ increased over time. This did not correlate with the asphaltenes content.

Wasan et al. (163) used a deep-channel viscometer in studying the interfacial shear viscosity of Salem crudes/water with and without the addition of petroleum sulfonate and salts, as well as Illinois crude/brine with pentadecyl benzenesulfonate. They showed a decreased coalescence time with decreased shear viscosity. A viscous traction shear viscometer was used for fractionated crude oil/brine by Pasquarelli and Wasan (164), who showed that increased interfacial shear viscosity is correlated with decreased coalescence. Later, Wasan correlated interfacial shear viscosity with film-drainage time to determine effective demulsifiers (178).

Taylor (165) observed that surfactants used as destabilizers retard the build up of material at the oil/water interface for aging Ninian, Romaskino, and Kuwait crudes. Mohammed et al. (144) studied the rheological behavior of the North Sea Buchan crude oil/double-distilled water interface by oscillatory interfacial shear rheometry, using the biconical bob rheometer in the oscillation and creep modes. They used creep measurements to obtain relaxation data. This crude is known to produce very stable emulsions, yet its heptane-precipitated asphaltenes were only 0.058 g/mL. They attributed Newtonian viscosity to the adsorption of low-molecular-weight components of crude, and the build up of the interfacial films to high-molecular-weight components forming a network structure with viscoelastic properties. They investigated the effects of temperature and of demulsifiers Unidem 120 and BJ18 in xylene on the film viscoelasticity. They found that at 45°C the interfacial film viscoelasticity became more liquid-like. The increased viscosity with increase in temperature was a result of loss of light ends. The Unidem 120 caused a change in the aged film character from solid-like to liquid-like, and BJ18 prevented the film buildup. They traced these effects through the relaxation time from creep curves with and without demulsifiers, as well as observing a decreased viscous component in the oscillatory measurements.

Chen et al. (179) used optical microscopy to study the same crude-oil emulsions in an electric field. They modeled the system with a computer simulation based on a hard-sphere model describing the droplets as stabilized by a rigid asphaltene film. They identified two different types of coalescence. The mobile interfacial films led to low emulsion conductivities owing to immediate droplet/droplet coalescence, and the incompressible interfacial film led to low emulsion conductivities due to droplet-chain formation and bridging between the electrodes.

Malhotra (180), in a parametric study, showed that interfacial mobility and the rate of film drainage depend on the interfacial shear viscosity when the viscosity is in the range $10^{-5}$ to $10^{-3}$ Pa.ms. Outside this range the interfacial viscosity had no effect. Interfacial viscosity as low as $< 10^{-5}$ Pa.ms can lead to stable emulsions (181).

The reported information on film rigidity, compressibility, and other aspects of interfacial rheology indicates that the highly viscous and elastic films are stable. In most of the compressibility studies on crude oil films, the films were made with the various surface-active components of crude oil and/or the crude oil itself. The drop-retraction, Langmuir-type trough, and pendant drops pulsed at various frequencies of expansions and contractions were tools for studying the mechanics and fluid dynamics of the dilating interfacial films. In addition, sophisticated shear rheometers, including oscillatory and creep measurements, were used to obtain shear viscosity, elasticity, and film relaxation data. Typically, these measurements were made with and without demulsifiers, with changes in water quality, pH, or solvent properties when demulsifiication was investigated.

Wasan and coworkers (63, 65, 174) extended techniques for studying film rheology of the foam lamella to studies of crude-oil emulsion lamella. Using a capillary balance technique and light interferometry, the film thinning of foams was studied with and without chemical demulsifiers, with solvent properties changed, etc. (182). They confirmed that there were two contributions to emulsion stability – a structural component that originates from the nature of the bulk phase, and an adsorbed-layer contribution to film stability (170). This will be covered in another chapter in this series.

## F. Destabilization and Interfacial Tensions

The equilibrium interfacial tension between water and oil is a measure of the adsorption of surface-active components to the interface and can be related to surface excess by the Gibbs equation (183). However, in crude oil systems the activity/concentrations of the surface-active components are not easily determined. Indirect measures are applied. In most process conditions with short resident times, it is the dynamic interfacial tension gradient that is important. Interfacial tension also tells whether or not the demulsifier is surface active, and as will be shown later, this is important for demulsification. The interfacial tension gradient is the important factor for rendering the interface immobile. The interfacial tension gradient is used in a measure of Gibbs surface elasticity, $E$, or as an indication of changes in free energy. When the interface is expanded or contracted, then $E = -d\gamma/d\ln\Gamma$, where $\gamma$ is the measured interfacial tension, and $\Gamma$ is the interfacial concentration or Gibbs surface excess. This elasticity is an indication of the ability of the interface to adjust the interfacial tension when stressed.

The interfacial tension, $\gamma$, in the Gibbs adsorption equation is used for equilibrium conditions as bitumen components are adsorbed. Measurement techniques available are extensive. Some of these methods are: duNouy ring, maximum bubble pressure, drop volume, Wilmhelmy plate, sessile drop, spinning drop, pendant drop, capillary rise, oscillating jet, and capillary ripples. These and many others are referenced extensively by Malhotra and Wasan (153). These authors also showed that there is no correlation between emulsion stability and interfacial tension. The nature of the film dominates stability. Some relationships between interfacial tensions and crude oil properties follow.

The interfacial tension values of crude oil/water or asphaltenes–resins/water are strongly dependent on pH and salt content of the aqueous phase. With toluene as the solvent, bell-shaped $\gamma$–pH curves were observed for heavy Venezuelan crude/brine systems (125), distilled-water washings of diluted bitumen/air (11), and light crudes (126). Luthy et al. (184) showed similar results for heavy southern Californian, light Arabian, and waste petroleum oil/brine systems. Figure 15 illustrates the interfacial tension pH effect of crude oil/water systems (184). A bell-shaped $\gamma$–pH curve with a maximum at pH 6 was observed for all crudes. Similar findings were observed for Zakum, Murban, and Tia Juana medium crude oil/water interfaces, with pH changes showing a maximum interfacial tension at pH 6 by Cairns et al. (185). They suggested that increased emulsion stability is expected to occur at very high and low pH only because of adsorption of surface-active components at these pH values. The interfacial tension in these cases reflects only the adsorption of interfacially active components at the interface. Acevedo et al. (186) found similar results for Venezuelan crudes and explained that, at low pH (< 7), the adsorption of basic groups is dominant. At

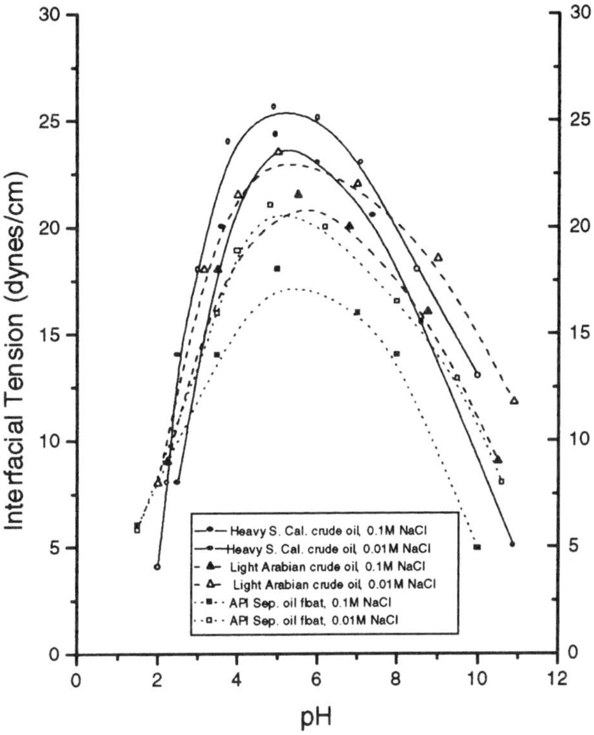


**Figure 15** Interfacial tension changes with pH, showing the maximum for various crudes. (From Ref. 184.)

high pH (> 8), ionized carboxylic acids integrated in the adsorbed asphaltenes micelles are dominant, and an acid–base pair is adsorbed in the neutral region (pH 6–8). Xu (187) showed that the dynamic interfacial tension of the bitumen/water interface decreased with increased caustic concentrations as the interface aged. Similar decreases were noted as the bitumen concentration increased, indicating that adsorption of natural surfactants after saponification is time dependent. Similar observations were made for Ratawi asphaltenes (188). Through interfacial tension measurements the critical micelle concentrations of light and heavy fractions of Ratawi asphaltenes and desorption kinetics in organic solvents (189) were studied.

In all cases adsorption at the W/O interface is very slow and time dependent. The rate is lower for acidic and neutral pH. At basic pH, adsorption is much faster. These are often reported as aging effects, especially as changes at the interfaces go on for days or weeks. Malhotra (180) discussed detailed adsorption models and related this to film-drainage rates.

One can thus offer the conjecture that demulsification may be enhanced when films are weaker, which may or may not correspond with a high interfacial tension, since the film properties/architecture are dependent on many factors that in turn depend on both the crude oil and the water. Eley et al. (72) and others discussed earlier have shown that, at acidic and neutral pH, the films are solid-like and highly viscous. Large increases in interfacial viscosity and elasticity were observed for crude oil/water with a change in pH from neutral to alkaline (73).

The times for development of an adsorbed layer or the conditions by which adsorption is minimized are factors which can be used to improve chemical demulsification. There is no correlation between decrease in interfacial tension and increased coalescence rates (190), notwithstanding the data showing decreased shear viscosity and increased coalescence rates (163).

Low interfacial tension is important for decreased energy requirements in the emulsion formation stage, but it does not appear to play such a role in demulsification. At an interfacial tension of 5 dyne/cm, in order to overcome LaPlace pressure and break up of a drop of 10 $\mu$m, one requires an external pressure gradient of 100 psi/cm, and for a 0.1-$\mu$m drop, a gradient of $10^6$ psi/cm. The increase in interfacial energy for changing 10 ml water into droplets of 10 $\mu$m at interfacial tensions of 5 dyne/cm is $3\times10^4$ dyne/cm. If we compare this energy with that required to raise the temperature of the same water by 1°C at an energy cost of $4.2\times10^8$ dyne/cm, then the emulsification process of producing droplets from a bulk uses significantly less energy than the energy needed for temperature change or the energy for breakup of an already formed emulsion droplet. The smaller droplets behave as hard spheres. Thus, an easier means of destabilization would be by chemical demulsification augmented by changes in internal natural surfactant composition.

## G. Effects of Temperature and Heat on Destabilization

Heat has its own advantages and disadvantages in the demulsification process (135, 191). Heat may also contribute to the disruption of emulsion film material by the differential expansion of the water inside the droplets, perhaps creating higher burst pressures. With increasing temperature the viscosity and density differences between water and heavy crude oils begin to diverge. The density of the oil is reduced faster than the density of the water as temperature increases, showing a larger divergence below 150°C. Since heat increases the density differences between bulk oil and water droplets it leaves a greater chance of accelerated settling, between 50°C and 125°C.

Also, as heating reduces the bulk viscosity of the oil, it also reduces the fluid resistance (106). The settling of water droplets is faster in less viscous oils. Because heating puts increased energy into a system there is a higher frequency of droplet collisions.

The increased temperature of the system may lower the interfacial shear viscosity, which may lead to an improved rate of film drainage between adjacent drops. Menon and Wasan (62) demonstrated the decrease in interfacial viscosity of shale oil/water at increased temperatures and showed analogous behavior at increased demulsifier concentration. Thompson et al. (133) reported that, at increased temperatures, some incompressible nonrelaxing films tend to relax and the rate of buildup of the resistance to film compression increases. Demulsifiers then reduce the kinetic barrier to coalescence.

Sometimes heat will deactivate the demulsifier if its solution chemistry is sensitive to heat. The cloud points of the nonionics occur at specific temperatures based on the chemical structure and oil-phase chemistry. In aromatic oils the phase-inversion temperature (PIT) is lower than the cloud point; in n-paraffins and cycloparaffins the PIT is higher than the cloud point. Phase inversion may result in some cases (192, 193) as the temperature approaches the cloud point.

Heating may augment destabilization of stable crude emulsions in which waxes play a large role. This is more likely with highly paraffinic crudes found in the North Sea (20) and paraffin wax crystals melted by heat. With asphaltenic or highly aromatic crudes, heating has less effect on the stability of the emulsions unless the viscosity and density of the interfacial barrier and continuous phase are reduced considerably. Stockwell et al. showed that the heating prehistory affected the emulsion stability of North Sea crudes but not the Canadian asphaltenic crudes (17). This was related to the crystallization sizes of the waxes in the North Sea crudes and the absence of wax crystals in the asphaltenic Canadian crudes. Neumann and Paczynka-Lahme (120, 194) held that, at increased temperatures, the interfacial films with asphaltenes remain intact but with the resins, stability increases. At high temperatures, resins form liquid-crystalline lyotropic mesophases for several crudes tested at 90°C. Stable films are suggested to have considerable viscoelasticity (195). Breaking emulsions with high resin contents is not easy because of the formation of either hexagonal or lamella liquid crystals at the interface at high temperatures.

Sometimes heating enhances the transport of the demulsifier chemicals to the interface as well. A combination of heating and demulsifiers was found to be synergistic in breaking stable emulsions formed with Alaskan North Slope, Bonny Light, and BCF-17 crudes in a simulated spills study conducted by Stroem-Kristainsen et al. (196). This synergism with heat was also observed for breaking other North Sea (197) crudes' W/O emulsions (66, 198) and Nigerian (199) crude oil emulsions, when heat alone was ineffective. Nonionic surfactants performed better as demulsifiers at elevated temperatures for Hungarian (135) Algyo crude oil emulsions. If heat solubilizes the stabilizing surfactants into either the oil or water phase, the interfacial film will be weakened, leading to destabilization of the emulsion.

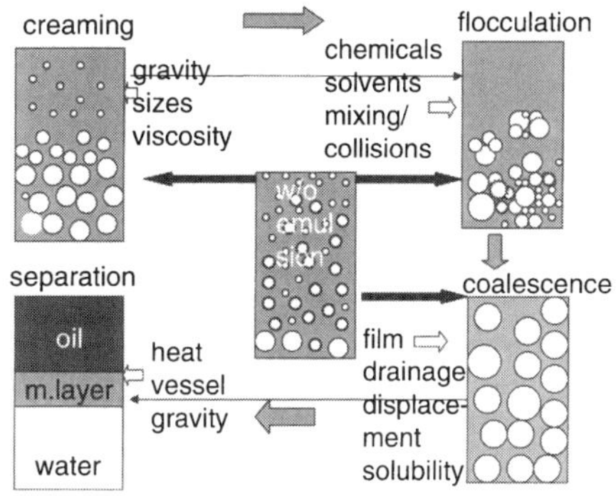


Figure 16 Schematic illustrating the various stages of demulsification process.

## IV. CHEMICAL DEMULSIFICATION PROCESS

Emulsification and demulsification are both complex processes. However, as noted earlier, demulsification is by no means the opposite of emulsification (200, 201). This is especially the case in the petroleum industry. In order to demulsify a crude W/O emulsion efficiently, it has been emphasized that it is advantageous to understand first the characteristics of the emulsions, the nature of interfacial films, and hence the causes of stability. Accordingly, in choosing a demulsification protocol, one would first identify key factors responsible for the stability, find the target properties to modify toward destabilization, introduce sufficient energy to promote coalescence, and find the best conditions to allow phase separation.

Demulsification of emulsions can be summarized by four phenomena occurring either sequentially or simultaneously. These are flocculation and/or aggregation, creaming or sedimentation, coalescence, and phase separation. The efficiency depends upon the matching of the demulsifier with the process residence time, the concentration and stability of the emulsion, the temperature, the process vessel, and the mixing, all of which affect the aggregates before coalescence occurs. These various stages are illustrated in Figure 16. The figure shows a schematic describing four stages of demulsification. It summarizes the changes as the emulsion is influenced by chemicals, solvents, mixing conditions, size distributions of the droplets, heat, and the vessel used for separation.

Figure 17 is a photomicrograph of flocculation and coalescence of a bitumen-stabilized W/O emulsion after treatment with an oil-soluble demulsifier. The

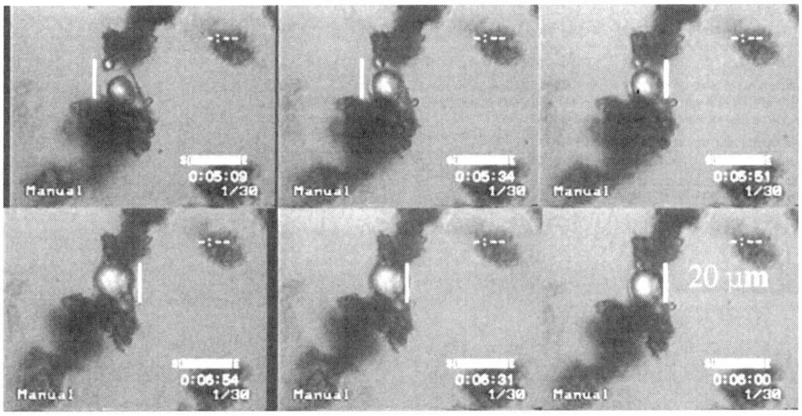


Figure 17 Photomicrograph of flocculation and coalescence of a bitumen-stabilized W/O emulsion after treatment with an oil-soluble demulsifier. (From Ref. 82.)

large droplet approaches the smaller droplet and they eventually coalesce. In this case, coalescence is the rate-determining step. Figure 18 shows bitumen W/O emulsions droplets before and after treatment with a demulsifier in the oil phase. The aggregates of larger droplets, after treatment, indicate that coalescence occurs during or before flocculation. Note the change in the distribution of finer droplets from 10 s to 24 h after treatment.

In industry, if crude oil emulsions do not coalesce and/or phase separate in a given time frame, and persist throughout the process, the emulsion is deemed stable or tight. Demulsification can be monitored by bulk phase separation over time and/or by a more fundamental approach of examining the interfacial dynamics which provides some understanding of the demulsification mechanisms. The first is used in the field and/or as part of a laboratory study. Here, efficiency is defined as the rate of coalescence and water resolution, by tracing the fraction of water resolved over time and using an initial slope of the curve as an efficiency factor (202). The second approach is more complex and is most often used in research. It includes flocculation, coalescence, and film-drainage phenomena. In the second approach, monitoring the growth of droplets over time or the disappearance of droplets into the bulk phase provides data that have been used in the development of many models on coalescence kinetics (203–205). However, obtaining experimental data to test these models relies on techniques such as microscopy and image analysis, light scattering, or turbidimetry (62, 76, 206, 207). Many theories (208, 209) have been developed for the droplet growth or disappearance over time as a measure of emulsion stability (90, 210) and instability (211). Droplet growth through water attraction by holes developed in the interface, or by Oswald ripening mechanisms, are a few of the reasons behind coalescence. The lamella drainage approach involves the interplay of both surface/interfacial forces and hydro-

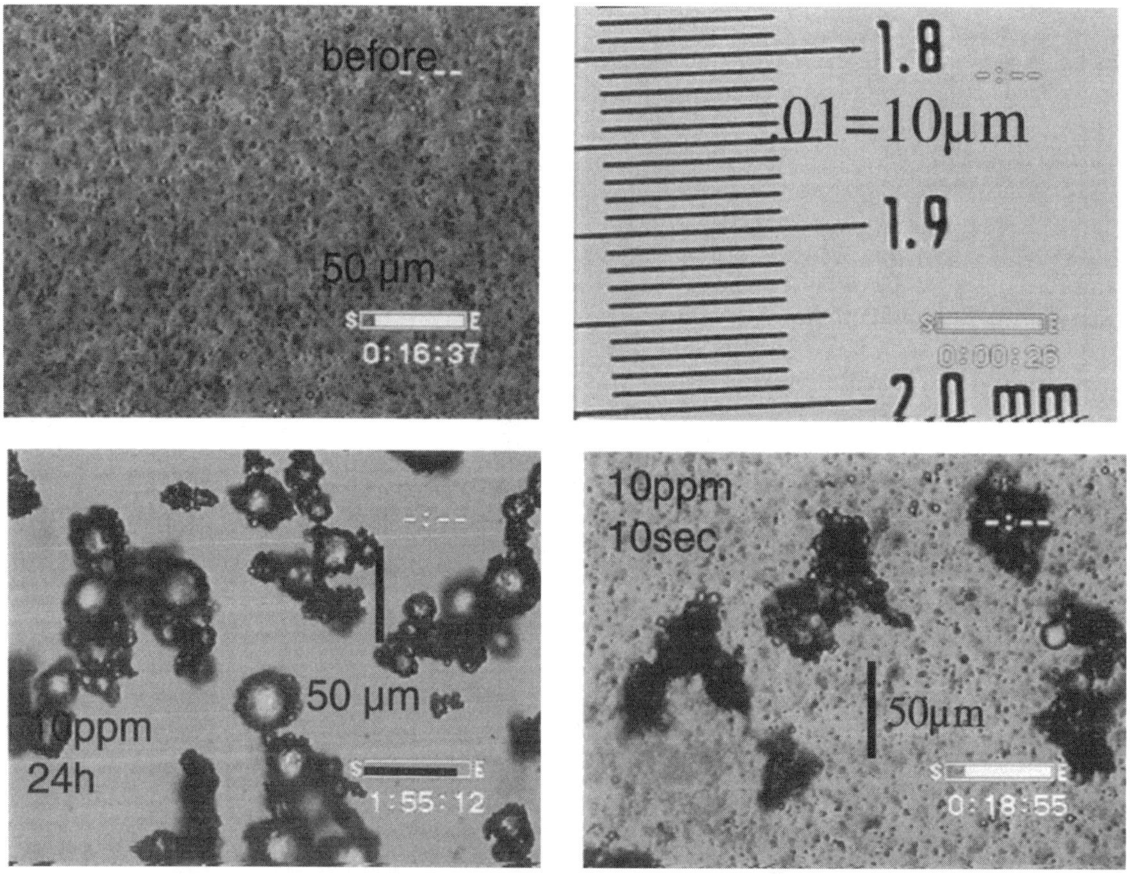


**Figure 18** Photomicrograph of bitumen W/O emulsion before and after treatment (after 10 s and 24 h clockwise) with a demulsifier showing the aggregates of larger droplets with smaller droplets attached. (From Ref. 82.)

dynamic forces (212). Interfacial rheology is important in the latter.

## A. Creaming and Coalescence

For a simple two-phase system consisting of an upper organic phase and a lower oil/water emulsion phase, Ostrovsky and Good (213) distinguished between the kinetic and aggregative instability of macroemulsions. The kinetic instability was identified as sedimentation or creaming, which was distinguished from aggregative instability. They developed a model in which the system coalescence occurred under agitation, and then traced coalescence and sedimentation times. The latter arose out of their studies on drop size versus agitation relationships, the former through low interfacial tension (range 0.024–0.33 dyne/cm) and coalescence–time correlations.

However, according to Vold and Vold (214), creaming is the separation of an emulsion into a concentrated and a dilute fraction through centrifugation or gravitational settling. The concentrated part is rich in the dispersed phase and the remaining dilute phase has finer droplets. The rate of creaming of noninteracting particles depends on density differences and the square of the droplet radius. If particles are interacting and held in a confined space, the rate of creaming involves complex hydrodynamics, wall effects, and hindrances. If the particles are aggregated the creaming rates are higher than the rate with the primary particles owing to the larger aggregate radius and the relative porosity (tightness or loose nature) of the aggregate. Flocculation may help to accelerate creaming, as particle concentration increases. The floc which is sedimenting more rapidly also sweeps or intercepts smaller droplets during the mass movement. Visually, creaming appears as a moving boundary of highly turbid material away from a lesser turbid material. This boundary is used to trace separation rates.

If coalescence or rupture of the interdroplet films common to two contacting droplets occurs during or before creaming, the destabilization rates are measured by tracing the changes in droplet size distributions with time, or by counting the number of droplets of specific diameters over the same time. In the case where a droplet merges with the bulk, the times of both approach and merge are measured.

One of the first theoretical interpretations of the coagulation process for hydrophobic sols was developed by Smoluchowski (215). This was later extended to flocculation and coalescence kinetics (216). Becher (89, 100, 117, 211) describes these theories in some depth. He stated that demulsification is the most important and the most complete example of emulsion stability. There is a distinction between a dilute emulsion system and a concentrated system. In the dilute system it is expected that the rate of flocculation is much lower than the rate of coalescence. As the droplets increase in number (or volume fraction), there is a much faster increase in the rate of flocculation and a slower increase in the rate of coalescence. In highly concentrated emulsions, coalescence can be rate determining. Over a certain range of concentrations the two processes can be the same order of magnitude.

One can thus add chemicals in a dilute system such that the rate of flocculation is unaffected but coalescence is inhibited. In some cases, creaming can be ruled out if the densities of the two phases are adjusted closely as in bitumen emulsions. Becher has noted that tracing the number of particles as a function of time is more sensitive than the specific interface method, as a 10% decrease in interfacial area is accompanied by a 27% decrease in the number of particles for a fixed size distribution.

Smoluchowski (215) modeled the decrease in particle numbers over time. In the model there was no distinction between single droplets, primary particles, and aggregates. The number of particles diffusing through a sphere surrounding a given particle in a unit time is equal to the number of particles adhering to a single particle. Therefore, for fast, irreversible flocculation alone in a dilute dispersion Smoluchowski's expression is:

$$N = N_0/(1 + a N_0 t) \tag{6}$$

where $N_0$ = the number of particles at time zero, $N$ = the number of particles at time $t$, $a$ = rate-determining constant = $8\pi Dr$, $D$ = diffusion coefficient = $kT/6\pi\eta r$, $r$ = droplet diameter, $k$ = Boltzmann constant, $T$ = temperature, $\eta$ = the viscosity, and the half-life of an emulsion becomes $1/aN_0$. A plot of $1/N$ versus $t$ gives the rate constant $a$, which approximates to $a = 10^{-11}\text{cm}^3\text{s}^{-1}$.

In a flocculating concentrated emulsion, $aN_0 \gg K$, and the contribution of unreacted primary particles is negligible. A rapid flocculation rate relative to the rate of coalescence is given by Van den Tempel (216):

$$N = N_0\,[1 - \exp(-Kt)]/KT \tag{7}$$

Here, $K$ = coalescence rate constant, and $T$ = absolute temperature. When the rates are equal for both coalescence and flocculation, $a = 10^{-12}$ to $30 \times 10^{-11}$ cm$^3$s$^{-1}$. When coalescence is slow the collision frequency and the duration of collisions are more important. In this case, mixing enhances collision frequency.

Smoluchowski's theory assumed that coagulation has been going on for a long time and the system is near steady state. Flocculation is considered irreversible. Van den Tempel (216) assumed that the rate of coalescence is proportional to the number of points of contact between the particles in an aggregate. He considers flocculation and coalescence to occur simultaneously. The number of primary particles not yet combined into aggregates is:

$$N_1 = N_0\,(1 + aN_0 t)^{-2} \tag{8}$$

the number of aggregates is $N_v$:

$$N_v = aN_0^{\ 2}\, t\,(1 + aN_0 t)^{-2} \tag{9}$$

and the number of primary particles in unit volume associated into all aggregates is:

$$N_0 - N_1 = N_0\,[1 - 1/(1 + aN_0 t)^2] \tag{10}$$

The average number of primary particles in an aggregate is:

$$N_a = N_0 - N_1 / N_v = 2 + aN_0\, t \text{ at time } t_1 \tag{11}$$

because coalescence has taken place. If $M$ is the average number of separate particles existing in an aggregate at time $t$, then $M$ can be unity if coalescence is fast, and slightly less than $N_a$ if coalescence is slow. The rate of coalescence is then proportional to $M-1$, i.e., the number of contacts between particles in an aggregate. Van den Tempel claimed that in dilute emulsions a small aggregate consists of one large particle with one or two smaller ones and these build up linearly. $M$ increases by adherence of new particles and the rate of increase in $M$ is caused by flocculation:

$$\mathrm{d}M/\mathrm{d}t = aN_0 - k(M-1) = aN_0 - kM + k \tag{12}$$

where $K$ is the rate of coalescence. Integrating for $M=2$ when $t=0$,

$$M - 1 = aN_0/k + (1 - aN_0/k)\exp(-kt) \tag{13}$$

The number of primary particles, whether flocculated or not, was found by adding the number of unreacted primary particles to the number of particles in an aggregate:

$$N = N_1 + N_v M \tag{14}$$

$$= N_0/(1 + aN_0 t) + aN_0^2\, t/(1 + aN_0 t)^2 \{aN_0/K + (1 - aN_0/K)\exp(-Kt)\} \tag{15}$$

The first term in Eq. (15) is the number of particles found if each aggregate has been counted as a single particle; the second describes the number of particles that enter the aggregate and is not found in the classical Smoluchowski treatment. Here, the composition of the aggregate is taken into account. When $K = \infty$, coalescence is immediate, and the expression is reduced to the first term. When $K=0$ no coalescence occurs, and $N = N_0$ for all times $t$. For $0 < K < \infty$, the rate of aggregation changes the particle concentration.

Further reviews of coalescence and flocculation kinetics were reported by Becher (211), Tadros and Vincent (90), and Hartland (217). For all practical purposes the above treatments usually suffice in crude-oil studies. Extensive treatments of coalescence and flocculation kinetics were modeled as required for various other emulsion applications. Borwanker et al. (218) developed a mathematical model to account for flocculation and coalescence kinetics occurring simultaneously. They modified Van den Tempel's treatment for coalescence to include coalescence occurring in small flocs. They showed how the rate-controlling mechanism could change from coalescence-rate controlling to flocculation-rate controlling during an emulsion lifetime. They further extended the model for concentrated emulsions.

The disappearance of droplets by counting numbers of particles in a given field of view is modeled kinetically for most experimental data. Bhardwaj and Hartland (206) have shown that, with their demulsifier and crude oil emulsions, coalescence occurred in the first few seconds by a binary coalescence mechanism, then, after a lag of 7 min, the coalescence time was several minutes. The coalescence was enhanced by gentle mixing to improve collision frequency (with solids this is called orthokinetic coagulation), in contrast to the quiescent approach which relies on Brownian motion or thermal convection currents for collisions. Menon and Wasan (62) traced the number of droplets per unit volume as a function of time for water-in-shale oil emulsions. They fitted their data to

$$N = N_o \exp(-Kt) \tag{16}$$

where $N$ is the number of droplets per unit volume of aqueous phase, $N_o$ is the number of droplets per unit volume at initial time, $K$ is the coalescence rate constant, and $t$ is the time in seconds. After the addition of demulsifier, the plot of number of droplets versus time was not linear and could not be represented by a first-order rate equation. They used a rate expression containing both coalescence and flocculation rate constants (211) for the system treated with demulsifier. Here, $a$ is the flocculation rate constant:

$$N = N_o/(1+aN_o t) + aN_o^2 t/(1+aN_o t)^2 \; [aN_o/K + (1-aN_o/K)\, e^{-kt}] \qquad (17)$$

They showed that the coalescence rate constant, $K$, increases while the flocculation rate constant decreases with increased demulsifier concentration. Flocculation is high at low demulsifier concentration. At increased concentration it breaks the interfacial film and promotes coalescence. A plot of initial coalescence rate constant versus dosage indicates that the demulsification of this system was in a flocculation-rate controlling state, within its environment. Aggregation is reversible and the drop identity is not lost.

Mixing or agitation has been shown to augment coalescence by enhancing the rates of collisions. Menon and Wasan (62) have shown that the flocculation rate constant increases to a maximum with increasing speed of mixing. In order to promote coalescence there is an optimum mixing speed for every system. Redispersion occurs with excessive mixing or high rates of mixing.

Mason et al. (219) concurred on system specificity for mixing, when they showed that aged crude emulsions had less droplet growth during mixing and that separation was slow. However, the aged emulsion required increased demulsifier concentrations and a longer mixing time after demulsifier addition. This led to larger droplet size and faster separation. An emulsion mixed for 45 min had a mean size of 61 $\mu$m in contrast with 28 $\mu$m for 15 min of mixing at the same speed. Larger droplet size could promote boundary coalescence instead of binary coalescence. They used the half-lives of oil and water (i.e., time required to generate a clear layer containing one half of the oil or water initially present in the emulsion) for slowly separating systems as a measure. They also showed that for the size range between 25 and 30 $\mu$m, separation was a function of age, demulsifier concentration, and mixing time, and that mixing time could be optimized with lower demulsifier concentration, or could provide a measure of demulsifier efficiency.

Bhardwaj and Hartland (206) showed that binary coalescence improved with demulsifier and with mixing. Increasing temperature from 20° to 40°C was significant in producing increased droplet sizes over time with 100 mg/L demulsifier. The higher dosages reduced the coalescence time from 5.2 s (50 mg/L) to 4.2 s (100 mg/L). They found that initial rapid coalescence was followed by slow coalescence. They traced coalescence rates by plotting the natural logarithm of interfacial area per gram of dispersed phase against time, using the slope as a measure of coalescence rate (fractional decrease of surface area per minute). As a measure of binary coalescence rate, the droplet diameter growth was traced with time. Fast coalescence corresponds to a few seconds of binary coalescence time. Sjöblom et al. (136) showed that increased speed of mixing over time for North Sea crude oil emulsion produced decreased droplet sizes. Thus, the experimenter should be aware that if the objective of the work is emulsification, high speeds are desirable; if demulsification is the objective, then mixing must be done gently and with great care.

Tracing the resolved volume fraction of the collected bulk free-water layer over time is also a common means of measuring destabilization (220). Centrifugal forces cause the droplets to flocculate and cream faster, facilitating the drainage of thin liquid films formed between them (221–230). Vold and Maletic (231) indicated an Arrhenius type of relationship between centrifugal forces and dosage of demulsifiers for demulsification of an ideal O/W emulsion.

Before coalescence occurs between the primary drop and the bulk separated phase, or between two or three interacting droplets, increased hydrostatic pressures are developed in the creamed layer, regardless of whether the creaming is achieved by gravity or centrifugation. These pressures are the driving forces for drainage. Not every collision results in coalescence. This is because coalescence time depends on the rate of film drainage between the droplets. Allan and Mason (232) and Hartland (233) predicted that the film thinning was inversely proportional to the droplet diameter.

In another view, Vold and Vold (214) suggest that holes are formed in the interfacial film and this allows the droplets to merge. Ivanov and Dimitrov (234) indicated that holes are due to surfactant depletion at the interface. However, extensive studies conducted to understand the mechanism of destabilization of the thin/thick films formed between two droplets, or between droplet and bulk phase, indicate that the process is much more complex and may involve more than one mechanism. These are not all fully understood as yet for crude oil and bitumen systems.

### B. Film (Lamella) Drainage – Model of Coalescence

A model of coalescence via film-drainage phenomena and flow dynamics has been discussed theoretically and

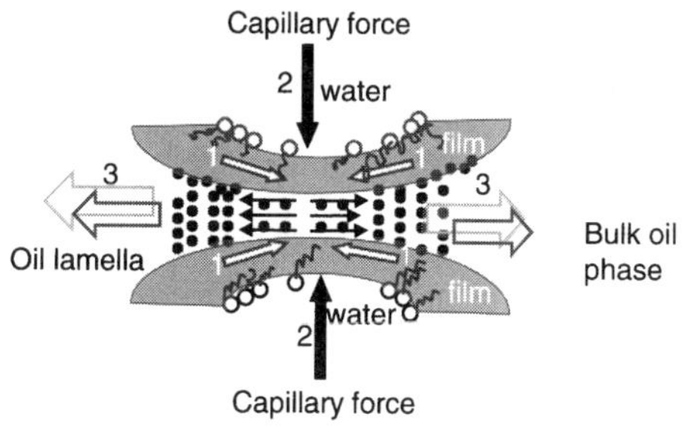


**Figure 19** Schematic of a dynamic bitumen film lamella between two approaching water droplets which are stabilized by an interfacial skin.

experimentally by several investigators over the past decades (235). From this body of work the belief is that the coalescence process of two or more droplets, drop/drop, drop/bulk, can be divided into three steps: approach, film drainage, and rupture. Film drainage is a function of bulk and interfacial fluid rheology in a balance of hydrodynamic and thermodynamic (sum of all surface forces) interactions. The hydrodynamic interactions increase rapidly as the gap width between the droplets decreases. The flow results from hydrostatic pressures normal to the surface due to the nature of the interfacial fluid in a given space. The interfacial fluid is affected by the tangential mobility and deformation of the droplets' interfaces. Figure 19 is a schematic which shows the lamella of two approaching water droplets stabilized by a bitumen film. Some of the forces for the draining process are illustrated.

Surface forces include both long and short range. The long-range forces are van der Waals attraction, steric, and, lately, structural forces (151, 236). The short-range forces include the chemical bonding to surface groups, dipole interaction, hydrophobic bonding, and Born repulsion. The short-range forces determine the interfacial structure, and the long-range forces determine whether the emulsion droplets are aggregated/flocculated. The surface forces are lumped together and are also called disjoining pressures/forces – a term coined by Derjaguin (237). These are important in very thin films of thicknesses less than 100 nm. A positive disjoining pressure gradient is required to impart resistance to film thinning, and a negative disjoining pressure has the opposite effect and increases in magnitude as the film thins. However, in simpler systems such as soap-film experiments, Scheludko and Exerowa (238, 239) showed that the negative disjoining pressure depends on the inverse third power of film thickness.

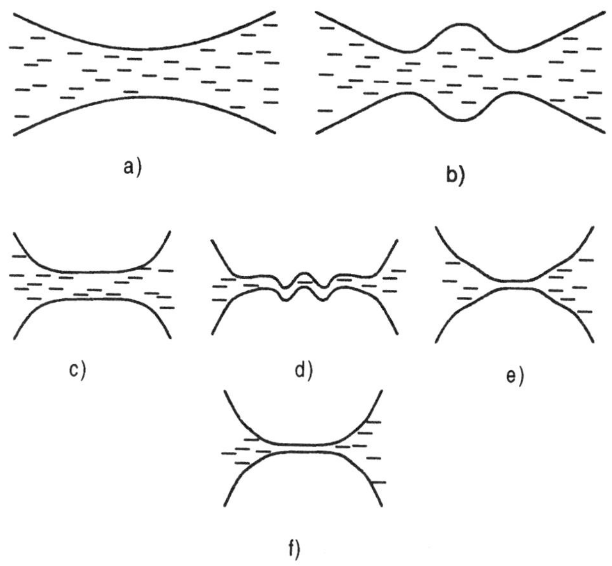


**Figure 20** Schematic of the evolution of a thin liquid lamella between two approaching droplets (147,151): (a) droplets' mutual approach with slight deformation of interfaces; (b) dimple formation on surfaces; (c) near plane-parallel film; (d) thermal or mechanical fluctuations at interface; (e) black (common) film formation; (f) growth of black film or Newton film to equilibrium radius.

The stages of thinning for a simple emulsion system can be described as follows (see Fig. 20 for a schematic example of an ideal foam system for comparison):

1. When two droplets are approaching, the thickness, $\delta$, decreases rapidly with time, and dimpling (also corrugations or oscillations) precedes the formation of a plane parallel film.
2. As the viscous and interfacial resistance forces in the film increase the film is slowly thinned to a critical thickness, $\delta_{cr}$, for rupture.
3. Rupture occurs when a hole is formed.

Step 2 has the slowest rate of thinning and is thus a rate-limiting step that determines the film lifetime. The thinning rates of steps 1 and 3 are fast. If we examine this in terms of flow we can explain the simple drainage process by what has been observed under a microscope.

Increased capillary forces at the Plateau borders allow film drainage to occur until the films have thinned to an upper limit of about 100 nm. The film reaches a metastable state and may rupture suddenly due to dust particles, thermal or mechanical shocks, or may just reach a critical thickness that it can no longer sustain. A hole in the film may form as a result of thermal fluctuations of the two expanding droplet surfaces, and at the same time there are forces working to maintain an equilibrium film thickness. These forces originate from the surfactant monolayer which is undergoing dilating and shearing deformations, causing stresses and opposing flow. The tangential bulk stress from the film liquid causes surface-layer flow. The droplets are pushed together by external pressures such as buoyant forces, or other applied forces such as dynamic pressure gradients in the continuous medium.

Thick films ($\approx$500 nm) before drainage appear as colorful interference fringes created by passing a monochromatic light beam through the film, and this can be measured from interferometric patterns (240). The thicknesses would correspond to the wavelength of the color. As the film thins down to thicknesses below the wavelength of visible light (10–100 nm), the film appears to be black. During thinning, liquid flows out in a radial flow pattern to the Plateau borders pulled by osmotic or capillary pressures. Very thin films with negative disjoining pressures, and with low viscosity will follow Reynold's law for radial flow, where the change in thickness, $D$, in time, $t$, is $d(1/D^2)/dt = a$P, where $P$ is the hydrostatic pressure, $a$ is a constant $(=4/3\eta r^2)$, $r$ is the radius of the film, and $\eta$ is the viscosity of the fluid. Reynold's flow expresses the motion of a fluid being squeezed between two approaching solid surfaces with fixed interfaces. Detailed treatments of Reynold's flow are given by Hunter (122). Emulsion films approximate to type C in the Hunter categorization of film types.

Film lifetime is taken as the time taken to reach a given thickness plus the breaking time at that thickness. When this critical thickness, $\delta_{cr}$, is reached, and if there is an equilibrium black film, it only persists at an equilibrium or stable state provided that the barrier of potential energy is high enough. In foams the first black films, known as the common films (10–100 nm thickness, and gray or silvery), precede the second black films known as Newton black films (5 nm thickness, black). The Newton films are seen to appear after the colors of the interferometric pattern (created by the film's dispersion of the monochromatic light beam) change into gray, silver, or black. The Newton films may form spots or holes.

Drainage rates are low for rigid films and high for mobile films. The stability of the films is dependent on the rheological properties of the interfacial layer, the adsorbed layer of surface-active materials, the temperature, and the composition of the fluid and its dynamics inside the droplets. Most theoretical analyses of film drainage describe the relationship between the coalescing forces (which are suction forces at the Plateau borders and which promote drainage), the bulk phase, and the interfacial effects that resist drainage. Since the film-drainage step is rate determining, investigations have been focussed on the kinetics through the many drainage models, the important effects being the critical collapse thickness and time, bulk viscous effects, and the interfacial viscosity and elasticity. Various hydrodynamic models of the film-drainage process have been developed (241–248).

Several generalized models that account for mobility of the surface, kinetics of adsorption – desorption of surfactants, surface and bulk diffusion, surface rheological properties, and flow in both film and bulk phases were developed by Wasan (236) and Nikolov et al. (249). They introduced the concept of structural forces resulting from the narrow size distribution of micelles or colloids forced into the restricted volume of the film. The thinning process for these films becomes stepwise through various stratification stages as each micellar layer flows out. The approximate sizes of micelles has been be determined from these steps as a photocurrent detects the change in the light transmitted (240). There were correlations found between the number of film-thickness transitions and the increased chain length of simple surfactants such as $N$-alkyl sulfates. Stepwise thinning depends on effective micellar volume, polydispersity, film size, and film thickness. The driving force for this drainage is the gradient of the chemical potential of the micelles at the film's periphery.

Nikolov et al. (250) derived an expression for multilayering of the micelles, relating them to the interaction free energy. They also integrated disjoining pressures with respect to film thickness in their theoretical model. The reader is referred to the references for more detail. According to Nikolov et al., an increase in surface viscosity means a decrease in mobility and a longer drainage time. A low surface viscosity means that the Gibbs–Marangoni effect has more impact on drainage time and coalescence rate. They reported that the estimated drainage time for a mobile surface with no surfactant is small in comparison with an immobile surface having large surface rheological stresses. Thin-film drainage times of foams were shown to increase with the increased dilatational modulus and with increased surfactant carbon chain length (248, 251, 252). According to these authors, drainage is very important in preceding coalescence, and is affected by film viscosity, film thickness, surface diffusion, surfactant adsorption, both surface shear and dilational viscous properties, surface shear, and dilational elastic properties, as discussed earlier.

Thus, a demulsifier that would enhance drainage rates and film thinning may also counteract Marangoni flows through the demulsifier's competitive adsorption. Drainage time is long when the interfacial tension gradient is high, shear elasticity is high, and both bulk and surface diffusion of demulsifier cannot counteract the tension gradient. Demulsifiers can reduce the drainage time by inducing decreased interfacial viscosity and lower interfacial film elasticity while promoting high interfacial mobility.

When the interfacial shear elasticity is moderate, at moderate surface viscosity, the thinning velocity will be greater than the Reynolds velocity. An increased surface viscosity means decreased surface mobility and a longer drainage time. These are all factors to be considered in decisions toward positive steps of destabilization.

Although stable diluted bitumen films are more complex than soap films, they appear to follow a classical film-drainage pattern without demulsifier as shown in a study by Angle et al. (82) and reproduced in Fig 21, 22.

Figure 21 shows actual droplets of water surrounded by a film of bitumen. The fine particles forming the structural barrier component of the bitumen film can be seen around the bright water droplet. Other stabilizers preventing coalescence are in the film lamella. The drainage of an actual film of bitumen between two water droplets is depicted in

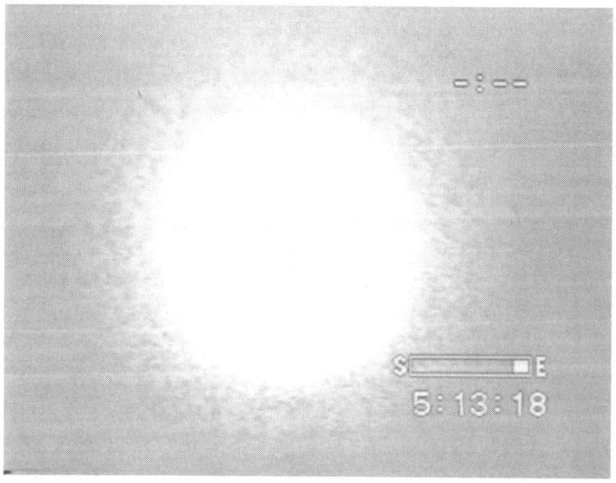


**Figure 21** Photomicrograph of stabilizing asphaltenes particles in a 300-$\mu$m diameter bitumen film structure between two bitumen-stabilized water droplets.

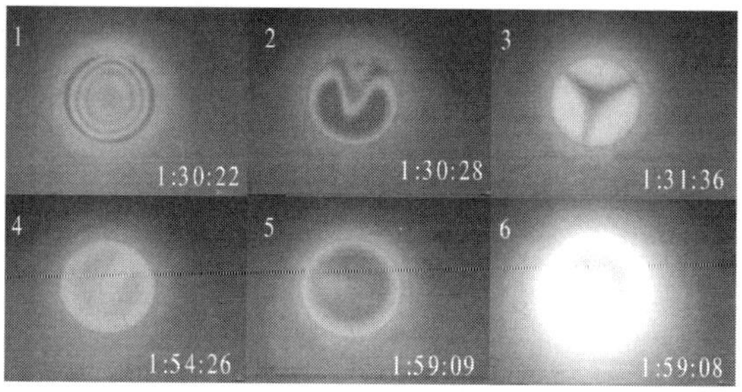


**Figure 22** Photomicrograph of film drainage and thinning of a 300-$\mu$m diameter bitumen film (lamella) over time, as measured by videomicrography, capillary balance techniques, and interferometry. (From Ref 82.)

Fig. 22 in frames 1–5. It was observed by Angle et al. that, generally, for a stable bitumen film, a typical drainage time to a common film was 25 min and to arrive at a Newton film was approximately 30 min in a 300-$\mu$m film diameter. Film thicknesses were measured by interferometry, using a capillary balance technique for plane parallel films (A.D. Nikolov and D.T. Wasan, personal communication, 1998) (253).

In Fig. 22, the multiple colors (frame 1 – dark gray) depict a thick film before drainage. The plane white/gray depicts a drained stable bitumen film closely resembling a common film (frame 4) or stable Newton black film (frame 5). A close-up photomicrograph shows the uniformly sized and distributed particles of asphaltenes and resins at the interface of the water droplet, but held in the confined space within the lamella (frame 6). These uniformly sized aggregates are responsible for providing a structural component of disjoining pressure in a coherent stable bitumen film according to Wasan's structural stabilization model. Angle et al. (82) showed that an oil-soluble demulsifier increased film-thinning rates up to a critical film radius and for an optimum concentration of demulsifier. Not all emulsion systems would follow this drainage process before coalescence. Although stepwise thinning may occur, the diluted bitumen and crude oil/water interfaces are more networked and require chemicals to achieve demulsification. Other effects of demulsifier blends on crude-oil film properties, rheology, drainage, and, lately, film thickness stability for Louisiana crudes are reported by Kim et al. (65).

The choice of demulsifiers is not an easy process as there are thousands of patents published on various formulations.

## V. PERFORMANCE DEMANDS ON DEMULSIFIERS

### A. Basic Behaviors Expected of Demulsifiers

From this review, it would appear that the basic demands on demulsifiers are the abilities to have one or more of the following behaviors: (1) strong attraction to the oil/water interface with the ability to destabilize the protective film around the droplet and/or to change the contact angle of the solids which may be part of the interfacial film; (2) ability to flocculate the droplets; (3) ability to promote coalescence by opening pathways for water's natural attraction to water; and (4) promotion of film drainage and thinning of the interdroplet lamella by inducing changes to the interfacial rheology such as decreased interfacial viscosity and increased compressibility.

It has been shown by Berger et al. (64) and confirmd by Kim et al. (63) that equal partitioning of demulsifier from the oil into the water phase appears to be important for an effective demulsifier. The change in Gibbs free energy for transfer of surfactant from oil to water is related to the relative equilibrium solubilities or partitioning coefficient of the demulsifier in either phase. It would suffice to infer that partitioning would occur only if the interface barrier's pores are opened by adsorption of surfactant. Transfer would also entail diffusion through the film based on a strong attraction to the water phase. However, partitioning would not be a dominant factor when the other effects such as dissolution of the interfacial material or their flocculation by the demulsifier (82) occur. The demulsifier's relative solubility in oil is important for mass transport to the interface, and where this is inadequate, carrier solvents have been used. Demulsification mechanisms include displacement, disruption by adsorption, solubilization, and competition with the emulsifier for interfacial sites. The work of desorption of the emulsifier from the interface would be important for the process.

Strong attraction to the oil/water interface is often dependent on diffusibility and interfacial activity of the demulsifier. This also involves speed of migration to the interface and the ability to compete or interact with the emulsifier by one or more mechanisms. The demulsifier must be relatively soluble in the continuous phase yet not completely soluble, and able to transport itself to the interface. In some cases, if the interface is stretched, the demulsifier must get there before the emulsifier can readsorb. The presence of demulsifier in the dispersed water phase has been shown to increase the rate of film drainage (254). According to Mukerjee and Kushnick (167, 255, 256) an effective demulsifier lowers the interfacial tension gradient and enhances the coalescence rates by rapidly diffusing to the interface. For fast diffusion, the molecular weight of the demulsifier becomes important. It may also dampen the growth of surface waves which alternatively stretch and compress the film. Thus, as the gradient in the film is created, the demulsifier may counter the inward flows. If it counters the inward flows it may enhance drainage, depending on the flexibility of the film. If it makes the surface more rigid, the liquid in a film does not drain as rapidly, but drains more slowly than would a film with a flexible surface. This leads to both reduced thinning and rates of coalescence.

When the continuous phase is not compatible with the demulsifier, carrier solvents are used. Most carrier solvents are alcohols or benzene derivatives such as glycols and xylene. Carrier-solvent effects were emphasized by Neustadter et al. (166) with xylene being a good carrier for the demulsifier. Wasan and coworkers (63, 164, 182) showed that alcohol not only acted as a cosurfactant but was also a good carrier solvent. The relative solubility of the demulsifier is thus related to the organic groups, the polar groups, the configuration of the demulsifier, and the molecular weight. Most successful demulsifiers are of intermediate molecular weight as is shown in the next section. Temperature and solvent effects on the demulsification of North Sea crude oil emulsions were investigated in some detail by Sjöblom and coworkers (81, 136). Of the large number of solvents investigated only three were considered to be effective destabilizers.

A demulsifier with the ability to destabilize the protective film around the droplet can create changes to the film as well as to the natural stabilizers. The demulsifier may influence the droplet interfacial film material by displacement (98, 115, 257), complexation, changing the solubility in the continuous phase, changing the viscosity of the interfacial film, or through quick diffusivity (65, 258) and adsorption, thus inhibiting the Gibbs–Marangoni effect, which counteracts film drainage.

An example of film dissolution and displacement is shown in Fig. 23, which shows the effects of a fast-acting demulsifier on a diluted bitumen W/O emulsion. The action is traced clockwise as the demulsifier solution flows in a capillary in which the stable bitumen emulsion is held within the confined space but not deformed. At time 5 min:45 sec the system is untreated. At time 6 min:00 sec, the movement of the demulsifier solution into the emulsion is initiated as a destabilization front. The demulsifier contact front causes the

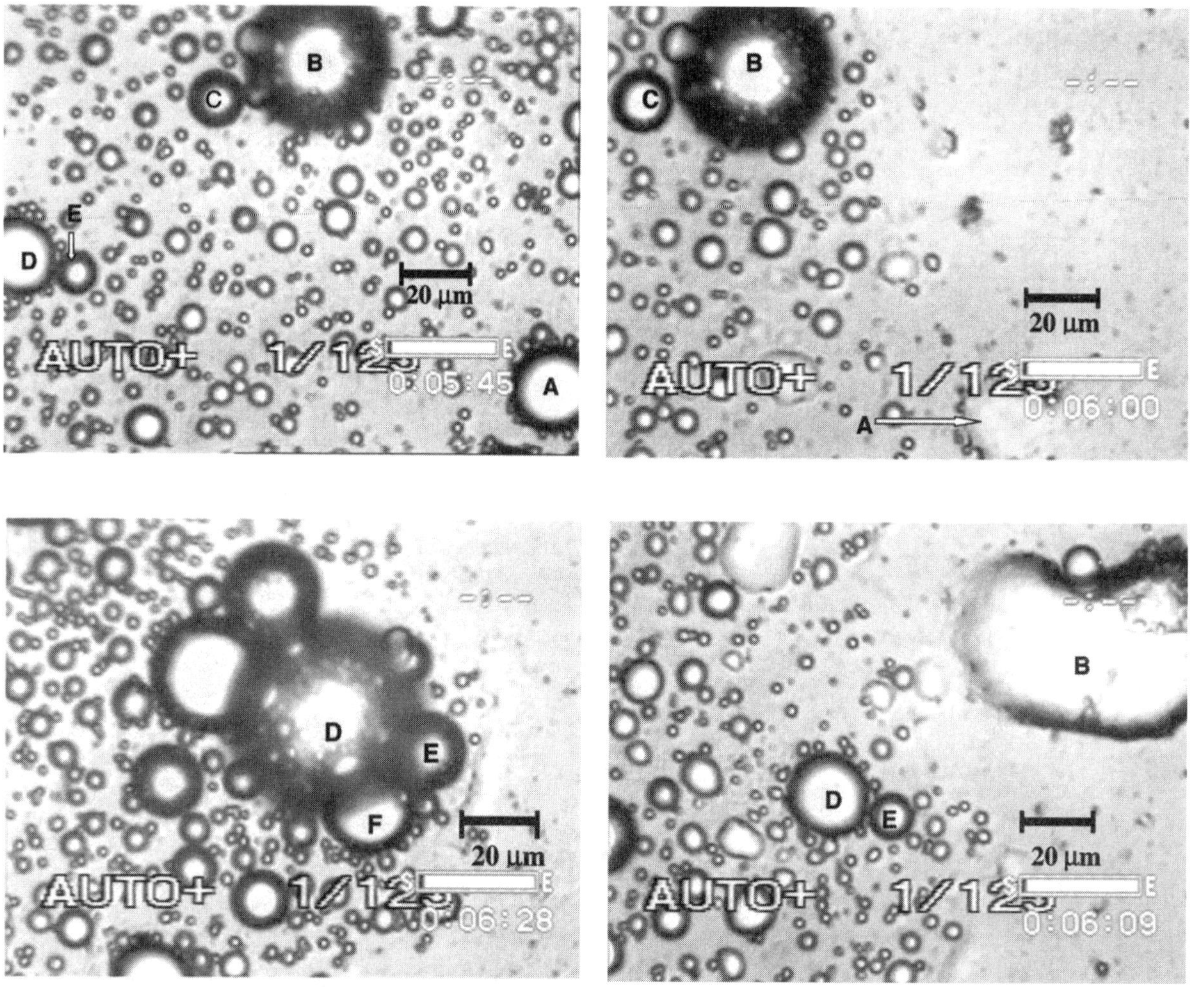


**Figure 23** Photomicrograph of a moving front of solvent plus a fast-acting demulsifier dissolving/disrupting the interfacial material on contact, causing instant demulsification. Measurements were in real time.

interface of the droplets to be broken by disruption or dissolution. There is also growth of the droplets with diffusion of the demulsifier solution into the sample, as seen in D (time 6 min:09 sec to time 6 min:28 sec). The large droplets are broken, leaving the smaller more rigid droplets in the solution. It appears that the interfacial material is being disrupted by the demulsifier. Note the relatively large droplet, B, with its attached colony of fine droplets.

## B. Demulsifier with Ability to Flocculate Droplets

Based on the interparticle distance or tightness of packing of droplets, the ability to flocculate may not be a necessary criterion for demulsifiers. When there is a high volume fraction of water droplets in the oil, as shown in Fig. 10, the emulsions are already in close contact and in a state of readiness for demulsification. In this case the added demulsifier must have mass-transport power to be integrated into the gaps between the droplets. High diffusibility helps with demulsification and heating may improve the transport. If the residual emulsions left in the oil are around 3% water and the droplets are very finely dispersed and widely distributed, the flocculating ability of the demulsifier is required to gather up the droplets. This happens more often in oily effluent treatment and in very dilute systems where droplet collisions are not frequent. Then, indeed, the high-molecular-weight and highly branched demulsifier molecules with an affinity for the water droplet provide some advantage. Here, gentle mixing will provide an additive effect.

### C. Demulsifier Creating Surface Changes to Solid Stabilizers

Lucassen–Reynder and Van den Tempel (259) emphasized that, in order to develop a strong interfacial film, solids as emulsifiers require a certain tightness of packing but this needs material and time to be built. The solids, which are W/O stabilizers (62, 260), have the correct wettability (261) or three-phase contact angle at the interface (174, 262–264). The role of solids as stabilizers is complex (174, 208, 265). The nature of the solids can be changed by a demulsifier. The demulsifier may adsorb on to the solids causing them to be more oil or water wettable. Often wetting agents of low molecular weights achieve this function. Thus, solids become more compatible with either the hydrophobic or hydrophilic phase, and are easily transported into the continuous phase away from the interface.

The nature of the solids may be variable, and may be one or more of the following species of minerals encountered in the petroleum production process: aluminum sulfates, calcium carbonates, iron sulfides, clays, drilling muds, crystallized paraffins, and asphaltenes. Most often after association with petroleum, inorganic solids adsorb the organics and form complexes with improved surface activity and thus influence the bulk rheology (266). It is thus desirable for a demulsifier to remove minerals to the water phase and the paraffins and organics to the oil phase (which can be treated easily by the refiners), while leaving a clean, sharp mirror-like interface of oil and water. Asphaltenes as partially solubilized solids are best recommended to be solubilized into demulsifier micelles in the oil phase. In fact, Little (267) has suggested that the demulsifiers must be such that they form micelles in the oil phase and do not themselves become stabilizers to the droplets by creating a more rigid interface stronger than the original natural emulsifiers.

### D. Demulsifier Inhibition of Film Forming Before Emulsification

In some situations demulsifiers have been used to inhibit emulsification (97). Demulsifiers used as inhibitors (1) in emulsification was considered in the prevention of "chocolate mousse" formation for oil spills (1, 268, 269). This topic has received considerable attention as increased environmental concerns demanded clean up of oil spills. Consequently, technology and chemical demulsifiers have been developed to address sea spills. However, discussion of this topic warrants a separate section, as a large part of the understanding of paraffinic W/O emulsions and demulsifier chemistry was stimulated by North Sea operations and oil spills at sea.

In downhole applications where mixing is very thorough and the temperature is high, reduced viscosity of the oil decreases the high lift pressure requirements. The use of demulsifiers as emulsion inhibitors has been indicated as an advantage. Sometimes a high concentration of fine micrometer-sized droplets would produce higher viscosity in some cases than the oil alone. The fine emulsions produced would consequently counteract the decrease in oil viscosity caused by temperature increases, and thus inhibition of emulsification with chemicals is advantageous.

### E. Demulsifier Working Together with Process Equipment

One or more of the above demulsification mechanisms may be required for fast or slow resolution, depending on the production process. Generally high-throughput processes demand more complete chemical treatment and fast resolution. In contrast, low-volume throughput such as vertical treaters, or systems where the residence time is longer as the material travels through pipes in a turbulent environment and mixing is continuous, demands a slower-acting demulsifier. If heat is encountered the demulsifier must be capable of performing at the higher temperature. Synergy is often desirable to reduce costs of chemicals or heat.

## VI. CHEMICAL NATURE OF DEMULSIFIERS

### A. The Users' Dilemma with Available Information

No single chemical by itself has been found to perform every destabilization and resolution function required in a process. This is the main reason why blends of chemicals are formulated. Each component in the blend addresses the change of a specific emulsion characteristic.

The most common problems encountered in any application of macromolecular surfactants are the matching of the commercially available products to a required end effect, according to Hancock (270). Manufacturers have assembled product combinations and provided end-use categories such as desalting chemicals, oil-slick dispersants, oil-well water-flooding viscosity improver, oil-well wettability improver, demulsifiers for W/O emulsions stabilized by heavy, light or high-acid-number oils, etc. Most chemical suppliers, based upon their experience, have put together application kits consisting of possible combinations of products that may work. Demulsifiers are labeled fast droppers, desalters, dehydrators, etc. A guide for selection is usually the relative solubility number (RSN), defined as the amount of water in milliliters required to reach cloud point at 25°C for 1 g of demulsifier dissolved in 30 mL of a solvent system containing 4% xylene in dioxane. The RSN is normally provided with the product specifications. It is rare that all basic systematic knowledge of the manufacture, composition, and performance trends are disclosed to the user.

Faced with this applications knowledge gap, the practitioner would first perform conventional bottle tests (271) to decide which are the most effective "chemical" products for the application in mind, working with samples that represent field samples as closely as possible. The screening is usually done on the spot and its success is aimed at convincing the engineers in the production application. Past experience then becomes an asset to the chemical-service representative. Although bottle tests are used for estimating the ranges of treating temperature, retention times, and settling times, it must be borne in mind that it is only a guide. It is estimated that 6–8 h of separation time in a bottle test is equivalent to 24 h in a process (272).

However, to date, it is generally agreed, by both researchers and practitioners that bottle testing is still a good guide (257). The bottle test is static and does not model closely the dynamic effects of water droplets dispersed or coalescing in the actual equipment such as control valves, pipes, inlet delivery, baffles, water wash, etc. If the point of injection of chemicals is upstream of the settler, then the test approximates the situation better. It is, however, still crucial that the characteristics of the emulsion be understood before the treatment system is selected (273, 274).

A typical laboratory bottle test is indicated in Fig 24 and 25, which illustrate the separations of a model 30% W/O bitumen (Athabasca, Alberta) emulsion at 50°C with two demulsifiers, A and B. In Fig. 24 (top photograph) the emulsion was allowed to resolve over 24 h by gravity. This is compared with the bottom photograph which shows the same treated emulsion assisted in separation by centrifugation immediately after demulsifier addition. The difference illustrates that enhancement of separation with a process aid such as centrifugation is useful for this system if time is a factor. The demulsifiers are considered to

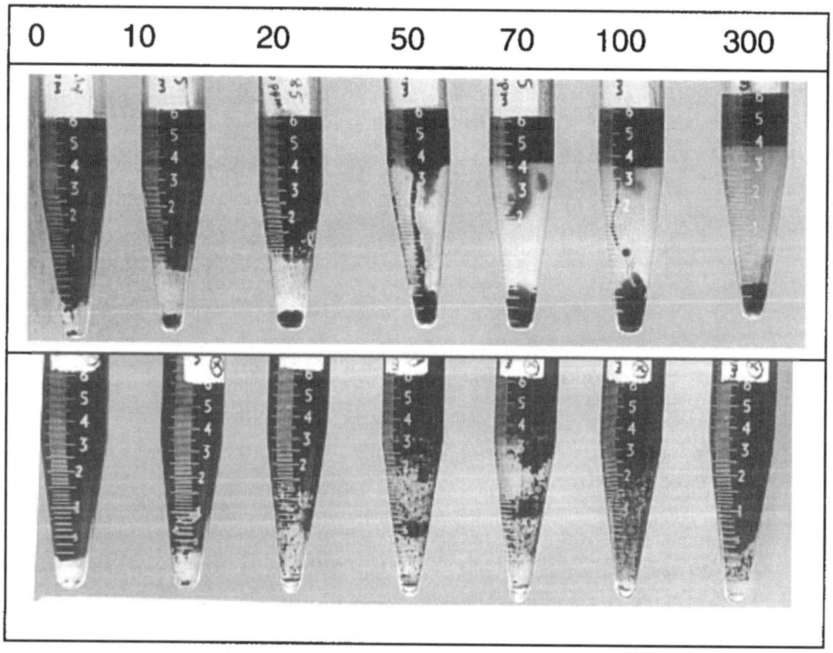


**Figure 24** Separation of bitumen W/O emulsion in tubes after demulsifier A was added: (top) by centrifugation; (bottom) by gravity separation. Dosages are indicated at the top.

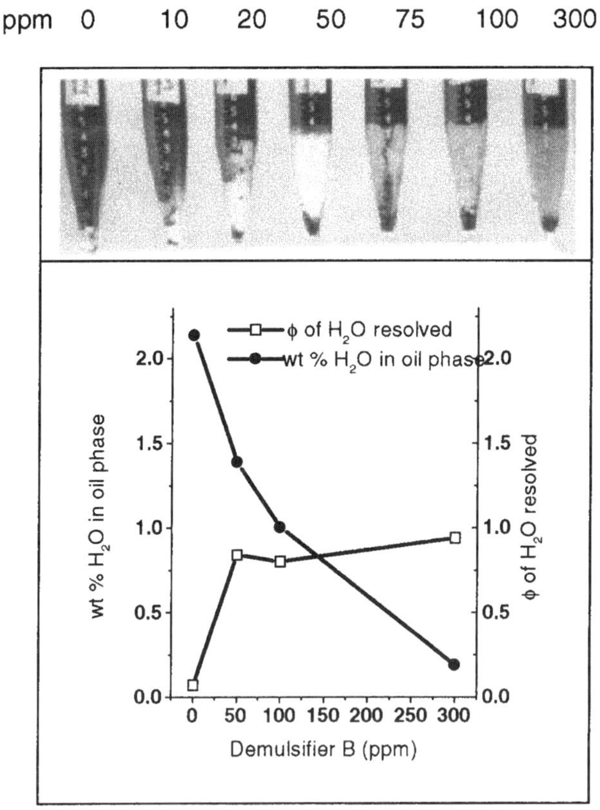


**Figure 25** Resolution of water and decrease in oil-phase moisture traced by changes in each phase of a bitumen W/O emulsion after demulsifier B was added at 50°C; middle phase is not visible. Top–photograph of its separation, showing clarity and optimum dosage.

be effectively fast droppers. The dosages are indicated at the top. A dosage dependence is shown as separation has an optimum dosage at 70 ppm in Fig. 24. Figure 25 illustrates an improvement in emulsion resolution at 50 ppm, showing water clarity. Overdosing is observed between 70 and 300 ppm, indicated by a turbid water phase, which suggests inversion to an O/W emulsion. The resolution is represented graphically by a plot of volume fraction of water resolved, on the right *Y* axis and the moisture reduction of the oil phase on the left *Y* axis against dosage showing effectiveness. The interface pad or middle phase is not represented (275) as it was not apparent.

Researchers first used the bottle tests preliminarily to select effective potential chemical demulsifiers, then would carry out further investigative studies on mechanisms for understanding demulsification phenomena for the crude at hand. Some studies are not concerned with the chemical identity of demulsifiers, but only with the effects or the demonstration of phenomena for developing the theory. It is important to demonstrate demulsifier relative effectiveness for these studies.

## B. Choosing Demulsifiers

Most of the research laboratories (IIT, University of Bergen/NTU, Energy Technology of CANMET, Environment Canada, British Petroleum, and Indian, Egyptian, Petroleum Institutes, among others) characterize and classify both the emulsions and the demulsifiers to determine some common factors that may be used in the extraction process. The test results would act as a prescreening step to the refiners when evaluating the crudes, or as a means of preparing to deal with specific demulsification problems.

Recently, there have been many studies to explore the selection of demulsifiers by more scientific and/or empirical means for matching the emulsion or oil with the demulsifier properties. Some advances as to what is useful and what is not have been made. Demulsifiers are required to have intermediate solubility in the crude oil or bitumen and not to form strong associations with other components of the crude. Sometimes the matching of RSN and equivalent alkane carbon numbers (EACNs) with BS&W are used as selection tools.

The determination of EACN is based on the minimum interfacial tension derived from a test surfactant in a series of hydrocarbon solvents, and then in the crude oil. The alkane carbon numbers are assigned to the solvent. In its determination for the crude oil/component, the minimum interfacial tension (IFT) against the reference surfactant may match one of the reference hydrocarbon solvent. The n-alkane with this minimum would be the EACN for the crude. The demulsifiers are tested in a similar way to the surfactants and are assigned preferred alkane carbon numbers (PACN) in the crude. This appeared to be a way of predicting their behavior in crude oils from their behavior in the n-alkanes. However, this process is long and labor intensive.

Cash et al. (276) and Cayais et al. (277) used EACNs of crudes and matched the numbers with PACNs of demulsifiers.

De Silva et al. (278) correlated the RSN with performance information and crude-oil properties to arrive at a predictive tool for choosing demulsifiers. They attempted to derive an alternative to the bottle test by considering the oil, its acidity, associated water, salinity, and demulsifier RSN and EACN relating the demulsifier affinity for oil and water. An unstable system was defined as having equal affinities for oil and water. This surfactant affinity difference was empirically related to solubility, EACN, and RSN. Their conclusion was that a less polar demulsifier of RSN = 8 would be suitable for a more polar crude oil. A nonpolar paraffinic oil required a polar demulsifier of RSN = 12. However, most often these empirical correlations become specific to the laboratory samples, and cannot be universally applied unless the models are tested globally.

Berger et al. (64) used 2400 field samples in bottle tests to correlate RSN with BS&W (water drop). From their studies they claimed no correlations between hydrophilic/lyophilic balance (HLB), RSN, and demulsifier performance for crude oil emulsions. They suggest that in the choice of demulsifiers the demulsifiers can be first characterized by their PACN and paired with the EACN of crude. In bitumen or crude oil systems the matching of EACN of the crude to the actual carbon numbers of the demulsifiers may be important for compatibility. To date, extensive studies have not been conducted to validate this.

In the earlier studies, a method of demulsification which was believed to be reliable was the reversal or inversion of the emulsion types. High concentrations of hydrophilic soaps accomplished this. The anionics were among the first to be used for this purpose. The nonionics were an improvement over the years. Later, a guide to the selection of demulsifiers was also sought in the HLB originally established for emulsifiers.

HLB is a number assigned originally to nonionic surfactants based on the hydrophilic head groups and hydrocarbon tails on an empirical scale, developed as a method for selecting simple emulsifiers for O/W or W/O systems. This was based on the surfactant dispersibility in water at ambient temperature. For demulsification, the HLB scale suggested that demulsifiers should possess HLBs in the intermediate range (8–11), which is neither oil soluble ($O < HLB < 6$), nor water soluble ($HLB > 12$) (68). Emulsifiers for O/W systems have high HLB numbers, wetting agents have intermediate HLB numbers, and emulsifiers of W/O emulsions have low HLB numbers. The HLB scale was originated by Griffin (279) and was later modified to include linear ethoxylated polymeric surfactants (280). Both solubility and HLB changed with changes in the hydrophilic portion in a series of ethoxylated and propoxylated compounds such as nonyl phenol formaldehyde resins. However, predicting an emulsifier type from molecular structure alone was not possible.

Shinoda's HLB temperature or phase-inversion temperature (PIT) became an important property of the surfactant–oil–water system (281). For nonionics, the PIT occurs below the temperature at which the surfactant preferentially partitions into the water phases as oil-soluble micelles and above the temperature at which it partitions preferentially into the oil phase as water-swollen inverted micelles. The PITs are affected by salinity, alcohol, temperature, and type of oil. The PIT indicated the point of emulsion inversion from oil/water to water/oil type. This was especially true for nonionic ethoxylated hydrocarbon surfactants, which obeyed Bancroft's rule for water solubility at low temperature and oil solubility at high temperature. This behaviour was illustrated in a linear relationship between HLB and PIT for ethoxylated hydrocarbon surfactants in cyclohexane and water. A lower HLB of 10 corresponded to a low PIT of 30°C (282) and a high HLB corresponded to higher PITs. Thus, an HLB balance in the molecules may be important where solubilities are concerned, but this property still has conflicting connotations in demulsification, especially for polymeric surfactants where branched configuration and molecular weight of the chemicals are more important in the changing interfacial environment. Crude oil and bitumen complexity would present new difficulties for using this parameter as the sole criterion for selection of demulsifiers.

Hayes et al. (283) related the $EACN_{min}$ with the HLB of ethoxylated alcohols (dinonyl phenols, tridecanols), in which HLB is the ethylene oxide percentage divided by 5. They found a simple linear relationship between HLB and $EACN_{min}$. The EACN values of 5–20 were in the range corresponding to HLBs of 11–12, for which the emulsion inverts or there is a transition region between oil and water solubility. However, HLB is still not simple to apply to demulsification especially with highly branched hydrophobes in complex systems such as crudes and anionic surfactants. Walker et al. (284) emphasized that HLBs of 7–9 are effective for demulsifiers of crude oil emulsions, as they represent emulsifiers of neither emulsion type, but are effective wetting agents. However, Cooper et al. (285) suggest that HLB is important in demulsifier efficiency, showing that heavy oil emulsions have optimum demulsification at HLBs 4–6. On the other hand, Berger et al. (64) found no correlations.

Aveyard et al. systematically studied phase behavior of the specific nonionics with varied HLBs to illustrate demulsification (202). They emphasized that changing the HLB of the system and not the HLB number of the surfactant can promote the demulsification of crude or model emulsions. The HLB of the system of Forties crude oil emulsion with ethoxylated phenol–formaldehyde resins was changed by increasing salt concentrations in the water phase. Inversion occurred close to the critical aggregation concentration. Anionic AOT in water/nonane emulsion showed a similar response, but at lower salt concentrations in the water. They indicated that the preferred curvature of the surfactant at the interface is modified by the HLB of the system.

Israelachvili (286) presented an overview on the phase inversion of emulsions based on the geometry of adsorbed surfactants, the sizes of the heads and tails, and the volume of the surfactants at the interface. W/O emulsions were favored by smaller less hydrated headgroups, higher ionic strength, smaller interfacial space occupied by heads, higher pH for cationics, lower pH for anionics, and higher temperatures for nonionics. W/O emulsions are favored also by a shorter length, $L$, of tails with surfactants occupying a larger volume, $v$, into the interface. The tail group would be either branched unsaturated chains or double chains. Nonionics at higher temperatures have greater oil penetration, and often are aided by cosurfactant addition. On the other hand, O/W emulsions were favored by large, more hydrated headgroups, lower ionic strength, larger head diameter/chain length ratio, lower pH for cationics, higher pH for anionics, and lower temperatures for nonionics, the single saturated chains, shorter chains, less oil penetration, and higher-molecular-weight oils. Intermediate to this measure is the bicontinous lamella phase where the surfactant's volume/area × length = 1 ($V/A \times L$), the occupied space is square in geometry at the interface and it corresponds to an HLB of 10.

Sharma et al. (71) used a concept of deformation of closely clustered W/O crude oil emulsions to explain demulsification by inversion. This explanation bore some similarity to the wedge concept used for changes in head and tail geometry of surface-active agents that adsorb on the interface of the droplet while promoting demulsification.

Marshall (287) investigated the emulsion inversion point (EIP) as a function of HLB for simple nonylphenol ethoxylates ($NPE_x$) where $x$ ranged from 3 to 25, in a paraffin-oil base W/O emulsion. The state of orientation of the $NPE_x$ was modeled to show the change in configuration of the polyethylene oxide chains, from zigzag or fully extended to meandering in the water phase with changes in $x$ between 9 and 12. The EIP occurred with eight ethylene oxide (EO) units at HLB = 12.3. The HLBs in these systems were also related to the curvature at the interface. A W/O interface was convex toward oil and concave toward water. This curvature changed with the stabilizer. This suggested that a change in the stabilizer property can induce demulsification.

Some of these studies indicate that HLB is not the only property of the chemical which determines the demulsifier power. Cooper et al. (285) indicated that water reduction was dependent on the chemical structure of the surfactant when two surfactants with similar HLBs gave opposite results. The effects of the interaction of the chemical structure with emulsion interfaces are the more important factors in demulsification, as these influence the film rheology of the system.

## C. Types of Chemicals

Chemicals used as demulsifiers may be simple surfactants. These may be cationic such as quaternary amines $(NR_1R_2R_3R_4)^+$, where R can be any alkyl or aryl group; anionic such as sodium dodecybenzenesulfonates (R-$PhSO_3Na$), petroleum sulfonates $(RSO_3^-M^+)$ and sodium di-iso-octylsulfonosuccinates [$ROOCC(CH_2COOR)H$ $SO_3$-$Na^+$, trade name Aerosol OT]; nonionic such as fatty alcohol ethers [$CH_3(CH_2)_{10}CH_2O(C_2H_4O)_nH$], fatty esters [$(CH_3(CH_2)_{10}COO$ $(C_2H_4O)_nH$], alkyl phenol ethers [R-Ph-O-$(C_2H_4O)_nH$], polyoxypropylene glycol ethers, and fatty amides; and zwitterionic such as alkylbetaine derivatives [$RCH_2COO^-N^+(CH_3)_2$], which are pH dependent. Simple copolymers of EO and propylene oxide (PO) may be used alone or in combination with a surfactant.

Staiss et al. (288) have summarized the developments in chemical demulsifiers and their effective dosages used until 1991. They indicated then that the most recent developments in poly(ester amines) at very low dosages were most efficacious for crude oils (Table 4). Table 5 extends this development to encompass some of the demulsifiers used by research groups globally to date. Table 5 summarizes some of the chosen chemistries or products used in published studies on demulsification of a variety of crude W/O emulsions world wide. Aerosol OT is still used today in formulations and is one of the few demulsifiers approved by the Norwegian environmental authorities (284). Aerosol OT still appears to be successful in demulsifying conventional crude emulsions. However, it easily partitions into the water phase and cannot be available for a long time.

**Table 4** Chemicals Used as Demulsifiers of Crude Oil Emulsions

| Up to year | Dosage (ppm) | Chemicals as demulsifiers |
|---|---|---|
| 1920 | 1000 | Soaps, salts of naphthenic acids, aromatic and alkylaromatic sulfonates Turkish red oil, and sulfated castor |
| 1930s | 1000 | Petroleum sulfonates, "mahogany soaps," oxidized castor oil, and sulfosuccinic acids (Aerosol OT) |
| 1935 | 500–100 | Ethoxylates of fatty acids, fatty alcohols, and alkylphenols |
| 1950 | 100 | EO/PO copolymers, *p*-alkylphenol – formaldyhyde resins + EO/PO and modifications |
| 1965 | 30–50 | Amine alkoxylates |
| 1986 | 5–20 | Polyesteramines and blends |

*Source*: Ref. 288.

However, there are many more combinations of chemicals synthesized to reduce the effective dosages (77, 89). The patent literature covers thousands of chemicals. A few examples for bitumen and heavy oils are indicated in Refs 83–88. Other simpler compounds have also been used and some of these appear in the footnote.*

Most of the bases or intermediates used in the synthesis of demulsifiers are derived from either agricultural or petroleum sources. Table 6 shows examples of the bases as published by Hancock's schematic (270). All large-scale production of EOs and POs are synthesized from naphtha (a crude-oil distillate feedstock) or natural gas. Benzene is obtained from naphtha and, from benzene, phenol is obtained.

* Chemicals: Selection is of extreme importance in tailoring to the crudes as well as operating conditions. Examples are: polyoxyalkylene (ethylene, propylene) substituted phenols, alcohols, esters, ketones, aldehydes, amines, nitrocompounds, organometallic salts, solvents.
Physicochemical: combinations of chemical and physical, e.g., demulsifier, mixing and heating to high *P* or Temp. increasing dispersed phase by 10–20% by adding very dilute demulsifier.

**Table 5** Chemicals Used as Demulsifiers for W/O Emulsions in Crude Oils from Literature Survey to Date

| Year | Dosage (ppm) | Molecular weight | Successful demulsifier chemicals published |
|---|---|---|---|
| 1948 | Varied | Very high | Polyoxethylene |
| 1974 to 1980 | | Low | Aerosol OT, alkoxylated phenol + alkoxylated ester + soap of carboxylic acids<br>Petroleum sulfonate and n-hexanol, pentadecyl benzenesulfonate |
| 1981–1990 | Varied | Varied | Nonylphenol ethoxylate (NPE), Octylphenol ethoxylates (OPE), and formaldehyde resins; polyethers, polyols, blends; Oxyalkylated phenol resins where alkyl = amyl, butyl, nonyl oxyalkylated polyols with EO varied<br>Aerosol OT and Triton X – R EO (9–30); straight chain and branched chain; butanols, hexyl to decylamines, NPE (4–20) |
| 1991–1999 | 25 | 3500–4500 | Varied molecular weights of EO/PO; phenol – formaldehyde glycidyl ether<br>Bisphenol A glycidyl ether with EO/PO |
| | 20–40 | | Iminobis (polyoxyalkylene) polyester |
| | 32 | 2500 | EO/PO copolymer hexanetriol ether blend with poly(thioalkylene oxide) of polyethers |
| | 100–1000 | 25,000, 4000 | Alkoxylated polyol |
| 1991–1999 | 20–300 | 7500,3700 | EO/PO block copolymers; straight chains and cross-linked, branched |
| | | 8400 | complexes of EO/PO with polyfunctional amine |
| | | 13,500 | Sorbitan – monolaurate, mono-oleate |
| | | 400,000 | Polyoxyphenol (NP, DP 1–9), polyalkylenepolyamine formaldehyde ethoxylate (11–25)<br>Trade names, such as Dissolvan 4455<br>Corexit 95000, Breaxit 126,<br>Novolak resols, polarchem, tetronics |

**Table 6** Intermediates for Demulsifiers and Their Feedstocks

| Agriculture sources of intermediates | Petroleum feedstocks source for intermediates |
|---|---|
| Alcohol ethoxylate (AE) | Alcohol ethoxylates |
| $R[CH_2CH_2O]_nOH$ | $R[CH_2CH_2O]_nOH$ |
| Fatty acid ethoxylates | Nonyl phenol ethoxylate (NPE) |
| $R'COO[CH_2CH_2O]_nH$ | Dodecyl phenol ethoxylate (DPE) $R'\phi$ $[CH_2CH_2O]_nOH$ where |
| $R''COO[CH_2CH_2O]_nOCR''$ | $R'$ is $C_9H_{19}$ or $C_{12}$ $H_{25}$ or linear or branched |
| Sorbitol ester | EO/PO copolymers (block, random), e.g., |
| $R''COOCH_2[CHCOH]_4CH_2OH$ | EO/PO/EO |
| | $OH[CH_2CH_2O]_p[CH_3CHCH_2O]_q[H_2CH_2O]_p$ |
| Fatty alkanolamides | Amine ethoxylate (from fatty acids, ethylene, alcohol) |
| $R''CON[[CH_2CH_2O]_nH]_2$ | $R''$ $N[[CH_2CH_2O]_nH]_2$ |

$R''$ = linear alkyl, alkenyl, C12–C18 even carbon only; $n = 1$–40, $p = 2$–100, $q = 15$–50.
*Source*: Ref. 270.

Any product that includes EO and PO as copolymers are generally highly surface active. Their oil solubilities are determined by, not only the molecular weight, but also the EO (hydrophilic) content as well as the PO (hydrophobic) components. Generally, experience has shown that for the solution behavior of nonyphenol ethoxylates, the lower the EO content the lower the cmc in an aqueous phase, and the lower the surface tension (290).

Recently, demulsifier chemicals supplied have been polymerized surfactants containing EO and PO as linear blocks or random copolymer chains ($EO_x/PO_y/EO_z$; $PO_y/EO_x/PO_y$), added to various polyglycols (289) whose molecular weight may be varied. Starting with poly(propylene glycols) of selected molecular weight, the EO is polymerized sequentially or randomly. Compounds with more EO groups are more water soluble, and with more PO groups are more oil soluble. The lower molecular weights are more water soluble than the high molecular weights, which are more oil soluble. The solution phase behavior of these compounds is affected by salt content, increased temperatures, and solvent type at these temperatures. For example poly(ethythylene oxide) (PEO) is more readily soluble in toluene at high temperatures than at room temperatures (291). Lower consolute temperatures depend on the molecular weights and polymer concentrations.

For demulsifiers of crude W/O emulsions, a low EO content is preferred, at low molecular weights of 1500–3500. The random copolymers are usually of lower molecular weights. However, factors such as the concentration, solvent, and temperature affect the phase behaviors of these demulsifiers. Because of the wide applications and versatility of these compounds there have been extensive studies on their chemistries (291). Their properties are as diverse as the many core molecules with which they can be copolymerized.

Table 6 illustrates some of the simple intermediates for demulsifiers derived from either agriculture or petroleum feedstocks. In some respects all of the above may be used as demulsifiers. In addition to these are the final reaction products of block copolymers of EO/PO polymerized with initiators such as glycerol, phenol–formaldehyde resins, melamine–formaldehyde resins, polyamines, siloxanes, and polyols. The flexibilities of the backbone structures are designed for function as emulsifying agents, and the molecular weights range from 3000 to 100,000, and in some cases higher. Demulsifiers for O/W emulsions will have typically 80% PEO, while for W/O emulsions 20–50% PEO. The limited W/O solubility is the driving force for attraction to the interface. These are commonly formulated into multicomponent solutions in aromatic solvents and used at levels from 5 to 100 ppm. These, however, are inferior as wetting agents or for detergency. The simpler linear surfactants are more suitable.

To date there are thousands of products appearing in directories and patent literature. Generally, the compounds with EO/PO copolymers are exceptionally surface active and they migrate and spread readily at the interface. The fatty amines and quaternary cationics adhere to all surfaces including asphaltenes, resins, naphthenic acids, paraffin waxes, inorganic clays, carbons, and silica (288).

As the chemistries improved and became more complex over the years, there was also a decrease in active concentrations of demulsifiers required to demulsify crude W/O emulsions. The efficacious demulsifiers such as the poly(ester amines) or combinations of bases with EO and poly(propylene oxide) copolymers (289) fit into this category as is shown in Table 4. Some published cases of successful outcomes for the various demulsifiers follow in Table 5.

Taylor (68) reported success in the demulsification of Kuwait crude oil emulsion with ethoxylated nonylphenol–formaldehyde (NPE) resins in which the EO content varied from 0 to 20 mol. They found optimum demulsification with $n = 5$ mol of EO per phenol group and strong evidence of NPE interacting with the asphaltenes of the bulk. Here, water solubility was changed with change in the length of the hydrophobic portion of the NPE resins.

Aveyard et al. (257) studied the demulsification rates of North Sea Forties crude oil emulsions using a homologous series of octyl phenol polyethoxylates [Triton X series – $C_8H_{17}\phi(EO)_n$], where $n$ varied from 3, 5, 9–10, 16, to 30. Crude oil was solubilized in the water phase for $n = 12$ and greater. Maximum water resolution was obtained for compounds with $n = 9$–10 at 500 and 2000 ppm. The compounds with lower $n$ values partitioned in the oil phase as monomers at dosages below the critical aggregation concentration (cac). Maximum resolution rates were found to coincide with the cac of each demulsifier in the homologous series. Of course, in most homologous series of this type it is expected that not only temperature, but also alkane chain length, cosurfactant, and molecular structure would change the solubility.

The demulsification of the crude oil emulsions with Aerosol OT, an anionic surfactant, was also studied with change in HLB of the system via increasing the salt concentrations. Similarly, the cac coincided with maximum rate of resolution at different salt concentrations. At low salt concentrations AOT is water soluble, stabilizing O/W emulsions, which invert to W/O emulsions with increased salt concentration. At high salt concentrations (0.7–1 M) 10 molecules of AOT solubilized one molecule of water in crude oil. Salt has a drastic effect on the distribution of aggregated AOT, especially at concentrations greater than the cac.

Sharma et al. (71) chose polyoxyethylene alkylphenols, their sulfonates, and sodium sulfonates in various combinations as demulsifiers for W/O emulsions from the Assam fields of India. They found xylene to be a better solvent than water for the effective demulsifiers. Successful demulsifiers were those of the nonylphenol type with 30 molecules of EO per mole, and the octylphenols with 40 molecules of EO per mole, followed by treatment with polyvalent cations. All demulsifiers studied were of HLB between 13 and 18 in various combinations. The W/O emulsions were from crudes of relatively low asphaltene (0.27, 0.35, and 0.25%) and resin (0.1, 0.15, and 0.1%) and high wax (11, 5, and 15%) contents. The emulsions were between 15 and 25% water content and of high salt content. They introduced a new concept to select the demulsifier by using the ratio of the number of EO units in the surfactant over the number of carbon atoms in the surfactant, instead of the HLB for demulsifier selection. They found no interaction between the demulsifier and natural emulsifiers from the crude oil emulsions.

Amarvathi and Pandey (289) synthesized and tested several demulsifying agents of increasing chemical complexity starting from the alkoxylated alkyl phenol–formaldehyde resins. The other compounds were bis(glycidyl ether) EO/PO copolymers, EO/PO copolymers of amino compounds, and EO/PO copolymers of sulfur compounds. In using the EO/PO block copolymers for dewatering W/O emulsions of 34% water in Ramashkino naphtha, they found that the copolymers of MWt 3500–4500 were best. These polymeric surfactants were made with 60% EO condensed with poly(-propylene glycol) of MWt 1400–1800. If the EO content was above 60% the product showed decreased demulsifying power. For the copolymers of phenolic resin group, the EO/PO block copolymer of phenol–formaldehyde resin glycidyl ether was best at removing 100% water and salt from the crude oil emulsions at dosages as low as 25 ppm. A slightly modified bisphenol A – bis(glycidyl ether) EO/PO copolymer at 20–40 ppm was effective for 42–63% water removal at 30–45°C. The Saudi Arabian crude-oil emulsions containing 36% water and 9% salt was dewatered by 94% and desalted to 0.9% within 30 min by 32 ppm of imino-bis(polyalkylene)polyalkylene polyester. At an even lower dosage of 20 ppm, the EO/PO copolymers of hexanetriol ether and the poly(thioalkylene oxides) of polyethers were effective as demulsifiers. For the aromatic amine derivatives of EO/PO copolymers, the effectiveness of the diamines were not only dependent on the ratios of EO/PO but also on the isomer. The *para*-substituted diamine having 28 units of EO to 80 units of PO was best. The *ortho*- and *meta*-substituted diamines did not perform as well. The esters of PPO glycols and PE glycols were effective for not only the Indian crude oil emulsions but also West African and North Sea emulsions at 20 ppm. Thus, it appears that greater effectiveness can be achieved by improved design of demulsifiers.

Formulation can be specific to the crude oil emulsions such as was shown for Buchan crude by Mohammed et al. (67). They used nonionic surfactants from the Pluronic (PE) and Tetronic series, which differed basically in the degree of EO/PO copolymers added to straight chains or branches. In combination with wetting agents and octy- or nonyl-phenol formaldehyde condensates containing EO/PO copolymers, they formulated effective demulsifier blends. They indicated that there is some degree of functional specificity in the demulsifier molecules in the interaction with the interfacial films.

Cooper et al. (285) demulsified Alberta's Cold Lake heavy oil emulsions with PEs (block copolymers of EO/PO) of HLBs of 4–6 or 14. While Tweens (esters of sorbitan and fatty acids of various lengths) and Brij (polymers of EO with terminal fatty alcohols of various lengths) were poor as demulsifiers. Kim and Wasan (63) used demulsifier blends of EO/PO as copolymers. They used EO/PO diepoxides and EO/PO with phenolic resins of varying MWts from 3700 to 8000 in their partitioning studies of destabilized emulsion films. They also tested polyamine glycols, alkyl aryl sulfonates, phenolic resins, and polyamines. These showed different performances, with blends having the same partition coefficients between water and oil phases. Sjöblom and Coworkers (81, 136, 292–294) varied the EO content from 4, 10, 20, to 30 molecules per unit of nonylphenol ethoxylate; together with solvents, the demulsification of Norwegian crude oil emulsions was thereby optimized. The medium-chain alcohols and amines speeded up the separation. They concluded that amines showed a strong specific interaction with the emulsion film which rendered the film more hydrophilic.

In a series of studies, Zaki (106) treated Geisum crude oil emulsions which was additionally stabilized by the anionic surfactant dodecylbenzenesulfonic acid. They found that 60 ppm of EO/PO alkylated alkylphenol formaldehyde resins at 50°C was successful. Zaki et al. (78) then used EO/PO block copolymers of MWt 5000 and 7000 to destabilize model asphaltene-stabilized water-in-benzene emulsions. They found that the efficiency increased for the higher-molecular-weight (7000) polymer. Temperatures in the range 50°–70°C caused increased efficiency. They also synthesized more complex demulsifiers by making further changes to the basic alkylphenol–formaldehyde resins. Zaki and Al-Sabagh (77) recently published some successes with polyalkylphenols–polyalkylenepolyamines formaldehyde ethoxylate in which the molecular weights of the polyethylene glycol used in the synthesis were varied. The alkyl groups in the phenols were either nonyl or dodecyl, and amines were varied from triamine to pentamine. The HLBs consequently were varied between 10 and 15. They found HLBs of 12–13.5 to be optimum for demulsification. Increasing the temperature from 50°C to 70°C enhanced resolution. They demulsified the W/O emulsions of the two crude oils, MB and LB, which were of high (8.8%) and low asphaltene (1.4%) contents, respectively. The salt content of the water was varied. The higher-asphaltenes, MB crude oil emulsions demulsified with an efficiency greater than that of the low-asphaltenes crude oil, LB. The nonylphenols were the best in all cases. The demulsifiers with the shorter alkyl groups were better than those with the longer alkyl groups, and efficiency increased with increasing number of amino groups. Molecular weights were within a narrow range for all the synthesized demulsifiers tested. They also showed a variation of demulsifier performance with water quality and temperature.

It would appear that, from fundamental tenets of physical chemistry, the demulsifier structure and chemistry determine the degree of interfacial activity and the HLB. In addition, its molecular weight governs viscosity and diffusivity as well as its solubility. The overall structural configuration determines the elasticity and viscosity at an interface or in solution. However, carrier solvents play a major role in the demulsifier configuration as well as in compatibility with the continuous oil phase and at the oil/water interface. These factors all apply in the demulsification of crude oils and bitumen emulsions.

## D. Solvent Effects

Solvents and cosolvents as demulsifiers or as carriers influence demulsification. The solvent not only influences the natural emulsifier components as was shown earlier, but also affects the micellization of the surfactant. Good solvents are those in which the demulsifier can dissolve but remains surface active. Poor solvents do not allow demulsifier dissolution or transport to the interface. For emulsions from North Sea crudes, Graham and coworkers (295, 296) showed that xylene was the best carrier and did not cause aggregation, while 1-propanol and 2-propanol were poor solvents. Sjöblom et al. (294) conducted an extensive survey of solvent effects as carriers and demulsifiers. They found that *t*-butanols and hexylamine were good solvents. Walker et al. (284) used glycol ethers as good solvents for coupling liquids of different polarity to provide a stable demulsifier mixture. Aerosol OT functions better in alcohol/water or propylene glycol ethers. Sometimes when solvency is reduced there is precipitation of components of crudes.

According to Zaki et al. (78) not only does a good solvent assist in solubility of the demulsifiers, but it also assists in depressing the pour point of the crude to effect demulsification at low temperatures. Water; water-miscible hydroxy compounds such as n-butanols, isopropanol, and monoethylene glycols; and aromatic hydrocarbons such as benzene, toluene, and xylene are commonly used solvents. They found that the efficiency of EO/PO block copolymers as demulsifiers decreased as the number of methyl groups increased in the aromatic hydrocarbon solvents. The efficiencies were related to the solvation power. However, they concluded that, for breaking W/O emulsions, the solvents for EO/PO block copolymers should be preferentially compatible with the dispersed water phase rather than with the oil phase. Oil-soluble demulsifiers may be better at demulsification, depending on the mechanism, as was illustrated earlier. However, demulsifier action can be augmented by the pH of the water around 7.0, low salinity, and increased temperature, as these factors affect the interface surface properties as well as the solution behavior of the demulsifier. The solution properties of the surfactants play a large role in the efficiency of demulsification.

Since the early 1980s, it was believed that demulsifiers had to be high-molecular-weight polymers. These include polymerized alkoxylated polyglycols, polyglycol esters, polymerized oils, alkanolamine condensates such as oxyalkoxylated phenols, and polyamides. Today, the increasing trends are toward lower dosages and more highly surface-active products which are more often polymeric surfactants.

However, it is understood that the selection of a group of chemicals for a specific application takes a great deal of skill and knowledge to arrive at a solution. There are high costs and hence profits associated with a successful formulation, because of high-volume throughput. This market becomes competitive. There are thousands of patents on product formulations for these applications. These are the main reasons behind the lack of detail in formulation and nondisclosure to researchers.

For these reasons the basic understanding of crude oil and bitumen emulsions and demulsification has received considerable attention. Advances in the knowledge of the physicochemical–mechanical structure of the stable emulsions and their films are being made in our laboratory and elsewhere. The films' response to demulsifiers are studied in order to understand the detailed mechanisms of demulsification. This knowledge will provide a more scientific basis for formulating products. The knowledge of surfactant chemistry, the behavior in solutions, the behavior at interfaces, and interactions with the crude-oil solvent base involve a wide interplay of complex processes in designing formulations.

## E. Demulsifier Selection and Petroleum Recovery

Field experience with use of specific chemical groups of demulsifiers on conventional crudes is indicated in Table 7 (274). A demulsifier with a low-molecular-

**Table 7** Field Observations of Behaviors of Chemical Types of Demulsifiers

| Chemical type | Behaviors |
|---|---|
| Polyglycol esters | Quick brightening of emulsions<br>Slow water drop, sludging, problems associated with overtreating |
| Low-molecular-weight resin derivatives | Rapid water drop, fair to good overall demulsification, overtreatment a problem with high API° emulsions |
| High-molecular-weight resin derivatives | Strong wetting tendency, fair brightening and water drop ability, can be used in combination with other materials. |
| Sulfonates | Fair to good wetting, and water drop, brightens the oil only sometimes, little tendency to overtreatment, best with high gravity emulsions |
| Polymerized oils and esters | Poor for widespread applications and are always used with other materials |
| Alkanolamine condensates | Promotes water drop, some brightening, used with other materials |
| Oxyalkylated phenols | Blended to improve performance, poor demulsifier properties, some wetting, and brightening |
| Polyamine derivatives | Poor water drop, good brightening, good with blending |

*Source*: Ref. 273.

weight resin base which would perform on a 35 API° oil emulsion with a rapid water drop would be ineffective for heavy oil. Generally, it is easier to treat high API° crude oil emulsions than heavy oils and bitumen because of the many differences in physical/chemical properties. Thus, in the study of emulsions for demulsification purposes and for the selection and design of demulsifiers it is imperative to know the production operations and the chemical prehistory of the emulsions to be treated. Chemical demulsification is the most economical and commonly used method of dehydration of crudes.

In the first stages of crude oil production, after crude is extracted or drilled, the fluids are under high pressures and temperatures. It is essential to recognize that the first-stage treatment involves the removal of excess of free water and gas. If there are high salt concentrations in the water, a fresh-water wash is conducted. This wash is then followed by addition of demulsifiers and defoamers. The chemical demulsifiers assist in the dropout of the major amounts of water. Further dehydration is achieved by use of hydrotreaters or electrostatic coalescers. This latter cannot tolerate greater than 6% water or solids.

The final criteria for dry oils rest in the treatment process in the plant and the changes in fluid quality that chemical demulsification would require. These factors would influence the chemical selection. The selected treating vessels are also crucial in chemical choices. For instance, if a large settling vessel is used it is desirable to employ a longer-acting demulsifier for slower adsorption. Gun barrels, although they have a high throughput, have a longer time available for chemicals to perform and fast resolution is not required. The need for augmentation by electrocoalescers, which speed resolution, dehydrating filters, and a variety of other physical methods* are indicated by production demands. The mixing requirements and the operations temperature would all be included in the design and would determine the choice of demulsifier.

The mixing, which is crucial for adsorption, may occur in static mixers or during flow conditions in the piping. The major limitations to the equipment and methods used for demulsification would be the small platform space for offshore operations, and the short residence time. For heavy oil or bitumen, or drilling on land, it is easier to use large settling tanks, heater treaters, and dispose of free-water. Retention times may be longer, up to 40 min, and are taken into account in the choice of slower-adsorbing demulsifiers.

After treatment, it is also desirable to have no accumulations at the layer between the oil and the water which forms a middle phase or interface pad. A large middle phase (interface pad) would have to undergo treatment separately as it tends to collect solids, emulsifiers, asphaltenes, and other assorted surface-active impurities. On the one hand, an interface pad presents no problem as the resolved water is drawn out from the bottom of a setting tank, or the pad may act as a filter for the mass transfer of solids into the oil phase. This interface can be removed and treated separately, after oil dehydration. On the other hand, a large pad creates slop oil and presents its own separation and disposal problems. For this reason the user is cautioned against overdosing in treatment. The choice of chemicals used is thus crucial to avoid overdosing or incompatibilities.

These all suggest that a demulsification program must be streamlined to the process in the majority of cases. A very good general review of this streamlining is discussed by Svetgoff (191), in which the economics and retention times are emphasized along with equipment choices (297). Most processes use a combination of chemical addition, heat, electrical methods (electrotreaters, dual-polarity dehydrators), and settling (272). The final results thus depend on the choices of injection points, mixing, temperature, process equipment, demulsifiers, and the characteristics of the crude emulsions. In the majority of cases many combinations of methods are used for complete dehydration. Each type of equipment comes with performance limitations. Economics determines the path in the end.

* Physical methods of treatment include the following: heating; centrifuging (used in oil sand extraction): washing through a water column; filtering through porous media passing through high pressure jets, which involves shaving off the interfacial layer as found to be successful on emulsions from Indian crudes; and the application of a sudden pressure drop. The increase of dispersed phase volume by dilution with water only works when soluble inorganics are the stabilizers. Other sophisticated methods are: application of high magnetic fields; application of high electrical fields – the most widely used 6–36 kV-induced polarity on the dispersed phase causes the water droplets to form pearlized strings of oppositely charged ends; application of ultrasonic waves. It is not unusual to use combinations of heat and high pressure.

### F. Other Chemicals for Consideration

If the crude is to be transported either by pipeline or tanker, chemicals such as wax inhibitors for highly paraffinic crudes, corrosion inhibitors, and lubricants are added. The excess of fluids produced in extraction of crude oils are also transported for further treatment to remove solids, water, and impurities before disposal. Demulsification treatment can also be complicated by crudes that have already been subjected to many other chemicals.

In typical offshore production the fluids produced are gathered into a collection line or common manifold. All the lines that feed into this manifold have been subjected to chemicals such as corrosion inhibitors, scale inhibitors, wax inhibitors, and hydrate inhibitors. In enhanced oil recovery of conventional crudes, surfactant micellar floods have been used in the past to promote recovery. For heavy oil, fluid flow is enhanced by viscosity-reducing agents and or diluents. The reality is a more complex system for upgrading or refining.

## VII. RECOMMENDATIONS

Oil clean-up is a necessary requirement to meet environmental specifications before water is discharged into disposal streams or natural lakes or oceans. The producer considering a separation strategy must consider what will be done with what is left behind. It is generally agreed that the job of demulsification is not complete until one has found a place for each fraction of the emulsion system. After the oil is separated from the water, and is recovered, what happens to water-soluble oils, water, solids, and surfactants? The choice of demulsifers can address these problems. If the mineral solids are transported to the water, and the asphaltenes, waxes, and resins are transported into the oil phase, clean-up can be better dealt with. The "reverse" emulsion, which is the O/W emulsion, is mostly found in waste effluent streams and may be a result of overdosing. The demulsification of the latter involves different demulsifier chemistries that are compatible with the continuous water phase. These systems are often very dilute, and flocculants, together with coalescing filter beds or membranes, are chosen for clean-up.

The published literature on demulsification of crude oil emulsions is often based on studies of samples derived from the author's main research sponsors and so the sources of samples are often from the closest geographical location of a production facility. Thus, some collaborative effort between the producer and the researcher is required to ensure that samples are representative for study and actually simulate the real situations. These studies are often only undertaken when expertise is sought by the producer or is identified and offered by the performer after considerable discussions/negotiations.

It is not very often that service companies, which supply technology to the producers, seek external support for the fundamental research to be performed. The chemical service companies have their core competencies based on their knowledge of chemistry and some knowledge of oil production technology. Dessemination of their knowledge does not serve their interest. Therefore, most chemical products supplied by service companies are protected formulations which are coded. These are not identified, chemically, to the producers. Demulsifier formulations and demulsification even to this day still appear to be an art more than a science. However, there are some new developments in the understanding of demulsification. In this paper some of these developments have been highlighted.

### A. Research Needs in This Industry

Research on petroleum emulsion formation, stabilization, and destabilization, and especially interactions with chemical demulsifiers is very necessary for optimizing production. However, this is expensive in comparison with other industries' requirements for emulsion studies. The field of emulsion technology would not have arrived at the present knowledge base had it not been for the initiative of the petroleum industry in finding answers. However, to date there are only a handful of research facilities worldwide that are involved in this type of combined fundamental and applied investigation aimed at understanding crude oil emulsions. In addition to this, few research facilities involved in crude oil emulsions publish their findings. The oil producers keep process and emulsions research results in-house to ensure their competitive edge, and if specialized facilities are engaged, information becomes classified. There is a need for a consistent reporting of the properties of crude oils studied in addition to the chemical identification of surfactants in order to contribute more to the science, as well as compare notes across the broad scientific community.

Thus, the number of articles published on the topic does not reflect the degree of research effort in this

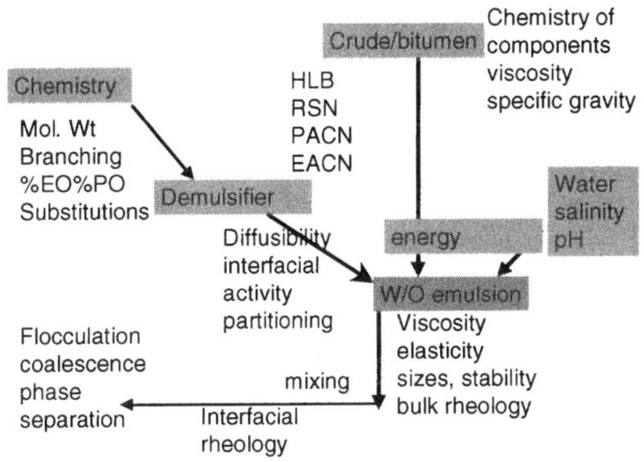


**Figure 26** Schematic summarizing factors that affect demulsification of crude oil and bitumen emulsions as discussed in this chapter

field. Crude oil prices often dictate the incentives for funding research.

## VIII. CONCLUSIONS

We have discussed the nature and origins of crude oils and bitumens and compared field and bench emulsions first for developing the theme of demulsification. The nature of the crude oil and bitumen emulsions, film architecture, and the developments in their understanding for destabilization have been presented. The influence of the crude oil components on the interfacial skin strength and stability/instability was discussed. Demulsification was discussed in terms of film drainage, compression, coalescence, and water resolution. The demulsification process, chemical choices, and chemistry of the demulsifiers involved in this process were also discussed. Figure 26 provides a summary flow diagram of the factors impacting demulsification, which have been touched upon in this review.

## ACKNOWLEDGMENTS

Support in funding and time was provided partially by CANMET, Natural Resources Canada, and the Federal Panel on Energy Research and Development – the energy sector of Natural Resources Canada.

## REFERENCES

1. AL Bridie, TH Wanders, W Zegveld, HB Van Der Heijde. Mar Pollut Bull 11: 343–348, 1980.
2. ASTM Designations D-96, 1983; DA-96, 1983.
3. S Mitchell, DM Smith. Aquametry, Part III. 2nd ed. New York: John Wiley, 1980.
4. JG Speight. The Chemistry and Technology of Petroleums. New York: Marcel Dekker, 1991, Ch 11, p 14.
5. CW Angle, N Berkowitz. Fuel 70: 891–896, 1991.
6. CW Angle. MSc thesis, University of Alberta, Edmonton, 1989.
7. L Blom, L Edelhausen, DW Van Krevelen. Fuel 36: 135–322, 1957.
8. DW Van Krevelen. Coal: Functional Group analysis, Typology Chemistry, Physics, Constitution. Amsterdam: Elsevier, 1961, pp 160–176.
9. RB Long, JG Speight. In: JG Speight, ed. Petroleum Chemistry and Refining. Washington, DC: Taylor & Francis, 1998, Ch 1.
10. JW Bunger, NC Li, eds. Chemistry of Asphaltenes. Advances in Chemistry Series 195. Washington, DC: American Chemical Society, 1981.
11. CW Bowman. Molecular and interfacial properties of Athabasca tar sands. Proceedings of Seventh World Petroleum Congress, Houston, TX. Vol 3, 1967, p 583.
12. JA Koots, JC Speight. Fuel 54: 179–184, 1975.
13. MM Sharma, GV Chilingarian, TF Yen. In: TF Yen, GV Chillingarian, eds. Asphaltenes and Asphalts. Developments in Petroleum Science Series 40A. New York: Elsevier, 1994, Ch 12.
14. JP Pfeiffer, RNJ Saal. J Phys Chem, 44, 139, 1940.
15. D MacKay. Formation and stability of water-in-crude oil emulsions. Report TU-EE93, MICROLOG-88-02079, 1987 (Abstr no. 157, 257, Vol 110, no. 18).
16. RN Tuttle. J Petrol Technol (June): 1192–1196, 1983.
17. A Stockwell, AS Taylor, DG Thompson. The rheological properties of water-in-crude oil emulsions. Presented at Fifth International Symposium: Surfactants in Solution, Bordeaux, France, 1984.
18. AJ McMahon. In: J Sjöblom, ed. Emulsions – A Fundamental and Practical Approach. NATO AST Series 363. Dordrecht: Kluwer Academic, 1992, pp 135–136.
19. KJ Leontaritis. Fuel Sci Technol Int 14: 13–39, 1996.
20. PA Wheeler. The significance of wax crystal formation in water-in-crude oil emulsion stability. Proceedings of Second Norwegian Institute of Technology Lerkendal Petroleum Engineering Workshop, Trondheim, Norway, 1992, pp 183–192.
21. JD McLean, PK Kilpatrick. J Colloid Interface Sci 189: 242–253, 1997.
22. S Puskas, J Balazs, T Haraszti, L Turi, I Dekany. The influence of paraffinic deposits and their fractions on the stability of crude oil emulsions. Petroleum abstracts: 1st SPE Brazil Sect et al., Colloid Chemistry in Oil Production, Ashphaltenes & Wax Deposition International Symposium, Proceedings 230–235, Abstract no. 627, 639, Vol 36, no. 28, 1996.

23. MB Manek. Interface control for heavy crude oils. Petroluem Abstracts: 1st SPE Brazil sect et al., Colloid Chemistry in Oil Production, Ashphaltenes & Wax Deposition International Symposium, Proceedings 136–141, Abstract no. 627, 638, Vol 36, no. 28, 1996.
24. HN Dunning, JW Moore, MO Denekas. Ind Eng Chem 45: 1759, 1952.
25. CG Dodd, JW Moore, MO Denekas. Ind Eng Chem 44: 2285, 1952.
26. V Hornoff, JK Liu. Dynamic interfacial tension effects of oil sands processing. Proceedings of Fourth World Congress, Chemical Engineering, Karlsruhe, 1991.
27. RL Piehl. Mater Perform 27: 1: 37, 1988.
28. HY Jennings. Soc Petrol Eng J (June) 1975.
29. RJ Parker, ESN Chung. J Can Pet Technol (July–August): 72, 1986.
30. EE Isaacs, KF Smolek. Can J Chem Eng 61: 233, 1983.
31. ST Dubey, PH Doe. Base numbers and wetting properties of crude oil. Proceedings of 66th Annual Technical Conference Exhibition, SPE, Dallas, TX 1991.
32. RL Marquez-Silva, S Key, J Marino, C Guzman, S Buitriago. Chemical dehydration: correlations between crude oil, associated water and demulsifier characteristics in real systems. Proceedings of SPE Oilfield Chemical International Symposium, Houston, TX, 1997, pp 601–607.
33. IMRA Bruning. Oil Gas J 9: 31: 38, 1991.
34. IMRA Bruning. Fuel 69: 646f, 1990.
35. Data from Environment Canada – Environmental Technology Centre – www.etcentre.org.
36. Syncrude Analytical Methods for Oil Sands and Bitumen Processing. Edmonton: Syncrude Research, 1979, p 123.
37. KA Ferworn. Thermodynamic and Kinetic Modelling of Asphaltene Precipitation from Heavy Oils and Bitumens. PhD thesis, University of Calgary, Alberta, 1995.
38. KA Ferworn. Asphaltene Flocculation and Precipitation. MSc thesis, University of Calgary, Alberta, 1992.
39. KA Ferworn, WY Svrcek, AK Mehrotra. Ind Eng Chem Res 32: 955, 1993.
40. KA Ferworn, AK Mehrotra, WY Svrcek. Can J Chem Eng 71: 699, 1993.
41. EW Funk, Can J Chem Eng 57: 333, 1979.
42. JG Speight. In: TF Yen, GV Chilingarian, eds. Asphaltenes and Asphalts. Developments in Petroleum Science Series 40A. New York: Elsevier, 1994, Ch 2.
43. RB Long. The Concept of Asphaltenes. Advances in Chemistry Series 195, Washington, DC: American Chemical Society, 1981, p 17.
44. TF Yen, GV Chilingarian. In: TF Yen, GV Chilingarian, eds. Asphaltenes and Asphalts. Developments in Petroleum Science Series 40A. New York: Elsevier, 1994, Ch 1.
45. TF Yen. In: TF Yen, GV Chilingarian, eds. Asphaltenes and Asphalts. Developments in Petroleum Science Series, 40A. New York: Elsevier, 1994, Ch 5.
46. NF Carnahan, L Quintero. In: RF Meyer, ed. Fueling for a Clear and Safe Environment, World Petroleum Congress, Houston, TX, 1995, pp 237–250.
47. FJ Nellensteyn. The Colloidal Structure of Bitumens. The Science of Petroleum. Vol 4. London: Oxford University Press, 1938, p 2760.
48. EY Sheu, DA Storm. In: EY Sheu, OC Mullins, eds. Asphaltenes – Fundamentals and Applications, New York: Plenum Press, 1995, Ch 1.
49. SE Moschopedis, JF Freyer, JG Speight. Fuel 55: 227, 1976.
50. JG Speight, SC Moschopedis. In: JW Bunger, NC Li, eds. Chemistry of Asphaltenes. Washington, DC: American Chemical Society, 1981.
51. CH Bardon, L Barrie, D Espinat, V Guille, Min Hui Li, J Lambard, JC Ravey, E Rosenberg, T Zemb. Fuel Sci Technol Int 14: 203–242, 1996.
52. S Acevedo, MA Ranaudo, G Escobar, LB Gutierrez, X Gutierrez. In: EY Sheu, OC Mullins, eds. Asphaltenes – Fundamentals and Applications. New York: Plenum Press, 1995, Ch 4.
53. R Cimino, C Correra, A Del Bianco. In: EY Sheu, OC Mullins, eds. Asphaltenes – Fundamentals and Applications. New York: Plenum Press, 1995, Ch 3.
54. DA Storm, EY Sheu. In: TF Yen, GV Chilingarian, eds. Colloidal Nature of Petroleum Asphaltenes, Asphaltenes and Asphalts 1. Developments in Petroleum Science Series 40A. New York: Elsevier, 1994, Ch 6.
55. CW Wiggins. J Phys Chem 69: 3500, 1965.
56. DA Storm, EY Sheu. Fuel 74: 1140, 1995.
57. SJ Park, GJ Mansoori. Organic deposition of heavy petroleum crudes. Proceedings Fourth International Symposium on Heavy Crudes and Tar Sands, Edmonton, Alberta, 1988, Preprint paper 225.
58. H Pan, A Firoozabadi. SPE production and facilities, AICHE J 46: 416–426, 2000.
59. TJ Jones, EL Neustadter, KP Whittingham, JCPT (April–June): 100–108, 1978.
60. CW Angle. CANMET Division Report WRC 93-21 (CF). Natural Resources Canada, 1993.
61. Fine Tails Fundamentals Consortium. Advances in Oil Sands Tailings Research, Vol 2, Ch 11. Alberta Department of Energy, Oil Sands and Research Division, 1995.
62. VB Menon, DT Wasan. Sep Sci Technol 19: 555–574, 1984.

63. YH Kim, DT Wasan. Ind Eng Chem Res 35: 4: 1141–1149, 1996.
64. PD Berger, C Hsu, JP Arendell. Designing and selecting demulsifiers for optimum field performance based on the production fluid characteristics. SPE publication of AIME. Proceedings of International Symposium on Oilfield Chemistry, San Antonio, TX, 1987, Paper SPE 16285, pp 457–464.
65. Y Kim, AD Nikolov, DT Wasan, H Diaz-Arauzo, CS Shetty. J Dispers Sci Technol 17: 33–53, 1996.
66. T Stroem-Kristiansen, PS Daling, AB Nordvik. Demulsification of Water-in-Oil (W/O) Emulsions resulting from Offshore Crude Oil Spills by use of Heat and Emulsion Breaker. Phase 1. Marine Spill Response Corporation Technical Report Series 93-026, 1993, pp. 51.
67. RA Mohammed, AI Bailey, PF Luckham, SE Taylor. Colloids Surfaces A: Physiochem Eng Aspects 83: 261–271, 1994.
68. SE Taylor. Chem Ind 20: 770–773, 1992.
69. BP Singh, BP Pandey. Indian J Technol 29: 443–447, 1991.
70. IC Sharma, I Haque, SN Srivastava. Indian J Technol 20: 175–178, 1982.
71. IC Sharma, I Haque, SN Srivastava. J Colloid Polym Sci 260: 616–622, 1982.
72. DD Eley, MJ Hey, MA Lee. Colloids Surfaces 24: 173–182, 1987.
73. S Acevedo, G Escobar, LB Gutierrez, H Rivas, X Gutierrez. Colloids Surfaces A: 71: 65, 1993.
74. A Bhardwaj, S Hartland. Ind Eng Chem Res 33: 1271–1279, 1994.
75. A Bhardwaj, S Hartland. J Dispers Sci Technol 14: 5: 541–557, 1993 (Abstr. no. 229604, Vol 119, no. 22).
76. DJ Miller, R Bohm. J Petrol Sci Eng 9: 1–8, 1993.
77. NN Zaki, AM Al-Sabagh. Tenside Surfactants Deterg 24: 12–14, 16–17, 1997.
78. NN Zaki, ME Abdel-Raouf, AAA Abdel-Azim. Monatsh Chem 127: 1239–1245, 1996.
79. J Sjöblom, L Mingyuan, H Hoiland, E Johnsen. Colloids Surfaces 46: 127–139, 1990.
80. E Johnsen, M Skjarvo, T Lund, J Sjöblom, H Soderlund, G Bostrom. Colloids Surfaces 34: 353–370, 1989.
81. J Sjöblom, H Soederlund, S Lindblad, EJ Johansen, IM Skaervoe. Colloid Polym Sci 268: 389–398, 1990.
82. CW Angle, AD Nikolov, DT Wasan, HA Hamza. Demulsification of water-in-bitumen emulsions: Role of emulsion film thickness stability. Twelth International Symposium on Surfactant in Solution Conference, Stockholm, 1998, Paper B05.
83. DR McCoy, EE McEntire, RM Gipson. Method for Demulsification of Bitumen Emulsions. US Patent 4: 457 371, 1984.
84. DR McCoy, RM Gipson, KB Young. Demulsification of Bitumen Emulsions with a High Molecular Weight Mixed Alkylene Oxide Polyol. Can Patent 1 160 172, 1984.
85. DR McCoy. Method for Demulsification of Bitumen Emulsions using Polyalkylene Polyamine Salts. US Patent 4 439 450, 1984.
86. RB Duke. Demulsification of a Crude Oil Middle Phase Emulsion. US Patent 4 439 450, 1984.
87. DR McCoy, CL Laberge. Demulsification of Bitumen Emulsions with a High Molecular Weight Polyol Containing Discrete Blocks of Ethylene and Propylene Oxide. Can Patent 1 153 717, 1983.
88. EE McEntire, DR McCoy. Demulsification of Bitumen Emulsions using Polymers of Diquaternary Ammonium Monomers Containing Hydroxyl Groups. US Patent 4 387 017, 1983.
89. P Becher. Theory of Emulsions: Creaming, Inversion, and Demulsification. 2nd ed. ACS no. 62, New York: Krieger, 1977, pp 1–149.
90. TF Tadros, B Vincent. In: P Becher, ed. Encyclopedia of Emulsion Technology. Vol 1. New York: Marcel Dekker, 1983, pp 1–102.
91. ESR Gopal. In: P Sherman, ed. Emulsion Science. New York: Academic Press, 1968, pp 1–72.
92. J Sjöblom, ed. Emulsions – A Fundamental and Practical Approach. NATO AST Series 363. Dordrecht: Kluwer, 1992.
93. SL Mason, K May, S Hartland. Colloids Surfaces A, 96: 85–92, 1995.
94. CM Blair. Chem Ind 20: 538–544, 1960.
95. HP Ronningsen, J Sjöblom, L Mingyuan. Colloids Surfaces A: Physiochem Eng Aspects 97: 119–128, 1995.
96. A Barnes. Colloids Surfaces A: Physicochem Eng Aspects 91: 89–95, 1994.
97. GP Canevari. Mar Pollut Bull 13: 49–54, 1982.
98. A Hayward-Walker, DL Ducey, JR Gould, AB Nordvik. Formation and Breaking of Water-in-Oil Emulsions: Workshop Proceedings. Marine Spill Response Corp. (Kananaskis AB 06/14–15/93), Tech. Rep. Series 93–018.
99. TF Tadros. In: J Sjöblom, ed. Emulsions – A Fundamental and Practical Approach, NATO AST Series 363. Dordrecht: Kluwer Academic, 1992, pp 173–188.
100. P Becher. Emulsions Theory and Practice. 2nd ed. ACS no. 62, New York: Krieger, 1977, pp 111–116.
101. KJ Lissant. Demulsification – Industrial Application. Surfactant Science Series 13, New York: Marcel Dekker, 1983.
102. TH Plegue, SF Fruman, JL Zakin. J Colloid Interface Sci 114: 88, 1983.
103. JR Brenan. Pipelining Oil Emulsion. Company Rep., IMO Industries Inc., NC, USA.

104. LA Burkholder, JM Rashan, MH Auerbach, RP Gill. Production and pipeline transportation of heavy crude oils by use of specially formulated emulsions. Third Zulia University et al. Enhanced Oil Recovery International Symposium, Maracaibo, Venezuela, Proceedings 2, 1989, pp 765, 767–777. (Abst. no. 469, 279, Vol 29, 1989).
105. L Berero, A DiLullo, A Lentini, L Terzi. A (Patented) innovative way to produce and transport heavy (and very viscous) oil through dispersion in water (using sodium naphthalene sulfonate condensed with formaldehyde): Laboratory study and field test results. Proceedings of SPE Annual Technical Conference/Production Operations and Engineering. Vol 2, New Orleans, 1994, pp 283–295.
106. NN Zaki. Colloids Surfaces Chem Abstr 125: 19–25, 1997.
107. VA Adewusi, AO Ogunsola. Chem Eng Res Design 71(A1): 62–68, 1993.
108. KS Pedersen, P Skovborg, HP Ronningsen. Energy Fuels 5: 924f, 1991.
109. WD Bancroft. J Phys Chem. 17: 501, 1913.
110. M Van der Waarden. Kolloid Z 156: 116, 1958.
111. JA Bichard. Oilsands Composition and Behaviour Research. AOSTRA Technical Publication Series 44, 1987, Ch 2, p 27.
112. JP Pfeiffer, RNJ Saal. Asphaltic bitumen as a colloidal system. Proceedings of 16th Colloid Symposium, Stanford University, CA, 1939, p139.
113. JE Strassner. Effect of pH on interfacial films and stability of crude oil–water emulsions. Proceedings of 42nd Annual Spec of AIME Fall Meeting, 1969, Preprint no. SPE-1969.
114. JE Strassner, J Petrol Technol 243: 303, 1968.
115. A Lewis, M Walker. Formation and Breaking of Water-in-Oil Emulsions Workshop (Kananaskis, Alberta 6/14–15/93). Marine Spill Response Corp. Tech. Rep. Series N 93-018, 1993, pp 223–238.
116. A Lewis, M Walker, K Colcomb-Heiliger. Formation and Breaking of Water-in-Oil Emulsions Workshop (Kananaskis, Alberta 6/14–15/93). Marine Spill Response Corp. Tech. Rep. Series N.93-018, 1993, pp 203–220.
117. P Becher. Emulsions: Theory and Practice. 2nd ed. ACS no. 62, New York: Krieger, 1977, Ch 5.
118. P Sherman, ed. Emulsion Science. New York: Academic Press, 1969, pp 131–216.
119. GD MacKay, SG Mason. Can J Chem Eng 41: 203, 1963.
120. HJ Neumann, B Paczynska-Lahme. Progr Colloid Polym Sci 101: 101–104, 1996.
121. FF Bartell, DO Niederhauser. Film forming constituents of petroleum oils. API Research Project 27. Fundamental Research on the Occurrence and Recovery of Petroleum, API 1964–47, 1949, p 45.
122. RJ Hunter. Foundations of Colloid Science. Oxford Science Publication Series, Vol II, 1989, Ch 15.
123. CHM Roberts. J Phys Chem 36: 3102, 1932.
124. ASC Lawrence, WJ Killner, J Inst Petrol 34: 821, 1948.
125. C Blakey, ASC Lawrence. J Inst Petrol 40: 203, 1954.
126. J Reisberg, TM Doscher. Prod Mon 21: 43, 1956.
127. JE Strassner. J Petrol Technol 20: 203f, 1968.
128. H Foerdedal, Y Schildberg, J Sjöblom, JL Volle. Colloids Surfaces 124:14: 106: 33–47, 1996.
129. H Foerdedal, J Sjöblom. J Colloid Interface Sci 181: 589–594, 1996.
130. PL Cottingham, FA Birkholz, LG Nickerson. Am Chem Soc Div Fuel Chem 23: 4, 1978, preprint.
131. S Puskas, J Balazs, A Farkas, I Regdon, O Berkesi, I Dekany. Colloids Surfaces A: Physicochem Eng Aspects 113: 279–293, 1996.
132. J Oren, GDM MacKay. Fuel 56: 382–384, 1977.
133. DG Thompson, AS Taylor, DE Graham. Colloids Surfaces 15: 175f, 1985.
134. DG Thompson, AS Taylor, DE Graham. Emulsification and demulsification related to crude oil production. Proceedings of SCI/Institute of Chemical Engineering Symposium, London, 1984.
135. B Felian, J Balazs, J Lakatos-Szabo, I Lakatos. Acta Phys Chem 30: 183–192, 1984.
136. J Sjoblom, O Urdahl, H Hoiland, AA Christy, EJ Johansen. Progr Colloid Polym Sci 82: 131–139, 1990.
137. RM Murzakov, SA Sabanenkov, AA Gureev. Chem Technol Fuels Oils 16: 69–71, 1980.
138. RZ Syunyaeva. Chem Technol Fuels Oils 16: 407–409, 1980.
139. R Murzakov, SA Sabanenkov, ZI Syunyaev. Chem Technol Fuels Oils 16: 674–677, 1980.
140. OK Kimbler, RL Reed, IH Silberberg. Soc Petrol Eng J 6: 153, 1966.
141. DJ Shaw. Introduction to Colloid and Surface Chemistry. 2nd ed. New York: Butterworth, 1970, p 91.
142. AG Taylor, J Mingins. J Chem Soc Faraday Trans I 5: 1161, 1975.
143. BP Singh. Energy Sources 16: 377–385, 1994.
144. RA Mohammed, AI Bailey, PF Luckham, SE Taylor. Colloids Surfaces A: Physicochem Eng Aspects 80: 237–242, 1993.
145. RA Mohammed, AI Bailey, PF Luckham, SE Taylor. Colloids Surfaces 80: 223–235, 1993.
146. KG Nordli, J Sjöblom, J Kizling, P Stenius. Colloids Surfaces 57: 83–98, 1991.
147. IB Ivanov, ed. Thin Liquid Films – Fundamentals and Applications. Surfactant Science Series, Vol 29. New York: Marcel Dekker, 1988.
148. JA de Feijter. In: IB Ivanov, ed. Thin Liquid Films – Fundamentals and Applications. Surfactant Science Series, Vol 29. New York: Marcel Dekker, 1988, Ch 7.

149. GJ Hirasaki. In: NR Morrow, ed. Interfacial Phenomena in Petroleum Recovery. New York: Marcel Dekker, 1991, Ch 2.
150. C Maldarelli, RK Jain. In: IB Ivanov, ed. Thin Liquid Films – Fundamentals and Applications. Surfactant Science Series, Vol 29. New York: Marcel Dekker, 1988, Ch 8.
151. DA Edwards, H Brenner, DT Wasan. Interfacial Transport Processes and Rheology. Boston: Butterworth-Heinmann, 1991.
152. TF Tadros. Colloids Surfaces: Physicochem Eng Aspects 91: 39–55, 1991.
153. AK Malhotra, DT Wasan. In: IB Ivanov, ed. Thin Liquid Films – Fundamentals and Applications. Surfactant Science Series v29. New York: Marcel Dekker, 1988, Ch 12.
154. CI Kristov, L Ting, DT Wasan. J Colloid Interface Sci 85: 362–373, 1982.
155. G Loglio, U Tesser, R Miller, R Cini. Colloids Surfaces 61: 219–226, 1991.
156. DA Edwards, DT Wasan. J Colloid Interface Sci 139: 479–486, 1990.
157. J Lucassen-Reynders, M Van den Temple. Chem Eng Sci 27: 1283–1291, 1972.
158. A Yeung, T Dabros, J Masliyah. Langmuir 13: 6597–6606, 1998.
159. J Boyd, P Sherman. J Colloid Interface Sci 34: 76f, 1970.
160. RJR Cairns, DM Grist, EL Neudstadter. In: AL Smith, ed. Theory and Practice of Emulsion Technology. New York: Academic Press, 1976, pp 135–151.
161. DM Grist, EL Neustadter, KP Whittingham. J Can Petrol Technol (June): 74–78, 1981.
162. DE Graham, EL Neudstadter, A Stockwell, KP Whittingham, RJR Cairns. Symposium on Surface Active Agents. Society of Chemical Industries Colloids Surfaces Chemistry Group, London, 1979, pp 127–.
163. DT Wasan, JJ McNamara, SM Sampath, S Aderangi. J Rheol 23: 181f, 1979.
164. CH Pasquarelli, DT Wasan. In DO Shah, ed. Surface Phenomena in Enhanced Oil Recovery. New York: Plenum Press, 1979, p 237f.
165. SE Taylor. Colloids Surfaces 29: 29f, 1988.
166. EL Neustadter, KP Whittingham, DE Graham. In: DO Shah, ed. Surface phenomena in Enhanced Oil Recovery. New York: Plenum Press, 1981, pp 307–326.
167. Mukherjee, AP Kushnick. Effect of demulsifiers on interfacial properties governing crude oil demulsification. Proceedings of 195th ACS National Meeting, Toronto, 1988, ACS Div Petrol Chem Preprint 33; pp 205–210.
168. JH Clint, EL Neustadter, TJ Jones. Proceedings of the Third European Symposium on Enhanced Oil Recovery, Bournemouth, UK, 1981.
169. IC Callaghan, EL Neustadter. Chem Ind 17: 53–57, 1981.
170. AD Nikolov, DT Wasan, M Randie, CS Shetty. Chem Eng Commun 152/153: 337–350, 1996.
171. D Tambe, J Paulis, MM Sharma. J Colloid Interface Sci 171: 163–469, 1995; 147: 137f, 1991.
172. DT Wasan. SPE J 409–417, 1978.
173. VB Menon, DT Wasan, Colloids Surfaces 19: 7–27, 1988.
174. VB Menon, AD Nikolov, DT Wasan. J Dispers Sci Technol 9: 575–593, 1988–89.
175. RA Mohammed, AI Bailey, PF Luckman, SE Taylor. Colloids Surfaces A 91: 129–139, 1994.
176. D Tambe, J Paulis, MM Sharma. J Colloid Interface Sci 171: 463–469, 1995.
177. RJR Cairns, DM Grist, EL Neustadter. In: DO Shah, ed. Theory and Practice of Emulsion Technology. New York: Academic Press, 1976, p 135.
178. DT Wasan. In: J Sjöblom, ed. Emulsions – A Fundamental and Practical Approach. Dordrecht: Kluwer, 1991, pp 283–295.
179. TY Chen, RA Mohammed, AI Bailey, PF Luckham, SE Taylor. Colloids Surfaces A: Physiochem Engin Aspects 83: 273–284, 1994.
180. AK Malhotra. PhD thesis, Illinois Institute of Technology, Chicago, 1984.
181. DT Wasan, AK Malhotra, N Aderangi. Int J Multiphase Flow 9: 105, 1983.
182. PJ Breen, DT Wasan, YH Kim, AD Nikolov. In: J Sjoblom, ed. Emulsions and Emulsion Stability. Surfactant Science Series, Vol 61. New York: Marcel Dekker, 1996, pp 237–285.
183. SR Ross, ID Morrison. Colloidal Systems and Interfaces. New York: John Wiley, 1988, pp 180–189.
184. R Luthy, R Selleck, T Galloway. Environ Sci Technol 11: 1211–1217, 1977.
185. RJR Cairns, DM Grist, EL Neustadter. Effects of crude oil – water interfacial properties on water – crude oil emulsion stability. Proceedings of SCI Symposium Theory and Practice of Emulsion Technology, Brunel University, London, 1974, Ch 8.
186. S Acevedo, G Escobar, LB Gutierrez, H Rivas. Fuel 71: 619, 1992.
187. Y Xu. Energy Fuels 9: 148–154, 1995.
188. EY Sheu, MM De Tar, DA Storm. Fuel 71: 1277–1280, 1992.
189. EY Sheu, MM De Tar, DA Storm, SJ DeCanio. Fuel 71: 299–302, 1992.
190. DT Wasan, K Sampath, N Aderangi. AIChE Symp Ser 76: 93, 1980.
191. J Svetgoff. Petrol Eng Int 61: 48–50, 1989.
192. K Shinoda, H Arai. J Phys Chem 68: 3485, 1964.
193. K Shinoda, H Sagitani. J Colloid Interface Sci 255: 172, 1977.

194. B Paczynska-Lahme. Progr Colloid Polym Sci 83: 196f, 1990.
195. S Friberg. J Colloid Interface Sci 40: 291–295, 1971; 37: 291, 1971.
196. T Stroem-Kristainsen, AB Nordvik, A Lewis, PS Daling. Demulsification by Use of Heat and Emulsion Breaker. Phase 2. Marine Spill Response Corp. Tech. Rep. Series, N94–012, 1994.
197. RJ Fiocco, MA Walsh, JN Hokstad, PS Daling, A Lewis, KW Becker. Improved laboratory demulsification tests for oil spill response. API-EPA-USCG et al. Proceedings of International Oil Spill Achieving and Maintaining Preparedness Conference, Long Beach, CA, API Publication no. 4620, 1995, pp 165–170.
198. T Stroem-Kristiansen, A Lewis, PS Daling, JN Hokstad, I Singsaas. Weathering and dispersion of naphthenic, ashphaltenic, waxy crude oils. Proceedings of International Oil Spill Conference, Fort Lauderdale, FL, 1997, pp 631–636.
199. B Obah. Nigeria Fed Univ Technol Erdoel Kohle-Erdgas-Petrochem 41: 71–74, 1988.
200. BJ Carroll. In: E Matijevic, ed. Surface and Colloid Science. Vol 9. New York: John Wiley, 1976, pp 1–66.
201. KJ Lissant. Demulsification. New York: Marcel Dekker, 1982.
202. R Aveyard, BP Binks, PD Fletcher, X Ye, JR Lu. In: J Sjöblom, ed. Emulsions – A Fundamental and Practical Approach. Dondrecht: Kluwer, 1992.
203. SR Reddy, HS Fogler. J Colloid Interface Sci 82: 128–135, 1981.
204. SR Reddy, S Fogler. J Colloid Interface Sci 79: 101–104, 1981.
205. SR Reddy, HS Fogler. J Colloid Interface Sci 79: 98–104, 1981.
206. A Bhardwaj, S Hartland. J Dispers Sci Technol 15: 133–146, 1994.
207. SL Mason, K May, S Hartland. Colloids Surfaces 96: 85–92, 1996.
208. VB Menon, DT Wasan. In: P Becher, ed. Encyclopedia of Emulsion Technology. Vol 2. New York: Marcel Dekker, 1985, pp 1–76.
209. S Hartland. In IB Ivanov, ed. Thin Liquid Films – Fundamentals and Applications. Surfactant Science Series, Vol 29. New York: Marcel Dekker, 1988, Ch 10.
210. E Nakache, P-Y Longaive, S Aeillo. Colloids Surfaces A: Physicochem Eng Aspects 96: 69–76, 1995.
211. P Becher. Emulsions: Theory and Practice. 2nd ed. ACS No. 62, New York: Krieger, 1977, pp 367–380.
212. DA Edwards, H Brenner, DT Wasan. Interfacial Transport Processes and Rheology. Boston, MA: Butterworth-Heinmann, 1991, Ch 11.
213. MV Ostrovsky, RJ Good. J Disper Sci Technol 7: 95–125, 1986.
214. RD Vold, MJ Vold. Colloid Interface Chemistry. Boston, MA: Addison Wesley, 1983, pp 415–439.
215. M von Smoluchowski. Z Phys Chem 92: 129, 1917.
216. M Van den Temple. Rec Trav Chim 72: 433,442, 1953.
217. S Hartland. In: IB Ivanov, ed. Thin Liquid Films – Fundamentals and Applications. Surfactant Science Series, Vol 29. New York: Marcel Dekker, 1988, Ch 10.
218. RP Borwanker, LA Lobo, DT Wasan. Colloids Surfaces 69: 135–146, 1992.
219. SL Mason, K May, S Hartland. Colloids Surfaces A: 96: 85–92, 1994.
220. CW Angle, HA Hamza. Evaluation of some biological surfactants as emulsion destabilizers. Proceedings of 59th International Surface and Colloid and 5th International Colloid Science Conference, Clarkson University, Potsdam, NY, 1985; CANMET Div. Rep. ERP/ERL 81–99(CF), 1981.
221. R Vold, K Mittal, A. Hahn. Surface Colloid Science 10. New York: Plenum Press, 1978, pp 45–97.
222. MJ Vold. Langmuir 1: 74–78, 1985.
223. AU Hahn, RD Vold. J Colloid Interface Sci 51: 133–142, 1975.
224. CW Angle, JL Picard, HA Hamza. The effects of ultracentrifugal stress on the stability of crude oil/water emulsions. Proceedings of 38th Canadian Chemical Engineering Conference, Edmonton, Alberta, 1988.
225. RD Vold, KL Mittal. J Soc Cosmet Chem 23: 171–188, 1972.
226. AU Hahn, KL Mittal. Colloid Polym Sci 9: 959–967, 1979.
227. MK Sharma. Curr Sci 46: 131–132, 1977.
228. MK Sharma. Acta Cienca Indica 3: 139–141, 1977.
229. MK Sharma. Progr Colloid Polym Sci 63: 75–77, 1978.
230. AU Hahn, KL Mittal. Colloid Poly Sci 257: 959–967, 1979.
231. RD Vold, M Maletic. J Colloid Interface Sci 65: 390–393, 1978.
232. RS Allan, SG Mason. J Colloid Sci 17: 383, 1962.
233. S Hartland. Chem Eng Sci 22: 1675, 1967.
234. IB Ivanov, DS Dimitrov. Thin Film Drainage. In: IB Ivanov, ed. Thin Liquid Films – Fundamentals and Applications. Surfactant Science Series, Vol 29. New York: Marcel Dekker, 1988, pp 379–485.
235. JA Kitchener, PR Musselwhite. In: P Sherman, ed. Emulsion Science. New York: Academic Press, 1968, Ch 2.
236. DT Wasan. Interfacial transport processes and rheology – structure and dynamics of thin liquid films. ASEE Chemical Engineering Education, 1992, pp 104–112.
237. BV Derjaguin. Theory of Stability of Colloids and Thin Films. New York: Consultants Bureau, 1989.
238. A Scheludko, D Exerowa. Kolloid Z 165: 148–51, 1960; 168: 24–8, 1959. (Taken from RJ Hunter.

Foundations in Colloid Science. Vol II. Oxford: Oxford Science Publications; Clarendon Press, 1989, Chs 15 and 16.)
239. A Scheludko. Colloid Chemistry. Amsterdam: Elsevier, 1966, p 277.
240. AD Nikolov, DT Wasan, SE Friberg. Colloids Surfaces A: 118: 221–243, 1996.
241. RW Flummerfelt, AB Catalano, CH Tong. Proceedings: Surface Phenomena in Enhanced Oil Recovery, 1981, pp 571–594.
242. XB Reed, E Riolo Jr, S Hartland. Int J Multiphase Flow 3: 485, 1977.
243. IB Ivanov, TT Traykov. Int J Multiphase Flow 2: 397, 1976.
244. TT Traykov, ED Manev, IB Ivanov. Int J Multiphase Flow 3: 485, 1977.
245. AD Barber, S Hartland. Can J Chem Eng 54: 279, 1976.
246. AF Jones, SDR Wilson. J Fluid Mech 87: 263, 1978.
247. DT Wasan, SM Shah, M Chan, JJ McNamara. Soc Petrol Eng J 18: 409, 1978.
248. DT Wasan, ME Ginn, DO Shah, eds. Surfactants in Chemical and Process Engineering, Surfactant Science Series 28, 1988.
249. AD Nikolov, DT Wasan, PA Kralchevsky, IB Ivanov. In: N Ike, I Sogami, eds. Proceedings of Yamada Conference XIX. Singapore: World Scientific Publications, 1988.
250. AD Nikolov, PA Kralchevsky, IB Ivanov, DT Wasan. J Colloid Interface Sci 133: 1, 1989; J Colloid Interface Sci 133: 13, 1989.
251. AD Nikolov, DT Wasan, ND Denkov, PA Kralchevsky, IB Ivanov. Progr Colloid Polym Sci 82: 1, 1990.
252. DT Wasan, AD Nikolov, L Lobo, K Koczo, DA Edwards. Progr Surface Sci 39: 2, 1992.
253. AD Nikolov, DT Wasan. Langmuir 8: 2985–2994, 1992.
254. MA Krawczyk, DT Wasan, CS Shetty. Ind Eng Chem Res. 30: 367–375, 1991.
255. S Mukherjee, AP Kushnick. Effect of demulsifiers on interfacial properties governing crude oil demulsification. Proceedings of 195th ACS National Meeting, Toronto, 1988, ACS Div. Petrol. Chem. Preprint, Vol. 33, No. 1, pp 205–210.
256. S Mukherjee, AP Kushnick. In: JK Borchardt, TF Yen, eds. Oil Field Chemistry – Enhanced Recovery and Production Stimulation. Washington, DC: American Chemical Society, 1989, pp 364–374.
257. R Aveyard, BP Binks, PDI Fletcher, JR Lu. J Colloid Interface Sci 139: 128–138, 1990.
258. YH Kim, DT Wasan, PJ Breen. Colloids Surfaces A: Physicochem Eng Aspects 95: 235–247, 1995.
259. J Lucassen-Reynders, M Van den Temple. J Phys 67: 731, 1963.
260. VB Menon, DT Wasan. Colloids Surfaces 29: 7–27, 1988.
261. VB Menon. Characterization of dispersed three-phase systems with applications to solids-stabilized emulsions. Dissertation Abstracts International: Section B Science & Engineering, Vol 47, No. 5 2077-B, 1986.
262. A Gelot, W Friesen, HA Hamza. Colloids Surfaces 12: 271f, 1984.
263. AB Taubman, AF Koretskii. Colloid J Wash 20: 631, 1958.
264. VB Menon, DT Wasan. Colloids Surfaces 29: 7–27, 1988.
265. VB Menon, AD Nikolov, DT Wasan. J Colloid Interface Sci 124: 317–327, 1988.
266. CW Angle, R Zrobok, HA Hamza. J Appl Clay Sci 7: 455–470, 1993.
267. RC Little. Environ Sci Technol 15: 1184–1190, 1981.
268. IA Buist, SL Ross. Emulsion inhibitors: a new concept in oil spill treatment. Proceedings of Oil Spill Conference. American Petroleum Institute, Washington, DC, 1987, p 217.
269. SL Ross. An experimental study of oil spill treating agents that inhibit emulsification and promote dispersion. Report EE87. Ottawa: Environment Canada, 1986.
270. RI Hancock. In: TF Tadros, ed. Surfactants. New York: Academic Press, 1983, pp 287–321.
271. G Leopold. In: LL Schramm, ed. Emulsions Fundamentals and Applications in the Petroleum Industry. Advances in Chemistry Series 231, Washington, DC: American Chemical Society, 1992, Ch 10.
272. R Grace. In: LL Schramm, ed. Emulsions Fundamentals and Applications in the Petroleum Industry. Advances in Chemistry Series 231, Washington, DC: American Chemical Society, 1992, Ch 9.
273. VH Smith, KE Arnold. In: HB Bradley, ed. Petroleum Engineering Handbook. SPE, 1992, Ch 19.
274. B Rowan. The Use of Chemicals in Oil Field Demulsification. Spec Publ Royal Soc Chem 107 (Ind Appl Surfactants III), 1992, pp 242–251.
275. CW Angle, T Dabrus, H Hamza. Mechanism of destabilization of water-in-bitumen emulsions by polymeric surfactants. Paper presented at 13th International Symposium on Surfactants in Solution, SIS-2000, Gainesville, FL, June 11–16, 2000, p 25.
276. L Cash, JL Cayais, RS Schechter, G Fournier, D Macallister, T Schares, RS Schechter, WH Wade. J Colloid Interface Sci 59: 39–44, 1977.
277. JL Cayais, RS Schechter, WH Wade. Soc Petrol Eng J (Dec.): 351: 357, 1976.
278. RL De Silva, S Key, J Marino, C Guzman, S Buitriago. Chemical dehydration: correlations

between crude oil, associated water and demulsifier characteristics in real systems. Proceedings of SPE Oilfield Chemical International Symposium, Houston, TX 1997, pp 601–607.
279. WC Griffin. J Soc Cosmet Chem 1: 311, 1949; 5: 249, 1954.
280. HT Davis. Colloids Surfaces A: 9: 9–34, 1994.
281. K Shinoda, H Saito. J Colloid Interface Sci 26: 70, 1968.
282. K Shinoda, H Kuneida. In: P Becher, ed. Encyclopedia of Emulsion Technology. Vol 1. New York: Marcel Dekker, 1983.
283. ME Hayes, M El-Emary, RS Schechter, WH Wade. J Colloid Interface Sci 68: 591–592, 1979.
284. AH Walker, DL Ducey, JR Gould, AB Nordvik. Proceedings of Formation of Water-in-Oil Emulsions. Marine Spill Corp. Tech. Rep. Series 93–018, Kananaskis, Alberta, 1993.
285. DG Cooper, JE Zajic, EJ Cannel, JW Wood. Can J Chem Eng 58: 576–579, 1980.
286. J Israelachvili. Colloids Surfaces A: 91: 1–8, 1994.
287. L Marshall. Cosmet Perfum 90: 37, 1975.
288. F Staiss, R Bohm, R Kupfer. Improved demulsifier chemistry: A novel approach in dehydration of crude oil. SPE Production Engineering, 1991, pp 334–338.
289. M Amarvathi, BP Pandey. Res Ind 36: 198–202, 1991.
290. K Shinoda. J Colloid Interface Sci 24: 4–9, 1967.
291. FE Bailey Jr, JV Koleske, eds. Alkylene Oxides and their Polymers. Surfactant Science Series, Vol 35. New York: Marcel Dekker, 1990, Ch 6.
292. O Urdahl, J Sjöblom. J Dispers Sci Technol 16: 557–574, 1995 (Abstr. no. 37002, Vol 124, No 4)
293. O Urdahl, AE Moevik, J Sjöblom. Colloids Surfaces 74: 293–302, 1993.
294. J Sjöblom, O Urdahl, H Hoeiland, AA Christy, EJ Johansen. Progr Colloid Polymer Sci 82: 131–139, 1990.
295. DE Graham, A Stockwell, DG Thompson. Chemical Demulsification of Produced Crude Oil Emulsions. Spec Publ Royal Soc Chem 45: 73–91, 1983 (Abstract no. 70810).
296. DE Graham, A Stockwell. Selection of demulsifiers for produced crude oil emulsions. Proceedings of Europe Offshore Petroleum Conference, London, 1980, Vol 1, pp 453–458.
297. J Svetgoff. Petrol Eng Int 62(1): 1990 (Abstract no. 476 918, Vol 30, No. 7).

# 25

# Demulsifiers in the Oil Industry

**Johan Sjöblom, Einar Eng Johnsen, and Arild Westvik**
*Statoil A/S, Trondheim, Norway*

**Marit-Helen Ese and Jostein Djuve**
*University of Bergen, Bergen, Norway*

**Inge H. Auflem and Harald Kallevik**
*Norwegian University of Science and Technology, Trondheim, Norway*

## I. INTRODUCTION

According to our traditional understanding of emulsion formation and stabilization, there is a need to introduce mechanical energy and stabilizing agents to a water/oil mixture in order to create stable emulsions. In bench experiments the energy input is typically controlled by means of different kinds of rotors and homogenizers. Under real oil-production conditions, pressure gradients over chokes and valves will guarantee that there will be a sufficiently high mechanical energy input in order to rupture original solution structures and to form new fresh W/O interfaces. Obviously the magnitude of the pressure gradient over the choke/valve will be decisive for the droplet size distribution in the fresh emulsions. Since the transport over a choke/valve will mean the creation of a new W/O interface the nature of the emulsion before and after the valve/choke may differ significantly (1–5).

The lifetime of the emulsion (and the retention time in the full-scale separator) depends on the kind of stability mechanisms involved. There exist several possibilities of finding stabilizing agents (or solid fines) in either the crude oil itself or in added production chemicals. Among the indigenous stabilizers, asphaltenes/resins/porphyrins are mentioned as possible candidates for the stabilization of W/O emulsions. In most cases it is, however, not possible to pick out one single component responsible for the stability of the dispersed droplets, but several components/fractions in parallel are the source for the stability of the emulsion. The components/fractions in the crude oil show a large range of molecular weights. Lighter components like the resins can act as individual monomers in a similar manner to traditional surfactants. The driving force for their action is the presence of water (and the existence of a W/O interface). Usually the low molecular weight resins have a tendency to be the most interfacially active, i.e., to reach first and cover a fresh W/O interface. However, this is mostly a necessary requirement but not a sufficient one for the formation of stable W/O emulsions. The next step in the stabilization process involves interaction with the heavier crude oil components, i.e., the asphaltenes. Depending on the production history and the fluid properties these molecules can be either in a monomeric or associated state. In the latter case small particles are formed. The formation of these is normally a result of the stacking tendency of the individual asphaltene molecules. These nanosized particles will have a strong tendency to accumulate at the W/O interfaces, if the solution conditions or changes herein so favor. Obviously the final particle size or the flocculation of these nanosized particles is critical with regard to the stabilizing capacity of these entities (6, 7).

It is of crucial importance to understand the stabilizing mechanisms when discussing demulsifiers and the efficiency of these. In this chapter we are going to discuss different types of demulsifiers, i.e., from simple solvents to intriguing macromolecules. It is also our intention to view how new instrumentation can reveal important and to some extent unexpected properties of these chemicals. In this chapter we introduce, in addition to conventional techniques, the use of Langmuir and Langmuir–Blodgett techniques, atomic-force microscopy (AFM) and near-infrared spectroscopy (NIR), when analyzing the effects of the demulsifying chemicals. We can also for the first time report on destabilization experiments (with demulsifiers) at elevated pressures. These experiments have been carried out in a special separation rig constructed for Statoil.

## II. EXPERIMENTAL TECHNIQUES

### A. The Langmuir Technique

The Langmuir technique is used in order to characterize monolayer properties of surface-active materials. The instrumentation consists of a shallow rectangular container (trough) in which a liquid subphase is added until a meniscus appears above the rim, whereupon the film is spread. The barrier for manipulation of the film rests across the edges of the container. For a more thorough description of the experimental setup, Petty and Barlow (8) are recommended. The surface pressure is measured by means of the Wilhelmy method (8–10). Modification of the trough design has made it possible to carry out the same kind of experiments on a liquid/liquid interface, i.e., the oil/water interface, instead of on the liquid surface. A prototype trough has been designed by KSV Chemicals in collaboration with the University of Compiègne, France (11). The trough, entirely made of Delrin, is a "double" trough (Fig. 1) where the barriers contain holes to allow the flow of the light phase as the compression of the interface proceeds. The Wilhelmy plate is first placed in the aqueous phase, then the oil phase is added until the plate is totally immersed.

The most common and adequate way of presenting the results obtained from the Langmuir technique is a plot of surface pressure as a function of the area of surface available to each molecule, i.e., the mean molecular area. The measurements are carried out at a constant temperature and are known as surface pressure/area isotherms (Fig. 2). The film is compressed at a constant rate by the moving barriers while the surface pressure is continuously monitored. Generally, a number of distinct regions are apparent on examining the isotherm. As the surface area is reduced from its initial high value, there is a gradual onset of surface pressure until an approximately horizontal region is reached. In this region the hydrophobic parts of the molecules, originally distributed near the water surface, are being lifted away. However, this part of the isotherm is often not resolved by the apparatus, because the surface pressure at which this occurs is usually quite small (< 1 mN/m) due to the weakness of interaction between water and the tailgroups. This region is followed by a second abrupt

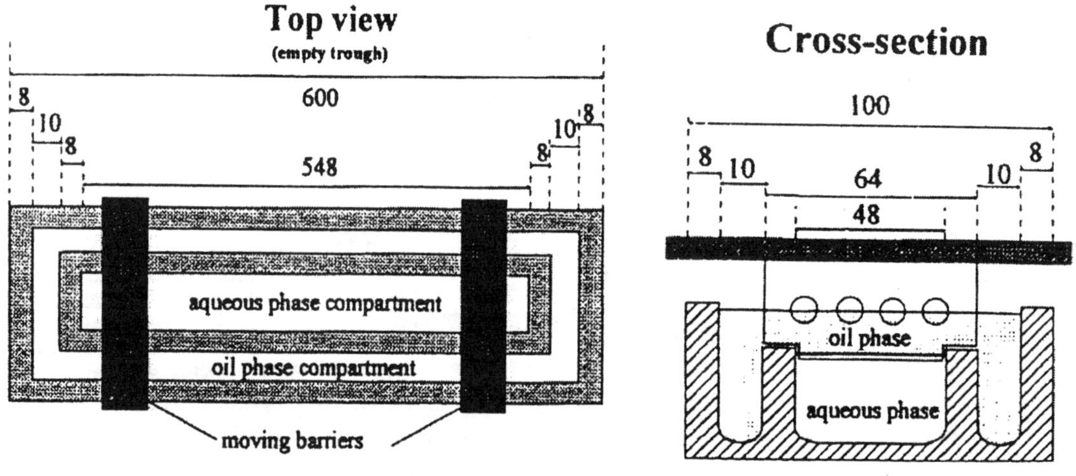


**Figure 1** Schematic drawing of the liquid–liquid interfacial trough (size in mm).

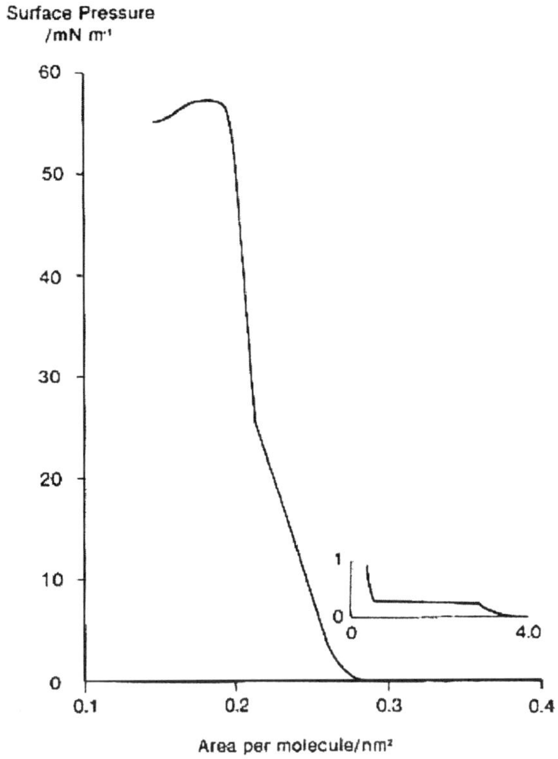


**Figure 2** Typical surface pressure/area isotherm of stearic acid on an acidified water subphase.

transition to a steeply linear region, with an approximately constant compressibility.

Further reduction in surface area results in an abrupt increase of slope, and hence reduced compressibility. All these different regions indicate different states of the monolayer, and analogously to bulk matter these are characterized as gas-, liquid- and solid-like (12–14). Additional techniques (15) such as X-ray scattering, electron diffraction, fluorescence, polarized fluorescence, atomic-force microscopy, and Brewster angle microscopy have proved the existence of mesophases in Langmuir films. The complexity of behavior is a result of the differerent intermolecular interactions in the film (alkyl group/alkyl group and polar group/polar group interactions) and between the film and the subphase (polar group/subphase interactions). Hence, the interaction forces would undergo certain changes, which would be related to the packing of the molecules in the two-dimensional plane. More details may be found in the books by Gains (9) and Birdi (10).

A sharp break at small areas in the $\Pi$–$A$ isotherm is attributed to the collapse of the monolayer under the given experimental conditions. In general, the collapse pressure is the highest surface pressure to which a monolayer can be compressed without a detectable movement of the molecules in the film to form a new phase.

In order to investigate the stability of a monolayer, the area loss at constant surface pressure or the decrease in surface pressure at constant area is measured. Different destabilization mechanisms are illustrated by the shape of the relative area relaxation isotherms in Fig. 3. The compression is stopped at a predetermined surface pressure and the relative area loss is plotted as a function of time. Curve (a) in Fig. 3 shows the behavior of a totally stable film, with no area reduction. Isotherm (b) shows an initial area loss, which is attributed to structural rearrangements in the monolayer to form a coherent close-packed film (16–19). This process depends on the rate of compression. Fast barrier movement creates a higher degree of disorder in the monolayer, and the initial area loss is increased. A continuous decrease in area, as illustrated by curve (c), is the result of a slow dissolution of the film-forming material into the subphase or evaporation of the monolayer. This film loss increases with surface pressure, and is a consequence of the huge volume difference between the material in the film and the subphase liquid and gas phase to which the monolayer is exposed. Even a low solubility or vapor pressure may lead to destabilization of the film due to solution or evaporation (20). Isotherm (d) in Fig. 3 is characterized by a relaxation rate that increases with time. This kind of behavior is observed for systems that undergo nucleation and further growth of the film material into bulk fragments (19, 21).

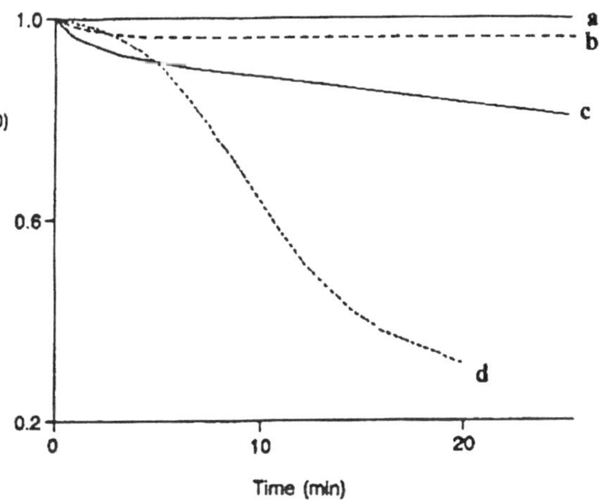


**Figure 3** Relative relaxation curves at constant surface pressure for monolayers showing different stabilities: (a) stable monolayer; (b) rearrangements of the film molecules; (c) dissolution of film molecules into the subphase; (d) collapse by nucleation and growth of bulk solid fragments.

## B. Langmuir–Blodgett Deposition

The most commonly used process of transferring a floating insoluble monolayer to a solid surface is Langmuir's original method (22). A clean wettable solid is placed in the subphase before a monolayer is spread, and then drawn up through the surface after formation of the film. The transfer process is critically dependent on the surface pressure, so it is desirable to maintain a constant pressure as the film is removed from the surface. Using this kind of deposition technique, both the film and a thin layer of water are transferred to the solid substrate. The water is later removed by drainage or evaporation, leaving a monolayer on the solid surface.

The rate of deposition is an important factor. Optimal values of this parameter depend partly on the rate of drainage of the intervening liquid film from the monolayer/slide interface and partly on the dynamic properties of the monolayer on the liquid surface, i.e., the film viscosity.

When conducting structural studies of Langmuir–Blodgett (LB) films, careful consideration has to be taken of possible effects that might arise during film deposition. Irregularity in the dipping motion may result in formation of striations in the deposited layer. Trapped water droplets between the film and the solid surface is another possible reason for imperfections in the monolayer. Circumstances such as low surface pressure or weak interactions between the film-forming molecules, i.e., the monolayer is not coherent, may result in deposition of irregular films. This is probably a result of expansion, contraction, or flow during the transfer process.

A solid surface, which is not smooth on a molecular scale, may lead to problems when transferring a monomolecular film on to it. At the moment of deposition the film may bridge over the surface roughness, especially if the film is closely packed and under high surface pressure. This kind of bridging is often supported by the intervening water layer, so when this layer is removed the film may collapse.

## C. Atomic-force Microscopy

Atomic-force microscopy (AFM) (23) is a nearly ideal, high-resolution method in providing a molecular-scale topographic view of a variety of solid surfaces, organic, inorganic, or biomolecular. Under optimal conditions this microscopic technique is capable of producing images showing details of molecular resolution in LB films.

The AFM technique exploits the forces that exist between atoms and molecules. The force exerted upon a tip mounted on to a cantilever (with a known spring constant, weaker than the equivalent spring constant between atoms) is monitored as the tip passes over the surface. Measurements of the cantilever deflection, which is proportional to the magnitude of the force, during the scan make it possible to obtain images of the surface topography. All types of materials exert these forces, so there is no restriction regarding composition of the analyzed components. Another important advantage with AFM is that it can operate in a variety of environments. It is especially convenient that the measurements may be performed in air at atmospheric pressure. In addition, AFM may also be operated in liquids.

There are three scanning modes for AFM, i.e., contact mode, noncontact mode, and tapping mode (24). In the contact mode the tip is touching the sample surface where the repulsive forces dominate, while the attractive forces dominate in the noncontact mode. The tapping mode represents a compromise between these two, giving better resolution than the noncontact mode and is not as damaging for the sample as the contact mode. A tip in contact with the surface may generate extremely high pressures on the small contact area between the tip and the sample, which may result in indentation of the tip in soft materials. In tapping mode the cantilever is oscillating near its resonance frequency, with a high enough amplitude to allow the tip to dip periodically in the contamination layer. Measuring the change in amplitude or phase for the oscillating cantilever provides images of the surface (Fig. 4).

For more information regarding the use of AFM and related probe techniques for imaging LB films, review articles by Zasadzinski et al. (25) and DeRose and Leblanc (26) are recommended.

## D. Test Procedures for Demulsification

It is commonly known that the administration of chemicals is very essential. Depending on the administration procedure one can expect different efficiencies of the demulsifiers. Different administration procedures are reviewed in the following.

The traditional testing of demulsifiers is to undertake bottle-shake tests. In these tests one has a premixed emulsion and the chemical under study is applied. After this the bottle is gently shaken in order to distribute the chemical evenly into the emul-

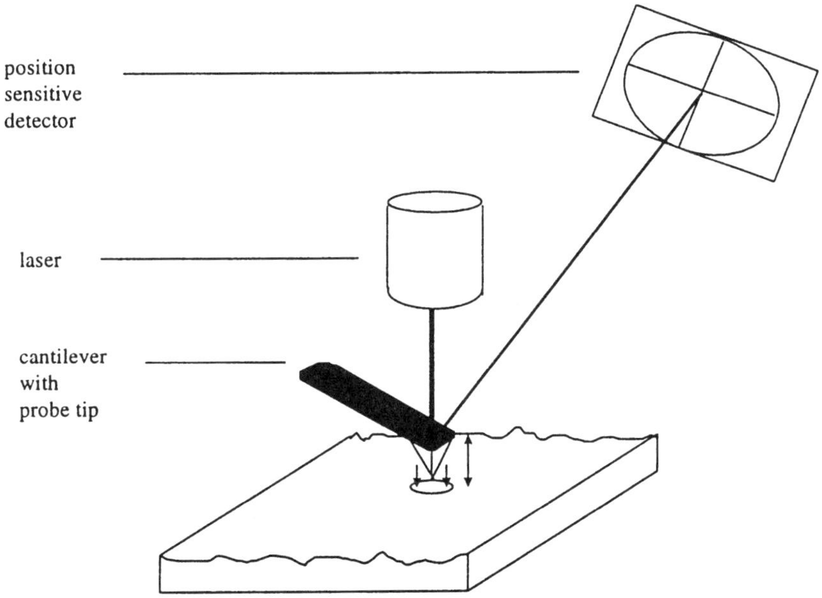


Figure 4 Schematic representation of an atomic-force microscope.

sified system. The efficiency of the chemical applied is read from the resolution of the dispersed phase in volume as a function of time. Normally the size and shape are such that area effects can be neglected. The demulsification normally progresses under stagnant conditions with no mechanical energy input. When the coalescence is in progress there will be a resolution of the dispersed phase as a function of time, i.e., so-called volume–time plots. If the administration of the chemical only results in an accelerated creaming/sedimentation, there will be an increased concentration of droplets on the top (or the bottom) of the otherwise clear bulk phase. This situation should not be misinterpreted as a breaking of the emulsion, since gentle shaking will redistribute the droplets again.

The weakness of the bottle tests is that a true process is not reproduced. This fact has been accounted for in a variety of test rigs simulating true flow conditions, process kits, and separation conditions. The Statoil R&D Center has recently constructed a special rig for simulation of high-pressure processes. In this rig (Fig. 5) emulsions can be formed at different pressures before being brought into the separation chamber. The final droplet sizes and size distributions are determined by the pressure drop over the chokes and valves. The separation can be performed at elevated pressures if so wanted. The rig is described in detail below.

## E. Separation Rig

A separation rig, as illustrated in Fig. 5, is used to prepare W/O emulsions and monitor the separation of them. The principle is that two pressurized fluids meet just before a choke valve (VD1) and flow through the valve into the separation cell. The two fluids can either be oil and water or for instance two premixed oil/water dispersions. Through the choke valve the fluid mixture undergoes a pressure drop. The low-pressure side of the choke valve equals the separator pressure. The pressure drop through the choke leads to the creation of more interface between the oil and water, i.e., water droplets are formed and dispersed into the oil. At the same time the light end of the oil undergoes a phase change from liquid to gas. (The gas evolved may form a foam and may influence the sedimentation and coalescence of the water droplets.) After the cell has been filled, the amount of the different phases (foam, oil, emulsion/dispersion, and water) is recorded as a function of time. Upstream of the choke valve VD1 there are two other choke valves; one on each line (VD2 and VD3). Through these choke valves the same processes take place as described for VD1. Through VD2 and VD3 dispersions of oil and water can be made. In this way the dispersion which enters the separation cell can be a mixture of the two dispersions made through VD2 and VD3.

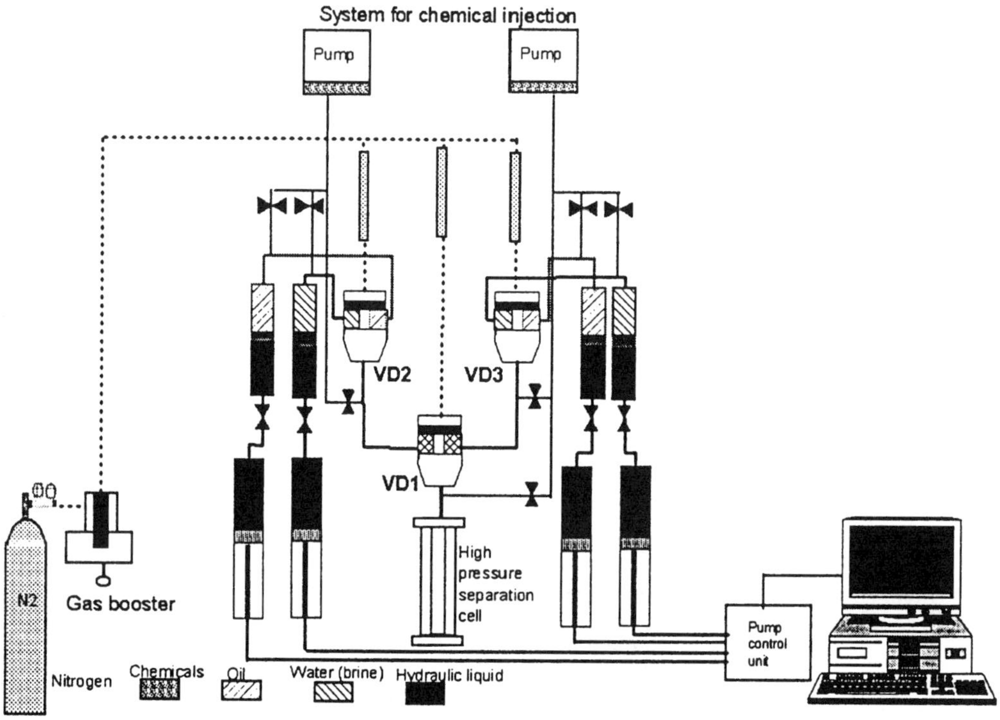


**Figure 5** The high-pressure separation rig.

Chemicals can be injected into any of the flow lines; at the high pressure upstream of VD2 and VD3, at the medium pressure upstream of VD1 or at the lowest pressure downstream of VD1 just before the fluid enters the separation cell. Chemicals can also be injected into the bottom of the cell where a stirrer can be used to distribute the chemicals. The injection of for instance demulsifiers takes place through 1/16-inch tubes with very small inner diameters. The chemical injection pumps deliver volumes down to 0.03 ml/h.

The rig consists of four 600-ml high-pressure sample cylinders. Usually, two are filled with water (brine), and two are filled with oil. With the aid of four motor-driven high-capacity piston pumps, water and oil are pumped through the choke valves. The four pumps are independent of each other, but in most of the experimental series the total flow has been kept constant. The pressure drops through the choke valves are back-pressure controlled. The pressure in the separation cell is regulated by a back-pressure controlled valve. The maximum pressure in the cell is 200 bar. The cell is filled with gas, inert or natural gas, to the desired pressure before the filling of water and oil into the separation cell starts. The separation cell (450 ml) is made of sapphire, assuring full visibility of the separation process. Video cameras are installed to follow the separation process.

All parts of the rig are thermostated. Temperature, cell pressures, valves, pumps, and video cameras are computer controlled.

The above-mentioned techniques all have in common that they are of macroscopic scale, often enabling diagrams of separated volume as a function of time to be displayed.

## III. STABILITY OF WATER-IN-CRUDE OIL EMULSIONS

It is well recognized that these emulsions are stabilized by means of an interplay between different heavy components, organic and inorganic particles, respectively. Heavy components cover asphaltenes, resins, etc. In a depressurized anhydrous crude oil the asphaltenes are normally in a particulate form. The role of the resins (and lighter polar components) is to stabilize the asphaltene dispersion (suspension) by adsorption mechanisms. Owing to this strong interaction the asphaltene particles are prevented from concomitant coagulation and precipitation. The stability will also put some restriction with regard to particle sizes since the largest particles are supposed to show the highest rate of sedimentation. When water is mixed with the crude oil, the situation will drastically change. The system will reach an energetically higher level, where the energy difference is proportional to the interfacial area created during the mixing process. This fresh interfacial area will attract components in the system. The molecules possessing the highest interfacial activity will try to cover the fresh W/O interface and hence minimize the energy level of the system. This category of indigenous components is normally covered by the lighter polar fraction, i.e., the resins. As a consequence a competition situation between resin molecules at the W/O interface and on the solid asphaltene particles will occur. Decisive factors determining the final position of the resins are the hydrophilic/lipophilic balance of these molecules and the corresponding properties of the solid surface. One could imagine that a very hydrophobic particle surface and a very polar W/O interface would extract different types of resins for the different activities. However, as pointed out, highly interfacially active resins will show preference for the W/O interface over not only less hydrophobic resin molecules but also over asphaltenes. As a consequence the solubility conditions for the asphaltenes will drastically change and a particulate precipitation will take place. With aqueous droplets coated by an interfacial resin film as closest neighbors the asphaltene particles will precipitate and accumulate at the droplet surface. The resulting interfacial properties will be much more rigidified and the stability of the corresponding emulsions profoundly improved. Central mechanisms involved in the stabilization process will hence be both steric and particle stabilization.

The mechanical properties of the protecting interfacial film are essential for the final stability level of the W/O emulsions. Concentrated polymeric interfacial films may display either elastic or viscous properties that make the destabilization process difficult and time consuming. The aromatic asphaltene molecules will normally undergo a stacking into sandwich-like structures as a consequence of the molecular association. The presence of other nanosized-particles like organic wax particles and inorganic clay particles will further enhance the stability level. However, these compounds are not further dealt with in the present chapter.

The interfacial conditions are reflected in the level of the interfacial pressure ($\pi$). Sjöblom et al. (27) showed that there is a correlation between the level of $\pi$ and the macroscopic emulsion stability. Preferably the interfacial pressure should be above 10–14 mN/m for stable emulsions. Aromatic molecules such as benzene will substantially lower the level of $\pi$. With increasing content of aromatic molecules the interfacial activity of the indigenous surfactants will be canceled and hence the emulsion stability will vanish. The dilution with aromatic solvents is in practical use in many places in the world where heavy crude oils create transport and emulsion problems.

## A. Coalescence

Coalescence is defined as the combination of two or more droplets to form a larger drop. When these droplets approach each other, a thin film of the continuous phase will therefore be trapped between the droplets, and it is obvious that the properties of this film will determine the stability of the emulsion (28). The mechanism of coalescence occurs in two stages: film thinning and film rupture. In order to have film thinning there must be a flow of fluid in the film, and a pressure gradient present. It is obvious that the rate of film thinning is affected by the properties of the colloidal system. Some of the most important parameters (29) are defined as viscosity and density of the two phases present, interfacial tension and its gradient, interfacial shear and dilational viscosities and elasticities, drop size, concentration and type of surfactant present at the interface, and forces acting between the interfaces. Considerable effort has been made to develop models for prediction of the rate of film thinning and critical film thickness. Reynolds (30) made the first mathematical analysis of parallel disks. He assumed the bounding interfaces to be solid and the film to be of uniform thickness. Frank and Mysels (31) investigated dimple formation and drainage through the dimple. Later models of film thinning are those of Zapryanov *et al* (32) and Lin and Slattery (33, 34). Zapryanov investigated surfactant partitioning at the interface using the parallel-disk model. This model has later been extended to account for the adsorption/desorption kinetics of surfactants (35). Film rupture is a nonequilibrium process that may occur as a result of flow instabilities, temperature fluctuations, electric fields, or Marangoni effects (36). Investigations by de Vries (37) and Lang (38) showed that there exists a critical film thickness. Above this thickness the probability of rupture is zero, and below it the probability of rupture increases with decreasing film thickness. Scheludko and Manner (39) investigated the rupture of thin liquid films between two droplets in relation to fluctuations at the interface. He also developed an expression for the critical film thickness with only van der Waals forces acting: $d_c = [A\pi/32K^2\gamma_0]^{0.25}$, where $A$ is the Hamaker constant, $\gamma_0$ is the interfacial tension between the continuous and dispersed phase, and $K$ is the wavenumber of the surface fluctuations. Vrij (40) has derived an alternative expression for $d_c$; for larger thicknesses: $d_c = 0.268[A^2R^2/\gamma_0\pi f]^{0.14}$, where $R$ is thc droplet radius, and $f$ is dependent on $d$. For small thicknesses: $d_c = 0.22[AR^2/\gamma_0 f]^{0.25}$. According to the first equation $d_c \to \infty$ when $\gamma \to 0$, i.e., the film should spontaneously rupture at large $d$ values. However, this is not the case since emulsion droplets become highly stable when $\gamma \to 0$. Also, the first equation predicts that as $R \to 0$, $d_c \to 0$, i.e., small emulsion droplets would never rupture. Sonntag and Strenge (41) showed that $d_c$ will not change when the contact area is varied. This is due to the fact that the lamella formed between two droplets, at nonequilibrium separations, does not have an idealized planar interface between them. Sonntag and Strenge (41) also showed that emulsion films of octane/water droplets stabilized by a nonylphenol ethoxylated surfactant plus an oil-soluble surfactant had a $d_c$ independent of $\gamma_0$.

## IV. GENERAL THEORY OF DEMULSIFIER ACTION

Commercially available demulsifiers can generally be described as "chemical cocktails." The terminology is introduced in order to describe the fact that one expects to find a synergistic effect of one or two (or more) active components that are dissolved in an active solvent. The solvent should be so hydrophobic that the active components can be readily dissolved in the crude oil. From this one can see that here are some general rules of thumb for the demulsifiers. Below, we summarize some of the most pertinent features. We can briefly classify the demulsifiers according to their molecular weight, as high molecular weight (HMW) and low molecular weight (LMW) demulsifiers and pure solvents.

The HMW molecules include different kinds of polymers and macromolecules (block copolymers, etc.), together with polyelectrolytes. Typical HMWs should be > 5000 g/mol. The LMW demulsifiers are in most cases some types of oil-soluble surfactants with co-operativity between the molecules. The solvents in use can be classified according to the polarity. Simple examples on increasing polarity would be pure paraffinic hydrocarbons < aromatic hydrocarbons < alcohols < diols, etc.

### A. Low Molecular Weight (LMW) Demulsifiers

Basically, the functionality of this category of demulsifiers is based on two specific mechanisms, i.e., increased interfacial activity and changed wettability of stabilizing components, respectively. The increased interfacial activity results in a suppression of the interfacial tension. Hence, these molecules tend to replace other, already existing molecules at the interface. This is a thermodynamical result, but in practice it can be difficult for these surfactant-like molecules to reach the W/O interface. A common retention mechanism is the adsorption on to solid material, primarily asphaltenes, but also inorganic oxides and organic waxes. The adsorption process can change the wettability of the solid particles. In order to complete the adsorption process the LMW adsorbent must complete with naturally occurring dispersants like resins. The final equilibrium conditions on the surface of the solid particles will hence reflect a balance between attraction to the surface, interaction with resin-like molecules on the surface, and retention mechanisms in the bulk phase.

### B. High Molecular Weight (HMW) Demulsifiers

These molecules are actually supposed to penetrate the interfacial film surrounding the water droplets and hereby to alter the rheological properties of the film material. From the low dosage levels used, i.e., 5–20 ppm, one can conclude that these molecules are extremely efficient as film modifiers. A critical and decisive step for the HMW demulsifiers to perform optimally is the time requirement for the diffusion to the interfacial membrane and for the reorientation movement inside the film until local equilibrium is attained.

### C. Solvents

The action of the solvent can be manifold. However, the commonly used aromatics efficiently dissolve the aromatic particles in a swelling process leaving behind oligomeric and monomeric asphaltenes. As shown before, an increasing aromatic content will gradually decrease and finally eliminate the interfacial activity of the indigenous crude oil components. By experimentally following the interfacial pressure as a function of the aromatic concentration one can conclude qualitatively if the level of emulsion stability is high or low.

Most chemical agents used for demulsification are preferentially oil-soluble blends consisting of HMW polymers. These blends commonly consist of: (1) flocculants (large, slow acting polymers); (2) coalescers (LMW polyethers); (3) wetting agents; and (4) solvents/cosolvents. Some chemical structures of demulsifiers used for breaking crude oil emulsions have been listed by Jones et al. (42). Much work has been carried out in order to identify and understand the mechanisms behind chemical demulsification. Fiocco (43) concluded that the interfacial viscosity was kept at a low level when demulsifiers were present. Later on it was realized that the interfacial shear viscosity of crude oil emulsions does not have to be very low in order to ensure accelerated water separation (44).

Wasan and coworkers (45, 46) investigated the coalescence of systems containing petroleum sulfonates. They concluded that the coalescence rates correlated well with the interfacial shear viscosity, while no correlation was observed with the interfacial tension. Aveyard and coworkers (47, 48) investigated the correlation between surfactant interfacial behavior, surfactant association, and the destabilization efficiency. They observed a clear correlation between the demulsifier concentration at optimal demulsification efficiency and the critical micellization concentration (CMC) of the demulsifier in the crude oil system as long as simple surfactants were used. This means that the monomer activity of the surfactants is crucial the for destabilization of the emulsion system. Wasan and coworkers (49, 50) investigated in detail the processes taking place at the O/W interface during a destabilization process with a LMW amphiphilic compound. From studies of different additives they concluded that oil-soluble destabilizers should be able to partition into the aqueous droplets in order to act as destabilizers. The concentration of the demulsifier inside the droplets should be high enough to ensure a diffusion flux to the O/W interface. In order to be efficient as destabilizers the additives must show a high rate of adsorption to the interface. Wasan and coworkers also emphasized the importance of sufficiently high interfacial activity of the demulsifier to suppress the interfacial tension gradient. In this way the film drainage will be accelerated and droplet coalescence will be promoted. Little (51) suggested that the sequence of steps leading to demulsification of peteroleum emulsions involves the displacement of asphaltic material from the interface by the demulsifier followed by the formation of demulsifier micelles which solubilize and/or stabilize the asphaltene compounds in the oil.

Krawczyk (44) investigated the influence of different demulsifiers on the stability of water-in-crude oil emulsions. He defined a partitioning coefficient, $KP = c_a/c_o$, where $c_a$ refers to the demulsifier concentration in the aqueous phase, and $c_o$ to the concentration in the oil phase. He concluded that demulsifiers with $K = 1$ gives the best results. He also concluded that the interfacial activity and adsorption kinetics of the demulsifier are important parameters. The interfacial region can be expected to be more dynamic, and considerable interfacial fluctuations may occur in the presence of medium-chain alcohols.

The mechanism behind a destabilization with surfactants is probably an interfacial competition. In this situation the indigenous crude oil film will be totally or partially replaced by a surfactant layer which cannot stabilize the crude oil emulsion. When comparing two different hydrophobic surfactants, tetraoxyethylene–nonylphenol ether (Triton N-42) and sodium bis-(2-ethylhexyl)sulfosuccinate (AOT), it was found that the ionic surfactant, AOT, was more efficient than the nonionic analogue. Three different hydrophilic, fluorinated surfactants were also investigated in Ref. 52. They were all very efficient as destabilizers, probably because of their high interfacial activity (53, 54). As mentioned earlier, Wasan and coworkers (49, 50) have analyzed in detail the processes taking place at the O/W interface during destabilization. The results for the hydrophobic surfactants are in direct agreement with their conclusions. Also, in the case of common solvents where we found medium-chain alcohols to be efficient as destabilizers, the results also correspond with their conclusions. The medium-chain alcohols are soluble in all three pseudophases and will therefore partition between these. The hydrophobic surfactant AOT is soluble in water up to a few per cent, and will therefore also be present in the aqueous phase, whereas Triton N-42 is completely water insoluble. This will most likely contribute to the differences between the surfactants.

Aveyard and coworkers (47, 48) have stressed the importance of monomer activity when simple surfactants are used as demulsifiers. For a commercial demulsifier the interfacial tension between oil/water seems to pass through a minimum for NaCl concentrations between zero and 1 M. According to Menon and Wasan (55), AOT has been found to have a cmc at approximately 300 ppm in a water/oil system with asphaltenes present. This means that in our destabilization tests, where the concentration of AOT is up to 100 ppm, the results correspond with the conclusions from Aveyard (47, 48). Fluorinated surfactants have been investigated for systems containing both distilled water and synthetic formation water (52). The results showed that the resolution of water was faster when synthetic formation water was used as the dispersed phase. The explanation of this might be in accordance with Aveyard's conclusions (47, 48). In the case of Triton N-42 the influence of salt is not believed to be significant since this is a nonionic surfactant, and its phase behavior is not so sensitive to the addition of salt.

The mechanism behind destabilization with macromolecules is very dependent on the size of the molecule. Polymers of lower molecular mass can show a strong affinity to the oil/water interface, adsorb irreversibly and destabilize in this way. Another route of destabilization is flocculation. Flocculation is an aggregation process in which droplets form three-dimensional clusters, each droplet retaining its individual identity. In order to model the importance of flocculation in the destabilization of model systems, one can investigate $\alpha$-alumina dispersions (52).

## V. EXPERIMENTAL DEMULSIFICATION

In this section we compile information about demulsifiers active in W/O emulsions (or added prior to the emulsification) and their performance in Langmuir and Langmuir–Blodgett films. We have also performed an AFM study on the demulsifiers in order to visualize the interactions taking place between indigenous crude oil surfactants and the LMW/HMW demulsifiers.

### A. Crude Oil Matrix

The crudes span geographically over large areas: North Sea, European continent, Africa, Asia, etc. This is a necessity since if the crude oils in the test matrix are interrelated one cannot universalize the results. Table 1 lists the crude oils and their origin. To start with we determined the inversion point (or alternatively, the maximum content of water that can be introduced into the oil without a phase separation). We have chosen to study emulsions that are 10% below the inversion point. Exceptions in this respect are the two European crudes with 5% water stabilized. The crude oils were characterized by means of density, surface tension, and viscosity measurements. The results are summarized in Table 2. All experiments involving emulsions were carried out at 50°C. The reason for working at elevated temperature is to melt the wax in the oils and thereby prevent the influence of the wax on emulsion stability. The elevated temperature is also more closely related to the real working temperature used in the processes in the field.

**Table 1** Crude Oils and Their Origin. Types 1–3 Refer to Increasing Emulsion Stability; Model Emulsions are Based on Extracted Asphaltenes

| Emulsion | Crude | Ratio of W/O used | Inversion point | Crude based emulsion | Model oil based emulsion |
|---|---|---|---|---|---|
| 1 | Venezuela | 40/60 | 50/50 | Type 3 | Type 2 (3) |
| 2 | Nigeria | 50/50 | 60/40 | Type 2 | Type 2 |
| 3 | Nigeria | 30/70 | 40/60 | Type 2 | Type 2 |
| 4 | North Sea | 50/50 | 60/40 | Type 2 | Type 2 |
| 5 | North Sea | 40/60 | 50/50 | Type 2 (3) | Type 3 |
| 6 | European continent | 40/60 | 50/50 | Type 2 | Type 2 |
| 7 | North Sea | 5/95 | 5/95 | Type 1 | [a] |
| 8 | North Sea | 5/95 | 5/95 | Type 1 | [a] |
| 9 | North Sea | 40/60 | 50/50 | Type 2 (3) | Type 2 (3) |
| 10 | North Sea | 30/70 | 30/70 | Type 1 | [a] |
| 11 | North Sea | 20/80 | 30/70 | Type 1 | [a] |
| 12 | North Sea | 20/80 | 30/70 | Type 2 | [a] |
| 13 | North Sea | 20/80 | 30/70 | Type 2 (1) | Type 2 |

[a]Model oils based on asphaltene that did not stabilize any water.
Type 1 – least stable; Type 3 – most stable.

**Table 2** Physical Characteristics of Crude Oils

| Crude nr. | Surface tension, $\gamma_{ao}$(mN/m) at 25°C | Density (g/ml), $\delta$ 25°C | Interfacial tension with 3.5% NaCl (mN/m) | Viscosity (cP), 25°C | Viscosity (cP), 50°C | Asphaltene content (%) |
|---|---|---|---|---|---|---|
| 1 | 27.0 | 0.88 | 23.95 | 35.7 | 13.8 | 7.82 |
| 2 | 26.5 | 0.84 | 23.06 | 8.95 | 4.34 | 2.58 |
| 3 | 27.8 | 0.87 | 16.20 | 70.8 | 20.05 | 5.16 |
| 4 | 27.6 | 0.87 | 27.79 | 14.3 | 2.12 | 3.63 |
| 5 | 25.9 | 0.86 | 27.66 | 15.6 | 6.97 | 11.69 |
| 6 | 26.9 | 0.83 | 28.00 | 12.0 | 5.96 | 1.86 |
| 7 | 24.8 | 0.84 | 19.75 | 2.93 | 1.93 | Negligible |
| 8 | 24.7 | 0.84 | 21.31 | 3.05 | 2.35 | Negligible |
| 9 | 29.9 | 0.92 | 25.08 | Above critical viscosity | 50.95 | 2.59 |
| 10 | 24.5 | 0.80 | 23.5 | 3.95 | 2.37 | Negligible |
| 11 | 25.4 | 0.82 | 23.86 | 5.38 | 2.69 | 0.5 |
| 12 | 26.8 | 0.84 | 31.80 | 12.7 | 5.43 | 1.99 |
| 13 | 24.8 | 0.81 | 25.45 | 3.27 | 2.09 | 2.19 |

**Table 3** Interfacial Tensions of Low Molecular Weight Chemical Additives

| Nr. | Chemical | Interfacial tension (25 ppm), $\gamma_{wo}$(mN/m) | Interfacial tension (50 ppm), $\gamma_{wo}$(mN/m) | Interfacial tension (100 ppm), $\gamma_{wo}$(mN/m) |
|---|---|---|---|---|
| 1 | Fatty alcohol ethoxylate | 27.8 | 25.4 | 23.6 |
| 2 | Vinylidene alcohol | 40.2 | 40.0 | 39.4 |
| 3 | Sulfosuccinate | 12.4 | 9.3 | 5.3 |
| 4 | Linear monoisopropylamine alkylbenzene sulfonate | 5.3 | 2.6 | 1.3 |
| 5 | Emulsifier for hydrophobic system | 14.9 | 11.7 | 7.5 |
| 6 | Coconut fatty acid diethanolamide | 27.5 | 23.7 | 19.8 |
| 7 | Plasticizer | 36.5 | 36.2 | 36.5 |
| 8 | Rape oil fatty acid diethanol | 26.5 | 23.2 | 19.3 |
| 9 | Anionic/nonionic blend | 4.5 | 2.6 | 1.6 |
| 10 | Emulsifier for polymerization processes | 39.3 | 39.4 | 38.6 |
| 11 | Fatty alcohol | 39.2 | 38.5 | 38.4 |
| 12 | Long-chain fatty alcohol | 38.3 | 35.4 | 31.5 |
| 13 | long-chain $\alpha$-olefin | 39.4 | 38.8 | 38.6 |

Oil phase = 30/70 toluene/decane.
Aqueous phase = 3.5% NaCl.

**Table 4** Interfacial Tensions of High Molecular Weight Chemical Additives

| Nr. | Chemical | Interfacial tension (25 ppm), $\gamma_{wo}$(mN/m) | Interfacial tension (50 ppm), $\gamma_{wo}$(mN/m) | Interfacial tension (100 ppm), $\gamma_{wo}$(mN/m) |
|---|---|---|---|---|
| A | Complex block polymer | 3.4 | 0.7 | 0.6 |
| B | Complex block polymer | 10.5 | 7.7 | 7.3 |
| C | Phenolic resin alkoxylate | 1.1 | 0.6 | 0.4 |
| D | Phenolic resin alkoxylate | 0.2 | 0.2 | 0.2 |
| E | Phenolic resin alkoxylate | 5.0 | 3.6 | 2.4 |
| F | Complex block polymer | 17.4 | 16.2 | 15.1 |
| G | Complex block polymer | 10.4 | 9.2 | 7.7 |
| H | Mildly cationic; high molecular weight block polymer | 15.6 | 15.5 | 15.8 |
| I | Polyester | 9.2 | 8.0 | 6.3 |
| J | Polyol dispoxide reaction product | 16.2 | 14.6 | 12.5 |

Oil phase = 30/70 toluene/decane.
Brine = 3.5% NaCl.

## B. Chemical Additives

The chemicals added represent two of the categories mentioned above, i.e., the LMW and HMW demulsifiers. The definition of these additives (see Tables 3 and 4) is very general. However, from a functional point of view their interfacial activity under operational conditions is of interest. The interfacial tension between an organic phase (30/70 toluene/decane) and an aqueous phase (with 3.5% NaCl) was determined upon addition of 25, 50, and 100 ppm of additives; $\gamma_0$ without additives is 36.3 mN/m.

## C. Destabilization

Experimental conditions are found in the paper by Djuve et al. (56).

The stability of the different crude oil-based emulsions varies a lot. The water cuts range from 5 to 60%. These values can be compared with the stability for the model emulsions containing only dissolved asphaltene residues (see Table 1). This large difference in the water cut dispersed cannot be explained by the differences in asphaltene content alone (Table 2). The large difference in stabilization ability reflects the wide and different distribution in size, state, structure, polarity, and mass that exists between the asphaltenes found in each oil. It is generally believed that it is mostly the state of the asphaltene and not the amount that controls the stability in an W/O emulsion, i.e., whether the asphaltenes are in a particulate form or not.

The effect of adding a demulsifier is presented in Tables 5 and 6. Table 5 contains the test results for LMW additives, and Table 6 gives the test results when the HMW additives have been used. Both tables refer to the percentage of water separated after 30 min. From the tables, some interesting features are revealed.

Naturally, the addition of demulsifiers affect the rate of separation. If we compare the separation rate of water from crude oil emulsions with the addition of chemicals compared with blank samples, it is clear that the addition can enhance separation. This is not true for all chemicals added. Some demulsifiers have no apparent effect on separation and some demulsifiers even make the emulsions more stable. For the last case the separation of water after addition of demulsifier is lower than without demulsifier. The average separation without addition of any chemicals is 21.5% of water after 30 min and particularly for the LMW chemicals a reduced separation is observed. The HMW chemicals on the other hand seem to enhance separation.

The difference found in the performance between the HMW and LMW chemical activites should be traced back to the interfacial film and the added species. It should be noted that the interfacial tension measurements for the demulsifiers refer to a pure W/O interface, while the demulsifier action actually refers to a W/O interface covered by indigenous components, like asphaltenes, resins, waxes, etc. With a weak reversible adsorption on to the film material, the destabilizing effect of the LMW additives (e.g., oil-soluble surfactants) will be rather limited, if penetration into

**Table 5** Effect of Low Molecular Weight Chemicals on Crude Oil Based W/O Emulsions[a]

| Emulsion sample | 1 | 2 | 3 | 4 | 5 | 6 | 7 | 8 | 9 | 10 | 11 | 12 | 13 | $\sum sep/n_{oil}$ |
|---|---|---|---|---|---|---|---|---|---|---|---|---|---|---|
| 1 | 0 | 0 | 0 | 0 | 0 | 0 | 50 | 20 | 0 | 40 | 0 | 0 | 20 | 10 |
| 2 | 0 | 0 | 0 | 0 | 0 | 0 | 40 | 0 | 0 | 73 | 0 | 0 | 0 | 8.6 |
| 3 | 0 | 0 | 0 | 100 | 0 | 20 | 100 | 60 | 0 | 56 | 0 | 0 | 0 | 25.8 |
| 4 | 0 | 0 | 0 | 16 | 0 | 0 | 100 | 40 | 0 | 65 | 0 | 0 | 0 | 17 |
| 5 | 0 | 0 | 0 | 4 | 0 | 0 | 80 | 60 | 0 | 70 | 0 | 0 | 0 | 16.5 |
| 6 | 0 | 0 | 0 | 0 | 0 | 0 | 60 | 40 | 0 | 83 | 0 | 0 | 0 | 14 |
| 7 | 0 | 0 | 0 | 0 | 0 | 0 | 40 | 0 | 0 | 93 | 0 | 0 | 0 | 10.3 |
| 8 | 0 | 0 | 0 | 0 | 0 | 0 | 60 | 10 | 0 | 100 | 0 | 0 | 0 | 13 |
| 9 | 0 | 8 | 0 | 0 | 0 | 5 | 80 | 60 | 0 | 100 | 0 | 0 | 0 | 19.5 |
| 10 | 0 | 0 | 0 | 0 | 0 | 0 | 0 | 80 | 0 | 100 | 0 | 0 | 0 | 13.8 |
| 11 | 0 | 0 | 0 | 0 | 0 | 0 | 0 | 80 | 0 | 67 | 0 | 0 | 0 | 11.3 |
| 12 | 0 | 0 | 0 | 0 | 0 | 0 | 80 | 75 | 0 | 67 | 0 | 0 | 0 | 17.0 |
| 13 | **0** | 0 | 0 | 0 | 0 | 0 | 80 | 55 | 0 | 83 | 0 | 0 | 0 | 16.5 |
| Blank | 0 | 0 | 0 | 0 | 0 | 0 | 80 | 100 | 0 | 100 | 0 | 0 | 0 | 21.5 |

[a]Separation is given in percentage of water separated after 30 min.

the film material is obstructed. From Langmuir studies of asphaltene films it is known that polymeric demulsifiers can penetrate the films and strongly modify the film properties (57).

Each demulsifier also behaves differently, depending on which oil the emulsion is based on (Tables 5 and 6). That was expected since most of the crude oils are not interrelated.

Based on the results from the bottle tests outlined above a selection of crudes for the next experiments was made, omitting the crude oils that caused either spontaneous separation or a complete separation within 30 min.

### D. Model Emulsions Based on Asphaltene

Generally, for water-in-crude oil emulsions the indigenous component thought to have the largest effect on stability is the asphaltenes. Therefore, model systems based on asphaltenes were prepared for selected oils. From Table 1 one can observe that not all precipitated asphaltene fractions could stabilize a model emulsion. However, many model emulsions had a similar stability as the original crude oil-based emulsion, indicating that the fraction extracted from the oil plays a central role in the stabilization.

**Table 6** Effect of High Molecular Weight Chemicals on Crude Oil-based W/O Emulsions[a]

| Emulsion sample | 1 | 2 | 3 | 4 | 5 | 6 | 7 | 8 | 9 | 10 | 11 | 12 | 13 | $\sum sep/n_{oil}$ |
|---|---|---|---|---|---|---|---|---|---|---|---|---|---|---|
| A | 0 | 64 | 0 | 100 | 0 | 15 | 60 | 84 | 0 | 67 | 98 | 98 | 58 | 49.5 |
| B | 0 | 20 | 0 | 0 | 0 | 0 | 80 | 100 | 0 | 100 | 85 | 5 | 50 | 33 |
| C | 0 | 80 | 0 | 38 | 0 | 47.5 | 80 | 80 | 0 | 56.7 | 93 | 100 | 95 | 51.6 |
| D | 0 | 63 | 0 | 100 | 0 | 45 | 80 | 82 | 0 | 80 | 87 | 95 | 40 | 51.7 |
| E | 0 | 63.8 | 0 | 18 | 0 | 65 | 100 | 100 | 0 | 100 | 100 | 100 | 85 | 49.7 |
| F | 0 | 10 | 0 | 0 | 0 | 62.5 | 100 | 100 | 0 | 100 | 100 | 100 | 0 | 44.0 |
| G | 0 | 67 | 0 | 84 | 0 | 54 | 60 | 94 | 0 | 93 | 85 | 100 | 68 | 54.2 |
| H | 0 | 10 | 0 | 100 | 0 | 70 | 80 | 60 | 0 | 70 | 100 | 0 | 93 | 44.8 |
| I | 0 | 61 | 0 | 0 | 0 | 18 | 60 | 70 | 0 | 93 | 85 | 0 | 48 | 33.5 |
| J | 0 | 0 | 0 | 0 | 0 | 54 | 40 | 60 | 0 | 88 | 100 | 0 | 48 | 30 |
| **Blank** | 0 | 0 | 0 | 0 | 0 | 0 | 80 | 100 | 0 | 100 | 0 | 0 | 0 | 21.5 |

[a]Separation is given in percentage of water separated after 30 min.

The representability of such a "model emulsion" vis-à-vis the original emulsions has been debated. This brings forward the question of stabilization mechanisms and the state of asphaltenes. Basically we try to mimick interfacial conditions from true crude oil-based emulsions to model emulsions. In order to do so it is essential that the model oil used (heptol) can promote particle formation in the asphaltenes. In this way there should be a similarity in asphaltene-based nanoparticles located at the W/O interface in both types of emulsions under study. Since the asphaltene particles will hinder an efficient coalescence of the aqueous droplets, one can expect approximately the same level of stability against coalescence and similar actions of the demulsifiers. Addition of demulsifiers to the asphaltene-stabilized model emulsions accelerated in some cases the resolution of water. In particular three demulsifiers seemed to be most efficient, i.e., A, C, and G. It is obvious when comparing Tables 6 and 7 that the same demulsifiers are effective in both model emulsion systems and true crude oil-based W/O emulsions. This means that we can trace back the destabilization effect to an interaction between the demulsifying agent and the asphaltene fraction in the crude, and that this interaction is the most significant one. Other possibilities for interactions leading to destabilization would be demulsifier/wax particles, demulsifier/resins, demulsifier/solid inorganic particles, etc. An investigation of asphaltene-stabilized W/O emulsions has obviously shed light on fundamental destabilization mechanisms in the crude oil-based emulsions.

Table 4 shows that the demulsifiers A and C exhibit a low interfacial tension, i.e., of the order of a couple of units. In the case of G, $\gamma_{w/o}$ is about 10 times higher and reaches a value of $\simeq$ 10 mN m$^{-1}$. Obviously, pure displacement processes, where demulsifiers A and C (owing to a lower interfacial tension toward water) can create a new interface and in this way destabilize an emulsion, play a role in the destabilization process. Also, G has a $\gamma_{w/o}$ value lower than that of pure asphaltenes at the interface. In the latter case $\gamma_{w/o}$ is around 20–25 mN/m.

### E. Demulsifiers Used as Inhibitors

We have also added the destabilizing agents directly to the oil *before* the emulsification. The result as revealed from Tables 8 and 9 is very encouraging. In most cases there is a substantial enhancement in the efficiency of the action of the added chemicals, also for the low molecular species. The reason for this can be two-fold, i.e., either an interfacial competition or a strong bulk interacation between the demulsifier added and the stabilizing crude oil species. The interfacial competition can be traced back to the $\gamma_{w/o}$ values. Although the concentrations of the demulsifier added are small ($\approx$ 50 ppm) and there is most likely not enough molecules to create a stable emulsion with all the water molecules dispersed, some molecules of A, B, and G present at the interface can cancel the stabilizing properties of asphaltene particles at the W/O interface. A strong bulk interaction between the demulsifiers and the asphaltenes must change the state of the asphaltene particles in order to cancel their stabilizing effects. Obviously, one should anticipate the demulsifying agents dissolving the asphaltene particles to substantially smaller units

**Table 7** Effect of High Molecular Weight Chemicals on Asphaltene-based Model Emulsions[a]

| Emulsion chemical | 1 | 2 | 3 | 4 | 5 | 6 | 9 | 13 | $\sum$ sep/$n_{oil}$ |
|---|---|---|---|---|---|---|---|---|---|
| A | 0 | 92.5 | 0 | 80 | 0 | 92.5 | 0 | 0 | 33.1 |
| B | 0 | 0 | 3 | 43 | 0 | 43 | 0 | 0 | 11 |
| C | 58 | 88 | 77 | 46 | 0 | 75 | 0 | 0 | 43 |
| D | 0 | 0 | 0 | 0 | 0 | 62.5 | 0 | 0 | 7.8 |
| E | 15 | 87.5 | 6.6 | 5 | 0 | 40 | 0 | 0 | 19.3 |
| F | 0 | 0 | 0 | 0 | 0 | 27.5 | 0 | 0 | 3.4 |
| G | 90 | 90 | 3.3 | 12.5 | 0 | 90 | 0 | 75 | 45.1 |
| H | 0 | 0 | 3 | 0 | 0 | 10 | 0 | 0 | 2 |
| I | 0 | 0 | 0 | 0 | 0 | 80 | 0 | 0 | 10 |
| J | 0 | 0 | 3 | 0 | 0 | 18 | 0 | 0 | 3 |
| Blank | 0 | 0 | 0 | 0 | 0 | 0 | 0 | 0 | 0 |

[a]Separation is given in percentage water separated after 30 min.

Table 8 Effect of High Molecular Weight Chemicals as Inhibitors on Crude Oil Emulsions[a,b]

| Emulsion chemical | 1 | 2 | 3 | 4 | 5 | 6 | 9 | 13 | $\sum sep/n_{oil}$ |
|---|---|---|---|---|---|---|---|---|---|
| A | 75 | 92 | 7 | 100 | 55 | 80 | 100 | 35 | 68.0 |
| B | 40 | 44 | 77 | 80 | 85 | 78 | 85 | 25 | 64.3 |
| C | 0 | 94 | 53 | 78 | 65 | 93 | 63 | 100 | 68.3 |
| D | 35 | 94 | 0 | 100 | 60 | 83 | 45 | 65 | 60.3 |
| E | 43 | 76 | 67 | 78 | 0 | 60 | 88 | 85 | 62.9 |
| F | 50 | 84 | 0 | 80 | 63 | 73 | 53 | 5 | 51.0 |
| G | 0 | 98 | 73 | 100 | 75 | 83 | 75 | 95 | 74.9 |
| H | 65 | 98 | 70 | 100 | 78 | 93 | 83 | 100 | 85.9 |
| I | 100 | 92 | 0 | 76 | 60 | 88 | 80 | 100 | 74.5 |
| J | 78 | 92 | 47 | 82 | 83 | 80 | 70 | 100 | 79.0 |
| Blank | 0 | 0 | 0 | 0 | 0 | 0 | 0 | 0 | 0 |

[a]Chemicals added before emulsification.
[b]Separation in percentage water separated after 30 min.

*not* possessing stabilizing effects. There are some indications that this might be the case (58).

## VI. DEMULSIFICATION UNDER LABORATORY AND FIELD CONDITIONS

Normally, the bottle-shake tests with depressurized crude oils are upscaled to real separation conditions topside. However, it has been constantly pointed out that the samples in use at the laboratory are not representative of the samples from the same field. The main reason for this is that the laboratory samples have undergone oxidation upon storage. Another essential deviation in sample representability is due to the time delay in sampling of the crude oil samples to be compared. It is well known that the crude oil characteristic from a field consisting of several wells (up to 30–40 wells) will change over time. The best way to overcome the classical difficulties with representative samples is to work with pressurized samples. The separation rig presented in Fig. 5 has the great advantage of permitting this and preventing the crude oils under study to contact air. In addition to this the mixing conditions (the magnitude of $\Delta P$ over the chokes) can be adjusted to real process conditions.

Table 9 Effect of Low Molecular Weight Chemicals as Inhibitors on Crude Oil Emulsions[a] (Chemicals added before emulsification)

| Emulsion chemical | 1 | 2 | 3 | 4 | 5 | 6 | 9 | 13 | $\sum sep/n_{oil}$ |
|---|---|---|---|---|---|---|---|---|---|
| 1 | 1 | 46 | 0 | 44 | 0 | 25 | 0 | 0 | 13 |
| 2 | 0 | 72 | 0 | 0 | 0 | 0 | 0 | 0 | 8 |
| 3 | 0 | 0 | 0 | 100 | 73 | 73 | 0 | 0 | 27 |
| 4 | 0 | 52 | 0 | 70 | 40 | 80 | 0 | 0 | 27 |
| 5 | 0 | 80 | 0 | 90 | 53 | 70 | 23 | 45 | 40 |
| 6 | 65 | 7 | 0 | 0 | 4 | 0 | 0 | 10 | 22 |
| 7 | 0 | 80 | 0 | 0 | 0 | 0 | 0 | 0 | 9 |
| 8 | 0 | 70 | 0 | 0 | 83 | 0 | 3 | 5 | 21 |
| 9 | 0 | 72 | 0 | 0 | 0 | 48 | 58 | 20 | 22 |
| 10 | 0 | 66 | 0 | 0 | 0 | 0 | 0 | 10 | 8 |
| 11 | 0 | 40 | 0 | 0 | 0 | 38 | 0 | 8 | 10 |
| 12 | 0 | 0 | 0 | 0 | 0 | 0 | 0 | 0 | 0 |
| 13 | 0 | 4 | 0 | 0 | 0 | 0 | 0 | 0 | 5 |
| Blank | 0 | 0 | 0 | 0 | 0 | 0 | 0 | 0 | 0 |

[a]Separation in percent water separated after 30 min.

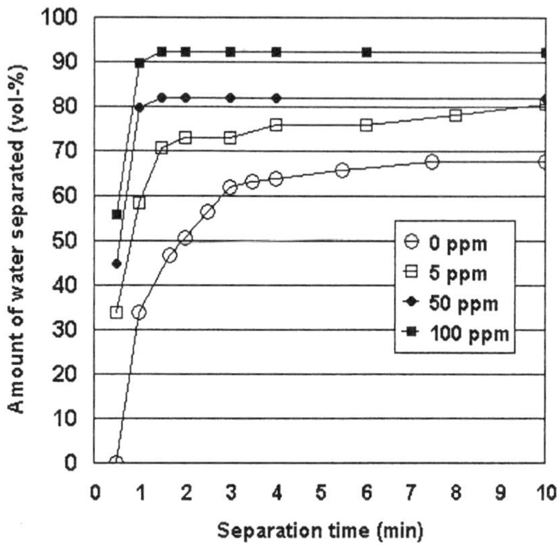


**Figure 6** Examples of separation as function of time. Oil 1 with 20% water cut; Demulsifier B.

## A. Tests of Demulsifiers – Comparison with Field Tests

The laboratory tests were conducted to qualify the separation rig by performing tests as tsimilar as possible to the field tests done previously. An important difference between the tests performed offshore and in the laboratory is the type of separator. The field tests were performed in a horizontal continuous gravity separator whereas the separation in the laboratory rig took place in a vertical batch separator. The oils and brine used in the laboratory were sampled offshore and kept under pressure until the tests were performed.

In the laboratory tests only one module of the separation rig was used. Only one oil was tested at a time, and there was a pressure drop through only one of the choke valves (VD1 in Figure 5). Oil and water were mixed upstream of VD2. There was no pressure drop through VD2. The demulsifier was mixed into the flow line just downstream of VD2. The pressure drop in the system was through VD1 just ahead of the separation cell. The oils tested were at their bubble

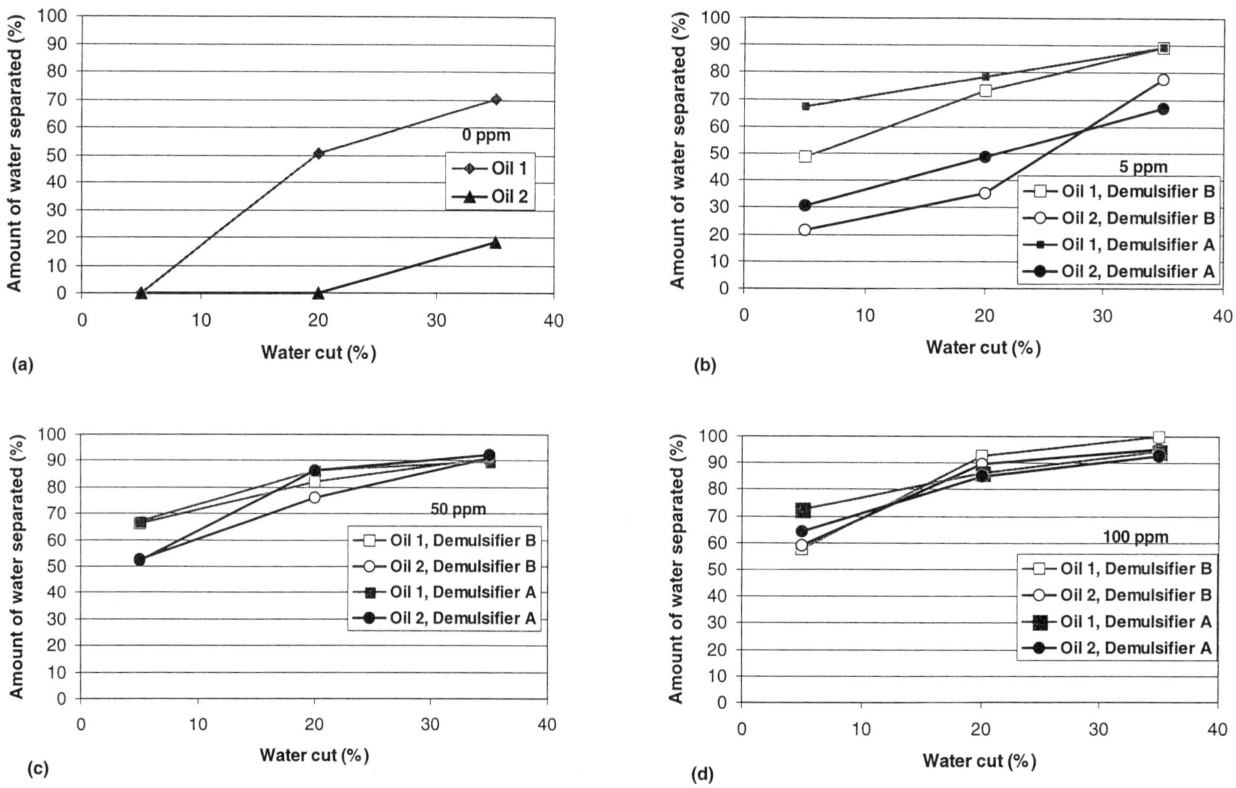


**Figure 7** Amount of water separated after 1.5 min for the two oils 1 and 2 at various water cuts and for the two demulsifiers A and B at various concentrations (1.5 min separation time corresponds to 2 min separation time in Fig. 6. In Fig. 6 the filling time of 30 s is included in the separation time).

points at 11 bar and 60°C. The experiments were performed at 60°C and with a pressure drop through VD1 from 11 to 7 bar. The separation took place under 7 bar pressure. They were typical North Sea crude oils with density and viscosity values for stabilized oils at 60°C at ≃ 0.8 g/ml and ≃ 3.5 mPa. There were small differences in the characteristics of the oils. The more dense oil was also the more viscous oil. The compositions of the two tested demulsifiers were totally different from each other. Three concentrations of the demulsifiers were tested: 5, 50, and 100 ppm. In addition, tests without demulsifier were performed. The water cut values were 5, 20, and 35 vol. %.

Some of the results are shown in Figs 6 and 7. The main results in the laboratory tests were:

1. Oil 1 had better separation characteristics than Oil 2. (Oil 1 was the lighter of the two oils.)
2. Demulsifier A performed better than Demulsifier B at concentrations of 5 and 50 ppm (for water cuts of 20 and 35%).
3. Demulsifier B performed better than Demulsifier A at 100 ppm (for water cuts of 20 and 35%).
4. No increase in separation efficiency was observed when the concentration of Demulsifier A was increased from 50 to 100 ppm.
5. Demulsifier B increased the separation efficiency with increasing concentrations up to 100 ppm.
6. Foam was never any problem (stable for a maximum of 30 s).

All these results confirmed the offshore field-test results. In addition, one could observe visually in the laboratory tests how the demulsifiers affected the system. Without demulsifier in the system an emulsion layer always formed between the oil and the water phase. When the demulsifier was added, no such separate emulsion layer was observed (except for 5 ppm demulsifier in the system with 5% water cut in the more viscous oil).

Results from the laboratory separation rig have also been verified with results from other field tests.

As a conclusion to this section one can say that a laboratory test kit has been constructed which can be used to test oils and chemicals in pressurized systems. The results are consistent with results achieved under offshore field conditions. The results obtained in the laboratory are based on correct sampling and handling of the fluids. The oil samples are kept under pressure and are never exposed to air during storing. Further, the results show that we can dose with chemicals down to concentrations as low as 5 ppm. The advantages of laboratory studies are smaller volumes, cheaper tests, more parameter variations can be performed within

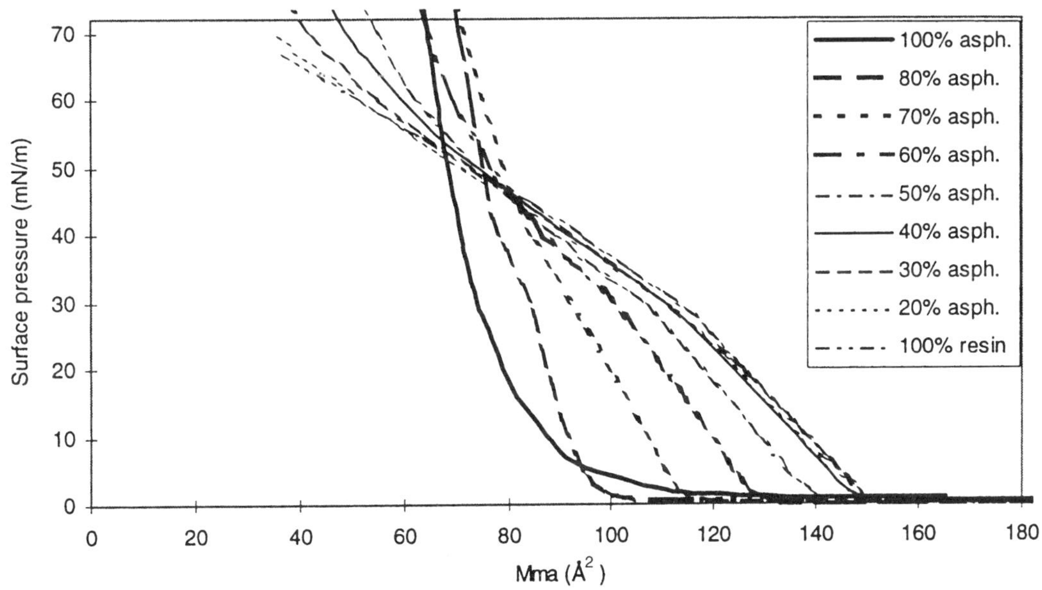


**Figure 8** Π-*A* is isotherms of asphaltene/resin mixtures spread from pure toluene on pure water (bulk concentration = 4 mg/ml).

short time limits, and access to more advanced characterization systems (e.g., drop size measurements) is available.

The separation rig has also been used to show the influence of an internal separator pressure up to 180 bar on the separation characteristics and efficiency.

## B. Langmuir Films

In order to obtain a better understanding of the mechanisms behind the effect of asphaltenes and resins on emulsion stability, we chose to investigate the film properties of these components. Such studies provide information on the rigidity and stability of films consisting of indigenous surface-active material. The rigidity of the interfacial film is important for the stability of emulsions, in as much as a rigid film on the emulsion droplets prevents coalescence, while a highly compressible film is more easily ruptured, leaving the droplets free to coalesce.

By means of the Langmuir technique, asphaltenes are found to build up close-packed rigid films, which give rise to quite high surface pressures. Resin films, on the other hand, are considerably more compressible (Fig. 8). This may explain the experimental observations showing that asphaltenes are able to stabilize crude oil-based emulsions, while resins alone fail to do so. Singh and Pandey (59) also concluded that a high interfacial pressure correlated with high W/O emulsion stability. On adding asphaltenes and resins together to a mixed film, the properties gradually change from a rigid to a compressible structure as the resin content is increased. The resins start to dominate the film properties when the amount of this lighter fraction exceeds 40 wt% (Fig. 8). The more hydrophilic resin fraction starts to dominate the film properties owing to the higher affinity towards the surface.

The influence of chemical additives on asphaltene films on the water surface and at the oil/water interface have also been studied by means of the Langmuir technique. This was done in order to view the interaction between demulsifiers added and asphaltenes, and to show the importance of this on emulsion stability.

The film properties of pure demulsifiers of high molecular weight are shown by the isotherms in Fig. 9. The shape of some of these isotherms, especially that of Demulsifier G and to some extent those of H and I, resembles pure resin films. The others, especially Compound A, give more rigid films, characteristic of the pure asphaltene film.

Compressible resin films will not alone stabilize a crude oil emulsion. Related to this, demulsifiers, which form films of low rigidity and high compressibility, should be the most efficient. When used as demulsifiers, the efficiency depends on the ability of the chemicals to interact with and modify the film built up by asphaltene particles.

Addition of demulsifiers of high molecular weight to the asphaltene film gave the isotherms in Fig. 10. The influence of the chemicals G, H, and I is most pronounced with respect to an increased compressibility, together with a reduced rigidity. The effect of this kind of manipulation of the asphaltene film is similar to the effects observed when resins are mixed together with

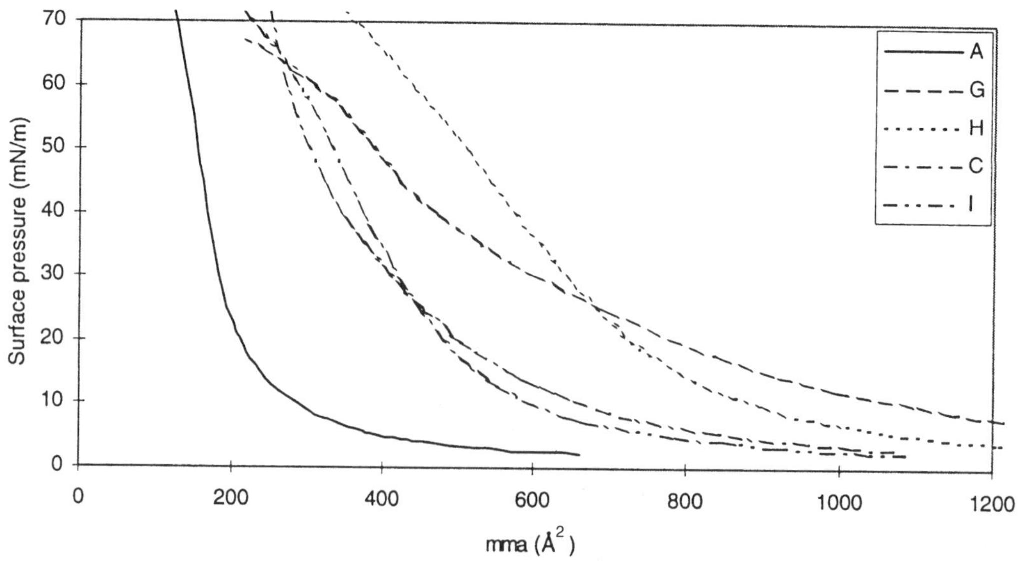


**Figure 9** Π-*A* isotherms of high molecular weight demulsifiers on pure water.

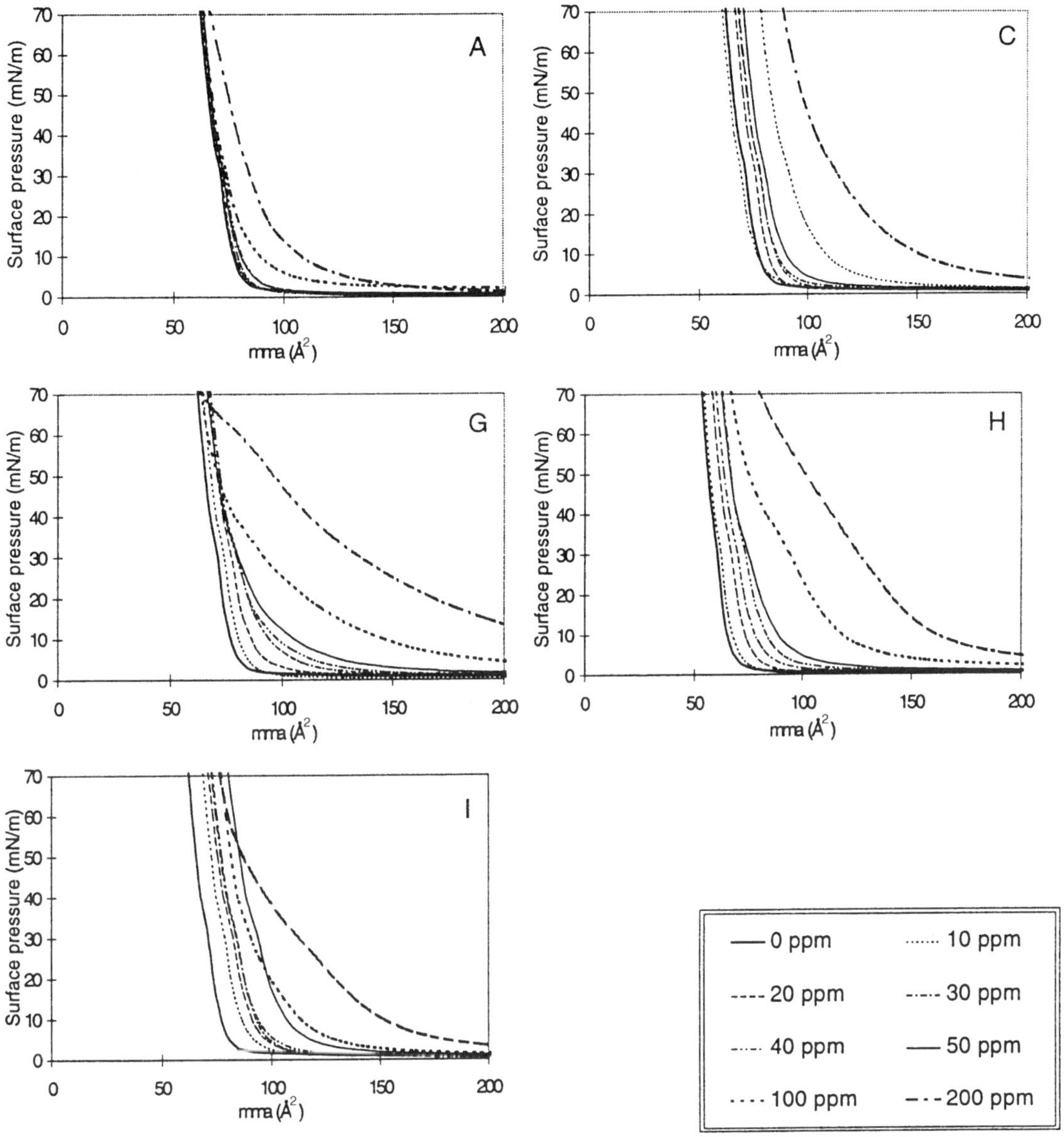


**Figure 10** Π-*A* isotherms of mixed monolayers of asphaltenes and varying concentrations of different demulsifiers on pure water.

asphaltenes (Fig. 8). However, the concentration needed to achieve the same effects is considerably lower when demulsifiers are used instead of resins. Demulsifier A has a quite small influence on a film of asphaltenes. A comparison with Fig. 9 shows that chemical A is the component with the most rigid and asphaltene-like film behavior of all the tested HMW demulsifiers.

From the film studies outlined above one can conclude that the best candidates for emulsion breaking should be G, H, and I. However, the efficiency depends not only on the direct influence of chemical additives within the film, but also on the ability of demulsifiers to reach the W/O interface in an emulsion (diffusion through the fluid). This is a critical step regarding the effective concentration of demulsifiers at the interface. These aspects make it difficult to undertake a direct comparison between the influence of demulsifier on Langmuir surface films, where all demulsifier molecules are implanted in the film, and on real emulsions.

In order to represent more realistic emulsion conditions, Langmuir interfacial films adsorbed at the O/W interface were analyzed. The isotherms depicted in Fig. 11 illustrate some of the film properties of naturally

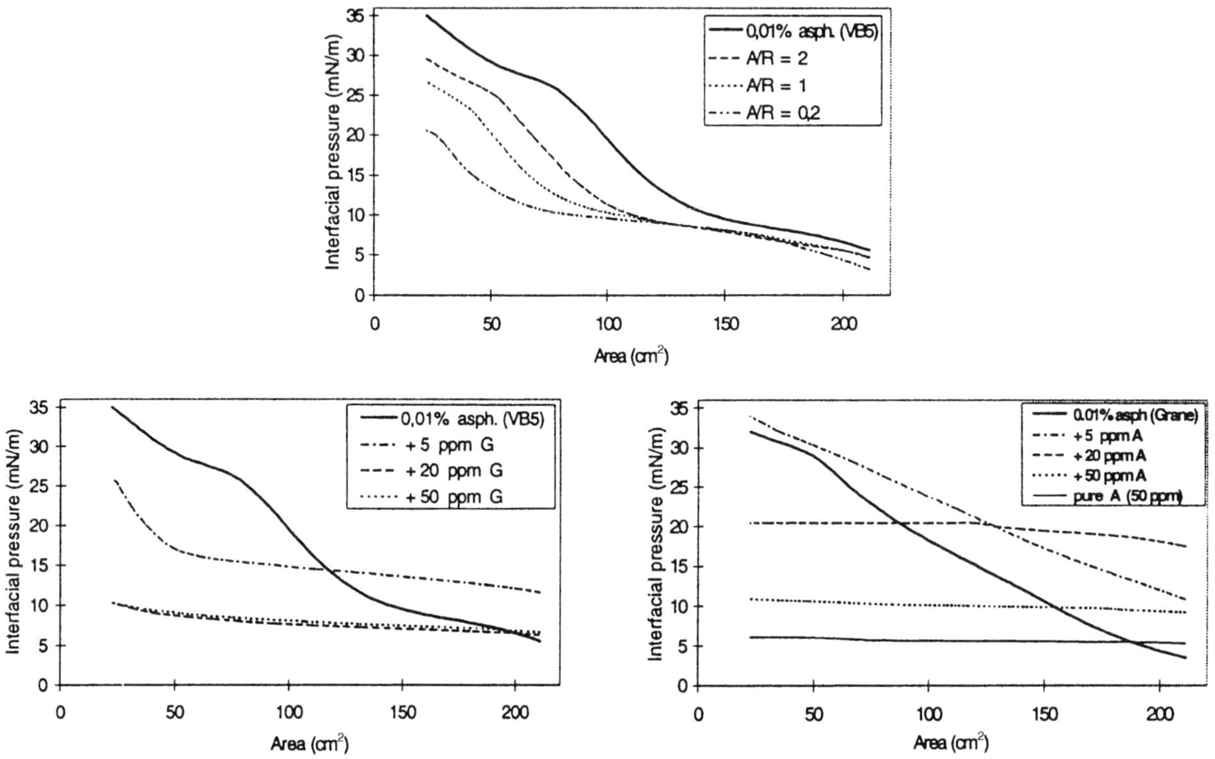


**Figure 11** Interfacial pressure isotherms of films formed between water and oil containing different ratios of asphaltenes and resins or different amounts of added chemicals.

occurring crude oil components adsorbed at the W/O interface.

The oil phase containing only 0.01 wt % asphaltene gives rise to a less rigid interfacial film than observed at the water surface (Fig. 8). This is most likely due to the possibility of the hydrocarbon tails of the asphaltenes to orient toward the highly aliphatic oil phase, making the interactions between the film material and, hence, the pressure increase during film compression, less extensive. In general, interactions between the bulk phase and interfacial components are different from the water/air case.

Addition of resins to 0.01 wt % asphaltene solutions further reduces the adsorption of interfacially active components on to the O/W interface, even if the total amount of naturally occurring surfactants is considerably higher in these oil phases. The reduction is seen as reduced pressure at constant interfacial area. These changes may be attributed to the ability of resins to disperse asphaltenes in the bulk oil phase, and thus prevent this heavy fraction from building up a stabilizing film between oil and water.

Introducing chemical additives together with asphaltenes into the oil phase may highlight the ability of these chemicals to prevent formation of relatively rigid asphaltene films at the O/W interface. For concentrations higher than 20 ppm of chemical A there is no pressure increase during the compression. Hence, the film that is formed at the interface is highly compressible. So instead of increasing the pressure, the components will build up a multilayer, or the film may dissolve under the influence of compression. An increased inhibitor concentration reduces the interfacial pressure, but has no influence on the film behavior. The reduced pressure is probably as a result of a more complete cover of inhibitor at the interface. That is, fewer components from the asphaltene fraction are adsorbed together with the chemical additive when the inhibitor concentration becomes high enough.

The results obtained upon addition of Demulsifier G are similar to those of A. However, G clearly increases the compressibility of the film even at low concentration. The difference between 20 and 50 ppm is quite small, so it is reasonable to believe that maximum efficiency, resulting from the competing adsorption in a system like this, is already reached at a concentration of 20 ppm in

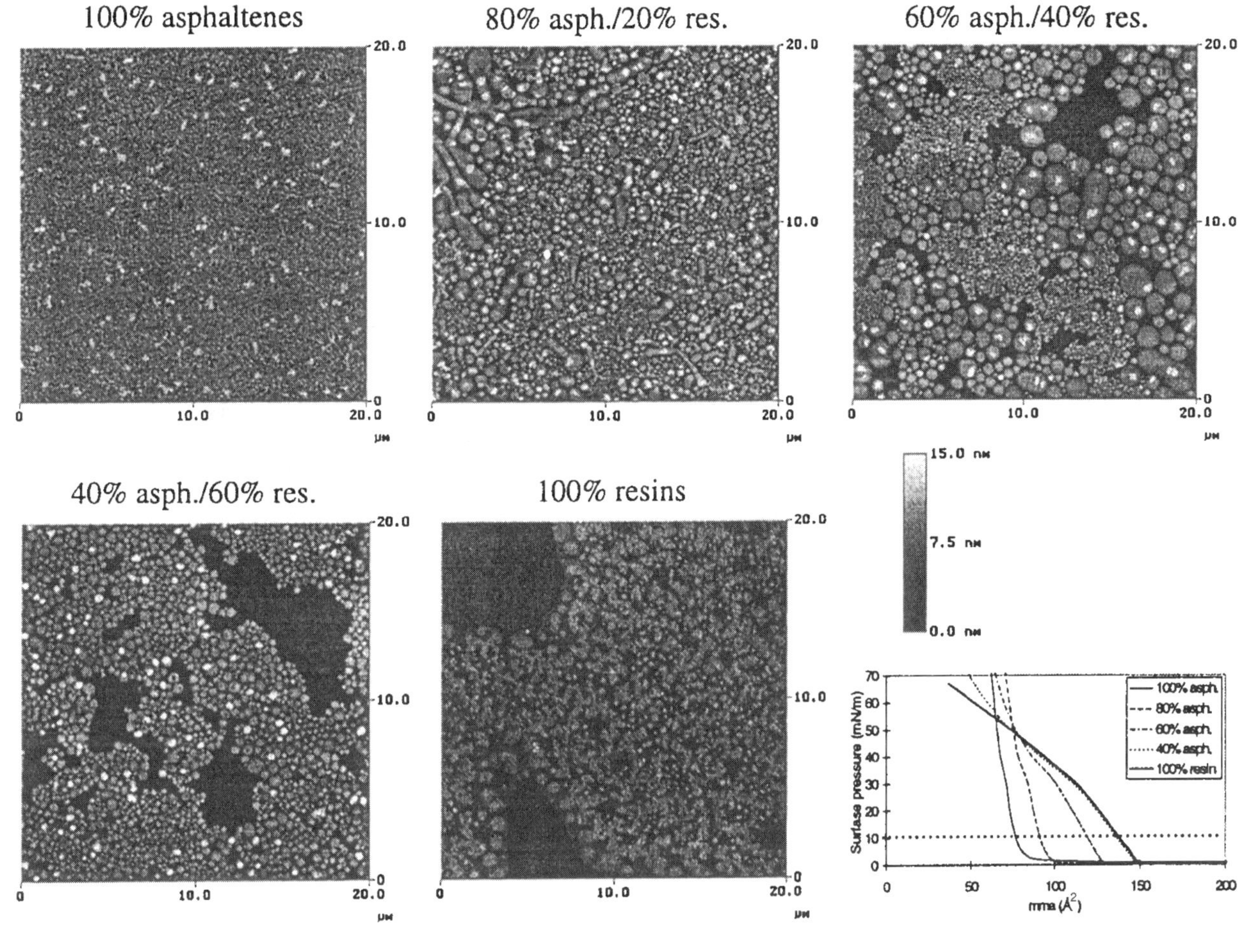


**Figure 12** AFM images (20 × 20 μm) of monolayers with increasing resin-to-asphaltene (*R/A*) ratio; LB film deposited onto mica substrates. The fractions are extracted from a crude from a production field in France (crude F).

the oil phase. With 20 ppm or more of G present, only small amounts of asphaltene will reach the interface.

The results obtained from the Langmuir interfacial film studies are important in explaining why certain chemicals are more effective as inhibitors than as demulsifiers. Obviously, the inhibitor/asphaltene interaction is so strong in the bulk oil phase that the interfacial structures being gradually built up will no longer possess properties required to stabilize W/O emulsions.

### C. Langmuir–Blodgett Films Studied by Means of AFM

Monolayers of asphaltenes and resins on the water surface were transferred at a surface pressure of 10 mN/m on to mica substrates by using the Langmuir–Blodgett technique. In order to visualize the earlier investigated film properties, AFM was used to examine the topography of these deposited layers.

The images shown in Figs 12 and 13 show the structural change in the monolayer at a surface pressure of 10 mN/m, when the composition of the film was gradually changed from pure asphaltenes to pure resins. Images of pure asphaltene show a closed-packed structure of nanosized particles. Addition of resins modifies this rigid structure toward an open structure with regions completely uncovered by film material. Pure resins build up a layer with an open fractal network.

The individual film units increase in size upon addition of resins. This indicates interactions between asphaltenes and resins, providing aggregates of larger dimensions than observed for the pure fractions. Small and moderate amounts of resins give rise to a more polydisperse distribution of the film material, while a further increase in the resin content (i.e., 60 wt % resins) reduces the polydispersity, i.e., the monolayer

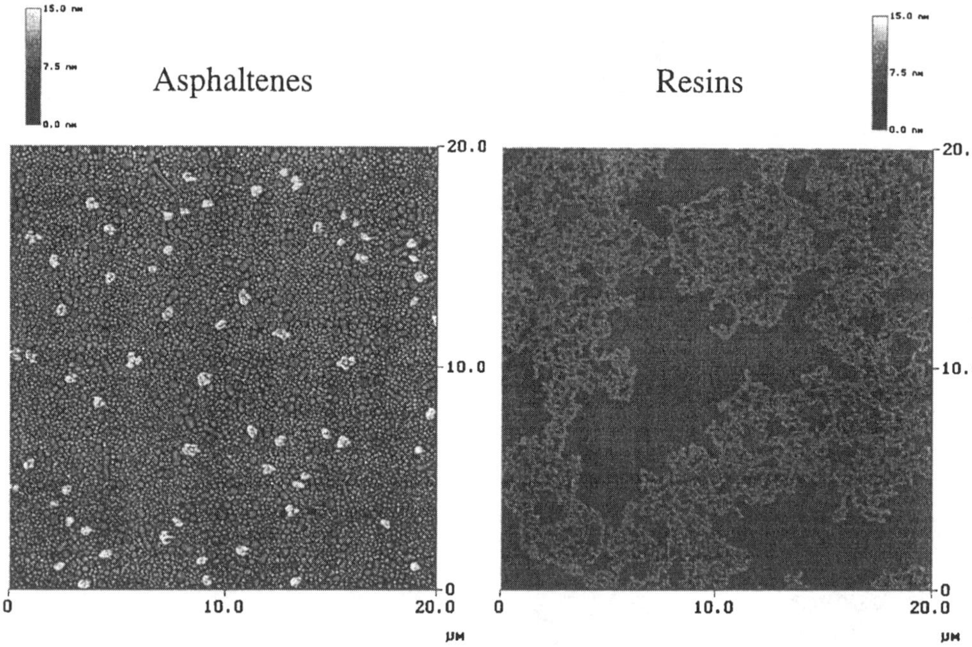


**Figure 13** AFM images (20 × 20 $\mu$m) of monolayers of pure components from a crude from a production field in the North Sea; LB film deposited on to mica substrates.

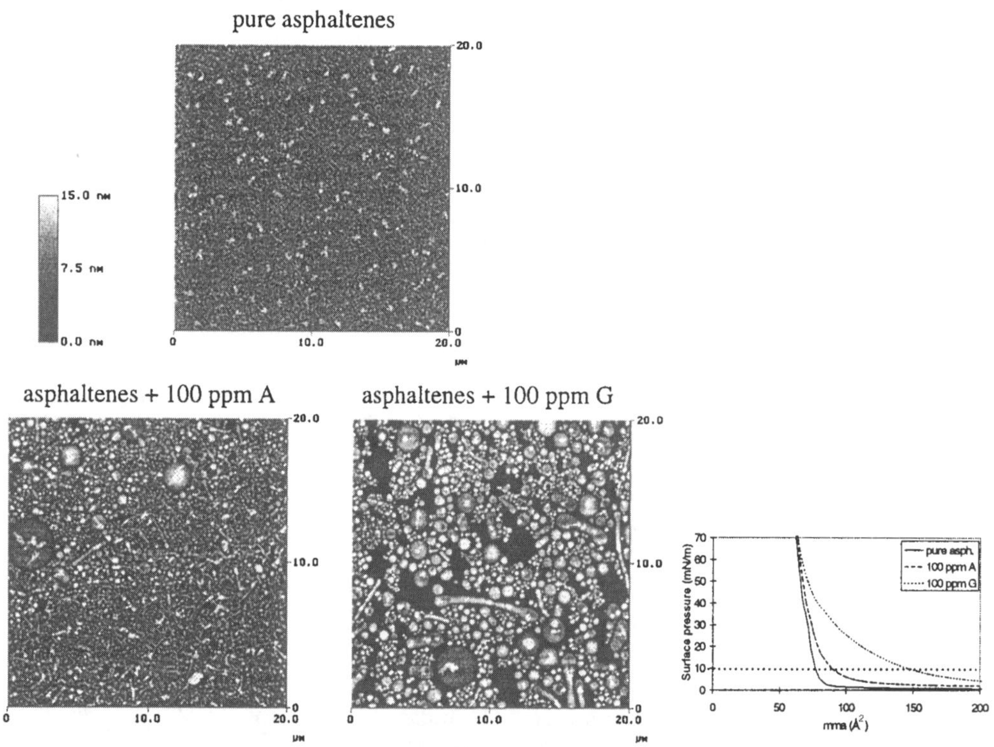


**Figure 14** AFM images (20 × 20 $\mu$m) of monolayers consisting of asphaltenes from crude F and 100 ppm high molecular weight demulsifiers/inhibitors; LB film deposited on to mica substrates.

becomes more uniform in component size when one of the pure fractions dominates the film properties.

The AFM images visualize why asphaltenes alone can stabilize emulsions while films dominated by the resin fraction do not. Hence, when the amount of resins present in the film is so large that the structure in the film changes toward a more open fractal network, the efficiency of film components as emulsifier is reduced.

The AFM images of asphaltene films containing 100 ppm of different HMW demulsifiers/inhibitors (Fig. 14) show that the effect of these components on the film is quite similar to the effect on structural changes brought about by the resins. These results indicate that the observed structural changes in the film are qualitatively essential in order to reduce the emulsion stability.

It is important to keep in mind that the AFM images visualize conditions in Langmuir films at the aqueous surface. Once again all interactions between an oil phase and interfacial components are lacking. In a real W/O emulsion there are no guaranteees that all these components will be present at the W/O interface due to solubility in the oil phase. Hence, results from an AFM study of LB films should not be too far-reach-

**Table 10** Composition of the Samples

| Sample name | Inhibitor concentration (ppm) | Toluene concentration (%) |
|---|---|---|
| Crude | 0 | 0 |
| Crude + Toluene | 0 | 1.25 |
| Inhibitor G (300 ppm) | 300 | 0.75 |
| Inhibitor A (300 ppm) | 300 | 0.75 |

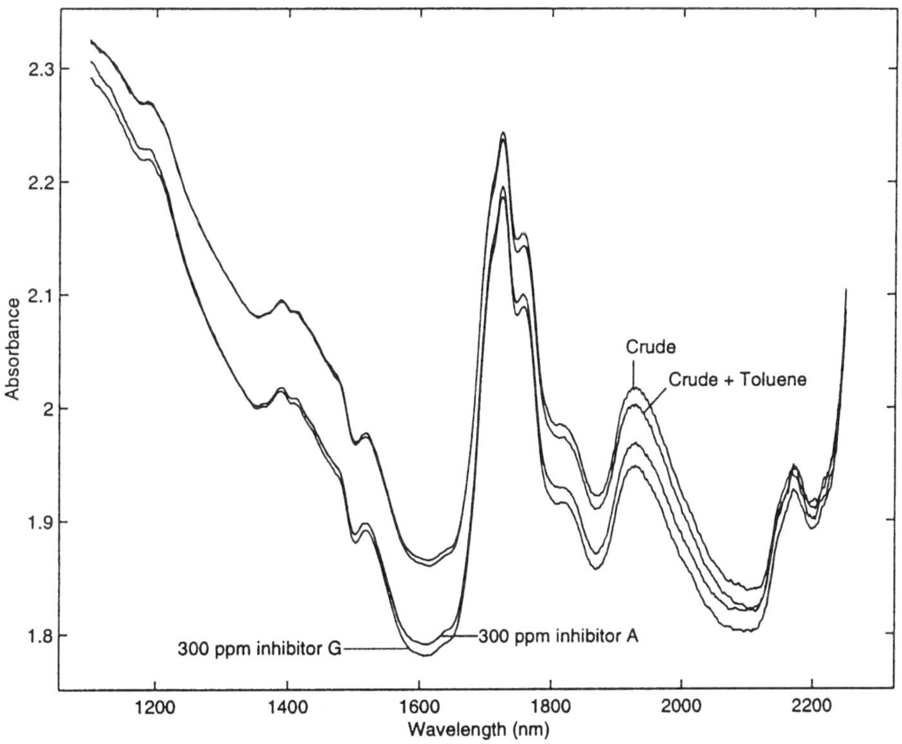


**Figure 15** Near infrared spectra of the Grane crude oil with no additives, and with the addition of 500 ppm toluene, 300 ppm inhibitor G, and 300 ppm inhibitor A.

ing when considering real conditions in W/O emulsions. However, the effect of demulsifiers on the film material remains indisputable.

### D. Near-infrared (NIR) Characterization of the Effect of Emulsion Inhibitors

Aggregates of colloidal size scatter near-infrared radiation ($\lambda = 700$–$2500$ nm) in accordance with Rayleigh theory (60). This is observed in the spectrum as a rise of the spectral baseline. The extinction of radiation increases with increasing radius of the scattering particles, and thus the spectrum yields information about the size of the aggregates. The effect of adding two different emulsion inhibitors to a crude oil was determined by means of near-infrared spectroscopy. In previous work (57) by our group it was stated that these inhibitors have a resin-like influence on the aggregation of asphaltenes, i.e., a solvating effect. Near-infrared spectroscopy should thus be able to detect the changes in the aggregation state by direct measurements on the crude oil. The near-infrared sampling of the crude oil was performed on a NirSystems 6500 spectrophotometer, equipped with a fiber-optic sampling probe for transflectance sampling. The wavelength region was set to 1100–2250 nm. The total pathlength was 2.5 mm. The total number of scans per spectra was set to 32 and the sampling was carried out at 25°C. The compositions of the four samples investigated are listed in Table 10.

The inhibitors were diluted in toluene because of their high viscosity. The effect of toluene alone was tested on one of the samples. Figure 15 shows the near-infrared spectra of the four samples.

The interpretation of Fig. 15 is that inhibitors have a solvating effect on the asphaltene aggregates. The reduction in aggregation size is observed as a decrease in the extinction of radiation due to scattering. It is shown that the effect of the inhibitors is more prominent than the effect of toluene alone. The findings suggest that near-infrared spectroscopy could be used for characterization of the effects of inhibitors on crude oils.

## ACKNOWLEDGMENTS

The technology programs Flucha I and II, financed by the oil industry and the Norwegian Research Council (NFR), are acknowledged for PhD grants to Marit-Helen Ese, Jostein Djuve, Harald Kallevik, and Inge H. Auflem. Statoil A/S is acknowledged for permission to publish results from the high-pressure separation rig.

## REFERENCES

1. P Becher, ed. Encyclopedia of Emulsion Technology. New York: Marcel Dekker, 1983.
2. K Larsson, S Friberg, Food Emulsions. New York: Marcel Dekker, 1990.
3. J Sjöblom, ed. Emulsions – A Fundamental and Practical Approach. NATO ASI Series, Vol 363. Dordrecht: Kluwer, 1991.
4. J Sjöblom, ed. Emulsions and Emulsion Stability. New York: Marcel Dekker, 1996.
5. OC Mullins, E Sheu, eds. Structures and Dynamics of Asphaltenes. New York: Plenum Press, 1998.
6. J Sjöblom, Ø Sæther, Ø Midttun, M-H Ese, O Urdahl, H Førdedal. In: OC Mullins, EY Sheu, eds. Structures and Dynamics of Asphaltenes. New York: Plenum Press. 1998, p 337.
7. JD McLean, PM Spiecker, AP Sullivan, PK Kilpatrick. In: OC Mullins, EY Sheu, eds. Structures and Dynamics of Asphaltenes. New York: Plenum Press, 1998, p 377.
8. MC Petty, WA Barlow. In: G Roberts, ed. Langmuir–Blodgett Films. New York: Plenum Press, 1990.
9. GL Gaines, Jr. Insoluble Monolayers at Liquid–Gas Interfaces. New York: John Wiley, 1966.
10. KS Birdi. Lipid and Biopolymer Monolayers at Liquid Interfaces. New York: Plenum Press, 1989.
11. L Galet, I Pezron, W Kunz, C Larpent J Zhu, C Lheveder. Colloids Surfaces 151: 85, 1999.
12. DG Dervichian. J Chem Phys 7: 931, 1939.
13. WD Harkins, D Boyd. J Phys Chem 45: 20, 1941.
14. NK Adam. Physics and Chemistry of Surfaces. 3rd ed. London: Oxford University Press, 1941.
15. GA Overbeck, D Möbius. J Phys Chem 97: 7999, 1993.
16. RD Smith, JC Berg. J Colloid Interface Sci 74: 273, 1980.
17. E Pezron, PM Claesson, D Vollard. J Colloid Interface Sci. 138: 245, 1990.
18. M Tomoaia-Cotisel, J Zsako, E Chifu, DA Cadenhead. Langmuir 6:191, 1990.
19. EP Honig, JHT Hengst, D den Engelsen. J Colloid Interface Sci 45: 92, 1973.
20. JH Brooks, AE Alexander. Proceedings of the Third International Congress of Surface Activity, Vol II, p 196.
21. ES Nikomaro. Langmuir 6: 410, 1990.
22. I Langmuir. Trans Faraday Soc 15:62, 1920.
23. G Binnig, CF Quate, C Gerber. Phys Rev Lett. 56: 930, 1986.
24. D Parrat, F Sommer, JM Solleti, Trans Minh Duc. J Trace Microprobe Tech 13: 343, 1995.

25. JA Zasadzinski, R Viswanathan, DK Schwartz, J Garnaes, L Madsen, S Chiruvolu. JT Woodward, ML Longo. Colloids Surfaces 93: 305, 1994.
26. JA DeRose, RM Leblanc. Surfactant Sci Rep 22: 73, 1995.
27. J Sjöblom, L Mingyuan, T Gu, AA Christy. Colloids Surfaces 66: 55, 1992.
28. AH Brown. Chem Ind 30:990, 1968.
29. AJS Liem, DR Woods. Review of Coalescence Phenomena. AIChE Symp Ser 70, 8, 1974.
30. O Reynolds. Phil Trans R Soc (London) A177:157, 1886.
31. SP Frank, KJ Mysels. J Phys Chem 66: 190, 1960.
32. Z Zapryanov, AK Malhorta, N Aderangi, DT Wasan. Int J Multiphase Flow 9: 105, 1983.
33. CY Lin, JC Slatter. AIChEJ 28: 147, 1987.
34. CY Lin, JC Slatter. AIChEJ 28: 786, 1982.
35. AK Malhorta. PhD thesis, Illinois Institute of Technology, Chicago, 1984.
36. CV Sternling, LE Scriven. AIChE J 5: 514, 1959.
37. AJ de Vries, Recl Trav Chim 77: 383, 441, 1958.
38. SS Lang. PhD thesis, University of California, 1962.
39. A Scheludko, E Manner. Trans Faraday Soc 64: 1123, 1968.
40. A Vrij. Disc Faraday Soc 42:23, 1966.
41. H Sonntag, K Strenge. Coagulation and Stability of Disperse Systems. New York: Halsteady-Wiley, 1969.
42. TJ Jones, EL Neustadter, KP Whittingham. J Can Petrol Technol 17: 100, 1978.
43. R Fiocco. US Patent 3 536 529, 1970.
44. MA Krawczyk. PhD thesis, Illinois Institute of Technology, Chicago, 1990.
45. DT Wasan, K Sampath, N Aderangi. AIChE Symp Ser 76: 93, 1980.
46. DT Wasan, NF Djabbarah, MK Vora, ST Shah. Lect Notes Phys 105: 205, 1979.
47. R Aveyard, PB Binks, PDI Fletscher, JR Lu. J Colloid Interface Sci 139: 128, 1990.
48. R Aveyard, PB Binks, PDI Fletscher, RYe, JR Lu. In: J Sjöblom, ed. Emulsions – A Fundamental and Practical Approach. NATO ASI Series, Dordrecht: Kluwer Academic 1992, p 97.
49. MA Krawczyk, DT Wasan, SS Chandrashekar. Ind Eng Chem Res 30: 367, 1991.
50. DT Wasan. In: J Sjöblom, ed. Emulsions – A Fundamental and Practical Approach. NATO ASI Series, Dordrecht: Kluwer Academic 1992, p 283.
51. RC Little. Environ Sci Technol 15: 1184, 1981.
52. O Urdahl, AE Møvik, J Sjöblom. Colloid Surface A 74: 293, 1993.
53. HB Clark, MT Pike, GL Rengel. Petrol Technol 7: 1565, 1982.
54. HB Clark, MT Pike, GL Rengel. Proceedings of the AIME International Symposium on Oilfield and Geothermal Chemistry, Houston, TX, 1979, SPE Paper 7894.
55. VB Menon, DT Wasan. Colloids Surfaces 19: 89, 1986.
56. J Djuve, X Yang, IJ Fjellanger, J Sjöblom, E Pelizzetti. Colloid Polym Sci. In press, 2000.
57. M-H Ese. Langmuir film properties of indigenous crude oil components: influence of demulsifiers, PhD thesis, University of Bergen, 1999.
58. H Kallevik. Characterisation of crude oil and model oil emulsions by means of near infrared spectroscopy and multivariate analysis. PhD thesis, University of Bergen, 1999.
59. BP Singh, BP Pandey. Indian J Technol 29: 443, 1991.
60. ‘OC Mullins. Analyy Chem 62: 508–514, 1990.

# 26

# Dynamic Processes in Surfactant-stabilized Emulsions

**Krassimir D. Danov, Peter A. Kralchevsky, and Ivan B. Ivanov**
*University of Sofia, Sofia, Bulgaria*

## I. INTRODUCTION

The process of emulsification usually takes place under essentially dynamic conditions. It is accompanied with the creation of new drops (new phase boundary) between the two liquids and with frequent collisions between the drops. Their instantaneous size distribution is the result of a competition between two oppositely directed processes: (1) breaking of the drops into smaller ones by the shear strain; and (2) coalescence of the newly formed drops into larger ones upon collision. If *surfactant* is present, it tends to adsorb at the surface of the drops and thus to protect them against coalescence. The rate of surfactant adsorption should be large enough to guarantee obtaining a sufficiently high coverage of the oil–water interface during the short period between two drop collisions. Therefore, an important parameter characterizing a given surfactant as emulsifier is its characteristic *adsorption time* $\tau_1$; the latter can vary by many orders of magnitude depending on the type of surfactant, its concentration, and the presence or absence of added nonamphiphilic electrolyte (salt) in the aqueous phase. In Sec. II.B we demonstrate how to quantify $\tau_1$ for both ionic and nonionic surfactants.

The adsorbed surfactant molecules counteract the drop coalescence in two ways (1, 2). The presence of surfactant gives rise to repulsive surface forces (of either electrostatic, steric, or oscillatory structural origin) between the drops, thus providing a *thermodynamic* stabilization of the emulsion; see also Refs 3 and 4. Moreover, the adsorbed surfactant reduces (or completely removes) the tangential mobility of the drop surfaces and in this way markedly decelerates the interdroplet collisions; this is known as *kinetic* stabilization (1). The latter is related to the Marangoni effect, i.e., to the appearance of gradients of adsorption and interfacial tension along the surfaces of two colliding droplets (see Fig. 1):

$$\nabla_s \sigma = -\frac{E_G}{\Gamma_1} \nabla_s \Gamma_1 \tag{1}$$

where $\nabla_s$ is the surface gradient operator, $\sigma$ is the surface tension, $\Gamma_1$ is the surfactant adsorption, and $E_G$ the Gibbs (surface) elasticity; expressions for estimating $E_G$ can be found in Sec. II.A.

In the case of low interfacial coverage with surfactant, the collision of two emulsion drops (step A→B in Fig. 2) usually terminates with their coalescence (step B→C in Fig. 2). The merging of the two drops occurs when a small critical distance between their surfaces, $h_c$ is reached. Sometimes, depending on the specific conditions (larger drop size, attractive surface forces, smaller surface tension, etc., – see, e.g., Ref. 2), the approach of the two drops could be accompanied with a deformation in the zone of their contact (step B→D in Fig. 2); in this way a *liquid film* of almost uniform thickness $h$ is formed in the contact zone. This film could also have a critical thickness $h_c$ of rupture; in fact, the film rupture is equivalent to drop coalescence (see step D→C in Fig. 2). The mechanisms

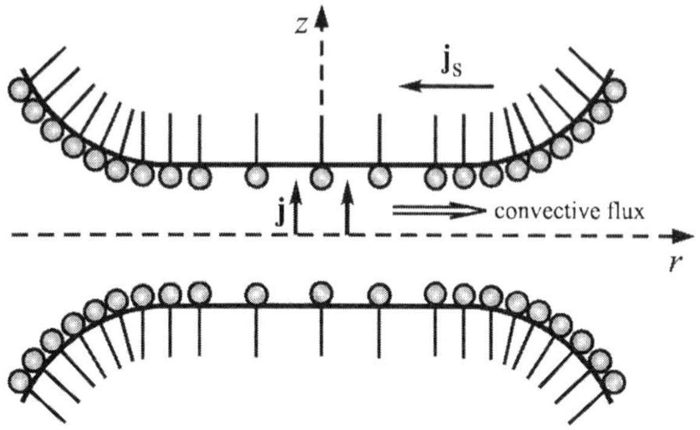


**Figure 1** Schematic presentation of the zone of contact between two approaching emulsion drops. The convective outflow of liquid from the gap between the drops drags the surfactant molecules along the two film surfaces: **j** and $\mathbf{j}_s$ denote the bulk- and surface-diffusion fluxes of surfactant.

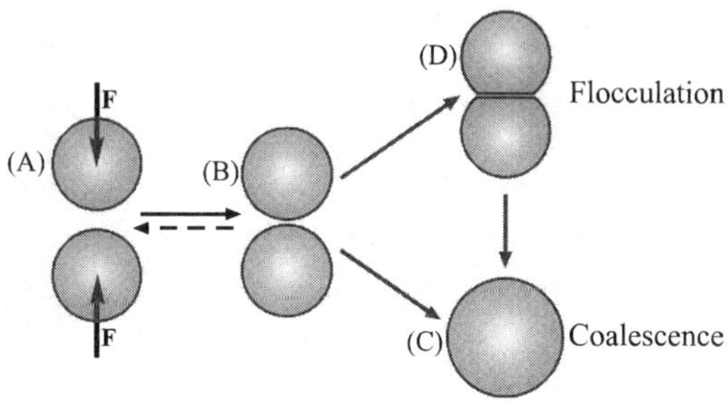


**Figure 2** Possible consequences from a collision between two emulsion drops. Step A → B: the two drops approach each other under the action of a driving force **F**; the viscous friction, accompanying the expulsion of liquid from the gap between the two drops, decelerates their approach. Step B → C: after reaching a given *critical distance* between the two drop surfaces coalescence takes place. Step B → D: after reaching a given threshold distance, $h_{inv}$, between the two drop surfaces, called the *inversion thickness*, the spherical drops deform and a film is formed in the zone of their contact. Step D → C: the film, intervening between the two drops, thins and eventually breaks after reaching a certain *critical thickness*, then the two drops coalesce.

of coalescence and the theoretical evaluation of $h_c$ are considered in Sec. III.

The driving force of the drop–drop collisions (**F** in Fig. 2) can be the Brownian stochastic force, the buoyancy force, or some attractive surface force (say, the van der Waals interaction); in stirred vessels an important role is played by the hydrodynamic (including turbulent) forces. The mutual approach of two emulsion drops (step A→B in Fig. 2) is decelerated by the *viscous friction* due to the expulsion of the liquid from the gap between the drops. If a doublet of two drops (Fig. 2D) is sufficiently stable, it can grow by attachment of additional drops; thus, aggregates of drops (flocs) are produced.

If the stirring of an emulsion is ceased, there is no longer generation of new droplets, but the opposite processes of drop flocculation and/or coalescence continue. After some period of time this will lead to the appearance of sufficiently large flocs and/or drops, for which the gravitational force is stronger than the Brownian force; this will lead to a directional motion of the drops/flocs upwards (creaming) or downwards (sedimentation), depending on whether the buoyancy force or the drop weight prevails.

As an illustrative example, Fig. 3 shows the occurrence of the creaming in an oil-in-water emulsion stabilized by the protein $\beta$-lactoglobulin – data from Ref. 5. The rise of the boundary between the lower transparent aqueous phase (serum) and the upper turbid emulsion phase is recorded as a function of time; in particular, the ratio of the volume of the serum to the total volume (turbid plus transparent phase) is plotted versus time. Two types of emulsions are used in these experiments: "coarse" and "fine" emulsion of average drop size 5 and 0.35 $\mu$m, respectively. In Fig. 3 one sees that, in the *fine* emulsion, creaming is not observed (the volume of the separated serum is zero). In contrast, there is creaming in the *coarse* emulsion, which starts some time after the initial moment (the ceasing of agitation); this period is necessary for "incubation" of

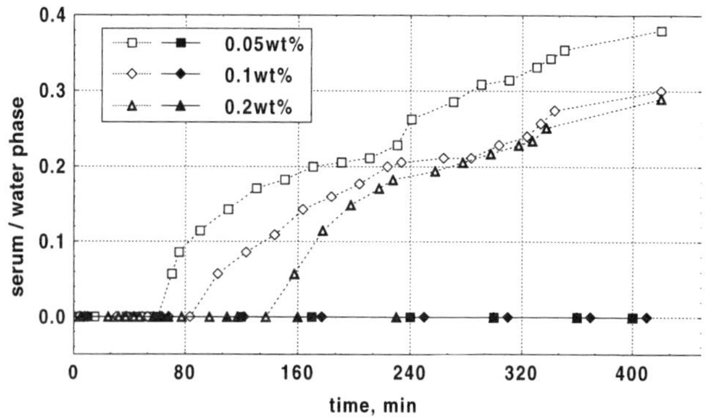


**Figure 3** Experimental data for creaming in xylene-in-water emulsions. The volume of the transparent "serum" left below the creaming emulsion, scaled with the total volume of the liquid mixture, is plotted against the time elapsed after ceasing the agitation. The emulsion is stabilized with $\beta$-lactoglobulin, whose concentrations, corresponding to the separate curves, are shown in the figure. The empty and full symbols denote, respectively, "coarse" emulsion (mean drop size 5$\mu$m) and "fine" emulsion (mean drop size 0.35 $\mu$m).

sufficiently large flocs, which are able to emerge under the action of the buoyancy force. The stabilizing effect of $\beta$-lactoglobulin is manifested as an increase in the "incubation period" with the rise of protein concentration.

The theoretical description of the mutual approach and coalescence of two emulsion drops is the subject of Sec. IV; the Bancroft rule on emulsification is interpreted and generalized in Sec. V; and the kinetics of flocculation is considered in Sec. VI, where the size of the aggregates needed for the creaming to start is estimated.

## II. DYNAMICS OF SURFACTANT ADSORPTION MONOLAYERS

### A. Gibbs (Surface) Elasticity

#### 1. Nonionic Surfactant Solutions

Let us consider the boundary between an aqueous solution of a nonionic surfactant and the oil phase. We choose the dividing surface to be the equimolecular dividing surface with respect to water. The Gibbs adsorption equation then takes the form (6, 7):

$$d\sigma = kT\,\Gamma_1\,d\ln c_1 \tag{2}$$

where the subscript "1" denotes the nonionic surfactant, $c_1$ and $\Gamma_1$ are its bulk concentration and adsorption, $k$ is the Boltzmann constant, and $T$ is the temperature. The surfactant adsorption isotherms, expressing the connection between $\Gamma_1$ and $c_1$, are usually obtained by means of some molecular model of the adsorption. The most popular is the Langmuir (8) adsorption isotherm;

$$\frac{\Gamma_1}{\Gamma_\infty} = \frac{Kc_1}{1+Kc_1} \tag{3}$$

which stems from a lattice model of localized adsorption of *noninteracting* molecules (9). In Eq. (3) $\Gamma_\infty$ is the maximum possible value of the adsorption ($\Gamma_1 \to \Gamma_\infty$ for $c_1 \to \infty$). On the other hand, for $c_1 \to 0$ one has $\Gamma_1 \approx Kc_1$; the adsorption parameter $K$ characterizes the surface activity of the surfactant: the greater $K$ the higher the surface activity.

Table 1 lists the six most popular surfactant adsorption isotherms, i.e., those of Henry, Freundlich, Langmuir, Volmer (10), Frumkin (11), and van der Waals (9). For $c_1 \to 0$ all other isotherms (except that of Freundlich) reduce to the Henry isotherm. The physical difference between the Langmuir and Volmer isotherms is that the former corresponds to a physical model of *localized* adsorption, whereas the latter corresponds to *nonlocalized* adsorption. The Frumkin and van der Waals isotherms generalize, respectively, the Langmuir and Volmer isotherms for the case when there is interaction between the adsorbed molecules; $\beta$ is a parameter which accounts for the interaction. In the case of the van der Waals interaction, $\beta$ can be expressed in the form (12, 13):

$$\beta = -\pi kT\int_{r_0}^{\infty}\left[1-\exp\left(-\frac{u(r)}{kT}\right)\right]dr \approx -\pi\int_{r_0}^{\infty} u(r)\,r\,dr \tag{4}$$

where $u(r)$ is the interaction energy between two adsorbed molecules, and $r_0$ is the distance between the centers of the molecules at close contact. The com-

**Table 1** The Most Popular Surfactant Adsorption Isotherms and the Respective Surface Tension Isotherms; $c_{1s}$ is the Subsurface Concentration of Surfactant Molecules

| | **Sufactant adsorption isotherms** (for nonionic surfactants: $Kc_{1s} \equiv c_1$) |
|---|---|
| Henry | $Kc_{1s} = \frac{\Gamma_1}{\Gamma_\infty}$ |
| Freundlich | $Kc_{1s} = \left(\frac{\Gamma_1}{\Gamma_F}\right)^{1/m}$ |
| Langmuir | $Kc_{1s} = \frac{\Gamma_1}{\Gamma_\infty-\Gamma_1}$ |
| Volmer | $Kc_{1s} = \frac{\Gamma_1}{\Gamma_\infty-\Gamma_1}\exp\left(\frac{\Gamma_1}{\Gamma_\infty-\Gamma_1}\right)$ |
| Frumkin | $Kc_{1s} = \frac{\Gamma_1}{\Gamma_\infty-\Gamma_1}\exp\left(-\frac{2\beta\Gamma_1}{kT}\right)$ |
| Van der Waals | $Kc_{1s} = \frac{\Gamma_1}{\Gamma_\infty-\Gamma_1}\exp\left(\frac{\Gamma_1}{\Gamma_\infty-\Gamma_1}-\frac{2\beta\Gamma_1}{kT}\right)$ |
| | **Surface tension isotherm** $\sigma = \sigma_0 - kTJ + \sigma_d$ (for nonionic surfactants: $\sigma_d \equiv 0$) |
| Henry | $J = \Gamma_1$ |
| Freundlich | $J = \frac{\Gamma_1}{m}$ |
| Langmuir | $J = -\Gamma_\infty \ln\left(1-\frac{\Gamma_1}{\Gamma_\infty}\right)$ |
| Volmer | $J = \frac{\Gamma_\infty\Gamma_1}{\Gamma_\infty-\Gamma_1}$ |
| Frumkin | $J = -\Gamma_\infty \ln\left((1-\frac{\Gamma_1}{\Gamma_\infty}\right) - \frac{\beta\Gamma_1^2}{kT}$ |
| van der Waals | $J = \frac{\Gamma_\infty\Gamma_1}{\Gamma_\infty-\Gamma_1} - \frac{\beta\Gamma_1^2}{kT}$ |

parison between theory and experiment shows that the interaction parameter $\beta$ is important for air–water interfaces, whereas for oil–water interfaces one can set $\beta = 0$ (14, 15). The latter fact, and the finding that $\beta > 0$ for air–water interfaces, leads to the conclusion that $\beta$ takes into account the van der Waals attraction between the hydrocarbon tails of the adsorbed surfactant molecules across air (such attraction is missing when the hydrophobic phase is oil). Note, however, that even for an oil–water interface one could have $\beta < 0$ if some nonelectrostatic repulsion between the adsorbed surfactant molecules takes place, say steric repulsion between some chain branches of amphiphilic molecules with a more complicated structure.

Concerning the parameter $K$ in Table 1, this is related to the standard free energy of adsorption, $\Delta f = \mu_1^{(0)} - \mu_{1s}^{(0)}$, which is the energy gain for bringing a molecule from the bulk of the water phase to a diluted adsorption layer (3, 16):

$$K = \frac{\delta_1}{\Gamma_\infty} \exp\left[\frac{\mu_1^{(0)} - \mu_{1s}^{(0)}}{kT}\right] \tag{5}$$

Here, $\delta_1$ is a parameter, characterizing the thickness of the adsorption layer, which can be set (approximately) equal to the length of the amphiphilic molecule. Let us consider the integral:

$$J \equiv \int_0^{c_1} \Gamma_1 \frac{dc_1}{c_1} = \int_0^{\Gamma_1} \Gamma_1 \frac{d \ln c_1}{d\Gamma_1} d\Gamma_1 \tag{6}$$

The derivative $d \ln c_1 / d\Gamma_1$ can be calculated for each adsorption isotherm in Table 1 and then the integration in Eq. (6) can be carried out analytically (17). The expressions for $J$ thus obtained are also listed in Table 1. The integration of the Gibbs adsorption isotherm, Eq. (2), along with Eq. (6), yields (17):

$$\sigma = \sigma_0 - kTJ \tag{7}$$

which in view of the expressions for $J$ in Table 1 presents the surfactant adsorption isotherm, or the two-dimensional (surface) equation of state.

As mentioned in the Sec. I, an important thermodynamic parameter of a surfactant adsorption monolayer is its Gibbs (surface) elasticity. The physical concept of surface elasticity is the most transparent for monolayers of *insoluble* surfactants, for which it was initially introduced by Gibbs (18, 19). The increments $\Delta\sigma$ and $\Delta\Gamma_1$ in the definition of Gibbs elasticity:

$$E_G = -\Gamma_1 \lim_{\Delta\Gamma_1 \to 0} \left(\frac{\Delta\sigma}{\Delta\Gamma_1}\right) = -\Gamma_1 \left(\frac{\partial\sigma}{\partial\Gamma_1}\right)_T \tag{8}$$

correspond to variations in surface tension and adsorption during a real process of interfacial dilatation. Expressions for $E_G$, corresponding to various adsorption isotherms, are shown in Table 2. As an example, let us consider the expression for $E_G$, corresponding to the Langmuir isotherm; combining the results from Tables 1 and 2 one obtains:

$$E_G = \Gamma_\infty kT\, Kc_1 \qquad \text{(for Langmuir isotherm)} \tag{9}$$

One sees that for Langmuirian adsorption the Gibbs elasticity grows linearly with the surfactant concentration $c_1$. Since the concentration of the monomeric surfactant cannot exceed the critical micellization concentration, $c_1 \le c_{CMC}$, then from Eq. (9) one obtains:

$$E_G \le (E_G)_{\max} = \Gamma_\infty kTKc_{CMC} \quad \text{(for Langmuir isotherm)} \tag{10}$$

Hence, one could expect higher elasticity $E_G$ for surfactants with higher $c_{CMC}$; this conclusion is consistent with the experimental results (20).

The Gibbs elasticity characterizes the lateral fluidity of the surfactant adsorption monolayer. For high values of the Gibbs elasticity the adsorption monolayer at a fluid interface behaves as tangentially immobile. Then, if two oil drops approach each other, the hydrodynamic flow pattern, and the hydrodynamic interaction as well, is the same as if the drops were solid particles, with the only difference that under some conditions they could deform in the zone of contact. For lower values of the Gibbs elasticity the Marangoni

**Table 2** Expressions for the Gibbs Elasticity of Adsorption Monolayers (Valid for Both Nonionic and Ionic Surfactants), Which Correspond to the Various Types of Isotherms in Table 1

| Type of surface tension isotherm | Gibbs elasticity $E_G$ |
|---|---|
| Henry | $E_G = kT\Gamma_1$ |
| Freundlich | $E_G = kT\frac{\Gamma_1}{m}$ |
| Langmuir | $E_G = kT\Gamma_1\frac{\Gamma_\infty}{\Gamma_\infty - \Gamma_1}$ |
| Volmer | $E_G = kT\Gamma_1\frac{\Gamma_\infty^2}{(\Gamma_\infty - \Gamma_1)^2}$ |
| Frumkin | $E_G = kT\Gamma_1\left[\frac{\Gamma_\infty}{\Gamma_\infty - \Gamma_1} - \frac{2\beta\Gamma_1}{kT}\right]$ |
| van der Waals | $E_G = kT\Gamma_1\left[\frac{\Gamma_\infty^2}{(\Gamma_\infty - \Gamma_1)^2} - \frac{2\beta\Gamma_1}{kT}\right]$ |

effect appears, see Eq. (1), which can considerably affect the approach of the two drops. These aspects of the hydrodynamic interactions between emulsion drops are considered in Sec. IV.

In the case of a soluble nonionic surfactant the detected increase in $\sigma$ in a real process of interfacial dilatation can be a pure manifestation of surface elasticity only if the period of dilatation, $\Delta t$, is much shorter than the characteristic *relaxation time of surface tension* $\tau_\sigma$, $\Delta t \ll \tau_\sigma$ (21). Otherwise, the adsorption and the surface tension would be affected by the diffusion supply of surfactant molecules from the bulk of solution toward the expanding interface. The diffusion transport tends to reduce the increase in surface tension upon dilatation, thus apparently rendering the interface less elastic and more fluid. The initial condition for the problem of adsorption kinetics involves an "instantaneous" ($\Delta t \ll \tau_\sigma$) dilatation of the interface. This "instantaneous" dilatation decreases the adsorptions $\Gamma_i$ and the subsurface concentrations $c_{is}$ of the species (the subsurface is presumed to be always in equilibrium with the surface), but the bulk concentrations $c_{i\infty}$ remain unaffected (22–24). This initially created difference between $c_{is}$ and $c_{i\infty}$ further triggers the diffusion process. Now, let us inspect closer how this approach is to be extended to the case of ionic surfactants.

## 2. Ionic Surfactant Solutions

The thermodynamics of adsorption of *ionic* surfactants is more complicated due to the presence of long-range electrostatic interactions in the system. Let us consider a boundary between two immiscible fluid phases (say, water and oil), which bears some electric charge owing to the adsorption of charged amphiphilic molecules (ionic surfactant). The charged surface repels the *coions*, i.e., ions having a charge of the same sign, but it attracts the *counterions*, which bear a charge of the opposite sign (Fig. 4). Thus, an electric double layer (EDL) appears, that is, a nonuniform distribution of the ionic species in the vicinity of the charged interface (25). The conventional model of the EDL stems from the works of Gouy (26), Chapman (27), and Stern (28). According to this model the EDL consists of two parts: (1) *adsorption layer*; and (2) *diffuse layer* (see Fig. 4). The adsorption layer includes surfactant molecules, which are immobilized (adsorbed) at the phase boundary, as well as bound counterions, which form the Stern layer. The diffuse layer consists of free ions in the aqueous phase, which are involved in Brownian motion and are influenced by the electric field of the charged interface. The boundary, separating the adsorption from the diffuse layer, called the *Gouy plane*, can be used as a Gibbs dividing surface between the two neighboring phases (15). The electric potential varies across the EDL: $\psi = \psi(x)$. The boundary values of $\psi(x)$ are $\psi(x=0) = \psi_s$ at the Gouy plane (at the interface) and $\psi(x\to\infty) = 0$ in the bulk of the solution. At equilibrium, the subsurface concentrations of the ionic species, $c_{is}$, are related to the respective bulk concentrations, $c_{i\infty}$, by means of the Boltzmann distribution (25):

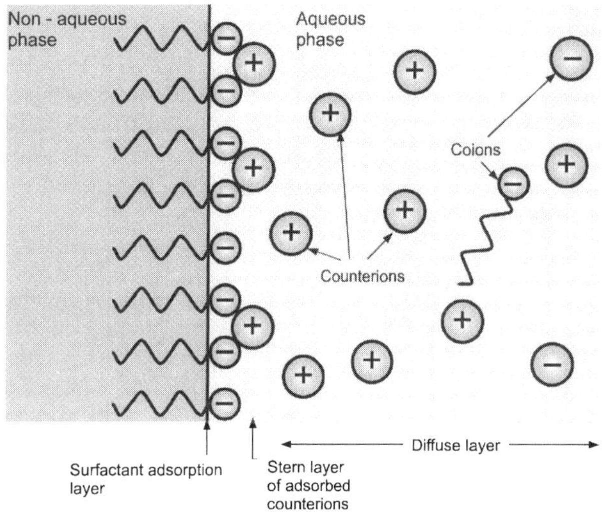


**Figure 4** EDL formed in the vicinity of an adsorption monolayer of ionic surfactant. The diffuse layer contains free ions involved in Brownian motion, while the Stern layer consists of adsorbed (immobilized) counterions. Near the charged surface there is an accumulation of counterions and a depletion of coions.

$$c_{is} \equiv c_i(x=0) = c_{i\infty} \exp\left\{-\frac{z_i e \psi_s}{kT}\right\} \qquad (11)$$

where $i = 1, 2, 3, \ldots, N$. Here, $e$ is the electronic charge, and $z_i$ is the valency of the $i$th ion. The Gibbs adsorption equation can be presented in the form (15, 17, 29–31):

$$d\sigma = -kT \sum_{i=1}^{N} \tilde{\Gamma}_i d \ln c_{i\infty} \qquad (T = \text{constant}) \quad (12)$$

Equations (11) and (12) are rigorous in terms of activities of the ionic species, rather than in terms of concentrations. For simplicity, here we set the activities equal to the concentrations, which is a good approximation for ionic strengths below 0.1 M; see Refs 14, 15 and 17 for details. In Eq. (12), $\tilde{\Gamma}_i$ denotes the adsorption of the $i$th component, and $\tilde{\Gamma}_i$ represents the surface excess of component "$i$" with respect to the *uniform* bulk solution. For an ionic species, $\tilde{\Gamma}_i$ is a total adsorption, which includes contributions $\Gamma_i$ and $\Lambda_i$, respectively, from the adsorption layer (adsorbed surfactant plus counterions in the Stern layer) and the diffuse layer, which are defined as follows (17, 29–31):

$$\Lambda_i \equiv \int_0^\infty [c_i(x) - c_{i\infty}]dx, \qquad \Gamma_i \equiv \tilde{\Gamma}_i - \Lambda_i \qquad (13)$$

Using the theory of EDL and Eq. (13) one can prove that the Gibbs adsorption equation, Eq. (12), can be represented in the following equivalent form (17):

$$d\sigma_a = -kT\sum_{i=1}^{N} \Gamma_i d\ln c_{is} \qquad (T = \text{constant}) \quad (14)$$

where $\sigma_a = \sigma - \sigma_d = \sigma_0 - kTJ$ is the contribution of the *adsorption* layer to the surface tension [$J$ is the same as in Eq. (6) and Table 1], and $\sigma_d$ is the contribution of the *diffuse* layer (17, 29):

$$\sigma_d = -\frac{\varepsilon}{4\pi}\int_0^\infty \left(\frac{\partial\psi}{\partial x}\right)^2 dx \qquad (15)$$

where $\varepsilon$ is the dielectric permittivity of the aqueous phase. The integrand in Eq. (15) represents the anisotropy of the Maxwell electric stress tensor, which contributes to the interfacial tension in accordance with the known Bakker formula (32–34). The comparison between Eqs (12) and (14) shows that the Gibbs adsorption equation can be expressed either in terms of $\sigma$, $\tilde{\Gamma}_i$, and $c_{i\infty}$, or in terms of $\sigma_a$, $\Gamma_i$, and $c_{is}$. The total surface tension is

$$\sigma = \sigma_a + \sigma_d \qquad (16)$$

Note that $\sigma_d$ represents a nonlocal, integral contribution of the whole diffuse EDL, whereas $\sigma_a$ is related to the two-dimensional state of the adsorbed surfactant ions and bound counterions (Fig. 4).

Let us consider a solution of ionic surfactant, which is a symmetric $z$:$z$ electrolyte, in the presence of additional nonamphiphilic $z$:$z$ electrolyte (salt); here, $z \equiv z_1 = -z_2 = z_3$. We assume that the counterions due to the surfactant and salt are identical. For example, this can be a solution of sodium dodecyl sulfate (SDS) in the presence of NaCl. We denote by $c_{1\infty}$, $c_{2\infty}$, and $c_{3\infty}$ the bulk concentrations of the surface active ions (1), counterions (2), and coions (3), respectively. For the special system of SDS with NaCl $c_{1\infty}$, $c_{2\infty}$, and $c_{3\infty}$ are the bulk concentration of the $DS^-$, $Na^+$, and $Cl^-$ ions, respectively. The requirement for the bulk solution to be electroneutral implies that $c_{2\infty} = c_{1\infty} + c_{3\infty}$. The binding of *coions* due to the non-amphiphilic salt is expected to be equal to zero, $\Gamma_3 = 0$, because they are repelled by the similarly charged interface (17). However, $\Lambda_3 \neq 0$; hence, $\tilde{\Gamma}_3 = \Lambda_3 \neq 0$. The difference between the adsorptions of surfactant ions and counterions determines the surface charge density, $\rho_s = ez(\Gamma_1 - \Gamma_2)$. For the considered system, Eq. (11) can be presented in the form:

$$c_{is} = c_{i\infty}\exp[(-1)^i \phi_s], \qquad \phi_s \equiv \frac{ze\psi_s}{kT} \qquad (17)$$

($i = 1, 2, 3$). Note that the dimensionless surface electric potential $\phi_s$ thus defined is always positive, irrespective of whether the surfactant is cationic or anionic.

Let us proceed with the definition of Gibbs elasticity for an adsorption monolayer from *ionic* surfactant. The main question is whether or not the electric field in the EDL should be affected by the "instantaneous" dilatation of the interface, $-\Delta\Gamma_1$, which is involved in the definition of $E_G$ – see Eq. (8). This problem has been examined in Ref. 35 and it has been established that a variation of the electric field during the initial instantaneous dilatation leads to results that are unacceptable from a theoretical viewpoint. The latter conclusion is related to the following facts: (1) the speed of propagation of the electric signals is much greater than the characteristic rate of diffusion; and (2) even a small initial variation in the surface charge density $\rho_s$ immediately gives rise to an electric potential, which is linearly increasing with the distance from the interface (potential of a planar wall). Consequently, a small initial perturbation of the interface would immediately affect the ions in the *whole* solution; of course, such an initial condition is physically unacceptable. In reality, a linearly growing electric field could not appear in an ionic solution, because a variation of the surface-charge density would be immediately suppressed by exchange of counterions, which are abundant in the subsurface layer of the solution. The theoretical equations suggest the same (35): to have a mathematically meaningful initial condition of *small* perturbation for the diffusion problem, the initial dilatation must be carried out at constant surface-charge density $\rho_s$; for details see the Appendix in Ref. 35. Thus, the following conclusion has been reached: the initial sudden interfacial dilatation, which is related to the definition of Gibbs elasticity of a soluble ionic surfactant, must be carried out at $\rho_s$ = constant. From Eq. (16) one obtains (36):

$$(d\sigma)_{\rho_s} = (d\sigma_a)_{\rho_s} + (d\sigma_d)_{\rho_s} \tag{18}$$

An interfacial dilatation at constant $\rho_s$ does not alter the diffuse part of the EDL, and consequently, $(d\sigma_d)_{\rho_s} \equiv 0$, see Eq. (15). Since (17),

$$\sigma_a = \sigma_0 - kTJ \tag{19}$$

the expressions for $J$ in Table 1 show that $\sigma_a$ depends only on $\Gamma_1$ at constant temperature. The definition of Gibbs elasticity of nonionic adsorption layers can then be extended to ionic adsorption layers in the following way (36):

$$E_G \equiv -\Gamma_1 \left(\frac{\partial \sigma}{\partial \Gamma_1}\right)_{T,\rho_s} = -\Gamma_1 \left(\frac{\partial \sigma_a}{\partial \Gamma_1}\right)_T \quad \text{(ionic surfactant)} \tag{20}$$

The definition of Gibbs elasticity given by Eq. (19) corresponds to an "instantaneous" ($\Delta t \ll \tau_\sigma$) dilatation of the *adsorption* layer (that contributes to $\sigma_a$) without affecting the *diffuse* layer and $\sigma_d$. The dependence of $\sigma$ on $\Gamma_1$ for nonionic surfactants is the same as the dependence of $\sigma_a$ on $\Gamma_1$ for ionic surfactants, cf. Eqs (7) and (19). Equations (8) and (20) then show that the expressions for $E_G$ in Table 2 are valid for *both* nonionic and ionic surfactants. The effect of the surface electric potential on the Gibbs elasticity $E_G$ of an ionic adsorption monolayer is implicit, through the equilibrium surfactant adsorption $\Gamma_1$, which depends on the electric properties of the interface. To illustrate this let us consider the case of Langmuir adsorption isotherm for an *ionic* surfactant (17):

$$Kc_{1s} = \frac{\Gamma_1}{\Gamma_\infty - \Gamma_1}, \qquad K \equiv K_1 + K_2 c_{2s} \tag{21}$$

where $K_1$ and $K_2$ are constants. Note that the above linear dependence of the adsorption parameter $K$ on the subsurface concentration of counterions, $c_{2s}$, can be deduced from the equilibrium exchange reactions, which describe the adsorption of surfactant ions and counterions (see Ref. 37). Combining the respective expression from Table 2 with Eq. (21) we obtain $E_G = \Gamma_\infty kTKc_{1s}$. Further, having in mind that $K = K_1 + K_2\, c_{2s}$, we substitute Eq. (17) to derive

$$E_G = \Gamma_\infty kTc_{1\infty}\,[K_1 \exp(-\phi_s) + K_2\, c_{2\infty}] \quad \text{(for Langmuir isotherm)} \tag{22}$$

Equation (22) reveals the effect of salt on $E_G$: when the salt concentration increases, $c_{2\infty}$ also increases, whereas the (dimensionless) surface potential $\phi_s$ decreases (see Fig. 5, from Ref. 17); then, Eq. (22) predicts an increase in $E_G$ with the rise in salt concentration.

A numerical illustration of the latter prediction is given in Fig. 4. The Gibbs elasticity is calculated with the help of Eq. (22), i.e., the Langmuir isotherm, using the values of $K_1$, $K_2$, and $\Gamma_\infty$ determined in Ref. 17 from the fit of experimental data due to Tajima and coworkers (38, 39) for sodium dodecyl sulfate (SDS). The surface potential $\phi_s$ is computed as a function of the surfactant and salt concentrations using steps 2–6 of the calculation procedure described in Sec. 9.2 of Ref. 17 with $\beta = 0$. As seen in Fig. 4, $E_G$ increases with the rise in surfactant (SDS) concentration. Moreover, for a fixed surfactant concentration one observes a strong increase in $E_G$ with increase in NaCl concentration. To understand this behavior of $E_G$ we notice that, according to Table 2, $E_G$ depends explicitly only on $\Gamma_1$ at fixed temperature $T$. Hence, the influence of surfactant and salt on the Gibbs elasticity $E_G$ can be interpreted as an increase in the surfactant adsorption $\Gamma_1$ with the rise in both surfactant and salt concentrations.

## B. Characteristic Time of Adsorption

### 1. Nonionic Surfactant Solutions

The characteristic time of surfactant adsorption at a fluid interface is an important parameter for surfactant-stabilized dynamic systems such as emulsions. Sutherland (22) derived an expression describing the relaxation of a *small* dilatation of an initially equili-

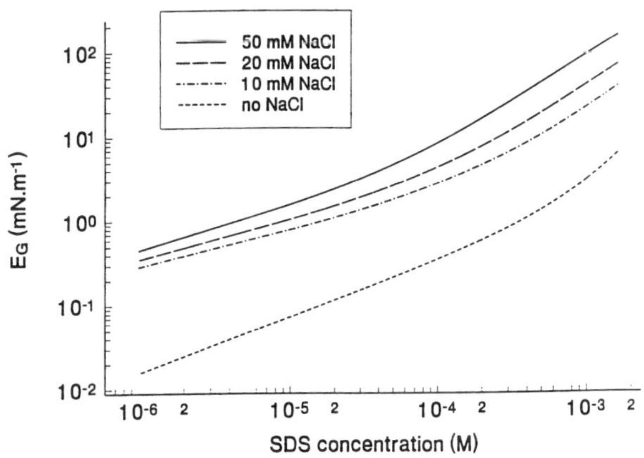


**Figure 5** Plot of the Giibs (surface) elasticity $E_G$ vs. the surfactant (SDS) concentration, $c_{1\infty}$. The four curves correspond to four fixed NaCl concentrations: 0, 20, 50, and 115 mM; $E_G$ is calculated by means of Eq. (22) using paraameters values determined from the best fit of experimental data in Ref. 17.

brium adsorption monolayer from soluble *nonionic* surfactant (diffusion control):

$$\frac{\sigma(t)-\sigma^{(e)}}{\sigma(0)-\sigma^{(e)}}=\frac{\Gamma_1(t)-\Gamma_1{}^{(e)}}{\Gamma_1(0)-\Gamma_1{}^{(e)}}=\exp\left\{\frac{t}{\tau_1}\right\}\operatorname{erfc}\left\{\sqrt{\frac{t}{\tau_1}}\right\} \tag{23}$$

where $t$ is time,

$$\tau_1\equiv\frac{1}{D_1}\left\{\frac{\partial\Gamma_1}{\partial c_1}\right\}^2 \tag{24}$$

is the characteristic relaxation time, and $D_1$ is the surfactant diffusivity; here and hereafter the superscript "$(e)$" denotes the equilibrium value of the respective parameter; erfc($x$) is the complementary error function (40–42). Using the asymptotics of the latter function for $x \gg 1$ one obtains

$$\frac{\sigma(t)-\sigma^{(e)}}{\sigma(0)-\sigma^{(e)}}=\frac{\Gamma_1(t)-\Gamma_1{}^{(e)}}{\Gamma_1(0)-\Gamma_1{}^{(e)}}=\sqrt{\frac{\tau_1}{\pi t}}+0(t^{-3/2}) \tag{25}$$
$$(t\gg\tau_1,\ \text{small perturbation})$$

Equation (25) is often used as a test to verify whether the adsorption process is under diffusion control: data for the dynamic surface tension $\sigma(t)$ are plotted versus $t^{-1/2}$ and it is checked if the plot complies with a straight line; the extrapolation of this line to $t^{-1/2}\to 0$ is used to determine the equilibrium surface tension $\sigma^{(e)}$ (23, 43).

In the experiment one often deals with *large* initial deviations from equilibrium; for example, such is the case when a new oil–water interface is formed by the breaking of larger emulsion drops during emulsification. In the case of large perturbation there is no general analytical expression for the dynamic surface tension $\sigma(t)$ since the adsorption isotherms (except that of Henry, see Table 1) are nonlinear. In this case one can use either a computer solution (44, 45) or apply the von Karman approximate approach (46, 47). Analytical asymptotic expressions for the long time ($t \gg \tau_1$) relaxation of surface tension of a *nonionic* surfactant solution was obtained by Hansen (48):

$$\sigma(t)-\sigma^{(e)}=(\Gamma_1^{(e)})^2\,\frac{kT}{c_1{}^{(e)}}\sqrt{\frac{1}{\pi D_1 t}}\qquad(t\to\infty) \tag{26}$$

When deriving Eq. (26), the surfactant adsorption at the initial moment was set to zero, $\Gamma_1(0) = 0$. Equation (26) has been verified, utilized, and generalized by many authors (24,49–53). With the help of Eqs (2), (8), and (24) one can represent Eq. (26) in the following equivalent form:

$$\sigma(t)-\sigma^{(e)}=E_G\sqrt{\frac{\tau_1}{\pi t}}+0(t^{-3/2}) \tag{27}$$
$$(t\gg\tau_1,\ \text{large perturbation})$$

where, as usual, $E_G$ denotes Gibbs elasticity. Comparison of Eqs (25) and (27) shows that the relaxation of surface tension is characterized by the *same* relaxation time $\tau_1$, irrespective of whether the interfacial perturbation is large or small. (The same conclusion is valid also for ionic surfactants, see below.) For that reason the relaxation time can be considered as a general kinetic property of the adsorption monolayer (36).

## 2. Ionic Surfactant Solutions

In the case of *ionic* surfactants the existence of a diffuse EDL essentially influences the kinetics of adsorption. The process of adsorption is accompanied by a progressive increase in the surface-charge density and electric potential. The charged surface repels the incoming surfactant molecules, which results in a deceleration of the adsorption process (54). Theoretical studies on the dynamics of adsorption encounter difficulties with the nonlinear set of partial differential equations, which describes the electrodiffusion process (55).

Another important effect, which adds to the complexity of the problem, is the *adsorption* (*binding*) *of counterions* at the conversely charged surfactant headgroups in the adsorption layer, see Fig. 4. The adsorbed (bound) counterions form the Stern layer, which strongly affects the adsorption kinetics of ionic surfactants insofar as up to 70–90% of the surface electric charge could be neutralized by the bound counterions (17, 56–58). The addition of nonamphiphilic electrolyte (salt) in the solution increases the occupancy of the Stern layer. It turns out that in the case of ionic surfactants (with or without salt) there are two adsorbing species: the surfactant ions and the counterions. The adsorption of counterions can be described by means of the Stern isotherm (6, 17, 28). It is worthwhile noting that the counterion binding enhances the adsorption of surfactant (17); formally, this appears as a linear increase in the surfactant adsorption parameter $K$ with the rise in the subsurface concentration of counterions, $c_{2s}$, see Eq. (21).

In recent papers (35, 36) the problem of the kinetics of adsorption from an ionic surfactant solution has been addressed in its full complexity, including the time evolution of the EDL, the effect of added salt, and the counterion binding. An analytical solution was found only in the asymptotic cases of *small* and *large* initial deviations from equilibrium and *long* times of adsorption. Thus, generalizations of Eqs (25) and (27) for the case of ionic surfactants was obtained (see below). An interesting result is that the electrostatic interaction leads to the appearance of three distinct characteristic relaxation times, those of surfactant adsorption $\tau_1$, of counterion adsorption (binding) $\tau_2$, and of surface-tension relaxation $\tau_\sigma$. In particular, the relaxation of surfactant and counterion adsorptions, $\Gamma_1$ and $\Gamma_2$, under electrodiffusion control, is described by the equation:

$$\frac{\Gamma_i(t) - \Gamma_i^{(e)}}{\Gamma_1(0) - \Gamma_1^{(e)}} = \sqrt{\frac{\tau_i}{\pi t}} + 0(t^{-3/2}) \qquad (t \to \infty, i = 1, 2) \tag{28}$$

where $\tau_1$ and $\tau_2$ are given by a generalized version of Eq. (24), which can be found in Refs 35 and 36 together with the procedure for calculations. The relaxation of interfacial tension of *ionic* surfactant solutions is given again by Eqs (25) and (27), in which $\tau_1$ is to be replaced by $\tau_\sigma$ defined as follows (36):

$$\sqrt{\tau_\sigma} = (1 + w)\sqrt{\tau_1} - w\sqrt{\tau_2} - \frac{2qw\lambda}{\kappa} \tanh \frac{\phi_s^{(e)}}{4} \tag{29}$$

where $\kappa$ is the Debye parameter,

$$\kappa^2 \equiv \frac{4\pi z^2 e^2 \sum_i c_{i\infty}}{\varepsilon kT}; \tag{30}$$

$$w \equiv \frac{2kT\Gamma_1{}^{(e)}}{E_G} \tanh \frac{\phi_s^{(e)}}{2}, \qquad \lambda = \frac{1 + \zeta\Gamma_2{}^{(e)}/\Gamma_1{}^{(e)}}{1 + \zeta}, \zeta = \exp(-\phi_s{}^{(e)}/2) \tag{31}$$

The latter expression for the parameter $\lambda$ corresponds to the case of large perturbations; for *small* perturbations one simply has $\lambda \equiv 1$ (36). The computations show that for large perturbations $\lambda$ is close to 1, and therefore the relaxation time is not sensitive to the magnitude of perturbation.

As an illustration, we show in Fig. 6 the calculated dependence of the relaxation times $\tau_1$, $\tau_2$, and $\tau_\sigma$ on the surfactant concentration. As in Fig 5, we have used the values of $K_1$, $K_2$, and $\Gamma_\infty$ determined in Ref. (17) from the fit of experimental data due to Tajima and coworkers (38, 39) for SDS. All necessary equations and the procedure of calculation are described in Ref. 36 for the case of large perturbations. The range of surfactant and salt concentrations correspond to the *nonmicellar* surfactant solutions studied experimentally in Refs 38 and 39. In Fig. 6a and b one notices the wide range of variation in relaxation times, which is from 2 to 6 orders of magnitude. For example, the relaxation time of surface tension, $\tau_\sigma$, drops from about 40 s for $10^{-5}$ M SDS down to $\approx 4 \times 10^{-5}$ s for $10^{-3}$ M SDS (see Fig. 6a). In addition, one sees that systematically $\tau_2 < \tau_1 < \tau_\sigma$; the difference between these three relaxation times can be greater than one order of magnitude for the lower surfactant concentrations, especially in the case without added electrolyte (Fig. 6b). One can conclude that the terms proportional to $w$ in Eq. (29), which give rise to the difference between $\tau_1$ and $\tau_\sigma$, play an important role, particularly for solutions of lower ionic strength. Figure 6 demonstrates

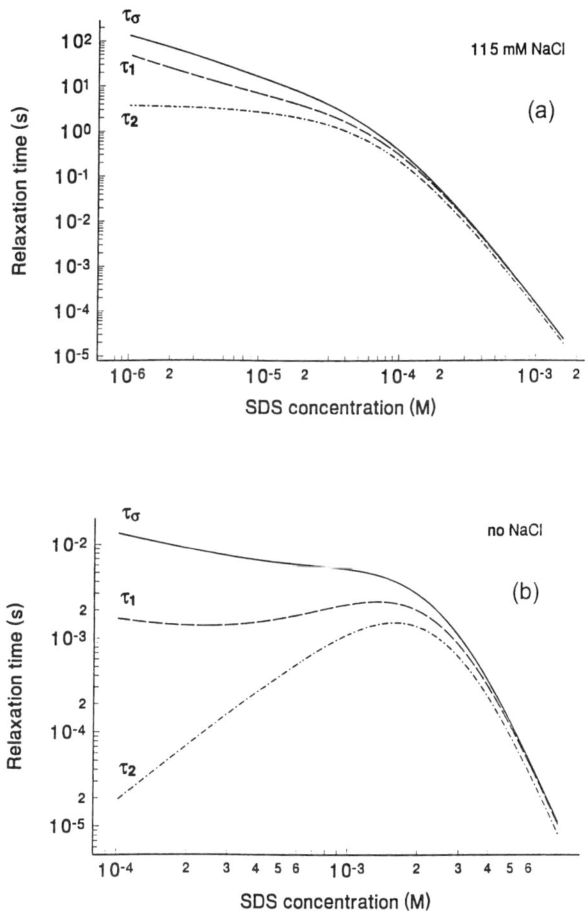


**Figure 6** Ionic surfactant solution: relaxation times of interfacial tension. $\tau_\sigma$, of surfactant adsorption, $\tau_1$, and of counterion adsorption (binding), $\tau_2$, calculated in Ref. 36 as functions of surfactant (SDS) concentration, $c_{1\infty}$, using parameters values determined from the best fit of experimental data in Ref. 17. (a) SDS solutions with 115 mM added NaCl; (b) SDS solutions without added NaCl.

that the approximation $\tau_\sigma \approx \tau_1$, which is widely used in the literature, is applicable only for the higher surfactant concentrations, for which $\tau_\sigma \to \tau_1$. Note also that for a given surfactant concentration $\tau_2$ is always smaller than $\tau_1$ and $\tau_\sigma$, that is, the adsorption of counterions relaxes faster than does the adsorption of surfactant ions and the surface tension.

The physical importance of these results is related to the fact that the coalescence of drops at the early highly dynamic stages of emulsion production is expected to be sensitive to the degree of saturation of the newly created interfaces with surfactant, and correspondingly, to the relaxation time of surfactant adsorption. The surfactant transport is especially important when the emulsion is prepared from non-pre-equilibrated liquid phases. In such cases one can observe dynamic phenomena like the *cyclic dimpling* (59, 60) and *osmotic swelling* (61), which bring about additional stabilization of the emulsions (see also Refs 1 and 62).

### 3. Micellar Surfactant Solutions

Emulsions are often prepared from micellar surfactant solutions. As known, above a given critical micelle concentration (cmc) surfactant aggregates (micelles) appear inside the surfactant solutions. At rest the micelles exist in equilibrium with the surfactants monomers in the solution. If the concentration of the monomers in the solution is suddenly decreased, the micelles release monomers until the equilibrium concentration, equal to cmc, is restored at the cost of disassembly of a part of the micelles (63, 64).

The dilatation of the surfactant adsorption layer leads to a transfer of monomers from the subsurface to the surface, which causes a transient decrease in the subsurface concentration of monomers. The latter is compensated for by disintegration of a part of the micelles in the subsurface layer. This process is accompanied by a diffusion transport of surfactant monomers and micelles due to the appearance of concentration gradients. In general, the micelles serve as a powerful source of monomers which is able to saturate quickly the surface of the newly created emulsion drops. In this way, the presence of surfactant micelles strongly accelerates the kinetics of adsorption.

The theoretical model developed by Aniansson and coworkers (65–68) describes the micelles as polydisperse aggregates, whose growth or decay happens by exchange of monomers. The general theoretical description of the diffusion in such a solution of polydisperse aggregates taking part in chemical reactions (exchange of monomers) is a heavy task; nevertheless, it has been addressed in several works (69–72). The relaxation of surface tension of a micellar solution at small initial deviation from equilibrium can be described by the following expression, derived in Ref. 70:

$$\frac{\sigma(t)-\sigma^{(e)}}{\sigma(0)-\sigma^{(e)}} = \frac{\Gamma_1(t) - \Gamma_1^{(e)}}{\Gamma_1(0) - \Gamma_1^{(e)}} = \frac{1}{g_1-g_2}[E(g_1,\tau)-E(g_2,\tau)]\exp\left\{-\frac{t}{\tau_m}\right\} \tag{32}$$

$$E(g,\tau) = g\exp(g^2\tau)\mathrm{erfc}(g\sqrt{\tau})$$
$$\tau = t/\tau d$$
$$g_{1,2} = [1 \pm \sqrt{1+4\tau_d/\tau_m}]/2$$

$$\tau_m = [K_d(1+m^2 c_m{}^{(e)}/c_1{}^{(e)})]^{-1}$$
$$\tau_d = (\partial\Gamma_1/\partial c_1)^2/D_1 \tag{33}$$

where $\tau_m$ and $\tau_d$ are the characteristic relaxation times of micellization and monomer diffusion (see Ref. 73). For the sake of estimates $\tau_d$ can be identified with $\tau_1$ as given by Eq. (24); $K_d$ is the rate constant of micelle decay; as earlier, the index "$(e)$" refers to the equilibrium state; and $m$ is the average micelle aggregation number. In the absence of micelles $\tau_d/\tau_m \to 0$; then, $g_1 = 1$, $g_2 = 0$, and Eq. (32) reduces to Eq. (23), as should be expected. One can estimate the characteristic time of relaxation in the presence of micelles by using the following combined expression:

$$\tau_\sigma \approx \frac{4\tau_d}{(1+\sqrt{1+4\tau_d/\tau_m})^2} \tag{34}$$

According to the latter $\tau\sigma$ expression $\tau_\sigma \approx \tau_m$ for $\tau_d \gg \tau_m$, and $\tau_\sigma \approx \tau_d$ for $\tau_d \ll \tau_m$.

Equation (32) is applicable only for small perturbations. An approximate analytical approach, which is applicable for both small and large deviations from equilibrium, is developed in Ref. 47.

## III. MECHANISMS OF COALESCENCE

### A. Mechanisms of Rupture of Emulsion Films

#### 1. Thermodynamic and Kinetic Factors Preventing Coalescence

Often the contact of two emulsion drops is accompanied by the formation of a liquid film between them. The rupture of this film is equivalent to coalescence of the drops, that is, step D→C in Fig 2. Figure 7 shows schematically the zone of contact between two emulsion drops of different radii, $R_1$ and $R_2$ ($R_1 < R_2$). For the sake of simplicity we assume that the two drops are composed of the same liquid and have the same surface tension $\sigma$. The film formed in the contact zone has radius $R$ and thickness $h$. The interaction of the two drops across the film leads to the appearance of an additional *disjoining pressure* $\Pi$ inside the film, which in general depends on the film thickness: $\Pi = \Pi(h)$ (see, e.g., Refs 2–4 and 62). Positive $\Pi$ corresponds to repulsion between the two film surfaces (and the two drops), whereas negative $\Pi$ corresponds to attraction between them. The presence of a disjoining pressure gives rise to a difference between the tension of the film surfaces, $\sigma^f$, and the interfacial tension $\sigma$ of the droplets. The force balance at the contact line reads (62, 74, 75):

$$\sigma^f = \sigma \cos\alpha; \tag{35}$$

where $\alpha$ is the contact angle, which is related to the disjoining pressure $\Pi$ as follows (62, 76):

$$\cos\alpha = 1 + \frac{1}{2\sigma}\int_h^\infty \Pi(h)\,dh \tag{36}$$

Since $\cos\alpha < 1$, a necessary condition to have a contact angle is for the integral in Eq. (36) to be negative; for emulsion drops this can be ensured by the long-range van der Waals attraction, which dominates $\Pi$ for the larger $h$ (see Fig. 8a). Geometrically, $\alpha$ appears as the angle subtended between the tangents to the film and drop surfaces at the contact line (Fig. 7). At *equilibrium* (no applied external force) the radius of the film between the two drops is determined by the equation:

$$R = \frac{2\sigma\sin\alpha}{\Pi}\cos\theta \approx \frac{2\sigma\sin\alpha}{\Pi}; \qquad \cos\theta \equiv (1 - R^2/R_f^2)^{1/2} \approx 1 \tag{37}$$

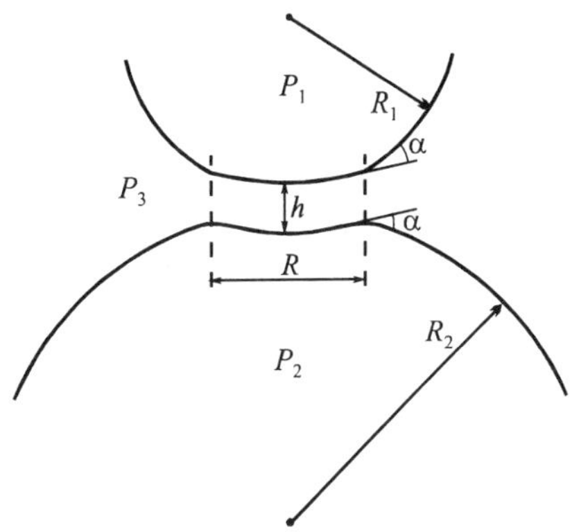


**Figure 7** Sketch of a film between two nonidentical emulsion drops of radii $R_1$ and $R_2$. The film thickness and radius are denoted by $h$ and $R$, respectively; $\alpha$ is the contact angle, and $P_1$, $P_2$, and $P_3$ denote the pressure in the respective liquid phases.

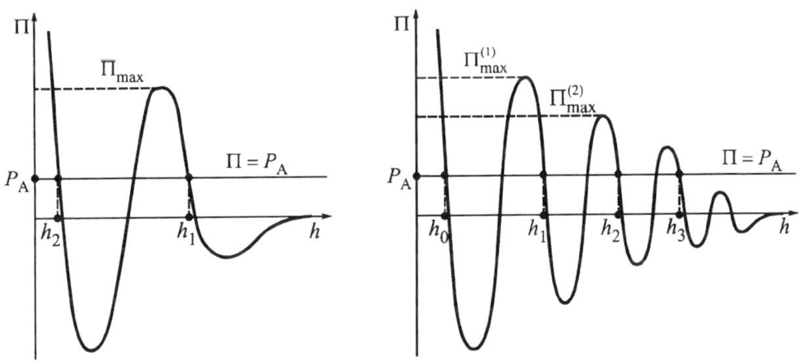


(a) Electrostatic DLVO Barrier (b) Multiple Barriers Due to Micelles

**Figure 8** Typical plots of disjoining pressure $\Pi$ vs. film thickness $h$; $P_A$ is the pressure difference applied across the film surface, see Eq (43); the equilibrium states of the liquid film correspond to the points in which $\Pi = P_A$. (a) DLVO-type disjoining-pressure isotherm $\Pi(h)$; the points at $h = h_1$ and $h_2$ correspond to primary and secondary films, respectively; $\Pi_{max}$ is the height of a barrier resulting from the electrostatic repulsion between the film surfaces. (b) Oscillatory structural force between the two film surfaces caused by the presence of surfactant micelles (or other monodisperse colloidal particles) in the continuous phase; $\Pi(h)$ exhibits multiple decaying oscillations; the stable equilibrium films with thickness $h_0$, $h_1$, $h_2$, and $h_3$ corresponds to stratifying films containing 0, 1, 2, and 3 layers of micelles, respectively (see Fig. 9).

where $R_f$ is the curvature radius of the film. Equation (37) follows from Eqs (144) and (179) in Ref. 75. One sees that the greater the contact angle $\alpha$, the larger the equilibrium film radius $R$. On the other hand, for $\alpha = 0$ the equilibrium film radius $R$ is also zero and there are no equilibrium doublets or larger aggregates (flocs) of emulsion droplets.

If the two drops have different radii, as in Fig. 7, the film between them is curved. The balance of the pressures applied per unit area of the two film surfaces can be expressed by means of versions of the Laplace equation (75):

$$\frac{2\sigma^f}{R_f} = P_1 - (P_3 + \Pi) \qquad \text{(upper film surface)} \quad (38)$$

$$\frac{2\sigma^f}{R_f} = (P_3 + \Pi) - P_2 \qquad \text{(lower film surface)} \quad (39)$$

where $P_1$ and $P_2$ are the pressures inside the respective drops (Fig. 7), and $P_3$ is the pressure in the continuous phase; the effect of disjoining pressure is equivalent to an increase in the pressure within the film, which is $P_3 + \Pi$. To obtain Eqs (38) and (39) we have neglected some very small terms, of the order of $h/R_f$ (see Ref. 75 for details). To determine $R_f$ we apply the Laplace equations for the two drops (Fig. 7):

$$\frac{2\sigma}{R_1} = P_1 - P_3, \qquad \frac{2\sigma}{R_2} = P_2 - P_3 \qquad (40)$$

Combining Eqs (36) and (38)–(40) one determines the curvature radius of the film:

$$\frac{1}{R_f} = \frac{1}{2\cos\alpha}\left\{\frac{1}{R_1} - \frac{1}{R_2}\right\} \qquad (41)$$

If the two drops have identical size ($R_1 = R_2$), then Eq. (41) yields $1/R_f \to 0$, i.e., the film between the drops is flat, as should be expected.

The disjoining pressure $\Pi$ is the major *thermodynamic* stabilizing factor against drop coalescence. A stable equilibrium state of a liquid film can exist only if the following two conditions are satisfied (3):

$$\Pi = P_A \qquad \text{and} \qquad \left\{\frac{\partial \Pi}{\partial h}\right\}_{\Pi = P_A} < 0 \qquad (42)$$

Here, $P_A$ is the pressure difference applied across the surface of the film, which in view of Eqs (38) and (39) can be expressed in the form:

$$P_A = P_1 - P_3 - \frac{2\sigma^f}{R_f} = \frac{2\sigma^f}{R_f} + P_2 - P_3 = \sigma\left\{\frac{1}{R_1} + \frac{1}{R_2}\right\} \qquad (43)$$

At the last step we have used also Eqs (35), (40), and (41). For two identical drops $R_f \to \infty$, and then $P_A$ reduces to the capillary pressure of the drops: $P_A = 2\sigma/R_1 = 2\sigma/R_2$. The condition $\Pi = P_A$, see Eq. (42), means that at equilibrium the disjoining pressure $\Pi$ counterbalances the pressure difference $P_A$ applied across the film surface. In addition, the condition $\partial\Pi/\partial h < 0$ guarantees that the equilibrium is *stable* (rather than unstable).

As an illustration, Fig. 8a shows a typical DLVO-type disjoining-pressure isotherm $\Pi(h)$ (see Refs 3, 4 and 62 for more details). There are two points, $h = h_1$ and $h = h_2$, at which the condition for stable equilibrium, Eq. (42), is satisfied. In particular, $h = h_1$ corresponds to the so-called *primary film*, which is stabilized by the electrostatic (double layer) repulsion. The addition of electrolyte to the solution may lead to a decrease in the height of the electrostatic barrier, $\Pi_{max}$ (3,4); at high electrolyte concentration it is possible to have $\Pi_{max} < P_A$, then the primary film does not exist. Note, however, that the increase in electrolyte concentration may lead also to a shift in the maximum toward smaller thicknesses and to an increase in the barrier $\Pi_{max}$. Therefore, primary films could be observed even at relatively high salt concentrations.

The equilibrium state at $h = h_2$ (Fig. 8a) corresponds to a very thin *secondary film*, which is stabilized by the short-range Born repulsion. The secondary film represents a bilayer of two adjacent surfactant monolayers; its thickness is usually about 5 nm (slightly greater than the doubled length of the surfactant molecule) (77). Secondary films can be observed in emulsion flocs and in creamed emulsions.

The situation is more complicated when the aqueous solution contains surfactant micelles, which is a common experimental and practical situation. In such a case the disjoining pressure isotherm $\Pi(h)$ can exhibit multiple decaying oscillations, whose period is close to the diameter of the micelles (Fig. 8b) (for details see, e.g., Ref. 78). The condition for equilibrium liquid film, Eq. (42), can be satisfied at several points, denoted by $h_0$, $h_1$, $h_2$, and $h_3$ in Fig. 8b; the corresponding films contain 0, 1, 2, and 3 layers of micelles, respectively. The transitions between these multiple equilibrium states represent the phenomenon *stratification* (see Fig. 9 and Refs 78–91). The presence of disjoining-pressure barriers, which result from either the electrostatic repulsion (Fig. 8a) or the oscillatory structural forces (Fig. 8b), has a stabilizing effect on liquid films and emulsions (2).

The existence of a stable equilibrium state (see Fig. 8) does not guarantee that a draining liquid film can "safely" reach this state. Indeed, hydrodynamic instabilities, accompanying the drainage of liquid, could rupture the film before it has reached its thermodynamic equilibrium state (1). There are several *kinetic stabilizing factors*, which suppress the hydrodynamic instabilities and decelerate the drainage of the film, thus increasing its lifetime. Such a factor is the *Gibbs* (*surface*) *elasticity*, $E_G$, of the surfactant adsorption monolayers (see Sec. II. A); it tends to eliminate the gradients in adsorption and surface tension and damps the fluctuation capillary waves. At higher surfactant and salt concentrations the Gibbs elasticity is also higher and it renders the interface tangentially immobile (see Fig. 5). The *surface viscosity* also impedes the drainage of water out of the films because of the dissipation of a part of the kinetic energy of the flow within the surfactant adsorption monolayers (see Sec. IV). The *surfactant adsorption relaxation time* (see Sec. II.B) is another important kinetic factor. If the adsorption relaxation time is short enough, a dense adsorption monolayer will cover the newly formed emulsion drops during the emulsification and will protect them against coalescence upon collision. In the opposite case (slow adsorption kinetics) the drops can merge upon collision and the emulsion will be rather unstable.

### 2. Mechanism of Film Breakage

The role of the emulsion stabilizing (or destabilizing) factors can be understood if the mechanism of film breakage is known. Several different mechanisms of rupture of liquid films have been proposed, which are briefly described below.

The *capillary-wave mechanism* has been proposed by de Vries (92) and extended in subsequent studies (2, 24, 93–98) (see Fig. 10a). The conventional version of this mechanism is developed for the case of monotonic attraction between the two surfaces of a liquid film (say, van der Waals attraction). Thermally excited fluctuation capillary waves are always present at the film surfaces. With the decrease in average film thickness, $h$, the attractive disjoining pressure enhances the amplitude of some modes of the fluctuation waves. At a given critical value of the film thickness, $h_c$, corrugations on the two opposite film surfaces can touch each other and then the film will break (97). The same mechanism takes place also in the case of slightly deformed emulsion drops. If the emulsion drops are quiescent, only the thermodynamic and geometric factors determine the critical thickness; indeed, the finite area of the drops (films) imposes limitation on the maximum length of the capillary waves (see Secs III.B and III.C). When the breakage happens during the drainage of the emulsion film (during the approach of the emulsion drops), then the critical thickness is also affected by various hydrodynamic factors (see Sec. IV for details).

The mechanism of *film rupture by nucleation of pores* has been proposed by Derjaguin and Gutop (99) to explain the breaking of very thin films, built up from two attached monolayers of amphiphilic molecules. Such are the secondary foam and emulsion films and the bilayer lipid membranes. This mechanism was further developed by Derjaguin and Prokhorov (3, 100, 101), Kashchiev and Exerowa (102–104), Chizmadzhev and coworkers (105–107), and Kabalnov and Wennerström (108). The formation of

**Figure 9** The spot of lower thickness in a stratifying liquid film corresponds to a local decrease in the number of micelle layers in the colloid-crystal-like structure of surfactant micelles formed inside the liquid film. The appearance of spots could be attributed to the condensation of vacancies in that structure. (From Ref. 82.)

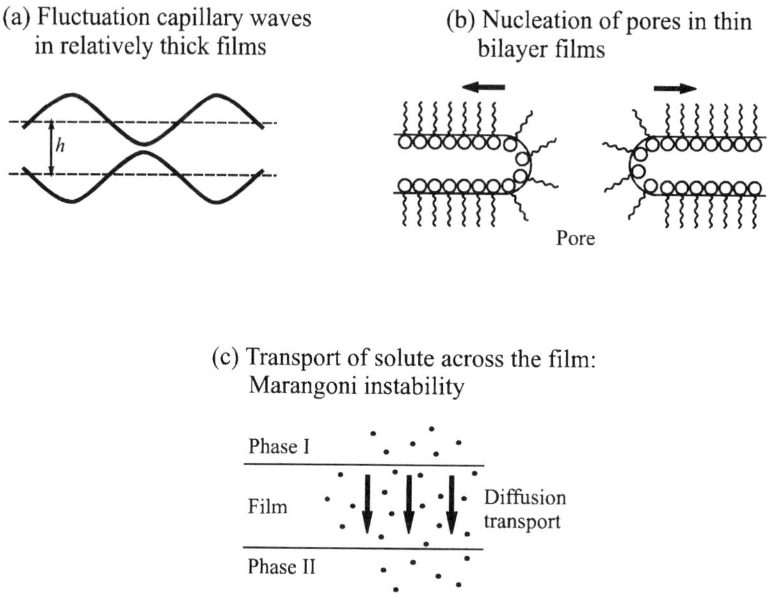


**Figure 10** Mechanisms of breakage of liquid films. (a) *Fluctuation-wave-mechanism*: the film rupture results from growth of capillary waves enhanced by attractive surface forces (92). (b) *Pore-nucleation mechanism*: it is expected to be operative in very thin films, virtually representing two attached monolayers of amphiphilic molecules (99). (c) *Solute-transport mechanism*: if a solute is transferred across the two surfaces of the liquid film due to gradients in the solute chemical potential, then Marangoni instability could appear and break the film (109).

a nucleus of a pore (Fig. 10b) is favored by the decrease in surface energy, but it is opposed by the edge energy of the pore periphery. The edge energy can be described (macroscopically) as a line tension (100–104) or (micromechanically) as an effect of the spontaneous curvature and bending elasticity of the amphiphilic monolayer (108). For small nuclei the edge energy is predominant, whereas for larger nuclei the surface energy gets the upper hand. Consequently, the energy of pore nucleation exhibits a maximum at a given *critical* pore size; the larger pores spontaneously grow and break the film, while the smaller pores shrink and disappear.

A third mechanism of liquid-film breakage is observed when there is a *transport of solute across the film* (see Fig. 10c). This mechanism, investigated experimentally and theoretically by Ivanov and coworkers (109–111), was observed with emulsion systems (transfer of alcohols, acetic acid, and acetone across liquid films), but it could appear also in some asymmetric oil–water–air films. The diffusion transport of some solute across the film leads to the development of Marangoni instability, which manifests itself as a forced growth of capillary waves at the film surfaces and eventual film rupture. Note that Marangoni instability can be caused by both mass and heat transfer (112–114).

A fourth mechanism of film rupture is the *barrier mechanism*. It is directly related to the physical interpretation of the equilibrium states depicted in Fig. 8. For example, let us consider an electrostatically stabilized film of thickness $h_1$ (Fig. 8a). Some processes in the system may lead to an increase in the applied capillary pressure $P_A$. For instance, if the height of the column of an emulsion cream increases from 1 to 10 cm, the capillary pressure in the upper part of the cream increases from 98 to 980 Pa owing to the hydrostatic effect. Thus, $P_A$ could become greater than the height of the barrier, $\Pi_{max}$, which would cause either film rupture (and coalescence) or transition to the stable state of secondary film at $h = h_2$ (Fig. 8a). The increase in the surfactant adsorption density stabilizes the secondary films. In addition, the decrease in $\Pi_{max}$ decreases the probability of the film rupturing after the barrier is overcome. Indeed, the overcoming of the barrier is accompanied by a violent release of mechanical energy accumulated during the increase in $P_A$. If the barrier is high enough, the released energy could break the liquid film. On the other hand, if the barrier is not too high, the film could survive the transition.

The overcoming of the barrier can be facilitated by various factors. Often the transition happens through the formation and expansion of spots of lower thickness within the film, rather than by a sudden decrease in the thickness of the *whole* film. Physically this is accomplished by a nucleation of spots of submicrometer size, which resembles a transition with a "tunnel effect," rather than a real overcoming of the barrier. A theoretical model of spot formation in stratifying films by condensation of vacancies in the structure of ordered micelles (vacancy mechanism) has been developed in Ref. 82 (see Figs 8b and 9). The nucleation of spots makes the transitions less violent and decreases the probability of film breakage. The increase in applied capillary pressure $P_A$ facilitates spot formation and the transition to a state with lower film thickness; this has been established by Bergeron and Radke (85), who experimentally obtained portions of the stable branches of the oscillatory disjoining-pressure curve (Fig. 8b) for foam films. Oscillatory disjoining-pressure curves resulting from reverse micelles in an oily phase were directly measured by Parker et al. (86) by using a version of the surface-force apparatus. Marinova et al. (91) investigated the stabilizing role of the oscillatory disjoining pressure in oil-in-water emulsions which contained surfactant micelles in the aqueous phase.

Below we present in more detail the predictions of the capillary-wave mechanism.

## B. Critical Thickness of Quiescent Emulsion Films

Let us first consider a quiescent emulsion film, say the film between two drops within a floc or cream. At a given sufficiently small thickness of the film, termed the *critical thickness* (92–97, 115), the attractive surface forces prevails and causes growth of the thermally excited capillary waves. This may lead to either film rupture or transition to a thinner secondary film. Two modes of film undulation have been distinguished: symmetric (squeezing, peristaltic) and antisymmetric (bending) modes; it is the symmetric mode which is related to the film breakage/transition. The critical thickness, $h = h_c$, of a film having area $\pi R^2$ can be estimated from the equation (94):

$$\frac{2R^2}{\sigma}\left(\frac{\partial \Pi}{\partial h}\right) = j_i^2 = 5.783\ldots, \tag{44}$$

where $j_1$ is the first zero of the Bessel function $J_0$; as usual, $\sigma$ denotes surface tension. Equation (44) has been derived in a different manner by Vrij (94), Ivanov et al. (95), and Malhotra and Wasan (116). It is obvious that Eq. (44) can be satisfied only for positive $\partial\Pi/\partial h$. If, in the special case of van der Waals interaction one is to substitute $\partial\Pi/\partial h$ by $A_H/(2\pi h^4)$, where $A_H$ is the Hamaker constant, then from Eq. (44) it follows that the critical thickness increases with increase in the film radius $R$, i.e., the films of larger area break more easily (at a greater thickness) than those of smaller area. Note that the effect of surfactant on the tangential mobility of the interface, which involves the surface elasticity, viscosity, and diffusion, does not affect the form of Eq. (44), and correspondingly, the critical thickness $h_c$. The surfactant affects Eq. (44) and $h_c$ only indirectly, through the values of $\sigma$ and $\partial\Pi/\partial h$. These conclusions are valid only for quiescent films, which do not thin during the development of instability.

When an aqueous film is stabilized by an ionic surfactant, then the stability problem becomes more complicated owing to the electrostatic interactions between the charged film surfaces (117). Electrolyte films surrounded by dielectric were initially studied by Felderhof (118), who examined the stability of an equilibrium infinite plane-parallel film surrounded by a vacuum. Sche and Fijnaut (119) extended Felderhof's analysis to account for the effect of surface shear viscosity and surface elasticity. In these studies the electrostatic (double-layer) component of disjoining pressure $\Pi$ was involved in the theory, and a quasistatic approximation was used to describe the electrostatic interaction (117–119). In other words, it has been assumed that the ions immediately acquire their equilibrium distribution for each instantaneous shape of the film. The electric field has been computed by solving the Poisson – Boltzmann equation for the respective instantaneous charge configuration. This quasistatic approximation, which neglects the electrodiffusion fluxes, leads to a counterpart of Eq. (44) in which the total disjoining pressure $\Pi$ includes an electrostatic component. The latter leads to $\partial\Pi/\partial h < 0$ at the equilibrium state ($h = h_1$ in Fig. 8a) and then Eq. (44) has no positive root for $h = h_c$; that is, the film should remain stable for an infinitely long time in agreement with the conventional DLVO theory (3). On the other hand, if electrolyte is added at sufficiently high concentration, the double-layer repulsion is suppressed and the liquid films rupture under the action of the van der Waals force [see Ref. 120 and Eq. (86)].

In reality, aqueous films stabilized with ionic surfactant, without electrolyte, also rupture, especially at surfactant concentrations below the cmc. The latter fact cannot be explained in the framework of the quasi-static approximation (117–119); this is still an open problem in the theory of liquid-film stability.

### C. Critical Distance Between Quiescent Emulsion Drops

Let us consider two emulsion drops of different radii, $R_1$ and $R_2$, like those depicted in Fig. 7 but without the formation of a film between them, i.e., $R = 0$. In this case the gap between the two drops represents a liquid film of uneven thickness. The frequently used lubrication approximation (121) is not applicable to a description of the fluctuation capillary waves on the drop surfaces because it presumes infinite interfacial area and does not impose the natural upper limits on the capillary wavelength, originating from the finite size of the drops. On the other hand, it is possible to solve the problem by means of the usual spherical co-ordinates, locating the co-ordinate origin at the center of one of the two drops. We consider the case in which effects of surface electric charge are negligible and the interaction between the drops (the disjoining pressure) is dominated by the van der Waals attraction. The critical distance between the two drops can be determined from a thermodynamic requirement, viz., the fluctuation of the local disjoining pressure in the narrowest zone of the gap to be equal to the fluctuation of the capillary pressure of the drops. (For shorter distance the fluctuation of the attractive disjoining pressure will prevail and will initiate film rupture.) This requirement leads to the following equation:

$$\frac{1}{\sin\theta}\frac{d}{d\theta}\left[\sin\theta\frac{d\zeta}{d\theta}\right] + \left\{2 + \frac{A_H a^2 \cos^2\theta}{\pi\sigma}[h + 2a(1-\cos\theta)]^{-4}\right\}\zeta = 0, \tag{45}$$

Here, $\zeta$ is the fluctuation in the drop shape, $\theta$ is the polar angle of the spherical coordinate system, $h$ is the shortest distance between the two drop surfaces, $A_H$ is the Hamaker constant and

$$a = 2R_1R_2/(R_1 + R_2) \tag{46}$$

is the mean drop radius. We used the following two boundary conditions: (1) $d\zeta/d\theta = 0$ at $\theta = 0$, i.e., at the narrowest region of the gap; and (2) $\zeta = 0$ for $\theta = \pi/2$, that is, far from the gap zone. The value $h = h_c$, corresponding to the *greatest* eigenvalue of the spectral problem, Eq. (45), gives the critical distance between the two drops. Note that the effect of surfactant on the tangential mobility of the interface, which involves the surface elasticity, viscosity, and diffusion, does not affect the form of Eq. (45), and correspondingly, the critical distance $h_c$. We found the greatest eigenvalue numerically. The results for the critical distance as a function of the drop radius $a$ and the Hamaker constant $A_H$ are shown in Fig. 11; for the interfacial tension we used the value $\sigma = 30$ mN/m. One sees that the critical distance is of the order of dozens of nanometers and that it increases with the rise of both $A_H$ and $a$.

Note, however, that if the two drops are not quiescent, but instead approach each other, the critical distance is influenced by the hydrodynamic interactions – see the next section.

## IV. HYDRODYNAMIC INTERACTIONS AND DROP COALESCENCE

First, we consider the hydrodynamic interactions between two emulsion drops, which remain spherical when the distance between them decreases (Sec. IV.A); this is the transition A→B in Fig. 2. Second, we consider the thinning of the film formed between two emulsion drops (Sec. IV.B): this is stage D in Fig. 2. In both cases the effect of surfactant is taken into account and the critical distance (thickness) for drop coalescence is quantified.

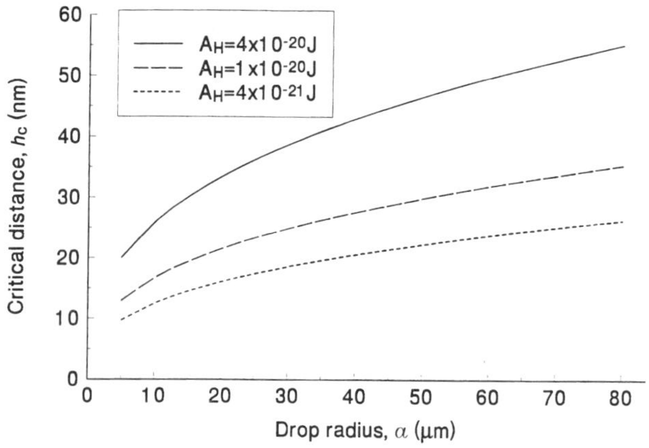


**Figure 11** Plot of the critical distance between two quiescent drops, $h_c$, vs the mean drop radius, $a$, calculated by means of Eq. (45) for three values of the Hamaker constant $A_H$.

## A. Interaction of Spherical Emulsion Drops

### 1. Limiting Cases of Low and High Surface Mobility

The solution to the problem of hydrodynamic interaction between two rigid spherical particles, approaching each other across a viscous fluid, was obtained by Taylor (122). Two spherical emulsion drops of *tangentially immobile* surfaces (due to the presence of dense surfactant adsorption monolayers) are hydrodynamically equivalent to the two rigid particles considered by Taylor. The hydrodynamic interaction is due to the dissipation of kinetic energy when the liquid is expelled from the gap between the two spheres. The resulting friction force decreases the velocity of the two spherical drops proportionally to the decrease in the surface-to-surface distance $h$ in accordance with the Taylor (122) equation:

$$V_{\mathrm{Ta}} = \frac{2h}{3\pi\eta\, a^2}(F - F_s) \quad \text{(tangentially immobile surfaces)} \tag{47}$$

Here, $a$ is the mean drop radius defined by Eq. (46), $F$ is the external force exerted on each drop, and $F_s$ is the surface force originating from the intermolecular interactions between the two drops across the liquid medium. When the range of the latter interactions is much smaller than the drop radii, then $F_s$ can be calculated by means of the Derjaguin approximation (3, 4):

$$F_s = \pi a \int_h^\infty \Pi(h)\, dh \tag{48}$$

where, as before, $\Pi$ is the disjoining pressure.

If the surface of an emulsion drop is mobile, it can transmit the motion of the outer fluid to the fluid within the drop. This leads to a circulation of the fluid inside the drop and influences the dissipation of energy in the system. The problem about the approach of two nondeformed spherical drops or bubbles in the *absence of surfactants* has been investigated by many authors (123–132). A number of solutions, generalizing the Taylor equation [Eq. (47)], have been obtained. In particular, the velocity of central approach of two spherical drops in pure liquid, $V_{\mathrm{p}}$, is related to the Taylor velocity $V_{\mathrm{Ta}}$, defined by means of a Padé-type expression derived by Davis et al. (131):

$$V_{\mathrm{p}} = V_{\mathrm{Ta}} \frac{1 + 1.711\xi + 0.461\xi^2}{1 + 0.402\xi} \quad \text{(no surfactant)} \tag{49}$$

$$\xi \equiv \frac{\eta_{out}}{\eta_{in}} \sqrt{\frac{a}{2h}} \tag{50}$$

where, as usual, $h$ is the closest surface-to-surface distance between the two drops, and $\eta_{in}$ and $\eta_{out}$ are the viscosities of the liquids inside and outside the drops. In the limiting case of solid particles one has $\eta_{in} \to \infty$, $\xi \to 0$ and then Eq. (49) reduces to the Taylor equation, Eq. (47). Note that in the case of a *close* approach of two drops ($h \to 0$, $\xi \gg 1$) the velocity $V_{\mathrm{p}}$ is proportional to $h^{1/2}$. This implies that the two drops can come into contact ($h = 0$) in a finite period of time ($\tau < \infty$) under the action of a given force, $F$, because the integral expressing the lifetime (97):

$$\tau = \int_{h_c}^{h_{in}} \frac{dh}{V} \tag{51}$$

(with $V = V_{\mathrm{p}}$) is convergent for $h_c = 0$; $h_{in}$ is the surface-to-surface distance at the initial moment $t = 0$. In contrast, in the case of immobile interfaces ($\xi \ll 1$) Eq. (47) gives $V_{\mathrm{Ta}} \propto h$ and $\tau \to \infty$ for $h_c \to 0$. Moreover, the counterbalancing of the external force by the surface force, i.e., $F - F_s = 0$, implies $V_{\mathrm{Ta}} = V = 0$ and $\tau \to \infty$ (equilibrium state) irrespective of whether the drop surfaces are tangentially mobile or immobile.

It has been established both theoretically and experimentally (133, 134) that, if the surfactant is dissolved only in the drop phase, the film formed between two emulsion drops (Fig. 2D) thins just as if surfactant is missing. Likewise, one can use Eq. (49) to estimate the velocity of approach of two emulsion drops when surfactant is contained only in the drop phase (2).

### 2. Effects of Surface Elasticity, Viscosity, and Diffusivity

When surfactant is present in the continuous phase at not too high concentration, then the surfactant adsorption monolayers, covering the emulsion drops, are tangentially mobile, rather than immobile. The adsorbed surfactant can be dragged along by the fluid flow in the gap between two colliding drops, thus affecting the hydrodynamic interaction between them. The appearance of gradients of surfactant adsorption are opposed by the Gibbs elasticity, surface viscosity, and surface and bulk diffusion. Below, we consider the role of the enumerated factors on the velocity of approach of two emulsion drops.

If the driving force $F$ (say, the Brownian or the buoyancy force) is small compared to the capillary pressure of the droplets, the deformation of two spherical droplets upon collision will be only a small perturbation in the zone of contact. The film thickness and the pressure within the gap can then be presented as a sum of a nonperturbed part and a small perturbation. Solving the resulting hydrodynamic problem for low (negligible) interfacial viscosity, an analytical formula for the velocity of drop approaching, $V = -dh/dt$, can be derived (121):

$$\frac{V}{V_{\mathrm{Ta}}} = \frac{h_s}{2h}\left[\frac{d/h+1}{d/h}\ln(d/h+1) - 1\right]^{-1} \tag{52}$$

where $a$ is the mean drop radius defined by Eq. (46), and $V_{\mathrm{Ta}}$ is the Taylor velocity, Eq. (47); the other parameters are defined as follows:

$$d \equiv \frac{h_s}{1+b} \tag{53}$$

$$h_s \equiv \frac{6\eta_{out} D_{1,s}}{E_G} \tag{54}$$

$$b \equiv \frac{3\eta_{out} D_1}{E_G}\left(\frac{\partial c_1}{\partial \Gamma_1}\right)^{(e)} = \frac{3\eta_{out} D_1}{\Gamma_1^{(e)}}\left(\frac{\partial c_1}{\partial \sigma}\right)^{(e)} \tag{55}$$

As usual, the superscript ($e$) denotes that the respective quantity should be estimated for the equilibrium state; the dimensionless parameter $b$ accounts for the effect of *bulk diffusion*, whereas $h_s$ has a dimension of length and takes into account the effect of *surface diffusion*. In the limiting case of very large *Gibbs elasticity* $E_G$ (tangentially immobile interface) the parameter $d$ tends to zero and then Eq. (52) yields $V \to V_{\mathrm{Ta}}$, as should be expected (121, 135, 136).

If the effect of *surface viscosity* is taken into account, then Eq. (52) can be expressed in the generalized form (137, 138):

$$\frac{V}{V_{\mathrm{Ta}}} = \Phi_{\mathrm{v}}(h; h_s, d, s_{\mathrm{v}}) \tag{56}$$

where $\Phi_{\mathrm{v}}$ is termed the *mobility factor* (*function*); the dimensionless parameter $s_{\mathrm{v}}$ takes into account the effect of surface viscosity:

$$s_{\mathrm{v}} \equiv \frac{\eta_s}{6a\,\eta_{out}}, \qquad \eta_s \equiv \eta_{\mathrm{sh}} + \eta_{\mathrm{dil}} \tag{57}$$

Here, $\eta_{\mathrm{sh}}$ and $\eta_{\mathrm{dil}}$ are, respectively, the interfacial shear and dilatational viscosities. In fact, Eq. (52) gives an analytical expression for the mobility factor $\Phi_{\mathrm{v}}$ in the case when $s_{\mathrm{v}} \ll 1$, i.e., the effect of surface viscosity can be neglected. However, if the effect of surface viscosity is essential, there is no analytical expression for $\Phi_{\mathrm{v}}$; in this case a numerical procedure for computation of $\Phi_{\mathrm{v}}$ has been developed (137, 138). Table 3 contains asymptotic expressions for $\Phi_{\mathrm{v}}$. A general property of $\Phi_{\mathrm{v}}$ is

**Table 3** Asymptotic Expressions for Mobility Factors

| Asymptotic cases | Mobility factors ($\Phi_{\mathrm{v}}$ and $\Phi_{\mathrm{p}}$) |
|---|---|
| $\frac{d}{h} \ll 1$ | $\Phi_{\mathrm{v}} = (b+1)/\left[1 - \frac{d}{3h} + \frac{8s_{\mathrm{v}}b}{3(b+1)}\left(b + \frac{d}{h}\right)\right]$ |
| | $\Phi_{\mathrm{p}} = \frac{\Phi_{\mathrm{v}}}{b+1}\left[1 - \frac{2d}{3h} + \frac{16s_{\mathrm{v}}b}{3(b+1)}\left(b + \frac{2d}{h}\right)\right]$ |
| $\frac{d}{h} \gg 1$ | $\Phi_{\mathrm{v}} = \frac{h_s}{2h}/\left[\frac{4s_{\mathrm{v}}h_s}{3h} - 1 - 8s_{\mathrm{v}} + \ln\frac{d}{h}\right]$ |
| | $\Phi_{\mathrm{p}} = \frac{2h\Phi_{\mathrm{v}}}{h_s}\left(\frac{1+8s_{\mathrm{v}}h_s}{3h}\right)$ |

*Source*: Ref. 138.

$$\Phi_{\mathrm{v}} = 1 \quad \text{for} \quad E_G \to \infty \quad \text{(tangentially immobile surfaces)} \tag{58}$$

It is important to note that the surface viscosity parameter $s_{\mathrm{v}}$ appears only in the combinations $s_{\mathrm{v}}h_s/h = \eta_s D_{1s}/(hE_G a)$ and $s_{\mathrm{v}}b$ (see Table 3). In view of Eqs (53)–(55), it then follows that the surface viscosity can influence the mobility factor $\Phi_{\mathrm{v}}$ only if either the Gibbs elasticity, $E_G$, or the drop radius, $a$, or the gap width, $h$, are small enough.

To illustrate the dependence of the mobility function $\Phi_{\mathrm{v}}$ on the concentration of surfactant in the continuous phase, in Fig. 12 we present theoretical curves, calculated in Ref. 138 for the nonionic surfactant Triton X-100, for the ionic surfactant SDS (+0.1 M NaCl) and for the protein bovine serum albumin (BSA). The parameter values, used to calculated the curves in Fig. 12, are listed in Table 4; $\Gamma_\infty$ and $K$ are parameters of the Langmuir adsorption isotherm used to describe the dependence of surfactant adsorption, surface tension, and Gibbs elasticity on the surfactant concentration (see Tables 1 and 2). As before, we have used the approximation $D_{1s} \approx D_1$ (surface diffusivity equal to the bulk diffusivity). The surfactant concentration in Fig. 12 is scaled with the reference concentration $c_o$, which is also given in Table 4; for Triton X-100 and SDS + 0.1 M NaCl, $c_o$ is chosen to coincide with the cmc. The driving force, $F$, was taken to be the buoyancy force for dodecane drops in water. The surface force $F_s$ is identified with the van der Waals attraction; the Hamaker function $A_{\mathrm{H}}(h)$ was calculated by means of Eq. (86) (see below). The mean drop radius in Fig. 12 is $a = 20\ \mu$m. As seen in the figure, for such small drops $\Phi_{\mathrm{v}} \approx 1$ for Triton X-100 and BSA, i.e., the

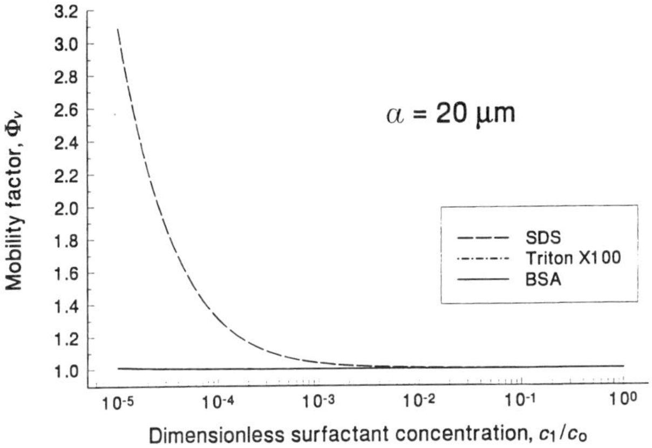


**Figure 12** Theoretical dependence of the mobility factor $\Phi_v$, on the surfactant concentration $c_1$, calculated in Ref. 138 for the nonionic surfactant Triton X-100, ionic surfactant SDS + 0.1 M NaCl, and the protein BSA; the curves for Triton X-100 and BSA coincide. The mean drop radius is $a =$ 20 $\mu$m and the film thickness is $h \approx 10$ nm; the other parameters values are listed in Table 4.

drop surfaces turn out to be tangentially immobile in the whole concentration range investigated. On the other hand, $\Phi_v$ becomes considerably greater than unity for the lowest SDS concentrations, which indicates increased mobility of the drop surfaces.

### 3. Formation of Pimple

Let us consider two spherical emulsion drops approaching each other, which interact through the van der Waals attractive surface force. Sooner or later interfacial deformation will occur in the zone of drop–drop contact. The calculations (138) show that, if the drop radius $a$ is greater than 80 $\mu$m, the drop interfaces bend inwards (under the action of the hydrodynamic pressure) and a *dimple* is formed in the contact zone; soon the dimple transforms into an almost plane-parallel film (Fig. 2D). In contrast, if the drop radius $a$ is less than 80 $\mu$m, then at a given surface-to-surface distance $h = h_p$ the drop surface in the contact zone bends outwards and a *pimple* forms due to the van der Waals attraction (see the inset in Fig. 13). Correspondingly, $h_p$ is called the pimpling distance. Since the size of the drops in an emulsion is usually markedly below 80 $\mu$m, we will consider here only the formation of a pimple.

The formation of pimples was discovered by Yanitsios and Davis (139) in computer calculations for emulsion drops from pure liquids, without any surfactant. Next, by means of numerical calculations, Cristini et al. (140) established the formation of a pimple for emulsion drops covered with insoluble surfactant in the case of negligible surface diffusion; their computations showed that rapid coalescence took place for $h < h_p$. A complete treatment of the problem for the formation of pimples was given in Ref. 138, where the effects of surface and bulk diffusion of surfactant, as well as the surface elasticity and viscosity, were taken into account, and analytical expressions were derived.

The origin of pimple formation is the fact that the van der Waals disjoining pressure, $\Pi \propto 1/h^3$, grows faster than the hydrodynamic pressure with decrease in $h$. For a certain distance, $h = h_p$, $\Pi$ counterbalances the hydrodynamic pressure (138):

$$\frac{F - F_s}{\pi a h_p} \Phi_p (h_p; b, d, s_v) + \Pi(h_p) = 0 \qquad (59)$$

where $\Phi_p$ is the mobility factor for the pressure. Further, for a shorter distance between the drops, $h < h_p$, the pimples spontaneously grow until the drop surfaces touch each other and the drops coalesce. The pimple formation at $h = h_p$ can be interpreted as an onset of instability *without* fluctuations.

Analytical asymptotic expressions for the pressure mobility factor, $\Phi_p$, can be found in Table 3. In general, $\Phi_p$ is to be calculated numerically. In the case of tangentially immobile surfaces of the drops Eq. (59) yields a very simple formula for the pimpling distance (138):

**Table 4** Parameters for the Solutions of Triton X-100, SDS + 0.1 M NaCl, and BSA

| | Triton X-100 | SDS + 0.1 M NaCl | BSA |
|---|---|---|---|
| $\Gamma_\infty$ | $1.75 \times 10^{-6}$ mol/m$^2$ | $3.75 \times 10^{-6}$ mol/m$^2$ | $0.38 \times 10^{-6}$ mol/m$^2$ |
| $1/K$ | $4.9 \times 10^{-6}$ kg/m$^3$ | $1.4 \times 10^{-3}$ kg/m$^3$ | $0.124 \times 10^{-6}$ kg/m$^3$ |
| $D_1$ | $2.6 \times 10^{-10}$ m$^2$/s | $6.0 \times 10^{-10}$ m$^2$/s | $1.0 \times 10^{-10}$ m$^2$/s |
| $c_0$ | 0.4 kg/m$^3$ (cmc) | 0.33 kg/m$^3$ (cmc) | 0.6 kg/m$^3$ |
| $\eta_s$ | $1.0 \times 10^{-6}$ mPa.s | $1.0 \times 10^{-6}$ mPa.s | $5.0 \times 10^{-3}$ mPa.s |

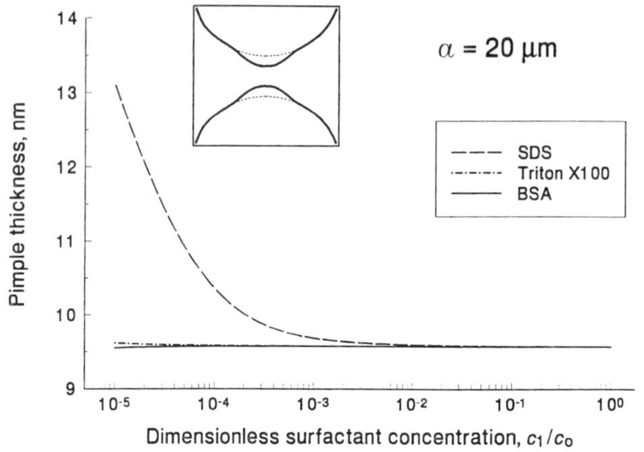


**Figure 13** Calculated dependence of the pimple thickness, $h_p$, on the surfactant concentration, $c_1$, for emulsion films formed from aqueous solutions of SDS + 0.1 M NaCl, Triton X-100, and BSA; the parameters values used are listed in Table 4. The inset illustrates the shape of the drop surfaces in the zone of contact.

$$h_p = \sqrt{\frac{aA_H}{12F}} \qquad \text{(tangentially immobile surfaces)} \tag{60}$$

In the more complicated case of mobile drop surfaces Eq. (59) has to be solved numerically. Figure 13 shows calculated curves for the dependence of $h_p$ versus the surfactant concentration; the parameter values used are the same as for Fig. 12 (see Table 4). Since the surfaces of the drops with BSA and Triton X-100 are tangentially immobile, the respective pimpling distance is practically constant (independent of surfactant concentration) and given by Eq. (60). The effect of surface mobility shows up for the emulsions with SDS + 0.1 M NaCl, for which the pimpling distance $h_p$ is greater (Fig. 13). These calculations demonstrate that $h_p$ is typically of the order of 10 nm.

If the pimpling distance is greater than the critical distance, $h_p > h_c$, then the pimpling will be the reason for coalescence. On the other hand, if $h_c > h_p$, then the coalescence will be caused by the fluctuation capillary waves (see the next subsection).

### 4. Transitional and Critical Distance

As already mentioned, when two emulsion drops approach each other, the attractive surface forces promote the growth of fluctuation capillary waves in the contact zone. At a given, sufficiently small surface-to-surface distance, called the transitional distance, $h_t$, the waves with a given length (usually the longest one) begin to grow; this is a transition from stability to instability. During the growth of the waves the gap width continues to decrease, which leads to destabilization and growth of waves with other lengths. Finally, the surfaces of the two drops touch each other owing to the enhanced interfacial undulations, and coalescence takes place. The latter act corresponds to a given mean surface-to-surface distance, called the critical thickness, $h_c$. The difference between the transitional and critical distance, $h_t > h_c$, is due to the fact that during the growth of the capillary waves the average film thickness continues to decrease, insofar as the drops are moving against each other driven by the force $F - F_s$. In the simpler case of immobile drops ($F - F_s = 0$), considered in Sec. III.C, one has $h_t = h_c$.

A general equation for determining $h_t$, which takes into account the effect of surface mobility, has been reported in Ref. 136:

$$[F - F_s]\Psi(\bar{d}) = \frac{\pi a^2 h_t^3}{32\sigma}\left[\left(\frac{\partial \Pi}{\partial h}\right)^2 - \frac{8\sigma}{a}\left(\frac{\partial^2 \Pi}{\partial h^2}\right)\right]_{h=h_t} \tag{61}$$

where

$$\bar{d} \equiv \frac{d}{h}, \qquad \Psi(\bar{d}) = \frac{(\bar{d}+2)\bar{d}^2}{4(\bar{d}+1)^2[(\bar{d}+1)\ln(\bar{d}+1) - \bar{d}]} \tag{62}$$

The function $\Psi(\bar{d})$ accounts for the effect of the surface mobility. For large interfacial elasticity one has $\bar{d} \to 0$, see Eq. (53); then $\Psi \to 1$ and Eq. (61) acquires a simpler form, corresponding to drops of tangentially immobile interfaces. In the other limit, small interfacial elasticity, one has $\bar{d} \gg 1$ and in such a case $\Psi \propto 1/\ln \bar{d}$, i.e., $\Psi$ decreases with the increase in $\bar{d}$, that is, with the decrease in $E_G$. A numerical solution to this problem is reported in Ref. 24. The effect of the interfacial viscosity on the transitional distance, which is neglected in Eq. (61), is examined in Ref. 141. It is established therein that the critical distance, $h_c$, can be with in about 10% smaller than $h_t$.

The dependence of the transitional distance $h_t$ on the surfactant concentration, calculated with the help of Eq. (61), is shown in Fig. 14; the three curves correspond to three fixed values of the mean drop radius $a$. The calculations are carried out for the system with SDS + 0.1 M NaCl in the aqueous phase (see Table 4); the oil phase is dodecane. One sees that the increase in surfactant concentration leads to a decrease in transitional thickness, which corresponds to a greater sta-

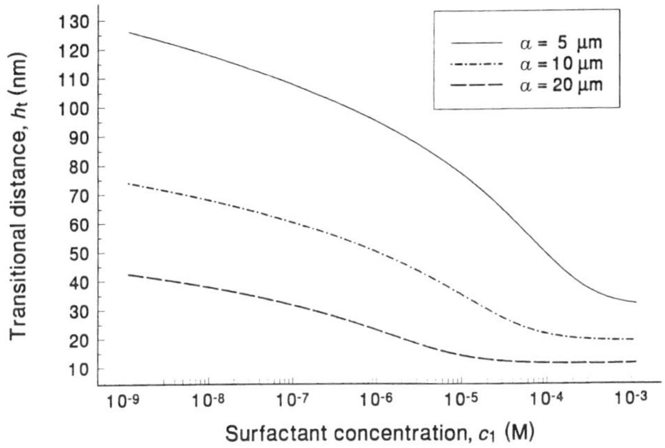


**Figure 14** Dependence of the transitional distance between two drops, $h_t$, on the surfactant concentration, $c_1$, calculated with use of Eq. (61) for three values of the mean drop radius $a$.

bility of the emulsion against coalescence. Physically this is related to the damping of the fluctuation capillary waves by the adsorbed surfactant. Moreover, the transitional thickness for two approaching drops increases with the decrease in drop radius $a$ (Fig. 14), which is exactly the opposite to the tendency for quiescent drops in Fig. 11 (we recall that $h_t = h_c$ for quiescent drops). The difference can be attributed to the strong dependence of the buoyancy force $F$ on the drop radius $a$ (such an effect is missing for the quiescent drops).

The comparison between Figs 13 and 14 shows that for the emulsion with SDS + 0.1 M NaCl one has $h_t > h_p$. In other words, the theory predicts that in this emulsion the drops will coalesce due to the fluctuation capillary waves, rather than owing to the pimpling.

If the coalescence is promoted by the van der Waals attractive surface force, from Eq. (61) one can deduce asymptotic expressions for $h_t$, corresponding to tangentially immobile drop surfaces ($\Psi = 1$) (136):

$$h_t = \left[a^2 A_H^2/(128\pi\sigma F)\right]^{1/5}, \qquad \text{for} F \gg F_a \tag{63}$$

$$h_t = [5aA_H/(12F)]^{1/2}, \qquad \text{for} F \ll F_a \tag{64}$$

where $F_a = 12.66(a\sigma^2 A_H)^{1/3}$. In particular, if $F$ is the buoyancy force, then $F \propto a^3$ and for small droplets ($F \ll F_a$) one obtains $h_t \propto 1/a$, i.e., the critical thickness markedly increases with the decrease in droplet radius.

## B. Interaction Between Deforming Emulsion Drops

### 1. Drops of Tangentially Immobile Surfaces

In this subsection we consider the case in which a liquid film is formed in the zone of contact between two emulsion drops (see Fig. 7). Such a configuration appears between drops in flocs and in concentrated emulsions, including creams.

In a first approximation, one can assume that the viscous dissipation of kinetic energy happens mostly in the thin liquid film intervening between two drops. (In reality, some energy dissipation happens also in the transition zone between the film and the bulk continuous phase.) If the drop interfaces are tangentially immobile (owing to adsorbed surfactant), then the velocity of approach of the two drops can be estimated by meanss of the Reynolds formula for the velocity of approach of two parallel solid disks of radius $R$, equal to the film radius (142):

$$V_{\text{Re}} = \frac{2h^3 F_{\text{tot}}}{3\pi\eta_{out} R^4} \tag{65}$$

As usual, here $h$ is the film thickness, and $F_{\text{tot}}$ is the total force exerted on a drop (2):

$$F_{\text{tot}} = F - F_s - F_{\text{def}} \tag{66}$$

As before, $F$ is the applied external force (buoyancy, centrifugal force, Brownian force, etc.); $F_s$ is the surface force of intermolecular origin, which for deformable drops can be expressed in the form (2, 143):

$$F_s = -\frac{dW}{dh}, \qquad W(R, h) \approx \pi R^2 f(h) + \pi a \int_h^\infty f(\hat{h})d\hat{h} \tag{67}$$

where

$$f(h) \equiv \int_h^\infty \Pi(\hat{h})d\hat{h} \tag{68}$$

is the interaction free energy per unit area of a plane-parallel liquid film, and $W$ is the drop – drop interaction energy due to surface forces, which is a sum of contributions from the planar film and the transition zone film – bulk liquid; for $R = 0$, Eq. (67) reduces to Eq. (48). Finally, $F_{\text{def}}$ is a force originating from the deformation of the drop interfaces (2):

$$F_{\text{def}} = -\frac{d(W_{\text{dil}} + W_{\text{bend}})}{dh} \tag{69}$$

where $W_{\text{dil}}$ is the work of interfacial dilatation (143–145),

$$W_{\text{dil}} = \sigma\frac{\pi R^4}{2a^2} + \frac{1}{2}E_G\left[\frac{\pi R^4}{2a^2}\right]^2 + \cdots, \text{ for } \left[\frac{R}{a}\right] \ll 1 \tag{70}$$

and $W_{\text{bend}}$ is the work of interfacial bending (146):

$$W_{\text{bend}} = -2\pi R^2 B_0/a, \quad (R/a)^2 \ll 1 \tag{71}$$

where $B_0 = -4k_c H_0$ is the interfacial bending moment; $H_0$ is the so-called spontaneous curvature, and $k_c$ is the interfacial curvature elastic modulus.

Initially, the two approaching drops are spherical. The deformation in the zone of contact begins when the surface-to-surface distance reaches a certain threshold value, called the inversion thickness, $h_{\text{inv}}$. One can estimate the inversion thickness from the simple expression $h_{\text{inv}} = F/(2\pi\sigma)$ (see, e.g., Refs 98 and 121). The generalized form of the latter equation, accounting for the contribution of the surface forces, reads (136):

$$h_{\text{inv}} = \frac{F}{2\pi\sigma} + \frac{a}{2\sigma}[h_{\text{inv}}\Pi(h_{\text{inv}}) - f(h_{\text{inv}})] \tag{72}$$

The inversion thickness can be determined by solving Eq. (72) numerically.

A generalized expression for the velocity $V = -dh/dt$, which takes into account the energy dissipation in both film and the transition zone film – bulk liquid, has been derived in Refs 2 and 147:

$$\frac{1}{V} = \frac{1}{V_{\text{Re}}} + \frac{1}{V_{\text{Ta}}} + \frac{1}{\sqrt{V_{\text{Re}}V_{\text{Ta}}}} \tag{73}$$

where the Taylor velocity, $V_{\text{Ta}}$, and the Reynolds velocity, $V_{\text{Re}}$, are defined by means of Eqs (47) and (65). For $R \to 0$ (nondeformed spherical drops), Eq. (73) reduces to $V = V_{Ta}$. On the other hand, for $h \to 0$ one has $1/V_{\text{Ta}} \ll 1/V_{\text{Re}}$, and then Eq. (73) yields $V \to V_{\text{Re}}$. Substituting Eqs (47) and (65), and assuming $F \gg (F_s + F_{\text{def}})$ one can bring Eq. (73) into the form (147):

$$\frac{V_{\text{Ta}}}{V} = 1 + \frac{R^2}{ha} + \frac{R^4}{h^2a^2} \tag{74}$$

One sees that $V \to V_{\text{Ta}}$ for $R^2/(ha) \ll 1$. If the external force $F$ is predominant, then $R^2 \approx aF/(2\pi\sigma)$, $h_{\text{inv}} \approx F/(2\pi\sigma)$ and it follows that $R^2/a \approx h_{\text{inv}}$ (97, 135); the substitution of the latter equation into Eq. (74) yields:

$$\frac{V_{\text{Ta}}}{V} \approx 1 + \frac{h_{\text{inv}}}{h} + \frac{h_{\text{inv}}^2}{h^2} \tag{75}$$

Equation (75) shows that for $h \leq h_{\text{inv}}$ the velocity $V$ becomes considerably smaller than $V_{\text{Ta}}$.

## 2. Effect of Surface Mobility

When the surfactant is soluble only in the *continuous* phase (we will call such a system "System I", see Fig. 15), turns out that the respective rate of film thinning $V_{\text{I}}$ is affected by the surface mobility mainly through the Gibbs elasticity $E_G$, just as it is for foam films (97, 121):

$$\frac{V_{\text{I}}}{V_{\text{Re}}} \approx 1 + \frac{1}{\varepsilon_f}, \qquad \frac{1}{\varepsilon_f} = \frac{6\eta_1 D_{1s}}{hE_G} + \frac{3\eta_1 D_1}{\Gamma_1}\left(\frac{\partial c_1}{\partial \sigma}\right)^{(e)} \tag{76}$$

Here, $\varepsilon_f$ is the so called *foam parameter*, and $\eta_1$ is the viscosity in the surfactant-containing phase (Liquid 1 in Fig. 15); the influence of the transition zone film – bulk liquid is not accounted for in Eq. (76). Note that the bulk and surface diffusion fluxes (see the terms with $D_{1s}$ and $D_1$ in the latter equation), which tend to damp the surface tension gradients and to restore the uniformity of the adsorption monolayers, accelerate the film thinning (Fig. 1). Moreover, since $D_{1s}$ in Eq. (76) is divided by the film thickness $h$, the effect of surface diffusion dominates that of bulk diffusion for small values of the film thickness. On the other hand, the Gibbs elasticity $E_G$ (the Marangoni effect) decelerates the thinning. Equation (76) predicts that the rate of thinning is not affected by the circulation of liquid

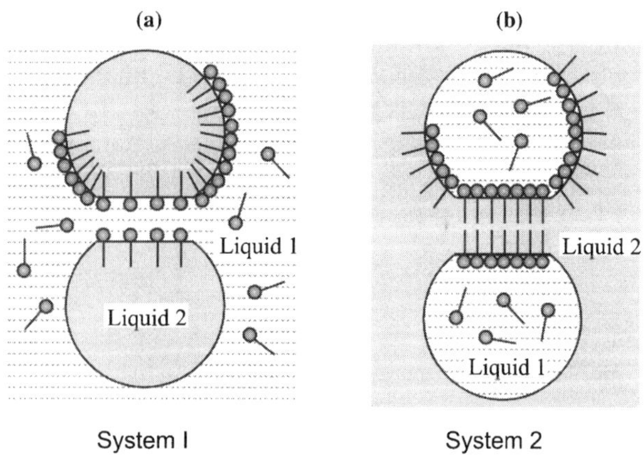


**Figure 15** Two complementary types of emulsion system obtained by a mere exchange of the continuous phase with the disperse phase. The surfactant is assumed to be soluble only in Liquid 1. (a) Liquid 2 is the disperse phase; (b) Liquid 2 is the continuous phase.

in the droplets, i.e., System I really behaves as a foam system.

It was established theoretically (97, 133) that when the surfactant is dissolved in the *drop* phase (System II in Fig. 15) it remains uniformly distributed throughout the drop surface during film thinning, and interfacial tension gradients do not appear. This is the result of a powerful supply of surfactant, which is driven by convective diffusion from the bulk of the drops toward their surfaces. For that reason, the drainage of the film surfaces is not opposed by surface-tension gradients, and the rate of film thinning, $V_{\text{II}}$, is the same as in the case of pure liquid phases (97, 133):

$$\frac{V_{\text{II}}}{V_{\text{Re}}} \approx \frac{1}{\varepsilon_e} \approx \frac{\eta_1 \delta}{\eta_2 h} \approx \left\{ \frac{108\pi\eta_1^3 R^4}{\rho_2 \eta_2 h^4 F} \right\}^{1/3} \tag{77}$$

Here, $\varepsilon_e$ is called the emulsion parameter, $\delta$ is the thickness of the hydrodynamic boundary layer inside the drops, and $\rho_2$ and $\eta_2$ are the mass density and dynamic viscosity of Liquid 2, which does not contain dissolved surfactant. The validity of Eq. (77) was confirmed experimentally (134).

The only difference between the two systems in Fig. 15 is the exchange of the continuous and drop phases. Assume for simplicity that $V_{\text{Re}}$ is the same for both systems. In addition, usually $\varepsilon_f \approx 0.1$ and $\varepsilon_e \approx 10^{-2}$ to $10^{-3}$. From Eqs (76) and (77) one then obtains (97, 121, 133):

$$\frac{V_{\text{II}}}{V_{\text{I}}} \approx \frac{\varepsilon_f}{\varepsilon_e} \approx \frac{0.1}{10^{-2}\ \text{to}\ 10^{-3}} \approx 10\ \text{to}\ 100 \quad \text{(deformed drops)} \tag{78}$$

Hence, the rate of film thinning in System II is much greater than that in System I. Therefore, the location of the surfactant has a dramatic effect on the thinning rate and, thereby, on the drop lifetime. Note also that the interfacial tension in both systems is the same. Hence, the mere phase inversion of an emulsion, from Liquid 1-in-Liquid 2 to Liquid 2-in-Liquid 1 (Fig. 15), could change the emulsion lifetime by orders of magnitude. As discussed in Sec. V, the situation with interaction in the Taylor regime (between spherical, nondeformed drops) is similar. These facts are closely related to the explanation of the Bancroft rule for the stability of emulsions (see Sec. V) and the process of chemical demulsification (1).

Equations (76) and (77) do not take into account the hydrodynamic interactions across the transition zone around the film, which can be essential if the film radius $R$ is relatively small. In the latter case the effect of surface viscosity becomes important for System I. Equation (76) can then be presented in a more general form (137).

$$\frac{V_{\text{I}}}{V_{\text{Re}}} = \Phi_{\text{v}}\,(h; b, d, R, s_{\text{v}}) \tag{79}$$

where $\Phi_{\text{v}}$ is a mobility function. In Ref. 137 a general, but voluminous, analytical expression for $\Phi_{\text{v}}$ is derived in the form of an infinite series expansion; it accounts for the effects of surface elasticity, surface viscosity, and bulk and surface diffusion. In some special cases this infinite series can be summed up and closed expressions for $\Phi_{\text{v}}$ can be obtained. Such is the case when the effect of the surface viscosity is negligible, $s_{\text{v}} \to 0$; the respective expression for $\Phi_{\text{v}}$ reads (137):

$$\frac{1}{\Phi_{\text{v}}} = \frac{1}{1+b+h_s/h} + \frac{2h}{N_R^4 h_s} \left\{ \left[ \frac{h}{h_s}(1+b)(1-N_R^2)+1 \right] \ln\left[ 1 + \frac{h_s}{h(1+b)} \right] + N_R^2 - 1 \right\} \tag{80}$$

where the dimensionless parameter $N_R \equiv R/(ah)^{1/2}$ accounts for the effect of the film radius. In the case of emulsion drops $N_R \approx 1$; however, if experiments with emulsion films are performed in the experimental cell of Scheludko and Exerowa (148, 149), which allows independent control of $R$, then one usually has $N_R \gg 1$. (The original experiments in Refs 148 and 149 have been carried out with foam films, but a similar technique can be appllied to investigate emulsion films, see, e.g., Refs 91 and 150–158.) In the limit of large plane-parallel film, $N_R \gg 1$, Eq. (80) reduces to the result of Radoev *et al.* (159): $V_{\text{I}}/V_{\text{Re}} = 1 + b + h_s/h$ (effect of the transition zone negligible). For insoluble surfactants the parameter $b$ in Eq. (80) must be set equal to zero.

Under certain experimental conditions, like those in Ref. 60, the motion of surfactant along an oil–water interface represents a flow of a two-dimensional *incompressible viscous* fluid. In such a case Eq. (79) acquires the following specific form (137):

$$\frac{1}{V_{\text{I}}} = \frac{1}{V_{Ta}} + \frac{1}{V_{\text{Re}}} + \frac{1}{\sqrt{V_{\text{Ta}} V_{\text{Re}}}} - \frac{1}{2s_{\text{v}}} \left[ \frac{1}{V_{\text{Ta}}} + \frac{1}{V_{\text{Re}}} + \frac{1}{\sqrt{V_{\text{Ta}} V_{\text{Re}}}} + \frac{N_R^2}{3\sqrt{V_{\text{Ta}} V_{\text{Re}}}} \right] \tag{81}$$

Equation (81) is a truncated power expansion for $s_{\text{v}} \gg 1$. In the limit of tangentially immobile interfaces ($s_{\text{v}} \to \infty$) Eq. (81) reduces to Eq. (73).

To illustrate the effects of various factors on the velocity of approach of two deforming emulsion drops (Fig. 15a) we used the general expression from Ref. 137 (the infinite series expansion) to calculate the mobility factor $\Phi_V$; the results are shown in Figs 16 and 17. First of all, in Fig. 16 we illustrate the effects of bulk and surface diffusion. For that reason $\Phi_V \equiv V_I/V_{Re}$ is plotted versus the parameter $b$, related to the bulk diffusivity, for various values of $h_s/h$; $h_s$ is related to the surface diffusivity, see Eq. (54). If the hydrodynamic interaction were operative only in the film, then one would obtain $V_I/V_{Re} \geq 1$. However, all calculated values of $V_I/V_{Re}$ are less than 0.51 (Fig. 16); this fact is evidence for a significant effect of the hydrodynamic interactions in the transition zone around the film. Moreover, in Fig. 16 one sees that for $b > 10$ the mobility factor $\Phi_V$ is independent of the surface diffusivity. On the other hand, for $b < 10$ a considerable effect of surface diffusivity shows up: the greater the surface diffusivity effect, $h_s/h$, the greater the interfacial mobility factor $\Phi_V$. For the upper curve in Fig. 16 the interfacial mobility is determined mostly by the effect of surface viscosity, $s_V$, which is set equal to unity for all curves in the figure.

To illustrate the effect of surface viscosity, $s_V$, in Fig. 17 we have plotted the mobility factor $\Phi_V \equiv V_I/V_{Re}$ versus $b$ for three different values of $S_V$. For the higher surface viscosities, $s_V = 1$ and 5, and the mobility factor is $V_I/V_{Re} < 1$, which again indicates a strong hydrodynamic interaction in the transition zone around the film. For the lowest surface viscosity, $s_V = 0.1$, the mobility factor is sensitive to the effect of bulk diffusion, characterized by $b$: for $b > 3$ we have $V_I/V_{Re} > 1$, i.e., we observe a considerable rise in the interfacial mobility (Fig. 17).

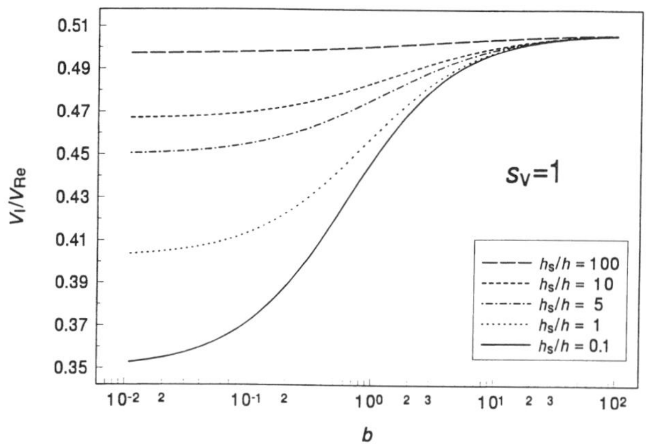


**Figure 16** Effect of the surface diffusion parameter, $h_s/h$, on the variation of the mobility factor, $\Phi_V \equiv V_I/V_{Re}$, with the bulk diffusion parameter, $b$, for fixed $s_V$ and $N_R = 1$.

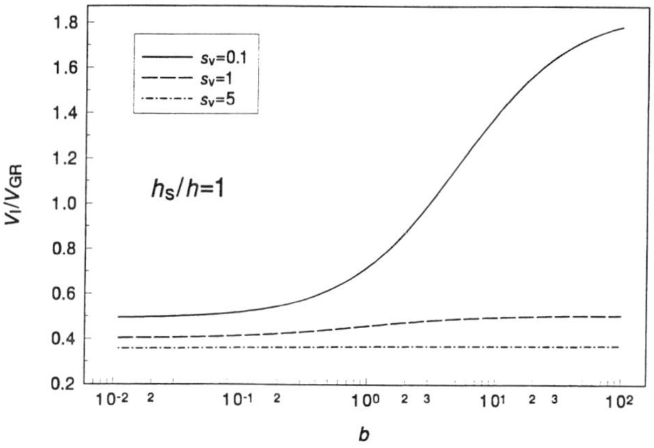


**Figure 17** Effect of the surface viscosity parameter, $s_V$, on the variation of the mobility factor, $\Phi_V \equiv V_I/V_{Re}$, with the bulk diffusion parameter, $b$, for fixed $h_s/h = 1$ and $N_R = 1$.

### 3. Critical Thickness of the Film Between Two Deforming Drops

As already mentioned, the transition from stability to instability occurs when the thickness of the gap between two colliding emulsion drops decreases down to a "transitional" thickness $h_t$. For $h_t > h > h_c$ the film continues to thin, while the instabilities grow, until the film ruptures at the critical thickness $h = h_c$.

Equation (61) determines the transitional distance between two spherical emulsion drops. An analog of this equation for the case of two deformed drops (Fig. 15a) has been obtained in the form of a transcendental equation (2, 136):

$$\frac{2h_t + d}{h_t + d} = \frac{h_t R^2[\Pi'(h_t)]^2}{8\sigma[2\sigma/a - \Pi(h_t)]}, \qquad \Pi' \equiv \frac{\partial \Pi}{\partial h} \tag{82}$$

Equation (82) shows that the disjoining pressure significantly influences the transitional thickness $h_t$. The effect of surface mobility is characterized by the parameter $d$, see Eq. (53); in particular, $d = 0$ for tangentially immobile interfaces. Equation (82) is valid for $\Pi < 2\sigma/a$, i.e., when the film thins and ruptures before reaching its equilibrium thickness, corresponding to $\Pi = 2\sigma/a$ [cf. Eqs (42), (43), and (59)].

The calculation of the transitional thickness $h_t$ is a prerequisite for computing the critical thickness $h_c$, which can be obtained as a solution to the equation (95, 96):

$$h_c^2 = \frac{2kT}{I(h_c, h_t)} \exp\left[\frac{I(h_c, h_t)}{4\sigma}\right] \tag{83}$$

where $I(h_t,h_c)$ represents the following function:

$$I(h_c, h_t) = \Pi'(h_t)R^2 \int_{h_c}^{h_t} \frac{\Pi' dh}{\Phi_v(h)[2\sigma/a - \Pi(h)]} \tag{84}$$

In the special case of tangentially immobile interfaces and large film (negligible effect of the transition zone) one has $\Phi_v(h) \equiv 1$, and the integration in Eq. (84) can be carried out (95):

$$I(h_c, h_t) = \Pi'(h_t)R^2 \ln\left[\frac{2\sigma/a - \Pi(h_c)}{2\sigma/a - \Pi(h_t)}\right] \tag{85}$$

Note that Eqs (82)–(85) hold not only for an emulsion film formed between two oil drops, but also for a foam film intervening between two gas bubbles. In Fig. 18 we compare the prediction of Eqs (82)–(84) with experimental data for $h_c$ versus $R$, obtained by Manev et al. (120) for foam films formed from an aqueous solution of 0.43 mM SDS + 0.1 M NaCl. The mobility factor $\Phi_v(h)$ was calculated by using the exact expression (the infinite series) from Ref. 137. Parameters such as $\sigma$, $E_G$, $\Gamma_1$, and $\partial\Gamma_1/\partial c_1$, see Eqs (53)–(55), are obtained from the experimental fit in Ref. 17, in the same way as the numerical data in Fig. 5 have been obtained (see Sec. II.A.2). The disjoining pressure was attributed to the van der Waals attraction: $\Pi = -A_H/(6\pi h^3)$. To account for the effect of the electromagnetic retardation on the dispersion

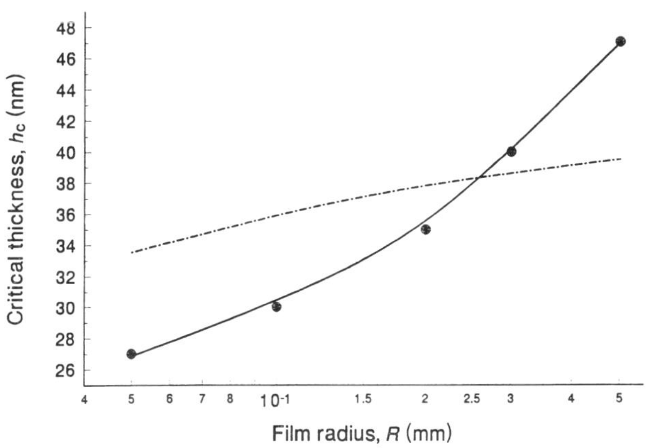


**Figure 18** Critical thickness, $h_c$, vs radius, $R$, of a foam film formed from aqueous solution of 0.43 mM SDS + 0.1 M NaCl: comparison between experimental points, measured by Manev et al. (120), with our theoretical model based on Eqs. (82)–(87) (the solid line) and the model by Malhotra and Wasan (116) (the dashed line).

interaction, we used an expression proposed by Russel et al. (160):

$$A_H(h) = \frac{3h_P\nu}{4\pi} \frac{(n_0^2 - n_w^2)^2}{(n_0^2 + n_w^2)^{3/2}} \int_0^\infty \frac{(1 + 2\tilde{h}z)\exp(-2\tilde{h}z)}{(1 + 2z^2)^2} dz \tag{86}$$

Here, $h_P = 6.63 \times 10^{-34}$ J.s is the Planck constant, $\nu \approx 3.0 \times 10^{15}$ Hz is the main electronic absorption frequency, and $n_o$ and $n_w$ are the refractive indices of the nonaqueous and aqueous phases; for a foam film $n_o = 1$ and $n_w = 1.333$. The dimensionless thickness $\tilde{h}$ is defined by the expression:

$$\tilde{h} = n_w(n_0^2 + n_w^2)^{1/2} \frac{2\pi\nu h}{c} \tag{87}$$

where $c = 3.0 \times 10^{10}$ cm/s is the speed of light. For small thickness $A_H$, as given by Eqs (86) and (87), is constant, whereas for large thickness $h$ one obtains $A_H \propto h^{-1}$. The solid line in Fig. 18 was calculated with the help of Eqs (82)–(87) *without* using any adjustable parameters; one sees that there is an excellent agreement between this theoretical model and the experiment.

The dot–dashed line in Fig. 18 shows the prediction of the theoretical model by Malhotra and Wasan (116). Our calculations showed that for the specific surfactant and salt concentrations (0.43 mM SDS + 0.1 M NaCl) the interfaces are almost tangentially immobile. Moreover, in these experiments the film radius $R$ is sufficiently large, which allows one to neglect effects of the transition zone, i.e., to ignore the last two terms in Eq. (73). Consequently, the difference between the model from Ref. 116 and the experimental data (Fig. 18) cannot be attributed to the latter two effects (interfacial mobility and transition zone), which have not been taken into account in Ref. 116. The main reasons for the difference between the output of Ref. 116 and the experiment are that (1) these authors have, in fact, calculated $h_t$ and identified it with $h_c$; and (2) a constant value of $A_H$ has been used, instead of Eq. (86), i.e., the electromagnetic retardation effect has been neglected. It is interesting to note that the retardation effect turns out to be important in the experimental range of critical thicknesses, in this specific case: 25 nm < $h_c$ < 50 nm.

## V. INTERPRETATION OF THE BANCROFT RULE

A simple rule connecting the emulsion stability with the surfactant properties was formulated by Bancroft

(161). The Bancroft rule states that "in order to have a stable emulsion the surfactant must be soluble in the continuous phase." Most of the emulsion systems obey this rule, but some exclusions have also been found (162). The results on drop–drop interactions, presented in Sec. IV, allow one to give a semiquantitative interpretation of the rule and the exclusions (1, 2, 163).

According to Davies and Rideal (6), both types of emulsions (water-in-oil and oil-in-water) are formed during the homogenization process, but only the one with lower coalescence rate survives. If the initial drop concentration for the two emulsions (Systems I and II, see Fig. 15) is the same, the corresponding coalescence rates for the two emulsions will be (approximately) proportional to the respective velocities of film thinning, $V_{\mathrm{I}}$ and $V_{\mathrm{II}}$ (163):

$$\frac{\text{Rate I}}{\text{Rate II}} \approx \frac{V_{\mathrm{I}}}{V_{\mathrm{II}}} \tag{88}$$

## A. Case of Deforming Drops

In the case of *deforming* drops, using Eqs (65), (76), and (77), one can represent Eq. (88) in the form (1, 163):

$$\frac{\text{Rate I}}{\text{Rate II}} \approx (486\rho_2 D_{1s}^3)^{1/3} \left[\frac{h_{c,\mathrm{I}}^3}{h_{c,\mathrm{II}}^2}\right]^{1/3} \left[\frac{\eta_2}{R^2}\right]^{1/3} \frac{2\sigma/a - \Pi_{\mathrm{I}}}{E_G(2\sigma/a - \Pi_{\mathrm{II}})^{2/3}} \tag{89}$$

where $h_{c,\mathrm{I}}$ and $h_{c,\mathrm{II}}$ denote the critical thickness of film rupture for the two emulsion systems in Fig. 15; $\Pi_{\mathrm{I}}$ and $\Pi_{\mathrm{II}}$ denote the disjoining pressure of the respective films. To obtain Eq. (89) we have also used the estimate $F_{\mathrm{tot}} \approx \pi\,(2\sigma/a - \Pi)R^2$ (see Ref. 149). The product of the first three multipliers on the right-hand side of Eq. (89), which are related to the *hydrodynamic* stability, is $8 \times 10^{-5}$ dyn$^{2/3}$cm$^{-1/3}$ for typical parameter values (1). The last multiplier in Eq. (89) accounts for the *thermodynamic* stability of the two types of emulsion film. Many conclusions regarding the type of emulsion formed can be drawn from Eq. (89) (1, 62, 163).

In thick films the disjoining pressures, $\Pi_{\mathrm{I}}$ and $\Pi_{\mathrm{II}}$, are zero, and then the ratio in Eq. (89) will be very small. Consequently, emulsion I (surfactant soluble in the continuous phase) will coalesce much more slowly than emulsion II; hence, emulsion I will survive. Thus, we obtain an explanation of the empirical Bancroft rule. The emulsion behavior in this case is controlled mostly by the hydrodynamic factors, i.e., the factors related to the *kinetic* stability.

The disjoining pressure, $\Pi$, can substantially change, and even reverse, the behavior of the system if it is comparable by magnitude with the capillary pressure, $2\sigma/a$. For example, if $(2\sigma/a - \Pi_{\mathrm{II}}) \to 0$ at a finite value of $2\sigma/a - \Pi_{\mathrm{I}}$, then the ratio in Eq. (89) may become much larger than unity, which means that System II will become *thermodynamically* stable. This fact can explain some exclusions from the Bancroft rule, like that established by Binks (162). Moreover, a large stabilizing disjoining pressure is operative in emulsions with a high volume fraction of the disperse phase, above 95% in some cases (164).

The Gibbs elasticity, $E_G$, favors the formation of emulsion I (Fig. 15a), because it slows down the film thinning. On the other hand, increased surface diffusivity, $D_{1s}$, decreases this effect, because it helps the interfacial-tension gradients to relax, thus facilitating the formation of emulsion II.

The film radius, $R$, increases, whereas the capillary pressure, $2\sigma/a$, decreases with the rise in drop radius, $a$. Therefore, larger drops will tend to form emulsion I, although the effect is not very pronounced, see Eq. (89). The difference between the critical thicknesses of the two emulsions affects only slightly the rate ratio in Eq. (89), although the value of $h_c$ itself is important.

The viscosity of the surfactant-containing phase, $\eta_1$, does not appear in Eq. (89); there is only a weak dependence on $\eta_2$. This fact is consistent with the experimental findings about a negligible effect of viscosity (see Ref. 6, p. 381 therein).

The interfacial tension, $\sigma$, affects directly the rate ratio in Eq. (89) through the capillary pressure, $2\sigma/a$. The addition of electrolyte would affect mostly the electrostatic component of the disjoining pressure (see Fig. 8a), which is suppressed by the electrolyte; the latter has a destabilizing effect on O/W emulsions. In the case of ionic surfactant solutions the addition of electrolyte rises the surfactant adsorption and the Gibbs elasticity (see Fig. 5), which favors the stability of emulsion I.

Surface-active additives (such as cosurfactants, demulsifiers, etc.) may affect the emulsifier partitioning between the phases and its adsorption, thereby changing the Gibbs elasticity and the interfacial tension. The surface-active additive may change also the surface charge (mainly through increasing the spacing among the emulsifier ionic headgroups), thus decreasing the electrostatic disjoining pressure and favoring the W/O emulsion. Polymeric surfactants and adsorbed proteins increase the steric repulsion between the film surfaces; they may favor either of the emulsions O/W or W/O, depending on their conformation at the interface and their surface activity.

The temperature affects strongly both the solubility and the surface activity of nonionic surfactants (165). It is well known that at higher temperatures nonionic surfactants become more oil soluble, which favors the W/O emulsion. These effects may change the type of emulsion formed at the phase-inversion temperature (166). The temperature effect has numerous implications, two of them being the change in the Gibbs elasticity, $E_G$, and the interfacial tension, $\sigma$.

## B. Case of Spherical Drops

Equation (89) was obtained for deforming emulsion drops, i.e., for drops which can approach each other at a surface-to-surface distance less than the inversion thickness $h_{inv}$, see Eq. (72). Another possibility is the drops to remain *spherical* during their collision, up to their eventual coalescence at $h = h_c$; in such a case the expressions for $V_I$ and $V_{II}$, which are to be substituted in Eq. (88), differ from Eqs (76) and (77).

Let us first consider the case of System II (surfactant inside the drops, Fig. 15b) in which case the two drops approach each other like drops from pure liquid phases (if only the surface viscosity effect is negligible). Therefore, to estimate the velocity of approach of such two aqueous droplets one can use the following approximate expression, which directly follows from Eq. (49) for $\xi \gg 1$:

$$V_{II}/V_{Ta}^{(II)} \approx 0.811 \frac{\eta_2}{\eta_1}\sqrt{\frac{a}{h}} \tag{90}$$

(For the system from Fig. 15b one is to set $\eta_{out} = \eta_2$ and $\eta_{in} = \eta_1$.) On the other hand, the velocity $V_I$ of droplet approach in System I can be expressed by means of Eq. (52). Note that the Taylor velocities for Systems I and II, $V_{Ta}^{(I)}$ and $V_{Ta}^{(II)}$, are different because of differences in viscosity and droplet–droplet interaction, see Eq. (47). By combining Eqs (47), (52), (88), and (90) we then arrive at the following criterion for formation of emulsions of type I or II (2):

$$\frac{\text{Rate I}}{\text{Rate II}} \approx 1.233 \frac{h_s}{2h}\sqrt{\frac{h}{a}}\left[\frac{\bar{d}+1}{\bar{d}}\ln(\bar{d}+1) - 1\right]^{-1} \frac{(F-F_s)_I}{(F-F_s)_{II}} \tag{91}$$

where $\bar{d} \equiv d/h$; see Eqs (53)–(55) for the definitions of $d$, $b$, and $h_s$. In the case of large surface (Gibbs) elasticity, $E_G \gg 1$, one has $\bar{d} \ll 1$; hence, one can expand the logarithm in Eq. (91) to obtain (2):

$$\frac{\text{Rate I}}{\text{Rate II}} \approx 1.233 \sqrt{\frac{h_c}{a}} \frac{1+b}{1-\bar{d}/3+O(\bar{d}^2)} \frac{(F-F_s)_I}{(F-F_s)_{II}} \tag{92}$$

Here, we have substituted $h_c$ for $h$, which is fulfilled at the moment of coalescence. For typical emulsion systems one has $a \gg h_c$, and then Eq.(92) yields Rate I/Rate II $\ll 1$; therefore, System I (with surfactant in the continuous phase, Fig. 15a) will survive. This prediction of Eq. (92) for spherical drops is analogous to the conclusion from Eq. (89) for deformable drops. Both these predictions essentially coincide with the Bancroft rule and are valid for cases in which the *hydrodynamic* stability factors prevail over the *thermodynamic* ones. The latter become significant close to the equilibrium state, $F_s \approx F$, and could bring about exclusions from the Bancroft rule, especially when $(F - F_s)_{II} \to 0$. The following conclusions, more specific for the case of *spherical* drops, can be also drawn from Eqs. (91) and (92).

For larger droplets (larger $a$) the transitional distance $h_t$ (and the critical distance $h_c$ as well) is smaller (see Fig. 14). It then follows from Eq. (91) that the difference between the coalescence rates in Systems I and II will become larger (2). On the contrary, the difference between Rates I and II decreases with the reduction in droplet size $a$, which is accompanied by an increase in the critical thickness $h_c$. Note that this effect of $a$ cannot be derived from the criterion for deforming drops, Eq. (89).

The effect of the bulk viscosity is not explicitly present in Eq. (92), although there could be some weak implicit dependence through the parameters $d$ and $b$ [see Eqs (53) and (55)]. This conclusion agrees with the experimental observations about a very weak dependence of the volume fraction of phase inversion on the viscosity of the continuous phase (6).

The increase in bulk and surface diffusivities, $D_1$ and $D_{1s}$, which tend to damp the surface-tension gradients, leads to an increase in the parameters $b$ and $d$, which decreases the difference between Rates I and II [see Eqs (53), (55), and (92)]. In contrast, the increase in the Gibbs elasticity, $E_G$, leads to a decrease in $d$ and thus favors the survival of System I. These are the same tendencies as for deforming drops (Sec. V. A). In the limit of tangentially *immobile* interfaces ($E_G \to \infty$) one has $d = 0$ and $b = 0$ and the criterion, Eq. (92), further simplifies (2):

$$\frac{\text{Rate I}}{\text{Rate II}} \approx \frac{V_{\text{I}}}{V_{\text{II}}} \approx 1.233\sqrt{\frac{h_c}{a}\frac{(F - F_s)_{\text{I}}}{(F - F)_{\text{II}}}} \qquad (E_G \to \infty) \tag{93}$$

The effect of surface viscosity, $\eta_s$, is neglected when deriving Eqs (91)–(93). Based on the hydrodynamic equations one can estimate that this effect is really negligible when (2)

$$\frac{\eta a^2}{\eta_s h} \gg 1 \tag{94}$$

where $\eta$ represents the bulk viscosity, which is assumed to be of the same order of magnitude for the liquids inside and outside the drops. If for a certain system, or under certain conditions, the criterion, Eq. (94), is not satisfied, one can expect that the surface viscosity will suppress the interfacial mobility for both Systems I and II. The difference between Rates I and II will be then determined mostly by thermodynamic factors, such as the surface force $F_s$.

Although Eqs (89) and (91) lead us to some more general conclusions than the original Bancroft rule (e.g., the possibility for inversion of the emulsion stability owing to disjoining pressure effects), we neither claim that the Bancroft rule, or its extension based on Eqs. (89) and (91), have general validity, nor that we have given a general explanation of the emulsion stability. The coagulation in emulsions is such a complex phenomenon, influenced by too many different factors, that according to us any attempt at formulating a *general* explanation (or criterion) is hopeless. Our treatment is theoretical and as every theory, it has limitations inherent to the model used and therefore is valid only under specific conditions. It should not be applied to a system where these conditions are not fulfilled. The main assumptions and limitations of the model are (2): the fluctuation-wave mechanism for coalescence is assumed to be operative (see Fig. 10); the surfactant transfer on to the surface is under diffusion or electro-diffusion control; parameter $b$ defined by Eq. (55) does not account for the demicellization kinetics for $c_1 >$ cmc; and the effect of surface viscosity is not taken into account in Eqs (89) and (91). Only *small* perturbations in the surfactant distribution, which are due to the flow, have been considered; however, under strongly nonequilibrium conditions (like turbulent flows) we could find that new effects come into play, which may significantly alter the trend of the phenomenon.

## VI. KINETICS OF COAGULATION IN EMULSIONS

### A. Types of Coagulation in Emulsions

The coagulation in an emulsion is a process in which the separate emulsion drops merge to form larger drops (coalescence) and/or assemble into flocs (flocculation), see Fig. 2. If the films intervening between the drops in a floc are unstable, their breakage is equivalent to coalescence, see step D→C in Fig. 2. In other words, the coagulation in an emulsion includes flocculation and coalescence, which could occur as parallel or consecutive processes.

Various *experimental methods* for monitoring the kinetics of coagulation in emulsions have been developed, such as the electroacoustic method (167), direct video-enhanced microscopic investigation (168), and ultrasonic attenuation spectroscopy (169).

To a great extent the occurrence of coagulation is determined by the energy, $W(R, h)$, of the interaction between two drops. Equation (67), which defines $W(R, h)$, can be applied to any type of surface force (irrespective of its physical origin) if only the range of action of this force is much smaller than the drop radius $a$. In Ref. 2 one can find theoretical expressions for the components of $W$ stemming from various surface forces: electrostatic, van der Waals, ionic, correlations, hydration repulsion, protrusion and steric interactions, oscillatory structural forces, etc.

If the two drops remain spherical during their interaction (i.e., there is no film in the contact zone and consequently $R = 0$), then $W$ depends only on a single parameter, $W = W(h)$; as usual, $h$ is the surface-to-surface distance between the two drops. When the approach of the two drops is accompanied by the formation and expansion of a film in the contact zone (Fig. 7), then one can characterize the interaction by $\bar{W}(h)$, which is obtained by averaging $W(R, h)$ over all configurations with various $R$ at fixed $h$ (see Ref. 143).

The shape of $W(h)$, or $\bar{W}(h)$, qualitatively resembles that of $\Pi(h)$ (see Fig. 6). In particular, if only electrostatic and van der Waals interactions are operative, the shape of the dependence $W = W(h)$ resembles Fig. 6a, where an electrostatic barrier is present. The coagulation is called *fast* or *slow*, depending on whether that electrostatic barrier is less than $kT$ or higher than $kT$. In addition, the flocculation is termed *reversible* or *irreversible*, depending on whether the depth of the primary minimum (that on the left from the barrier in Fig. 6a) is comparable with $kT$ or much greater than $kT$. The driving forces of coagulation can be the following:

1. The *body forces*, such as gravity and centrifugation, cause rising or sedimentation of the droplets, depending on whether their mass density is smaller or greater than that of the continuous phase. Since drops of different size move with different velocities, they are subjected to frequent collisions, leading to drop aggregation or coalescence, called *orthokinetic coagulation*.
2. The *Brownian stochastic force* dominates the gravitational body force for droplets, which are smaller than 1 $\mu$m. Thus, the Brownian collision of two droplets becomes a prerequisite for their flocculation and/or coalescence, which is termed *perikinetic coagulation*.
3. The heating of an emulsion produces *temperature gradients*, which in their own turn cause thermocapillary migration of the droplets driven by thermally excited gradients of surface tension (170–172):

$$\nabla_s\sigma = -\frac{E_T}{T}\nabla_s T \qquad E_T \equiv -\left(\frac{\partial\sigma}{\partial \ln T}\right)_{\Gamma_1} \tag{95}$$

Here, $\nabla_s$ is the surface gradient operator and $E_T$ is the coefficient of interfacial thermal elasticity, [cf. Eq. (1)]. The drops moving with different thermocapillary velocities can collide and flocculate or coalesce; this is the *thermal coagulation*.

## B. Kinetics of Irreversible Coagulation

### 1. Basic Equations

The kinetic theory of the fast irreversible coagulation was first developed by Smoluchowski (173, 174) and later extended to the case of slow and reversible coagulation. In any case of coagulation the general set of kinetic equations reads (175):

$$\frac{dn_k}{dt} = \frac{1}{2}\sum_{i=1}^{k-1} a_f^{i,k-i} n_i n_{k-i} - n_k \sum_{i=1}^{\infty} a_f^{k,i} n_i + q_k \qquad (k = 1, 2, \ldots) \tag{96}$$

where $t$ is time, $n_1$ denotes the number of single drops per unit volume, $n_k$ is the number of flocs of $k$ drops ($k$ = 2, 3, ...,) per unit volume, and $a_f^{i,j}$ ($i,j$ = 1, 2, 3, ...,) are rate constants of flocculation (see Fig. 19); $q_k$

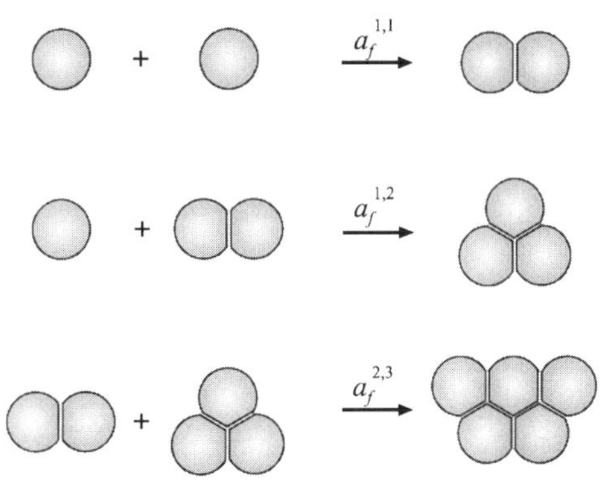


**Figure 19** Examples for elementary acts of flocculation according to the Smoluchowski scheme; $a_f^{i,j}$ ($i,j = 1, 2, 3, \ldots$,) denote the respective rate constants of flocculation.

denotes a flux of flocs of size $k$ which are products of other processes, different from the *flocculation* itself [say, the reverse process of floc disassembly, or the droplet coalescence, see Eqs (116) and (120)]. Analogously to flocculation, the *coalescence* in emulsions can be considered as a kind of irreversible coagulation (176–179).

In the special case of *irreversible* coagulation one has $q_k \equiv 0$. The first term on the right-hand side of Eq. (96) is the rate of formation of $k$ flocs by merging of two smaller flocs, whereas the second term expresses the rate of disappearance of $k$ flocs due to their incorporation into larger flocs. The total concentration of flocs (as kinetically independent units), $n$, and the total concentration of the constituent drops (including those in flocculated form), $n_{tot}$, are given by the expressions:

$$n = \sum_{k=1}^{\infty} n_k, \qquad n_{tot} = \sum_{k=1}^{\infty} k n_k \tag{97}$$

The rate constants in Eq. (96) can be expressed in the form:

$$a_f^{i,j} = 4\pi D_{i,j}^{(0)}(R_i + R_j)E_{i,j} \tag{98}$$

where $D^{(0)}{}_{i,j}$ is the relative diffusion coefficients for two flocs of radii $R_i$ and $R_j$, and aggregation number $i$ and $j$, respectively; and $E_{i,j}$ is the collision efficiency (180, 181). Below we give expressions for $D^{(0)}{}_{i,j}$ and $E_{i,j}$ applicable to the various types of coagulation.

The Einstein approach to the theory of diffusivity $D$ gives the following expression:

$$D = kT/B, \qquad B \equiv F/V \tag{99}$$

where $B$ is the friction coefficient, and $V$ is the velocity acquired by a given particle under the action of an applied net force $F$. For a solid sphere of radius $R_0$ one has $B = 6\pi\eta R_0$. For a liquid drop, $B$ is given by the equation of Rybczynski (182) and Hadamar (183):

$$B = 2\pi\eta R_0 \frac{3\eta_d + 2\eta}{\eta_d + \eta} \tag{100}$$

where $\eta_d$ is the viscosity inside the drop, and $\eta$ is the viscosity of the continuous phase. The combination of Eqs (99) and (100) yields the following expression for the relative diffusivity of two isolated Brownian droplets of radii $R_i$ and $R_j$.

$$D_{i,j}^{(0)} = \frac{kT}{2\pi\eta} \frac{\eta_d + \eta}{3\eta_d + 2\eta} \left( \frac{1}{R_i} + \frac{1}{R_j} \right) \tag{101}$$

(perikinetic coagulation)

The limiting case $\eta_d \to 0$ corresponds to two bubbles, whereas in the other limit, $\eta_d \to \infty$, Eq. (101) describes two solid particles or two liquid drops of tangentially immobile surfaces.

When the relative motion of the drop is driven by a *body force* or by *thermocapillary migration* (rather than by self-diffusion), Eq. (101) is no longer valid. Instead, in Eq. (98) one has formally to substitute the following expression for $D^{(0)}{}_{i,j}$, see Rogers and Davis (184):

$$D_{i,j}^{(0)} = \frac{1}{4}(R_i + R_j) \mid \mathbf{v}_i - \mathbf{v}_j \mid \tag{102}$$

(orthokinetic coagulation)

Here, $\mathbf{v}_j$ denotes the velocity of a floc of aggregation number $j$. Physically, Eq. (102) accounts for the fact that the drops/flocs of different size move with different velocities under the action of the body force. In the case of gravity-driven flocculation $\mathbf{v}_j$ is the velocity of a rising/sedimenting particle, which for a drop of tangentially immobile surface is given by the Stokes formula:

$$\mathrm{v}_j = \mathrm{v}_{\mathrm{bf}} \equiv 2g\,\Delta\rho R_j^2/(9\eta) \tag{103}$$

(sedimentation velocity)

see, e.g., Ref. 16; here, $g$ is the acceleration due to gravity, and $\Delta\rho$ is the density difference between the two liquid phases.

In the case of *thermal coagulation*, the drop velocity $\mathbf{v}_j$ is given by the expression (185):

$$\mathbf{v}_j = \mathbf{v}_{\mathrm{tm}} \equiv \frac{2R_j E_T \lambda}{(3\eta_d + 2\eta)(\lambda_d + 2\lambda)} \nabla(\ln T) \tag{104}$$

(thermocapillary velocity)

where the thermal conductivity of the continuous and disperse phases are denoted by $\lambda$ and $\lambda_d$; the interfacial thermal elasticity $E_T$ is defined by Eq. (95).

The *collision efficiency* $E_{i,j}$ in Eq. (98) accounts for the interactions (of both hydrodynamic and intermolecular origin) between two colliding drops. The inverse of $E_{i,j}$ is called the *stability ratio* or the *Fuchs factor* (186) and can be expressed in the following general form (3, 180):

$$\Phi_{i,j} = \frac{1}{E_{i,j}} = 2\int_0^\infty \frac{\beta(s)}{(s+2)^2} \exp\left[\frac{W_{i,j}(s)}{kT}\right] \mathrm{d}s, \qquad s \equiv \frac{2h}{R_i + R_j} \tag{105}$$

$$\beta \equiv \left( 2\pi\eta a \frac{3\eta_d + 2\eta}{\eta_d + \eta} \right)^{-1} B(s) \tag{106}$$

As usual, $h$ is the closest surface-to-surface distance between the two drops; $a$ is defined by Eq. (46); $W_{i,j}(s)$ is the energy of *non hydrodynamic* interactions between the drops, see Eq. (67); $\beta(s)$ accounts for the *hydrodynamic* interactions; and $B(s)$ is the drop friction coefficient. For $s \to \infty$ one obtains $\beta \to 1$, since for large separations the drops obey the Rybczynski – Hadamar equation (100). In the opposite limit, $s \ll 1$, i.e., close approach of the two drops, $B(s) \equiv F/V$ can be calculated from either Eq. (47), (49), (52), or (56), depending on the specific case. In particular, for $s \ll 1$ one has $\beta \propto s^{-1/2}$ for two spherical droplets of tangentially *mobile* surfaces, whereas $\beta \propto 1/s$ for two drops of tangentially *immobile* surfaces (or two solid particles). In the latter case the integral in Eq. (105) seems to be divergent. To overcome this problem it is usually accepted that for the smallest separations $W_{i,j}$ is dominated by the van der Waals attraction, i.e., $W_{i,j} \to -\infty$ for $s \to 0$, and consequently, the integrand in Eq. (105) tends to zero for $s \to 0$.

The Fuchs factor $\Phi_{i,j}$ is determined mainly by the values of the integrand in the vicinity of the electrostatic maximum (barrier) of $W_{i,j}$ (cf. Fig. 6a) since $W_{i,j}$ enters Eq. (105) as an exponent. By using the method of the saddle point, Derjaguin (3) estimated the integral in Eq. (105):

$$\Phi_{i,j} \equiv \frac{1}{E_{i,j}} \approx \left[\frac{8\pi kT}{-W''_{i,j}(s_m)}\right]^{1/2} \frac{\beta(s_m)}{(s_m+2)^2} \exp\left[\frac{W_{i,j}(s_m)}{kT}\right] \tag{107}$$

Here, $s_m$ denotes the value of $s$ corresponding to the maximum. One sees that the higher the barrier,

$W_{i,j}(s_m)$, the smaller the collision efficiency, $E_{i,j}$, and the slower the coagulation.

The infinite set of Smoluchowski equations [Eq. (96)] was solved by Bak and Heilmann (187) in the particular case when the flocs cannot grow larger than a given size; an explicit analytical solution was obtained by these authors.

## 2. Special Results

For imaginary drops, which experience neither long-range surface forces ($W_{i,j} = 0$) nor hydrodynamic interactions ($\beta = 1$), Eq. (105) yields a collision efficiency $E_{i,j} = 1$, and Eq. (98) reduces to the Smoluchowski (173, 174) expression for the rate constant of *fast irreversible* coagulation. In this particular case, Eq. (96) represents an infinite set of nonlinear differential equations. If all flocculation rate constants are the same and equal to $a_f$, the problem has an exact analytical solution (173, 174):

$$n = \frac{n_0}{1 + a_f n_0 t/2}, \qquad n_k = n_0 \frac{(a_f n_0 t/2)^{k-1}}{(1 + a_f n_0 t/2)^{k+1}} \quad (k = 1, 2 \ldots) \tag{108}$$

The total average concentration of the drops (in both singlet and flocculated form), $n_{\mathrm{tot}}$, does not change and is equal to the initial number of drops, $n_0$. Unlike $n_{\mathrm{tot}}$, the concentration of the flocs, $n$, decreases with time, while their size increases. Differentiating Eq. (108) one obtains:

$$\frac{dn}{dt} = -\frac{a_f}{2} n^2, \qquad \frac{d\bar{V}}{dt} = \frac{a_f}{2}\phi_0, \qquad \bar{V} \equiv \frac{\phi_0}{n} \tag{109}$$

where $\bar{V}$ is the *average volume per floc*, and $\phi_0$ is the initial volume fraction of the constituent drops. Combining Eqs (98) and (109) one obtains the following result for *perikinetic* (*Brownian*) coagulation:

$$\frac{\bar{V}}{V_0} = 1 + \frac{t}{t_{\mathrm{Br}}}, \qquad t_{\mathrm{Br}} = \frac{R_0^2}{3\phi_0 D_0 E_0} \tag{110}$$

where $V_0 = 4\pi R_0^{\,3}/3$ is the volume of a constituent drop of radius $R_0$, $t_{\mathrm{Br}}$ is the characteristic time of the coagulation process in this case, $E_0$ is an average collision efficiency, and $D_0$ is an average diffusion coefficient. Equation (110) shows that for fast irreversible coagulation, $\bar{V}$ increases linearly with time.

In contrast, $\bar{V}$ is not a linear function of time for *orthokinetic* coagulation, except in the limit of short times. When the flocculation is driven by a *body force*, i.e., in case of sedimentation or centrifugation, one obtains (181):

$$\frac{\bar{V}}{V_0} = \left[1 - \frac{t}{3t_{\mathrm{bf}}}\right]^{-3}, \qquad t_{\mathrm{bf}} = \frac{2R_0}{3\phi_0 v_{\mathrm{bf}} E_0} \tag{111}$$

where $t_{\mathrm{bf}}$ is the characteristic time in this case, and $v_{\mathrm{bf}}$ is an average velocity of floc motion, which can be expressed by means of Eq. (103) if the body force is the gravitational one.

If the orthokinetic coagulation is driven by *thermocapillary migration*, the counterpart of Eq. (111) reads (181):

$$\frac{\bar{V}}{V} = \exp\left[\frac{t}{t_{\mathrm{tm}}}\right] \qquad t_{\mathrm{tm}} = \frac{2R_0}{3\phi_0 v_{\mathrm{tm}} E_0} \tag{112}$$

where $v_{\mathrm{tm}}$ is an average velocity of thermocapillary migration, see Eq. (104), and $t_{\mathrm{tm}}$ is the respective characteristic time. Note that $D_0 \propto R_0^{\,-1}$, $v_{\mathrm{bf}} \propto R_0^{\,2}$, and $v_{\mathrm{tm}} \propto R_0$, cf. Eqs (99) and (104). From Eqs (110)–(112) it then follows that the three different characteristic times exhibit different dependencies on drop radius: $t_{\mathrm{Br}} \propto R_0^{\,3}$, $t_{\mathrm{bf}} \propto R_0^{\,-1}$, while $t_{\mathrm{tm}}$ is independent of $R_0$. Hence, the Brownian coagulation is faster for the smaller drops, and the body force-induced coagulation is more rapid for the larger drops, whereas the thermocapillary-driven coagulation is not sensitive to the drop size.

Using the Stokes–Einstein expression for the diffusivity $D_0$ and Eq. (110) one obtains:

$$t_{\mathrm{Br}} = \frac{2\pi\eta R_0^3}{\phi_0 k T E_0} \tag{113}$$

On the other hand, the combination of Eqs (103) and (111) yields:

$$t_{\mathrm{bf}} = \frac{3\eta}{\phi_0 R_0 E_0 g \Delta\rho} \tag{114}$$

Let us consider the quantity:

$$\chi(R_0) \equiv \frac{t_{\mathrm{bf}}}{t_{\mathrm{bf}} + t_{\mathrm{Br}}} = \frac{1}{1 + 3R_0^4/(5R_{\mathrm{cr}}^4)}, \qquad R_{\mathrm{cr}} = \left[\frac{9kT}{10\pi g \Delta\rho}\right]^{1/4} \tag{115}$$

For $R_0 < R_{\mathrm{cr}}$, Eq. (115) yields $\chi(R_0) \approx 1$, i.e., $t_{\mathrm{Br}} \ll t_{\mathrm{bf}}$, and the Brownian flocculation is much faster than the orthokinetic flocculation. In contrast, for $R_0 > R_{\mathrm{cr}}$, Eq. (115) yields $\chi(R_0) \approx 0$, i.e., $t_{\mathrm{bf}} \ll t_{\mathrm{Br}}$, and the orthokinetic flocculation is much more rapid than the Brownian flocculation. At $R_0 = R_{\mathrm{cr}}$, a sharp transition from Brownian to orthokinetic flocculation takes place; $R_{\mathrm{cr}}$ corresponds to the inflection point of the dependence $\chi = \chi(R_0)$. Since the orthokinetic flocculation happens through a directional motion of the particles, then $R_{cr}$ can be considered as a *threshold radius of the flocs* needed for the creaming (or sedimentation) to begin. With $\Delta\rho = 0.1$ g/cm$^3$ and $T = 298$ K from Eq. (115) one calculates $R_{cr} = 1.05$ $\mu$m. It turns out that the threshold size for creaming is around 1 $\mu$m. This conclusion is consistent with the experimental data in Fig. 3, which show that emulsions with $2R_0 = 5$ $\mu$m do cream, whereas those with $2R_0 = 0.35$ $\mu$m do not.

### C. Kinetics of Reversible Flocculation

If the depth of the primary minimum (that on the left from the maximum in Fig. 6a) is not so great, i.e., the attractive force which keeps the drops together is weaker, then the flocs formed are labile and can disassemble into smaller aggregates. This is the case of *reversible flocculation* (3). For example, a floc composed of $i+j$ drops can be split into two flocs containing $i$ and $j$ drops. We denote the rate constant of this *reverse* process by $a^{i,j}{}_r$ (see Fig. 20a). In the present case both the straight process of flocculation (Fig. 19) and the reverse process (Fig. 20a) take simultaneously place. The kinetics of aggregation in this more general and complex case is described by the Smoluchowski set of equations, Eq. (96), where one is to substitute:

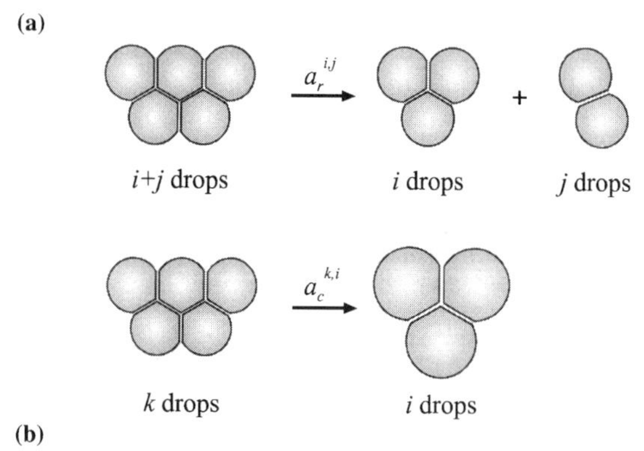


**Figure 20** (a) elementary act of splitting of a floc, containing $i+j$ constitutive drops, into two smaller flocs containing, respectively, $i$ and $j$ constitutive drops; (b) coalescence transforms a floc composed of $k$ drops into a floc containing $i$ drops ($i < k$). The rate constants of the respective processes are $a_r^{i,j}$ and $a_c^{k,i}$ $(i, j, k = 1, 2, 3, \ldots,)$.

$$q_1 = \sum_{i=1}^{\infty} a_r^{1,i} n_{i+1}, \quad q_k = \sum_{i=1}^{\infty} a_r^{k,i} n_{i+k} - \frac{1}{2} n_k \sum_{i=1}^{k-1} a_r^{i,k-i}$$
$$(k = 2, 3, \ldots) \tag{116}$$

Here, $q_k$ is the rate of formation of $k$ flocs in the process of disassembly of larger flocs minus the rate of decay of the $k$ flocs. As before, the total number of constituent drops, $n_{tot}$, does not change. However, the total number of the flocs, $n$, can either increase or decrease depending on whether the straight or the reverse process prevails. Summing up all equations in Eq. (96) and using Eq. (116) one derives the following equation for $n$:

$$\frac{dn}{dt} = \frac{1}{2} \sum_{i=1}^{\infty} \sum_{j=1}^{\infty} \left( a_r^{i,j} n_{i+j} - a_f^{i,j} n_i n_j \right) \tag{117}$$

A general expression for the rate constants of the reverse process was obtained by Martinov and Muller (188):

$$a_r^{i,j} = \frac{D_{i,j}^{(0)} E_{i,j}}{Z_{i,j}} \frac{1}{(R_i + R_j)^2} \tag{118}$$

Here, $Z_{i,j}$ is the so-called *irreversible factor*, which is defined as follows:

$$Z_{i,j} = \frac{1}{8} \int_{Wi,j<0} (s+2)^2 \exp\left[ -\frac{W_{i,j}(s)}{kT} \right] ds \tag{119}$$

The integration in Eq. (119) is carried out over the region around the primary minimum, where $W_{i,j}$ takes negative values (cf. Fig. 6a). In other words, $Z_{i,j}$ is determined by the values of $W_{i,j}$ in the region of the primary *minimum*, whereas $E_{i,j}$ is determined by the values of $W_{i,j}$ in the region of the electrostatic *maximum*, cf. Eqs (107) and (119). When the minimum is deeper, $Z_{i,j}$ is larger and the rate constant in Eq. (118) is smaller. Moreover, Eqs (107) and (118) show that the increase in the height of the barrier also decreases the rate of the reverse process. The physical interpretation of this fact is the following: to detach a drop from a floc, the drop has to first emerge from the well and then to "jump" over the barrier (cf. Fig. 6a).

As an illustration, in Fig. 21 we show theoretical curves for the rate of flocculation calculated in Ref. 62. The curves are computed by solving numerically the set of Eqs (96), (116), and (117). To simplify the problem the following assumptions have been used (62): (1) the Smoluchowski assumption that all rate constants of the *straight* process are equal to $a_f$, (2) flocs containing more than $M$ drops cannot decay;

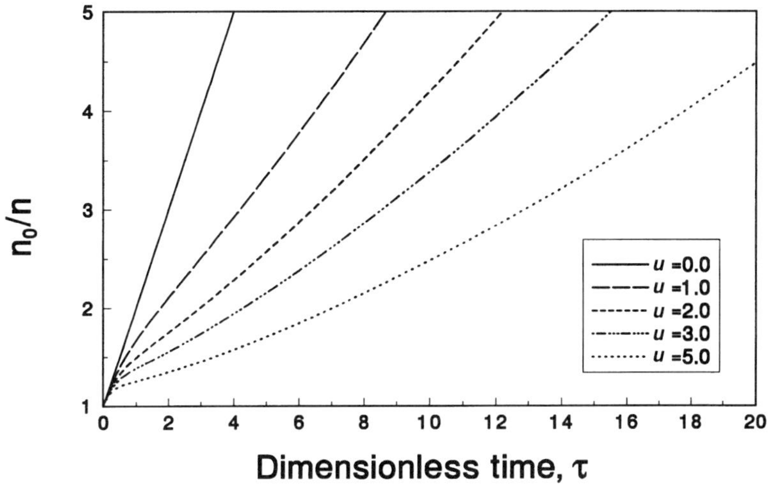


**Figure 21** Plot of the inverse dimensionless concentration of flocs, $n_0/n$, vs. the dimensionless time, $\tau = a_f n_0 t/2$, for $M = 4$ and various values of the dimensionless ratio $u = 2a_r/(n_0 a_f)$; *ar* and *af* are the rate constants for the reverse and straight processes. Theoretical curves for reversible flocculation from Ref. 62.

(3) all rate constants of the *reverse* process are equal to $a_r$; and (4) at the initial moment only single constituent drops of concentration $n_0$ are available. In Fig. 21 we present the calculated curves for $n_0/n$ versus the dimensionless time, $\tau = a_f n_0 t/2$, for a fixed value $M = 4$ and various values of the ratio of the rate constants of the straight and the reverse process, $u \equiv 2a_r/(n_0 a_f)$. Note that $n$ is defined by Eq. (97). The increase in $n_0/n$ with time means that the concentration $n$ of the flocs decreases; i.e., the emulsion contains a smaller number of flocs, but their size is larger. Consequently, a larger $n_0/n$ corresponds to a larger degree of flocculation. It is seen that for the short times of flocculation ($\tau \to 0$) all curves in Fig. 21 touch the Smoluchowski distribution (corresponding to $u = 0$), but for the longer times one observes a reduction in the degree of flocculation, which is smaller for the curves with larger values of $u$ (larger rate constants of the reverse process). The "S-shaped" curves in Fig. 21 are typical for the case of *reversible* flocculation; curves of similar shape have been obtained experimentally (3, 168, 189).

## D. Kinetics of Simultaneous Flocculation and Coalescence

In the case of *pure* flocculation considered above the total number of constituent drops, $n_{\text{tot}}$, does not change, see Eq. (97). In contrast, if coalescence is present, in addition to the flocculation, then $n_{\text{tot}}$ decreases with time (6). Hartland and Gakis (190) and Hartland and Vohra (191) developed a model of coalescence, which relates the lifetime of single films to the rate of phase separation in emulsions of comparatively large drops (> 1 mm) in the *absence* of surfactant. The effect of surfactant (emulsifier) was taken into account by Lobo et al. (192), who quantified the process of coalescence within an already creamed or settled emulsion containing drops of size less than 100 $\mu$m. Danov et al. (175) generalized the Smoluchowski scheme of flocculation to account for the fact that the droplets within the flocs can coalesce to give larger droplets, as illustrated in Fig. 20b. In this case, on the right-hand side of Eq. (96) one has to substitute (175):

$$q_1 = \sum_{i=2}^{\infty} a_c^{i,1} n_i\,,$$
$$q_k = \sum_{i=k+1}^{\infty} a_c^{i,k} n_i - n_k \sum_{i=1}^{k-1} a_c^{k,i} \qquad (k = 2, 3, \ldots) \tag{120}$$

where $a^{k,i}{}_c$ is the rate constant of transformation (by coalescence) of a floc containing $k$ droplets into a floc containing $i$ droplets (see Fig. 20b). The resulting floc is further involved in the flocculation scheme, which thus describes the interdependence of flocculation and coalescence. In this scheme the total coalescence rate, $a^i{}_{c,\text{tot}}$, and the total number of droplets, $n_{\text{tot}}$, are related as follows (175):

$$\frac{dn_{\text{tot}}}{dt} = -\sum_{i=2}^{\infty} a^i_{c,\text{tot}}\, n_i\,,$$
$$a^i_{c,\text{tot}} = \sum_{k=1}^{i-1} (i-k) a_c^{i,k} \qquad (i = 2, 3, \ldots) \tag{121}$$

To determine the rate constants of coalescence, $a^{k,i}_{c}$, Danov et al. (147) examined the effects of the droplet interactions and the Brownian motion on the coalescence rate in dilute emulsions of micrometer- and submicrometer-sized droplets. The processes of film formation, thinning, and rupture were included as consecutivc stages in the scheme of coalescence. Expressions for the interaction energy due to various DLVO and nonDLVO surface forces between two deformed droplets were obtained (143).

*Average models* for the total number of droplets have also been proposed (193, 194). The average model of van den Tempel (193) assumes a linear structure for the flocs. The coalescence rate is supposed to be proportional to the number of contacts within a floc. To simplify the problem van den Tempel used several assumptions, one of them being that the concentration of the single droplets, $n_1$, obeys the Smoluchowski distribution, Eq. (108), for $k=1$. The model of Borwankar et al. (194) employs some assumptions, which make it more applicable to cases in which the flocculation (rather than the coalescence) is slow and is the rate-determining stage. This is illustrated by the curves shown in Fig. 22, which are calculated for the same rate of *coalescence*, but for two different rates of *flocculation*. For relatively high rates of flocculation (Fig. 22a) the predictions of the three theories differ, but the model of Borwankar et al. (194) gives values closer to that of the more detailed model by Danov et al. (175). For very low values of the flocculation rate constant, $a_f$, for which the coalescence is not the rate-determining stage, all three theoretical models (175, 193, 194) give results for $n_{tot}/n_0$ versus time, which almost coincide numerically (Fig. 22b).

Finally, it is worthwhile noting that the simultaneous flocculation and coalescence in emulsions could be also accompanied with adsorption of amphiphilic molecules on the drop surfaces (195); this possibility should be kept in mind when interpreting experimental data.

## VII. SUMMARY

Surfactants play a crucial role in emulsification and emulsion stability. A first step in any quantitative study on emulsions should be to determine the equilibrium and dynamic properties of the oil–water interface, such as interfacial tension, Gibbs elasticity, surfactant adsorption, counterion binding, surface electric potential, adsorption relaxation time, etc. Useful theoretical concepts and expressions, which are applicable to ionic, nonionic, and micellar surfactant solutions, are summarized in Sec. II.

The emulsion drops in flocs and creams are separated with thin liquid films, whose rupture leads to coalescence and phase separation. At equilibrium the area of the films and their contact angle are determined by the surface forces (disjoining pressure) acting across the films (Sec. III.A.1). Several ways of breakage of these emulsion films have been established: capillary-wave mechanism, pore-nucleation mechanism, solute-transport mechanism, barrier mechanism, etc. (Sec. III.A.2).

Experimental and theoretical results show evidence that the capillary-wave mechanism is the most frequent reason for the coalescence of both deformed and spherical emulsion drops. For a certain critical thickness (width), $h_c$, of the film (gap) between two emulsion drops the amplitude of the thermally excited fluctuation capillary waves begins to grow, promoted by the surface forces, and causes film rupture. The capillary waves can bring about coalescence of two spherical emulsion drops, when the distance between them becomes smaller than a certain critical value, which is estimated to be about 10–50 nm (see Sec. III.C).

The interactions of two emulsion drops, and their theoretical description, become more complicated if the drops are moving against each other, instead of being quiescent. In such a case, which happens most frequently in practice, the hydrodynamic interactions come into play (Sec. IV). The velocity of approach of two drops and the critical distance (thickness) of drop coalescence are influenced by the drop size, disjoining pressure, bulk and surface diffusivity of surfactant, Gibbs elasticity, surface viscosity, etc. If attractive (negative) disjoining pressure prevails, then "pimples" appear on the opposite drop surfaces in the zone of contact; thus, the drop coalescence can be produced by the growth and merging of these "pimples" (Sec. IV.A.3). Alternatively, drop coalescence can be produced by the growth of fluctuation capillary waves; the theory of the respective critical thickness is found to agree excellently with available experimental data (Sec. IV.B.3).

The finding that the hydrodynamic velocity of mutual approach of two emulsion drops is much higher when the surfactant is dissolved in the drop phase (rather than in the continuous phase) provides a natural explanation of the Bancroft rule in emulsification (Sec. V). A generalized version of the Bancroft rule is proposed, Eqs (89) and (91), which takes into account the role of various thermody-

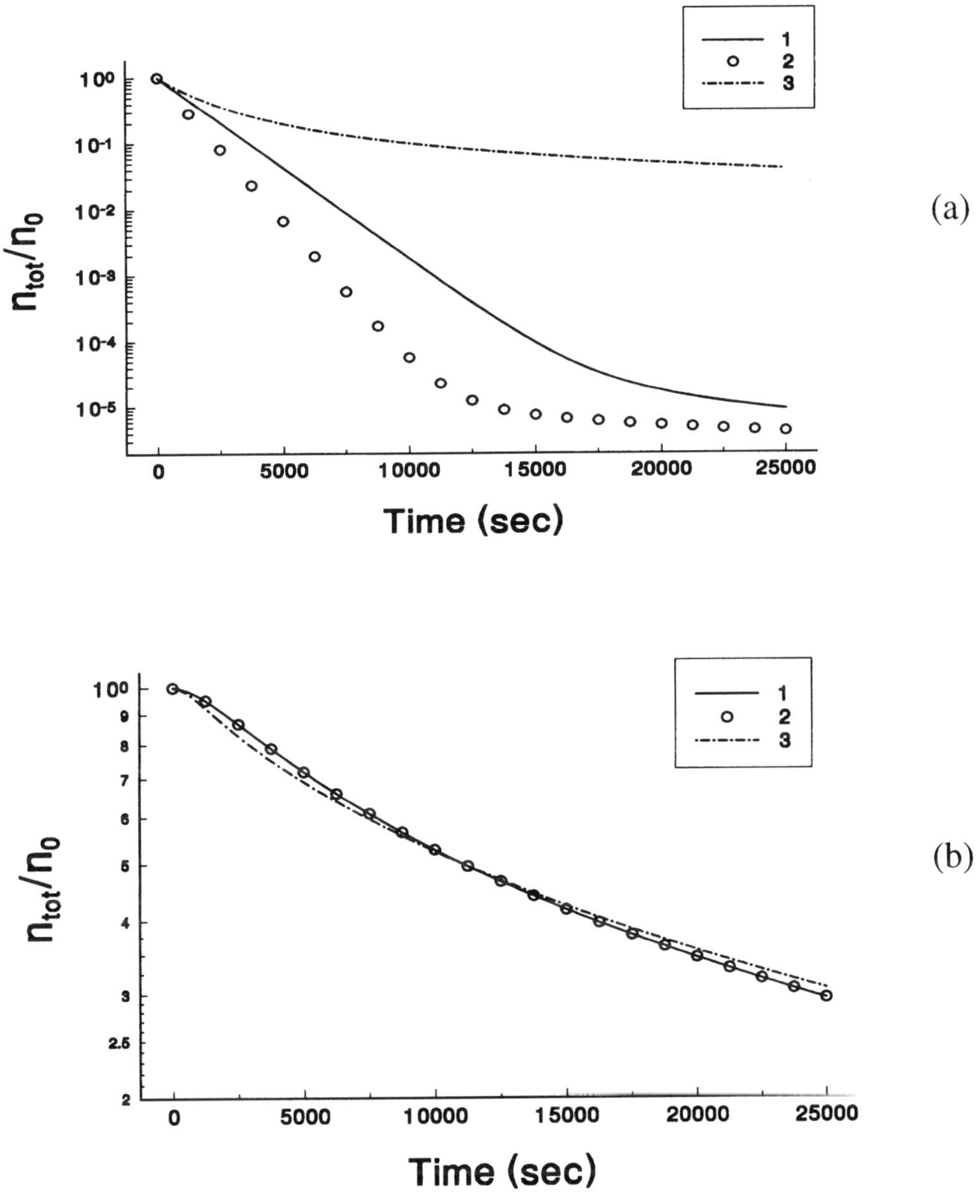


**Figure 22** The total number of constituent drops in a flocculating emulsion, $n_{tot}$, decreases with time, $t$, because of a parallel process of coalescence. The curves are calcualted for the following parameter values: initial number of constituent drops $n_0 = 10^{12} \text{cm}^{-3}$; coalescence rate constant $k^{2,1}{}_c = 10^{-3}\ \text{s}^{-1}$. Curve 1 is a numberical solution to Eq. (121); Curves 2 and 3 are the results predicted by the models of Borwankar *et al.* (194) and van den Tempel (193), respectively. The values of the flocculation rate constant are: (a) $a_f = 10^{-11}\ \text{cm}^3/\text{s}$; (b) $a_f = 10^{-16}\ \text{cm}^3/\text{s}$.

namic and hydrodynamic factors. For example, the existence of a considerable repulsive (positive) disjoining pressure may lead to exclusions from the conventional Bancroft rule, which are accounted for in its generalized version.

Knowledge concerning the individual acts of drop–drop collision is a prerequisite for development of a kinetic theory of such collective phenomena as flocculation/coalescence and phase separation. The cases of fast and slow, perikinetic and orthokinetic, and irreversible and reversible flocculation are considered in Sec. VI. Special attention is paid to the case of parallel flocculation and coalescence. Much work remains to be done in order to build up united theory including both individual drop interactions and collective phenomena in emulsions.

## ACKNOWLEDGMENTS

Financial Support from the Inco-Copernicus Project No. IC15 CT98 0911 of the European Commission is gratefully acknowledged. The authors are indebted to Ms Mariana Paraskova and Mr Vesselin Kolev for their help in the preparation of the figures.

## REFERENCES

1. IB Ivanov, PA Kralchevsky. Colloids Surfaces A 128: 155–175, 1997.
2. IB Ivanov, KD Danov, PA Kralchevsky. Colloids Surfaces A 152: 161–182, 1999.
3. BV Derjaguin. Theory of Stability of Colloids and Thin Films. New York: Plenum Press – Consultants Bureau, 1989.
4. JN Israelachvili. Intermolecular & Surface Forces. London: Academic Press, 1992.
5. VM Mikova. Investigation of Emulsions Stabilized by $\beta$-Lactoglobulin. MSc thesis, University of Sofia, 1999.
6. JT Davies, EK Rideal. Interfacial Phenomena. London: Academic Press, 1963.
7. PC Hiemenz, R Rajagopalan. Principles of Colloid and Surface Chemistry. New York: Marcel Dekker, 1997.
8. I Langmuir. J Am Chem Soc 15: 75–84, 1918.
9. TL Hill. An Introduction to Statistical Thermodynamics. Reading, MA: Addison-Wesley, 1962.
10. M Volmer. Z Phys Chem 115: 253–259, 1925.
11. A Frumkin. Z Phys Chem 116: 466–475, 1925.
12. LD Landau, EM Lifshitz. Statistical Physics. Part 1. Oxford: Pergamon Press, 1980.
13. TD Gurkov, PA Kralchevsky, K Nagayama. Colloid Polym Sci 274: 227–238, 1996.
14. EH Lucassen-Reynders. J Phys Chem 70: 1777–1785, 1966.
15. RP Borwankar, DT Wasan. Chem Eng Sci 43: 1323–1337, 1988.
16. ED Shchukin, AV Pertsov, EA Amelina. Colloid Chemistry. Moscow: Moscow University Press, 1982 (in Russian).
17. PA Kralchevsky, KD Danov, G Broze, A Mehreteab. Langmuir 15: 2351–2365, 1999.
18. JW Gibbs. The Scientific Papers of J. W. Gibbs. Vol 1. New York: Dover, 1961.
19. EH Lucassen-Reynders. In: P Becher, ed. Encyclopedia of Emulsion Technology. Vol 4. New York: Marcel Dekker, 1996, p 63.
20. Y Tian, RG Holt, R Apfel. J Colloid Interface Sci, 187: 1–10, 1997.
21. A Ozawa, A Minamisawa, K Sakai, K Takagi. Jpn J Appl Phys 33: L-1468, 1994.
22. K Sutherland. Austr J Sci Res Ser A 5: 683–689, 1952.
23. SS Dukhin, G Kretzschmar, R Miller. Dynamics of Adsorption at Liquid Interfaces. Amsterdam: Elsevier, 1995.
24. KD Danov, PA Kralchevsky, IB Ivanov. In: G Broze, ed. Handbook of Detergents. New York: Marcel Dekker, 1999, p 303.
25. JTG Overbeek. In: HR Kruyt, ed. Colloid Science. Vol 1. Amsterdam: Elsevier, 1953; J Colloid Sci 8: 420, 1953.
26. G Gouy. J. Phys Radium 9: 457–466, 1910.
27. DL Chapman. Phil Mag 25: 475–485, 1913.
28. O Stern. Z. Elektrochem 30: 508–512, 1924.
29. S Hachisu. J Colloid Interface Sci 33: 445–455, 1970.
30. DG Hall. In: DM Bloor, E Wyn-Jones, eds. The Structure, Dynamics and Equilibrium Properties of Colloidal Systems. Dordrecht: Kluwer, 1990, p 857.
31. DG Hall. Colloids Surfaces A 90: 285–293, 1994.
32. G Bakker. Handbuch der Experimentalphysik. Band 6. Leipzig: Akademische Verlagsgesellschaft, 1928.
33. S Ono, S Kondo. In: S Flügge, ed. Handbuch der Physik. Vol 10. Berlin: Springer, 1960, p 134.
34. JS Rowlinson, B Widom. Molecular Theory of Capillarity. Oxford: Clarendon Press, 1982.
35. KD Danov, PM Vlahovska, PA Kralchevsky, G Broze, A Mehreteab. Colloids Surfaces A 156: 389–411, 1999.
36. KD Danov, VL Kolev, PA Kralchevsky, G Broze, A Mehreteab. Langmuir 16: 2942–2956, 2000.
37. VV Kalinin, CJ Radke. Colloids Surfaces A 114: 337–350, 1996.
38. K Tajima, M Muramatsu, T Sasaki. Bull Chem Soc Jpn 43: 1991–1998, 1970.
39. K Tajima. Bull Chem Soc Jpn 43: 3063–3067, 1970.
40. E Janke, F Emde, F Lösch. Tables of Higher Functions. New York: McGraw-Hill, 1960.
41. M Abramowitz, IA Stegun. Handbook of Mathematical Functions. New York: Dover, 1965.
42. GA Korn, TM Korn. Mathematical Handbook. New York: McGraw-Hill, 1968.
43. G Loglio, E Rillaerts, P Joos. Colloid Polym Sci 259: 1221–1230, 1981.
44. R Miller. Colloid Polym Sci 259: 375–381, 1981.
45. YM Rakita, VB Fainerman, VM Zadara. Zh Fiz Khim 60: 376–388, 1986.
46. PA Kralchevsky, YS Radkov, ND Denkov. J Colloid Interface Sci 161: 361–365, 1993.
47. KD Danov, PM Vlahovska, TS Horozov, CD Dushkin, PA Kralchevsky, A Mehreteab, G Broze. J Colloid Interface Sci 183: 223–235, 1996.
48. RS Hansen. J Chem Phys 64: 637, 1960.

49. R van den Bogaert, P Joos. J Phys Chem 83: 2244–2251, 1979.
50. E Rillaerts, P Joos. J Chem Phys 96: 3471–3477, 1982.
51. R Miller, G Kretzchmar. Adv Colloid Interface Sci 37: 97–136, 1991.
52. VB Fainerman, AV Makievski, R Miller. Colloids Surfaces A 87: 61–75, 1994.
53. LK Filippov. J Colloid Interface Sci 164: 471–482, 1994.
54. A Bonfillon, F Sicoli, D Langevin J Colloid Interface Sci 168: 497–504, 1994.
55. C MacLeod, CJ Radke. Langmuir 10: 2965–2975, 1994.
56. AW Cross, GG Jayson. J Colloid Interface Sci 162: 45–53, 1994.
57. SB Johnson, CJ Drummond, PJ Scales, S Nishimura. Langmuir 11: 2367–2378, 1995.
58. RG Alargova, KD Danov, JT Petkov, PA Kralchevsky, G Broze, A Mehreteab. Langmuir 13: 5544–5551, 1997.
59. OD Velev, TD Gurkov, RP Borwankar. J Colloid Interface Sci 159: 497–501, 1993.
60. KD Danov, TD Gurkov, TD Dimitrova, D Smith. J Colloid Interface Sci 188: 313–324, 1997.
61. OD Velev, TD Gurkov, IB Ivanov, RP Borwankar. Phys Rev Lett 7–75: 264–267, 1995.
62. PA Kralchevsky, KD Danov, ND Denkov. In: KS Birdi, ed. Handbook of Surface and Colloid Chemistry. Boca Raton, L: CRC Press, 1997, p 333.
63. MJ Rosen. Surfactants and Interfacial Phenomena. New York: John Wiley, 1989.
64. J Clint. Surfactant Aggregation. London: Chapman & Hall, 1992.
65. EAG Aniansson, SN Wall. J Phys Chem 78: 1024–1030, 1974.
66. EAG Aniansson, SN Wall. J Phys Chem 79: 857–864, 1975.
67. EAG Aniansson, SN Wall. In: E Wyn-Jones, ed: Chemical and Biological Applications of Relaxation Spectrometry. Dordrecht: Reidel, 1975, p 223.
68. EAG Aniansson, SN Wall, M Almgren, H Hoffmann, I Kielmann, W Ulbricht, R Zana, J Lang, C Tondre. J Phys Chem 80: 905–922, 1976.
69. CD Dushkin, IB Ivanov. Colloids Surfaces 60: 213–233, 1991.
70. CD Dushkin, IB Ivanov, PA Kralchevsky. Colloids Surfaces 60: 235–261, 1991.
71. BA Noskov. Kolloidn Zh 52: 509–517, 1990.
72. BA Noskov. Kolloidn Zh 52: 796–805, 1990.
73. GC Kresheck, E Hamori, G Davenport, HA Scheraga. J Am Chem Soc 88: 264–279, 1966.
74. PA Kralchevsky, IB Ivanov. Chem Phys Lett 121: 111–116, 1985.
75. IB Ivanov, PA Kralchevsky. In: IB Ivanov, ed. Thin Liquid Films. New York: Marcel Dekker, 1988, p 49.
76. IB Ivanov, BV Toshev. Colloid Polym Sci 253: 593–599, 1975.
77. JA. de Feijter, A Vrij. J Colloid Interface Sci 70: 456–467, 1979.
78. PA Kralchevsky, ND Denkov. Chem Phys Lett 240: 385–392, 1995.
79. AD Nikolov, DT Wasan, PA Kralchevsky, IB Ivanov. In: N Ise, I Sogami, eds. Ordering and Organisation in Ionic Solutions. Singapore: World Scientific, 1988, p 302.
80. AD Nikolov, DT Wasan. J Colloid Interface Sci 133: 1–12, 1989.
81. AD Nikolov, PA Kralchevsky, IB Ivanov, DT Wasan. J Colloid Interface Sci 133: 13–22, 1989.
82. PA Kralchevsky, AD Nikolov, DT Wasan, IB Ivanov. Langmuir 6: 1180–1189, 1990.
83. AD Nikolov, DT Wasan, ND Denkov, PA Kralchevsky, IB Ivanov. Prog Colloid Polym Sci 82: 87–98, 1990.
84. DT Wasan, AD Nikolov, PA Kralchevsky, IB Ivanov. Colloids Surfaces 67: 139–145, 1992.
85. V Bergeron, CJ Radke. Langmuir 8: 3020–3026, 1992.
86. JL Parker, P Richetti, P Kékicheff, S Sarman. Phys Rev Lett 68: 1955–1958, 1992.
87. ML Pollard, CJ Radke. J Chem Phys 101: 6979–6991, 1994.
88. XL Chu, AD Nikolov, DT Wasan. Langmuir 10: 4403–4408, 1994.
89. XL Chu, AD Nikolov, DT Wasan. J Chem Phys 103: 6653–6661, 1995.
90. ES Basheva, KD Danov, PA Kralchevsky. Langmuir 13: 4342–4348, 1997.
91. KG Marinova, TD Gurkov, TD Dimitrova, RG Alargova, D Smith. Langmuir 14: 2011–2019, 1998.
92. AJ de Vries. Rec Trav Chim Pays-Bas 77: 44–53, 1958.
93. A Scheludko. Proc K Akad Wetensch B 65: 87–91, 1962.
94. A Vrij. Disc Faraday Soc 42: 23–31, 1966.
95. IB Ivanov, B Radoev, E Manev, A Scheludko. Trans Faraday Soc 66: 1262–1273, 1970.
96. IB Ivanov, DS Dimitrov. Colloid Polym Sci 252: 982–990, 1974.
97. IB Ivanov. Pure Appl Chem 52: 1241–1262, 1980.
98. PA Kralchevsky, KD Danov, IB Ivanov. In: RK Prud'homme, SA Khan, eds. Foams. New York: Marcel Dekker, 1995, p 1.
99. BV Derjaguin, YV Gutop. Kolloidn Zh 24: 431–438, 1962.
100. BV Derjaguin, AV Prokhorov. J Colloid Interface Sci 81: 108–115, 1981.
101. AV Prokhorov, BV Derjaguin. J Colloid Interface Sci 125: 111–123, 1988.
102. D Kashchiev, D Exerowa. J Colloid Interface Sci 77: 501–511, 1980.
103. D Kashchiev, D Exerowa. Biochim Biophys Acta 732: 133–142, 1983.

104. D Kashchiev. Colloid Polym Sci 265: 436–441, 1987.
105. YA: Chizmadzhev, VF Pastushenko. In: IB Ivanov, ed. Thin Liquid Films. New York: Marcel Dekker, 1988, p 1059.
106. LV Chernomordik, MM Kozlov, GB Melikyan, IG Abidor, VS Markin, YA Chizmadzhev. Biochim Biophys Acta 812: 643–655, 1985.
107. LV Chernomordik, GB Melikyan, YA Chizmadzhev. Biochim Biophys Acta 906: 309–352, 1987.
108. A Kabalnov, H Wennerström. Langmuir 12: 276–292, 1996.
109. IB Ivanov, SK Chakarova, BI Dimitrova. Colloids Surfaces 22: 311–316, 1987.
110. BI Dimitrova, IB Ivanov, E Nakache. J Dispers Sci Technol 9: 321–341, 1988.
111. KD, Danov, IB Ivanov, Z Zapryanov, E Nakache, S Raharimalala. In: MG Velarde, ed. Proceedings of the Conference of Synergetics, Order and Chaos. Singapore: World Scientific, 1988, p 178.
112. CV Sterling, LE Scriven. AIChE J 5: 514–520, 1959.
113. SP Lin, HJ Brenner. J Colloid Interface Sci 85: 59–74, 1982.
114. TD Gurkov, KD Danov, N Alleborn, H Raszillier, F Durst. J Colloid Interface Sci 198: 224–240, 1998.
115. J Lucassen, M van den Tempel, A Vrij, FT Hesselink. Proc Konkl Ned Akad Wet B 73: 109–113, 1970.
116. AK Malhotra, DT Wasan. Chem Eng Commun 48: 35–41, 1986.
117. C Maldarelli, RK Jain. In: IB Ivanov, ed. Thin Liquid Films: Fundamental and Applications. New York: Marcel Dekker, 1988, p 497.
118. BU Felderhof. J Chem Phys 49: 44–52, 1968.
119. S Sche, HM Fijnaut. Surface Sci 76: 186–198, 1978.
120. ED Manev, SV Sazdanova, DT Wasan. J Colloid Interface Sci 97: 591–604, 1984.
121. IB Ivanov, DS Dimitrov. In: IB Ivanov, ed. Thin Liquid Films. New York: Marcel Dekker, 1988, p. 379.
122. P Taylor. Proc Roy Soc (London) A108: 11–14, 1924.
123. E Rushton, GA Davies. Appl Sci Res 28: 37–42, 1973.
124. S Haber, G Hetsroni, A Solan. Int J Multiphase Flow 1: 57–66, 1973.
125. LD Reed, FA Morrison. Int J Multiphase Flow 1: 573–584, 1974.
126. G Hetsroni, S Haber. Int J Multiphase Flow 4: 1, 1978.
127. FA Morrison, LD Reed. Int J Multiphase Flow 4: 433–434, 1978.
128. VN Beshkov, BP Radoev, IB Ivanov. Int J Multiphase Flow 4: 563–570, 1978.
129. DJ Jeffrey, Y Onishi. J Fluid Mech 139: 261–276, 1984.
130. YO Fuentes, S Kim, DJ Jeffrey. Phys Fluids 31: 2445–2455, 1988.
131. RH Davis, JA Schonberg, JM Rallison. Phys Fluids A1: 77–81, 1989.
132. X Zhang, RH Davis. J Fluid Mech 230: 479–491, 1991.
133. TT Traykov, IB Ivanov. Int J Multiphase Flow 3: 471–483, 1977.
134. TT Traykov, ED Manev, IB Ivanov. Int J Multiphase Flow 3: 485–494, 1977.
135. IB Ivanov, BP Radoev, T Traykov, D Dimitrov, E Manev, C Vassilieff. In: E Wolfram, ed. Proceedings of the International Conference on Colloid and Surface Science. Vol 1. Budapest: Akademia Kiado, 1975, p 583.
136. KD Danov, IB Ivanov. Critical film thickness and coalescence in emulsions. Proceedings of the Second World Congress on Emulsions, Paris, 1997, Paper No. 2–3–154.
137. KD Danov, DS Valkovska, IB Ivanov. J Colloid Interface Sci 211: 291–303, 1999.
138. DS Valkovska, KD Danov, IB Ivanov. Colloids Surfaces A 156: 547–566, 1999.
139. SG Yanitsios, RH Davis. J Colloid Interface Sci 144: 412–432, 1991.
140. V Cristini, J Blawzdziewicz, M Loewenberg. J Fluid Mech 366: 259–273, 1998.
141. DS Valkovska, KD Danov, IB Ivanov. Colloids Surfaces A 175: 179–192, 2000.
142. O Reynolds. Phil Trans Roy Soc (London) A177: 157–234, 1886.
143. KD Danov, DN Petsev, ND Denkov, R Borwankar. J Chem Phys 99: 7179–7189, 1993.
144. ND Denkov, PA Kralchevsky, IB Ivanov, CS Vassilieff. J Colloid Interface Sci 143: 157–173, 1991.
145. ND Denkov, DN Petsev, KD Danov. Phys Rev Lett 71: 3226–3229, 1993.
146. DN Petsev, ND Denkov, PA Kralchevsky. J Colloid Interface Sci 176: 201–213 1995.
147. KD Danov, ND Denkov, DN Petsev, IB Ivanov, R Borwankar. Langmuir 9: 1731–1740, 1993.
148. A Scheludko, D Exerowa. Kolloid-Z 165: 148–157, 1959.
149. A Scheludko. Adv Colloid Interface Sci 1: 391–440, 1967.
150. OD Velev, AD Nikolov, ND Denkov, G Doxastakis, V Kiosseoglu, G Stalidis. Food Hydrocolloids 7: 55–71, 1993.
151. OD Velev, TD Gurkov, SK Chakarova, BI Dimitrova, IB Ivanov, RP Borwankar. Colloids Surfaces A 83: 43–55, 1994.
152. OD Velev, GN Constantinides, DG Avraam, AC Payatakes, RP Borwankar. J Colloid Interface Sci 175: 68–76, 1995.
153. KG Marinova, TD Gurkov, OD Velev, IB Ivanov, B Campbell RP Borwankar. Colloids Surfaces A 123/124: 155–167, 1997.
154. KP Velikov, OD Velev, KG Marinova, GN Constantinides. J Chem Soc Faraday Trans 93: 2069–2075, 1997.

155. OD Velev, KD Danov, IB Ivanov. J Dispers Sci Technol 18: 625–645, 1997.
156. KG Marinova, TD Gurkov, TD Dimitrova, RG Alargova, D Smith. Progr Colloid Polym Sci 110: 245–250, 1998.
157. OD Velev, BE Campbell, RP Borwankar. Langmuir 14: 4122–4130, 1998.
158. TD Gurkov, KG Marinova, A Zdravkov, C Oleksiak, B Campbell. Progr Colloid Polym Sci 110: 263–268, 1998.
159. BP Radoev, DS Dimitrov, IB Ivanov. Colloid Polym Sci 252: 50–55, 1974.
160. WB Russel, DA Saville, WR Schowalter. Colloidal Dispersions. Cambridge, England: Cambridge University Press, 1989, p 155.
161. WD Bancroft. J Phys Chem 17: 514–522, 1913.
162. BP Binks. Langmuir 9: 25–28, 1993.
163. KD Danov, OD Velev, IB Ivanov, RP Borwankar. Bancroft rule and hydrodynamic stability of thin films and emulsions. Proceedings of First World Congress on Emulsions, Paris, 1993.
164. H Kunieda, DF Evans, C Solans, M Yoshida. Colloids Surfaces A 47: 35–46, 1990.
165. MJ Schick. Nonionic Surfactants: Physical Chemistry. New York: Marcel Dekker 1986.
166. K Shinoda, S Friberg. Emulsions and Solubilization New York: John Wiley, 1986.
167. EE Isaacs, H Huang, AJ Babchin. Colloids Surfaces 46: 177–192, 1990; Colloids Surfaces A 123/124: 195–203, 1997.
168. O Holt, O Seather, J Sjöblom, SS Dukhin, NA Mishchuk. Colloids Surfaces A 123/124: 195–207, 1997.
169. K Demetreades, DJ McClements. Colloids Surfaces A 150: 45–54, 1999.
170. F Feuillebois. J Colloids Interface Sci 131: 267–275, 1989.
171. RM Merritt, RS Subramanian. J Colloid Interface Sci 131: 514–521, 1989.
172. KD Barton, RS Subramanian. J Colloid Interface Sci 133: 214–221, 1989.
173. M von Smoluchowski. Phys Z 17: 557–569, 1916.
174. M von Smoluchowski. Z Phys Chem 92: 129–143, 1917.
175. KD Danov, IB Ivanov, TD Gurkov, RP Borwankar. J Colloid Interface Sci 167: 8–17, 1994.
176. JL Klemaszewski, KP Das, JE Kinsella. J Food Sci 57: 366–371, 1992.
177. PT Jaeger, JJM Janssen, F Groneweg, WGM Agterof. Colloids Surfaces A, 85: 255–264, 1994.
178. E Dickinson, A Williams. Colloids Surfaces A 88: 317–326, 1994.
179. AJF Sing, A Graciaa, JL Salager. Colloids Surfaces A, 152: 31–39, 1999.
180. X Zhang, RH Davis. J Fluid Mech 230: 479–491, 1991.
181. H Wang, RH Davis. J Colloid Interface Sci 159: 108–118, 1993.
182. W Rybczynski. Bull Int Acad Sci (Cracovie) A, 40–49, 1911.
183. JS Hadamar. Comp Rend Acad Sci (Paris) 152: 1735–1740, 1911.
184. JR Rogers, RH Davis. Metal Trans A21: 59–68, 1990.
185. NO Young, JS Goldstein, MJ Block. J Fluid Mech 6: 350–365, 1959.
186. NA Fuchs. Z Phys 89: 736–742, 1934.
187. TA Bak, O Heilmann. J Phys A: Math Gen 24: 4889–4893, 1991.
188. GA Martinov, VM Muller. In: BV Derjaguin, ed. Research in Surface Forces. Vol 4. New York: Plenum Press – Consultants Bureau, 1975, p 3.
189. IM Elminyawi, S Gangopadhyay, CM Sorensen. J Colloid Interface Sci 144: 315–323, 1991.
190. S Hartland, N Gakis. Proc Roy Soc (London) A369: 137–148, 1979.
191. S Hartland, DK Vohra. J Colloid Interface Sci 77: 295–308, 1980.
192. LA Lobo, IB Ivanov, DT Wasan. AIChE J 39: 322–334, 1993.
193. M van den Tempel. Recueil 72: 419–461, 1953.
194. RP Borwankar, LA Lobo, DT Wasan. Colloids Surfaces 69: 135–144, 1992.
195. R Hogg. Colloids Surfaces A 146: 253–270, 1999.

# 27

# Three-phase Wellstream Gravity Separation

**Richard Arntzen**
*Kværner Process Systems a.s., Lysaker, Norway*
**Per Arild K. Andresen**
*Provida ASA, Oslo, Norway*

## I. INTRODUCTION

Production of petroleum is a stepwise process. The reservoir fluid is taken through a wellhead and transported to a production site by using the reservoir pressure, submersible pumps and/or gas/water injection. The production site takes the pressure stepwise down to shipping conditions (often atmospheric pressure) and removes water, gas, and solids from the oil.

The first process equipment the incoming reservoir fluid enters after the initial pressure reduction is the primary separator. This unit removes most of the water from the oil, which continues to a secondary and possibly ternary separator (which is often equipped with electrodes to enhance coalescence). Often the primary separator removes all the water, and the other stages are used for pressure reduction and gas removal only. The final goal is to match the specifications of the refineries and/or transport companies.

The water separated from the oil is postprocessed for dumping or reinjection by flotation-based or centrifugal equipment, according to the applicable specification. The gas is similarly dried and compressed for shipping or reinjection.

All water/oil separation processes utilize the immiscibility and density difference between the two phases (the electrostatic unit uses the difference in polarity as well).

The primary gravity separator is an important factor, especially offshore, in making the process cost effective. At an offshore platform where volume is an expensive resource, it is important to design the separator as small and light as possible. This is particularly important with regard to gravity separators since they are usually larger than other equipment and have a potential for size reduction.

In the petroleum production on the Norwegian continental shelf new trends will emerge during the next 3–5 years. First, the amount of water produced from offshore platform separators will increase mainly as a result of aging fields where water breakthrough has taken place, giving a concomitant production of injection water together with the oil. This high production rate of water will place high demands on separator efficiency and treatment of wastewater. In addition to this, many new fields to be explored in the future will be complicated to develop, as the crude oil produced will contain large amounts of heavy components such as asphaltenes and resins. These components strongly increase the capability of the crude oil to bind water, which increases the necessary retention time in the separator. These new types of crude oils will also most likely lead to an increase in the use of production chemicals in the separator and in the transport process.

The effects of these two trends have to be implemented into the design tools used for optimizing gravity separators. Tools in use today do not have a coalescence model for the dispersion entering the separator, but use only modified versions of Stokes' law when describing the settling/creaming of droplets and hence the separation. The influence of higher watercut and more stabilizing components increases the need for a coalescence model.

### A. Basic Principles of Gravity Separation

The two major phenomena recognized in phase separation are drop break-up and coalescence. Drop break-up is the process where one phase in an immiscible (multiphase) system forms an unstable, heterogeneous state of two or more distinct phases (drops) dispersed in a continuous phase. Coalescence is the reverse process where the system returns to the state of lowest total energy, i.e., separate homogeneous phases with a minimized common interface.

Of these two phenomena, break-up is by far the best understood. This is because turbulent forces, usually correlated with turbulent dissipation, often dominate break-up, and can be analyzed in terms of equilibrium states by considering the energy transport. Coalescence is often dominated by kinetics, and depends heavily on the chemical composition of the system.

The aim of a separator vessel is to give the coalescence process the necessary time and create conditions for satisfactory phase separation. The classical approach is to use an overall residence time criterion, which allows the drops to: (1) reach a bulk interface by sedimentation; and (2) coalesce with this bulk interface, forming a single homophase. One seeks to minimize any events contributing to drop break-up.

### B. Classical Design Philosophy–Sizing the Vessel

Correct modeling of the coalescence process for use in separator design is very difficult. The engineering solution has been to circumvent the problem by focusing on settling as described by Stokes' law, assuming that coalescence is sufficient. A standard developed throughout the years is to design the separator to handle a cut size of 200 $\mu$m in both the water and oil phases, assuming that there exists a sharp interface controlled by the interface control system. Thus, more residence time is allowed for viscous oils. In addition, the different engineering companies may have proprietary safety factors/cut-size relations based on various parameters available at the time of design.

Often the design based on separation (or settling) characteristics is overridden by the necessity of certain fill-up times between the different alarm settings. These are based on the operation of the outlet valves versus shutdown criteria for the process. Standard shutdown criteria range from 30 s to 1 min between the different alarm levels. Following the applicable API specification (1), the vessel size is selected and possibly increased until the 200-$\mu$m cut-size criterion is met. Proprietary factors and relations may modify the design slightly.

The separation involves settling and coalescence mechanisms. The settling velocity is a function of droplet size. The local velocity in the vessel and the settling regime (usually Stokian or Newtonian) determines the exact relationship between these two mechanisms (2). The coalescence of droplets within the dispersion and the dispersed continuous phase boundary is a complex function of droplet diameter as the gravity and surface forces that control the coalescence are both related to the droplet diameter, or more precisely the local curvature at the interface. It is known that when the droplet size reduces, so does the separation rate. When the droplets reach a size of about 30–60 $\mu$m the separation is found to be settling controlled (2). In these cases the settling time is usually longer than the residence time of the bulk phases. Systems of this type are classified as secondary dispersions. Although they are thermodynamically unstable they lead to poor outlet quality of the continuous phase. Knowledge of the characteristics of the dispersion (droplet size distribution) is obviously of great importance in designing separator equipment.

There are several obvious flaws with this technique. First, there is no reason to believe that Stokes' law will be valid for the problems it is applied to, especially in the continuous phase (normally the oil phase) where the high droplet concentration may lead to significant droplet/droplet interaction. Second, the 200-$\mu$m cut-size seems rather arbitrarily chosen (possibly from API specifications for refinery separators, following Ref. 3), and as far as the authors have been able to find out, no evidence exists that this necessarily resembles any actual droplet entering the separator. Third, the coalescence process is disregarded and the separator is assumed to be settling controlled. This is in reality a limiting case for real fluid systems, and by no means generally applicable. This also leads to the assumption of a sharp interface. This is generally incorrect, as the coalescence process normally creates a band of noncoalesced drops residing at the interface. This region has special characteristics, and will hereafter be referred to as the dispersion band.

These problems with classical design are the core of this chapter, and will be discussed below.

## II. FLUID PROPERTIES AND STABILIZING MECHANISMS

This section describes the various chemical properties of the fluids entering the separator, and how they interact with the coalescence rate. The intention is not to describe a coalescence model for design use, but merely to give an overview of the different stabilizing mechanisms that must be taken into account when trying to model coalescence.

The overall coalescence rate of a dispersion/emulsion in a separator is the most important design criterion. Unfortunately, this rate is a product of several complex mechanisms like binary coalescence, interfacial coalescence, and settling/creaming. Each of these mechanisms is further related to other even more complex processes/factors like hydrodynamic micro- and macro-motion, droplet size distribution, and interfacial components. In order to understand the overall coalescence rate one must also understand the interactions between these mechanisms. This makes it difficult to separate the overall rate into a sum of distinct rates, and is probably the reason why there exists no generalized coalescence model for concentrated dispersions with a sound theoretical foundation.

Coalescence is the process where two or more droplets combine and form a larger droplet. This is necessary to form a clear liquid layer from an initially dispersed phase. Droplets can coalesce owing to binary or interfacial coalescence.

Binary coalescence is coalescence between droplets that are settling/creaming or packed in the dispersion band, while interfacial coalescence is the coalescence of a droplet with its own phase (a droplet with infinite dimensions). In both cases a liquid film of continuous phase separates the dispersed droplets and this film has to be drained and broken in order to complete the coalescence process. Hartland describes this draining process in detail (4).

The following will mainly focus on the effect and mechanisms that are associated with the hindering of film drainage by the interfacial components.

When describing stabilizing mechanisms, one can distinguish between three types:

1. Steric stabilization;
2. Electrostatic stabilization;
3. Mechanical stabilization.

Steric stabilization occurs when the interfacial components have a long chained part of the molecule that stretches into the continuous phase. The term steric stabilization can be associated with several different mechanisms, but they are all related to an increase in the system free energy. The penetration mechanism is known to be the most important and can be described as a local increase in concentration of polymer segments in the film separating the droplets. If the continuous phase is a good solvent for the polymer segments, the local increase in concentration when the drops move closer will be thermodynamically unfavorable ($\Delta G > 0$). The chemical potential of the solvent in the area between the droplets will decrease. This creates an osmotic pressure, $\pi$, that will oppose the increase in concentration by making the continuous phase flow into to the film between the droplets. The mechanisms is depicted in Fig. 1.

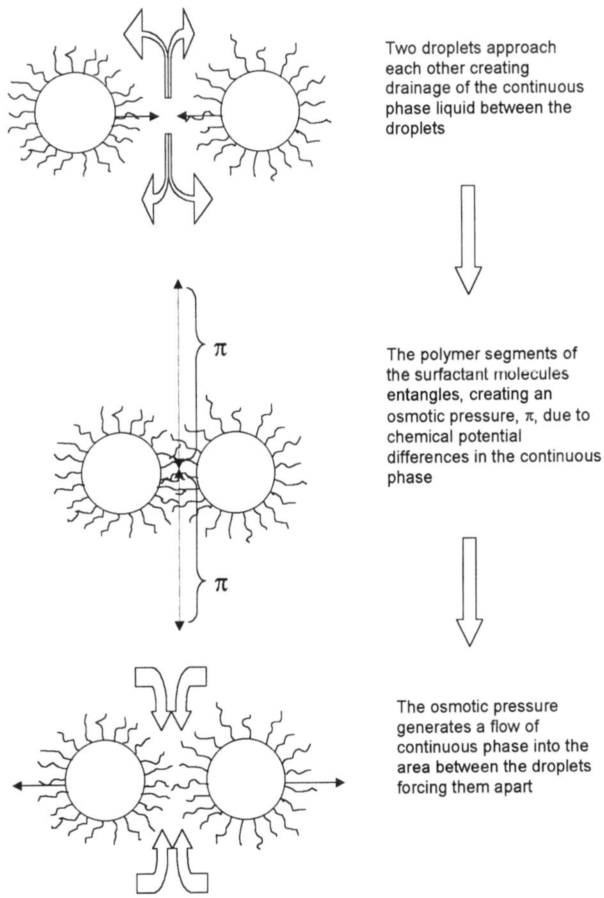


**Figure 1** Sketch of the steric stabilization of drops.

Electrostatic stabilization occurs when the interfacial components are charged and the electric double layer between two or more droplets overlap. The resulting repulsive force counteracts further drainage of the film. Authors have, however, disregarded this repulsion as a significant stabilizing factor in describing water-in-crude oil stability (5, 6).

Mechanical stabilization is a process where interfacial components act as particles, creating a mechanically stable film on the surface of the droplets. This film encapsulates the droplets and, due to its immobility and low solubility in both water and oil, creates a very stable emulsion. Asphaltenes, resins, wax particles, minerals, and clay are compounds believed to enhance the formation of mechanically stable films.

All these mechanisms can be present when an emulsion is formed (although some are more predominant than others). This makes it very difficult to model emulsion stability as a function of fluid properties and interfacial components.

A more quantitative theory based on interfacial gradients can also be used to describe how film drainage is retarded, thereby stabilizing the dispersion. When interface-active components adsorb on a water/oil interface the interfacial tension will decrease monotonically as a function of the surface concentration. As two droplets approach each other the resulting film drainage will carry interfacial components away in the drainage direction, creating a concentration gradient. This gradient results in an inward interfacial-tension gradient, creating a positive inward force counteracting the film drainage. This theory assumes mobile W/O interfaces. Figure 2 shows a simplified scenario of this theory.

Accepting these stabilizing mechanisms may give qualitative ideas on how to explain the known separation of a given system, but cannot be used to predict the separation in advance. Even if the composition of the probable stabilizing components (asphaltenes, resins, wax, minerals, and clay) is known, it is very difficult to predict the stability of the resulting emulsion. One of the reasons is that the history of the fluid greatly influences the stability of the emulsion. Factors such as temperature–pressure variation and well combinations affect the solubility and thereby the size distribution of the stabilizing particles, which is thought to be an important stabilizing factor. Aging of an emulsion is also known to greatly affect the emulsion character. It is currently impossible to control all these factors, and difficult to gain detailed information about them.

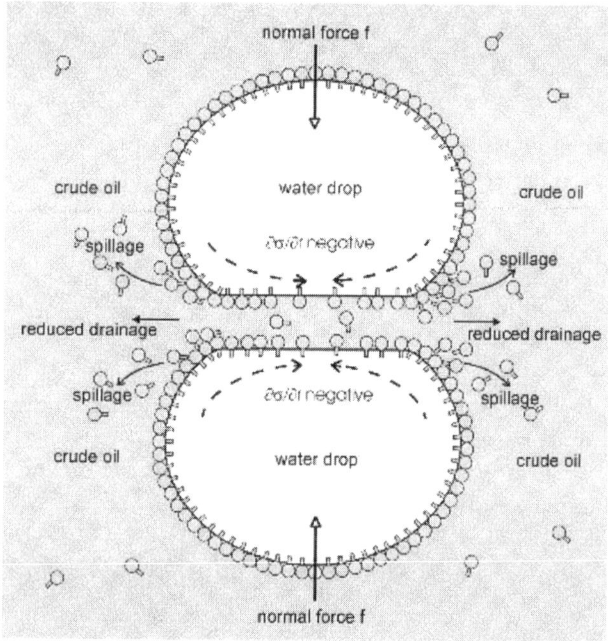


**Figure 2** Sketch of the marangoni effect. (From Ref. 22.)

The petroleum industry generally "solves" the emulsion problem by adding demulsifiers in an ad hoc manner, often based on simple bottle tests. There are many problems associated with this solution. First, the chemical composition of a given well changes with time and can in a worst-case scenario result in a composition totally incompatible with the given demulsifier. Second, little is known about the exact interaction between demulsifiers and other chemical additives (e.g., corrosion inhibitors and "flow enhancers"). One may, therefore, create a new problem by solving another.

Another well-known term used in the petroleum industry is "watercut curves". For a given crude oil they give the fraction of separated water as a function of the watercut at a given retention time (normally 4 min). A typical watercut curve is shown in Fig. 3. As can be seen from the figure, a higher watercut gives better separation. The lowest watercut that is associated with complete separation is called the critical watercut. A qualitative explanation of this could be that at this watercut, the total area of the dispersed droplets is larger than the maximum area that the interfacial components can cover for proper stabilization of the droplets. Operating a separator at this watercut solves the problem of a stable emulsion, but is not always cost effective as the water production increases at the expense of oil production.

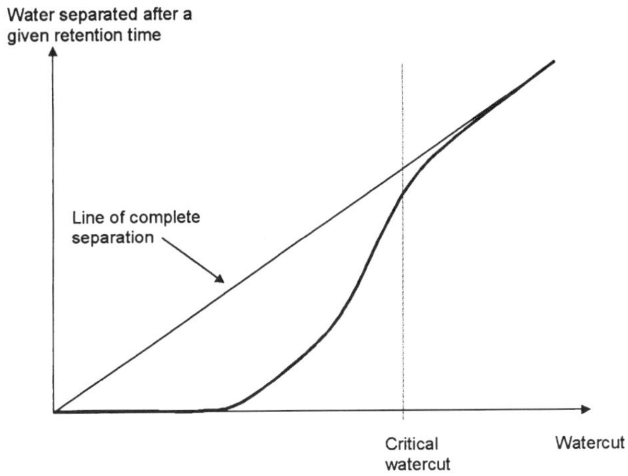


**Figure 3** Sketch of a watercut curve.

## III. DROP SIZES IN TURBULENT REGIMES

It is widely recognized that the size of the drops entering the separator is a parameter of great importance, since it will affect both the settling and coalescence mechanics inside the vessel. The drop size at a given point in the process is dependent on the turbulent fluctuations, the history of the fluids up to that point, and the physical properties of the mixture. Traditionally, the existence of a specific drop size equilibrium in any turbulent field is assumed. This has been investigated experimentally by several authors, particularly for stirred systems (7). Some authors have also looked more specifically into tube flow (8).

Turbulent modeling is a large field of research in itself. Various turbulence models have been developed with the evolution of computers, as this is a field of high computing intensity. The bulk of experimental work has, as mentioned above, been performed on simple geometries, and the isotropic turbulence model (see below) has been used with spatially averaged values to approximate break-up effects in these experiments. For the more complex geometries typically found in an oil-producing process, such as manifolds, various valves, bends, and inlets, the relevance of these relations is questionable. Additionally, when following a multiphase flow from the well to the separator, equilibrium cannot always be assumed at the various regions observed. Depending on the stability of the emulsion(s) formed, the approach toward equilibrium will depend on the residence time in the various regions. For highly stable systems, the drop size used for design basis should probably be in the regime of highest turbulent intensity, corresponding to the smallest drops. This also leads to the statement that, in order to make the separator inlet conditions more favorable, chemical destabilization should be performed as early as possible in the process. Turbulence theory is traditionally limited to single-phase flow, and the extension to multiphase behavior should be carried out with care. Kolev (9) gives a comprehensive review of various approaches and models. This chapter is, however, limited to the traditional approach of modeling the multiphase mixtures with single-phase turbulence relations.

Drop break-up is usually associated with turbulence and is most prominent in sections with high turbulent shear. Kolmogoroff [from Davies (7)] showed that the maximum diameter a drop can have in a local isotropic turbulent field is given by

$$d_{\max} = C\left\{\frac{\sigma}{\rho_c}\right\}^{3/5} \varepsilon^{-2/5} \tag{1}$$

Here, $C$ is a constant near unity, $\sigma$ is the interface tension, $\rho_c$ is the continuous phase density, and $\epsilon$ is the turbulent energy dissipation (energy input per mass unit). This equation has been verified experimentally for dilute, stirred systems of various types.

Davies (10) concluded that it is the turbulent fluctuation velocity $\nu\prime$ that is responsible for drop break-up. The dissipation term $\epsilon$ is given by the following relation:

$$\varepsilon = -\frac{1}{2}\left\{\frac{d}{dt}\right\}[(\tilde{\nu}'_x)^2 + (\tilde{\nu}'_y)^2 + (\tilde{\nu}'_z)^2] \tag{2}$$

The assumption of isotropic turbulence arises when one assumes that the various mean turbulent velocity fluctuations are equal, and thereby obtains a mean dissipation for a given mass (knowing the energy input):

$$\varepsilon = -\frac{3}{2}\frac{d(\tilde{\nu}')^2}{dt} \tag{3}$$

If the turbulent velocity fluctuations are known from simulations or measurements, the maximum drop sizes can also be calculated more accurately for anisotropic turbulence.

Davies (10) proposed a viscosity correction as shown in Eq. (4):

$$d_{\max} = C\left\{\frac{4\sigma + \mu_d\,\tilde{\nu}'}{4\rho_c}\right\}^{3/5} \varepsilon^{-2/5} \tag{4}$$

Here, $\mu_d$ is the dispersed phase viscosity.

For dilute pipe flow, Eq. (1) is expected to hold for the flow near the centre of the pipe where the turbulence can be regarded as isotropic. However, the shear in turbulent pipe flow is mainly located near the walls where isotropy is a poorer assumption. Karabelas (8) showed that for a dilute pipe-flow system the maximum diameter surviving in turbulent pipe flow is given by

$$d_{\max} = 4.0D\left\{\frac{D\rho_c\mu^2}{\sigma}\right\}^{-3/5} = 4.0\,D\,\mathrm{We}^{-3/5} \qquad (5)$$

It should be noted that later work (11) has shown that this (and similar) equations may not be entirely correct as the experiments may not have been performed under steady-state conditions. This, however, falls beyond the scope of this chapter.

Polderman et al. (12) suggested a water-cut dependency on droplet size distribution at the Draugen field:

$$d_{\max} \propto \Delta P^{-3/5}(1 + 20\phi_0^2) \qquad (6)$$

The pressure term represents the turbulent energy input across a valve. An interesting feature of this equation is that it shows a dependence of dispersed phase fraction. This equation is of a more empirical nature than the others, and the extra term could include both binary coalescence in the downstream region of the valve and the possibility that a larger dispersed mass would absorb and dissipate energy internally at larger eddy sizes. It addresses, however, the nondilute situation usually encountered in crude oil/water separation processes.

For the problem at hand, three regions are regarded as important for drop break-up: the choke valves, the tube from the choke to the inlet, and the inlet. For noncentrifugal inlets, the inlet momentum from the tube can be regarded as the energy input, and the length scale of the inlet can be used for scaling the dissipative volume. For cyclonic inlets, the dominating velocity is usually the tangential one, and the region of large dissipation is the liquid outlet region. Hence, inlet cyclone-related drop break-up is related to the cyclone liquid-outlet momentum and the liquid-outlet length scale. Depending on the stability of the system and the residence time in the tube, the engineer will have to estimate the size of the drop entering the separator, based on these regimes.

### A. Turbulence-induced Coalescence

Meijs and Mitchell (13) examined the possibilities of inducing coalescence by gentle turbulent mixing in tube flow, and found that droplets with an initial $d_{max} = 20\ \mu$m were coalesced to $d_{max} = 250\ \mu$m, by applying a turbulent mixing intensity $\epsilon$ of 2 $m^2/s^3$. This effect decreased linearly with log $\epsilon$ until $d_{max} = 39\ \mu$m for $\epsilon$ of $10^3$ $m^2/s^3$, with the same initial drop size distribution. They also indicated that a time span of the order of 10+ min was needed to reach these new equilibria, depending on the amount of dispersed phase and energy input. The most efficient energy input level for the tests was found to be 0.59 $m^2/s^3$, and the efficiency decreased with decreasing dispersed fraction. The energy input level was estimated by the relation:

$$\varepsilon = \frac{\mu^3 f}{D} \qquad (7)$$

where $u$ is the mean velocity in the tube, $D$ is the tube diameter, and $f$ is the friction factor defined as 16/Re. They also found that Eq. (1) overpredicted the drop sizes by a factor of 1.5–3 for the system investigated.

## IV. GRAVITY SEPARATOR MECHANISMS

This section covers the status of mechanistic understanding of the processes in a gravity separator. Recently, several new philosophies (12, 15, 20) have emerged which study the mechanisms inside a separator, using different approaches.

### A. Setting Laws

The classical approach to settling is Stokes' law, a balance between gravity forces and drag on a solid, spherical particle in infinite dilution and for creeping flow where the Reynolds number is $\ll 1$ (experiments have shown that the equation has validity for systems approaching Re = 1). The particle's terminal vertical velocity is given by

$$v_{st} = \frac{\Delta\rho g d^2}{18\,\mu_c} \qquad (8)$$

Stokes' law is an analytic solution of the Navier–Stokes equation for the simplified flow case with solid particles and creeping flow. If the particles are fluid and in the absence of surface-active components, internal circulation inside the particle will reduce the drag. (Note that this is not necessarily valid for small fluid particles, but these are irrelevant in gravity separation.) The viscosity correction term for this case is given in Eq. (9). From this equation it can be seen that, for large viscosity differences between the dispersed and continuous phases, the settling will approach the Stokes velocity or 3/2 Stokes velocity (the two limiting cases), depending on what is dispersed. Viscous liquid drops in a gas will approach Stokes' law (negligible circulation), while gas bubbles in viscous liquids will approach 3/2 Stokes' law (high degree of circulation inside the bubbles).

$$v_{St,\,visc} = \frac{\Delta\rho g d^2}{18\,\mu_c}\frac{\mu_c + \mu_d}{2/3\,\mu_c + \mu_d} \tag{9}$$

Kumar and Hartland (14) reviewed 14 different data sources published, with 998 results for 29 different liquid/liquid systems and correlated Eq. (10) by nonlinear fitting. This equation is different from Eqs (8) and (9) in that it includes the initial amount of dispersed phase $\phi_0$. This feature is called hindered settling, as it does not use infinite dilution as an initial assumption.

$$v_{KH} = \frac{12\mu_c}{0.53\,\rho_c\,d}\left[-1 + \sqrt{1 + \frac{0,53\,\rho_c\,\Delta\rho g d^3 (1-\phi_0)}{108\mu_c^2 (1 + 4.56\phi_0^{0.73})}}\right] \tag{10}$$

By calculating the mean residence time between two heights inside the separator, the smallest particle diameter that will traverse the vertical distance can be calculated. This is known as cut size. It is customary in gravity-separator design to calculate the cut size for an oil drop between the bottom of the vessel (BV) and the various oil/water interfaces (NIL, LIL, HIL). Likewise, the cut size for a water drop between the various liquid interfaces (NOL, LOL, HOL) and the interface is calculated. The vertical velocity criterion thus becomes:

$$v_c = \frac{h_1 - h_2}{\tau\eta} \tag{11}$$

where $h_1$ and $h_2$ are the chosen levels, $\tau$ is the residence time of the fluid between these levels, and $\eta$ is an efficiency depending on the flow regime; $\eta = 1$ corresponds to a plug-flow approximation. The cut sizes for the various settling equations by this relation become:

$$d_c^{\,\mathrm{St}} = \sqrt{\frac{18 v_c \mu_c}{\Delta\rho g}}$$

$$d_c^{\,\mathrm{St\,visc}} = \sqrt{\frac{18 v_c \mu_c (\mu_d + 2/3\,\mu_c)}{\Delta\rho g(\mu_d + \mu_c)}}$$

$$d_c^{\mathrm{KH}} = \frac{\frac{0.53\,\rho_c v_c^2}{12\mu_c} + \sqrt{\frac{0.53^2\rho_c^2 v_c^4}{12^2\mu_c^2} + \frac{\Delta\rho g(1-\phi_0)v_c}{\mu_c(1+4.56\phi_0^{0.73})}}}{\frac{\Delta\rho g(1-\varepsilon_0)}{4\mu_c(1+4.56\varepsilon_0^{0.73})}} \tag{12}$$

### 1. Plug Velocities and Retention Times

The retention-time calculations are based on the volume between the perforated plate and the weir plate. The cross-sectional area available to the phases (oil, water (or liquid), and gas) is calculated and the mean plug velocity and retention times are given. The cross-sectional area below a level $h$ in a horizontal cylinder is generally given by the geometric relation:

$$A_c = r^2 \arccos\left\{\frac{r-h}{r}\right\} - r(r-h)\sin\left[\arccos\left\{\frac{r-h}{r}\right\}\right] \tag{13}$$

The plug velocity of a phase is then found by dividing a phase flow by its respective cross-sectional area, and the retention time is given by the ratio between the effective separator length and this plug velocity.

Also, a dispersion retention time can be defined, as in Eq. (14), by calculating the available dispersion volume between the O/W interface setting for the dispersed [water] phase:

$$\tau_d = F\frac{A_{\mathrm{NIL}}\,(h_w - \mathrm{NIL})}{Q_w^{\mathrm{out}}} \tag{14}$$

here, $F$ is a currently unknown function depending on flow conditions [e.g., the concentration gradient within the dispersion layer, suggested as linear by Polderman et al. (12)]; $F > 1$. Setting $F = 1$ will give the volume available above the interface setting, thus being incorrect as the dispersion layer will extend below this level. The available region below NIL will depend on the suction from the water outlet, i.e., water outlet geometry and velocity, and weir height. If the concentration gradient through the dispersion layer is known, the correct volume occupied by the dispersion layer can be calculated. Note that this approach also puts stringent demands on knowing the absolute value of the interface setting which, depending on the control method, is often slightly inaccurate.

The splitting of the dispersed region into a (dense) dispersion layer and a (dilute) settling region has been developed for batch tests, where the only transport is parallel to gravity. These types of tests are common in laboratories for characterizing the separability of an oil/water system, are the origin of the "critical water cut" concept as depicted in Fig. 3, and are discussed briefly in Sec. II. In continuous systems there will also be a transport normal to gravity, and this will possibly create a concentration gradient within the dense region, as suggested by Polderman, et al. (12).

## B. Load Factors

The load factors attempt to compare different separator performances. Their origin is uncertain (possibly API). Three different load factors have been observed: the liquid (LLF), oil (OLF), and water (WLF) load factors. These are described in Eq. (15). Load factors have a unit of $(\mathrm{m}^3\mathrm{s}^2)^{-1}$).

$$\begin{aligned} LLF &= \frac{Q_1 \mu_c}{A_{\mathrm{NIL}}\, \Delta\rho} 1000 \\ OLF &= \frac{Q_o\, \mu_w}{A_{\mathrm{NIL}}\, \Delta\rho} 1000 \\ WLF &= \frac{Q_w \mu_o}{A_{\mathrm{NIL}}\, \Delta\rho} 1000 \end{aligned} \qquad (15)$$

The physical interpretation of the load factors is that a liquid flow should be transported through the interface, and this transport is augmented by the density difference and decreased by the continuous phase viscosity. The liquid load factor accounts for transport of both phases through the interface, while the other two only transport the applicable dispersed phase. Hence, the liquid load factor is in line with the new separator design philosophy proposed by Polderman et al. (12, 15), while the oil and water load factors are in line with the dispersion layer theory developed by Jeelani and Hartland (16–20) (for the corresponding dispersed phase).

## C. Binary Coalescence and Settling Hindered Systems

In systems with rapid coalescence the limiting parameter for separation will be the settling velocity (e.g., the droplet size distribution entering the separator). Hafskjold et al. (21) modeled the outlet oil quality from a continuous model separator by using batch data. The basis for their model was, in addition to Stokes' law [Eq. (8)] and the plug-flow approximation [Eq. (11)], the water concentration in the oil outlet as a function of the concentration gradient over the oil phase:

$$\phi_{\text{oil outlet}} = \frac{1}{h_{\text{oil}} - h_{\text{water}}} \int_{h_{\text{water}}}^{h_{\text{oil}}} \phi(z)\, dz \qquad (16)$$

here, $\phi$ is the fraction of dispersed phase (e.g., water as all systems investigated were oil continuous), $z$ is the height above the water level, and $\phi(z)$ is the local water concentration profile immediately upstream of the weir. This profile was modeled from batch tests as a Padé approximant [Eq.(17)], with $\phi_0$ as the separator inlet concentration and $\phi_{residual}$ as the concentration still in the oil phase after a long separation time; $t$ refers to a time in the batch tests corresponding to a retention time $\tau$ in Eq. (11).

$$\frac{\phi_0 - \phi}{\phi - \phi_{\text{residual}}} = \left[\frac{z}{h_{\text{oil}} - h_{\text{water}}} - z\right]^{a_1} \left[\frac{t}{a_3}\right]^{a_2} \qquad (17)$$

Here, $a_1$ and $a_2$ are characteristic parameters for the drop size distribution, and $a_3$ characteristic separation time. This model was tested on an offshore test separator (light crude), with good results.

## D. Dispersion Layer Theory

For the past 15 years, Hartland and coworkers have developed a theory referred to here as the dispersion layer theory (20). The theory has been developed for batch tests, and has the following assumptions:

1. The incoming fluid has a defined, pseudohomogeneous continuity. Thus, the incoming mixture consists of one defined continuous phase and one defined disperse phase. The total volume of drops entering the separator is equal to the incoming dispersed phase flow.
2. All of the dispersed phase (in drop form) has to be transported through the continuous phase layer and the interface in order to achieve separation. This is divided into steps:
   a. Transport through the continuous layer to the dispersion layer by settling.
   b. Transport through the dispersion layer by stackwise removal of the interfacial drop layer by coalescence. The dispersion layer is considered to be a packed layer with a fixed drop concentration.
   c. Transport through the interface by coalescence.
3. The settling through the continuous layer is hindered, and described by Eq (10).
4. The coalescence rate for a drop of diameter $d_0$ at the interface is either correlated from experiments or calculated by theory.

Following these assumptions, there will exist a point in time where the last drop settles at the top of the dispersion layer, the inflection point $t_i$. A mass (or rather a volume) balance is calculated from this point, describing the heights of the region boundaries:

$$\frac{dh_c}{dt} = \Psi = \Psi_i \left[\frac{\Delta h}{\Delta h_i}\right]^p \qquad (18)$$

Here, $h_c$ is the height of the coalescing interface [the interface between the two different liquid continuous regimes corresponding to $h_{water}$ in Eq. (16)], $\Delta h$ is the thickness of the dispersion layer, and $\Delta h_i$ is the thickness at the inflection point; $p$ is the interfacial coalescence index, a parameter describing the degree of packing within the dispersion layer, ranging from 0 to 1; $\psi_i$ (m/s) is the coalescence rate at the inflection point, the velocity with which the coalescing interface moves, given by

$$\Psi_i = \frac{2\,\phi_p\,d_0}{3\,\tau_0}\left[\frac{\Delta h_i}{d_0}\right]^p \tag{19}$$

Here, $d_0$ is the initial mean droplet size, $\phi_p$ the packed dispersed-phase hold-up, and $\tau_0$ the coalescence time for drops of size $d_0$.

A modeling of a light/medium crude oil in batch settling is shown in Fig. 4 (from Ref. 22).

Panoussopoulos (22) has studied several real and model systems, using this model, and reported values of $\epsilon_p$ ranging from 0.65 to 0.875 and $p$ ranging from 0.23 to 0.68. The current lack of a theory predicting these variations is considered to be the major weakness of this model.

## E. Design Philosophy for Dewatering Vessels (Developed by Shell)

In recent years, Shell has published a new design philosophy (12–15) based on extensive laboratory tests and field trials. The basis of this philosophy is close to that of Jeelani and Hartland (16–20) and looks at the transport of the dispersed phase through an interface. This philosophy has certain prerequisites, being:

1. The dispersed phase has to be appropriately destabilized (in order to make viscosity the only stabilizing factor)
2. The model includes the vertical transport of both continuous and dispersed phases.

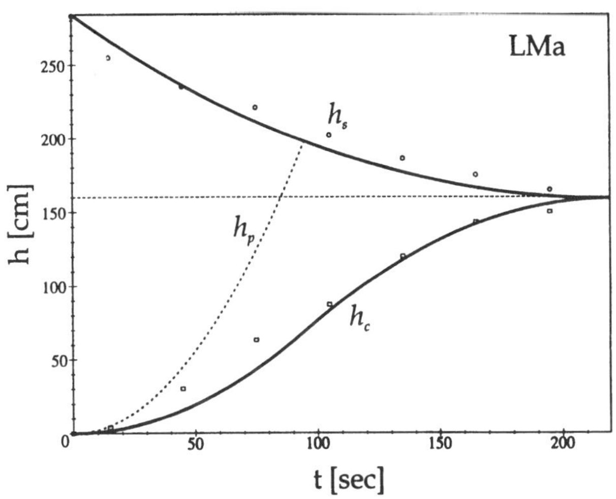


**Figure 4** Theoretical and experimental variation of the sedimenting and coalescing interface with time for a crude oil system – batch test (From Ref. 22.)

The theory is developed for vertical separators, with the oil flowing upward, opposing the settling velocity of the dispersed water. This may explain the inclusion of oil flow into the relations.

The main feature of the theory is the prediction of the dispersion layer thickness $h_d$ as a function of liquid flow rate and interface area:

$$\frac{Q_l}{A_{\text{interface}}} = \frac{h_d}{a + bh_d}$$

$$h_d = \frac{a\,Q_l}{A_{\text{interface}} + b\,Q_l} \tag{20}$$

here, $a$ and $b$ are constants depending on feed properties and operating conditions. These are determined by batch tests in the laboratory or in the field, where the decay of the dispersion band with time $dh_d/dt$ is as follows:

$$\left.\frac{dh_d}{dt}\right|_{h_d} \equiv \left.\frac{Q_l}{A_{\text{interface}}}\right|_{h_d} \tag{21}$$

Combining Eqs (20) and (21) yields the relationship (22), giving $a$ and $b$ from batch tests for a given system:

$$\frac{h_d}{\left[dh_{d/dt}\right]_{hd}} = a + bh_d \tag{22}$$

Combined with experimental and field data, Polderman et al. (15) have deduced generalized design windows for destabilized crude oils, as shown in Fig. 5. For comparison, the applicable API specification (1) suggests a design basis as shown in Table 1.

## F. Brief Discussion of Models Presented in Sections IV.C, IV.D, and IV.E

The main differences between the different models presented are as to where the coalescence processes take place, and whether coalescence or settling is the limiting factor. Hafskjold et al. (21) discusses a system where rapid coalescence takes place, and attributes the separation characteristics to binary coalescence (increasing the settling rate) and flow conditions.

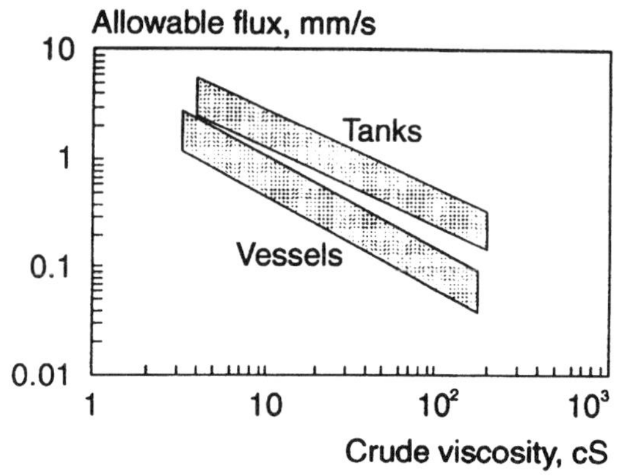


**Figure 5** Generalized design window for primary separators. (From Ref. 15, © Society of Petroleum Engineers.)

Jeelani and Hartland (20) neglect binary coalescence and focus on systems where coalescence is slow and takes place only at the interface. They distinguishes between a packed layer formed by the settled drops where coalescence takes place and a dilute layer where settling takes place. Polderman et al. (15) have a combined view, suggesting that binary coalescence is important in the inlet tube upstream of the separator and at the inlet (or that the actual break-up is reduced by increasing the dispersed phase), and that interfacial coalescence is the main parameter within the separator, together with flow conditions. Our view is that all the mentioned interpretations are valuable, and that the different mechanisms will appear in varying degrees for different systems and conditions.

However, one important shortcoming is that the quality of the dispersed-phase effluent is not accounted for by any of the models. In real operation this parameter may also be limiting, e.g., in oil continuous systems the water quality is often the limiting factor. The instrumentation used in batch tests may be inadequate for measuring dispersed-phase effluent quality, as the concentration levels are much lower in this phase. However, the various interpretations of continuous-phase behavior (in particular the degree of binary and bulk phase coalescence) would have an effect on the predicted dispersed-phase effluent quality.

**Table 1** Design Criteria for Liquid-Liquid Separation

| Oil gravities (°API) | Typical residence time, equal in both liquid phases (min) |
|---|---|
| Above 35 | 3–5 |
| Below 35, $T > 100$°F | 5–10 |
| Below 35, 80°F $< T$ $< 100$°F | 10–20 |
| Below 35, 60°F $< T$ $< 80$°F | 20–30 |

*Source*: Ref. 1.

## V. INTERNALS

Several internal components for improving the performance of gravity separators are available on the market. These address different problems that may occur within the process, and are here divided into foam handling, mist handling, flow distributing, and settling enhancing devices. Because of the costs associated with maintenance, the normal trend in design is to keep the number of internals at a minimum, and to maximize the simplicity of the process. It is therefore important to choose the correct configuration based on the actual problem at hand.

### A. Devices Used

#### 1. Foam Handling Devices

Mechanical foam handling is normally done at the inlet section, and seeks to separate the gas phase from the liquid phase by utilizing the inlet momentum. Two inlet types, traditionally regarded as efficient for foam handling, are the baffle type and the cyclonic inlets. Both types rely on a smooth reduction of the inlet momentum, to reduce the mixing energy at the inlet. The design of the baffles aims at using the difference in momentum between the gas and liquid phases by directing them separately into the separator and thereby avoiding foam generation. Cyclonic inlets force the incoming mixture to rotate, thus creating centrifugal forces enhancing the separation of the lighter gas phase from the liquids. The separated phases exit through separate outlets into the separator.

The obvious advantage of the cyclone type is that it physically separates the phases before the completion of momentum reduction; baffle-type inlets rely only on the capability to reduce the momentum smoothly. As such, it is difficult to design a baffle-type inlet that will perform satisfactorily for high loadings, especially if the mixture is highly susceptible to foaming. Cyclonic inlets may have a higher associated pressure drop than baffle-type inlets and/or occupy a larger volume. Note that these effects are strongly dependent on design.

### 2. Mist Handling Devices

Mist handling devices seek to remove liquid drops from the gas phase to meet the specifications of downstream equipment. Demisters are typically based on impingement and centrifugal separation mechanisms.

Mesh pads and filters have the highest efficiency with respect to the drop sizes they a are able to remove. They work on the impingement principle, guiding the gas through channels formed by the media, and making the liquid drops coalesce at solid surfaces. The pressure drop and efficiency are functions of the density of the mesh, and are often high. Unfortunately, mesh pads and filters have a low turndown ratio and are highly susceptible to fouling by liquid overload and clogging. The main drawback is, however, the onset of flooding which will occur at relatively low gas velocities (depending on system pressure). Flooding is characterized by a re-entrainment of the liquid resulting from high gas velocities through the mesh pad. This makes the mesh pads and filters relative large units compared to vane packs and axial flow cyclones since the gas velocity must be kept low.

Vane packs also operate on the impingement principle, but the channels are much wider, providing a pressure drop lower than that of mesh pads. These are, therefore, less susceptible to foaming, but do not have the same efficiency versus drop size (commercial claim: 10+ $\mu$m, depending on design). They are also limited with respect to liquid loading. When it comes to flooding, the vane packs are less sensitive than the mesh pads. The liquid drainage is arranged as slots to guard the liquid drain from the gas flow.

Axial cyclones are flow-through devices that set the entering fluids in rotation and separate the liquid from the gas by centrifugal forces. The liquid forms a film on the wall, and is drained through slots and flows back to the bulk liquid phase through a downcomer. These units have a turndown ratio and efficiency versus drop size (commercial claim: 4+ $\mu$m, depending on design) higher than those of vane packs. The pressure drop is the dimensional criterion for axial-flow cyclones. Axial-flow cyclones are characterized by a relatively high-flow throughput, which makes them the most compact alternative.

Finally, reversible-flow cyclones can be used for gas cleansing. These have the highest turndown ratio and good efficiency versus drop size (commercial claim: 3+ $\mu$m, depending on design). They do, however, have a pressure drop higher than that of vane packs and axial flow cyclones.

Generally, separation efficiency is a function of pressure drop – increasing the available pressure drop improves separation (until flooding occurs for the vane packs and the mesh pads). Pressure drop is often critical for demisting devices. The liquid removed from the gas will be at a lower pressure than the gravity separator as the pressure drop is the driving force of the separation. To reintroduce the liquid into the gravity separator, the static height between the demister and the liquid surface in a downcomer pipe is utilized. If the pressure drop across the device becomes larger than the liquid static height in this pipe, there will be no liquid transport in the pipe and the liquid will follow the gas.

### 3. Flow Distributing Devices

Separation is normally a function of time, as either the settling or the coalescence process is limiting. It is, therefore, imperative that the flow through a separator is controlled in some manner. The design equations presented in Sec. IV are all based on an even distribution of flow over the cross-section of the separator. A skewed inlet distribution will lead to a distribution in residence times and impaired separation.

The conventional method to avoid uneven inlet velocity distributions is to divide the separator into compartments with baffle plates that provide a low pressure drop in the flow direction. The effect of this will be briefly discussed in Sec. V.B.

### 4. Settling Enhancing Devices

Settling enhancing devices seek to improve the settling process by providing channels with reduced height. This has two effects: reducing the diameter available for the flow reduces the Reynolds number and the turbulence, and shortening the vertical space reduces the residence time required for a drop to reach its bulk phase. The bulk phase in this respect is the film that is formed on the channel surface. The channels are usually formed by several plates, which are inclined to allow the liquid film to drain vertically. Meon and Blass (23) studied the performance of inclined plates, and found that performance variation was a function of plate inclination, drop size, and flow regime. Similar structures are also used for foam suppression if the inlet does not perform satisfactorily.

## B. Modeled Flow Patterns in Gravity Separators – Impact of Various Models and Internals

The flow pattern inside a gravity separator is complex owing to the simultaneous transport of multiple, heterogeneous bulk phases, mass transfer between these phases, and the impact of various internals. This section seeks to give a qualitative description of understanding flow patterns, in the light of the different effects of these results, by case descriptions.

The simplest mechanical design basis possible is a separator with a homogeneous inflow in which gas and liquid are separated, a weir plate to provide suitable safety criteria for the oil phase, and three outlets for the respective (clean) water, oil, and gas phases. The traditional simplified view is to assume plug flow in the liquid and gas phase, and slip between the liquid and gas, thus neglecting inlet effects and the possible slip between oil and water. Furthermore, Stokes' law is used for mass transport between the bulk phases (assuming rapid coalescence). This view is depicted in Fig. 6.

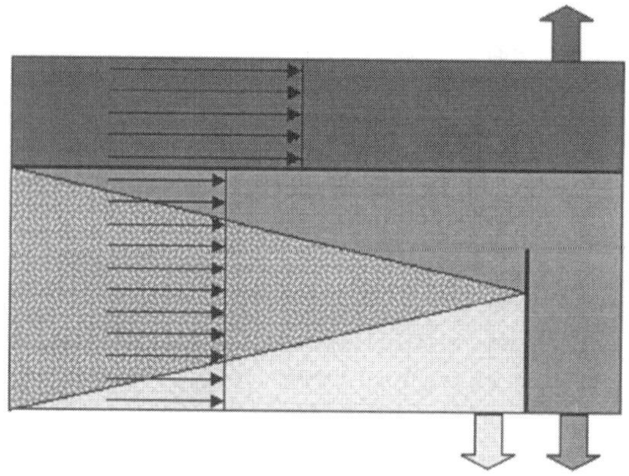

**Figure 6** Sketch of velocity profiles according to a plug-flow assumption, with cut-size calculation from Stokes' law.

This situation is idealized and outlet driven, in that plug flow is initiated at the inlet in some manner. This is certainly not the case for any inlet reported. As inlets have to handle an incoming liquid momentum from the inlet nozzle, the incoming fluid will enter the vessel in a defined region. Depending on the length and diameter of the vessel, the flow pattern will be partly driven by the inlet and outlet boundaries of the flow, as imposed by the inlet and outlet geometries/velocities. The best tool to look at these effects is computational fluid dynamics (CFD) codes. Figures 7 and 8 shows the velocity profile and velocity vectors of a CFD single-phase, three-dimensional simulation of a separator with a cyclonic inlet (thus only the liquid phase is shown, as the gas phase is assumed to be separated by the inlet). Even though the velocities are greatly reduced by the removal of the gas, the velocity profile downstream of the inlet is far from the plug-flow assumption

In CFD codes, coalescence models have not yet been implemented and multiphase solutions should be used with great care. As the hydrodynamics often are controlled by gravity and coalescence as well as by momentum, the absence of coalescence models will affect CFD results, and consequently a complete quantitative evaluation of two-phase flow through a gravity

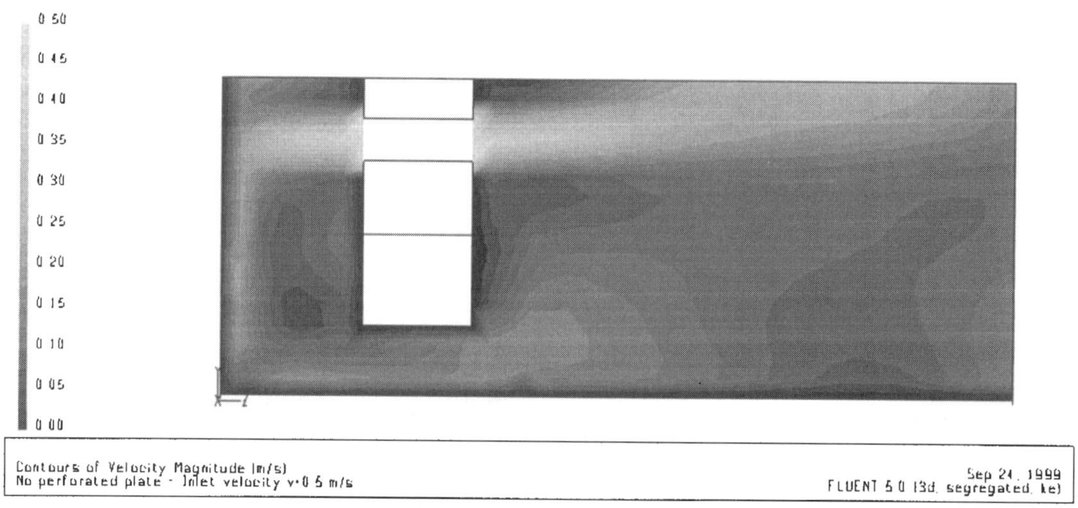


**Figure 7** Velocity profile in a simulated gravity separator inlet liquid section; inlet velocity 0.5 m/s.

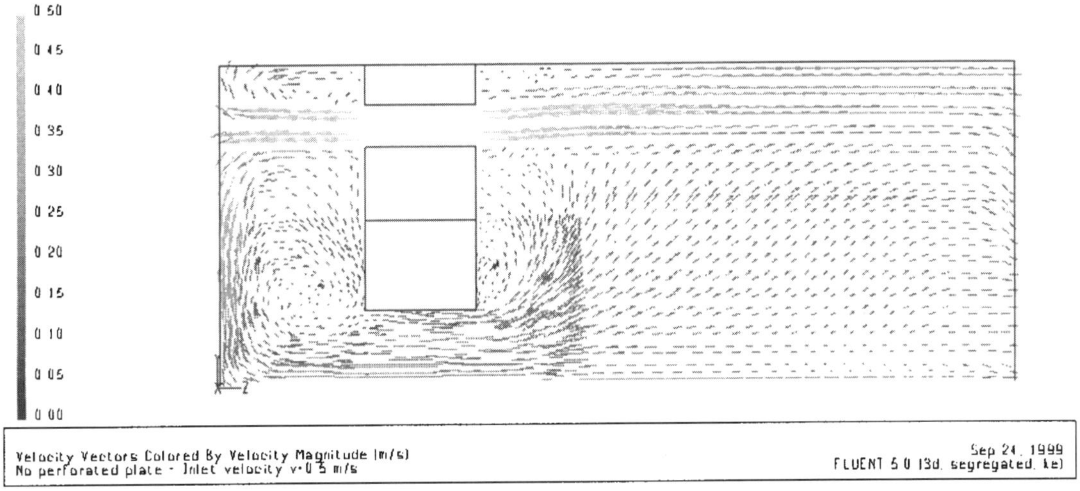


**Figure 8** Cross-sectional views of the velocity profile for the simulation in Fig. 7.

separator is still not within reach. However, sections of the separator are expected to be controlled by momentum (typically inlet and outlet regions), and CFD modeling of these parts can be expected to yield reasonable results.

The implementation of a flow distributor, such as a perforated plate, will greatly enhance the downstream conditions with regard to flow distribution. Figure 9 shows the same case as Figures 7 and 8, with a porous region resembling two baffle plates, each with 20% open area. The velocity distribution downstream of this region is rather unaffected by the inlet region, as can be seen by the figure. This is still the case when doubling the inlet velocity, as shown in Figs 10 and 11.

Moving toward the impact of the models mentioned in Secs IV.D and IV.E, these will also have significance on the modeling of velocity and phase distribution. Assuming a case where settling is adequately completed within the residence time in the oil phase, a packed layer may form near the oil/water interface. This layer will have a dispersed-phase fraction of approximately 0.7 according to Jeelani and Hartland (20).

Assuming further that the outlet-driven flow will not give local velocities large enough to draw from

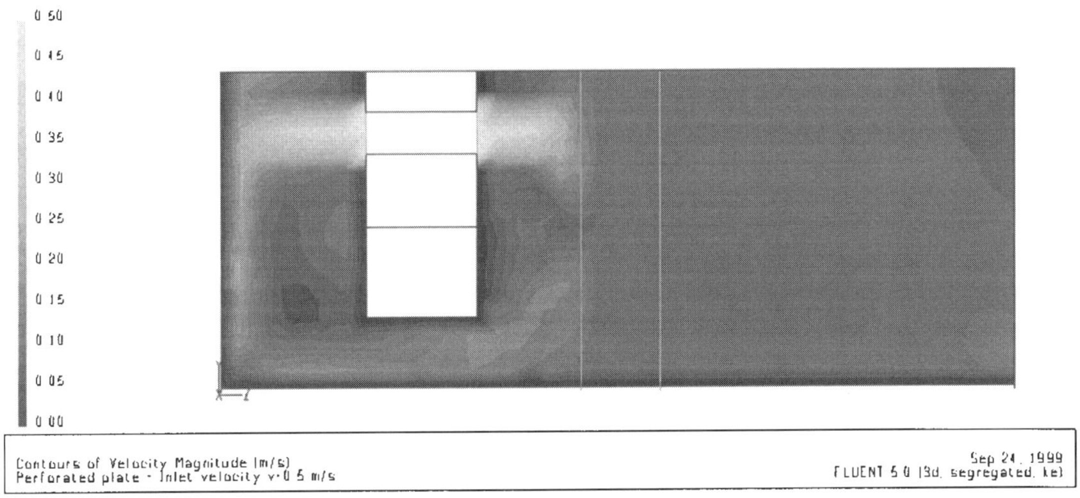


**Figure 9** Simulation of the case in Fig. 7, with porous section.

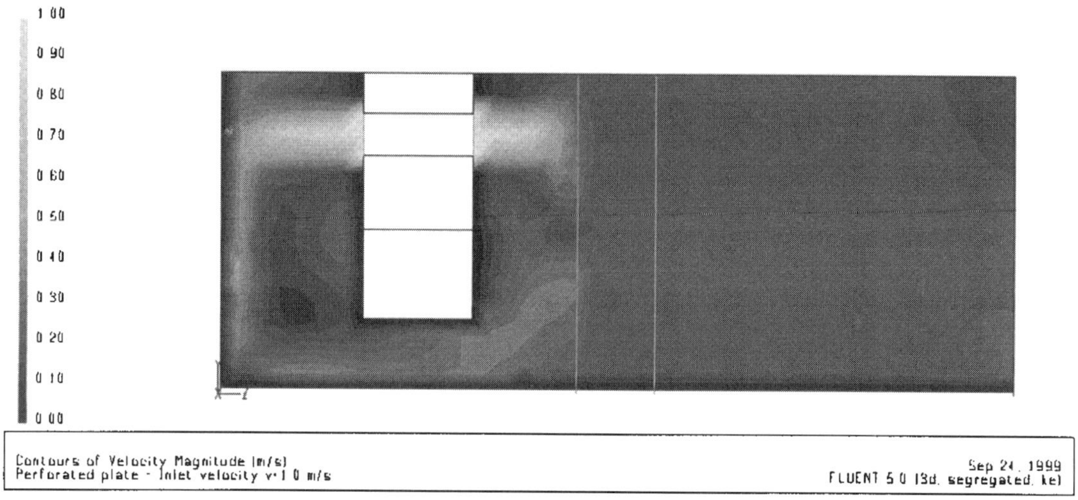


**Figure 10** Simulation of the case in Fig. 7, with added porous section; inlet velocity is doubled (1 m/s).

this layer, the source term of this layer will be the incoming dispersed phase, and the removal term will be coalescence. As the velocities found by the plug-flow assumption are very low, the dynamic head of this layer will be several orders of magnitude lower than the (vertical) force induced by gravity. The packed layer will therefore remain between two fixed horizontal planes as determined by the interface control setting.

This suggests that the dispersed layer may be treated as if it were stagnant. The momentum transfer between the "pure" bulk oil and water phase will be greatly reduced, implying that they will have independent flow behavior and may be treated separately. Also, the residence time within the packed layer will be independent of its horizontal velocity (which is assumed to be zero) and will vary only with the coalescence rate (which can be interpreted as a vertical velocity through the applicable interface). As the dispersed layer grows, it will eventually reach down to the region of suction from the water outlet or flow over the weir, and a step change in the respective outlet quality will occur.

## VI. DOWNSTREAM PROCESSING

The outlet specifications for a primary gravity separator are normally given by the downstream processing

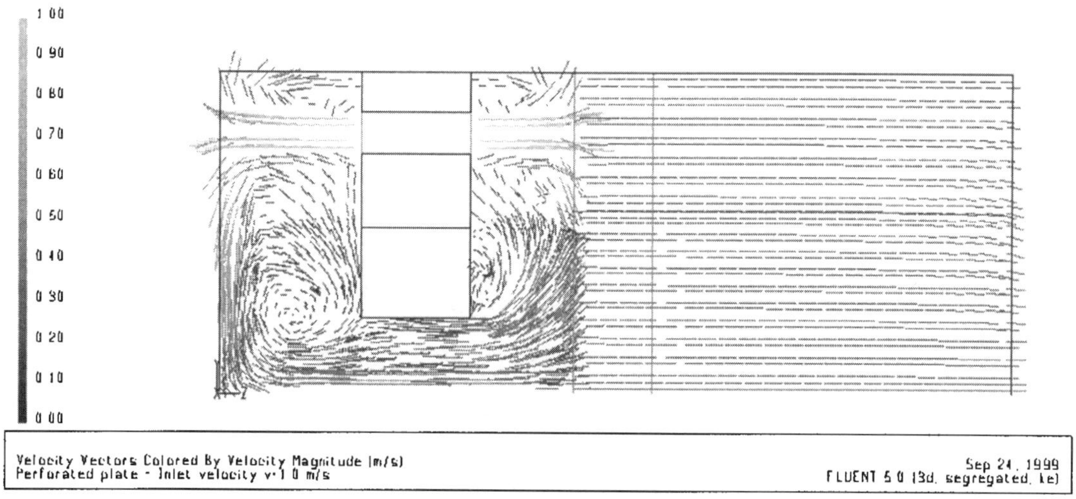


**Figure 11** Cross-sectional views of the velocity profile for the simulation in Fig. 10.

equipment, and a brief review of these conditions is in order.

## A. Water Downstream Processing

The standard downstream processing equipment on the produced water side is hydrocyclones, also known as deoiler cyclones. Standard specifications for this type of equipment are often given as parts per million levels, and normal commercial specifications are inlet qualities no higher than 1000 ppm. This should give an outlet quality no higher than 40 ppm. It is appropriate to mention that these qualities are with respect to saturated hydrocarbons. Aromatic and polar groups are not included. As such, crude oils high in aromatic and polar components will give outlet values lower than those of paraffinic crudes, and this is also reflected in government regulations in different oil-producing regions – the requirements for dumping water quality are often more restrictive in regions with traditionally aromatic crude oils.

Deoiler hydrocyclones separate oil from water by inducing a strong centrifugal field, of the order of 500–1000 *g*. As such, their performance is not necessarily a function of inlet concentration, but rather of drop size distribution and continuous phase viscosity. A high concentration will increase the coalescence rate (by increased collision frequency) and therefore give a slight improvement in performance. It is, however, imperative that the droplets are protected from excessive shear, particularly as the concentration of dispersed phase is low (1000 ppm) and binary coalescence will not usually prevail. For example, deoiler efficiency will suffer greatly if centrifugal pumps are used for upstream pressure boosting.

The deoiler hydrocyclone is considered to be at a mature technology level, and little improvement is expected in mechanical design. Current research on improving the performance tends to focus on upstream chemical additives, such as flocculants and low-viscosity solvents.

## B. Oil Downstream Processing

Downstream processing of the oil is more complicated than for produced water, as cost-effective pressure reduction versus compression work has to be considered alongside oil/water separation. The objective of the downstream oil processing is to provide the necessary quality required for shipping/transporting, usually known as "stock tank oil" or "dead oil." Normal requirements are less than 0.5 Wt % BS&W (basic sediment and water) as defined by a standard method, such as that in Ref. 24.

Downstream oil processing normally consists of subsequent gravity separator units, possibly enhancing the last step with electrostatic equipment. The water separation efficiency of these steps will normally be much lower than for the primary separator, and higher residence times are often required.

## C. Gas Processing

The gas is processed for either reinjection or sale. In both cases, the gas phase has to be dried and compressed to the relevant pressure. As mentioned above, this is an important feature of the process as compressors are among the most expensive and mechanically complex equipment at a production facility. The downstream processing of the oil is therefore often associated with minimizing the compressor work, and the number of stages and pressures at each stage is determined by the gas/liquid equilibrium. Compressors are highly sensitive to liquid following the gas, and there is often a sharp focus on removing the liquid from the gas prior to it entering the compressors.

# VII. EMERGING TECHNOLOGY

There is a strong drive in offshore processing to reduce cost and weight. This is done by exchanging large and bulky components on a platform with more compact units, or by moving them closer to the well (subsea) or indeed into the well itself (downhole). These possibilities will be briefly discussed below.

## A. Compact Separation Units

Compact separation equipment seeks to reduce weight and size, demanding less of the costly support structure. This is done primarily by reducing the retention time needed for separation, by increasing the separating forces. Typical examples are cyclones and centrifuges, creating a centrifugal force field several times the magnitude of gravity by inducing rotational flow and radial mass transfer. Also, the utilization of turbulent coalescence and electric fields has been implemented in a compact electrostatic coalescer unit. Recently, equipment of this type has become commercially available, customized for the different tasks needed. Typical compact components are:

1. Gas–liquid (liquid dominated) separation done by reverse-flow cyclonic devices operating at 5–20 *g*. These are already extensively used as inlets in ordinary gravity separators, to eliminate foaming.
2. Gas–liquid (gas dominated) separation done by axial- and reverse-flow cyclonic devices operating at 20–200 *g*. These are also used for polishing at gravity separator gas outlets.
3. Sand–liquid separation done by reverse-flow cyclonic devices operating at 10–500 *g*. These have been commercially available in the mining industry for a long time, and are extensively used for sand cleaning in the offshore industry
4. Oil/water (high oil content, water continuous) separation done by cyclonic devices operating at 50–200 *g*.
5. Oil/water (ppm oil content, "deoilers") separation done by cyclonic devices operating at 500–1000 *g*.
6. Oil/water (up to 10–20% water) separation done by compact coalescers (actually droplet growth promoters), reducing the size of downstream separators.
7. Oil/water (in principle any range) separation, done by centrifuges.

Items 1, 2, and 5 are units with a large number of applications within the oil-producing industry, while the others are at pilot unit level and will probably be installed commercially in the near future.

Cyclonic and centrifugal devices (all except item 6) work under the principle of setting a multiphase flow into rotational movement, thereby forcing the heavy phase radially outward and the light phase inward. For cyclones, the difference between reverse- and axial-flow mode is the mechanism of splitting the separated phases. The axial flow drains the heavy phase at the wall, while the reverse-flow types apply a back pressure forcing the light phase to exit countercurrently within the low-pressure core of the swirling flow. Reverse-flow types have a larger turndown, but also a larger pressure drop versus efficiency than that of axial-flow types.

All of these units have a large potential as individual components, but the major impact on cost/weight reduction would appear when using them throughout a truly compact process. The main concern of such a compact process is the lack of an accumulator volume, needed for safety during shutdown. The lack of an accumulator volume puts higher demand on the control system and makes the equipment more sensitive to variations in flow. Thus, the compactness cannot be exploited fully unless these restrictions can be modified. In addition, the lack of compact equipment managing high-pressure, oil continuous oil/water separation is also a drawback. The two available compact units for this separation regime are the compact coalescer and the centrifuge – the compact coalescer has an upward limit in water content while the centrifuge will be sensitive to operating pressure and gas content of the oil. Several attempts at producing (static) cyclonic equipment for oil continuous flows have been reported, but so far it has not been possible to prove the general handling of oil continuous flows.

## B. Subsea and Downhole Separation

Another trend is to move processing equipment to the seabed, known as subsea processing. This requires reliability with regard to process control and stability/durability. In particular, sand handling is often a concern. The first subsea separation unit is due for installation this summer, at Norsk Hydro's Troll field (25).

Downhole separation is perhaps the hottest subject in separation today. Experiments suggest that the multiphase fluids are easier to separate at conditions normally present in the well without free gas, at high pressures and temperatures (26). Prototype tests with two-stage cyclonic devices and hydraulic pumps for offshore wells up to 20,000 barrels/day have recently been performed, with promising results (27). Also, a concept involving gravity mechanisms in horizontally drilled wells is being patented (28) and is due for onshore prototype testing in the near future.

The purpose of performing separation downhole is to increase the production rate to the platform, by removing the main part of the water for reinjection. The lowering of the liquid volume will enhance the effect of gas lift, by lowering the slope of dynamic pressure versus gas flow rate, and the reduced produced volume will also reduce the strain on the existing process equipment. Cyclonic separation will, as mentioned earlier, often be constrained by the demand of a water continuous flow.

## ACKNOWLEDGMENTS

Richard Arntzen would like to thank the Norwegian Research Council (NFR) and Kvaerner Process

Systems (KPS) for a PhD grant. Per Arild Andresen would like to acknowledge the technology program Flucha financed by NFR and the oil industry. KPS is also thanked for the kind permission to publish the material underlying the chapter.

## NOMENCLATURE

Symbols, with units where applicable, are shown in Table 2.

**Table 2** Nomenclature

| Symbol | Unit | Definition |
|---|---|---|
| $A$ | $m^2$ | Area |
| $C$ | | Constant |
| $d$ | m | Droplet diameter |
| $D$ | m | Pipe diameter |
| $f$ | | Friction factor |
| $g$ | $m/s^2$ | Acceleration by gravity |
| $G$ | J/mol | Gibbs free energy |
| $h$ | m | Level height |
| $P$ | Pa | Pressure |
| $Q$ | $m^3/h$ | Volumetric flow rate |
| $r$ | m | Radius |
| $t$ | s | Time |
| $T$ | °F | Temperature |
| $u$ | m/s | Mean velocity |
| $v$ | m/s | Terminal (vertical) velocity |
| We | | Weber number |
| $\varepsilon$ | $m^2/s^3$ | Energy dissipation |
| $\mu$ | kg/(ms) | Kinematic viscosity |
| $\tilde{v}\prime$ | m/s | Turbulent fluctuation velocity |
| $\rho$ | $kg/m^3$ | Density |
| $\sigma$ | N/m | Interface tension |
| $\tau$ | s | Residence time |
| $\phi$ | | Fraction of internal phase |
| $\Psi$ | m/s | Coalescence rate |

## REFERENCES


1. Specification for Oil and Gas Separators. API Specification 12J. Washington, DC: American Petroleum Institute, 1989, p 20.
2. GA Davies, FP Nilsen, PE Gramme. The formation of stable dispersions of crude oil and produced water: The influence of oil type, wax and asphaltene content. SPE 36587. Denver, CO: Society of Petroleum Engineers, 1996.
3. DR Woods, E Diamadopoulos. In: DT Wasan, ME Ginn, DO Shah, eds. Surfactants in Chemical/Process Engineering. New York: Marcel Dekker, 1988, pp 369–539.
4. S Hartland. In: IB Ivanov, ed. Thin Liquid Films: Fundamentals and Applications. New York: Marcel Dekker, pp 663–763.
5. IC Sharma, I Haque, SN Srivatava. Colloid Polym Sci 260: 616–622, 1982.
6. TJ Jones, EL Neustadter, KP Wittingham. J Can Petrol Technol 17: 100, 1978.
7. JT Davies. Turbulence Phenomena. London: Academic Press 1972, p 363.
8. AJ Karabelas. AICLE J 24: 170–180, 1978.
9. NI Kolev. Exp Therm Fluid Sci 6: 211–251, 1993.
10. JT Davies. Chem Eng Sci 40: 839–842, 1985.
11. M Kostoglou, AJ Karabelas. Chem Eng Sci 53: 505–513, 1998.
12. HG Polderman, FA Hartog, WAI Knaepen, JS Bouma, KJ Li. Dehydration field tests on Draugen. Proceedings of SPE Annual Technical Conference and Exhibition, New Orleans, LA, 1998.
13. FH Meijs, RW Mitchell. J Petrol Technol 5: 563–570, 1974.
14. A Kumar, S Hartland. Can J Chem Eng 63: 368–376, 1985.
15. HG Polderman, JS Bouma, H van der Poel. Design rules for dehydration tanks and separator vessels. Proceedings of SPE Annual Technical Conference and Exhibition, San Antonio, TX, 1997.
16. SAK Jeelani, S Hartland. AICLE 31: 711–720, 1985.
17. SAK Jeelani, S Hartland. Chem Eng Sci 48: 239–254, 1993.
18. SAK Jeelani, S Hartland. Effect of Interfacial Mobility on thin Film Drainage. J Colloid Interface Sci 164: p 296–308, 1993.
19. SAK Jeelani, S Hartland. Chem Eng Sci 48: 239–254, 1993.
20. SAK Jeelani, S Hartland. Effect of dispersion properties on the separation of batch liquid-liquid dispersion. Ind Eng Chem Results 37:547, 1998.
21. B Hafskjold, TB Morrow, HKB Celius, DR Johnson. Drop–drop coalescence in oil/water separation. Proceedings of the 69th Annual Technical Conference and Exhibition of the Society of Petroleum Engineers. New Orleans, LA, 1994.
22. K Panoussopoulos. Chemical Engineering. Zurich: Swiss Federal Institute of Technology ETHZ, 1998, p 191.
23. W Meon, E Blass. Chem Eng Technol 14: 11–19, 1991.
24. D96–88(1998). Standard Test Methods for Water and Sediment in Crude Oil by Centrifuge Method (Field Procedure). West Conshocken, PA: American Society for Testing and Materials, 1998.
25. C Fougner. Production Separation Systems. Oslo: IBC, 1998.

26. PE Gramme. Production Separation Systems. Aberdeen, UK: Institute for International Research, 1999.
27. GI Olsen. Offshore Downhole Oil/Water Separation – DOWS – Humble Testing. Oslo: Kværner Oilfield Products, 1998, p 48.
28. T Søntvedt, H Kamps, PE Gramme, PM Almdahl. Fremgangsmåte og anordning for separering av et fluid omfattende flere fluidkomponenter, fortrinnsvis separering av et brønnfluid i forbindelse med et rør for produksjon av hydrokarboner/vann. Berg, Norway: André, 1998, p. 25.

# 28

# Compact Electrostatic Coalescer Technology

**Olav Urdahl**
*Veslefrikk Operations, Statoil, Sandsli, Norway*

**Nicholas J. Wayth**
*BP Amoco Exploration, Greenford, Scotland*

**Harald Førdedal**
*Statoil A/S, Trondheim, Norway*

**Trevor J. Williams and Adrian G. Bailey**
*University of Southampton, Southampton, Hampshire, England*

## I. INTRODUCTION

In the offshore production of petroleum, technical problems are sometimes encountered with emulsions which are formed at different stages of the production and transportation processes. These have to be taken into consideration at an early stage of the planning and construction of a platform. Enough space must be reserved for emulsion destabilization equipment such as coalescers and separators. With effective methods of emulsion separation, based on reliable information about crude oil and its tendency to form emulsions, much of this space could be reserved for other more useful purposes.

The stability of water-in-crude oil emulsions has been investigated thoroughly during the last 20 years, which has resulted in an increased understanding of the underlying mechanisms (1–17). This information could be utilized in order to develop more efficient chemical demulsifiers and, as a result, improve the separation efficiency of platforms. Another way of improving separation efficiency is to establish more refined or new methods of physical separation. In this chapter, the electrostatic destabilization of water-in-oil emulsions under flowing conditions is investigated.

## II. HISTORICAL OVERVIEW

In the petroleum industry, the first work on electrocoalescence dates back to the beginning of the 20th century when Cottrell applied external electric fields to crude-oil emulsions (18, 19). Subsequently, much effort has been made to gain a deeper understanding of the processes taking place during the breaking of oil-continuous emulsions in electric fields (20–27). Allan and Mason (21) examined the application of a d.c. electric field to a water–heptane system containing surfactant. They concluded that the rate of film drainage was significantly enhanced by the electric field, resulting in a reduced droplet lifetime. This conclusion was later confirmed by Brown and Hanson (22, 23) for water-in-kerosene emulsions subjected to an a.c. electric field.

Bailes and Larkai (24) used insulated electrodes energized by the application of a pulsed d.c. field. The main advantage of this approach is that short-circuiting, due to water slugs or droplet chains bridging the electrodes, can be avoided. Galvin (25), who also employed insulated electrodes, emphasized the importance of using a properly designed power supply.

Taylor (26) investigated the influence of high-voltage electric fields on the stability of water-in-crude oil emulsions. He concluded that two types of behavior (termed types I and II) occur, which are related to the rheological properties of the crude oil/water interface. For incompressible crude oil/water films, it was proposed that chains of water molecules formed which hindered droplet coalescence and increased the conductivity of the emulsion (type I behavior). However, efficient coalescence of water droplets was thought to be associated with a minimal increase in the electrical conductivity of the emulsion. It was suggested that this could be explained by the interfacial film being compressible (type II behavior). These findings were later verified by Chen et al. (27).

Gestblom et al. (28) used dielectric spectroscopy to investigate the behavior of concentrated water-in-oil emulsions stabilized by C9PhE4 and subjected to strong electric fields. They concluded that the breakdown process of emulsions built up gradually. The reason for this was believed to be an angular dependence of the membrane potential between closely packed droplets. This implies that droplet pairs aligned parallel to the applied electric field have the highest probability of coalescence. Further, the membrane potential was found to be directly dependent on droplet size. Thus, in a polydisperse emulsion, the electric field required to promote coalescence is inversely related to droplet size since, for a given applied field strength, the membrane potential increases with droplet size.

Conventional electrocoalescers are large vessels containing electrodes, between which a "treating space" exists where dispersed water droplets grow mainly by electrocoalescence, and a "settling zone" where phase separation takes place under laminar-flow conditions. A considerable residence time, typically 30–40 min, is required, hence the need for large vessels. This leads to problems offshore where, in order to economize on platform structure, an important issue is the reduction of weight and size of topside equipment. By decoupling the electrocoalescence and phase separation processes, it should be possible to obtain droplet growth in laminar or turbulent flow and subsequently separate the phases by using conventional equipment (centrifugal separation is also an option). This should make it feasible to reduce the size, weight, and residence time of separation equipment.

## III. THEORETICAL OVERVIEW

An understanding of the droplet size distribution created during flow mixing of immiscible fluids has long been of importance to the chemical engineering industry. The nature of both the size distribution of dispersed droplets and the mean droplet size affect many chemical processes and are of great significance to the study of coalescence and phase separation. In this section, the mechanisms which cause droplet break-up are examined and their effects compared.

### A. Droplet Break-up Processes

#### 1. Droplet Break-up due to Turbulent-flow Conditions

Kolmogorov (29) is believed to have been one of the first workers to investigate droplet break-up in dispersed systems. For turbulent flow, Kolmogorov determined the microscale eddy length to be:

$$\eta = \left[\frac{\nu_c^3}{\varepsilon}\right]^{0.25} \tag{1}$$

where $\epsilon$ is the turbulent energy dissipation rate per unit mass, and $\nu_c$ is the kinematic viscosity of the continuous phase. Kolmogorov also determined a time microscale by combining the two parameters $\epsilon$ and $\nu_c$ in a different way:

$$\tau = \left[\frac{\nu_c}{\varepsilon}\right]^{1/2} \tag{2}$$

For droplets of diameter smaller than the Kolmogorov microscale ($d < \eta$) and with a relaxation time less than the time microscale ($\tau_r < \tau$, where $\tau_r = d^2/18\nu_c$), local viscous stress forces dominate. However, for droplets larger than the Kolmogorov microscale ($d > \eta$), dynamic pressure effects dominate droplet and interdroplet processes. From calculations based on the experimental systems used in studies at Southampton University (which employed rectangular and annular ducts), estimates of the Kolmogorov microscale $\eta$ were made. In the rectangular ducts, these range from around 300 $\mu$m, at the onset of turbulence, down to around 60 $\mu$m in the outlet pipe-work at high flow rate. The droplet diameters examined ranged from around 1 $\mu$m to over 1000 $\mu$m, indicating that, under different conditions, droplet diameters may be below or above the Kolmogorov microscale and therefore that both regimes are relevant.

Hinze (30) investigated the splitting of globules under different flow regimes and identified three different types of droplet deformation: "lenticular," "cigar-shaped," and "bulgy." Lenticular deformation commences with a globule being flattened into an oblate ellipsoid. The subsequent stages leading to break-up depend on the magnitude and type of external forces causing the deformation. Hinze gave the example of a droplet deforming into a torus before breaking into many droplets. The cigar-shaped deformation is defined by elongation of a droplet into a long cylindrical thread which subsequently becomes unstable and breaks up into smaller droplets. Bulgy deformation occurs when the surface of a droplet is deformed locally; protuberances appear and the droplet becomes irregular in shape.

Hinze also discussed various well-defined flow forms and the types of droplet deformation associated with them. The flow patterns described are: parallel flow, plane hyperbolic flow, rotating flow, axisymmetric hyperbolic flow, Couette flow, and irregular flow (turbulent).

In the case of droplet deformation and break-up in dynamic pressure flow, Hinze (30) estimated the maximum stable droplet diameter ($d_{max}$) under turbulent shear conditions to be:

$$C = d_{\max}\left[\frac{\rho_c}{\gamma}\right]^{0.6}\varepsilon^{0.4} \tag{3}$$

where $\rho_c$ is the continuous phase density, and $\gamma$ is the interfacial tension between the two phases.

Hinze interpreted data from Clay (31) in order to determine a value of 0.725 for $C$ which allows the diameter $d_{95}$ in Couette flow to be deduced [see Eq. (4)]. Karabelas (32) questioned the assumptions made in Eq. (3), as turbulent flows are sometimes neither isotropic nor homogeneous. However, a number of workers have found the expression to be satisfactory. The $d_{95}$ diameter may also be expressed as a function of the Weber number:

$$\frac{d_{95}}{D} = 1.516\left[\frac{\gamma}{D\rho_0\,\bar{U}^2}\right]^{0.6}\mathrm{We}^{0.10} \tag{4}$$

where $D$ = pipe diameter, and $D\rho_c[dO]U[/dO]^2/\gamma$ = We (Weber number).

Sleicher (33) carried out experimental work using a pipe section, of length 14.6 m and internal diameter 38 mm, in which droplets of uniform size were accelerated. The diameter $d_{max}$ was arbitrarily defined as the initial drop diameter for which 20% of the droplets broke up. Taking into account viscous forces, Sleicher derived the following expression:

$$\left[\frac{d_{\max}\,\rho_c\,\bar{U}^2}{\gamma}\right]\left[\frac{\mu_c\,\bar{U}}{\gamma}\right]^{0.5} = 38\left[1 + 0.7\left\{\frac{\mu_d\,\bar{U}}{\gamma}\right\}^{0.7}\right] \tag{5}$$

This differs from Eq. (4) in that it does not account for the pipe diameter $D$. However, later work by Paul and Sleicher (34) indicated a small dependency on pipe diameter given by $d_{max} \propto D^{0.1}$. Collins (35) questioned whether the flow length was sufficient for the droplets to have reached a stable maximum size.

Following experimental and theoretical work on water-in-oil emulsions, Karabelas found the Hinze expression [Eq. (3)] more accurate than that of Sleicher [Eq. (5)]. Further, Karabelas developed the following empirical expression for $d_{max}$:

$$\frac{d_{95}}{D} = 4.0\,\mathrm{We}^{-0.6} \tag{6}$$

where again, We = $D\rho_c[dO]U[/dO]^2/\gamma$ (Weber number).

Based on experimental results, Karabelas found Eq. (6) to be superior to both the Hinze and Sleicher expressions since it offers reliability throughout the range of practical Reynolds numbers (Re). However, caution was recommended for very viscous systems where inertial forces also influence droplet break-up.

### 2. Droplet Break-up Due to Laminar-flow Conditions

Droplet break-up under laminar-flow conditions is less prevalent than under turbulent flow owing to the lower energy dissipation in the fluid. However, shear stress may still cause the break-up of larger droplets, particularly if the fluid flow rate is at a fairly high level, approaching the turbulent regime. The fluid shear, for a given geometry, is greatest at the boundaries and vanishes along the line of maximum flow velocity. Rumscheidt and Mason (36) suggested that droplets undergoing shear from a velocity gradient disintegrate when the dimensionless velocity gradient group ($N$ defined below) reaches a critical value:

$$N = \tau_0\,G\,\mu_c/\gamma \tag{7}$$

where $r_0$ is the maximum, stable, undistorted droplet radius (initial), and $G$ is the shear rate.

The critical velocity gradient $N_{crit}$ is dependent on the viscosities of the two phases and is defined as:

$$N_{crit} = \frac{1}{2}\left\{\frac{1+k}{1+19k/16}\right\} \tag{8}$$

where $\kappa$ is the ratio of dispersed-phase viscosity to continuous-phase viscosity $\mu_d/\mu_c$. $N_{crit}$ varies between limits of 0.50 and 0.42 for the entire range of viscosity ratios. For a typical water-in-oil system, $\kappa$ is at the higher end of this scale.

### 3. Droplet Break-up Due to Electrostatic Forces

An electrostatic field applied across an emulsion will place a limit on the maximum stable droplet size. The electrostatic field has a polarizing effect which creates charges of opposite polarity at opposing sides of the droplet. This elongates the droplet, in the direction of the applied electric field, and may result in break-up if the disruptive electrostatic force exceeds the cohesive interfacial force.

Rosenkilde (37) described the shape of a droplet subjected to an electrostatic field as a prolate ellipsoid (rugby ball shaped). The relationship between electrostatic forces and interfacial forces acting on droplets is described by the electrostatic Weber number defined as follows:

$$\frac{d\varepsilon\, c\, E_0^2}{\gamma} = We_{erit} \tag{9}$$

The value of the critical Weber number $We_{crit}$ was found by Rosenkilde to be 0.409 which compares well with the value of 0.41 derived by Wilson and Taylor (38). At the point of droplet break-up, Rosenkilde deduced also that the ratio of the semimajor to semiminor axes was 1.8391.

Williams (39) reviewed other electrostatic phenomena which hinder or counteract the coalescence of emulsion droplets under the action of an applied electrostatic field. These include: Taylor-cone formation, contact–separation charging, and droplet disruption due to the possession of charge. The formation of Taylor cones occurs when a conducting droplet is subjected to such a strong electric field that a conical protrusion is formed. This may subsequently lead to a jet which sprays many tiny droplets towards an oppositely charged or grounded object. These effects are well illustrated in work by Zeleny (40). Taylor (41) proved theoretically that the conical interface between two fluids can only exist at a semivertical angle of 49.29°. Experimentally, Taylor demonstrated that deviations beyond this angle result in instability.

A criterion for stability was shown to be:

$$\frac{V^2}{r_0}\gamma = C = 140 \tag{10}$$

where $r_0$ is the undistorted droplet radius, and $V$ is the applied potential in electrostatic units (1 V $\equiv$ 1/300 esu and 1 Coulomb $\equiv 3 \times 10^9$ esu).

Using a surface tension of $\gamma = 37$ mN m$^{-1}$ and a relative permitivity of $\epsilon_r = 2.2$, Taylor found the critical potential for transformer oil in his experimental apparatus to be:

$$\begin{aligned} V_{\text{crit}} &= 1.432 \times 10^3 \times (37)^{0.5} \times (1.25)^{0.5} \times (2.2)^{-0.5} \\ &= 6.5 \times 10^3\ V \end{aligned} \tag{11}$$

where $1.432 \times 10^3$ is a constant associated with the experimental geometry, and 1.25 is the height in centimeters of the upper disk electrode above the conical tip.

Experimentally, however, Taylor found the critical potential to vary slightly, two typical values being $7.2 \times 10^3$ V and $7.6 \times 10^3$ V. He explained the discrepancy in terms of molecular surface effects and imperfect insulation.

## B. Droplet Coalescence Processes

The process of droplet coalescence may involve many different mechanisms. In simple terms, there are four stages in the coalescence of a droplet pair. First, the droplets must come into very close proximity, under long-range flocculation forces, before the second stage of film thinning occurs. The rate of thinning of the continuous-phase membrane, between two droplets, is dependent on droplet size, the level of deformation of the droplets, and the force between them. At the end of the film-thinning phase, the two droplets come into contact as the fourth and final stage of film rupture, and droplet coalescence occurs.

It is appropriate to consider first the influences of long-range flocculation or collision mechanisms. Depending on the size and movement of dispersed droplets, different mechanisms will play different roles in the collision process. In a compact electrostatic coalescer (CEC), Brownian motion, sedimentation, laminar shear, turbulent shear, or turbulent inertia may play a role in droplet movement owing to hydrodynamic effects. Additionally, electrophoretic and dielectrophoretic forces, arising from the applied electric field, may act on dispersed droplets.

### 1. Brownian Motion

The bombardment of suspended water droplets, by molecules in the surrounding oil phase, will impart forces on the droplets causing them to move (Brownian motion). Small droplets are more susceptible to this effect than larger ones and this may result in collisions between neighboring droplets. Friedlander and Wang (42) investigated the effect of Brownian motion on dispersions, and the droplet size distribution was found to be self-preserving. In a CEC, the dynamic forces created by laminar and turbulent shear are likely to dominate the effects of Brownian movement.

### 2. Sedimentation

Sedimentation occurs when droplets are allowed to settle under the effect of gravity. According to Stokes' law, settling velocity is proportional to the droplet diameter squared. Larger droplets will therefore settle at greater velocities than the smaller ones, resulting in collisions between droplets of different sizes in a polydisperse system. This process is commonplace in large, conventional settling tanks. In a rapid through-flow unit, such as a CEC, sedimentary coalescence is more likely to occur at the lower flow rates, particularly for the larger water droplets. Sedimentary coalescence also occurs if a tangential inlet configuration is used in a CEC. Such an inlet design would accelerate the larger droplets to greater velocities than the smaller ones and result in collisions between the droplets. Any increase in the centripetal acceleration over gravity would produce a proportional increase in collision frequency. Even a relatively low centripetal acceleration, of say 100 m $s^{-2}$ (many times lower than that produced by a centrifugal device such as hydrocyclone or centrifuge), would still give an order of magnitude increase in collision frequency.

### 3. Laminar Shear

Droplet collision under laminar shear can occur due to velocity differences between droplets in different streamlines. The collision rate is therefore proportional to the shear rate of the fluid and this implies that droplet collisions are more likely near the walls of a duct where the velocity gradient is greatest. The work of Allan and Mason (21) is of particular interest in this respect. They investigated the coalescence of droplets subjected to laminar shear with and without an imposed electric field. Silicone oil was used as the continuous phase and water (distilled water plus 0.07% KC1) for the dispersed phase. Two counter-rotating cylinders were used to create Couette flow of the oil, into which was placed a pair of charged or uncharged water droplets. The collision and coalescence mechanisms were observed, under a microscope, with and without an applied d.c. electric field. The coalescence of droplets without shear or an electric field, due to van der Waals' forces, was also observed (although these tests were aborted owing to erratic behavior probably caused by convection currents).

The injected droplets had diameters of 750 ± 25 $\mu$m. This represents the higher end of the droplet size distribution one would expect to find in a CEC, but the results are still of interest. The injected droplets were so large that deformation resulted, which led to asymmetrical paths of approach and recession. The path of recession was found to be at a smaller angle than that of the approach. This contrasts with the behavior of rigid droplets, investigated using 500-$\mu$m polystyrene spheres, where symmetry was observed in the approach and recession paths. With uncharged droplets, and an applied electrostatic field, the coalescence angle $\theta_c$ of the droplets, shown in Fig. 1, was found to decrease as the applied potential was increased. There was also a reduction in the coalescence time of droplets as the rate of film-thinning increased. However, at the highest field strength of 1000 V/cm, the droplets were seen to approach one another before suddenly moving apart. This was thought to be due to charge exchange, whereby the droplets were left with an equal and opposite charge, causing them to move apart in the applied electric field. When the applied field was removed, the droplets were once again attracted to one another, which resulted in coalescence.

With no applied electric field, Allan and Mason (21) made comparisons between doublet interactions with both droplets charged or one charged and the other uncharged. When both droplets were charged at the same level, repulsive effects were seen which increased as the potential was raised to +250 V. Collision could be induced by increasing the shear rate from 0.15 to 4 $s^{-1}$. With one droplet charged, the paths of the droplets were the same as for uncharged droplets. However, at the highest applied potential, charge exchange was observed between the charged and uncharged droplets on contact. This left both droplets positively charged and caused repulsion between them. Coalescence could only then be induced by increasing the shear rate.

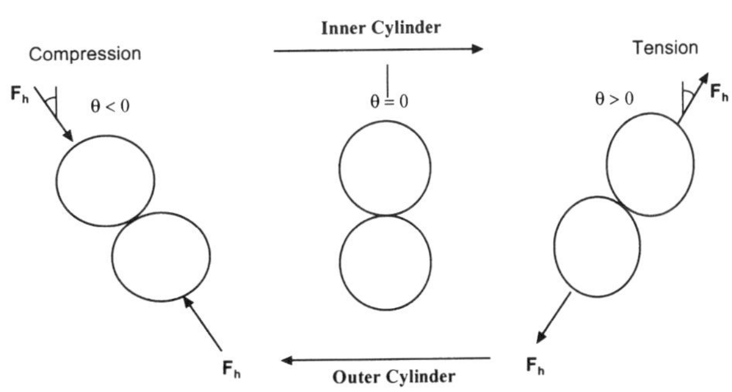


**Figure 1** Collision of droplets under laminar shear.

## 4. Turbulent Collisions

Saffman and Turner (43) considered collisions between droplets due to turbulence in rain clouds. Under turbulent conditions, droplet collision is governed by two different mechanisms: isotropic turbulent shear and turbulent inertia. The choice of regime applicable to a droplet is determined by its size in relation to the Kolmogorov microscale $\eta$ defined earlier. Droplets of diameter $d > \eta$ are subjected to the former of these processes (small-scale motion). Spatial variations in the flow give neighboring droplets different velocities and this result in collisions. Droplets of diameter $d > \eta$ are subjected to turbulent inertia. In this case, collisions result from the relative movement of droplets in the surrounding fluid. Droplets of different diameter will have different inertias and this results in collisions. Droplets of equal diameter, however, will not collide under this mechanism as they have the same inertia.

Saffman and Turner (43) produced collision expressions for droplet collisions governed by both small-scale motion and turbulent inertia. The first of these is shown below:

$$N = 1.30\,(r_1 + r_2)^3\, n_1\, n_2 \left\{\frac{\varepsilon}{\nu}\right\}^{1/2} \tag{12}$$

where $\epsilon$ and $\nu$ are as defined earlier and $n_1$ and $n_2$ are, respectively, the number densities of droplets having radii $r_1$ and $r_2$. This expression is valid for values of the ratio $r_1/r_2$ between 1 and 2. The multiplication factor 1.3 in Eq. (12) was later found to be incorrect by a factor of $\pi^{1/2}$ owing to an algebraic mistake. This was pointed out by Pearson et al. (44) who deduced a new factor of 2.3. In the case of turbulence inertia, the droplet collision rate is given by:

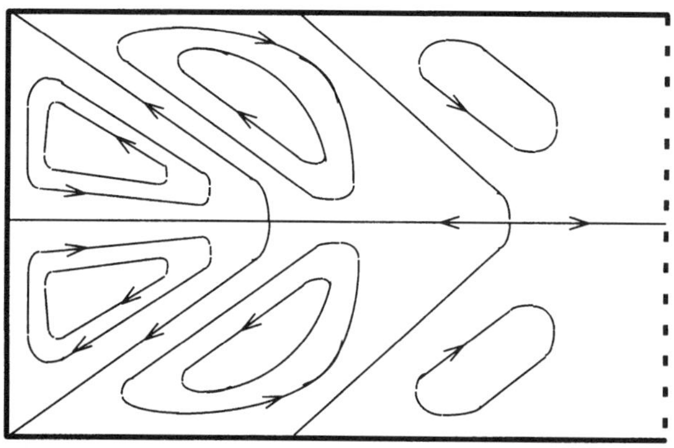

**Figure 2** Secondary flow patterns in a rectangular duct based on Prandtl (45); cross sectional view of flow.

$$N = \frac{1.27(\rho_p - p_f)}{\mu} \left\{\frac{\varepsilon^3}{\nu}\right\}^{1/2} (r_1 + r_2)^2 \mid r_1^2 - r_2^2 \mid n_1\, n_2 \tag{13}$$

The efficiency of turbulence-induced collisions was found to be equivalent to that of gravity at a turbulent energy dissipation rate per unit mass of approximately $\epsilon = 2000\ \mathrm{cm^2\ s^{-3}}$, which is equivalent to "vigorous turbulence." This shows that the turbulent growth of droplets in cumulus clouds might be sufficient to induce the formation of rain drops, but that in high-level stratified clouds would be too low to initiate rainfall.

## 5. Secondary Flow Effects

The existence of secondary flow in a duct of rectangular cross-section was deduced by Prandtl (45) following measurements made by Nikuradse (46). There is a tendency for the liquid to flow toward the corners of the duct before returning to the center (as shown in Fig. 2). In the corners corners of the duct, where the shearing stress is less, flow moves from the inside to the duct wall. Where the shearing stress of the boundary is greatest, the flow is forced to the center of the duct due to turbulence.

The effects of secondary flow on droplet collision and coalescence mechanisms have not been considered in the literature currently reviewed. The scale of secondary flow is much larger than the Kolmogorov microscale $\eta$ and it will, therefore, only affect droplets of diameter d $> \eta$ (those subject to turbulent inertia). Secondary flow will be most prevalent following changes in duct geometry, particularly where there is some form of duct divergence.

## 6. Comparison Between Collision Mechanisms

Pearson et al. (44) compared the collision functions of various collision mechanisms, shown in Table 1. It is interesting to note that, although all the mechanisms shown are dependent on the continuous-phase properties, the droplet sizes, and the flow conditions, only sedimentation and turbulent inertia are dependent on the density difference between the dispersed and continuous phases. These two mechanisms only occur where the droplets are of different size, and their collision functions tend to zero as the droplet sizes become closer. From the collision functions of these two mechanisms it is seen that turbulent inertia will only dominate sedimentary coalescence when the characteristic acceleration is greater than that of gravity:

**Table 1** Collision Functions and Dimensional Parameters for Various Processes

| Mechanism | Collision function ($\beta$) | Source | Dimensional parameter |
|---|---|---|---|
| Brownian motion | $\frac{2kT}{3\mu}\frac{(r_i+r_j)^2}{r_i r_j} = 4\pi(D_i+D_j)(r_i+r_j)$ | Smoluchowski (1916) | $K_B = \frac{kT}{\mu}$ |
| Laminar shear | $1.33\,G(r_i+r_j)^3$ | Smoluchowski (1917) | $G$ |
| Pure strain (extension) | $4.89\dot{\gamma}(r_i+r_j)^3$ | Zeichner and Schowalter (1977) | $\dot{\gamma}$ |
| Isotropic turbulent shear | $2.3(r_i+r_j)^3\left(\frac{\varepsilon}{\nu}\right)^{0.5}$ | Saffman and Turner (1956) | |
| Turbulent inertia | $1.27\frac{(\rho_p-\rho_f)}{\mu}\left(\frac{\varepsilon^3}{\nu}\right)^{0.25}(r_i+r_j)^2\lvert r_i^2+r_j^2\rvert$ | Saffman and Turner (1956) | $\frac{\rho_p-\rho_f}{\mu}\left(\frac{\varepsilon^3}{\nu}\right)^{0.25}$ |
| Differential sedimentation | $\frac{0.7g(\rho_p-\rho_f)}{\mu}(r_i+r_j)^2\lvert r_i^2+r_j^2\rvert$ | Findeisen (1939) | $K_{ds} = \frac{g(\rho_p-\rho_f)}{\mu}$ |

$$1.27\left[\frac{\varepsilon^3}{\nu}\right]^{0.25} > 0.7g \tag{14}$$

The comparison between different collision mechanisms is examined further in the following section.

## 7. Collisions Due to Electrostatic Forces

Under an applied electric field, a droplet may be subject to two different electrostatic forces, depending on whether the drop is charged or neutral. Electrophoresis is the motion arising from the force exerted on a charged drop by the applied field:

$$F = \frac{q_1 q_2}{4\pi\varepsilon_0 r^2} \tag{15}$$

where $q$ is the droplet charge. Clearly, the direction of motion is dependent on polarity of the charge and the applied field.

For a droplet charged by direct contact with an electrode, the predominant means by which a charging is likely to occur, the resultant force may be rewritten as:

$$F = \left[\frac{\pi^2}{6}\right]4\pi r^2\varepsilon_c E^2 \tag{16}$$

This force will cause a charged droplet to migrate toward an oppositely charged electrode. In doing so, it is likely to come into contact with other droplets, leading to coalescence. In a d.c. electric field, the droplet will migrate in a continuous path with its velocity determined by the viscosity of the continuous phase. The droplet will gradually lose its charge, depending on the relaxation time $\epsilon/\sigma$ of the continuous phase, and the driving force will diminish. In the case of an a.c. electric field, a charged droplet will tend to oscillate about its mean position between the electrodes.

A droplet may become charged by other mechanisms such as: ionization, preferential adsorption of ions at the interface (electric double layer), and droplet disintegration.

Neutral droplets can also be made to collide by inducing a dielectrophoretic force of interaction between neighboring droplets which arises from the polarization of the droplets in the applied electric field. The local electric field must be nonuniform, and the presence of the droplets will distort the field even if it is uniformly applied. The force, which is independent of field polarity, depends on the permittivity $\epsilon_c$ of the continuous phase and the volumes of the droplets. At larger separations, dielectrophoretic forces tend to be small in comparison with electrophoretic ones. However, at very close proximity, dielectrophoretic forces will dominate. The dielectrophoretic force acting on a droplet is given by:

$$F = 24\pi\epsilon_c E^2 r^2\left(\frac{r}{\delta}\right)^4 \text{ (large droplet separation)} \tag{17}$$

and at small separations ($\delta << r$) by:

$$F = \frac{2}{9}\pi^5 \varepsilon_c E^2 r^2 \frac{(r/h)}{\ln^2(r/h)} \quad \text{(small droplet separation)} \tag{18}$$

The different mechanisms described above have been compared graphically in Fig. 3. The factor $\epsilon_c E^2 r^2$ has been set to unity for all three mechanisms, and a relative force is plotted as a function of $r/h$.

The coalescence force $F$ between two aligned droplets of equal size (radius $r$) in an applied electric field $E$ was given, in electrostatic units, by Waterman (47) as:

$$F = 6\varepsilon_c E^2 \frac{r^6}{\delta^4} = 6\varepsilon_c E^2 r^2 \left\{\frac{r}{\delta}\right\}^4 \tag{19}$$

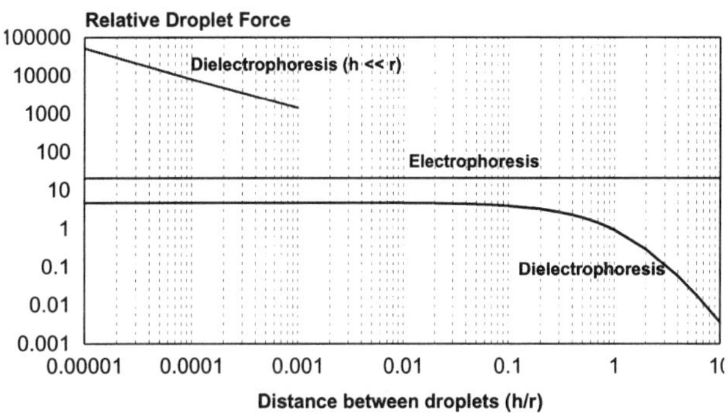


**Figure 3** Comparison between droplet forces under electrophoresis and dielectrophoresis.

where $\epsilon_c$ is the dielectric constant of the continuous phase, and $\delta$ is the distance between droplet centers. Note that this differs from Eq. (17), which is presented in SI units, only by the factor $4\pi\epsilon_0$, where $\epsilon_0$ is the permittivity of free space. The difference only arises because of the different conventions used for the units. The $r/\delta$ part of Eq. (19) is proportional to the cube root of the water cut, and is therefore independent of average droplet size, unless sedimentation and phase separation occur.

The angle between two polarized equisized droplets, in relation to the applied electric field, plays a large role in the resultant force between them and therefore in their chance of collision. Work by Krasny-Ergen (48) gives zones of dipolar attraction and repulsion, as shown in Fig. 4. At large droplet separations, an attractive force exists for angles between $\theta = \pm 54.7°$ from the direction of the applied electric field $E_0$. For droplets in contact, Krasny-Ergen gave the equivalent angle as $\theta = \pm 75.1°$ (the limiting angle must vary as a function of the droplet separation). In both cases, the force between neighboring droplets is greatest when they are aligned with the electric field ($\theta = 0°$). The presence of regions of repulsion is significant as it will hinder the collision and coalescence of droplets if they are outside the regions of attraction. However, a torque is established for droplets which initially repel one another. This rotates the droplets relative to one another so that the angle between them reduces, and attractive forces result. Fluid forces may also rotate droplet pairs into different angular orientations.

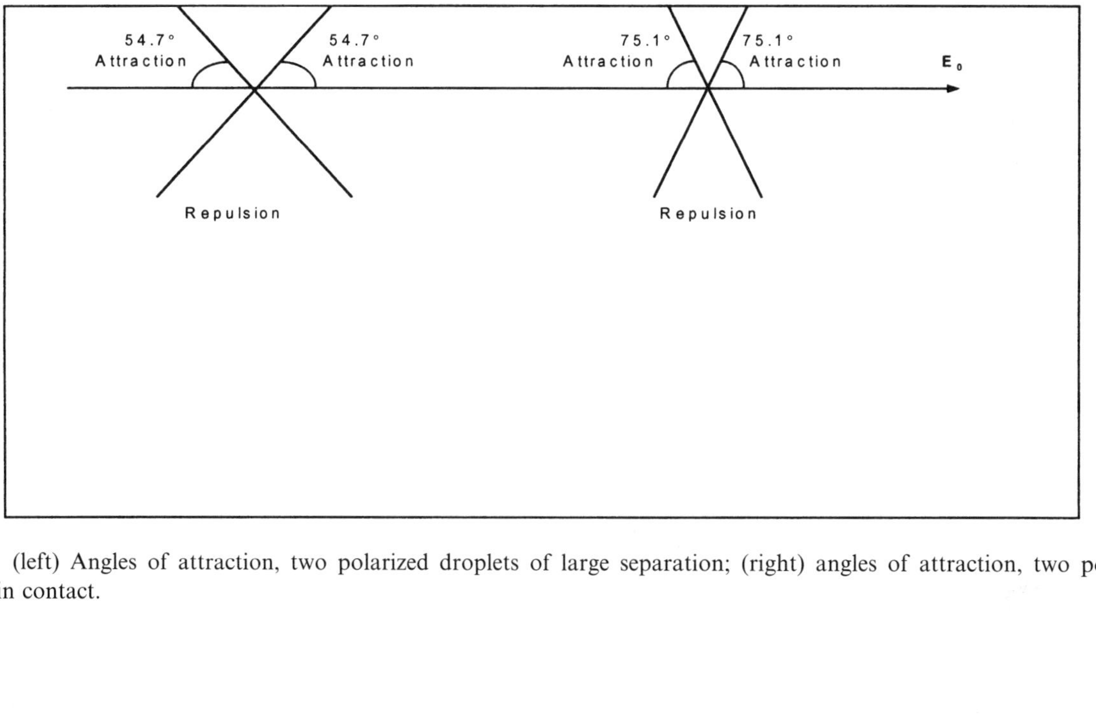


**Figure 4** (left) Angles of attraction, two polarized droplets of large separation; (right) angles of attraction, two polarized droplets in contact.

### 8. Film Thinning and Droplet Coalescence

Once long-range flocculation of droplets has taken place, due to whatever hydrodynamic and electrostatic forces are present, film thinning occurs between adjacent droplets. The chances of coalescence will depend on the rate of film thinning and the forces holding the droplets together. The film-thinning rates depend on whether or not droplet deformation occurs and were considered by Williams (39):

$$\frac{dh}{dt} = \frac{-8\pi\gamma^2 h^3}{3\mu_c r^2 F} \quad \text{(deformable droplets)} \quad (20)$$

$$\frac{dh}{dt} = \frac{-2Fh}{3\pi\,\mu_c\,r^2} \quad \text{(nondeformable droplets)} \quad (21)$$

Equations (20) and (21) may be considered in terms of the electric field strength applied across a dispersion:

$$\frac{dh}{dt} = \frac{-12\gamma^2 h^4 \ln^2(r/h)}{\pi^4 \mu_c\, r^5 \varepsilon_c\, E_0^2} \quad \text{(deformable droplets)} \quad (22)$$

$$\frac{dh}{dt} = \frac{-4\pi^4 r\varepsilon_c E_0^2}{27\,\mu_c \ln^2(r/h)} \quad \text{(nondeformable droplets)} \quad (23)$$

Williams plotted the film-thinning time for deformable and nondeformable droplets against droplet radius. While an increase in droplet size increases the time required for thinning of a deformable droplet, nondeformable droplets experience a reduction in film thinning time as their size increases. It is interesting also to note the square relationship on thinning rate with nondeformable droplets and an inverse square relationship for deformable droplets. Clearly, increasing the applied field across a system with deformable droplets could result in a reduction in coalescence efficiency.

Oweberg et al. (49) looked at droplet coalescence mechanisms. By pressing together two droplets suspended on platinum wires (using a rack and pinion arrangement) and applying an electrical potential, the mechanisms of coalescence were studied using a high-speed camera. As water drops were held together the interface between them was seen to flatten and a "lens" appeared. This eventually disappeared and the two droplets coalesced. Oweberg described coalescence as the formation of intermolecular bonds across the interface between the drops. Two mechanisms were then described, by which bonds could be switched from within the droplets to across the interface. In the first mechanism, bonds were assumed to be broken then reformed in a process equivalent to evaporation followed by condensation. This mechanism occurred above a threshold potential difference (typically 6 V), and the rate of coalescence was found to be proportional to $\epsilon_d V^2$ (where $\epsilon_d$ is the relative permittivity of the water forming the droplets). In the case of the second mechanism, at potential differences below the threshold level, bonds were assumed to be gradually rearranged rather than broken in a process equivalent to diffusion. The rate of coalescence in this instance is given by $(\epsilon_d - 1)^{1/2} V$. From Oweberg's work it is clear that the rate of coalescence is increased as the potential difference between two adjacent droplets is increased.

Kitchener and Musselwhite (50), following work by Mason et al. (21, 36), examined the approach of two dispersed droplets. Three situations were discussed, the first for large drops where the inertial forces outweigh the surface forces. Here, concave dimpling occurs (Fig. 5a) and liquid is trapped between the deflections. Coalescence will occur on the ring of the dimple, which is the thinnest area. If the droplets are smaller, they are depressed by contact but remain convex (Fig. 5b). Coalescence takes place on the center line of the two droplets, the closest point of contact, as film drainage occurs. Slowly moving larger droplets also coalesce in this way. In the third situation (Fig. 5c) a thin liquid lamella forms between the droplets. This tends to occur in the presence of surfactants.

## C. Electrostatic Separation of Water-in-Oil Emulsions

A multitude of different methods have been used to separate oil from water and water from oil. These techniques include gravity differential (settling and centri-

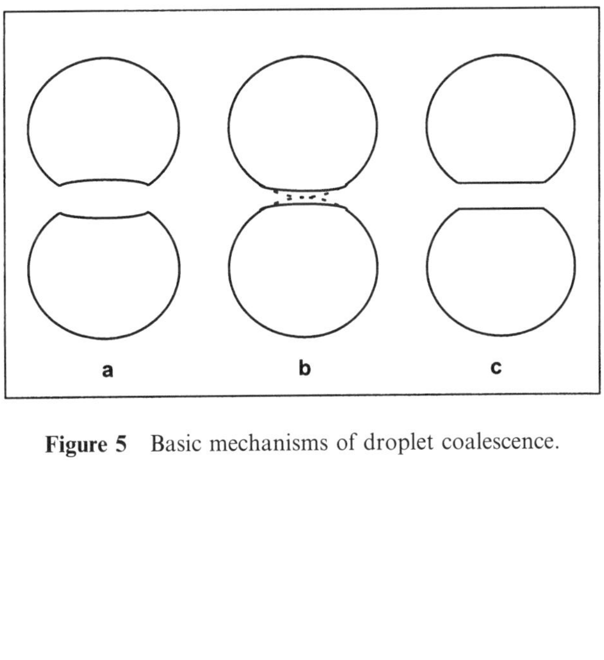


**Figure 5** Basic mechanisms of droplet coalescence.

fugal), as well as filtration, membrane, ultrasonic, thermal, adsorption, electromagnetic, viscosity actuated, and Coanda techniques. Despite the myriad techniques available, both novel and conventional, the main technique employed to separate water from oil continues to be gravity separation using settling tanks, often enhanced by an electrostatic field, increased temperature, or destabilizing chemicals.

### 1. Conventional Electrostatic Dehydrators

As already reviewed, there are many papers on certain aspects of coalescence, but the literature available specifically on electrostatic coalescers is mainly in the form of patents. One of the most comprehensive papers is by Waterman (47), which discusses both commercial and scientific aspects (a rarity). Waterman explains the role of electrostatic coalescers in removing salts such as those of sodium, iron, and arsenic. Two coalescence mechanisms are explained: first, induced-dipole coalescence, which occurs in both a.c. and d.c. electric fields, and second, the coalescence resulting from the force produced by a unidirectional (d.c.) electric field acting on a charged droplet. The latter process is ineffective in an a.c. field. Dipole coalescence has been shown to be the dominant force, as coalescence occurs at least as efficiently in an a.c. electric field. Waterman developed one of the first models for electrocoalescence.

Sadek and Hendricks (51) were also responsible for developing a model for the electrical forces on suspended droplets. Taylor (26), again within an oil-industry context, carried out tests in which an electric field was applied to water-in-crude oil samples under a microscope. Three crude oils were used: Ninian, Kuwait, and Romashkino. Tests were carried out with 5% water at an applied voltage of 1 kV and with two additives. Two types of coalescence behavior were observed as discussed earlier. Type I behavior was defined as being related to droplet-chain formation. This caused an increase in emulsion conductivity and occurred in oils with incompressible interfacial films. Type II behavior was observed with low emulsion conductivities in high-strength electric fields, where droplets coalesced too quickly to form droplet chains. Taylor's joint work with Mohammed et al. (52–54) and Chen et al. (27) looked at many of the fundamental surface-chemistry topics relating to the dewatering of crude oil. Taylor (55) provides a comprehensive review of work in this area from both an industrial and academic viewpoint.

Mori et al. (56) carried out tests to break W/O emulsions in a small sample cell. Kerosene and 50 mol/$m^3$ hydrochloric acid were emulsified by using Rheodol SP-O10 surfactant (equivalent to ICI's Span 80). Tests were carried out at frequencies between 40 and 2000 Hz at potentials of up to 8 kV. Coalescence was found to be enhanced with increase in frequency. Taking into account power requirements, 1000 Hz was found to be the most effective operating frequency. Phase separation was found to be faster for a smaller initial hold up of water but, with an aqueous content of less than 40%, coagulation occurred before coalescence and this slowed the process.

Wang et al. (57) investigated the demulsification of W/O emulsion by using an intense a.c. electric field. Their laboratory test cell consisted of an acrylic tube (7 cm in diameter and 10 cm high) with a metal plate attached to the base which acted as a grounded electrode. The energized electrode, which was insulated, was rather elaborate. It was formed by suspending an insulating beaker in the cell 2 to 6 cm above the grounded electrode. A copper wire was passed into the beaker to make contact with conductive aqueous sodium chloride solution contained inside. Silicone oil was floated on top of the liquid electrode to insulate the operator from electric shock. An emulsion was formed by suspending a mixture of an electrolyte (sulfuric acid) and deionized water (which formed the aqueous phase) in an organic phase of paraffin. Span 80 and ECA4360, both commercially available surfactants, were used to stabilize the emulsion. A mechanical homogenizer was used to shear the dispersed aqueous phase and vary the droplet size. Although not clearly stated, it would appear that all the tests were carried out at an aqueous phase concentration of 50% by volume. Measurements were made of the resolution time for the emulsion under varying conditions of electric field strength, initial droplet size, electrolyte concentration, and surfactant type and concentration.

The demulsification rate ($k_w$) was found to increase as a function of electric field strength as follows:

$$k_w \propto E^{0.626} \tag{24}$$

Similarly, $k_w$ was found to increase as the initial droplet size was increased from a mean of 14.4 to 27.0 $\mu$m:

$$k_w \propto d^{2.21} \tag{25}$$

The exponent 2.21 determined by Wang et al. (57) was slightly lower than found by other workers; Hano et al. (58) and Fujinawa et al. (59) deduced $k_w \, \alpha \, d^{3.5}$ and $k_w \, \alpha \, d^3$, respectively.

The increase in aqueous-phase electrolyte level was found to reduce $k_w$, despite increasing the density of the water and therefore the density difference between the two phases. Wang et al. claim that the reduction in $k_w$, at higher electrolyte concentrations, is due to an electric shielding effect which results in a reduction of the electrostatic force. However, the increase in electrolyte level will clearly affect both the physical and electrostatic properties of the aqueous phase, and this may explain the reduction in performance. A number of physical changes are likely. First, the overall conductivity of the emulsion increases, causing a reduction in the effective electric field strength across the emulsion. The nature of the electrical double layer may also change, perhaps increasing the repulsion between neighboring droplets. Additionally, the interracial tension between the two phases will be affected and this may further enhance emulsion stability.

## 2. Pulsed d.c. Waveform Systems

Bailes and Larkai (24) first experimented with the use of a pulsed d.c. waveform applied to a (W/O) emulsion. Early trials by these workers discovered problems with the use of d.c. and bare electrodes (in contact with the emulsion). Conducting droplets eventually created a short circuit from one electrode to the other or from an energized electrode to a nearby ground. These obstacles were overcome by the use of insulated electrodes and pulsed d.c. energization. Tests were carried out with acrylic insulation thickness of 3, 6, 10, and 13 mm. Coalescence was found to be optimized when the d.c. applied voltage was modulated at low frequency. With a steady d.c. field, interfacial polarization occurs. This is a process whereby the insulation is charged to the opposite polarity of the adjacent electrode, and the electric field across the actual emulsion is greatly reduced, effectively ending electrostatically enhanced coalescence processes. Bailes and Larkai carried out experiments with two W/O systems. System A was based on Escaid 100, a kerosene-type hydrocarbon, with cyclohexane as the organic phase and water as the aqueous phase. System B was based on Escaid 100 with LIX 64N as the organic phase, and sulfuric acid in water as the aqueous phase. Tests were carried out with square, triangular, and semisinusoidal waveforms. Performance of the electrocoalescer was assessed in terms of the dispersion band depth in a subsequent gravity settling tank (a small dispersion band depth corresponds to efficient coalescence and vice versa).

This was extended by further work (60), which led to the formation of theoretical and experimental optimum frequencies for the pulsed d.c. system. An experimental optimum frequency for the system was found to lie between 8 and 10 Hz. At higher frequencies it was suggested that coalescence-enhancing droplet chains cannot form while, at lower frequencies, the droplet chains produce a current leakage path. A model was developed using a term for average droplet spacing and a function for the work done per collision, the force being produced by the applied electric field. The average number of collisions $N$ was given as:

$$N = \frac{d \cdot I_m}{2.89\varepsilon\varepsilon_0 \cdot v_c \cdot E_{\max}\,\phi[(0.74/x)^{1/3} - 1]} \tag{26}$$

where $d$ = distance between electrodes, $I_m$ = mean conduction current, $E_{max}$ = peak electric field strength, and $\phi$ = fractional water hold-up.

For the system used by Bailes and Larkai (60), with $\phi = 0.5$ and an electrode area of $A$ in contact with the emulsion, this becomes:

$$N = \frac{5}{\varepsilon\varepsilon_0 A} \bullet \frac{I_m}{E_{\max}} \tag{27}$$

The Bailes and Larkai model incorporates a number of assumptions such as the use of a monodispersion and uniform interdroplet spacing. However, developing a model incorporating a typical droplet distribution with random droplet spacing would be significantly more complicated. No attempt is made, either, to incorporate the effects of flow velocity or regime, and the experimental results do not indicate whether tests were carried out in laminar or turbulent flow (though laminar flow can be deduced). These parameters would also have had an effect on droplet collision frequency, and therefore the rate of coalescence.

Bailes and Larkai (61) investigated the effects of dispersed-phase hold-up. The optimum applied pulsed d.c. frequency was found not to be affected by the level of dispersed water hold-up. However, a minimum threshold level for water content (25%) was found, above which the best coalescence performance was produced. This was explained in terms of the drop size, which increases with rise in water cut, and the effective electric field, which reduces with rise in water cut. The optimum frequency for efficient coalescence was in the range 4–5.5 Hz. This is lower than the earlier value (8 Hz) as an acrylic insulation thickness of 3 mm rather than 6 mm was used.

Joos and Snaddon (62) did not agree with the ideas put forward by Bailes and Larkai (24, 60) to explain their experimental results. They argued that coalescence performance was an average field-strength effect and was not dependent on field frequency. The high-voltage pulsed d.c. power supply, used by Bailes and Larkai, incorporated a 100-Ω current-limiting resistor which connected the stabilized d.c. supply to the switching circuit. As the operational frequency was increased, the effective electric field applied to the emulsion reduced with consequent reduction in coalescence performance. If a smaller current-limiting resistance value had been chosen, the "optimum frequency" would have been increased. At frequencies below the "optimum frequency," interfacial polarization reduced the effective field strength across the emulsion by causing charge to build up at the emulsion/insulation interface (interfacial polarization). Joos and Snaddon produced a model based on Bailes and Larkai's work and argued that coalescence is a function of the mean value of the square of the effective electric field. Using a model based on Bailes and Lankai's work, they found an optimum frequency of 22 Hz, somewhat higher than the empirically derived 8 Hz. This is illustrated in Fig. 6 which shows an effective electric field applied to the emulsion as a function of operational frequency. It can be seen that at low frequencies the effective field strength is small owing to interfacial polarization. As the frequency is increased the effective field strength rapidly increases, reaching a maximum before it decreases due to the current-limiting resistor. Joos and Snaddon pointed out that their optimum frequency would be reduced from 22 to 8 Hz if the effective emulsion capacitance or resistance were to be increased by a factor of about 5.

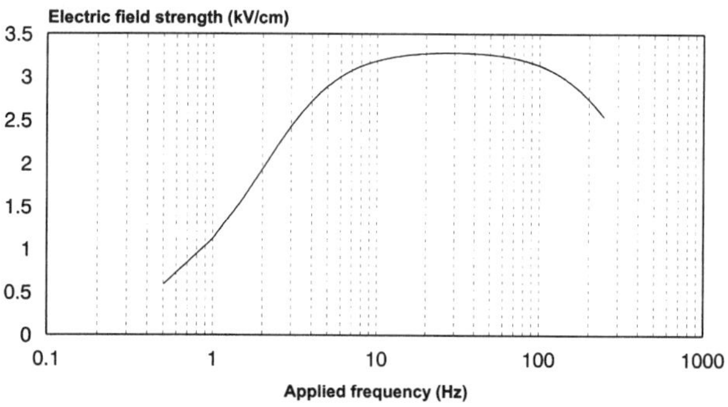


**Figure 6** Frequency effects on effective electric field strength. The declining influence of interfacial polarization and the increasing influence of the current limiting resistor (10 MΩ) with frequency on effective field strength. Based on $\phi = 0.15$.

Bailes (64) has developed a mathematical model to explain previous experimental findings (24, 60). Taylor (55) provided further explanations of why there should be an optimum frequency. Gomis (65) investigated the work of Bailes and also of Joos and Snaddon. He pointed out that the model produced by Joos and Snaddon did not predict other trends found. For example, it did not explain why the optimum frequency is less critical at higher voltages, and why the optimum frequency is less for a thinner layer of insulation. Gomis extended Joos and Snaddon's model to include these parameters. The Gomis model is dynamic and therefore takes account of the applied electric field at all times. This is opposed to the Joos and Snaddon model which uses a time-averaged mean electric field value.

Drelich et al. (66) also performed tests on a laboratory rig to investigate the optimum frequency of a pulsed d.c. electric field on W/O emulsion separation efficiency. A mixture of 0.08–0.2 wt% distilled water and an aromatic extraction solvent were emulsified. The resulting emulsion was allowed to settle for 40 min to remove any large droplets. The viscous nature of the organic phase ensured that complete separation did not occur in this time. The emulsion was then pumped through an electrostatic cell of dimensions 150 mm (length) × 100 mm (width) × 70 mm (height). A bare cathode was fitted to the base of the cell and an insulated anode was fitted at the top of the cell. Epoxy resin was used to provide insulation thicknesses of 0.2 and 2.0 mm. A high-voltage pulse generator was used to apply an electric field across the emulsion at potentials of up to 20 kV and at frequencies between 5 and 25 Hz. The emulsion was then passed through a settler, and the separation efficiency was determined from the expression:

$$\text{Separation efficiency} = \frac{\phi_{inlet} - \phi_{outlet}}{\phi_{inlet}} \tag{28}$$

A sharp increase in separation efficiency, from about 20 to 60%, was reported when the electrostatic field strength was increased from 0.32 to 1.33 kV/cm. When the field strength was increased further, up to a value of 10.6 kV/cm, only a small increase in separation efficiency was seen. This implied the presence of an optimum field strength. Additional measured values between electric field strengths of 0.32 and 1.33 kV/cm would have been useful since the critical value may have been significantly lower than 1.33 kV/cm. Bailes and Larkai (61) reported critical field strengths of 0.3 kV/cm for concentrated emulsions and about 1 kV/cm for emulsions with a water hold-up of $\phi < 0.09$.

Drelich et al. (66) suggested that separation performance was optimized with pulsation frequencies of between 8 and 11 Hz, though not by more than 5–7%. They concluded that this improvement is of little practical significance. The paper fails to give details of the power supply and electric circuit used. Thus, it is not clear whether factors other than coalescence processes may have been influenced by the variations in frequency.

The question arises, therefore, whether an optimum frequency exists beyond that defined by the power-supply circuitry at high frequencies and the effects of interfacial polarisation at low frequencies.

## IV. TECHNOLOGY STATUS

The only compact electrostatic coalescer that is commercially available at present is the Electro-Pulsed Inductive Coalescence (EPIC) (made by the National Tank Company (NATCO)). The EPIC device has a number of patents filed at this time; Ref. 67 shows a single-annulus down-flow unit. The W/O emulsion is injected tangentially at the inlet and swirls between the inside of the outer tubular vessel and an insulated inner electrode tube. An electric field is applied across the emulsion and that provided by a pulsed d.c. voltage is said to be preferable. It is claimed that this unit can improve water separation rates by as much as 1250% over conventional methods. Patents (68, 69) again relate to the EPIC device described in Ref.67 and, in addition, a double-annulus unit is described. As before, this incorporates a tangential inlet which causes the emulsion to swirl in the applied electric field. The emulsion first moves in down-flow, in the outer annular region, before its axial direction is reversed and it passes up into the inner annular region. The outer and inner annular regions are separated by an additional concentric electrode which allows an electric field to be applied to the emulsion before it passes upward out of the unit from the inner annular region. To facilitate the removal of any free water, which would be more likely in the outer down-flow region, an outlet is fitted at the bottom of the vessel. The use of a pulsed d.c. field, and an optimum frequency, is again mentioned in these patents but the use of other types of electrostatic field is not excluded.

During the first half of year 2000, Kvaerner Process Systems had planned to market a CEC. The theoretical framework for this design, for which a patent application was filed in 1998 (70), is based on work by Urdahl and coworkers (71, 72), Harpur et al. (73), and Wayth et al. (74).

This system, which has no inherent limitations with regard to water cut, is based on the use of a regular a.c. field (50–60 Hz) and insulated electrodes. The system has been shown to have a dramatic effect on the droplet growth in laboratory experiments (71, 73, 74) and in prototype testing it significantly improved the water/oil separation rate of a downstream gravity settler (72).

Another type of CEC has been patented by Provost and Rojey (75, 76) but does not appear to be available commercially. This system is based on a combined centrifugal/electrocoalescer device for separating water from the oil. These two patents show a wide variety of compact configurations in which W/O emulsions are subjected to centripetal accelerations of up to 500 *g* in combination with applied electrostatic fields of strength up to 6 kV/cm. It is stated that the applied frequency of the a.c. electric field should preferably be between 50 and 60 Hz. The level of development of these devices in unknown but an efficient commercial version would certainly be of great interest to operators. The benefit of this type of approach is that larger droplets are separated immediately and there is less of a problem with droplet break-up in downstream pipework. However, such a CEC is necessarily larger and more complicated as it must incorporate a quiescent settling zone and apparatus for removing excess of water. Additionally, since this type of CEC contains a water/oil interface, it will be more susceptible to platform orientation and motion.

## V. APPLICATIONS OF COMPACT COALESCERS

As more satellite fields are developed and connected to distant existing installations, efficient pipeline transport of multiphase, unprocessed well fluids is of increasing importance. The well fluid can contain large amounts of water which becomes emulsified in turbulent flow over several kilometers. As the well flow reaches the processing facilities, the system is choked, leading to further emulsification of the fluid system.

High water cuts often lead to a bottle-neck in the production process whereby the rate of oil production is constrained by large, undesired volumes of water. The problem is further compounded for emulsions which require longer residence times for separation. The stability of the emulsion formed depends on the properties of the oil. Heavy oils and oils which are acidic are more prone to forming stable emulsions. The viscosity of a W/O emulsion tends to be far higher than that of the oil itself, which, as a consequence, increases pressure drop and reduces transport capacity.

The separation of water from oil is a major challenge in the processing of hydrocarbon fluids. There is a continual demand to improve the quality of crude oil before it is exported in pipelines or tankers to refineries. Stringent criteria restrict the maximum water content allowed in the export oil (normally 0.5% maximum) and the oil content of the effluent water (normally 40 ppm maximum). The separation of water from oil depends on several fluid- and system-dependent factors. Water not only leads to a threat of corrosion scale and hydrates, but can also dramatically increase pumping costs. First, the pumps must deal with a larger volume of fluid, and, second, the formation of a W/O emulsion can significantly increase fluid viscosity and thereby pressure drop in the pipeline (as mentioned above).

As demonstrated in other parts of this chapter, some advantages of the compact coalescer unit are: short residence times (seconds rather than many minutes), order-of-magnitude droplet growth, and effectiveness over a large range of water cut (1–30%).

### A. Debottle-necking

The mixing of water and oil during production can cause very stable water-in-crude-oil emulsions. In addition to the mixing of water and oil in turbulent, multiphase flow, the fluid system is further mixed as the well stream is choked topside, ahead of the first-stage separator. In particular, heavy oils form stable W/O emulsions and, since the density difference is less, they are more difficult to separate. The location of a coalescer unit, upstream of the first-stage separator, increases the mean droplet size of the dispersed water. The consequences of this are: more effective phase separation, reduced residence time, a direct saving in chemical costs, and savings in heating costs if the process temperature can be reduced. Assuming that the level of demulsifier dosage can be reduced by 40 to 50%, it should be possible to save several million dollars in large oilfields.

### B. Between Separator Stages

Large separators are needed to process water-in-crude-oil systems which require long residence times in the separation process. A series of three separators is often used for the purpose. Additional equipment, such as heaters and coalescers, as well as process plant for the treatment of produced water, may be connected to the separation train. This is especially true when processing heavy crude oils since it is possible for several per cent water to be left in the oil after the first stage of separation. Techniques for removing the remaining water may involve heating the oil between the first and second stages of separation. Alternatively, demulsifier or combinations of demulsifying chemicals may be added.

### C. Alternative to Traditional Coalescer for Removing Remaining Water After Final Separation Stage

A traditional coalescer is the same size as a separator and hence is a large and heavy unit. If such a unit were to be replaced by a compact coalescer, a direct investment saving would result. Additionally, as the unit is smaller and lighter, a weight reduction in the production platform or ship on which it is mounted is possible.

### D. Desalter at Refineries

So far, only the potential use of the compact coalescer in upstream processes has been considered. However, there is also potential for using the unit at refineries. In order to remove salt from a crude oil, fresh water is added to the oil and intimately mixed with it. In some cases, this water may stay in the oil for a long time. In order to remove this water, the oil must again be heated or treated with chemicals or both. The installation of a compact coalescer here can, therefore, provide a more effective desalting process.

## VI. SUMMARY

This chapter has covered different physical phenomena and processes, ranging from bulk-fluid dynamics to microscopic interdroplet surface chemistry. All of these topics play a role in the electrostatic separation of W/O emulsions and the development and construction of an optimal, compact electrostatic coalescer. In some areas, such as turbulent droplet break-up, the understanding is well developed. In other fields there are still many questions to be answered. It is interesting to note that various authors have performed experimental assessments of W/O emulsion separation by using electrostatic fields. There is agreement on some aspects such as the general improvement in coalescer performance as the electric field strength is increased, to which the law of diminishing returns applies. In other areas, such as the existence of an optimum frequency for the applied electric field, there is still disagreement between researchers.

It is apparent that there are many different mechanisms working simultaneously when a W/O emulsion is treated in an electrostatic coalescing device. The overall growth of droplets is a balance of numerous hydrodynamic, electrostatic, chemical, and physical properties of the emulsion being treated. Some of these factors are double-edged swords, with both beneficial and detrimental effects on droplet growth. While high levels of turbulence or electric field strength promote the collision and coalescence of the smaller droplets, both mechanisms increase the chances of larger droplet break-up. Optimal droplet growth is therefore a careful balancing act of all of the factors, which must be carefully incorporated into the design of CECs.

## ACKNOWLEDGMENT

Statoil is acknowledged for giving permission to publish the results.

## REFERENCES


1. TJ Jones, EL Neustadter, KP Whittingham. J Can Petrol Technol 17: 100, 1978.
2. DG Thompson, AS Taylor, DE Graham. Colloids Surfaces 15: 103, 1985.
3. VB Menon, DT Wasan. Colloids Surfaces 19: 89, 1986.
4. VB Menon, DT Wasan. Colloids Surfaces 19: 107, 1986.
5. VB Menon, DT Wasan. Cooloids Surfaces 23: 353, 1987.
6. VB Menon, AD Nikolov, DT Wasan. J Colloid Int Sci 124: 317, 1988.
7. VB Menon, DT Wasan. Colloids Surfaces 29: 7, 1988.
8. EJ Johansen, IM Skjærvø, T Lund, J Sjøblom, H Søderlund, G Bostrøm. Colloids Surfaces 34: 353, 1988/89.
9. J Sjøblom, H Søderlund, S Lindblad, EJ Johansen, IM Skjærvø. Colloid Polym Sci 268: 389, 1990.
10. J Sjøblom, L Mingyuan, H Høiland, EJ Johansen. Colloids Surfaces 46: 127, 1990.
11. J Sjøblom, O Urdahl, H Høiland, AA Christy, EJ Johansen. Progr Colloid Polym Sci 82: 131, 1990.
12. KG Nordli, J Sjøblom, J Kizling, P Stenius. Colloids Surfaces 57: 83, 1991.
13. J Sjøblom. ed. Emulsions – A Fundamental and Practical Approach. NATO ASI Series. Dordrecht: Kluwer Academic, 1992.
14. L Mingyuan. PhD thesis, University of Bergen, Norway, 1993.
15. O Urdahl. PhD thesis, University of Bergen, Norway, 1993.
16. O Urdahl, AE Møvik, J Sjøblom. Colloids Surfaces A 74: 293, 1993.
17. B Balinov, O Urdahl, O Søderman, J Sjøblom. Colloids Surfaces A 82: 173, 1994.
18. FG Cottrell. US Patent 987 114, 1911.
19. FG Cottrell, JB Speed. US Patent 987 115, 1911.
20. TJ Williams, AG Bailey. IEEE Trans Ind Appl IA-22: 536, 1986.
21. RS Allan, SL Mason. Trans Faraday Soc 57: 2027, 1961.
22. AH Brown, C Hanson. Trans Faraday Soc 61: 1754, 1965.
23. AH Brown, C Hanson. Chem Eng Sci 23: 841, 1968.
24. PJ Bailes, SKL Larkai. Trans Inst Chem Eng 59: 115, 1981.
25. CP Galvin. Inst Chem Eng Symp Ser (No. 88): 101, 1984.
26. SE Taylor. Colloids Surfaces 29: 29, 1988.
27. TY Chen, RA Mohammed, AI Bailey, PF Luckham, SE Taylor. Colloids Surfaces A 83: 273, 1994.
28. B Gestblom, H Førdedal, J Sjøblom. J Dispers Sci Technol 15: 449, 1994.
29. AN Kolmogorov. Dokl Akad Nauk USSR 66: 825, 1949.
30. JO Hinze. AIChE J 1: 289, 1955.
31. PH Clay. Proc Roy Acad Sci (Amsterdam) 43: 852 and 979, 1940.
32. AJ Karabelas. AIChE J 24: 170, 1978.
33. CA Sleicher. Jr. AIChE J 8: 471.
34. HI Paul, CA Sleicher Jr. Chem Eng Sci 20: 57, 1965.
35. SB Collins. PhD thesis, Oregon State University, Corvallis, OR, 1967.
36. FD Rumscheidt, SG Mason. J Colloid Sci 16: 238, 1961.
37. CE Rosenkilde. Proc Roy Soc A312: 473, 1969.
38. CTR Wilson, GI Taylor. Proc Camb Phil Soc 22: 782, 1925.
39. TJ Williams. PhD thesis, University of Southampton, UK, 1989.
40. J Zeleny. Phys Rev 10: 1, 1917.
41. G Taylor. Proc Roy Soc A280: 383, 1964.
42. SK Friedlander, CS Wang. J Colloid Int Sci 22: 126, 1966.
43. PG Saffman, JS Turner. J Fluid Mech 1: 16, 1956.
44. HJ Pearson, IA Valioulis, EJ List. J Fluid Mech 143: 367, 1984.
45. L Prandtl. The Essentials of Fluid Dynamics, Blackie and Son Ltd. 1952.
46. J Nikuradse. Ing Arch 1: 306, 1930.

47. LC Waterman. Chem Eng Progr 61: 51, 1965.
48. W Krasny-Ergen. Ann Phys 27: 459, 1936.
59. TG Oweberg, GC Fernish, TA Gaukler, J Atmos Sci 20: 153, 1963.
50. JA Kitchener, PE Musselwhite. The Theory of Stability in Emulsions. Emulsion Science. New York: Academic Press, 1968.
51. SE Sadek, CD Hendricks. Ind Eng Chem Fundam 13: 139, 1974.
52. RA Mohammed, AI Bailey, PF Luckham, SE Taylor. Colloids Surfaces A 80: 223, 1993.
53. RA Mohammed, AI Bailey, PF Luckham, SE Taylor. Colloids Surfaces A 80: 237, 1993.
54. RA Mohammed, AI Bailey, PF Luckham, SE Taylor. Colloids Surfaces A 83: 261, 1994.
55. SE Taylor. Trans I Chem E 74 (part A): 526 1996.
56. Y Mori, M Tanigaki, N Maehara, W Eguchi. J Chem Eng Jpn 27: 340, 1994.
57. SS Wang, CJ Lee, CC Chan. Sep Sci Technol 29: 159, 1994.
58. T Hano, T Ohtake, K Takagi, J Chem Eng Jpn 21: 345, 1988.
59. K Fujinawa, M Morishite, M Hozawa, N Imaishi, H Ino. J Chem Eng Jpn 17: 632, 1984.
60. PJ Bailes, SKL Larkai. Trans I Chem E 60: 115, 1982.
61. PJ Bailes, SKL Larkai. Chem Eng Res Design 62: 33, 1984.
62. FM Joos, RWL Snaddon. Chem Eng Res Design 63: 305, 1985.
63. PJ Bailes, SKL Larkai. Chem Eng Res Design 65: 445, 1987.
64. PJ. Bailes. Trans I Chem E 73 Part A: 559, 1995.
65. V Gomis. Trans I Chem E 71 (part A): 85 1993.
66. J Drelich, G Bryll, J Kapczynski, J Hupka, JD Miller, FV Hanson. Fuel Process Technol 31: 105, 1992.
67. GW Sams, FL Prestridge, MB Inman. US Patent 5 565 078, 1996.
68. GW Sams, FL Prestridge, MB Inman, KD Manning. US Patent 5 575 896, 1996.
69. US Patent 5 645 451.
70. JP Berry, SJ Mulvey, O Urdahl, AG Bailey, MT Thew, NJ Wayth, TJ Williams. US Patent Application 09/090 060, 1998.
71. O Urdahl, TJ Williams, AG Bailey, MT Thew. Chem Eng Res Design 74 (A2): 158, 1996.
72. O Urdahl, P Berry, NJ Wayth, TJ Williams, KH Nordstad, AG Bailey, MT Thew. Proceedings of the 73rd SPE Annual Technical Conference and Exibition, New Orleans, 1998, SPE Paper 48990, p 111.
73. IG Harpur, NJ Wayth, AG Bailey, MT Thew, TJ Williams, O Urdahl. Electrostatics 40/41: 135, 1997.
74. NJ Wayth, TJ Williams, AG Bailey, MT Thew, O Urdahl. Proceedings of Electrostatics 99 Conference, Cambridge, UK, 1999.
75. I Provost, A Rojey. US Patent 5 643 469, 1997.
76. I Provost, A Rojey. US Patent 5 647 981, 1997.
77. M Smoluchowski. PhysZ 17: 557, 1916.
78. M Smoluchowski. Phys Chem 92: 129, 1917.
79. GR Zeichner, WR Schowalter. AI ChE J 23: 243, 1977.
80. W Findeisen. Meteorol Z 56: 365, 1939.

# 29

# Formation of Gas Hydrates in Stationary and Flowing W/O Emulsions

**Tore Skodvin**
*University of Bergen, Bergen, Norway*

## I. INTRODUCTION

Whenever gas and liquid water is brought together the formation of so-called gas hydrates may take place provided that the pressure and temperature conditions are in favor of this process, and that the gaseous molecules are able to stabilize the hydrates. The hydrates are solid ice-like compounds in which the gas molecules are trapped in cavities in a lattice formed by hydrogen-bounded water molecules.

Hydrates bring about a great deal of concern within the oil- and gas-producing industry in as much as all through the production chain problems related to the formation of gas hydrates may be experienced. A complete plugging of pipelines and processing equipment may be the result if the hydrates are allowed to grow without limitations. Apart from the obvious economical consequences connected to the closing down of a pipeline, hydrate plugging also imposes a threat to the safety of both workers and equipment.

In this chapter we focus on one of the areas where the formation of gas hydrates imposes a great challenge, namely, during long-distance transport (of untreated multiphase fluids) in pipelines situated on the sea floor. With the development and exploitation of increasingly more marginal oil and gas fields in, for instance, the North Sea one may foresee that this kind of multiphase flow will be increasingly more common. Also, multiphase transport from offshore installations to onshore processing plants is highly probable. The multiphase fluid contain more often than not liquid water, crude oil, and natural gas. The depths (and thus the pressure) and temperatures at the sea floor in many instances favor the formation of gas hydrates. An inevitable mixing of the phases during transport also increases the possibility of build up of hydrates.

The most common means of avoiding the hydrate problem is to inject large amounts of chemicals (often methanol or glycols) into the fluid, thus preventing the hydrates from forming. From both environmental and economical points of view the addition of these inhibitors is not desirable; thus, great efforts are made to find more effective and, at the same time, more acceptable inhibitors. This chapter reports on an alternative approach to hydrate inhibition, where the capability of crude oils to form stable water-in-oil (W/O) emulsions is exploited (1–3). The idea is that, as long as the water is present fincly dispersed in the oil phase, the probability of forming flow-obstructing hydrate structures will be minimized, since, in order to form gas hydrates, the gas molecules must diffuse across a film separating the oil and aqueous phases. Even if hydrates are formed within the water droplets the same mechanisms stabilizing the emulsion droplets may stabilize hydrate particles, thus preventing the agglomeration to larger structures.

Following a very brief introduction to the field of gas hydrates, this chapter gives details on laboratory-scale experiments where the formation of gas hydrates in W/O emulsions can be followed, as well as flow-loop experiments where the transportability of small hydrate particles is demonstrated.

## II. GAS HYDRATES

Gas hydrates are crystalline inclusion compounds where "guest" molecules (gas molecules) are trapped in the cavities of a "host" lattice. The physical appearance resembles that of ice; in some instances the hydrates are in the form of a "slush" while at other times (in different conditions) the hydrates appear to be dry, more like snow (4). The hydrates may be formed at temperatures well above the freezing point of pure water.

The main brick in the making of the host lattice is the pentagonal dodecahedron (Fig. 1a) formed by hydrate-bonded water molecules (5). It is not possible to fill a space with pentagonal dodecahedra without creating voids, and in the various kinds of hydrates the shape of these voids will be different. For the best known hydrate structures the voids have the shape of tetrakaidecahedra (Fig. 1b) (as found in structure I hydrates) or hexakaidecahedra (Fig. 1c) as in the structure II hydrates. The guest molecules may occupy either the small cavity enclosed by the pentagonal dodecahedron or the larger cavities depicted in Fig. 1b and 1c. The unit cell of structure I hydrates (formed from 46 water molecules) consists of two pentagonal dodecahedra and six tetrakaidecahedra, while the structure II hydrate unit cells (formed from 136 $H_2O$ molecules) contain 16 pentagonal dodecahedra and eight hexakaidecahedra. When all the voids in a structure II hydrate are filled, the gas:water ratio is 3:17, when only the large voids are filled the ratio is 1:17. The van der Waals interactions between the enclosed gas molecules and the water molecules are absolutely necessary for the stability of the gas hydrates; the host lattice will collapse without the gas present.

### A. Structure

The size and shape of the gas molecules are the most important features decisive for what the structure of a gas hydrate will be like. For instance, both methane and ethane form hydrates of structure I, while propane forms a structure II hydrate. Neither ethane or propane are able to enter the smallest cavities in the hydrates; these cavities may thus be empty. In the case of gas mixtures, small guest molecules may enter the smallest cavities while the larger guests are restricted to the larger ones.

### B. Hydrate-forming Species

The hydrate-forming gases include light alkanes (methane to isobutane), carbon dioxide, hydrogen sulfide, nitrogen, and oxygen. As hydrate-forming species in laboratory experiments it may be convenient to use substances such as tetrahydrofuran or short-chained CFCs, since gas hydrates are readily formed at temperatures between 3° and 8°C, at 1 atmos.

## III. GAS HYDRATE FORMATION IN W/O EMULSIONS

### A. Hydrate Formation at Normal Pressure – Experimental Techniques

Time-domain dielectric spectroscopy (TDS) may be used to follow hydrate formation in emulsified systems, and by means of this technique the role that different fractions of the crude play in gas hydrate formation in water-in-crude oil emulsions may be investigated (6, 7).

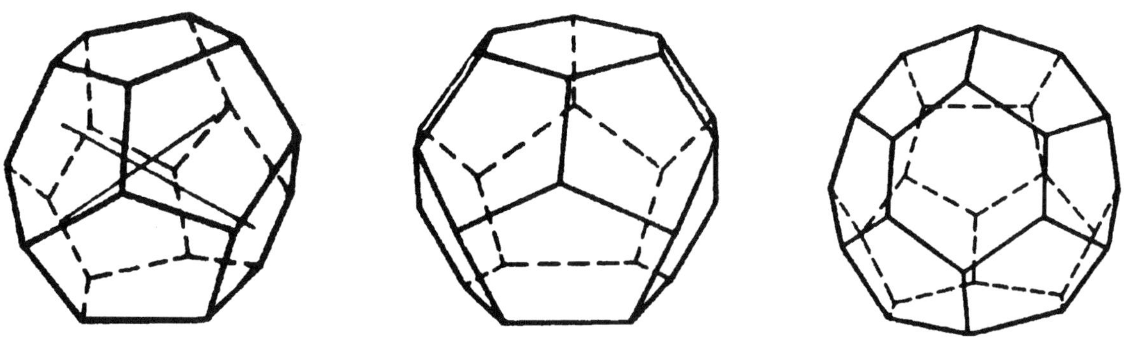

(a) (b) (c)

**Figure 1** Different cavities found in gas hydrates: (a) pentagonal dodecahedron; (b) tetrakaidecahedron; (c) hexakaidecahedron.

## B. Detection of Hydrate Formation in W/O Emulsions by Use of Dielectric Spectroscopy

The so-called Hanai equation (8, 9) gives the complex (frequency dependent) permittivity $\epsilon^*$ of an emulsion as

$$\left(\frac{\varepsilon^* - \varepsilon_2^*}{\varepsilon_1^* - \varepsilon_2^*}\right)\left(\frac{\varepsilon_2^*}{\varepsilon^*}\right)^{\frac{1}{3}} = 1 - \phi$$

where $\epsilon_1$* and $\epsilon_2$* are the permittivities of the continuous and the dispersed phase, respectively, and $\phi$ is the volume fraction of the dispersed phase. For a W/O emulsion the permittivity will thus be given by the volume fraction occupied by the water droplets and the dielectric properties of the aqueous phase (static permittivity $\approx$ 80–90) and the oil (static permittivity $\approx$ 3–4). *However, if the dielectric properties of either phase change by some process, then the dielectric properties of the total emulsified system must change also.* In the case of hydrate formation the permittivity of the aqueous phase will decrease since the permittivity of gas hydrates is substantially lower than that of liquid water (10). (At frequencies in the megahertz region the hydrate permittivity normally is in the range 5–10). The permittivity of the system at large will thus decrease to a level determined by the amount of water converted from the liquid state into immobilized hydrate water.

## C. Sample Preparation and Experimental Procedure

A dielectric sensor is attached to the inner wall of a cylindrical sample container made of aluminum. A stirring rod is inserted through a lid at the top, and the sample container is immersed in a thermostated waterbath. This experimental set-up (Fig. 2) allow dielectric measurements during hydrate formation (11).

Before an experiment the different components (i.e., the hydrate-forming compound, the aqueous phase, the oil phase, or, as in some cases, a premixed W/O emulsion) were thermostated in a waterbath at the same temperature as the sample container. The components were then transferred to the sample container where constant stirring assured proper mixing of the components. Dielectric data were recorded from the beginning of mixing and thereafter at even intervals.

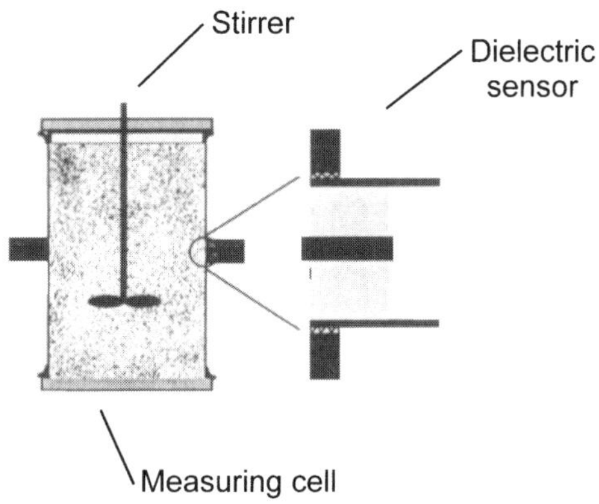


**Figure 2** Measuring cell used for dielectric monitoring of gas hydrate formation in W/O emulsions. (Adapted from Ref. 11.)

## D. Illustrative Results from Experiments on Model Systems

It has been found (12) that $CCl_3F$ forms clathrate hydrates at atmospheric pressure and 8.4°C. In a system consisting of a 60/40 W/O emulsion prepared from Exxol D-80, $CCl_3F$, 1% NaCl solution, and the surfactant Berol 26 (4% of total volume), no hydrate formation was detected at the reported equilibrium temperature even after 5 h of constant stirring (13). The onset of hydrate formation was found to be 3.0°C and the equilibrium temperature 4.4°C. The high degree of supercooling in this system may be analogous to the freezing of W/O emulsions, where it is found that small liquid droplets can undergo large supercooling while bulk samples do not (14, 15). Also, the presence of sodium chloride in the aqueous phase will lower the equilibrium temperature of clathrate hydrate.

TDS measurements were performed continuously during the course of the experiment. The dielectric spectra were fitted to the Cole–Cole model (this volume, Chapter 6). Figure 3 shows the changes in the dielectric parameters due to hydrate formation.

The static permittivity ($\epsilon_s$) and high-frequency permittivity ($\epsilon$) show the same behavior with regard to hydrate formation (Fig. 3a). When hydrate formation is taking place, free water is converted into hydrate water, and the volume fraction of free water in the emulsion falls. As a consequence, the low-frequency permittivity decreases, and the level of this permittivity may thus be taken as a direct measure of the amount of water converted into hydrate.

Electrolytes are not incorporated in the clathrate structures, leading to an increasing electrolyte concen-

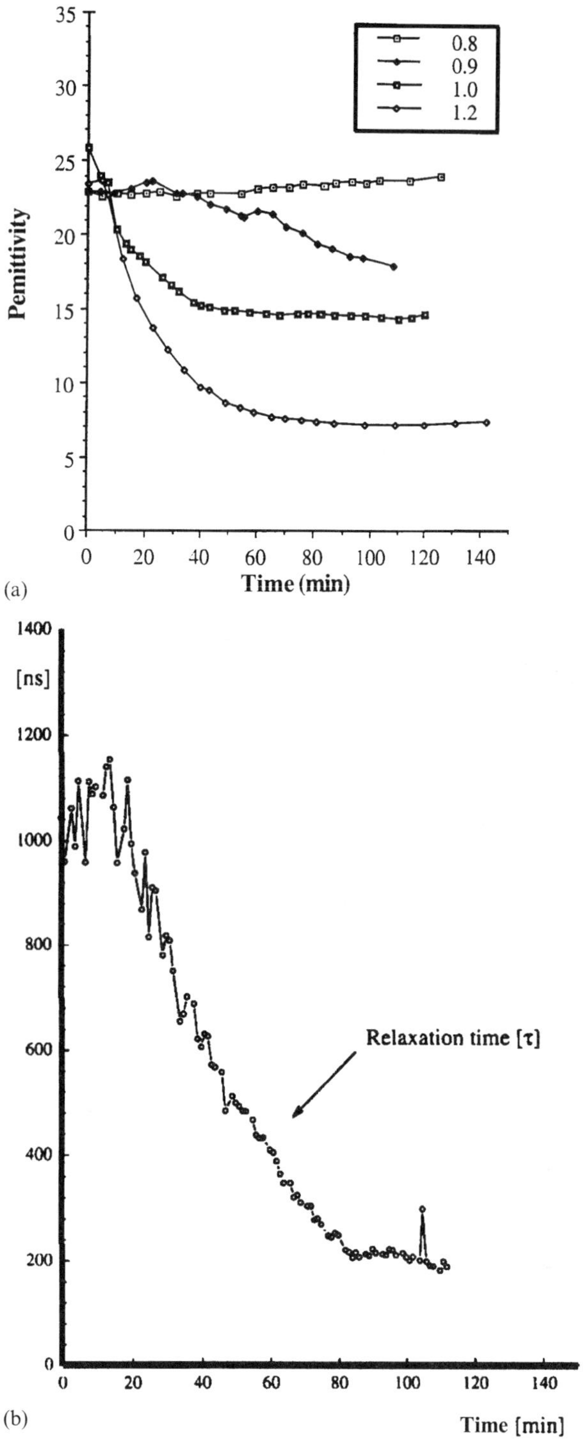


**Figure 3** Dielectric parameters as a function of hydrate formation in W/O emulsions. (a) Static permittivity vs. time, during hydrate formation; the different symbols indicate different mole fractions between the hydrate-forming compound and water relative to the 17:1 water:guest ratio expected in hydrates of structure II. (b) Dielectric relaxation time vs. time, during hydrate formation. (Adapted from Ref. 6.)

tration in the remaining free water as more hydrates are built up. This also influences the dielectric relaxation modes, as can be seen from Fig. 3b. By using the dielectric shell model initially proposed by Hanai et al. (16; this volume, Chapter 6) it is possible to convert the measured change in permittivity into a change in water concentration. In this model a shell with dielectric properties different from that of the core and the continuous phase is included in the dispersed droplets (Fig. 4).

In this model it is assumed that the $CCl_3F$ hydrate formation starts at the droplet interface. As the clathrate hydrate grows to the center of the droplets, a shell (with a permittivity typical of hydrates) forms. By inserting the dielectric parameters (Table 1) the experimentally obtained spectra may be fitted to model spectra for emulsion systems having different relative thicknesses of the shell. Hence, it is possible to calculate the amount of free water converted into clathrate hydrates. By using such a procedure it has been possible to evaluate the kinetics of hydrate formation in W/O emulsions (13).

### E. Hydrate Formation in W/O Emulsions Using Tetrahydrofuran

In another set of experiments (7) the influence on hydrate formation of various surface-active compounds extracted from crude oils was investigated. Tetrahydrofuran was used as the hydrate forming compound. In Fig. 5 the high-frequency permittivity of emulsions stabilized by asphaltenes from crude A (see Table 2) is displayed as a function of time. The oil-phase solvent was a mixture of toluene and decane (1:4 or 3:2 in volumetric ratio, respectively). The aqueous phase constituted 40 v/v% of the emulsion. The asphaltene contents were 2 or 5 w/w%, respectively (w/w with regard to the oil phase). The water bath temperature was kept constant at -1.3°C through the course of the experiment.

Initially, the high-frequency permittivity is steadily reduced from an initial value of $\simeq 8$ down to values in the range 3–5, reflecting the formation of hydrates. After $\simeq 0.5$ h the permittivity has attained a fairly constant level, indicating that no more hydrates are formed. The final permittivity levels for the samples containing 5% asphaltenes seem to be somewhat higher than for the samples with 2% asphaltenes. However, the composition of the oil phase (i.e., the ratio between toluene and decane) also seems to have some influence on the final permittivity levels, and thus the amount of hydrates formed.

**Table 1** Dielectric Values Used in Calculation of the Shell Model

| Medium | Static permittivity ($\epsilon_s$) | High-frequency permittivity ($\epsilon$) | Relaxation time ($\tau$) | Conductivity (S pdm$^{-1}$) |
|---|---|---|---|---|
| Water | 86.5 | 5.6 | 15 ps | 1.5 |
| Oil | 2.0 | 2.0 | 0 | 0 |
| $CCl_3F$ | 2.0 | 2.0 | 0 | 0 |
| $CCl_3F$ hydrate | 58.0 | 4.5 | 5 $\mu$s | 0 |
| Emulsion | 26.0 | 15.0 | 1 ns | 0 |

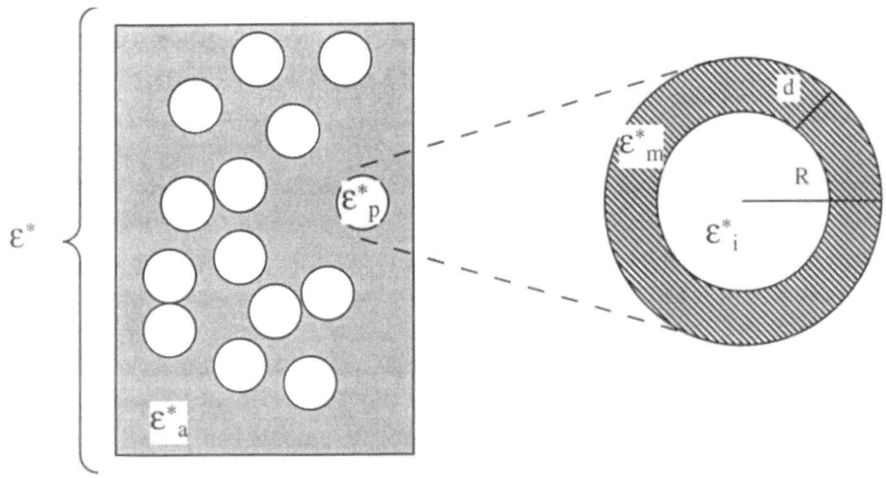


**Figure 4** Dielectric shell model. The total permittivity $\varepsilon^*$ will be a function of the permittivity of the continuous phase, $\varepsilon_a^*$, the volume fraction of particles, and the permittivity of the particles, $\varepsilon_p^*$. The particle permittivity depends on the core and shell permittivities, $\varepsilon_c^*$ and $\varepsilon_m^*$, and the ratio between the particle radius $R$ and the shell thickness $d$.

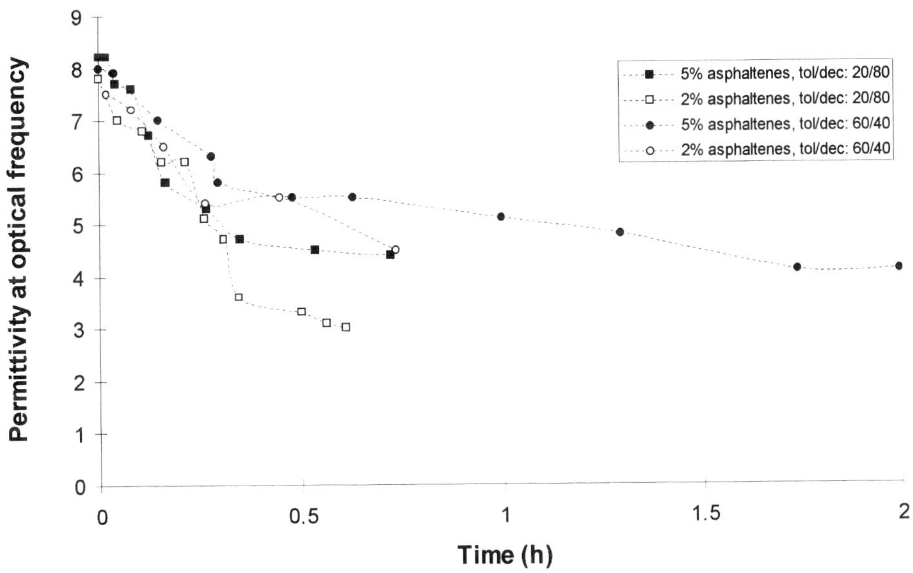


**Figure 5** High-frequency permittivity vs. time, during hydrate formation; influence of asphaltenes. (Adapted from Ref. 7)

**Table 2** Selected Characteristics of Crude Oils Used in the Studies

| | Density (kg/m$^3$) | Viscosity (cp) | Wax appearance Temp. (°C) | Asphaltene (wt%, precipitated in pentane) | Resins (adsorbed on silica) | Paraffins (wt% $C_{10}+$) |
|---|---|---|---|---|---|---|
| Crude A | 890 | 70 at 20°C<br>310 at 3°C | | 5 | 20 | 5 |
| Crude B | 838 | 15 at 20°C<br>200 at 3°C | 35 | 0.5 | 15 | 12 |
| Crude C | 861 | | | 0.5 | 7 | 6.5 |
| Crude D | 900 | | | 0.3 | 19 | 0 |
| Crude E | 922 | | | 18 | 12 | 4 |

In Fig. 6 we compare the time needed for hydrate formation at 1.2°C, with or without wax particles present in the continuous phase (decane). The inclusion of 10% wax (extracted from crude B) in the continuous phase drastically shortens the time needed before the onset of hydrate formation.

## F. Role Played by the Asphaltenes

From Fig. 5 we note that when the asphaltene content is increased from 2 to 5% the amount of hydrates formed is reduced. The influence of the asphaltenes on the initial rate at which the hydrates form seems to be relatively small.

The oil-phase composition has an effect on the solvation state of the asphaltenes and thus an effect on their ability to stabilize emulsions. However, based on these experiments we cannot be conclusive about the impact of the asphaltenes' solvation state on hydrate formation.

## G. Role Played by the Waxes

From Fig. 6 it is seen that the addition of 10 w/w% waxes (extracted from crude B) to the decane phase considerably reduces the induction time. However, the amount of gas hydrate formed does not change upon the addition of wax. One explanation of the reduced induction time would be that wax particles act as nuclei for the hydrates, thus speeding up the initial stage of the reaction.

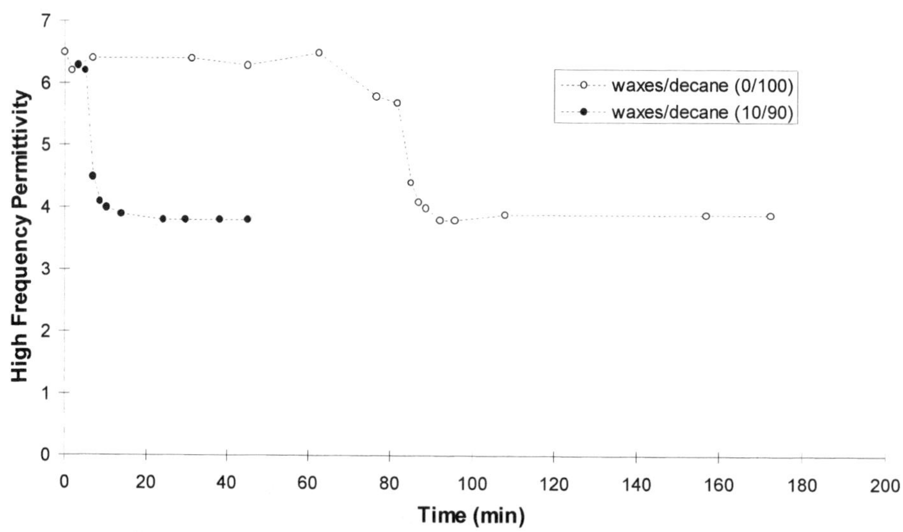


**Figure 6** High-frequency permittivity vs. time during hydrate formation; Influence of waxes. (Adapted from Ref. 7.)

### H. Conclusions to Be Drawn from Experiments on Model Systems

First of all, the experiments on the model systems confirm that hydrate formation can take place in W/O emulsions. Thus, the possibility of transporting the hydrates in emulsion droplets is definitely present.

Further, we have seen that the amount of water converted into hydrates depends somewhat on the asphaltene content of the emulsion systems. However, in the systems investigated the composition of the continuous phase also influences the hydrate formation. We have not been able to be conclusive about how the state of the asphaltenes (monomeric or aggregated) relates to the hydrate conversion. In the investigated systems, waxes reduce the induction time for the gas hydrate formation, probably because the waxes act as crystallization nuclei. The conversion rate was not influenced by the presence of waxes. To what extent the results from the model systems can be applied to real systems is not clear at present.

When hydrate formation takes place in a W/O emulsion, the hydrates seem to be formed inside the emulsion droplet. Hydrates in systems which are not emulsified have a tendency to form solid plugs. The present studies have shown that hydrates in emulsions have a stronger tendency to agglomerate than do emulsions without hydrate present. However, by selecting the appropriate surfactant this agglomeration may be delayed or completely prevented and thus plugging can be avoided.

## IV. FLOW LOOP TESTS

In the following, some of the results obtained with Elf's test rig are reported. A complete and more detailed description is given in Ref. 17.

Tests were performed at different water cuts with different crudes enriched or not with natural surfactants.

### A. The Elf Test Rig

The test rig includes a 22 m long loop of 1 inch diameter and a 90-liter tank used to store adequate amounts of liquid hydrocarbon, deionized water, and gas. The liquid phase is circulated through the loop by means of a Moineau pump. The loop is further equipped with a sapphire window located at the inlet of the pipe, a Coriolis flowmeter and a thermostatic device to adjust the temperature of the loop. A computerized data-acquisition system is used to monitor the test rig and to store and process the experimental results.

About 40 kg of liquid are needed to operate properly. Standard commercial gas (G20: $CH_4$ 96.2%, $C_2H_6$ 3.2%, $C_3H_8$ 0.1%, $N_2$ 0.5%) is admitted at a pressure of 75 bar, resulting in a hydrate thermodynamic equilibrium temperature of 13°C. The liquid velocity is fixed at 1 or 2 m/s during an experiment, and the temperature is initially set at 20°C.

Gas–liquid equilibrium and stabilized operating conditions are reached by pumping the liquid mixture through the loop and the tank for several hours. The maximum differential pressure drop acceptable for the Moineau pump is 5 bar.

After equilibrium is reached the tank is set off-stream and the liquid phase (saturated with gas) is recirculated through the loop only. The loop temperature is then decreased at a constant rate of 10°C/h from 20°C down to 4°C. This temperature is then kept constant during the whole test period, i.e., about 20 h if pipe plugging has not occurred. At the end of each test the loop is heated to 25°C and maintained at this temperature for several hours in order to ensure complete hydrate dissociation. The loop is then ready for the next test after a new stabilization period at 20°C.

During an experiment the mass flow rate is controlled by means of the Coriolis flowmeter, and the pressure drop in the loop is measured by a differential pressure cell. Make-up gas is added to the loop in order to keep the static pressure constant when the hydrates form. Measurement of the make-up gas flow rate is a way to observe the hydrate growth kinetics and amount formed. Knowing the amount of gas necessary to saturate the oil phase at 75 bar, when the temperature is decreased from 20°C to 4°C, and assuming that crystals of hydrates form in stoichiometric conditions, it is possible to evaluate the rate of conversion and the mass of water consumed.

Hydrate formation is detected by the monitoring of three variables:

- Sharp increase in gas consumption as it goes from thermodynamic saturation of the gas to compensation of the gas trapped in the hydrates.
- Increase in temperature profile (exothermicity of the reaction).
- Abrupt increase and irregularities in pressure loss caused by appearance of solid particles in the flow, viscosity change, deposition, and plugging.

It is possible to qualify each test in terms of the transportability of the hydrate slurry. The slurry is considered "untransportable" if one of the following occur:

- The pipe is plugged (i.e., the pressure drop through the loop exceeds 5 bar).
- Hydrates are crushed by the pump (i.e., a significant increase in pressure drop followed by a slight decrease, and stabilization is observed).
- Accumulation and/or deposition of hydrates. (The criterion is an increase in the static pressure owing to partial dissociation.)

The slurry is considered "transportable" if during the course of the test every parameter remains unchanged for several hours.

## B. Experimental Procedure

Crudes were tested with increasing water cuts in order to establish their capability of transporting hydrate slurry (maximum acceptable water cut). Crudes with low potential for hydrate transportation were then doped with surfactants (in order to enhance their hydrate transportability properties by increasing their natural surfactant content). The addition of surfactants could be accomplished in two ways.

### 1. Addition of Extracted Natural Surfactants

Natural surfactants (mainly asphaltenes and resins) extracted from one crude are added to a lighter crude or a condensate.

### 2. Mixtures of Crudes

Different crudes are mixed together, or one crude and a condensate are mixed. This second procedure was selected for practical reasons and was mainly used in this study.

The results for different mixtures are summarized in Figs 7–9, where the water cut is plotted versus the weight percentage of asphaltenes in the oil phase. The hydrate transportability is indicated by symbols. In the right part of the figures the tests have failed not necessarily because of hydrate formation but due to wax deposition or other modifications of the rheological properties of the fluid: this is ascribed to the experimental limitation of the pilot loop.

## C. Properties of Tested Crudes and Condensate

The main physicochemical properties of the five fluids reported on in this study are presented in Table 2.

The black oil (crude A) has relatively high amounts of heavy components and exhibits nonNewtonian behavior with viscosity, depending on the temperature. In spite of the low asphaltene content, crude B from the North Sea was selected as a suitable candidate

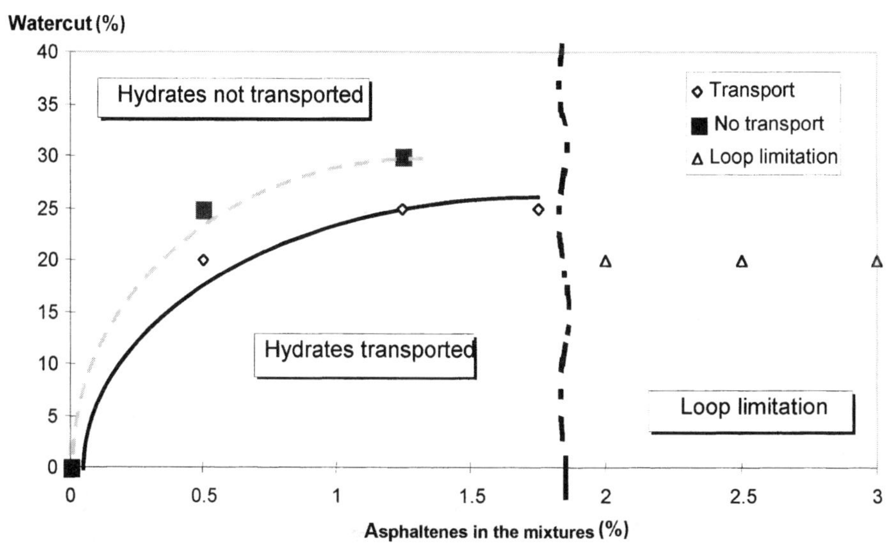


**Figure 7** Hydrate transportability of different mixtures between crude A and a condensate (Adapted from Ref. 17.)

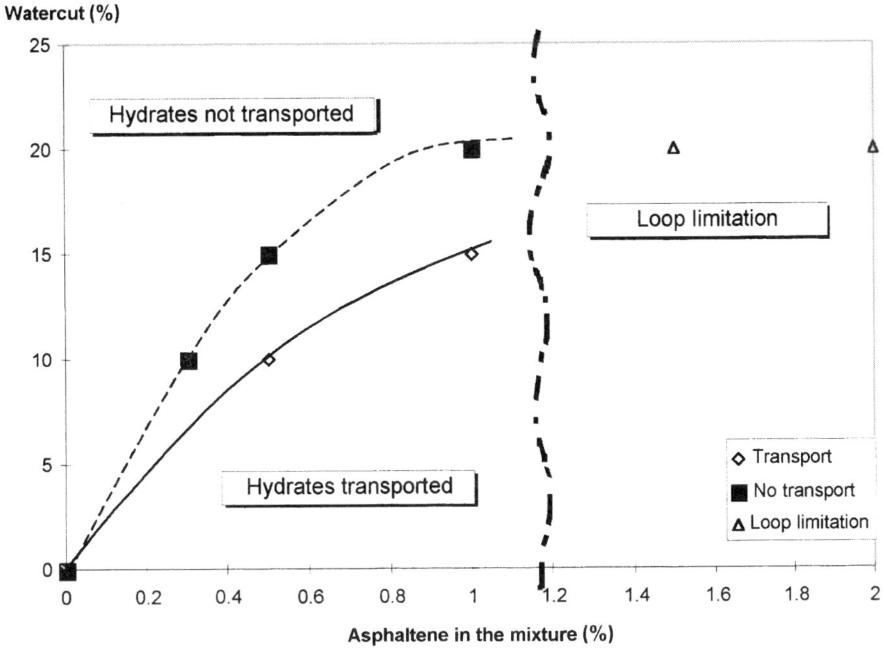


**Figure 8** Hydrate transportability for mixtures of the nonwaxy crude D and crude E. (Adapted from Ref. 17.)

because of previous good results in terms of hydrate transportation. As it is a waxy crude, this North Sea crude exhibits highly nonNewtonian behavior, with a viscosity which depends a lot on temperature and shear rate. The West African crude C is classified as a light paraffinic fluid. Its wax content is relatively low; nevertheless, the wax components are heavy. The West African crude D is biodegraded and contains no wax. The North African crude E has a high asphaltene content and is very viscous. The condensate from a North Sea gas field has a density of 820 kg/m$^3$.

## D. Tests Results

### 1. Loop Tests on Individual Crudes

*a. Crude A*

The viscosity of this crude increases drastically as the temperature decreases (Table 2), and the pressure drop exceeds 5 bar in the loop (which is the maximum acceptable) when the W/O emulsion (20% water cut) turns into a dispersion of hydrate crystals. It was, therefore, not possible to discriminate between the

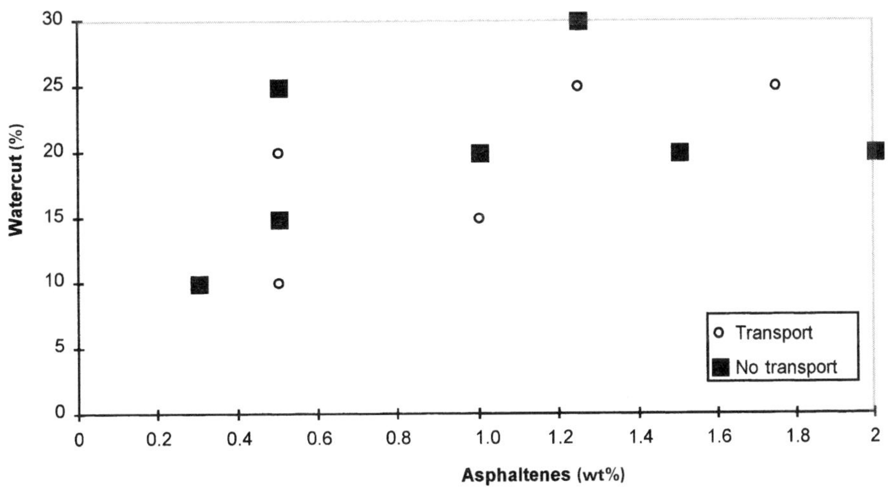


**Figure 9** Hydrate transportability vs. asphaltene content. (Adapted from Ref. 17.)

effect of the viscosity and hydrate plugging. Anyway, it was clear that hydrates transportation with this crude alone was not feasible under the testing conditions.

In order to lower the viscosity and in turn the pressure drop, vol 10% of a 1:1 mixture of xylene and Exxsol D60 was added to the crude. This improved the behavior of the emulsion but did not significantly modify the result of the test, as the natural surfactants from crude A were not able to ensure the transport of a hydrate slurry.

*b. Crude B*

At 10 and 20% water cuts the hydrate slurry was transportable. At 30% water cut the transportability test was considered as "failed" even if the circulation did not stop for 16 h.

The other "pure" crude oil systems showed varying and limited capability of transporting a hydrate slurry.

### 2. Condensates and Crudes Enriched with Natural Surfactants

*a. North Sea Condensate with Extracted Natural Surfactants*

Extracted surfactants were added to the condensate, but no substantial hydrate transportability was observed.

*b. Mixtures of Crude A and North Sea Condensate*

Since extraction and resolubilization of large quantities of natural surfactants is difficult and therefore limits the range of concentration available it was decided to mix the North Sea condensate and crude A in order to increase the contents of resins and asphaltenes in the condensate. Hence, in these experiments two parameters were evaluated, i.e., the concentration of crude A in the oil phase (from 5 to 50%) and the water cut.

A test with pure North Sea condensate and 20% water cut failed as soon as the hydrates formed. Complete plugging did not occur (owing to a high shear rate) but hydrates started to agglomerate and were crushed by the pump until overpressure was reached. This test showed that the condensate was not self-inhibited.

If 5% of crude A were added to the condensate, the same amount of hydrate formed (same gas consumption), but no agglomeration or deposition was observed for 4 h. However, the overpressure observed after 6 h indicated that the hydrate slurry was not totally stable.

When the amount of crude A was increased in the condensate to 10, 25, and 35%, respectively, the hydrate slurry was transportable for at least 16 h and at water cuts of 20 and 25% (Fig. 7). For water cuts above 30% the tests failed due to overpressure.

Thus, it was demonstrated that by adding natural surfactants to this North Sea condensate, transportation of a hydrate slurry was made possible. A similar result had previously been observed by using a selected commercial surfactant.

An additional test on the waxy crudes indicated that the amount of wax in the crude may play a role on the transportability of hydrates, i.e., relatively high amounts of wax lead to an enhanced transportability.

A summary of all the performed tests is presented in Table 3.

## E. Conclusions from Loop Tests

The loop tests have led to a better understanding of the mechanisms and better identification of some key parameters. It has been noticed that waxes may have positive interactions with the natural surfactants of the oil to stabilize the slurry. It was further demonstrated that a crude oil, up to a certain water cut, can be self-inhibited against hydrate blockage.

**Table 3** Summary of the Flow Experiments with Different Water Cuts and Different Oil Phases

| Water cut (%) | Crude A | Crude B | Crude A + condensate | Crude A + crude B | Crude C | Crude D + crude E |
|---|---|---|---|---|---|---|
| 10 | | + | | + | + | + |
| 15 | | | | | | + |
| 20 | - | + | + | + | +/- | - |
| 25 | | | + | - | | |
| 30 | | - | - | - | - | |
| 40 | | - | - | | | |

(+) Denotes that the test passed (i.e., that hydrates were transportable); (-) indicates a nonsuccessful test.

From the tests based on mixtures of crude A and a low asphaltenic crude (such as crude B or a condensate), it has been shown that if the amount of asphaltenes in the mixture is sufficiently high (1.5–2.0% by weight) the system can transport up to 30% of water under hydrate conditions. This indicates that an appropriate mixture of a condensate and a crude can be protected against hydrate blockage.

Unlike commercial surfactants, natural surfactants represent a class of components rather that a pure substance. Natural surfactants have different compositions and properties from one crude to an other. The *nature* of the asphaltenes is probably as important as the concentration of these surface-active agent when it comes to hydrate transportability, and other factors such as resins, waxes, aromatics, water quality, etc., have to be assessed in order to have a complete understanding of the transportability. An enhanced hydrate transportability of a crude as a result of increasing the amount of natural surfactants was demonstrated for several systems. In all cases it seems likely that the hydrate-slurry transportability is limited to 30% water cut.

Both the experiments on model systems and the flow-loop systems gave encouraging results, showing that hydrate formation really can take place inside emulsion droplets, and that the hydrates formed are transportable for a prolonged period. Nevertheless, further work is needed before this can be applied in a real production environment.

## ACKNOWLEDGMENTS

The oil companies Elf and TOTAL are acknowledged for their financial and experimental support. The dielectric spectroscopy equipment was financed by The Norwegian Research Council.

## REFERENCES

1. Y Schildberg, J Sjöblom, AA Christy, JL Volle, O Rambeau. J Dispers Sci Technol 16: 575–605, 1995.
2. H Førdedal, Y Schildberg, J Sjöblom, JL Volle. Colloids Surfaces A: Physicochem Eng Aspects 106: 33–47, 1996.
3. O Mouraille, T Skodvin, J Sjöblom, J-L Peytavy. J Dispers Sci Technol 19: 339–367, 1998.
4. T Austvik. Hydrate Formation and Behaviour in Pipes. Thesis, Norwegian Polytechnic University, Trondheim, 1992.
5. YF Makogon. Hydrates of Natural Gas. Tulsa, OK: Penn Well, 1981.
6. T Jakobsen. Chlathrate Hydrates Studied by Means of Time-domain Dielectric Spectroscopy. Thesis, University of Bergen, Norway, 1996.
7. O Mouraille, T Skodvin, J Sjöblom, J-M Fourest. J Dispers Sci Technol, in press.
8. T Hanai. Kolloid Z 171: 23–31, 1960.
9. T Hanai. Kolloid Z 175: 61–62, 1961.
10. NE Hill, WE Vaughan, AH Price, M Davies. Dielectric Properties and Molecular Behaviour. London: Van Nostrand Reinhold, 1969.
11. T Jakobsen, K Folgerø. Measure Sci Technol 8: 1006–1015, 1997.
12. TA Wittstruck, WS Brey, AM Buswell, WH Rodebush. J Chem Eng Data 6: 343, 1961.
13. T Jakobsen, J Sjöblom, P Ruoff. Colloids Surfaces A: Physiochem Eng Aspects 112: 73–84, 1996.
14. M Clausse. In: P Becher, ed. Encyclopedia of Emulsion Technology. Vol 1. Basic Theory. New York: Marcel Dekker, 1983.
15. ID Chapman. J Phys Chem 72: 33–38, 1968.
16. T Hanai, K Asami, N Koizumi. Bull Inst Chem Res 57: 197, 1979.
17. EM Leporcher, JL Peytavy, Y Mollier, J Sjöblom, C Labes-Carrier. Proceedings of SPE Annual Technical Conference and Exhibition, New Orleans, 1998, SPE 49172.

# 30

# Asphaltene Emulsions

**Peter K. Kilpatrick and P. Matthew Spiecker**
*North Carolina State University, Raleigh, North Carolina*

## I. INTRODUCTION

An emulsion is a thermodynamically unstable dispersion of two immiscible liquids. Surface tension requires that the dispersed phase forms spherical droplets in the continuous phase provided that the dispersed phase volume fraction is less than that corresponding to close droplet packing. The droplets, when stable, are slow to flocculate and coalesce. Emulsions are formed quite often in industrial processes and can be either desirable or undesirable. Examples of useful emulsions abound, as in foods, cosmetics, pharmaceuticals, agricultural products, and a host of other areas of technology such as can be found in this encyclopedia. Emulsions are also found in the petroleum industry where they are typically undesirable and can result in high pumping costs, reduced throughput, and special handling equipment (1). There are, however, examples from the petroleum industry in which emulsions may be desirable, as in the generation of oil-in-water (O/W) emulsions for transportation and in the generation of water-in-oil (W/O) emulsions for gas hydrate inhibition.

Crude oil, or petroleum, is found in reservoirs along with water or brine and is typically produced as an emulsion (2–5). Water is also injected into the crude during processing to wash out contaminants or is used as steam to improve fractionation (6). While contamination of water when processing crude oil often leads to emulsions of the O/W type (7), W/O emulsions are much more prevalent in the petroleum industry (8–16). Figure 1 illustrates the various types of water–crude oil emulsion systems. Crude oil typically exists and is produced as a W/O emulsion. When crude oil is spilled on the sea and agitated, high viscosity, stable W/O emulsions are formed (17,18). Petroleum emulsions of the W/O variety are almost exclusively stabilized by asphaltenes, at least in part, and that is the subject of this chapter. Asphaltene-stabilized W/O emulsions occur during production, as a result of spills, in the desalting operation, and in downstream wastewater handling. While W/O petroleum emulsions often contain considerable quantities of inorganic solids (e.g., calcium and iron oxides and hydroxides) or other organic solids (e.g., waxes), the dominant contributor to the stabilizing film is the asphaltene fraction from the crude.

Crude oils have markedly differing abilities to stabilize W/O emulsions, but a considerable proportion of these emulsions are very stable and can lead to significant sludge generation (10). An important issue that we will touch upon in this chapter is the identification of those chemical and structural delimiters of petroleum that determine its ability to form W/O asphaltene emulsions. A central goal of research in this area is the fundamental understanding of the relationships among asphaltene molecular structure, thermodynamic variables such as solvent composition and temperature, and the resulting W/O emulsion stability. An important first step in developing this type of correlative understanding is elucidating the mechanism of asphaltene stabilization of W/O emulsions.

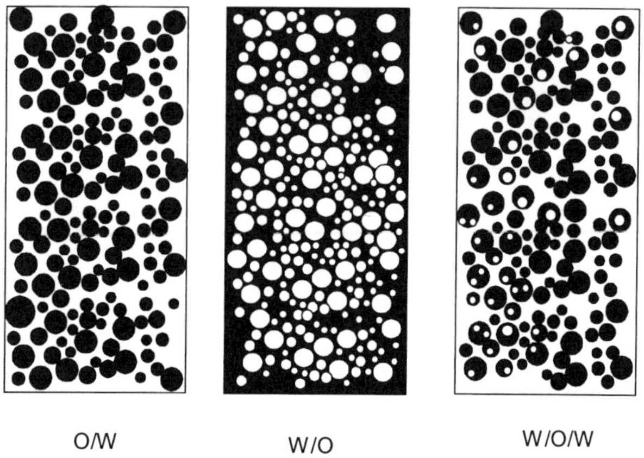


**Figure 1** Schematic drawing of emulsion types.

The formation of an interfacial layer consisting of surface-active material present in crude oil (asphaltenes and resins) produces a physical barrier for droplet–droplet coalescence. Numerous researchers have noted the presence of an interfacial "skin" in oil–water systems with these surface-active components present (3,5,9,19–26). Mohammed et al. (21), using a Langmuir film balance, found the interfacial dilatational modulus to be dependent on the presence of asphaltenes and resins. Førdedal et al. (20) have shown that these components are responsible for stabilizing emulsions. From observations of interfacial tension measurements, they reasoned that the ratio of resins to asphaltenes is an important factor for determining emulsion stability. As we will show, the evidence is compelling that the primary mechanism of asphaltene stabilization of W/O emulsions is through the formation of a viscous, crosslinked three-dimensional network with high mechanical rigidity (Fig. 2). Shown on the left side of Fig. 2 is a schematic depiction of asphaltenic aggregates interacting through donor–acceptor interactions (either of the proton or electron type) and solvated on the edge by resins. On the right side of the figure is shown the adsorption of these aggregates at an oil–water interface and accompanied by interaggregate interactions to form a viscous, mechanically rigid film.

The schematic of Fig. 2 is obviously oversimplified; nonetheless, there are several salient features illustrated which likely capture the essence of asphaltene-stabilized films. First, surface adsorption of asphaltene molecules is probably driven by hydration of polar functional groups in the aromatic core of an individual asphaltene molecule. Second, resin molecules probably serve to solvate primary aggregates (asphaltene micelles) in the bulk phase, but these resins are likely shed and do not appreciably participate in the actual stabilizing film. In fact, as we will show later, resins are totally unnecessary in the stabilization of asphaltenic films. A missing detail in Fig. 2 is the means whereby individual asphaltene molecules crosslink to form a networked interfacial film. In the sche-

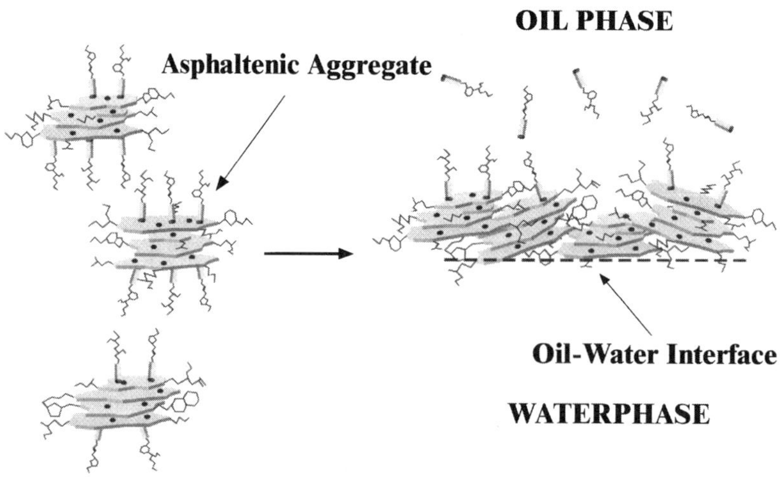


**Figure 2** Depiction of asphaltenic aggregates, shown as cofacial stacks of individual asphaltene molecules and solvated on edges by resin molecules. Aliphatic side chains and moieties are shown as jagged, stick-like groups with rings; fused aromatic moieties are shown as flat shaded groups with edges; polar functional groups are shown as dark dots. On right-hand edge of the drawing is a depiction of the self-assembly of these asphaltenic aggregates at an oil–water interface.

matic, the aliphatic side-chains on the aromatic core are shown to commingle intimately. The specific forces which crosslink asphaltenic molecules at oil–water interfaces are likely much stronger than simple dispersion forces, and the likeliest candidates are H-bonds or electron donor–acceptor interactions. However, this is still an active area of research and conflicting opinions abound (27). A contributory element in the stabilization of W/O emulsions may be the presence of organic (wax) or inorganic solid particles or aggregates in the thinning films between water droplets that raise film viscosity and reduce film drainage. We will, however, not be able to review these solids-based contributions to emulsion stability here.

Years of research and observation have clearly established that a rigid and protective asphaltenic film surrounding the water droplets governs the long-term stability of crude W/O emulsions. The detailed properties of this asphaltenic film are still an active area of research. Stabilization by asphaltenes is brought about by the formation of an interfacial skin at the water droplet interface that prevents disperse-phase coalescence. The conditions of asphaltene, resin, and solvent composition that promote the formation of this interfacial skin are a focus of our research and of this chapter. Central to the development of our understanding is a view of asphaltene solution behavior in which surface activity, propensity to adsorb, and ability to crosslink and form a strongly viscoelastic film or "skin" is immediately related to the size and cohesive energy of asphaltenic aggregates in solution. Accordingly, much of what will be described in this chapter is the literature and current understanding of the relationship between asphaltene molecular structure, molecular aggregation, and solvent composition. After reviewing this literature, the role of aggregate size and lability will be described in dictating adsorption and viscoelastic film formation at model oil–water interfaces and at crude oil–water interfaces. Finally, the relationship between film properties and emulsion stability will be discussed.

## II. REVIEW OF ASPHALTENE CHEMISTRY, AGGREGATION, FLOCCULATION, AND SOLUBILITY

### A. Introduction

The stabilization of emulsions by asphaltenes is strongly mediated by their interfacial activity, state of aggregation in solution, and lability at the oil–water interface. These properties, in turn, are strongly dependent on asphaltene solubility and nearness of the crude oil or asphaltene solution to the phase boundary at which asphaltenes precipitate. The role of molecular chemistry in dictating these properties is not totally understood, but much is known and will be summarized in this section.

Asphaltenes are defined as the toluene-soluble and *n*-heptane- or *n*-pentane-insoluble fraction of crude oil, while maltenes or petrolenes are the alkane-soluble portion. The soluble maltene fraction consists of saturated hydrocarbons (including waxes), aromatics, and a polar fraction called resins (defined as that fraction of maltenes that elutes from silica gel with polar solvent mixtures including acetone, methylene chloride, tetrahydrofuran, and ethyl acetate). While asphaltenes are recognized to be remarkably polydisperse – in heteroatomic functionality, molecular weight, and carbon backbone structure – much chemical structure elucidation with asphaltenes seems to suggest some common features. Specifically, the aromatic carbon content of asphaltenes falls typically between 40 and 60%, with a corresponding H/C atomic ratio of 1.0–1.2. A high percentage of these aromatic carbon rings are interconnected in the molecular structure and, consequently, the typical asphaltene molecule is flat or planar. This has a significant impact on asphaltene physical chemistry, aggregation, solubility, and interfacial film formation, as we discuss below.

### B. Chemistry of Asphaltenes

Many hundreds of studies have reported on the chemical composition of asphaltenes (28–44) and excellent summaries exist. We will only attempt to summarize some key findings here as they relate to aggregation, solubility, and interfacial film formation.

#### 1. Elemental Analysis

The range and typical values of key elements found in asphaltene fractions are summarized in Table 1. One primary delimiter of asphaltene structure is the atomic H/C ratio, which correlates directly with aromaticity (40). An even more telling indicator of asphaltene structure is the distribution of H/C ratio, or equivalently, the Jurkiewicz $N$ factor (41), over the range of molecular weights, which characterizes the aromaticity and polydispersity of asphaltenes. As we will discuss below, the tendency of asphaltenes to associate and aggregate certainly increases with acidity, but may also increase with molecular weight or aromaticity as well. This seems sensible, as in the first case intermo-

**Table 1** Range and Typical Values of Elemental Compositions in Asphaltenes

| Element (in wt%) | Range | Typical |
|---|---|---|
| Carbon | 78–90 | 82–84 |
| Hydrogen | 6.1–10.3 | 6.5–7.5 |
| Nitrogen | 0.5–3.0 | 1.0–2.0 |
| Sulfur | 1.9–10.8 | 2.0–6.0 |
| Oxygen | 0.7–6.6 | 0.8–2.0 |
| Vanadium (ppm) | 0–1200 | 100–300 |
| H/C | 0.8–1.5 | 1.0–1.2 |

lecular H-bonding is increased, while in the latter cases, $\pi$-bonding between delocalized aromatic electrons is increased. Both likely play a role in viscoelastic film formation.

## 2. Molecular Weight

The literature is full of molecular weights reported for asphaltene in a variety of solvents – unfortunately, it is difficult to determine whether reported values accurately reflect the molecular weights of monomers, aggregates, or some volatile subfractions. Based on "typical" chemical structures determined from elemental analyses and functional groups from spectroscopy (described below), average molecular structures of the type shown in Fig. 3 can be assembled. These structures are invariably flat or planar (as mentioned above), and one can easily envision cofacial stacking interactions through $\pi$-bonding, dispersion forces, and/or H-bonding, resulting in the molecular aggregates depicted in Fig. 2. Under what solvent conditions would one expect to measure monomolecular weights?

Boduszynski et al. (28) used field ionization mass spectrometry on asphaltenes isolated from a crude oil from the then Soviet Union and obtained molecular masses ranging from 500 to 1500 Da with number averages around 900–1000 (the molecular structures shown in Fig. 3 are in this range). Although this group was very careful, the challenge with MS methods is uncertainty as to whether one has adequately volatilized the sample with no bond breaking or oligomer formation. Vapor-pressure osmometry has been applied to many asphaltene solutions (36). The best data taken under extreme conditions (90–120°C) in strong solvents (chloro- and nitro-benzene) seem to indicate again low molecular weight ranges (500–1500 Da) with number averages of around 900–1000 [see McLean et al. (45) for further discussion].

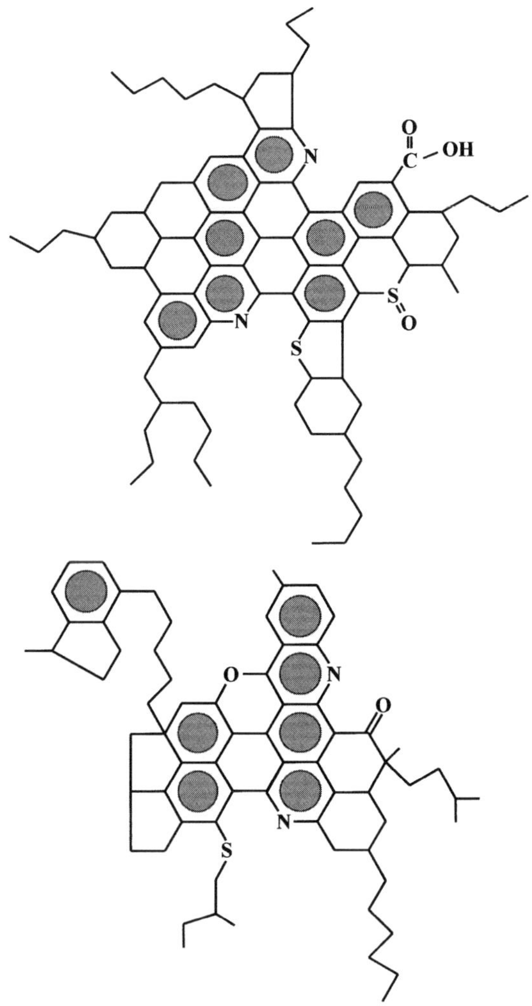


**Figure 3** Molecular structures of two plausible asphaltene molecules. The top drawing has a structural formula of $C_{84}H_{98}N_2S_2O_3$, a molecular weight of 1248 amu, and an H/Cratio of 1.18. The lower drawing has a structural formula of $C_{78}H_{87}N_2S_1O_2$, a molecular weight of 1045 amu, and an H/C ratio of 1.21. Both are flat molecules.

## 3. Functional Group Analysis

Infrared spectroscopy, NMR spectroscopy, X-ray methods such as X-ray absorption near-edge structure spectroscopy and ESR spectroscopy have been used primarily to probe the detailed chemistry of heteroatom speciation, polar functional group determination, and hydrogen and carbon types in asphaltenes. The consensus seems to indicate that most asphaltene molecules have one to three heteroatoms (S, N, and O) per molecule. Sulfur exists predominantly as thiophenic heterocycles (typically 65–85%) with the remainder as sulfidic groups (46,47). Thiophenic moieties are not particularly polar and thus this dominant form of sulfur species likely plays little role in aggregation, H-bonding, or viscoelastic film formation. Only in highly biodegraded crudes does there appear to be a large amount of sulfoxide. Thus, if sulfur-based functional groups contribute any significant way to emulsion film formation, it is probably through sulfoxide moieties. Nitrogen occurs in pyrrolic, pyridinic, and quinolinic groups, the dominant portion being pyrrolic. Interestingly, relatively small amounts of porphyrin complexes appear to exist in asphaltenes. However, when asphaltenes are extracted with warm acetone, what little porphyrin material present is extracted. The film-forming capability of the asphaltene fraction, and the shear strength of that film, appear to be diminished when the porphyrin fraction is removed (48), although this result does not appear to have been duplicated and confirmed. Oxygen species are predominantly hydroxylic, carbonyl, carboxylic, and ether (41,49). Acidic functional groups appear to play a critical role in asphaltene films which stabilize emulsions (50). By fractionating asphaltenes and resins from North Sea crudes using a solvent extraction procedure, Sjoblom's group has shown that model emulsions were strongest when stabilized by asphaltene fractions richest in open-chain carbonyl functional groups. They hypothesized that the hydrogen bonding afforded by flexible carboxyl groups in asphaltenes yields a rigid mechanical barrier film to water droplet coalescence. This view is consistent with recent data we have collected which suggest that acidic asphaltenes are considerably stronger viscoelastic film formers than their basic and neutral counterparts. We will describe this in Sec. III. $^{13}C$ NMR methods have been applied to asphaltenes to gage the degree of aromatic ring condensation (35). This fused-ring character is also reflected in the H/C ratio and aromaticity, and probably plays an important role in the aggregation properties of asphaltenes (51).

## C. Aggregation Studies

Based on their planar molecular structure, it is not surprising that asphaltenes associate through stacking interactions to form supramolecular aggregates [see Sheu and coworkers (52–55)]. A substantial number of data from small-angle X-ray and neutron scattering (SANS) techniques confirm that this aggregation is ubiquitous in good solvents – such as toluene and pyridine – and results in small clusters of 6–10 nm hydrodynamic diameter. Recent SANS data we have obtained on a Californian offshore crude asphaltene sample in varying proportions of *n*-heptane and toluene are consistent with this picture of smallish aggregates in the soluble region (see Fig. 4).

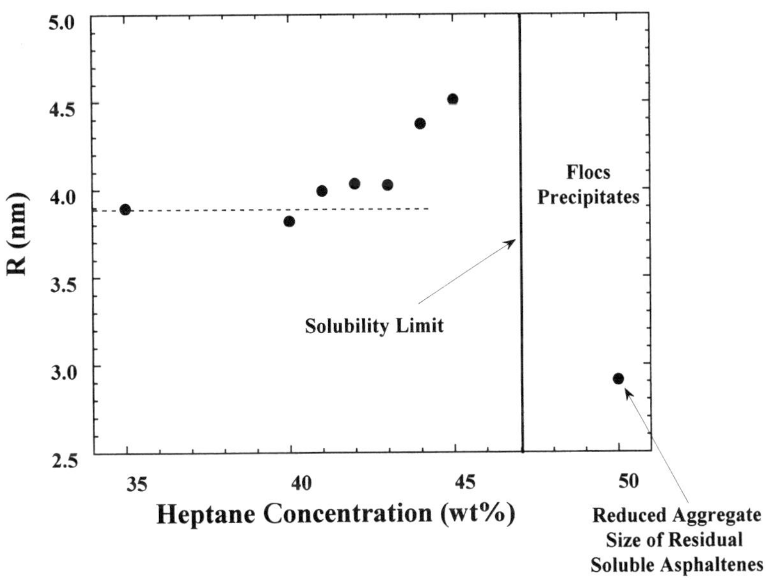


**Figure 4** Radii of gyration of asphaltenic aggregates obtained from SANS data. The samples were 2% (w/w) of asphaltenes obtained from a California crude and dissolved in heptane–toluene mixtures of different proportions. The heptane and toluene were deuterated to obtain adequate scattering contrast.

Interestingly, there appears to be a growth in aggregate size close to the solubility limit. This may reflect strong dispersion forces which ultimately drive the phase separation. Beyond the precipitation boundary, the asphaltenes which remain in solution and do not sediment appear to have a smaller aggregate dimension, possibly as a result of weaker interactions between these residually soluble species. As we will discuss later, this may very well be because acidic asphaltenes dominate the aggregation story in the soluble regime, but are the first subfraction to precipitate. The remaining soluble basic, neutral, and more weakly acidic asphaltenes apparently do not associate as strongly as the more strongly acidic fraction. Should this be true, it could explain which fraction indeed dominates the viscoelastic film-forming properties of asphaltenes.

Beyond the solubility boundary, the small several nanometer-sized asphaltenic aggregates apparently flocculate through diffusion and reaction-limited mechanisms to form classical colloidal flocs of fractal dimension (56–59). Over fairly narrow ranges of solvent composition, the precipitation occurs to form micrometer-sized particles. Clearly, the chemistry of these flocs hold important clues to the interactions which drive aggregation in the soluble regime. Efforts to discriminate the chemistry of these flocs on the edge of the solubility regime are ongoing (60–66).

## D. Solubility Studies

Asphaltenes are defined as a solubility class: soluble in toluene and insoluble in alkane solvents, either *n*-pentane or *n*-heptane. It is thus not surprising that *n*-alkanes would be used as antisolvents in differential solubility experiments to fractionate asphaltenes further. Fogler's group has performed a series of studies in which asphaltenes have been resolubilized in chloroform and then differentially fractionated by precipitation with *n*-pentane (62,63). The result is a series of fractions termed F1–F6 which should, in principle, have decreasing polarity and increasing kinetics of redissolution in heptane solutions of surfactants (resin-like solvaters). While the latter is definitely true, multiple efforts to demonstrate decreasing polarity though some chemical measurement have been more difficult.

We have performed experiments in which asphaltenes isolated from Safaniya due (also known as Arab Heavy) have been subjected to ion-exchange chromatography (66). The neutral fraction in our experiments – i.e., that fraction which binds neither to cationic or anionic resin – is clearly different chemically from the other fractions. Specifically, it has a higher H/C ratio (lower aromaticity), a much lower nitrogen content (by a factor of 3–4), and a lower carbonyl content. However, the differences between the "acidic" (those binding to the cation column) and the "basic" (those binding to the anion column) asphaltenes were much more subtle. Functionally, however, the acidic and basic asphaltenes were very different. The acidic asphaltenes were noticeably less soluble in mixtures of heptane and toluene than the original whole fraction of asphaltenes, while the basic asphaltenes were considerably more soluble. Interface rheology experiments indicated that the acidic asphaltenes also formed much stronger films. Thus, while demonstrating chemical differences between fractions of asphaltenes of varying solubility and film-forming strength may be challenging, there is no doubt that such chemical differences must exist. One possible explanation is that the acidic functional groups in the "acidic" asphaltenes lie on the periphery of the molecular structure and hence are accessible to binding both other asphaltenic molecules as well as the ion-exchange matrix, while such acidic functional groups exist in "basic" asphaltenes but are simply inaccessible.

Independent evidence of the least soluble fractions also being the most prone to associate comes from recent work by Yarranton and coworkers (64,65). In their studies, asphaltenes were precipitated from *n*-hexane–toluene mixtures as a function of increasing *n*-hexane content. They measured the number average molecular weight (MW) from vapor-pressure osmometry and deduced that apparent MW increases with decreasing solubility, i.e., the least soluble fractions which precipitated at the highest toluene concentration also gave the highest apparent MW. These values of osmotically determined MW were as high as 7000–8000 Da, certainly not monomers. Again, a stack of four to six monomers, each of perhaps 1200–1400 Da, is consisent with their reported MW as well as with an aggregate of 5–10 nm diameter.

## E. Emerging Picture of Asphaltene Aggregation

While there is a great deal of experimental data detailing a variety of structural properties of asphaltenes, much of the molecular chemical picture as it relates to film-forming structure remains unclear. The early Nellensteyn (67) and Pfeiffer and Saal (68) model of a locally structured solution consisting of a graded interfacial zone between asphaltenic species and the crude solvent (saturated and aromatic hydrocarbons) seems to have some legitimacy in the light of detailed scattering data suggesting spherical aggregates. Moreover, the molecular structural distribution data of Bestougeff and coworkers (34,38) suggest polycondensed structures connected by flexible linkers, which further implies a relatively disordered aggregate interior. On a very local level [≃10 Å], the interaction of fused aromatic ring systems suggests a local director axis similar to discotic mesogens. Specific scattering evidence of sedimented asphaltenic films indicates discotic lamellar structures (69). Thus, a balanced picture of the literature suggests that asphaltenes may aggregate to form locally directed anisotropic structures connected through space to form a three-dimensional network. This picture is in fact very close to Yen's model (70,71). It seems clear that strong, directed intermolecular forces must hold asphaltene molecules together in supramolecular aggregates, and the likeliest forces are $\pi$-bonds, hydrogen bonds, and electron donor–acceptor bonds. We also know that primary aggregates of asphaltenes agglomerate further in "poor" solvents, i.e., $n$-alkanes, to form larger colloidal particles. This subsequent agglomeration may also be attributable to the aforementioned types of forces as it is driven by increasingly aliphatic solvents which would promote all three types of interactions. In "good" solvents, such as toluene and pyridine, asphaltenes are known to micellize or aggregate (52–55), but the extent of aggregation is modest (small spherical aggregates of radius 30–40 Å). Thus, it is the subsequent larger scale aggregation or self-assembly at interfaces that ultimately gives rise to three-dimensional network formation, viscoelastic film formation, and emulsion stabilization, as we will describe below.

## III. RHEOLOGICAL STUDIES OF PLANAR OIL–WATER INTERFACES WITH ADSORBED ASPHALTENES

### A. Introduction

The mechanical properties of asphaltene films at interfaces can be probed by a variety of rheological techniques. These methods provide valuable insight into the origins of stability of asphaltene emulsions and into the role of concentration, and solvation by resins and aromatic solvents on the adsorption and self-assembly of asphaltenes. Miller et al. provide a comprehensive review of methods for probing interfacial dilational and shear properties of adsorption layers at liquid interfaces (72). They describe devices that measure surface velocity profiles (indirect methods) or determine torsional stress values (direct methods). Indirect techniques such as canal surface viscometers, deep-channel surface viscometers and rotating wall knife-edge surface viscometers require measurements of fluid flow using visible inert particles from which surface viscosities can be determined. In direct methods, a pendulum is placed at the interface containing adsorbed material, and the torque generated is measured after the application of an oscillatory stress (73).

The biconical disk or bob interfacial viscometer is a modification of the flat-disk viscometer. In the typical use of the biconical bob rheometer, a stress is applied to the interface by oscillating the cylindrical cup containing the fluids. This stress in turn confers motion to the bob that is generally monitored through deflection of incident light on a torsion wire. Oh and Slattery have analyzed the biconical bob interfacial viscometer and developed expressions for the torque required to hold the bob stationary when rotating the dish at constant velocity (74). Wibberley applied biconical bob viscometry to the study of aqueous potassium arabate–air and –liquid paraffin interfaces (75). They found these films were pseudoplastic in nature and became more rigid with time. Recently, liquid–liquid interfaces containing various polymers were studied by both the biconical bob technique and by using a deep-channel rheometer (76). Comparable measurements of interfacial viscoelasticity were achieved using both rheometers. The biconical bob device appeared to provide enhanced sensitivity when working with very elastic interfaces as compared to the deep-channel rheometer.

The biconical bob rheometer has also received attention for its use at crude oil–water interfaces (12,77 79). The most important findings include (1) high values of interfacial viscosity and elastic moduli under conditions conducive to W/O emulsion formation; and (2) solid-like behavior of the interfacial films, i.e., the films were observed to prevent the motion of the bob or were difficult to deform. It was also seen that increased aging time allowed stronger and more elastic films to form. Mohammed et al. (12) found that aging times approaching 8 h yielded films with viscosities 400 times greater than those for films aged for 6 h. Beyond 8 h the films were too rigid to deform when subjected to the same applied stress.

In our laboratory, we have applied biconical bob rheometry to highly elastic films formed at asphaltene-containing, model oil–water interfaces (80). Instead of using a torsion wire, however, we have utilized a commercially available, highly sensitive dynamic stress rheometer in conjunction with a serrated edge stainless steel bob. As we will show, the results of the rheometry method compare well with measures of emulsion stability under comparable conditions. Thus, this method provides useful insights into the mechanical origins of emulsion stability in asphaltenic films and provides a platform from which to ask and answer relevant questions about the dependence of stabilized films on asphaltene chemistry and solution properties.

## B. Results and Discussion

### 1. Typical Stress and Frequency Sweeps

The basic experiment utilizing a biconical bob and stress rheometer returns elastic ($G'$) and viscous ($G''$) moduli as functions of oscillatory stress frequency and stress amplitude, reported in radians per second and pascals, respectively. Initial experiments were performed at the limit of asphaltene solubility, 50% toluene and 50% *n*-heptane (v/v), with the asphaltene samples reported here (B6, an offshore California crude). The properties of the crude oil used (B6) and the asphaltenes from this crude are shown in Table 2. Representative frequency and stress sweeps of a well-formed asphaltene interface are shown in Fig. 5. The oil phase containing 0.75% (w/v) B6 asphaltenes in 35 ml of 50% heptane-50% toluene was aged in the presence of water for 8 h. The clear-glass circulating water bath allowed the interface to be viewed beyond the edges of the serrated tooth biconical bob. From below, the interface was visible through the aqueous phase and initially appeared to be the black color of the asphaltenic oil phase. After 8 h, the asphaltene-laden interface became light brown and extended from the glass wall to the bob. At the edge of the bob, the film grew within the spaces between the serrated teeth. First, a frequency sweep was performed at 1 mPa from 0.1 to 3 rad/s (within the linear viscoelastic regime) followed by a stress sweep at 1.0 rad/s from the minimum applied stress of 0.664 mPa to the rupture stress. It should be noted that the minimum applied stress is governed by instrument and stress/strain factors. After 8 h of aging the interface exhibited rheological behavior typical of a gel; note the relative independence of the elastic modulus with frequency,

**Table 2:** Properties of B6 Crude Oil and Asphaltenes

| | |
|---|---|
| B6 Crude Oil | |
| Density (g/ml @ 60° F) | 0.935 |
| Viscosity (cP @ 100°F) | 2030 |
| Asphaltenes (wt %) | 13.1 |
| Resins (wt %) | 12.0 |
| R/A[a] | 0.92 |
| H/C | 1.55 |
| B6 Asphaltenes | |
| H/C | 1.22 |
| Nityrogen (wt %) | 1.91 |
| Sulfur (wt %) | 7.32 |
| Oxygen (wt %) | 1.90 |
| Vanadium (ppm) | 460 |

[a]R/A, mass ratio of resins to asphaltenes

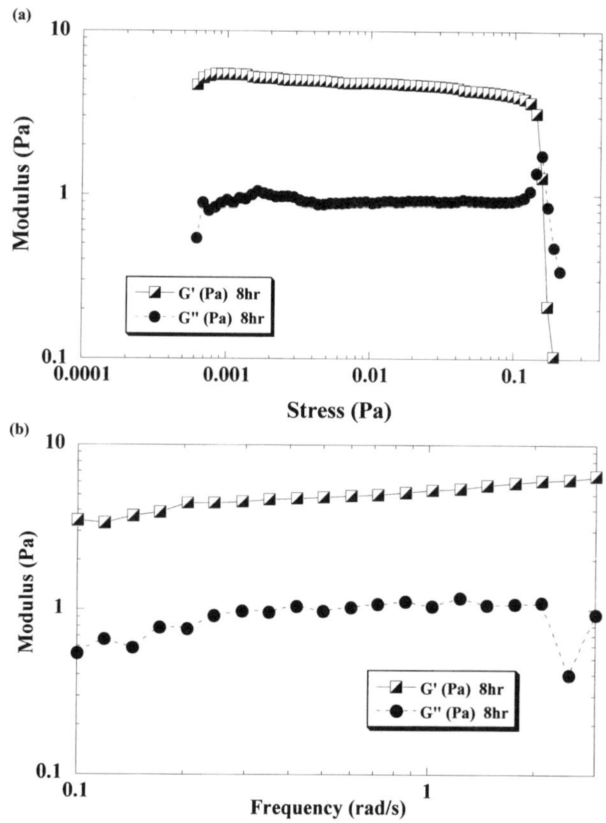


**Figure 5** (a) Elastic and viscous moduli as functions of shear stress obtained from oscillatory shear dynamic rheometry (see text and Ref. 80 for details) performed at the oil–water interface. The system consisted of 40 ml of water on to which 35 ml of an oil phase was layered, consisting of 0.75% (w/v) B6 asphaltenes dissolved in 50% (v/v) toluene in heptane. The asphaltenes were allowed to adsorb for 8 h at which point the rheometrical measurements were performed. The frequency of oscillation of the biconical bob was 1 rad/s. (b) Elastic and viscous moduli as functions of *frequency*. The same system and experiment as described for Fig. 5a were used with the exception that the shear stress was fixed at 1 mPa and the frequency was varied. Asphaltenes were allowed to adsorb for a period of 8 h.

and its value was six to seven times that of the viscous modulus. Above 2 rad/s the response to the applied oscillation became clouded by the effects of bob inertia. From the stress sweep (Fig. 5a), the viscous and elastic moduli were unchanging over two decades of applied stress. As the rupture stress was reached, the film began to break down as witnessed by the sharp decrease in elastic modulus. At the same time, the viscous modulus rose to a maximum when the film ruptured. This peak in $G''$ is likely caused when the initial solid-like character of the interface is transferred to liquid-like behavior upon yielding. Further details of the experiment are provided elsewhere (80).

### 2. Kinetics of Film Formation

In Fig. 5 we saw evidence of appreciable interfacial film formation after 8 h of aging. In Fig. 6, film growth was monitored over a 24-h period beginning at 2 h. Even though there was evidence of asphaltene adsorption within minutes of the oil phase addition, there was not sufficient coverage or consolidation to impart a measurable signal when probed by the rheometer. The most significant film growth occurred in the first several hours of aging in which $G'$ increased roughly two-fold from 2 to 4 h. As the aging time lengthened, the rate at which $G'$ increased tapered off. The important stages of film growth were highlighted by the initial film formation in the first hours when asphaltenic aggregates diffused to the interface and were adsorbed. As time progressed, the adsorbed asphaltenes likely consolidated and imparted a higher elastic modulus; in addition, the diffusive process continued and more asphaltenic aggregates migrated and were adsorbed at the interface.

Figure 7 shows the result of 8 and 24 h stress sweeps on two interfaces formed under similar conditions. The 8-h experiment is replotted from Fig. 5a. A two-fold increase in both $G'$ and the rupture stress is observed after 24 h as compared to 8 h. In both films the linear viscoelastic region spanned the entire spectrum of applied stress until rupture occurred.

### 3. Effect of B6 Concentration (0.25, 0.75, 1.5% w/v) and Kinetics

Figure 8 illustrates the effects of varying the bulk asphaltene concentration on the elastic moduli of oil–water asphaltenic films as a function of frequency and aging time. Even though the asphaltene concentrations were increased 3-fold from 0.25 to 0.75% and 6-fold from 0.25 to 1.5% the corresponding increase in elastic modulus was not as great. Values of $G'$ taken at 8 h and 1 rad/s increased by a factor of 2.24 from 0.25 to 0.75% (w/v) and a factor of 3.1 from 0.25 to 1.5%. When working at very low bulk concentrations, incremental increases in asphaltene content lead to large jumps in film magnitude. There is likely a minimum

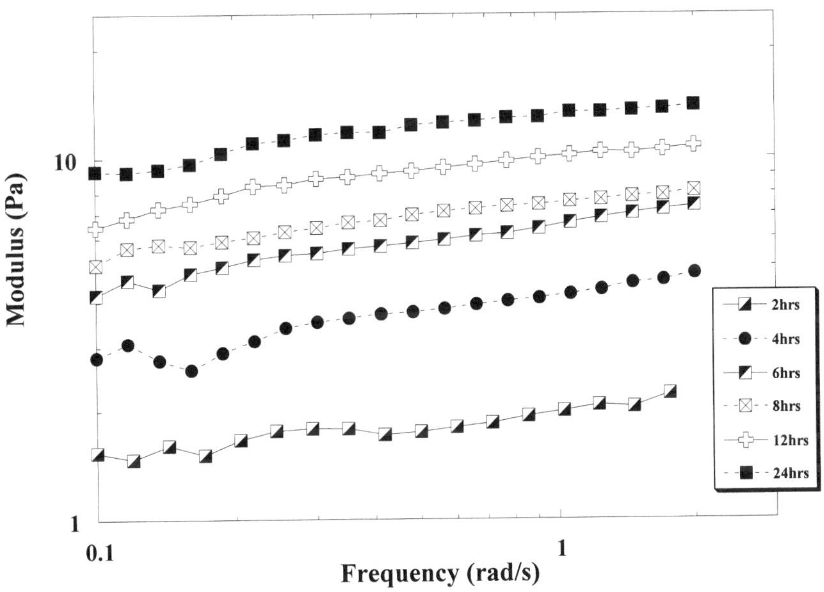


**Figure 6** Elastic modulus as a function of shear stress for the system described in Fig. 5. Aging time was varied from 2–24 h and the shear stress was fixed in 1 mPa.

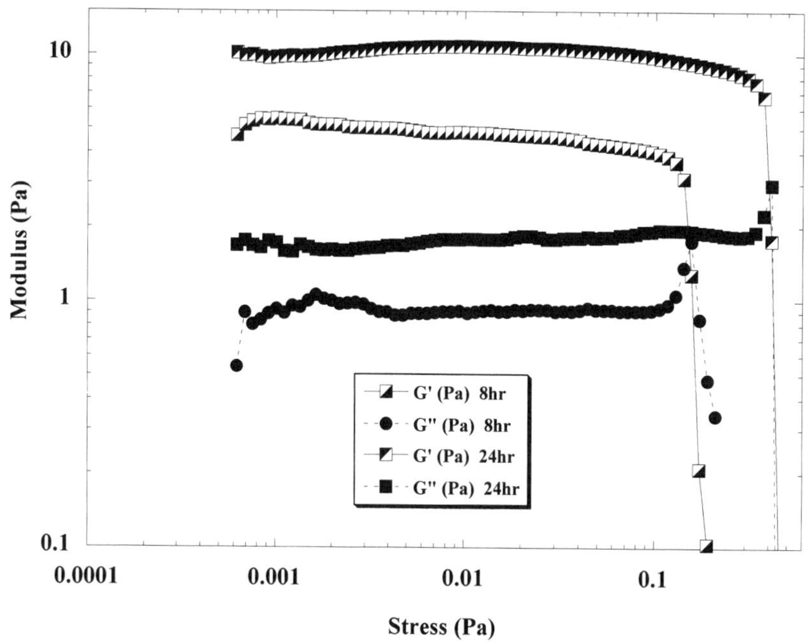


**Figure 7** Elastic and viscous moduli as functions of shear stress for the system described in Fig. 5. The result of aging from 8 to 24 h is shown.

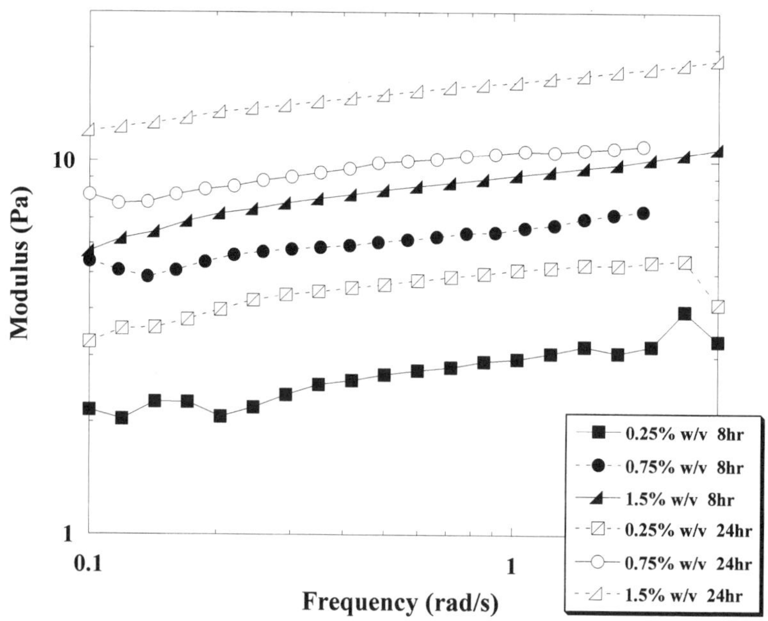


**Figure 8** Elastic modulus as a function of frequency obtained from oscillatory shear dynamic rheometry (see text) performed at the oil–water interface. The proportions of oleic and aqueous phase are the same as described for Fig. 5. Concentrations of 0.25, 0.75, and 1.5% (w/v) B6 asphaltenes were dissolved in the oleic phase (50% v/v toluene in heptane) before adsorption. Samples were aged for 8 or 24 h. Shear stress was held constant at 1 mPa.

bulk concentration necessary to generate a film with sufficient cohesive properties to confer a signal with the stress rheometer.

The stress sweeps corresponding to the 24-h frequency sweeps of Fig. 8 are shown in Fig. 9 for all three asphaltene concentrations tested. In each case the films demonstrated linear viscoelastic behavior and ruptured cleanly upon reaching their critical stress. In addition to the increase in $G'$ from 0.25 to 1.5%, we saw a similar trend in rupture stress. From 0.25 to 0.75% the yield stress increased by a factor of 2.2 and increased by a factor of 3.56 from 0.25 to 1.5%. These factors of increase closely followed the values obtained for the elastic moduli, but were not as large as the three- and six-fold differences in asphaltene concentration. The adhesion of the film to the bob was critical for accurate and reproducible measurement of modulus and yield stress. Different geometries of the biconical bob and its edge resulted in similar elastic moduli, but yield stress values corresponded to their ability to "grip" the films. A smooth, knife-edged bob would not maintain sufficient contact with the films to obtain yield stresses as high as the serrated tooth bob. If adhesive forces exceeded the internal film yield stress, the various bob geometries would return similar values for yield stress regardless of geometry. In our case, the film yield stresses clearly exceed the tangential adhesive force between the film and the bob. In a few cases where the films were exceptionally high in modulus and yield stress, such as the 1.5% (w/v), 24-h experiments, the film and bob remained in contact, but rupture occurred at the glass surface.

The $G'$ data were selected from 2–24 h frequency sweeps at 1 rad/s and plotted for each asphaltene concentration as a function of time. The moduli at short times tended to increase quickly and then shifted to a region of more gradual increase. The most rapid asphaltene adsorption likely occurred in the short period up to 6 or 8 h. The additional accumulation of asphaltenes was hindered by the material present at the interface and so the adsorption rate decreased.

## 4. Effects of Asphaltene State of Aggregation and the Role of Aromaticity

The solubility of asphaltenes is highly dependent on the medium in which they are placed (81). The presence of dissolved asphaltenes in crude oil is mediated by a combination of crude aromaticity and petroleum resins that act to solvate asphaltene aggregates. Adding an excess of aliphatic solvent, namely *n*-heptane, sufficiently reduces the solubility of asphaltenes in crude oil and causes precipitation. To perform subsequent film experiments these precipitated asphaltenes were then redissolved in toluene. As *n*-heptane was added to the asphaltene–toluene solutions, the

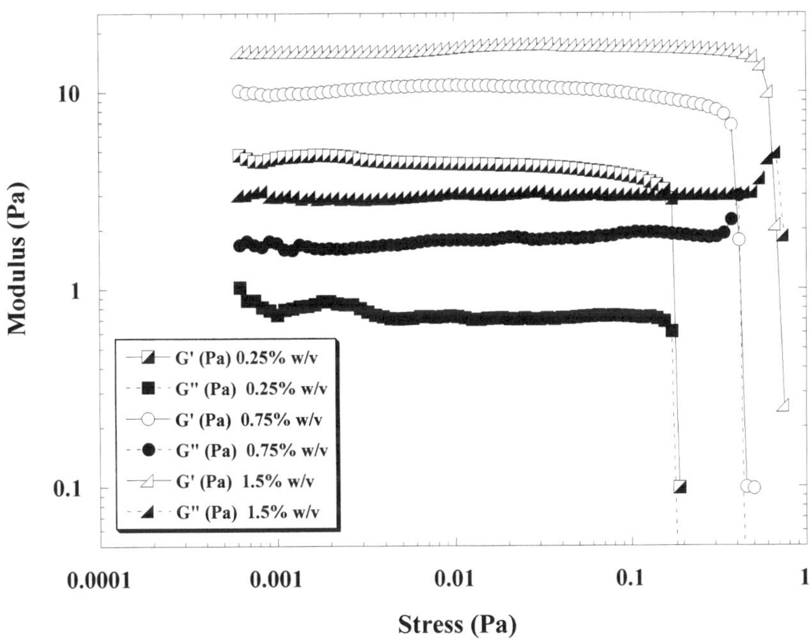


**Figure 9** Elastic and viscous moduli as functions of *shear stress* for the system described for Fig. 8. Samples were aged for 8 h and the frequency of oscillation of the biconical bob was 1 rad/s.

state of aggregation began to increase. At a high toluene concentration (above 60% v/v) negligible precipitated material remained after filtration through a 1.6-μm syringe filter, as seen in Fig. 10. Decreasing the aromaticity of the solution to 50% (v/v) toluene began to increase the size of large aggregates. This is the point of incipient flocculation which is characterized by asphaltene aggregates whose size approaches that necessary for precipitation. Further reduction in aromaticity caused greater amounts of asphaltenes to precipitate, increasing to 100% precipitation in pure *n*-heptane.

The asphaltene solubility experiments provided the locations of important regions to perform subsequent film rheology experiments. Above 50% (v/v) toluene the asphaltene aggregates are increasingly soluble and we would expect the driving force for interfacial adsorption to be low. A low driving force would result in correspondingly weak films because the asphaltene aggregates remain in solution and fewer asphaltenes adsorb. As the percentage of toluene in the solvent approaches 50, the asphaltene aggregates become less soluble in the heptane–toluene mixture. This situation arises because the intermolecular $\pi$- and hydrogen-bond forces between individual and stacks of asphaltenes begin to overcome the solvating ability of the toluene. We expect enhanced film adsorption and modulus as the aggregates become less soluble in the heptane–toluene mixture and are driven to the oil–water interface where hydrogen bonding interactions dominate. Once the aromaticity of the solvent drops below 50% (v/v) toluene, some fraction of the asphaltene aggregates become unstable thermodynamically and precipitate. The fraction of material that precipitates grows as the percentage of toluene decreases until the solvent is purely aliphatic, whereupon all of the asphaltenes precipitate.

In Fig. 11 we see the effect of increasing aromaticity on asphaltenes at 0.75% (w/v) and 24 h aging. At 50% toluene the asphaltene aggregates are at the point of incipient flocculation and form the strongest films. Above 50% toluene the aggregates become well solvated and the elastic modulus of the film correspondingly decreases. Clearly, the enhanced aromaticity has solvated the asphaltene aggregates well and rendered them less interfacially active. The resulting films at higher toluene concentration are weaker.

Below 50% toluene, a fraction of the asphaltenes precipitate. At 40% toluene this fraction approaches 20%. These precipitates are drawn by gravity to the interface and compete with soluble aggregates for the interface. The mixture of soluble and insoluble asphaltenes leads to much weaker films, as shown in Fig. 12. These films were allowed to age for 8 h and are best compared to the 8-h films created at 50% toluene. At 50% toluene the interfacial film is an elastic gel with a distinct yield stress of $\simeq 0.15$ Pa.

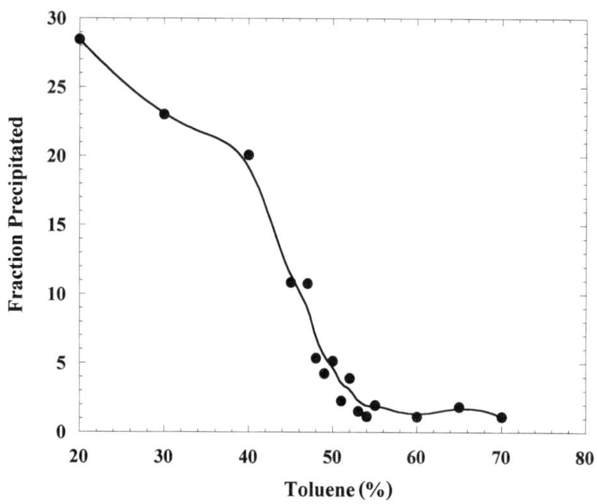


**Figure 10** Fraction of precipitated B6 asphaltenes as a function of toluene content (v/v) in heptane. Experiments were performed gravimetrically by weighing the % of precipitated asphaltenes which were retained by a 1.6-μm filter.

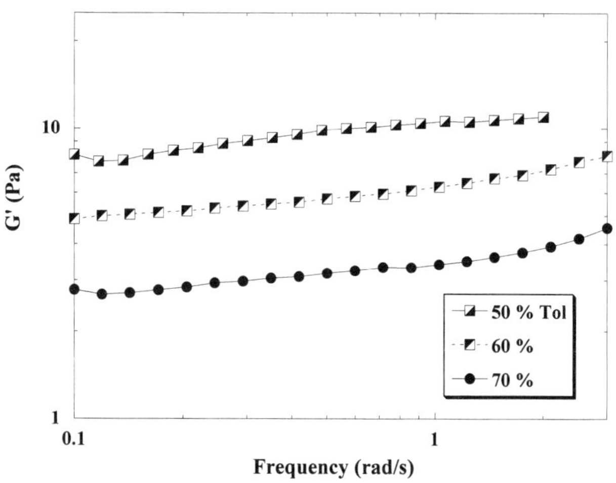


**Figure 11** Elastic modulus as a function of frequency and toluene concentration obtained from oscillatory shear dynamic rheometry (see text) performed at the oil–water interface. The proportions of oleic and aqueous phases are the same as described for Fig. 5. A concentration of 0.75% (w/v) B6 asphaltenes was dissolved in the oleic phase [varying % (v/v)] toluene in heptane before adsorption. Samples were aged for 8 h. Shear stress was held constant at 1 mPa.

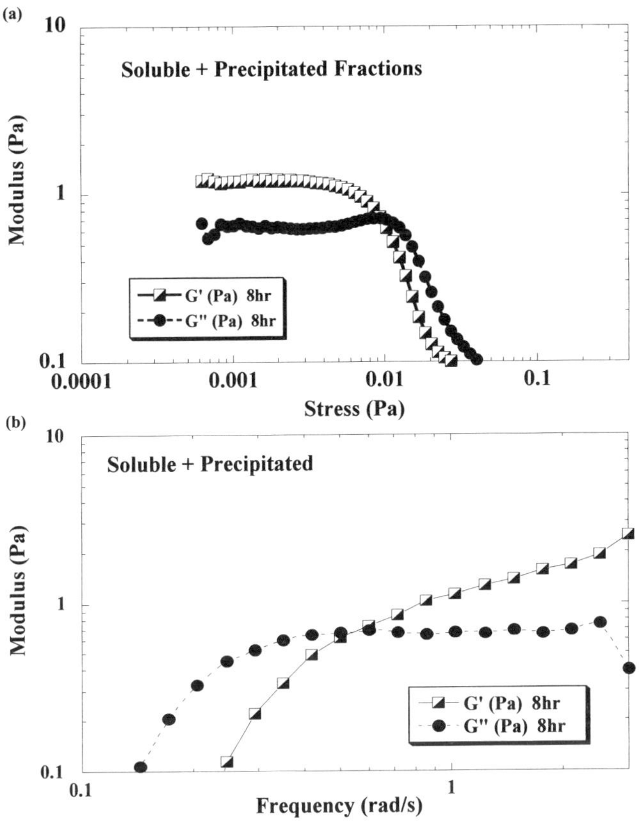


**Figure 12** (a) Elastic and viscous moduli as functions of shear stress. The proportions of oleic and aqueous phase are the same as described for Fig. 5. A concentration of 0.75% (w/v) B6 asphaltenes was dissolved in the oleic phase [40% (v/v)] toluene in heptane before adsorption. Samples were aged for 8 h. The frequency was held constant at 1 rad/s. (b) Elastic and viscous moduli as a function of *frequency*. The same system and experiment as described for Fig. 12a were used with the exception that the shear stress was fixed at 1 mPa and the frequency was varied. Asphaltenes were allowed to adsorb for a period of 8 h

In the presence of precipitated material (40% v/v toluene) the interface is viscoelastic ($G'$, $G''$ crossover between 0.5 and 0.6 rad/s, see Fig. 12b) with moduli that are considerably more frequency dependent than at 50% toluene. In addition, the plateau elastic modulus falls nearly five-fold below the soluble condition and the yield stress is less than ten times that of the soluble condition. Thus, the films formed at 40% (v/v) toluene are in every respect weaker, less solid-like, and less gel-like than those formed at 50% (v/v) toluene, despite the fact that considerably more material resides at the interface in the 40% (v/v) toluene experiment. It is clear that precipitated asphaltenes form far weaker and less elastic interfaces than those of soluble asphaltenes. This may be due to a high concentration of defects and grain boundaries in the precipitated interface case. In the soluble asphaltene case, one can speculate that the lability of the small asphaltenic aggregates adsorbing at the interface enables them to knit themselves into a more cohesive film with fewer weak spots.

The evidence of varied film characteristics in the presence of precipitated material led us to test the effectiveness of the remaining soluble material to form a film. One of two scenarios likely caused the dramatic reduction in film strength: either (1) precipitates greatly outnumbered soluble asphaltenic aggregates which were not effective at enhancing film properties in the absence of precipitated material; or (2) precipitates hindered the formation of a stable film even though there was an adequate supply of interfacially active soluble aggregates. After removing the precipitated material from a stock solution of 0.75% (w/w) asphaltenes in 60:40 heptane:toluene, the remaining soluble asphaltenes were isolated and redissolved in 50:50 heptane:toluene. Under these conditions a weak film would indicate that the interfacially active material was primarily located in the fraction removed as precipitate. In Fig. 13 we see that the soluble material was marginally capable of forming a measurable film. The film formed was weakly viscoelastic and had a plateau $G'$ less than half that of the precipitated film and more than ten times lower than that of the whole asphaltene solution.

### C. Summary of Conclusions Drawn from Interfacial Rheological Studies

The results of the interfacial rheological studies on asphaltene adsorption at oil–water interfaces teach us a great deal about the behavior of asphaltenes and their role in emulsion stabilization. The conclusions drawn are based largely on the assumption that the rheological properties measured, namely the elastic film modulus $G'$, are directly related to the surface excess concentration of asphaltenes. Γ. It is understood that the elastic modulus actually depends on both the surface excess concentration and the relative conformation (i.e., connectivity) of the adsorbed asphaltenes. It is further understood that a minimum adsorbed level is required to observe a finite value of $G'$ and that the relationship between $G'$ and $G$ is not linear. With these caveats in mind, we can conclude the following:

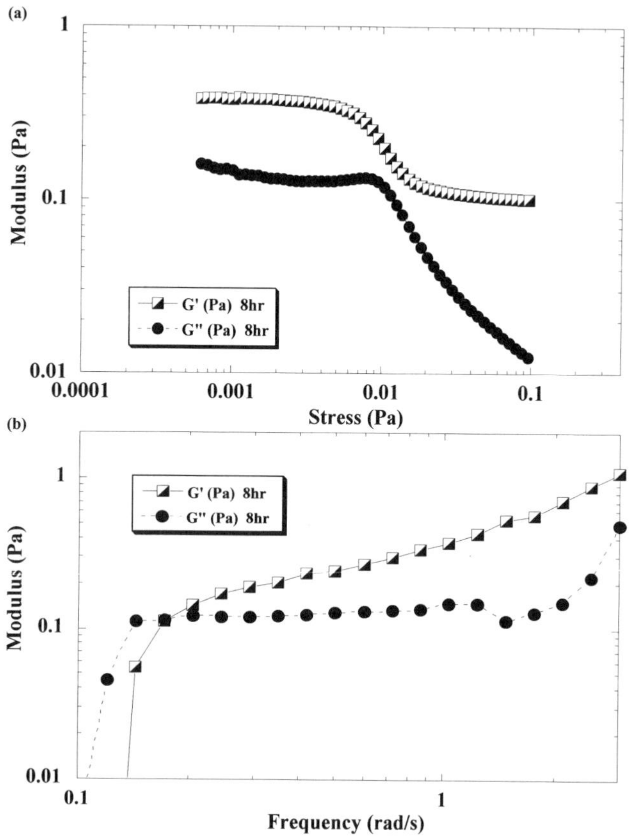


**Figure 13** (a) Elastic and viscous moduli as functions of shear stress. The proportions of oleic and aqueous phase are the same as described for Fig. 5. A concentration of 0.5% (w/v) *soluble* B6 asphaltenes was dissolved in the oleic phase (40% v/v toluene in heptane) before adsorption. These asphaltenes differed from the original *whole* sample of B6 asphaltenes in that the fraction which precipitated in 40% toluene in heptane was removed. Samples were aged for 8 h. The frequency was held constant at 1 rad/s. (b) Elastic and viscous moduli as functions of *frequency*. The same system and experiment as described for Fig. 13a were used with exception that the shear stress was fixed at 1 mPa and the frequency was varied. Asphaltenes were allowed to adsorb for a period of 8 h

1. Asphaltenes adsorb in a manner limited by the diffusion of soluble asphaltenic aggregates to the oil–water interface, whereupon they self-assemble to form an elastic, rigid stable film of high mechanical strength. It is this film which is primarily responsible for the stability of asphaltenic W/O emulsions.
2. The adsorption process appears to proceed indefinitely, provided that there is an adequate inventory of asphaltenes in the bulk phase and there is sufficient time for adsoption. What this implies, of course, is that the asphaltenes can always lower their free energy (relative to remaining in a solvated slate in the bulk) by adsorbing to a structured, self-assembled third-phase film at the oil–water interface. Clearly, in emulsion systems, the oil–water interfacial area per unit mass of asphaltenes is considerably larger than in the case of a bulk planar interface. Thus, in the former case (emulsions), adsorption is more likely to be asphaltene limited, while in the latter case (bulk planar interface), adsorption is more likely to proceed indefinitely.
3. The bulk concentration of asphaltenes appears to be an important variable driving asphaltene adsorption, higher concentrations leading to greater adsorption.
4. The value of $G'$ measured by the biconical bob method appears to be independent of bob material or geometry. The yield stress of asphaltenic interfacial films appears to depend on bob material and peripheral area of contact of the bob with the interface.
5. Solvent plays a critically important role in driving asphaltene adsorption, presumably related to the solvation of individual asphaltenic aggregates and molecules and the overall solubility parameter of the solvent versus that of the asphaltenes. Asphaltenes have their greatest tendency to adsorb and make the strongest interfacial films per unit mass at their limit of solubility.
6. Asphaltenes clearly consist of a distribution of molecular weights, types, and polarity, which can be conveniently subdivided according to their solubility behavior in mixtures of a "good" solvent (e.g., toluene or methylene chloride) and a 'poor' or nonsolvent (e.g., heptane). The most polar or least soluble fraction of asphaltenes (as gaged by this solvent–nonsolvent fractionation method) appears to form the strongest interfacial films, all other thermodynamic parameters being equal.

## IV. STUDIES OF ASPHALTENE EMULSIONS AND THEIR STABILITY TO COALESCENCE

### A. Introduction

As discussed throughout this chapter, asphaltene emulsions are definitely stabilized by a rigid, elastic, crosslinked network of asphaltenic aggregates adsorbed from the bulk oil phase at the oil–water interface. This picture of emulsion stabilization has been described in some detail elsewhere [McLean et al. (45)]. Moreover, that previous review also summarized in its Tables 2 and 4 at least 50 citations of previous work performed on either crude oil–water emulsions or asphaltene–oil–water emulsions, or of studies on planar interfaces of oil and water at which asphaltenes had been adsorbed. Again, the consensus picture from all of these studies is the self-assembly of asphaltene at the oil–water interface to form a rigid, elastic interfacial film, often characterized as a "skin" or "plastic film." Many thermodynamic and molecular structural variables attenuate or mediate this adsorptive self-assembly process. Several of these were alluded to in the previous section summarizing interfacial rheological studies peformed in our laboratory.

## B. Factors Affecting Asphaltene Adsorption and Self-Assembly

As seen in Sec. III.B.3, the *state of solvation* of asphaltenic aggregates affects critically (1) their tendency to adsorb at oil–water interfaces; (2) their solubility; and (3) the mechanical strength of the film formed. Three key factors govern this state of solvation: (1) the aromaticity of the solvent (i.e., the balance between aliphatic and aromatic hydrocarbons in the solvent medium); (2) the relative proportions of asphaltenes and resin molecules in the mixture; and (3) the details of the structural chemistry of the asphaltene and resin fractions. This last factor is probably the least well understood at this stage because of the challenges associated with fractionating and analyzing the chemical functionality of asphaltenes and resins in a standardized fashion, which clearly delineates the details that govern aggregation and film formation. The role of aromatic and aliphatic solvents, and of resins, are better understood at this stage.

## C. Methods for Probing Stability: % Resolved and Centrifugation Tests

Most of the early methods developed for probing the relative stability of asphaltene and crude emulsions were centered on monitoring the amount or percentage of an emulsion which will phase separate on a macroscopic scale into a bulk continuous or disperse phase under the influence of gravity or a centrifugal field. These experiments are usually performed in vials or bottles under ambient conditions and are often referred to as "bottle tests." While the methods clearly are an oversimplification of the more complex situations found in reservoirs or refineries, in which the crude experiences high pressures and higher temperatures, these methods are nonetheless still quite useful in understanding factors which influence emulsion stability and relative trends. In the sections below, we will review some of the key results uncovered using these methods with respect to the primary factors that influence asphaltene emulsion stability. We will then discuss the results of a more recently developed method, the critical electric field for W/O emulsion breakdown.

### 1. Role of Solvent Aromaticity in Asphaltene Emulsion Stability

Many researchers have demonstrated that varying solvent aromaticity has a profound effect on asphaltene solubility, asphaltene aggregation, and adsorption at interfaces and resulting emulsion stability. In Sec. II, recent SANS data on asphaltene–heptane–toluene solutions indicate that asphaltenes aggregate to form clusters of approximate diameter 80 Å ($R_g = 40$ Å) in solutions of high toluene content. Near the solubility limit (which is approximately 50% heptane in these solutions), the aggregate size reaches a maximum. This certainly suggests that the intermolecular forces between asphaltene molecules (relative to solvent–asphaltene forces), which give rise to aggregation, are most attractive right at the solubility limit. Beyond the solubility limit, those asphaltene molecules that possessed the strongest attractive forces for each other have presumably precipitated and the remaining soluble asphaltene molecules form aggregates of smaller dimension (see Fig. 3). This trend appears to hold in interfacial film rheology and in emulsion stability, as we will show. The dominant theme suggested by these and other experiments is that asphaltenes are most prone to aggregation and surface adsorption at their limit of solubility. Aromaticity and resin:asphaltene ratio are both key thermodynamic variables in dictating this solubility, as we will discuss.

McLean and Kilpatrick (82) described in detail the role of aromatic solvent in controlling emulsion stability in both crude oil–water and model asphaltene–heptane–toluene–water mixtures. Maximum stability was observed in a host of model systems at or near the limit of solubility, as discussed above. This typically occurs at a heptane–toluene fraction of 50–60% heptane (50–40% toluene). Figure 14 illustrates a typical profile of percentage water resolved in a

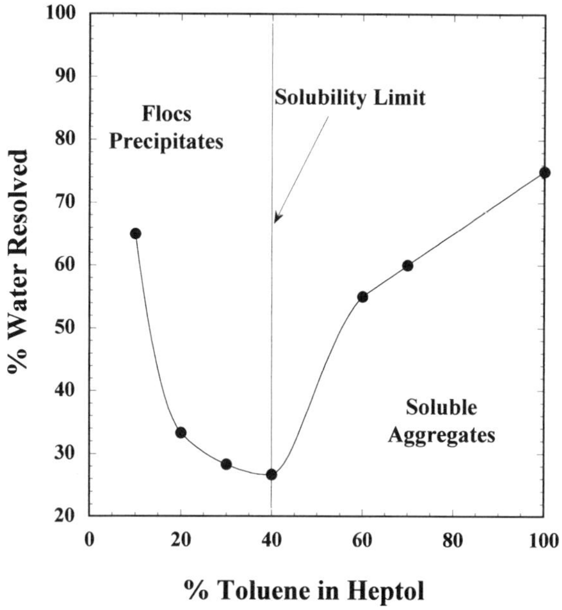


**Figure 14** Emulsion stability as gaged by % water resolved following centrifugation (see Ref. 82 for details) versus % (v/v) toluene in heptane for 0.5% Arab Heavy (AH) asphaltenes dissolved in oleic phase and mixed with water in a 40:60 (v/v) proportion. The AH asphaltenes are partially precipitated below 40% toluene.

model system of asphaltenes–heptane–toluene–water as percentage toluene in the oleic solvent is varied. The asphaltenes utilized were those from Arab Heavy crude oil (or Safaniya), although quite similar results were observed with other crude asphaltenes under these conditions, such as Alaska North Slope, Arab Berri, and San Joaquin Valley (a California crude). Annotated on Fig. 14 is the location of the solubility limit. Upon comparison of this limit with differing asphaltene systems, it is observed to vary somewhat depending on the chemistry of the asphaltene fraction (83). As can be seen, the evidence from rheological data on asphaltene adsorption at planar oil–water interfaces and the "bottle tests" on percentage water resolved in a centrifugation test both indicate that film strength and emulsion stability are maximized at the solubility limit. This suggests that the driving force for adsorption of asphaltenes increases as the solubility limit is approached. Moreover, the lability of the asphaltenic aggregates appears to be a very important factor in dictating the ease with which asphaltenic films self-assemble. It is this facile self-assembly and the strength of interasphaltene attraction (i.e., H-bonding versus $\pi$-bonding or dispersion forces) which leads to integral films with high elastic and mechanical strength.

### 2. Role of Resins and Resin: Asphaltene Ratio in Asphaltene Emulsion Stability

The role of resins in solvation (also called peptization) of asphaltenes has been presumed for many decades. It is believed that resins solvate asphaltenes and enhance their solubility by interacting with the polar and aromatic groups through comparable moieties in the resins (illustrated schematically in Fig. 2). In model systems of asphaltenes dissolved in heptane and toluene near or at the limit of solubility, addition of resins can initially enhance emulsion stability at resin:asphaltene mass ratios of 2:1 or less. This can be understood in the light of a molecular model in which the resins assist the asphaltenic aggregates by dynamically solvating them (i.e., resin monomers are freely exchanging between the surface of the aggregates and the surrounding solvent and creating adsorbable species, which are more labile than the resin-free asphaltenic aggregates and can then self-assemble at the oil–water interface). Thus, one can understand the asphaltenes in the absence of resins to be somewhat static (non-exchanging) aggregates which have a limit of solubility and which adsorb at the interface at this limit as maximally sized aggregates (80–100 Å diameter) and self-assemble, but without complete freedom of structural rearrangement at the interface to knit themselves into an elastic film in the shortest possible times. With the addition of resins at modest levels (R:A $<$ 2:1), the asphaltenic aggregates are solvated and now somewhat smaller and more labile in self-assembling. Presumably, this can lead to elastic films as strong or stronger than purely asphaltenic films, all other things being equal (such as the total amount of adsorbed asphaltenes). Of course, with the addition of resins and the concomitant solvation of the aggregates, the asphaltenes are now somewhat less surface active and it is this balance of lability and surface activity which governs the emulsion stability of the system. As seen in Figs 15 and 16, this can produce a local maximum or plateau in emulsion stability (as gaged by a local minimum in percentage water resolved in a centrifugation experiment), or it can simply lead to decreasing stability monotonically as resins are added. Clearly, the chemistry and the resulting molecular specifics of resin : asphaltene interactions are what ultimately govern details of the relative surface activity and film stability (82). This is an active area of research in our laboratory and others.

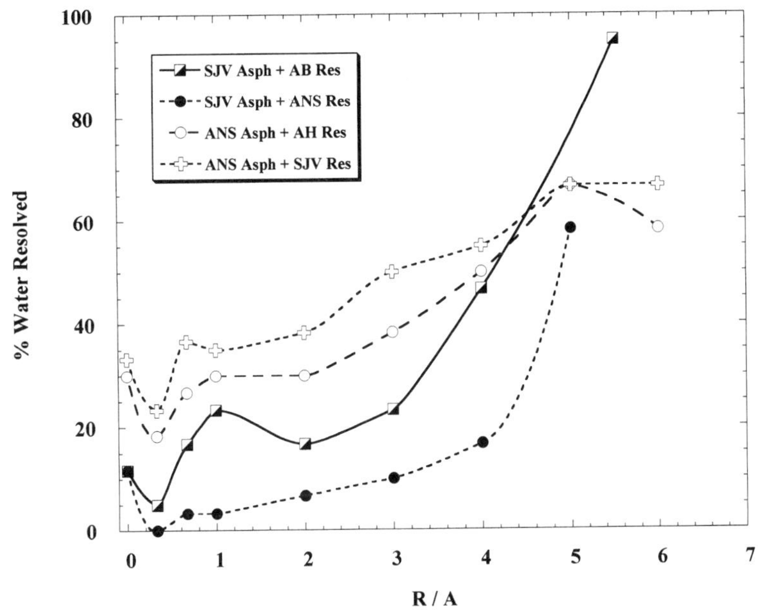


**Figure 15** Emulsion stability as gaged by % water resolved following centrifugation (see Ref. 82 for details) versus % (v/v) toluene in heptane for 0.5% (w/w) San Joaquin Valley (SJV) asphaltenes and 0.5% (w/w) Alaska North Slope (ANS) asphaltenes dissolved in 30% (v/v) toluene in heptane and mixed with water in a 40:60 (v/v) proportion. Varying amounts of resins from AH, ANS, SJV, and AB (Arab Berri) crude oils were added to the oleic solutions to give ratios (w/w) of resins to asphaltenes ranging from 0–6. In these experiments, it should be noted that a large range of R/A exists for which the emulsion stability is fairly invariant.

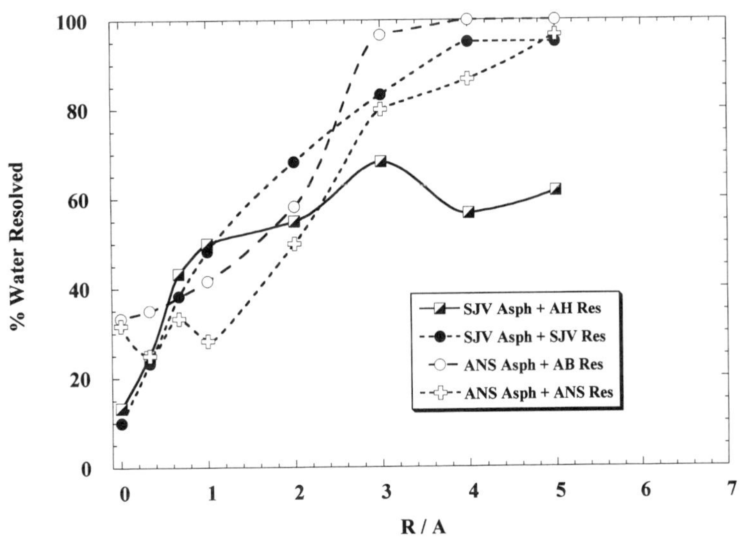


**Figure 16** Emulsion stability as gaged by % water resolved following centrifugation (see Ref. 82 for details) versus % (v/v) toluene in heptane for 0.5% (w/w) San Joaquin Valley (SJV) asphaltenes and 0.5% (w/w) Alaska North Slope (ANS) asphaltenes dissolved in 30% (v/v) toluene in heptane and mixed with water in a 40:60 (v/v) proportion. Varying amounts of resins from AH, ANS, SJV, and AB (Arab Berri) crude oils were added to the oleic solutions to give ratios (w/w) of resins to asphaltenes ranging from 0–6. In these experiments, it should be noted that the emulsion stability is sensitive to R/A and the resins destabilize the emulsions at even the lowest ratios. Please note that the asphaltene/resin combinations are different from those in Fig. 15.

### 3. Role of Fused Aromatic Ring Additives in Asphaltene Emulsion Stability

Much of our understanding of asphaltene emulsion stability and its relative dependence on aromaticity and R/A is based on our molecular picture of asphaltene aggregate structure and how this should depend on aromatic and resinous additives. Those molecular additives which solvate most strongly the asphaltenes will have a profound effect on asphaltene aggregate size and emulsion stability (as well as surface activity). One set of experiments performed to illustrate this were those in which fused-ring aromatic additives were combined with a crude oil system known to form strong emulsions (84). The resulting emulsion stability of the "doped" crude oil when blended with water was monitored and is shown in Fig. 17. As seen here, toluene initially stabilizes the crude emulsion and, after sufficient addition of 20–25 vol %, destabilizes the emulsion. Again, as described above, this is understood in terms of the solubility and surface activity of the asphaltenes as the solvent becomes increasingly aromatic. Addition of methylnaphthalene has an even more profound effect on emulsion stability, destabilizing the emulsion after addition of 10–12 vol %. Finally, a polar, fused-ring aromatic compound phenanthridine (three fused rings with a basic nitrogen) was blended into methylnaphthalene and added, and at very modest phenanthridine concentrations (< 1–2 vol %) dramatically destabilized the water-in-Arab Heavy crude oil emulsion. We explain this on the basis of the molecular model depicted in Fig. 2; the phenanthridine behaves somewhat like an asphaltene mimic and inserts itself into the cofacial, stacked aggregates. This molecule is highly aromatic as well as being an electron donor (proton acceptor). If one of the primary modes of interaction between asphaltene molecules is H-bonding among H-bond donors and acceptors (such as carboxyl, phenolic, amine, pyridinic, etc.), one can envision these phenanthridine molecules breaking up H-bond mediated crosslinks between the asphaltenes. This in turn would reduce the elasticity of the interfacial films formed and reduce emulsion stability. We have proposed that this is precisely what happens in experiments embodied by Fig. 17 (84). If this is indeed the case, this has profound implications for inventing methods of minimizing emulsion formation during production and refining and, possibly, for managing asphaltene deposits.

## D. Methods for Probing Emulsion Stability: Critical Electric Field Measurements

Sjöblom and coworkers have developed and applied an electrical method for assessing the stability of W/O

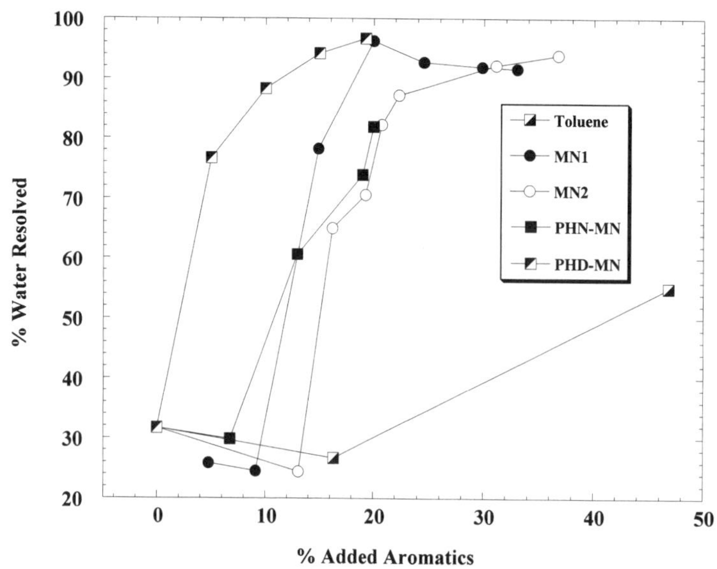


**Figure 17** Emulsion stability as gaged by % water resolved following centrifugation (see Ref. 84 for details) versus % (v/v) aromatic solvent modifier added. The emulsions consisted of oil and water mixed in a 40:60 ratio for 0.5% (w/w) Arab Heavy asphaltenes dissolved in oleic phase and mixed with water in a 40:60 (v/v) proportion. The AH asphaltenes are partially precipitated below 40% toluene.

emulsions, called the critical electric field (cef) method (20,85–88). The technique relies on the very low ionic conductivity of oleic phases and the polarizability of dispersed water droplets into which a modest (< 1% w/w) amount of electrolyte has been dissolved. By slowly increasing the d.c. voltage applied across a narrow gap between two electrodes in which a W/O emulsion has been placed, the droplets become polarized and align themselves in the field created between the electrodes. As the field strength is increased the droplets begin to chain together (see Fig. 18) and become distorted (89–92). When the field strength is sufficiently large, the electromotive force on the ions and the distortion of the droplets become sufficient for the interfacial films stabilizing the droplets to rupture. When this occurs and a sufficient number of droplets have ruptured, the conductivity measured across the gap width increases markedly (by orders of magnitude). The field strength at which this occurs is termed the "critical electric field" and is typically of the order of 0.01–10 kV/cm in asphaltenic emulsions. The dynamic range of three orders of magnitude provides a useful gage of ordering the relative strength of W/O emulsions. The technique is somewhat subtle and requires great care and patience in sampling the system, performing repeated measurements, carefully accounting for the dynamics of film formation and droplet coalescence as a function of time in the samples, droplet sedimentation, and other factors confounding the reproducibility of the results. Nonetheless, in the hands of a skilled and experienced practitioner, the method is extremely useful and, as we shall see, agrees quite well with results from rheological studies and bottle tests.

### 1. Kinetics of Critical Electric Field Development

The time scale over which asphaltenic aggregates adsorb at water droplet–oil interfaces and begin to stabilize emulsions is extremely fast, of the order of seconds after the droplets are created by shear. This is not surprising when one considers the extremely short diffusion times required over short length scales. Typical droplet sizes produced in strong shear are of the order of micrometer or less when the Weber numbers are high (> 100+). Thus, the diffusion times of asphaltenic aggregates with diffusivities of $10^{-6}\,\text{cm}^2/\text{s}$ over interdroplet separation distances of $\simeq 1\,\mu\text{m}$ should be:

$$D = l^2/t;\; t = l^2/D = (1\,\mu\text{m})^2/10^{-7}\,\text{cm}^2/\text{s} = 0.1\,\text{s}$$

As should be apparent from the above simple calculation, within a matter of seconds in a dispersion of fine water droplets in crude oil (or model oils containing asphaltenes), an appreciable inventory of adsorbed asphaltenic aggregates begin to self-assemble and stabilize the oil–water interface. This is in stark contrast to the bulk oil–water interface discussed in the section on rheology, in which the relevant diffusion length is in

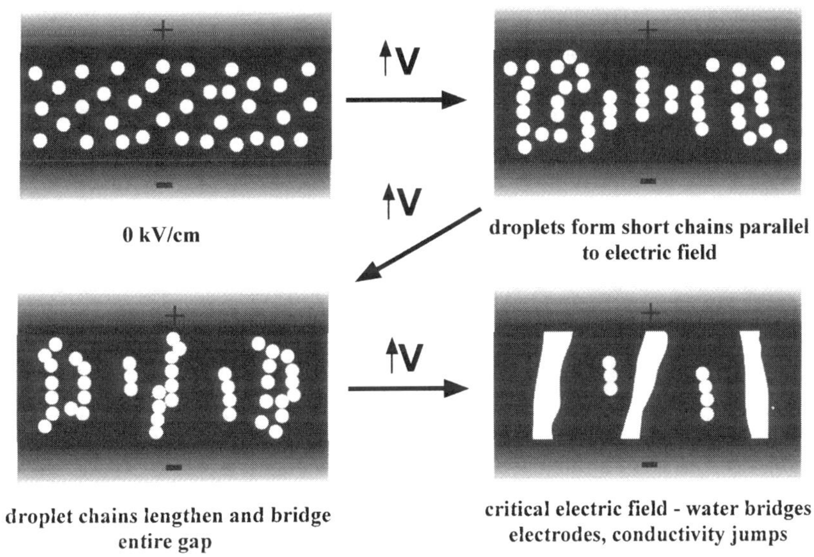


**Figure 18** Schematic illustration of the putative process of emulsion breakdown during the measurement of cef in W/O emulsions (see Refs 20 and 85–92 for detailed discussion of the method and phenomenon; Ref. 93 describes our application).

millimeters. In that experiment, the diffusion times are at least $10^6$ times longer, or several hours, precisely the time scale we observe in those experiments (see discussion in Sec. III).

The time scale for the dynamic development of a cef in model asphaltene–heptane–toluene emulsions with water is illustrated in Fig. 19, in which solutions of Arab Heavy (AH) asphaltenes (0.5–1.0% w/w) in 40–50% toluene-in-heptol are emulsified with 30% water, and the cef is monitored as a function of time. As should be apparent, there is a characteristic time scale for the cef to rise markedly towards its steady-state value which varies with asphaltene concentration and solvation state of the asphaltenes. The time scale is most rapid at the limit of solubility (40% toluene) and with the higher concentration of asphaltenes (1% versus 0.5%). As concentration is reduced, the time required to reach the near-steady value decreases. Interestingly, it appears to be the same value, regardless of concentration ($\simeq 1.2$ kV/cm). Also, as the toluene concentration is increased from 40 to 50%, the time scale increases and the long-term value of the cef decreases. This is consistent with a reduced driving force for adsorption of aggregates and with a reduction in total adsorbed material.

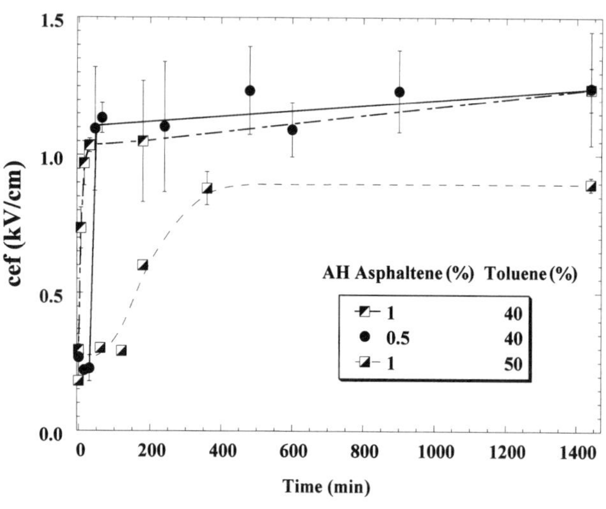


**Figure 19** Variation in model oil emulsion stability with time. Emulsions were stabilized with 0.5 and 1% (w/w) AH asphaltenes in 40% (v/v) toluene in heptane, which is at the solubility limit (Fig. 14) and 50% (v/v) toluene in heptane, a solvent concentration at which the AH asphaltenes are less surface active. Emulsion stabilization occurs in a two-step process: rapid interfacial adsorption, followed by long-term film consolidation. The associated rate constants in this process are clearly a function of solvent conditions, as seen here in comparing the 40 and 50% toluene results. Details of these experiments are described in Ref. 93.

## 2. Role of Aromatic Solvency in Asphaltene Film Stabilization

Experiments have been performed to complement the "bottle tests" performed on model emulsions of AH asphaltenes dissolved in heptol solutions and presented above (Fig. 14). The cef was monitored at 24 h for emulsions prepared with model oils emulsified to contain 30% water, varying amounts of AH asphaltenes from 0.25–3.0% (w/w), and varying toluene in heptol amounts from 30–60% (see Fig. 20). What is striking about the data in this figure is that the cef appears to be relatively independent of asphaltene concentration at long times, above an AH concentration of 0.5%. As mentioned above (Fig. 19), however, the cef does depend strongly on the percentage of toluene. Moreover, the cef appears to reach a local extreme at about 40% toluene, which is the limit of solubility of the asphaltenes and the point at which the asphaltenic aggregates would be expected to most surface active. When the data in Fig. 20 are converted into calculated interfacial film thickness (94) by assuming that 10% of the asphaltenes have adsorbed at 24 h (a value consistent with weighed film mass experiments we have performed in our laboratory), we observe that the cef appears to plateau beyond a critical interfacial cover-

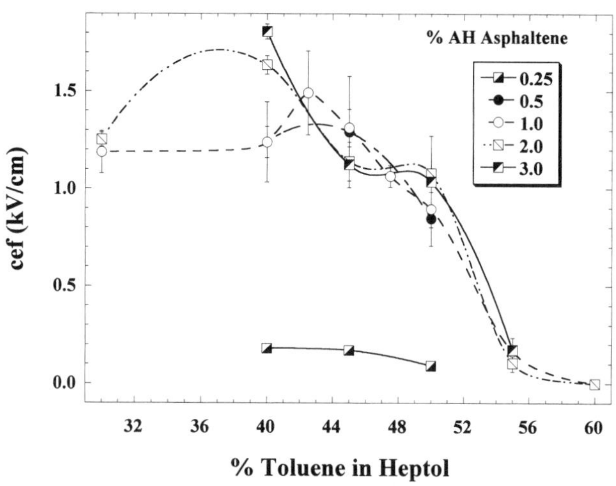


**Figure 20** Critical electric field measurements on W/O emulsions prepared by mixing oil phase containing AH asphaltenes in toluene–heptane mixtures in concentrations ranging from 0.25 to 3.0% (w/w) asphaltenes with water. The W/O emulsion consisted of 30% water and 70% oil (continuous phase).

age corresponding to 1.5 mg/m$^2$, which is equivalent to four to five molecules thick at the interface, assuming reasonable values for the molecular mass (1000 amu) and density (1.2 g/ml). Such a film thickness is remarkably consistent with the dimensions of an individual asphaltenic aggregate of roughly 4 nm radius corresponding to the primary adsorbing species and the critical film thickness equaling a monolayer of these adsorbed primary aggregates. Thus, it appears from these types of experiments that once a monolayer of aggregates has adsorbed and knitted itself into a coherent film, the droplet stability to coalescence in an electric field has achieved a large percentage of its maximum value.

### 3. Critical Electric Field Studies of Crude Oil Emulsions

The time dependence of the cef in emulsions prepared with water and crude oils is very similar to the time dependence in model systems (Fig. 19) with some important exceptions. In certain instances, local maxima are observed in these cef measurements which appear to be reproducible. The primary difference between the model systems described above and crude oil emulsions is the presence of resins in the crude oil. As we have mentioned above and demonstrated using percentage water resolved methods, resins can actually stabilize "pure" asphaltene emulsions in model solvents by possibly making the asphaltenic aggregates which adsorb smaller and more labile. Accordingly, we have performed preliminary experiments studying the cef of asphaltenes and resins in varying proportions in heptol solutions. These experiments indicate that with ratios of resins to asphaltenes of 0.25–0.75 (w/w), local maxima can be observed in the cef profile as a function of time. This maximum can be quite large relative to (1) the magnitudes of the cef that are observed in asphaltene alone systems; and (2) the long-term value of the cef. For example, with 1% B6 asphaltenes (a California offshore crude) in heptol in the absence of resins, the highest cef observed at 50% toluene is $\simeq$ 1.0 kV/cm. With B6 resins in an amount to give mass ratios of 0.75–1:1 (relative to the asphaltenes), the cef is observed to pass through a local maximum of $\simeq$ 1.9 kV/cm at adsorption times of 50–60 min, while at long times (several hours), the cef decreases to $<$ 1.0 kV/cm. This observation has been reproduced many times in our laboratory with different asphaltene–resin combinations. Clearly, a phenomenon is occurring at short times which leads to very stable emulsions and, presumably, highly ordered self-assembled asphaltenic films, which diminish over time in either film thickness or film elasticity per unit mass. In this regard, we believe resins may serve to solvate asphaltenes as smaller aggregates, making them more labile and enabling larger amounts of asphaltene to self-assemble at the interface over short times. At longer times, however, thermodynamics drives increasing amounts of resin to the interface which interacts with the asphaltenic film and either "loosens" the asphaltenic crosslinks in the interface (reduction in elasticity per unit mass) or "dissolves" the asphaltenes and removes them from the interface (reduction in interface thickness). This local maximum in cef versus time has also been observed in a variety of crude oil emulsions, suggesting that the above cited phenomenon may be somewhat general. This is currently an active area of research in our laboratory.

As a final point of discussion, we have measured the cef of W/O emulsions for a large array of crudes. The systems studied were 30% water in whole crudes, emulsified and measured at 60°C, a temperature at which the effects of wax might be expected to be somewhat minimized or eliminated. The crudes represent a very broad range of asphaltene content (0.8–15% w/w), resin-to-asphaltene ratios (0.9–6.2), wax or paraffin content (0.5–32% w/w), and viscosities (4–2300 cP at 100°F). The cefs measured correlate remarkably well with the asphaltene content of the crudes when the asphaltene concentration is multiplied by an effective driving force for adsorption to the interface (namely, the difference between the H/C ratio of the asphaltenes and the H/C ratio of the maltenes, i.e., the crude solvent). This correlation is shown in Fig. 21, and the only significant deviation from the correlation is for the crude SCS (South China Sea), a crude with a remarkably high paraffin content (32% w/w). Thus, it appears that the cef may be a very useful method for studying the relative and absolute stabilities of W/O emulsions and for unraveling essential scientific questions about the nature of asphaltene adsorption to interfaces.

## V. CONCLUSIONS

In this chapter, we have attempted to review and summarize the state of knowledge of the mechanisms of stabilization of W/O emulsions by asphaltenes. This has necessarily taken us into the realm of asphaltene chemistry, physical chemistry, adsorption at interfaces, film structure, and film rheology. The self-assembly of

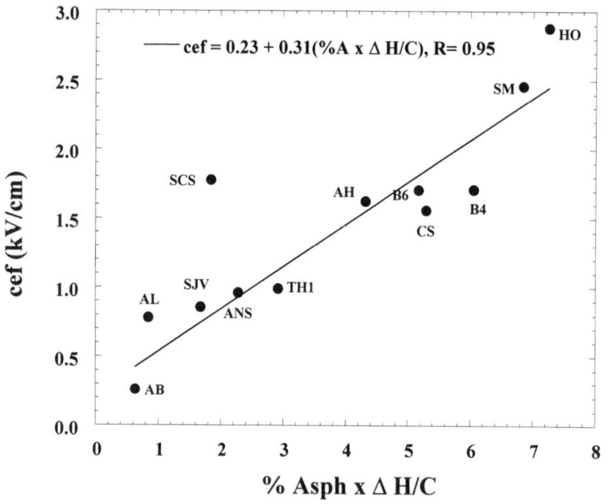


**Figure 21** Petroleum emulsion stability (30% water) as a function of %A × ΔH/C. Twelve different crude oils are included in the correlation and represent crudes varying in asphaltene content from 0.8% to 15% (w/w), resin content (3.2–20% w/w), and wax content (0.5–32% w/w), and in viscosity at 100°F (4–2300 cP). Details are described in Ref. 9.

asphaltenes at oil–water interfaces into a physically crosslinked, mechanically strong, viscoelastic network is clearly the means by which asphaltenes stabilize water droplets in oil. The kinetics of this process, the magnitude of the film strength, and the consequent stability of W/O emulsions stabilized by asphaltenes all depend strongly on thermodynamic variables, particularly (1) the aromaticity of the oleic medium; (2) the concentration and chemistry of the resin fraction and other specific solvating species in the crude; and (3) the polarity and specific chemistry of the asphaltenes. While item (1) is well understood and is described thoroughly in this chapter and in the cited bibliography, items (2) and (3) are less well understood and are active areas of research. With the continued high demand for petroleum and petroleum products, and the increasing heaviness of crudes produced around the world, the detailed mechanistic understanding of asphaltene-stabilized emulsions will continue to be of high interest to the scientific and engineering community.

## ACKNOWLEDGMENTS

The authors are pleased to acknowledge the encouragement, support, and technical guidance of Professor Johan Sjöblom. The research described here was supported in part by grants from Shell Oil Co. Nalco-Exxon Energy Chemicals, L.P., Texaco, ExxonMobil, and the National Science Foundation (CTS-9817127).

## REFERENCES

1. LL Schramm, ed. Emulsions: Fundamentals and Applications in the Petroleum Industry. Washington, DC: American Chemical Society, 1992, pp 1–49.
2. F Stalss, R Bohm, R Kupfer. Soc Petrol Eng – Product Eng 334–338, 1991.
3. OK Kimbler, RL Reed, IH Silberberg. Soc Petrol Eng 153–165, 1966.
4. JE Strassner. Petrol Technol 20: 303–312, 1968.
5. CM Blair. Chem Ind 538–544, 1960.
6. R Grace. In: LL Schramm, ed. Emulsions: Fundamentals and Applications in the Petroleum Industry. Washington, DC: American Chemical Society, 1992, pp 313–339.
7. GE Jackson. Chem Oil Ind 45: 92–107, 1983.
8. H Neumann. Petrochemie, 18: 776–779, 1965.
9. DD Eley. MJ Hey, JD Symonds, JHM Willison. J Colloid Interface Sci 54: 462–466, 1976.
10. B Obah. Erdohl, Kohle, Erdgas Petrochem 41: 71–74, 1988.
11. SE Taylor. Chem Ind 20: 770–773, 1992.
12. RA Mohammed, AI Bailey, PF Luckham, SE Taylor. Colloids Surfaces A: Physicochem Eng Aspects 80: 223–235, 1993.
13. RA Mohammed, AI Bailey, PF Luckham, SE Taylor. Colloids Surfaces A: Physicochem Eng Aspects 91: 129–139, 1994.
14. AJS Liem, DR Woods. AIChE Symp Ser 70: 8–23, 1974.
15. EJ Johansen, M Skjarvo, T Lund, J Sjöblom, H Soderlund, G Bostrom. Colloids Surfaces 34: 353–370, 1989.
16. J Sjöblom, H Soderlund, S Lindblad, EJ Johansen, IM Skjarvo. Colloid Polym Sci 268: 389–398, 1990.
17. AL Bridie, TH Wanders, W Zegveld, HB van der Heijde. Mar Pollut Bull 11: 343–348, 1980.
18. M Bobra, M Fingas, E Tennyson. Chemtech 236–241, 1992.
19. DT Wasan. In: J Sjöblom, ed. Emulsions – A Fundamental and Practical Approach, Dordrecht: Kluwer Academic, 1992, pp 283–295.
20. H Førdedal, Y Schildberg, J Sjöblom, JL Volle. Colloids Surfaces A: Physicochem Eng Aspects 106: 33–47, 1996.
21. RA Mohammed, AI Bailey, PF Luckham, SE Taylor. Colloids Surfaces A: Physicochem Eng Aspects 80: 237–242, 1993.
22. CH Pasquarelli, DT Wasan. In: DO Shah, ed. Surface Phenomena in Enhanced Oil Recovery. New York: Plenum Press, 1981, pp 237–248.

23. GDM Mackay, AY McLean, OJ Betancourt, BD Johnson. Petrol 59: 164–172, 1973.
24. CG Dodd. Phys Chem 64: 544–550, 1960.
25. EE Isaacs, H Huang, AJ Babchin, RS Chow. Colloids Surfaces, 46: 177–192, 1990.
26. M van der Waarden. Kolloid ZZ Polym 156:116–122, 1958.
27. E Rogel. Energy Fuels 14:566–574, 2000.
28. MM Boduszynski, JF McKay, DR Latham. Proc Assoc Asphalt Paving Technologists 49: 123–143, 1980.
29. M Boduszynski, BR Chadha, T Szkuta-Pochopien. Fuel 56: 432, 1977.
30. JF McKay, PJ Amend, TE Cogswell, PM Harnsberger, RB Erickson, DR Latham. In: PC Uden, S Siggia, HB Jensen, eds. Analytical Chemistry of Liquid Fuel Sources, Tar Sands, Oil Shale, Coal and Petroleum. Washington, DC: American Chemical Society, 1978, pp 128–142.
31. M Boduszynski. In: Preprints Division of Petroleum Chemisty. Washington, DC: American Chemical Society, 1979, p 935.
32. JG Speight. Preprints – ACS Div Petrol Chem 34: 321–328, 1989.
33. V Kiselev. Petrol Chem, USSR 19: 714, 1979.
34. MA Bestougeff, P Gendrel. Study on the structure of asphaltenic constituents by combined physical and chemical methods. Proceedings of 147th American Chemical Society Meeting, Division of Petroleum Chemistry, Philadelphia, 1964.
35. V Calemma, P Iwanski, M Nali, R Scotti, L Montanari. Energy Fuels 9: 225–230, 1995.
36. MMH Al-Jarrah, AH Al-Dujaili. Fuel Sci Technol Int 7: 69–88, 1989.
37. TA Filimonova, LV Gorbunova, VF Kam'yanov. Petrol Chem, USSR 27: 210–220, 1987.
38. MA Bestougeff. Bulletin Chem Soc Fr 12: 4773, 1967.
39. JD McLean, PK Kilpatrick. Energy Fuels 11: 570–585, 1997.
40. J Jurkiewicz, S Rosinski. Koks, Smola, Gaz 1: 143–152, 1956.
41. MA Bestougeff, RJ Byramjee. In: TF Yen, GV Chilingarian, eds. Asphaltenes and Asphalts, 1. Developments in Petroleum Science, 40. Elsevier Science, pp 67–94, 1994.
42. JW Bunger, KP Thomas, SM Dorrence. Fuel 58: 183–195, 1979.
43. TA Filimonova, LV Gorbunova, VF Kam'yanov. Petrol Chem, USSR 27: 257–265, 1987.
44. JG Speight, SE Moschopedis. In: JW Bunger, NC Li, eds. Chemistry of Asphaltenes. Washington, DC: American Chemical Society, pp 1–15, 1981.
45. JD McLean, PM Spiecker, AP Sullivan, PK Kilpatrick. In: OC Mullins, EY Sheu, eds. Structures and Dynamics of Asphaltenes. New York: Plenum Press, 1998, pp. 377–422.
46. J Lamathe. Comptes Rend Seances, Acad Sci, Ser C 2663: 862–874, 1966.
47. OC Mullins. In: EY Sheu, OC Mullins, eds. Asphaltenes: Fundamentals and Applications, New York: Plenum Press, 1995, pp 53–96.
48. IR Mansurov, EZ Il'yasova, VP Vygovskoi. Chem Technol Fuels Oils 23: 96–98, 1987.
49. TF Yen. Energy Sources 1: 447–463, 1974.
50. L Mingyuan, A Christy, J Sjöblom. In: J Sjöblom, ed. Emulsions – A Fundamental and Practical Approach. Dordrecht: Kluwer Academic, 1992, pp 157–172.
51. AN Khryashchev, NA Popov, NA Posadov, DA Rosendal. Petrol Chem, USSR 31: 601–604, 1991.
52. EY Sheu, DA Storm, MMD Tar. Non-Cryst Solids 131–133: 341–347, 1991.
53. EY Sheu, MM DeTar, DA Storm, SJ DeCanio. Fuel 71: 299–302, 1992.
54. EY Sheu, DA Storm. In: EY Sheu, OC Mullins, eds. Asphaltenes: Fundamentals and Applications. New York: Plenum Press, 1995, pp 1–52.
55. EY Sheu. In: OC Mullins, EY Sheu, ed. Structures and Dynamics of Asphaltenes. New York: Plenum Press, 1998, pp 115–143.
56. IK Yudin, GL Nikolaenko, EE Gorodetskii, EL Markhashov, MA Anisimov, JV Sengers. Physica A 251: 235–244, 1998.
57. IK Yudin, GL Nikolaenko, EE Gorodetskii, EL Markhashov, D Frot, Y Briolant, VA Agayan, MA Anisimov. Petrol Sci Technol 16: 395–414, 1998.
58. H Rassamdana, B Dabir, M Nematy, M Farhani, M Sahimi. AIChE J 42: 10–22, 1996.
59. H Rassamdana, M Sahimi. AIChE J 42: 3318–3332, 1996.
60. CL Chang, HS Fogler. Langmuir 10: 1749–1757, 1994.
61. CL Chang, HS Fogler. Langmuir 10: 1758–1766, 1994.
62. P Permsukarome, CL Chang, HS Fogler. Ind Eng Chem Res 36: 3960–3967, 1997.
63. TJ Kaminski, HS Fogler, N Wolf, P Wattana, A Mairal. Energy Fuels 14: 25–30, 2000.
64. HW Yarranton, JH Masliyah. AIChE J 42: 3533–3543, 1996.
65. HW Yarranton, H Hussein, JH Masliyah. J Colloid Interface Sci 228: 52–63, 2000.
66. PM Spiecker, PK Kilpatrick. Ion exchange fractionation of asphaltenes and their resulting interfacial properties, to be submitted.
67. FJ Nellensteyn. J Inst Petrol Technol 10: 311–325, 1924.
68. JP Pfeiffer, RNJ Saal. Phys Chem 44: 139–149, 1940.
69. E Papirer, C Bourgeois, B Siffert, H Balard. Fuel 61: 732–734, 1982.
70. JP Dickie, TF Yen. Analyt Chem 39: 1847–1852, 1967.
71. TF Yen. In: TF Yen, RD Gilbert, JH Fendler, eds. Advances in the Applications of Membrane-Mimetic Chemistry. New York: Plenum Press, 1994, pp 255–279.

72. R Miller, R Wüstneck, J Krägel, G Kretzschamr. Colloids Surfaces A: Physicochem Eng Aspects 111: 75–118, 1996.
73. I Langmuir. Science 84: 303, 1936.
74. S-G Oh, JC Slattery. J Colloid Interface Sci 67: 516–525, 1978.
75. K Wibberley. Pharmacy Pharmacol 14: 87T, 1962.
76. R Nagarajan, SI Chung, DT Wasan. J Colloid Interface Sci 204: 53–60, 1998.
77. RJR Cairns, DM Grist, EL Neustadter. The effect of crude oil–water interfacial properties on water–crude oil emulsion stability. In: AL Smith, ed. Theory and Practice of Emulsion Technology. London: Academic Press, 1974.
78. EL Neustadter, KP Whittingham, DE Graham. In: DO Shah, ed. Surface Phenomena in Enhanced Oil Recovery. New York: Plenum Press, 1981, 307–326.
79. S Acevedo, G Escobar, LB Gutierrez, H Rivas, X Gutierrez. Colloids Surfaces A: Physicochem Eng Aspects 71: 65–71, 1993.
80. PM Spiecker, PK Kilpatrick. Rheological properties of adsorbed asphaltenic films at oil–water interfaces. Langmuir, to be submitted.
81. R Cimino, S Correra, A Del Bianco, TP Lockhart. In: EY Sheu, OC Mullins, eds. Asphaltenes: Fundamentals and Applications. New York: Plenum Press, 1995, pp 97–130.
82. JD McLean, PK Kilpatrick. J Colloid Interface Sci 196: 23–34, 1997.
83. PM Spiecker, PK Kilpatrick. A comparison of asphaltenes from differing sources and the impact on solubility and rheological properties of adsorbed films. Energy and Fuels, to be submitted.
84. S Singh, JD McLean, PK Kilpatrick. J Dispers Sci Technol 20: 279–293, 1999.
85. J Sjöblom, H Fordedal, T Jakobsen, T Skodvin. In: KS Birdi, ed. Dielectric Spectroscopic Characterization of Emulsions. Boca Raton, FL: CRC Press, 1997, pp 217–237.
86. T Skodvin, J Sjöblom, JO Saeten, O Urdahl, B Gestblom. J Colloid Interface Sci 166: 43–50, 1994.
87. T Skodvin, J Sjöblom. Colloid Polym Sci 274: 754–762, 1996.
88. H Førdedal, E Nodland, J Sjöblom, OM Kvalheim. J Colloid Interface Sci 173: 396–405, 1995.
89. PJ Bailes, SKL Larkai, Trans Inst Chem Eng 60: 115–121, 1982.
90. TY Chen, RA Mohammed, AI Bailey, PF Luckham, SE Taylor. Colloids Surfaces A: Physicochem Eng Aspects 83: 273–284, 1994.
91. O Urdahl, TJ Williams, AG Bailey, MT Thew. Trans Inst Chem Eng 74: 158–165, 1996.
92. L Wang, Z Yan. J Dispers Sci Technol 15: 35–58, 1994.
93. AP Sullivan, NN Zaki, PK Kilpatrick. The stability of water-in-crude and model oil emulsions. Colloids Surfaces A: Physicochem Eng Aspects, to be submitted.

# Index